D1567475

Giant Oil and Gas Fields of the Decade 1968-1978

Giant Oil and Gas Fields of the Decade 1968-1978

Edited by

Michel T. Halbouty

Published by

The American Association of Petroleum Geologists

Tulsa, Oklahoma 74101, U.S.A.

Published December 1980

Library of Congress Catalog Card Number: 80-69780
ISBN: 0-89181-306-3

The Association gratefully acknowledges
assistance from the Halbouty Fund
of the
American Association of Petroleum Geologists Foundation

Table of Contents

Introduction

This volume is the result of a symposium of worldwide participation on "Giant Discoveries of the Past Decade" sessions which were presented at the 64th annual meeting of The American Association of Petroleum Geologists, in Houston, Texas, April 1-4, 1979. The papers represented are distinctly international. I know of no other collection of papers in the annals of our Association, or any other society, which has brought together such a wealth of information which gives the petroleum geologist an opportunity to significantly analyze the anatomy of what constitutes a giant field.

The areas represented in this volume are China, Malaysia, Indonesia, Middle East, Africa, Europe, including the North Sea, Brazil, Trinidad, Mexico, United States, Canada and Alaska. Knowledge can be gained from these papers so that we, as searchers for petroleum, may more effectively serve our country and mankind by finding many more giants throughout the world.

The papers in this volume discuss the accumulations of vast reserves of petroleum in, on, and around anticlines, arched monoclines, faulted anticlines, folds, faulted structures, reefs, salt domes, stratigraphic traps, paleogeomorphic traps, and a combination of two or more of these conditions. Most importantly, these papers represent a synthesis of a mass of data obtained on the geology and conditions under which oil and/or gas were formed and accumulated in reservoirs as young as Pleistocene in age and as old as Cambrian.

Considerable information will be presented which we, as geologists, should mentally record and use in our future thinking as explorationists. For that reason, I will not attempt to summarize the contents. Instead, I want to stress the importance of looking for the one or more factors which cause such vast accumulations of petroleum, so that those same factors may be applied to find the "Giants" which are waiting to be found—not only in the United States, but in other parts of the world.

We should make it a point to learn as much as we can concerning accumulations occurring in each of the giant fields, the type trap, how the trap was formed and found, the age of the reservoir rocks, and the significance of all these factors to one another. We should ask ourselves: What is usual? What is unusual? And we should concentrate our study on the unusual factors more than on those which are common. Prejudiced ideas should be discarded, as it is those hard-nosed prejudices which are destructive in petroleum exploration and which should not be tolerated in our future thinking.

We should be encouraged and comforted in our knowledge that giant fields do occur, and as they have been found in reasonable numbers in the past, it is logical to assume there are many other undiscovered giants just waiting for the proper scientific determination to find them—or even the intestinal fortitude of the rank wildcatter, who, without science, drills at random in areas where science condemns or fears to tread.

I wonder if anyone in 1968 would have had the audacity to remotely predict that 288 giants would be found worldwide in the coming decade. That is the number which were discovered. So, what about the next decade? Will we find more or less? It would be interesting ten years from now to have a symposium on the giants discovered in that decade. However, I can safely predict without any reservation whatsoever that many giants *will* be discovered. Just how many remains to be seen.

Thirty papers on selected giants and areas are included in this special publication with case histories of how and why conditions existed to permit large accumulations of petroleum. The estimated oil reserves of the fields covered by these papers total over 50 billion barrels and the gas reserves total over 100 trillion cubic feet. Therefore, it is obvious that with such large reserves, the world's giant fields are of great importance to the economy and strength of not only the petroleum industry but of the nations in which these fields occur.

The great majority of the discoveries of the giants of the world are the results of exploration which was oriented purposely toward finding and exploiting the structural trap. In some instances, as a result of drilling, what was mapped as a geophysical structure turned out to be a subtle trap—stratigraphic, paleogeomorphic, and/or unconformity. In other words, the structure did not exist at all; however, inadvertently and certainly unintentionally, a giant field was discovered. Call it blind luck if you wish. I am sure that the operator did not care whether the structure existed or not as long as he had to his credit a giant discovery.

It is most important to note that although only a relatively few of these subtle giant traps have been found to date, and most if not all of these accidentally, they appear fairly persistently in the discovery records throughout the years. It is certainly logical to assume that this persistent, though accidental, discovery history of subtle trap accumulations suggests that many more must occur in the basins of the world. If this conclusion is valid, just think of what might happen in future exploration if our search efforts were directed purposely toward the subtle and obscure trap.

There are more subtle trap giants, such as East Texas, to be found somewhere in this country and in other parts of the world. They will certainly be found from time to time—hopefully by design rather than by accident. It is hoped that a few will be found in the United States, and that they will be discovered by some enterprising company which is blessed by a bold explorationist supervisor, or by some independent geologist who has the courage to back his convictions.

Furthermore, we must recognize that the future exploration position of the United States is not the same as that of foreign countries. Since exploration in most foreign areas has not been so extensive and thorough as it has been in the United States, it is apparent that further exploration in these foreign countries will discover many more structural giants. It is most logical to assume that the search for petroleum in foreign regions will continue to be focused on the obvious type of structural traps which can readily be found by today's exploratory tools. In the United States, however, possibilities for finding the obvious structural type giant fields are not as favorable onshore as they are in foreign lands because of the intensive exploration which has already been conducted in all known petroliferous provinces in this country.

Since it is imperative that the exploratory effort in the United States continue to find giant fields to meet the nation's future needs for petroleum, and since most of the obvious easy-to-find giant structural traps probably already have been found in this country, it is time that the American oil industry focus its attention on the subtle traps, and expend a large portion of its exploratory effort and funds toward searching for them.

I firmly believe that, in this country alone, there are many more giant fields to be found, but it is evident that we must change our exploration methods and re-orient our thinking to the accumulations which surely must have occured in traps other than structural. We must find these hidden giants in the United States. We need the reserves to meet our own future needs.

The contents of this volume should add immeasurably to the art of discovering giant fields wherever they are located.

Michel T. Halbouty
Houston, Texas

Giant Field Discoveries 1968-1978: An Overview[1]

By T. A. Fitzgerald[2]

A giant oil field is defined in this collection as being a contiguous surface area beneath which one or more reservoirs either has produced or is expected to produce 100 million bbl or more of conventional oil, except in the Middle East, North Africa, and Asiatic Russia, where 500 million bbl or more is required for giant status.

A giant gas field is similary defined as a contiguous surface area beneath which one or more reservoirs either has produced or is expected to produce 1 Tcf or more of combustible gas; again with the exception of the Middle East, North Africa, and Asiatic Russia, where 3 Tcf or more is required. Natural gas liquids are included with gas.

These definitions exclude heavy oil sands and tar deposits, as well as pervasive ultra-tight and other nonconventional gas accumulations. No credit is given for possible tertiary recovery, nor are fields included which may qualify as giants only on a combined oil plus gas basis.

There are some arbitrary features about these definitions which make for difficult statistical analysis. These are the energy equivalent differences between the minimum sizes of oil fields versus gas fields in most of the areas, and the size change between geographic subdivisions themselves. The period covered in this paper is 11 years, which also is statistically cumbersome.[3]

The general statistics regarding past oil and gas discovery are shown in Figure 1. Through the end of 1977, some 1.1 trillion bbl of oil and 4,100 Tcf of natural gas plus gas liquid equivalent have been found. About 85% of this is in giant fields as per definition. Keep in mind this is the sum of geologic estimates of all fields found—and as such exceeds the sum of the commonly accepted proved and probable reserves plus cumulative production by about 100 billion bbl and 500 Tcf. Also note that 6,000 cu ft of gas equals 1 bbl oil for all oil-equivalent numbers.

The location of these 288 fields also are shown by major geographic area. The dashed line indicates the definitional size change. Also shown is the approximate contribution of the giants in billions of oil-equivalent barrels and (in parentheses) the average size of the giant fields for each geographic area. About 80 of these are gas fields only; in others which qualify as oil giants, the gas fraction is the larger on a BTU basis. About 39 giants were found in the U.S. and Canada. This area had one very large giant field, Prudhoe Bay, accounting for 60% of the oil equivalent discovered in giants. The rest were in fields at the lower end of the range, and gas fields were predominant.

Nearly half of the 43 giant fields found in Latin America were located in Mexico. These Mexican discoveries contributed the bulk of the hydrocarbons in the giant category and may still be understated. The high, 1 billion bbl (equivalent) average-field sizes reflect the large size of the Mexican finds.

About 58 giant fields were found in Europe and, as might be expected, the North Sea fields predominate.

[1] Manuscript received, June 1, 1979; accepted for publication, August 24, 1979.

[2] Exxon Corporation, New York, New York 10020.

[3] The data in this paper are taken from Exxon internal files and modified by outside sources only to that extent which is normal for any large organization.

Article Identification Number:
0065-731X/80/M030-0001/0.

GENERAL STATISTICS

APPROXIMATE OIL AND GAS DISCOVERED IN WORLD TO 12/31/77

OIL B BBL	GAS TCF	OIL EQUIV. B BBL	GIANTS % OF TOTAL
1120	4080	1800	85%

APPROXIMATE OIL AND GAS DISCOVERED IN WORLD 1/1/67 – 12/31/77

OIL B BBL	GAS TCF	OIL EQUIV. B BBL	GIANTS % OF TOTAL
235	1250	445	75%

GIANT DISCOVERIES BY GEOGRAPHIC AREA 1967–77

AREA	APPROX. NO.	APPROX DISC. (AVG. SIZE) B BBL OIL EQUIV.	
U.S. AND CANADA	39	25	(0.6)
LATIN AMERICA	43	45	(1.0)
EUROPE (INCL. WEST U.S.S.R.)	58	40	(0.7)
AFRICA (WEST AND SOUTH)	33	10	(0.3)
FAR EAST	50	35	(0.7)
MIDDLE EAST AND NO. AFRICA	38	85	(2.2)
ASIATIC U.S.S.R.	27	90	(3.3)
TOTAL	288	330	(1.1)

FIG. 1—General statistics table relating past oil and gas discovery to that during the 1967-1977 period. Note that giants accounted for 85% of the total discovery prior to 1967, but only 75% of the discovery since then.

AREAS OR BASINS
WITH GIANT OIL AND GAS DISCOVERIES
DURING THE PERIOD 1967–77

AREAS WITH PRIOR GIANT FIELDS	
ARABIAN GULF	(32)
ASIATIC U.S.S.R.	(27)
NORTH AFRICA	(6)
NIGER DELTA	(25)
GULF COAST (U.S.)	(9)
EUROPEAN U.S.S.R.	(7)
NORTH EUROPE (INCL. SO. NORTH SEA)	(7)
GIPPSLAND (AUSTRALIA)	(5)
CHINA	(4?)
INDONESIA (SUMATRA)	(4)
BRUNEI	(4)
MISSISSIPPI SALT (U.S.)	(3)
ALL OTHER	(19)
TOTAL	(152)

AREAS WITH NO PRIOR GIANT FIELDS	
NO. AND CENT. NORTH SEA	(43)
REFORMA (MEXICO)	(15)
ECUADOR/PERU (ORIENTE/MONTANA)	(11)
GABON AND CONGO COASTAL	(8)
MALAY	(7)
INDONESIA (OFFSHORE JAVA)	(7)
BRAZIL (OFFSHORE)	(6)
INDONESIA (KALIMANTAN)	(5)
TRINIDAD (ORINOCO)	(5)
SVERDRUP (CANADA)	(5)
ALL OTHER	(24)
TOTAL	(136)

( ) APPROXIMATE NO. OF GIANTS

FIG. 2—Discoveries categorized by basin area rather than by broad geographic region. Note that nearly half the giants discovered were found in basins with no prior giant discoveries. Giants listed here were discovered since 1967.

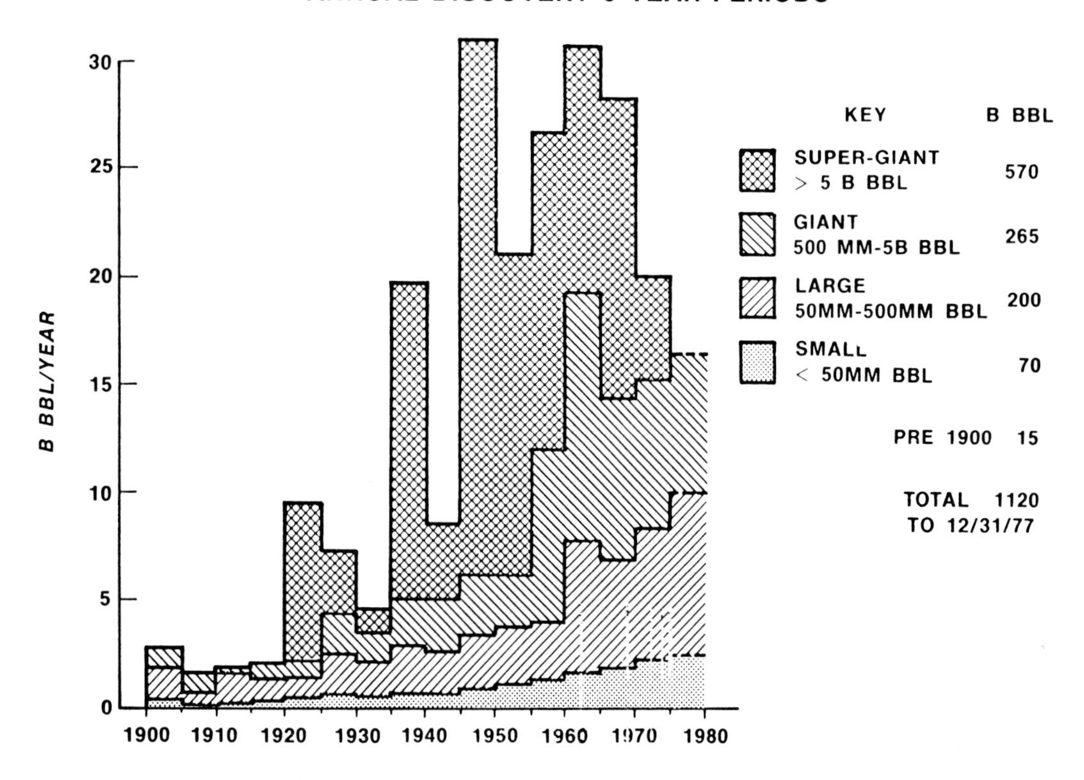

FIG. 3—Discovery history of oil since 1900. Note that discoveries were on the increase until 1960, but the general trend has declined since then. The decline has not been due to finding fewer fields, rather it has been due to not finding more supergiants.

During the 11-year period, January 1967 through December 1977, it is presently estimated that some 235 billion bbl of oil and 1,250 Tcf of gas were discovered and that roughly 75% of this reservoir is in giant fields as defined. It appears therefore that the giants' contribution to the total may have declined somewhat during the subject period. Nevertheless, these 288 fields account for most of the hydrocarbons found. It should be emphasized that there are a number of fields near the minimum sizes and the estimated number of fields may range from about 5% too many to 10% too few.

In West and South Africa, approximately 33 giants were found principally in the Niger delta onshore and offshore, and the Gabon and Congo coastal basins. None of these discoveries were exceptionally large.

Some 50 or so fields were found in the Asia/Pacific region. The largest discoveries in this area were gas.

In the larger size definition, about 38 giant fields were discovered in the Middle East and North Africa, with the Middle East contributing both the bulk of the fields and the bulk of the hydrocarbons.

About 27 fields were found in Asiatic Russia with the super giant gas fields in the northern part of the West Siberia basin being the most important.

Figure 2 shows giant discoveries by basin area and is perhaps more meaningful than broad geographic subdivisions.

Here the 288 giant discoveries have been divided into two basin categories; those that have had giant discoveries prior to 1967 and those that had giant discoveries for the first time since 1967. Most importantly, it may be noted that they are split almost evenly—152 in those basins which already had giant fields and 136 in those basins which did not.

Above the dashed line in the left column (areas with prior giant fields) are those giant discoveries in areas where the 500 MM bbl or 3 Tcf of gas is the minimum size. There were 65 of these. Had these areas been categorized on the same basis as the others, the left-hand column would have been proportionally larger. Also in the left column, note the question mark (?) by China, indicating the difficulty in analyzing this area properly. The scanty data coming out of China make it difficult to separate fields from groups of fields or even provinces.

In the right-hand column, areas with no prior giant fields, the Central and Northern (or Viking) area of the North Sea leads with 43 giant discoveries. The Reforma area of Mexico follows with 15, but this number also is somewhat suspect because some of these fields are being joined by development drilling.

At the bottom of this column, I have noted 24 giants in the category "all other." There are some extremely important discoveries covered in this category—the Alaska North Slope, the Bombay High offshore India, the Tarragona area offshore

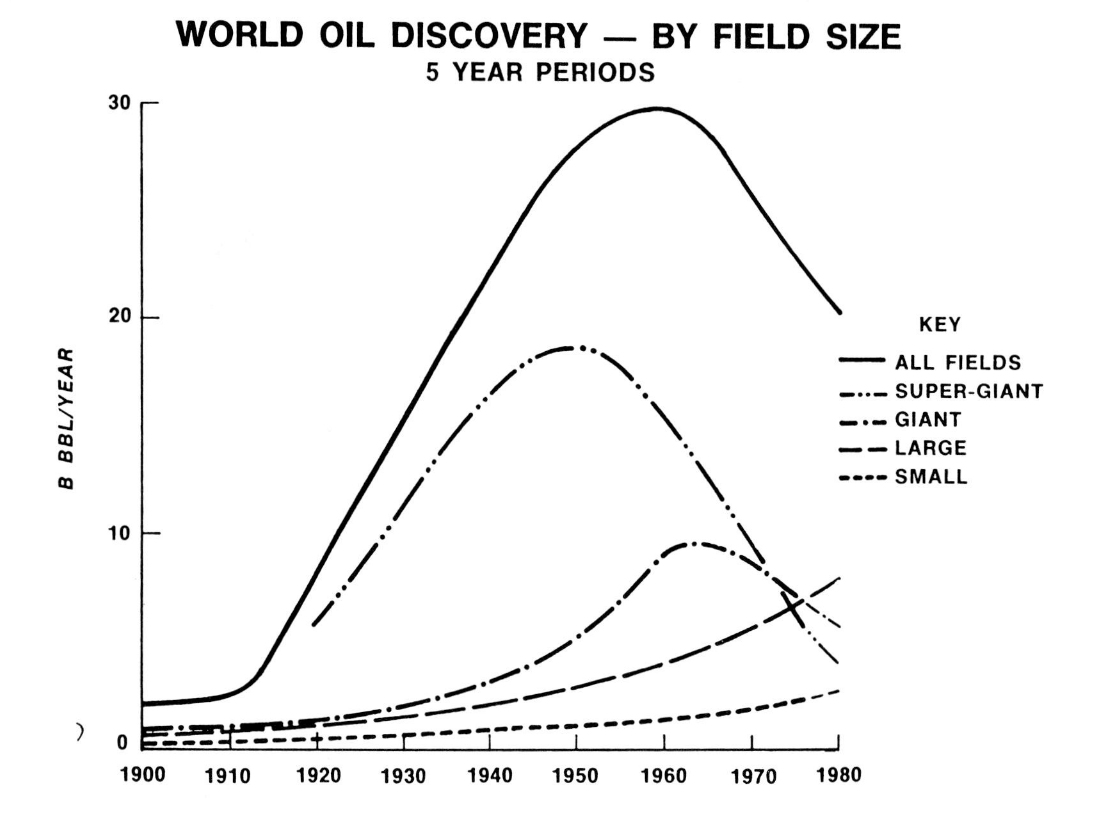

FIG. 4—Smoothed graph of discovery history since 1900 (in billion bbl/year). Note that whereas the number of small and large field discoveries is still on the increase, the number of giant and supergiant discoveries has been declining since a peak in the 1950s and 1960s.

Spain, offshore Taranaki in New Zealand, the Campeche Shelf of Mexico, and offshore Palawan in the Philippines among them.

I think it is fairly obvious that exploration is finding new oil producing basins—roughly 45 were discovered in the 11 year period—and some 20 odd of these already have giant fields. Recall that this is a current estimate. The number of 136 giant fields will likely increase as additional information becomes available about the more recent finds.

Perhaps some longer range historic perspective on worldwide oil discovery may provide additional background. Figure 3 represents the discovery history for oil since the year 1900. This graph displays oil discovery per year averaged over 5-year periods since that time in order to smooth year-to-year variations. Discoveries are further divided into 4 field-size categories: those called super giants with 5 B bbl or more of expected ultimate recovery; giants in the 500 million to 5 B bbl range; large fields in the 50 to 500 million bbl range; and small fields being those with ultimate recovery of less than 50 million bbl. The key on the right also indicates the approximate amounts of oil in each category, totalling the 1,120 billion bbl referred to previously.[4]

The rate of discovery increased rapidly until the early 1960s. The depression of the 1930s, World War II, and the excess producing capacity developed during the 1950s are reflected as interruptions in the upward trend. Since the early 1960s, the rate of discovery has been in apparent decline even though this has been the period of greatest exploration activity.

The decline has not been caused by finding fewer fields. On the contrary, the number of discoveries is continuing to increase along with the wildcat drilling and related activity. The decline in the amount discovered has been the direct result of finding fewer super giant fields. Note that about 50% of all the oil discovered to date has been in the super giants. There are perhaps only 32 super giant oil fields in the world which account for the 570 billion barrels. Of these, 24 are in the Middle East, 3 in the Soviet Union, 2 in the United States, and one each in Lybia, Mexico, and Venezuela. Three super giant oil fields have been found since 1967. These include Prudhoe Bay, Bermudez in Mexico, and Shaybah/Zarrara in Saudi Arabia/Abu Dhabi.

The 265 billion bbl discovered in giant fields is accounted for by perhaps 200 fields. The discovery rate for this class of field appears to be in decline also.

It is the two lower categories, large and small, which are continuing to increase in significance.

Figure 4 is a smoothed plot of the discovery rate for each of the aforementioned field size classes in billion barrels per year. The total smoothed curve for all discoveries also is shown.

Super giant field contributions to past discovery clearly peaked around 1950 and have recently declined to the point

[4] Note that data shown on Figure 3 and Figure 4 use somewhat different size categories than those used for the 11-year period.

where on average they contribute less than 5 billion bbl per year. This reflects an estimate of the recent contribution of the Bermudez and potential additional super giants in the Reforma area of Mexico.

Giant field contributions to discovery appear to have peaked in the middle 1960s, although this observation may be a bit premature. However, the contribution by this group is at least not growing as it did in the 1950s.

The large field category is making an increasing contribution as is the small size category.

Although not illustrated, gas discovery history is similar in most repsects except one, and that is the overall discovery peak occurs in the late 1960s, some 10 years after that for oil, and in fact the peak in gas discovery may be still in the making. Five gas super giants (30 Tcf or more) were found in 1967 to 1977, four in Russia and one in the Middle East.

While it would be misleading to read too much into these discovery patterns, they do show a certain progression and provide some food for thought.

In summary, several points can be made. The number of giant discoveries is not declining and, in fact, there are a sizable number in basins which did not previously contain giants. In terms of hydrocarbons discovered, giants in the past 11 years account for more than one-fifth of all giants to date. In spite of this, the size of the average giant discovery is declining (fewer super giants) as is the oil discovery rate for the world in total. Gas discovery (while not specifically dealt with) does not yet show an established decline.

However, it should be noted that: (1) statistics such as these are based solely on conventional fields with historic economic conditions; and (2) exploration is in the midst of a period of transition resulting from a reduction of spare producing capacity and price escalation. Thus, the whole statistical picture is in a state of flux, and many of the giants found in the past decade are not yet economical to develop, even at current prices, due to high costs involved with physical location and/or government economic and regulatory policies. Also, very large nonconventional oil and gas accumulations are beginning to enter the picture—even discovery date is not germane, since most of these large-but-low-grade deposits were found many years ago. These would include those such as Lloydminster and Cold Lake in Canada, Venezuela "tar" belt, and perhaps the recent Mexican disclosure of Chicontepec, among others. Finally, enhanced recovery techniques are starting to appear which will require substantial revisions of much of this data in the future.

Geologic Significance of Landsat Data for 15 Giant Oil and Gas Fields[1]

By Michel T. Halbouty[2]

Abstract If land satellite data had been available and applied to areas over which giant fields were found, how effective would the data have been in the exploration effort, and what kind of useful geologic information would have been generated from the satellite images? This question obsessed the writer and prompted the effort to find the answer by obtaining satellite images of fifteen existing giant fields in various parts of the world and interpreting whatever geological data the images provided. The studies clearly proved that the images would have been of considerable value in exploring for, and pinpointing the locations of, most of the giants under study. Such a conclusion would indicate that land satellite images and remote sensing data should be a top priority in the search for the future giants to be found in the remaining prospective areas of the earth.

INTRODUCTION

This paper is designed to estimate just how helpful land satellite images would have been for exploration in areas where giant fields exist. In other words, if land satellite data had been available and applied to areas over which giant fields were found, how effective would have been the use of the data in the exploration effort, and what kind of useful geological information would have been obtained from the satellite images?

This question prompted the study of the satellite images of fifteen existing giant fields picked at random in various parts of the world. It was reasoned that if such images would have been useful, then it would be only logical to conclude that land satellite images and remote sensing data should be a top priority in the search for the future giants which exist in the remaining petroleum-prospective areas of the earth.

The reader should constantly bear in mind that comments made on the interpretation of the images are stated in terms of "how helpful" or "significant" an observation or conclusion could have been to the explorer *prior* to any surface or seismic exploration within the image area. This premise also included that no oil or gas field existed in the image area. It was essential to think in these terms to study the images in a strictly unbiased manner.

For example, on the Ghawar image, the writer was aware that Dhahran and Bahrain are producing fields, but the text states "The anticlines at Dhahran and Bahrain clearly are visible on the image, as well as the small dome to the north. These are

[1]Modified from a paper presented at the AAPG Annual Meeting, April 3, 1979 (Houston). This paper previously was published in the AAPG Bulletin, v. 64, p. 8-36. Received and accepted for publication, August 8, 1979.

[2]Consulting geologist and petroleum engineer, independent producer and operator, Houston, Texas.

The writer expresses his sincere appreciation to Cynthia Sheehan, associate applications scientist—geology, at the EROS Data Center, who contributed much to the interpretation of the images and also made timely suggestions which were incorporated in the paper.

Thanks also are expressed to James V. Taranik, former principal remote sensing scientist at EROS Data Center and now Chief of the Branch of Non-Renewable Resources, NASA; Charles M. Trautwein, geological remote sensing instructor, EROS Data Center; G. Bryan Bailey, applications scientist—geology, EROS Data Center; and Patrick D. Anderson, Jr., applications scientist—geology, EROS Data Center, for their many recommendations on the applications and interpretations of the images. And the writer thanks James J. Halbouty, geologist; James Radtke, chief draftsperson; and Harlene Marcotte, secretary, (all of whom are associated with the writer) for their assistance in the overall preparation and completion of the manuscript.

Article Identification Number:
0065-731X/80/M030-0002/$03.00/0.

significant features." Remember that the study of the images was predicated on the assumption that each image was available *before* any kind of exploration was conducted in the area; consequently, the geologic interpretation and comments pertinent thereto were made accordingly.

After the images were interpreted, annotated, and made ready for publication, the general information on each field was written so as to provide the reader with a very brief sketch of some of the pertinent data on each field.

The fifteen fields are Ghawar, Saudi Arabia; Taching (Daqing), China; Kangan, Iran; Buzurgan, Iraq; Burgan, Kuwait; Messla, Libya; El Borma, Tunisia; Prudhoe Bay, Alaska; Samotlor (Samotlorskoye), U.S.S.R.; Kettleman Hills, California; Swanson River, Alaska; Strachan and Ricinus, Alberta; Jay, Florida; Malossa, Po Basin, Italy; and Groningen, The Netherlands and West Germany.

It is noteworthy that the estimated ultimate recoverable oil reserves of the fields covered by this study total approximately 200 billion bbl of oil and 180 Tcf of gas.

intermittent stream that drains them.

Figure 2 shows the geologic interpretation of the image. Most of the linear features on the image strike north-northwest to south-southeast. The folded strata in the southwestern corner of the image, and at Dhahran and Bahrain, parallel this trend. Linear segments of the coastline exhibit a definite preference for this direction. These may indicate faults or flexures.

The one major exception to this north-northwest to south-southeast trend is the north-northeast to south-southwest axis of a large anticlinal structure, visible on the image and very subtly recognizable on the ground, that corresponds to the location of the Ghawar field. This feature is more obvious on the lower area of the image where the outcrop can be traced around the southern nose of the fold. The field is situated within this feature, a broad, low-amplitude fold. A smaller feature just west of the southern nose of the fold is similar in appearance and may be of exploration interest. The anticlines at Dhahran and Bahrain clearly are visible on the image, as is the small dome to the north. These are significant features. There also is a weak circular anomaly in the southeastern corner of the lower image, which also is of interest.

Had the image of the area been available before conducting ground exploration, it would have probably pinpointed the Ghawar feature and other prospects of interest, thus, exploration costs would have been materially reduced and the drilling of the prospective targets advanced.

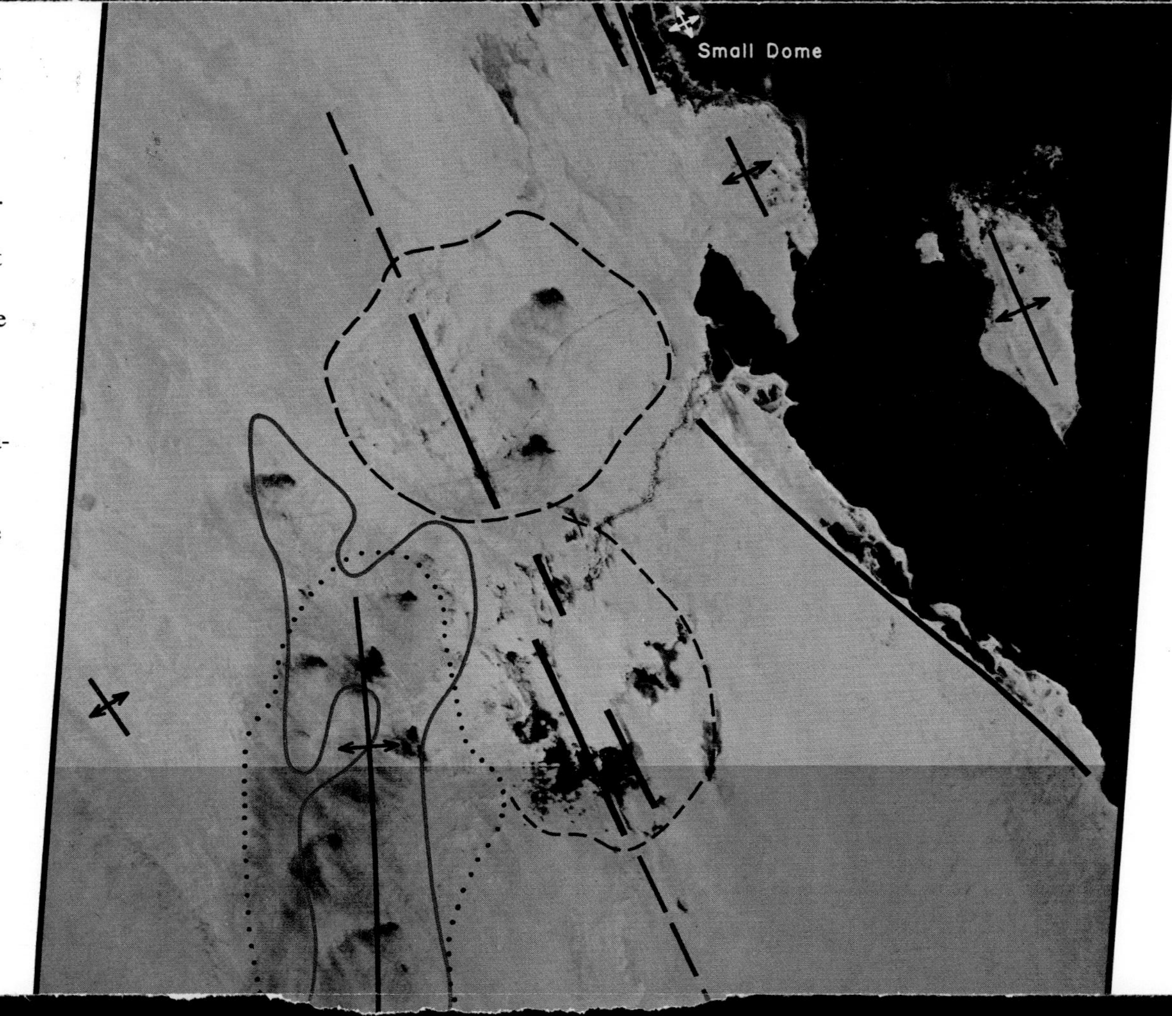

ching 80 million bbl annually in 1972 and over 328 million bbl annually in 1977. Proved and probable reserves in 1977 were estimated at 5 billion bbl with an additional 5 billion bbl estimated as potential reserves.

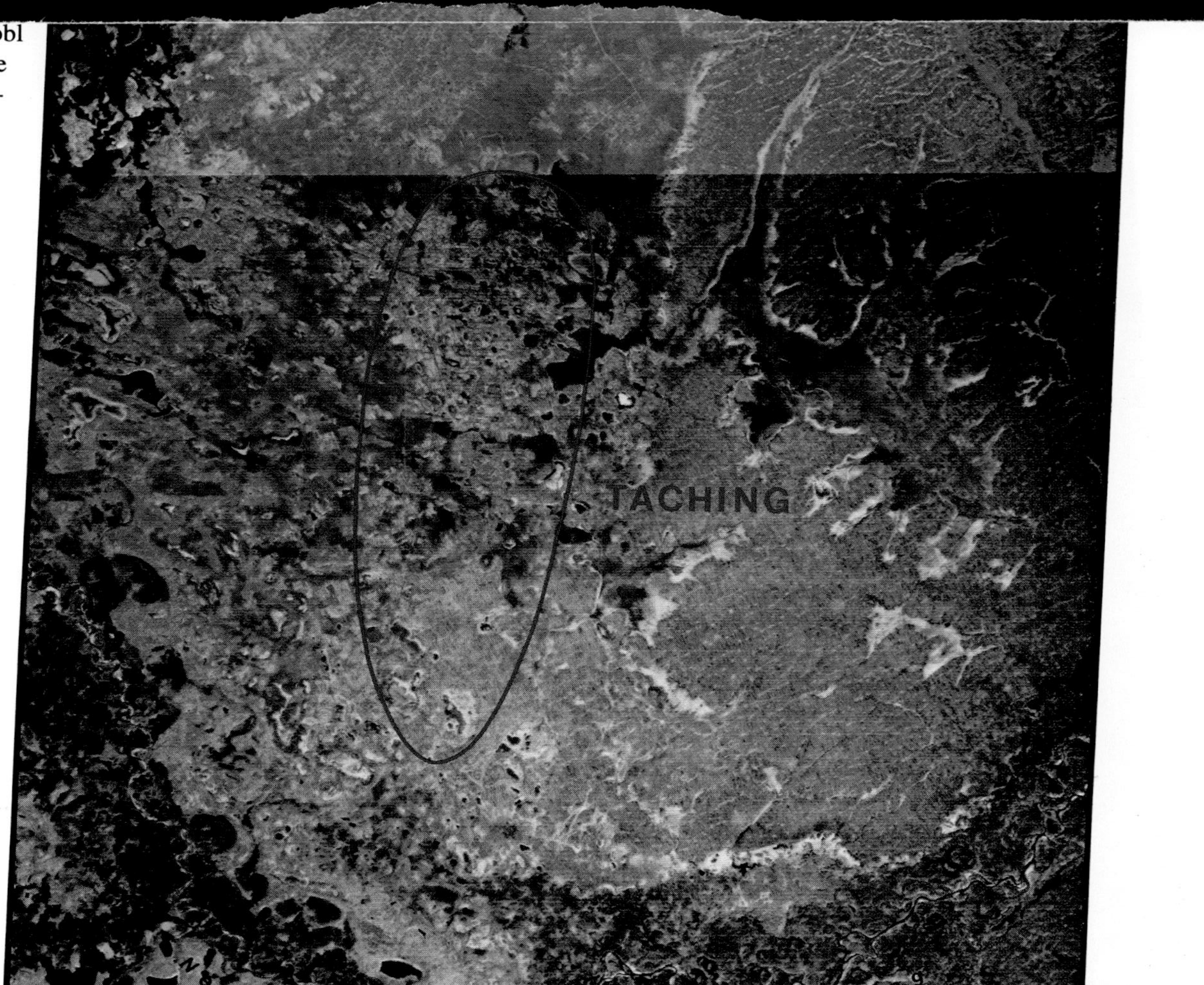

TACHING (DAQING), CHINA

General Field Information

The Taching (Daqing) field is located in the Songliao (Sung-Liao) basin of China. The basin, over 96,526 sq mi (250,000 sq km) in areal extent, was formed by Mesozoic rifting followed by a broader subsidence extending to the present, and contains 19,250 ft (6,000 m) of Mesozoic and Tertiary sediments. Local structures are controlled by deeper horsts and grabens but at productive depth are broad, gently-dipping arches and swells.

The field was discovered in September, 1959, by seismic mapping and a core-hole program. Taching field is a complex of ten separate productive areas, with one having the major part of the reserves. The productive area is 31 to 47 mi (50 to 75 km) long by 12 to 16 mi (20 to 25 km) wide, with 2,000 wells in 1975 and a reported 4,500 wells in 1977. The structure has flank dips of 2° to 5°. The reservoirs consist of 22 separate zones which are composed of fine to coarse-grained Lower Cretaceous lacustrine/fluviatile (nonmarine) sand, and occur between 963 ft (300 m) and 8,021 ft (2,500 m) in depth. Production began in 1961 when 3 million bbl of oil were produced. Production has climbed tremendously since then, rea-

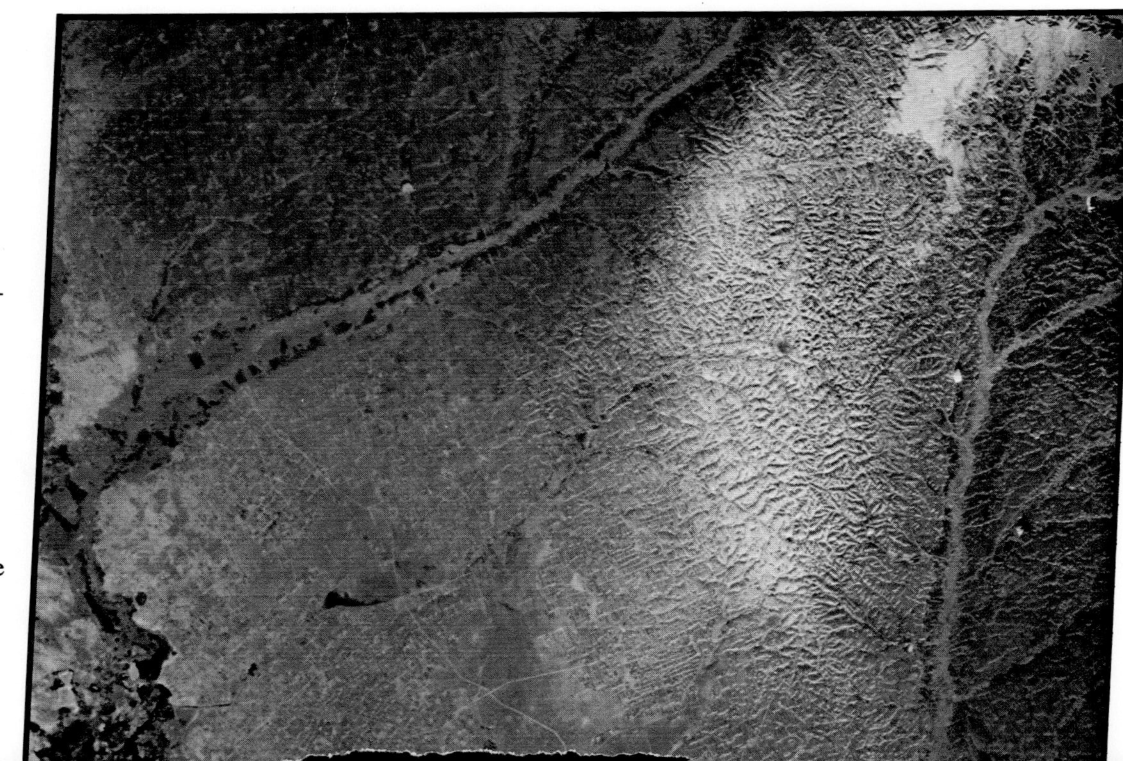

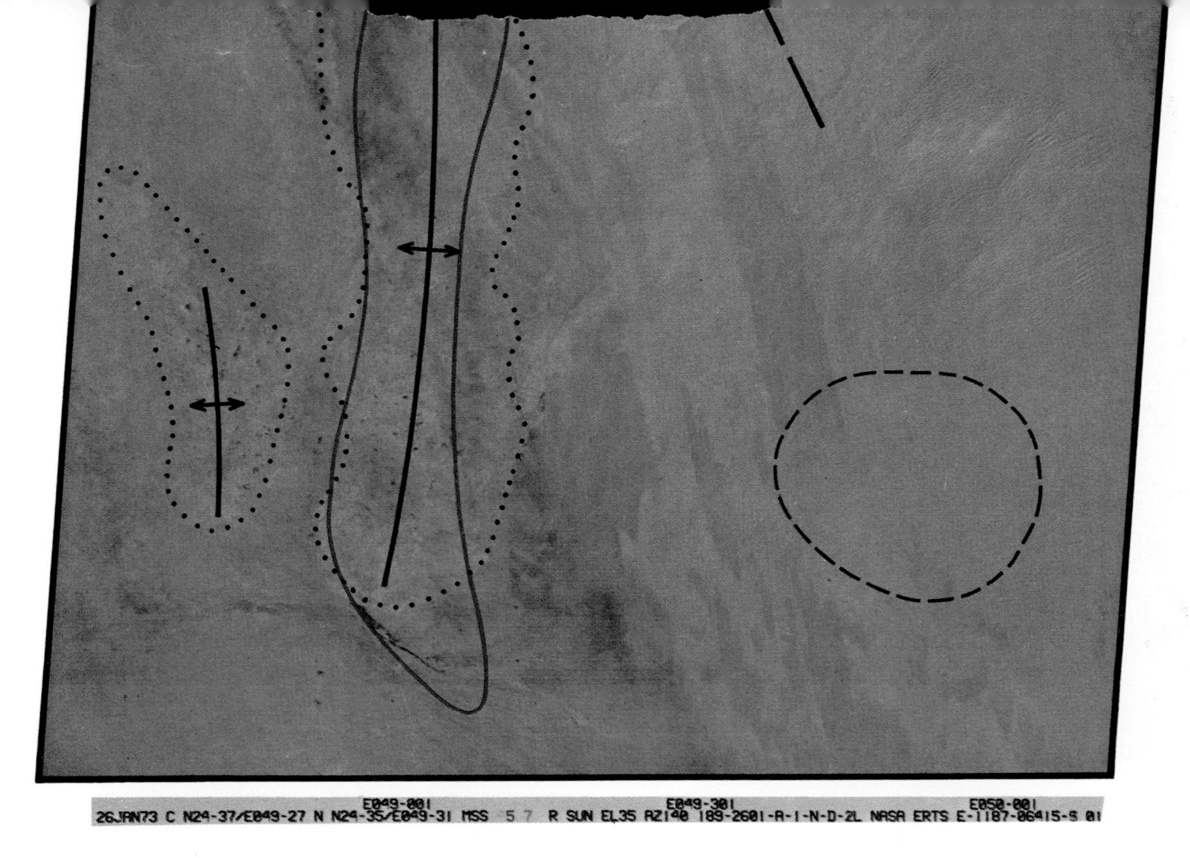
E049-001
E049-301
E050-001
26JAN73 C N24-37/E049-27 N N24-35/E049-31 MSS 5 7 R SUN EL35 AZ140 189-2601-A-1-N-D-2L NASA ERTS E-1187-06415-5 01

FIGURE 2

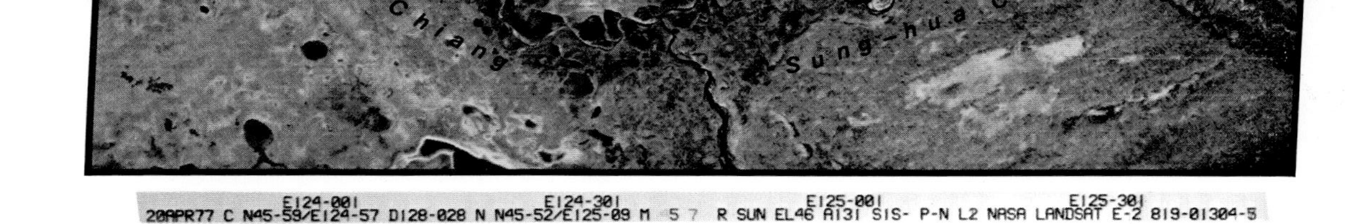

FIGURE 3

Landsat Image Interpretation

The large Taching oil field is located on relatively flat, partly cultivated, plateau as shown in Figure 3. The geologic intepretation of this Lower Cretaceous nonmarine field is shown in Figure 4.

The outstanding geologic feature on the Taching plateau is an elongated arch trending northeast to southwest. The prominent rivers, which are distinct on the image, curve around and almost enclose the area. The lakes are circular on the flat crest of the structure, indicating no preferred direction of slope in contrast to the oriented elongated lakes on the flanks. It is noteworthy that most of the lineaments parallel the axis of the arch, thus indicating a possible graben within the area of the arch with angular or perpendicular lineaments to the flanks of the axis.

A circular anomaly in the center of the arch suggests structural closure within the arch and could be the trap for the field. This area also is the location of the circular lakes. The circular anomaly shown to the north and the two anomalies to the

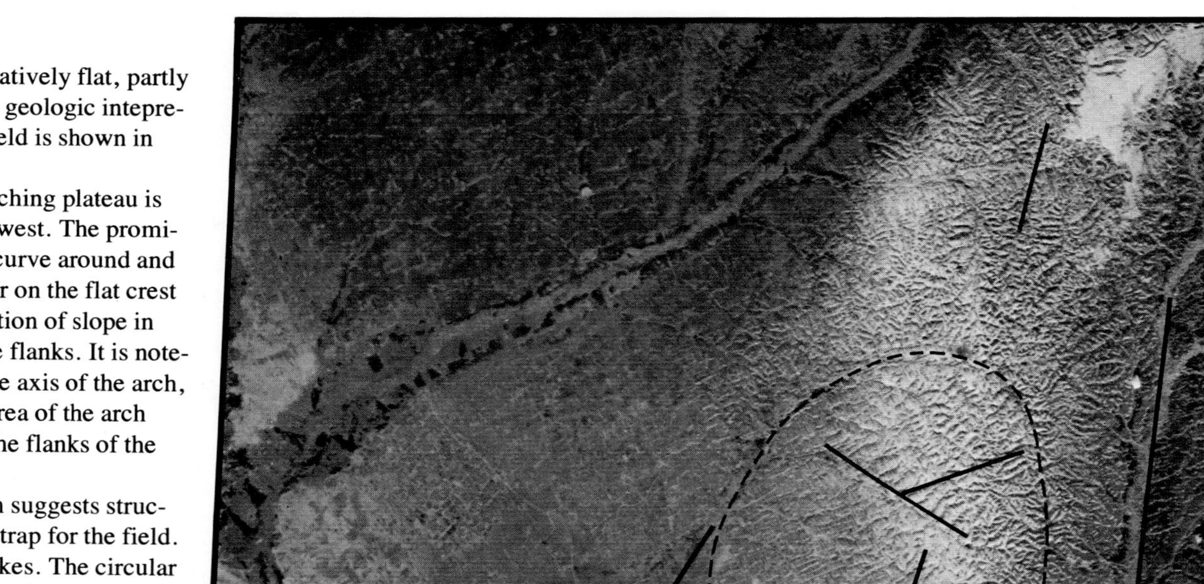

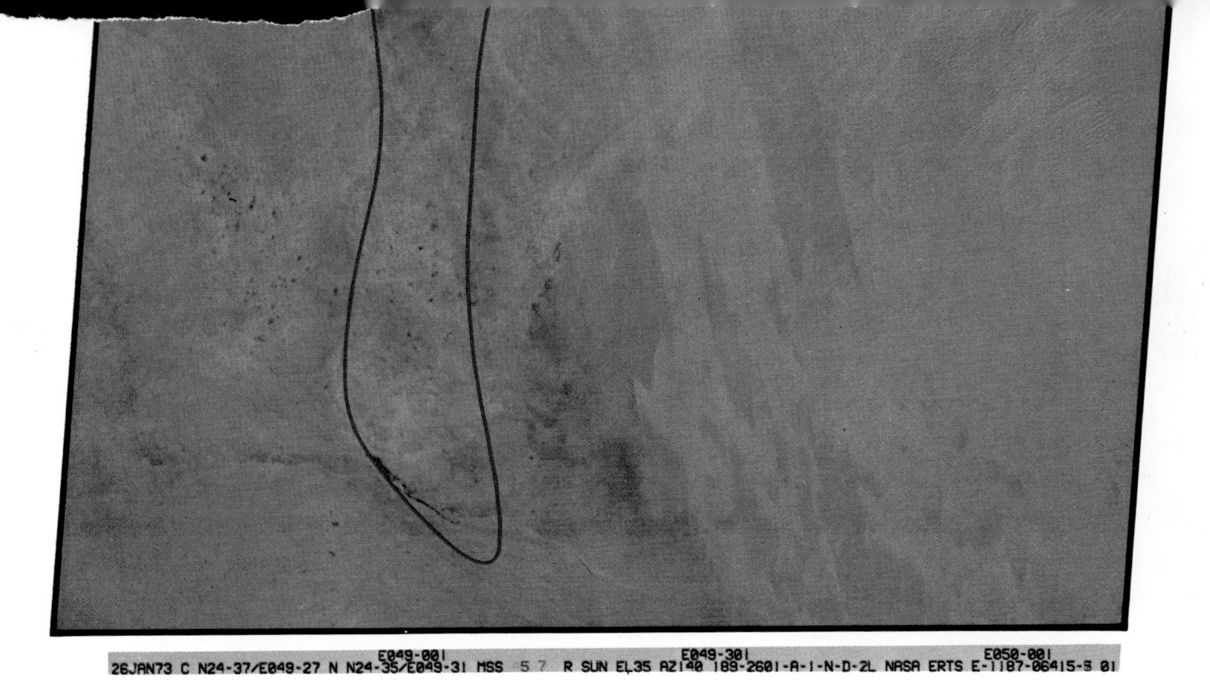

FIGURE 1

Landsat Image Interpretation

The oil field, as shown in the image (Fig. 1), is located on the eastern coast of the Saudi Arabian desert. Vegetation in the area is concentrated within two circular basins and along the

GHAWAR, SAUDI ARABIA

General Field Information

The Ghawar field is located approximately 50 mi (80 km) inland from the western shore of the Arabian Gulf, in Hasa Province, Saudi Arabia. It was discovered in 1948 and covers an area 150 mi (241 km) long and 22 mi (35 km) wide. The oil field is a structural accumulation along the north-south trending, north-plunging, compound (in part), En Nala anticlinal axis. The fold is simple in the south, develops two low marginal crestal closures in the center, and is subdivided into two adjacent anticlines in the north. Seven crestal closures have been found but the oil accumulation is continuous along the entire length of the field. The oil is Upper Jurassic, occurring in the shallow-water carbonate sediments of the upper Jubaila and lower Arab formations. The most prolific production is from calcarenites in which much original pore space survives (Arabian American Oil Company Staff, 1959). The ultimate recoverable reserves are estimated at 75 billion bbl of oil.

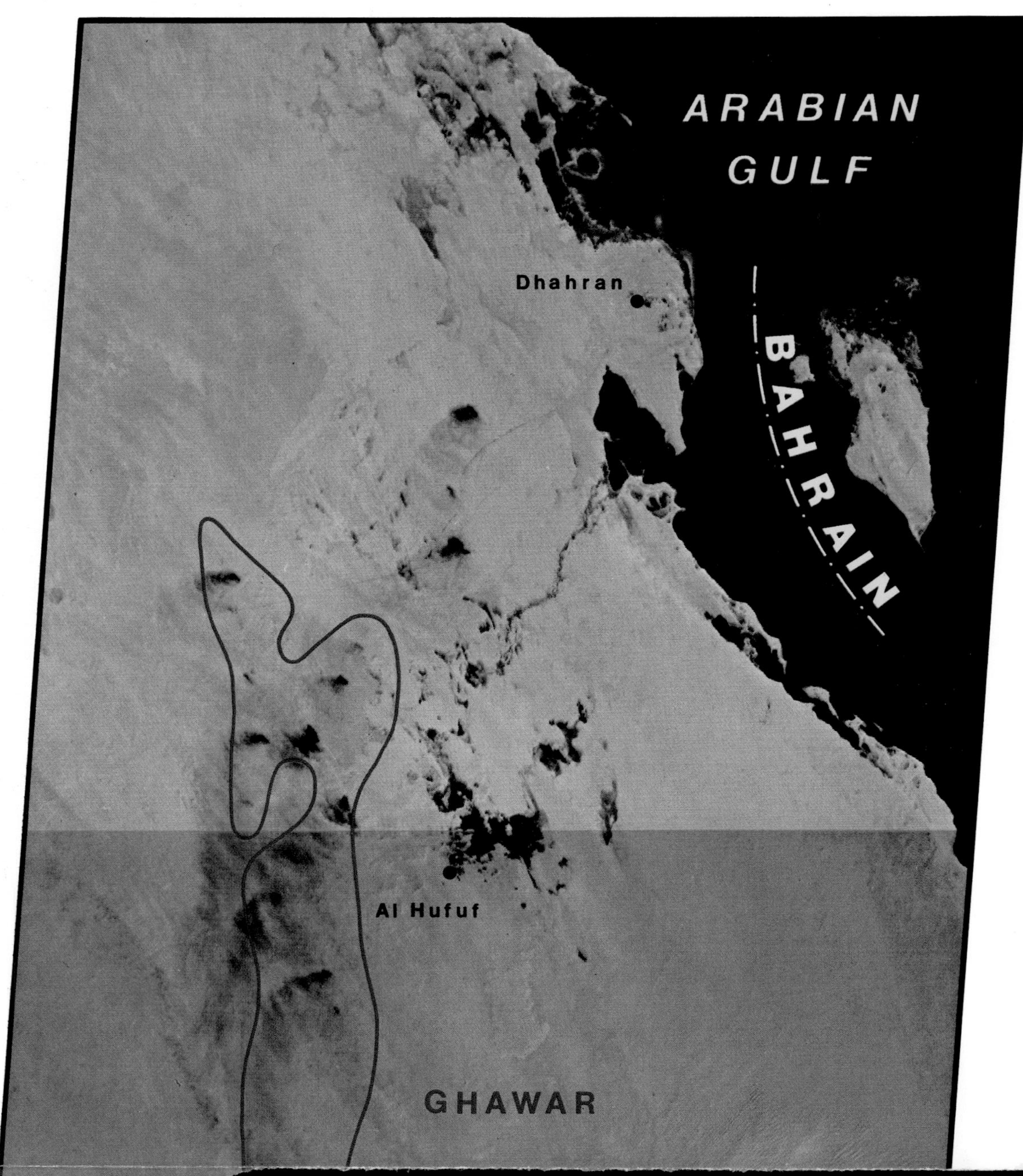

southeast are suspect areas.

If the eastern large lineament is a fault there could be a possible structural trap formed at the juxtaposition of the circular anomalies and the fault.

In summary, this image showing the plateau with its arch and circular features outlined by the rivers would have pinpointed the Taching area for exploration. Therefore, the image would have been of prime assistance to the explorationist had it been available prior to the discovery.

FIGURE 4

LEGEND

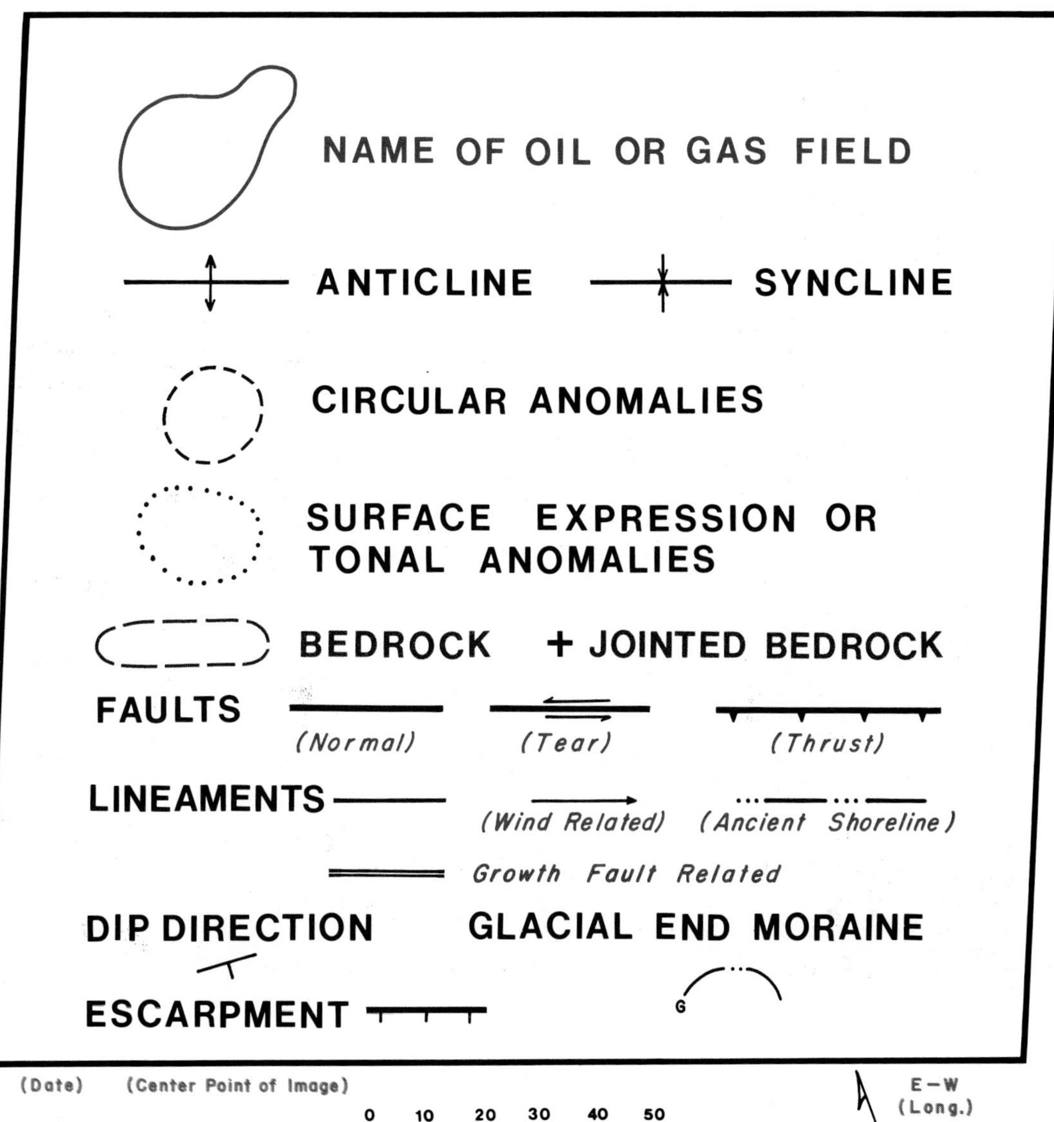
NAME OF OIL OR GAS FIELD
ANTICLINE
SYNCLINE
CIRCULAR ANOMALIES
SURFACE EXPRESSION OR TONAL ANOMALIES
BEDROCK
+ JOINTED BEDROCK
FAULTS
(Normal)
(Tear)
(Thrust)
LINEAMENTS
(Wind Related)
(Ancient Shoreline)
Growth Fault Related
DIP DIRECTION
GLACIAL END MORAINE
G
ESCARPMENT
(Date)
(Center Point of Image)
E-W
(Long.)
0 10 20 30 40 50
km
N

FIGURE 5

KANGAN GAS FIELD, IRAN

General Field Information

The Kangan area is located in the Fars Province of Iran and is within the Zagros foothills thrust belt. It differs, from the northwestwards Disful embayment which holds the main oil reserves of Iran, by its situation on the Fars High which together with Qatar arch have separated the Mesopotamian basin from the Rub Al Khali basin during all Mesozoic time. The discovery in Kangan was made in 1973 by Societe Francaise Des Petroles D'Iran. They acted as the operator for a European consortium (Elf Aquitaine, AGIP, Hispanoil, Fina, OEMV) in a service contract for National Iranian Oil Company. Since then, several other discoveries were made in Fars Province, in structures such as Pars, Dalan, Nar, Varavi, and Aghar, among which Pars and Nar are under development drilling.

The gas-bearing zones in the Kangan field are in the Lower Triassic to Permian Khuff Formation (Kangan and Dalan Formations according to NIOC nomenclature) constituted mainly of heavily fractured dolomitic and oolitic carbonates of intertidal environment; thin layers of anhydrite form the caprock. The Kangan field structure is a compressional folded anticline and has a gas column of approximately 3,000 ft (914 m). The ultimate recoverable gas reserves are estimated at 100 Tcf.

FIGURE 6

Landsat Image Interpretation

The field is located on the western margin of the Zagros Mountains, an intensely folded region in southeastern Iran (Fig. 5). The geologic interpretation of the image is shown in Figure 6.

The producing structure is a large anticline that parallels a major fault to the west. Near the northern nose of this fold a pair of faults dissect the structure and appear to extend westward to the neighboring refolded anticline. The arrow on Figure 6 points to a light-toned outcrop, an exposed salt diapir, between the two faults, which can be clearly seen in Figure 5. An unexposed diapir would account for the disruption of the strata on the northern end of the refolded anticline.

The image clearly indicates that the entire land area has been subjected to compression and folding and that erosion has exposed the upper parts of the folds. Displaced anticlinal flanks mark the location of other faults in the image area. Neighboring images indicate faulting increases to the east. Further image analysis indicates that there is an abundance of possible petroleum traps such as anticlines, diapirs, and faults.

Landsat would have been most useful in the initial exploration of this area. It would have aided the field geologist in tracing the resistant strata over long distances; the geophysicist in locating lines to cut across all the major structures with the least amount of topographic interference; and would have been encouraging to the wildcatter to drill on the Kangan structure, which, after all, is the largest anticline on the scene.

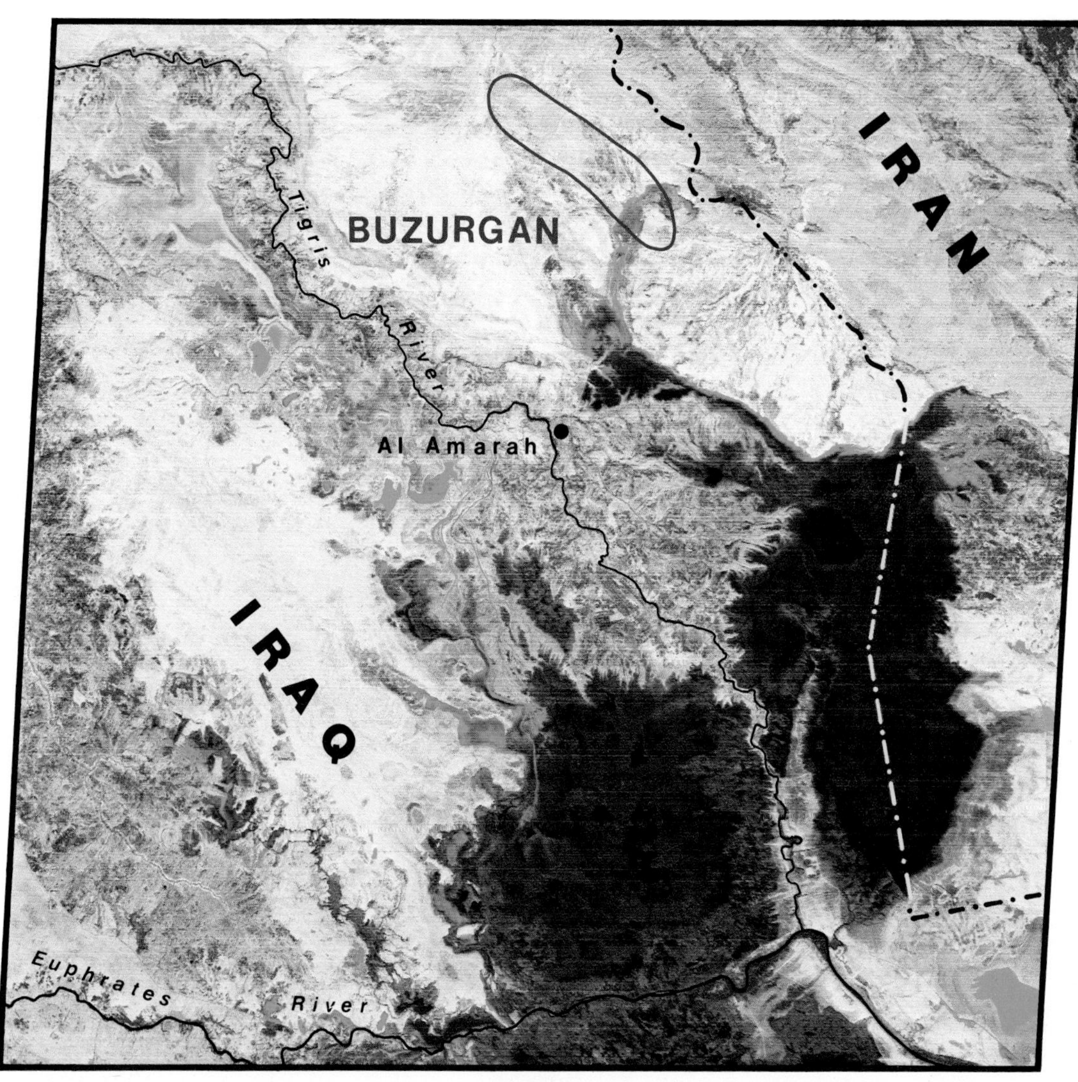

FIGURE 7

BUZURGAN, IRAQ

General Field Information

The Buzurgan oil field, Iraq, was discovered in January 1970. The field, which covers an area 31 mi (50 km) long and 5 mi (8 km) wide, is part of the Zagros foothills thrust belt. The field is located on a prominent anticline. The areal closure is 155 sq mi (246 sq km) and the vertical closure is 400 ft (122 m). The oil column is estimated to be approximately 500 ft (152 m). The major oil production is from the Mishrif limestone formation, middle Cretaceous, at depths of approximately 10,500 ft (3,200 m) and the minor oil production is from the Zubair sandstone formation, Lower Cretaceous, at approximately 13,000 ft (3,962 m). The depositional environment is upper shelf for the middle Cretaceous formation and shelf for the Lower Cretaceous formation. Ultimate recoverable reserves are placed at 1.1 billion bbl of oil.

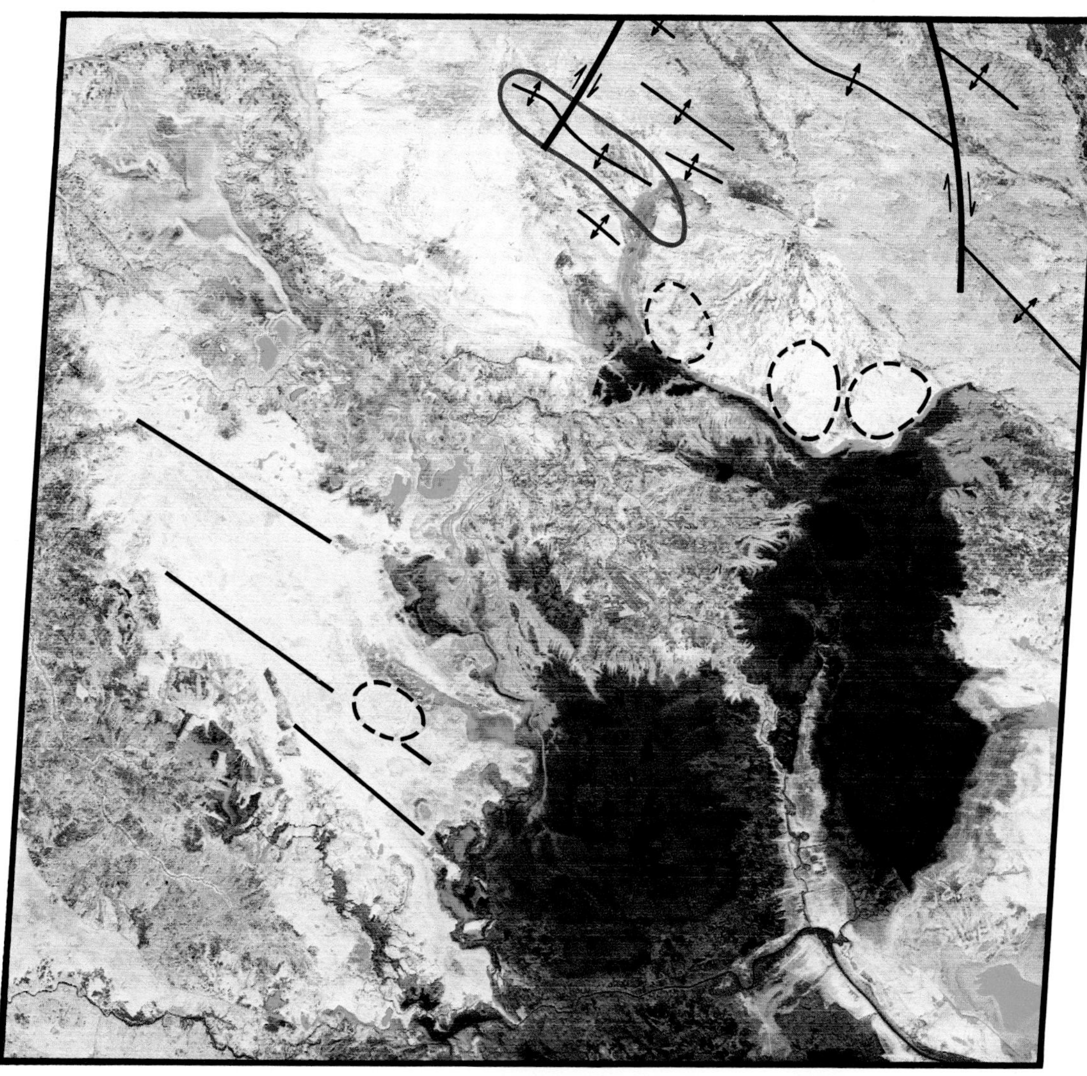

E046-301 E047-001 E047-301
15FEB75 C N31-40/E047-11 N N31-40/E047-10 MSS 5 7 R SUN EL34 AZ137 190-0328-N-1-N-D-2L NASA ERTS E-2024-06453-5 01

FIGURE 8

Landsat Image Interpretation

The Buzurgan field (Fig. 7) is located on the western margin of the Zagros Mountains. The delta of the Tigris River covers most of the image area, but the foothills of the Zagros are visible in the northeastern corner. The geologic intepretation is shown in Figure 8. Folds in the bedrock area trend north-northwest to south-southeast. Offset within the folds suggests a later faulting with right-lateral strike-slip displacement. This strike-slip displacement may be a fault within a thrust sheet whose leading edge has been covered by alluvial deposits.

The anticline that provides the structural trap for the Buzurgan field is only one of many anticlines in this area that would be of interest to the explorationist. The anticline is cut by a fault with apparent right-lateral displacement. Determining the location of faults like this one, that forms a seal on one side of reservoir, is of vital importance to the production and development geologist. Landsat imagery is useful in the exploration stage in locating these potential traps.

The circular anomalies are tonal features which indicate some structural relationship. The lineaments shown in the west-central area of the figure probably are structurally influenced. In the center of this image, just west of Al Amarah, there is a major stream deflection and anomalous drainage pattern related to the Tigris River. The image affords an excellent overview of the geomorphology which would be significant in initial exploratory efforts in this area.

Had the image been available, it would have been a useful tool in the exploration and discovery of the Buzurgan field and in delineating the many other anticlines and geologic features in the image scene.

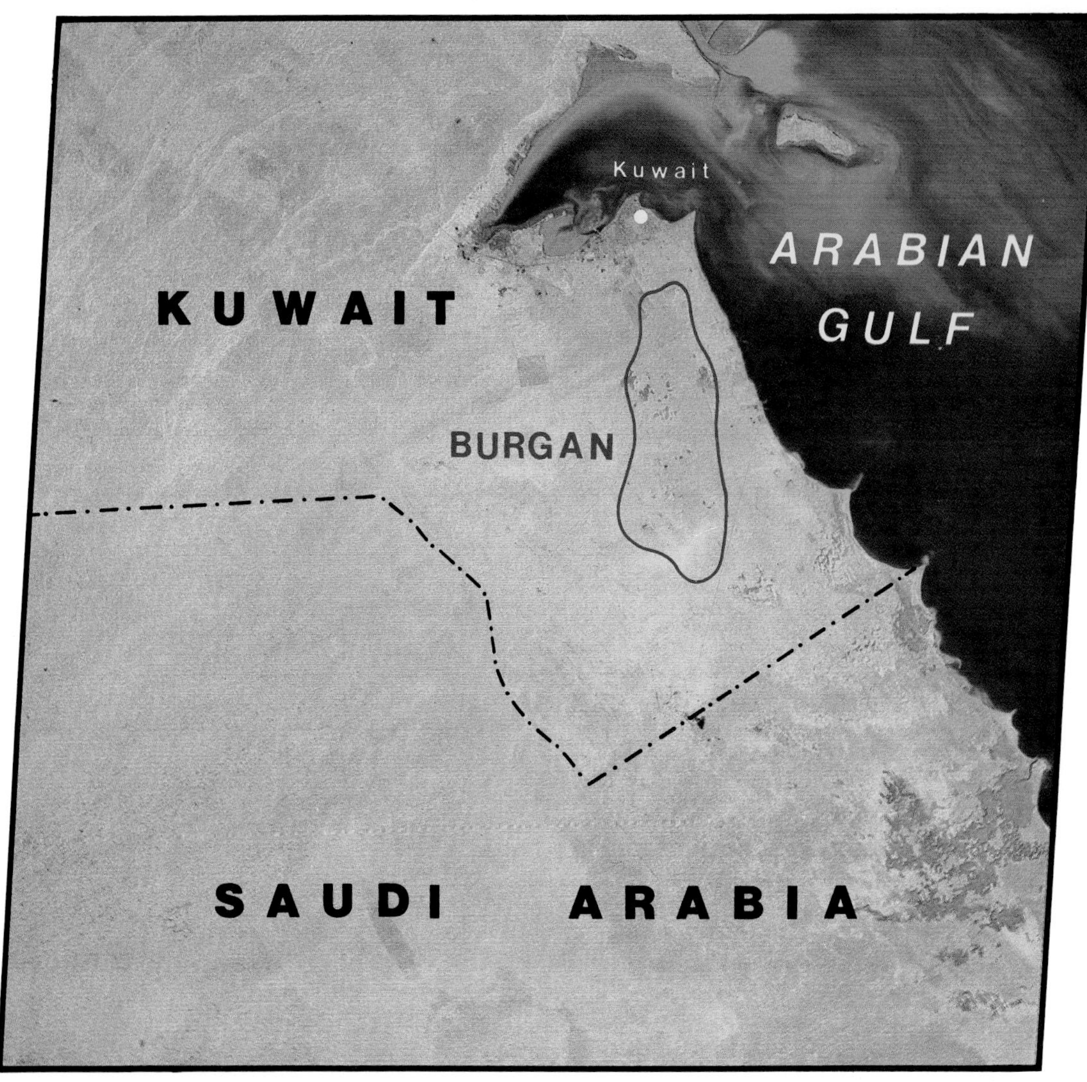

FIGURE 9

BURGAN (GREATER BURGAN), KUWAIT

General Field Information

Burgan is located approximately 28 mi (45 km) to the south of Kuwait Town and 14 mi (23 km) west of the coast of the Arabian Gulf. The Burgan structure is anticlinal. The crustal depth is approximately 3,600 ft (1,097 m). Three oil fields, Burgan, Magwa, and Ahmadi, all are interconnected by continuous oil-bearing zones and collectively are referred to as the Greater Burgan field (Davis, 1970). The producing sand sequence is of middle Cretaceous age. The net thickness of the sequence is more than 1,000 ft (305 m). The producing zones are highly porous and permeable. The depositional environment is considered to be shallow-water shelf, probably on the edge of a delta. Ultimate recoverable reserves are estimated at 66 billion bbl of oil.

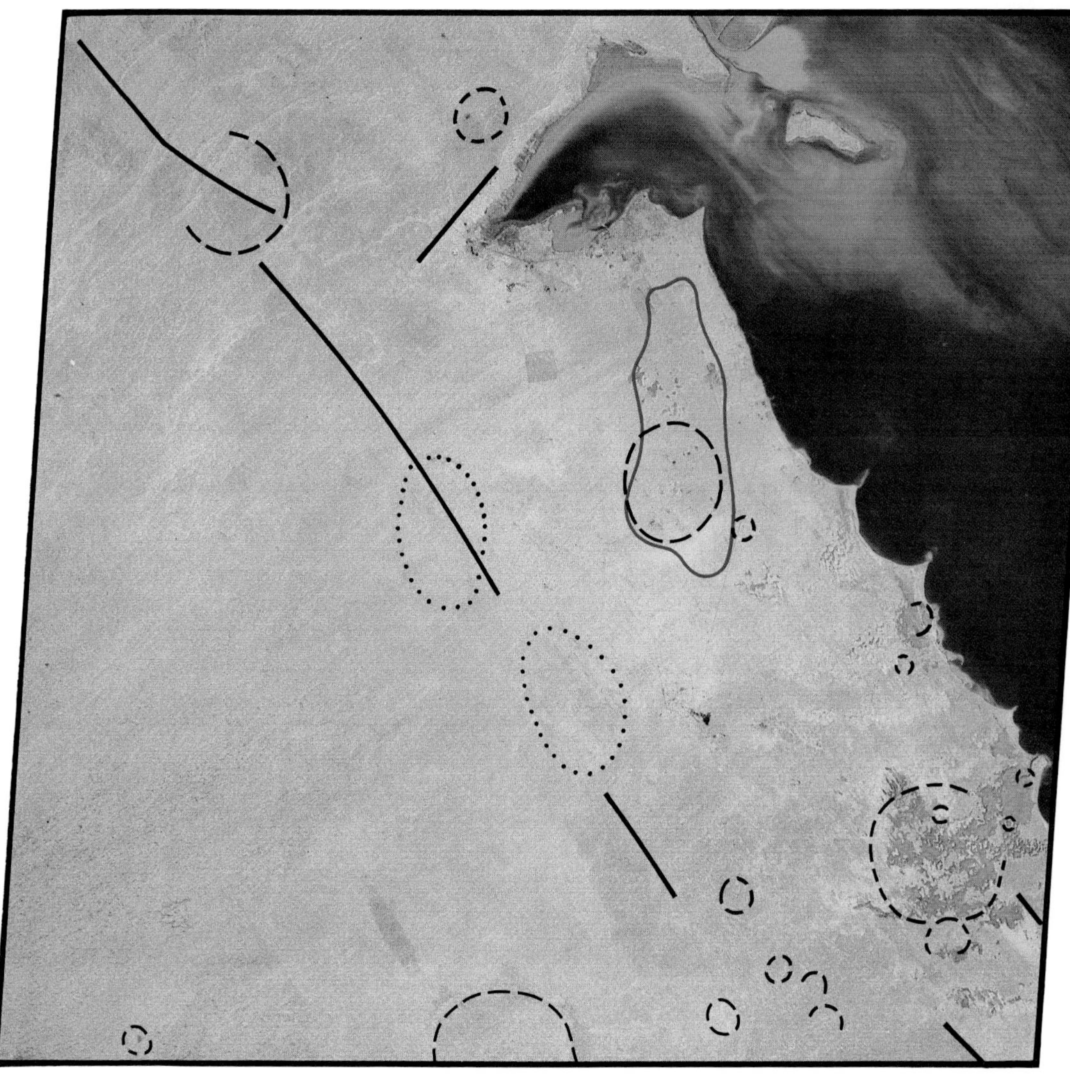

FIGURE 10

Landsat Image Interpretation

The location of the oil field is shown in Figure 9. The surface area on which the Burgan field is located lacks the image surface expression that highlighted the Ghawar structure.

Figure 10 is the geologic interpretation of the image. The north-northwest to south-southeast linear trend observed in the Ghawar image is somewhat similar in this scene, as evidenced by tonal changes related to sand thickness and moisture differences, and abrupt changes in drainage direction. These suggest that in the Burgan area the linear trend is a low, gentle, east-facing escarpment.

There are four types of anomalies visible on the image: (1) The smallest anomalies are similar in size to salt domes found elsewhere in the Persian Gulf area. With one exception, they are located east of the escarpment in the relatively sand-poor coastal zone. An intensively cultivated lowland adjacent to the Gulf forms a larger anomaly. (2) Two anomalies have a subtle mottling that may indicate a slight topographic high. However, one of these is located within the Burgan field and may be due to industrial land use. (3) The rectangular area in the south-central area of the image appears to be a land-use pattern. The boundaries appear too symmetrical, in contrast to the rest of the scene, to be natural. The circular features within this area may be land-use and/or topographically related. (4) A lineament is offset where it meets the northernmost circular anomaly. Although color variations in the sand tones delineate the anomaly, the offset suggests possible structural origins.

There is a tonal variation on the image, which reflects the outline of the existing field, which initially would have been observed as a feature to be explored. There are two other tonal, or surface expression, anomalies along this north-northwest to south-southeast trend which are outlined with a dot pattern: one to the west of Burgan and the other to the south-southwest of the field.

If used prior to exploration, this image would have helped pinpoint features for geophysical surveys.

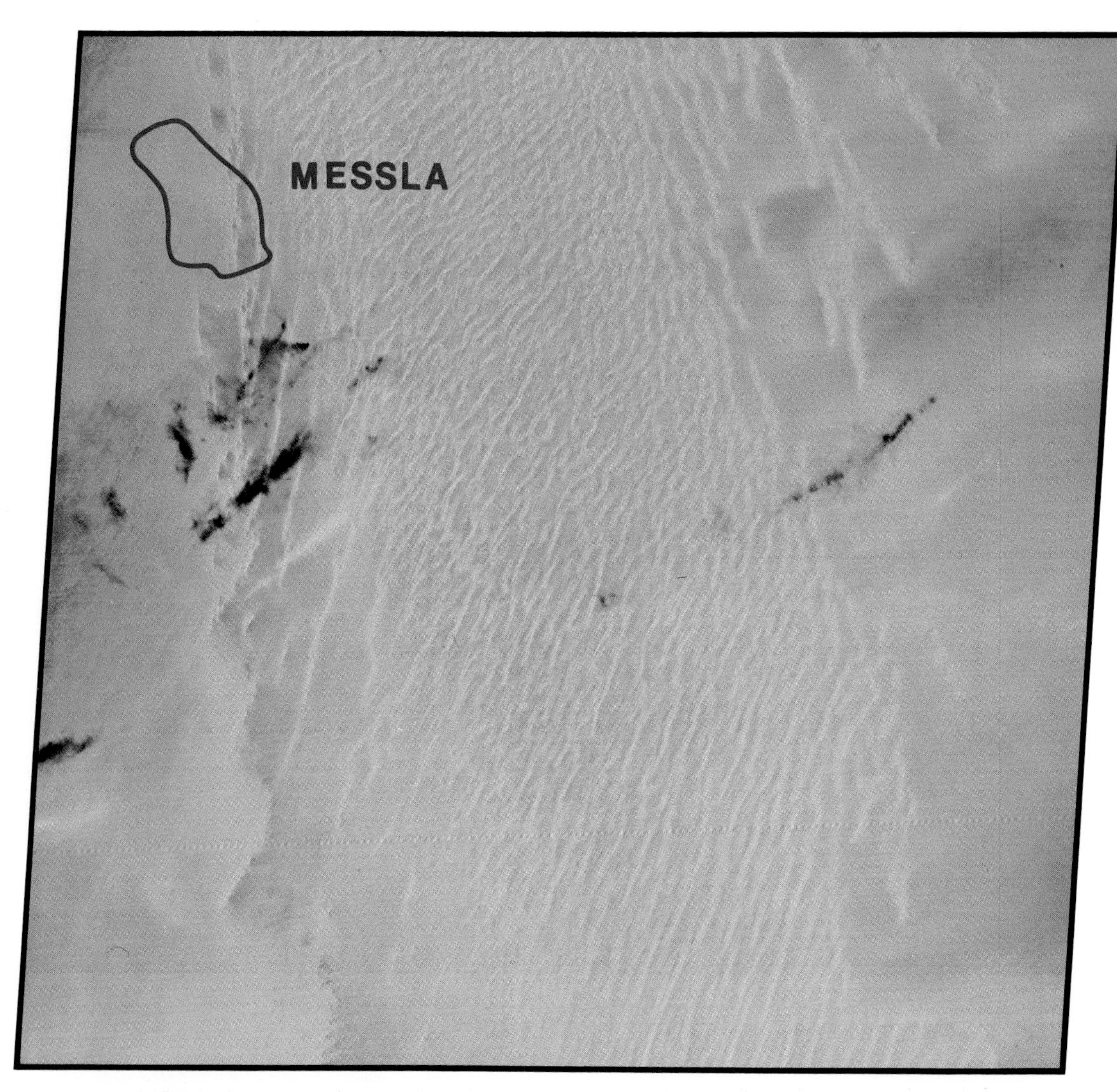

FIGURE 11

MESSLA, LIBYA

General Field Information

The giant Messla oil field, discovered in 1971, is located in the southeastern part of the Sirte basin, approximately 311 mi (500 km) southeast of Benghazi, Libya, in which 20 giant oil fields have been found to date. The field is a huge stratigraphic trap. Other giant fields in the basin produce from closed structures or carbonate "buildups."

The discovery well was located on a seismic-controlled south-easterly plunging nose. The location was based on the geologic concept that the Lower Cretaceous Sarir Sandstone wedged out toward the west and northwest. In this direction a prominent Sarir bald basement high was encountered by earlier wildcats. The field covers over 97 sq mi (250 sq km), with average productive oil-bearing sandstone thickness of 100 ft (30 m) and average productive depth of 8,800 ft (2680 m; Clifford et al, 1979).

The Messla trap, caused by a westward wedge-out of the Sarir Sandstone (apparently due to truncation by a basin-wide unconformity at the base of the capping Upper Cretaceous marine shales), is similar to the East Texas field trap. The East Texas field, containing approximately 9 billion bbl of oil in place, was formed by a wedge-out of the Cretaceous Woodbine sandstone, due to truncation by a regional unconformity, on the westward-dipping west flank of the Sabine uplift. Ultimate recoverable oil reserves are estimated at 3 billion bbl.

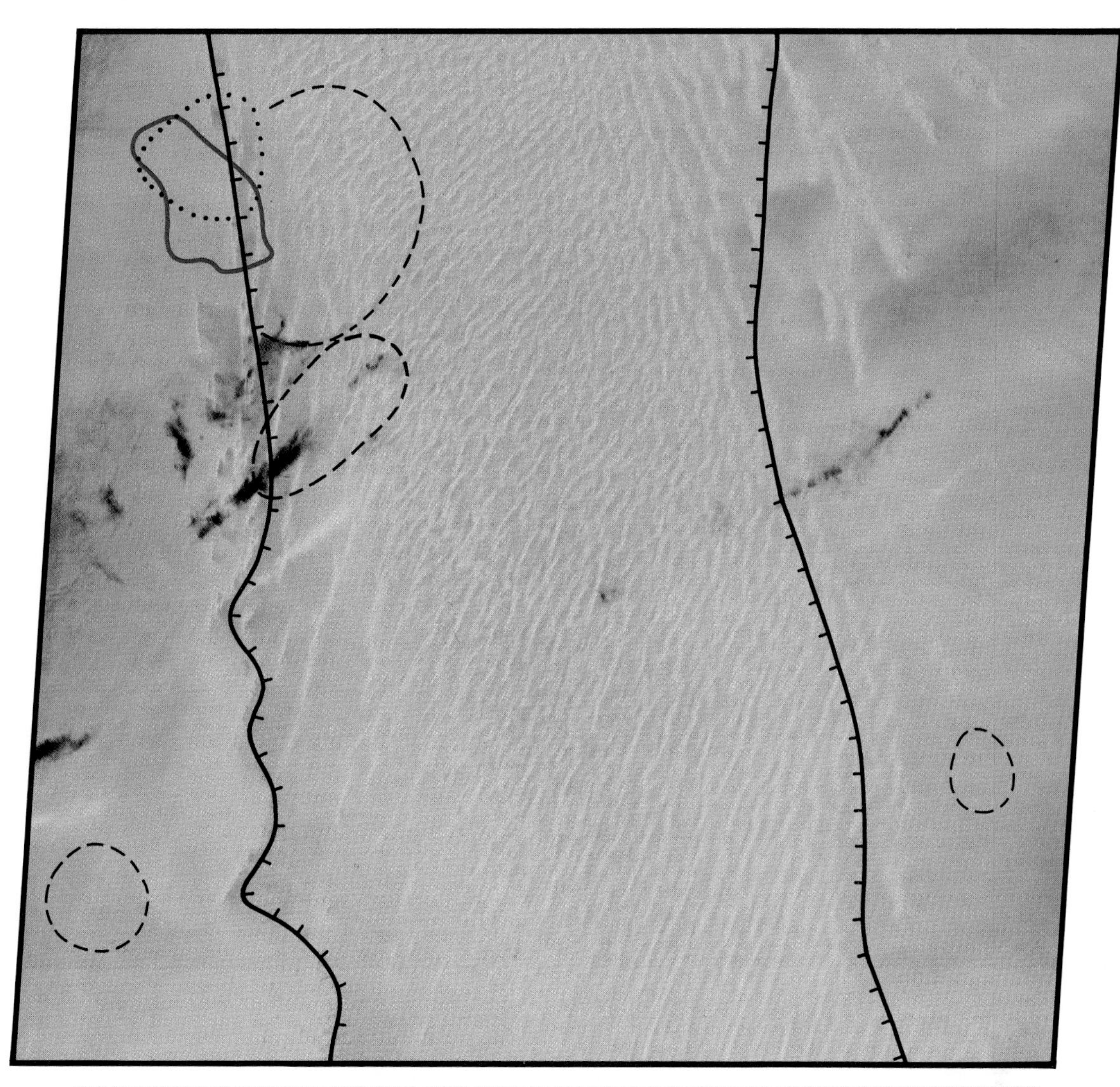

FIGURE 12

Landsat Image Interpretaton

This oil field is located on the western margin of the Sand Field of the Libyan Desert (Fig. 11). Eolian landforms dominate the scene. The central area contains longitudinal dunes, indicating a relatively sand-rich area. To the east and west, flat sand sheets indicate relatively sand-poor areas.

The boundaries between these eolian landforms follow a north to south trend which may be related to bedrock structure. In the southern area of the western boundary, subtle shadowing highlights a small east-facing escarpment. The Messla field occurs along this western escarpment. If the escarpment represents the surface epxression of a fault, migrating oil trapped at the fault plane could be the producing medium for the field.

A geologic analysis of this Landsat image (Fig. 12) suggests that the central area is a topographic low, perhaps formed by a graben structure. If the Messla field is controlled structurally by faulting along a graben margin, other fields also may exist along the margin. Geophysical lines should be run perpendicular to this trend to determine actual structure in other locations.

The small tonal anomaly located (in part) on the Messla field is possibly a land-use area related to its development. Careful examination of the image reveals a very subtle circular anomaly immediately east of the oil field. The size of this circular anomaly compares with the one to the north of the Ghawar field in Saudi Arabia—very large. There is a circular anomaly south of Messla which looks significant. To the southwest and southeast there are two more small circular anomalies which may be due to land usage, but which should not be overlooked from an exploration standpoint. The dark elongated pattern south of the field has been determined to be a cloud shadow, as is the dark elongated pattern to the east of the escarpment.

It is highly doubtful that the image alone would have indicated the location of the Messla field. However, it would have prominently shown the boundaries between the eolian landforms and the pronounced western escarpment. An explorationist could have used the image to plan geophysical activity along the edge of the escarpment, to determine subsurface structural significance, which could eventually have detected the anomaly of the field. The entire central longitudinal dune area also would be considered suspect for intensive geophysical exploration.

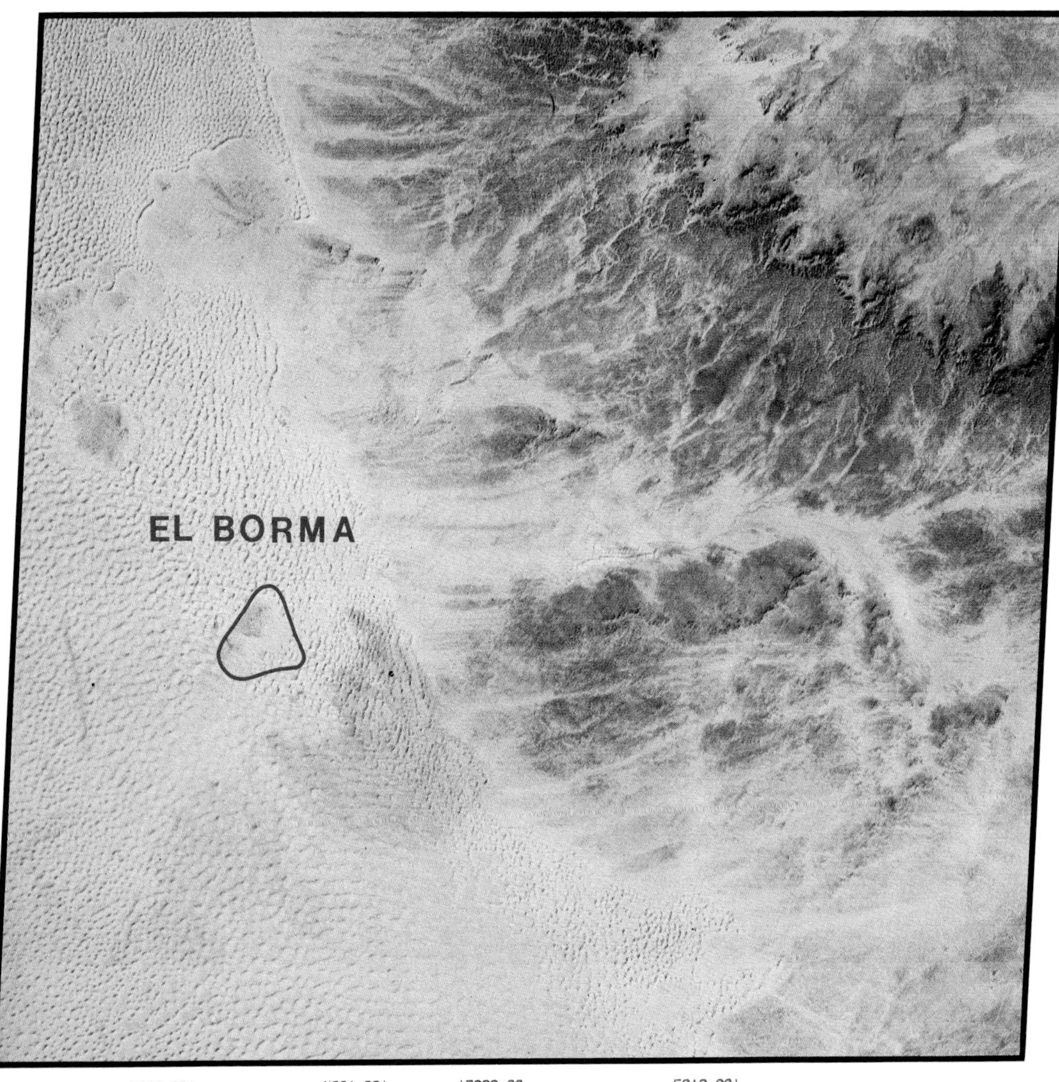

FIGURE 13

EL BORMA, TUNISIA

General Field Information

The El Borma field is located on the boundary between Tunisia and Algeria in northern Africa on the Saharan platform, Ghadames basin. Its discovery in October 1964 followed a seismic survey of the gently-dipping Upper Cretaceous surface structure. The discovery well was drilled to a total depth of 13,634 ft (4,248 m) in the Cambrian. The field is a faulted anticlinal structure with the producing section in the Lower Triassic Kirchaou sandstone formation. The top of the pay interval is 7,700 ft (2,400 m) with a net pay of 148 ft (46 m) in a gross interval of 241 ft (75 m). The sandstone has an average porosity of 19% and permeability of 500 md.

El Borma field began production in May of 1966 at 20,000 b/d and increased to 100,600 b/d in 1972 (75,600 b/d in Tunisia and 25,000 b/d in Algeria). Peak production was achieved in 1970 with 88,000 b/d from Tunisia only, declining to 62,700 b/d (Tunisia and Algeria) in 1978.

The combined total reserves are estimated at 750 million bbl of ultimately recoverable oil plus additional natural gas liquids.

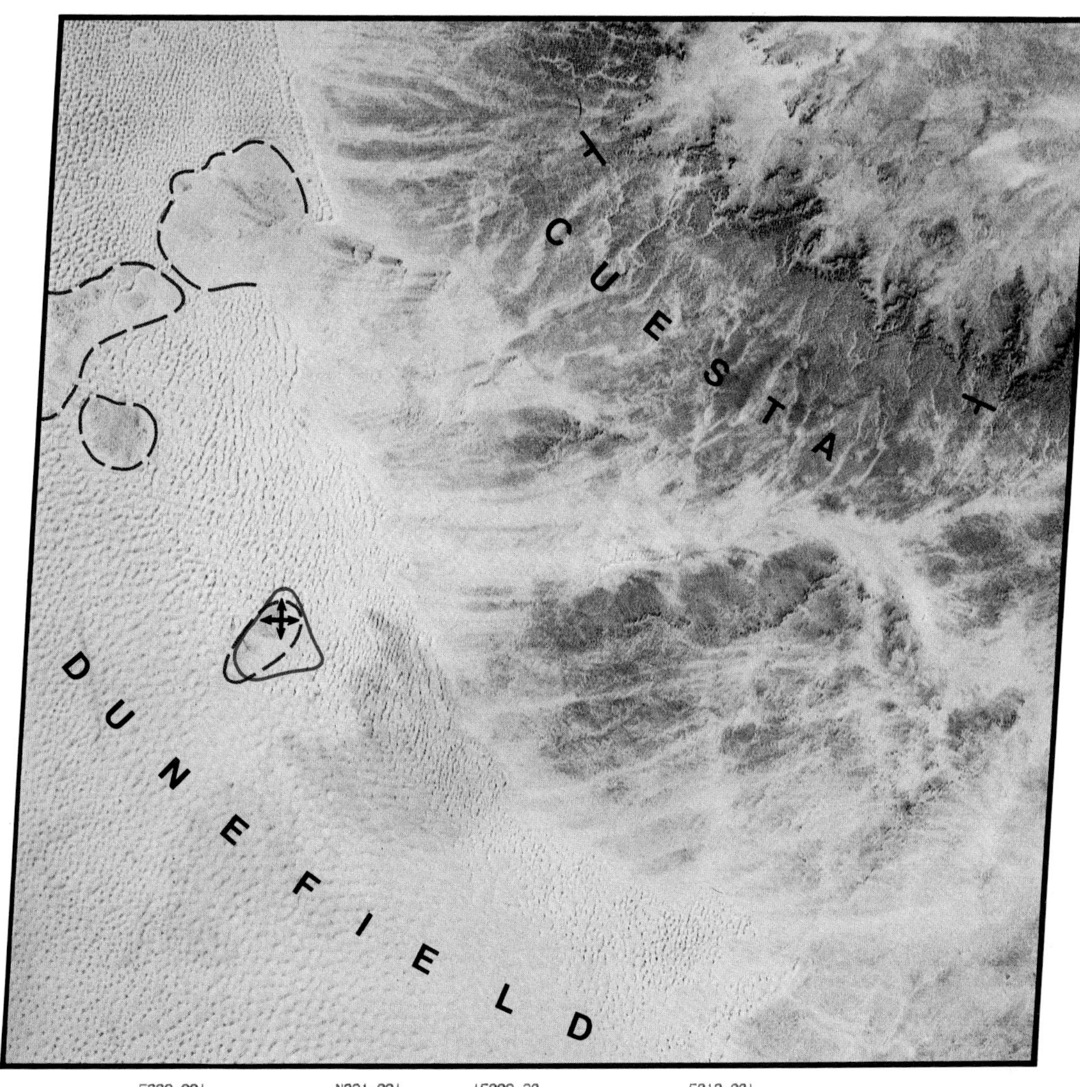

FIGURE 14

Landsat Image Interpretation

The Eastern Great Erg, a vast sand field in the north-central Sahara Desert, covers all but a few isolated bedrock outcrops in the western third of the image (Fig. 13). The El Borma field is located approximately in the center of the image's dune field.

Gently-dipping resistant sedimentary strata form a dissected cuesta to the east of the image (Fig. 14). The zig-zag appearance of the cuesta escarpment was produced by headward erosion of streams flowing to the northeast. The shifting direction of the cuesta's dip-slope drainage from northeast to southwest across the image area suggests a structural high, such as a dome or a low amplitude fold to the northeast.

Inspection of neighboring images shows that the cuesta resulted from erosion along a curved area of the Mediterranean coast that appears to be of regional fault or flexure, rather than fold, origin. The El Borma field is associated with one of the isolated bedrock outcrops in the dune area. Although the outcrop's elevation may be strictly due to a local, lithologically related erosional resistance, a circular pattern at its crest may indicate a small structural dome, as shown by the small anticlinal symbol, within the dotted area. If this is the case, the other bedrock outcrops to the northwest could be promising wildcat areas.

This image would have highlighted the isolated bedrock outcrops in the dune area which would have warranted detailed geophysical surveys and the resulting mapping and drilling of the Borma anomaly.

FIGURE 15

PRUDHOE BAY, ALASKA

General Field Information

The Prudhoe Bay field, the largest oil field in the United States, is located on the eastern extension of the Barrow arch and at the eastern end of the North Slope, Alaska, Prudhoe Bay area. The discovery of the field in April of 1968 resulted from geologic surface work and seismic surveys. The geology of the North Slope is characterized by complex structural and stratigraphic relationships. The sedimentary section contains many zones with reservoir potential and ranges in age from pre-Devonian through Tertiary. Productive zones range from Mississippian to Jurassic in age and the most important reservoirs are sandstones belonging to the Sadlerochit Formation of early Triassic age.

Hydrocarbon accumulations are controlled by both a westward-plunging faulted anticline and the unconformable Early Cretaceous Unnamed shale which truncates the Sadlerochit Formation in the eastern part of the field (Jamison et al, 1979). These accumulations are encountered in the Sadlerochit reservoir at subsea depths ranging from 8,000 ft (2,438 m) to 9,000 ft (2,742 m). The oil column reaches a maximum of 460 ft (140 m) within the Sadlerochit sandstones. The reservoir, with an areal extent of approximately seven townships, contains hydrocarbon volumes of 21.2 Tcf (standard) of gas-cap gas, 13.9 Tcf (standard) of solution gas, 729 million bbl (stock tank) of gas-cap condensate, and 19.1 billion bbl (stock tank) of oil (Alaska, Division of Oil and Gas, 1974).

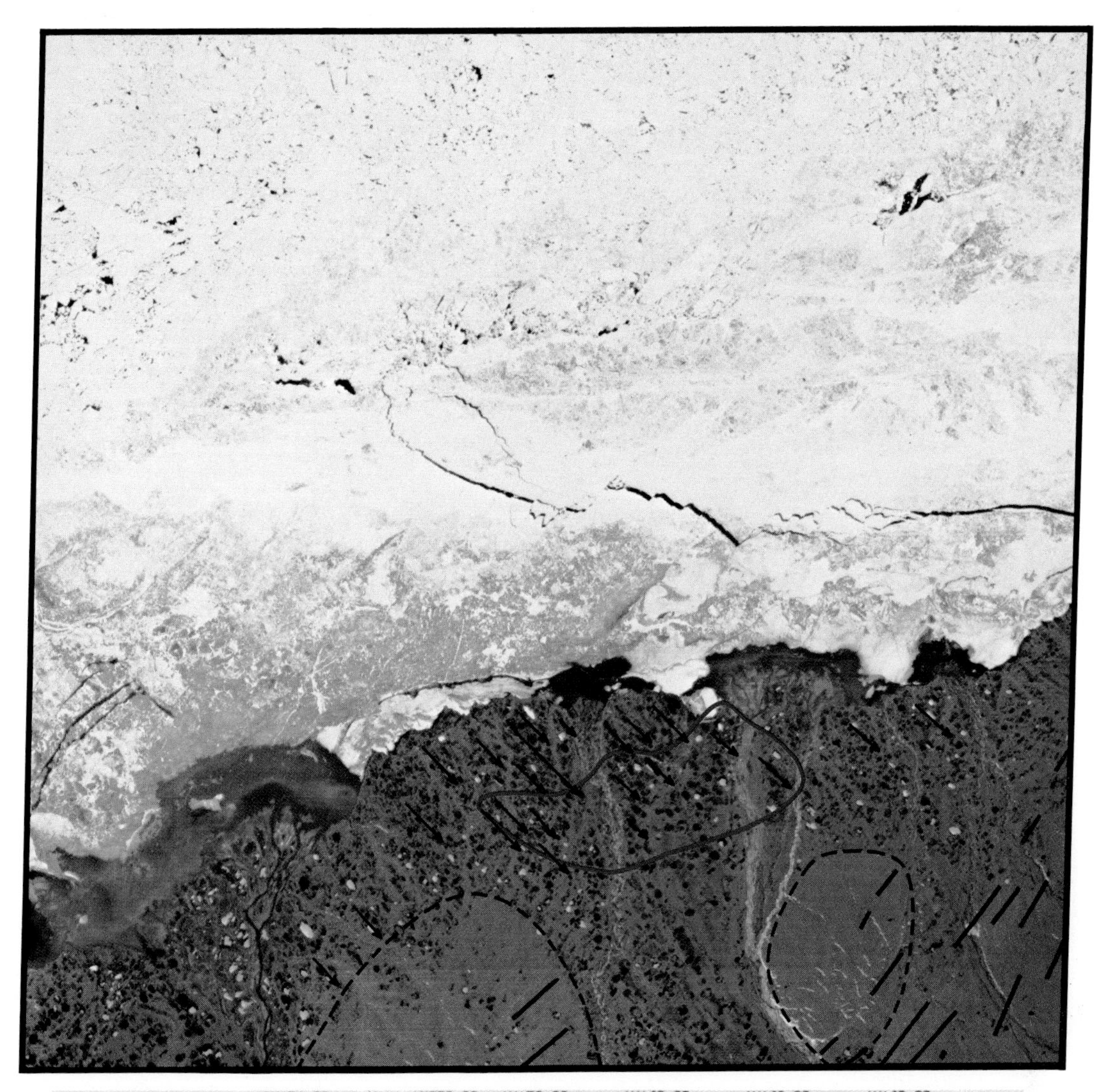

FIGURE 16

Landsat Image Interpretation

In contrast to previous images, the image of Prudhoe Bay field is located in an arctic, or high latitude desert (Fig. 15). The ground in the Prudhoe Bay area is permanently frozen to a depth of approximately 2,133 ft (650 m); this is the thickest known permafrost in Alaska. High reflectance in the infrared part of the spectrum from tundra grasses and sedges give the land area a uniform red tone on the false-color composite. Two lighter-toned hill areas exhibit a homogeneous texture and better drainage than surrounding river delta areas. Therefore, the surficial material is probably a uniform sand or silt deposit. Eolian dunes to the west of the image area tend to support this theory. Two major linear trends are visible in the image (Fig. 16). In the northeast to southwest trend, subsequent drainage developed parallel with bedrock strike. One trend continues southwest for over 200 mi (322 km). In the northwest to southeast trend, elongation of thaw lakes is perpendicular to the prevailing wind direction.

Studies of the image area have suggested that growth faults along the North Slope are reflected in the alignment of permafrost lakes (location rather than elongation). However, in the Prudhoe Bay area no alignments were noted.

Although there are no obvious surface indications of the oil field, Landsat imagery would have been useful in this area for synoptic geographical mapping, base mapping for setting geophysical lines, and monitoring ice movements. Also, the fact that the field lies (relatively) between the two lighter-toned areas might have been of some significance to the explorer had the image been available prior to geophysical survey. The image could have been useful in laying out seismic lines and other surface work.

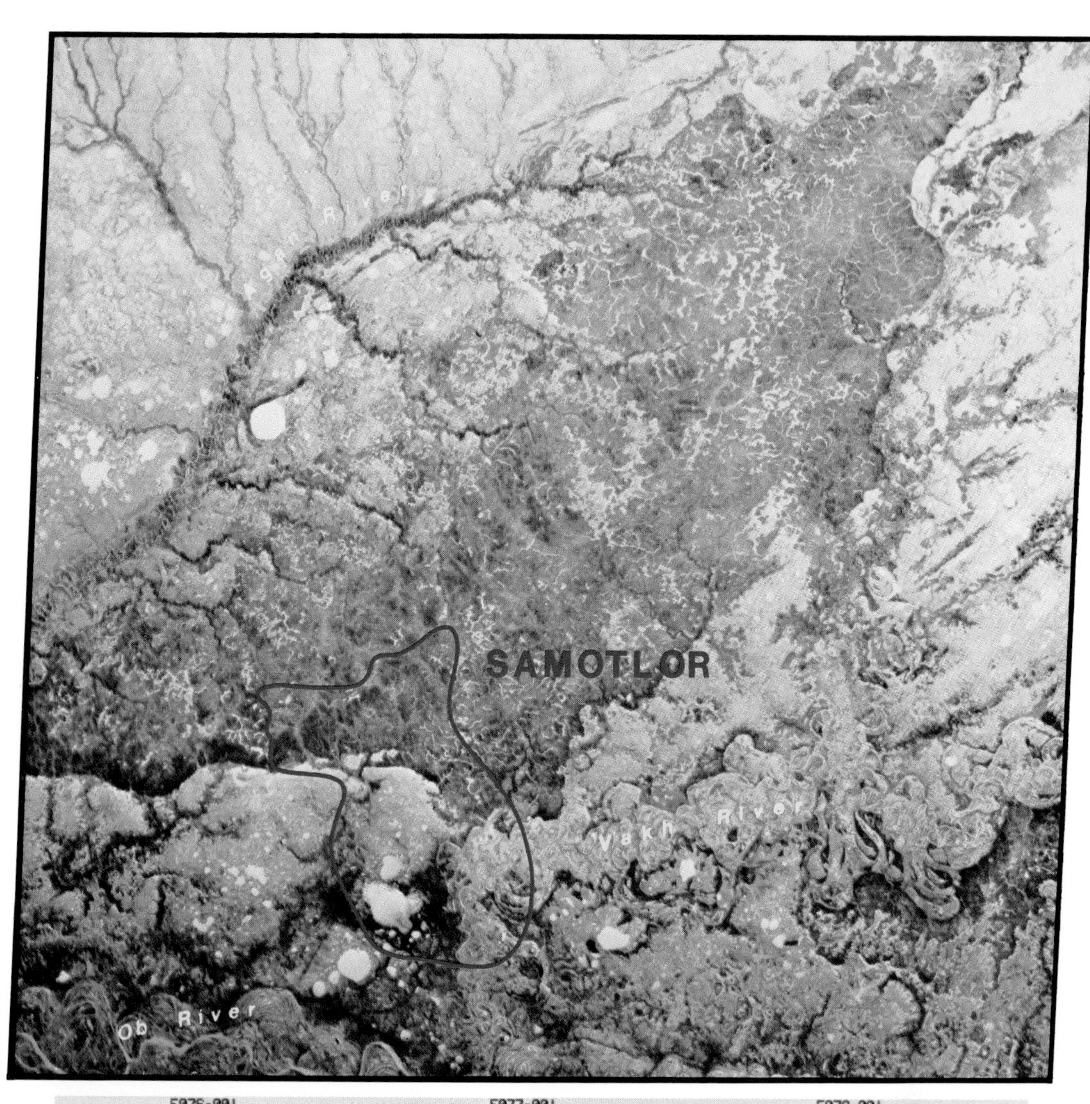

FIGURE 17

SAMOTLOR (SAMOTLORSKOYE), USSR

General Field Information

The Samotlor (Samotlorskoye) oil field is located in the Middle Ob region in the West Siberian basin of the USSR. The basin is considered to be one of the largest in the world. The field is the largest giant in the USSR and was discovered in 1965 by the first exploratory well which was located from seismic surveys. The field is on a large anticlinal structural closure. Productive zones occur within a section 1,925 ft (600 m) thick, ranging in age from Upper Jurassic to Lower Cretaceous-Cenomanian. The reservoir rocks are shallow-marine shelf sandstones interbedded with shale existing at depths between 5,000 ft (1,524 m) to 7,000 ft (2,134 m). Production began in 1969 with an average of 27,000 b/d. In late 1978, production was averaging 2.9 million b/d with an indication that the field was reaching its peak production limits. The total recoverable reserves are estimated at 15 billion bbl.

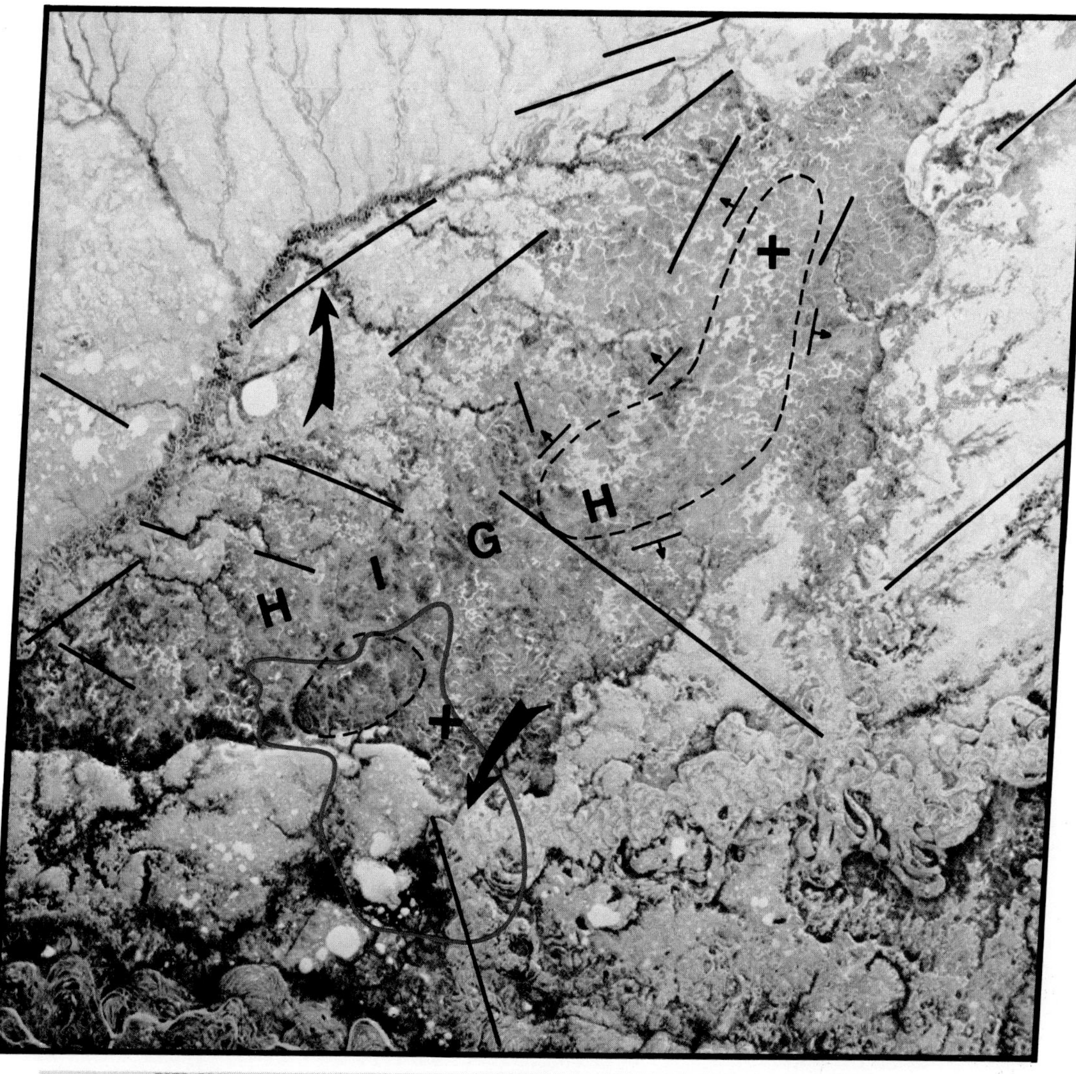

FIGURE 18

Landsat Image Interpretation

The location of the field is shown in Figure 17. The geologic interpretation indicates that the feature in this image (Fig. 18) appears similar to the Taching image. In both, a large gently-sloping plateau is encircled by two river systems. The drainage has been diverted around a topographic high which may reflect a bedrock high. Two arrows point to abrupt changes in the rivers' courses that support the idea of structural influences. In the northern valley the river makes a bend north of the oil field. In the southern valley, the river flows into the oil field area and then makes a 90° turn to continue in a direction parallel with the elongation of the field. The crosses on the geologic interpretation (Fig. 18) indicate areas of blocky or rectangular drainage which suggest a relatively flat-lying, jointed bedrock, possibly limestone.

In the northeast area of the image, there is an interesting geomorphic feature. The dip symbols show apparent dip of the bedrock and drainage direction.

Most of the linear features on the image parallel the northeast to southwest stream direction. However, a series of northwest to southeast linears occurs near the field and parallel its direction of elongation. The structural origins of these northwest to southeast linears could yield valuable information for the development of the Samotlor field and for exploration in the surrounding area, making it a primary target for seismic crews.

Had the image been available prior to exploration, the large gently sloping plateau encircled by two river systems would have delineated the area of interest for exploration which would have surely helped in the discovery of the 15 billion bbl Samotlor oil field.

FIGURE 19

KETTLEMAN HILLS, CALIFORNIA

General Field Information

The Kettleman Hills, California, form a part of a general zone of folding and faulting which extends from the middle part of the Diablo Range into the western margin of the San Joaquin Valley (Arnold and Anderson, 1910). The fields, discovered in 1928, 30 mi (48 km) long and 4 to 5 mi (6 to 8 km) wide, consist of three northwest-trending anticlines—North Dome, Middle Dome, and South Dome—which lie in echelon alignment along the western side of the southern San Joaquin Valley. Although a part of the foothills of the Coast Ranges, they are separated from the mountains by the synclinal Kettleman Plains.

South Dome is in the northern part of the anticline that is overlapped by the alluvium of San Joaquin Valley. North Dome is much longer and 400 ft (122 m) higher stucturally than Middle Dome. South Dome evidently is part of an anticline that is wider than North and Middle Domes. As the structure of South Dome is interpreted, it is 800 ft (244 m) higher structurally than North Dome. The southward extent of South Dome is uncertain owing to the overlap of alluvium, but South Dome may be the north end of the Lost Hills anticline, structurally the highest part of which is about 10 mi (16 km) south of South Dome (Woodring et al, 1940). The ultimate productive reserves are estimated at more than 500 million bbl of oil.

FIGURE 20

Landsat Image Interpretation

The location of the Kettleman Hills field is shown in Figure 19. The geologic interpretation indicates that the southern end of the San Joaquin or Great Valley dominates the image area (Fig. 20). To the east lie the block-faulted Sierra Nevada Mountains; to the west, the faulted and folded Coast Ranges.

The strike valley of the right-lateral San Andreas fault can be traced across the southwestern corner of the image. Between Kettleman Hills and the San Andreas fault, the offset of a fold suggests the presence of another fault with similar displacement. The image contains many linear features which are associated with the field. The anticlinal structure that forms the trap for the Kettleman Hills oil field is in a 30 mi (48 km) wide folded zone that borders the western edge of the Great Valley trough. Several of the lineaments on the slopes of the Sierra Nevadas appear to converge in the area of the oil field.

Pre-exploratory intensive geologic analysis of the image would have produced a photogeologic map from which the optimum seismic line positions could be determined. Although image interpretation can pinpoint structurally favorable sites, the extremely rapid facies changes in the stratigraphic units in this geosynclinal trough make subsurface data imperative for field development. Landsat provides one of the tools needed in such areas of the world. If an exploration team had this Landsat image available to them when first entering this area, its members would have immediately recognized the folded zone as structurally favorable for petroleum accumulation.

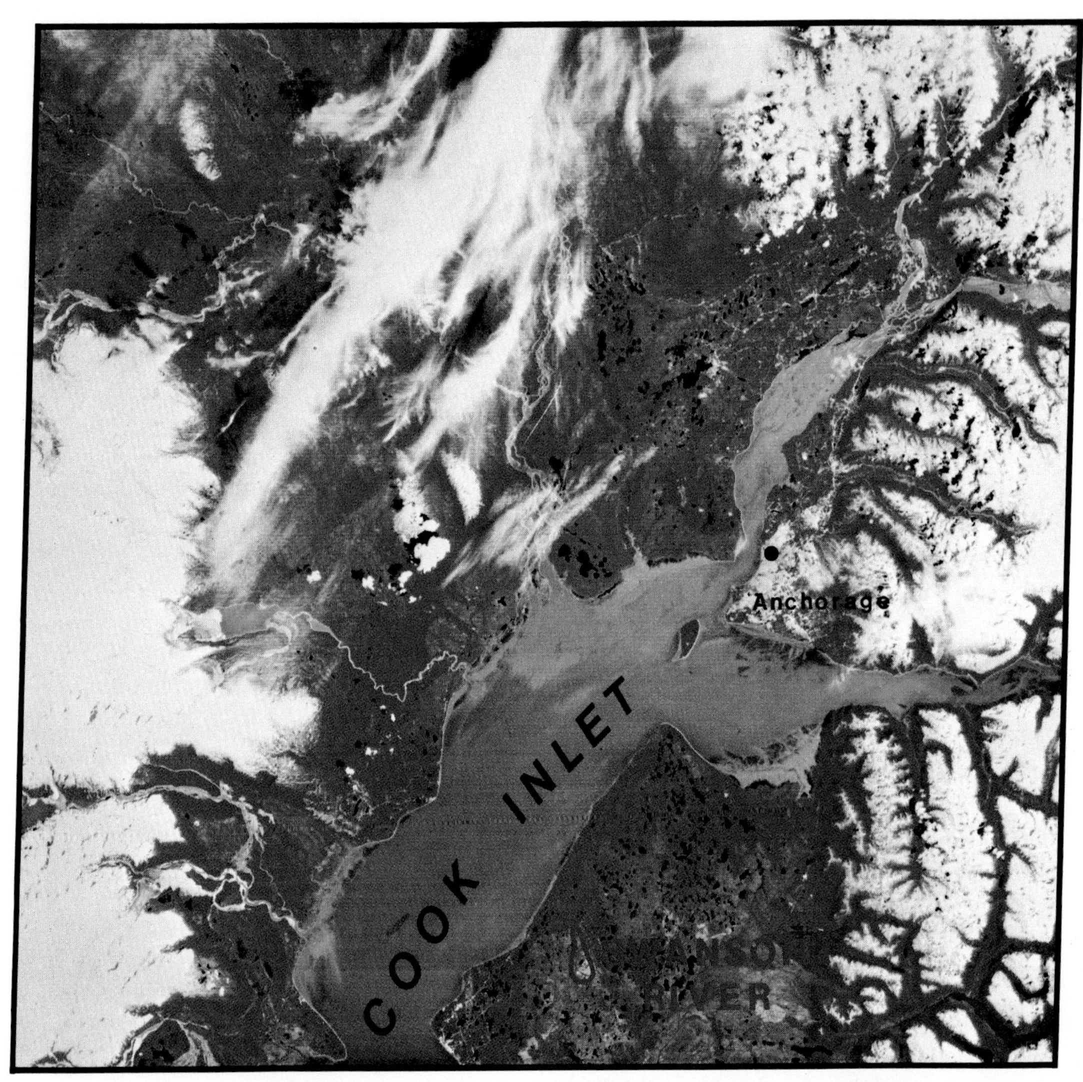

FIGURE 21

SWANSON RIVER, ALASKA

General Field Information

The Swanson River field, located in the Kenai National Moose Range approximately 50 mi (80 km) southwest of Anchorage, was the first commercial find in Alaska. The field was discovered in 1957 by the Richfield Oil Corporation. Standard Oil of California became the operator after reaching a land pooling agreement with Richfield that same year.

The hydrocarbon accumulations occur on a north to south trending elongated anticlinal structure. The anticlinal fold is cut by a series of east to west trending normal faults, dipping 45° to 55° northward. The field is approximmately 7 mi (11 km) long and 2 mi (3.2 km) wide; the hydrocarbon-bearing zones lie within the Tertiary Kenai formations which in this area are approximately 12,000 ft (3,657 m) thick and nonmarine. The Hemlock oil-bearing formation averages 650 to 750 ft (198 to 228 m) in thickness in the field and lies at depths between 10,000 to 11,500 ft (3,048 to 3,505 m). It is composed of a conglomeritic sandstone and fine- to coarse-grained sandstone with interbedded coal and carbonaceous siltstone. Net productive pay within the field ranges between 8 to 300 ft (2.4 to 91 m). Original oil-in-place is estimated at 437 million bbl and ultimate recovery at approximately 230 million bbl.

FIGURE 22

Landsat Image Interpretation

The location of the Swanson River field is shown in Figure 21. The linearity of the mountain fronts, especially the one to the east of Cook Inlet, indicates that these mountains are fault-related (Fig. 22). The study of this and other images shows that the movement appears to be primarily vertical, with the Cook Inlet lowland forming a downdropped graben. However, the geometry of the lineaments in this area suggests synthetic faulting and/or faulting at an angle (approximately 30°) to the major faults, indicating a strike-slip component of movement. Inspection of a mosiac of the surrounding area supports this combined movement picture.

Several curvilinear features can be seen in the lowland area near the Swanson River field. The larger curvilinears are glacial end moraines deposited during the last glaciation and should not be confused with structural (fold or dome) related curvilinears. The smaller circular anomalies are drainage and cover type breaks that may reflect small structural highs. A lineament intersects one of these small circular anomalies just north of the Swanson River field.

Cook Inlet has a rectangular, boxy shape that may be fault-controlled. The inlet contains producing fields along its western flanks that may be related to this faulting and associated folding.

If a simple shear system of faulting exists in the Kenai Mountain fault zone, right lateral movement along the north-northeast to south-southwest trending major fault and synthetic minor faults would produce folds oriented orthogonally. Had this image been available at the time of exploration, it would have assisted in estimating the geologic significance of the area, which would have led to the seismic delineation of the Swanson River structure.

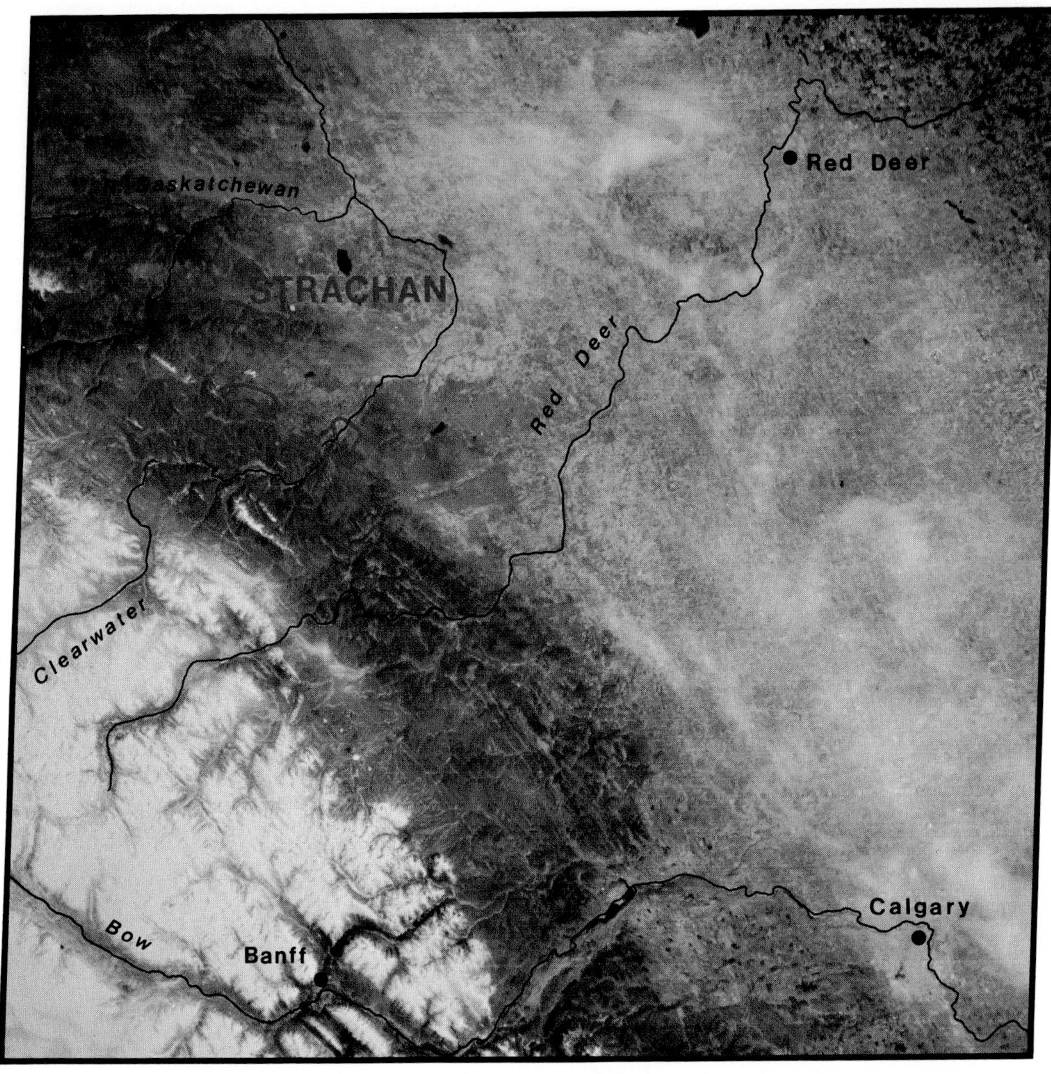

FIGURE 23

STRACHAN AND RICINUS FIELDS, ALBERTA

General Field Information

The Strachan and Ricinus gas fields are productive from two separate reefs of Late Devonian age. The Strachan field also is gas productive from Cretaceous sandstones above the reef; the Ricinus field also is productive of shallow oil and gas in Cretaceous sandstones above its reef. The Strachan field was discovered in 1967 and the Ricinus field in 1969. The reefs were delineated by adapting the seismic common-depth-point (CDP) techniques of data aquisition and the latest advances in data processing that were available. Such seismic surveys were most significant in the exploration for reef reservoirs in the deep basin of Alberta during the mid-60s. The results in the basin were considered to be very successful. In 1968, Banff and Aquitaine drilled a well into the Strachan reef which penetrated a full reef buildup of 900 ft (274 m) with a pay section of 540 ft (152 m). The discovery well in the Ricinus field was drilled by Banff and Aquitaine and had a reef buildup of 800 ft (244 m) with a maximum pay of 690 ft (210 m). Ultimate recoverable gas reserves in the two fields are estimated at approximately 1.75 Tcf. The recoverable oil reserves in the Ricinus field from the Cretaceous sandstones are estimated at 33 million bbl.

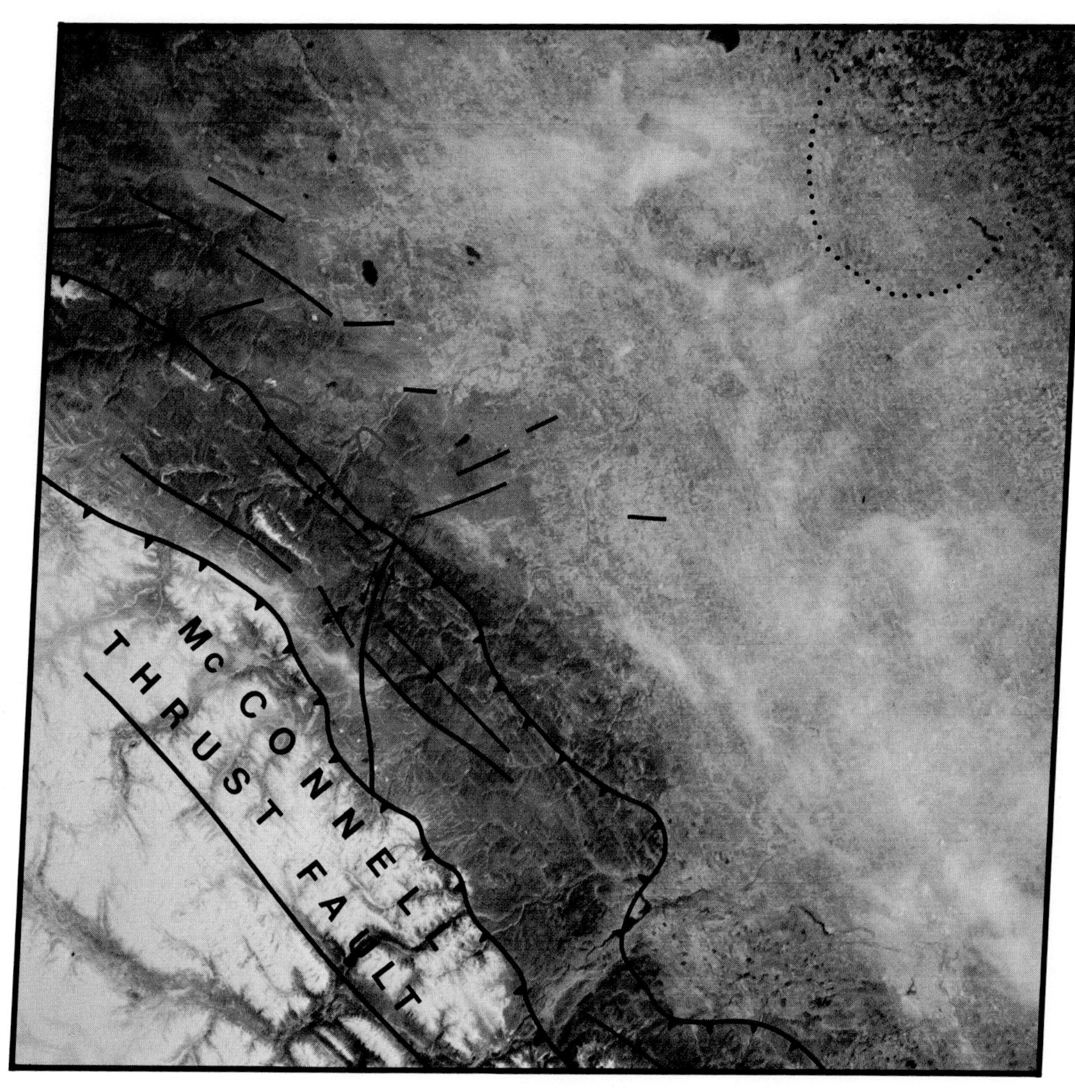

FIGURE 24

Landsat Image Interpretation

The location of the Strachan and Ricinus fields is shown in Figure 23. The geologic interpretation is shown in Figure 24. From the study of this and other adjacent images, the area can be divided into four structural units: (1) To the east are some gentle folding horizontal sedimentary units over which are flat-lying wheatfields. (2) Although lacking the ridges found to the west, the linear features of the foothills region suggest that faulting has occurred in this area but has been partly obscured by outwash debris from the mountains. (3) Folding and faulting in the lower overthrust plate has produced a series of parallel ridges. With the information provided by a detailed drainage analysis, the structure in this plate could be mapped from the image. Three of the major folds near the oil fields are easily spotted. Ridges in this area appear to be offset by a strike-slip fault, or unequal thrusting. (4) Structure in the upper overthrust plate is obscured. However, the major fault-controlled valleys still are visible.

Gravel terraces blanket the area of the two fields. Features observed on the image would have indicated that seismic lines should be located perpendicular to the faulting. By orienting the seismic lines northeast to southwest across the linear features, faults that form the traps for the fields in the foothills would have been readily detected. Drainage analysis would have given additional information as to the location of possible anticlinal and/or fault traps in the lower overthrust plate.

An examination of the abrupt change of the Red Deer River's course at the town of Red Deer and a large circular anomaly, which may be related to land usage or a topographic high, could have led to more geologic and geophysical exploration in this area. This image would have been helpful if it were available prior to geological investigation.

N029-30I IW087-30 W087-00I W086-30I
25FEB77 C N30-15/W086-52 D021-039 N N30-12/W086-39 M D SUN EL35 A129 SIS- P-N L NASA LANDSAT E-2 765-15240-7

FIGURE 25

JAY FIELD, FLORIDA

General Field Information

Jay Field, Florida, discovered in 1970, is a combination structural-stratigraphic trap with 420 ft (128 m) of oil column extending over an area 7 mi (11 km) long and 3 mi (5 km) wide, the productive area covering roughly 14,000 acres (5,670 ha.). Production is from the Jurassic Smackover Formation at depths between approximately 15,000 to 15,500 ft (4,572 to 4,724 m).

The field is located on a nose which plunges off the southern flank of the Flomaton structure, a large adjacent anticline immediately to the north. The Jay trap is formed by a facies change in the Smackover limestone in which dolomitized limestone in the productive area becomes a dense micritic limestone to the north of production. The porosity change is practically perpendicular to the axis of the nose. Estimated recoverable reserves exceed 300 million bbl of oil and 300 Bcf of gas (Ottoman et al, 1973).

FIGURE 26

Landsat Image Interpretation

The location of Jay field is indicated in Figure 25. This image over the western panhandle of Florida shows vegetation in the form of dense pine forests or cultivated fields which obscure much of the geology.

Drainage and cover-type contrasts reveal two lineament trends (Fig. 26). The northwest to southeast lineaments follow the strike of the Mississippi embayment structure and may be related to down-to-the-coast faulting, providing structural traps. The east to west lineaments parallel the present coastline and reflect ancient shorelines.

The courses of the rivers are interrupted by sharp or right angle bends. The Escambia River, in the western area of the image abruptly changes its course as it intersects the field. This could be coincidence or a reflection of a slight structural and surface high. The lateral extent of the field probably has been influenced by faulting. Even though the area is relatively flat and covered with vegetation, subtle drainage and cover changes yield some interesting geomorphic information. There are no prominent geomorphic features on the Jay area image which would indicate that the area overlying the Jay field is suspect, but the subtle features are of geologic interest for future exploration in the area.

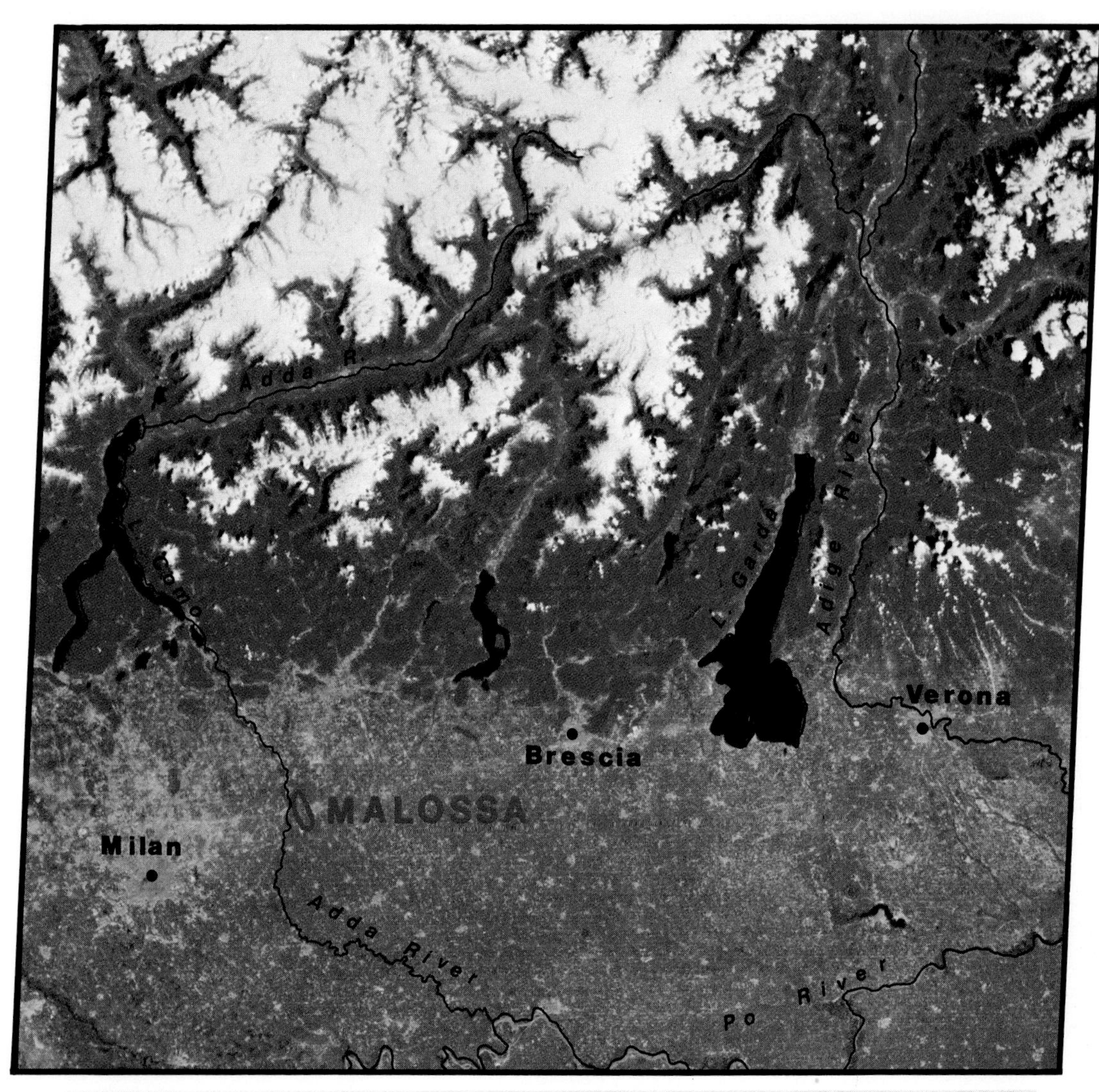

FIGURE 27

MALOSSA FIELD, ITALY

General Field Information

Malossa, a giant gas-condensate field discovered in 1973, is located in northern Italy, about 15 mi (24 km) east of Milan. It is in the northern part of the Po basin which covers an area approximately 19,305 sq mi (50,000 sq km). The Malossa structure, discovered through seismic exploration, is a faulted and overthrusted anticline about 3 mi (5 km) long and 2 mi (3 km) wide bordered to the southwest by a reverse fault and on the east by a normal fault. The discovery (and first) well at Malossa, spudded in 1972, was drilled to a depth of 18,891 ft (5,545 m) in February 1973.

Deepest test in the field to date, Malossa No. 2, reached 21,230 ft (6,470 m) and was bottomed in the Middle Triassic dolomites of the Esino formation. Principal pay zones in the field consist of the Dolomia Principate, an Upper Triassic dolomite and the Zandobbio dolomite of Early Jurassic age. The productive interval encompasses beds which extend from Upper Triassic (Dolomia Principate) upward through Lower Cretaceous (Maiolica limestone). The pay section in the gas-bearing beds totals 2,700 ft (821 m). The principal reserves, the Zandobbio and Dolomia Principate dolomites, are the most porous and fractured of the productive beds and are connected vertically through macro and micro fracture systems. The ultimate recoverable reserves of Malossa are estimated at 284 million bbl of oil and 2 Tcf of gas (Errico et al, 1979).

FIGURE 28

Landsat Image Interpretation

This is a most interesting image (Fig. 27). Malossa is located in the Po River Valley at the base of the Italian Alps. As in several of the other images in this study, intensive cultivation of the valley plain obscures the surface expression of geologic features. However, two image patterns of possible anomalous importance can be seen (Fig. 28).

Malossa field is located within a large circular anomaly in which a smaller one exists, defined by relatively high reflectance (darker color) around which the Adda River is diverted. A similar light area with curvilinear drainage segments occurs near Verona to the east.

The Po River contains several anomalously straight segments of drainage oriented approximately east to west. The Po River Valley contains a series of east to west faults, one of which forms the structural trap for the Malossa field. The strong east to west lineament, shown by a green dashed line, could have been influential in the trapping of the hydrocarbons.

Although its cause is unknown, the large tonal anomaly within which exists the Malossa field (Fig. 27), is the most obvious feature in the Po River Valley. In Figure 28, note that the similarity of the large Malossa anomaly to the one to the east establishes the latter as a potential exploratory target. Also of prospective exploratory interest is the north to south anomaly near Verona.

If this image was available and properly interpreted prior to exploration in this area, the anomalous areas could have been pinpointed for seismic surveys.

FIGURE 29

GRONINGEN, THE NETHERLANDS AND WEST GERMANY

General Field Information

The giant Groningen gas field is located in the northeastern part of The Netherlands. The discovery by the N. V. Nederlandse Aardolie Maatschappij (NAM) in 1959 triggered an intensive search for gas in The Netherlands, the adjoining countries, and on the continental shelf bordering the United Kingdom, The Netherlands, Germany, Denmark, and Norway. The field, which covers 180,000 acres (72,900 ha.), is on a culmination of the large, regional northern Netherlands high, a formation of the late Kimmerian tectonic phase (Late Jurassic–Early Cretaceous).

The reservoir overlies unconformably the truncated and strongly faulted coal-bearing Pennsylvanian strata, which are considered to be the main source of the gas. The reservoir, consisting of fluviatile and eolian sandstone and conglomerate of the Rotliegendes Formation (Lower Permian), is 300 to 700 ft (91 m to 213 m) thick. These coarse clastic beds are overlain by a few thousand feet of Permian Zechstein evaporites, notably rock salt and to a lesser amount anhydrite and dolomite, which constitute the reservoir seal. The thickness of the overlying Mesozoic and Cenozoic strata ranges from 3,000 ft (914 m) to more than 6,500 ft (1,981 m; Stauble and Milius, 1970). Ultimate recoverable reserves are estimated at 66 Tcf.

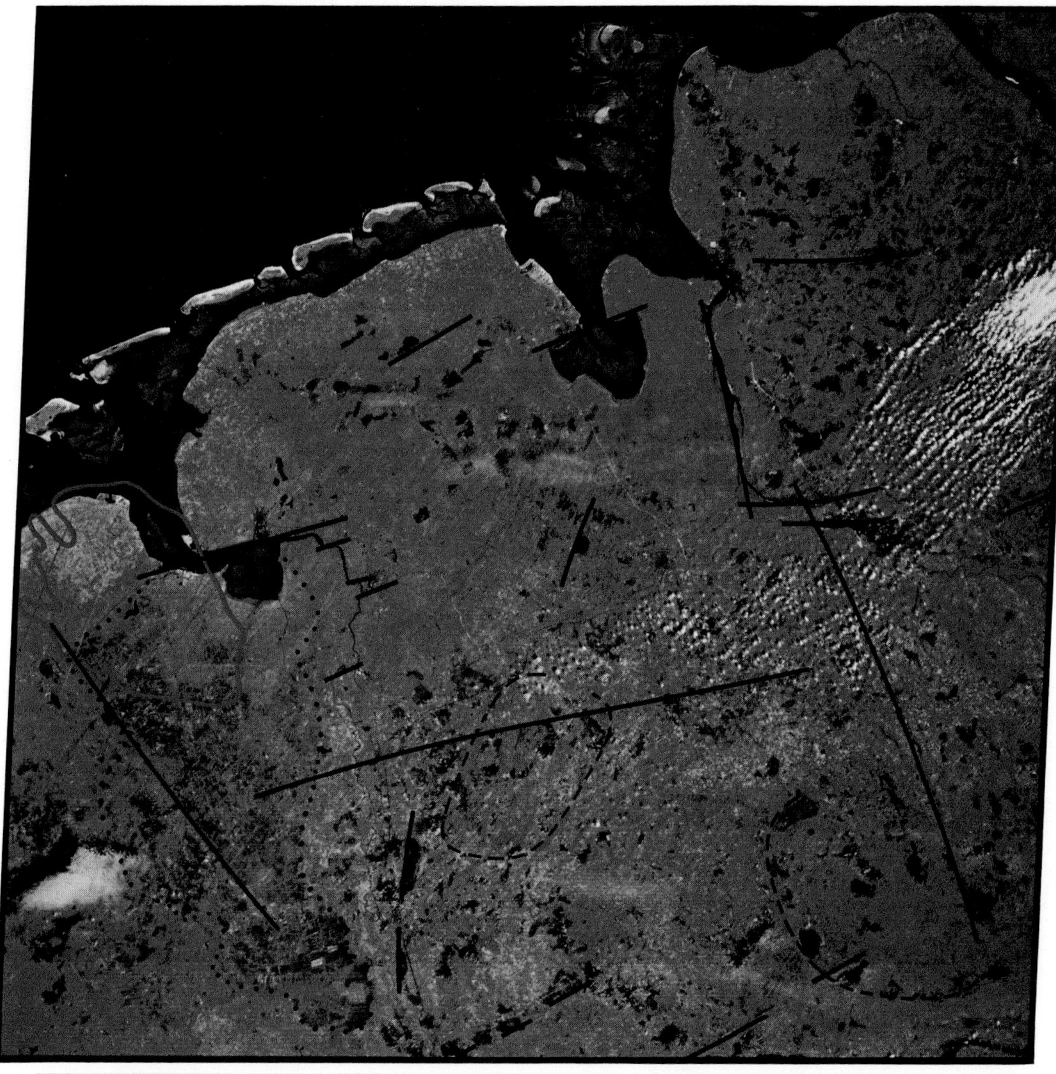

FIGURE 30

Landsat Image Interpretation

The location of the Groningen field is shown in Figure 29. Intensive cultivation of the rich soils of the coastal plain has obscured any prominent surface features in the area of this large gas field, however some interesting subtle surface expressions are observed. The dark soils south of the field are reclaimed wetlands that are below sea level. The estuary of the Ems River may originally have extended much farther inland than at present and may indicate a structural trough and/or a basin susceptible for the accumulation of hydrocarbons (Fig. 30).

The major producing area of the North Sea basin is located to the northwest of the image area. Linear features trending northeast to southwest may be the surface expression of down-to-the-coast faulting. Linear features parallel with the coast may reflect old shoreline positions and/or faulting. Other linear features and circular anomalies are of exploratory significance.

There are no prominent geomorphic features on the Landsat image which indicate the area overlying the Groningen field is suspect. However, once the location of the field is related to the image, the two major lineaments that intersect the field, as well as the circular anomalies and smaller linear features, become important in planning its development and for further exploration of the area.

The use of Landsat imagery in exploration is an interactive procedure. As each new piece of information is acquired, the imagery should be reevaluated and used in planning the next step of exploration, development, and production. Although this image lacks indications of the offlapping sequence of strata, geophysical lines should be run in the Ems River basin to check its correlation with the petroleum-rich Mississippi embayment.

SUMMARY AND CONCLUSION

In summary, there is no question that Landsat data would have been valuable to the initial exploration program in each but two (Groningen and Jay) of these areas had they been available; some being more significant than others. The obvious conclusion is that there are large explored and unexplored areas of the world which should be thoroughly and repeatedly studied with Landsat images. By doing so, anomalous areas could be outlined and then pinpointed for intensive geologic and geophysical investigation. Such a program would drastically cut exploration costs and, in turn, surely result in the discovery of more giants.

REFERENCES CITED

Alaska Division of Oil and Gas, 1974, In place volumetric determination of reservoir fluids, Sadlerochit Formation, Prudhoe Bay field, *in* The Trans-Alaska pipeline and west coast petroleum supply, 1977-1982: U.S. Senate Comm. Insular Aff., p. 126-175.

Arabian American Oil Company Staff, 1959, Ghawar oil field, Saudi Arabia: AAPG Bull., v. 43, p. 434-454.

Arnold, R., and R. Anderson, 1910, Geology and oil resources of the Coalinga district, California: U.S. Geol. Survey Bull. 398, 354 p.

Clifford, H. J., R. Grund, and H. Musrati, 1979, Geology of a stratigraphic giant—Messla oil field, Libya (abs.): AAPG Bull., v. 63, p. 433.

Davis, J. A., 1970, Burgan giant outproduces all others: Oil and Gas International, v. 10, no. 10, p. 100-114.

Errico, G., et al, 1979, Malossa field, deep discovery in Po Valley, Italy (abs.): AAPG Bull., v. 63, p. 446-447.

Halbouty, M. T., 1976, Application of Landsat imagery to petroleum and mineral exploration: AAPG Bull., v. 60, p. 745-793.

——— 1978, The impact of Landsat on scientific and technical orientation: OPEC Rev., v. 2, no. 3, p. 22-30.

Jamison, H. C., L. D. Brockett, and R. A. McIntosh, 1979, Prudhoe Bay, a ten-year perspective (abs.): AAPG Bull., v. 63, p. 699.

Ottman, R. D., P. L. Keyes, and M. A. Ziegler, 1973, Jay field—Jurassic stratigraphic trap: Gulf Coast Assoc. Geol. Socs. Trans., v. 23, p. 146-157.

Stauble, A. J., and G. Milius, 1970, Geology of Groningen gas field, Netherlands, *in* Geology of giant petroleum fields: AAPG Mem. 14, p. 359-369.

Woodring, W. P., R. B. Stewart, and R. W. Richards, 1940, Geology of the Kettleman Hills oil field, California: U.S. Geol. Survey Prof. Paper 195, 170 p.

A Review of the Viking Gas Field[1]

By Michael Gage[2]

Abstract The Viking Gas Field produces natural gas from the Leman Sandstone Formation in the Lower Permian Rotliegendes Group. The area is difficult to map seismically because of complexities introduced by the overlying Upper Permian Zechstein, mobile salt. Nine independent accumulations have been developed. Exploration and development programs must both take into account the rapidly changing facies of the desert sandstone reservoir. A fluvial-wadi course is outlined bisecting the field, accounting for poor reservoir performance in certain areas of the field. Petrographic work shows the deleterious influence of diagenesis of the varied and mixed facies. Integration of detailed facies analyses and well productivity data should lead to methods for attaining more efficient recoveries.

INTRODUCTION

The Viking gas field lies in blocks 49/12, 49/16, and 49/17 of the United Kingdom Southern North Sea. The field area encloses a number of pools (Fig. 1) which are grouped into the North and South Viking complexes. The single largest group, the North Viking complex, contains 1.6 Tcf of gas in place. The South Viking complex includes a wider geographical spread of pools containing a further 2 Tcf of gas for a combined field total of 3.6 Tcf of gas. The reservoir rock is sandstone of the Lower Permian Rotliegendes Group. The Viking field (i.e., the North and South complexes) fall withing the Viking Contract area (Fig. 1) and, as such, supplies approximately 12.5% of the U.K. natural gas requirements to the British Gas Corporation at rates of up to 950 MMcf/day.

Viking is operated by Conoco for the Conoco/British National Oil Corporation Group. The latter corporation is the successor of the National Coal Board, an early partner in the exploration combine. The first Conoco/National Coal Board well, 49/17-1, was completed in December, 1965, and found 43 ft (13.1 m) of gas-bearing Permian Rotliegendes Sandstone, at 9,044 ft 2,756 subsea. Its location is on the margin of the present South Viking C pool. Subsequent exploration drilling located the North Viking A, H, F, and FS pools and South Viking B, D, E, G, AND GN. A typical gas composition for the field is given in Table 1.

The geologic sequence is typical of the Southern North Sea (Fig. 2). The Lower Permian continental Rotliegendes Sandstone is the only productive reservoir in the Viking field. Normal faulting has created traps with an effective seal provided by overlying Upper Permian Zechstein evaporites. The underlying Carboniferous Westphalian Coal Measures are the source of the gas. The Lower Permain play owes its existence quite simply to a good reservoir that overlies coal source beds and possesses an ideal salt seal. Where the seal is breached, gas can escape, but given appropriate geologic circumstances, still can be trapped in the Lower Triassic Bunter Sandstones, a condition which prevails in Hewitt field to the west.

The margin of the Rotliegendes deposition lies to the south and west, flanking the London-Brabant Massif. Conglomeratic

[1] Manuscript received, May 15, 1979; accepted for publication, July 6, 1979.

[2] Conoco, Inc., The Woodlands, Texas 77380. The writer would like to thank his many colleagues within Conoco North Sea, Inc. and the British National Oil Corporation who assisted and contributed to this paper. In particular, he wants to thank Richard Schroder of Conoco North Sea, Inc. I also thank both corporations for their permission to publish.

Article Identification Number:
0065-731X/80/MO30-0003/$03.00/0.

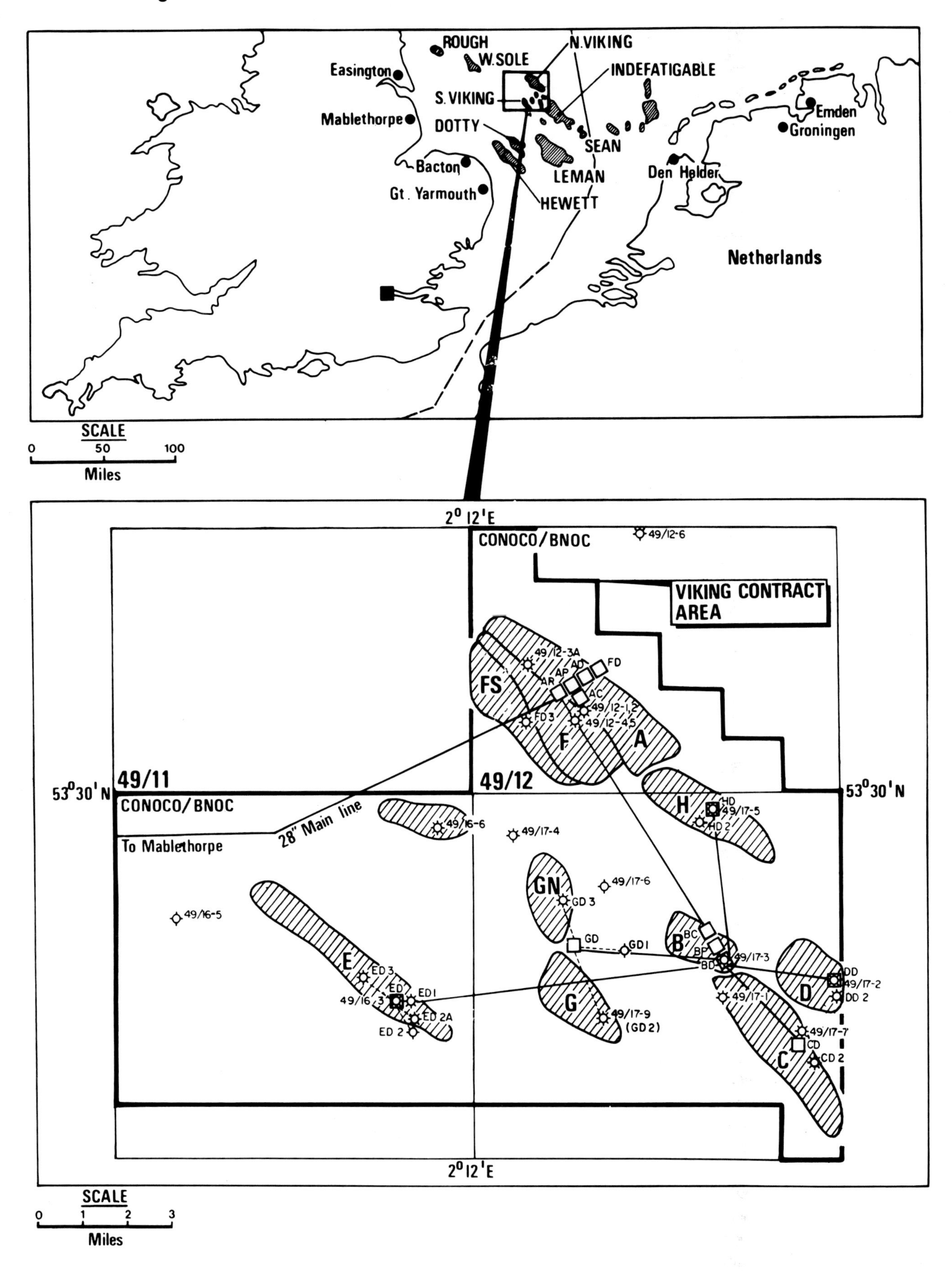

FIG. 1—Viking gas field, location maps.

sheet flood sands and silts spread from this upland area introducing sediments to the region and merging into the major east-west dune belts which cross Quadrangles 48 and 49.

The northwest-southeast structural grain of the Carboniferous persists through the Rotliegendes where major normal faults trend in the same direction. The overlying Zechstein salt has accommodated fault movement of the Rotliegendes and salt flow has caused flexure and faulting of the post-Zechstein sequence. Similar movements dated as Cimmerian and Laramian in age are identified from outside the salt area. Seismic records indicate that salt in the field area first moved during the Jurassic.

Exploration history has demonstrated that any suitably closed and sealed structure can be gas-bearing. Conoco's policy has been to test such structures by drilling but with development priority depending on the reservoir quality and geologic evaluation of specific areas.

During the current pause in exploration and development the opportunity has been taken to generally review the geology of the Viking gas field. The structure of the North Viking area previously has been the subject of a paper by Gray (1975).

THE ROTLIEGENDES SANDSTONE

Stratigraphy

Rotliegendes sandstones of early Permian age lie unconformably on the eroded surface of the Carboniferous which was gently folded in the late Hercynian orogeny. A large regional Carboniferous anticlinal structure extends northwest-southeast through blocks 49/16 and 49/17 exposing a Namurian subcrop along its crest and successively younger Westphalian units on its flanks.

In the Viking area, the Rotliegendes is represented by the Leman Sandstone Formation (Rhys, 1974), consisting mostly of eolian sandstones which subordinate interdune sabkha and wadi-fluvial deposition. North Viking is close to the northern limit of the east-west eolian belt. To the north and northeast, increasing interfingering of silts and shales of the Silver Pit Formation mark proximity to the North German basin. The Leman Sandstone thickens regionally southwestward over the field from 450 ft (137 m) in North Viking to more than 800 ft (244 m) in the E pool (Fig. 3).

Drilling shows significant local variations in thickness and facies of the Rotliegendes. After initial identification of the facies (Table 2) by sedimentary analyses of cores, studies were extended relating log and dipmeter characteristics to the uncored Rotliegendes section. Similar methods were described in some detail by van Veen (1975) for the Leman field. Of the two basic facies, eolian sandstones provide the optimum reservoir and wadi-fluvial, the poorest. Figure 4 indicates the facies distribution which was mapped in the Viking field. Three principal zones are identified: (1) the marginal sabkha formation (Silver Pit Formation), (2) the main area in which eolian sandstone prevails, and (3) the northwest-southeast trending zone characterized by an anomalously thick section which appears to mark the course of a wadi. Both of the latter facies are a part

Table 1. Typical gas composition, Viking gas field, North Sea.

COMPONENT	MOL PER CENT
Hydrogen Sulfide	0.00
Carbon Dioxide	0.38
Nitrogen	2.51
Methane	91.22
Ethane	4.07
Propane	0.97
iso-Butane	0.20
n-Butane	0.22
iso-Pentane	0.07
n-Pentane	0.06
Hexanes	0.06
Heptanes plus	0.24
Total	100.00

Calculated gas gravity = 0.614

of the Leman Sandstone Formation.

In the Leman and Indefatigable fields, the upper section of the Leman Sandstone Formation is a gray-colored, partially homogeneous, water-laid sandstone (van Veen, 1975; France, 1975). Its origin is attributed to the reworking of a pre-existing Rotliegendes deposits during the Zechstein transgression. However, its presence is not always apparent and it is possible absent from certain wells such as 49/12-2, the North Viking discovery well. Throughout Viking it has been referred to as the "tight zone," a compound of reworked and secondarily cemented standstones. Cores through the section show a horizontally bedded, normally gray sandstone. This zone is identified on field sections A-A' and B-B' whose locations are given on Figure 5, and are illustrated in Figures 6 and 7, respectively.

Typically the Leman Sandstone is difficult to correlate over any distance. An exception is over North Viking where intermixed fine-grained sandstone, siltstone, and mudstone beds ranging to 30 ft (9 m) in thickness may be correlated in an east-west direction and divide the reservoir into three zones. The intervals are thought to represent either temporary establishment of an interdune sabkha-lacustrine environment or periodic fluctuation in the level of the arm of the north Terman Lake to the north, causing the shoreline to advance and recede. The zonation has been extended basinwards to the Conoco/BNOC well 49/12-6, four miles north of Viking, though greatly increased mudstone and siltstone content (representative of the Silver Pit Formation) makes this correlation somewhat tenuous, and even impossible farther to the north.

The sabkha zone only impinges on the northern limit of the field as already mentioned, and was not a consideration in the exploration of Viking. Subordinate amounts of interdune sabkha, fluvial, and wadi sediments are interbedded with the dominant eolian sandstones. Throughout the Viking field, grain sorting in the depositional environment is the prime factor affecting the present porosity and permeability (Fig. 8). The petrographic analysis of two well, 49/16-3 from the eolian environment and 49/17-4 from the wadi environment, are discussed below.

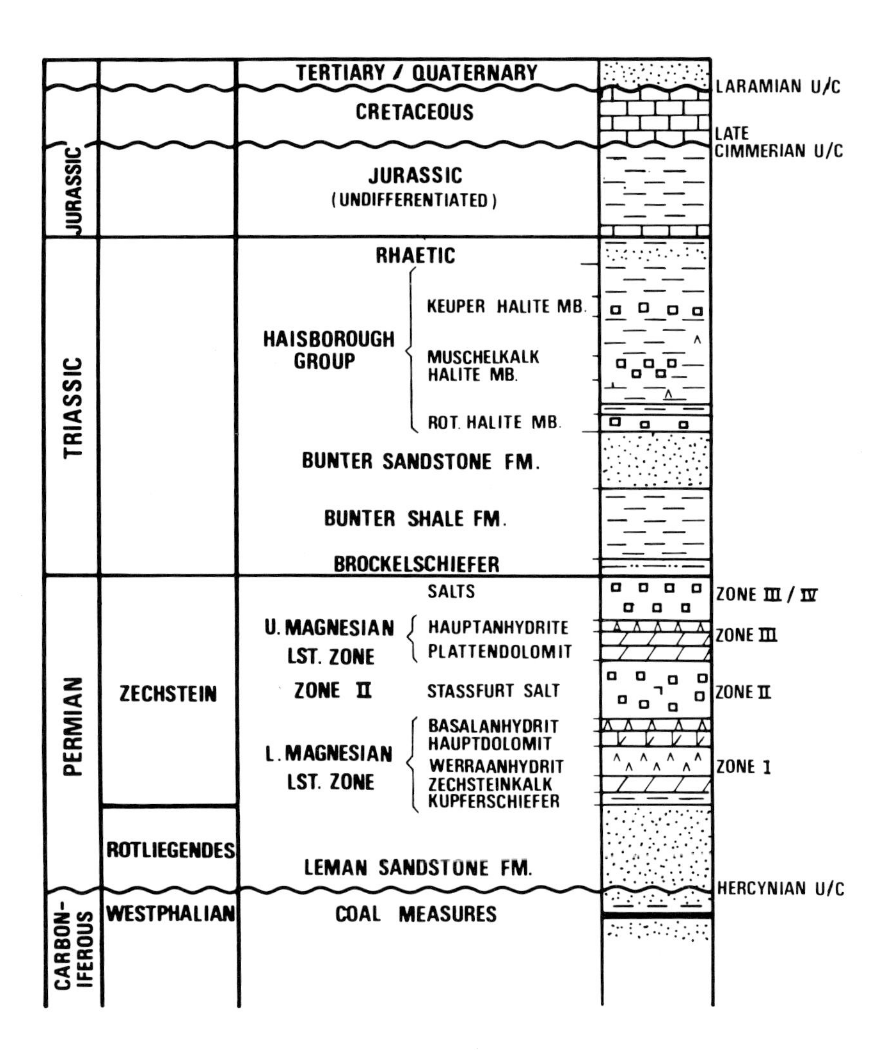

FIG. 2—Viking gas field, generalized stratigraphic column.

Well 49/16-3 (Fig. 9) is located in the E pool, which is one of the best producers in the Viking field. The section consists predominantly of eolian dune sands with interbedded sabkha and fluvial depostis. Distribution of porosity and permeability is largely facies controlled. The clean, well-sorted, eolian sands retain their porosity, whereas the poor sorting and interstitial clay matrix of the other facies result in much reduced porosity and permeability. Cementation is largely by carbonate, concentrated in those zones which have experienced some aqueous influence. Minor amounts of authigenic quartz and occasional clay are found in the eilian sands; otherwise pore space remains open.

By contrast, the 49/17-4 well (Fig. 9) is located within the wadi trent. In this well, both depositional fabric and diagenesis have determined present porosity and permeability. From a section of over 800 ft (244 m), only the upper 312 ft (95 m) were cored and therefore available for analysis. Eolian sands in this well are common and have not retained their original open pore structure. A largely fluvial environment has deposited clays and silts with the sands, Laolinite, filling interstitial pore space, is noted in almost every sample—it has even been introduced into the pores of the sparse eolian sandstones. Dolomite is the most common cementing agent in the well. Anhydrite and quartz cements also arepresent in minor quantities. The eolian sands tend to be the most heavily cemented.

An observation from these two wells, which holds true elsewhere in the Viking field, is that where eolian facies predominate, the section can survive the adverse affects of diagenesis fairly successfully. Where finer sediment fractions predominate, diagenetic products can migrate and occupy the pore space of the originally more porous section destroying the fabric of the entire section.

The proposed course and distribution of wadi sedimentation is illustrated by Figure 10. Superimposed on the isopach of gross thickness, pie diagrams give a breakdown of general facies from wells and demonstrate an increase in thickness correlated with an increasing proportion of wadi sediments. Whereas the wadi probably was confined in its upper course, it deposited sediment in a depression extending north from 49/17-4 to the F and FS pools of North Viking. Reworking of sediment between periods of flood may account for the high proportion of eolian sediments in 49/17-4 where compared to the well 49/17-6.

The presence of the wadi suggests some structural or topo-

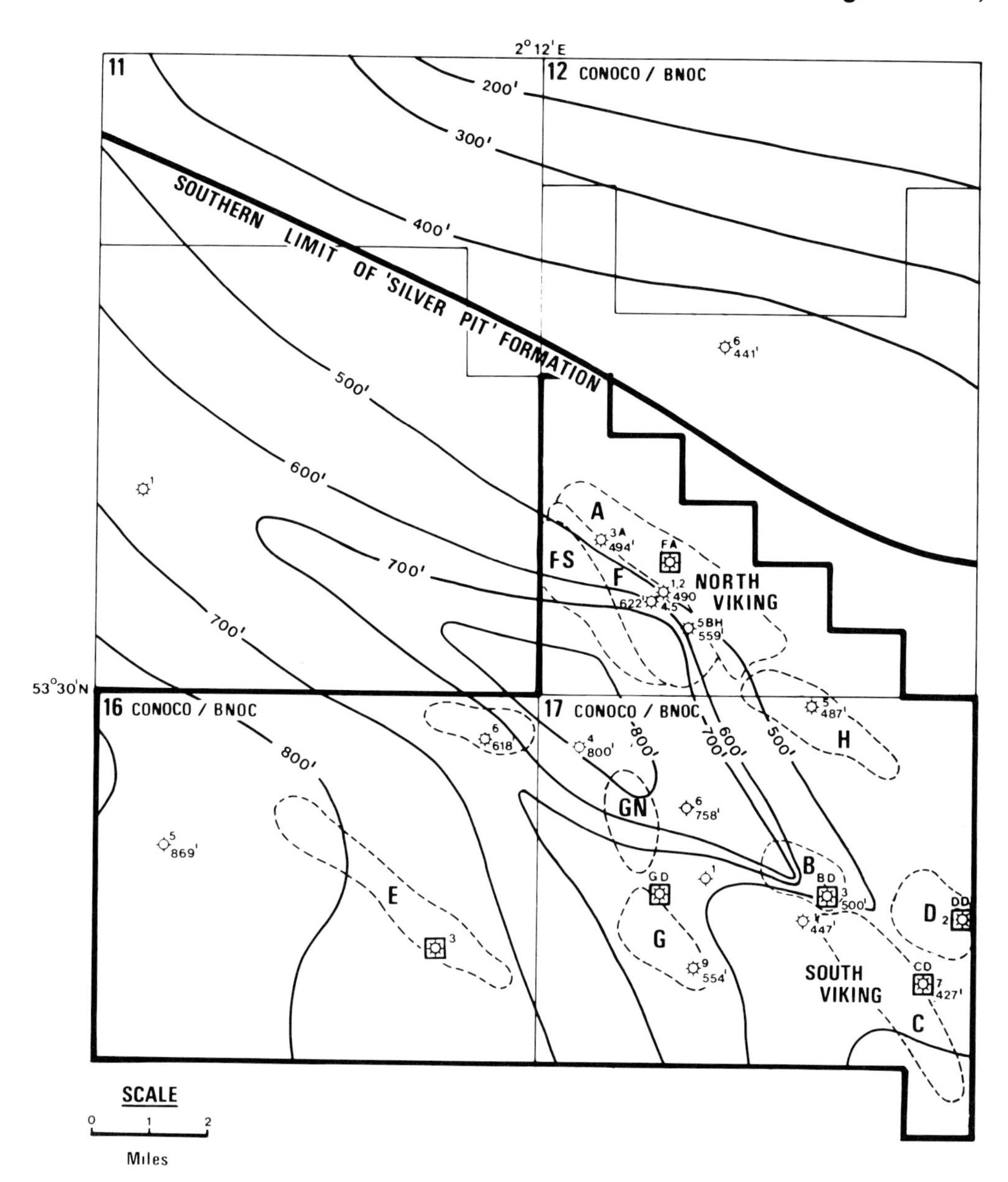

FIG. 3—Viking gas field, Rotliegendes sandstone gross isopach.

graphic cause for its inception. Because it trends northwest-southeast, it probably is linked to the structural grain which has the same orientation. It is notable that a regional isopach of the Leman Sandstone shows thinning eastwards. From Viking to the southern limits of Indefatigable field the section thins from 800 ft (244 m) to less than 200 ft (61 m). If this difference is due to the presence of an upland area during the Rotliegendes time, which on well correlation seems likely, then the Indefatigable area acted as an early sediment source for the channel before the topography was buried.

The North Viking pools normally are completed within three established zones. Pressure depletion data show that they are interconnected; however, recent production data indicate a selective contribution from tighter sands.

Well 49/17-GD2 (a South Viking G pool well) was drilled and completed in 1976. Three different lithologic zones were found. These were a reworked "tight" zone, an eolian zone, and an inland sabkha zone. Production logging indicated the differences between the zones as shown in Figure 11. Zone 1a is equated with the reworked "tight" zone at the top of the Leman Sandstone Formation and has low permeability. Zone 1b is an 8-ft (2.5 m) interval of high permeability dune sand beneath the reworked zone. Zone 1c is interpreted as a mixture ofinterbedded eilian and sabkha sandstones with a corresponding overall intermediate permeability. A break of siltstones and shales 13 ft thick separates zone 1 from zone 2, the latter being a more homogenous, predominantly aeolian section.

Skin damage is responsible for thelack of production over the interval 11,920 to 11,960 ft (3,633 to 3,645 m). It raises the question of whether other wells in the Viking field are producing effectively from the total perforated interval—so that poor production may not necessarily be a reflection of the Rot-

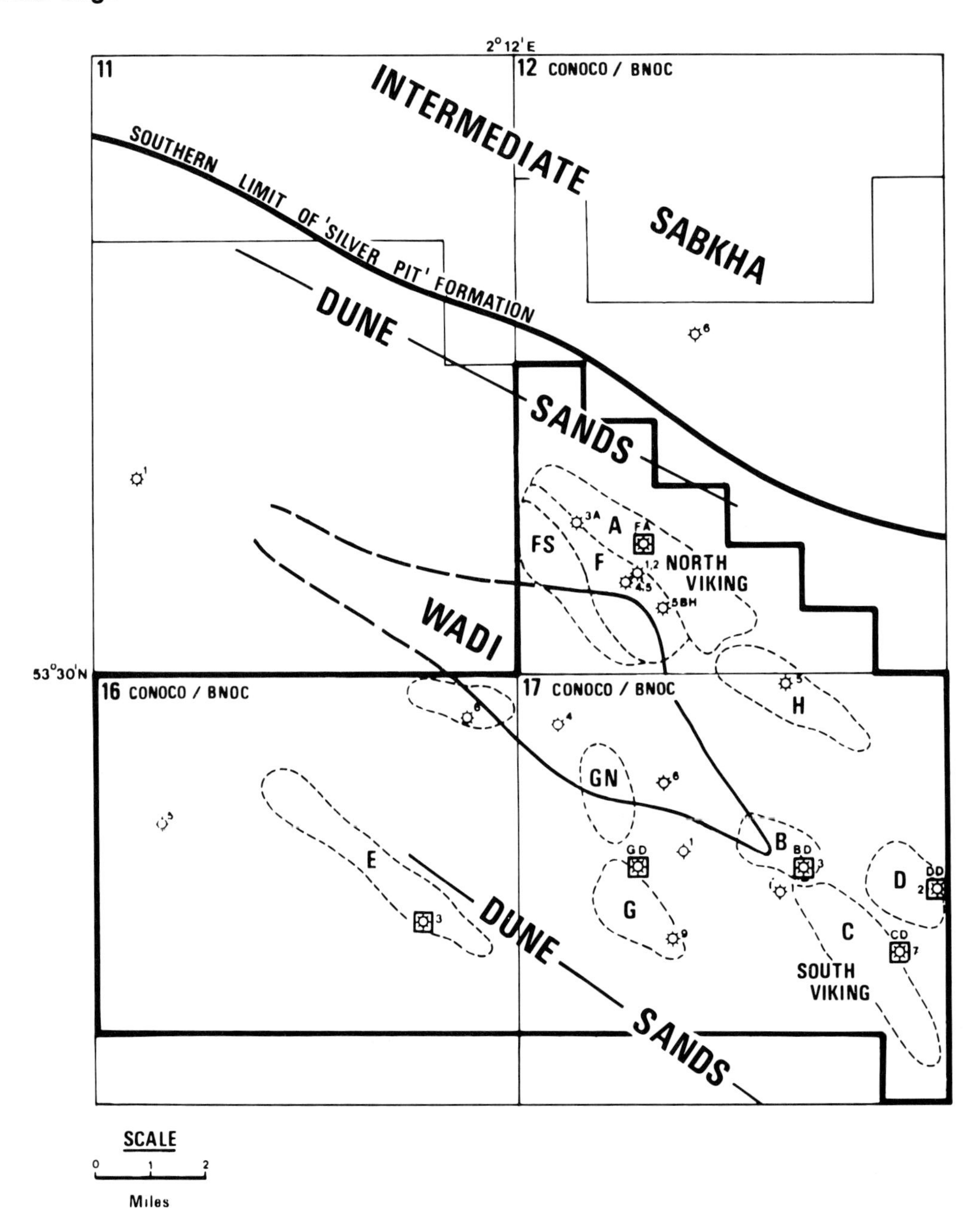

FIG. 4—Facies distribution of the Rotliegendes sandstone in the Viking gas field area. Note gradation from intermediate sabkha, through dune sands, fluvial-wadi, and back to dune sands.

liegendes facies encountered in the well. The integration of geologic data with production techniques should ensure that wells are produced efficiently.

Structure

A variety of interpretation problems exist within the United Kingdom's southern gas fields. The most significant are those related to identifying and mapping the base Zechstein carbonate seismic marker which is the unit just above the reservoir. In Conoco's terminology, this is the Lower Magnesian Limestone zone. It is overlain by beds which change in composition from predominantly mobile salt in the north to a thick carbonate and anhydrite sequence in the south. The Viking gas field is overlain by variable thicknesses of Zechstein Salt within which are layers of carbonate and anhydrite sometimes detached as raftlike bodies. On seismic sections these layers within the salt can sometimes be confused with the objective base Zechstein event, as demonstrated by the reflector at 1.73 sec S.P. 2811, on Figure 12. However, for most of the Viking field area the base Zechstein can be adequately identified and mapped. Only where extreme salt movement has occurred, as illustrated in Figure 13, does mapping become impossible. The salt wall shown on this section lies northwest of the E pool.

Base Zechstein time maps of the Viking field include many

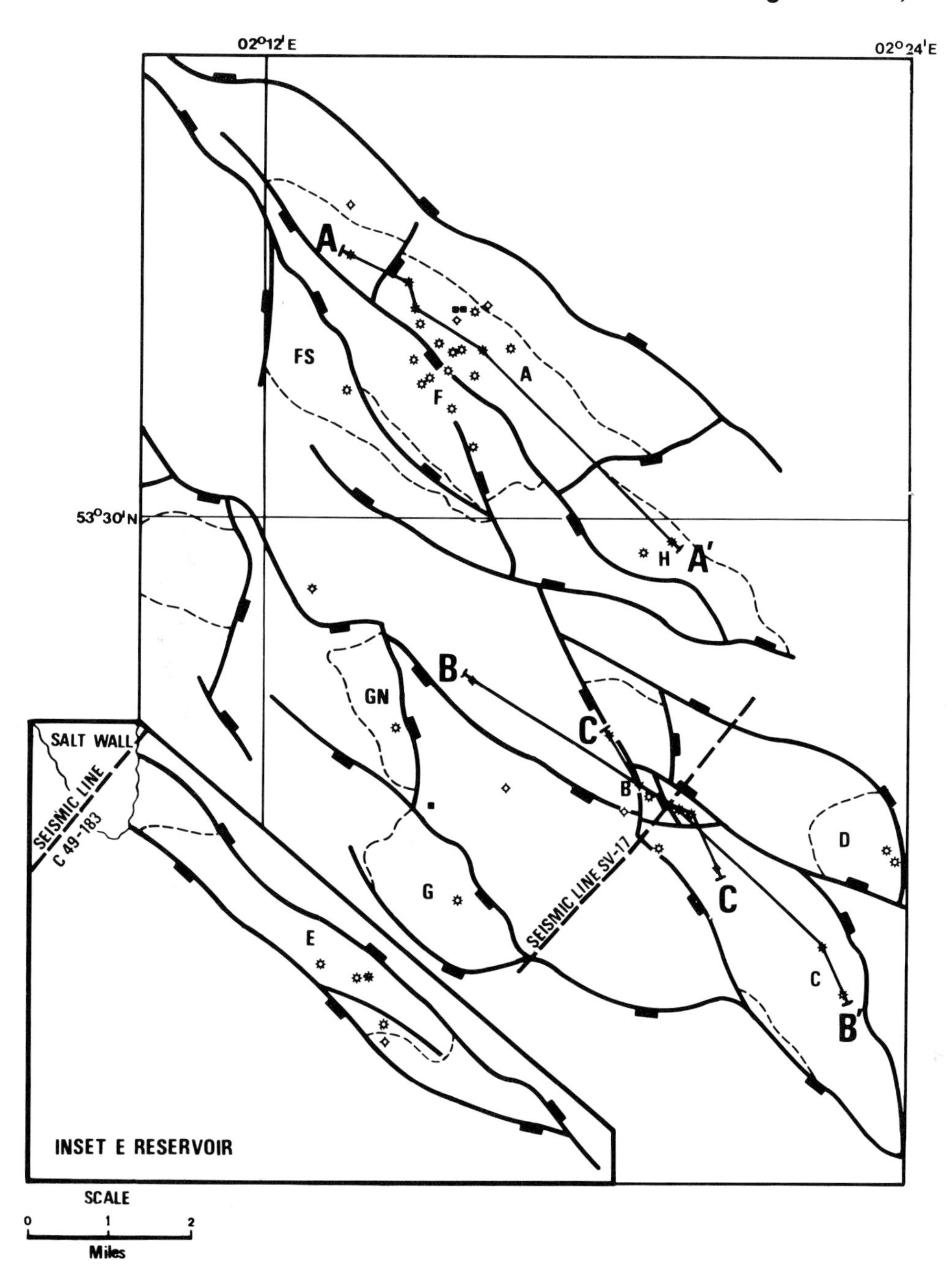

FIG. 5—Location map of wells in relation to structural features (not cross-section line locations) in Viking gas field, North Sea.

anomalies directly or indirectly related to the salt. The anomalies were removed by accurate depth conversion. The direct salt effect was removed by isopaching the salt interval based on seismic times and constant velocity.

The indirect salt effect is more subtle and more difficult to remove. Because top Zechstein highs bear no relationship to base Zechstein high and because there has been deep burial and subsequent inversion of the pre-Tertiary section, centered in the Sole Pit trough, lateral gradients exist which also require compensation. The trough lies westwards of Viking and has not been a major influence; however, gradients derived from well control are utilized (Fig. 14). These relate depth of burial of the Lower Triassic section in terms of two-way seismic time versus interval velocity. Base Zechstein depth maps can then be generated by a "layer cake" method as described by Gray (1974). The Lower Triassic is plotted because it is an extremely uniform shale section.

An analysis of faulting in the Viking field leads to the conclusion that faults are nonsealing where Leman Sandstone is juxtaposed against a similar sandstone. Thus, structures spill to form a staircase of Rotliegendes pools. Although faults may not be completely sealing, secondary cementation by carbonate- and sulphate-bearing ground waters moving preferentially along a fault plane does create permeability barriers.

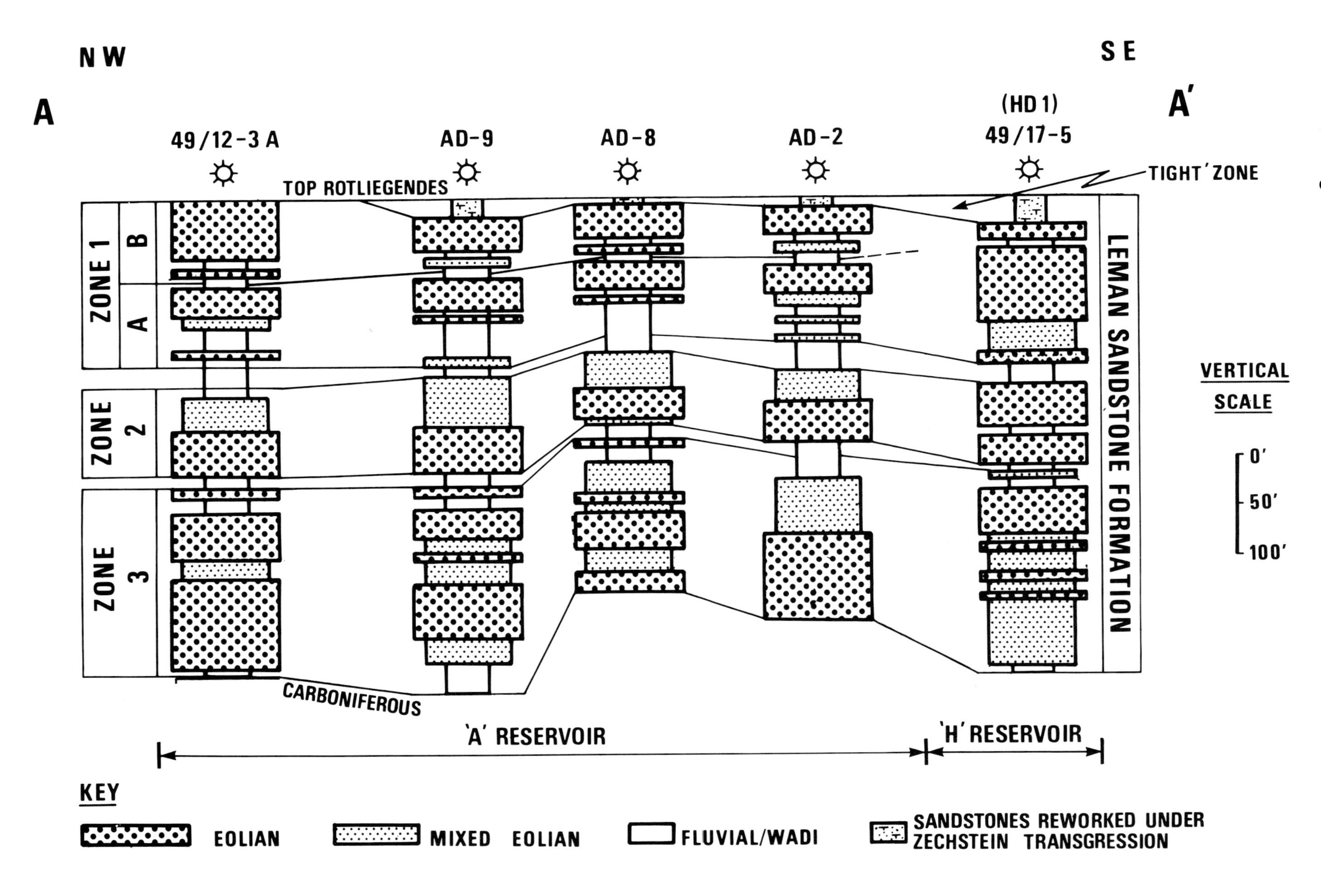

FIG. 6—Facies variations in the Rotliegendes, North Viking gas field, A and H reservoirs. Cross section along A-A' (see Fig. 5 for location).

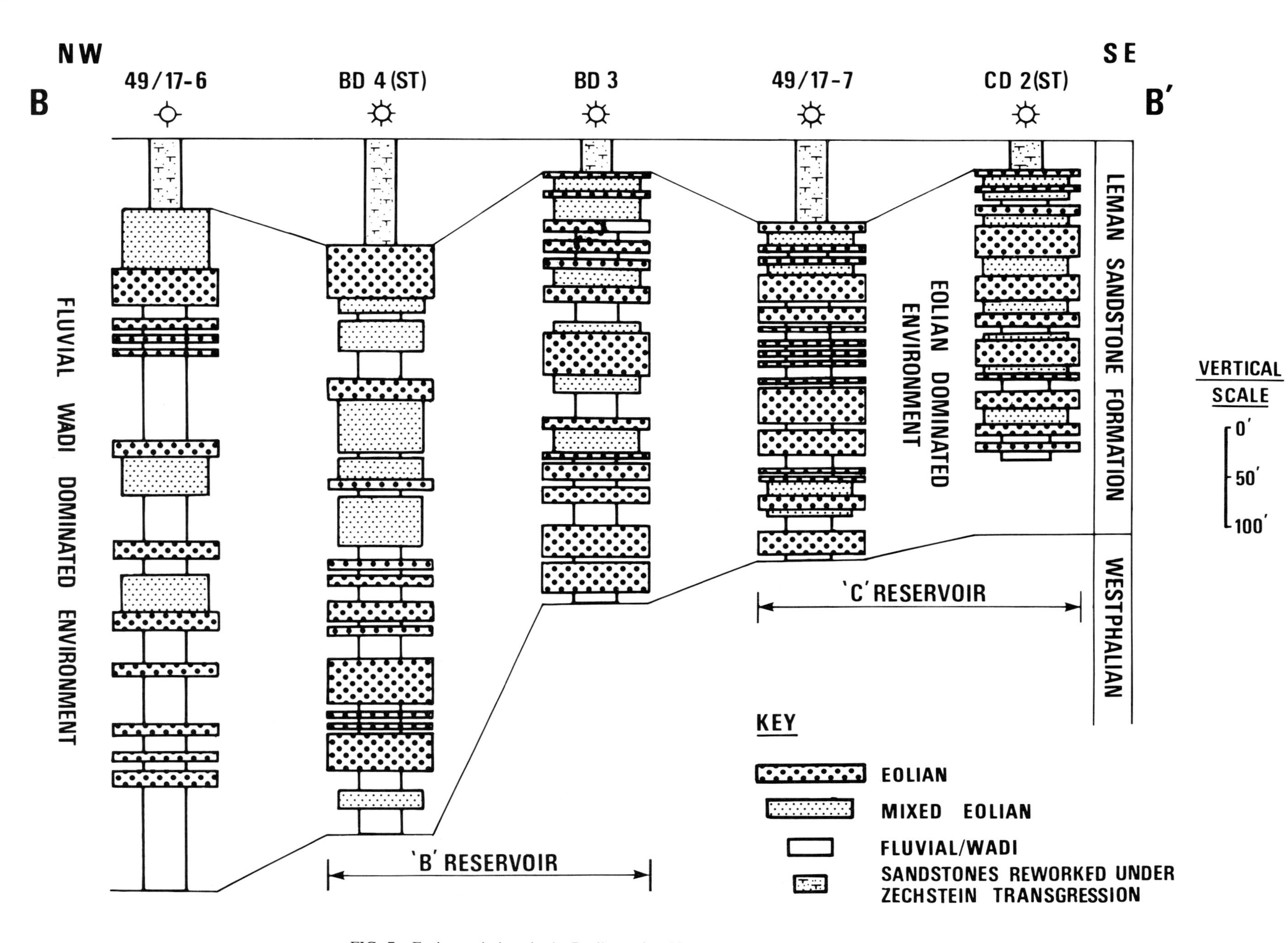

FIG. 7—Facies variations in the Rotliegendes, North Viking gas field, B and C reservoirs. Cross section along B-B' (see Fig. 5 for location).

Table 2. Facies Classification (after Glennie, 1972) & Relation to Permeability.

DUNE SANDS: Well sorted sandstone, fine-medium grained, massive or cross bedded with sedimentary dip of around 27°. Best reservoir, consistently good porosity and permeability.

FLUVIAL SANDS: Argillaceous sandstone, very fine-medium to coarse grained sub horizontal or shallow dips, often with thin shale interbeds. Tops commonly reworked by wind. Fair reservoir, depending on reworking and burial.

WADI & FANGLOMERATIC SANDS: Poorly-sorted course, pebbly sandstones spreading out from the paleo-highs. Typically braided deposits are unpredictable in exact location. Eolian reworking improves reservoir quality towards distal margins. Poor reservoir, improved by eolian reworking and redistribution.

INLAND SABKHA & DESERT LAKE: The inland sabkhas are characterized by fine adhesion - ripple sands and thin clay laminae. Towards the desert margin they become subordinate to shales and silts which merge into the major shale and evaporitic deposits of the desert lake. The silts and shales offer no reservoir potential.

REWORKED GRAY SANDSTONE: A gray sandstone is commonly found occupying the immediate top of the Rotliegendes and varies to over 100 ft (30m) in thickness. It appears to be a marine reworking of the Rotliegendes by the initial Zechstein transgression, an equivalent of the Weissliegendes. Some degree of in-situ percolation of solutions from Zechstein has also contributed to reducing poroperms by the introduction of dolomitic and anhydritic cements.

Several wells, paticularly 49/17-BD4 and 49/17-BD4A, lie within faulted zones and simply low production capabilities. Recent data from the producing pools now indicate that minor faults are similarly acting as local permeability barriers. Well data and observations by Glennie et al (1978) indicate that anhydrite is the most common fracture-filling mineral in faulted zones.

INDEPENDENT POOLS OF THE VIKING COMPLEX

In the development of the Viking, the individual pools have been given letters, A to H. Development wells bear the letter D prefixed by the pool letters. The addition of N as in GN and S as in FS merely refer to new pools established north of the G pool or south of the F pool, respectively. In the following account, the alphabetical designation is used without necessarily being qualified by "pool."

North Viking

The North Viking complex was discovered in March 1969 with the completion of the 49/12-2 well in the Rotliegendes Leman Sandstone Formation. It is a northwest-southeast trending periclinal fault structure, plunging to the northwest at 7° and more gently to the southeast at 1½°. The crest of the structure is at 8,433 ft (2,570 m) subsea in a well AD 10. A major fault along the axis of the structure is downthrown to the southeast between 100 to 1,200 ft (30 to 366 m), separating the A and H from the F pool. The latter occupies a graben whose southwest margin is downthrown around 400 ft (122 m) against the FS pool which lies farther southwest (see Fig. 15).

H and A Pools

The H pool is downthrown approximately 400 ft (122 m) relative to the A pool. This fault appears to completely offset the Rotliegendes at this level, although throw diminishes to the northeast. The fault is effective as shown by pressure measurements. There is no noticeable production across the fault, even though a common gas-water contact has been proved in the 49/12-AD6 and 49/17-5 wells at ± 9,680 ft (2,950 m) subsea.

The Rotliegendes sandstone increases in thickness from 450 ft (137 m) on the northeast margin of A to 600 ft (183 m) in the FS pool. The net sand quality deteriorates in the same direction so that the best reservoir sands are located in A and H. The three porous zones recognized in North Viking are demonstrated in Figure 6. The uppermost zone (I) in the H and A pools is further subdivided into "a" and "b" except in 49/17-5 (H area) where the sand is continuous.

FS and F Pools

Deterioration in net sand quality is especially noticeable in the FS pool and to a lesser extent in the F pool. This area is thought to be on the periphery of a structural low in which the thick fluvial wadi sands (identified in 49/17-4 and 17-6) accumulated.

Associated with the reduction in net sand from A to FS, 2 mi apart, is similar reduction in porosity and permeability:

	A	FS
Average porosity	14%	10%
Average permeability	30 to 80 md	0.1 md

It is possible that porosity in sands may locally improve to the northwest. However, at the present the potential of the FS pool is limited.

The porosity given for the reservoir in A is an average. Typically, figures for a clear eolian sandstone in the Viking field can range from 11 to 25%. The reduction in primary porosity is mostly due to compaction and moderate to locally high pressure solutioning of the quartz grains. The most severe affects of subsidence in the Sole Pit trough are south and west of Viking. Profound reductions in poroperms are due to cementation and diagenetic changes in the clay minerals of the "dirtier" sandstones, as found in the reservoir of FS pool.

According to seismic mapping, F and FS pools should be in communication. However, F shows a gas-water contact at 10,110 ft (3,082 m) subsea and FS at 10,187 ft (3,105 m) subsea. Engineering data indicate significant differences in reservoir and aquifer pressure suggesting two separate reservoirs with a fault separation. The structural relationship between the

FS and F to the A and H pools has been described previously and is illustrated in Figure 15.

South Viking Complex

The B, C, and D structures are adjacent to one another approximately 2½ miles (4 km) southeast of North Viking and share the same northwest-southeast alignment. G, GN, and E pools lie to the west of the B, C, and D cluster.

The structure at the top Rotliegendes is shown in Figure 15. A generalized structural section (Fig. 16) demonstrates the strong faulting of the Rotliegendes. The lithological section shown in Figure 7 demonstrated the changes in reservoir quality, which occur from C and B northwestward into the thick fluvial section in the area of the 49/17-6 well.

The B pool, discovered in the 49/17-3 well in December 1968, is an upthrown triangular fault-bounded block. The southern boundary is upthrown almost 1,000 ft (305 m) against C and similarly the northeastern boundary is substantially upthrown from the D structure.

The elevation at the crest, 8,045 ft (2,452 m) subsea in 49/17-BD3 falls away to the west due to a series of cross-faults which are downthrown to the west. The reservoir is sealed in all directions against Zechstein salt. The Rotliegendes sandstone is sub-horizontal and does not provide structural closure.

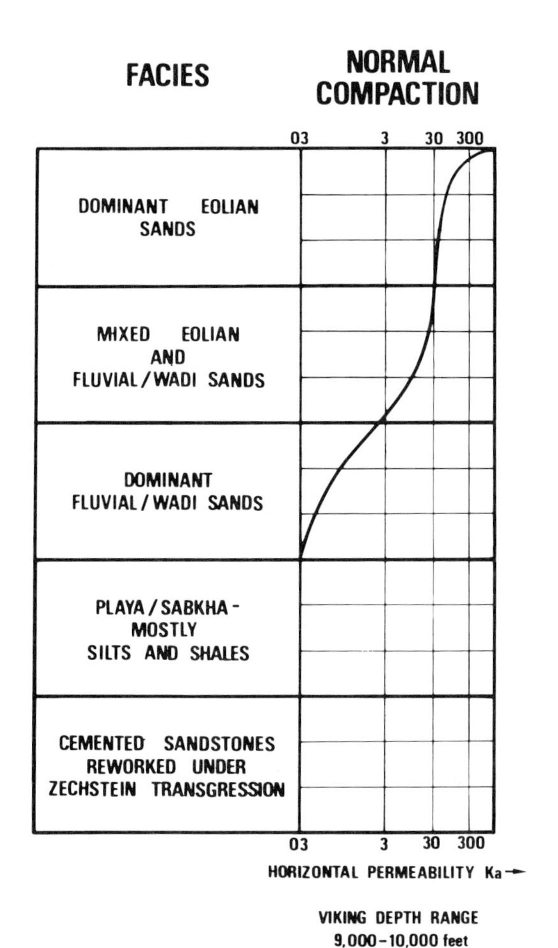

FIG. 8—Generalized facies/permeability relationship in the Rotliegendes of the Viking gas field.

Current interpretation shows that the BD 7 well drilled a separate gas accumulation but shares a common gas-water contact at 9,206 ft (2,806 m) subsea with B structure. The main fault block of B is completely gas-filled and sufficiently high such that potentially productive Carboniferous sandstones also are raised above the gas-water contact.

The fault block which forms B pool is remarkable for its small area (200 acres) and sharp relief, being approximately 1,000 ft (305 m) above surrounding Rotliegendes sandstone. Development drilling from the BD platform revealed how poorly the structure was defined seismically. The current interpretation is based largely on geologic data. Confidence in the present structure is given by a close match of mapped and material balance gas-in-place calculations.

The zones of North Viking are not recognized in B or any other South Viking pool. Wells BD4 and BD4 (sidetrack) were both notably poor producing wells. Their logs suggest a predominantly eolian section so it seems that the poor reservoir performance is caused by secondary cementation, principally by anhydrite along the fault planes which are in proximity to these wells. The top Rotliegendes section has been faulted out in well BD4.

The C pool is elongated in a northwest-southeast direction and measures approximately 5 by 1¾ mi (8 by 2.8 km). It is a horst block tilted southwest at 2½ to 3° at top Rotligendes. The throw of the faults ranges on both the northeast and southwest flanks between 500 and 800 ft (152 to 244 m), so that the Rotliegendes, averaging between 400 and 500 ft (122 and 152 m) in thickness, is completely offset. The northern boundary of the pool is faulted against the upthrown B structure. The C structure at top Rotliegendes rises from 9,100 ft subsea to 8,730 ft subsea (2,774 to 2,661 m) at the crest. A gas-water contact was identified at 9,073 ft (2,765 m) subsea, seen in 49/17-1, the initial discovery well, and confirmed by 49/17-7 located close to the crest of the structure.

The thickness of the sandstone appears similar to that in the 49/17-3 well of B pool, although correlation is tenuous. Some similarity in character exists between the wells but pronounced breaks are not present. Sandstones are interpreted as dominantly eolian. Erosion of the uppermost Rotliegendes is suggested in a correlation of wells CD2 (sidetrack) and 49/17-7.

The D pool is confined to the culmination of a north-northwest–east-southeast horst dipping 3° west-northwest. Spillpoint is reached at 9,060 ft (2,761 m) subsea. The structure is separated from Indefatigable field to the southeast by a fault with a throw of 200 to 250 ft (61 to 76 m). Fault displacement of 600 ft (183 m) forms the south boundary of D pool.

The discovery well, 49/17-2, was completed in May 1968. Rotliegendes sandstone thickness is 485 ft (148 m). It has a progressively mixed lithology towards its base, though it is basically comprised of eolian sandstones. Silty breaks are common in the lower section, providing markers for correlation.

The E pool shares northwest-southeast regional alignment with the other Viking accumulations. It occupies a horst with its long axis bounded northeast and southwest by major faults

<u>EOLIAN SANDSTONE</u> from well 49/16-3 at 9423 feet. This is a well sorted sandstone with good porosity and permeability. Cements are restricted to thin clay films on the quartz grains and some patchy carbonate

(x50 plane polarised light)

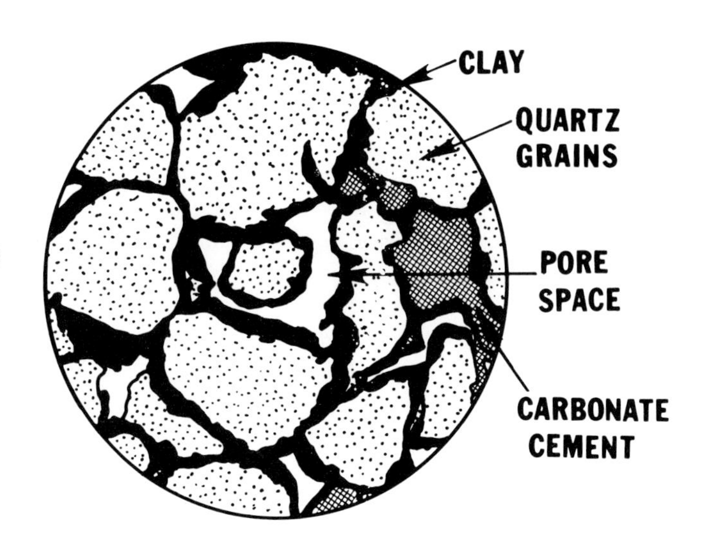

<u>EOLIAN SANDSTONE</u> from well 49/17-4 at 9302 feet. This is a bimodally sorted sandstone where any visible porosity has been destroyed by the presence of abundant authigenic kaolinite and some patchy carbonate cement

(x50 plane polarised light)

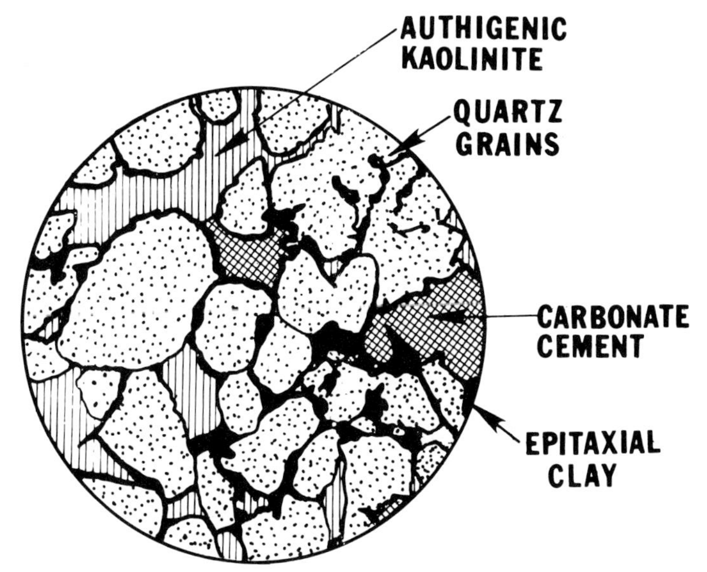

<u>EOLIAN SANDSTONE</u> from well 49/17-4 at 9335 feet . A well sorted sandstone in which the original porosity has been destroyed by anhydrite and rare quartz cement

(x40 plane polarised light)

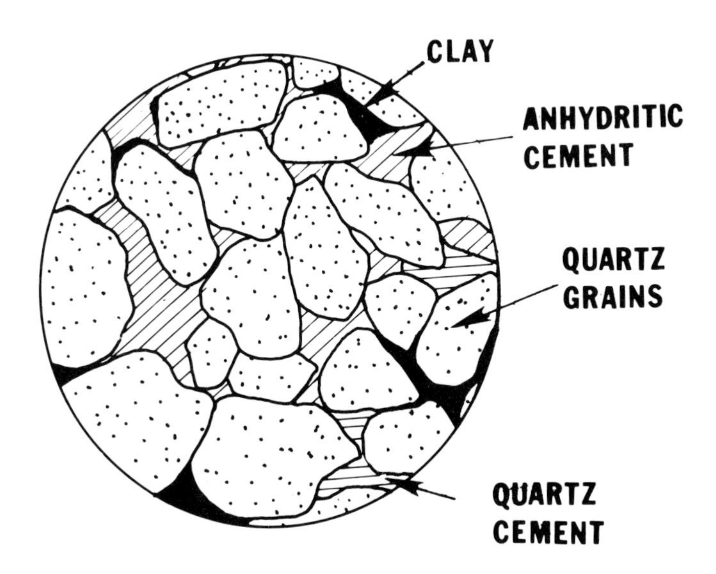

FIG. 9—Sketches of thin sections repesentative of eolian sandstones from two Viking wells. Samples from the Rotliegendes sandstone.

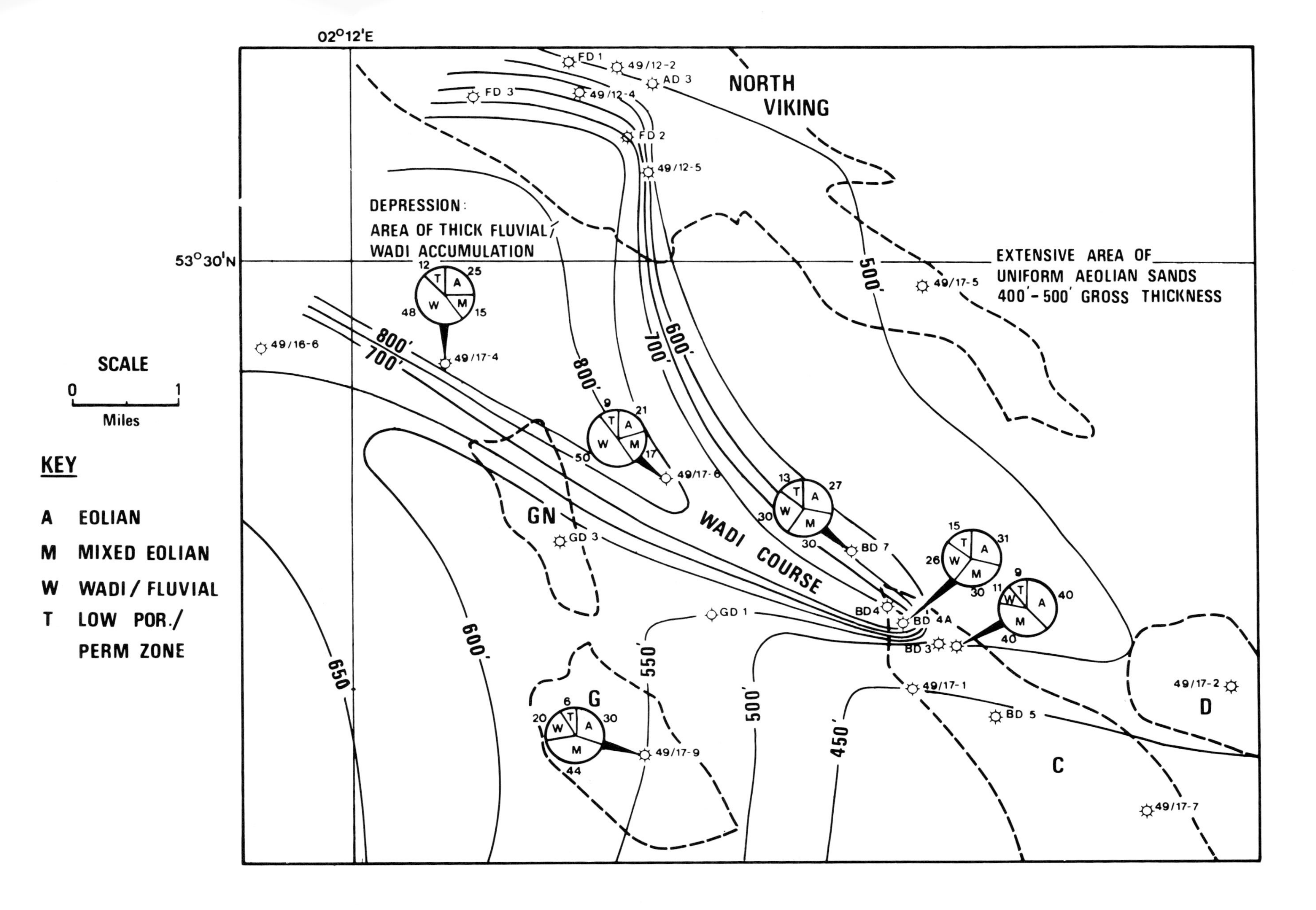

FIG. 10—Facies isopach of Rotliegendes sandstones in the Viking field.

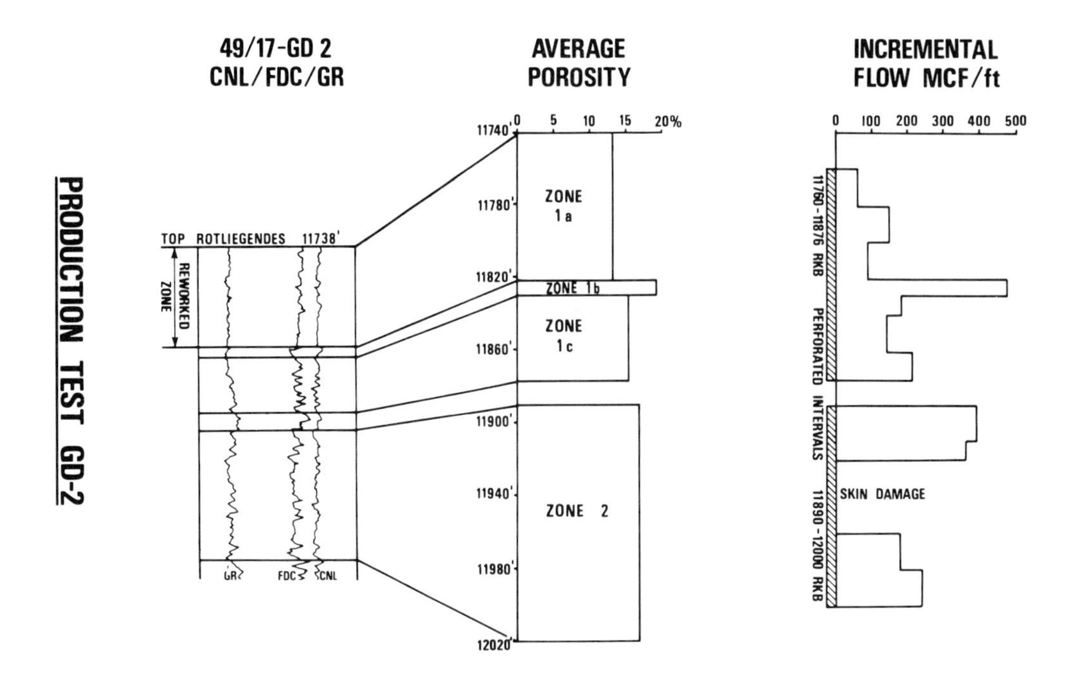

FIG. 11—Production test for well GD-2, Viking field. Porosity of zones within the Rotliegendes sandstone as compared to flow.

downthrown as much as 900 ft (274 m). Reservoir dimensions are approximately 4½ mi by ½ mi (7.2 km by 0.8 km), with 360 ft (110 m) maximum net vertical pay. Structural dip of 3° closes the structure in the southeast. To the northwest, the structure spills at 9,091 ft (2,771 m) subsea (gas-water contact). Minor cross-faults interrupt the reservoir along its axis.

The Rotliegendes sandstone thickens from 700 to 800 ft (213 to 244 m) in a northwesterly direction through the block. Well 49/16-ED3 found the Rotliegendes at 8,687 ft (2,748 m) subsea, close to its structural culmination. Wells ED2 and 2A, drilled on the southeastern limits of the structure, revealed that a branch of the principal southwest-bounding fault had downthrown the Rotliegendes sandstone 250 ft (76 m), thus limiting the gas-bearing reservoir in this direction. ED2A found only 26 ft (8 m) of Rotliegendes sandstone above the gas-water contact.

The E pool is one of the best producers of the Viking field. With limited exception, the sands are interpreted to be within the dune facies, units being 30 to 60 ft (9 to 18 m) in thickness and separated by thin, tight, bottom-set beds. Only wells 49/16-3 and ED3 completely penetrated the Rotliegendes section and both display facies deterioration in the lowest part of the section.

The G pool was discovered by the drilling of well 49/17-9, which was completed in February, 1973, and which shows 201 ft (61.3 m) of gas-bearing sandstone in a gross Rotliegendes thickness of 554 ft (169 m). In an attempt to prove an extension of the reservoir, well 49/17-GD1 was subsequently drilled into a structure to the north, and well 49/17-GD3 into a structure to the northwest. 49/17-GD1 was dry; it probably was located low, on a depressed structure south of the B pool. 49/17-GD3 found gas-bearing Rotliegendes and is now designated the *GN pool*.

The two accumulations demonstrate the spilling mechanism in the area. Gas in G appears to have spilled towards GN and from there into the 49/17-4 area. This area coincides with a thickening Rotliegendes section and the interpreted wadi-fluvial trend through 49/17-6. As such, this additional area offers little potential as a reservoir.

The G structure is a horst, regionally aligned northwest-southeast measuring approximately 2 mi by 1 mi (3.2 km by 1.6 km) and dipping 2½° to the northwest. The structure is fault-bounded except to the northwest and is contained in that direction by the structural dip. It spills at 9,530 ft (2,905 m) subsea across the northeastern fault into the GN pool. Maximum elevation of the structure is approximately 9,200 ft (2,804 m) subsea. Gross sandstone thickness averages 550 ft (168 m) and gradually thickens westward.

The GN pool occupies a similar horst position. Its structural dip to the west at 4½° confines the pool against the eastern bounding fault, which is downthrown around 600 ft (183 m). Dimensions are approximately 2 mi by 3/4 mi (3.2 km by 1.2 km). Gross Rotliegendes sandstone thickness in GN averages 540 ft (165 m) with maximum net vertical pay of 124 ft (37.8 m). The reservoir spills at 9,350 ft (2,849 m) subsea across the fault towards 49/17-6 well.

The Rotliegendes sandstone in both G and GN is comprised of similar dune sands. Each 10- to 20-ft (3 to 6 m) thick unit is isolated by sharp bottom-set breaks or thin fluvial horizons. The uppermost 40 ft (12 m) of sand have low porosity, probably caused by reworking and cementation by percolating solutions from the Zechstein. These wells, together with the GD1 well drilled into the northern prospect establish the presence of dune sands to the southwest of 49/17-4 and 49/17-6.

A tentative correlation is carried into the 49/17-4 well from GN pool. It seems that the major fluvial breaks of the two

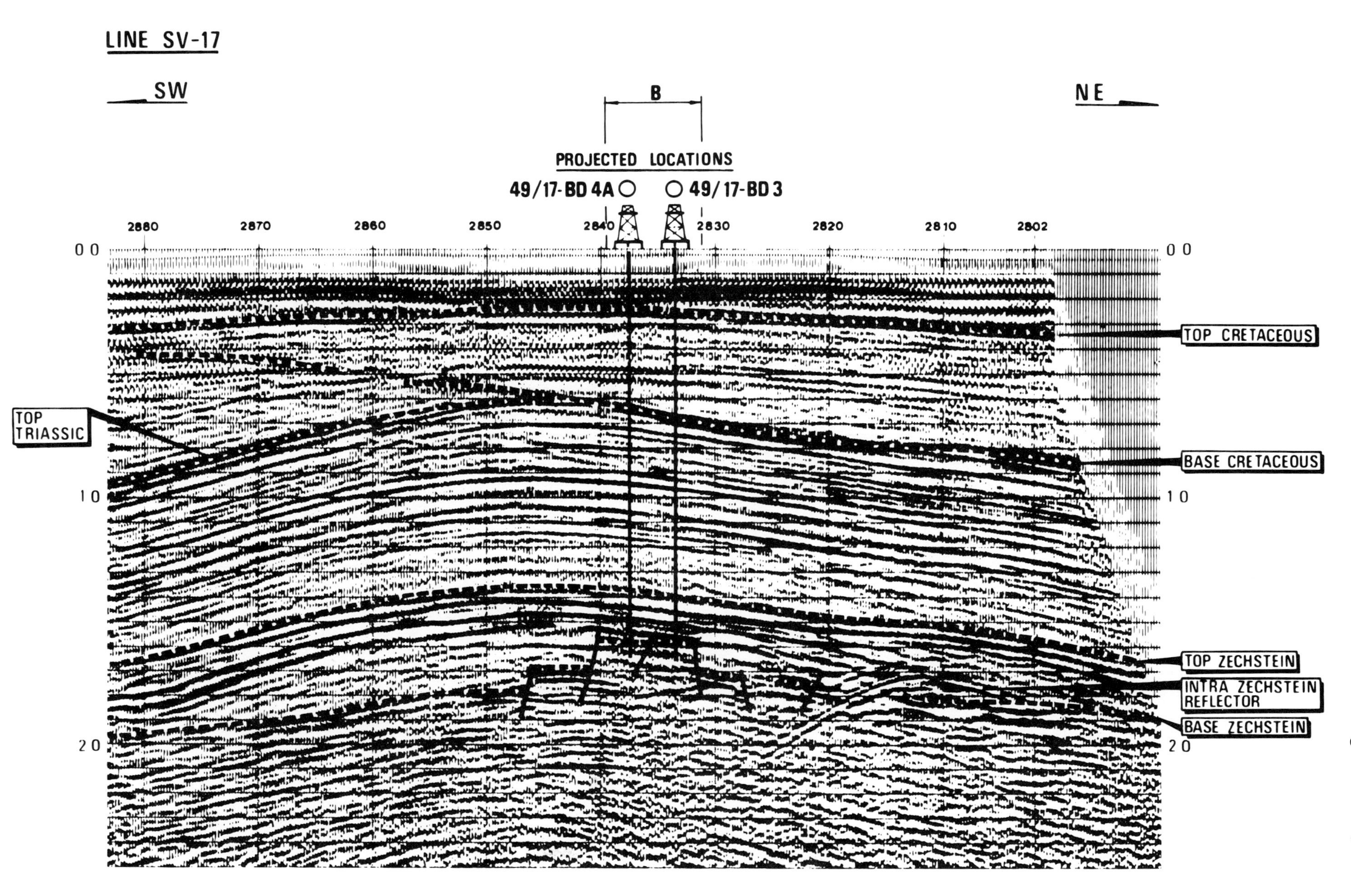

FIG. 12—Seismic line shot along line SV-17, showing stratigraphy through Viking B horst block.

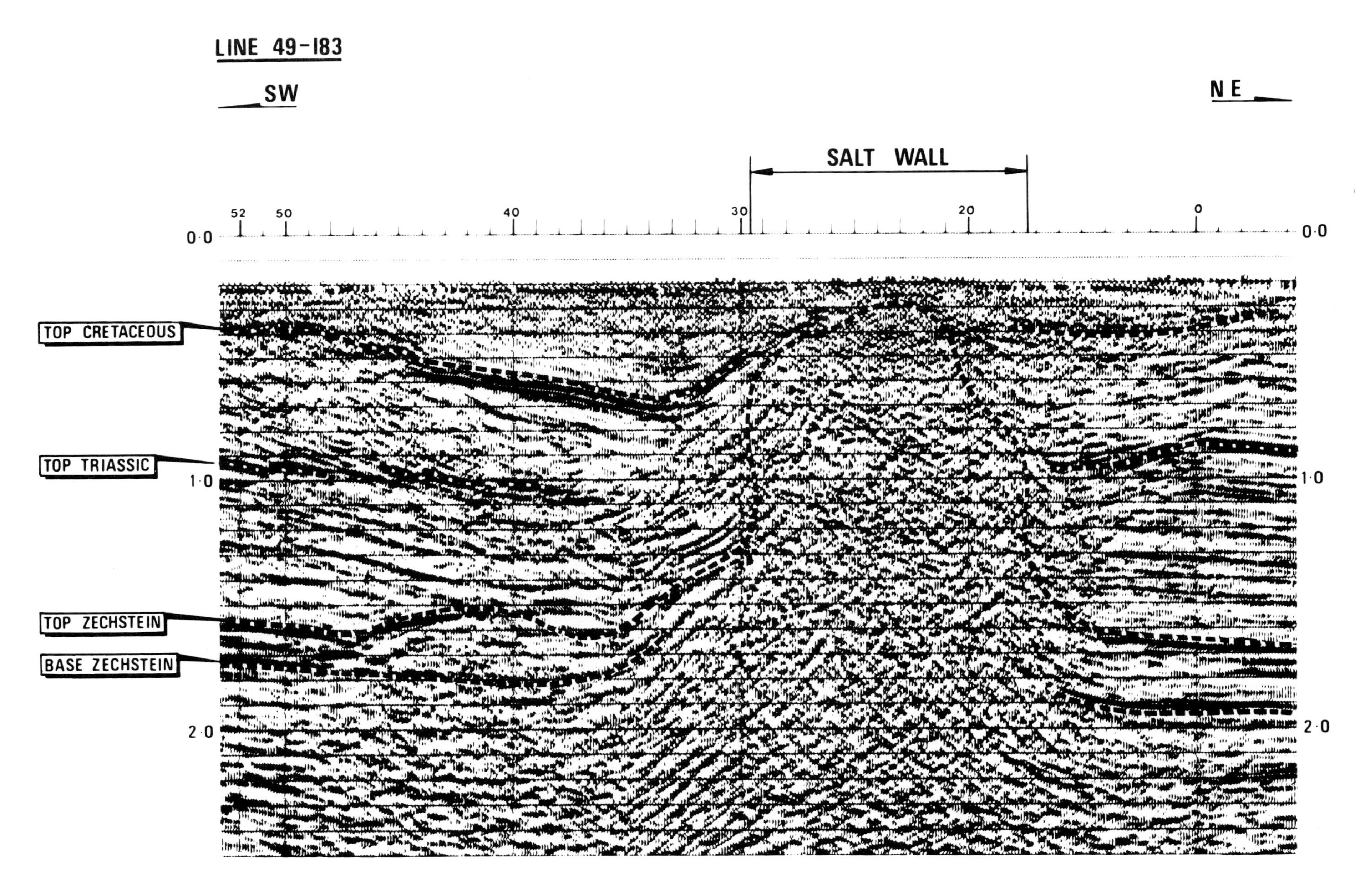

FIG. 13—Seismic line shot along line 49-183, showing stratigraphy through salt wall.

wells may correspond, though the 49/17-4 section is broken up by fluvial-wadi sediments.

To the north of the E reservoir, well 49/16-6 discovered a small gas accumulation which is uneconomical. The gas is contained in a structure aligned northwest-southeast and measures approximately 1 mi by 4 mi (1.6 km by 6.4 km). It is partly offset by cross-faults downthrown to the west. Faults downthrown northward define the pool margin in the southeast and northwest.

A combination of factors make this reservoir uneconomical. The structure is fairly flat, with a crest at approximately 9,350 ft (2,850 m) subsea and a gas-water contact at 9,438 ft (2,877 m) subsea. The facies of the complete Rotliegendes section is an alternation of dune and fluvial sands and has a conglomeratic wadi deposit at its base. The uppermost 120 ft (42 m) contains a high degree of anhydritic and dolomitic cements. Due to low structural relief, the better sands (beneath the tight zone) are below the gas-water contact.

CONCLUSIONS

The Viking field comprises 10 commercial gas pools producing from Lower Permian Rotliegendes desert sandstone; 3.6 Tcf of gas in place is estimated. This has been generated from the underlying Carboniferous Westphalian coal measures and trapped in the Rotliegendes by the Zechstein evaporite seal.

The Rotliegendes has been faulted into a number of northwest-southeast orientated horsts and (provided that a seal is present) it is reasonable to expect each structure to contain gas.

The quality of the reservoir sandstone varies considerably and is difficult to predict. Original porosity has been reduced in the eolian sandstones by pressure solution whereas cementation and clay minerals reduce both porosity and permeability. Introduction of clays into eolian sands can virtually eliminate permeability in a once good reservoir.

A wadi deposit is proposed, bisecting the north and south sections of the Viking field, and could account for an area of poor reservoir sandstones.

The structures are mapped by a layer cake construction on the base Zechstein, but must consider varying thickenss of Zechstein salt, anhydrite, and carbonate layers within the Zechstein, and a laterally changing velocity gradient in the overlying section.

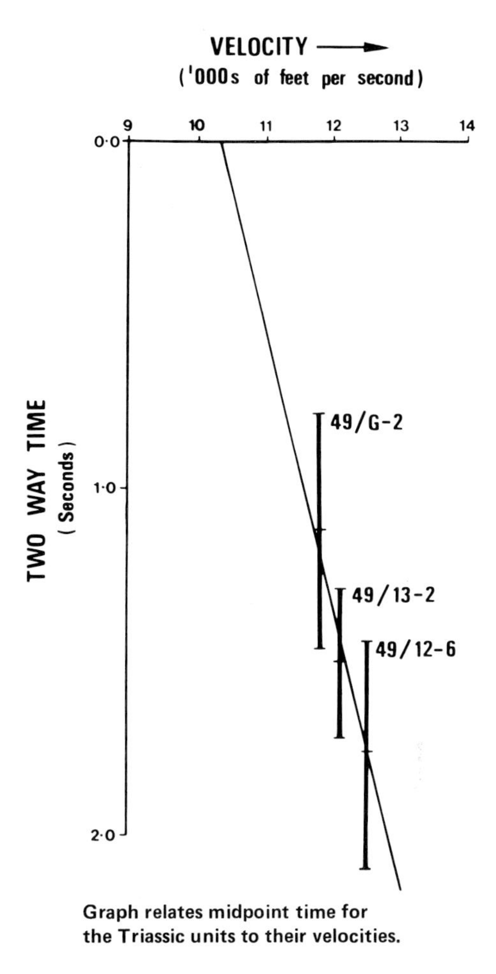

FIG. 14—Time/velocity chart plotted for Triassic units in Viking field.

REFERENCES CITED

France, D. S., 1975, The geology of the Indefatigable gas field, *in* A. W. Woodland, Petroleum and the continental shelf of northwest Europe, vol. 1: Barking, Essex, Applied Science Publishers, p. 233-239.

Glennie, K. W., 1972, Permian Rotliegendes of North-West Europe interpreted in light of modern desert sedimentation studies: AAPG Bull., v. 56, p. 1048-1071.

———G. C. Mudd, and P. J. C. Nagtegaal, 1978, Depositional environment and diagenesis of Permian Rotliegendes sandstone in Leman Bank and Sole Pit areas of the U.K. southern North Sea: Geol. Soc., London Jour., v. 135, p. 25-34.

Gray, I., 1975, Viking gas field, *in* A. W. Woodland, Petroleum and the continental shelf of northwest Europe, vol. 1: Barking, Essex, Applied Science Publishers, p. 241-247.

Marie, J. P. P., 1975, Rotliegendes statigraphy and diagenesis, *in* A. W. Woodland, Petroleum and the continental shelf of northwest Europe, vol. 1: Barking, Essex, Applied Science Publishers, p. 205-210.

Rhys, G. H., (comp.), 1974, A proposed standard litho-stratigraphic nomenclature for the southern North Sea and an outline structural nomenclature for the whole of the (U.K.) North Sea. A report of the Joint Oil Industry—Institute of Geological Sciences Committee on North Sea Nomenclature: Inst. Geol. Sci. Ann. Rept. 74/8.

Schlumberger, 1974, Well evaluation conference—North Sea.

van Veen, F. R., 1975, Geology of the Leman gas field, *in* A. W. Woodland, Petroleum and the continental shelf of northwest Europe, vol 1: Barking, Essex, Applied Science Publishers, p. 223-231.

Woodland, A. W., ed., 1975, Petroleum and the continental shelf of northwest Europe, vol. 1: Barking, Essex, Applied Science Publishers.

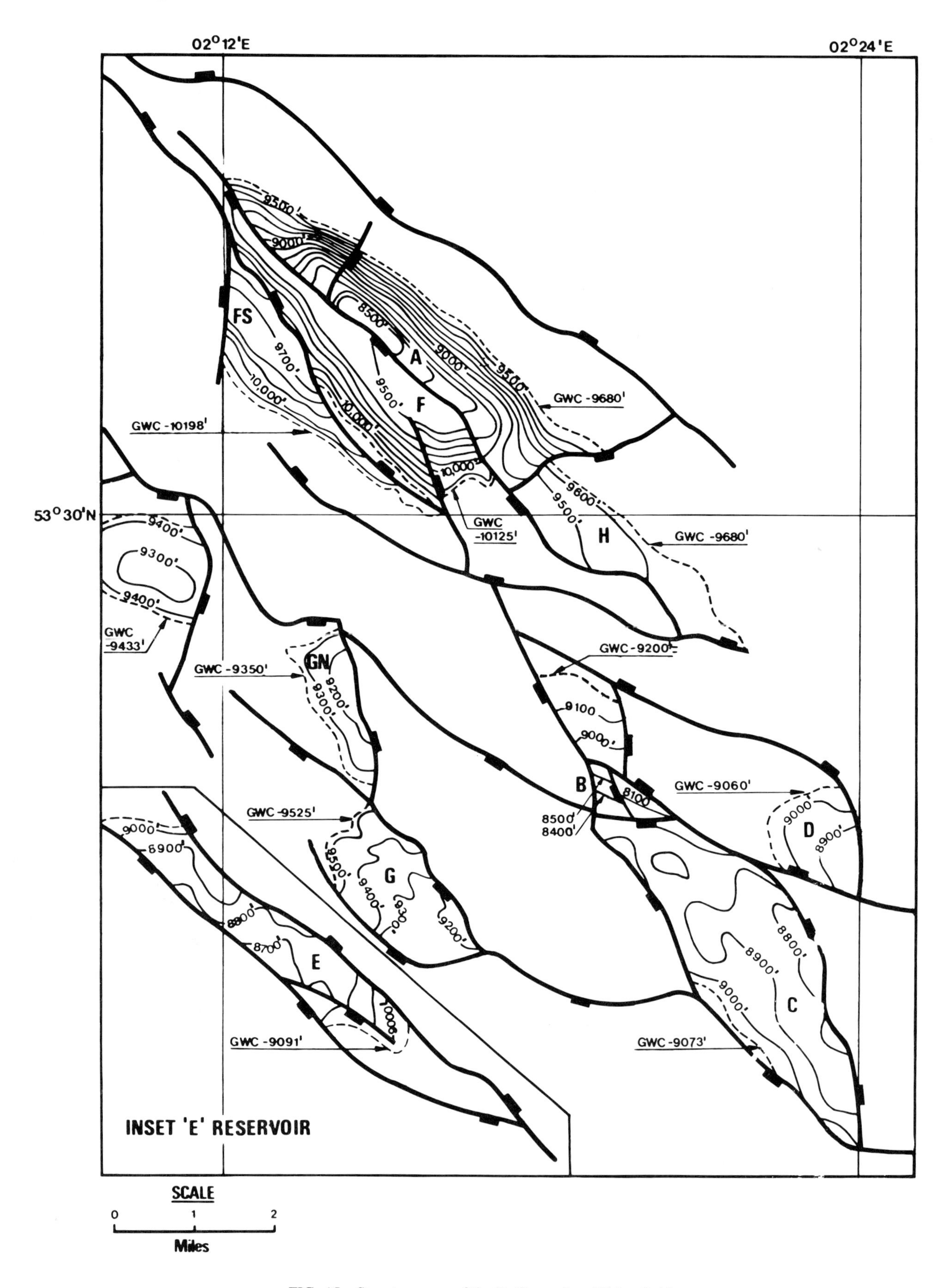

FIG. 15—Structure map of the Rotliegendes, Viking field.

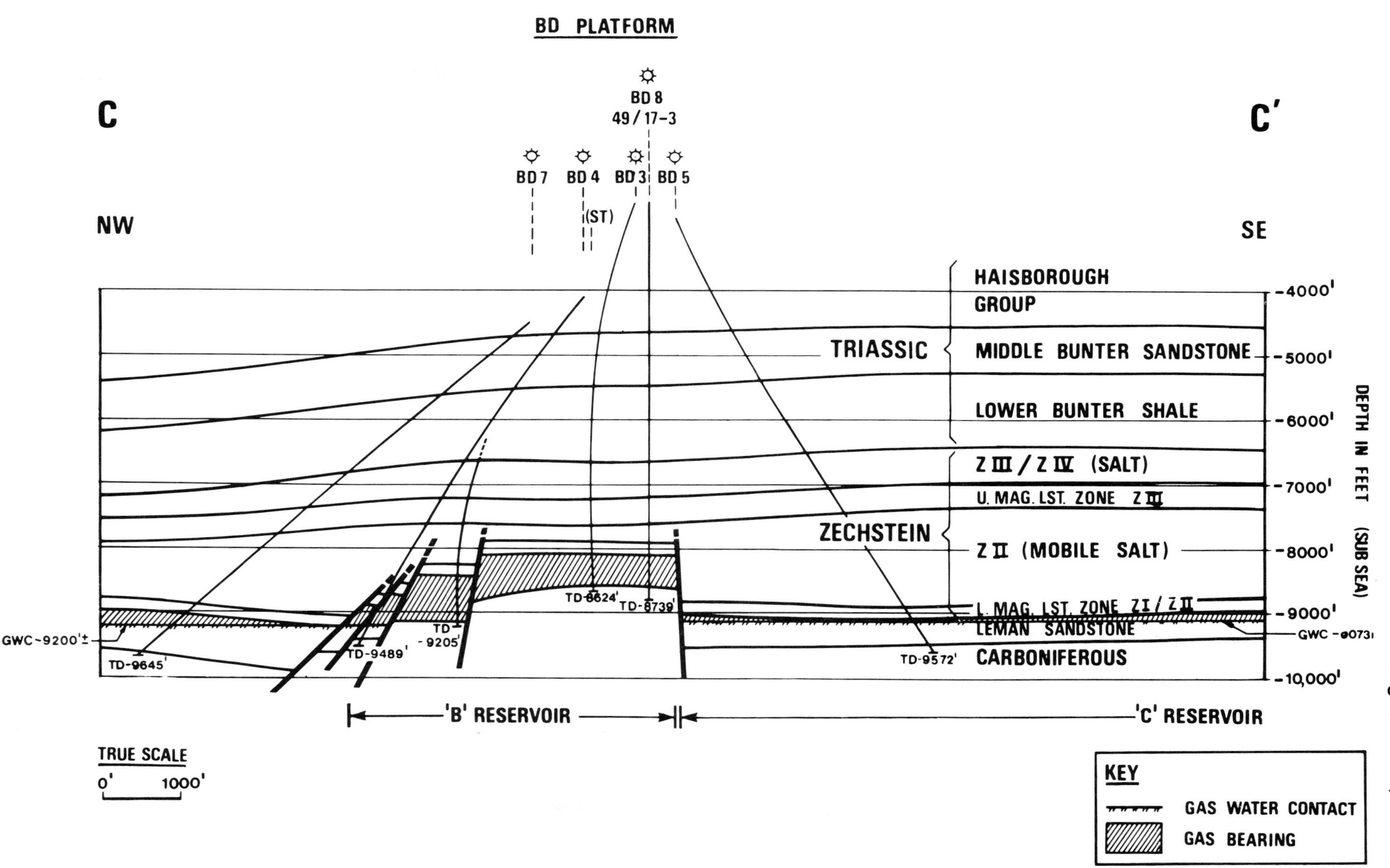

FIG. 16—Structure section of B and C reservoirs, South Viking field, compiled from well logs at the BD platform.

Frigg Field—Large Submarine-Fan Trap in Lower Eocene Rocks of the Viking Graben, North Sea[1]

F. E. Heritier[2], P. Lossel[3]
and E. Wathne[4]

Abstract In the deepest, axial part of the Viking subbasin of the North Sea, the Frigg Field, one of the world's largest offshore gas fields, straddles the border of the British and Norwegian continental shelf at 60° N lat. The discovery well was drilled in 1971 on Norwegian block 25/1 in 100 m of water. Gas was discovered at a depth of 1,850 m in a lobate submarine fan representing the ultimate phase of a thick Paleocene deposit.

Sealed by middle Eocene open-marine shales, the structure is mainly submarine-fan depositional topography enhanced by draping and differential compaction of sands. The area of structural closure is underlain by a typical "flat spot" on seismic sections, and the gas column lies on a heavy oil disc. Chromatographic analysis shows that source of both the oil and gas could be the underlying Jurassic section.

Recoverable gas reserves are estimated to be about 200 billion cu m (7 Tcf). Production began September 15, 1977; the gas is brought ashore at St. Fergus in Scotland by a 360-km pipeline.

INTRODUCTION

The Frigg field straddles the line between the Norwegian and the United Kingdom sectors of the North Sea at lat. 59°50′N and long. 2°E. It is located approximately 190 km west-northwest of Haugesund, Norway, 180 km east of the Shetland Islands, and 390 km northeast of Aberdeen, Scotland (Fig. 1).

On the Norwegian side, the Frigg field lies within the Petronord group Block 25/1, with only a small part extending into Esso Block 30/10. On the United Kingdom side, the field lies mostly on Total Oil Marine and Elf Aquitaine License Blocks 10/1, 10/6, and 9/10a. A small part of the field extends into BP License Block 9/5.

Average water depth throughout the field is approximately 100 m.

Frigg was named after a goddess of the Norwegian pantheon.

HISTORY OF DISCOVERY

In 1965, the Petronord group, created by Elf Aquitaine, Total Oil Marine Norsk, and Norsk Hydro to search for oil on the Norwegian continental shelf south of 62°N lat., commenced an active program of seismic mapping which eventually led to the discovery of the Frigg field in July 1971.

[1] Manuscript received April 26, 1978; accepted May 29, 1979.

[2] Societe Nationale Elf Aquitaine Norge (Production), Paris, France.

[3] Total Oil Marine Norsk, Paris, France.

[4] Norsk Hydro, Sandvika, Norway.

The writers thank many colleagues for valuable contributions and discussion, in particular, F. C. Duffaud, for his help in preparing this paper. The writers also thank their respective companies, as well as Statoil, for permission to publish this paper.

 This manuscript also has been published in the AAPG Bulletin, v. 63, p. 1999-2020 (November 1979).

Article Identification Number:
0065-731X/80/MO30-0004/$03.00/.

In 1966, the first interpretation of a very loose seismic grid 15 × 20 km showed the Frigg structure on a mapping horizon which, at the time, was considered to be the top of the Upper Cretaceous chalk.

After the 1968 discovery by the Phillips Petronord group of the Cod field in Paleocene sands, more seismic lines were shot in the Tertiary basin along the border of the Norwegian and United Kingdom waters to make a selection of blocks in the second round of licensing.

In 1969, four of the eight blocks applied for were granted to the Petronord group under conditions whereby the Norwegian State could enter as a partner if a commercial discovery were found.[5] Block 25/1 and 25/2 were among the blocks granted. At that time a new 5 × 5 km seismic grid was shot. The interpretation of this survey permitted the definition of the Frigg structure, particularly at the top of the basal Tertiary sands.

On June 8, 1970, the United Kingdom Blocks 10/1 and 10/6 on the western flank of the structure were granted by the British authorities to the Total Oil Marine and Elf Aquitaine group.

In 1971, the official agreement between the Petronord group and the Norwegian State made possible the drilling of the first well on the Frigg structure. The Petronord group spudded its first well in Block 25/1 using the *Neptune P 81* semisubmersible rig on March 30, 1971.

Well 25/1-1 was located at the crest of a very large but relatively low-relief structure with an area of closure on the seismic marker, then equated with the top of the Paleocene, of about 300 sq km.

At 1,812 m subsea, the well entered gas-bearing sands of early Eocene age. After penetrating a 135-m gas column, it entered an oil-bearing zone about 10 m thick. Gas-oil contact was established at 1,947 m subsea, and gas was subsequently tested at a rate of 24 MMCFD (675,000 cu m/day) with a ⅞-in. (2.2 cm) choke. A major gas field had been discovered.

[5] Den Norsk Stats Oljeselskap A/S (Statoil) became a partner of the Norweigian group in May, 1973.

FIG. 1—Index map of central and northern North Sea.

After that discovery, and to supplement the earlier work, two seismic surveys, totaling 550 line-km, were shot in both United Kingdom and Norwegian waters, and simultaneously a four-well appraisal program was started.

The first appraisal well, 25/1-2, was spudded in July 1971 about 5.5 km north of the discovery well and found a gas column of 49 m and an oil column of 10 m which confirmed the original gas discovery and proved an extension of the structure toward the north.

The second appraisal well, 25/1-3, was spudded in November 1971 about 5 km east-northeast of 25/1-1 and found a gas column of 17 m and an oil column of 10 m which proved a large extension of the structure toward the east.

The results of reinterpretation based on the 1971 seismic surveys, combined with data obtained from the first three wells, indicated a smaller and more complicated structure than had been envisaged originally.

The third appraisal well, 10/1-1 was spudded on the United Kingdom sector about 7 km southwest of 25/1-1, in a crestal position, and found a gas column of 92 m overlying 11 m of oil.

The fourth appraisal well, 10/1-2, was spudded about 4 km west of 25/1-1 to define the westward development and thickness of the Frigg sand facies. At a depth of 1,951 m, the well entered the oil zone in the lateral equivalent of the Frigg sand. This formation showed a facies of thin sand layers interbedded with shales, thus defining the western extension of the field.

At the same time, Esso drilled a stepout well in its Block 30/10, about 13 km north of 25/1-1, which found 2.5 m of gas-bearing sand and 7.5 m of oil-bearing sand that proved a small extension of the field into Esso 30/10 Block.

As the previous seismic grid was considered insufficient, a new detailed seismic survey consisting of 1,880 km of profiles was shot during the summer of 1973, along a regular grid of 1 × 1 km on the structure and 2 × 2 km off structure. The interpretation of this detailed survey and the results of the first six wells have thus confirmed a major gas field with a maximum gas column of 170 m in lower Eocene sands covering an area of about 115 sq km for the main accumulation. The decision to develop the find was made at that time.

REGIONAL SETTING

The Viking basin is a thick sedimentary domain which lies between the Norwegian shield on the east and the Shetland platform on the west, and between lat. 58 and 62°N. It is a collapsed area typical of rifts, which in terms of plate tectonics is the result of an aborted opening of the northern European continental shield (Fig. 2).

The basin's history involves two major geologic periods separated by the Cimmerian orogeny. The first period was a positive epicontinental sedimentary cycle which began with deposition of the red continental Triassic beds and ended with

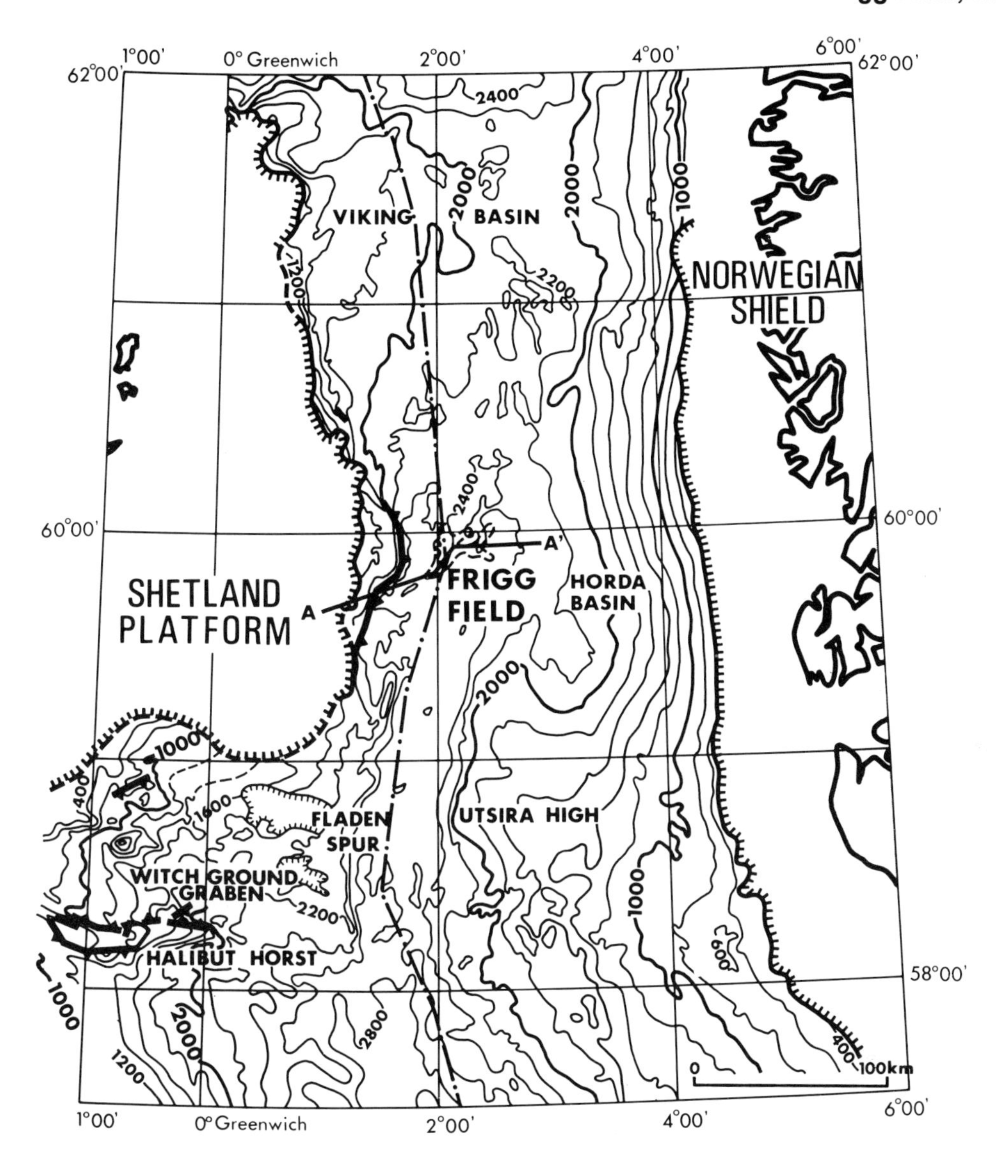

FIG. 2—Viking basin, structure map of top of Cretaceous. C.I. = 200 m. AA' is location of cross section shown in Figure 3.

Oxfordian marine shales. The second was a negative open-marine sedimentary cycle, which began with deposition of deep-marine black shales of Kimmeridgian age and ended in the present shallow-marine conditions.

Pre-Cimmerian Period

The first cycle comprises successively the continental Triassic red clastics, the lower to middle Lias fluviodeltaic sandy deposits, middle to upper Lias Marine shales, Dogger regressive deltaic sandstones, and the Callovian-Oxfordian marine shales. The source of the clastics was mainly the Norwegian shield in the northeast. Oxfordian clastics in the southern part of the basin were derived from the Shetland platform.

The major facies and thickness changes of the second geologic period were related to a synsedimentary breakup of the basin into individual subbasins, or tectonic trends, bounded by large antithetic normal faults.

In the Frigg area, two of these subbasins (the Beryl embayment and the Frigg subbasin) are present. Figure 3, a geologic cross section, shows the presence beneath the Cimmerian unconformity of three major fault blocks—the Bruce, Frigg, and East Frigg structures.

Post-Cimmerian Period

After the Cimmerian orogeny, which was a phase of intense tectonic activity, erosion, and relief inversion at the end of the middle Oxfordian, the basin received much wider transgressive infilling. Sedimentation started in late Oxfordian-Kimmeridgian time with the deposition of richly organic, black radioactive shales characteristic of an euxinic deep-water marine environment.

During Early Cretaceous time a stronger connnection into the Boreal sea became established; a Lower and middle Cretaceous shaly sequence onlaps the fault-block relief. The pres-

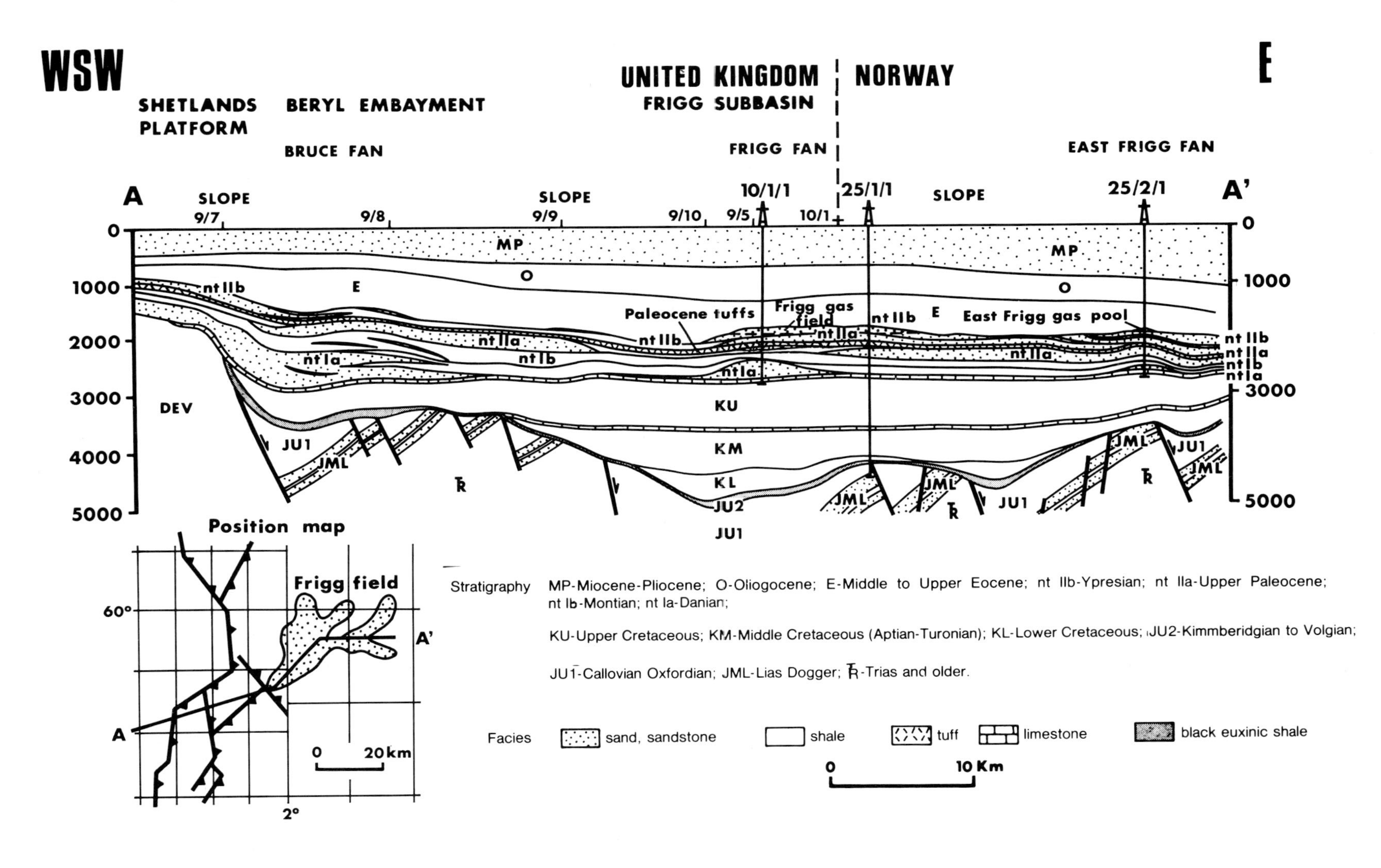

FIG. 3—Geologic cross section from Shetland platform to axis of Viking basin shows relations of pre-Cimmerian structures to successive Paleocene and Eocene fans. Depths are in meters. Location of section shown in Figure 2.

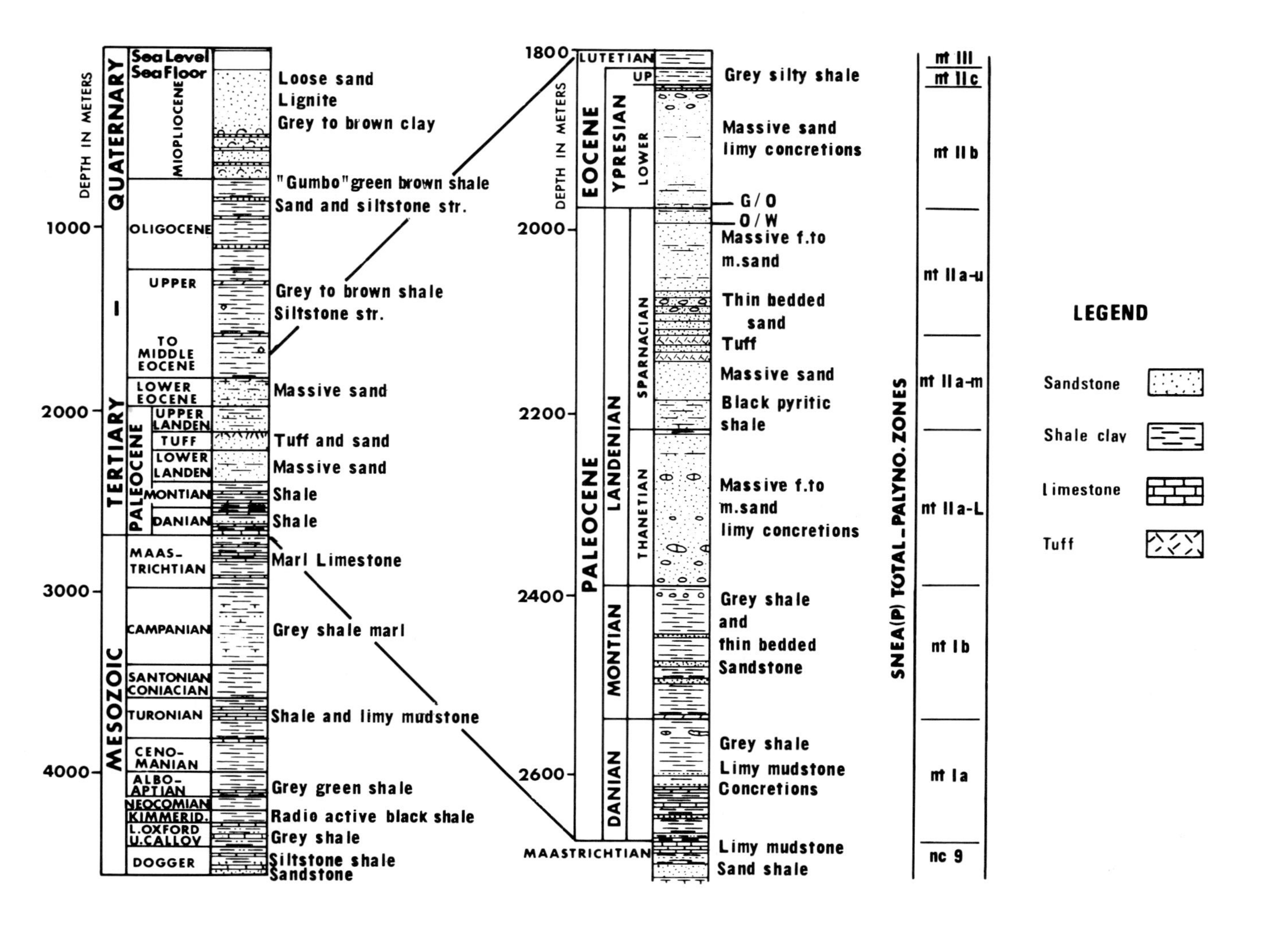

FIG. 4—Composite stratigrapic section of Frigg field showing palynologic zones of Paleocene and Eocene.

Table 1. Early Tertiary Faunal Assemblages, Frigg Field

Zones	Dinoflagellates	Associated Microflora	Proposed Age
nt III	*Areosphaeridium dictyoplokus*	Gymnosperms	Middle Eocene Lutetian
nt IIc	*Membranilarcia ursulae* *Wetzeliella edwardsii*	Gymnosperms Angiosperms	Early Eocene Ypresian
nt IIb	*Wetzeliella coleothrypta*	Angiosperms Gymnosperms	
nt IIa upper	*Detlandrea phosphoritica* *Wetzeliella pachyderma*	*Taxodiaceaepollenites hiatus* *Sequoiapollenites sp.*	Late Paleocene (Landenian) Sparnacian (to Ypresian)
nt IIa middle	*Wetzeliella meckenfeldensis* *Wetzeliella cf. lunaris*	*Taxodiaceaepollenites hiatus* *Sequoiapollenites sp.*	Thanentian (to Sparnacian)
nt IIa lower	*Wetzeliella hyperacantha* *Wetzeliella homomorpha*	*Caryapollenites sp.* *Platycaryapollenites sp.*	
nt Ib	*Areoligera senonensis* *Deflandrea speciosa*	Gymnosperms Pteridophytes, angiosperms	Early Paleocene Montian (to Thanetian)
nt Ia	*Eisenackia crassitabulata* *Paleoperidinium phrophorum*	Gymnosperms Pteridophytes	Danian

ence of a few limestone beds, of regional extent, emphasizes the eustatic low stand of the sea at the end of the Aptian, Cenomanian, and Turonian Stages.

The Upper Cretaceous in the Frigg area consists mostly of shale but includes chalky limestone beds of Campanian and Maestrichtian ages. The Upper Cretaceous became mostly pure chalk in the southern North Sea.

In the Frigg area, inversion of structural relief related to the collapse of Utsira high occurred at the end of the Cretaceous. That event, together with a correlative rejuvenation of the Shetlands-Orcadian belt on the west, was the main cause of the strong offlap of sediments from west to east which characterizes the Tertiary of the Frigg area.

The Paleocene sediments include a large amount of clastic material that originated in the west and was brought into the basin by turbidity currents creating fan complexes at the foot of both the Shetland escarpment and the Fladen Group spur.

After the Paleocene regression, a new phase of marine shaly sedimentation started during the Eocene. It began with a very important clastic influx: The Ypresian Frigg sands.

The Oligocene is also represented by predominantly shaly marine sedimentation, but sand deposition became more frequent at that time and finally predominated during the last regressive Miocene-Pliocene period.

STRATIGRAPHY

Frigg 25/1-1

The Frigg discovery well 25/1-1, whose stratigraphic column is typical of the entire field, was drilled through a predominantly sandy Pliocene and Miocene section of pronounced continental character (Fig. 4). Under that section was a greenish-brown, soft, silty mudstone of late middle Oligocene age underlain by a monotonous section of brown, soft gumbo clay of early Oligocene to middle Eocene age.

Below a 58-m (190 ft) section of apple-green, soft, pyritic clay followed by brown-red silty shales, the drill reached the top of the Frigg sand at 1,836 m. This sand is of early Eocene age and constitutes the gas reservoir.

The top of the Paleocene, identified on the basis of palynologic evidence, was found at 1,976 m, about 175 m above the tuffaceous layers which are a prominent seismic marker over most of the North Sea. The total thickness of the Paleocene, which is predominantly sand, is 593 m.

Beneath the Paleocene sands are 104 m of greenish, silty shales with dolomite stringers of Danian age that overlie a very thick 1,322-m shaly section of Late Cretaceous age. The uppermost part of this sequence, consisting of alternations of chalky limestones and shales, is Maestrichtian; the more clacareous lower part is Turonian. Below lie 208 m of dark-gray undercompacted shales of Early Cretaceous age.

At this point the drill reached the strong deep seismic reflector which corresponds to the late Cimmerian unconformity and went through 77 m of highly radioactive (250 API) oil shale of Kimmeridgian age.

Below the Kimmeridgian shales are 282 m of dark-gray pyritic and calcareous shales of Callovian age, with numerous dolomite stringers in the upper part, then finally the water-bearing Dogger sandstone. Total depth is 4,570 m.

Paleocene-Lower Eocene—Frigg Field

The formations of the lower Tertiary are almost devoid of fauna and microfauna. Though the Maestrichtian is characterized by an abundant marine microfauna including both pelagic and benthonic forms, only some Globigerinidae are present in the Danian, and the later series provides only common arenaceous forms. Thus, the stratigraphy is mainly based on palynologic assemblages of both marine microplankton (dinoflagellates) and continental microflora (pollens and spores) which permit very accurate dating. These assemblages allow us to define eight zones, for which an age attribution is proposed as in-

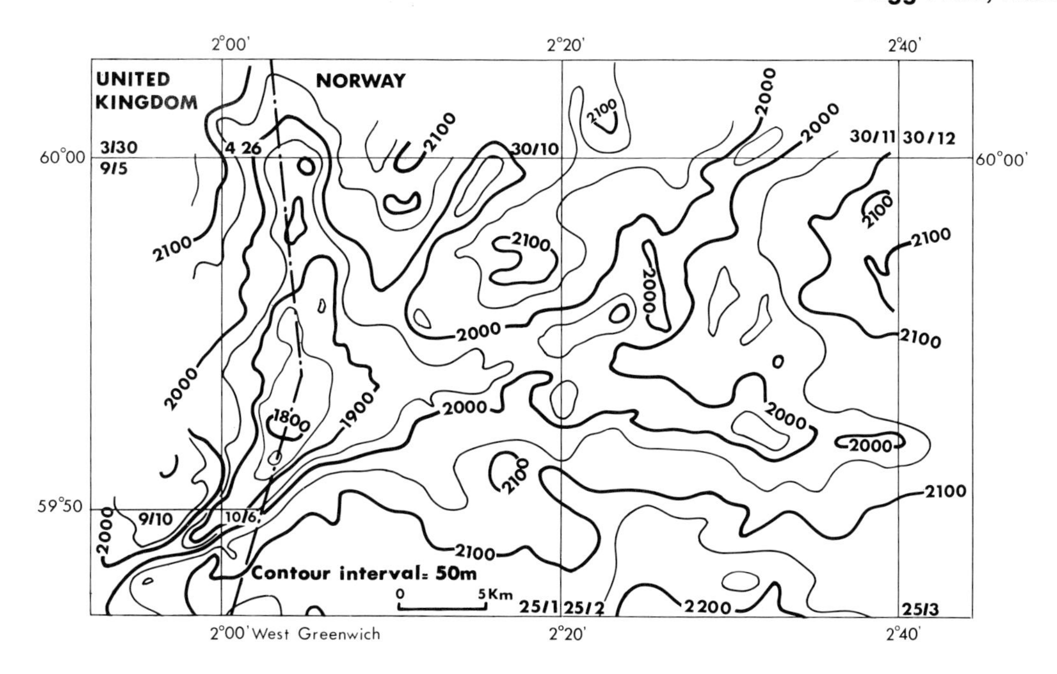

FIG. 5—Seismic structure of Frigg field at top of Frigg sand.

dicated on Table 1.

This correlation supports the idea that an initial period of marine sedimentation in the early Paleocene was followed first by a major regression in the late Paleocene and later by a new marine transgression in the early Eocene. These are regional events observed in other areas including the Paris basin. Nevertheless, an exact correlation with the classic stages of northwest Europe is difficult and, as indicated, the boundaries of some of the zones identified may not exactly coincide with the stage boundaries. The Frigg sequence probably represents a more complete and continuous stratotype than those of France, England, Belgium, or Denmark.

STRUCTURE

Structure of the Viking basin mapped on the top of the Cretaceous (Fig. 2) provides a regional model of the framework at the time of Paleocene deposition. The bordering platforms appear as shallow areas gradually deepening toward the central basin. A major feature, the Shetland escarpment, which is downthrown to the east from 500 to 1,000 m., represents an important structural discontinuity. This escarpment is considered to be a former shelf edge between the Shetland platform and the central deep-marine basin. In the Frigg area this escarpment has a maximum throw of 1,000 m.

The Frigg field is located on the western flank, not far from the deepest part of the Viking Tertiary embayment, in an area of low structural relief at the Tertiary base level where tectonism was of very mild intensity.

On seismic maps, the structure at the top of the Eocene appears as a low-amplitude, lobate, fan-shaped anticline, with a southwestern apex and three main lobes trending east, northeast, and north (Figs. 5, 6). Seismic maps of deeper horizons show that the Frigg field overlies a late Cimmerian faulted anticline which in turn overlies a complex Jurassic faulted block (Fig. 7). The migrated-depth contour map on the seismic marker very near the top of the reservoir indicates a closed area of 115 sq km with a vertical closure of about 170 m.

A well-defined "flat spot" can be recognized on most of the seismic sections which have been shot across the structure. This phenomenon, which is the seismic reflection due to the density contrast of the gas-liquid contact within a good reservoir, perfectly underlines the structural closed area (Fig. 8).

ORIGIN OF FRIGG STRUCTURE DEEP-SEA-FAN HYPOTHESIS

The location of the Frigg structure, in the deepest part of the basin but not very far away from the Shetland platform escarpment, and its particular shape suggest a deep-sea fan. Enhancing this hypothesis is the monotonous character of the facies, which show no major variations through time.

The shales are gray to green, in places very pyritic. The sands are either massive or thin layered, generally fine to medium, coarser in the proximal areas, finer in the distal; they commonly show a bimodal distribution—a few coarser supported grains in a finer homogeneous sand. They generally contain shale clasts and are characterized by association of glauconite and carbonaceous detritus.

All these criteria, and paticularly the association of glauconite and carbonaceous detritus, are typical of submarine sedimentation (Selley, 1976).

In such sedimentation, the differential compaction of sands and muds is an important factor in the depositional arrangement of the next clastic sediments, which have a tendency to be deposited above the shaly section on the flank of former thick sandy deposits (Figs. 9 to 12).

It can thus be inferred that the Frigg fan has been preserved to the present nearly in its final condition of deposition and that the structure is related mainly to the submarine-fan deposi-

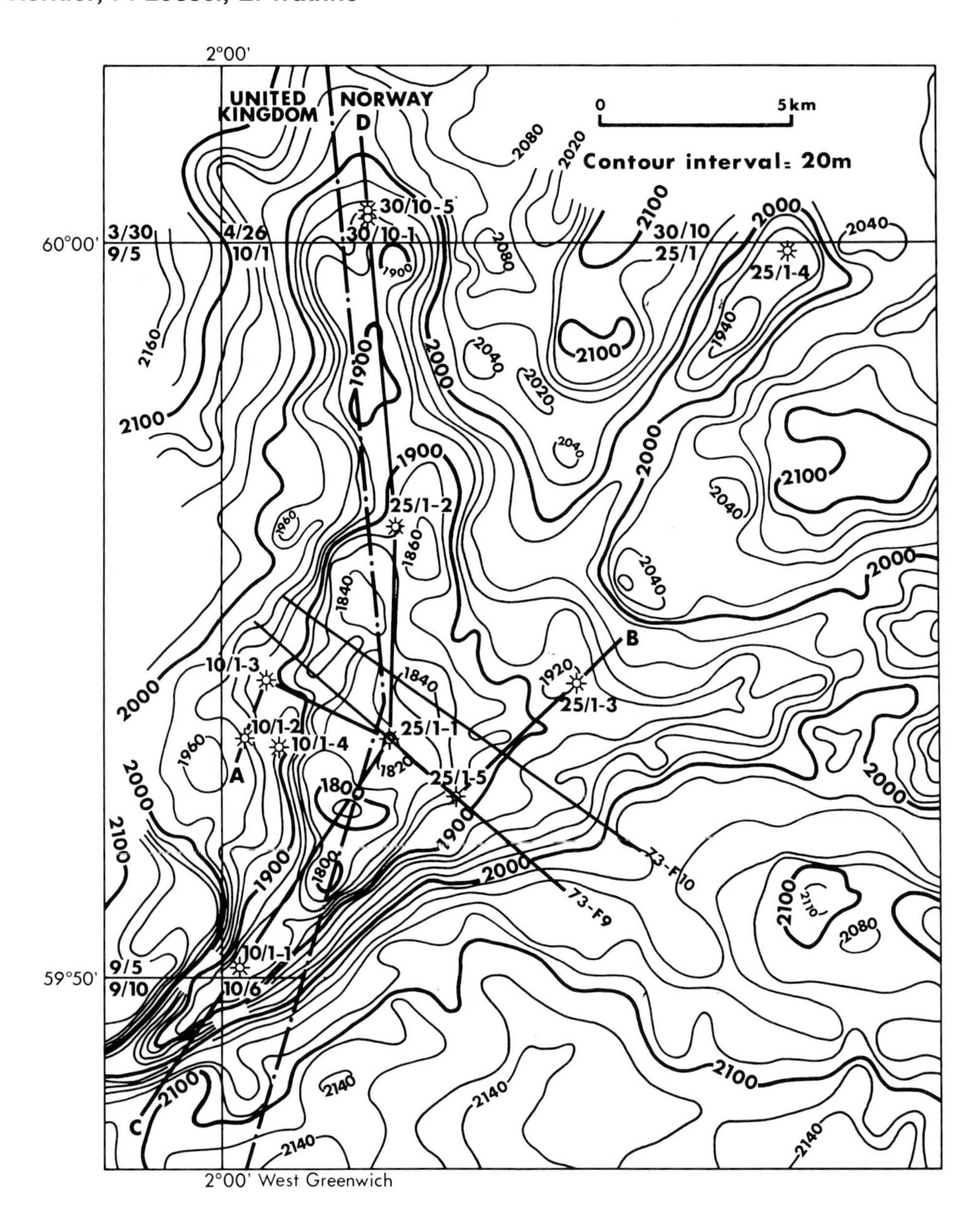

FIG. 6—Detailed structure map of top of Frigg sand in main Frigg field. AB, CD are locations of cross sections shown on Figures 9 and 10.

tional topography enhanced by differential compaction of sands and muds (Fig. 13). In this respect the main structure represents the upper part of the fan and the apex its feeder channel. Lobes represent outer channels and levees in the middle part of the fan whereas the low areas between lobes are due to the compaction of the more shaly beds between the channels (Figs. 5, 6).

South of Frigg, the Beryl embayment, which is a zone of weakness of the Shetland escarpment, provides the fundamental control for the deposition of clastics in this part of the basin. In particular, in providing routes of transport, it is responsible for the position of several fan complexes. Detailed study of the thick Tertiary sediments deposited along the Shetland escarpment has shown a very important deep-sea-fan sedimentation.

Along the eastern Shetland escarpment, widely differing fan complexes are present. The type of slope, the amount of clastic material locally available, the rate of subsidence, and the variations of the sea level, together with the overriding structural influences already discussed, define the precise nature of each fan.

The general mode of deposition throughout the area is progradation, or offlap, of the deposits to the east or southeast. The maximum thickness of the series is generally in the proximal areas, such as the foot of the escarpment (Fig. 3). Farther east, differentiation is more important and results in the development of local fan complexes, which are much thicker and more sandy than the surrounding embayments. Such differentiation is typical of deep-sea-fan sedimentation where no important reworking of initial deposits occurs.

The fan complexes of the Viking basin form part of a regional system, which implicates on its eastern border a final progradation slope that is apparent on the seismic lines.

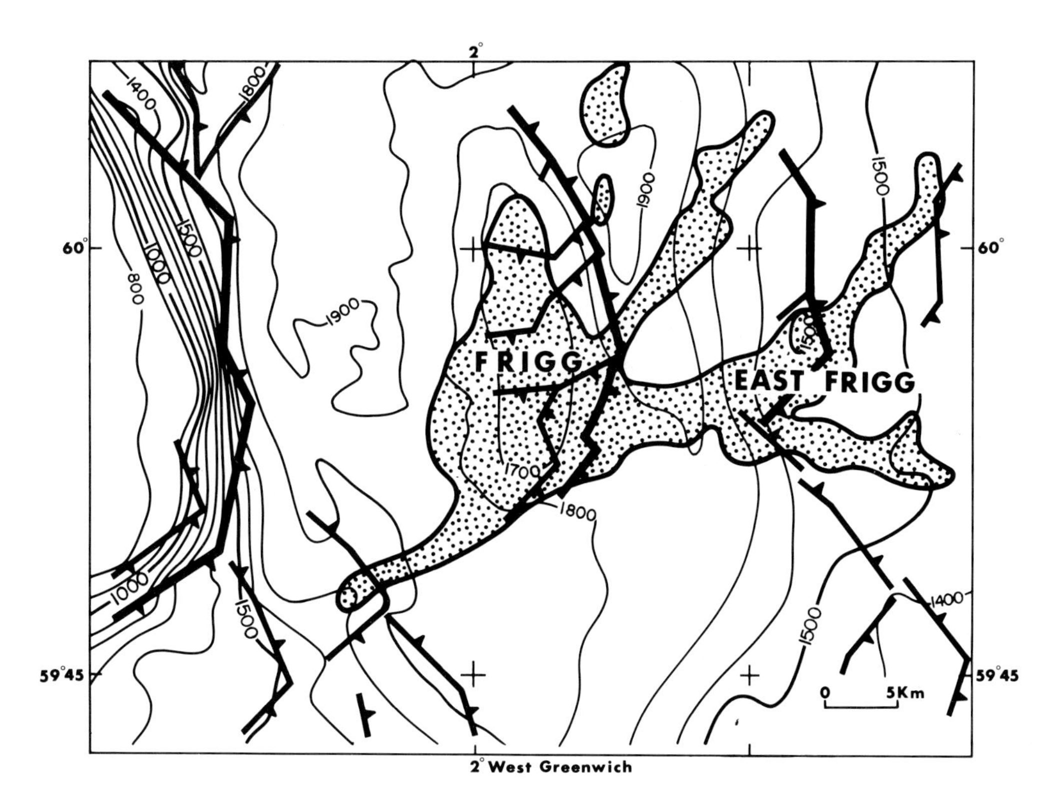

FIG. 7—Pre-Cimmerian structure (C.I. = 100 msec one-way time) and extent of Frigg sand (dotted). Axis of main Frigg field is parallel with axis of deep structure. Feeder channel follows deep fault zone which parallels old northeast-southwest Caledonian faults.

HISTORY OF TERTIARY SEDIMENTATION AND PALEOGEOGRAPHY

Early Paleocene (Fig. 14)

The first phase of sedimentation occurred in the early Paleocene and was characterized by rapid infilling of the basin under a general marine environment.

Danian—During the Danian stage (Table 1 nt Ia), in the southern area south of 59°, including the Witch Ground graben and the Viking basin, an initial phase of sedimentation 200 m thick is characterized by shale, marl, and limestone without detrital material (Fig. 14a).

The first important influx of clastic material occurred only in the late Danian and, at that time, two marine areas of thick sandy progradation were located in the Witch Ground graben and in the Beryl embayment. In the latter area, the Bruce fan shows a maximum thickness of 600 m of massive sands. After the first downslope infilling, progradation extended farther east and additional massive sand bodies were deposited farther into the basin in the Frigg area. These sand bodies (400 m thick at Frigg, 100 m thick at East Frigg) represent the first elements of the Frigg and East Frigg fans. They were deposited on the southeastern flank of both the Frigg and East Frigg Late Cretaceous anticlines. The relation is probably not random, but results from the antagonism between the turbidite flow and the structural obstacles. These low-amplitude structural obstacles created a velocity loss which allowed the coarser fraction of the flow to be deposited there. Apart from these events, thick distal shaly sedimentation occurred everywhere in the basin as well as on the slope areas between the Bruce, Frigg, and East Frigg fans.

At the end of the Danian, a new topography developed as a result of differential compaction of the sand mounds and the surrounding shales. The new topography modified to a great extent the location of later sand bodies.

Montian—The Montian Stage (Table 1, nt Ib) was also a period of rapid basinal infilling (Fig. 14b). In the Witch Ground graben, offlap continued to the southeast, with deposition of thick downslope deposits. At the same time, new fan complexes appeared in the Viking basin (Sleipner, Heimdal, and Ninian fans), where only distal Danian shales had been previously deposited. Other fans continued to grow, including the Frigg and East Frigg fans where new sand bodies flank on the southeast the former Danian mounds. The Bruce fan complex, which had been infilled during the Danian, became a wide slope area with only shaly sedimentation.

Late Paleocene (Fig. 15a)

The late Paleocene regression brought important changes in the mode of deposition of the fan complexes, mainly because of a decrease in the amount of clastic material available and progressive establishment of a lower sea level.

Thanetian—The Thanetian was characterized by deposition of a more distal fan complex. In Witch Ground graben, thick lower Paleocene sediments are overlain by thin shaly marine beds probably more representative of a prodeltaic environment. At the same time, the very distal fans of Forties, Montrose, and Cod fields were deposited in the Ekofisk basin. The

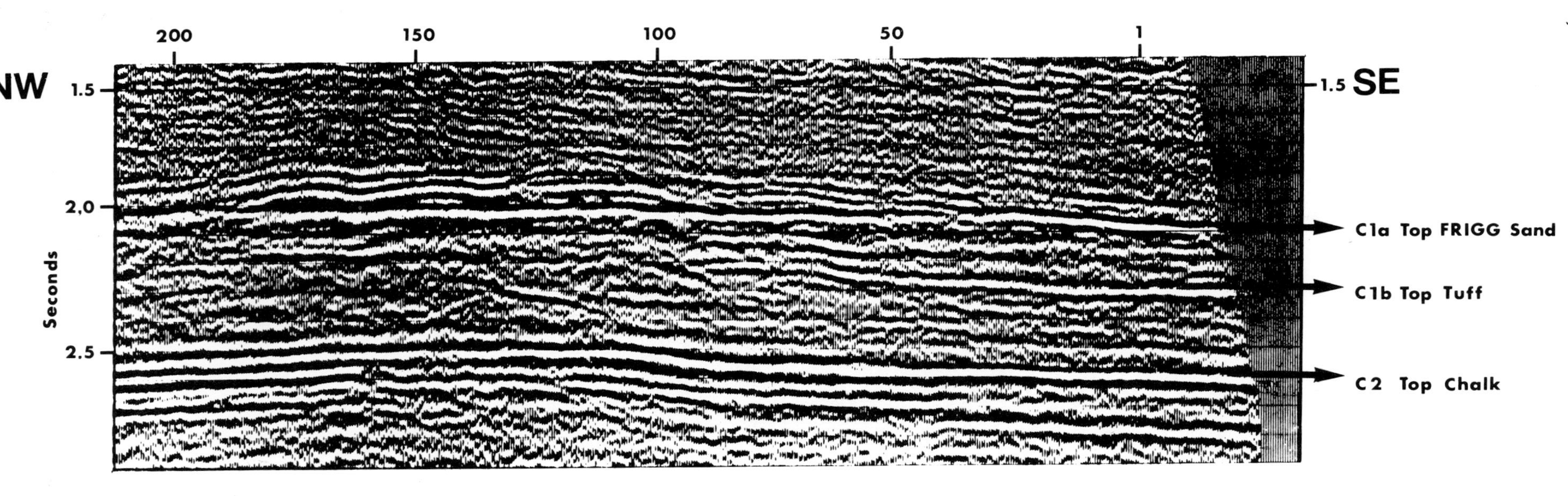

FIG. 8—Typical seismic section across Frigg field shows very clear "flat spot" at about 2 sec. This "flat spot" is real seismic reflection caused by density contrast of gas-liquid contact.

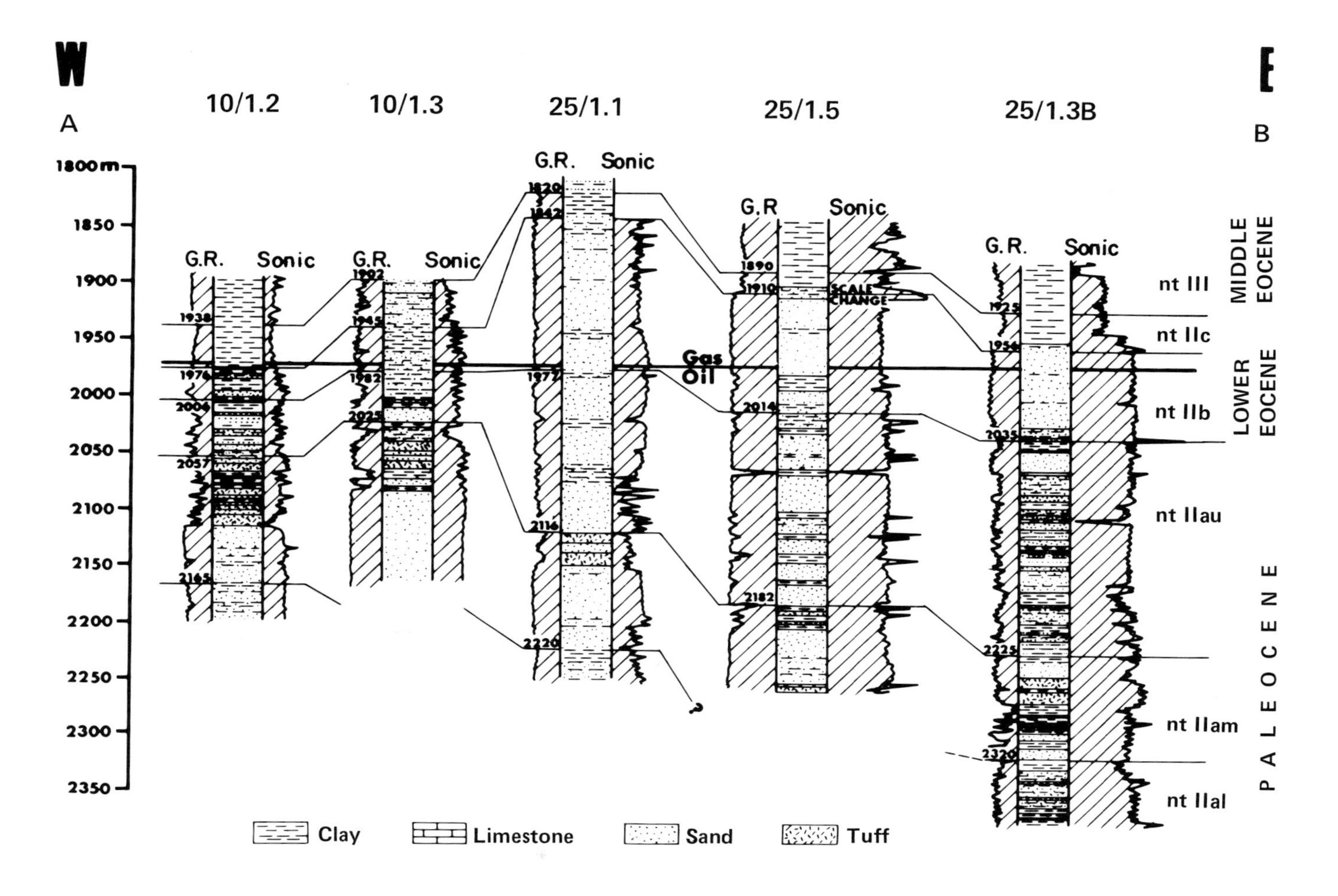

FIG. 9—West-east geologic section across Frigg structure shows eastward thickening of sand. For location see Figure 6.

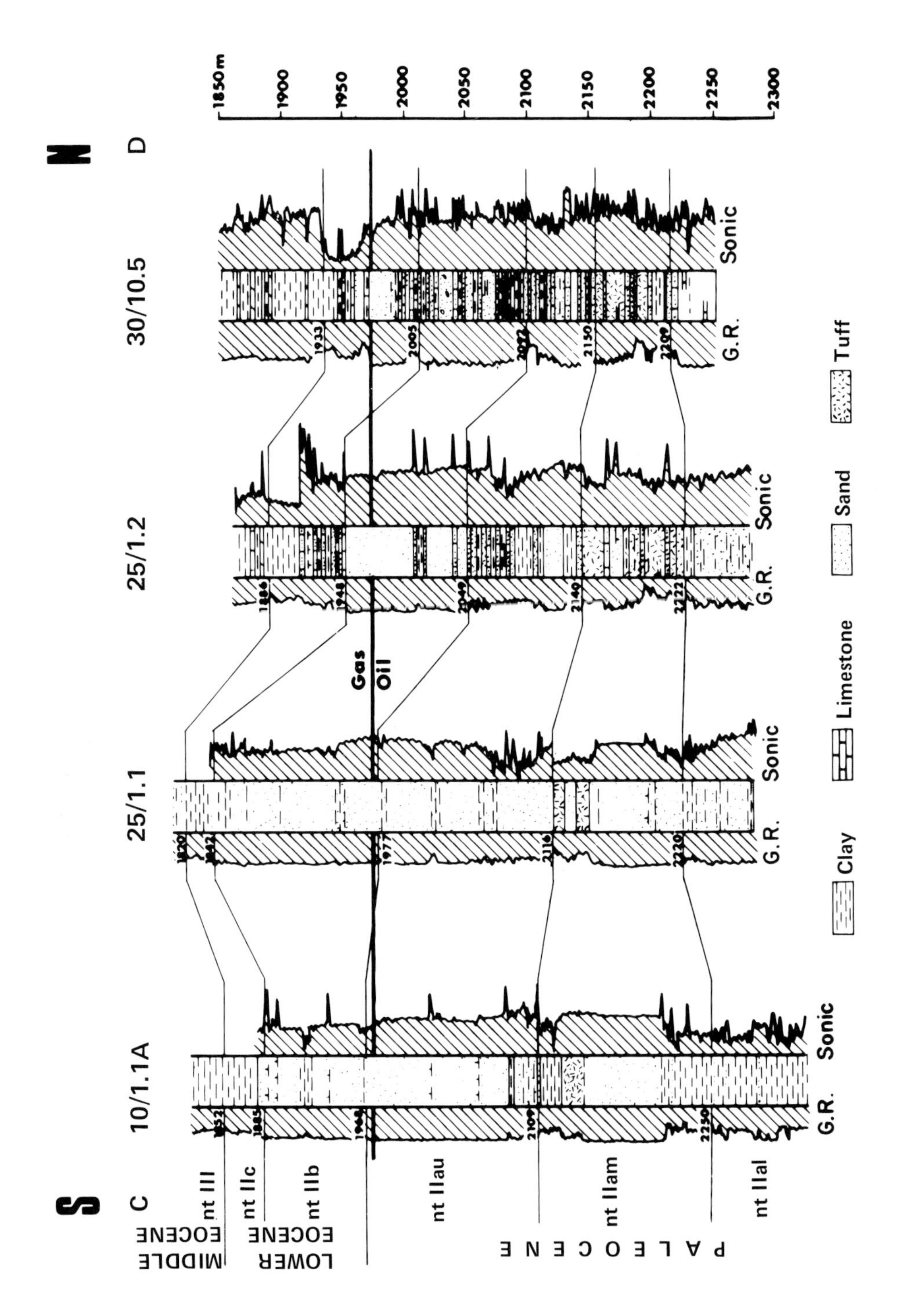

FIG. 10—North-south geologic section across Frigg structure shows northward thinning of sand. For location see Figure 6.

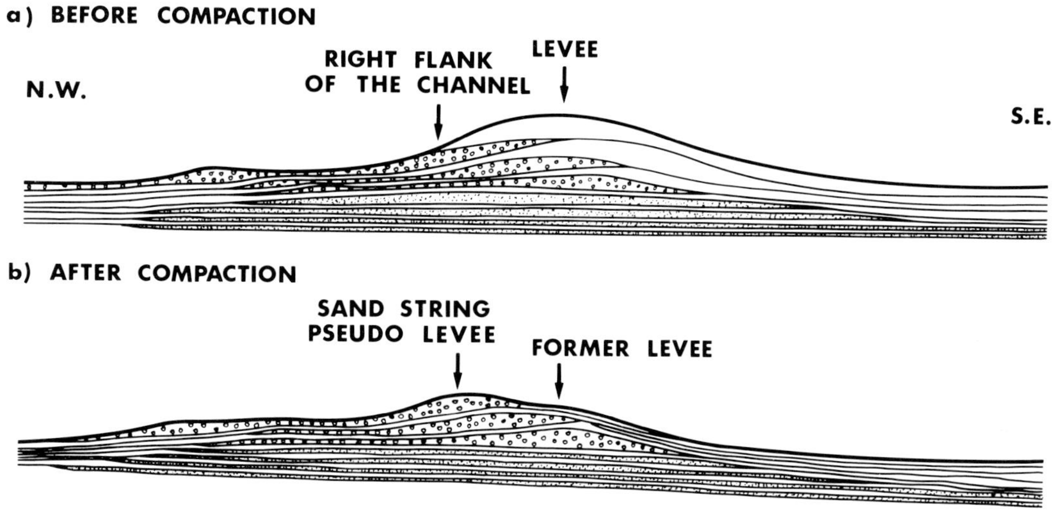

FIG. 11—Compaction of deepsea-fan channel. **A,** levee at time of deposit. **B,** same levee after compaction; sand deposits on flanks of original levee are now topographic highs.

same tendency is observed in the Viking basin, where new fans were deposited in the remaining low areas on the northern flanks of the original fans. The Frigg fan was developing in an area previously characterized by mainly shaly sedimentation between the Heimdal fan on the south and the South Alwyn fan on the north.

Sparnacian—The Sparnacian was the main period of regression. In the lower Sparnacian (Table 1, nt IIa middle) only local sand bodies are present. The sequence is widespread and relatively uniform in facies and thickness. Pyritic black shales are present at the base, indicating deposition in a more euxinic environment. The sequence also includes, near the top, a persistent layer of volcanic tuff.

The tuff very probably originated from the western Hebridean volcanic belt and is related to a major episode of volcanic activity resulting from the opening of the North Atlantic (Jacque and Thouvenin, 1975). At the same time, a thin deltaic complex prograding from the west developed in the Witch Ground graben and replaced the former sea, thus capping the very thick, early Paleocene fan system.

The early Paleocene fan complex, comprising a delta-front and a deltaic-plain facies, is well defined by the alternation of sand, shale, and coal layers. The present depth of the Sparnacian coals ranges from 900 to 1,200 m subsea, which may be interpreted as a regional paleo-sea level, representative of the late Paleocene. This paleo-sea level, which can be related in the Frigg area with the top of the escarpment, is probably the best direct evidence for the deep-sea character of the sedimentation, considering especially that it was a period of low stand of the sea.

The predominantly continental character of the "nt IIa middle" palynologic assemblage must be interpreted as due to the relative proximity of the deltaic plains.

The late Sparnacian (Table 1, nt IIa upper) was an episode of very local development represented by a persistent deltaic-plain facies in the northern Witch Ground graben. In the Frigg fan, the interval is represented by mainly sandy deposits, which constitute the base of the productive Frigg formation. The new sand lobes deposited at this time are on the southern flank of the former Thanetian fan.

Early Eocene (Fig. 15b)

A new marine transgression of Ypresian age (Table 1, nt IIb) occurred in the basin at the beginning of the Eocene and was characterized by a final influx of clastic material into the basin. However, at that time thick sand was deposited only in the Bruce, Frigg, East Frigg, and South Alwyn fans. These final sand lobes were deposited on the southern flank of the previous structures because of differential compaction of the Thanetian and Sparnacian fan deposits (cf. Fig. 12).

The shaly lateral facies of the sands is either very thin or absent, which means that Frigg sands were originally deposited in mounds and that the relief observed today mainly reflects this original deposition rather than the differential compaction.

These sands were subsequently sealed by a thin upper Ypresian marine shale (nt IIc) which is replaced by red oxidized shales on the top of the structure. A very distal sand lobe of this phase has been identified on the northernmost part of the Frigg field in Esso 30/10-5 well. The extent of sand deposits and the sedimentologic landscape at the end of the early Eocene time are shown in Figures 16 to 18.

HYDROCARBONS: COMPOSITION, DIAGENESIS, AND ORIGIN

Composition

The gas of the main Frigg field is a dry gas, containing only 3.7 g/cu m of condensate. Its percentage composition is as follows: N^2, 0.4; CO_2, 0.3; C^1, 95.5; C^2, 3.5; C^3, 0.04; C^4, 0.01; $>C^4$ . . . . The associated condensate is composed of 86.5% of C^{11} .

The oil beneath the gas accumulation has a gravity of 23 to 24° API and is of a naphthenic compositon. It is characterized

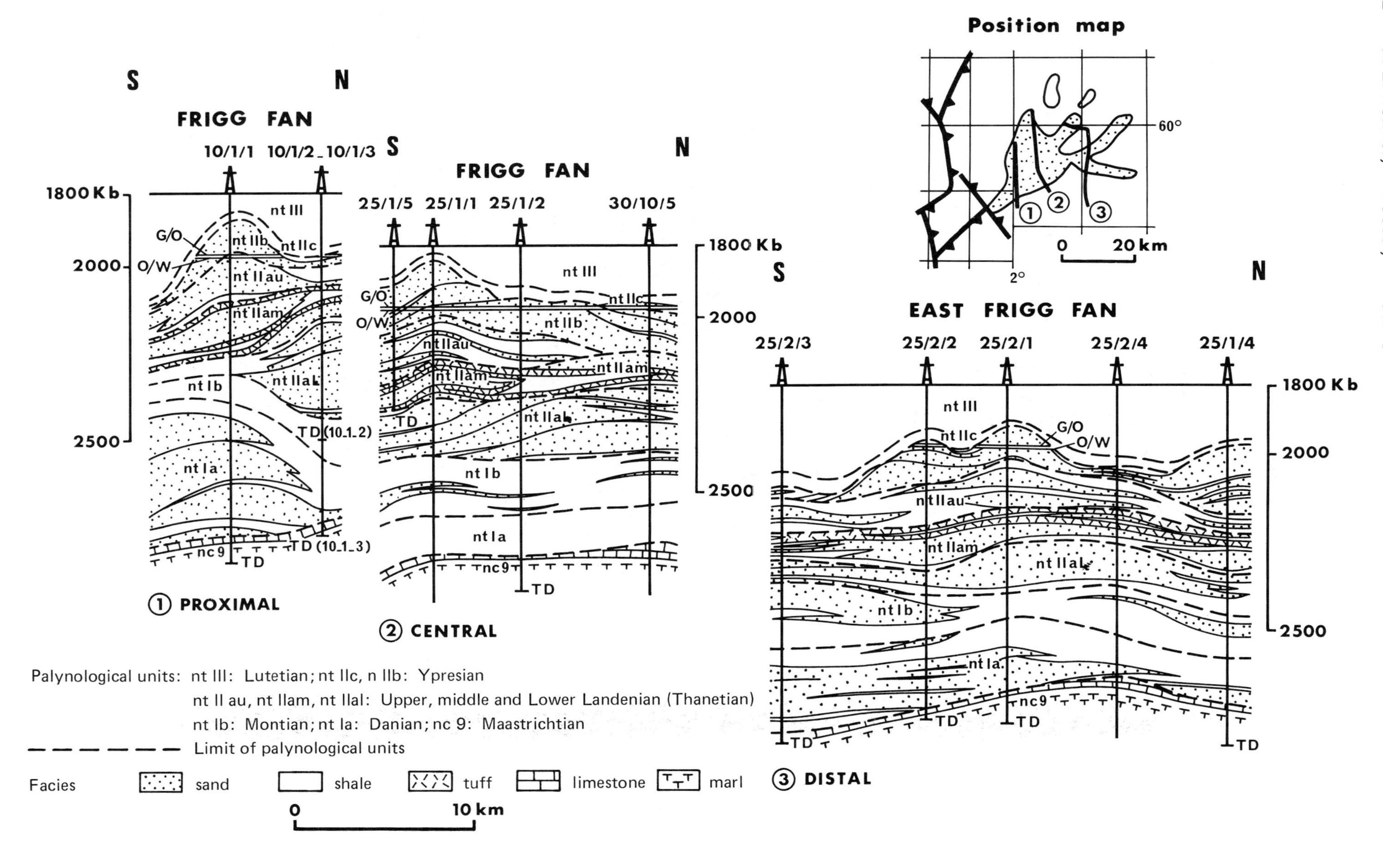

FIG. 12—Cross sections of Frigg field show deposition of thick sands (channels) on flanks of earlier structural features.

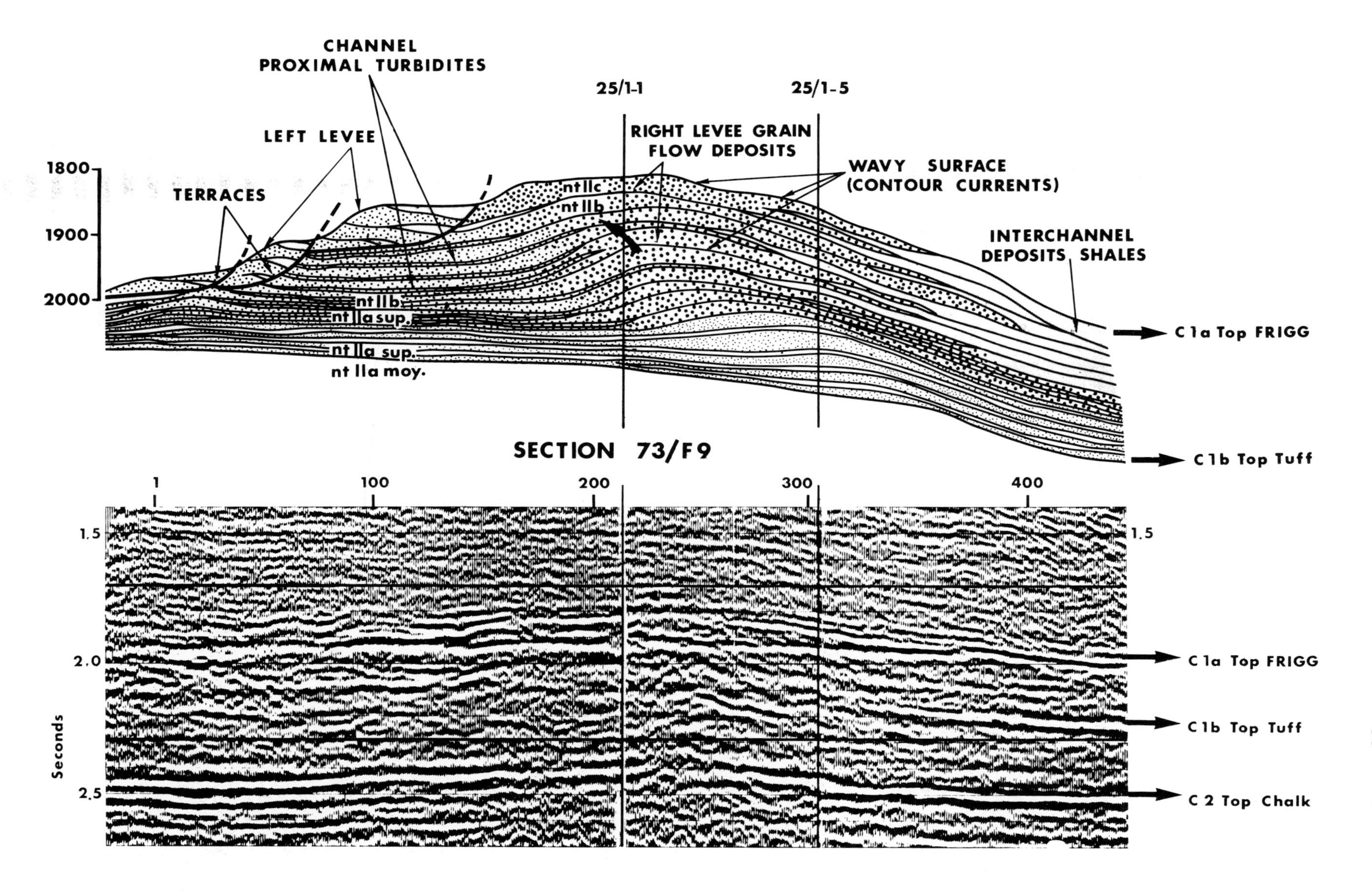

FIG. 13—Sedimentologic interpretation of typical Frigg seismic section crossing wells 25/1-1 and 25/1-5. Vertical exaggeration is 2×. Seismic "flat spot" is well defined.

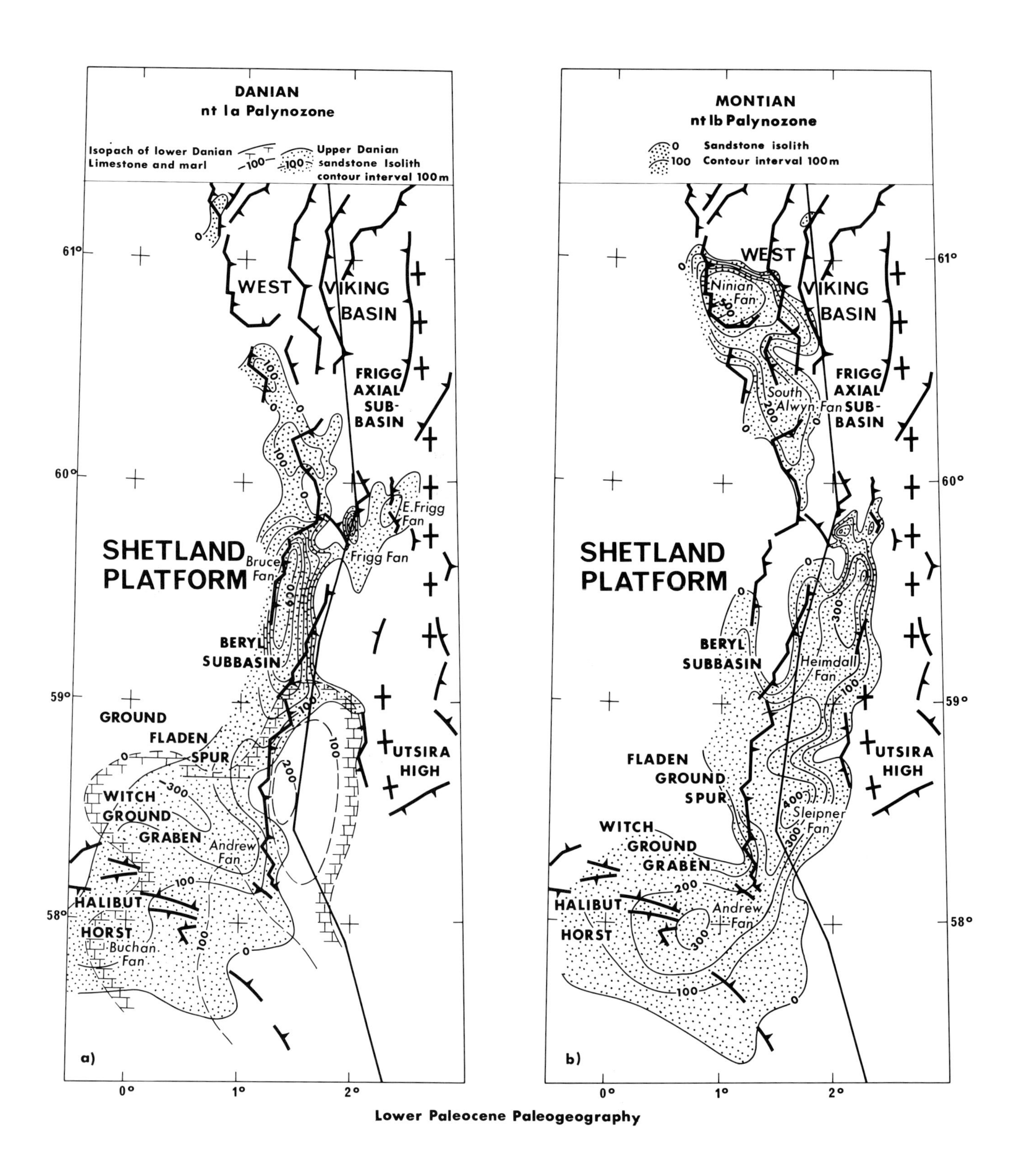

FIG. 14—Early Paleocene paleogeography. **A,** Danian. Structural framework of faults bordering Shetland platform and main pre-Cimmerian blocks of Viking basin, shows Danian episode of nonclastic sedimentation, south of 59°N lat., with maximum thickness of 200 m of limestone and marl. **B,** Montian sandstone isolith map shows spread of fan sedimentation over entire area.

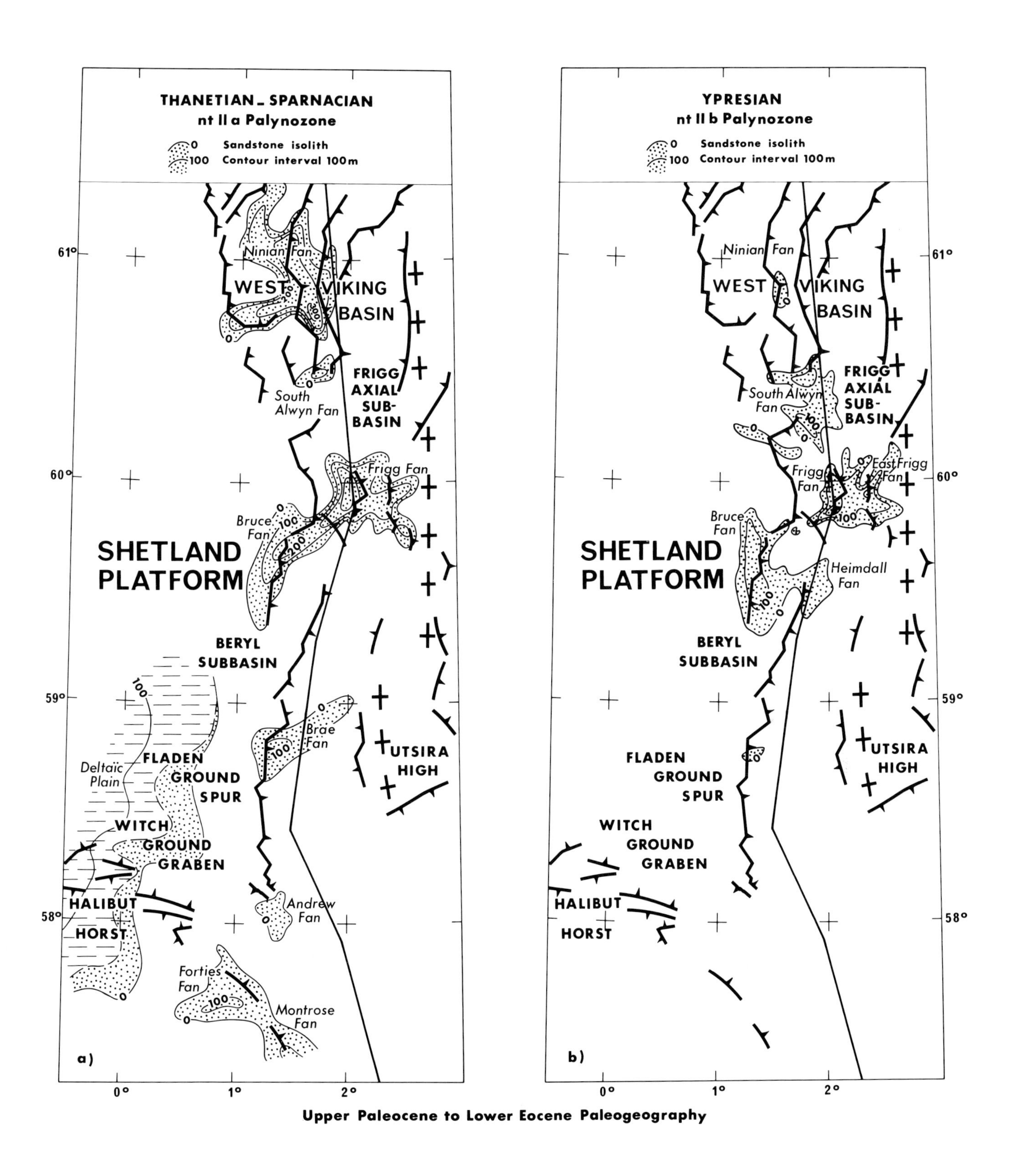

FIG. 15—Late Paleocene and early Eocene paleogeography. **A,** Late Paleocene distal-plain complex is invading Witch Ground graben; all Thanetian sand bodies are very distal fans completely separated from deltaic belt by slope area of thin shaly marine sedimentation. **B,** Early Eocene marine transgression characterized by final influx of sands mainly localized in Frigg area, repeating initial distribution of Danian sands.

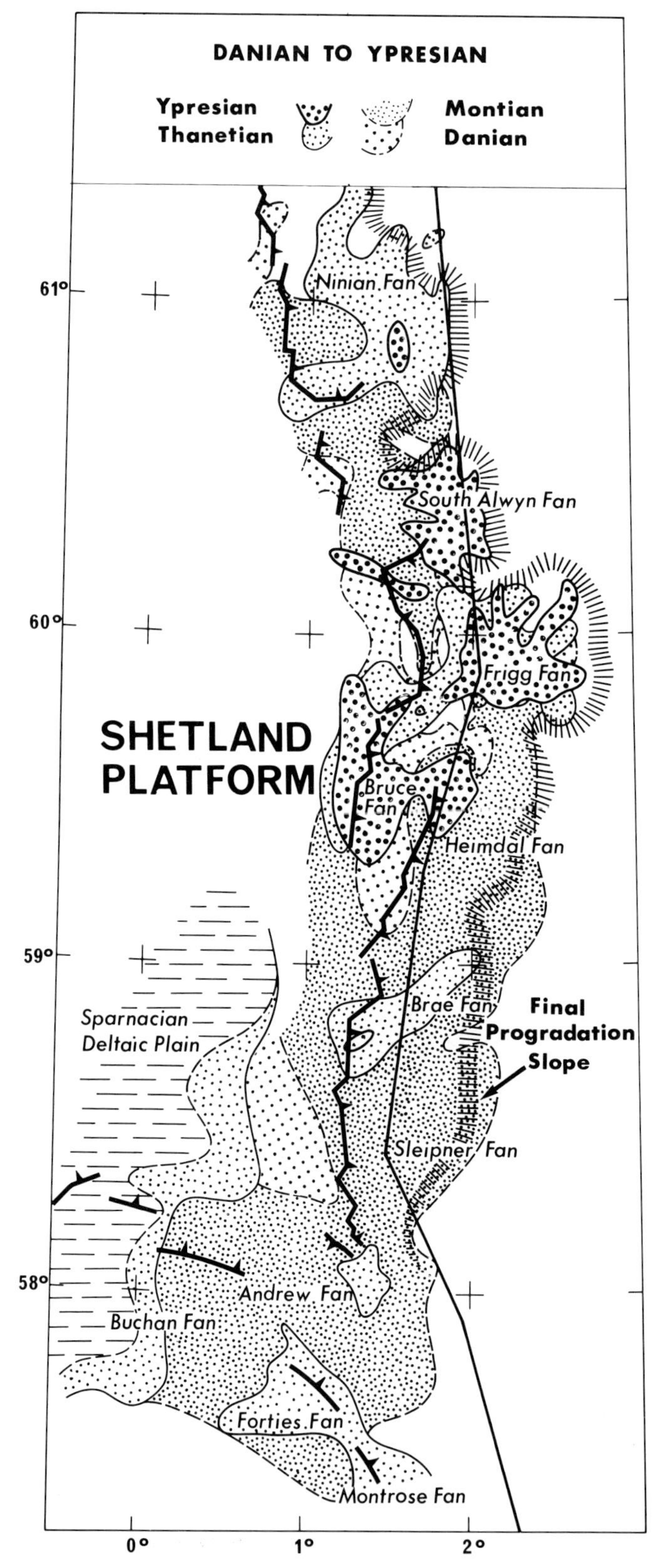

FIG. 16—Fan complex at end of early Eocene. Cumulative map shows nearly 50 km wide area of thick clastic sediments at foot of Shetland platform. Eastern limit of clastic deposition is final progradation slope, missing on south where basin deepens. Perspective views are shown in Figures 17 and 18.

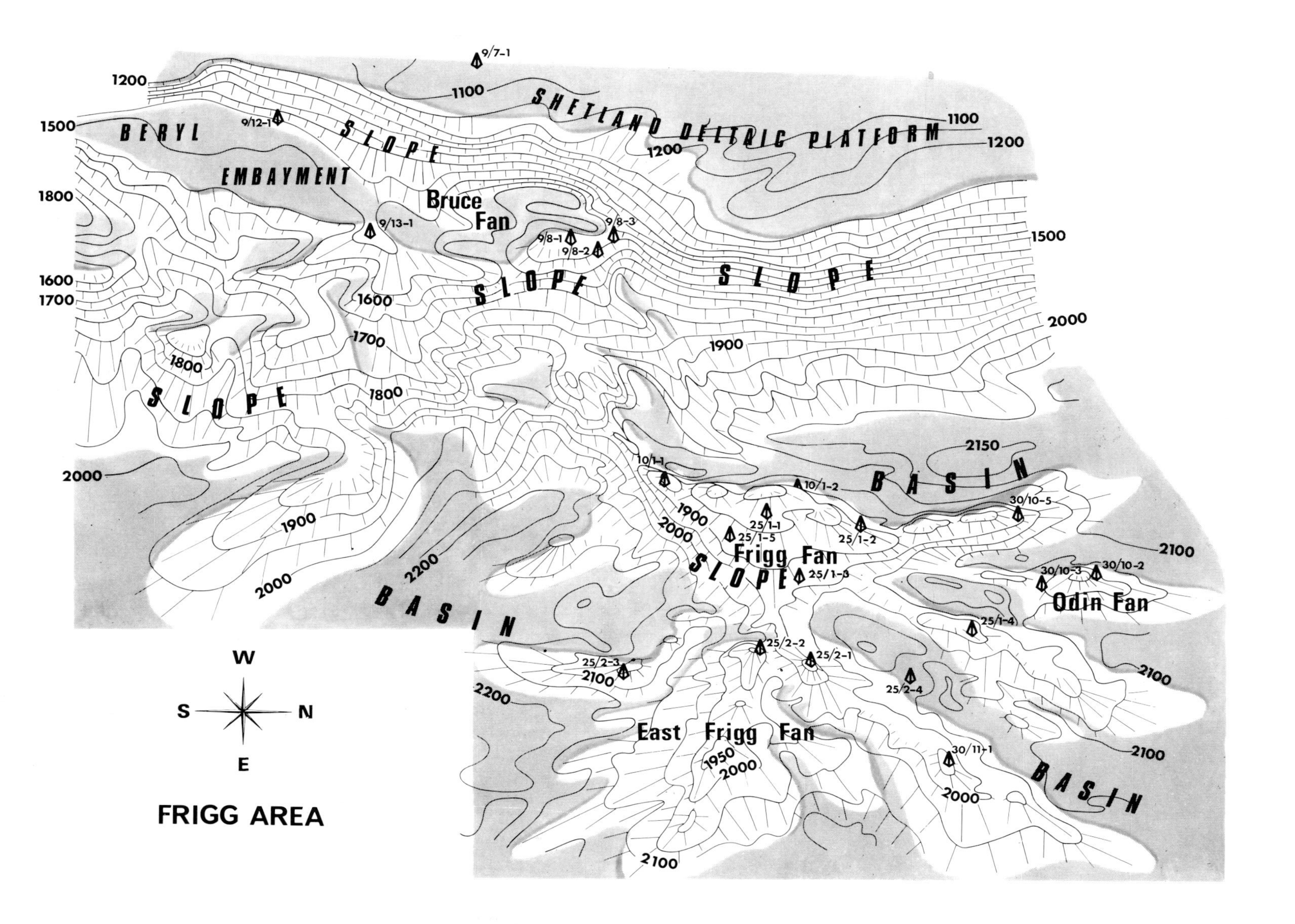

FIG. 17—Perspective view from east of Frigg field area shows landscape at end of lower Eocene which shows, in isobaths below sea bed, final shape of Frigg and East Frigg fan areas.

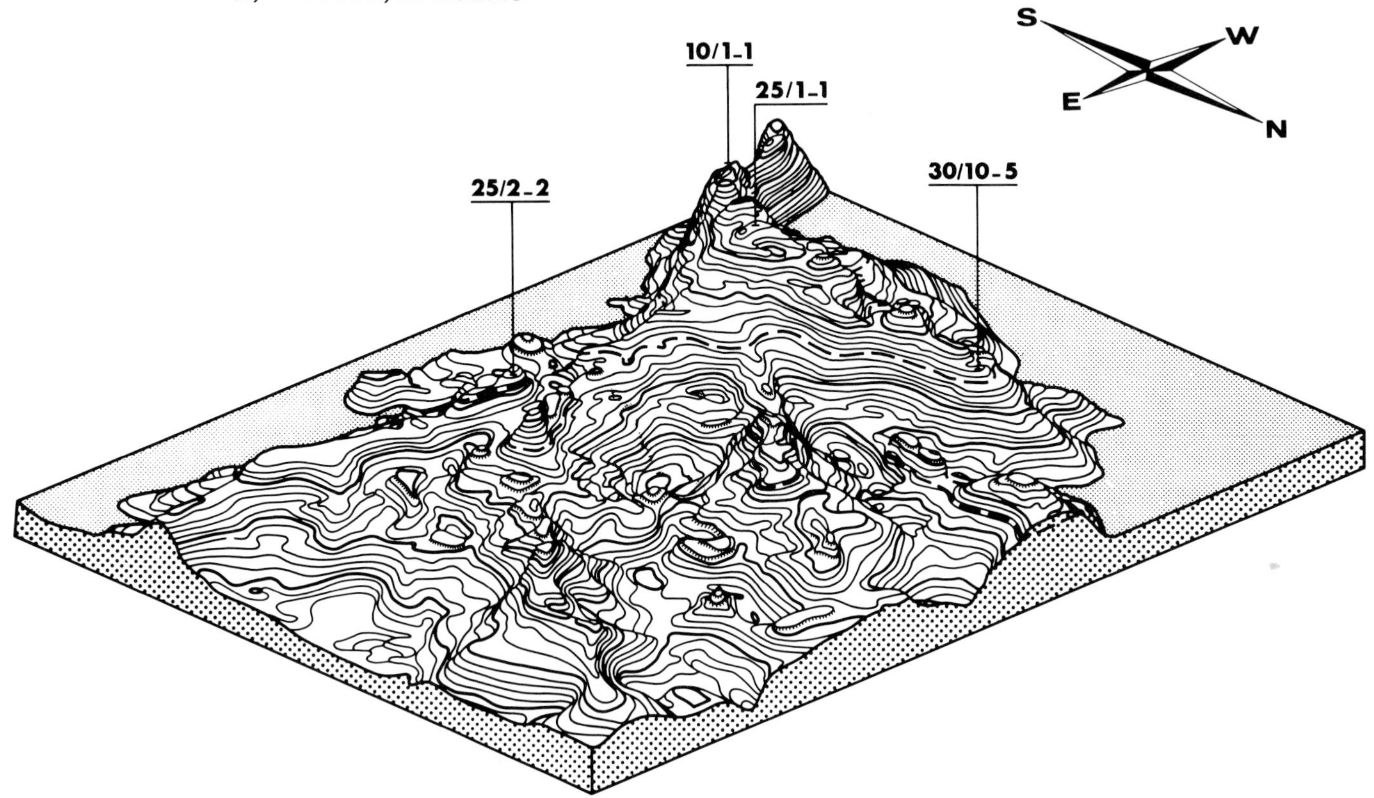

FIG. 18—Perspective view from northeast of Frigg field shows landscape of field and satellites. Drawn by computer from isobath map.

by a saturated/aromatic hydrocarbon ratio of 3:76. The light compounds with a carbon number of ≤C [17] represent a very minor fraction of the oil, and the N alkanes are almost absent. Thus, the saturated hydrocarbons are mainly represented by cyclanes and isoalkanes.

Diagenesis

The type of oil characteristic of the main Frigg field may be considered as anomalous, suggesting biodegradation caused by bacteria. Other Paleocene or Eocene pools of the area do not show the same anomaly. For example, the East Frigg pool (well 25/2-1) produces gas with condensate and shows an underlying oil containing a significant fraction of N alkanes.

Origin

The less altered oil of East Frigg permitted a measurement of the ratio of pristane/n C^{17} to phytane/n C^{18}.

A value of 2 for that ratio corresponds with the readings obtained from Dogger and Lias source rocks in the area (values of 2 to 2.4), but differs significantly from those measured on Kimmeridgian radioactive shales (1.4). Such a correlation suggests that the Frigg oil probably was generated and subjected to an early phase of migration from the pre-Cimmerian, deeply buried Dogger and Lias source rocks. A similar correlation between hydrocarbons in lower Tertiary reservoirs and pre-Cimmerian source rocks has been clearly established by analysis of the condensate gas and oil of the Paleocene Heimdal field, only 35 km south of Frigg.

The Frigg gas probably results form a late phase of the diagenesis of the Jurassic source rocks, which show vitrinite-reflectance values ranging from 1.5 to 1.8. These values are characteristic of gas-zone diagenesis. By contrast, lower Tertiary and Cretaceous source rocks of the Frigg area show only a low degree of diagenesis. The vitrinite-reflectance value of 1.0 is reached only at a depth of 4,000 m in the deep Frigg wells. The deep origin of the Frigg gas is also mainly supported by its carbon-isotope composition which shows a δ C^{13} value of −43.3% which is characteristic of a deep gas.

The absence of propane and butane in the Frigg gas, which could be interpreted as being indicative of a shallow origin of the gas, probably results from the same biodegradation as that of the oil.

RESERVOIR CHARACTERISTICS, RESERVES, AND PRODUCTION PLANS

The gas-bearing sand in the discovery well 25/1-1 was tested after perforating the interval between 1,920 and 1,928 m. With a ⅞-in. (2.2 cm) choke, a gas flow of 673,000 cu m/day (24 MMcf/day) was measured at 60°F (16°C) and 14.7 psi (101 kPa). The buildup of the bottom-hole shut-in pressure was instantaneous and became steady at 2,835 psi (19,547 kPa) at 1,880 m. The maximum static wellhead pressure is 2,465 psi (17,000 kPa). These tests have confirmed the very good permeability of the reservoir which had been measured from cores.

At an initial absolute pressure of 199.25 bars (19,925 kPa; 2,890 psi) at a subsea depth of 1,912.5 m, the gas viscosity is 0.018 cp at a reservoir temperature of 60°C. The dimensionless gas compressibility factor (Z) is 0.865.

The reservoir is predominantly composed of clean, fine, and unconsolidated sand with good characteristics: porosities range from 25 to 32% and average permeabilities from 1,200 to 1,600 md.

Gas-in-place and recoverable-gas-reserve calculations are based on the total accumulation in both Norway and the United Kingdom. The consulting firm of DeGolyer and MacNaughton estimated gas in place to be about 269 billion cu m (9.5 Tcf). Of this total, 60.82% is in the Norwegian sector and 39.18% is in the United Kingdom sector.

The probable recoverable gas depends on the effectiveness of the Frigg and Cod sand aquifers and is estimated at 227 billion cu m (8.0 Tcf).

The field is equipped with two drilling platforms (CDP1, DP2), two production platforms (TP1, TCP2) and one quarter platform. On each of the two drilling platforms, 24 production wells have been drilled and completed with 7 5/8-in. tubing and with 6 5/8-in. screens to prevent the wells from producing sand.

The behavior of these wells is excellent, with maximum potential flow rate of 2.3 million cu m/day (81 MMcf/day) per well with no skin or turbulence effects being noticed so far. Frigg production started in September 1977, and the first delivery from TP1 into the pipeline started on September 11, 1977. Gas sales at St. Fergus began on September 13, 1977 at a rate of 4 million cu m/day (140 MMcf/day), only 6 1/2 years after discovery.

For the present, production from the 47 completed wells varies from 30 to 60 million cu m/day (1059 to 2119 MMcf/day) depending on customer demand.

The gas deposit is underlain by an oil zone 10 m thick. Extensive tests in well 25/1-3 have proved the impossiblity of economically producing this heavy (24° API), naphthenic oil. Reserves in place amount to 125 million cu m (790 million bbl).

SELECTED REFERENCES

Blair, D., 1975, Structural styles in North Sea oil and gas fields, *in* A. W. Woodland, ed., Petroleum and the continental shelf of northwest Europe, v. 1, geology: New York, John Wiley & Sons, p. 327-337.

Fowler, C., 1975, The geology of the Montrose field, *in* A. W. Woodland, ed., Petroleum and the continental shelf of north-west Europe, v. 1, geology: New York, John Wiley & Sons, p. 467-476.

Jacque, M., and J. Thouvenin, 1975, Lower Tertiary tuffs and volcanic activity in the North Sea, *in* A. W. Woodland, ed., Petroleum and the continental shelf of north-west Europe, v. 1, geology: New York, John Wiley & Sons, p. 455-465.

Parker, J. R., 1975, Lower Tertiary sand development in the central North Sea, *in* A. W. Woodland, ed., Petroleum and the continental shelf of north-west Europe, v. 1, geology: New York, John Wiley & Sons, p. 447-453.

Selley, R. C., 1976, Subsurface environmental analysis of North Sea sediments: AAPG Bull., v. 60, p. 184-195.

Walker, R. G., 1978, Deep-water sandstone facies and submarine fans: models for exploration for stratigraphic traps: AAPG Bull., v. 62, p. 932-966.

Walmsley, P. J., 1975, The Forties field, *in* A. W. Woodland, ed., Petroleum and the continental shelf of north-west Europe, v. 1, geology: New York, John Wiley & Sons, p. 477-485.

Geology of the Forties Field, U.K. Continental Shelf, North Sea[1]

By P.J. Hill and G.V. Wood[2]

Abstract The Forties field is located nearly 180 km east-northeast of Aberdeen mainly in UK Licence Block 21/10, and was discovered in October 1970 in Paleocene sandstones of the Forties Formation. Four appraisal wells drilled during 1971-1972 proved the existence of a giant oil field with an area of some 90 sq km and (on current evidence) an estimated oil-in-place reserve of 4 billion barrels.

Additional geologic data from the 50 development wells drilled to date show rapid facies variations over the field with some sand bodies having a cross section less than the prime well spacing of 700 m. A large, partly isolated sand body, the Charlie Sand, is recognized in the upper, western part of the reservoir.

The sandstones and shales of the Forties Formation are considered to have been deposited in a middle and lower submarine fan environment. A mixture of sedimentary processes including grain flow, debris flow, and turbidity currents are identified and four broad facies were defined from cores. The facies types indicate significant vertical associations that are related to specific depositional environments and recognized by petrophysical log patterns.

Detailed lithofacies studies, together with pressure-decline data and log-pattern analyses, provide a practical means of correlating and mapping the complex sand geometry. These techniques provide a method for good well location and a better understanding of the reservoir performance.

INTRODUCTION

The Forties field, located in the North Sea nearly 180 km east-northeast of Aberdeen, lies mainly in the UK Licence Block 21/10 and was discovered in October 1970 by British Petroleum (BP) in Paleocene sandstones of the Forties Formation (Fig. 1). Four appraisal wells drilled during 1971-1972 proved the existence of this giant oil field and defined an area of some 90 sq km, a vertical closure of 155 m, and an estimated oil-in-place reserve of 4 billion stock tank barrels. A recovery factor of 45% (1.8 billion bbl) is expected.

The development plan was devised and completed during 1972-1975 and at the same time many studies were undertaken to assess the likely performance of the reservoir (Hillier et al, 1978). The development plan required four virtually identical drilling/production platforms to be installed with facility for 108 wells and a total system capable of producing at rates of

[1] Manuscript received, April 4, 1979; accepted for publication, June 11, 1979.

[2] BP Petroleum Development, Ltd., Dyce, Aberdeen, Scotland.
This manuscript was developed from studies undertaken by staff at BP Petroleum Development, Ltd., as part of the ongoing development program of the Forties field. The writers are indebted to many colleages for their help, comments, and suggestions in the preparation of this paper. They also thank the Chairman and the Board of Directors of British Petroleum Co., Ltd., for permission to publish this paper.

Article Identification Number:
0065-731X/80/M030-0005/$03.00/0.

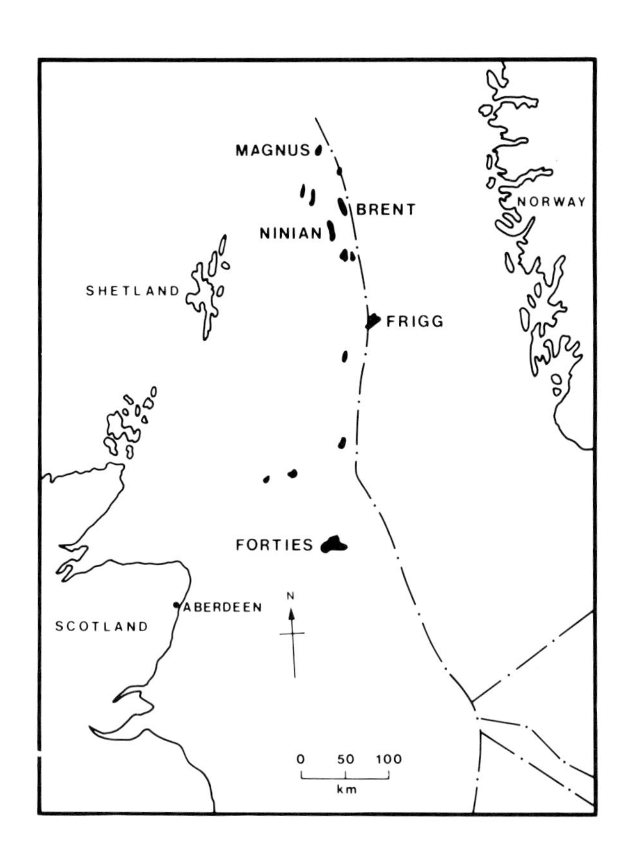

FIG. 1—Location map of the Forties field, U.K. continental shelf.

around 500,000 b/d (Fig. 2). Additionally, this inherently low energy reservoir was felt to require some form of external energy, either natural water influx or fluid injection, or both, for optimum recovery of reserves. Incorporated in the development plan therefore was a "complete replacement" seawater injection system with a capacity of up to 600,000 b/d. Development drilling began in June 1975 and production started in September 1975. First-stage separated crude is pumped via a 32-in. line, 169 km to land at Cruden Bay. A 36-in. landline carries the crude 208 km to Kinneil for final processing prior to refining or export (Fig. 3).

The order in which early wells were drilled on each platform was governed by the need to obtain a rapid production buildup, and the need to maximize information on the reservoir and well performance, together with the constraints on all operations caused by the offshore environment. The early drilling, with the above considerations in mind, highlighted the complex sand body geometry and variable nature of the reservoir lithologies.

The geologic data from the 50 development wells drilled to date are used to describe the abrupt facies variations seen across the field (Fig. 4). This new well data permits amplification of the earlier Forties field papers of Thomas et al (1974) and Walmsley (1975). The facies identified from cores and log data are related to specific depositional environments and show how detailed geology and well correlations provide a better understanding of the reservoir and future drainage patterns.

LITHOSTRATIGRAPHY AND STRUCTURAL SETTING

Earlier papers used an informal lithostratigraphic nomenclature to subdivide the Paleocene/Eocene sandstones and shales of the field (Thomas et al, 1974; Walmsley 1975). However, recent work by the Institute of Geological Sciences and a number of committees from the oil industry has resulted in a standard lithostratigraphic nomenclature for the central and northern North Sea area (Deegan and Scull, 1977). The Forties Formation across the field is informally divided into a lower interbedded sandstone, shale, and limestone sequence—the Shale Member, and an upper predominantly sandstone sequence—the Sandstone Member. In addition, the Sandstone Member is further subdivided into two sand units, the Main Sand unit and the Charlie Sand unit, separated by the Charlie Shale unit. It is the Sandstone Member which contains most of the Forties oil. A detailed description of the lithostratigraphy of the Forties area is given in Thomas et al (1974) and Deegan and Scull (1977) and is summarized in Figure 5.

The easterly regional dip at the base of the Tertiary sequence in the Forties area is interrupted by a large east-southeast trending nose, the Forties-Montrose ridge (Fig. 6). Reversal of dip on this nose provides the anticlinal closure in the overlying Paleocene sandstones of the field, in which no significant faulting has been observed.

CORE DATA

Studies of core material from the appraisal and early development wells resulted in the contrasting lithologies being grouped into four broad facies types referred to as Facies A, B, C, and D (Thomas et al, 1974). With the availability of more well data these facies types are redescribed and interpreted to define specific environments of deposition.

Facies A

The sediments are fine to medium grained, occasionally silty, quartzose sandstones, generally with a grain contact fabric. The sandstones contain detrital lignite and mica and are interbedded with laminated siltstones and shales as graded units commonly less than 1.5 m thick. The sandstones exhibit a vari-

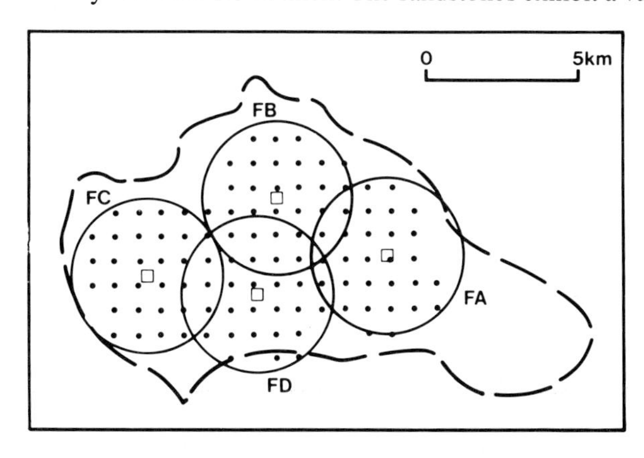

FIG. 2—Development plan of the Forties field showing well locations and drainage area of each platform.

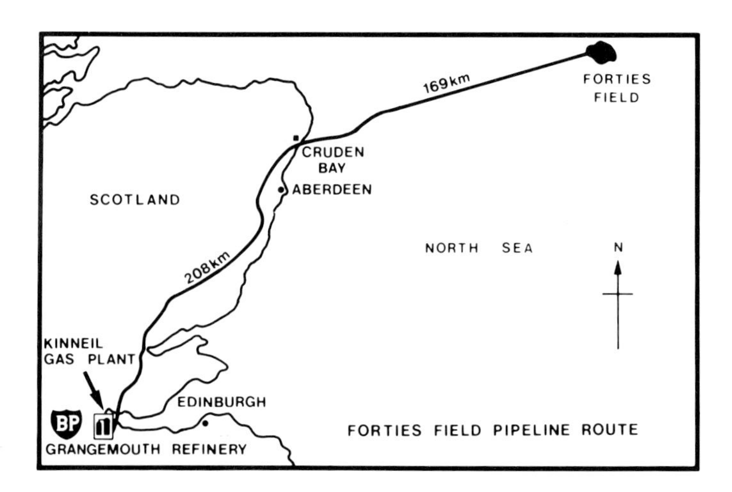

FIG. 3—Location of the Forties field and the pipeline route.

ety of structures including graded bedding, parallel lamination, sole marks and load structures, ripple cross-lamination, and contorted bedding. The graded beds, with the above associated structures and repetitive alternations of sandstones and shales are consistent with deposition by turbidity currents and can be best described by the Bouma model (Bouma 1962). Facies A sediments are interpreted to result from deposition by waning flows and, because of the predominance of sequences beginning with Bouma A/B divisions, are termed *proximal turbidites* (Walker 1978).

Facies B

The sediments are poorly sorted, coarse to medium/fine grained quartzose sandstones with a generally clay-free grain contact fabric. Thin pebbly or granule zones occur at the bases of many sandstone beds but normal and reverse grading generally are absent. Bed thickness ranges from tens of centimeters to 2 m, but units of sandstones up to 50 m thick occur with only a few interbedded shales. These units are believed to be composed of many amalgamated sandstone beds. Sedimentary structures are scarce and most sandstones appear to be massive. However, of those sedimentary structures which can be seen, coarse-grained laminae are the most common, with graded bedding, dish structures, vertical fluid escape pipes, convolute bedding, sole marks and flow casts only occasionally developed. Rare chaotic intervals occur up to 0.5 m thick, consisting of intraformational dark gray shale or sandstone clasts, set in a contorted matrix.

The thick, amalgamated, massive sandstone sequences up to 50 m thick are interpreted as channel sediments. However the thinner Facies B intervals up to 2 or 3 m thick are here considered as either minor channel fills or as nonchannelized proximal sheet flows, gradational between the proximal turbidites and the massive sandstones of Walker (1978).

Facies C

This facies is dominated by gray kaolinitic shales and graded siltstone/shale couplets 1 to 5 mm thick. Thicker siltstones up to 70 cm thick occur with ripple cross lamination and lenticular bedding. Rare fine-grained sandstones up to 80 cm thick occur and are occasionally graded with sharp erosional bases. Penecontemporaneous deformation structures are common. Facies C lithologies are consistent with deposition by low density turbidite currents (Ricci-Lucchi, 1975) and are best described as base-absent Bouma sequences. It is uncertain whether the kaolinitic shales represent turbiditic and/or hemipelagic sedimentation. Facies C sediments would be termed *thin bedded*

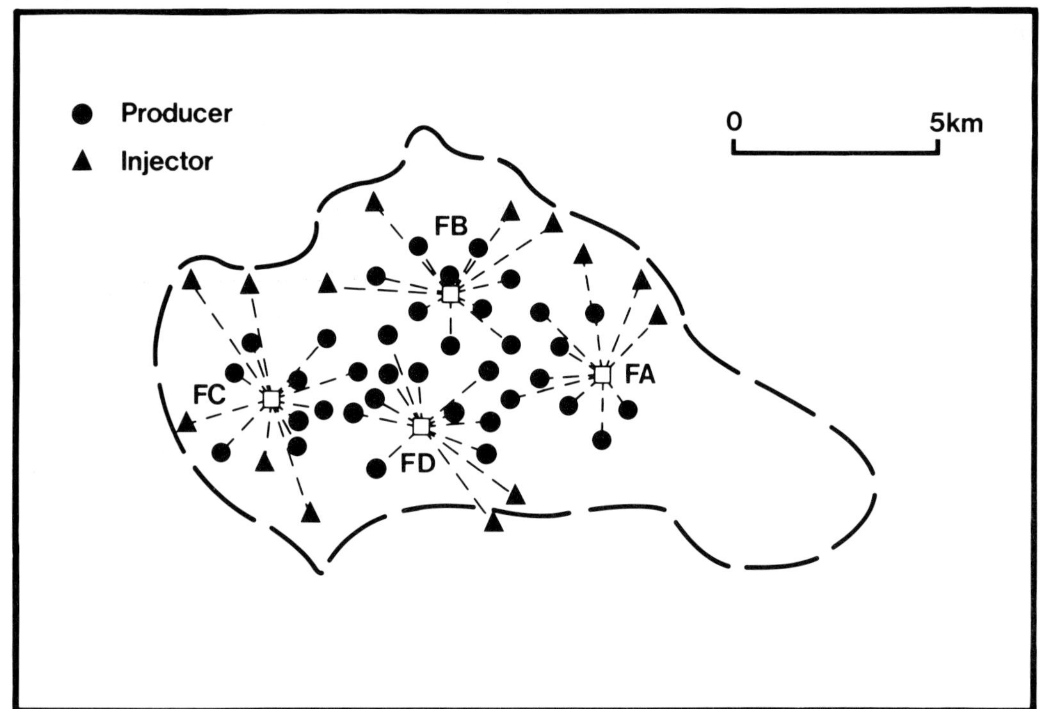

FIG. 4—Structure contour map of top of the Forties reservoir with present well locations, June 1978.

		Generalised Lithology	Thomas et al (1974)	Walmsley (1975)	This paper after Deegan and Scull (1977)		
Tertiary	Palaeocene		Unit IV	Unit IV	Balder Fm		Rogaland Gp
			Unit III	Unit III	Sele Fm		
		Charlie Sand Unit	Unit II	Unit II	Sandstone Mbr	Forties Fm	Montrose Group
		Charlie Shale Unit					
		Main Sand Unit					
			Unit I		Shale Mbr		
				Unit I	Andrew Fm		
					Maureen Fm		
			Danian Limestones	Danian Limestones	Ekofisk Fm		Chalk Gp
Cret.			Maastrichtian (U. Cretaceous)	Maastrichtian (U. Cretaceous)	Tor Fm		

FIG. 5—Early Paleogene lithostratigraphic nomenclature, Forties field.

turbidites.

Facies D

The sediments are burrowed, waxy green shales which have an abundant marine fauna and flora. The shales are associated with thin dolomites, limestones, and calcareous sandstones. In contrast with the shales of Facies A, B, and C which are kaolinitic, the shales of Facies D are montmorillonitic. This facies most commonly occurs in association with Facies C and is restricted to the southeastern part of the field. Facies D sediments are considered to be hemipelagic from the evidence of their contained fauna.

The porosity of sandstones within Facies A, B, and C are broadly similar. The mean porosity based on core data for Facies A and B is 24 to 29%. Facies C sandstones have slightly lower values, 20 to 22%. Each facies has a distinct horizontal permeability distributions. Facies A has geometric mean values of 9 to 68 md, Facies B has values of 214 to 753 md, and Facies C has values of 1 to 2.5 md. Facies D sandstones are insufficiently sampled but have porosities and permeabilities lower than those of the other facies.

Facies Transitions

The four facies described above represent a simplification of the spectrum of the lithologies seen in the Forties field. It is apparent from the core data that facies boundaries cannot be rigorously defined. There is a complete facies transition from Facies A through to Facies D. However, the recognition of preferred transitions between facies is important in outlining the general relationships and determining a depositional model to best describe the Forties Formation over the field area. It is recognized that many more facies types could be defined from the available core data. But as the value of these facies depends on their ability to be recognized from electric logs, the number of facies types has been purposely restricted.

Using all the available core data and tabulating the number of times all vertical facies transitions occur, preferred transitions have been determined by embedded Markov chain analysis (Krumbein and Dacey, 1969). In this type of study the transitions are tabulated regardless of the thickness of an individual bed. The nature of any preferred depositional process that is present is derived by following through the highest values of the the "probability matrix" modified after Miall (1973). These values reflect, although they do not prove, the presence of any Markovian dependency relationship. When applied to the three principle facies identified in core data, the following preferred depositional transitions are seen:

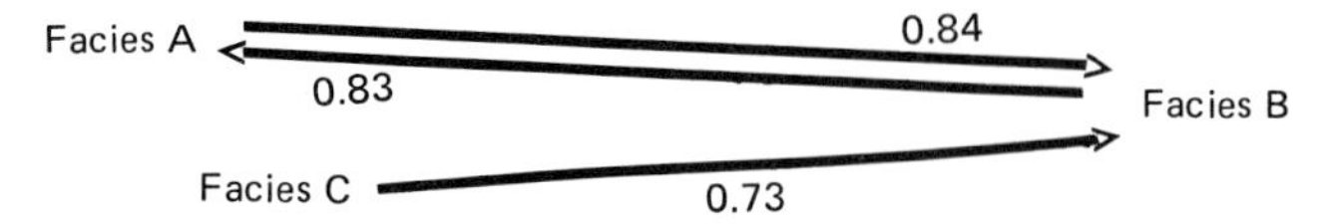

The proximal turbidites of Facies A are most commonly seen to pass into the sandstones of Facies B. This is seen as a thickening of the sandstone beds and the loss of the regular sand/shale interbeds. The thin bedded turbidites of Facies C also most commonly grade into Facies B and commonly are seen as a sudden thickening of the sandstone beds with loss of shales. Facies D is only found in association with Facies C sediments in the southeastern area of the field and is only recorded in the cores of one well. It has insufficient transitions to be included in the analysis. It is significant to note that Facies A and C show no apparent preference, indicating limited transitions between proximal and thin bedded turbidites.

Depositional Environment of the Forties Formation

The preferred facies relationships, the predominance of Facies B, the lack of association of Facies C and A sediments, and the internal structures shown by the sediments themselves, characterize deposition in the middle to lower fan area of a submarine fan environment (Fig. 7). In Figure 8 the various facies of the resedimented family (Walker, 1978) are shown in their interpreted positions in an idealized submarine fan model. Also shown are the interpreted positions where Facies A, B, C, and D are preferentially developed. The interpretation is based on the detail morphological studies of modern submarine fans (Normark, 1978) and on the extensive literature describing ancient submarine fans (see review by Walker, 1978). As can be seen in Figures 7 and 8, a mid-to-lower fan environment of deposition would account for the observed sedimentary features of the reservoir sandstones.

LOG DATA

Sequential Facies Associations

Early studies of redeposited sediments emphasized internal subdivisions of single beds of turbidite sequences according to the Bouma model (Bouma, 1962). More recently, nonrandom vertical variation in lithology of "flysch" successions many hundreds of meters thick has been recognized (see review in Walker, 1970). Ricci-Lucci (1975) applied the concept of sequential vertical facies analysis to the description and identification of ancient submarine fans. We now emphasize lithofacies and bed thickness relationships in determining depositional environments and correlations. Facies outlined in this paper are based on core data, including bed thickness, but the majority of oil industry geologic information is based on electric-log data. Bed thickness cannot readily be identified from logs, and correlation of individual sandstones and shales is extremely difficult and unreliable. To compensate for this, core data over the Forties field has been related to characteristic log shapes and patterns which correspond with the four facies (The four facies are considered to be responses to specific depositional environments.) The patterns are then correlated and reliable lithologic correlations are developed.

Five types of log patterns can be recognized in the Forties Formation by specific shape or pattern on a combination of the gamma-ray, sonic, density, and resistivity logs (Fig. 9). These patterns are described over cored intervals and then extrapolated to uncored intervals and other wells over the field. Many specific depositional environments other than those described by the log patterns occur in a submarine fan complex. However, the essence of this approach is correlation and, unless a log pattern is distinctive, it will not significantly help to understand sand body geometry and reservoir performance.

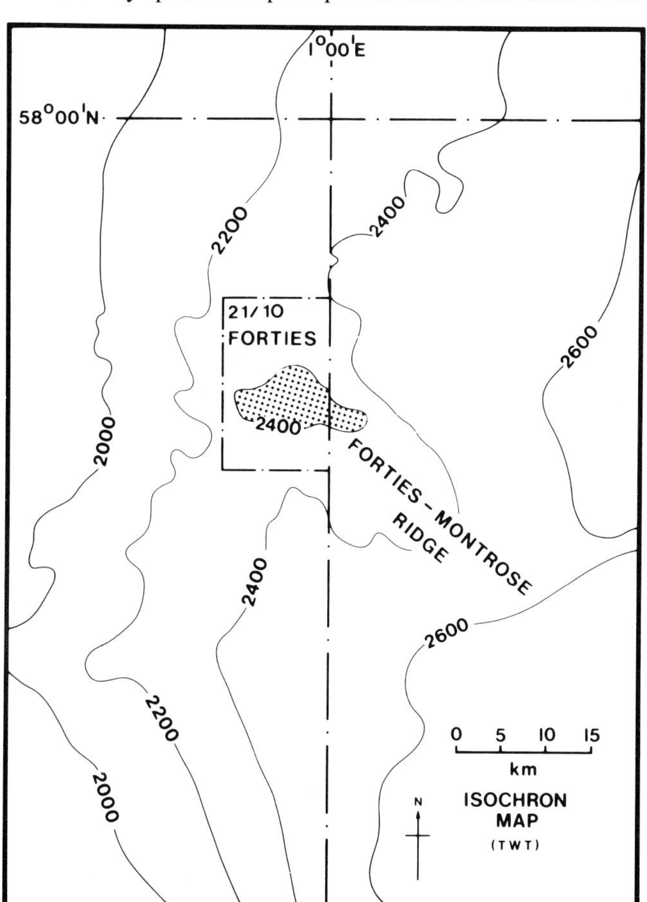

FIG. 6—Isochron map of the top Paleocene seismic horizon showing the structure of the Forties field. Isochron interval = 200 millisecs (two-way time). Map shown with UK Licence Block 21/10.

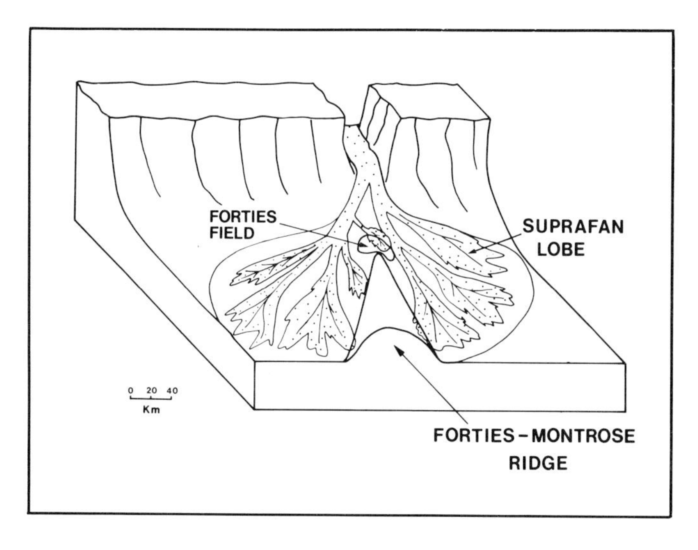

FIG. 7—Schematic representation of the regional depositional environment of the Forties Formation—a submarine fan emphasizing the depositional sandstone lobes or suprafans.

1. *Constant (High Gamma) Pattern*—This pattern 5 to 25 m thick is represented by a stable, high gamma-ray, sonic, density, and low-resistivity log response. The sequence is seen to consist of Facies C and D sediments—light to dark gray, occasionally waxy green shales with rare, thin interbeds of sandstone and limestone. The pattern is thought to be a response to deposition on the lower fan, or on the basin plain itself. The environment is characterized by hemipelagic deposition, interrupted periodically by low energy turbidity flows.

2. *Erratic Pattern*—An erratic log pattern refers to a sequence upward of 2 m thick of thinly interbedded sandstones and shales. This produces an irregular motif which predominantly contains thin bedded turbidites of Facies C with occasionally some thicker more proximal turbidites of Facies A. The pattern is considered typical of many environments in a submarine fan sequence including levee, channel margin, abandoned channel fill, and interchannel areas of the upper, middle, and indeed lower fan areas. In the absence of core data (and even with it), it is difficult to categorically interpret the erratic log motif; Mutti (1977) defined five thin-bedded turbidite facies. In the context of the Forties field it is important to take into account the associated facies both vertically, and where possible, laterally. Hence if it is found that the erratic pattern is predominantly associated with thick Facies B channelized sandstones then the motif probably represents an interchannel/levee environment. Conversely if this pattern is associated with Facies C and D sediments then the motif probably represents a lower fan environment.

3. *Upwards Decreasing Gamma Pattern*—This pattern,

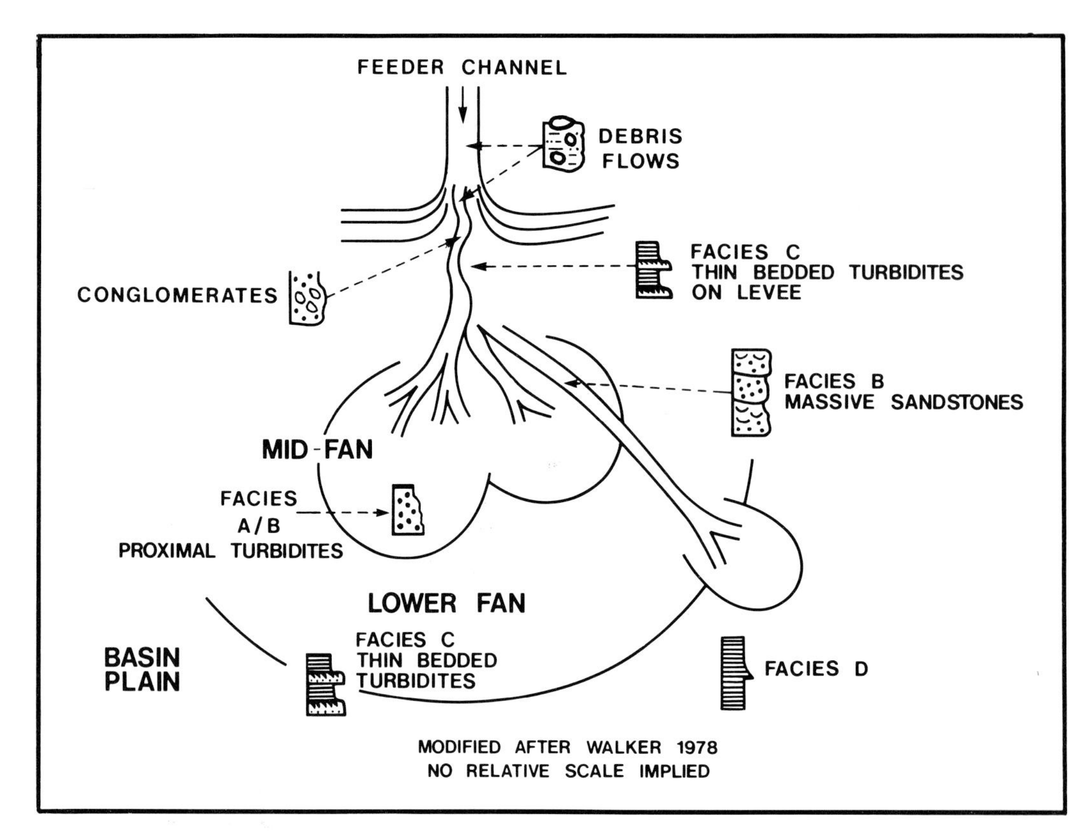

FIG. 8—A model of submarine fan deposition relating the four facies types in the Forties field to fan morphology and depositional environment. Facies A-D are compared to the resedimented course clastic family of Walker (1978).

generally 5 to 10 m thick, is easily recognized as an upwards-decreasing gamma ray pattern; density response and interval transit time all indicate decreasing shaliness and increasing porosity and permeability. The resistivity log (Deep Laterolog) shows an upwards increase in the oil column and this further suggests an increase in permeability. The pattern generally starts with mature turbidites of Facies A and passes gradually into more proximal, thicker bedded turbidites with a higher sand/shale ratio (Facies B). Such coarsening and upward-thickening sequences are interpreted in a submarine fan environment as prograding depositional lobes at the mouths of distributary channels (Walker and Mutti, 1973; Parker, 1975) and characterize the nonchannelized area of the middle fan (Mutti 1977).

4. *Constant (Low Gamma) Pattern*—This log pattern represents a fairly uniform sandstone sequence. Structureless conglomerates, pebbly sandstones, and occasional coarsely laminated sandstones are the most common features. Amalgamation of these sandstones and conglomerates form continuous bodies up to 20 m thick and may be separated from a succeeding thick unit of amalgamated sandstones by only a thin bed of shale or siltstone couplet of Facies C, or some occasional graded sandstones of Facies A. Taken as a whole, low gamma-ray log patterns with thin, interspersed shaley sequences are up to 50 m thick over the field area. This log pattern represents channel deposition, with sands forming thick, amalgamated units. The sequence terminates where the channels were suddenly choked and deposition switched to another channel. During abandonment, thin bedded sandstones, siltstones, and shales were deposited from the tails of passing turbidity currents giving thin units of Facies C or A.

5. *Upwards Increasing Gamma Pattern*—This log pattern is easily recognized as an upwards increase in gamma-ray and sonic-log response with a corresponding decrease in density and resistivity (Deep Laterlog) reading. Channel sandstones of Facies B, often with an eroded base, invariably underlie this log motif. The grain size and bed thickness decrease upwards and Facies A passes into Facies C sediments. It is considered that these upward-fining sequences 5 and 40 m thick reflect the progressive abandonment of a channel (Ricci-Lucchi 1975).

Facies Distribution over the Forties Field

The four facies described from core data are the starting point in the interpretation of the vertical changes seen in the log of the 50 development wells drilled to date (Fig. 4). The close spacing of wells provides a unique opportunity in the North Sea to recognize and correlate sequential and specific depositional events described by log patterns. Log pattern analysis extends the traditional and sometimes misleading "log pick" type correlation which assumes that the termination of a given lithology in one area, as reflected by a certain log response, is correlatable to a similar log response in another area. However, sequential log pattern analysis may require dissimilar log picks to be correlated. The results are meaningful in terms of defining sand body geometries and likely reservoir drainage patterns.

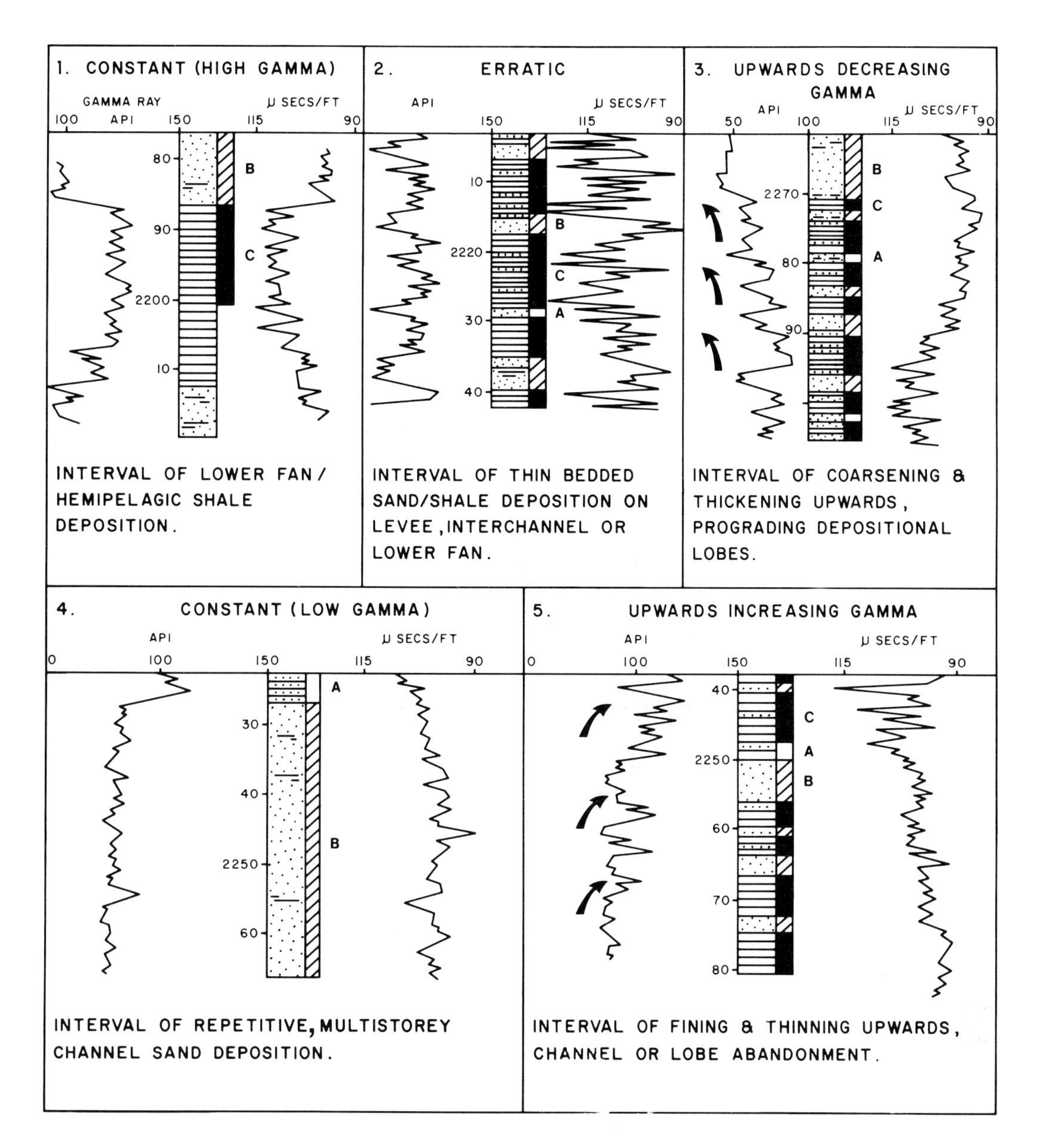

FIG. 9—Characteristic log patterns seen in the Forties Formation, Forties field. An interpretation of the preferred depositional location within a submarine fan complex is shown below each pattern. Facies types seen in the cores are shown to the right of the lithology. Depths are in meters (true vertical depth below rotary table).

The present interpretation of facies distribution of the Forties Formation over the field is shown in Figures 10 and 11. The 50% net sand line is an arbitrary line that broadly highlights the main channelized regions of the field. In the northeastern area of the field, continuous sand sequences of Facies B sediments are developed with a predominantly Constant (low gamma) log pattern with occasional large scale Upwards-Increasing gamma patterns and rare Upwards-Decreasing gamma patterns.

In the central and southeastern areas of the field, slightly lower net sand ratios are evident with Upwards-Decreasing gamma and Erratic patterns more commonly developed. This is interpreted as an essentially nonchannelized area of the middle fan with only infrequent channel progradations. Sedimentation in this region is considered to have been influenced by the Forties-Montrose ridge (Fig. 6). Because the channel sands of Facies B are essentially gravity flows, they are believed to have preferentially followed topographic lows and by-passed the ridge area.

In the western part of the field, thick channel sand sequences of Facies B are developed. However in this part of the field a major shale barrier some 20 m thick (the Charlie Shale unit), separates the sands of the Sandstone Member into a lower Main Sand unit and an upper Charlie Sand unit (Fig. 5). The Charlie Sand unit was only fully recognized and delineated by development drilling and its separate nature is evident from the pressure production history of the field (Hillier et al, 1978; see Fig. 12). The Charlie Sand unit is used here as an example of how sequential log pattern analysis is used to better understand the effect of sand geometry on reservoir drainage and pressure support requirements.

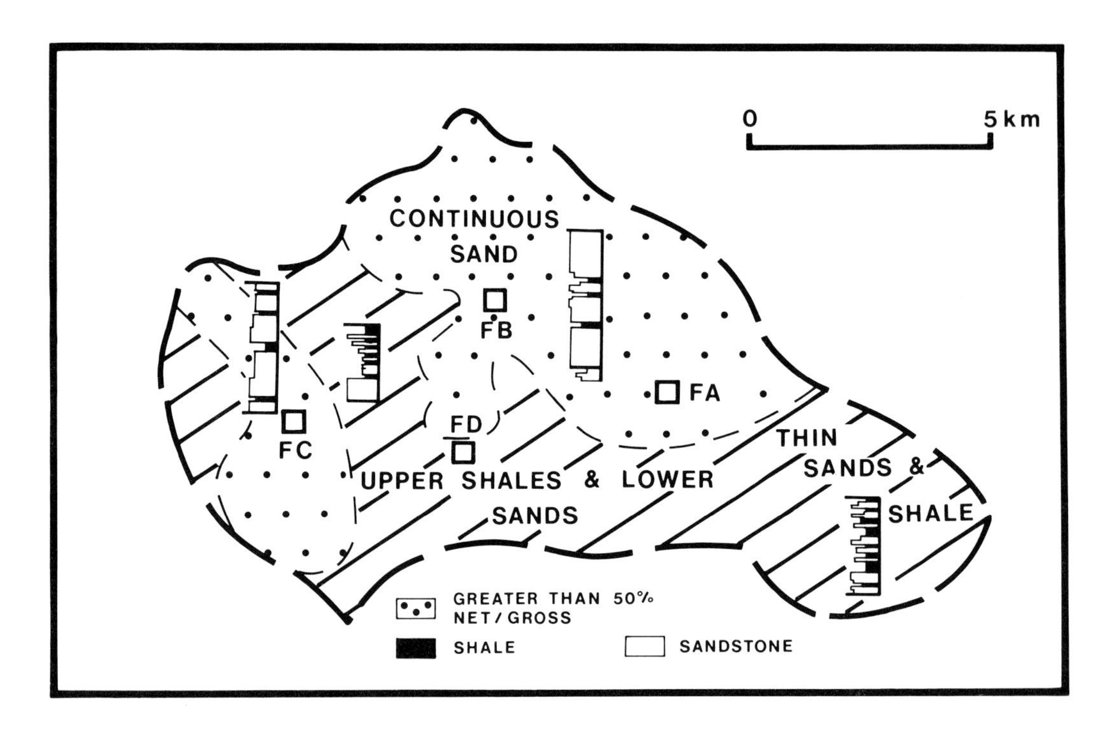

FIG. 10—The post-development interpretation of the facies distribution over the field area, using log pattern analysis.

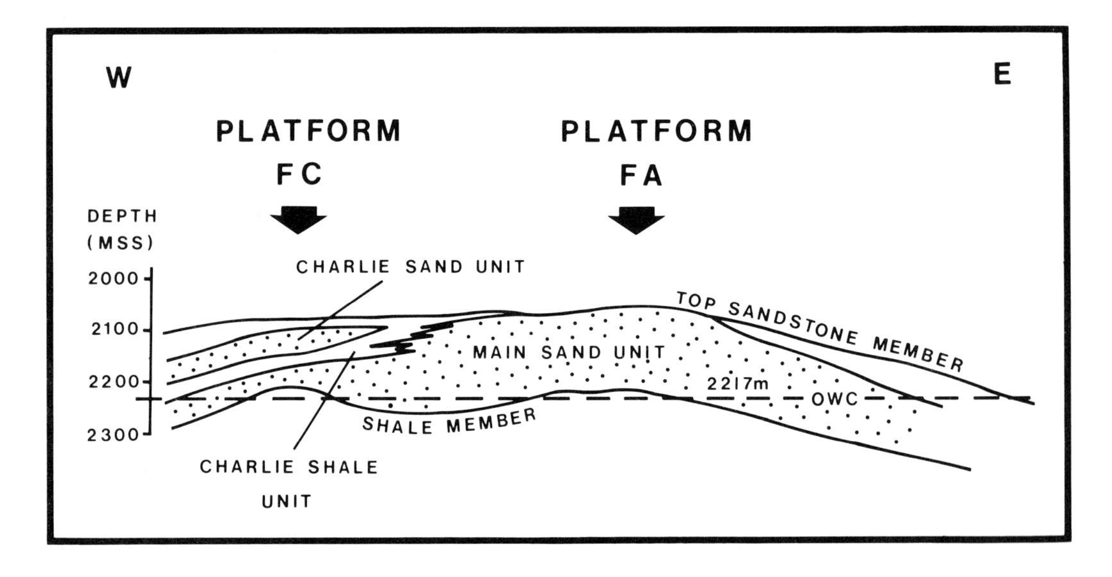

FIG. 11—A diagrammatic west-east cross section of the Forties field highlighting the separation of the Charlie Sand unit from the Main Sand unit by the Charlie Shale unit.

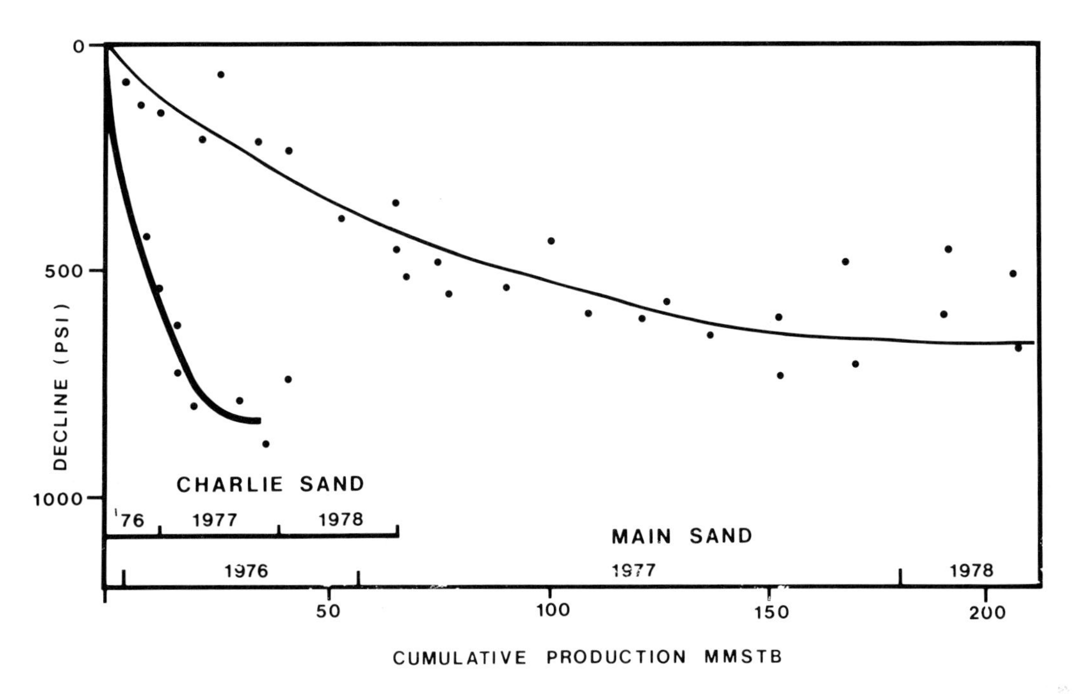

FIG. 12—Well pressure declines vs. cumulative oil production, Forties field. Since production start-up, the Charlie Sand behavior is marked by a greater pressure decline when compared with the Main Sand.

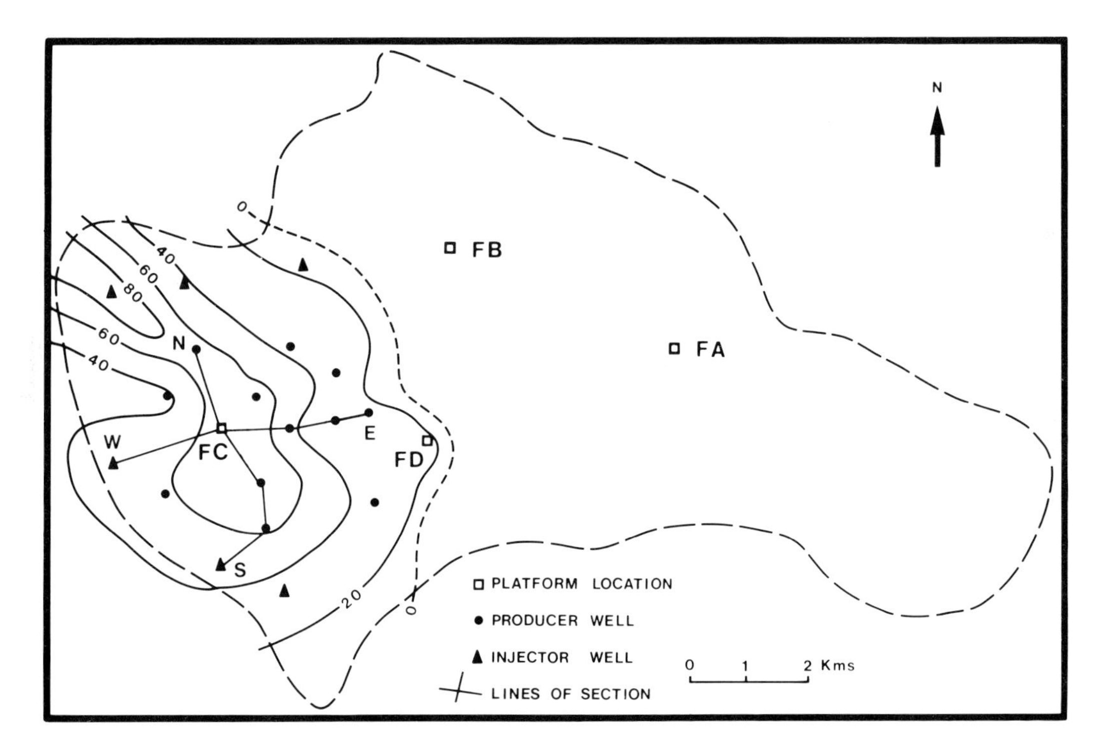

FIG. 13—An isopach map of the Charlie Sand unit based on well data and showing the line of section used in Figures 14 and 15.

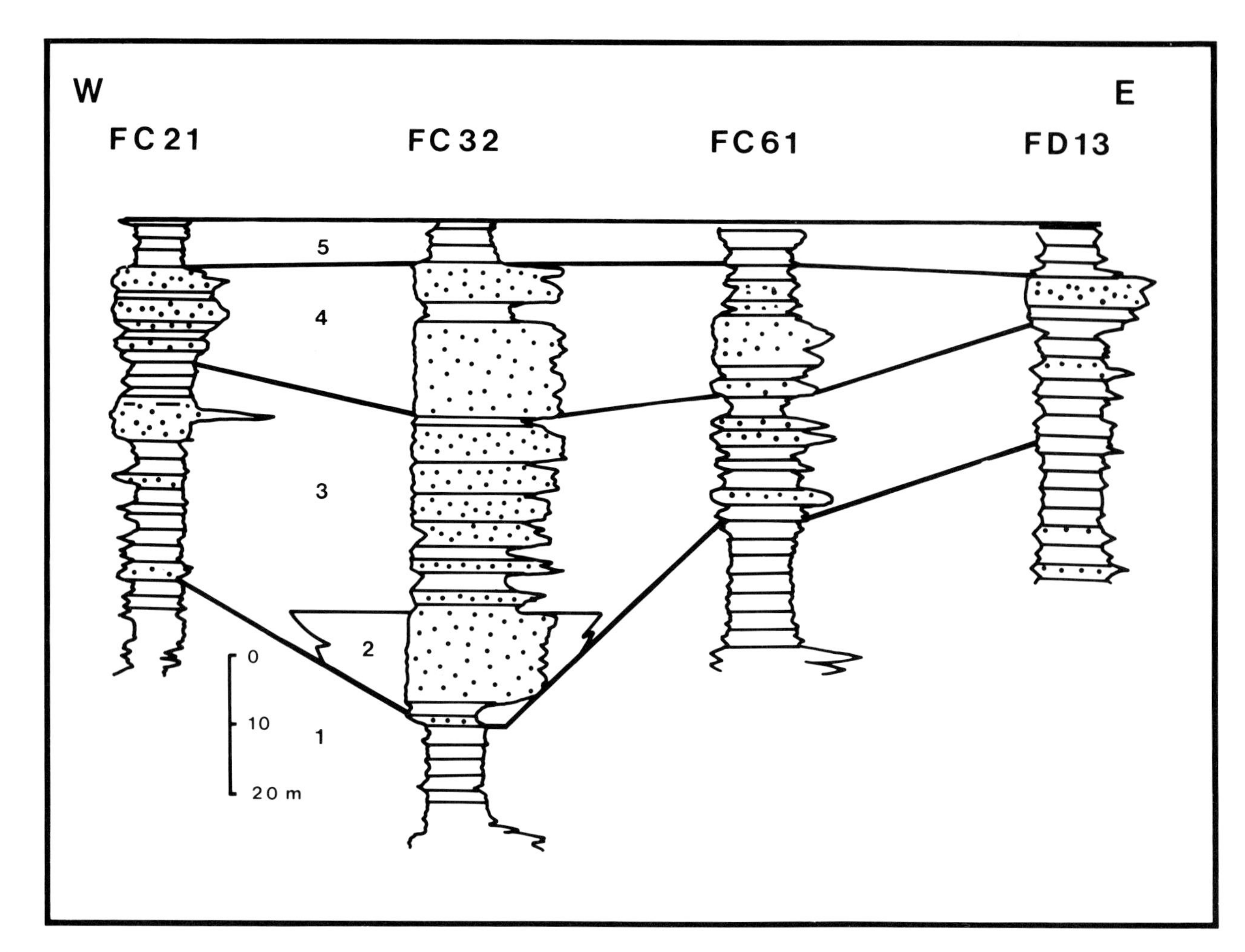

FIG. 14—West-east petrophysical log correlation across the Charlie Sand unit. The units 1-5 are described in the text. Gamma ray log is shown on the left and the deep resistivity log (DDL) on the right.

SEQUENTIAL LOG PATTERN ANALYSIS OF THE CHARLIE SAND UNIT

An isopach map of the Charlie Sand unit (Fig. 13)shows that the axis of the sand body trends in a north-northeast–south-southeast direction. By taking two cross sections, one west to east across the axis and one approximately north-northwest–south-southeast down the axis, a detailed morphological subdivision of the Charlie Sand is possible based on the identification of the log patterns defined above.

To emphasize the importance of using all available log data, and for comparison with Figure 9 the west to east section (Fig. 14) and the north-northwest–south-southeast section (Fig. 15) uses a gamma ray/resistivity (Deep Laterolog) display. Care was taken with the Laterolog to ensure that the effects of deep mud filtrate invasion or increasing water saturation with depth were not influencing the interpretation. Five informal intervals can be distinguished, starting with the Charlie Shale unit.

1. *Lower Fan Shales*—This is an interval of gray shales with occasional thin sandstones 20 m thick with a Constant (high gamma) log pattern. Predominantly made up of shales with occasional thin bedded turbidites of Facies C, the Charlie Shale unit is considered to reflect deposition on the lower fan area.

2. *Channel Sands*—This is the basal interval of the Charlie Sand unit, about 16 m thick, and is comprised of coarse to medium grained sandstones of Facies B with only infrequent, thin shale interbeds. The basal contact of the interval appears sharp on the logs and is possibly erosional. A Constant (low gamma) log pattern predominates in well FC32 with minor Upwards-Increasing gamma patterns suggesting multiple layers of channel sands (Fig. 14). Passing westwards and eastwards from well FC32 the sand quality decreases above the Charlie Shale unit and the logs change to an Erratic pattern indicating more levee or interchannel deposition with Facies C sediments predominant (Fig. 14). South of FC32, sand quality again decreases and the log pattern changes to an Upwards-Decreasing gamma pattern in wells FC51 and FC41 that appears to be related to the overlying section (Fig. 15). This pattern change is interpreted as a basinwards trend from a channel facies in FC32 to a nonchannelized progradational lobe facies in FC51, FC41, and FC36.

3. *Prograding Lobes*—This is an interval that is made up of many Upwards-Decreasing gamma patterns, 5 to 10 m thick, that makeup an overall sequence of Upwards-Decreasing gamma patterns about 35 m thick. The interval consists of Facies B sands generally between 2 to 3 m thick with subordinate

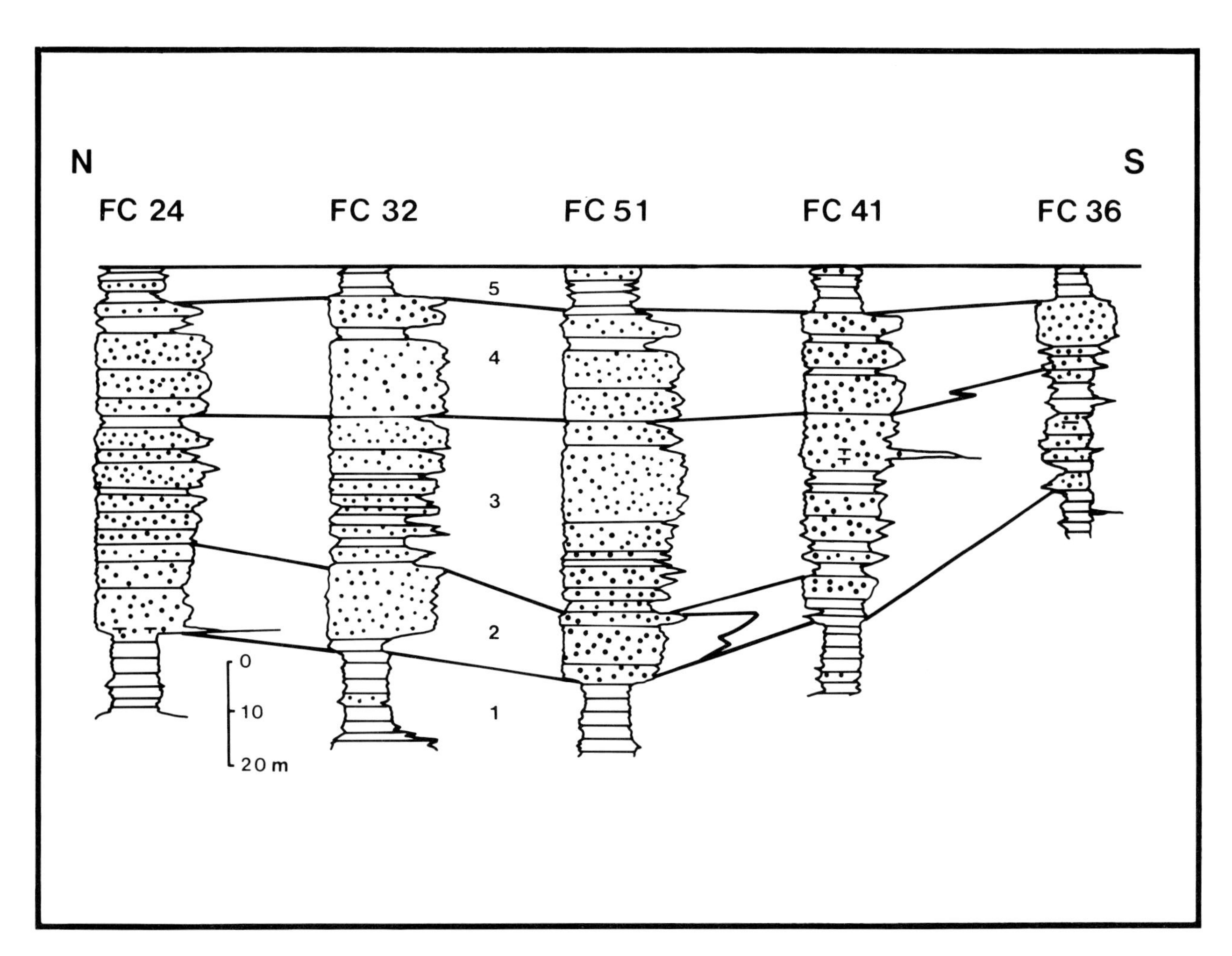

FIG. 15 — A north-northwest to south-southeast petrophysical log correlation across the Charlie Sand Unit. Units 1 to 5 are described in the text. Gamma Ray log is shown on the left and the deep resistivity log (DLL) on the right.

interbeds of sands and shales of Facies A and C. The small scale Upwards-Decreasing gamma patterns are interpreted as minor progadational lobes that were deposited at the mouths of constantly switching distributary channels. They can be compared with the thickening-upward, complex second order cycles of Ricci-Lucchi (1975). The overall coarsening and thickening upwards of the interval is interpreted as indicating the progradation of a major suprafan lobe. To the west, east, and south of FC32 the interval becomes shalier (Erratic log patterns) with an increase of Facies C sediments suggesting levee/ interchannel and/or lower fan deposition (unit 3b, Fig. 16).

4. *Channel Sands*—This is an interval of about 22 m of massive sandstones of Facies B with occasional shale interbeds. Constant (low gamma) patterns with occasional Upwards-Increasing gamma patterns are developed, indicating multiple-layered channel sequences. To the west and east of FC32 the massive sandstones thin and become shalier with a more channel-margin aspect. To the south of FC32, Upwards-Increasing gamma patterns pass into Upper-Decreasing gamma patterns which indicates a continuation of the previous interval, (3), and the predominance of prograding lobe sequences at the seaward edge of the suprafan lobe.

5. *Lower Fan Shales and Sands*—This is an interval about 8 m thick of mainly gray shales with occasional thin bedded sandstones of Facies C that give an overall Constant (high gamma) log pattern. However, there is one last prograding lobe sequence giving an Upwards-Decreasing gamma pattern as shown in wells FC51 and FC61. This interval marks the abandonment of the suprafan lobe and a return to nonchannelized, essentially lower-fan deposition in the western area of the field.

As a result of vertical log pattern analysis, the western part of the field is recognized as a prograding turbidite suite characterized by five depositional cycles that allow an interpretation of the depositional history (Fig. 16). Following the deposition of the lower fan shales and thin sands, (1), an active, rapidly advancing distributary channel system moved into the area, (2). The channels were then gradually choked and eventually abandoned, indicating a retrogradational phase. This was followed by a sequence of prograding depositional lobes, (3), deposited at the mouths of a more slowly advancing distributary system. Progradation continued and was followed by the appearance of middle-fan channels overlying and possibly scouring the topmost lobe deposits, (4). The abandonment of the distributary system feeding the suprafan lobe cut off the sand supply and allowed the deposition of essentially lower fan

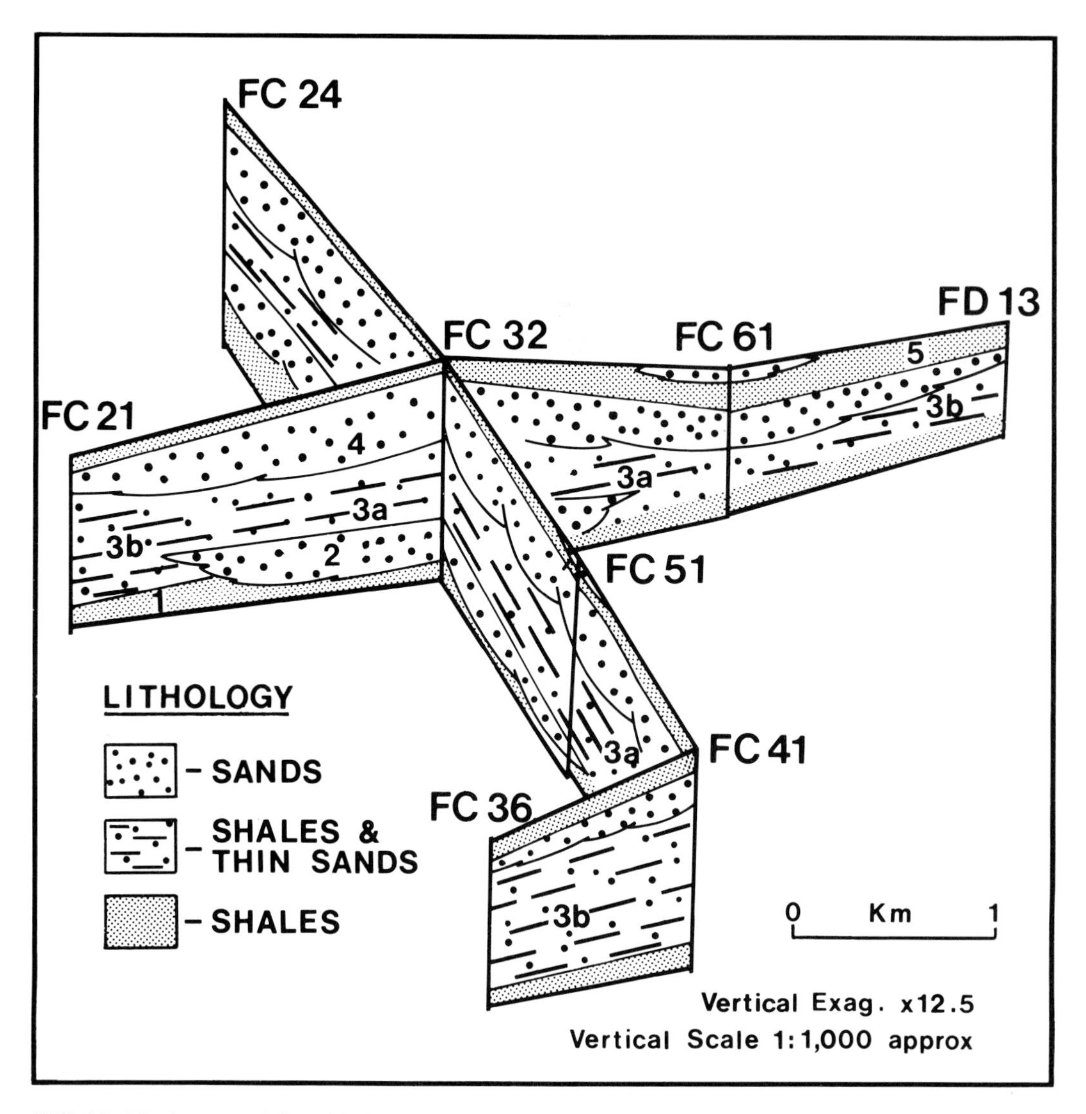

FIG. 16—The interpreted depositional history of the Charlie Sand unit based on sequential log pattern analysis. Units 1-5 are as described in the text. (1) Lower fan shales; (2) middle fan channel sands; (3a) Prograding lobes on the suprafan; (3b) Lower fan/interchannel deposition; (4) Middle fan channel sands; and (5) Lower fan shales with rare lobe sands.

shales and sands, (5). The relatively abrupt nature of the contact of (4) and (5) suggests a top cut-out cycle (Ricci-Lucchi, 1975) and a sudden avulsion of the distributary system feeding the suprafan.

The separate nature of the Charlie Sand unit has already been noted from pressure production data (Fig. 13) and this can in part be explained by the geometry and complexity of the depositional history of the western part of the field. Laterally to the west and east of well FC32, the Charlie Sand unit thins and shales out into what is considered an interchannel/levee environment with a decreased net/gross ratio and lower porosities and permeabilities. Thick channel sands continue to the north and dip below the oil/water contact. To the south the sands pass into thin-bedded turbidites of the lower fan, again with poorer reservoir characteristics (Fig. 16). Beneath the Charlie Sand unit lies the Charlie Shale unit; a sequence of shales and thin bedded turbidites that act as a vertical pressure barrier.

Log pattern analysis of the Charlie Sand unit allows the construction of a depositional model of the sand body. Such a model is predictive and is used when selecting down dip well locations, particularly water injection wells, which are important in the pressure support program of the Charlie Sand unit.

CONCLUSIONS

The Forties field is a giant oil field in an interbedded sandstone and shale sequence of the Forties Formation of Paleocene age. Based on core data, four broad facies types have been identified that show nonrandom vertical relationships. The internal sedimentary structures and facies relationships of these sandstones and shales are considered to characterize deposition in the middle to lower fan area of a submarine fan environment (Walker, 1978).

Because correlation of individual sands and shales is unreliable in this complex reservoir, five electric log patterns have been identified that are considered to be responses to specific depositional environments within a submarine fan complex.

Using the vertical changes seen in the logs of the 50 development wells drilled to date, detailed correlations are possible that greatly assist in the determination of sand-body geometry and drainage patterns.

The facies distribution over the field is described and shows thick multiple layered sandstones sequences in the northeastern part of the field that are interpreted as a channelized area of the middle fan. In the central and southeastern areas, lower net sand ratios suggest an essentially nonchannelized area of the middle to lower fan area. The western area of the field again shows thick multiple-layered channelized sandstones of the middle fan area.

As an example of sequential log pattern analysis, the western part of the field is described in detail. Five informal intervals are distinguished that suggest that the Charlie Sand unit developed as a prograding suprafan lobe sequence. The relatively abrupt contact with the overlying shales indicates rapid abandonment of the distributary channel feeding the lobe. The separate nature of the Charlie Sand unit reservoir can in part be explained by the geometry and complexity of this sand body. The depositional model of the Charlie Sand is a predictive tool and is used to aid the selection of downdip well locations.

REFERENCES CITED

Bouma, A. H., 1962, Sedimentology of some flysch deposits: Amsterdam, Elsevier, 168 p.

Deegan, C. E., and B. J. Scull, 1977, A standard lithostratigraphic nomenclature for the central and northern North Sea: Report 77/25 Bull. 1, HMSO, London.

Hillier, G. R. K., R. M. Cobb, and P. A. Dimmock, 1978, Reservoir development planning for the Forties field: European Offshore Petroleum Conference and Exhibition, Proc., v. 2, p. 325-335.

Krumbein, W. C., and M. F. Dacey, 1969, Markov chains and embedded chains in geology: Int. Assoc. Math. Geol., Jour., v. 1, p. 79-96.

Maill, A. D., 1973, Markov Chain analysis to an ancient alluvial plain succession: Sedimentology, v. 20, p. 347-364.

Mutti, E., 1977, Distinctive thin-bedded turbidites facies and related depositional environments in the Eocene Hecho Group (south-central Pyrenees, Spain): Sedimentology, v. 24, p. 107-31.

Normark, W. R., 1978, Fan valleys and depositional lobes on modern submarine fans: characters for recognition of sandy turbidite environments: AAPG Bull., v. 62, p. 912-931.

Parker, J. R., 1975, Lower Tertiary sand development in the central North Sea, *in* A. W. Woodland, ed., Petroleum and the continental shelf of north-west Europe, 1, Geology: London, Applied Sci. Pubs., p. 447-452

Ricci-Lucchi, F., 1975, Depositional cycles in two turbidite formations of northern Appennines (Italy): Jour. Sed. Petrology, v. 45, p. 3-43.

Thomas, A. N., P. J. Walmlsey, and D. A. L. Jenkins, 1974, Forties field, North Sea: AAPG Bull., v. 58, p. 396-405.

Walker, R. G., 1970, Review of the geometry and facies organization of turbidite bearing basins: *in* Flysh sedimentology in North America: Geol. Assoc. Canada Spec. Paper 7, 219-251.

———1978, Deep-water sandstone facies and ancient submarine fans: models for exploration for stratigraphic traps: AAPG Bull., v. 62, p. 932-966.

——— and E. Mutti, 1973, Turbidite facies and facies associations, *in* G. V. Middleton and A. H. Bouma, eds., Turbidites and deep water sedimentation: SEPM Pacific Sec. Short Course (Anaheim), p. 119-157.

Walmsley, P. J., 1975, The Forties field, *in* A. W. Woodland, ed., Petroleum and the continental shelf of north-west Eruope, 1, Geology: London, Applied Sci. Pubs, p. 477-485.

Statfjord Field—A North Sea Giant[1]

By R. H. Kirk[2]

Abstract Statfjord, largest single oil field in the North Sea, is located on the U.K.-Norwegian boundary between 61° and 61°30′ N lat. Initial estimates are that about 11% of the field lies in U.K. waters. Its discovery, in March 1974, was based on interpretation of seismic reflection surveys and extrapolation of a productive regional trend. Two principle sandstone reservoirs, Middle Jurassic Brent and Lower Jurassic–Upper Triassic Statfjord, contain reserves in the order of 3 billion bbl within a productive area of approximately 20,000 acres (580 sq km). Reservoir properties are excellent, with permeabilities in darcies. The field extends northeasterly 15.5 mi (24.8 km) and averages 2.5 mi in width (4 km).

Tilted Jurassic fault blocks form the primary hydrocarbon trap at Statfjord as throughout the East Shetland basin. Statfjord field is a structural-stratigraphic trap formed by westward tilting and erosion of a major fault block. Brent deltaic sands and underlying Statfjord continental (fluvial) sands are truncated by middle to late Kimmerian unconformities on the crest and east flank of the structure which is marked by a major fault system. Overlying and onlapping Jurassic and Cretaceous shales seal the trap. Organically rich Upper Jurassic shales provide an excellent oil source. Reservoirs have separate oil-water contacts. Normal faulting separates Statfjord field from Brent field to the southwest.

Joint development by Norway and the U.K. utilizes "condeep" type gravity platforms and initial offshore loading. Development drilling from Statfjord 'A' platform (towed to location in May 1977) began in late 1978. First production is expected late in 1979.

INTRODUCTION

As shown in Figure 1, the Statfjord field is located along the Norwegian-U.K. boundary in the northern part of the North Sea, approximately 120 mi (200 kms) northwest of the city of Bergen on the Norwegian mainland and 310 mi (500 kms) northeast of St. Fergus on the Scottish coast. The field lies primarily within Norwegian Petroleum Production License 037, Blocks 33/9 and 33/12, but extends into U.K. offshore blocks 211/24 and 211/25, both licensed to Conoco North Sea Inc. (operator), Gulf Oil Corporation, and BNOC (Exploration) Ltd.. The general Statfjord area is the most prolific part of the East Shetland basin (Fig. 2).

On August 10, 1973, the Norwegian government granted License 037 to a consortium of oil companies. Mobil Explora-

[1] Manuscript received, April 4, 1979; accepted for publication, July 23, 1979. This paper was presented at the Houston Annual Meeting of AAPG, but is an updated and enlarged manuscript from one presented by E. L. Jones, H. P. Raveling, and H. R. Taylor at a local Jurassic Northern North Sea symposium held in 1975 at Stavanger, Norway.

[2] Mobil Exploration Norway, Inc., Stavanger 4001, Norway.

The writer is grateful to Statoil, Mobil, and all of the Statfjord Unit Norwegian and U.K. partners for permission to present this paper. A number of restricted company and Statfjord Unit reports, not listed in the references, form the basis of the presentation. Reports by R. J. Moiola and J. W. Stinnett of Mobil Field Reserach Lab, Dallas, were used in the preparation of the sections on Statfjord and Brent stratigraphy and source rocks. The writer also gratefully acknowledges the help and assistance of Mobil Exploration Norway staff, in particular A. L. Chauvin, C. Killip, L. J. Reimer, L. Z. Valachi, and R. L. Whitney for their invaluable assistance and criticisms.

Article Identification Number:
0165-731X/80/M030-0006/$03.00/0.

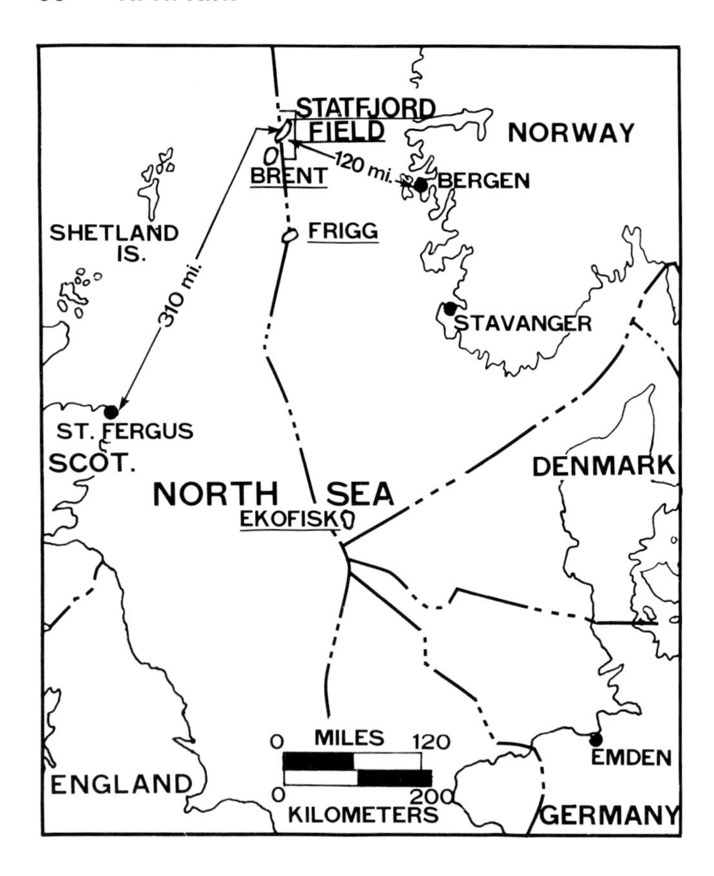

FIG. 1—Index map showing location of Statfjord field in North Sea.

tion Norway Inc., with an equity of 15%, was designated as operator for the group which included Den norske stats oljeselskap a.s. (Statoil), 50%; Conoco Norway Inc., 10%; Esso Exploration Norway Inc., 10%; A/S Norske Shell, 10%; and the Saga—Amoco—Amerada Hess—Texas—Eastern Group, 5%.

The License 037 area covers 143,300 acres (580 sq km) with water depths ranging from about 425 ft (130 m) in the south to 1,000 ft (305 m) in the northeast (Fig. 3). From the time of acquisition of the license to the end of 1978 some 2,237 mi (3,600 kms) of reflection-seismic record was gathered and 11 wildcats plus 4 appraisal wells drilled. This resulted in discovery of the Statfjord field in March 1974 and sufficient delineation of field limits to initiate development. In addition, two smaller fields were discovered in the license area.

REGIONAL SETTING

License 037 lies in the northern part of the Viking graben. This area is defined as the East Shetland basin and Viking trough following Norwegian terminology (Ronnevik et al, 1975; and Deegan and Scull, 1977, page 2). The basin is bounded on the west by the East Shetland platform, to the southeast by the Vestland arch and Horda platform, and to the south it narrows into the Viking trough (Fig. 2).

After establishment of an intracratonic basin during the Permian, rifting (during the Permo-Triassic) established the East Shetland basin and began an extensional stress regime which influenced northern North Sea (and License 037) tectonics and sedimentation throughout the Mesozoic (Ziegler, 1975).

Sedimentation in this basin probably began during Permo-Triassic time and continued through the Mesozoic and Tertiary except for interruptions during the Jurassic and Early Cretaceous. Periods of Kimmerian tectonic activity can be identified throughout the Jurassic. The first major period of block faulting, tilting, and erosion, (near the Middle/Upper Jurassic boundary) controlled deposition of onlapping late Jurassic sediments. Primarily erosional phases occurred during Late Jurassic—Jurassic-Early Cretaceous and near the end of the Early Cretaceous. By Late Cretaceous, basin subsidence became the main factor controlling sedimentation in this area. A schematic west to east cross section (Fig. 4) shows the regional configuration of East Shetland basin sediments.

Within the region of the East Shetland basin and the License 037 area, the most significant variations in stratigraphy occur primarily at, or near, the contact between the Jurassic and overlying Cretaceous where erosion surfaces truncate Jurassic and older sections. Thickness variations in Cretaceous sediments, particularly the Lower Cretaceous, are related primarily to onlap of Jurassic structures.

Figure 5 is a schematic, dip-oriented, structure section across the major westerly tilted Jurassic fault-block trend on which the Statfjord field is located. A typical configuration of Jurassic and older strata on Statfjord and adjacent fault blocks is illustrated.

STATFJORD AREA GEOLOGIC HISTORY

Reconstruction of the geologic history of the Statfjord field area is difficult, particularly on the east flank of the Statfjord structure, with available data. Here, truncation of Jurassic and uppermost Triassic units by successive periods of tilting and/or erosion has resulted in a complex system of preserved, reworked, or eroded Jurassic and uppermost Triassic sediments. Probable movements on secondary (gravity-type) faulting adjacent to the bounding fault system during onlap and erosion further complicate the geology.

The influence of tectonic activity on Jurassic sedimentation is more spectacular near the margins of the East Shetland basin, but its effects can be seen in the Statfjord field area as a series of transgressions and regressions with discontinuities commonly being restricted to local highs. Major unconformities are more easily recognized.

Reactivation of older (Caledonian?) northeasterly trending major fault systems, in combination with easterly basinward subsidence, probably was the prime control of Early Jurassic–Late Triassic sedimentation, both regionally and within the license. Kimmerian tectonic activity apparently recurred along these older fault trends and developed subsidiary cross faults, which often trend in a northwesterly direction (Fig. 6). The effect of Kimmerian block faulting, tilting, and erosion on Early Jurassic and Triassic sedimentation in License 037 is not well understood but it was certainly active, to various degrees, during much of the Jurassic. Indeed, it was one of the major influences on later Jurassic sedimentation and the general distribution of Jurassic-Triassic sediments in the Statfjord region.

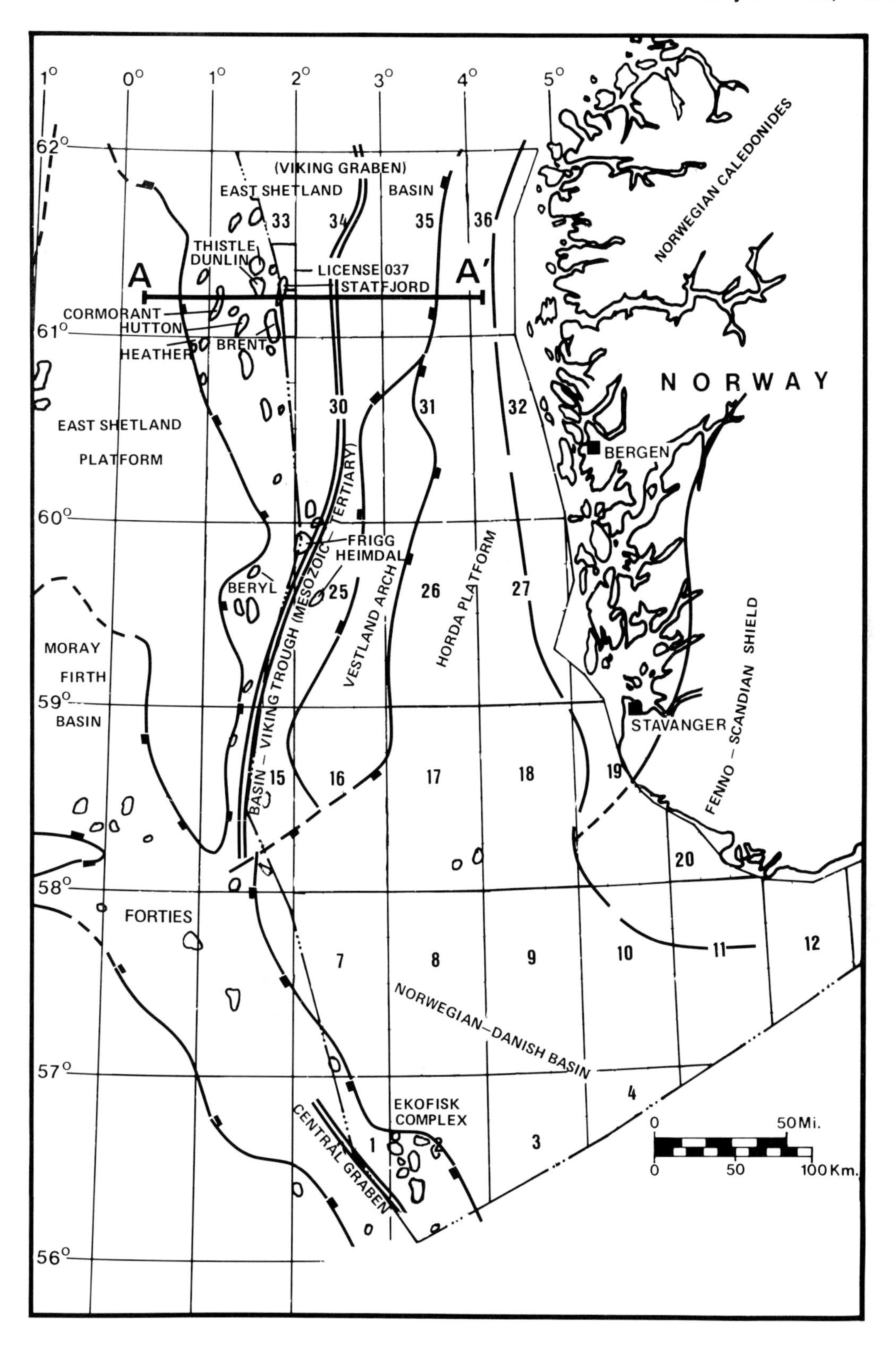

FIG. 2—Generalized northern North Sea major tectonic features (after Ronnevik et al, 1975).

Significant periods of tectonic activity, accompanied by varying degrees of growth faulting, depositional thinning and/or erosion of fault block highs, occurred during the later stages of Early Jurassic Dunlin deposition (Toarcian/Aalenian) and, following a rise in sea level, again near the end of the Middle Jurassic (Callovian/Oxfordian time) during deposition of the Middle to Late Jurassic Heather Formation. This latter time is interpreted as the main period of rotational block faulting and erosion responsible for the first major uplift of the Statfjord field and much of the erosion on what is now its east flank. Continuing syndepositional fault block rotation and gradual basin silling are interpreted as the main factors controlling distribution and progressive onlap of organically rich, Upper Jurassic Kimmeridge Clay Formation shales. Primarily erosional tectonic phases occurred at the end of the Jurassic, during Late Jurassic–Early Cretaceous time, and at the end of the Early Cretaceous, mainly affecting structurally higher areas within the license. During the Late Cretaceous, normal basin subsidence became the main influence on sedimentation.

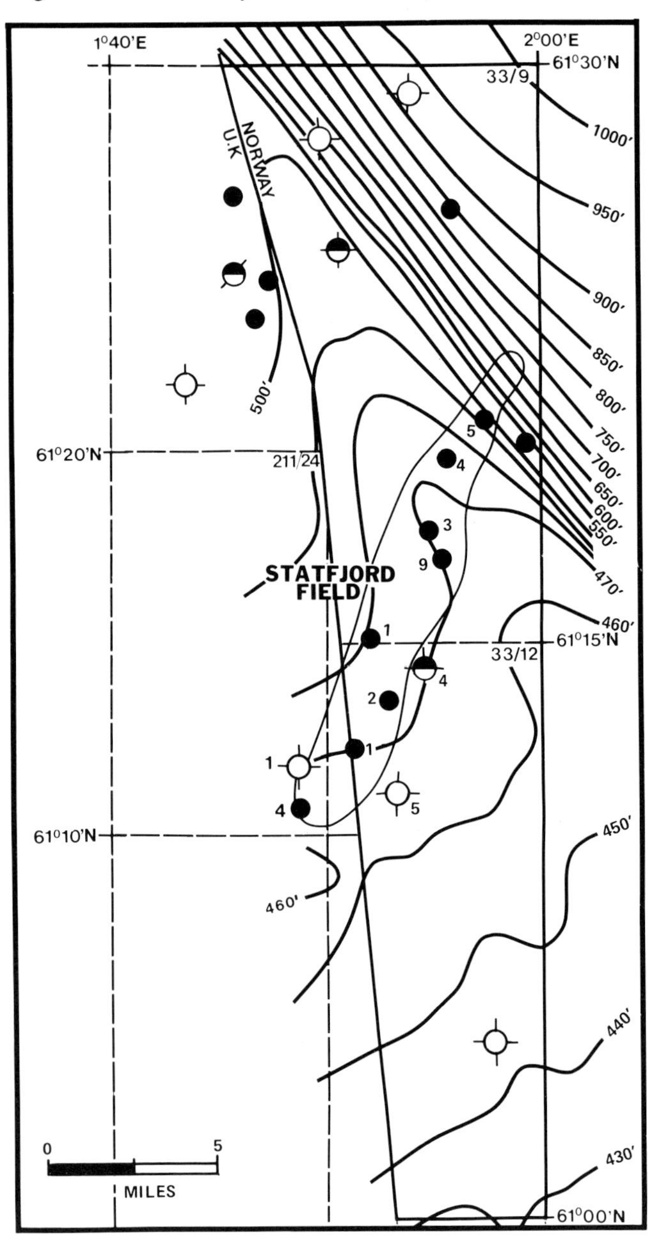

FIG. 3—Water depth map for License 037 area.

One of the primary tectonic phases in the formation of the Statfjord field structure is believed to have begun near the end of the Middle Jurassic with fault block rotation and with gravity faulting developed along the leading edge. Along this eastern edge of the Statfjord fault block, minor movement is interpreted to have continued periodically into the Late Cretaceous, with splinter fault blocks being downfaulted and eroded to form the Statfjord field's eastern flank. The Statfjord block remained a positive feature later than many other features in the region, until finally buried by Late Cretaceous sediments. Periods of tectonic activity are illustrated graphically on the License 037 lithologic column shown in Figure 7. Most major features in the area were affected by the same significant periods of tectonic activity, though there are differences in the magnitude and duration of the influence on individual fault blocks.

EXPLORATION HISTORY

Prior to the License 037 award, a number of hydrocarbon discoveries were already made in nearby U.K. blocks. The Brent field, the first major discovery in this part of the northern North Sea, was found by Shell/Esso in 1971. Five additional fields—Cormorant, Thistle, Dunlin, Heather, and Hutton—were discovered in the following two years (Fig. 2). In the Brent field, both Brent Formation and Statfjord Formation sandstones are productive. In all other fields only the Brent sandstones were known to be productive.

Therefore, when the License 037 area was awarded, primary exploration objectives in the area already were defined. Cretaceous and Paleocene reservoirs, found farther south, were not developed in this area.

Discovery and Delineation of Statfjord Field

Since the license award in 1973 approximately 2,237 mi (3,600 km) of reflection seismic data have been gathered from yearly surveys through the end of 1978. This has resulted in a detailed grid of seismic data, particularly over the Statfjord field.

A complete reshoot of the field was done in 1977 as a basis for remapping by the Statfjord Unit Group. Initial major problem areas, weak reflectors over the crest of the field and poor resolution of intra-Jurassic seismic reflectors above the top Statfjord Formation, remain a problem, as does the truncated east flank of the field.

Eleven exploratory wells have been drilled on the Statfjord field structure, 2 in the U.K. sector and 9 in the Norwegian sector (Fig. 6). Development drilling began in late 1978 and the first well was drilled in March 1979.

Initial mapping using an older regional grid of seismic data indicated the hydrocarbon-bearing "Brent" structural trend extended from U.K. waters across the border into the License 037 area. This large structural feature was mapped at top Jurassic level (late Kimmerian) and was interpreted as a large north-

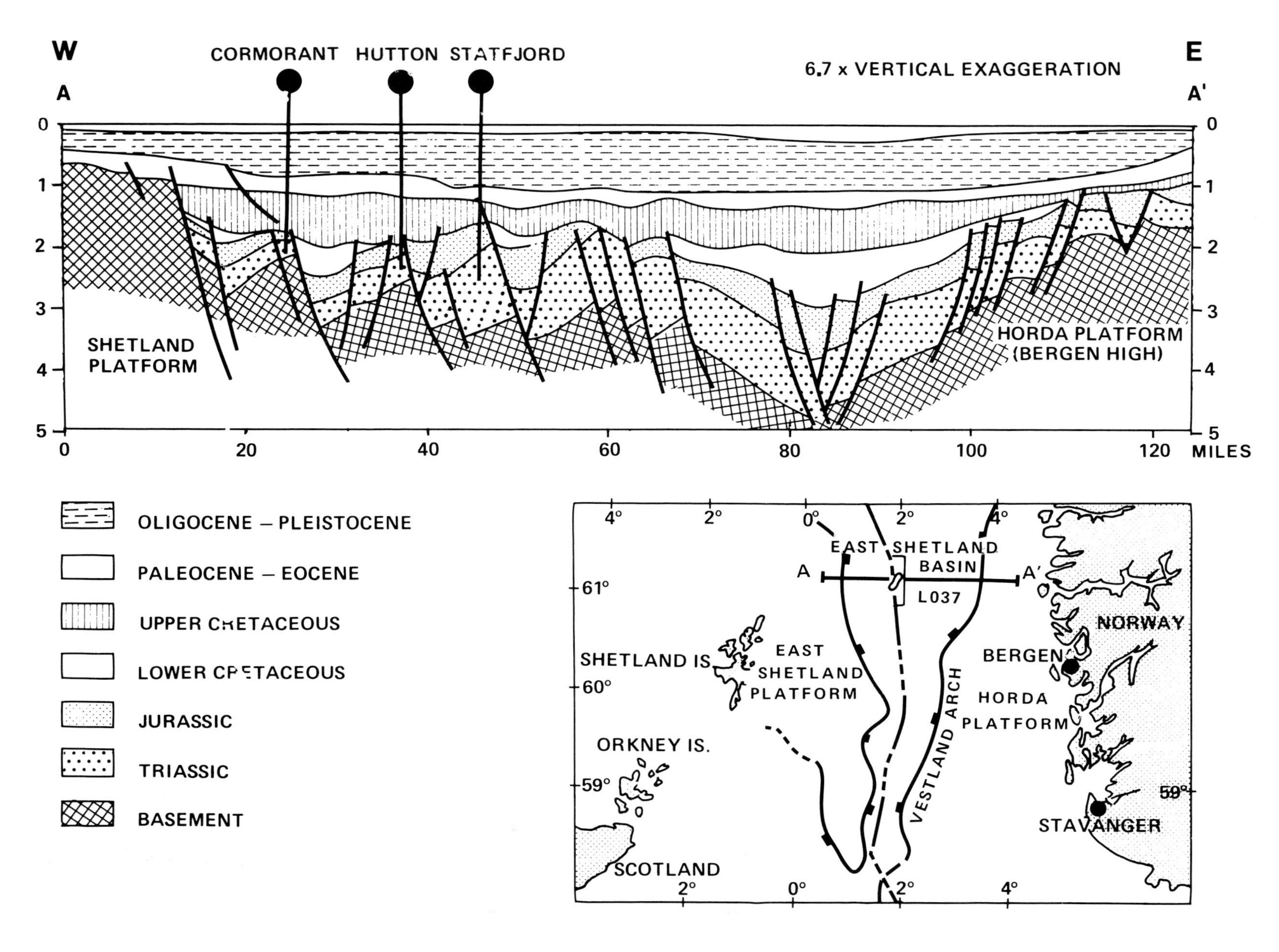

FIG. 4—Generalized west to east structural cross section, A-A' across the East Shetland basin.

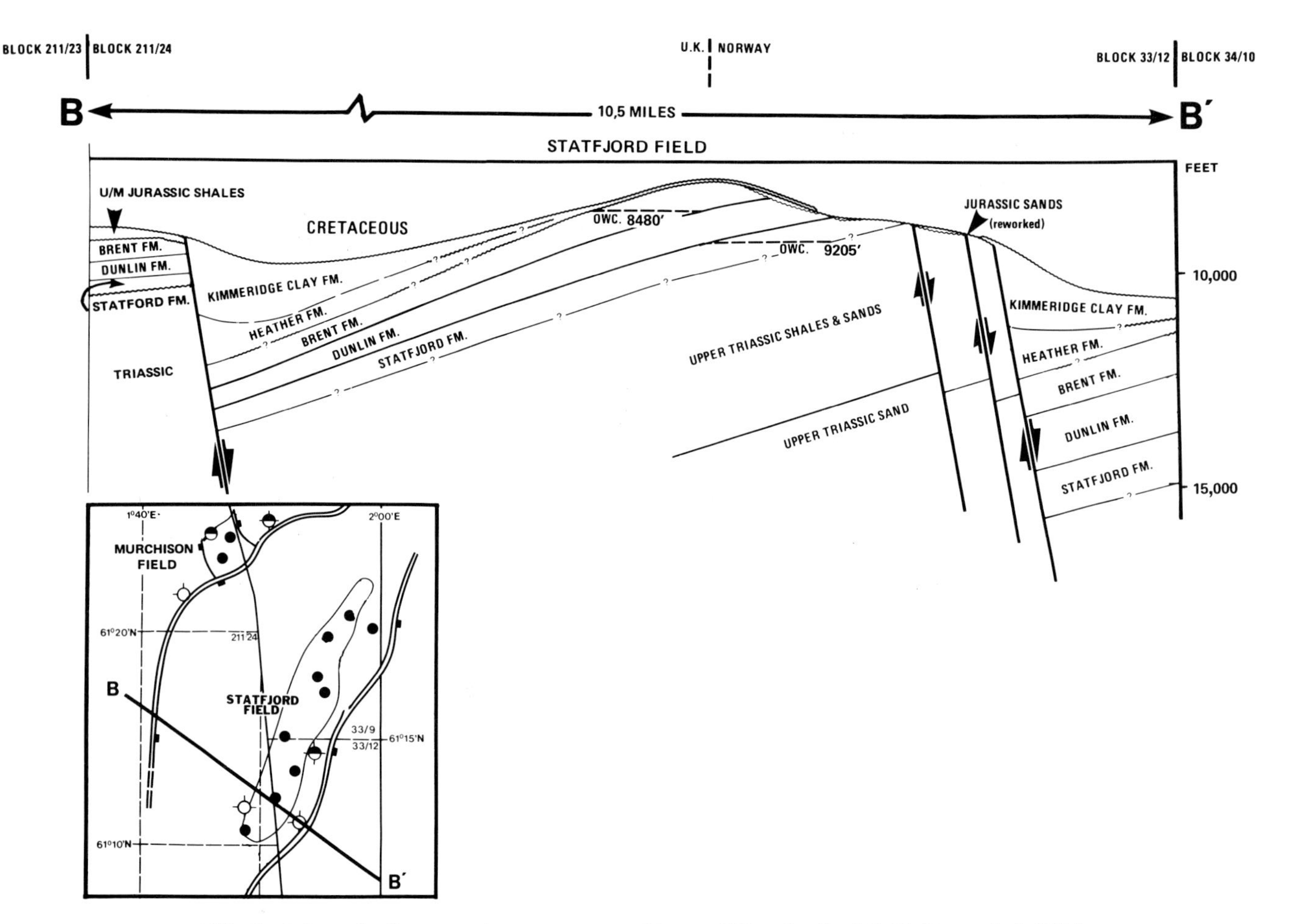

FIG. 5—Schematic dip-oriented structure across License 037 at Statfjord field. Datum is 7,380 ft subsea.

east-trending and northwesterly tilted, partly eroded, fault block. Water depth over the structure averages 475 ft (145 m). It was felt both the Brent and Statfjord sandstones could trap hydrocarbons as in the Brent field, some 12 mi (20 km) to the southwest in U.K. waters. The organically rich Upper Jurassic Kimmeridge Clay Formation shales were expected to provide both a source and a seal.

The first well on the Statfjord structure, 211/24-1, was drilled in U.K. waters by Conoco North Sea during late 1972 and early 1973. The test was located on the southwestern flank of the structure (Fig. 8) downdip from a seismic event which may have been interpreted as representing truncation of Jurassic reservoirs. The well found thick, porous, but water-bearing Brent and Statfjord Formation reservoir sections. The top of the Brent was encountered 66 ft (20 m) low to the eventual field oil-water contact (Fig. 9). The apparent truncation now is believed to represent thinning of Upper and Middle Jurassic shale units onlapping the Statfjord structure.

Mobil, as operator for the License 037 group, spudded the first well on the Norwegian side (33/12-1) less than 4 months after the license award. The well was located on the west flank of the structure where a thick prospective section was expected to be preserved updip from the 211/24-1 well. The entire Jurassic sequence was penetrated and the well was plugged and abandoned as an oil discovery in April 1974. Statfjord sandstones were wet but the entire 526 ft (160 m) of Brent Formation was oil bearing. The Brent sandstones were tested at a rate of 10,000 b/d (maximum capacity of the separator) of 37° API oil with a GOR of 952.

A second well, 33/9-1, located slightly downdip to the north, penetrated 478 ft (146 m) of Brent oil-bearing sandstone and established an oil-water contact for the Brent at 8,480 ft (2,584 m) subsea. No gas cap was found in either 33/9-1 or 33/12-1 wells. After confirmation of an accumulation separate from the Brent field by virtue of its different oil-water contact, a commercial accumulation was declared and the field was named *Statfjord*.

The third well, 33/12-2, was drilled to test Statfjord sandstones in a higher structural position. Younger, reworked (?), Upper to Middle Jurassic sandstones, 55 ft (16.8 m) thick were found overlying partly eroded Lower Jurassic shales and these sandstones (as well as Statfjord sandstones) were oil-bearing. The Statfjord reservoir contained an oil column of 417 ft (127 m) to a reservoir-shale contact and tested 12,200 b/d of 39.5° API oil. The oil-water contact for the Statfjord reservoir was subsequently placed by Unit agreement at 9,205 ft (2,806 m) subsea (Fig. 9).

Further appraisal drilling in 1974 and 1975 delineated the

field limits. They were extended into U.K. waters by Conoco North Sea's 211/24-4 well in 1975. This well found oil only in the Brent Formation.

An additional appraisal, 33/9-9, was drilled in 1977 in the northeastern part of the Statfjord field to provide control on the structure and to further appraise the Statfjord reservoir along the structural crest. Both the Brent and Statfjord reservoirs were oil bearing as anticipated. However, some 50 ft (15 m) of net oil-bearing sandstone, a new pay, was found in the Dunlin shales and tested at the rate of 8,314 b/d. This reservoir sandstone is interpreted to be in communication with the Brent and of limited areal distribution.

STRATIGRAPHY

The lithostratigraphic nomenclature used in this report (Fig. 7) is based on recommendations of a joint Norwegian U.K. Lithostratigraphic Nomenclature Committee (Deegan and Scull, 1977). As mentioned previously, sedimentation in the License 037 area was generally continuous through the Mesozoic and Tertiary except for known interruptions during the Jurassic and Early Cretaceous.

Triassic

The oldest section penetrated in the license area, at a depth of over 15,000 ft (4,572 m), was still of Late Triassic Carnian age. Triassic lithology typically consists of interbedded, continental, red to variegated claystones, siltstones, shales, and sandstones. Reservoir properties generally are poor due to calcareous cement and a clay matrix. Triassic sediments commonly are identified by their characteristic "red beds." However, in this area, red-colored clastics also occur in the Statfjord Formation, which spans the nebulous Triassic-Jurassic boundary. No hydrocarbon shows have been encountered except in the Statfjord reservoir.

Jurassic

Statfjord Formation—The Statfjord Formation is well developed in the Statfjord and Brent field areas. Regionally, it appears to unconformably overlie or onlap Triassic sediments to the west and northwest across the Murchison field bounding fault system, where only a thin upper calcareous member is present. To date, only the Brent and Statfjord fields contain commercial hydrocarbons in the Statfjord Formation, the second most important reservoir in the Statfjord field.

Except for truncation and possible erosion on the eastern leading edge of the Statfjord field structure, the base of the Statfjord Formation in the field area appears to be conformable with the underlying Triassic Cormorant Formation, and ranges from Late Triassic to Early Jurassic in age. The base of the formation commonly is difficult to determine due to paucity of fauna and flora and poor log correlations due to nondiagnostic lithologic changes. A continuous depositional sequence is normally present into the overlying Dunlin shales. Within the license area, thicknesses range from zero (where truncated on the east flank) to approximately 1,015 ft (0 to 310 m).

Gray and, in part, reddish colored shales, claystones, and siltstones are interbedded with thin sands at the base, which tend to coarsen upward into thicker, cleaner, more massive sandstones predominating in the upper section. These upper sandstones are white to gray, fine to very coarse and conglomeratic, and possess good to excellent reservoir properties. The sandstones are feldspathic to arkosic and contain both a kaolinitic matrix and calcite cement. Calcite cement predominates in the uppermost unit, which may locally grade into thin sandy limestones.

The Statfjord Formation is divided into three members on the basis of sand percent and thickness. It also has been subdivided on the basis of depositional environment as shown on Figure 10. The uppermost massive sandstone may, in part, represent a shallow-marine environment, as suggested by occa-

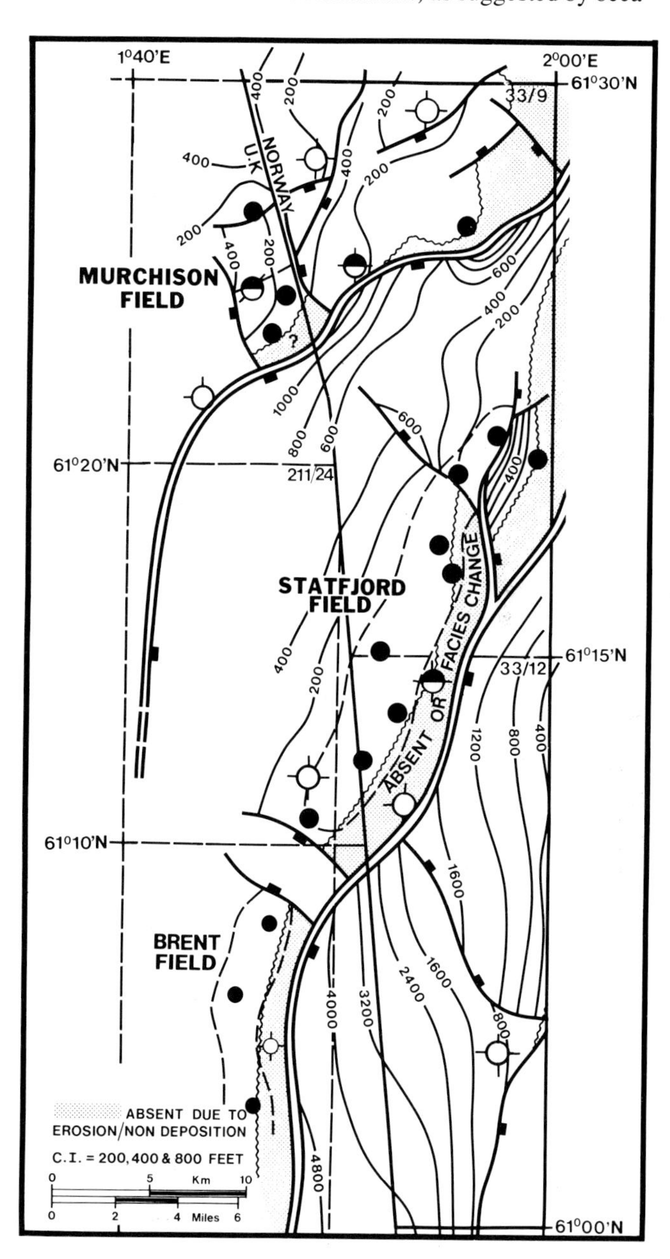

FIG. 6—Isopach map of Upper/Middle Jurassic shales (Humber Group) showing major Jurassic fault block trends and truncation areas.

sional glauconite and marine fossils. Regionally, a westerly source area is indicated (Deegan and Scull, 1977).

Dunlin Formation—The Lower Jurassic Dunlin Formation is composed essentially of shallow marine gray, brown, and black shales, mudstones, and siltstones. Minor thin, calcareous, silty sands and thin limestones are locally common near the base and within an upper Dunlin unit. The formation is divided into four members based primarily on shale to silt/sand ratios and resultant log character. Except for a local, areally restricted, oil bearing sandstone within the Cook Member on the crest of the Statfjord structure, no potential reservoir sandstones have been found within the license area.

The base of the Dunlin is normally conformable in the field and license area but shows disconformity to the west (Deegan and Scull, 1977). The contact with the overlying Brent shows local disconformity with some depositional thinning and/or erosion occurring over the crest of the Statfjord structure prior to later truncation on what is now the east flank. Other local, emergent features in the license area were similary affected and show thinning of upper Dunlin units.

Brent Formation—The Middle Jurassic Brent Formation is the primary oil reservoir of the northern North Sea. In Statfjord field the main reserves are contained within Brent deltaic sands.

A number of Brent subdivisions are currently in use in the East Shetland basin region. Bowen (1975) originally named the formation and divided it into five units. These five units, based on lithological variations and log character, are the Brent members proposed by Deegan and Scull (1977) and currently used in License 037. This subdivision permits separation of the lower, nondeltaic fluvial member from the primarily deltaic members, and allows a reasonable division of the various deltaic facies.

For reservoir purposes the Statfjord Field Unit subdivides the Brent into six zones based on rock properties. Zones 1, 4, and 5 primarily are massive sandstones and have the best reservoir properties. Zones 2 and 3 are interbedded sandstones, shales, and coals, and Zone 6 is a thin, poorly sorted, silty to conglomeratic sandstone; these exhibit poorer reservoir properties. Figure 11 compares the reservoir zones defined by the Statfjord Unit, the stratigraphic nomenclature, and an environmental subdivision based on core analysis by R. J. Moiola of Mobil's Field Research Laboratory. Despite the different criteria used, the main variance is primarily in the number of subdivisions. Figure 11 also defines pertinent lithologic and log correlation criteria as well as facies environments of the various Brent members.

In general, Brent lithology consists of massive to thin bedded, white to light gray and brown, fine to coarse sandstones, conglomeratic and arkosic in parts with common-to-abundant mica in the lower part and normally subordinate thin shales, coals, and siltstones in the upper part of the formation. Mica, kaolinite, and feldspar occur in minor amounts throughout the reservoir sandstones but appear more common in the lower two members. Lithologic descriptions of individual members are on Figure 11. Overall reservoir properties for the Brent Formation are excellent (Table 2).

Micaceous intervals, if the mica is sufficiently concentrated, can adversely affect log-derived porosity measurements and also permeability analysis. Recent log analysis work in the license area suggests mica effects can be reasonably accounted for. However, this may be because mica concentrations have been overestimated. Nyberg et al (1977) suggested that other radioactive, and/or heavy minerals besides mica, could have a significant effect on log-derived porosity measurements. Further studies will be required to properly quantify the mica problem.

LITHOSTRATIGRAPHIC COLUMN

NORWAY OFFSHORE LICENCE 037

Time Stratigraphic Units		Series/Stage	Group or Formation	Formation or Member	Thickness Range (ft)	Orogenic Events
CRETACEOUS	UPPER	MAASTRICHTIAN, CAMPANIAN, SANTONIAN, CONIACIAN, TURONIAN, CENOMANIAN	SHETLAND GROUP	UNDIFFERENTIATED	1758-3998	AUSTRIAN
CRETACEOUS	LOWER	ALBIAN, APTIAN, BARREMIAN, HAUTERIVIAN, VALANGINIAN, BERRIASIAN	CROMER KNOLL GROUP	(ALBIAN/APTIAN MARL/CLAYSTONE); BARREMIAN LS. UNIT	0-482	
JURASSIC	UPPER	VOLGIAN/PORTLANDIAN, KIMMERIDGIAN, OXFORDIAN	HUMBER GROUP	KIMMERIDGE CLAY FM.	0-298	LATE KIMMERIAN
JURASSIC		CALLOVIAN		HEATHER FM.	0-994 101	MAIN KIMMERIAN
JURASSIC	MIDDLE	BATHONIAN, BAJOCIAN	BRENT FORMATION	5 MBRS.	0-1023	
JURASSIC		AALENIAN (NOT RECOGNIZED)				
JURASSIC	LOWER	TOARCIAN, PLIENSBACHIAN	DUNLIN FORMATION	4 MBRS.	0-1430	
JURASSIC	LOWER	SINEMURIAN, HETTANGIAN	STATFJORD FORMATION	3 MBRS.	0-1017	EARLY KIMMERIAN
TRIASSIC	UPPER	RHAETIAN, NORIAN, CARNIAN	CORMORANT FORMATION	UPPER TRIASSIC SAND UNIT	UPPER TRIASSIC 6032 + (OLDEST SECTION PENETRATED IN L037. WELL 33/12-5)	
TRIASSIC	LOWER / MIDDLE	LADINIAN, ANISIAN, SCYTHIAN	?			
PALEOZOIC						

LEGEND ● OIL RESERVOIR

FIG. 7—Stratigraphic column for License 037 area.

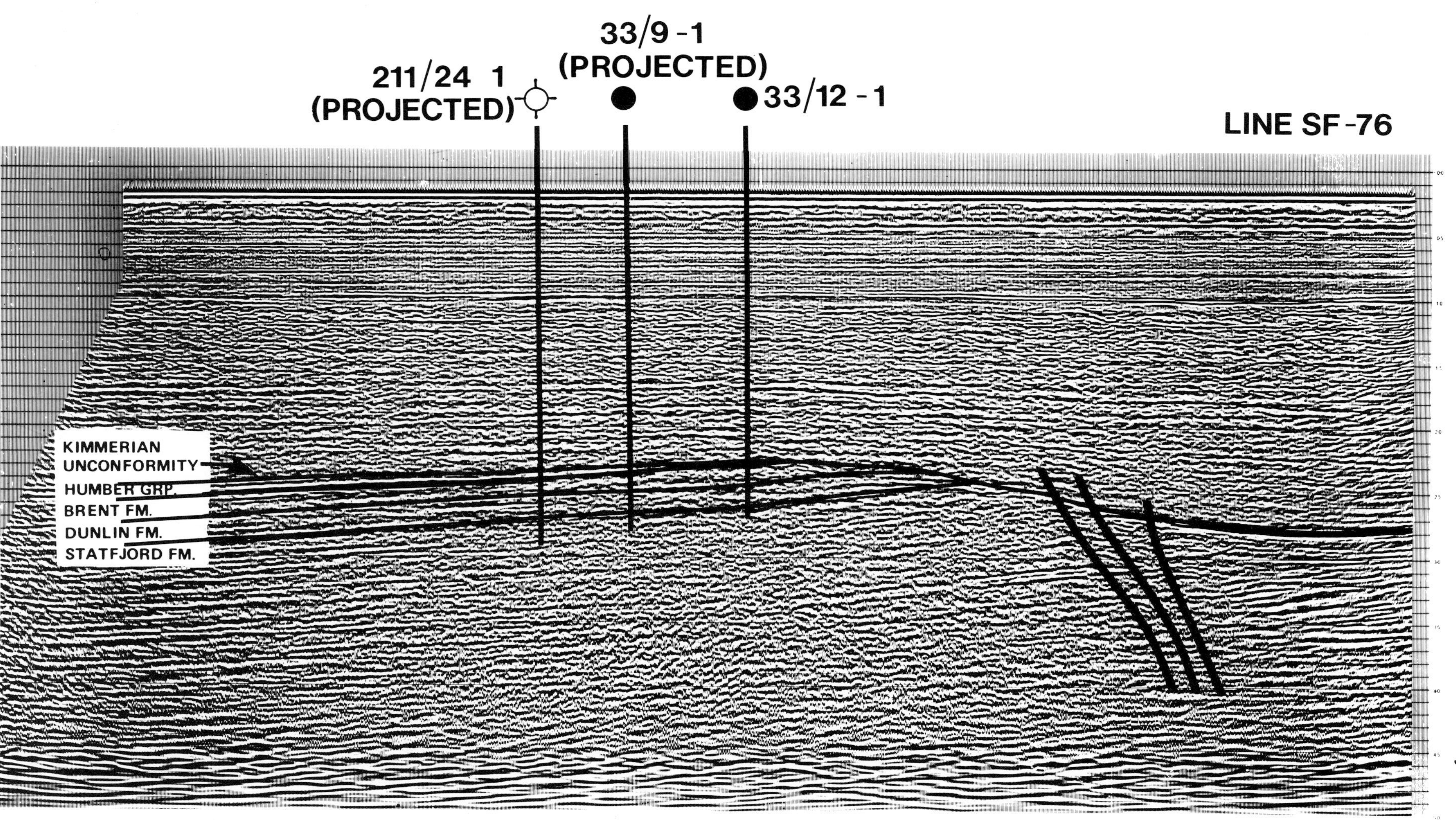

FIG. 8—Dip-oriented seismic line SF 76 through discovery well 33/12-1.

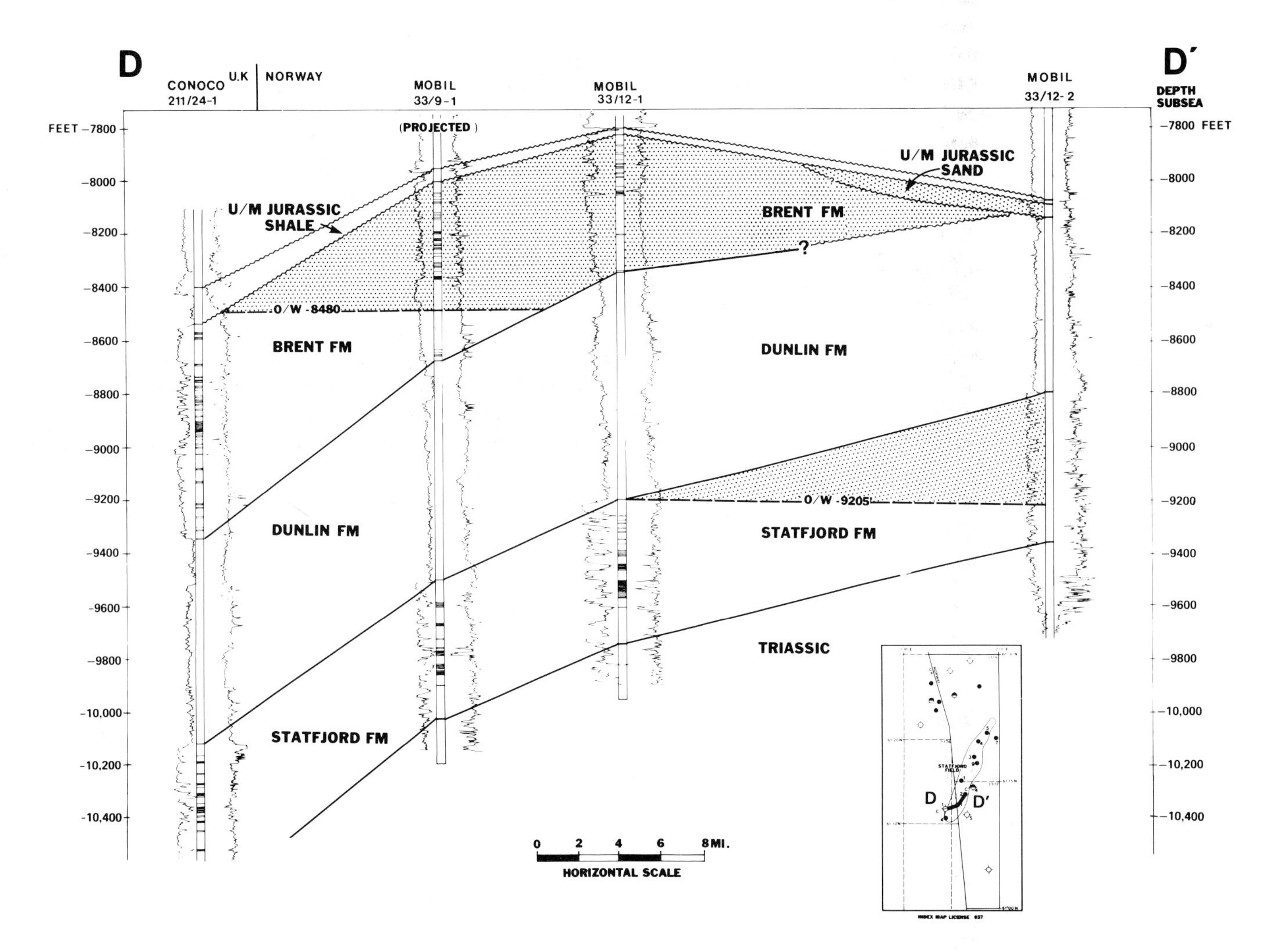

FIG. 9—Generalized electric-log structural section across Statfjord field. Section shows correlation of initial four wells on Statfjord structure and space relationship of Brent and Statfjord reservoirs. Datum is 7,730 ft subsea.

Within the license area, the base of the Brent Formation shows only local disconformity over crests of former topographic highs such as at Statfjord field. Except for crests of tilted Jurassic fault blocks, where erosion and/or truncation of the Brent occurs, the unconformity normally indicated at the top of the Brent need not be of major significance. In many areas, such as the west flank of the Statfjord feature, passage from Brent to overlying Heather shales of a similar age is believed to be conformable or nearly so. Though definitive well control is limited, depositional thinning of the Brent apparently occurs on emergent Jurassic features such as Statfjord field.

The base of the uppermost massive sandstone unit may also show disconformity in parts of the field but probably does not represent a widespread, regional erosion surface. The occurrence and distribution of reworked Upper to Middle Jurassic sandstones (e.g. 33/12-2, Fig. 9) in wells on the crest and east flank of the field is not well understood at this time. They probably are related to major tectonism initiated near the end of the Middle Jurassic.

In a typical deltaic facies distribution, prodelta and delta-front facies thicken away from the source area and delta-plain sediments thicken toward the source. Orientation of these facies units is normally more or less perpendicular to the distributary system and parallel with the shoreline trend. From data in License 037, as illustrated on the Brent facies tract (Fig. 12), it was concluded that the source of the Brent delta was from the southeast of the license area.

Regional distribution of the Brent Formation is not well known, especially to the east, but it attains its maximum known thickness in the area southeast of the Statfjord field. This may be related to both a more rapid accumulation of sediments toward the delta source to the southeast, and eastward thickening across major fault trends. Probable periodic syndepositional movements of some fault blocks and later truncation over individual Jurassic highs exerted local control on Brent distribution at Statfjord field and on other features.

On the Statfjord field structure, gross Brent thickness ranges from 810 ft (247 m) on the west flank to zero where truncated on the east flank.

Humber Group—The Humber Group includes both the Middle to Upper Jurassic Heather Formation and the Upper Jurassic Kimmeridge Clay Formation. Figure 6 is an isopach map of these Upper and Middle Jurassic shales. In Statfjord field, the shallow marine Heather shales normally are shown to unconformably overlie Brent sandstones (Fig. 9). Lithology consists primarily of gray to dark gray silty claystone sand shales. No reservoir development occurs in the area unless reworked sands on the east flank of the field are considered as age equivalent. Heather thicknesses in the field range from 157 ft (48 m) in the northern, downfaulted area to zero on the eroded east flank. Regionally the contact between the Bathonian to Oxfordian Heather and Oxfordian to Volgian Kimmeridge Clay commonly is considered to be diachronous (Deegan and Scull, 1977). However, on the Statfjord field fault block, this contact is interpreted as representing the first major period of tilting and erosion, with Kimmeridge Clay Formation sediments as young as Volgian overlying Bathonian/Callovian age Heather shales. On the east flank of the field this unconformity may have been responsible for much of the initial truncation of Jurassic and uppermost Triassic sediments.

Continued syndepositional fault-block movement and gradual basin silling are interpreted as the main factors controlling distribution and onlap of restricted, shallow marine sediments of the Kimmeridge Clay formation at Statfjord field. These dark gray to black, organically rich claystones and shales contain no known reservoirs within the Statfjord field and range in thickness from a known maximum of 75 ft (23 m) on the southwestern flank to zero on the truncated eastern flank. The top of the Humber Group represents the regional late Kimmerian Unconformity surface.

Cretaceous

Cromer Knoll Group—Lower Cretaceous Cromer Knoll sediments onlap, and are thin or absent over older Jurassic topographic highs such as Statfjord field. Both the lower and upper boundary of the Lower Cretaceous are unconformable, particularly in structurally high areas. In the Statfjord field, where the thickness ranges from less than 125 ft (38 m) to zero, the unit is primarily represented by a basal, shallow marine, time-transgressive, white to gray, argillaceous limestone. It sometimes is overlain by a section of deeper water, interbedded marly claystones and shales, often reddish brown near the top. This overlying sequence thickens off structure. No potential reservoir sandstones have been encountered within this sequence in the license area.

Complex secondary faulting, erosion and onlap may have continued along the leading edge of the Statfjord fault block into early Early Cretaceous before tectonic adjustments finally ceased. Sedimentation in this area, now the east flank of the Statfjord structure, is difficult to interpret.

Shetland Group—By Late Cretaceous time, normal basin subsidence became the primary control on sedimentation and the Statfjord field structure was finally buried by Late Cretaceous Shetland Group deep water claystones, shales, and subordinate siltstones. Thicknesses over the field area range from 2,610 ft (795 m) to 1,760 ft (536 m). Although scattered oil shows have been noted, no significant reservoirs are present in this area. Upper Cretaceous and Danian chalk reservoirs of the Ekofisk area in the Central North Sea are absent due to facies change. Sandy intervals are thin and poorly developed.

A less significant unconformity or hiatus is indicated at the Cretaceous/Paleocene boundary by the common absence of Danian-aged sediments.

Tertiary

Paleocene—The Paleocene Rogaland Group consists primarily of deep water claystones and mudstones with minor thin siltstones and sandstones identified as the Lista Formation. A regionally recognized mixed volcanic tuff sequence (ash fall deposits), known as the Balder Formation, is preserved over much of the North Sea basin at the top of the Paleocene (Deegan and Scull, 1977). This Formation is present in the License 037 area.

No significant reservoir development is present within the Paleocene at Statfjord field, although occasional oil shows have been noted. Sandstone reservoirs of the Frigg, Heimdal,

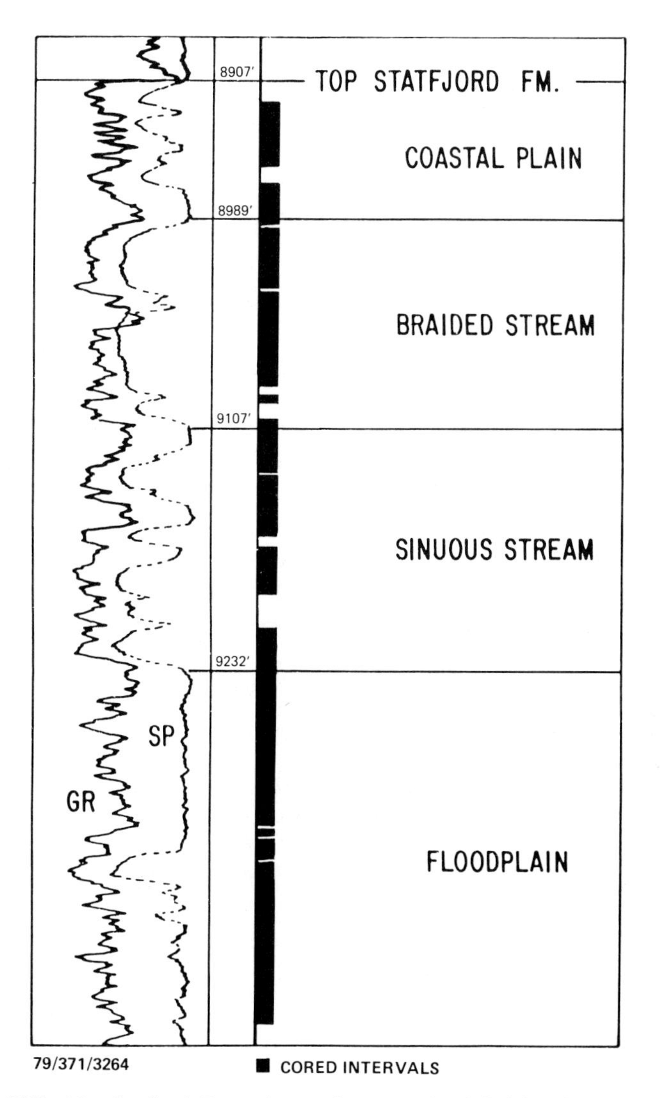

FIG. 10—Statfjord Formation environmental subdivision from core data (R. J. Moiola).

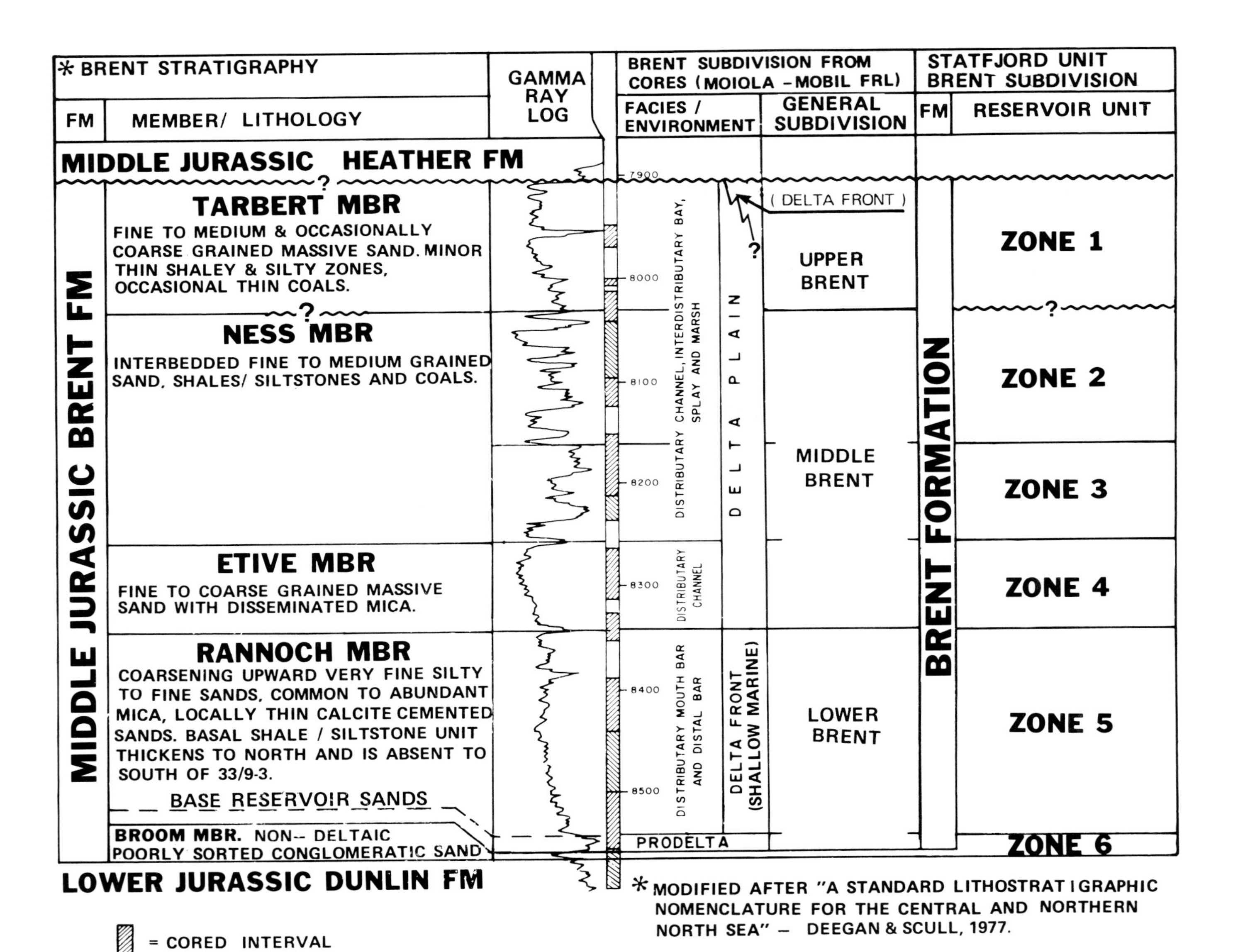

FIG. 11—A comparison of Brent Formation stratigraphy and general lithologies with Statfjord Unit reservoir zones and an environmental subdivision based on cores. Main variance is primarily in the number of subdivisions.

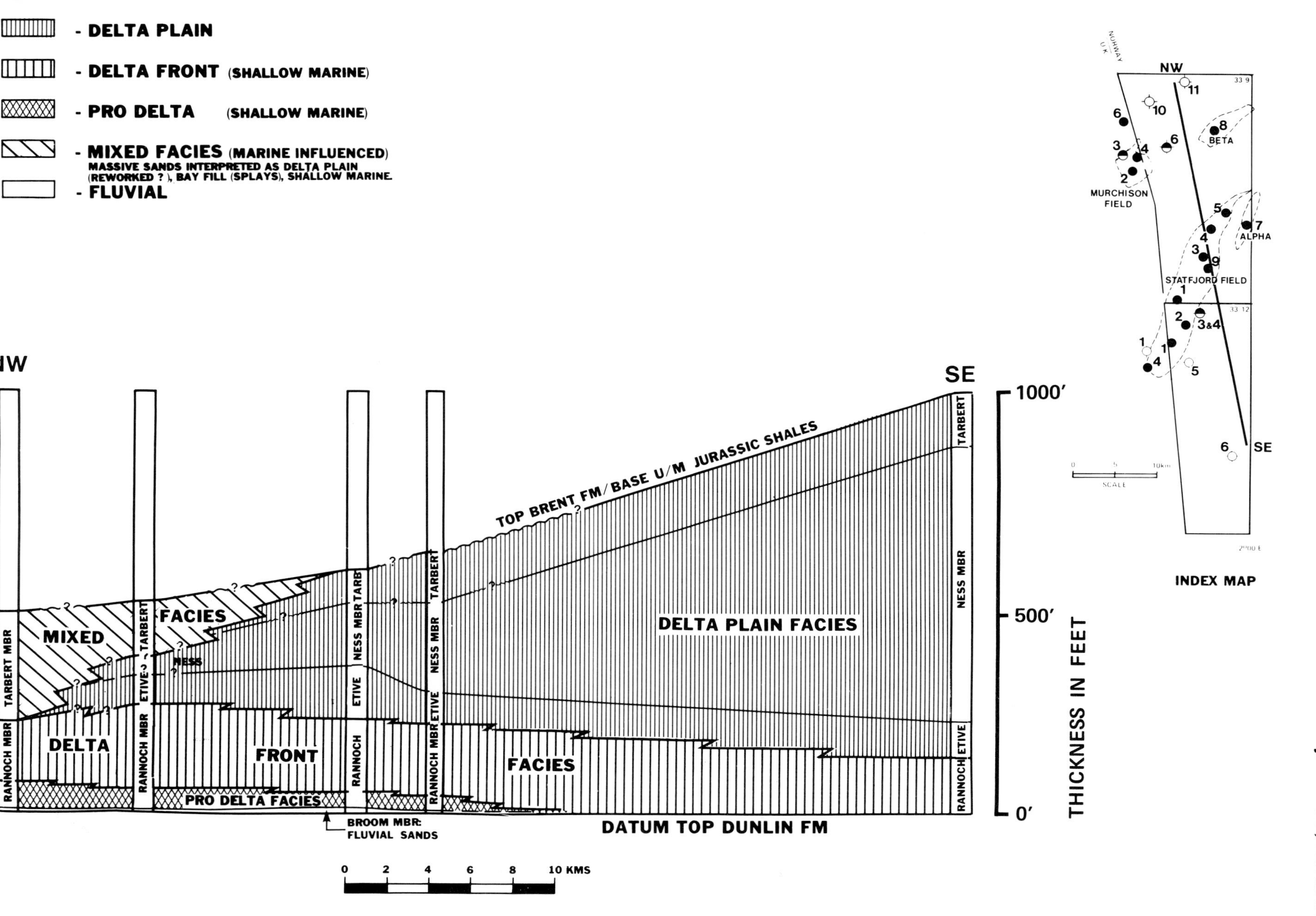

FIG. 12—License 037 Brent Formation facies tract from south to north showing general distribution of delta facies. This figure suggests a deltaic source from a southeasterly direction.

Table 1. License 037 source evaluation data averaged by stratigraphic unit.

PERIOD / SERIES GROUP / FM	LITHOLOGY	Thickness range (ft)	TOC [1]	EOM [2]	EOM/TOC	CPI [3]	Ro [4]
PALEOCENE ROGALAND GROUP		548-820	0.43 (64)	141 (3)	3.3 (3)	1.45 (3)	0.41 (3)
UPPER CRETACEOUS SHETLAND GROUP		1758-3998	0.78 (201)	172 (40)	2.64 (40)	1.71 (20)	0.43 (12)
LOWER CRETACEOUS CROMER KNOLL GROUP		0-482	1.04 (4)	78 (2)	1.41 (2)	—	0.71 (2)
UPPER JURASSIC KIMMERIDGE CLAY FM (HOT SHALE)	●	0-298	4.58 (4)	2160 (4)	4.00 (4)	—	—
MIDDLE JURASSIC HEATHER FM		0-994	2.24 (16)	350 (9)	1.71 (9)	1.68 (9)	0.40 (2)
MIDDLE JURASSIC BRENT FM	●	0-1023	—	—	—	—	0.44 (5)
LOWER JURASSIC DUNLIN FM	●	0-1430	1.47 (86)	234 (38)	1.95 (38)	1.68 (36)	0.59 (6)
LOWER JURASSIC / TRIASSIC STATFJORD FM	●	0-1016	0.44 (19)	—	—	—	0.68 (4)
TRIASSIC CORMORANT FM		Unknown. Max. Pen. 6032+ ft	0.35 (56)	50 (6)	1.71 (6)	1.37 (3)	0.73 (8)

LEGEND: ( ) NUMBER OF VALUES; ● OIL RESERVOIR

1. Total organic carbon (% rock weight)
2. Extractable organic material (ppm)
3. Carbon Preference Index
4. Vitrinite Reflectance

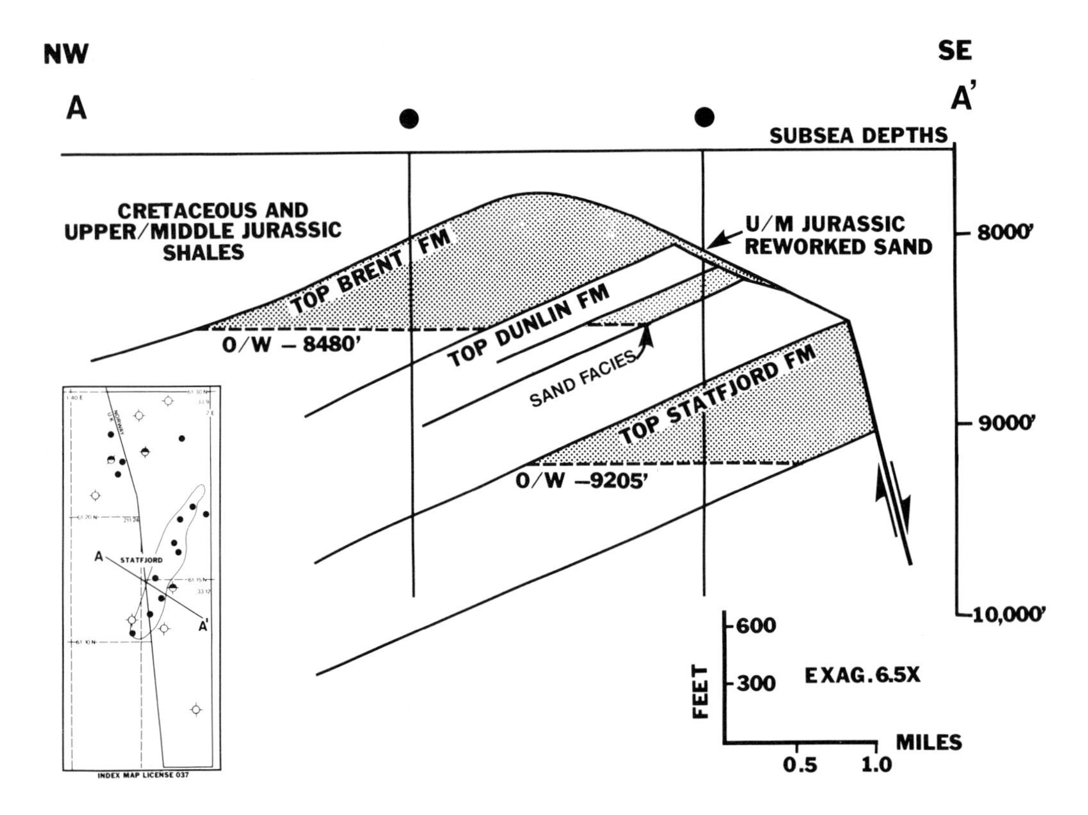

FIG. 13—Generalized dip-oriented geologic section across Statfjord. Location of section also is shown on Figures 15 and 16. Datum is 7,550 ft subsea.

and Forties fields farther south (Fig. 2) are not developed. Paleocene thicknesses over the Statfjord field area range from 730 ft (223 m) to 600 ft (183 m) and do not reflect the underlying paleostructure.

Eocene to Quaternary—The Eocene through Quaternary section is undifferentiated in the License 037 area.

Within the Statfjord field area these sediments are composed of unconsolidated siltstones and claystones, often with considerable fine to very coarse sandstones. No significant hydrocarbon shows have been found above the Paleocene.

SOURCE ROCKS

Source rock evaluations indicate that the Upper Jurassic Kimmeridge Clay Formation possesses the most favorable source characteristics of any stratigraphic unit within License 037 and the surrounding area (Table 1). These organically rich shales, overlying the Brent and Statfjord reservoirs, are interpreted to be the major source of oil for the Statfjord field. There generally is agreement that this unit, also known as the "Hot Shale," is the primary oil source in the East Shetland basin.

The Kimmeridge Clay Formation is thin or absent in the structurally high Statfjord field wells. Evaluation of its source potential was supported by data from other East Shetland basin wells containing thicker sections. The shale is extremely rich in organic matter and is mature in downdip sections. *Maturity in the wells on the crest of the Statfjord field is difficult to assess because of lack of definitive samples from the thin interval.*

STRUCTURE

General

Figures 13 and 14 are generalized dip and strike structural cross sections through the Statfjord field. The trapping mechanism is both structural and stratigraphic. Structurally it is a tilted fault block with Jurassic beds dipping westward at 6 to 8° and truncating on the faulted and eroded east flank. To the southeast the major down-to-the-east bounding fault system has a total displacement of over 5,900 ft (1,800 m) at Statfjord Formation level. Onlapping Upper Jurassic and Cretaceous shales bury the eroded fault block and provide the stratigraphic seal for subcropping Brent and Statfjord reservoirs. To the south normal faulting and a structural saddle provide separation from the Brent field, some 12 mi (20 km) to the southwest on trend with the Statfjord structure (Fig. 6). The western and northern limits of Statfjord field are defined by structural dip.

Productive closure defined by these structural limits and the oil-water contact is approximately 15.5 mi (25 km) long and 2.5 mi (4 km) wide. Total areal extent of the field is some 20,000 acres (81 sq km). The Brent reservoir covers approximately 15,000 acres (61 sq km) and the Statfjord reservoir about 9,000 acres (36 sq km).

Brent Structure

The structural configuration of the Brent reservoir is illustrated on Figure 15 and the dip and strike cross sections (Figs. 13, 14) which also show the configuration of the underlying Statfjord reservoir. Two culminations are shown on the Brent structure map, one in the northeast and one in the southwest area of the field.

The Brent reservoir subcrops and pinches out to the east and the oil accumulation is bounded to the west by structural closure and an oil-water contact at 8,480 ft (2,584 m) subsea. Northern and southern limits are controlled by a combination of faulting and structural dip. The main part of the oil accumulation is separated from a small northeastern extension by a northwest-trending cross fault near the 33/9-4 well.

Statfjord Structure

The underlying Statfjord Formation has a somewhat simpler northwest-dipping structural configuration (Fig. 16). The reservoir is limited to the north, northeast, and south by faulting and to the southeast by an unconformity surface. To the northwest the oil accumulation is controlled by an oil-water contact placed at 9,205 ft (2,806 m) subsea.

RESERVOIRS AND HYDROCARBONS

The Statfjord field has two major reservoirs, the Middle Jurassic Brent Formation and the Lower Jurassic/Triassic Statfjord Formation (Fig. 13). For purposes of reservoir and reserve discussion, Upper/Middle Jurassic sandstones are considered to be part of the Brent Formation. Oil has recently been found in a Lower Jurassic Dunlin Formation sand development on the crest of the Statfjord structure, located stratigraphically between the Brent and Statfjord Formations. Communication with the Brent reservoir is indicated. Currently, only one well defines the productive reservoir in the Dunlin and potential reserves are not included in the Statfjord field totals.

Brent and Statfjord sandstones have excellent reservoir properties. Brent gross pay thickness ranges up to 640 ft (195 m) with an average of 304 ft (93 m) in wells drilled in the productive area of the field. Porosities range up to 31% and average 29%, with permeabilities up to 8 darcies and averaging 1,500 millidarcies. The reservoir is approximately 50% overpressured and individual well potentials of 30,000 b/d are anticipated. Fluid injection is planned to maintain pressures above the bubble point in each reservoir and to sustain high production rates.

Statfjord field crude oil has a gravity of 38°/41° API, is low in sulfur content and without aggressive components such as H_2S. The pour point (+40°F) is relatively high, but the oil is of good quality.

Only two Statfjord field wells have encountered oil in the Statfjord Formation, though reservoir quality downdip below the oil column is reasonably well defined by a number of other wells. The characteristics of the oil productive zone are not as well defined as for the Brent, but general extent of the zone is known.

The quality of the Statfjord reservoir is lower than that of the Brent Formation and the extent of the oil accumulation is smaller. However, reservoir properties are still considered ex-

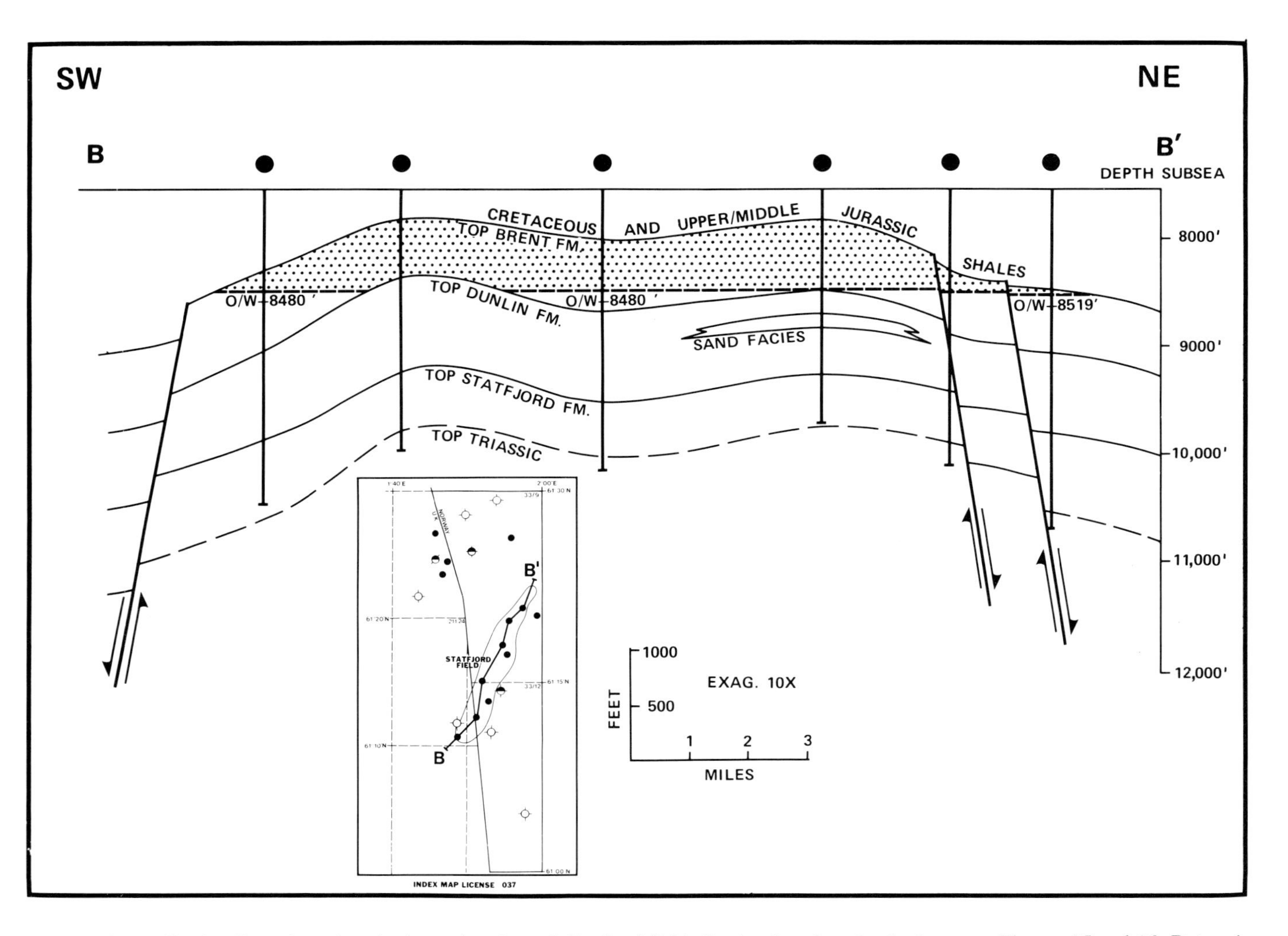

FIG. 14—Generalized strike-oriented geologic section through Statfjord field. Section location also is shown on Figures 15 and 16. Datum is 7,550 ft subsea.

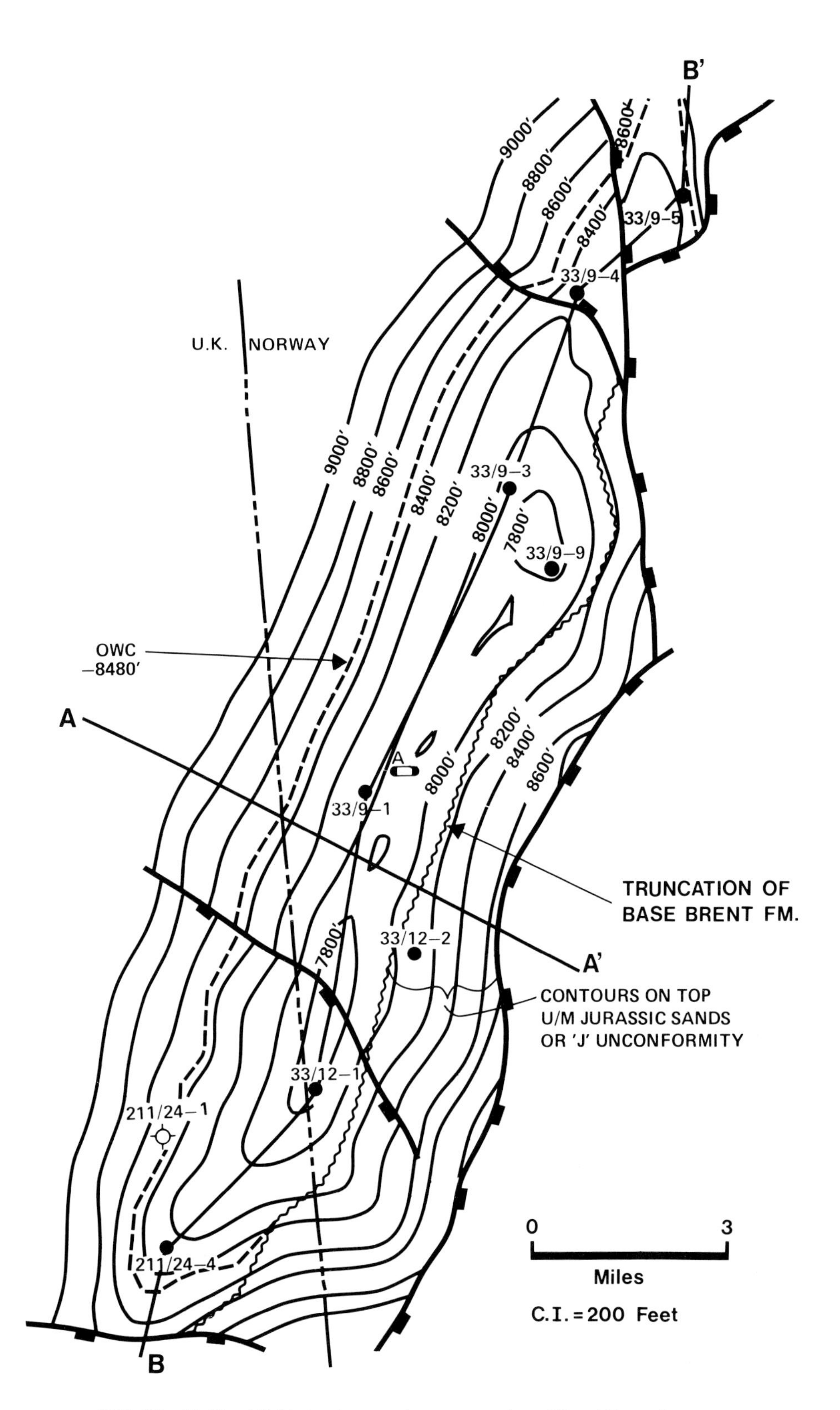

FIG. 15—Statfjord field structure contour map on top of Brent Formation.

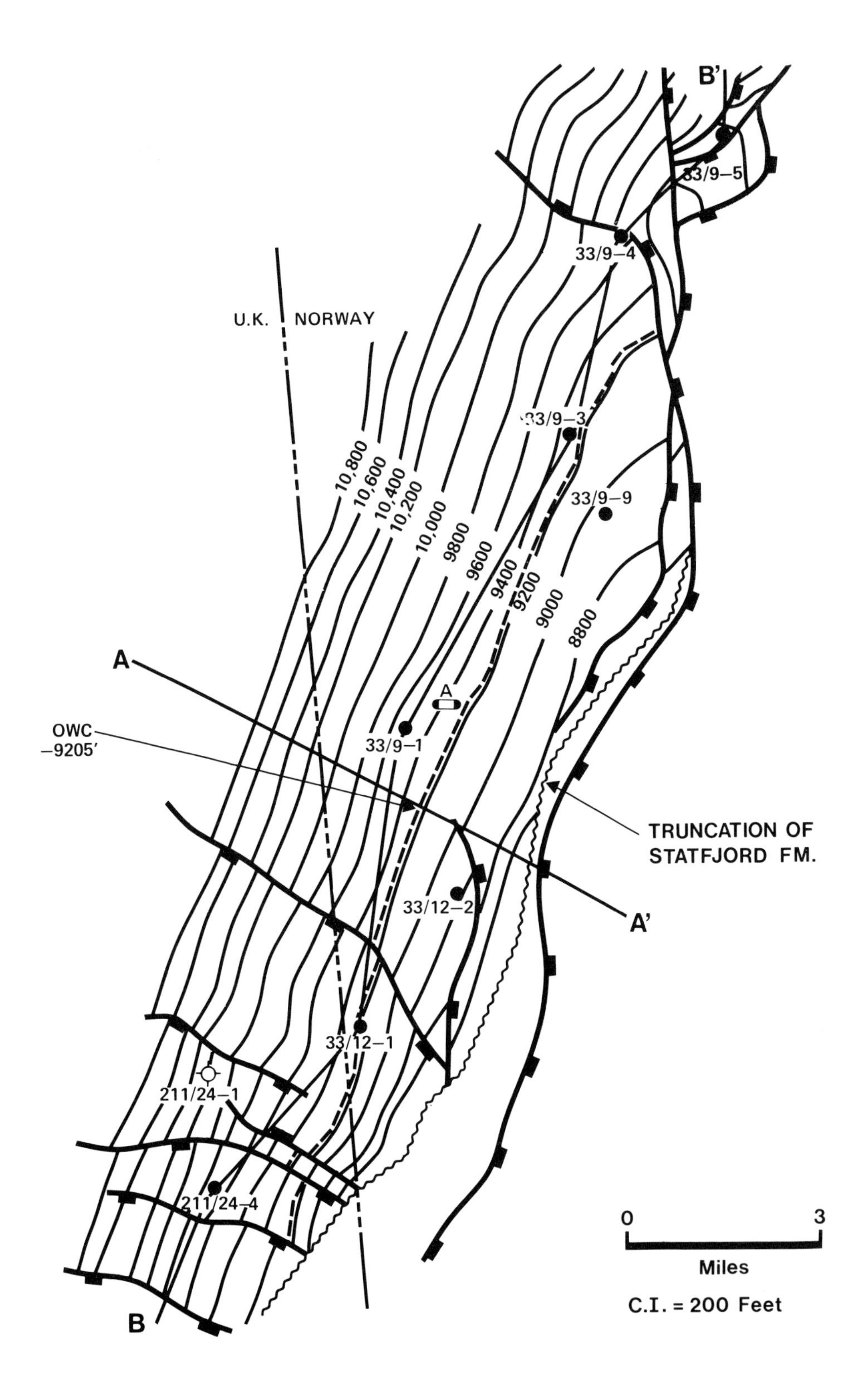

FIG. 16—Statfjord field structure contour map on top of Statfjord Formation.

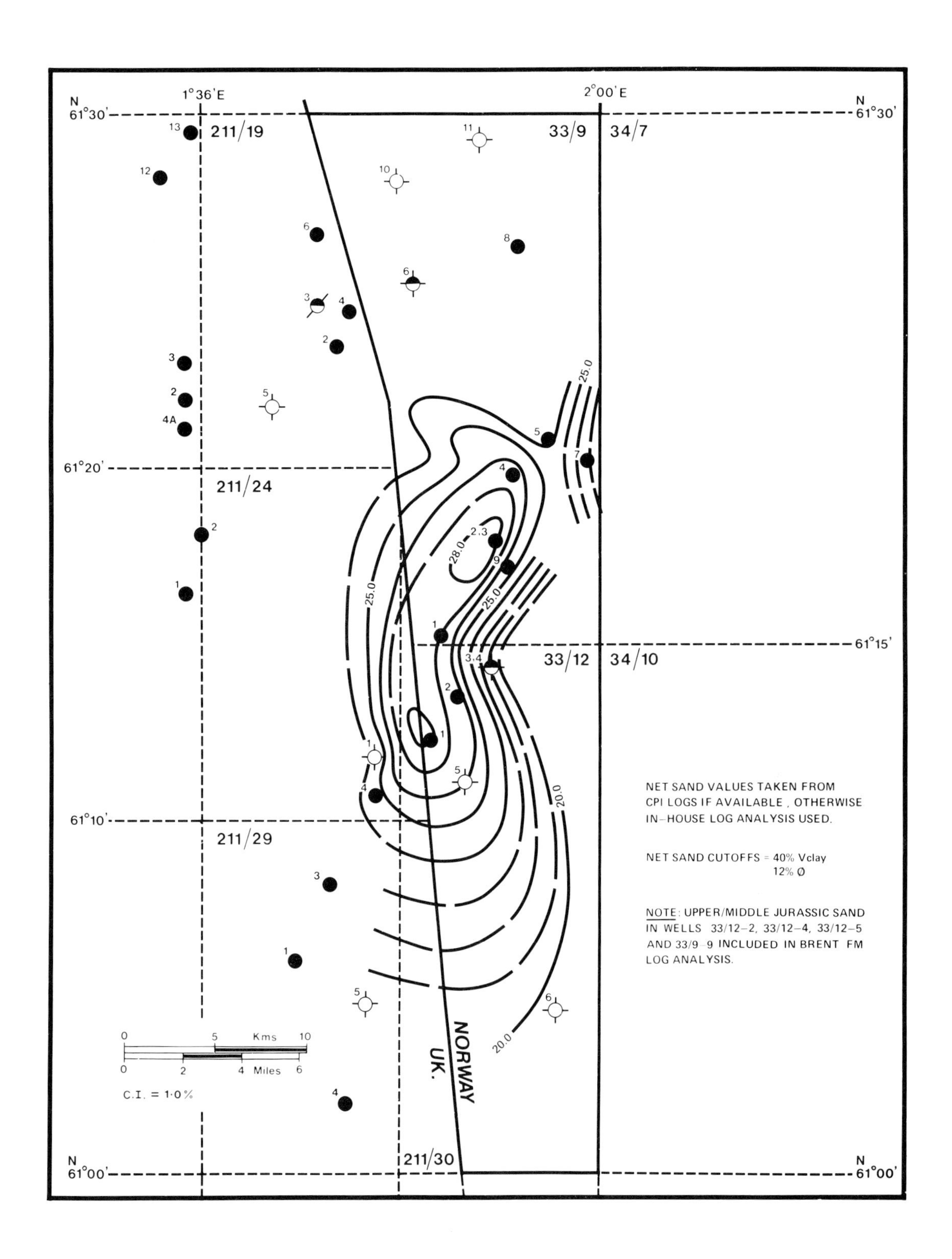

FIG. 17—Generalized isoporosity map of Brent Formation, Statfjord field area.

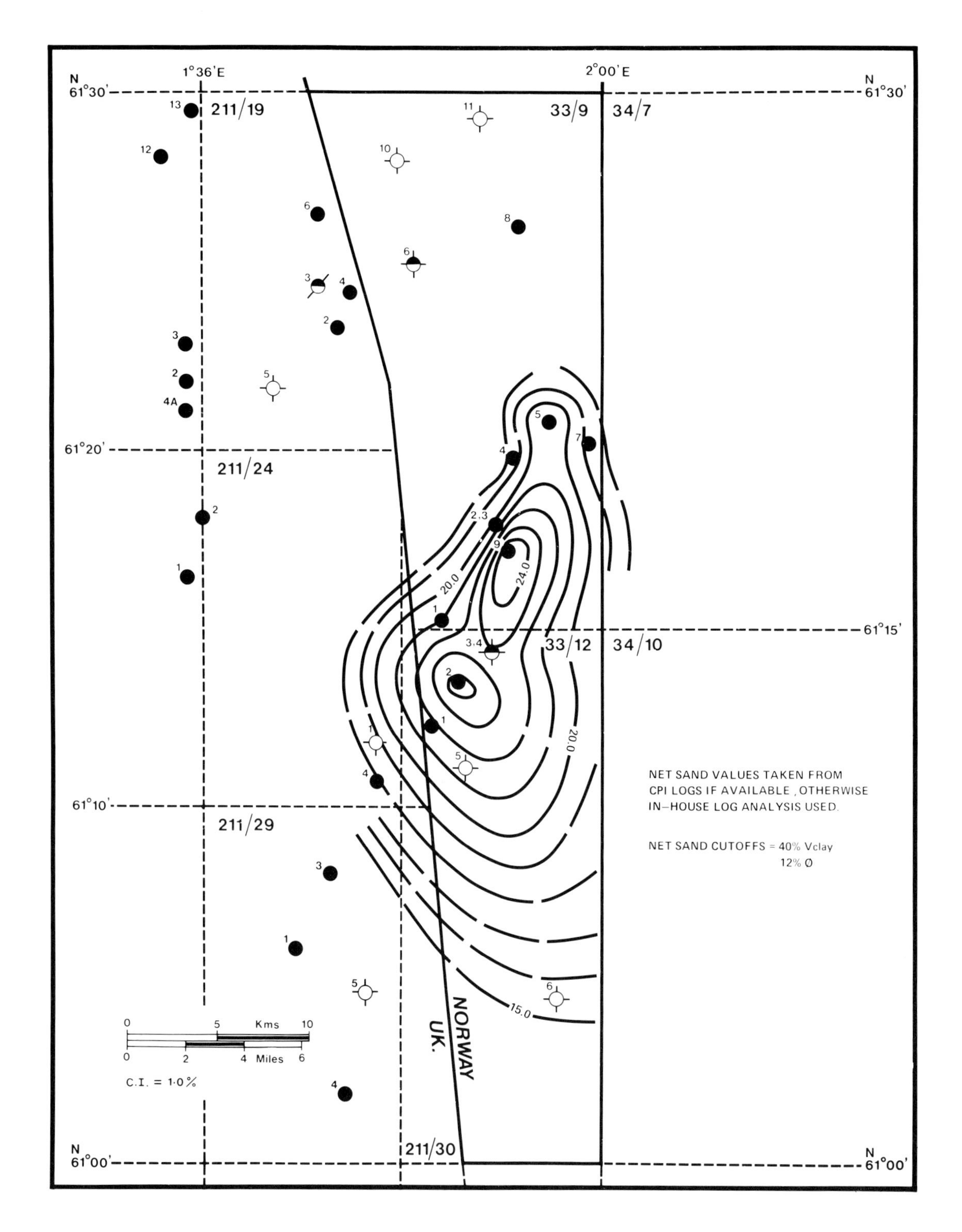

FIG. 18—Generalized isoporosity map of Statfjord Formation, Statfjord field area (Nansen and Eriksson Members).

cellent. Maximum gross porous-permeable reservoir thickness is 416 ft (127 m), porosities average 22% and permeabilities up to 12 darcies have been measured, though average permeability is 250 millidarcies.

Table 2 is a reservoir and reserve summary for the Brent and Statfjord Formations. Total combined Statfjord field reserves are estimated on the order of 3 billion bbl with the majority of the reserves contained within the Brent Formation.

A number of authors have commented on the uncommonly high porosities found in the Brent Formation at Statfjord, not normally expected in Mesozoic-age sandstones. Selected explanations include: (1) early complete cementation and much later decementation and etching (Schmidt et al, 1977); (2) early overpressuring of the reservoir resulting in preservation of primary porosity and little compaction since deposition (Blanche and Whitaker, 1978); and (3) possible leaching action of meteoric waters plus Statfjord's relatively higher structural position, which may have reduced effects of later diagenesis by pore fluids.

Table 2. Statfjord Field Reservoir Data.

Reservoir[1]	Brent	Statfjord[2]
Basic Data		
Average Porosity	29%	22%
Average Permeability	1,500 md	250 md
Average Water Saturation	17%	31%
Average Net/ Gross Ratio	88%	63%
Gross Pay Thickness:		
Maximum	640 ft (195m)	416 ft (127m)
Average	305 ft (92.6m)	372 ft (113.8m)
Oil-Water Contact (Subsea)	8,480 ft (2,585m)	9,205 ft (2,805m)
Productive Closure	15,000 ac (6,069 ha)	9,000 ac (3,641 ha)
Estimated Oil In Place	4.2 billion bbl	1.4 billion bbl
OTHER		
Gravity of Oil (°API)	38/39 (low sulphur)	38/41 (low sulphur)
Gas Cap	No	No
GOR	±1,000 cf/bbl	±850 cf/bbl
Water Resistivity (Rw)	0.11 at 171°F	0.128 at 181°F
Overpressured Reservoir	Yes (±50%)	Yes (±50%)
Drive Mechanism	Pressure Depletion (Pressure to be Maintained by Water Injection)	Pressure Depletion Pressure to be Maintained by Gas Injection)

[1] Only one well has penetrated the Dunlin sand in a porous reservoir facies. Communication with Brent reservoir is indicated. Reservoir data is limited and reserves are not included in Statfjord Field reserves.

[2] Data is limited as only two wells penetrated a Statfjord oil column.

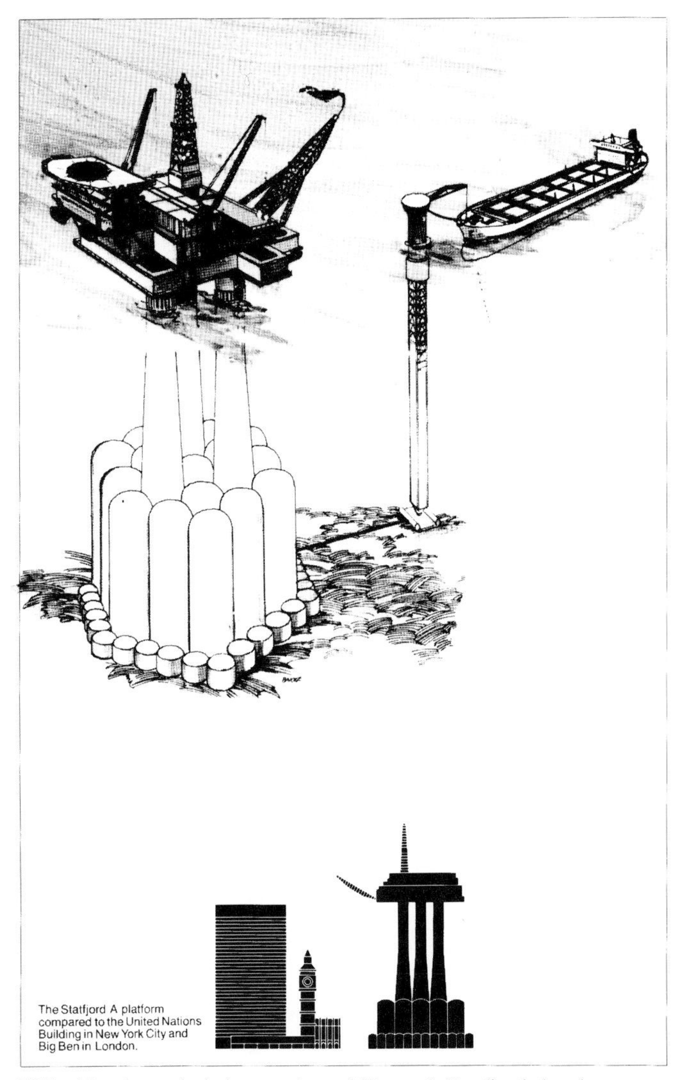

FIG. 19—An artist's impression of Phase 1 Statfjord development. Note the scale comparison of the Statfjord 'A' platform to the United Nations Building in New York and to Big Ben in London.

A general relationship of porosity to structure can be seen in Figures 17 and 18 which show the highest gross porosities in both the Brent and Statfjord reservoirs over the crestal area of Statfjord field.

DEVELOPMENT

The Statfjord field is currently being developed jointly by Norwegian and U.K. license holders. The first "Condeep" type platform (Statfjord 'A') was placed on location in May 1977. Development drilling was initiated in December 1978 and one development well has been drilled to date.

Figure 19 is an artist's conception of the Stafjord 'A' platform on location with a tanker loading crude from the storage cells via an ALP/SPM system. It also compares the size of the platform with the United Nations Building in New York and

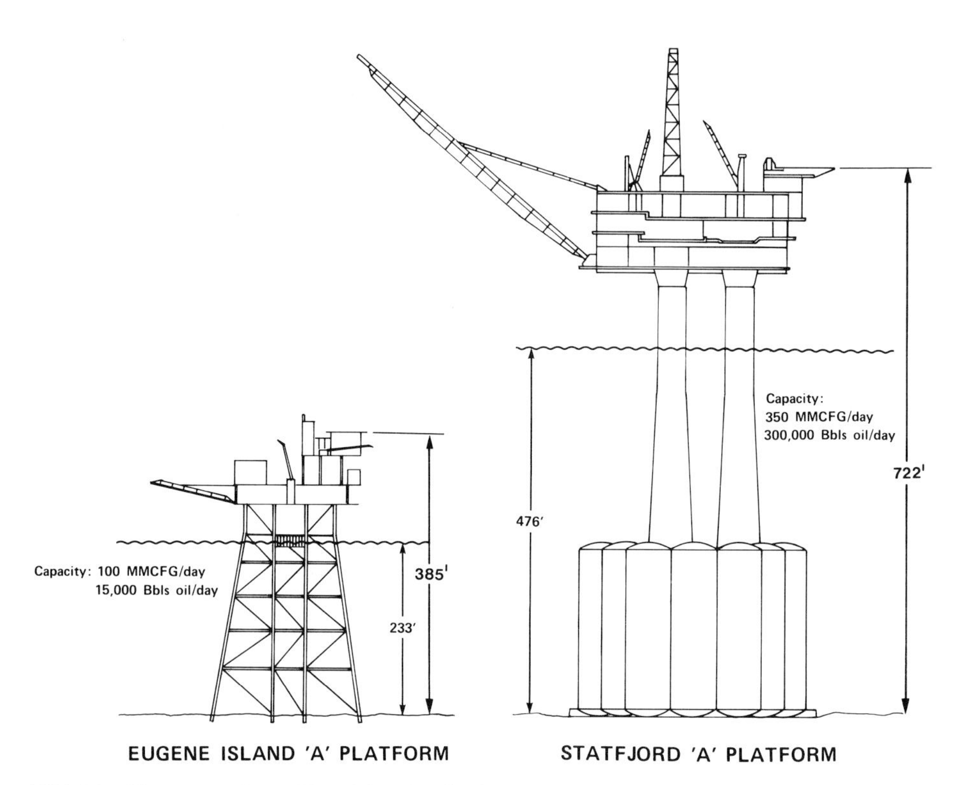

FIG. 20—Size comparison of North Sea Statfjord 'A' platform with a Gulf Coast production platform (Eugene Island 'A' platform).

Big Ben in London.

Figure 20 matches Statfjord 'A' against a typical Gulf Coast production platform to further demonstrate its scale.

CONCLUSION

The accumulation of major reserves at Statfjord field is believed to result from one of the larger traps in the basin being coincident with the region's thickest Brent reservoir which still maintains a high net-to-gross sand ratio. Reservoir properties of the Brent are related to depositional facies and diagenesis. However, the relatively higher structural position of the Statfjord area in relation to many other fields in the region may have played an important role in preservation of the excellent porosities and permeabilities at Statfjord field today.

REFERENCES CITED

Blanche, J. B., and J. H. M. Whitaker, 1978, Diagenesis of part of the Brent sand formation (Middle Jurassic) of the northern North Sea basin: Jour. Geol. Soc. London, v. 135, p. 73-82.

Bowen, J. M., 1975, The Brent oil-field, *in* A. W. Woodward, ed., Petroleum and the continental shelf of north-west Europe, v. 1: New York, John Wiley and Sons, p. 353-361.

Deegan, C. E., and B. J. Scull (compilers), 1977, A proposed standard lithostratigraphic nomenclature for the Central and Northern North Sea: Bull. Norwegian Petroleum Directorate No. 1; Inst. Geol. Sci. Rept. 77/25.

Jones, E. L., H. P. Raveling, and H. R. Taylor, 1975, Statfjord field: Norwegian Petroleum Soc., Jurassic Northern North Sea Symp., Proc., JNNSS/20, p. 1-19.

Nyberg, O., et al, 1978, Mineral composition—an aid in classical log analysis used in Jurassic sandstones of the Northern North Sea: Soc. Prof. Well Log Analysts Symp., Logging Sym. Trans. (El Paso), Sect. M, 35 p.

Ronnevik, H. C., W. Van den Bosch, and E. H. Bandlien, 1975, A proposed nomenclature for the main structure features in the Norwegian North Sea: Norwegian Petroleum Soc., Jurassic Northern North Sea Symp., Proc., JNNSS/18, p. 1-16.

Schmidt, V., D. A. McDonald, and R. L. Platt, 1977, Pore geometry and reservoir aspects of secondary porosity in sandstones: Bull. Canadian Petroleum Geology, v. 25, p. 271-290.

Ziegler, P. A., 1975, Geologic evolution of North Sea and its tectonic framework: AAPG Bull., v. 59, p. 1073-1097.

The Beatrice Field, Inner Moray Firth, U.K. North Sea[1]

By Philip N. Linsley[2,3], Henry C. Potter[4]
Greg McNab[2,5], and David Racher[2,5]

Abstract The Beatrice Field is located in Block 11/30 in the U.K. sector of the North Sea. The field is 14 mi (22 km) from the Scottish coast. The water depth is 160 ft (49 m).

In August 1976, well 11/30-1, the seventh wildcat well in the subbasin struck oil at about 6,000 ft (1,829 m) in an 831 ft (253 m) gross-pay column, at a time when most companies had written off the inner Moray Firth as a major oil province. The well produced an aggregate of 6,060 b/d (38° API) with a low GOR. The crude, though light and sweet, has a high wax content (17%) and high pour point (65°F).

Four more wells, three productive and one dry, have delineated the 4,271 acre field in which there are an estimated 476 million bbl of oil in place, 162 million bbl (34%) recoverable.

The stratigraphy and petrology of the field is described with emphasis on the alluvial to marine Jurassic (Sinemurian-Callovian) sandstone/shale reservoir sequence. Stratigraphic markers within the sequence are identified and related to the outcrops fringing the Moray Firth and an outline of the growth of the elongate fault bounded anticlinal trap is given. A summary of the reservoir characteristics and a review of the present and future field development are included.

INTRODUCTION

The Beatrice field is located in Block 11/30a, a partly relinquished block in the U.K. sector of the North Sea (Fig. 1). The field was discovered by a consortium consisting of Mesa Petroleum Co. (Operator), Kerr-McGee Corp., Creslenn Chelsea Company, P. & O. Oil Limited, and Hunt Oil Co. and Exploration Holdings. In early 1979 the British National Oil Corp. purchased half of the Hunt share in the field, and signed Heads of Agreement to purchase, with Deminex, the Mesa holding.

There are a number of significant features which set the Beatrice field apart from the other North Sea fields. It is in a fairly sheltered location, and in shallower water than any other oil field in the North Sea. The pay is shallow, and the crude has a high wax content and high pour point, although it is otherwise typical of most North Sea crudes.

However, to the explorationist, the significant features are as follows. First, it is 100 mi (160 km) from the nearest production (Claymore field) and is, to date, the only field in the inner Moray Firth basin (Fig. 1). The inner Moray Firth basin is

[1] Manuscript received, April 4, 1979; accepted for publication, May 30, 1979.

[2] Mesa (U.K.) Ltd., Aberdeen, Scotland, AB1 2FD.

[3] Presently with Mesa (Australia) Ltd., Perth, Western Australia, 6001.

[4] P & O Oil Ltd., London, England.

[5] Presently with London & Scottish Marine Oil Company Limited, Bastion House, 140 London Wall, London EC2Y 5DN.

The writers wish to thank the drafting department of Mesa (U.K.) Ltd., for preparing the illustrations, and Mrs. Kathleen Leiper for typing the manuscript. Acknowledgement also is made for help received from colleagues within the partnership, and from Tony King and Deryck Baylis of Palaeoservices, Ltd., who carried out paleontological determinations.

This paper is published by permission of the license holders in P. 187.

Article Identification Number:
0065-731X/80/M030-0007/$03.00/0.

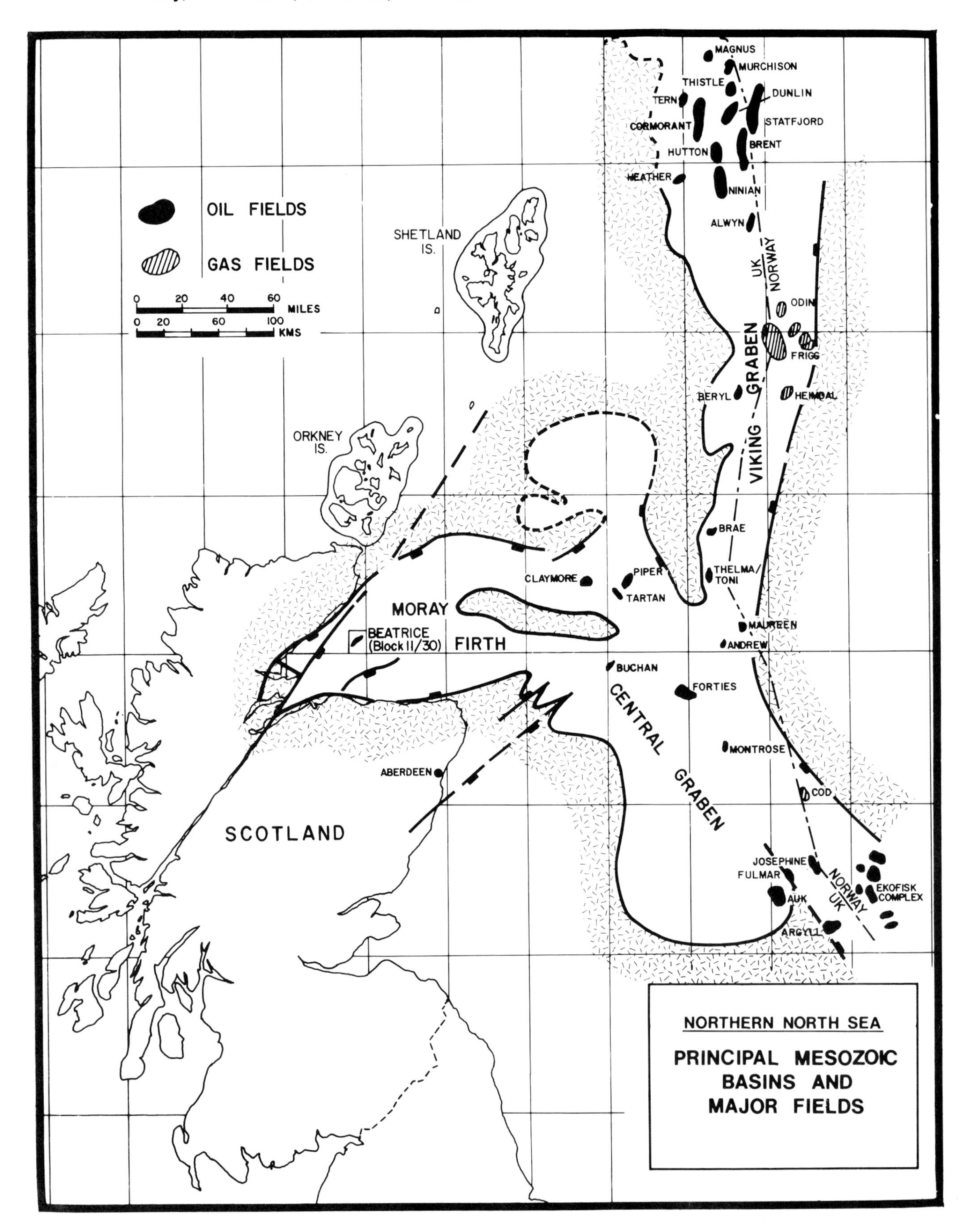

FIG. 1—Generalized location map of Mesozoic basins and major oil fields in the northern North Sea.

roughly that part of the Moray Firth basin west of 1°36′ W long. (see later section on Regional Structure; also Fig. 6).

Secondly, the play is different from that in any other North Sea field in that the subsurface geology can be related directly to the nearby surface geology. Halbouty (1978) recognized the importance of this and quoted it as an example of a "new" exploration play in a developing basin. In fact, it is only the success which is new; the play is a development of Jurassic outcrops 20 mi (32 km) away and were of interest for a long time, although elusive offshore. Many geologists thought that it was thoroughly tested by the first wells in the inner Moray Firth basin, although with hindsight the early errors in the biostratigraphic work are apparent. Thus, the Beatrice discovery well, 11/30-1, was the first to adequately test this play.

Lastly, the discovery has revived interest in the area and a large sedimentary basin has been reopened to exploration.

HISTORY OF EXPLORATION

The presence of prospective Jurassic rocks fringing the inner Moray Firth (Fig. 2) attracted interest to the basin the earliest days of North Sea oil exploration. The first well, Hamilton 12/26-1, was drilled in 1967 and like the next three, Total 12/23-1, 12/21-1, and 12/22-1, it was a dry hole. In spite of this discouragement, a group of companies which include most of those in the present Mesa group, applied for and was awarded a third-round license (1970) covering blocks 12/25 and 19/1 (Fig. 2). These blocks eventually were relinquished in August 1976 without drilling because subsequent seismic work did not delineate any viable prospects.

Following the award of the third-round license, efforts were concentrated in the innermost part of the Moray Firth, close to the line of the Great Glen fault. Inspection of the regional gravity map suggested that the gravity "low" in the area coincided with a thick Mesozoic sedimentary section in a basin immediately east of the fault. Regional seismic data supported this idea as did the presence of the thick onshore Jurassic section.

At that stage it was not clear which blocks along the Great Glen fault might actually be prospective, so in 1971 a regional seismic program was shot. The work indicated a prospective structure at a suitable depth in Block 11/30. In 1972, in the fourth-round of licensing, the Mesa group was awarded the block under License P.187. At the end of the initial six-year period, the size of the licensed area was reduced by half to about 27,000 acres.

Following the award of the license, additional seismic was acquired and integrated with earlier data, and the evaluation of the block and structure were continued. A drilling location was chosen at exactly the shot point where 11/30-1 was finally drilled (Fig. 3), but unfortunately in the summer of 1974 a suitable rig was not available and the proposed well had to be postponed. Meanwhile, two more dry holes, Petroswede 12/24-1 and BP 12/30-1, were drilled in the basin making a total of six, none of which confirmed the presence of the thick prospective onshore sequence. Consequently, most of the industry had discounted the inner Moray Firth as an area with major oil potential by the time the Mesa group prepared to drill its first well.

Nine years after the first well in the basin had been drilled, the Mesa group's first test was spudded on July 15, 1976; the well was located high on the mapped structure, close to the bounding fault (Fig. 3). After finding signs of hydrocarbons at shallow depth, major shows were encountered on August 5, 1976, at a depth of 5,953 ft (1,815 m) subsea, which proved to be the top of the Upper Jurassic Callovian stage. Well 11/30-1 proved a gross oil columnof 831 ft (253 m) with 304 ft (92.6 m) of net pay. Drill stem tests of four intervals produced 38° API, low sulfur, low gas to oil ratio, waxy oil, at an aggregate flow rate of 6,066 b/d.

Immediately after the discovery, a detailed seismic program on a kilometer-square grid was planned and shot over the block and the surrounding areas. Mesa group's second well was spudded on November 2, 1976, at a location 2¼ mi (3.6 km) southwest of the discovery well (Fig. 3).

The well, 11/30-2, confirmed the stratigraphy and the thickness of the oil column found in the first well and flowed an aggregate of 6,380 b/d from four intervals, in a pay section of 133 ft (40.5 m). The same rig was then used to drill 11/30-3 located 1.9 mi (3 km) northeast of the discovery well, and 11/30-4, 4 mi (6.4 km) southwest of 11/30-2 (Fig. 3). Well 11/30-3 produced 2,130 b/d on test from 92 ft (28 m) of net pay, but 11/30-4 was a dry hole.

At the time 11/30-4 was drilled, the closure (mapped both in two-way travel time and in depth) extended to the southwest, far beyond the location of 11/30-2. The depth conversion was based on the uniform seismic velocities determined by surveys in each of the first three wells. In 11/30-4 the section down to the top of the Callovian sandstone proved to be faster than in the first three wells and thus the saddle (mapped in two-way time) between 11/30-2 and 11/30-4 proved to be deeper than the oil-water contact. As migrated sections of the newly shot seismic data became available and were integrated with the earlier data, it became clear that the lowest point of the saddle is the spill point of the structure (Figs. 3, 4).

After the success of three of the first four wells, Mesa and its partners began field development studies which led to the submission of an Annexe 'B' to the Department of Energy in June 1977. The Department turned down the development plans in December 1977 because of the proximity of the planned offshore oil storage and tanker loading to shore and local fishing grounds.

Well 11/30-5, drilled in November 1977, was a successful flank delineation well, 1.5 mi (2.4 km) west of the discovery (Fig. 3). It flowed 5,200 b/d from a single interval. A revised development program based on a pipeline and shore terminal was submitted to the Department of Energy in May 1978 and was approved in August 1978. The first oil production is anticipated in May 1981.

REGIONAL STRATIGRAPHY

The oldest sediments found in the inner Moray Firth area lie with marked angular unconformity on the metamorphic and strongly deformed Caledonian basement (Fig. 5). North of the Great Glen fault this basement consists of the predominantly

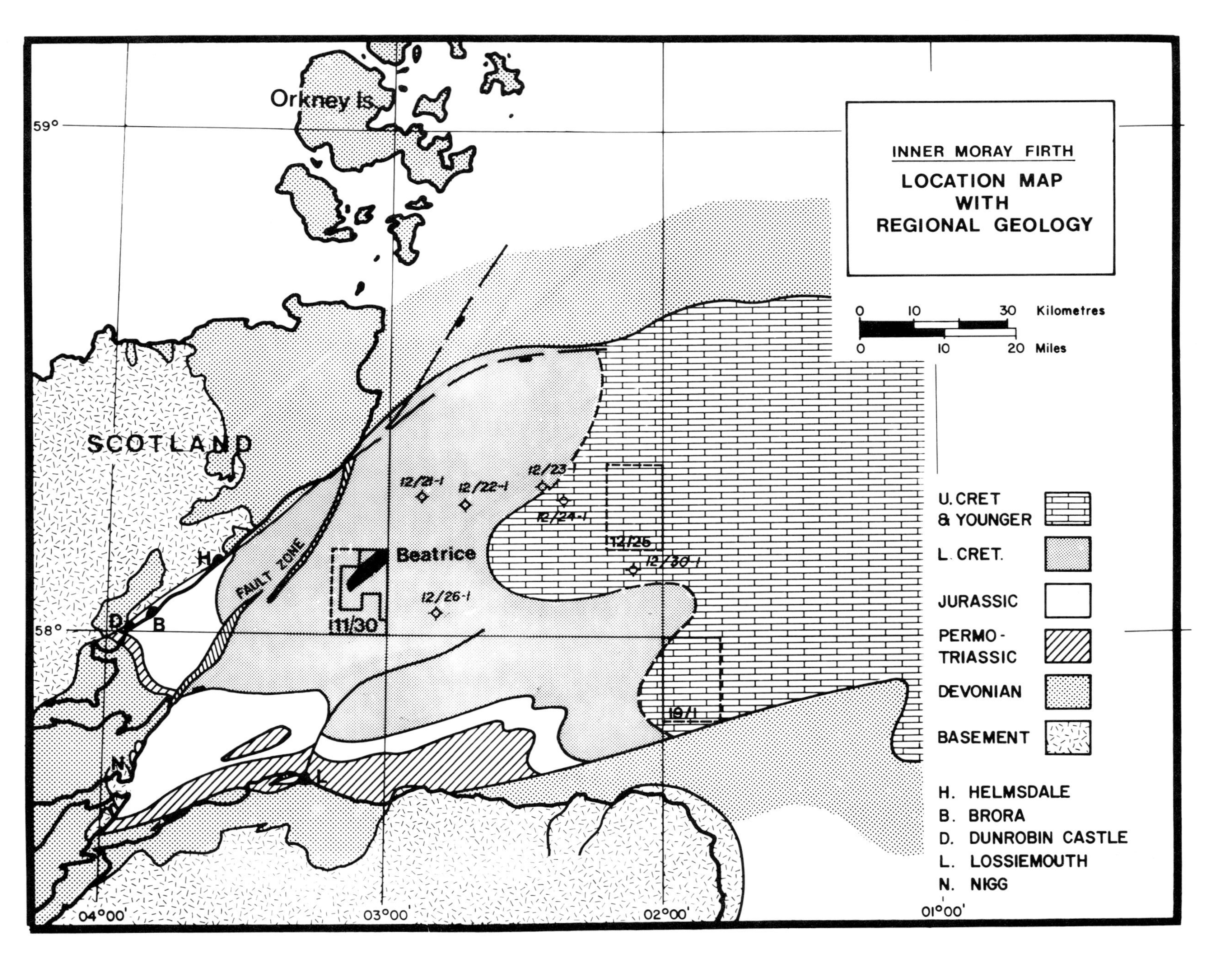

FIG. 2—Location map with regional geology of the inner Moray Firth area of the northern North Sea.

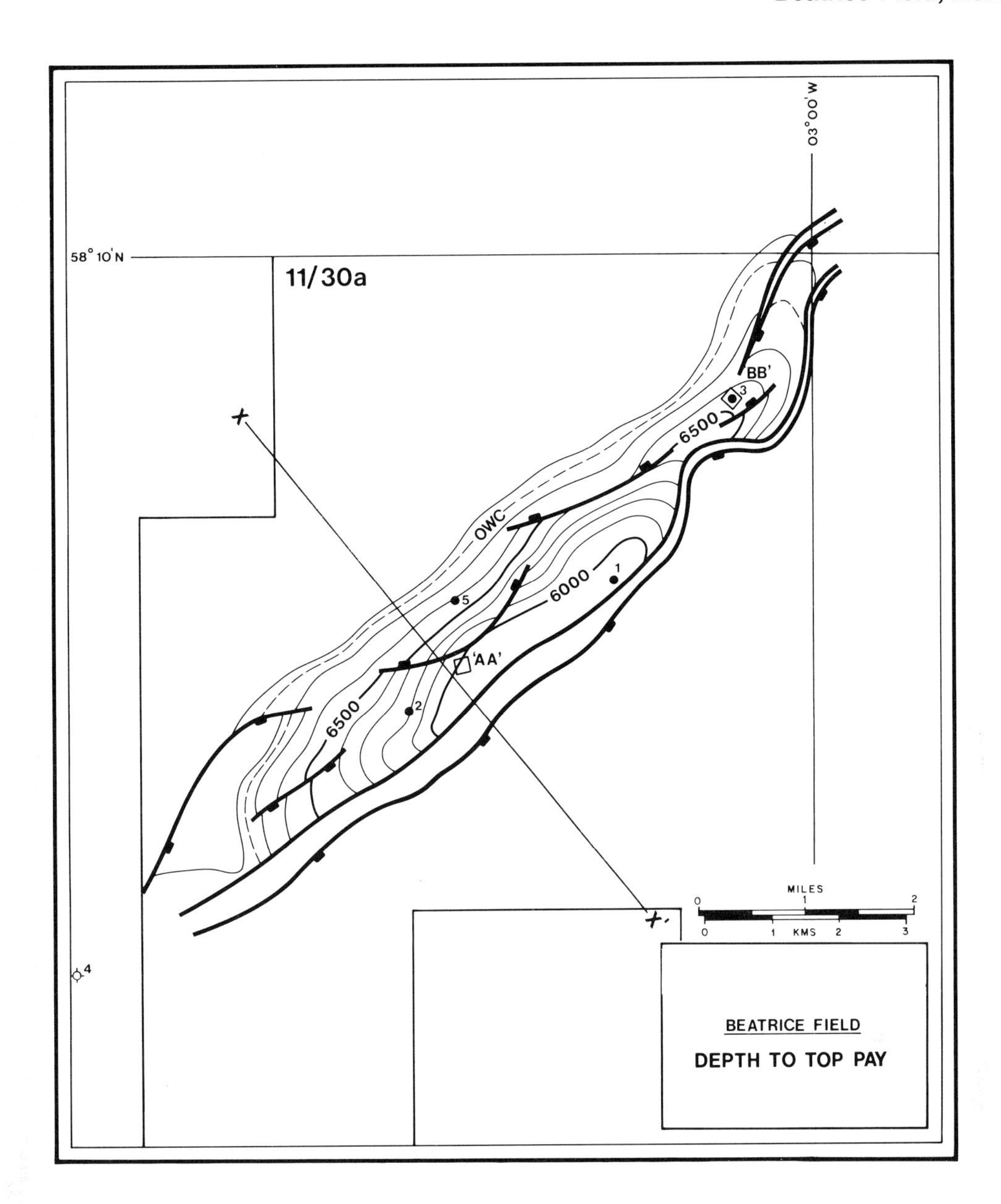

FIG. 3—Structure map indicating depth to top of pay, Beatrice field, North Sea.

gneissic Moinian Series, whereas south of the fault the Highland Granulites with a Dalradian section of lower metamorphic grade and/or Caledonian intrusives form the basement. Offshore, basement was reached only in well 12/23-1 (Fig. 2) which penetrated 115 ft (35 m) of quartz biotite schist and 55 ft (16.7 m) of granitic breccia.

Lying unconformably on the basement is the Devonian Old Red Sandstone Series, a thick sequence of continental clastics. They are the products of the erosion of the Caledonian Highlands which were formed at the end of the early Paleozoic orogenic cycle. The top of the sequence is also an unconformity in outcrop and in the western and central parts of the inner Moray Firth basin; the only Carboniferous rocks presently known are 119 ft (36.2 m) of basic volcanics found in well 12/23-1 (Fig. 2). Reworked Carboniferous microflora are widespread, but these could have been derived from sediments to the east.

Permo-Triassic sediments lie on the Devonian over most of the inner Moray Firth basin. The predominantly continental clastics were the first sediments to be deposited in the developing basin. They were derived from the thick Devonian section and the Caledonian Highlands which were rejuvenated by the Hercynian movements.

In outcrop, the top of the Permo-Triassic continental sequence is the Stotfield Cherty Rock, a calcrete with chert which is believed to be a fossil soil. The formation is a widespread stratigraphic marker in the inner Moray Firth basin, and is recognizable in well 12/26-1 and in the Beatrice field, where it is the base of the reservoir section. It is not present in well 12/23-1, probably due to erosion.

Overlying the Cherty Rock is a series of red-brown multicolored shales and mudstones of Rhaetian-Hettangian (early Liassic) age, which were deposited in an alluvial-lagoonal environ-

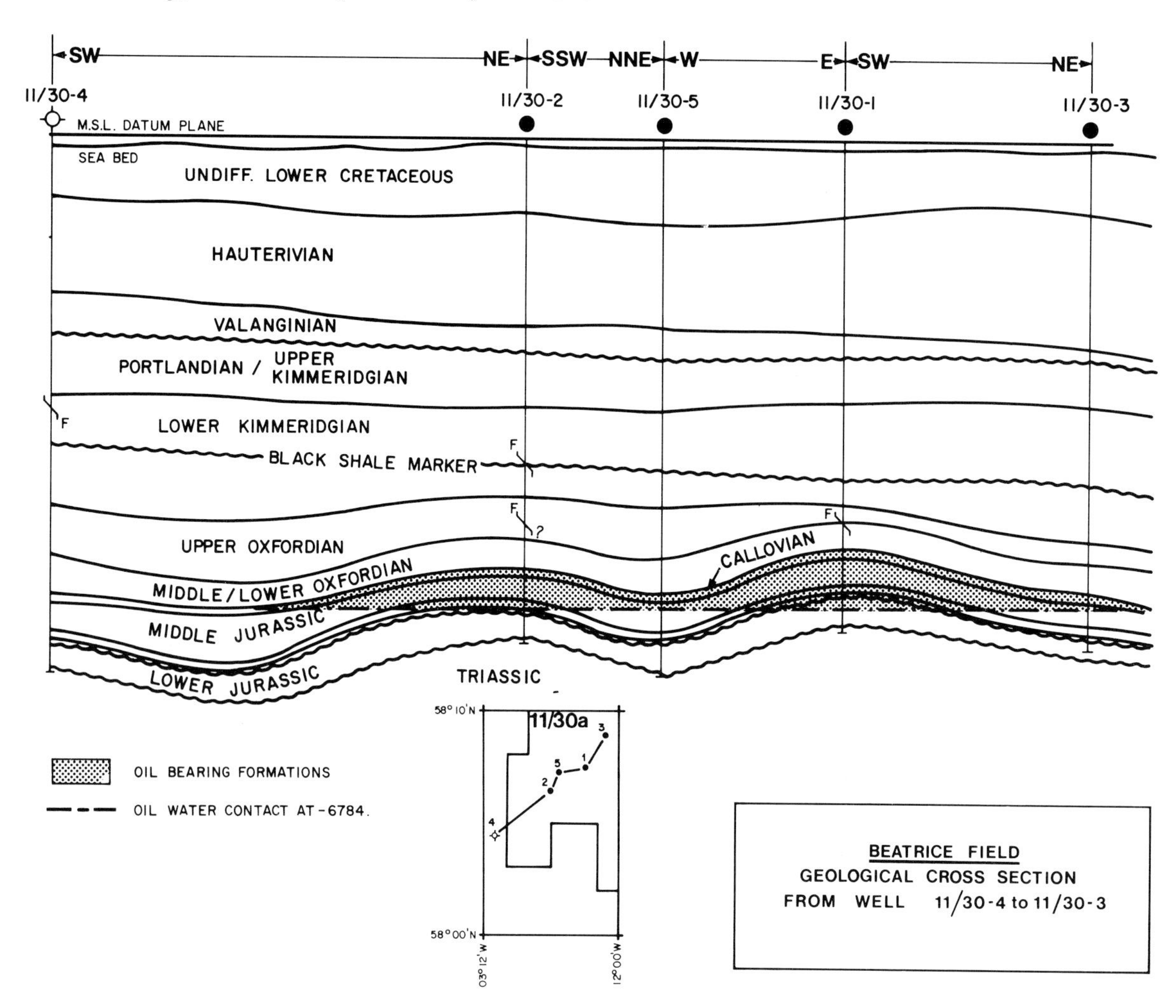

FIG. 4—Geological cross section, southwest to northeast through wells 11/30-4 to 11/30-3. Cross section trends the structural high depicted in Figure 3. Note locations of wells.

ment, the first phase of the developing Lower Jurassic marine cycle.

The most complete Liassic sequence in outcrop is at Dunrobin Castle, (Fig. 2) on the western side of the basin. On the foreshore, the Hettangian to early Pliensbachian sediments lying on the Triassic become progressively more marine upwards, a transition which is marked in the equivalent Beatrice field section by the incoming of deltaic sands overlain by marine clays. In Lossiemouth borehole, on the southern side of the basin, the equivalent section is less marine, presumably as a result of local uplift.

The top of the Lias, which is only represented offshore, is a thin sequence of Toarcian shaly sandstones/silty shales unconformably overlying the early Pliensbachian. These sediments, of a marginal marine origin, are the start of a regressive sequence of restricted marine delta-front sands, the youngest part of which is Bajocian (Middle Jurassic) in age.

A thick fluvial sand/shale sequence of Bajocian-Bathonian age, capped by a very thin coal unit, overlies the Toarcian-Bajocian deltaic sands. The coal was deposited in lagoonal-swamp conditions and is the equivalent of the Brora Coal exposed onshore on the western edge of the basin. The coal marks the start of a return to marine conditions at the end of the fluvial cycle. It is overlain by the cyclical, coarsening-upward, marginal marine Lower Callovian sediments.

West of the basin, the transgression led to the deposition of the Brora Argillaceous and Brora Arenaceous formations which can be seen in outcrop. The top of the sequence is the Brora Sandstone, a sand bar deposit. Offshore, its diachronous equivalent is the top of the Beatrice reservoir section (Fig. 5).

Renewed transgression in the Oxfordian led to the deposition of a thick marine shale sequence over the reservoir sands. The shales are occasionally phosphatic and also siliceous due to the presence of sponge spicules *(Rhaxella)* which also are common in the Ardassie Limestone exposed onshore at Brora.

Kimmeridgian shales, typically rich in organic carbon, overlie the Oxfordian, although not all of the sequence is radioactively "hot." Within the shales are thin limestones and thin beds of poorly sorted deepwater sands deposited in a distal channel/fan environment. The spectacular boulder-bed facies

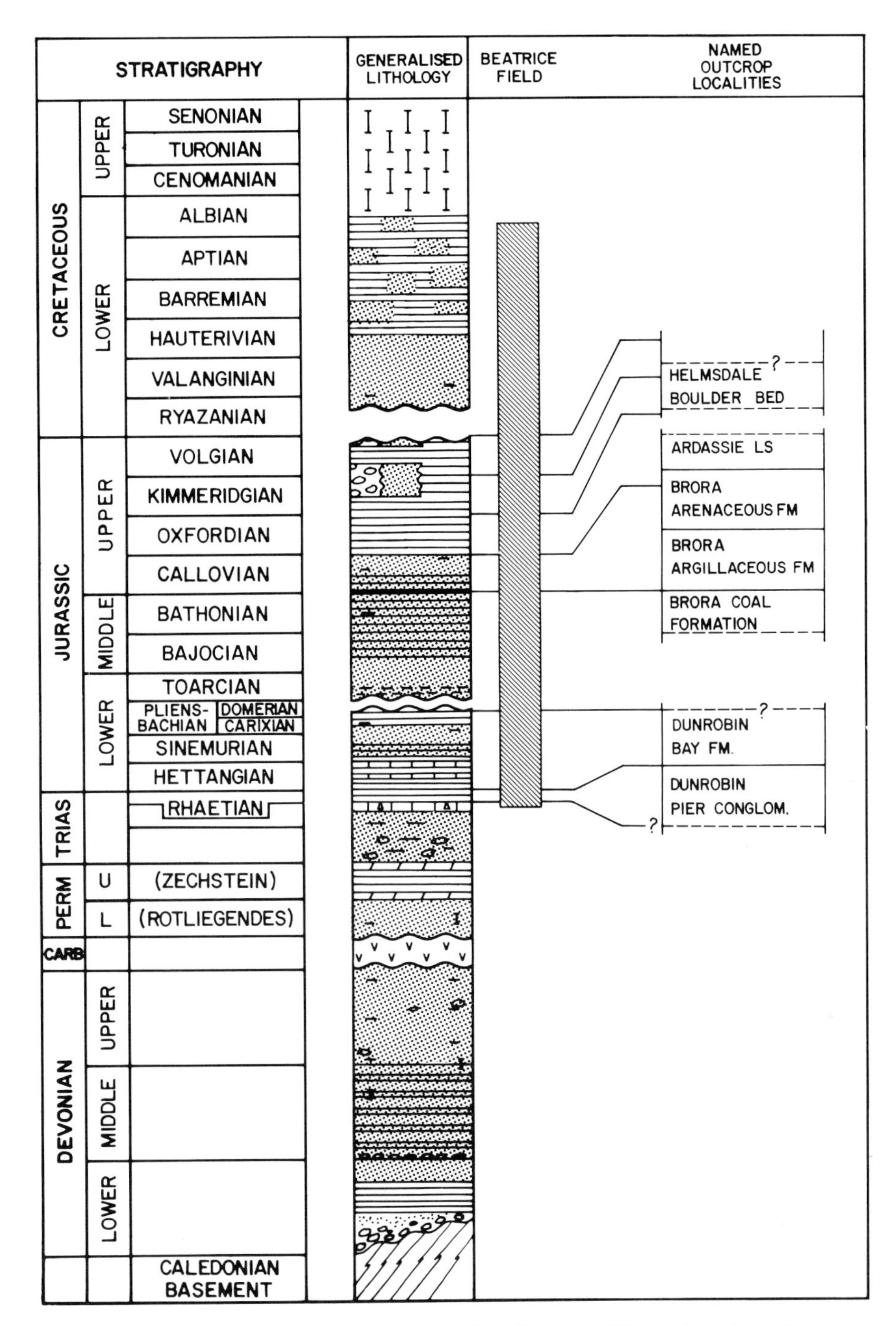

FIG. 5—Generalized stratigraphic column of the inner Moray Firth, North Sea. The total stratigraphic section encountered in the field is indicated by the hatched zone in the third column.

which abuts the Helmsdale fault has not been found offshore.

A period of uplift occurred before the deposition of the Valanginian (Lower Cretaceous) sands which lie unconformably on the Kimmeridgian shales. Over high blocks, such as the 12/23-1 structure, the unconformity is erosional, whereas in the basinal areas it is less clearly marked. Based on the gamma-ray log character and the benthic fauna of the interbedded clays, the medium to coarse grained gravelly and pyritic sandstones were deposited in moderately deepwater channels.

The Lower Cretaceous sands are overlain by marine clays of Hauterivian to Aptian/Albian age, which subcrop at the seafloor except where there is a thin cover of drift. The Upper Cretaceous chalk is present only in the easternmost part of the inner Moray Firth basin.

REGIONAL STRUCTURE

The Moray Firth basin is a sedimentary basin which was formed in the late Paleozoic. It is oriented roughly east-west, and runs from the Moray Firth area to the central North Sea where it joins the Viking and Central Grabens in a trilete trough system (Fig. 1).

Whiteman et al (1975) suggested that the trilete system was formed as the Laurasian supercontinent began to break up.

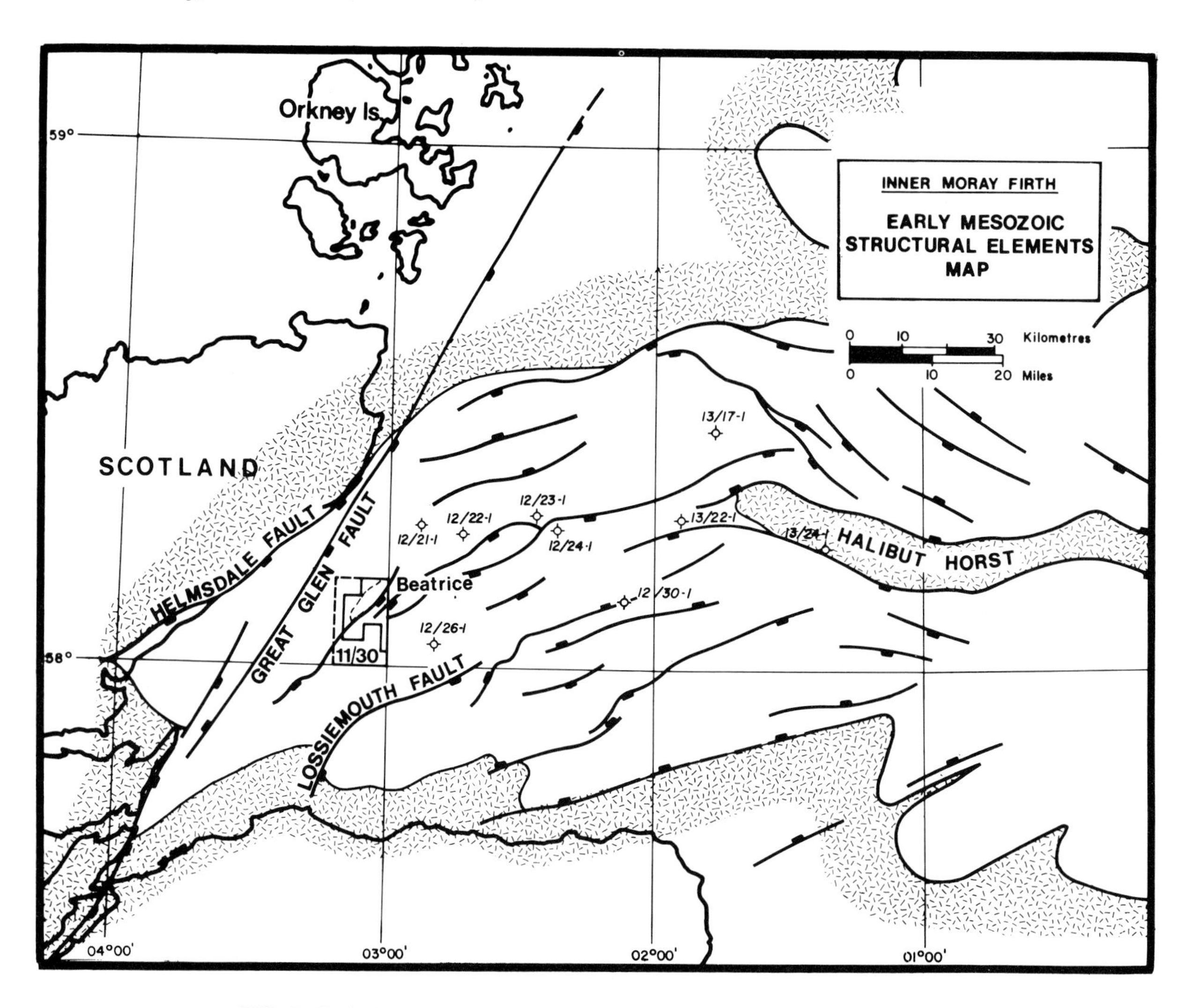

FIG. 6—Early Mesozoic structural elements map, inner Moray Firth, North Sea.

Each trough is described as the failed arm of a plume-generated crestal uplift, which cuts across the old lineations. In the outer Moray Firth, this theory may be valid, although the northwest to southeast trend is parallel with the Fair Isle Elbe lineation (Ziegler, 1975). However, in the inner Moray Firth the predominant tectonic trend is northeast to southwest, parallel with the Great Glen wrench-fault system which forms the western edge of the basin (Fig. 6). The change in tectonic trend, the boundary between the basins, occurs at about 1°36′W long.

The direction of the movement(s) on the Great Glen fault has been and is controversial, although the present consensus is that dextral movement only has occurred. Donovan et al (1976) demonstrated a post-middle Old Red Sandstone (Devonian) displacement of 19 mi (30 km), based on a facies reconstruction in the Moray Firth area. In Argyll, 90 mi (144 km) to the southwest, Speight and Mitchell (1979) demonstrated a 5-mi (8-km) displacement of a Permo-Carboniferous dyke swarm. The significance of the latter is that based on stratigraphy, the inner Moray Firth basin was formed in Permian time. Thus the movements on the Great Glen fault which (with the associated faulting along the northern and southern edges) formed the sedimentary basin, may have been initiated by the same forces which displaced the dykes.

The foregoing would seem to discount the theories of Whiteman et al (1975). However, the lack of Carboniferous sediments in the inner Moray Firth (due to basement arching?), and the presence of basic volcanics of the same age in well 12/23-1, could point to the forces involved having originated in the manner the authors suggest.

Throughout the Permian and Mesozoic, the main control on sedimentation in the inner Moray Firth basin was normal faulting associated with the Kimmerian movements. From the Permian up to Callovian/early Oxfordian time the basin in gross aspect was a half-graben controlled initially by the Great Glen fault, and from Rhaetian time onwards by the Helmsdale fault (Fig. 6). The basin had a single depocenter adjacent to the controlling fault, and through time it migrated towards the northeast. The movements on the bounding Helmsdale fault also controlled the Rhaetian to Callovian transgressive-regressive sedimentary cycles, whereas most of the faults within the basin had only a local effect on sediment thickness and facies.

From middle-late Oxfordian time the gentle Kimmerian movements became progressively stronger. As a result, several

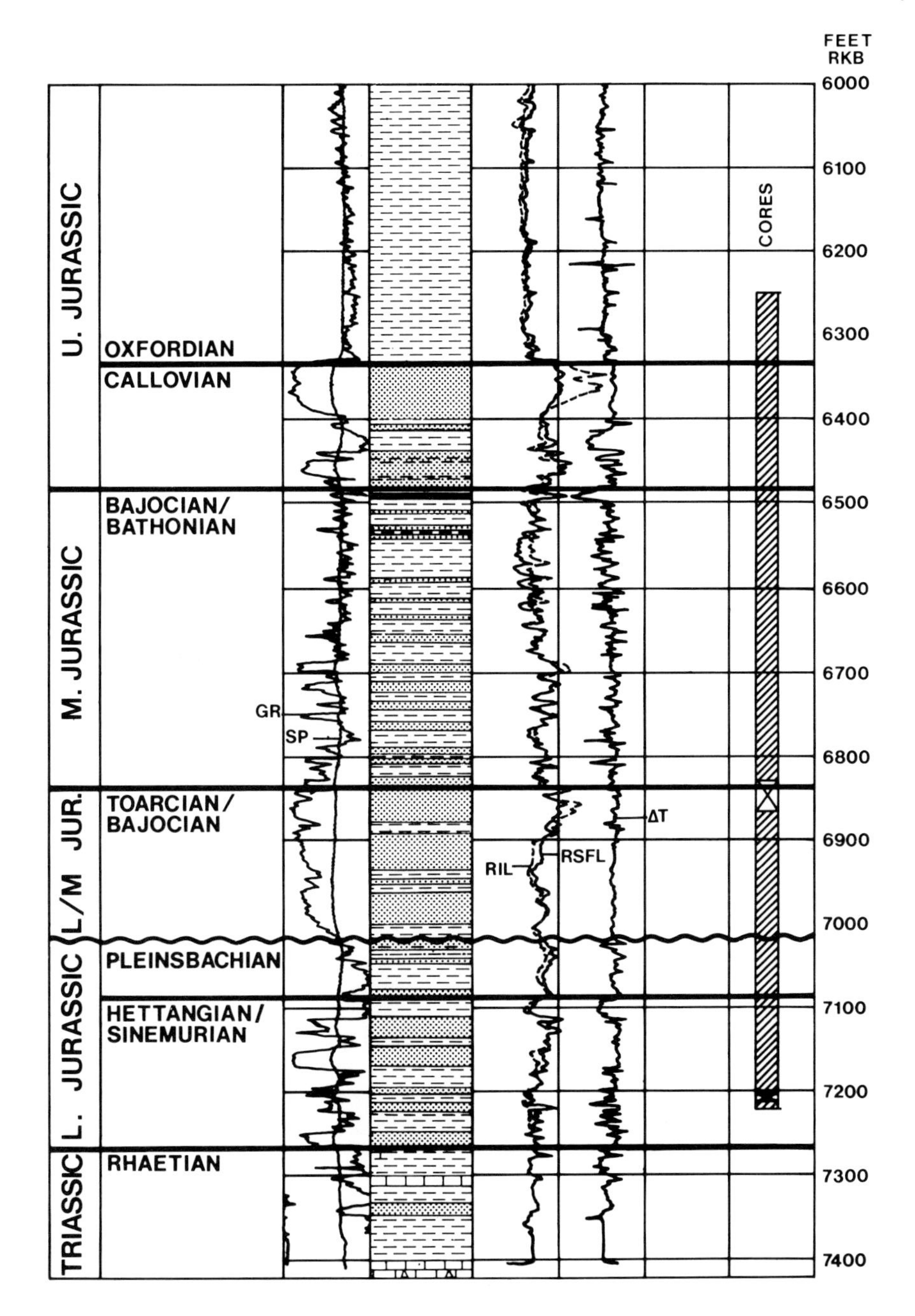

FIG. 7—Core section of reservoir in the 11/30-2 well, Beatrice field, North Sea. Note comparison to wireline logs, depth of section, lithology, and time scale.

major faults developed within the basin, producing a series of tilted blocks. In the depocenters which developed on the downthrown sides of the faults, thick clastic sequences accumulated, whereas on the dip slopes of the tilted blocks the section is greatly reduced. Based on the seismic sections the reduction is due to nondeposition rather than erosion, as the loss of section is at the base of each interval and the younger beds clearly onlap the older.

At the end of the Jurassic a general uplift occurred which resulted in the marked basal Lower Cretaceous unconformity over the high blocks. The major faults and depocenters, which originated in the Jurassic, continued to develop and deepen throughout the Lower Cretaceous. Eventually, tectonic equilibrium was attained and a blanket of Upper Cretaceous Chalk was deposited over the basin.

Renewed faulting occurred in post base Chalk times, and is assumed to be Late Cretaceous-Tertiary (Laramide) in age. The movements were on the major faults only, were in all cases slight, and in most cases were normal. Only on some of the major faults bounding the basin are there small drag faults which suggest any degree of transcurrent movement. This is in stark contrast to Holgate's (1969) suggestion of an 18-mi (29-km) dextral shift in the Great Glen fault in Tertiary time. In fact, the Great Glen fault is discontinuous beneath the Cretaceous cover, which is the reason Bacon and Chesher (1975) rejected Holgate's theory. Flinn (1975), however, suggested that the post Early Cretaceous movements on the Great Glen fault took place in the basement and led only to the "folding of the

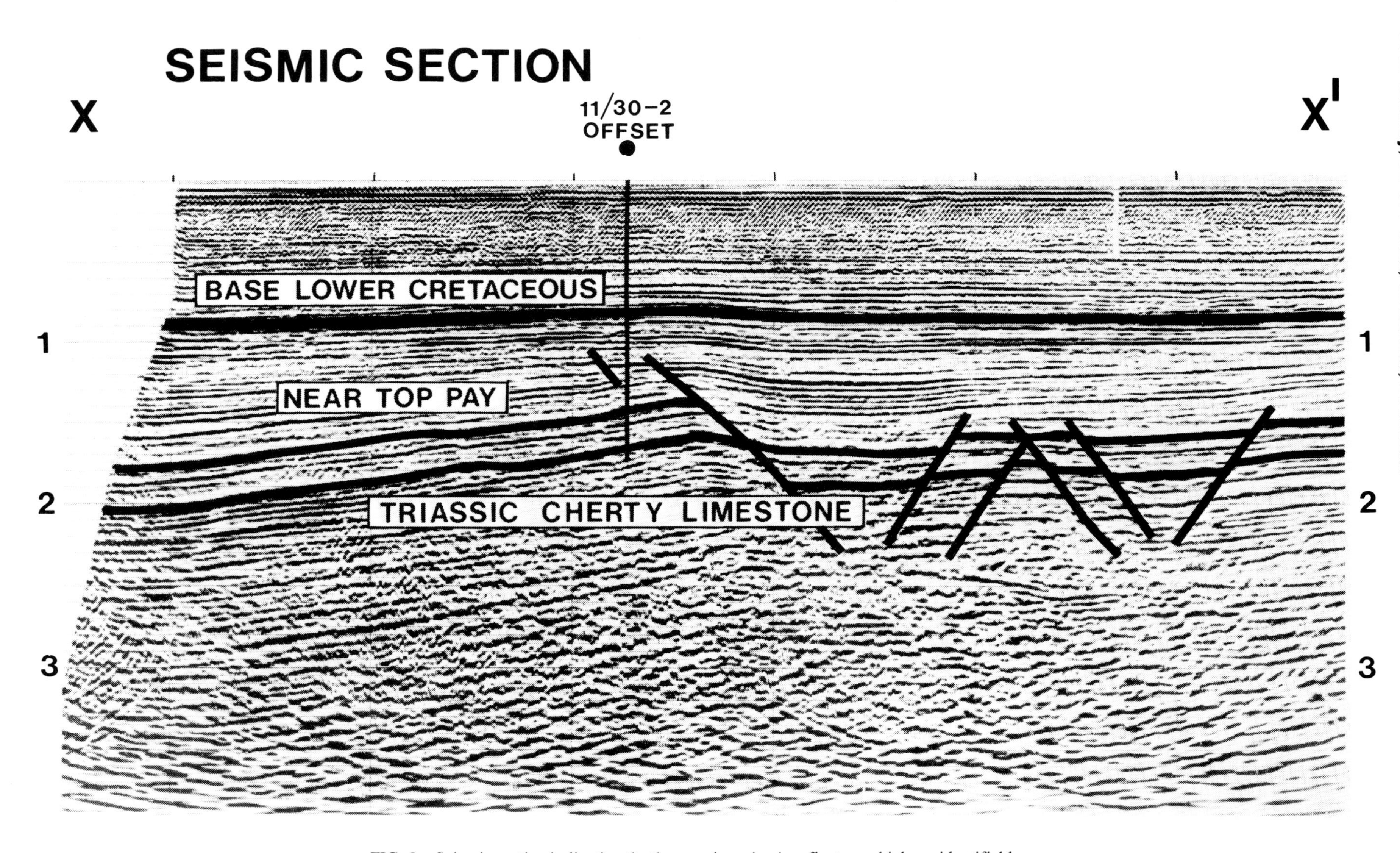

FIG. 8—Seismic section indicating the three major seismic reflectors which are identifiable over the Beatrice field, North Sea.

sedimentary cover and the discontinuous faulting along the general line of the underlying fault." This idea appears to be the most acceptable, based on the modern seismic data.

These transcurrent movements were related to the forces which caused a general tilt of the basin towards the northeast in post Chalk deposition, producing the present outcrop pattern at the seabed.

STRATIGRAPHY OF THE RESERVOIR

The Beatrice field reservoir has been divided into four lithostratigraphic units. These divisions are based on the wireline log data, the drill cuttings, paleontology, and the detailed lithologies of the conventional cores from well 11/30-2, in which almost all the reservoir was cored (Fig. 7). Short cores were also cut in wells 11/30-1 and 11/30-5.

Rhaetian-Lower Lias

The cherty limestone, which is the base of the Beatrice reservoir section, is overlain by a sequence of sediments which reflect a transition from a continental to a marine environment.

The lowest part of the sequence consists of alluvial flood basin and levee-crevasse deposits. These are fine grained, moderately sorted, and occasionally calcareous silty sandstones and sandy siltstones, interbedded with and also grading to mottled and veined gray-green claystones. Each bed is rarely more than 5 ft (1.5 m) thick. The sandstones are moderately- to well-cemented by authigenic silica, contain rare shell and ostracod fragments, and display a variety of planar and ripple cross-bedding structures. Rootlet horizons, rare plant material, and desiccation cracks occur in the interbedded claystones.

Higher in the sequence the claystones are replaced by gray to black organic shales with an abundant flora and fauna reflecting a change from a brackish water to a more marine environment. At the same time the sandstone bodies become thicker, coarser in grain, and more like distributary channel sands although in parts they are so calcareous-dolomitic that they become sandy limestones.

A series of transgressive calcareous shelf muds and mottled siltstones overlie the deltaic sandstones. The siltstones form a fining-up sequence with burrowed basal contacts whereas the claystone/mudstones contain a varied fauna of thin shelled pectens, crinoid fragments, and ammonites of early Pleinsbachian age.

In 11/30-2, two deltaic sandstone bodies are the main reservoirs in the 238-ft (73-m) thick Lower Lias unit, which is about the average thickness for the field. Approximately 35 ft (11 m) of the unit is net sand in which the average measured porosity and permeability are 15% and about 400 md respectively.

Toarcian-Bajocian

A distinct non-sequence occurs between the early Pleinsbachian and the overlying sandy silts of Toarcian age. In well 11/30-2, as in the other wells, there is sonic log marker (Fig. 7) at the non-sequence, although the dipmeter does not indicate an angular unconformity.

The regional movements which caused the late Pleinsbachian non-sequence also resulted in a change of environment, because the Toarcian sediments represent the start of a regressive sequence of high energy deltaic sands.

These massive sandstones are generally fine to medium grained and contain 10 to 12% feldspar and 15 to 18% clay, but vary greatly in sorting and degree of cementation. Quartz is the main cement, and a typical diagenetic cycle starts with syntaxial growths of authigenic quartz followed by authigenic kaolinite and amorphous silica. In parts of the cored reservoir below the oil-water contact this has resulted in a severe loss of pore space. However, in the oil zone, the diagenetic processes appear to have been arrested and the porosity and permeability preserved.

Based on gamma-ray character, the gross sandstone body is formed by two major coarsening-upward cycles, overlain by thin lagoonal-delta-top clays followed by a major distributary channel-sand. The claystones of the marsh-lagoonal phase contain a varied fauna of brackish water algae as well as marine acritarchs-dinocysts and thin coal and rootlet horizons.

The average thickness of the Toarcian-Bajocian unit in the first four wells is 250 ft (76 m) and the average net sand thickness is 122 ft (37 m). At the top of the section, the major distributary channel sand the top of the deltaic sand beneath form a zone of improved reservoir quality. In this zone the average porosity is 17%, which is at least 2% higher than in the zone below. However, the greatest change is in the permeabilities which improve from a range of 1 to 20 md to a range of 500 to 2,000 md. At present the exact reasons for the improved reservoir quality are not fully understood but they probably are a combination of sand type and diagenetic history.

Bajocian-Bathonian

The major channel sand is overlain by a cyclical sequence of alluvial flood plain deposits of Bajocian-Bathonian age. The five gross sandstone units presently recognized are based on the thicker productive sandstone bodies and wireline log markers. Generally the units contain more than one depositional cycle, a complete cycle being a sequence from an alluvial point-bar sandstone to a flood basin-levee-crevasse claystone. A typical cycle starts at the sharp base of a well sorted sandstone which fines upward into a mottled and veined siltstone. At the top of the cyclothem is a silty claystone with rootlet and coaly zones. The sandstones are typically cross-bedded and ripple laminated, and are cemented with authigenic silica. The siltstones commonly display convolute bedding but bioturbation is uncommon.

At the top of the Bajocian-Bathonian sequence is a coaly zone, the equivalent of the Brora Coal. Because the facies lies above the alluvial flood plain deposits and below the overlying marine Callovian, it is assumed to have been formed in an extensive lagoonal-coastal swamp environment.

The average thickness of the Bajocian-Bathonian in the first four wells is 350 ft (107 m) of which up to 80 ft (24 m) is net sand with an average porosity of 16.5%. The spread of the total sand-count values between the wells is small, although different sandstone units contribute different proportions of the total in each well. Where the sandstones are very thin the effective permeability is zero, however in the thicker units, permeabilities of up to 320 md have been measured.

Callovian

The Callovian sequence which overlies the Brora Coal equivalent is an alternating sequence of marine claystones and sideritic sandstones. Each transgressive-regressive cycle is a coarsening-upwards sequence following the initial transgression. A full cycle starts with silty fine grained bioturbated shelly sandstones with *Rhaxella* spicules which grade to silty shelly claystones-shelf muds. Fine-grained flasar-bedded and ripple-laminated silty sandstones overlie the claystones and coarsen up into clean, fine to medium grained, cross-bedded or bioturbated sandstones which were deposited in a beach or barrier environment.

The lower part of the Callovian includes three cycles in an average section of 50 ft (15 m). As a result the sandstone bodies, although continuous, are relatively thin and the reservoir properties suffer accordingly, particularly where calcite cement is common.

The upper part of the Callovian is the main Beatrice reservoir. It represents a single depositional cycle of the type described previously and is on average 100 ft (30 m) thick. In the Beatrice field the sandstone is overlain by a 2 ft (0.6 m) sandstone which includes numerous belemnite guards, some showing current orientation. The sandstone is a transitional zone to the shelf muds above, the lowest part of which has yielded a Callovian ammonite fauna. The arenaceous part of this cycle is therefore not the exact equivalent of the onshore Brora Arenaceous formation, which is Upper Callovian to Lower Oxfordian in age (Figs. 2, 5).

STRUCTURE OF THE FIELD

The origin and growth of the Beatrice structure has been deduced from maps on the three major seismic reflectors which are identifiable over the field area (Fig. 8). The shallowest is the "Base Lower Cretaceous" event which is mapped throughout the North Sea oil province; the intermediate reflector is near the top of the pay; and the deepest is at the top of the Triassic cherty limestone which is the base of the reservoir section (Fig. 7).

The map (Fig. 3) shows that the Beatrice structure is a faulted anticline. The dip reversal, which is only slight, is due to drape over the underlying block and also partly to fault drag. Although slight thinning occurred across the structure in pre-Oxfordian time, the structure really began to form with the development of the bounding normal fault in middle Oxfordian time, and most of the growth took place during the late Oxfordian.

The last movements on the fault occurred in early Kimmeridgian time, although strong movements continued on major faults elsewhere in the basin. The structure grew very slightly from then on as there is only a very slight closure at the "Base Lower Cretaceous" horizons.

DEVELOPMENT

The Beatrice field is being developed from two sites, AA and BB (Fig. 3). At AA, over the central part of the field, where the oil column thickness is at a maximum and the productivity good, a 32-slot drilling platform (linked by a bridge to a production platform) will be installed. The drilling platform will have a main and a workover rig while the production platform will have facilities to handle over 80,000 b/d of oil and 96,000 b/d of treated seawater for injection purposes.

At Site BB, where the oil column is thinner and productivity lower, a converted jack-up and free standing 9-slot conductor support will be installed. The production capacity of the jack-up will be 16,000 b/d. The oil will be pumped to the main production platform in a line running parallel with a return line carrying injection water.

The development of the field started in September 1978 with the location of an 8-well subsea template at Site AA. To date, (April 1979), the 8 conductors have been set and 3 production wells drilled. All template wells drilled prior to the installation of the drilling platform jacket will be suspended at the mud line. Development drilling should recommence at Site AA in 1980 using the main platform rig, while the workover rig will be used to tie back and complete the wells already drilled. Development drilling should commence at site BB, over 11/30-3 in 1979.

Meanwhile construction of the production platform and the shore facilities at Nigg (Fig. 2) in the Cromarty Firth, is in hand. These will be linked by a 49-mi (78-km) pipeline and should become fully operational in 1981. By that time as many as 22 wells may have been drilled, and it will then be possible to start production at or near the peak average rate of 80,000 b/d.

Because of the low energy of the reservoir system, wells will be pumped from the day they go on stream. For the same reason early water injection also is anticipated and up to 10 injectors are planned.

REFERENCES CITED

Anonymous, 1977, The Moray Firth area geological studies: Inverness Field Club, Scotland.

Bacon, M., and J. Chesher, 1975, Evidence against post-Hercynian transcurrent movement on the Great Glen fault: Scottish Jour. Geol., v. 11, p. 79-82.

Donovan, R. N., et al, 1976, Devonian paleogeography of the Orcadian basin and the Great Glen fault: Nature, v. 259, (February), p. 550-551.

Flinn, D., 1975, Evidence for post-Hercynian transcurrent movement on the Great Glen fault in the Moray Firth: Scottish Jour. Geol., v. 11, p. 266-267.

Halbouty, M. T., 1978, Acceleration in global exploration requirement for survival: AAPG Bull., v. 62, p. 739-751.

Holgate, N., 1969, Paleozoic and Tertiary transcurrent movements on the Great Glen fault: Scottish Jour. Geol., v. 5, p. 97-139.

Robertson Research, 1978, The Moray Firth area of Scotland—The stratigraphy, reservoir rocks, and source rock potential of the Devonian to Lower Cretaceous sediments—a non-exclusive industry report: Robertson Research, Ltd.

Speight, J. M., and J. G. Mitchell, 1979, The Permo-Carboniferous dyke swarm of northern Argyll and its bearing on the dextral displacement of the Great Glen fault: Jour. Geol. Soc. (London), v. 136, pt. 1 (January).

Whiteman, A., et al, 1975, North Sea troughs and plate tectonics: Tectonophysics, v. 26, p. 39-54.

Ziegler, W. H., 1975, Outline of the geological history of the North Sea, *in* A. W. Woodland, ed., Petroleum and the continental shelf of north-west Europe, vol. 1: London, Applied Science Pubs., p. 165-187.

Piper Oil Field[1]

By C. E. Maher[2]

Abstract Piper oil field lies in UK block 15/17, near the eastern end of the Moray Firth basin, 125 mi (200 km) northeast of Aberdeen, Scotland. The field was discovered in January, 1973 and confirmed as a major oil field in 1973 with five appraisal wells and one exploratory well. A steel platform with 36 well slots and space for two drilling rigs was centrally located over the field in 474 ft (144 m) of water in June, 1975, and made ready for production drilling by October 1976.

Production is from the Upper Jurassic Piper Sandstone, a high energy, marginal marine and shallow-marine, shelf sandstone with gross thickness ranging from 160 to 465 ft (50 to 142 m); net sand ranges from 131 to 378 ft (40 to 115 m); average porosity is 24%; permeabilities range from 200 to 1,200 md in lower energy bioturbated sandstones, and range from 2,000 to 10,000 md in higher energy sandstones. Kimmeridge Shale is the caprock over most of the field, but Upper Cretaceous marlstones provide the seal along some fault scarps where the Kimmeridge was removed during the Cenomanian.

The field is comprised of three folded, tilted blocks on the northern edge of the Witch Ground graben. Block I, a gently folded fault block dipping 5° northeast, has a common oil-water contact of 8,512 ft (2,594 m) subsea with Block II. Block II, a downthrown, northwest to southeast trending fault block lies southwest of Block I. Blocks I and II cover 7,149 productive acres (2,894 ha.) and have a gross reservoir column of 1,312 ft (399 m), from 7,200 ft (2,195 m) subsea to the oil-water contact. Block III (to the southwest) borders the Witch Ground graben and contains a small accumulation with a separate oil-water contact at 9,199 ft (2,804 m) subsea.

The P1 production well spudded October 10, 1976, established commercial production December 7, 1976, at more than 30,000 b/d, restricted by 5 1/2 in. (14 cm) tubing. The P7 well completed in April 1977 produced more than 50,000 b/d, restricted by 7 in. (17.8 cm) tubing.

Twenty-five wells have been drilled, four as water injectors to support a natural water drive of 250,000 b/d. Production is 280,000± b/d of 37° API, low sulfur oil and original recoverable reserves are estimated to be 618 million bbl.

Piper sand stratigraphy, reservoir performance, and development from a centrally located platform are discussed in detail.

INTRODUCTION

Piper field production began on December 7, 1976, when

[1]Manuscript recieved, May 29, 1979; accepted for publication, July 24, 1979.

[2]Occidental Petroleum (UK) Ltd., Aberdeen, Scotland, AB2 3TP.

The writer thanks the Occidental management, and partners in the consortium, Allied Chemical (North Sea) Ltd., Getty Oil Co., Thomson North Sea, Ltd., and British National Oil Corp., for permission to publish this paper. The writer also is thankful for the efforts of all persons working in the Occidental offices at Aberdeen, London, and Bakersfield who helped in the preparation of this manuscript.

It is especially noted that the writer received excellent cooperation from the Department of Energy, British National Oil Corp., other United Kingdom regulatory bodies, and partners of the Consortium who allowed maximum use of new data to improve subsequent development.

Article Identification Number:
0065-731X/80/M030-0008/$03.00/0.

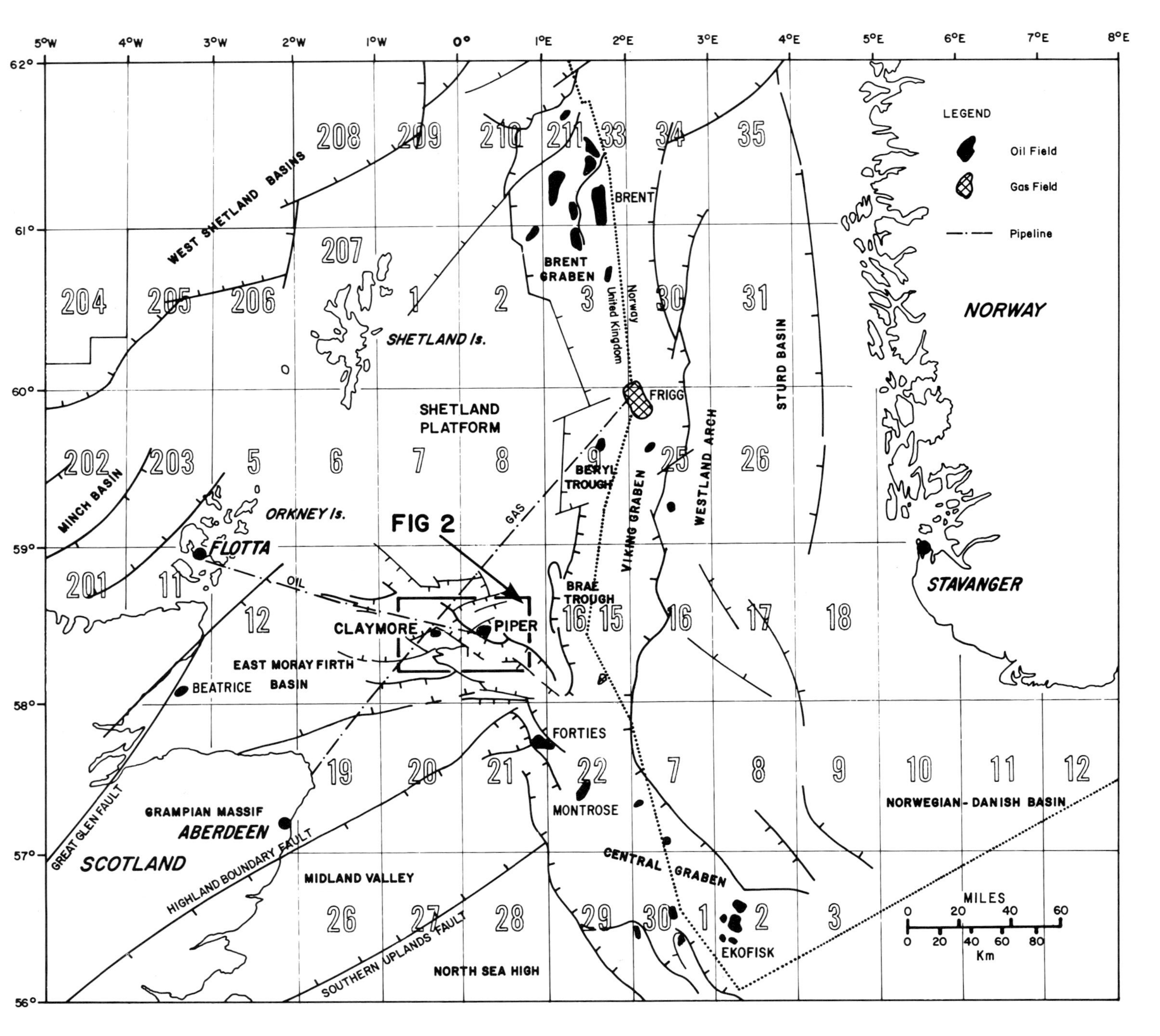

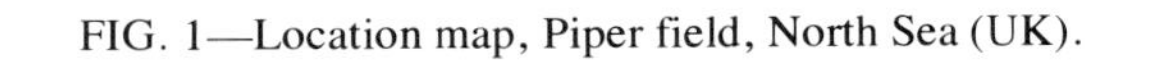
FIG. 1—Location map, Piper field, North Sea (UK).

the P1 well was placed on production at more than 30,000 b/d (rate restricted by 5 1/2 in. [14 cm] tubing) of 37° API, low sulfur oil. Production is from predominantly marine, Upper Jurassic sandstones. Current daily and cumulative production as of March 8, 1979 is summarized in Table 1.

Piper is located in UK block 15/17 in the eastern end of the Moray Firth basin and was discovered by drilling a seismically mapped structure in December 1972 (Figs. 1, 2). A combination of favorable geology and good reservoir parameters allows high production rates and large recoverable reserves to be fully developed from one platform. In June, 1975, a single steel platform was centrally located over the field in 474 ft (144 m) of water, secured with 24 piles extending 380 ft (116 m) beneath the seafloor and equipped with two drilling rigs, computers, and other equipment necessary to allow simultaneous drilling and production (Hulme, 1979; Fig. 3A).

In October, 1976, approval was granted by regulatory bodies of the British Government to commence drilling, and production well P1 was spudded October 10, 1976. Oil and gas liquids are shipped to Flotta Terminal in the Orkney Islands along a 130 mi (209 km), 30 in. (76 cm) submarine pipeline. Gas is shipped to the midcompressor platform O1 (MCP O1) of the Frigg gas pipeline through an 18 in. (46 cm), 30 mi (50 km) submarine pipeline (Fig. 3C).

Occidental of Britian, Inc., a subsidiary of Occidental Petroleum Corp. has a 36.5% interest and is operator for a group including Getty Oil (Britain) Ltd., 23.5%; Thomson North Sea Ltd., 20%; and Allied Chemical (North Sea) Ltd., 20%.

Williams et al, 1975, discussed in detail the exploration history leading to discovery of Piper field by the Oxy Group and the appraisal drilling which confirmed discovery of a major field.

GEOLOGY

General, Geographic and Tectonic Setting

Piper field is on a shelf in the eastern Moray Firth basin, south of the East Shetland platform, north of the Witch Ground graben and west of the Fladen Ground spur, which separates the Piper shelf from the Viking graben. The productive area covers 7,437 acres (30.1 sq km) and is comprised of three tilted, folded fault blocks parallel with the Witch Ground graben (Figs. 2, 3B, 4). The largest block, Block I, covers 5,115 productive acres, extends through fault D, and is separated from Block II along fault C, but has a common oil-water contact of 8,512 ft (2,594 m) subsea. Block I was discovered with the 15/17-1A well in December 1972 and appraised with the 15/17-2, 4, 5, and 15/12-1 wells in 1973. Block II contains 2,034 productive acres (823 ha.) and was appraised by the 15/17-3 and 15/17-6 wells in 1973. Block III, southwest of fault A, was discovered with the 15/17-7 well in January 1974 and contains 288 productive acres (117 ha.). Major movement on fault A (post-Upper Jurassic) resulted in a separate oil-water contact in Block III of 9,199 ft (2,804 m) subsea. Equalization of the common oil-water contact between Blocks I and II is through complex faulting between and north of wells P15 and P25 where there is thought to be sand-on-sand contact and/or

Table 1. Summary of Piper Production.

	DAILY	CUMULATIVE
OIL	280,000± b/d	175,954,566 bbl.
GAS	105,000 mcf/d	65,706,015 mcf
GAS LIQUIDS	4,000 b/d	507,012 bbl.

Table 2. Piper Field Productive Area and Volume.

BLOCK	AREA	VOLUME
	ACRES (SQ KM)	ACRE-FEET
I (west of D fault)	3,860 (15.6)	590,410
I (north of E fault)	98 (0.4)	4,672
I (east of D fault)	1,157 (4.7)	209.648
	5,115 (20.7)	804,730
II	1,542 (6.2)	185,287
II (P25 area)	492 (2.0)	32,829
	2,034 (8.2)	218,116
III	288 (1.2)	15,984
TOTALS	7,437 (30.1)	1,038,830

along the E fault north of P18 where there is sand-on-sand contact from this point to where the E fault dies out to the northeast. Productive area and oil-bearing sandstone volumes are summarized in Table 2.

The common oil-water contact and large volume of oil-bearing sandstone in Blocks I and II have a special significance. Upper Cretaceous, Campanian marl onlaps the Piper Sandstone and forms the caprock along the C fault scarp in Block I. The Maastrichtian chalk is not a caprock in the Piper area and if it had onlapped Piper Sandstone, no oil would have been trapped in Blocks I and II.

Rifting

There is a wide acceptance of the North Sea grabens having been formed as a result of rifting (Kent, 1975; P.A. Ziegler, 1975; Kahle, 1974). However, W. H. Ziegler (1975), Kahle (1975), Meyerhoff and Meyerhoff (1974), and Beloussov (1974) provided good reasons for these grabens forming largely as a result of near-vertical displacement along faults in response to local stresses generated by more major changes within the earth's interior and crust. Observed fault displacement is largely vertical, faults are predominantly vertical or near vertical, and laterally die out abruptly to be replaced nearby with other faults often having the opposite throw and forming a relay pattern (Lowell, 1976). It is difficult to recognize lateral displacement because seismic data indicating the faults do not indicate lateral movement, and well data are too sparse to define lateral movement.

Illies (1977) discussed ancient and recent rifting in the Rhinegraben which shows a remarkable similarity to the simplified fault patterns for the Piper and Claymore area (Figs. 2, 4).

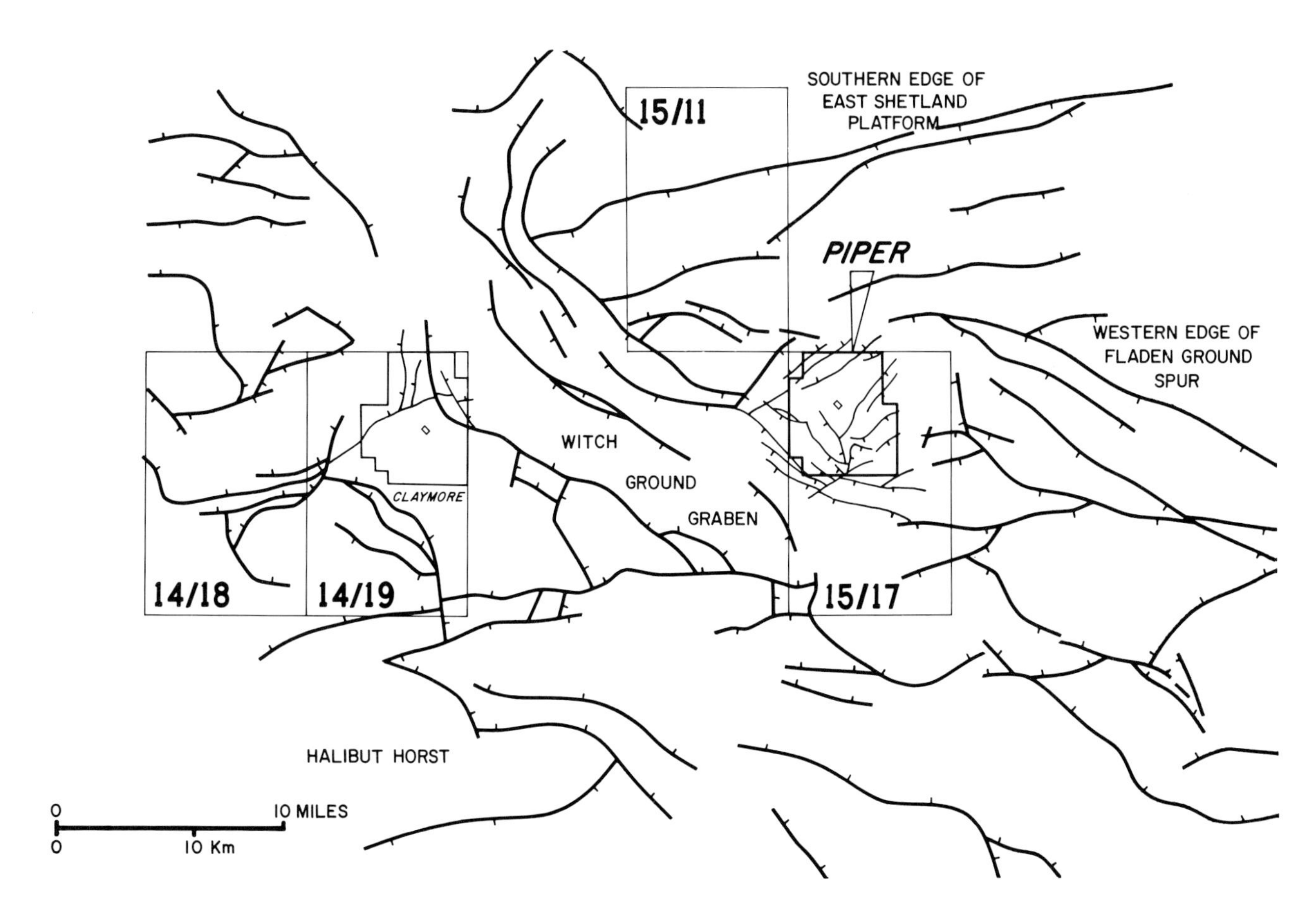

FIG. 2—Simplified base of Cretaceous fault pattern, Piper and Claymore area, North Sea (UK).

MESOZOIC HISTORY

General

Piper wells are drilled to about 140 ft (43 m) below the base of the Piper Sandstone. The only new data pertaining to the deeper formations since Williams et al, 1975 are 150 mi (242 km) of seismic data shot in 1977 for better delineation of the main faults and to aid in interpretation of more complexly faulted areas.

Figures 4 through 10 illustrate major structural elements and formations in Piper field and the onlap of Upper and Lower Cretaceous from the Witch Ground graben.

Permian

The North Sea and northwest Europe were part of the landmass until the Late Permian (Ramsbottom, 1978) when a shallow epicontinental gulf covering most of the North Sea basin area resulted in widespread evaporite deposition. A hot, arid climate resulted in thick salt deposits in the southern North Sea basin in the Zechstein formation, whereas at Piper field 100 ft (30 m) of interbedded dolomite and shale containing an assemblage of Late Permian palynomorphs was overlain by 300 ft (90 m) of massive anhydrite (Fig. 5A). Evaporites in the Piper area are indicative, in part, of a sabkha environment and lie unconformably on the Middle/Lower Carboniferous. The massive anhydrites are one of the best seismic reflectors in this area and provide the most consistent data for locating faults in the Piper field. This seismic reflector is not folded or bent anywhere over the Piper structure, except possibly southeast of the P25 well. Where the seismic quality is best, the massive anhydrite reflector is always seen in straight line segments, and changes in elevation are accomplished by numerous, small, complex fault displacements, and tilting of fault blocks. Overlying sediments are not offset by most of the small faults and commonly have less displacement along more significant faults. The anhydrites must have been too brittle to respond by bending under conditions which existed from Triassic through Early Cretaceous time.

Triassic

The Triassic of the North Sea and northwest Europe is predominantly nonmarine red sandstones, shales, and claystone (Brennand, 1975, Sellwood, 1978a).

At Piper, only 200 to 400 ft (61 to 122 m) of red shales (with no diagnostic fauna) are present and they thin over the structure (Williams et al, 1975). Thinning is interpreted to be caused by erosion on a structure initiated after deposition of the red shale and prior to deposition of Middle Jurassic sediments. The D fault may have formed at this time.

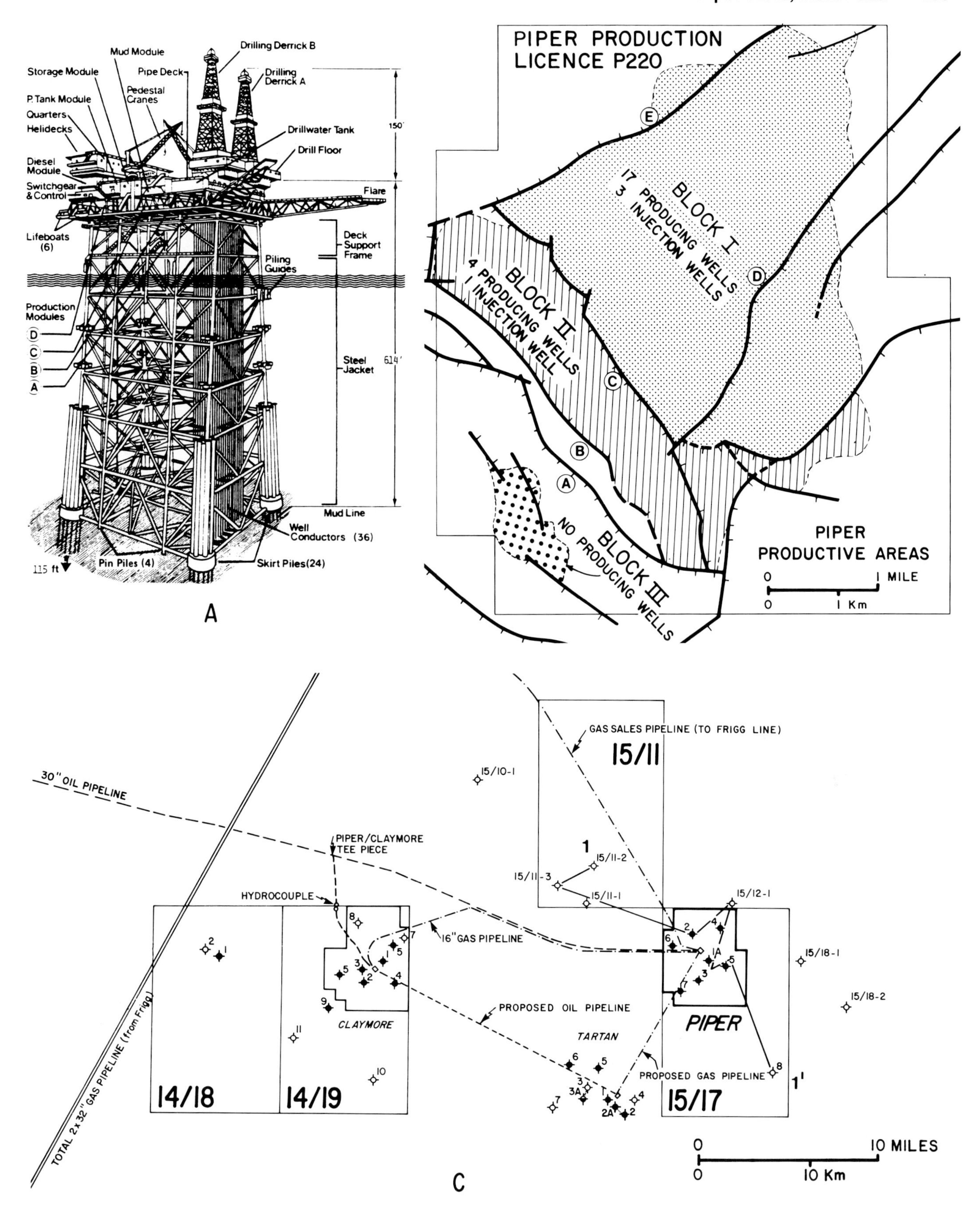

FIG. 3—**A:** schematic drawing, Piper platform; **B:** Piper productive areas; **C:** Oxy group blocks, pipeline system, location map for correlation section 1-1′.

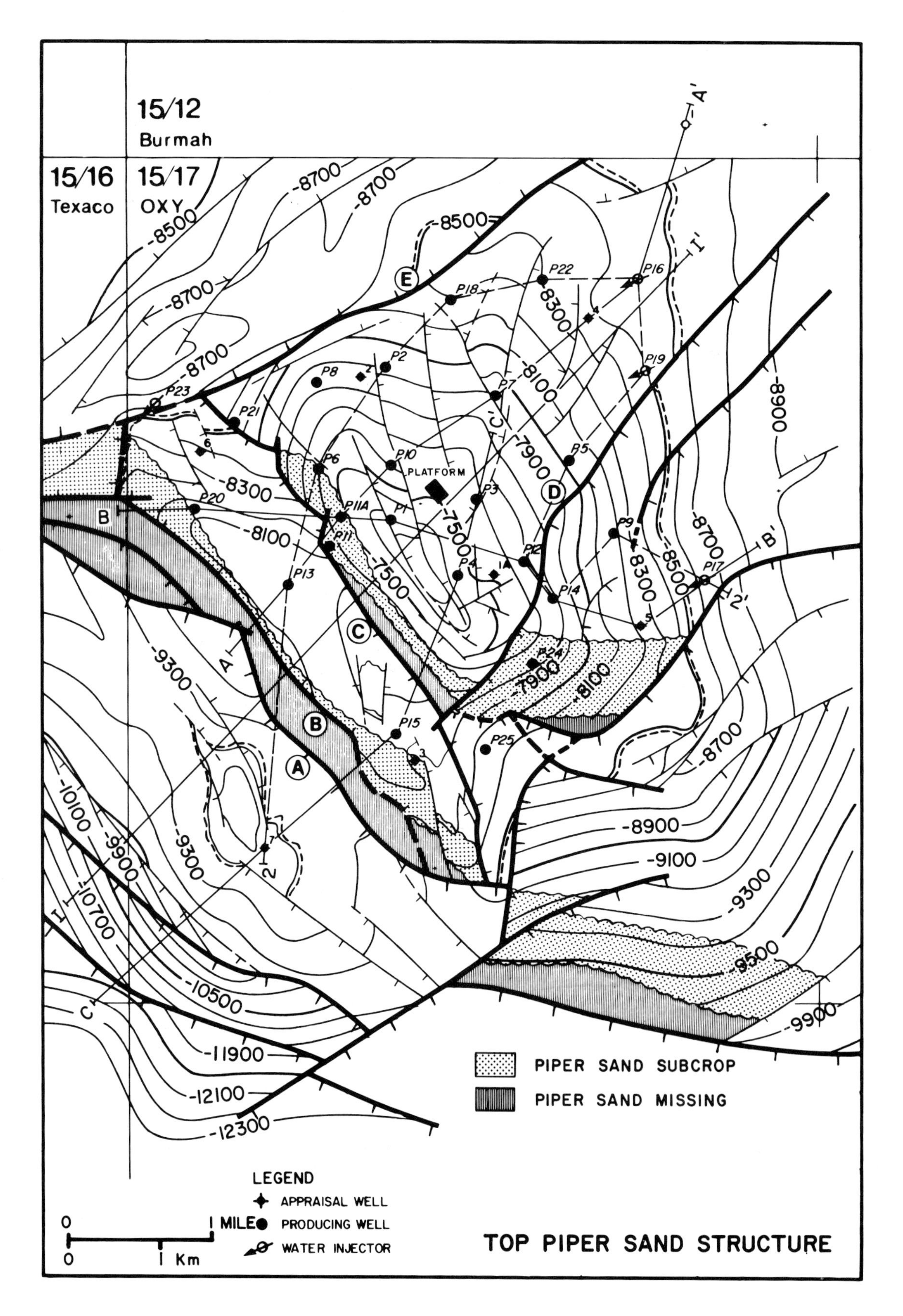

FIG. 4—Structure map top of Piper Sandstone showing location for correlation section 2-2, cross-sections A-A′, B-B′, C-C′, and seismic section I-I′

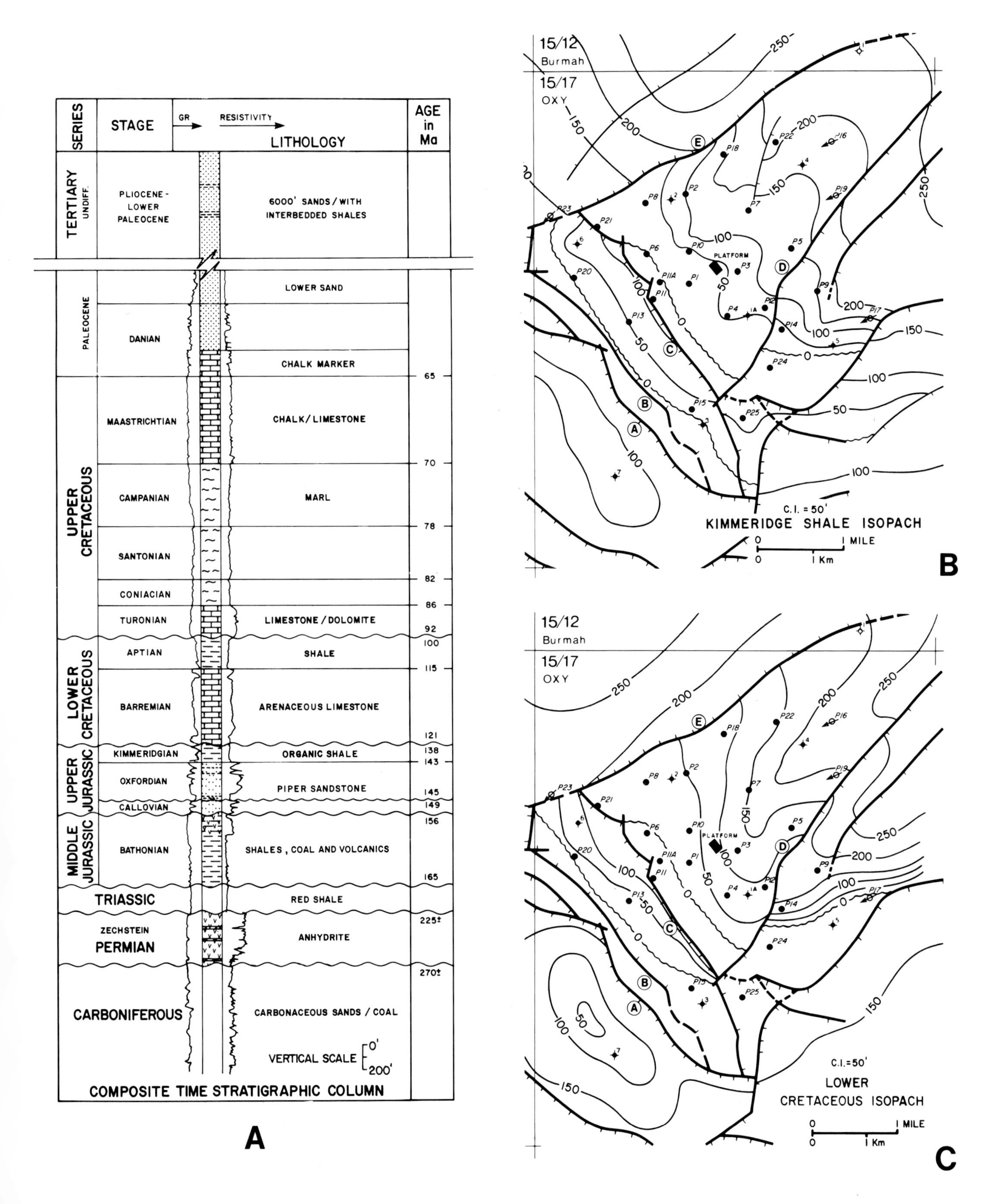

FIG. 5—**A:** Composite time stratigraphic column; **B:** Kimmeridge Shale isopach; **C:** Lower Cretaceous isopach.

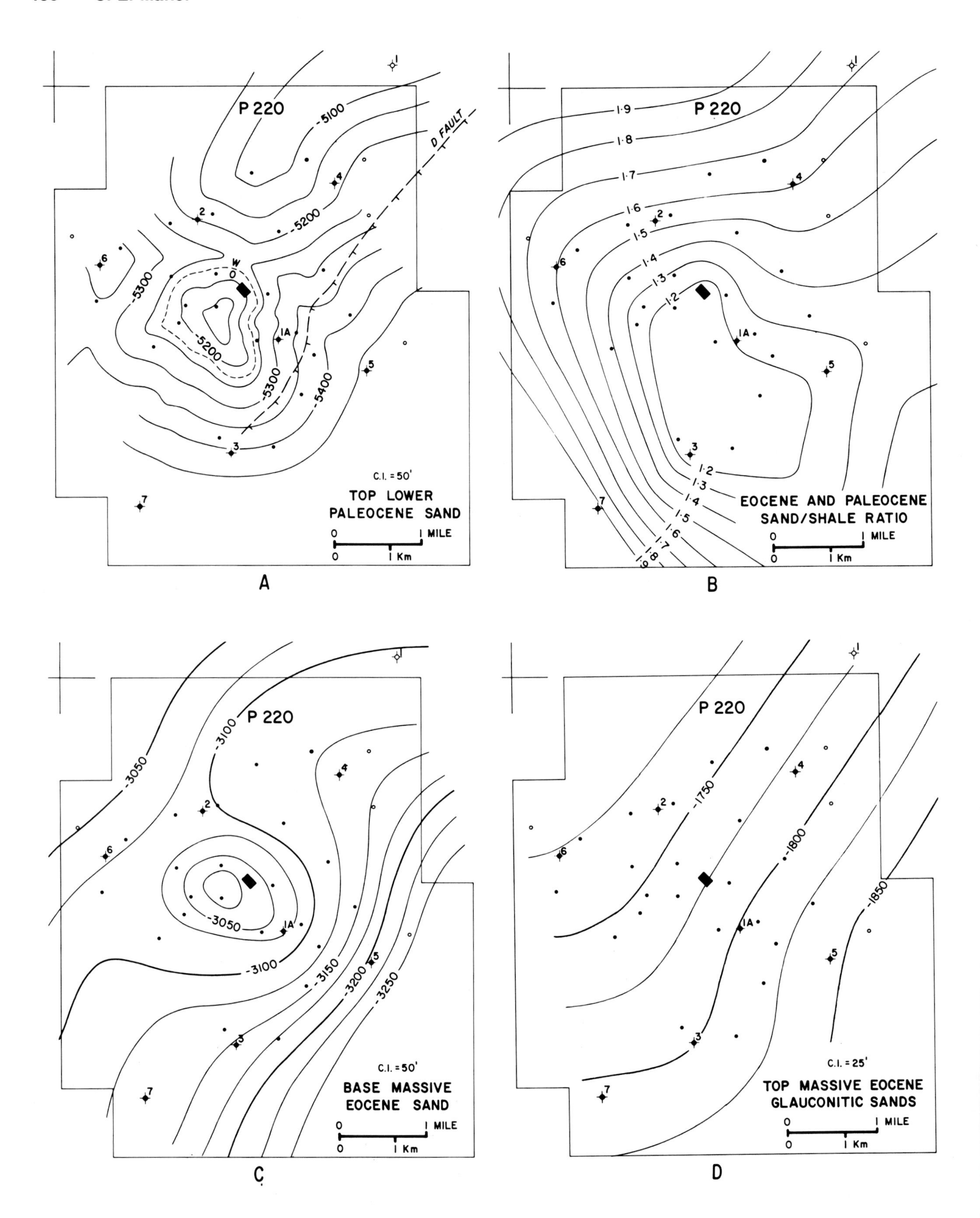

FIG. 6—**A:** Top of lower Paleocene sandstone; **B:** Eocene and Paleocene sand/shale ratio; **C:** base of massive Eocene sandstones; **D;** top of massive Eocene sandstone.

Middle Jurassic

Triassic red shale is overlain by a nonmarine sequence consisting of siltstone, shale, thin-bedded freshwater limestone, highly weathered volcanic sediments, coals interbedded with shale and sandstone, basalt, and coals interbedded with shales, weathered volcanic sediments, and sandstones. Tectonic activity at the end of the Middle Jurassic was accompanied by gentle folding, faulting, and tilting of fault blocks. Movement was renewed on fault D with 800 ft (244 m) of vertical displacement and tilting to the northwest on both the downthrown and upthrown blocks. Faults A, C, and major faults south of Piper were initiated at this time and were accompanied by minor northeasterly tilting of Blocks I, II, and III. Fault E may have formed at this time, but there is no evidence of movement until the post-Middle Jurassic (Fig. 4).

Erosion preceding deposition of Callovian removed the basalt along the D fault and in the central part of Block I in the area bounded by P3, P2, and P16. Most of the sediments above the basalt in Blocks II and III and along the C fault southwest of wells P1 and P4 were eroded away at the same time. Erosion may have resulted from a thinner and/or less resistant material in this area. Another explanation is for the central part of Block I to be in a surf or wave zone while the area southwest of P1 and 4, and Blocks II and III, were subaerial and subjected to a much slower rate of erosion.

Upper Jurassic

Mild tectonic activity and erosion at the end of the Middle Jurassic preceded deposition of a lower deltaic plain interdistributary sequence of Callovian age coals, interbedded silt, sand, and clays on an upper Bathonian unconformity surface. This was followed by a widespread marine transgression which deposited upper Oxfordian, micaceous, sandy, silty shale, containing abundant pyrite, plant fragments, coal clasts, and a well preserved molluscan fauna over the whole of the Piper area. Sedimentation in a shallow marine, shelf environment continued with gradual building up of offshore bars, interrupted by several minor transgressions. After the more significant transgressions, thin silty shales 2 to 10 ft thick (0.5 to 3 m) were deposited.

These grade to silts and silty sands over what were the highest structural positions during the Oxfordian. The transgressions are interpreted as being associated with increased tectonic activity which caused the Piper area to subside more rapidly than sediment supply and/or temporarily diverted the sediment source. There was intermittent movement on the D fault and gentle northeast tilting of all three Piper blocks during the Callovian-Oxfordian resulting in thinning of Piper sands toward faults C, A, and major faults to the south. More than 400 ft (122 m) of sand was deposited in downdip and structurally low areas along the northeast side of the field, in the 15/11 block and the 15/17-8A area.

The upper Oxfordian sandstones in the P10 well are overlain by a 6-ft (2-m) interval of highly bioturbated, poorly sorted sandstones containing abundant shell material and glauconite which grade upwards into dark gray to black silty shales of probable early Kimmeridgian age. This highly bioturbated interval probably formed during a transgression and may represent a long period of essentially no deposition at higher structural positions in Piper. There was also minor erosion and reworking of Piper sands from structurally high areas and redeposition downdip.

Kimmeridge silt and shale represent the last major transgression over Piper in the Jurassic. The Kimmeridge Shale thins rapidly to the southwest and the thinnest full section is near the C fault, indicating this to be the highest structural feature at Piper in the Late Jurassic (Fig. 5B).

Dipmeter data often show a moderate angular unconformity between Piper sandstones and overlying Kimmeridge Shale. Detailed stratigraphy of the Piper sandstones is discussed in a later section.

Lower Cretaceous

Barremian limestone and a thin Aptian-Albian shale were deposited subparallel with Kimmeridge Shale. The Lower Cretaceous thins toward the southwest in Blocks I and II. There is only one well for control in Block III, but the Lower Cretaceous (from seismic data) appears to be thinnest over the present structural high (Fig. 5C).

A thin Aptian shale was deposited over all the Piper field except along the northeast side of fault C.

The Lower Cretaceous thins dramatically from the Witch Ground graben on to the Piper structure. (Figs. 9, 10).

Upper Cretaceous

In the Witch Ground graben the 15/17-8A well penetrated 30 ft (9 m) of Turonian shale overlain by 140 ft (43 m) of hard, dense Turonian limestone. A moderate angular unconformity was discovered between the shale and the underlying Albian marls. This dates the last major fault movement and tilting of the Piper blocks as Cenomanian. Erosion of Lower Cretaceous and Upper Jurassic sediments along parts of the upthrown side of fault C, on the horst block between faults A and B, and along the major faults bordering the north side of the Witch Ground graben followed this movement (Figs. 2, 4). The Turonian onlaps Bock III to fault A and onlaps to just south of P25 on the east side of the field. Successively younger Upper Cretaceous stages onlap the Piper structure with 185 ft (56 m) of Campanian lying on truncated Piper sandstone in P6, and 180 ft (55 m) of Campanian lying on Callovian coal in P11 (Figs. 4, 7-9).

By late Campanian time, fault movement ceased and previous displacement had been leveled by sediment fill. Maastrichtian chalk represents uniform sedimentation with thickness ranging from 507 ft (155 m) in P1 to 618 ft (188 m) in 15/17-8A in the Witch Ground graben. Only occasional small faults appear to displace the Top Maastrichtian.

TERTIARY HISTORY

A change from chalk to predominantly sandstone and interbedded chalks which occurred in the Danian may be asso-

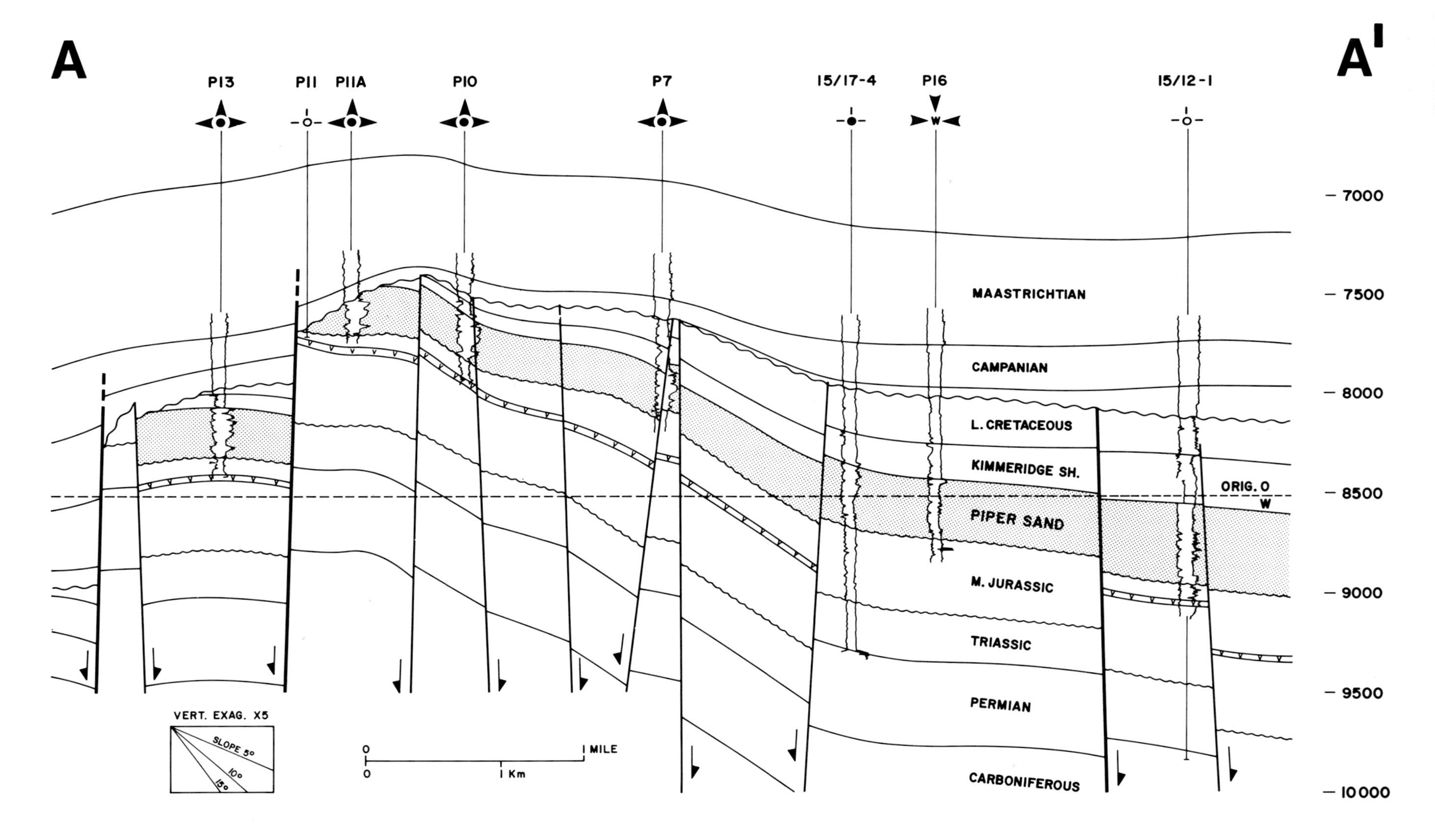

FIG. 7—Structural cross-section A-A'. See Figure 4 for location.

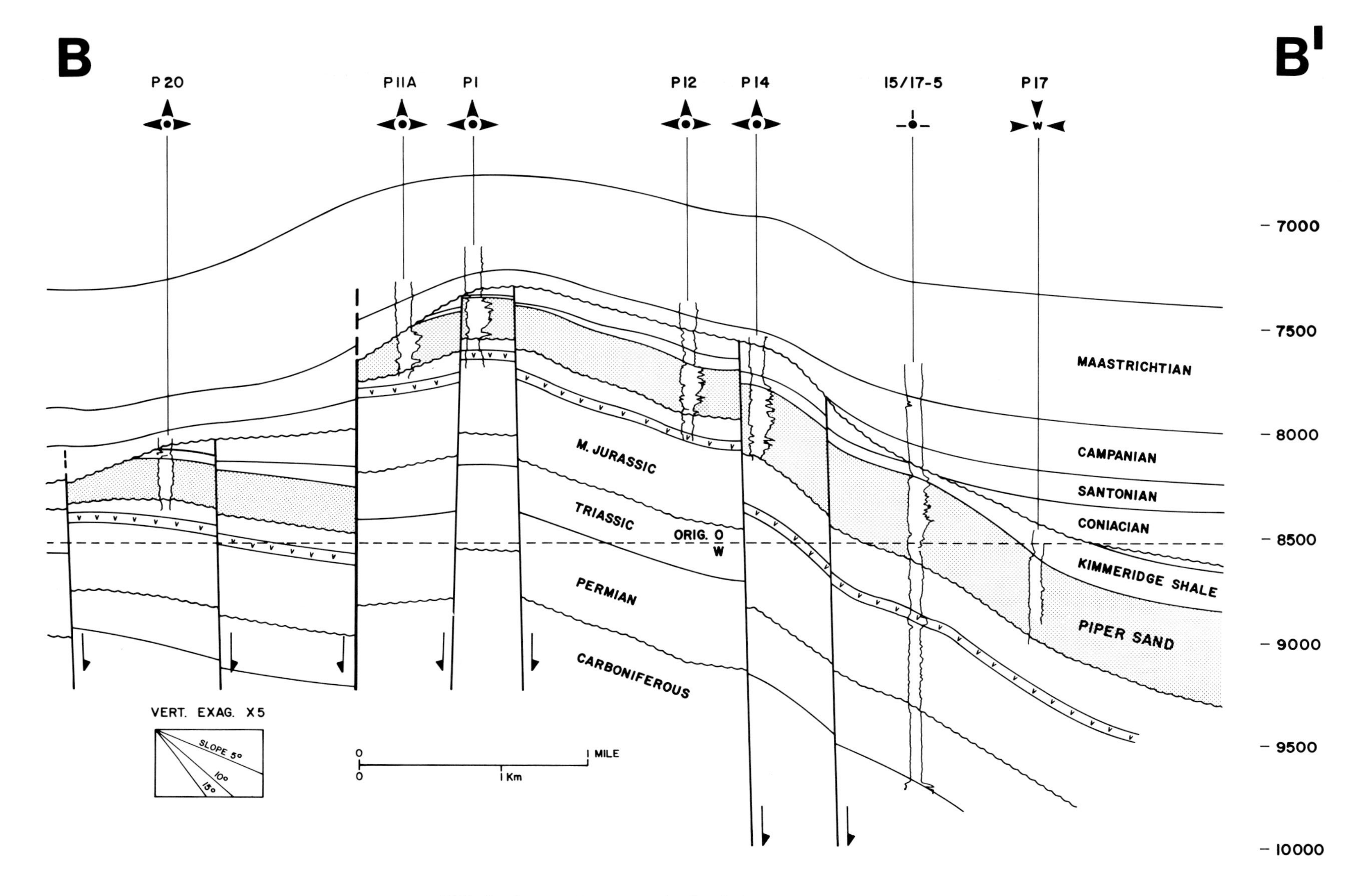

FIG. 8—Structural cross-section B-B'. See Figure 4 for location.

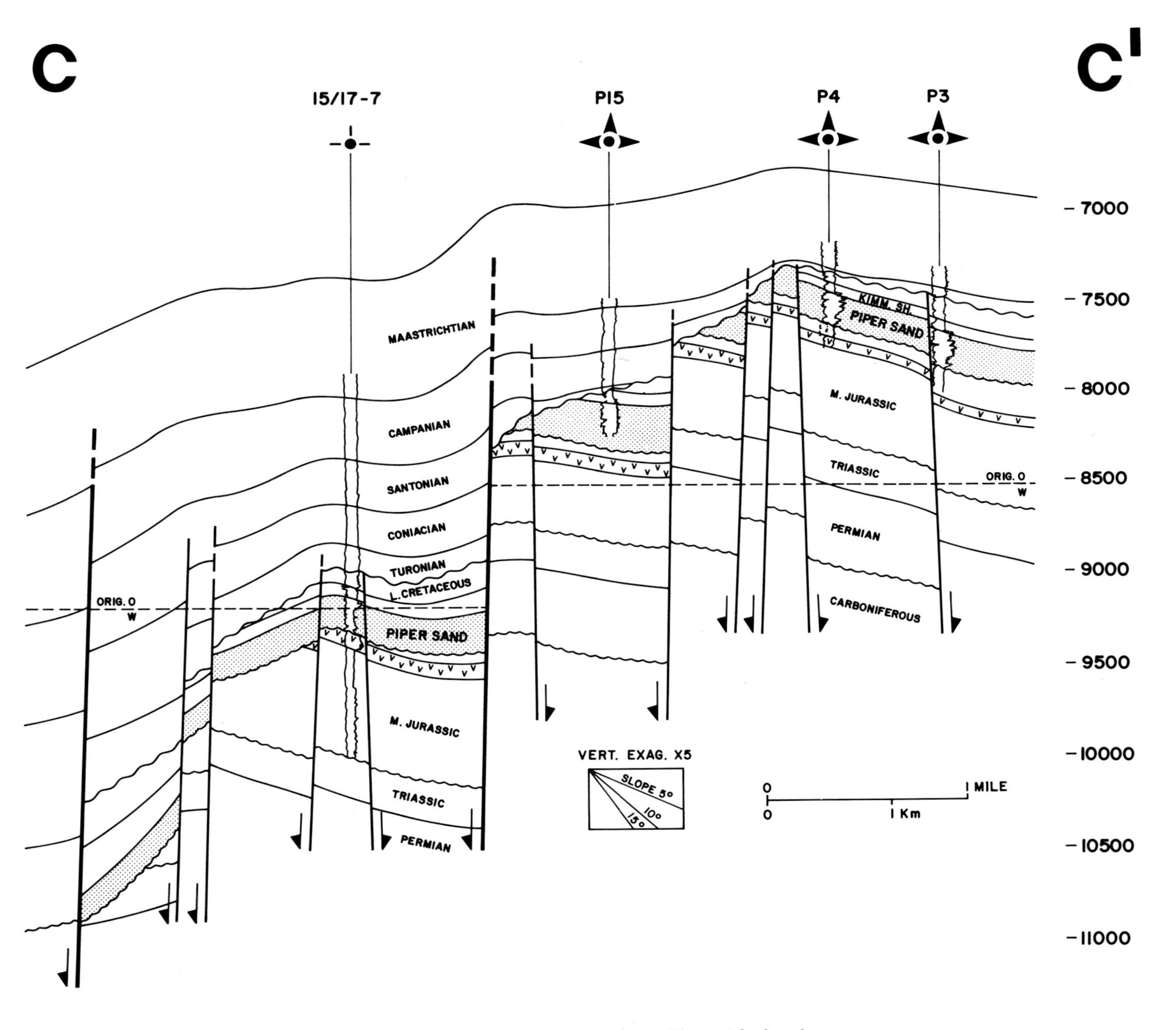

FIG. 9—Structural cross-section C-C'. See Figure 4 for location.

ciated with rifting. Individual chalk beds vary abruptly in thickness and do not correlate well at Piper. These massive chalks and sandstones grade upwards into thin limestones interbedded with sandstones and some thin shales. This interbedded sequence of discontinuous limestones, sandstones and shales makes it very difficult to make a reliable seismic pick for top of the Cretaceous over Piper. The Danian is overlain by 6,500 ft (1,980 m) of Paleocene, Eocene, and Oligocene interbedded sandstones and shales. There is 100 ft (30 m) of closure on the lower Paleocene sandstone which is overlain by a thick shale containing some swelling clays.

The shale forms a caprock for a small accumulation of heavy oil covering 419 acres (170 ha.) and containing 8,770 acre feet of oil sand (Fig. 6A). This small reservoir has been tested and found 8.7° API oil. Foreset deposits on a delta lobe to the northeast limit the closure and cause the lower Paleocene shale to thin from 647 ft (197 m) at 15/17-1A to 103 ft (31 m) at 15/12-1.

A sand/shale ratio map from the top of the Eocene to the base Tertiary shows less sand over the crest of Piper structure and is thought to reflect bypassing of sands deposited under deep-water conditions during the Paleocene (Fig. 6B).

The last closure over Piper is at the base of massive Eocene sandstones (Fig. 6C) and results from differential compaction of underlying shales. These sands thin from 550 ft (168 m) on the northwest to 300 ft (91 m) on the southeast of Piper and completely mask the underlying structure (Fig. 6D).

STRATIGRAPHY OF THE PIPER SANDS

Stratigraphy has been worked out with the use of log correlations, supplemented by nearly full core sections of the Piper Sandstone interval in the 15/17-4, 15/17-5, 15/17-6, 15/17-7 and P10 wells, additional core in 15/17-2, and sidewall samples. A total of 1,335 ft (407 m) of core was cut and 892 ft(272 m) was recovered. Macro- and micropaleontology of core material and sidewall samples (together with an excellent, detailed description of cores, supplemented by core photographs, petrographic and SEM studies, and porosity/permeability measurements) are extrapolated on the basis of log correlations.

In discussing thickness of Piper sandstone members and Kimmeridge Shale, minimum thickness is taken as the thinnest complete section. Where later erosion has partly or completely removed a member, the value is disregarded for this discussion (Table 3).

CALLOVIAN

The Callovian is interpreted to be a lower deltaic plain, interdistributary sequence deposited in 3 cycles which correlate from Claymore through the 15/11 blocks and into Piper field (Figs. 3C, 11-14). Log correlations with the 15/17-8A well are tentative, but east of the D fault are considered good.

Callovian deposition was on a surface of very low relief where small fault displacements and local topography greatly influenced sedimentation.

There is considerable variation at Piper with respect to the P2 well with an abrupt increase in thickness and percentage of sand on the downthrown side of the D fault. Topographic influence is clearly seen in the thinning of the Callovian from 15/11-2 to 15/11-3 (Fig. 11), and the influence of fault movement in wells north of Claymore and east of the D fault at Piper were thicker and possibly more marine deposits are preserved on the downthrown sides of faults (Figs. 8, 12).

Each cycle is composed of a prograding deltaic sand unit 20 to 60 ft (6 to 18 m) thick overlain by highly bioturbated silts and silty shales. Coals commonly are found in the lowermost cycle at Piper and the 15/11 block, whereas coals commonly are present in the second cycle at Claymore.

In the 15/17 well, the Callovian overlies Bathonian shales and clays containing rootlets with textures typical of *Equisetites* and in which macrofossils and bioturbation are absent. In P10, Callovian sandstones lie directly on highly weathered volcanics.

Callovian lithology at Piper consists of coal, silty, bioturbated shales, planar and trough cross-bedded, rippled, medium to coarse, poor to well sorted sandstones with occasional bioturbated surfaces west of fault D, and medium to very coarse, cross-bedded, moderate to poorly sorted sandstones with numerous bioturbated surfaces overlain by coarser grained sandstones east of the D fault. There are abundant plant and coal fragments and some carbonate cement and pyrite.

Body fossils are absent from the sandstones, but the silts and silty shales contain a sparse fauna including heterodonts and oyster fragments. There is only 10 ft (3 m) of Callovian sandstone in 15/17-4 and the occurance of *I depressa* to the exclusion of everything except *Procerithium* in underlying micaceous, glauconitic silt is considered highly suggestive of a brackish water or freshwater influence (Palmer, 1977).

Isopachs of the Callovian members of Basal Coal in M, M, L, J2, and J1 are shown in Figures 15 and 16A.

Van Hinte (1976a) showed a 7 m.y. time span for the Callovian. With a maximum of 100 ft (30 m) of Callovian sediments, average sedimentation rate at Piper was only 4.4 μm/yr.

OXFORDIAN

Callovian sandstones were cored in 15/17-5 and P10 and the upper surfaces were highly bioturbated. In P10 the upper surface of a 2-ft (0.6 m) bed of very coarse, poorly sorted sand lies on a scour surface. The top surface of this 2-ft bed (0.6 m) is burrowed and overlain by a few inches of medium to pebbly, bioturbated sandstone containing abundant belemnite fragments and large pelecypod shells. This is overlain by a bioturbated micaceous, sandy silt (I shale) containing wood fragments and a well preserved molluscan fauna, gastropods, bivalves, ammonites and belmnites. The molluscan fauna indicate marine conditions on a stable, well oxygenated seafloor. The lower half of the I shale contains ammonite forms which correspond to the middle zone of upper Oxfordian, and the upper half contains forms indicative of the upper part of upper Oxfordian. This indicates a hiatus of several million years between the end of Callovian and deposition of I shale.

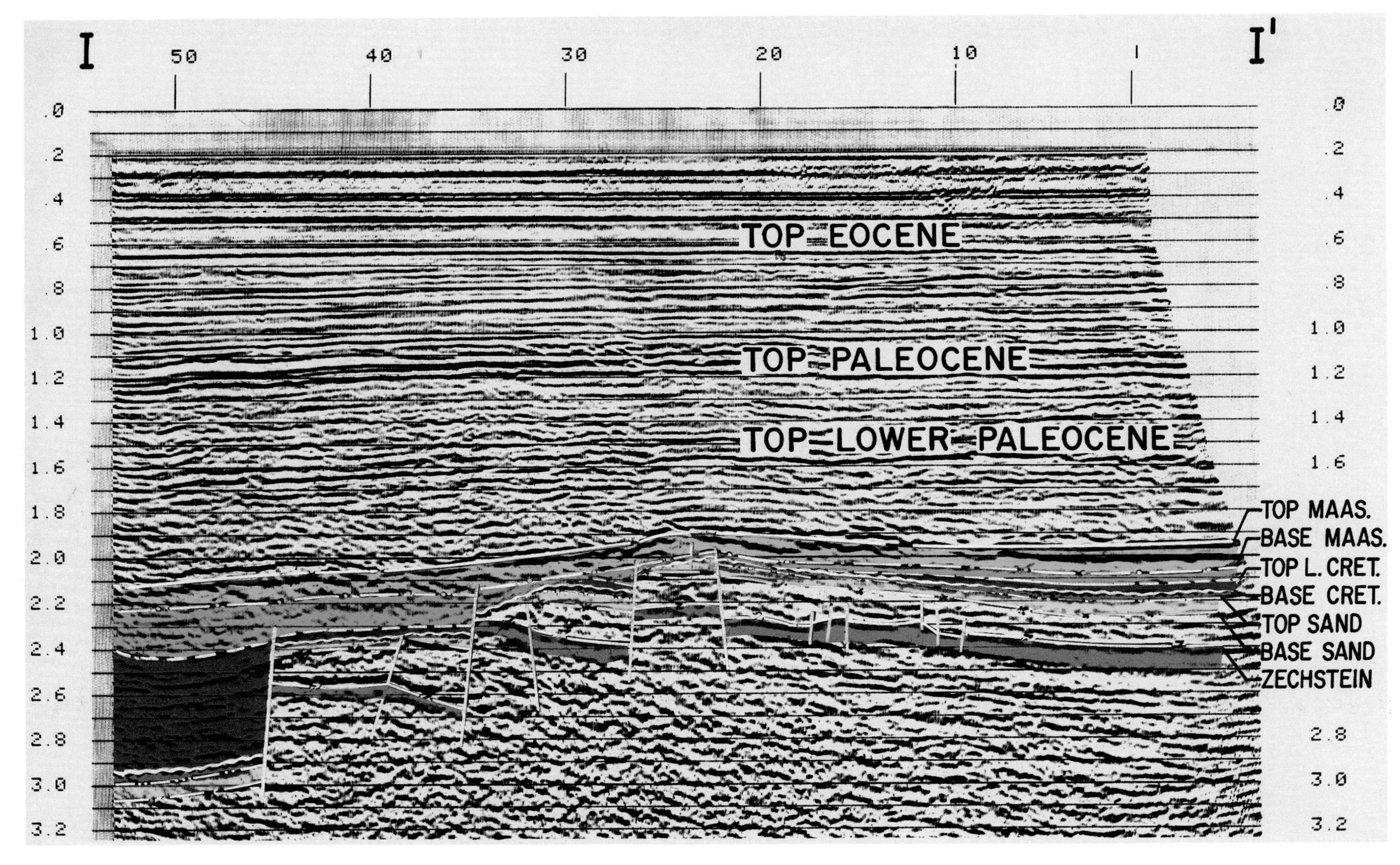

FIG. 10—Seismic section I-I′ from Witch Ground graben through Piper field. See Figure 4 for location.

Table 3. True Vertical Thickness (Ft.)

WELL NO.	LOWER CRET.	KIMM. SHALE	UNIT 4	A	B	C	D	E	F	G	H	I	J1	J2	L	M	TOTAL
P01	18	9	0	0	0	0	10	30	25	24	43	22	0	24	6	0	184
P02	49	110	3	0	0	0	31	30	27	30	54	36	8	34	22	16	291
P03	0	65	7	0	0	0	26	37	36	24	21	10	8	18	4	0	191
P04	74	58	0	0	0	0	0	28	36	18	30	12	6	22	8	0	160
P05	111	130	4	0	0	0	35	59	57	20	31	8	30	29	14	23	310
P06	0	0	0	0	0	0	13	32	21	27	45	27	8	50	4	8	235
P07	150	126	9	0	0	0	17	40	10	21	27	14	10	20	9	22	199
P08	21	40	10	0	0	0	22	22	16	15	32	16	6	20	5	6	170
P09	179	145	0	0	18	35	47	70	81	14	25	7	52	50	16	10	425
P10	38	72	0	0	0	0	21	35	24	28	42	27	2	22	8	0	209
P11	151	0	0	0	0	0	0	0	0	0	0	0	0	0	9	11	20
P11A	0	0	0	0	0	0	0	10	21	27	45	24	6	34	11	11	189
P12	94	83	0	0	0	0	6	43	32	16	28	5	16	26	20	27	219
P13	31	62	9	0	0	0	24	36	26	33	43	26	14	37	6	4	258
P14	144	46	0	0	0	31	47	56	66	21	22	8	34	42	20	0	347
P15	0	42	16	0	0	0	26	29	30	25	49	16	3	16	7	10	227
P16	225	168	0	0	30	41	42	53	51	13	18	6	7	16	12	7	296
P17	0	175	15	91	59	24	22	44	49	9	26	8	64	39	10	5	465
P18	81	152	13	0	0	0	26	33	24	24	62	23	5	21	14	6	251
P19	0	172	0	0	0	16	52	36	47	12	19	5	4	11	8	8	218
P20	43	15	13	0	0	15	27	22	9	21	36	23	5	20	6	6	203
P21	25	11	34	0	0	0	20	25	16	25	12	20	28	47	8	14	249
P22	134	205	3	0	0	29	42	27	56	20	39	5	22	39	11	10	303
P23	208	90	208	0	0	8	16	25	21	24	29	20	0	0	14	13	378
P24	0	0	0	0	0	0	0	21	56	12	28	7	34	48	0	0	206
P25	0	60	22	0	0	0	32	49	52	25	35	6	15	67	10	35	348
15/17-1A	0	49	0	0	0	0	7	26	40	29	22	7	12	26	6	10	185
15/17-2	53	69	8	0	0	0	29	30	19	28	50	26	7	46	10	14	267
15/17-3	0	0	0	0	0	0	0	0	11	28	29	17	7	15	10	0	117
15/17-4	218	154	2	0	0	29	45	54	48	18	23	5	14	13	12	11	273
15/17-5	0	50	7	0	21	36	36	47	64	15	24	13	52	32	16	10	373
15/17-6	75	73	32	0	0	16	12	20	21	37	36	23	6	32	22	10	267
15/17-7	79	65	0	0	0	17	14	19	22	22	36	16	6	20	7	0	179
15/12-1	192	232	0	62	91	28	46	35	62	15	20	6	0	10	14	6	395

The I shale, 5 to 36 ft (1.5 to 11 m) thick (Fig. 16B), represents the most significant marine transgression over Piper in Oxfordian time, and grades upwards from silt, with plant fragments, coal clasts, pyrite, and occasional sandy lenses (15/17-5) into the fine grained, moderately well sorted, bioturbated H sandstone containing carbonaceous laminae, shell debris, and coal clasts. The H sandstone ranges in thickness from 12 to 62 ft (3.7 to 17 m) and is shown in Figure 16C. Correlations across the D fault indicate no movement during the deposition of the I shale and the H sandstone. Bioturbation is continuous from the base of the I shale through the H sandstone, and very few bedding surfaces were preserved until the H sand body had built into a higher energy zone where bioturbated scour surfaces occur.

These surfaces are overlain by very coarse sandstone which infills burrows. In wells which have not been cored, very coarse sandstones at the top of the H sandstone are recognized by very high resistivity readings corresponding to high permeabilities and low water saturations.

Isopach maps of the I shale and H sandstone indicate that the H sandstone built as a classical offshore bar under moderate energy conditions, and alignment of the bar suggests, influence of the D fault although it was not active during this time.

A minor transgression at the end of the deposition of the H sandstone was followed by deposition of G silt and shale west of the H bar (Fig. 16D). This suggests a continued influence of the D fault and a predominant current direction from the east.

The G sandstone is 9 to 37 ft (2.7 to 11 m) thick and was deposited on the G silt, G shale, and H sandstone (Fig. 17A). Alignment of the isopach indicates the C fault to be active during deposition. It is fine to medium grained, moderately well sorted, bioturbated with plant fragments, coals clasts, carbonaceous streaks, and pyrite. There is essentially no difference between the H and G sandstones except for coarser sand at the top of H. There are no recognizable body fossils in these sandstones but burrows incude *nodosa, sueivica, planolites* and *chondrites*. Like the H sandstone, G sandstone is an offshore bar, and alignment of the ispach of G shows the influence of

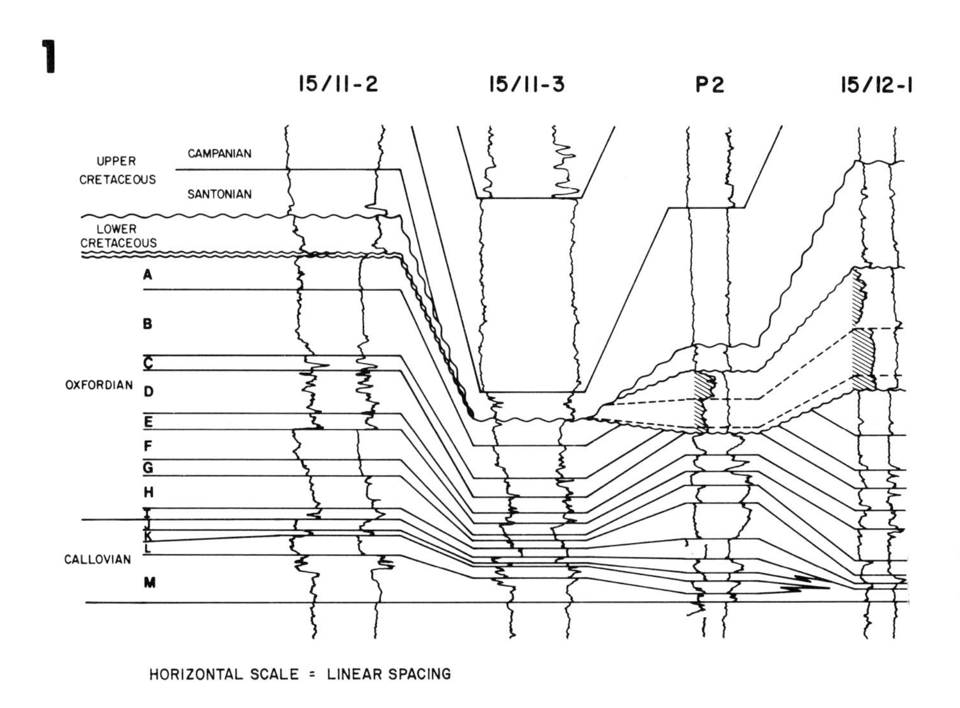

FIG. 11—Correlation section 1-1′ from 15/11 block through Piper field to

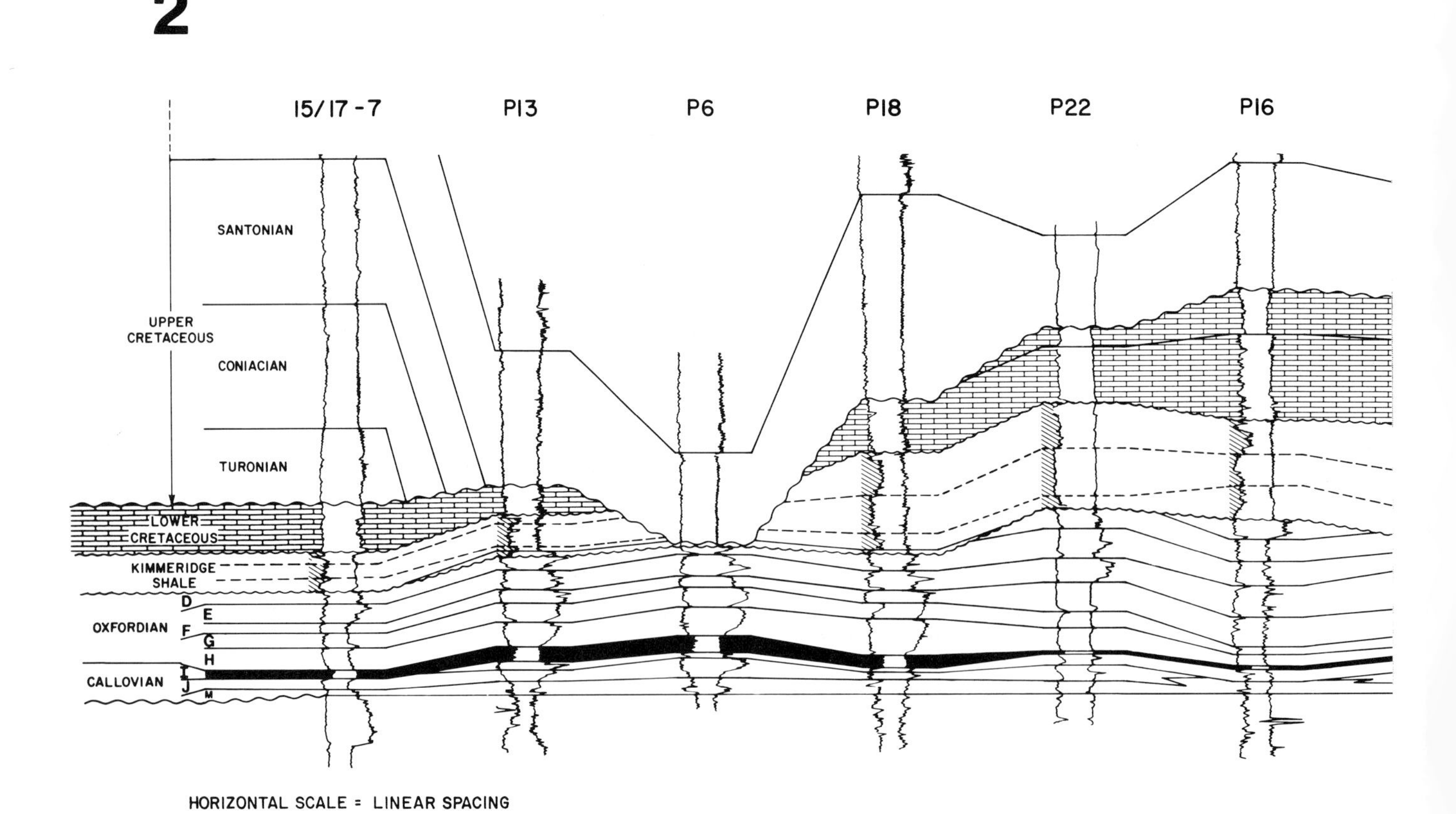

FIG. 12—Correlation section 2-2′ through Piper field. See Figure 4 for location.

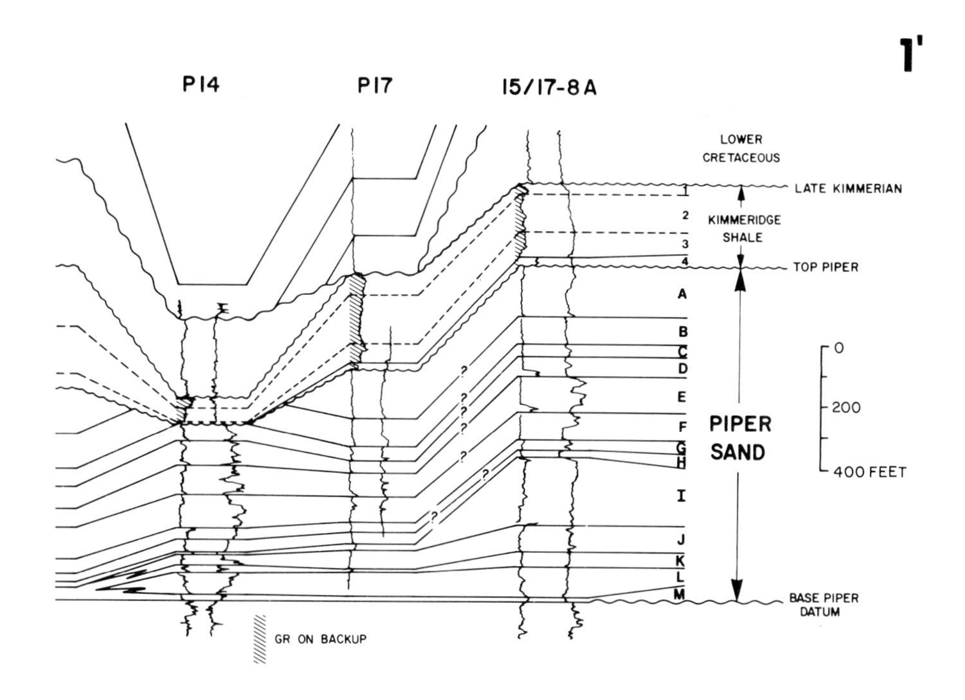

the 15/17-8A well in the Witch Ground graben. See Figure 3C for location.

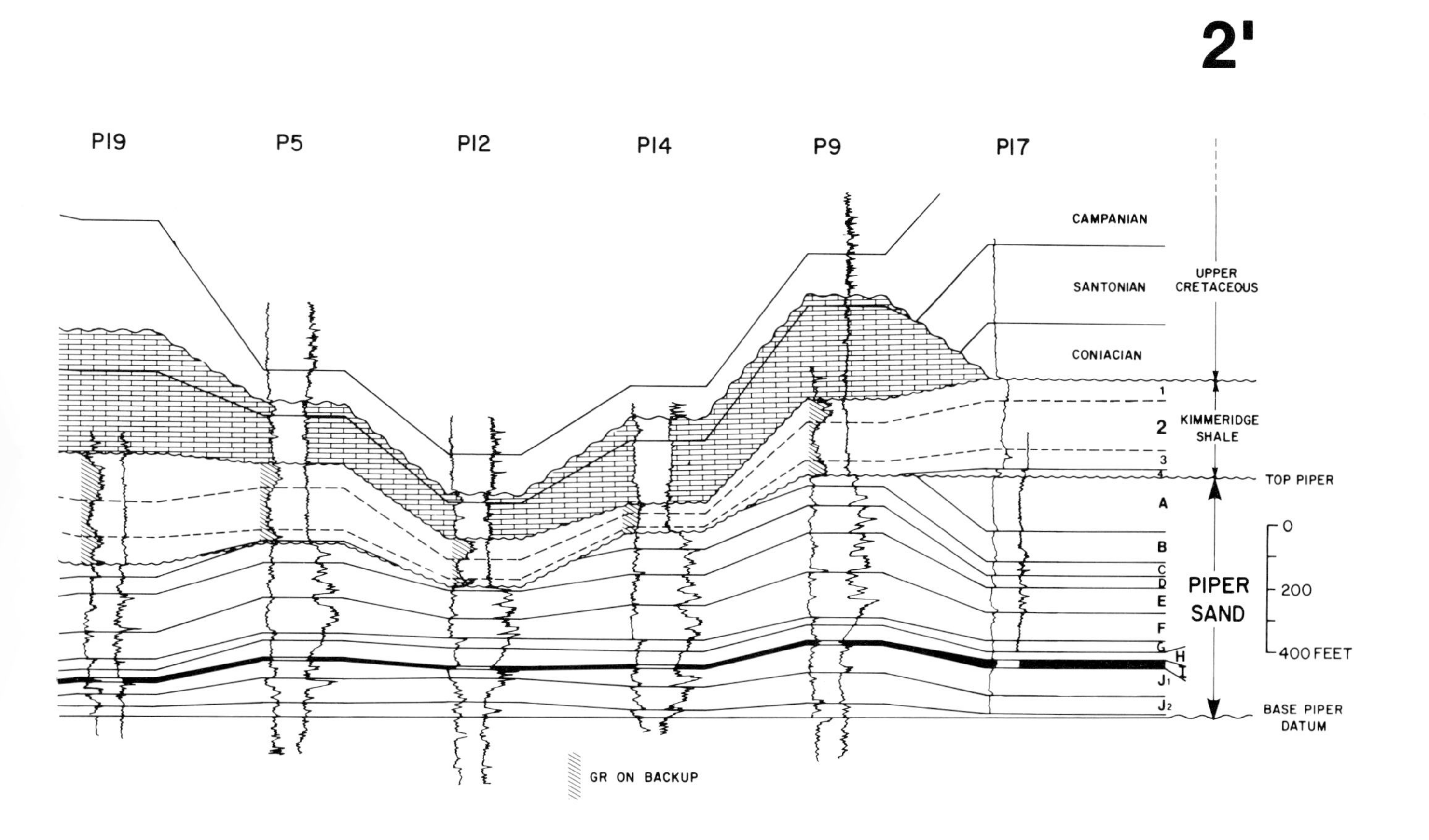

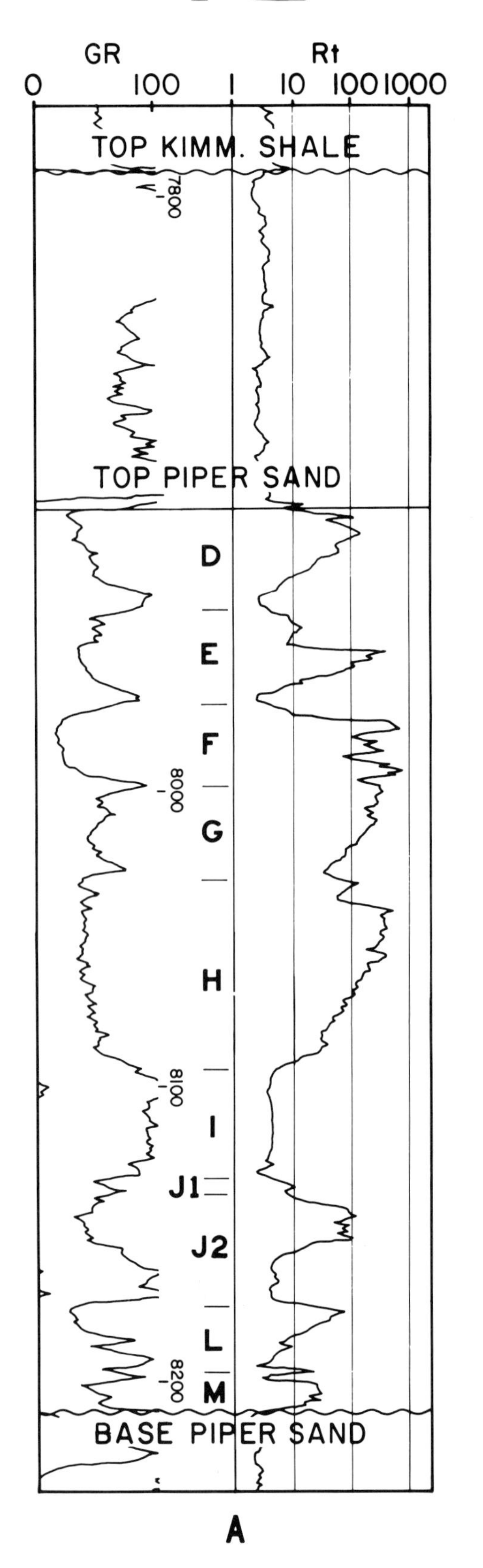

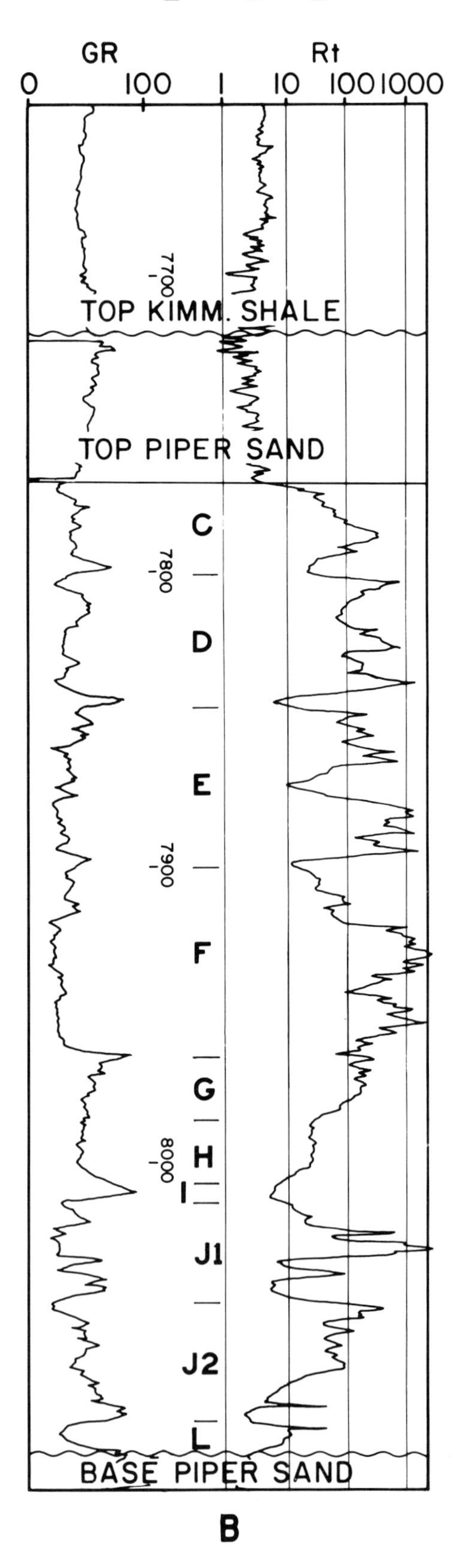

FIG. 13—**A:** Type R_t log for west side of Piper field. **B:** Type R_t log for east side of Piper field.

the C and D faults.

The G sandstone is overlain by 9 to 81 ft (2.7 to 25 m), medium to very coarse, moderate to poorly sorted, cross laminated F sandstone (Fig. 17 B). In the F sandstone, numerous scour surfaces and abrupt grain size changes occur from one to a few centimeters thick. Bioturbation is essentially absent in this sandstone, and it is coarsest in 15/17-7 and 15/17-6, decreasing in grain size towards the east, where the grain size is predominalty medium and where there also are fewer cross laminations.

In 15/17-7 the F sandstone is below the oil-water contact for Block III, contains a considerable amount of pyrite, and is a remarkably, clean, white, quartz sandstone.

The base of the F sandstone is marked by a very thin, high-gamma-ray sandstone (40 to 100 API units) which has little or no effect on porosity measurements, and also gives a sharp definitive reading on the thermal decay time (TDT) log.

Correlations across the D fault show that it was active during deposition of the F sandstone. Tectonic activity which led to the reviewed movement on the D fault was accompanied by a marked increase in depositional energy over Piper field.

The F sandstone is interpreted as an upper shoreface and beach deposit (Elliot, 1978; Reading 1978; Visher, 1969; Glaister and Nelson, 1974). The thin radioactive sandstone at the base is considered to be a heavy mineral lag (Houston and Murphy, 1977), deposited as a result of increased shear stress which preferentially removed quartz grains from the underlying fine grained sand (Middleton and Southard, 1978). Figure 17C shows some typical grain size probability plots for the F and H sandstones.

The upper part of the F sandstone, the "poor" F, shows increasing amounts of silt and clay, a corresponding decrease in porosity and permeability, but a decrease rather than an increase in the gamma-ray readings on the west side of the field (Figs. 13, 14). This is interpreted as lower energy due to subsidence exceeding sediment supply, and eventually a minor transgression over the field was followed by deposition of the E shale (Fig. 17D).

The E shale, 2 to 10 ft (0.6 to 3 m) thick, is a thin, silty, sandy shale commonly associated with one or two ft (0.3 to 0.6 m) of limestone at the top and base.

Well P10 found abundant shell material in the E shale and confirmed that shell material provides the nucleus for limestone doggers. These doggers became more numerous above the E shale and are occasionally up to 10 ft (3 m) thick (Fig. 25C). Core and dipmeter calipers show that the doggers are of different thickness on opposite side of the borehole and thus discontinuous. Cores and log correlations show that they parallel bedding planes and are like the doggers which occur in Jurassic sandstone outcrops on the Isle of Skye. The E sandstone 19 to 70 ft (5.8 to 21 m) thick, is much like the G and H sandstones except for some occassional beds with diffuse mottling which show weak oil stain and some occasional pebble beds in 15/17-5 (Fig. 18A).

There was little or no movement on the D fault during deposition of the E sandstone, but it influenced deposition.

The E sandstone is overlain by D shale (Fig. 18B), a highly bioturbated sandy, silty, claystone and dirty siltstone. Northerly tilting of Piper fault blocks is apparent from this isopach.

The D sandstone 10 to 52 ft (3 to 15 m) thick, overlies the D shale (Fig. 18C) and is much like the E sandstone around the 15/17-4 well, but is more fine-grained and contains fewer limestone doggers around P10. There are some scour contacts, bioturbated surfaces, and pebbles present at 15/17-5. D and C faults were both active during deposition of D sandstone, and there was northeasterly tilting in the Piper area.

After deposition of the D sandstone, a minor transgression introduced some clay which was thoroughly mixed with the sand by bioturbation. Although very sandy, there is no oil stain on the core in these 2- to 4-ft (0.6 to 2.4 m) intervals.

The A, C, and D faults were active before and during deposition of the overlying C and B and A sandstones (Figs. 18D, 19A). There was also northeasterly tilting of Piper blocks.

The C sandstone, 0 to 36 ft (0 to 11 m) thick, and B sandstone, 0 to 91 ft (0 to 28 m) thick, are very much like the G sandstone and, while not cored, a sandstone 0 to 91 ft (0 to 28 m) thick, appears (from log characteristics) to be very similar to B sandstone. It is thought that these sandstones were deposited around the flanks of the existing structure in water deep enough to prevent significant contemporaneous erosion on the higher areas of Piper.

Prior to deposition of the Kimmeridge Shale there was a transgression, probably accompanied by a long period of very slow sedimentation as indicated by a 6-ft (2-m) zone of intensely bioturbated sandstone grading to silt and containing abundant shell material and glauconite in P10. This transgression was accompanied by minor erosion of Piper sands and redeposition as Unit IV. Unit IV is 0 to 101 ft (0 to 31 m) thick, argillaceous, silty sandstone and occurs around the present crest of Piper (Fig. 19B).

Gross sandstone and net oil-bearing sandstone isopachs (Fig. 19D) show increasing sand thickness to the northeast in each of the blocks; most of the oil is in Block I, and Block II. Block III contains only a (relatively) small amount (less than 10 million bbl) of oil.

Glauconitic sand and silt of the Kimmeridge Shale in P10 contains a fauna representative of marine conditions and grade upwards into a dark gray to black shale, laminated, with no bioturbation and with silt in discrete laminae. Fish scales, fish debris, wood fragments, belemnites, and a lack of bioturbation indicate deeper, more stagnant marine conditions. Although ammonites indicate an early Kimmeridgian age for the Kimmeridge Shale, it is more likely on the basis of log correlations, palynology, and paleontology that the Kimmeridge Shale at Piper and Claymore is Volgian in age.

A proposed model for deposition of the Piper sandstones is shown in Figure 20. This model shows the Piper area at the end of F sandstone deposition with Piper field higher than the 15/11 block and deposition influenced by the D fault. There are beach, barrier bar, tidal flat, fluvial, and deep water depositional environments with active longshore, shelf, and slope processes (Gorsline, 1978). The Callovian sandstones below the I shale at Piper extend through Claymore and cover a minimum of 600 sq mi (1,660 sq km) and the Oxfordian sandstones

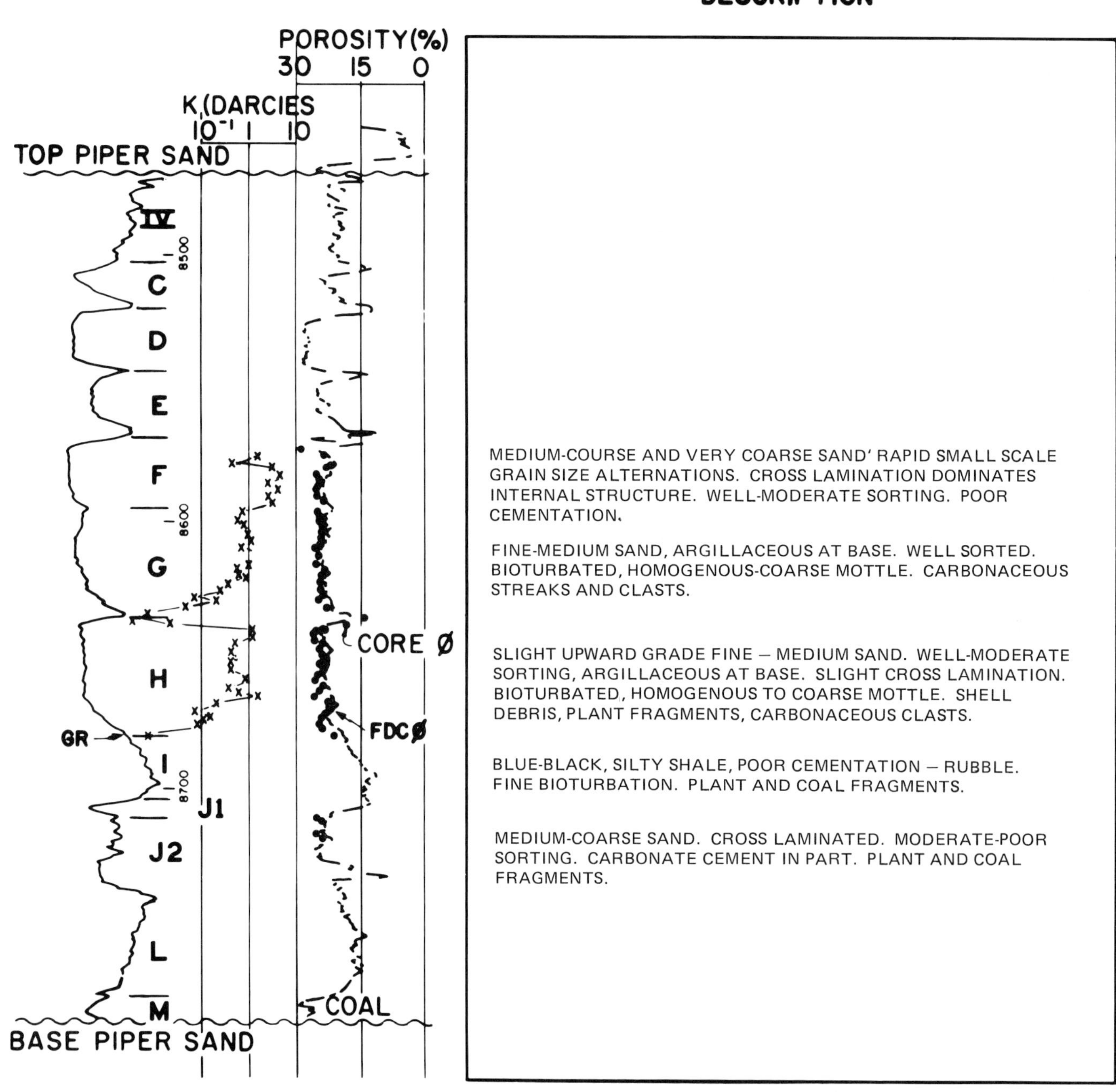

FIG. 14—**A:** Type porosity and permeability log for west Piper. **B:** Type porosity and permeability log for east Piper.

15/17 – 5

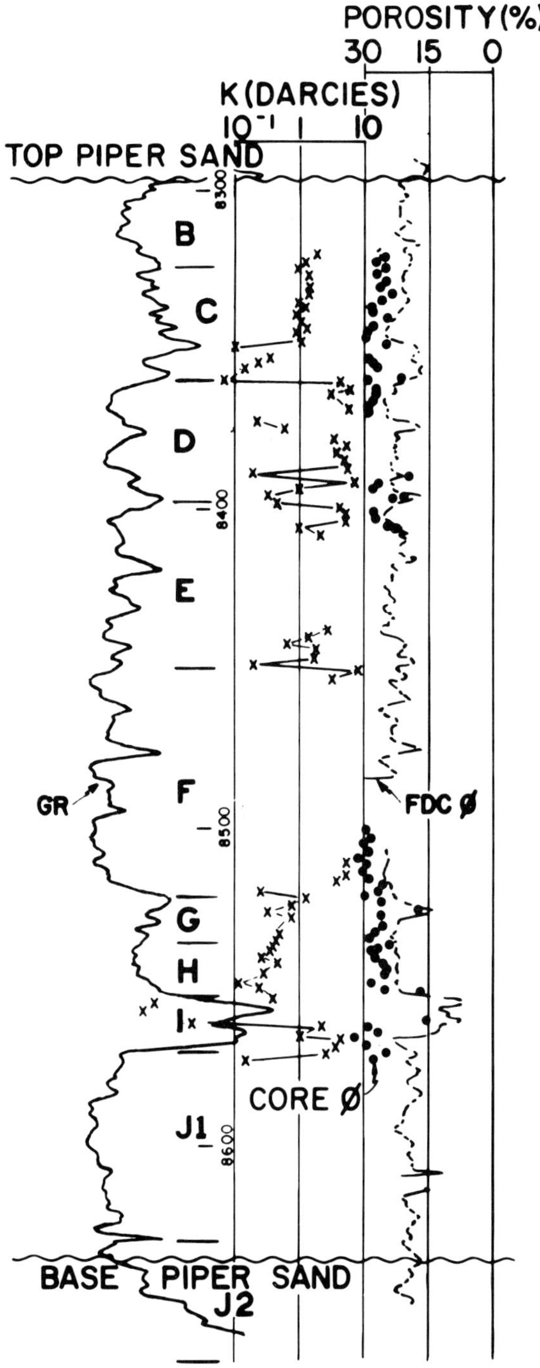

POROSITY (%)
30 15 0

K (DARCIES)
10⁻¹ 1 10

TOP PIPER SAND

8300 8400 8500 8600

B C D E F G H I J1

GR FDC ∅ CORE ∅

BASE PIPER SAND
J2

DESCRIPTION

FINE – MEDIUM SAND. WELL SORTED. BIOTURBATED, HOMOGENOUS – SLIGHT MOTTLE.

FINE-MEDIUM SAND, SCATTERED COARSE GRAINS. SILTY IN PART. WELL – MODERATE AND POOR SORTING. BIOTURBATED, HOMOGENOUS – MOTTLED. POOR CEMENT IN PART.

INTERBEDDED MEDIUM – COARSE SAND, PEBBLY IN PART ON SMALL SCALE SCOURS. ARGILLACEOUS IN PART. WELL – MODERATE AND POOR SORTING. BIOTURBATED, FINE – COARSE, MOTTLED. SMALL SCALE CROSS LAMINATION.

MEDIUM SAND, COARSE AND PEBBLY ON SMALL SCALE SCOURS. WELL – POOR SORTING. POORLY CEMENTED. COARSE BIOTURBATION, MOTTLED. DIAGNETIC CARBONATE CEMENTS.

RAPID SMALL SCALE MEDIUM AND COARSE GRAIN SIZE ALTERNATIONS, INDIVIDUALLY WELL SORTED. SMALL SCALE CROSS LAMINATION DOMINATES INTERNAL STRUCTURE. OCCASIONAL BIOTURBATION. MODERATE – POOR CEMENTATION.

MEDIUM – COARSE SAND, WELL – MODERATE SORTING. BIOTURBATED, HOMOGENOUS AND MOTTLED.

MEDIUM – FINE SAND, ARGILLACEOUS AT BASE. WELL-MODERATE SORTING, CARBONACEOUS, PLANT AND COAL FRAGMENTS.

SHALE, SILTY. BIOTURBATED, FINE – COARSE MOTTLING. PLANT FRAGMENTS, COAL CLASTS, BELEMNITES, AMMONITES.

MEDIUM – COARSE SANDS, MODERATE – POOR SORTING, CROSS LAMINATED. BIOTURBATED UPPER SURFACE, COARSE MOTTLE.

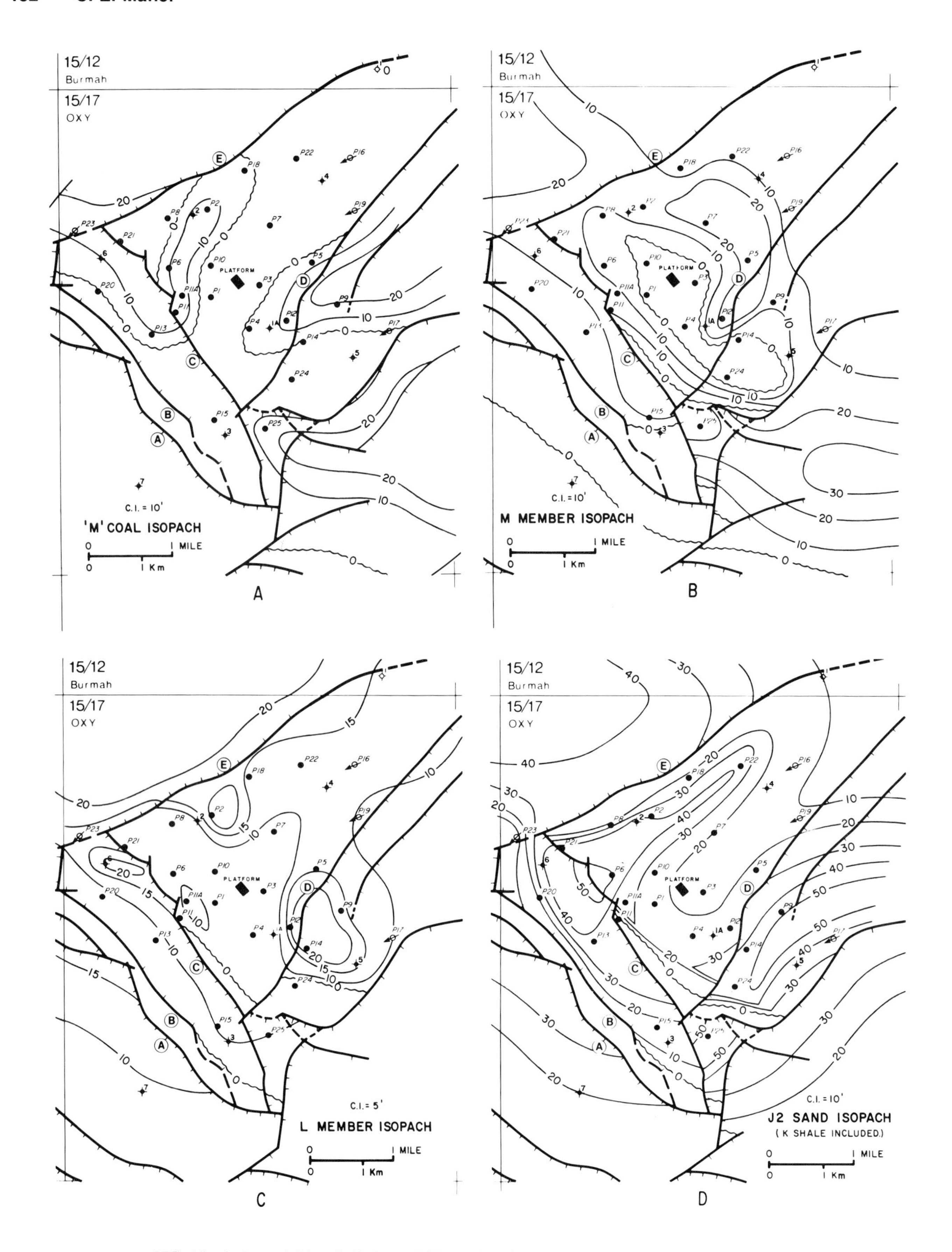

FIG. 15—**A:** Isopach M coals; **B:** isopach M member; **C:** isopach L member; **D:** isopach J2 member.

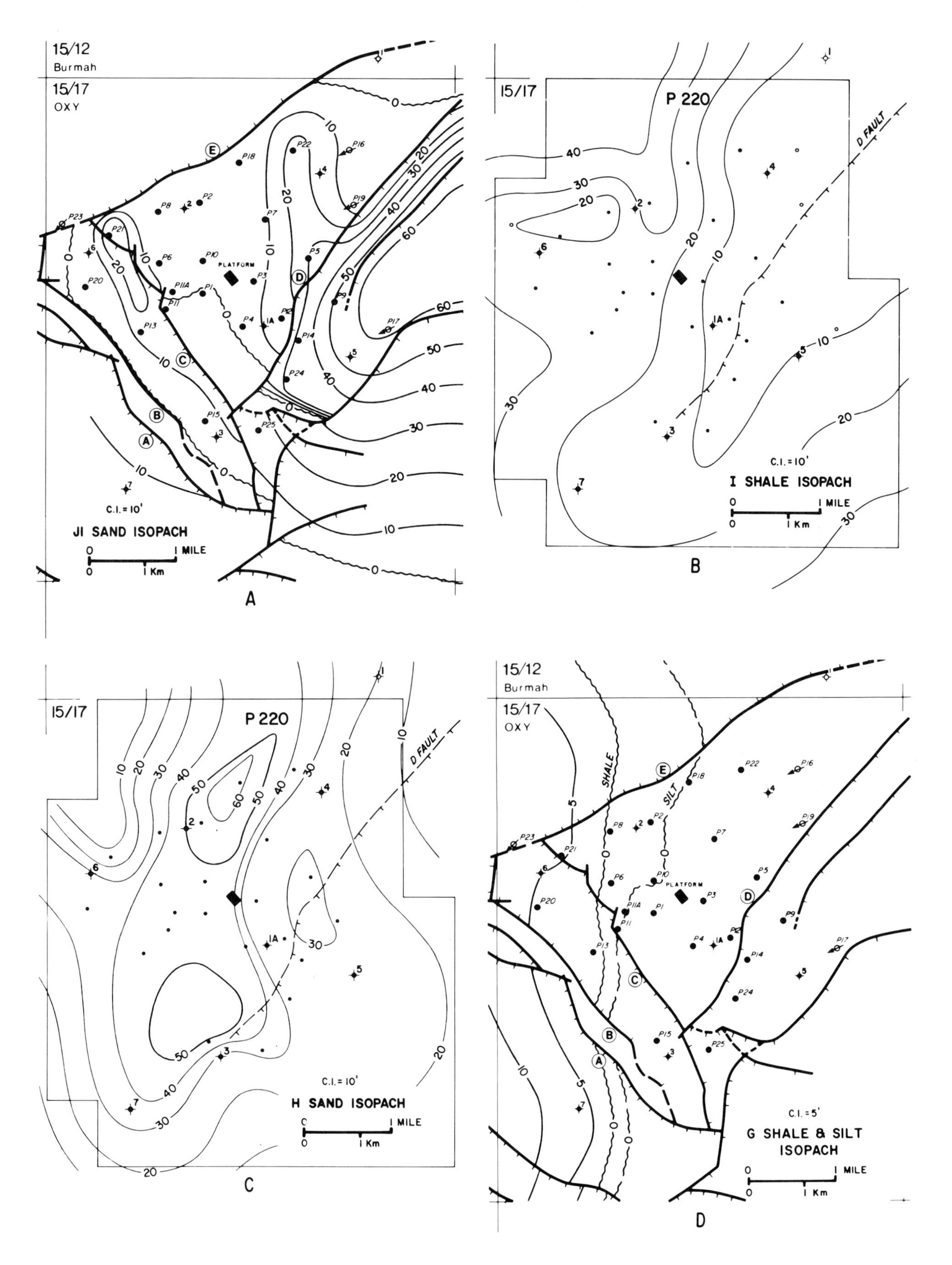

FIG. 16—**A:** Isopach J1 member; **B:** pre-Cretaceous isopach I shale; **C:** pre-Cretaceous isopach H sandstone; **D:** isopach G shale and siltstone.

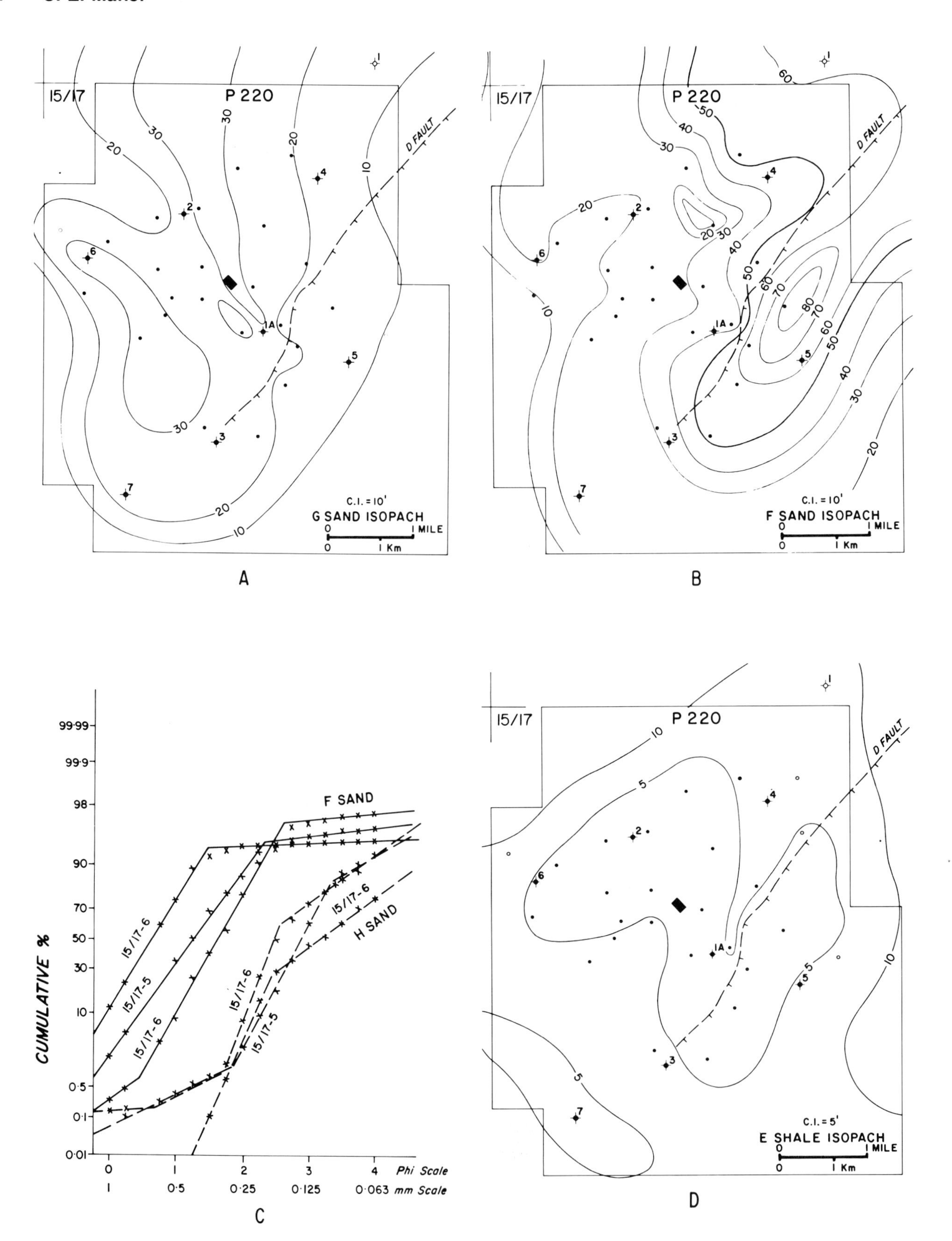

FIG. 17—**A:** Pre-Cretaceous isopach G sandstone; **B:** pre-Cretaceous isopach F sandstone; **C:** Grain-size probability plots for F and H sandstones; **D:** pre-Cretaceous isopach E shale.

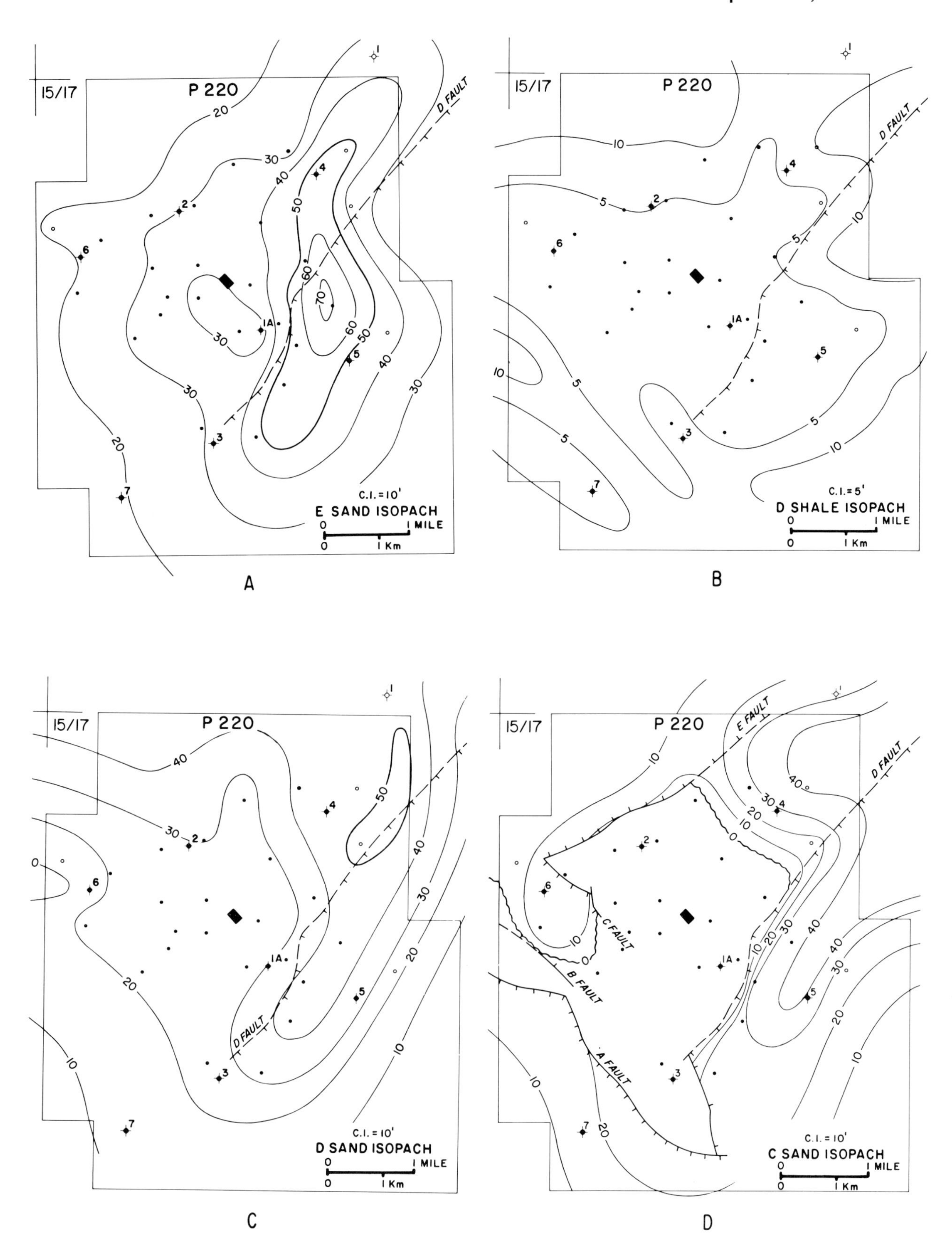

FIG. 18—**A:** Pre-Cretaceous isopach E sandstone; **B:** pre-Cretaceous isopach D shale; **C:** pre-Cretaceous isopach D sandstone; **D:** pre-Cretaceous isopach C sandstone.

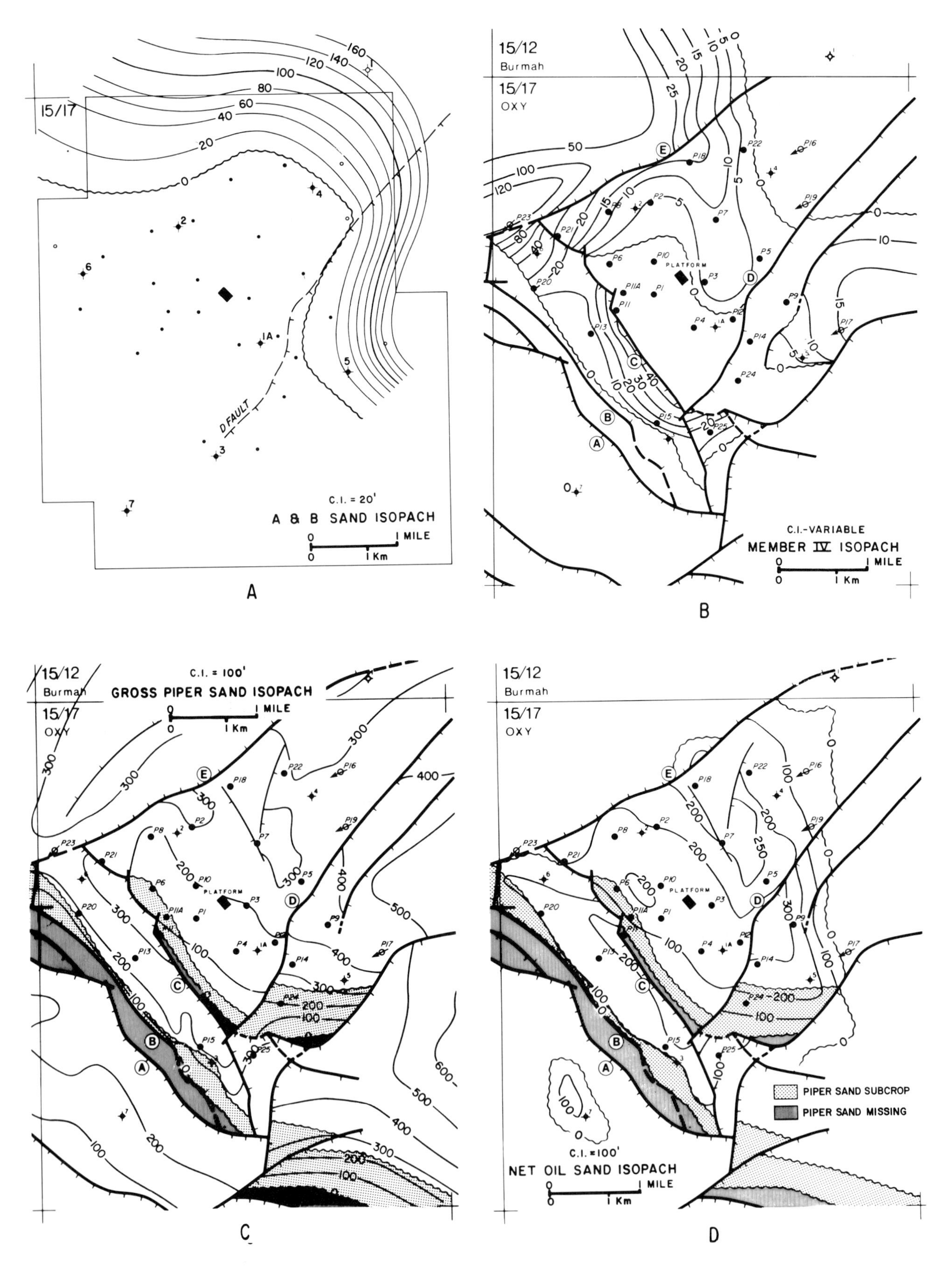

FIG. 19—**A:** Isopach A + B sandstone; **B:** isopach member IV; **C:** isopach gross Piper sandstone; **D:** isopach net oil-bearing sandstone.

above the I shale cover a minimum of 400 sq mi (1,000 sq km). Therefore, the Piper sandstones must be viewed a part of a widespread, moderate to high energy, marginal marine and marine sand sequence which reflects local topography and fault movement during deposition.

Gross thickness of Oxfordian sandstones at Piper ranges from 165 ft (50 m) at P8 to 349 ft (106 m) at P17. If the age of the sandstones are from middle late Oxfordian to the end of Oxfordian they would have been deposited in 1.5 million years, at an average rate of less than 70 μm/yr.

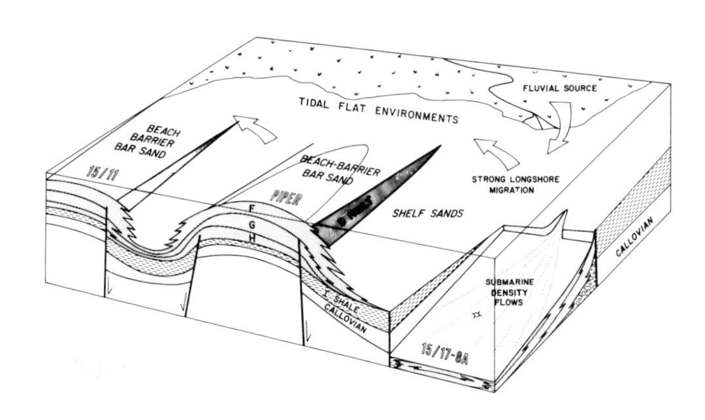

FIG. 20—Depositional model for Piper area at end of F sandstone deposition.

PETROGRAPHY OF PIPER SANDSTONES

Grain Size Analysis

Grain size analysis was performed on numerous samples. Selected samples were disaggregated and agitated through a nest of 21 British Standard sieves ranging from 14 to 400 (very coarse sand to silt and clay).

Examined sands were coarse to medium, fine or very fine grained having graphic means (Mz) of 3.5 to 0.7 phi.

Sorting varies from poor to very well sorted; moderate to well sorted was most common. Piper sands generally are positive-skewed or symmetrical with only isolated negative-skewed samples.

Grain size distributions plotted as cumulative percent probability curves are illustrated in Figure 17C for the F and H sandstone members of 15/17-6 and 15/17-5. The F sandstone is the coarsest, highest energy (upper shoreface–beach) sandstone in Piper field, and is cross-laminated, whereas H sandstone is predominantely a fine-grained homogeneous and bioturbated (offshore marine bar) sandstone.

Texture

Sphericity and Roundness

Grains range from subangular to well rounded according to Powers' (1953) visual charts. Coarse and very coarse grains have a distinct tendency to be more rounded than smaller material, and often are very well rounded.

Packing

Piper sandstones are grain supported with high porosity values from 20 to 28%. Grain contacts generally are tangential or concave/convex; the number of sutured contacts are minor. Where bimodal sands occur, the degree of packing increases with a greater number of grain contacts per grain; however, extensive suturing/pressure solution does not occur.

Recent discussions (Volkmar Schmidt and William Almon, personal commun.), have led to a detailed study of secondary porosity which is not complete, but preliminary results show some secondary porosity in the Piper sandstones.

Detrital Mineralogy

Within Piper sandstones there is little variation in the detrital mineralogy. Piper sandstones consist generally of 90%+ quartz most of which is a dominantly unstrained or simple strained variety. Metamorphic polycrystalline quartz varieties occur but are relatively rare. Other rock fragments are sometimes found; granitic fragments occur in 15/17-4.

Feldspar occurs in minor amounts in the sands; plagioclase perhaps is more abundant; microcline and orthoclase also are present. Detrital micas are uncommon, occurring only in argillaceous zones which generally are associated with silty shales. Other detrital grains may include trace amounts of tourmaline, chert, zircon, and (more common) glauconite pellets.

Clay Content

Piper sandstones are generally clean and nonargillaceous, and have high porosity/permeability values (Figs. 21A, B, 22). Argillaceous sandstones occur in association with silty shale zones, or where dirty sandstones form lateral equivalents of shale zones downdip.

In thin section and SEM, detrital clay appears as a pore-filling, structureless material which may be squeezed around and between detrital grains, and results in reduced permeability and reduced effective porosity.

Diagenesis/Cementation/Authigenesis

Quartz—Although statistically insignificant, 1 to 2% authigenic quartz occurs in many thin sections as small scale rims or needlelike projections in optical continuity with the parent grain (Fig. 21B, C, D, E, F).

Feldspar—Occuring as authigenic rims on detrital grains, feldspar forms less than 1% of the total mass, although its occurrence is not infrequent. The composition appears to be that of potassium feldspar.

Carbonate—Carbonate cements occur in areas throughout the sandstones, and generally form concretions or "doggers." Detrital grains are extensively replaced with irregular boundaries showing the effects of corrosion and replacement. Carbonate cements occur both in association with primary carbonate shell debris and where no primary material has been observed.

Authigenic Clays—Kaolinite, perhaps the most common diagenetic clay, occurs in Piper sandstones either as a pore pre-

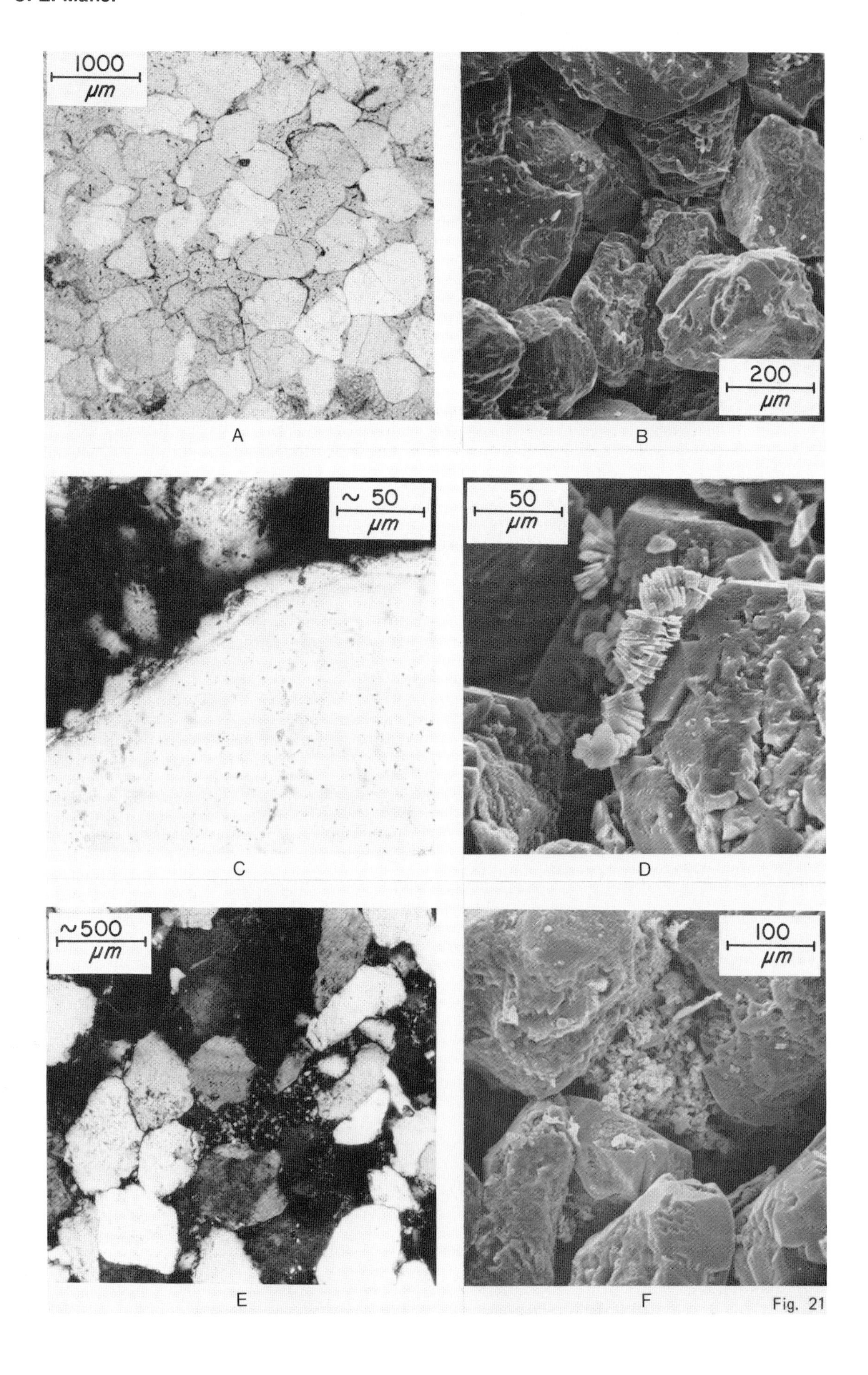

Fig. 21

cipitate or as a feldspar replacement product. The kaolinite forms well-crystallized books which may plug pores or occur as vermiform "worms" (Fig. 21B, D, F). There appears to be a slight increase in the amount of kaolinite present in the sandstones at depth. No major change occurs across the current oil-water contact, as reported in other North Sea Jurassic reservoirs in the Viking graben. Illite occurs in trace amounts as reconstituted detrital clay, and very occasionally as an alteration product of kaolinite.

Diagenetic Sequence

The following sequence is tentative and is based on replacement relationships observed in thin sections: (1) quartz and feldspar overgrowths; (2) kaolinite and illite; and (3) carbonate and pyrite. The proposed sequence of alteration is not mutually exclusive, and more than one process can occur at around the same time in the same rock.

Porosity Versus Depth

To achieve a better understanding of porosity versus depth, available formation density log (FDC) porosity for clean F sandstone was plotted for Piper, 15/11 Block, and 15/7-8A wells (Fig. 23).

Figure 23 assumes a pore fluid density of 1.0 g/cc. It is immediately apparent that wet sandstones in Piper and the 15/11 Block fall on a parallel but lower porosity trend to sandstones within the oil zone. This "shift" indicates a better porosity in the oil saturated Piper sandstones than would be expected from depth extrapolation of porosity in wet sandstones.

There are inadequate shallow formation resistivity (Rxo) measurements in Piper sandstones and it is difficult to make a reliable correction for lower density hydrocarbons within the zone measured by the formation density log (FDC). Enough data are available to suggest that about half of the shift be subtracted as hydrocarbon effect, and that the remainder is due to porosity preservation. This is confirmed by firmer, more compact, less friable, and better core recovery in the F sandstone from 15/17-6 and 15/17-7, where the sandstone was below the oil-water contact. Where the F sandstone was cored in the oil zone, 15/17-4, 15/17-5, P10, there was very little core recovery except as loose and very friable sand which crumbled under the slightest pressure. There is no evidence of additional clay in the wet sandstones, and the decrease in porosity is interpreted to be the result of greater dissolution (pressure solution) and tighter packing. There is no apparent increase in quartz cementation or overgrowths, but Fuchtbauer (1967) showed that the dissolved rock volume to achieve 50% compaction can be as little as 1.5%, depending on the angle of contact between grains. Thus, only minor amounts of dissolved rock and a minor change in quartz overgrowths might account for the shift. In 15/17-8A, where there is abundant evidence of pressure solution and recrystallization (Robin, 1978), the well plots are to the left of the trend for those wet sandstones showing only minor effects of dissolution and recrystallization.

Helium-derived core porosities are generally about 4% greater than formation density log (FDC) porosities. This requires a correction of core-derived porosity measurement of about minus one porosity unit, from 26% to 25% porosity to correct for decompaction.

OIL SOURCE

Piper oil has almost certainly originated in the Kimmeridge Shale. The formation of oil within the shale and subsequent release to carrier beds through discrete silt and sand laminae, or through a Kerogen network (McAuliffe, 1979) as well as fracturing of the shale due to pressure build-up during physical-chemical processes of generation (Tissot and Welte, 1978), fits the subsurface geologic setting in the Piper area. Piper oil contains 1% sulfur, has a relatively low gas/oil ratio (GOR), 446 cf/bbl, and a relatively high gravity, 37° API. There are significant differences in the crude oils in the Piper area, but there are insufficient data available to say whether this is simply due to variations in the local source, or in part due to differential entrapment or other unknown factors. All fields in this area appear to have been filled to a spill point during the main phase of oil migration.

Time of Migration

Oil migration could not have preceded deposition of the caprock, the Campanian marls, some 75 m.y. BP, and more than 65 m.y. after deposition of the Kimmeridge Shale, (Van Hinte, 1976a).

Campanian marl forms the caprock along the C fault where the Cenomanian erosion has cut through the Lower Cretaceous, Kimmeridge Shale, and part of the Piper Sandstone. Nearly 20 m.y. elapsed after the Cenomanian erosion before Campanian sediments onlapped the Piper sands and provided a seal to the reservoir. This was critical, because the Maastrichtian does not act as a caprock in this area and if Campanian marls had not covered the highest exposure of the Piper Sandstone, oil in Blocks I and II would have been lost.

It was also critical that erosion preceding deposition of the caprock had sufficiently lowered the highest point on the sand body so that it could be onlapped by Campanian sediments.

FIG. 21—**A:** Well sorted Piper sandstone, very clean, little or no clay, 26.7% intergranular porosity, permeability of 4,340 md. PPL.

B: Well sorted Piper sandstone showing trace amounts of kaolinite and authigenic quartz but essentially unobstructed pore space and pore throats. SEM × 230.

C: Small scale authigenic quartz developed on detrital grain from which it is distinguished by the presence of a dust rim. XPL.

D: Numerous authigenic quartz growths and well developed vermicular kaolinite. Close up of "B". SEM × 1,150.

E: Well sorted Piper sandstone, small scale quartz overgrowths, and pore plugging kaolinite precipitate. XPL.

F: Well developed authigeneic quartz on detrital grains surrounding authigenic kaolinite precipitate. Possible solution effects on the authigenic quartz. SEM × 50.

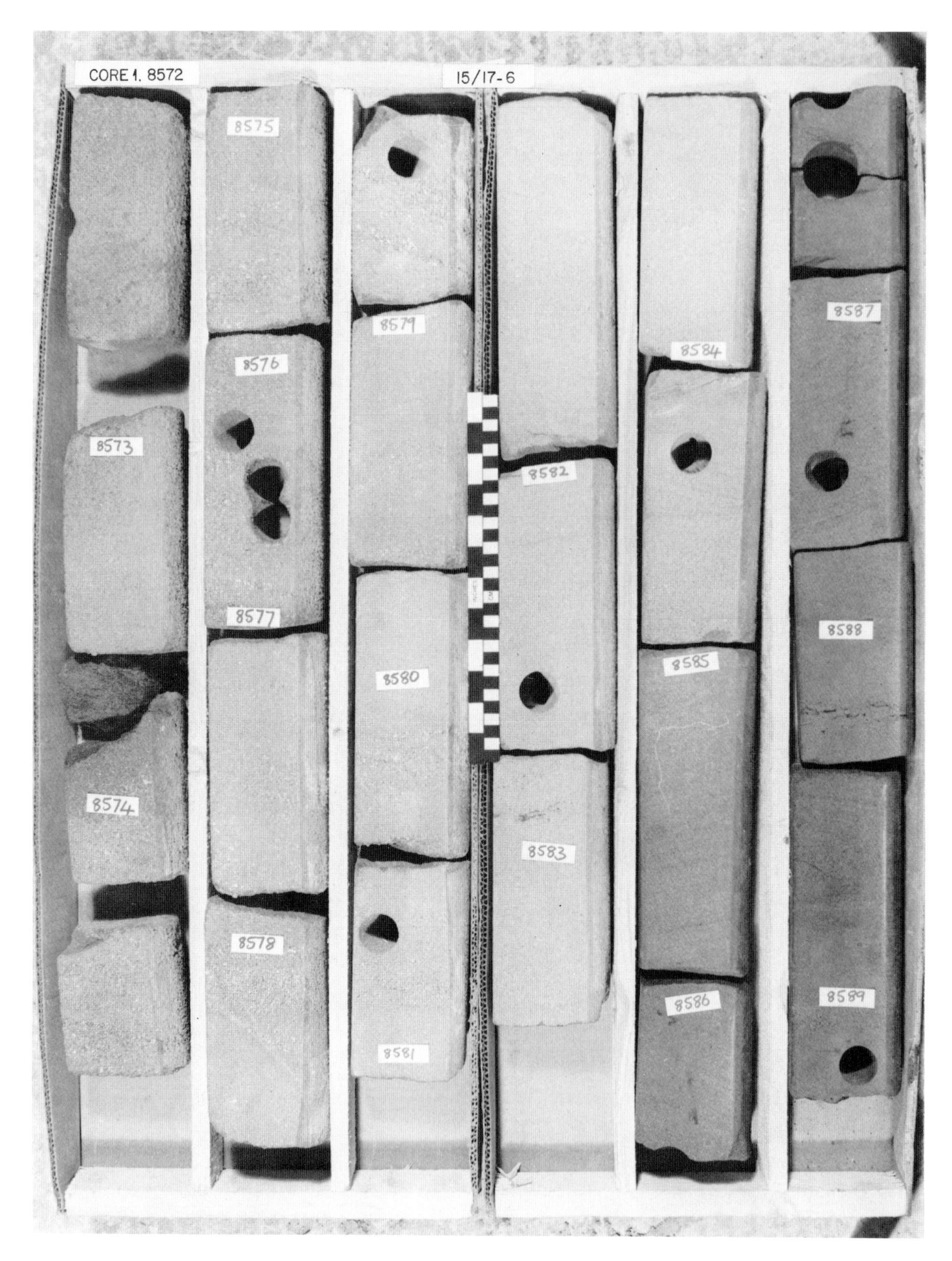

FIG. 22—Core photographs showing the coarse grained cross-bedded F sandstone, oil-water contact, and well preserved burrows in the underlying fine grained G sandstone.

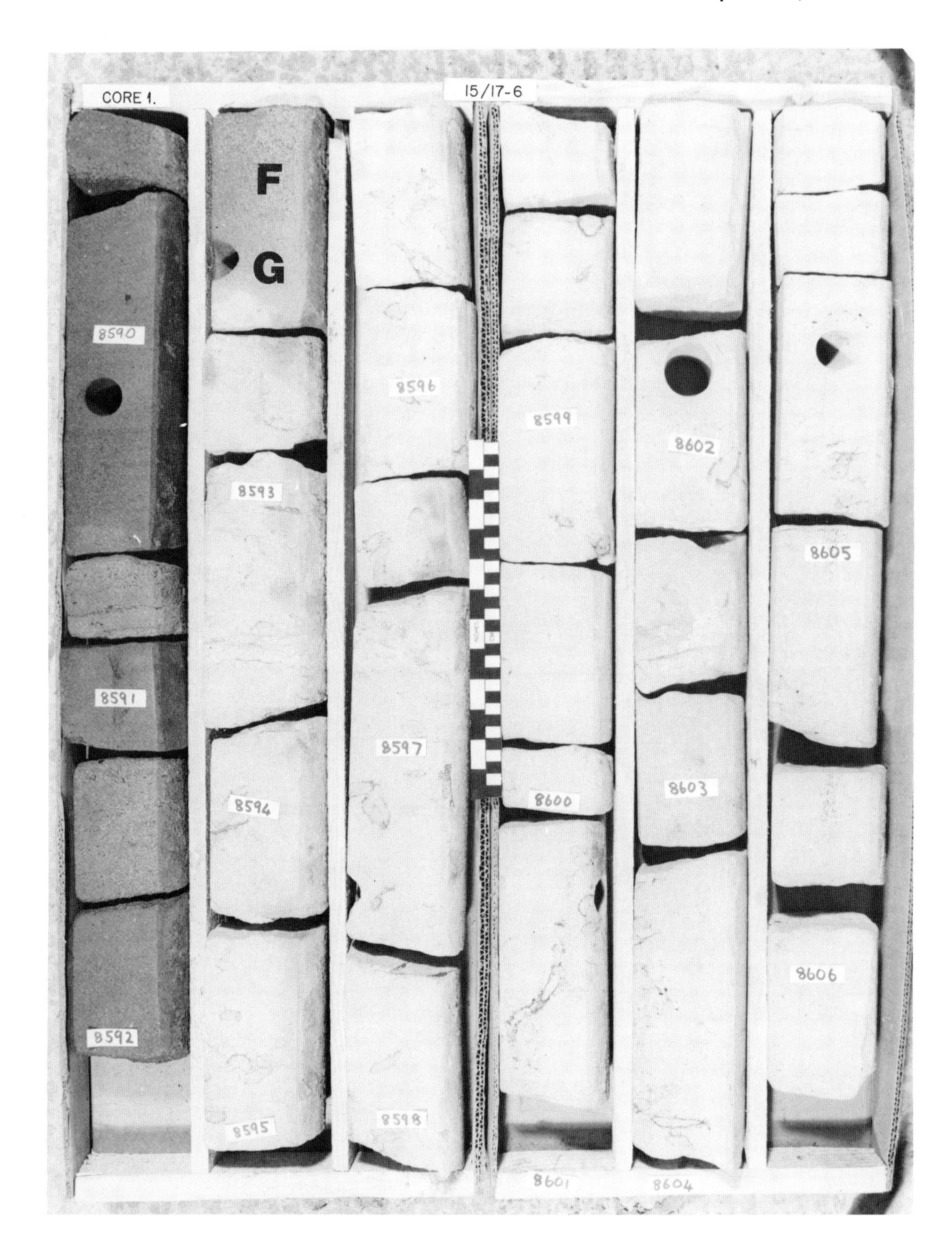
CORE 1.
15/17-6
F
G
8590
8591
8592
8593
8594
8595
8596
8597
8598
8599
8600
8601
8602
8603
8604
8605
8606

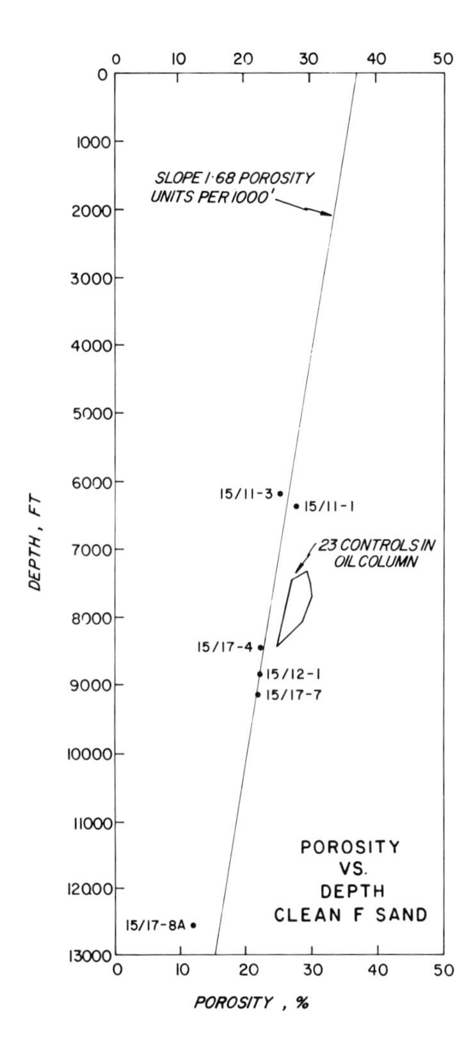

FIG. 23—Porosity trend clean F sandstone, Piper field, 15/11 block and 15/17-8A.

Lack of oil in the 15/11 block structure is attributed to lack of a caprock, with permeable Maastrichtian chalk lying directly atop upper Jurassic sandstones.

Before Campanian marl had onlapped Piper sandstone and formed a seal, Kimmeridge Shale had already been buried to more than 5,000 ft (1,525 m) in the adjacent Witch Ground graben (Figs. 9, 10). By the end of early Paleocene, depth of burial exceeded 7,000 ft (2,135 m), and it is quite possible that oil generation in the Witch Ground graben and accumulation in the Piper structure had commenced. Significant increases in resistivity and bulk density which occur in Kimmeridge Shale at about 9,000 (2,745 m) in the Piper area may be associated with diagenesis and initial migration of hydrocarbons.

There are no shows in Carboniferous sandstones or Zechstein dolomites at Piper, although there is more than 1,000 ft (300 m) of closure in these beds. However, the top of the Zechstein closure is below the oil-water contact, 8,512 ft (2,594 m) subsea, for Blocks I and II. The Piper sandstone is in contact with the Zechstein and Carboniferous along the complex fault system at Piper (probably along the C fault) and no oil could be trapped in these formations because they were below the spill point (Figs. 8, 9).

RESERVOIR PERFORMANCE

Production

Reservoir performance for Piper field closely follows what would be expected with our current geologic knowledge of Piper sandstone stratigraphy and petrography, fault displacement, and structural configuration. Seven wells have produced more than 10 million bbl each in the short time they have been on production, and P7 (completed with 7 in. [17.8 cm] tubing) has produced more than 18 million bbl in less than 2 years. Significant information and field parammeters are listed in Tables 4 through 6.

Liquid Extraction and Gas Processing

During 1977 and 1978, liquid extraction and gas processing facilities have been constructed on Piper platform. Residual dry gas is being transported to shore using the Frigg field pipeline. Excess condensate which (currently) cannot be handled at the Flotta terminal is being injected temporarily into well P4.

Permeability

Appraisal-well drill-stem tests and core data indicated that well production potential would be high. Core permeabilities averaged 790 md geometrically and 1,500 md arithmetically. However, the first development well, P1, exceeded expectations with an initial production rate greater than 30,000 b/d, restricted by 5 1/2 in. (14 cm) tubing.

High horizontal and vertical permeability in Piper required accurate pressure gauges to obtain necessary pressure data (de Boer, 1978), and Hewlett-Packard gauges are used to conduct

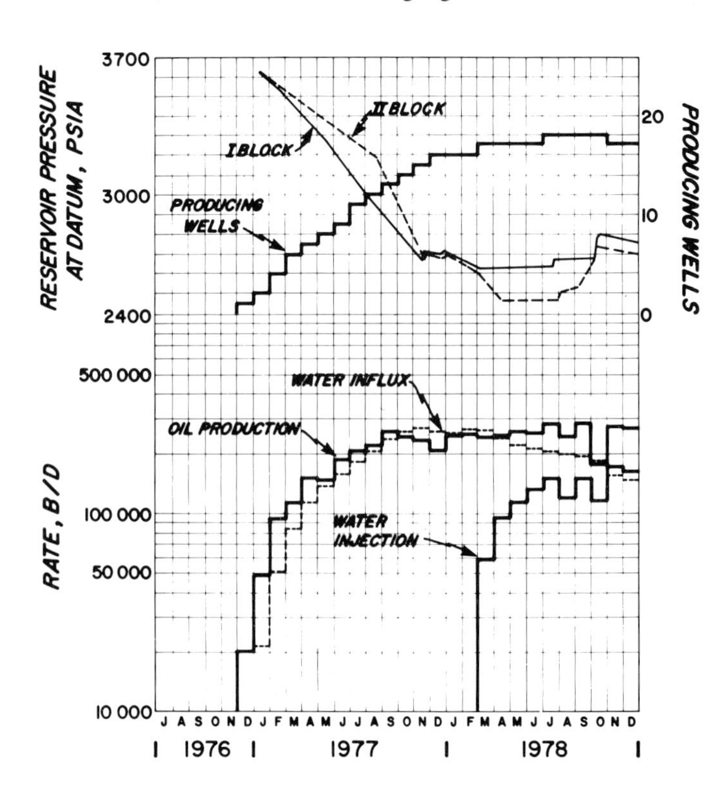

FIG. 24—Piper field performance.

Table 4. Piper Well Data.

WELL	SPUD DATE (DAY/MO/YR)	COMPLETION DATE (DAY/MO/YR)	GROSS PIPER SAND (FEET)	NET PIPER SAND (FEET)	NET PIPER OIL SAND (FEET)	I. P. (B/D)	CUMULATIVE OIL (bbl)[1] INJECTION (bbl)
15/17-1A	25.11.72	22.1.73	176	151	151	9,949	
15/17-2	27.1.73	18.3.73	258	198	198	32,129	
15/17-3	19.3.73	24.7.73	115	98	98	15,509	
15/17-4	8.7.73	9.8.73	263	225	123	No Test	
15/17-5	11.8.73	15.9.73	355	332	255	No Test	
15/17-6	16.9.73	17.10.73	264	210	72	No Test	
15/17-7	22.11.73	22.1.73	170	131	47	No Test	
P1	10.10.76	24.11.76	178	149	149	30,000	14,428,185
P2	24.11.76	5.1.77	289	196	196	30,000	12,483,780
P3	24.12.76	27.1.77	195	174	174	26,000	12,386,124
P4	7.1.77	5.2.77	159	141	141	26,200	10,802,995
P5	28.1.77	12.3.77	278	233	233	26,400	12,283,518
P6	6.2.77	18.3.77	227	176	176	27,000	10,407,346
P7	13.2.77	15.4.77	196	148	148	50,000	17,975,080
P8	19.3.77	25.4.77	165	132	132	22,400	9,216,356
P9	16.4.77	22.6.77	422	352	299	18,678	4,162,549
P10	24.4.77	4.6.77	210	157	157	20,903	9,606,583
P11	5.6.77	Sidetracked	Absent	--	--	--	---
P11A	5.6.77	21.7.77	192	126	126	14,761	8,303,945
P12	22.6.77	28.7.77	192	153	153	15,746	8,775,365
P13	23.7.77	24.8.77	254	198	198	17,991	7,386,214
P14	29.7.77	7.9.77	341	304	304	15,151	8,471,898
P15	29.8.77	25.10.77	231	186	186	16,159	6,423,722
P16 I[2]	8.9.77	8.11.77	294	205	27	51,360	16,133,813
P17 I	27.10.77	30.12.77	453	378	0	40,693	14,644,396
P18	22.10.77	6.1.78	238	176	176	23,129	10,024,133
P19 I	9.12.77	31.3.78	240	188		40,030	11,136,323
P20	1.1.78	6.3.78	198	140	140	10,500	4,060,396
P21	8.3.78	5.5.78	244	201	201	12,841	3,745,962
P22	2.4.78	12.7.78	275	200	85	11,140	790,171
P23 I	9.5.78	24.8.78	264	216	0	27,174	5,830,371
P24	21.10.78	2.12.78	191	171	171	17,000	1,277,003
P25	2.12.78	17.2.79	317	255	112	8,000	13,814
P26	17.2.79						

[1] February 28, 1979
[2] I - Injection well

pressure surveys. Pressure buildup and drawdown tests showed that the average well permeability varied from 2 to 4 Darcies, which is considerably higher than estimates based on core data.

The difference between core permeability and well test permeability resulted from a biased sampling of core data. Permeability measurements were made on core plugs, and very few of these were obtained in the highly permeable F sandstone. The F sandstone is medium to very coarse and quite friable. As a result, there was poor core recovery in the more permeable F sandstone, and little success in cutting plugs where there was core recovery. This explanation is supported by the observation that higher well test permeabilities are found in those wells with thicker F sandstone.

It also was found from well pressure test data that well interference often became significant only 1 to 2 hours after the start of a buildup test. This confirms the geologic description of a highly permeable and laterally continuous reservoir.

Pressure Trends for Blocks I and II

Reservoir performance to date (Fig. 24) shows different pressure trends for wells in Block I and Block II. Where more wells have been drilled and more pressure data are available, variation in static pressure about the pressure trend is very small, 30± psi. This also shows the Piper reservoir to be highly permeable and laterally continuous.

The pressure trend shown for Block II for the period from

Table 5. Piper Data and Reservoir Parameters

Reserves originally in place	618 MM bbl, 216 MMMscf, 13.4 MM bbl cond.
Productive area	7,437 acres (30.1 km^2)
Oil sand volume	1,038,830 Acre-ft
Bo	1.25 RB/bbl
Porosity:	19-30%, average 24%
Permeability	0.8-12 D, average 4 D
Water saturation (average)	Approx. 5%
API gravity	37°
Gas/oil ratio (GOR)	446 cf/bbl
Bubble point	1,600 psi
Sulfur content	1.0%
Original oil in place (OOIP)	1,368 MM bbl
Pore pressure gradient	0.43 psi/ft
Reservoir temperature	8,000 ft (2,469 m): 175°F (79°C)
Oil-water contact (Blocks I and II):	8,512 ft (2,594 M) subsea
Oil-water contact (Block III):	9,199 ft (2,803 m) subsea
Oil column (Blocks 1 and II):	7,200 to 8,512 ft (2,194 to 3,594 m) subsea
Height of oil column (Blocks I and II):	1,312 ft (400 m)
Piper sand: gross	165 to 453 ft (50 to 138 m)
net	131 to 378 ft (40 to 115 m)
net oil sand (maximum)	304 ft (93 m)
net oil sand (average)	140 ft (43 m)
R_w = 0.045 @ 175°F	m = 1.8 (measured)
R_o = 0.5 @ 175°F	n = 1.8 (measured)

December 1976 to August 1977 is inferred. Well P13, the first development well to be drilled in Block II, found a depleted pressure of 500 psi below the initial pressure, but 240 psi above Block I pressure. This indicated Block II was pressure-connected to Block I, but through a low transmissibility or longer path. Areal model simulations have indicated that more than one path exists. Interpretations in which a single path connecting the two blocks was postulated, failed to match measured pressure performance.

The best pressure match was obtained where simultaneous pressure communication between Blocks I and II was modeled by connecting: (1) the aquifer to the north of the E fault to Block I; (2) the oil zone in Block II, around P25, to the oil zone in the remainder of Block II and to Block I; and (3) the aquifer to the southeast of P25 to Block II.

Natural Water Drive

From field characteristics and availability of accurate pressure data, it became apparent that natural water influx was occurring in Piper field. This aquifer influx (Fig. 24) was calculated by material balance using a volume-weighted average pressure of Blocks I and II and areal model simulations. The main influx occurs at the eastern side of Block I.

Pressure Maintenance

Although aquifer influx is approximately 250,000 b/d, it was apparent by mid-1977 that reservoir pressure could not be sufficiently maintained at projected reservoir production rates of 250,000 to 300,000 b/d. In late 1977, the first water-injection wells, P16 and P17, were drilled to supplement natural water influx; injection began in early 1978. Two additional injection wells, P19 and P23, were drilled during 1978. Injection in P16, P19, and to a lesser degree P17 and P23, reduced aquifer influx, but as can be seen from Figure 24, reservoir pressure decline was halted even at an increased production rate. P25 found the upper Piper sandstones oil-bearing, and will be converted to a water-injection well when these sandstones water out due to natural water influx.

Water Breakthrough

Water breakthrough (WBT) has occurred in four production wells; P9, P22, P5, and P20. P9 penetrated the oil-water contact and was completed in the oil sands above in June 1977; however, WBToccurred in February 1978. A production log (PCT log) and thermal-decay time log (TDT log) showed that the C and D sandstones had become wet, but the stratigraphically deeper E, F, and G sandstones were still oil-bearing. By

October 1978 the F and G sandstones watered out as well, although these sandstones were not completed earlier.

Well P22 was completed in July, 1977. Open hole logs showed that the F and G sandstone were flushed by natural water influx to residual oil saturations of about 10%, and the uppermost sandstone was wet (residual oil saturation determination was difficult because of clay content in this sandstone). The well was selectively completed in the C, D, and E sandstones. Water breakthrough occurred only 2 months later through the bottom perforations of the E sandstone. This breakthrough could have been due to coning, a poor cement job, or the E sandstone becoming wet. Production logs and thermal-decay time logs in 1979 indicated that the E sandstone had become water saturated. A through-tubing bridge-plug that shut off the E sandstone perforations was successfully run, although the hole is inclined at 47°. This has resulted in dry oil production of 10,000 b/d.

P5 was completed in March 1977 and water breakthrough occurred in January, 1979. Production logs and thermal-decay time logs have shown that the F sandstone is becoming wet, probably due to injection in P19. Although P19 was selectively completed for injection into the sandstones below the F sandstone, a cement breakdown allows the injected water to enter the F sandstone almost exclusively.

P20 was completed in March 1978 and water production was first measured in March 1979. Production logs and thermal-decay time logs indicated entry in the lower part of the H sandstone.

Production of sand grains with the oil has been virtually nil at Piper and no increase has been noted with water breakthrough. Sorting and packing of Piper sand is apparently sufficient to prevent movement of fine grains when the wetting phase (connate water) begins to move (Muecke, 1979).

Water production performance of these wells show that Piper field has a layered reservoir behavior, even if there is a reasonable permeability between various sandstone members. However, within separate layers, sweep is efficient with advancing water at or near the base of the sandstone members, irrespective of permeability.

Water Injection

Piper formation water is incompatible with injected seawater. The formation water contains 1,990 mg/l Ca^{2+}, 620 mg/l Mg^{2+}, 80 mg/l BaU2+D, 590 mg/l St^{2+} and nil SO_4^{2-} while the injected seawater contains 485 mg/l Ca^{2+}, 1,810 mg/l Mg^{2+}, little to nil Ba^{2+}, and St^{2+}, but 2,850 mg/l SO_4^{2-}.

Gates and Caraway (1965) reported on scale formation from a similar injection program in Wilmington field, California. In addition to scale formation on injection water breakthrough, they reported the loss of 100 tons per day of magnesium from 500,000 b/d of injected water. It seems possible that dolomitic cement (Ca, Mg) CO_3 may be precipiated due to the excess SO_4^{2-}. If cements are formed they would tend to form preferentially along planes of reduced permeability or along thin beds containing clay minerals, and would further reduce permeability. The injection and produced water are frequently analyzed to determine if a similar situation develops at Piper.

An Eocene sandstone was tested and analyzed as a potential water source and was found to be much more compatible with Piper formation water. Should a problem develop, this sandstone could be used as a source for injection water.

Piper injection water is chemically treated to inhibit scale formation at the wellbore but high gamma-ray readings opposite normal thermal-decay time readings in zones taking water indicate scale formation containing a naturally radioactive substance. This could be radium associated with $BaSO_4$ scale (Gates and Caraway, 1965).

Extensive tests prior to commencement of injection determined that North Sea seawater had to be filtered to 5 μm to avoid core plugging. Plugging was due almost entirely to plugging at the inlet face by soft parts of nektonic and pelagic fauna living in North Sea waters. Plugging can be eliminated by reversing flow direction or by removing the inlet face.

DEVELOPMENT GEOLOGY

Method

Characteristics, lateral and vertical variations of different facies, and vertical sequences were observed and described. It was not considered important to classify these sandstone members (i.e., upper shoreface, interdistributary deltaic plain, etc.). Individual sandstone members were recognized as essentially time-rock units deposited under moderate to high energy, predominantly marine conditions interrupted by periodic minor transgressions. This understanding formed the basis for the next phase of the study.

Detailed core analysis logs were compiled for the discovery and appraisal wells; correlation of core and log porosity on these logs is excellent. After compilation of the core analysis logs, detailed correlation sections were constructed for Piper and Claymore.

Use of Seismic Data

Having established a time-rock framework, the real task of understanding the structural framework began. The available seismic data (approximately 325 mi [525 km]) was studied intensively and 11 working cross sections were constructed. This turned out to be very much more difficult than anticipated because of fault complexity, velocity gradient over Piper, and dif-

Table 6. Percent Oil Sand Volume by Sand Member.

Sand	Percent of Total Oil Sand Volume
B	2.2
C	7.1
D	13.7
E	13.8
F	13.6
G	18.8
H	15.7
J	15.1
	100

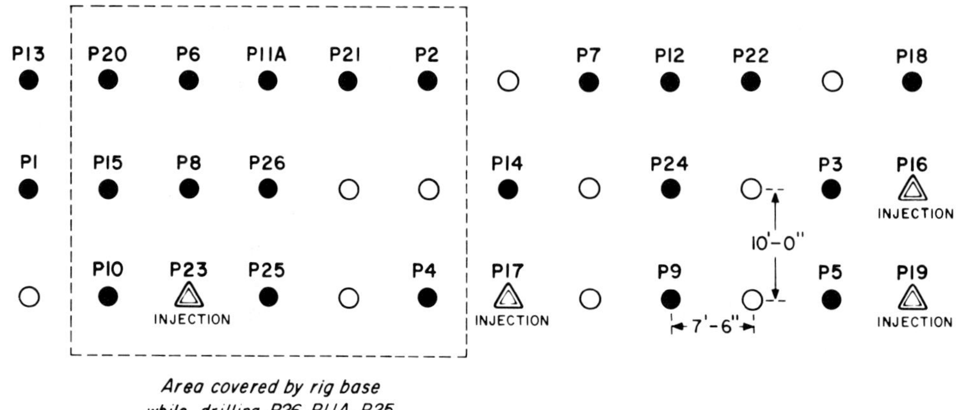

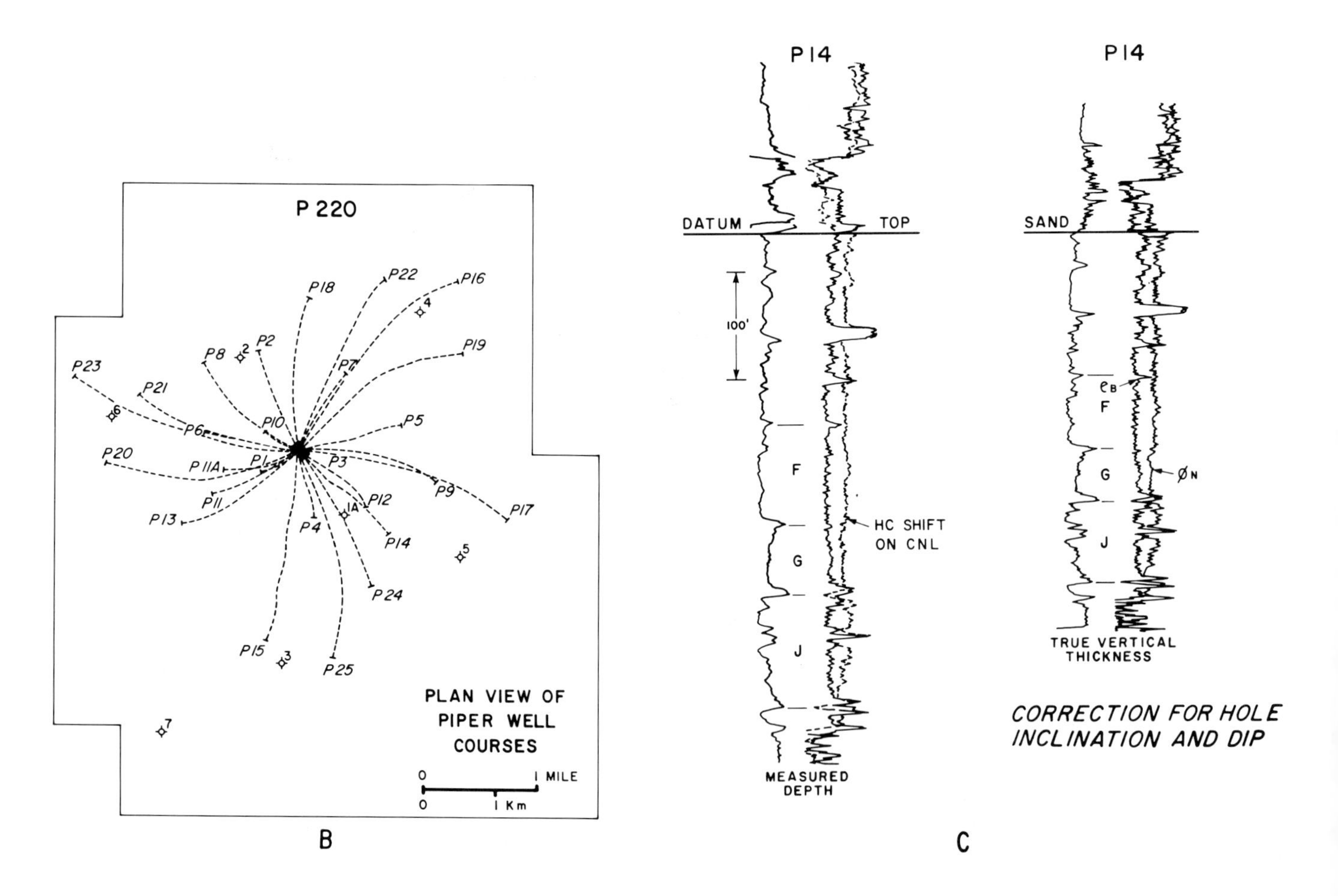

FIG. 25—**A:** Piper field cellar plan; **B:** plan view Piper well courses: **C:** correction for hole inclination and dip.

ficulty in identifying reflectors in fault blocks where there was no well control.

Seismic data are the most widely used source of data in understanding the geometry of the Piper field. All faults have been located with the use of seismic; only two wells, P21 and P23, have cut small faults. Neither of these minor faults can be seen on the seismic data, but because both wells are near major faults it is assumed that they have cut small antithetic faults.

Seismic quality is generally good, and where there is velocity control, velocity logs at the same vertical scale as the seismic sections make it relatively easy to identify reflectors (Payton, 1977). Base Piper Sandstone proved to be one of the most difficult reflectors to identify consistently because of variations in thickness of I shale and Callovian section, and the changing character of formations at the base of the Callovian (Sheriff, 1976). Basalt near fault C proved troublesome because it absorbed enough seismic energy to mask the deeper Zechstein formation and was much like the Zechstein reflector in appearance. Pre-1977 seismic data indicated a high probability of a horst block near fault C similar to that between faults A and B. To resolve this problem and obtain a better definition of fault alignment and the displacement at Piper, another 150 mi (242 km) of seismic lines were recorded in the summer of 1977. This, together with previous data enabled us to improve our fault control and to successfully locate several wells in very complexly faulted areas.

All pertinent seismic data were re-examined prior to recommending a production location, and reviewed immediately after each well was drilled to improve our understanding of individual reflectors. This technique enabled wells P6, P11A, and P24 to be successfully drilled in areas where Piper sandstone was partly truncated and provided confidence to locate wells very close to faults.

In summary, the level of geologic knowledge of the Piper field is based on more than 6,000 man-hours of seismic interpretation. Seismic data are used to determine time of movement on faults and all the changes in structure between wells. Without this data our understanding of Piper geology would not have progressed nearly as far and at a much slower and potentially costlier rate. An example of seismic interpretation is given in Figure 10, a northeast to southwest line showing faults and a tremendous change in thickness of Upper and Lower Cretaceous going from Block I across Blocks II and III into the Witch Ground graben.

DEVELOPMENT DRILLING

Development drilling called for as much long range planning as geologic knowledge of the structure could provide. A balance had to be struck between maximizing production, simultaneously drilling and the physical constraints imposed by a two-rig drilling operation (relative to cellar design and cellar orientation), and the need for geologic and reservoir information.

Orientation of the cellar (Fig. 25A) was fixed by platform design and weather conditions which required that the platform face 316°. Width of rig bases required 4 rows of slots between rigs. Thus, while P11A was drilling, the closest the second rig could be was the row containing P12, P24, and P9. It also was desirable to maintain maximum flexibility for future drilling and to avoid crossing under the cellar with any of the early wells (Fig. 25B). All adjacent production wells had to be closed in and temporarily plugged with a downhole safety valve during operations for drilling and setting 20-in (51-cm) casing, and until drilling for 13 3/8-in casing had reached a point where a rigidly adhered to program showed no possibility of intersecting an existing production string. For example, in the early stages of drilling P26, wells P25, P8, and P6, P11A, and P21 all had to be closed in and plugged.

Orientation of the cellar slots meant that the rig on the southwestern end of the cellar plan had to drill more wells with higher geologic risk.

Twenty-in. (51-cm) casing is set at 1,110± ft (335 m), or 500± ft (152 m) beneath the seabed. This section is composed of glacial till and, where boulders were encountered they tended to deflect the hole for the 20-in. (51-cm) casing which had to be maintained within 1/4° of vertical.

Programs for computing radius of error for wells being drilled were carried out on the platform with a dual computer system using the "Balanced Tangential Method" by Exploration Logging. The "Angle Averaging Method" was used by the Occidental direction drilling supervisor on the platform and the operations engineering group in the Aberdeen office. Because of the wide well spacing at Piper, a radius of 200 ft (61 m) was used for the bottom hole target at the midpoint of Piper Sandstone.

Significant differences between magnetic and gyro surveys were encountered and the gyro survey was chosen as being more accurate. The magnetic surveys are affected by the northern latitude, and in the Piper area by the presence of volcanics in Middle Jurasic beneath the Piper Sandstone. Because the gyro was run after casing, it was necessary to be able to predict what the error in the magnetic survey would be. It was most severe in an east to west direction but also varied with hole inclination. Our operations engineering group was able to use all of the data from early wells to successfully predict the magnitude of errors in magnetic surveys for subsequent wells.

From the first few wells it became apparent that a left-hand lead to the planned well course was required to compensate for a right-hand turn which developed in all wells (Fig. 25B). It also was learned that course corrections should be made above the Danian chalk. Course corrections made below the top of the Danian chalk were difficult and could lead to excessive torque. Those locations near a fault or fault intersection required especially careful planning to get everything right because it was necessary to be within the target at total depth. This was successfully accomplished in all cases.

Shale overlying the lower Paleocene sandstone contains some swelling clays which required mud weights of approximately 10.2 lb/gal. to control. These mud weights caused some differential sticking in Paleocene sandstones and also resulted in excessive torque and differential sticking while drilling the Piper Sandstone, in which the formation pressure had been drawn down to 2,600± psi compared to 4,300± psi in

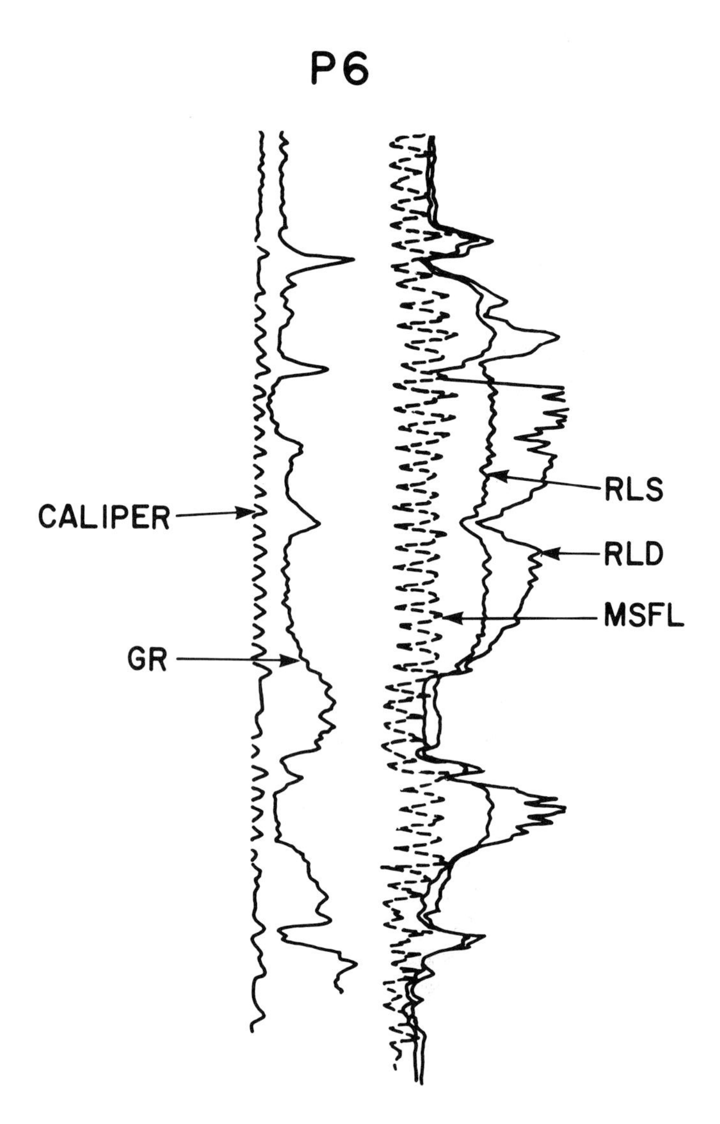

FIG. 26—Roller reamer effect.

the mud column. There also was some differential sticking of the logging tools and logging cable. This was overcome to some extent by running 9 5/8-in. (24.4-cm) casing above the Piper Sandstone. Mud weight could then be lowered enough to retain control of the Kimmeridge Shale and sandstones which might still be at virgin pressure, and at the same time reduce the pressure differential between the producing zone and the mud column to a more manageable 1,400± psi.

Severe sticking problems were encountered in wells P16 and P19 in Callovian silts and shales at the base of the Piper Sandstone and in the underlying Bathonian shales. The Bathonian shales contain the only pure clay minerals observed below the Kimmeridge Shale at Piper. There are scattered coal clasts in the Callovian section in these wells, but no coals beds. No sticking problems were encountered where relatively thick coal beds were drilled in the Callovian (Fig. 15A) and in the P5 and P9 where several coal beds were penetrated in the Middle Jurassic.

Roller reamers also were used to reduce torque while drilling. In holes which were in very good shape, no drag, nearly round, and a very smooth well course, the reamers cut a spiral groove up to 1/2 in. (1 cm) deep on the lower side of the borehole. This occurs while running in the hole when there is less or no weight on the hook and the pipe can rotate freely. The effect of the groove cannot be compensated for on pad type; open hole logging tools and the measurements from these tools are not usable (Fig. 26). Thus the only porosity log available is the sonic log.

In summary, many things had to be considered when determining drilling order for production wells. A need for geologic information had to be carefully considered relative to physical constraints, loss of production, reservoir performance, and relative geologic risk involved in a particular location.

LOGGING (OPEN HOLE)

With an average hole angle in excess of 45° and most wells exceeding 60° for some part of their course (Fig. 18), open hole logging encountered some unusual problems. Prior to production drilling, it was decided to use ISF-BHC Sonic-GR (Induction-Spherically Focused-Borehole Compensated-Compensated Neutron Log), HDT (High Resolution Dipmeter Tool), DST (Simultaneous Dual Laterolog run with Micro-Spherically Focused log) and ML-MLL (Microlog-Microlaterolog) in the initial production wells. Problems with high angle wells were anticipated and arranegments were made to have available slim hole SP-IES (Spontaneous Potential-Induction Electrical Survey), GR-BHC Sonic (Gamma Ray-Borehole Compensated Sonic). Later, slim hole GR-FDC-CNL became available. The diameters of these slim hole tools are approximately 1 1/2 in. (3.8 cm) and they can be run inside the drill pipe.

Very few problems were encountered in holes inclined less than 50°. The ISF-Sonic-GR was successfully run in holes where the inclination was up to 70° and where long sections of the borehole exceeded 65°. The ML-MLL collapsed in holes where the hole inclination exceeded about 35°. The DST (Simultaneous Dual Laterolog with Micro-Spherically Focused Log) and especially the GR-FDC-CNL proved troublesome to run in holes where the inclination approached 60°. These long, modern, combination tools with pads and especially the GR-FDC-CNL with the spring-loaded arm were not designed for high angle holes. When they were successfully run to bottom in high angle holes, they started off bottom with sufficient overpull to keyseat the line.

The logging problems were solved in several ways depending upon the specific need for data and borehole conditions:

1. Running through drill pipe logs after running and hanging off open-ended drill pipe to within 150± ft (46 m) of the zone of interest.

2. By reducing hole above the pay zone and drilling ahead to make sure the sandstones were present, attempting an induction spherically focused Sonic or through-drill-pipe log if deemed necessary, and then running 9 5/8 in. (24.4 cm) casing above the pay zone. After drilling ahead to the base of the sandstones, a normal logging operation could be run. When

this method was planned in advance, hole inclination was reduced above the pay zone to facilitate the logging operation.

3. Where severe problems were encountered and open hole logs were not considered as critical (water-injection wells), a through-drill-pipe log was run and the remainder of the logging was done in cased hole (gamma ray-formation density-compensated neutron, thermal decay time.)

After the initial five wells, the gamma ray induction spherically focused Sonic was dropped as a basic tool, but continued to be used where there were problems or danger of sticking the DSTor formation density compensated neutron log, and when these tools would not fall down the inclined hole.

High resolution dipmeter was run in most wells but proved very difficult to interpret in Piper because of high-energy sandstones or extensive bioturbation of sandstones and shales when energy was lower. The Schlumberger Mark IV program with horizontal reference and search angle equal to 85° minus the hole inclination proved best. Computer computations were checked by hand calculations from 1/20 film when irregular or suspect results were received. Several serious discrepancies were found which significantly changed net pay calculations. High resolution dipmeter results are used to calculate net vertical pay and to correct the measured depth logs to true vertical thickness for correlation and stratigraphic work (Fig. 25C)

The compensated neutron log shows a marked decrease of about 3 porosity units when passing from G sandstone into F sandstone. There is little or no decrease and often a slight increase in bulk density. This shift was so remarkable and occurred just above a high gamma ray peak that it led to considerable discussion as to its cause. It generally was agreed that it must be some form of hydrocarbon effect, but why in F sandstone and sandstones above, and not in G and H sandstones? A re-examination of cores and core results showed that this shift occurred when permeability of the sandstones was about 1.5 to 2 Darcies. Explanation of this observation is that in inclined boreholes with clean sandstones, at higher permeabilities, the filtrate drains away sufficiently to cause a significant hydrocarbon effect on the deeper reading CNL log (Fig. 25C). This effect is observed at about 1.5 Darcies for Piper sandstones and tends to make all of sandstones with permeabilities of this magnitude and above look alike on the formation density-compensated neutron log.

Study of this phenomenon and re-examination of the cores showed that open-hole logs give excellent and useable results, but much detailed information available from cores is not seen on logs and extreme caution should be used in interpreting sedimentary environments based largely on log data.

It also was interesting to note there was no increase and often a slight decrease in porosity going from G sandstone to F sandstone, but the permeability increased from less than one Darcy to more than six Darcies. Re-examination of the cores showed this to be a function of sorting and grain size. The G sandstone is a fine grained, well to very well sorted, bioturbated sandstone, whereas F sandstone is a medium to well sorted, coarse to very coarse grained sandstone. Poorer sorting reduces porosity, but coarser grains result in a dramatic increase in permeability of approximately a factor of 2 for each increase in grain size subclass (as was reported by Beard and Weyl, 1973).

Formation density readings give the most reliable results for effective porosity. They agree well with helium core porosities after core porosities have been reduced by approximately one porosity unit or 4% to correct for decompaction.

Water saturation calculations are straight forward as would be expected in a clean highly permeable sandstone. Lab derived values of 1.8 for both m and n are used.

Piper sandstones show a rapid decrease in water saturation in the first 200 ft (60 m) above the oil-water contact. Resistivities range from 0.5 ohm-meter from the induction log below the oil-water contact to more than 2,000 ohm-meter from the deep reading laterolog within 200 ft (60 m) above the oil-water contact. From 0 to 100 ft (0 to 30 m) above the oil water contact, water saturation decreased from 100% to less than 10% in the D, E, and F sandstones, and 15% in the G, H, and J sandstones. From 100 to 200 ft (30 to 60 m) above the oil-water contact, water saturation decreases to less than 2% in the F sandstone, 4% in the D sandstone and the H sandstone, 5% in the E and G sandstones, and 10% in the C sandstone. Water saturation continues to decrease to less than 2% for nearly all sandstones near the top of Piper reservoir. This is in complete agreement with the petrography of the Piper sandstones, predominantly pure quartz with little or no clay present and clean unaltered quartz surfaces.

There has been considerable discussion as to whether these low water saturations indicate an oil-wet reservoir. If the water absorbed on clean quartz surfaces is no more than a few molecular layers thick (Wyllie, 1962; Leversen, 1967; Tissot and Welte, 1978) and the irreducible water is held in the smallest pores and around grain contacts (Wyllie, 1962; Wilson and Pittman, 1977; Almon, 1979), it is completely reasonable to expect very low water saturations from 100 to 200 ft (30 to 60 m) above the oil-water contact.

Precise figures for low water saturations are very difficult to determine because the calculations depend upon ionic movement which requires a continuous thin film of water (Pingitore, 1976). A thin film of water, 1μm, on the surface of each grain and held in place by pore-lining authigenic clay, would be several thousand molecular layers thick and would increase the irreducible water saturation by approximately 1% for a medium to coarse sandstone. Clay lining pores are absent at Piper and the ionic movement must take place through a water-filled, interconnected network of small pores.

All of the sandstones show some conductivity and this suggests that the small pores form an interconnected network even for the very low water saturations of less than 2%.

All Piper sandstone members have water saturations in the first 100 ft (30 m) above the water table which fits all the available data and standard log calculations. It appears that bouyant forces were able to overcome capillary retentive forces (Berg, 1975) and reduce water saturations to uncommonly low values in a relatively short distance above the oil-water contact in the Piper sandstones (Fig. 27A, B).

Conner and Kelland (1974) reported the gamma ray as a good grain size indicator which is often true for Piper sand-

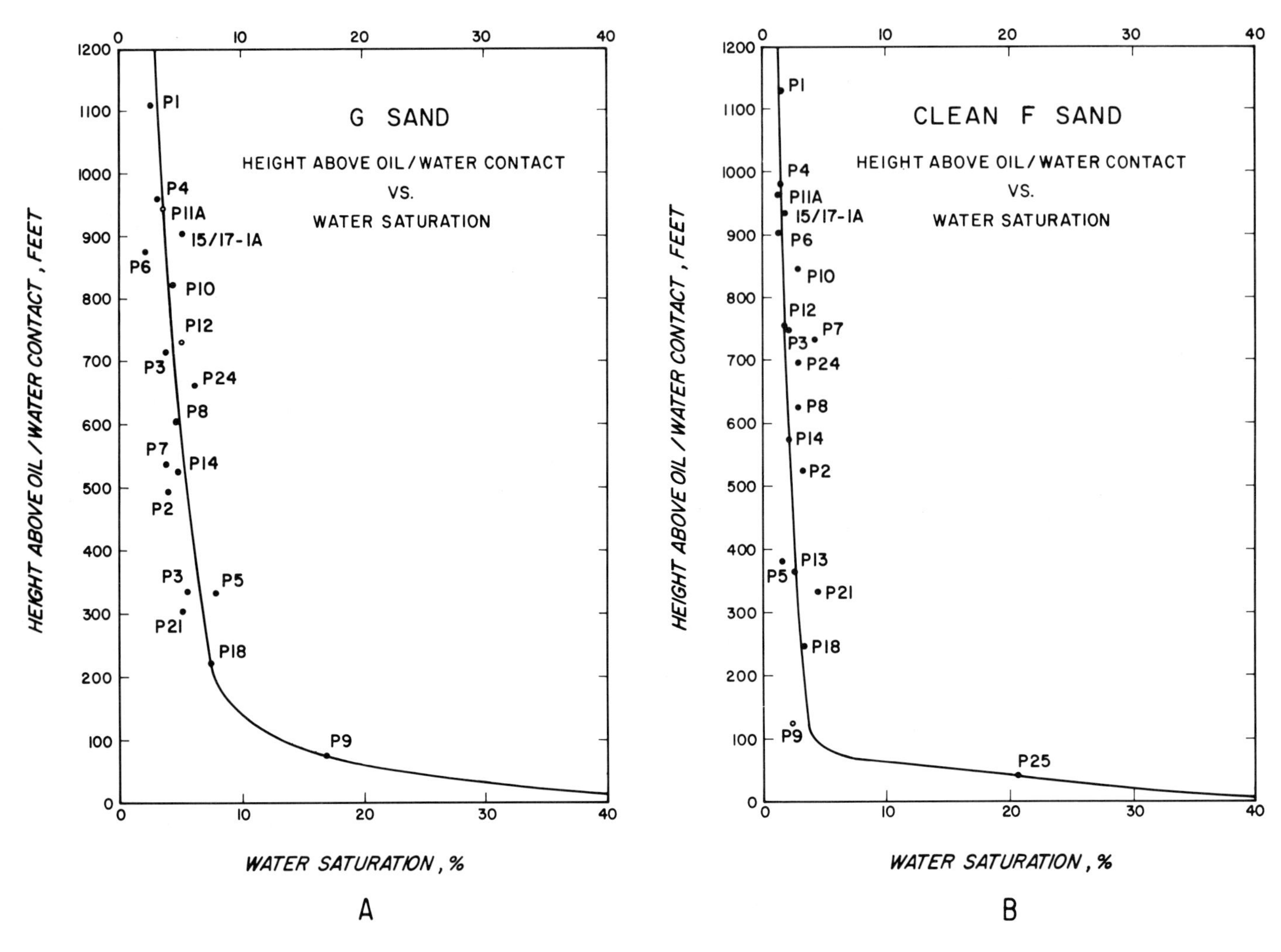

FIG. 27—**A:** Height above oil-water contact versus water saturation, G sandstone; **B:** height above oil-water contact versus water saturation, F sandstone.

stone, but there are several exceptions. The most important of these is shown on the type log 15/17-6 (Fig. 14) where upper F sandstones show decreasing gamma ray opposite decreased porosity and a marked decrease in permeability. Scanning electron microscopy (SEM) studies show a relatively large increase in authigenic kaolinite in this interval (Fig. 22).

A radioactive sandstone at base of F sandstone has been seen in all wells (See discussion under *Stratigraphy of the Piper Sandstones*). It was earlier reported to be caused by the mineral zircon (Conner and Kelland, 1974). They thought it was a random effect and would have no correlation value. The first production geology effort, before production drilling began, determined that this gamma ray peak was a correlatable event between 15/12-1, 15/17-4, and 15/17-5, and subsequent production drilling confirmed this correlation.

Thermal-decay time logging in a cased hole delineates this and other radioactive sandstones very well. Because the third gate is set to take out the background reading in the TDT, it was apparent that the radioactive sandstone also must have an element with a high-absorption cross section. This could be the element hafnium which generally is associated with zircon and has an absorption cross section of more than 100 barns. Uranium or thorium in the zircon would yield natural gamma rays.

This seems a reasonable explanation because hafnium was identified by analyzing X-ray spectra of Norwegian and Greenland zircons (Encyclopedia Britannica, 15th Edition).

Another source would be gadolinium which has the highest absorption cross section (49,000 barns) and has been reported in the North Sea (R. Simmond and C. Boyeldieu, Schlumberger, personal commun.). A likely mineral for the source of the gadolinium would be monazite, although gadolinium could be associated with many minerals. Most monazite contains thorium and thus both elements would be present to yield the responses seen on the gamma ray and TDT (Clavier and Hoyle, 1969). Material isolated within the core and sidewall samples by the use of radiation methods has been separated for microprobe analysis. Preliminary results indicate the mineral monazite as a source of the readings on the gamma-ray log and TDT. Zircon also is present, but appears to have little effect on the readings.

The repeat formation tester (RFT) was used to good advantage to measure pressures in individual sandstone members. It proved difficult to run in long open hole intervals, but where casing was run to just above the pay zone it worked well. Pressure measurements with the RFT in P25 showed all sandstone members plotting on the gradient of depleted pressure in the reservoir except for a basal 10-ft (3-m) sandstone which was at virgin pressure, approximately 1,000 psi greater than the sandstones above.

LOGGING (CASED HOLE)

The principal case hole logs are the formation interval tester (FIT), thermal-decay time log, production log (PCT; spinner survey), cement bond log, temperature log, and gamma ray-formation density log-compensated neutron log when no open hole log was obtained.

A thermal-decay time log is run in production holes shortly after they go on production as a base log. It has proved to be a very reliable and useful log at Piper. Because there is essentially no clay or shale in the sandstone, the 75,000 ppm formation water can be readily seen against the background log when water breaks through in a producing zone.

The formation interval testing log was successfully used to measure pressures in the separate sandstone members. It has the disadvantage of having to squeeze-cement a very small diameter hole, approximately 1/4 in. (0.6 cm), if the FIT is taken in a zone which is not to be left open during production or injection.

USE OF COMPUTER

Extensive use of the computer has been made to correct logs for dip and hole deviation. True vertical thickness (TVT) logs derived with the use of computer are used in all stratigraphic work (Fig. 25C). At present all logs, well courses, well data, and maps are stored in a fashion that enables us to draw profiles, have logs played back along well courses and to determine the volume of all sandstone members with respect to depth for any area of the field. This last feature is derived from surfaces of hand-contoured maps for each sandstone member which were stored in the computer.

The primary use for geologic data stored in the computer is integration of geologic information into a reservoir simulation study, but it also allows for a more rapid and efficient retrieval of data.

Figure 28 is an interesting three-dimensional view of Top Piper Sandstone with well courses which is derived from stored data.

CONTINUING STUDIES

Continuing studies include: (1) petrographic work to better understand the nature of the Piper sands; (2) cross section work to get all of the existing wells on a 1/2,000 scale cross section: (3) cross sections and re-examination of seismic data pertaining to water breakthrough problems; (4) integration of geologic data into a reservoir simulation study; and (5) studies to recommend infill and redrill locations.

FUTURE DRILLING AT PIPER

After the current location, the P26 well into Block II, there is only one planned location at Piper, a location to develop the small accumulation in Block III.

Any additional drilling will be largely dictated by reservoir performance.

CONCLUSIONS

1. Piper is a large and complex reservoir, containing 1.37 billion stock tank barrels of oil originally in place in three, tilted, folded, fault blocks.

2. Favorable geologic and reservoir conditions allow for complete development and high production rates from one centrally located platform.

3. Geologic understanding of the reservoir and the characteristics of the piper sandstones should enable Occidental to maximize the hydrocarbon recovery from Piper field.

REFERENCES CITED

Almon, W. R., 1979, Sandstone diagenesis—applications to exploration and exploitation: AAPG Clastic Diagenesis School, study notes, p. 1-94.

Bathurst, R. G. C., 1975, Carbonate sediments and their diagenesis: 2nd enlarged edition, Amsterdam, Elsevier, 658 p.

Beard, D. C., and P. K, Weyl, 1973, Influences of texture on porosity and permeability of unconsolidated sand: AAPG Bull., v. 57, p. 349-369.

Belsoussov, V. V., 1974, Seafloor spreading and geologic reality, *in* C. F. Kahle, ed., Plate tectronics—assessments and reassessments: AAPG Mem. 23, p. 155-166.

Berg, R. R., 1975, Capillary pressures in stratigraphic traps: AAPG Bull., v. 59, p. 939-956.

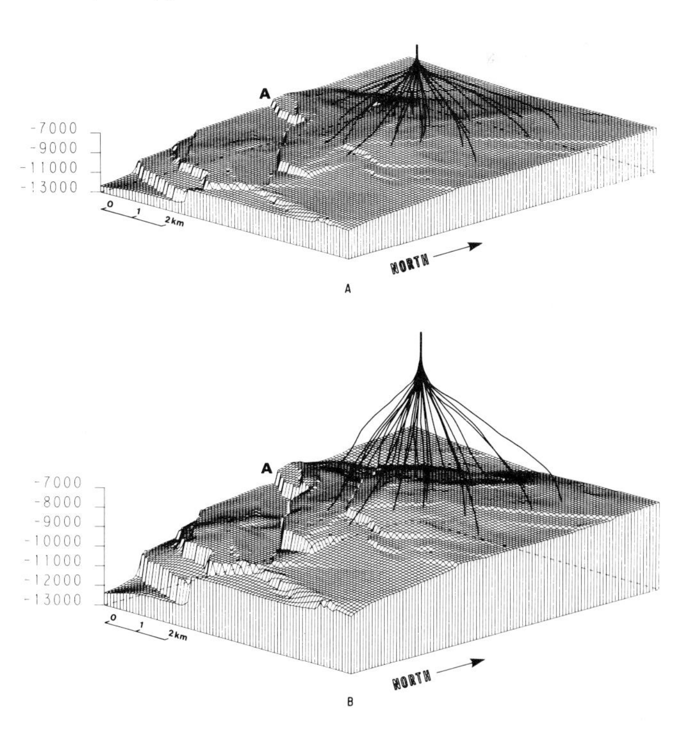

FIG. 28—**A:** True scale 3D view from SW corner of Piper field; **B:** Same view as Figure 28A with 2x vertical exaggeration.

Brennand, T. P., 1975, The Triassic of the North Sea, *in* A. W. Woodland, ed., Petroleum and the continental shelf of North West Europe, vol. 1, Geology: London, Applied Science Publ., p. 295-312.

Clavier, C., and W. R. Hoyle 1969, Quantitative interpretation of TDT logs, Parts I and II: Jour. Petroleum Techonology, p. 743-763.

Conner, D. C., and D. G. Kelland, 1974, Piper field, U. K. North Sea, interpretive log analysis and geologic factors, Jurassic reservoir sands, *in* 3rd European Formation Evaluation Symposium (London): Trans. Soc. Professional Well Log Analysts, p. 1-14.

de Boer, E. T., 1978, The use of accurate pressure data in a high permeability environment: Proc. European offshore Petroleum Conference (London), p. 387-396.

Elliot, T., 1978, Clastic shorelines, *in* H. G. Reading, ed., Sedimentary environments and facies: London, Blackwell Sci. Publ., p. 143-177.

Fuchtbauer, H., 1967, Influence of different types of diagenesis on sandstone porosity: Proc. 7th World Petroleum Cong.

Gates, G. L., and W. H. Caraway, 1965, Oil well scale formation in waterflood operations using ocean brines, Wilmington, California: U.S. Bureau of Mines. Rept. Inv. 6658.

Glaister, R. P., and H. W. Nelson, 1974, Grain-size distributions: an aid in facies identification: Bull. Canadian Petroleum Geology, v. 22, p. 203-240.

Gorsline, D. S., 1978, Anatomy of margin basins: Jour. Sedimentary Petrology, v. 48, p. 1055-1068.

Houston, R. S., and J. F. Murphy, 1977, Depositional environment of Upper Cretaceous black sandstones of the western interior: U.S. Geol. Survey Prof. Paper 994-A, 29 p.

Hulme, J. E., 1979, North Sea development drilling (preprint): 1979 Drilling Technology Conference, Int. Ass. Drilling Contractors (Denver).

Illies, J. H., 1977, Ancient and recent rifting in the Rhinegraben, *in* R. T. C. Frost and A. J. Dikkus, eds., Fault tectonics in N. W. Europe: Geol. en Mijnbouw, v. 56, p. 329-350.

Kahle, C. F., ed., 1974, Plate tectonics—assessments and reassessments: AAPG Mem. 23, 514 p.

Kent, P. E., 1975, The tectonic development of Great Britain and the surrounding Seas, *in* A. W. Woodland, ed., Petroleum and the continental shelf of North West Europe, vol., 1, Geology: London, Applied Science Publ., p. 3-28.

Leverson, A. I., 1967, Geology of petroleum (2nd edition): San Francisco, W. H. Freeman and Company, 724 p.

Lowell, J. D., 1976, Structural geology: Training reference manual by Oil and Gas Consultants International, Inc.

McAuliffe, C. D., 1979, Oil and gas migration—chemical and physical constraints: AAPG Bull., v. 63, p. 761-781.

Meyerhoff, A. A., and H. A. Meyerhoff, 1974, Ocean magnetic anomalies and their relations to continents, *in* C. F. Kahle, ed., Plate tectonics—assessments and reassessments: AAPG Mem. 23, p. 411-422.

——— and ——— 1974, Tests of plate tectonics, *in* C. F. Kahle, ed., Plate tectonics—assessments and reassessments: AAPG Mem. 23, p. 43-145.

Middleton, G. V., and J. B. Southard, 1977, Mechanics of sediment movement: SEPM Short Course Number 3.

Meucke, T. W., 1979, Formation fines and factors controlling their movement in porous media: Jour. Petroleum Technology, p. 144-150.

Palmer, T., 1977, Occidental Company report of macrofossil examination of shales in Piper wells 15/17-4 and P10.

Payton, C. E., ed., 1977, Seismic stratigraphy—applications to hydrocarbon exploration: AAPG Mem. 26, 516 p.

Pingitore, N. E., Jr., 1976, Vadose and phreatic diagenesis: processes, products and their recognition in corals: Jour. Sedimentary Petrology, v. 46, p. 985-1006.

Powers, C. M., 1953, A new roundness scale for sedimentary particles: Jour. Sedimentary Petrology, v. 23, p. 117-119.

Ramsbottom, W. H. C., 1978, Permian, *in* W. S. McKerrow, ed., The ecology of fossils; an illustrated guide; London, Gerald Ducksworth and Company, Ltd., p. 164-193.

Reading, H. G., 1978, Facies, *in* H. G. Reading, ed., Sedimentary environments and facies: London, Blackwell Sci. Publ., p. 4-14.

Robin, P. Y. F., 1978, Pressure solution at grain-to-grain contacts (Toronto Univ.): Geochim. et Cosmochim. Acta., v. 42, no. 9, p. 1383-1389.

Schlumberger, Ltd., 1976, The essentials of thermal-decay time logging, p. 31.

Sellwood, B. W., 1978a, Triassic, *in* W. S. McKerrow, ed., The ecology of fossils; an illustrated guide: London, Gerald Ducksworth and Company, Ltd., p. 194-203.

——— 1978b, Jurassic, *in* W. S. McKerrow, ed., The ecology of fossils; an illustrated guide: London, Gerald Ducksworth and Company, Ltd., p. 204-279.

Sheriff, R. E., 1976, Inferring stratigraphy from seismic data: AAPG Bull., v. 60, p. 528-542.

Tissot, B. P., and D. H. Welte, 1978, Petroleum formation and occurrence; New York, Springer-Verlag, 538 p.

Van Hinte, J. E., 1976a, A Jurassic time scale: AAPG Bull., v. 60, p. 489-497.

——— 1976b, A Cretaceous time scale: AAPG Bull., v. 60, p. 498-516.

Visher, G. S., 1969, Grain size distribution and depositional processes: Jour. Sedimentary Petrology, vol. 39, p. 1074-1106.

Weimer, R. J., 1976, Deltaic and shallow marine sandstones, sedimentation, tectonics, and petroleum occurrences: AAPG Education Course Notes Series No. 2, 169 p.

Williams, J. J., D. C. Conner, and K. E. Peterson, 1975, Piper oilfield, North Sea: fault-block structure with Upper Jurassic beach/bar reservoirs sands: AAPG Bull., v. 59, p. 1581-1601.

Wilson, M. D., and E. D. Pittman, 1977, Authigenic clays in sandstones: recognition and influences on reservoir properties and paleoenvironmental analysis: Jour. Sedimentary Petrology, v. 47, p. 3-31.

Wyllie, M. R. J., 1962, Relative permeability, *in* T. S. Frick, ed., Petroleum production handbook, vol. II, Reservoir Engineering: New York, McGraw-Hill Book Co. Inc.

Ziegler, P. A., 1975, North Sea basin history in the tectonics framework of North-Western Europe, *in* A. W. Woodland, ed., Petroleum and the continental shelf of North West Europe, vol. 1, Geology; London, Applied Science Publ., p. 131-150.

Ziegler, W. H., 1975, Outline of the geologic history of the North Sea, *in* A. W. Woodland, ed., Petroleum and the continental shelf of North West Europe, vol. 1, Geology: London, Applied Science Publ., p. 165-190.

Ninian Field, U.K. Sector, North Sea[1]

By W. A. Albright[2], W. L. Turner[3], and K. R. Williamson[2]

Abstract The Ninian oil field was discovered in January 1974 and went on production in December 1978. The reservoir is in fluvio-deltaic Middle Jurassic sand on a westward tilted horst block. The reservoir is limited on the various margins by a combination of erosional truncation faulting, and downdip by the oil water contact. Current estimates of reserves are 1.2 billion stock tank barrels of oil, 63 million stock tank barrels of natural gas liquids, and 230 BCF gas. As of June, 1979, field development has progressed to enable a daily production rate of approximately 225,000 BOPD.

INTRODUCTION

The Ninian oil field is located on the United Kingdom continental shelf in blocks 3/3 and 3/8, approximately 90 mi (144 km) east-northeast of the Shetland Islands. Approximate geographical coordinates are 60°50′N lat., and 01°28′E long. The field was discovered in January 1974 by the British Petroleum-Ranger 3/8-1 well and was named "Ninian" field after Saint Ninian, the patron saint of the Shetland Islands. (The location of the field relative to landmarks and other North Sea fields is illustrated in Figures 1 and 2.) Water depth in the field area varies from 440 to 490 ft (134 to 149 m).

Following the discovery, appraisal drilling continued without interruption during 1974 and 1975. A total of eight exploratory and appraisal wells were drilled to delineate the field, including six wells on block 3/3 and two wells on block 3/8. The rapid delineation of the field confirmed major oil reserves, and plans for development were well advanced by the end of 1974. Construction of facilities began in 1975, and currently (June 1979) the pipeline system and two platforms are installed. Development drilling is in progress and production began in December 1978. Nine wells are now on stream producing over 225,000 bbl of oil per day. The jacket for a third platform has been installed and will be completed during 1979.

HISTORY OF EXPLORATION AND DISCOVERY

General

The discovery of Gronigen gas field in the Netherlands in 1959 gave renewed impetus to petroleum exploration in northwestern Europe (Fig. 1). Following the trend of gas productive structures, the search moved offshore into the southern North Sea, resulting in the discovery of many large gas fields in the mid and late 1960s, primarily in the UK sector. Subsquently, progressive northward expansion of exploration led to a number of discoveries, including major oil fields in Danian-Maastrichtian chalk formations at Ekofisk field in 1969, in Lower Tertiary sandstones at Forties field in 1970, and in Jurassic sandstones at Brent field in mid-1971. With the Brent field discovery, nearly 600 mi (960 km) northwest of

[1]Manuscript received, July 9, 1979; accepted for publication, September 9, 1979.

[2]Chevron Overseas Petroleum, Inc., San Francisco, California 94105.
[3]Chevron Petroleum (U.K.) Ltd., London.

The writers want to acknowledge that the geology of the Ninian field, as described in this paper, reflects contributions from a number of colleagues both in Chevron and in partner companies. However, all interpretations presented in this paper are those of Chevron and do not necessarily reflect the views of the Ninian partners. We thank the management of Chevron and of the Ninian field partners for permission to publish this paper.

Article Identification Number:
0065-731X/80/M030-0009/$03.00/0.

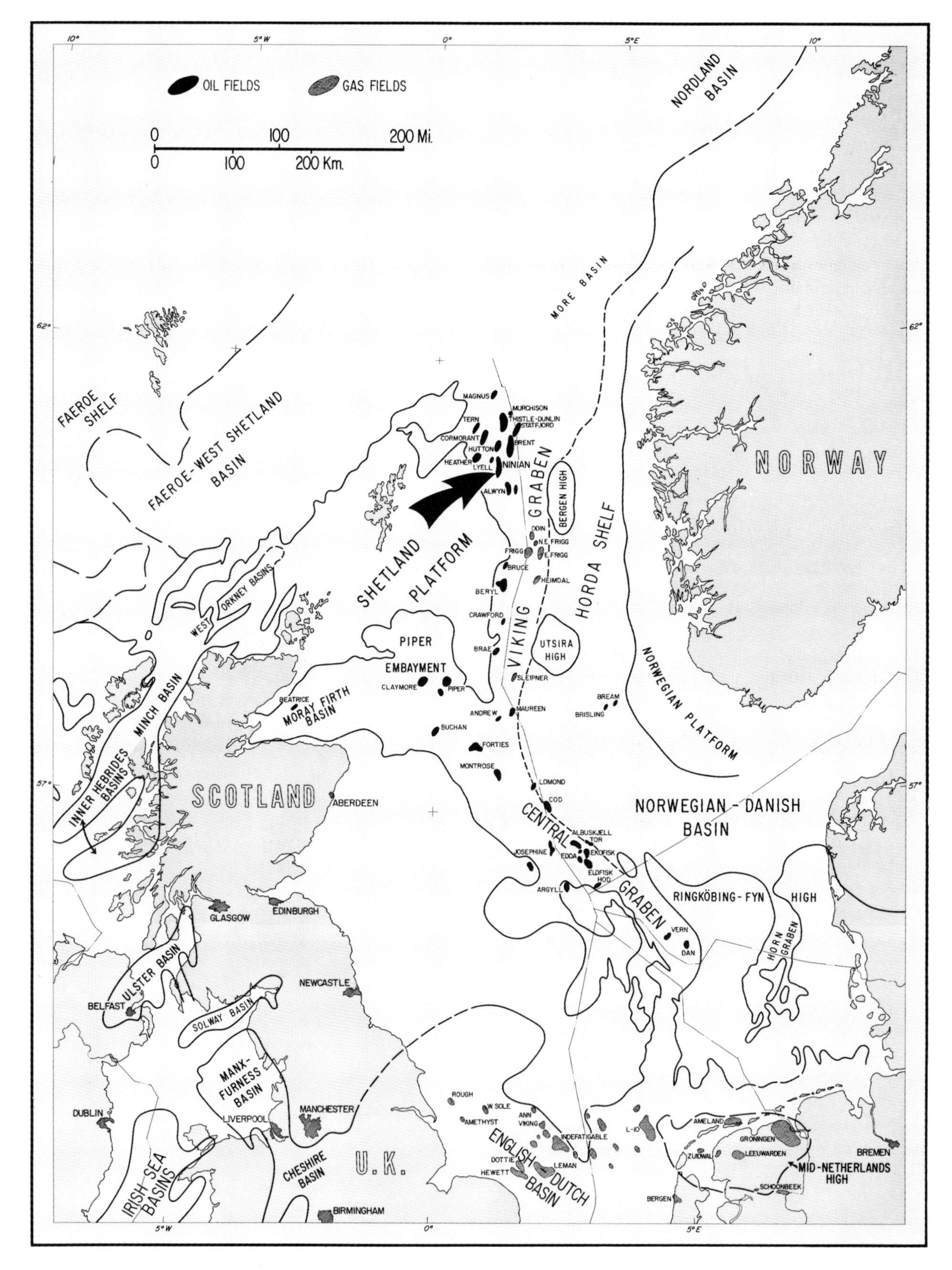

FIG. 1—Location of fields and salient features in the North Sea.

Groningen, the industry-wide exploration effort entered a major new phase. The geology of the major North Sea oil and gas accumulations and the criteria for successful exploration were generally well understood. The Brent discovery intensified exploration in the northern North Sea, and was followed in rapid succession by the discovery of a number of major oil fields in similar geological situations, including the Ninian field.

The licenses covering Ninian field blocks 3/3 and 3/8 were awarded by the British Government in the fourth round of awards in March, 1972. Participants and the interests in license P-202, which includes Block 3/3, were as follows: Burmah Oil North Sea, Ltd. (Burmah), 30%, ICI, Ltd. (ICI) (now ICI Petroleum, Ltd.), 26%; Chevron Petroleum (U.K.), Ltd. (Chevron), 24%; Murphy Petroleum, Ltd. (Murphy), 10%; and Ocean Exploration Co., Ltd. (Odeco), 10%. The Burmah license interest in Ninian field was subsequently purchased by the British National Oil Co. (BNOC). Participants and the interests in License P-199 which comprises block 3/8, are as follows: British Petroleum Development, Ltd. (BP), 50%; Ranger Oil (UK), Ltd. (Ranger), 20%; London and Scottish Marine Oil Co., Ltd. (LSMO), 23%; Scottish and Canadian Oil and Transportation Co., Ltd. (SCOT), 7%.

Drilling and Discovery History

Burmah served as operator for the P-202 license group and BP for the P-199 license group. Initial exploration on the blocks proceeded separately. However, it was recognized from the outset that a large structure was shared between blocks 3/3 and 3/8, and that unitization and joint development would be desirable if the structure proved productive. BP spudded the 3/8-1 well on September 16, 1973, and Burmah spudded the 3/3-1 well 4 mi (6.4 km) to the north on October 4, 1973, and both wells were drilling simultaneously in late 1973—early 1974. BP 3/8-1 was completed on March 12, 1974 as the Ninian discovery well. The Burmah 3/3-1 was completed on April 4, 1974, confirming the discovery and extension of Ninian field onto Block 3/3. Well and seismic data indicated that a larger part of the field reserves underlay block 3/3.

The license holders appointed a Ninian management committee with each participant represented, and delegated overall authority for a joint development project to this committee. Burmah was designated operator for field development and BP was designated operator for the construction and operation of pipeline and for securing terminal facilities. Chevron succeeded Burmah as Ninian field operator from March 1, 1975. Following approval of the Ninian Unit Agreement in January, 1979, Chevron was confirmed as operator of the Ninian Unit. Meanwhile BP succeeded Shell as operator of the terminal facility.

History of Geophysical Interpretation

By mid-1972 the two license groups in Ninian had each acquired and mapped a grid of seismic data of various origins. Line BB, Figure 3, is an example of 24-fold common-depth point stack (CDPS) data of that period. Figure 4 illustrates an early 1973 seismic map interpreted from those data. The mapped horizon is a regionally prominent seismic event which corresponds to the base of Cretaceous unconformity. A widespread velocity-density contrast occurs at the unconformity so reflection continuity at the unconformity is good. However, the target Jurassic section underlying the unconformity is structurally discordant with the unconformity surface, and critical pre-unconformity reflections are weak and difficult to resolve. Discovery wells 3/3-1 and 3/8-1 were positioned on the basis of data of this type and quality.

A new seismic survey was shot in 1974 with the lines spaced to produce a 1-km grid over the field area. These lines were also 24-fold CDPS data recorded on 48 channels with 50 m group spacing, half the spacing of Line BB. Line 34 (Fig. 3) shows the improvement in data quality obtained in 1974. The improvement is believed due to the closer group spacing, improved airgun arrays, and improved data processing. The maps included in this paper are based on a unitization subcommittee interpretation of the 1974 seismic data and a smaller survey shot in 1976.

In 1978 the area was resurveyed with lines spaced to provide a 0.5-km grid. This close control was planned to provide the best possible interpretation of structure and fault definition for development of the field. Lines 23 and 50 (Fig. 3) are examples from this survey. The lines were recorded 48-fold CDPS on 96 channel equipment with 25-m group spacing, using a 21-airgun energy sources. Processing included signature deconvolution and wave equation migration.

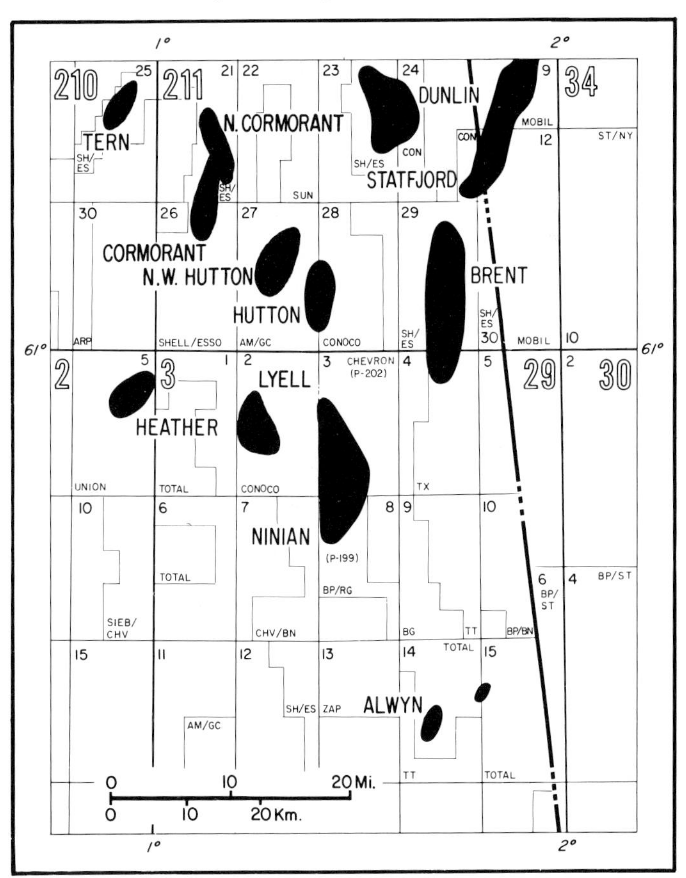

FIG. 2—Blocks and licenses in the Ninian field area, North Sea.

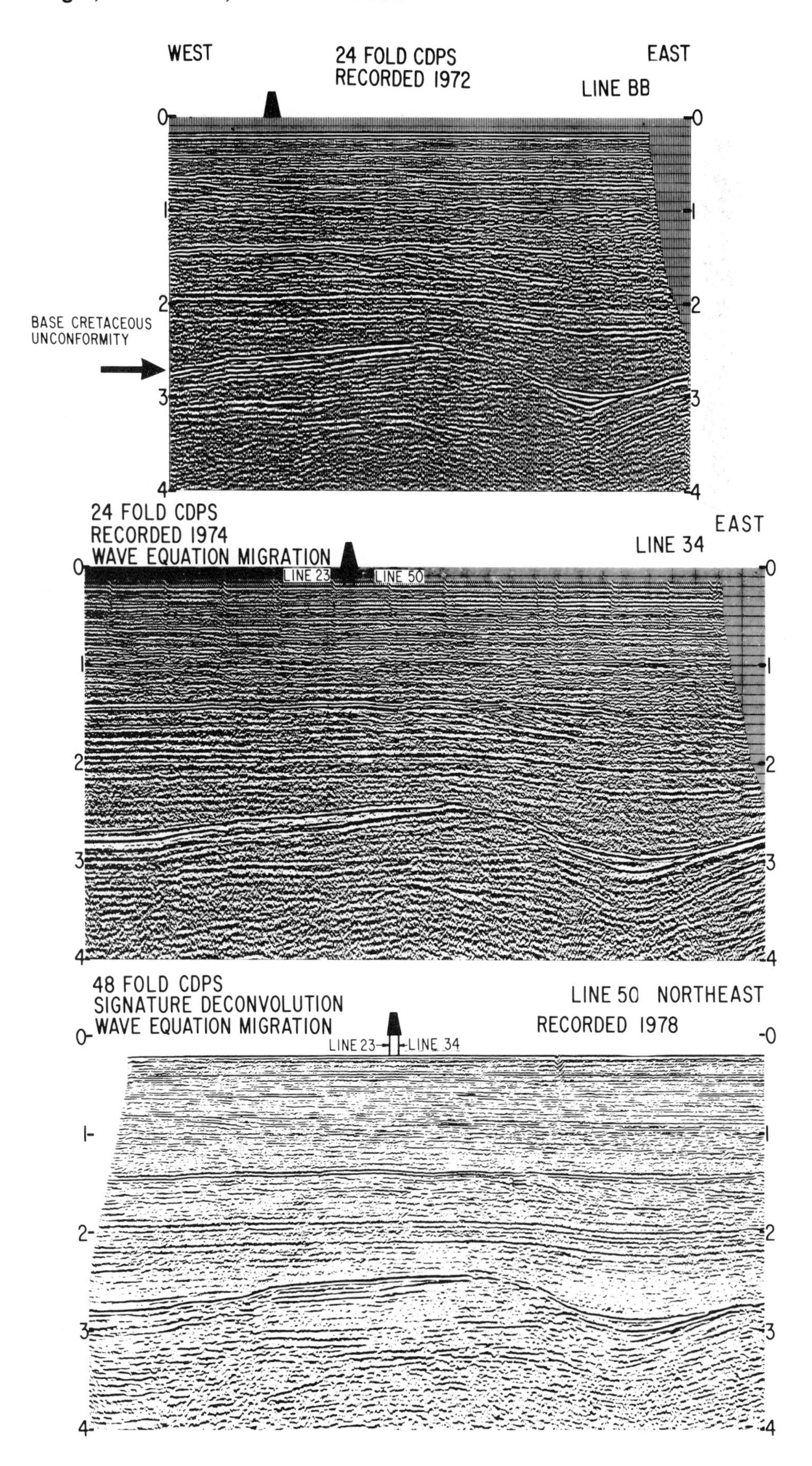
WEST
24 FOLD CDPS
RECORDED 1972
EAST
LINE BB
BASE CRETACEOUS UNCONFORMITY
24 FOLD CDPS
RECORDED 1974
WAVE EQUATION MIGRATION
EAST
LINE 34
LINE 23
LINE 50
48 FOLD CDPS
SIGNATURE DECONVOLUTION
WAVE EQUATION MIGRATION
LINE 50 NORTHEAST
RECORDED 1978
LINE 23
LINE 34
0
1
2
3
4

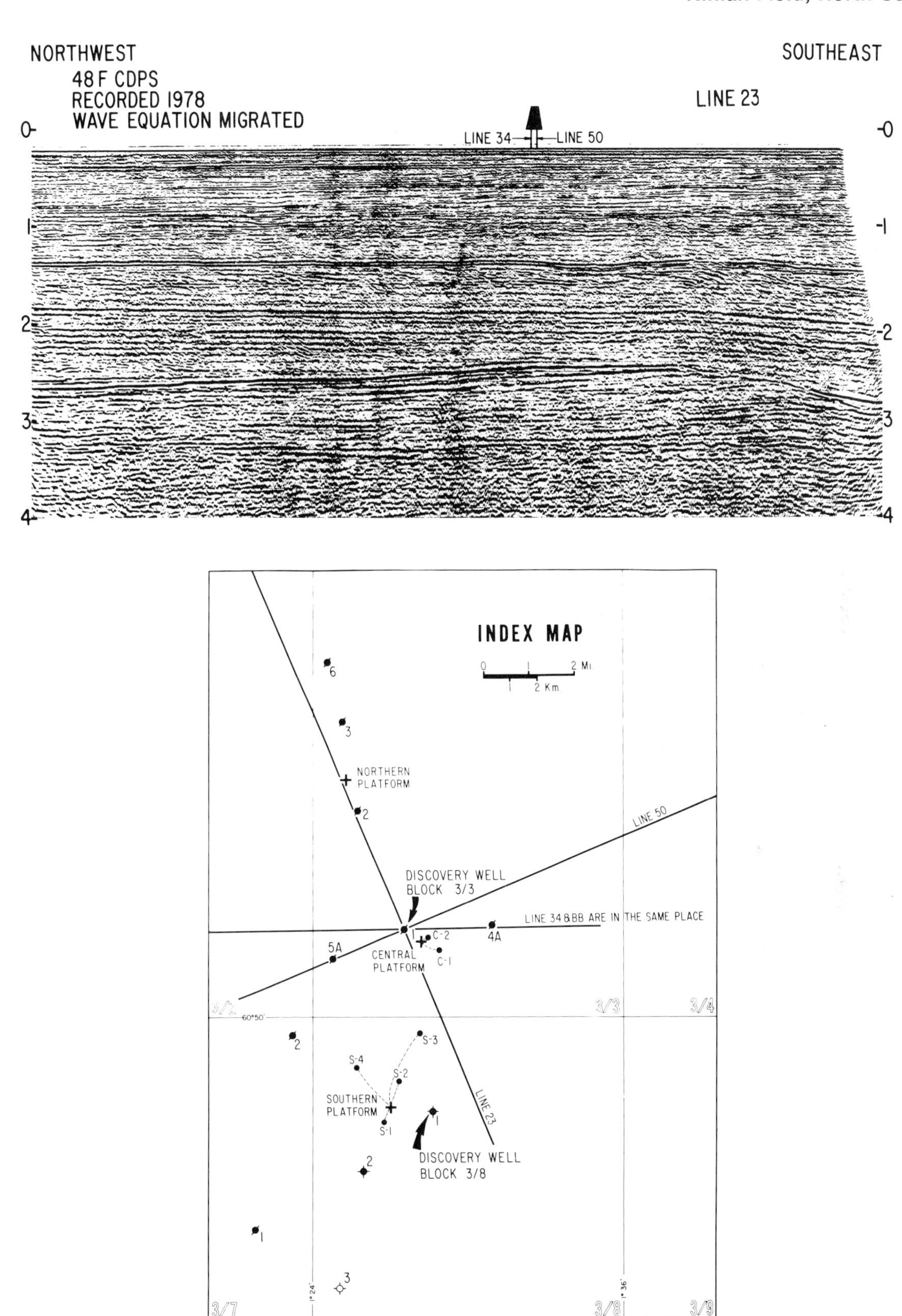

FIG. 3—Seismic profiles in the Ninian field, North Sea. Note different wavelet processing and common-depth-point stacking. Location of profiles on index.

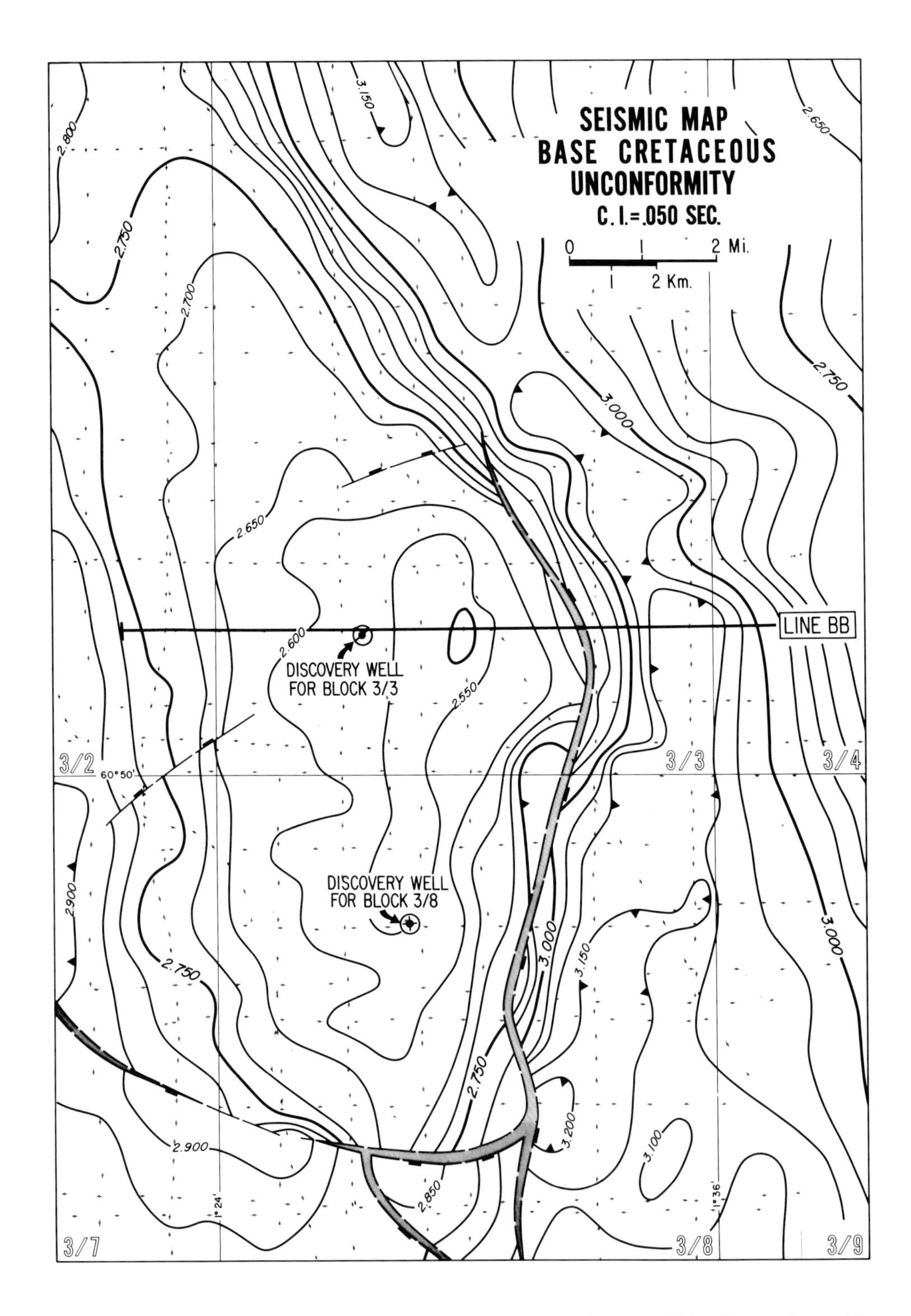

FIG. 4—Ninian seismic map of the base of the Cretaceous unconformity (1973). Contour interval is 0.5 seconds of two-way travel time.

GEOLOGY

General

Figure 1 depicts the major structural elements of the North Sea and the location of Ninian and other North Sea fields in the Northern Viking graben. The graben is part of a major rift system lying along a line of incipient crustal divergence between the British Isles and Norway. The Viking graben is bounded on the west by the Shetland platform and on the east by the Horda Shelf-Norwegian platform. The opening of the Viking graben may have begun as early as Late Permian time. Subsequent crustal extension resulted in major block faulting, downwarping, and initiation of subsidence of the North Sea basin. A thick section of Triassic and Jurassic sediment was deposited in the subsiding basin. Faulting reached a climax in the Middle and Late Jurassic and was accompanied by fault block rotation and erosion of structurally high areas on the tilted fault blocks. Several significant phases of erosion occurred. One of the major period is indicated by the regional base of Cretaceous (late Kimmerian) unconformity. Faulting strongly influenced the thickness of Jurassic sedimentation and wide variation in thickness occurs across the major faults. Movement on the faults had largely ceased by Early Cretaceous time, but broad basinal subsidence continued. Cretaceous and Tertiary sedimentation buried the block faulted pre-Cretaceous terrain. Characteristically, little or no fault displacement is evident in Cretaceous and younger beds. Poorly defined, low relief drape structure is characteristic of the Upper Cretaceous-Lower Tertiary section, due to differential compaction over the tilted Jurassic fault-block structures. These relationships are illustrated on Figure 5, a diagrammatic cross section extending from the Shetland platform eastward through Ninian field and the Brent-Stratfjord trend.

Ninian Field

Structure—The Ninian field oil accumulation occurs in a west dipping, Middle Jurassic sandstone on a basement-controlled westward-tiled horst block. The structural configuration of the "Top of Reservoir Sand" and probable oil-bearing area of the field are illustrated in Figure 12. The block is bounded on the east by a major, generally north to south trending fault system, downthrown to the east. Individual faults are segments in a zone of related fractures along which maximum displacement appears to transfer from one fault to another. In the Jurassic section, total displacement across the fault system exceeds 2,500 ft (762 m). Maximum displacement at basement level is probably about 5,000 ft (1,524 m). A second set of normal faults bounds the southern and western margins of the field. On the south these faults strike west-northwest and are downthrown southwestward. These faults turn northward along the western margin of the field and are downthrown toward the west. Although these faults exhibit several hundred feet of displacement of the Jurassic section, they are subordinate to the down-to-the-east system.

In the interior of the field only minor faults have been mapped from seismic data. Minor, normal faulting is interpreted in wells 3/3-4A and 3/8-2. A small horst block occupies the crestal area of the field near well 3/3-4A. This well has an atypically thin Jurassic sequence which results from a combination of depositional thinning, erosional truncation, and possible faulting. The attenuated Jurassic section in this well provides striking evidence that faults were active at Ninian during Jurassic deposition and strongly influenced bed thicknesses.

The eastern and southern limits of the reservoir are controlled by truncation of the Middle Jurassic sandstone by unconformities both in the Upper Jurassic and at the base of the Cretaceous. Truncation occurs at the base of the Callovian unconformity, base of the Kimmeridgian unconformity, and at the base of the Cretaceous unconformity. The western field limit is controlled by faulting and by the oil-water contact.

Mapping Procedure—The horizon picked for early mapping of the area was the reflection at the base of the Cretaceous/top of the Jurassic unconformity. Depth conversion was based on a simple time-to-depth relationship similar to that shown in the time/depth plot for well 3/3-1 (Fig. 6). Later, as well control was increased, it was found that stratigraphic units below the unconformity showed an areal variation in velocity that could be described by mapping the well velocity data. Check shot velocity surveys were obtained in all eight exploratory and appraisal wells. With the addition of computer methods in recent years it has been found advantageous to perform three dimensional migration in time using a time/depth function relationship in the migration equation. Depth conversion was based on mapped velocity values (Fig. 7). Maps of the base of the Cretaceous unconformity constructed in this manner typically tie well correlations to within 25 ft (7.6 m).

Precise mapping of geologic horizons below the base of the Cretaceous unconformity is more difficult because of the poorer seismic data quality and complex geology, but is essential to accurately define reservoir rock volume and oil in place, and to plan development well locations.

Figure 8 shows velocity and density traces and a synthetic seismogram for the 3/3-3 well. Figure 9 is a gamma ray-sonic log of the basal Cretaceous-Upper and Middle Jurassic section in this well. These figures show that the strongest seismic reflection occurs at the velocity-density interface at the base of the Cretaceous unconformity. This reflection is generated at the interface beween Cretaceous limestones, and the underlying lower velocity Kimmeridgian Shale (or older beds where Kimmeridgian is truncated). A second, but weaker, reflection occurs at the velocity-density interface at the contact between the Kimmeridgian shale and underlying Callovian shale or Middle Jurassic sandstone (or older formations). These unconformity relationships are shown in Figure 10 which is a diagrammatic east to west cross section of the Ninian field. The main problem in mapping the top of the Middle Jurassic sandstone reservoir is that there is not a well defined, persistent seismic reflection at that level. The nearest reflector is the top of the Callovian/pre-Callovian event which occurs as much as 90 ft (27 m) above the top of the reservoir.

At present, structure maps of deeper horizons are constructed by adding to the migrated base-of-the-Creteceous

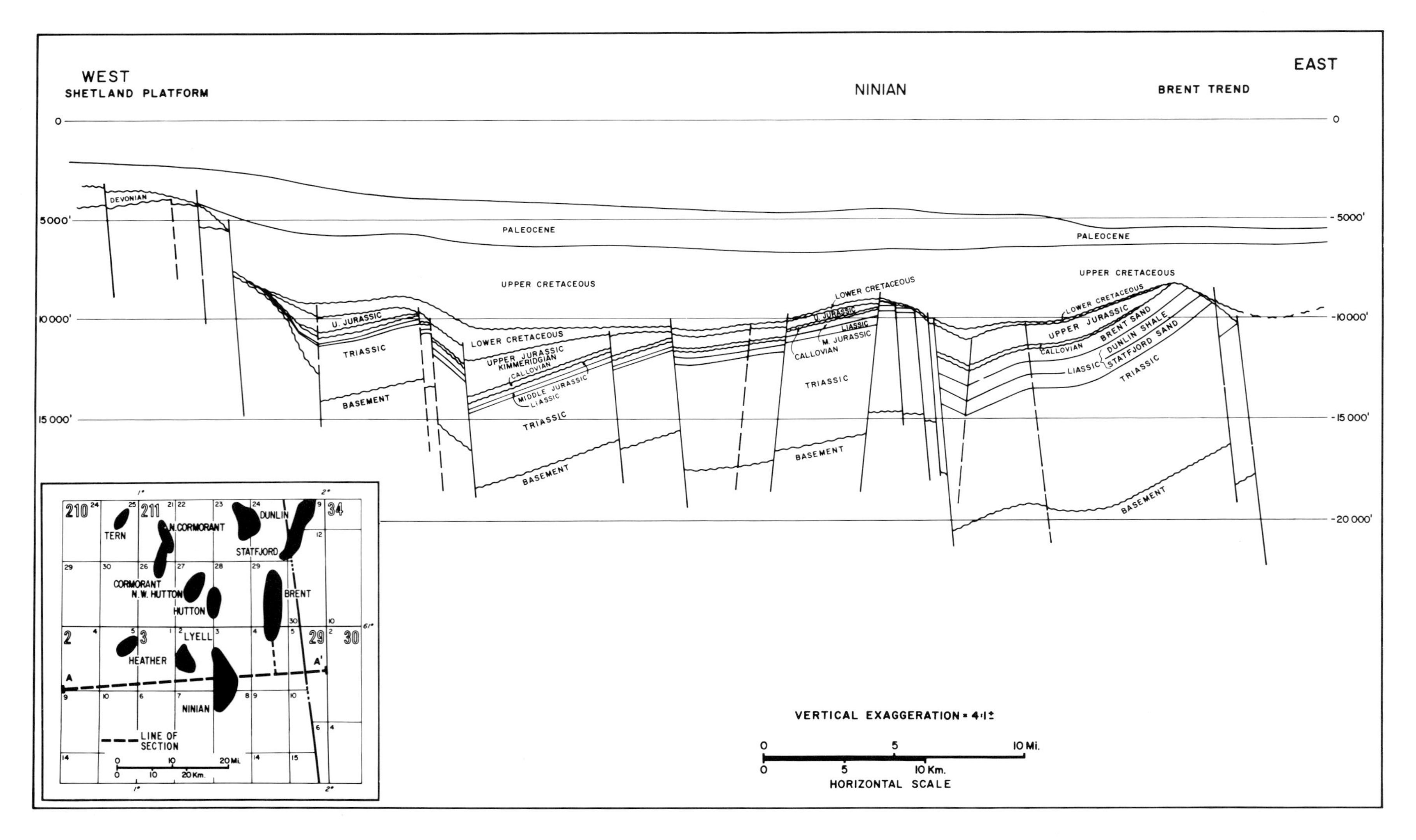

FIG. 5—Diagrammatic cross section of Ninian area, North Sea. Vertical exaggeration is 4:1. Line is west to east; location on index.

depth map the thicknesses of various underlying units as determined from combined seismic and well information. Accuracy of these maps is good. However, the accuracy of this method is dependent on an accurate knowledge of the unit thicknesses. Because these units are locally too thin to be defined seismically or are acoustically transparent (because of the lack of velocity or density contrast), good well control is needed for accurate mapping of the reservoir. Current interpretations of the structure at the base of the Cretaceous unconformity and top of the Middle Jurassic reservoir are illustrated on Figures 11 and 12. The gross isopay map was constructed by mapping the top of Liassic (base of the reservoir), subtracting the top of the reservoir and then superimposing the intersection with the oil-water contact (Fig. 15). The Liassic mapping was based on well data with consideration of seismic data.

Stratigraphy

A generalized stratigraphic column of the Ninian field section is depicted in Figure 13. Crystalline basement was encountered in the Block 3/3-4A well. A basement core recovered a dioritic gneiss similar in appearance to the Lewisian gneiss of northwestern Scotland. The minimum age, as determined by potassium-argon (K-Ar)) techniques is 350 m.y., which is a younger age than is usually associated with Lewisian rocks. It is interpreted that this dating may reflect a late Paleozoic tectonic overprinting, rather than the original age of emplacement of these rocks.

In the 3/3-4A well, basement is overlain by a redbed facies lacking diagnostic microflora or fauna. This redbed sequence is very similar to the Triassic section encountered elsewhere in the North Sea and is inferred to be Triassic in age. The possibility that the lower part of the section may be Permian in age cannot be entirely discounted. The redbed sequence is predominantly argillaceous and consists of red and various-colored claystone, with interbedded siltstone, sandstone, and occasional thin limestones. The redbeds represent a continental sediment infilling of the opening Viking graben.

The Lower Jurassic marks a transition from continental to marine sedimentation in this area and overlies the Triassic in an uncertain relationship. A minor unconformity may occur at this contact although no significant discordance is evident in Ninian field. The Ninian Jurassic section is somewhat attenuated as compared to adjacent synclinal areas. The thinner section is due primarily to depositional thinning on the structurally high block with some erosional thinning superimposed. The Lower Jurrasic is comprised of a basal Liassic sandstone overlain by a Liassic shale member. The Liassic sandstone is oil productive at Brent and Statfjord fields. Bowen (1975) proposed that the sandstone be recognized as the Statfjord Sand Formation, named after Statfjord field where the unit is best developed. The Statfjord Sand is considerably thinner at Ninian and is water saturated in all wells drilled to date.

The Liassic shale (proposed Dunlin Formation of Bowen) is overlain by a Middle Jurassic-Bajocian (and possibly Bathonian) sand sequence which is the Ninian oil reservoir. This unit also contains oil at Brent field, and Bowen has proposed that this sandstone be recognized as the Brent Sand Formation.

The Ninian Middle Jurassic sandstone reservoir is a progradational, fluviodeltaic sequence which reflects a dominant marine influence in the lower zones and becomes increasingly fluvial in character in the middle and upper zones.

The reservoir sequence has been subdivided into six zones based on log correlations with supporting sedimentological and paleoenvironmental evidence. The similarity and correlation to the Brent field type section is straightforward, with the possible exception of the upper two units of the Ninian zonation. Beginning at the base of the sequence, these zones are designated Zones I through VI. This zonation and correlation to the Brent field section is illustrated in Figure 14. Present correlation of Ninian Zones V and VI is tentative. Additional well data are needed to define this relationship.

Zones I, II, and III (Fig. 14) correspond to the Brent Lower Sand unit described by Bowen, who also recognized three sim-

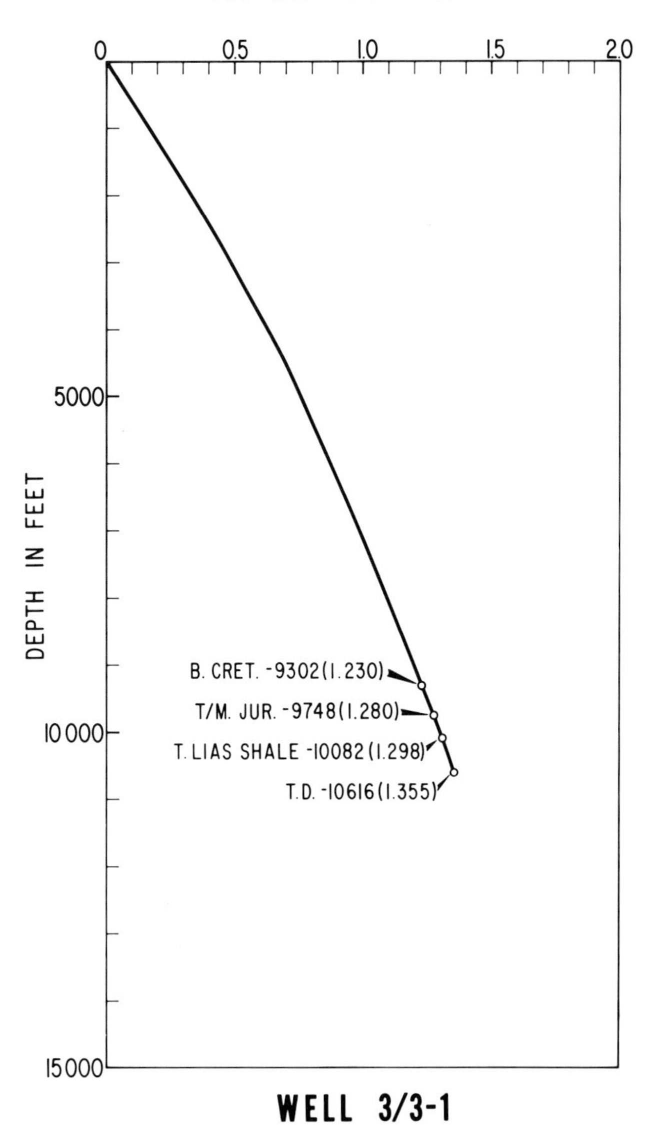

FIG. 6—Time/depth plot of well 3/3-1. Depth in feet; time in seconds of one-way travel time.

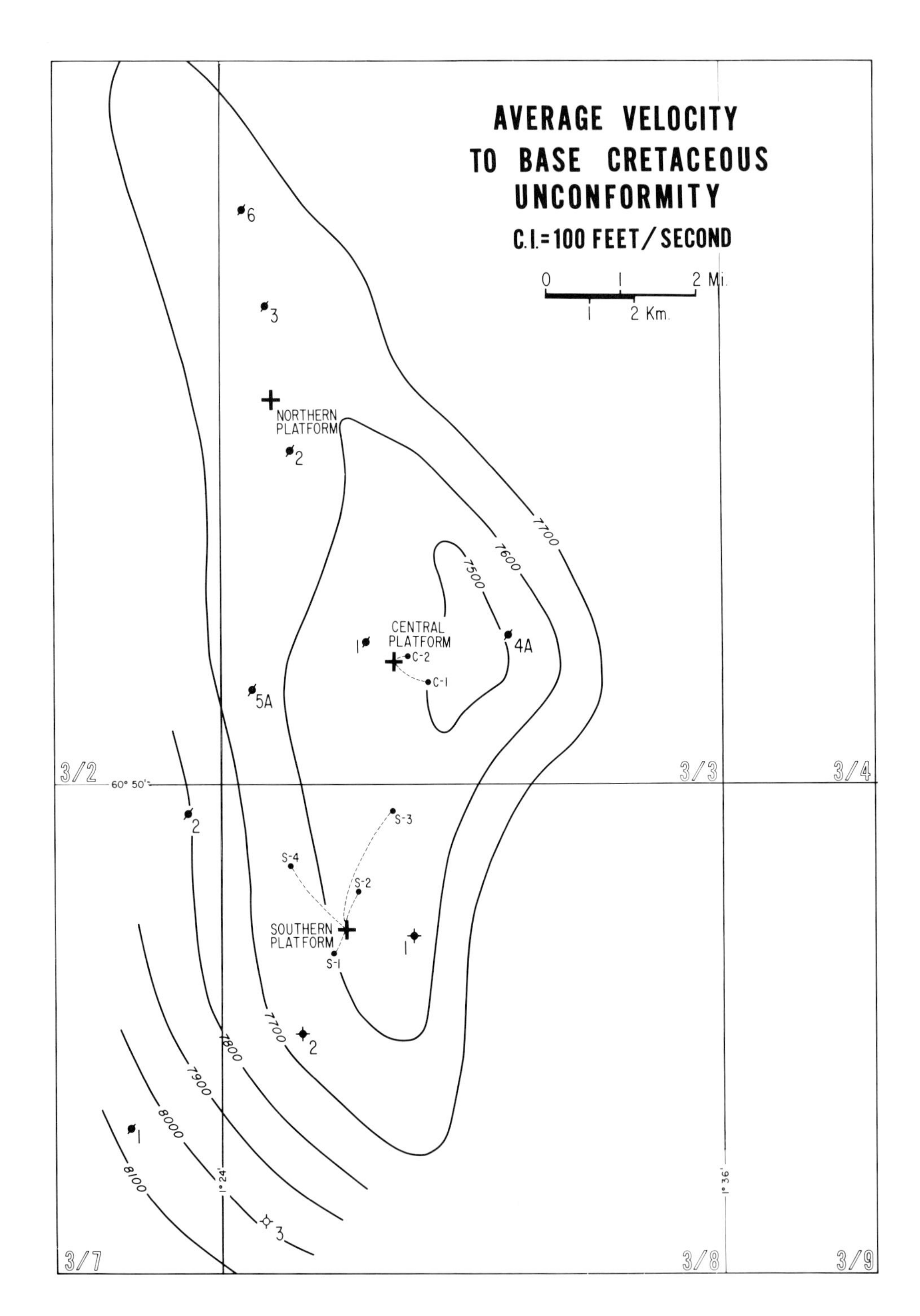

FIG. 7—Average velocity to base of the Cretaceous unconformity, Ninian field, North Sea. Note locations of northern, central, and southern platforms. Contour interval equals 100 ft/sec travel time.

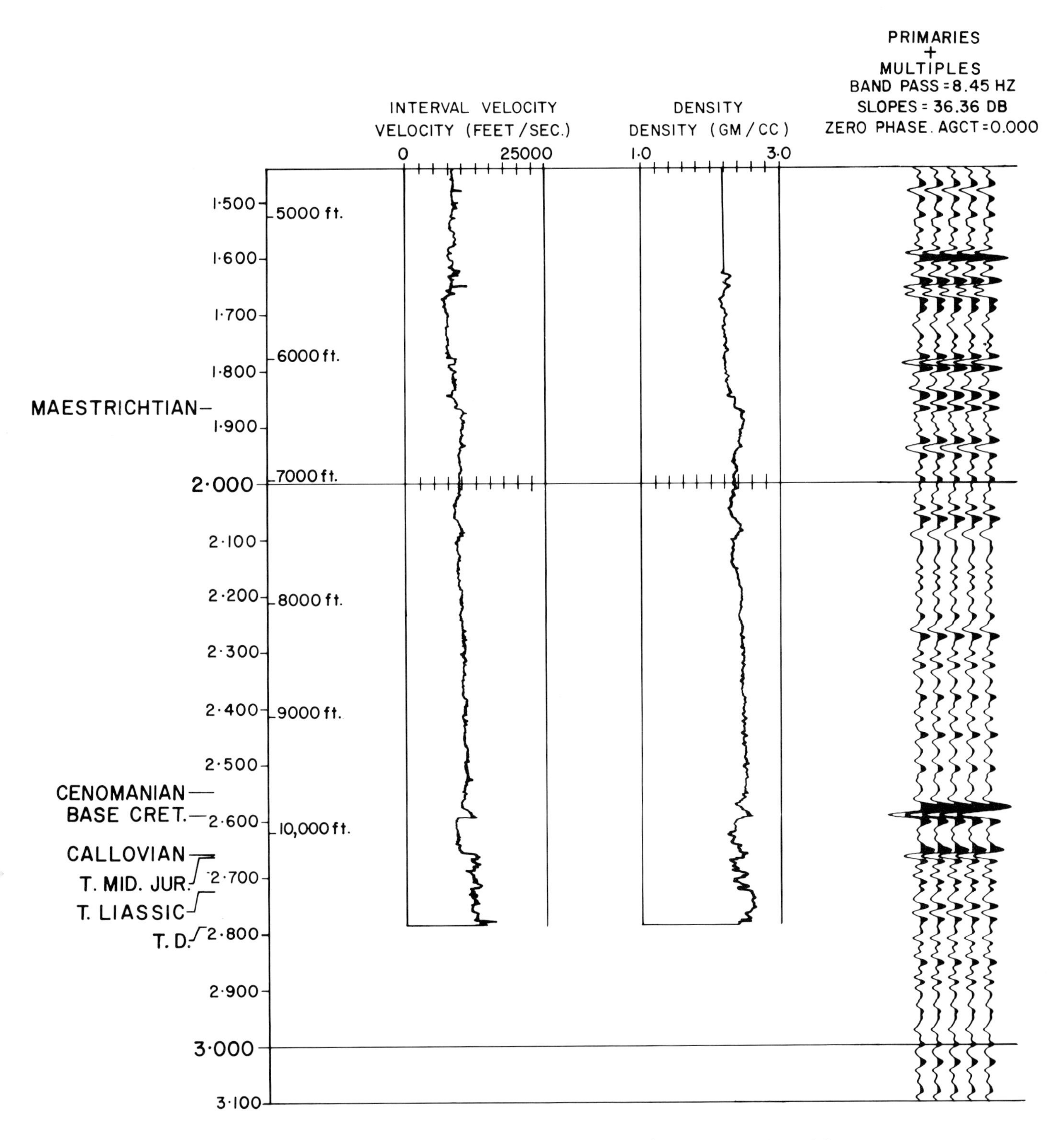

FIG. 8—Velocity-density and synthetic seismogram in well 3/3-3, Ninian field, North Sea.

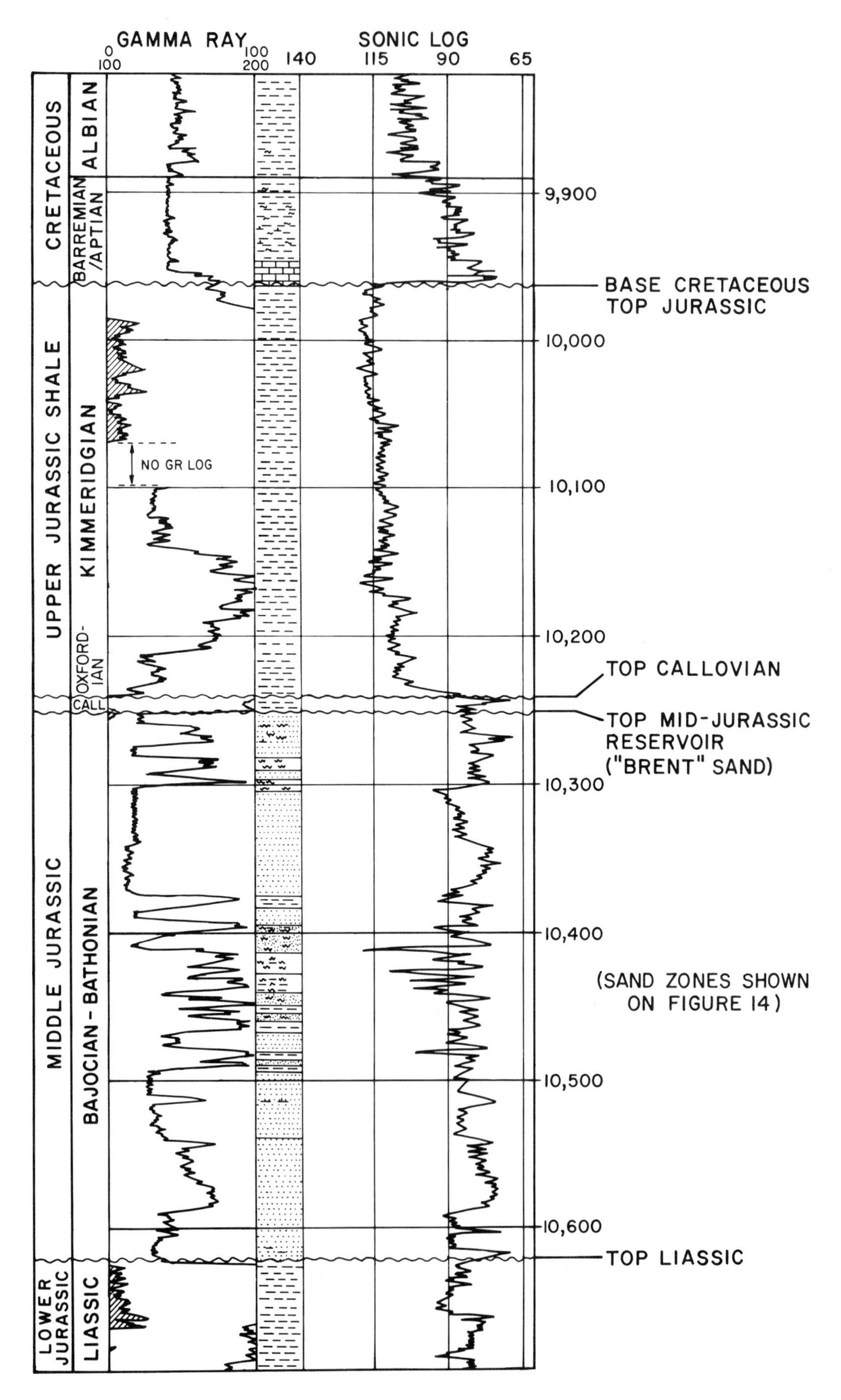

FIG. 9—Type log of well 3/3-3, Ninian field, North Sea.

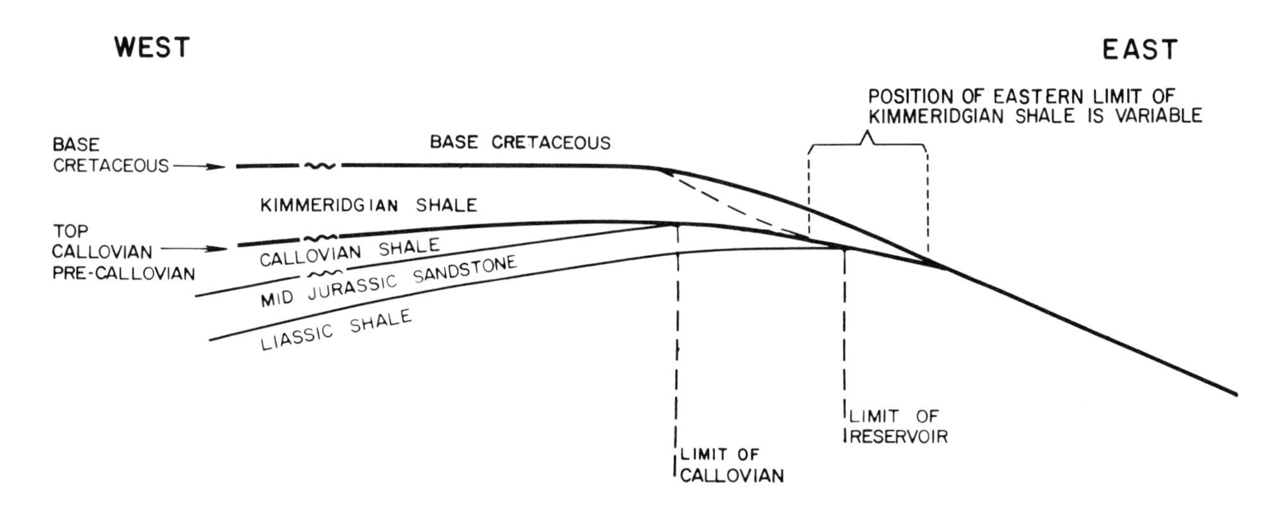

FIG. 10—Principal seismic reflectors across Ninian field, North Sea. See Figure 5 for location.

ilar subdivisions. Zones I, II, and III represent basal marine depositional units of a prograding deltaic sequence, and have blanket distribution throughout the field. Zone I is a sheet of medium to coarse grained massive sandstone, occasionally pebbly or conglomeratic. Zone II is a widespread sheet of fine grained micaceous sandstone with interlaminations of mica-rich shale. The lithofacies of the "mica sand" zone is remarkably similar over wide areas of the Viking graben. Zone III is a fine to coarse grained sandstone which typically coarsens upwards. Zone III is interpreted as a regressive sheet of coalesing river mouth bar sands, representing a prograding delta front. Zone IV correlates to the Brent Middle Shale unit, and is a complex of sandstones, siltstones, shales, and coal beds. Zone IV is a heterogeneous unit and has been subdivided into subzones IVA and IVB based on reservoir characteristics. Zone IV represents progradation of the delta plain and superposition of fluvial sands and interdistributary sediments over the basal marine facies. Zone V is a variable, usually massive, sand facies of uncertain distribution and debatable origin. Present well control does not permit a definitive interpretation of this zone, the correlations which define the zone are a subject of constant review as additional well data become available. The zone is interpreted as a channel deposit of the distributary system. Zone VI includes fluvial sandstones and interdistributary shale facies and represents a continuation of the delta-plain environment.

Zones V and VI are correlated to the Upper Brent Sand based on their similar stratigraphic position in the sequence and apparent contemporaneity of deposition. This interpretation is based on well-log correlations, but supporting biostratigraphic data are lacking. The correlation can be questioned since Ninian Zones V and VI are predominantly fluvial whereas the Upper Brent Sand is reported to be shallow marine. The differentiation of Zones V and VI dates from interpretation of early delineation wells, at which time Zone V was thought to be a distinctive widespread massive sandstone facies. Additional well data may require revision of the interpretation of the upper reservoir sequence of Ninian. The predominantly fluvial character of these zones at Ninian has been used to argue that they also may be correlative to the Middle Brent Member.

Correlation of the Ninian resevoir zones in this paper is referred to Bowen's (1975) convenient nomenclature. More recently the United Kingdom Institute of Geological Sciences (Deegen and Scull, 1977) proposed a formal stratigraphic nomenclature for the Middle Jurassic sandstone sequence. The nomenclature of Bowen, Deegen and Scull, and Ninian zonation are shown in Table 1.

The Middle Jurassic deltaic regime closed with a marine transgression. Some marine reworking and redistribution of sediment may have occurred at the top of the reservoir sequence but the evidence for this is not conclusive.

The Middle Jurassic reservoir sequence is unconformably overlain by the Callovian shale. This unconformity is relatively minor in a regional context, but is locally very important because it erodes part of the upper reservoir sequence. The "Kimmeridgian" shale, including possible but poorly defined Oxfordian shale, overlies the Callovian and is separated from it by another unconformity. This hiatus is again of minor regional importance, but is significant in the crestal area of Ninian field where, in some areas, it completely erodes the Callovian and bevels the reservoir section. The Kimmeridgian facies at Ninian is a dark gray to black, organic-rich shale, typical of the Viking graben.

The Cretaceous unconformably overlies the Jurassic succession and is separated from it by the late Kimmerian regional unconformity. In the Ninian area the Cretaceous system is predominantly a marine claystone facies with interbedded marls. The Lower Cretaceous is represented by a thin Barremian-Albian claystone overlying a basal marl/limestone unit. The Upper Cretaceous consists of a thick argillaceous se-

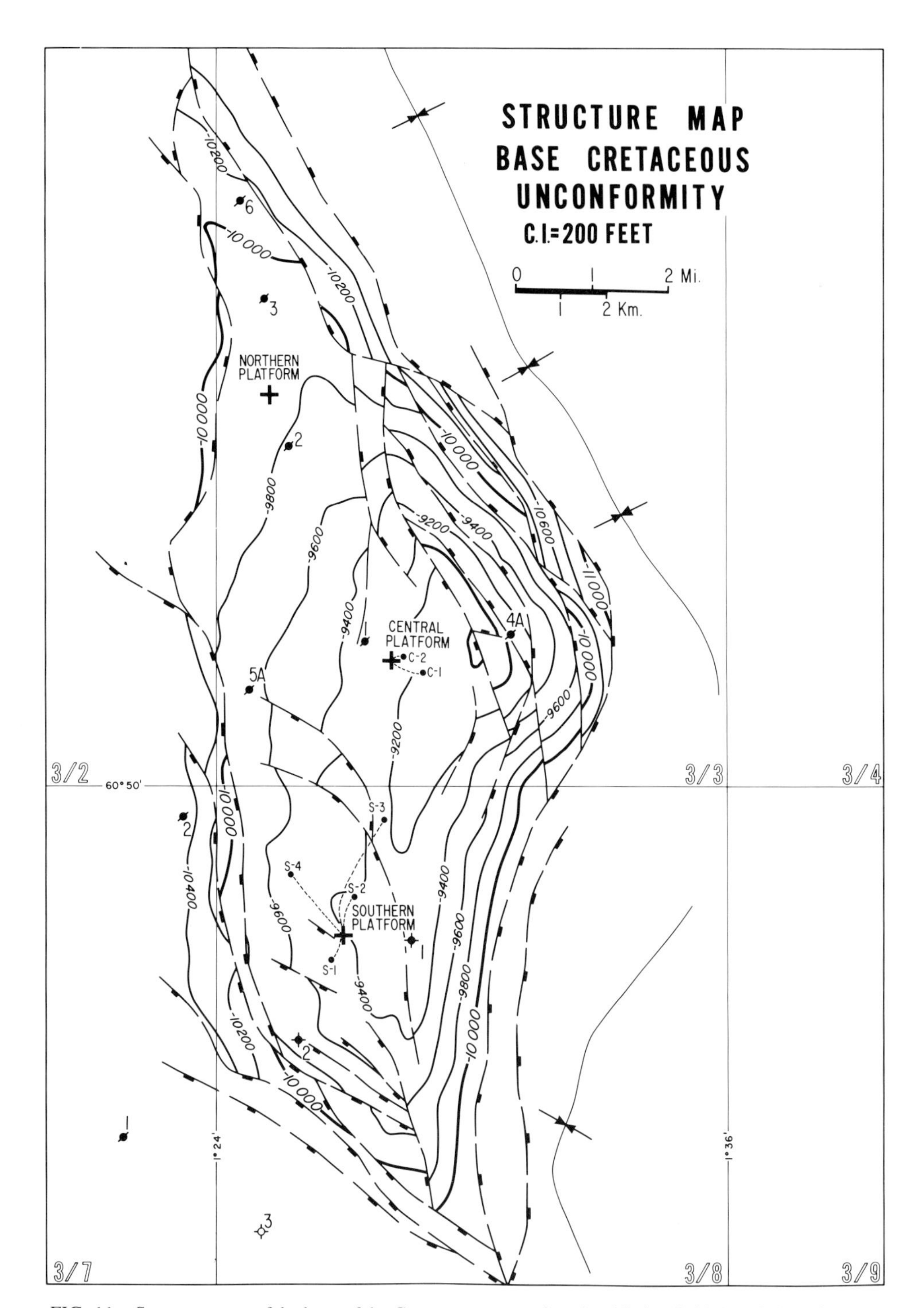

FIG. 11—Structure map of the base of the Cretaceous unconformity, Ninian field, North Sea. Contour interval is 200 ft.

quence with thin chalky limestone interbeds near the top. The Maastrichtian chalk facies typical of the central graben area of the southern North Sea is not developed in the northern Viking graben area. Cretaceous sedimentation continued to fill the North Sea basin as it subsided. By the close of the Cretaceous, the Jurassic tilted fault blocks were covered by nearly flat-lying Upper Cretaceous beds.

The onset of the Tertiary was accompanied by renewed uplift in Norwegian and Scottish sediment source areas and by renewal of coarse clastic sedimentation in the Viking graben. Paleocene sediments reflect a progradational sequence of sandstones, siltstones, and claystones. The Eocene, Oligocene, and Miocene are predominantly argillaceous with interbedded siltstone and thin sandstones. The Pliocene-Pleistocene consists of massive, generally coarse sandstones, gravel, and claystone.

Source Rock and Thermal History

Geochemical analyses indicate that the principal source of Ninian oil is the Kimmeridgian shale. The organic richness and

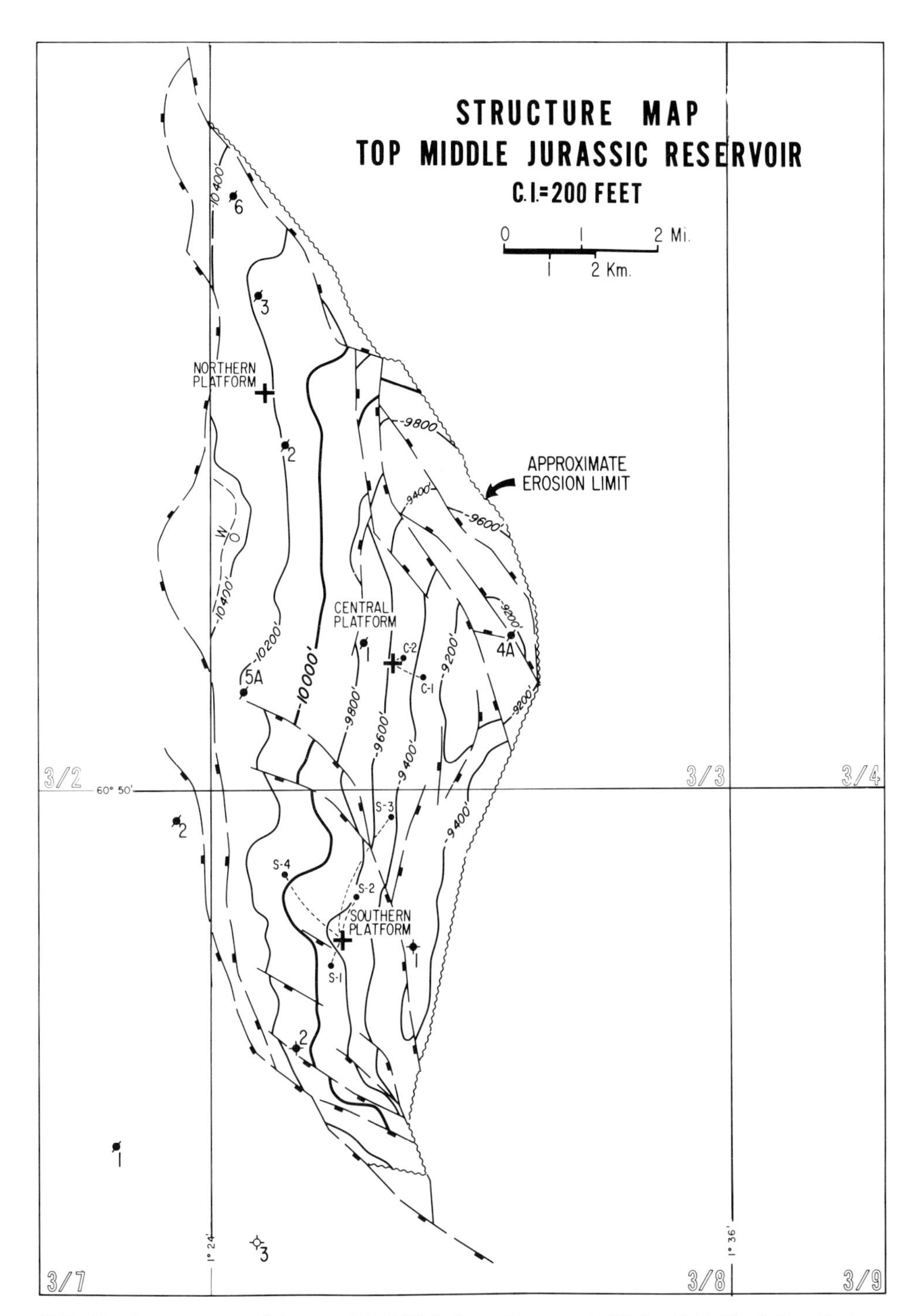

FIG. 12—Structure map of the top of the Middle Jurassic reservoir, Ninian field, North Sea. Contour interval is 200 ft.

prolific source potential of the Kimmeridgian in the Viking graben is well established. The Kimmeridgian at Ninian contains from 6 to 9% total organic carbon by weight and this material is strongly oil prone. Additional source potential occurs in the Callovian shale which has oil prone organic carbon content ranging up to about 4%. The Kimmeridgian-Callovian shale sequence forms both the cap rock and the source of oil for the underlying Middle Jurassic reservoir. Paleotemperatures, as interpreted from spore color (TAI) and vitrinite reflectance, indicate that the Kimmeridgian above the field is not sufficiently mature to have generated the oil in the reservoir. Both paleotemperature and present temperature are near the lower temperture limit of the oil generative window. This suggests that the bulk of the oil was generated in adjoining synclines which are 2,000 to 3,000 ft (610 to 914 m) lower and, consequently, more thermally mature. The present temperature gradient is about 1.7°F (0.94°C) per 100 ft (30 m) of depth. The compatibility between present reservoir temperature and paleotemperature indicators suggests that the Jurassic of Ninian is presently at maximum depth of burial.

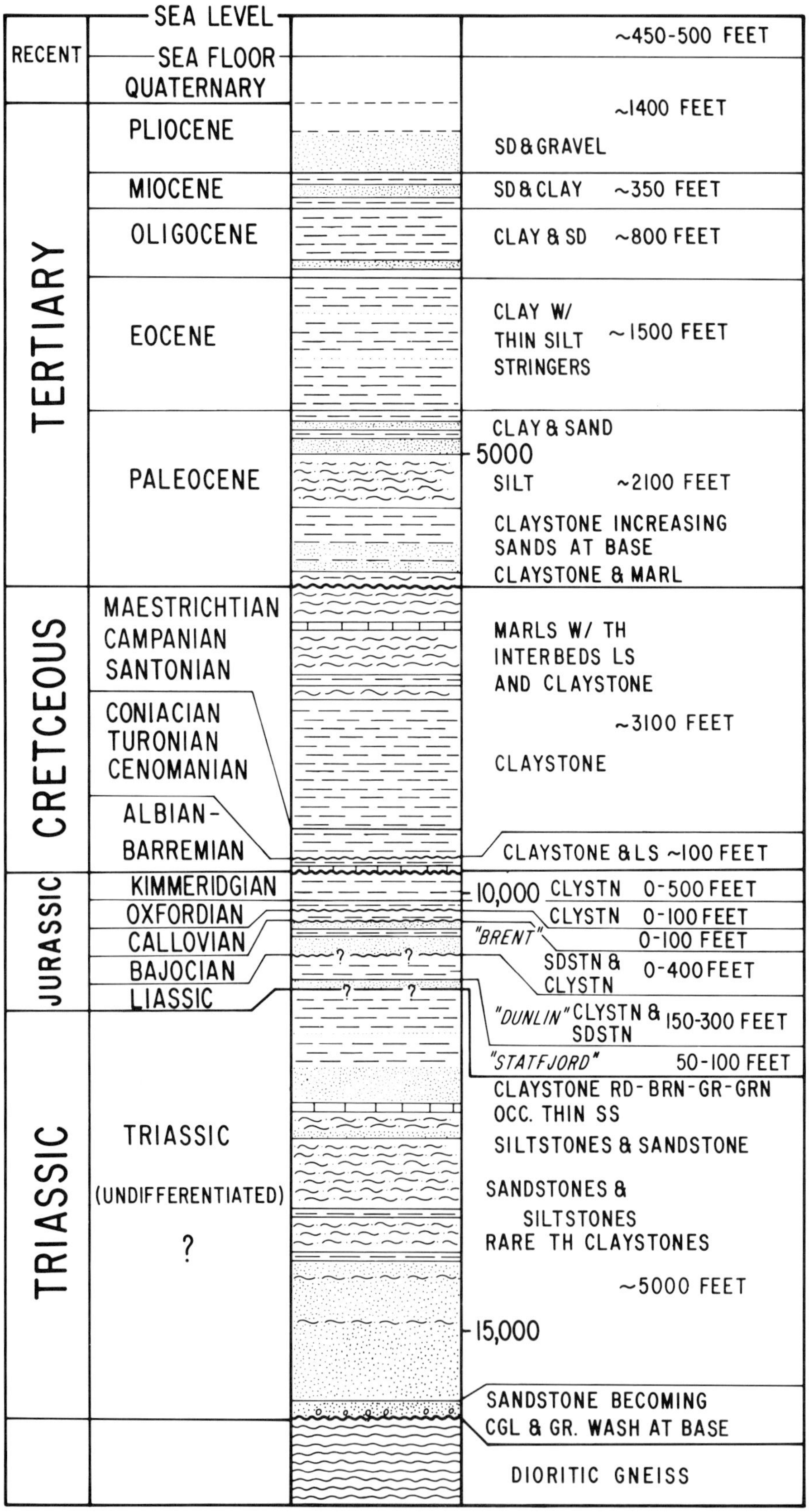

Era	Age	Depth	Lithology / Thickness
RECENT	SEA LEVEL		~450-500 FEET
	SEA FLOOR		
	QUATERNARY		~1400 FEET
TERTIARY	PLIOCENE		SD & GRAVEL
	MIOCENE		SD & CLAY ~350 FEET
	OLIGOCENE		CLAY & SD ~800 FEET
	EOCENE		CLAY W/ THIN SILT STRINGERS ~1500 FEET
	PALEOCENE	5000	CLAY & SAND; SILT ~2100 FEET; CLAYSTONE INCREASING SANDS AT BASE; CLAYSTONE & MARL
CRETCEOUS	MAESTRICHTIAN CAMPANIAN SANTONIAN		MARLS W/ TH INTERBEDS LS AND CLAYSTONE ~3100 FEET
	CONIACIAN TURONIAN CENOMANIAN		CLAYSTONE
	ALBIAN-BARREMIAN		CLAYSTONE & LS ~100 FEET
JURASSIC	KIMMERIDGIAN	10,000	CLYSTN 0-500 FEET
	OXFORDIAN		CLYSTN 0-100 FEET
	CALLOVIAN		"BRENT" 0-100 FEET
	BAJOCIAN		SDSTN & CLYSTN 0-400 FEET
	LIASSIC		"DUNLIN" CLYSTN & SDSTN 150-300 FEET
			"STATFJORD" 50-100 FEET
TRIASSIC	TRIASSIC (UNDIFFERENTIATED) ?		CLAYSTONE RD-BRN-GR-GRN OCC. THIN SS; SILTSTONES & SANDSTONE; SANDSTONES & SILTSTONES RARE TH CLAYSTONES ~5000 FEET
		15,000	SANDSTONE BECOMING CGL & GR. WASH AT BASE
			DIORITIC GNEISS

FIG. 13—Stratigraphic column, Ninian field area, North Sea. Based on wells 3/3 5-A and 3/3 4-A.

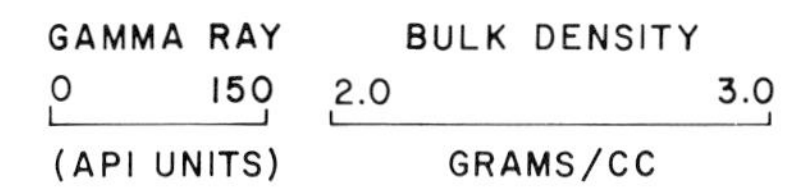

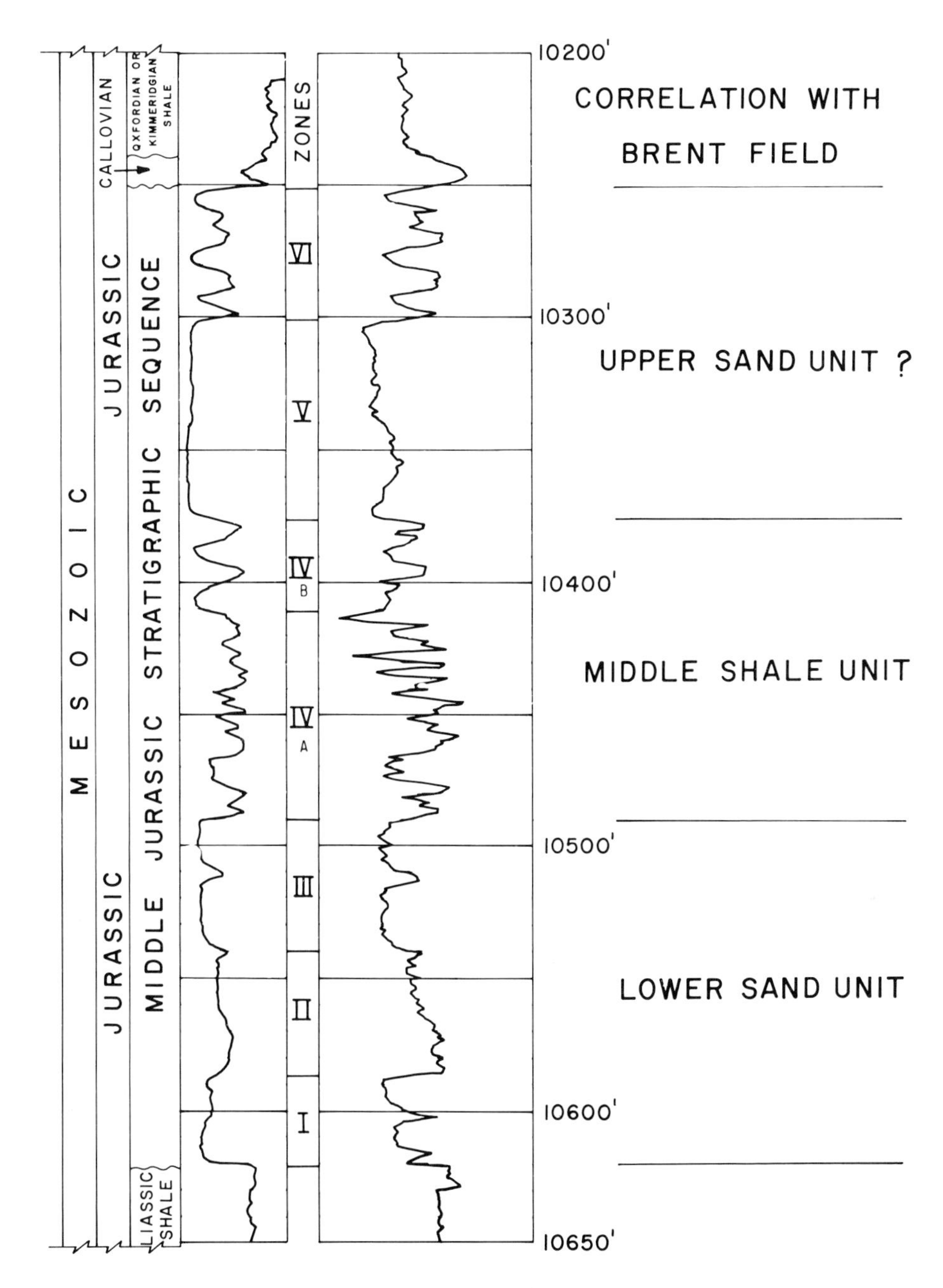

FIG. 14—Reservoir zonation type well 3/3-3, Ninian field, North Sea.

RESERVOIR CHARACTERISTICS AND RESERVES

The top of the reservoir section in the Ninian field has been found between 9,117 and 10,331 ft (2,779 and 3,149 m) subsea. Gross reservoir thicknesses encountered in wells range from 70 to 370 ft (21 to 113 m). The original oil-water contact occurs at depth of 10,430 ft (3,179 m) subsea, indicating an oil column of 1,313 ft (400 m). Based on presently available data, this appears to be the longest oil column in a single reservoir yet encountered in the Viking graben area.

The Middle Jurassic sandstone sequence has been subdivided into six sedimentological units. A consideration of reservoir criteria and probability of continuity between individual sandstone members has led to the further subdivision of Zone IV into two reservoir subzones, IVA and IVB. Seven reservoir zones (and subzones) are recognized in reservoir evaluation studies. Figure 14 shows the reservoir zonation in well 3/3-3, the type well in the field. Not all seven zones are present in all wells. Some higher zones may be absent due to truncation by unconformities at the top of the productive section.

Table 1. Nomenclature and Zonation, Ninian Field, North Sea

Age	Ninian Reservoir Zonation	Bowen (1975)		Deegan and Scull (1977)
Bajocian-Bathonian	Zone VI } Zone V }	Upper Brent Sand		Tarbert Formation
	Zone IV	Middle Brent Shale		Ness Formation
	Zone III	Lower Brent Sand	Massive Sand	Etive Formation
	Zone II		Micaceous Sand	Rannoch Formation
	Zone I		Basal Sand	Broom Formation

Zone I is a massive, medium to coarse grained, subangular to subrounded, moderately sorted, poorly to moderately well cemented quartzose sandstone. The sand includes minor feldspars, micas, clay (mostly kaolinite), carbonaceous material, and occasional calcite cemented streaks. This zone ranges from 14 to 71 ft (4 to 22 m) in thickness and averages 48 ft (15 m). Average porosity in the various wells ranges from 14.5 to 22%, and the ratio of net sandstone to gross zonal thickness ranges from 0.69 to 1.00. The zone has been divided into two subzones (I-A and I-B) for petrophysical analyses, principally based on different average grain densities in the two units.

Zone II is the micaceous zone. It consists of very fine to fine grained, silty, micaceous, quartzone sandstones interlaminated with very thin seams of carbonaceous and mica-rich claystones. The sandstones are subangular to subrounded, and fairly well sorted. Occasional low angle crossbedding is evident. This zone ranges in thickness from 21 to 49 ft (6 to 15 m) and averages 37 ft (11 m). Average porosity, by well, ranges from 10.8 to 18.9%. The net to gross ratio ranges from 0.33 to 1.00.

Zone III is a quartzose sandstone, fine to coarse grained but dominantly medium grained, friable to moderately well cemented but with occasional beds of very well cemented (with calcite) sandstone, subangular to subrounded, and moderately to well sorted. The zone contains minor amounts of mica, pyrite, carbonaceous material, and feldspars. The sandstone is predominantly massive but crossbedding is occasionally evident. The zone ranges in thickness from 15 to 75 ft (4.5 to 23 m), averaging 50 ft (15 m). Average porosity, by well, ranges from 14 to 24.6%, and the net-to-gross sand ratio ranges from 0.76 to 1.00.

Zone IV-A is composed of interbedded dark carbonaceous claystones, dark micaceous and carbonaceous shales with plant debris, hard micaceous siltstones, lustrous to clayey coal, and fine to coarse grained quartzose sandstones. The sandstones are commonly of two types; one is fine to medium grained, fairly tight, silty, and contains minor carbonaceous material and pyrite; the other type is medium to coarse grained, fairly well sorted, and poorly cemented. Beds are massive, but with some crossbedding, and are commonly 20 to 25 ft (6 to 7.6 m) thick. The zone ranges from 20 to 102 ft (6 to 31 m) in thickness with an average of 82 ft (25 m). Average porosity by well, ranges from 17.8 to 24.8% and net-to-gross sand ratio ranges from 0.14 to 0.46. The degree of continuity of individual sandstones is probably low in Zone IV-A. Only additional well control will determine the extent to which this problem will affect pressure maintainance and producing operations.

Zone IV-B consists of generally coarse sandstones which were originally though to have a more planar distribution than Zone IV-A, and to include less interbedded siltstone and claystone. Early development wells have shown that Zone IV-B is much more like Zone IV-A in distribution than originally supposed. Some sandstones appear to be continuous between wells while others are significantly isolated by claystone and siltstone interbeds. Nevertheless, the better sands in Zone IV-B are still a very good reservoir facies. The zone ranges from 19 to 50 ft (5.8 to 15 m) in thickness and averages 39 ft (12 m). Average porosity, by well, ranges from 16.1 to 21.3%, and net-to-gross ratio ranges 0.57 to 0.96. As more well data become available the subdivision of Zone IV may require revision.

Zone V has been encountered in only three delineation wells and two development wells. As currently correlated between wells, it shows a tremendous variability, ranging from a massive 75-ft (23 m) clean sandstone bed with excellent reservoir qualities to an eroded 8 ft (2.4 m) section of thin, fine grained sandstone and silty claystone. The massive clean sandstone facies is more typical in this zone and consists of quartzose sand, fine to coarse but predominantly medium grained, subangular to subrounded, and moderately sorted. Only very minor amounts of other materials occur in Zone V sandstones. The zonal thickness ranges from 0 to 75 ft (0 to 23 m) with an average of 38 ft (11.6 m). Average porosity, by well, ranges from 15.6% (from 2.5 ft [0.75 m] of net sand in the 8 ft [2.4 m] thick well) to 20.9 % and net-to-gross sand ratio ranges 0.31 to 1.00.

Zone VI was encountered in only two delineation wells and consists of a sequence of sandstones, siltstones, and claystones similar in character to Zone IV-B. The sandstones are fine to coarse grained, poorly to fairly well sorted, subangular to subrounded, and contain minor amounts of carbonaceous material. The wells containing this zone are 3 km apart, and show

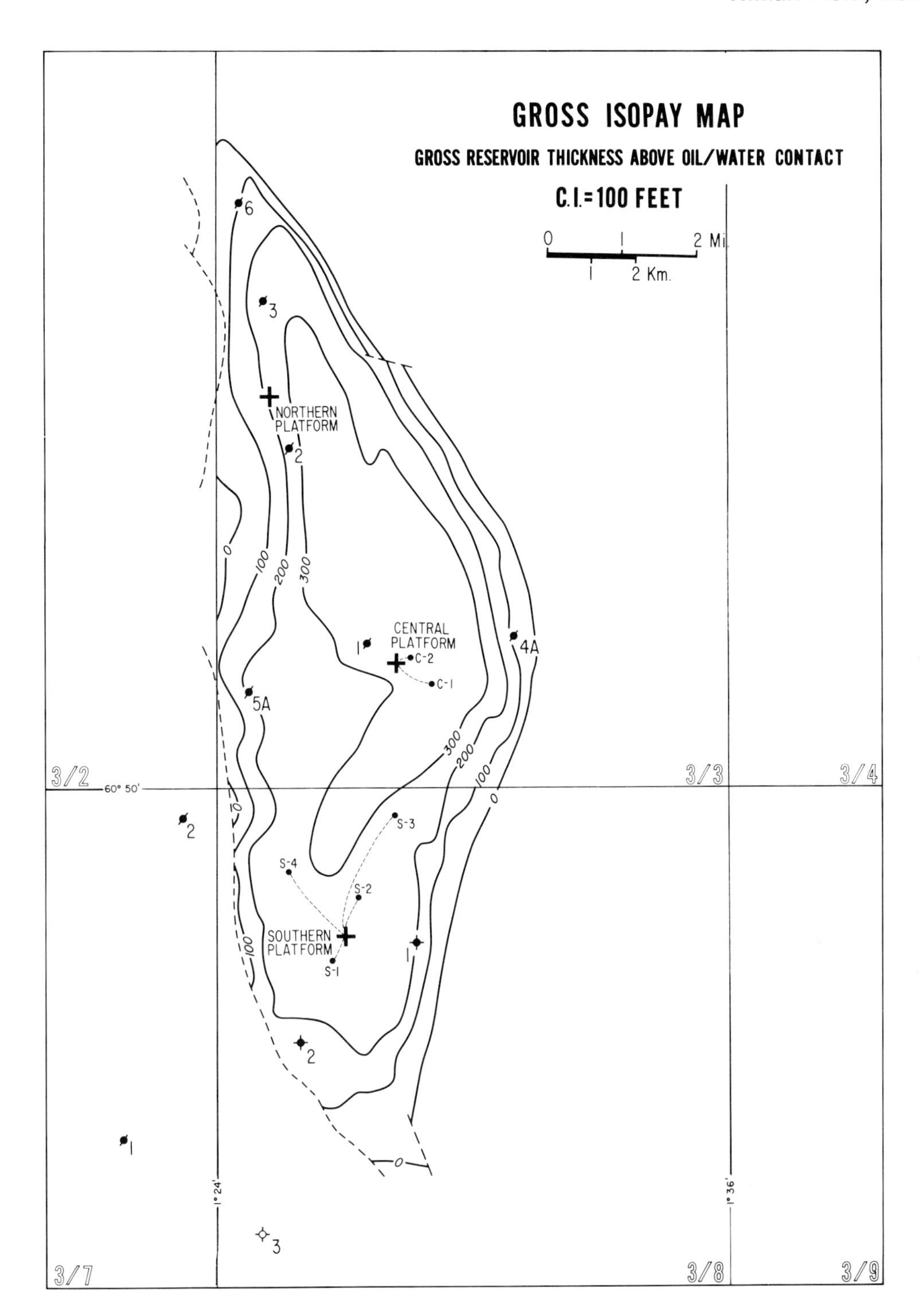

FIG. 15—Gross isopay map (gross reservoir thickness above oil-water contact), Ninian field, North Sea. Contour interval equals 100 ft.

38 and 50 ft (11.6 and 15 m) of the section present, so averages for this zone are subject to revision. On the present limited data, average porosity for the zone is 16.4% and average net-to-gross ratio is 0.67.

Permeability measurements of core samples from productive sandstones range from zero to several darcies. Permeability in Zone I is commonly in the 50 to 150 md range; Zone II is commonly less than 15 md and rarely as high as 35 md; Zone III ranges from 180 to 1,300 md but is normally a few hundred md; Zone IV-A ranges 20 to 800 md; Zone IV-B ranges 400 to 2,850 md and is commonly in the 800 to 1,000 md range. In Zone V, representative cores in two wells have averages of 1,550 and 1,800 md. The one well cored in Zone VI has sandstones with an average permeability of 225 md.

An isopay map of gross reservoir thickness above the oil-water contact is illustrated in Figure 15.

Reserves

At present, only eight delineation wells and nine development wells have been drilled in a productive area of approximately 20,000 acres (580 sq km), and estimates of oil in place and reserves must be considered as preliminary. At this early stage, reservoir performance data are of very limited value. Based on the data available, volumetric estimates of original oil and gas in place in Ninian field are 3.25 billion stock tank bbl of oil and 940 Bcf of solution gas. Currently accepted estimates of recovery are 1.2 billion bbl of stock tank oil, 63 million stock tank bbl of natural gas liquids, and 230 Bcf of gas. These reserve estimates are based on a development plan which includes early water injection and reservoir pressure maintenance.

Reservior Pressure and Temperature

Drill stem tests (DST) pressure build-up analyses from the 23 tests run in Ninian delineation wells indicate an initial reservoir pressure of 6,492 psi (absolute) at a datum of 9,750 ft (2,972 m) subsea. The average pressure gradient in the oil column of 0.335 psi/ft is consistent with the measured oil density from pressure-volume-temperature (PVT) analysis.

Ninian reservoir temperature is estimated at 218°F (103°C) at a datum of 9,750 ft 2,972 m) subsea, and the gradient through the reservoir section is 0.025°F (0.014°C) per ft (per 0.3 m).

Reservoir Fluid Characteristics

Ninian oil gravity is 35.4° API, and has a gas/oil ratio (GOR) of 320 scf/bbl and a mean viscosity of 1.32 centipoise at original reservoir conditions of temperature and pressure at the average depth of the reservoir. A formation volume factor value of 1.2 reservoir bbl per stock tank bbl has been agreed by Ninian partners.

Resistivity of formation water, measured from drill stem test (DST) samples is 0.37 ohm-metres at 60°F (15.5°C).

DEVELOPMENT PLAN

Development of Ninian field is being accomplished by the installation of three fixed platforms which have drilling slots for 109 wells. Platform hydrocarbon processing facilities are designed to handle up to 436,000 b/d of liquids. Mainline oil pumps are designed to deliver 400,000 b/d of Ninian crude to the shore terminal at Sullom Voe, Shetland Islands.

The Ninian central platform is a gravity base concrete structure. The southern and northern platforms are steel-piled jacket structures. The central and southern platforms each have two drilling rigs capable of drilling deviated wells to a maximum planned measured depth of 17,000 ft (5,488 m; 11,000 ft or 3,385 true vertical depth). Each of these platforms has 42 drilling slots and a limited number of spare risers and "J-tubes" which could accommodate subsea well completions and crude oil lines from nearby oil pools. The northern platform is in position but construction is not completed. It will have one drilling rig of comparable capability, 25 drilling slots, and two "J-tubes."

Pressure maintenance in Ninian field will be accomplished by water injection. The central and southern platforms will each have facilities to inject up to a total of 220,000 b/d of treated seawater at a pressure of 4,200 psi. The northern platform will initially have facilities for injecting 132,000 b/d.

The central plaform serves as the field junction with the 36-in. (92 cm) pipeline to the onshore terminal. The shore pipeline is connected to the central platform production train and also independently connected to the 24-in. (61-cm) feeder oil line riser from the southern platform and to the 16-in. (41-cm) oil line riser for the Heather field connecting line. Production from the northern platform will go into the shore pipeline through the central platform main oil line pumps.

Current plans are to drill 96 development wells, although slots are available on the northern platform for 13 more wells, if needed. Well completion intervals will be based on two separate reservoir completion units: an upper completion unit, consisting of Zone VI, V and IV-B; and a lower completion unit, consisting of Zones IV-A, III, II, and I. This concept of separate well completion units was planned because of the disparity in capacity (md-ft) shown between the reservoir zones in the delineation wells. The upper unit has a significantly higher capacity than the lower unit and if all zones were completed together, significant problems in the pressure maintenance program could occur.

The upper reservoir unit is planned to have 28 completions of the 96-well program. These consist of 20 producers, six injectors (two of which are currently optional wells), and two producers that will be subsequently converted to injection. A downdip, line drive flood is planned for this unit because of the high capacity observed in delineation wells and the apparent continuity of sandstones within the unit. This unit is estimated to contain approximately 30% of the effective oil in place in Ninian field.

The lower reservoir unit is planned to have 68 completions in the 96 well program—35 producers, 22 injectors, and 11 producers that will subsequently be converted to injectors. A five-spot water flood is planned in the southern and central platform areas because of the lower reservoir capacity of the unit. In the northern platform area a line drive type of flood is planned.

In the southern platform area a coarse grid of producers is planned with an injection well to complete a five-spot pattern of roughly 600 acres (243 ha.). When this stage of the development is completed, infill drilling on the sides of the coarse grid will convert to a nine-spot development. Subsequent conversion of the original producers to injectors will occur if a closer spacing than the nine-spot flood pattern is required. In the central platform area, the coarse grid of the five-spot is planned but no infill drilling will follow unless performance history dictates its necessity. In that event, subsea completions would be used.

Well 3/3-5A, originally drilled as a delineation well, is to be used as a prototype subsea completion. This well will be completed as a producer for a time but will ultimately be converted to an injector. Its most important function will be to determine the mechanical and economic feasibility of subsea injectors in the overall development program for Ninian field.

SELECTED REFERENCES

Bowen, J. M., 1975, The Brent oil field, *in* A. W. Woodward, ed., Petroleum and the continental shelf of North-west Europe, v. 1, geology: New York, John Wiley and Sons, p. 353-361.

Deegan, C. E., and B. J. Scull, 1977, A standard lithostratigraphic nomenclature for the central and northern North Sea: London, Inst. Geol. Sci. Rept. 77/25; also Stavanger, Norway, Norwegian Petroleum Directorate Bull. 1.

Jones, E. L., H. P. Raveling, and H. R. Taylor, 1975, Statfjord field: Norwegian Petroleum Soc., Jurassic Northern North Sea Symposium (Stavanger).

Ziegler, P. A., 1975, North Sea basin history in the tectonic framework of north-western Europe, *in* A. W. Woodward, ed., Petroleum and the continental shelf of North-west Europe, v. 1, geology: New York, John Wiley and Sons, p. 131-149.

Ekofisk: First of the Giant Oil Fields in Western Europe[1]

By Edwin Van den Bark and Owen D. Thomas[2]

Abstract Discovery of the giant Ekofisk field in block 2/4 in the Norwegian part of the North Sea in 1969 was a major turning point in the exploration for petroleum in Western Europe. Since that time, the North Sea has proven to be one of the best areas for exploration anywhere in the world. Current production is 1.5 million barrels of oil per day, and North Sea proven reserves total 18 billion barrels, with estimates of ultimate reserves as high as 40 billion barrels.

Ekofisk is located in the Central Graben in the southern part of the Norwegian sector of the North Sea. Although several periods of tectonism have affected this area, it has remained an intercratonic basin since Devonian time. The main elements of the tectonic fabric were established during the Caledonian and Hercynian orogenies, and later remained as controlling features for facies and sediment distribution.

A stratigraphic history reveals petrography of the main chalk group, environment of deposition, and diagenetic history of the area. A section outlines the preservation of porosity which has allowed the Ekofisk and Tor Formations to retain an average porosity of 30 to 40 percent which would not have been possible under normal circumstances. It is appropriate to state that the anomalously high porosity in Ekofisk field is probably due to a combination of: (1) overpressuring of the reservoir, (2) magnesium rich pore fluids, and (3) early introduction of hydrocarbons.

The six Greater Ekofisk fields now being developed were all located during the 1960s by reconnaisance seismic work. The history of the geophysical exploration is outlined and a study has been made on source rock analysis and geopressures. Also outlined is the three-phase, Greater Ekofisk development program.

INTRODUCTION

The discovery of the giant Ekofisk field in Block 2/4 in the Norwegian part of the North Sea in December 1969 was a major turning point in the exploration for petroleum in Western Europe, and rejuvenated the search for oil in the North Sea.

After this breakthrough, the North Sea has proved to be one of the best areas for petroleum exploration anywhere in the world and a number of giant fields have been found. Moreover, development of North Sea oil is one of the outstanding events in the economic development of Europe during this century. Current production is nearly 1.5 million b/d and North Sea proven reserves total 18 billion bbl with estimates of ultimate reserves ranging as high as 40 billion bbl. It is difficult to reconcile these figures with the pessimistic attitude that prevailed in the industry in 1969.

Whereas exploration in the North Sea was triggered by the discovery of the giant Groningen gas field in the Netherlands, exploration efforts were declining rapidly before the discovery

[1]Manuscript received, June 18, 1979; accepted for publication, November 27, 1979.

[2]Philips Petroleum Company, Bartlesville, Oklahoma 74004.

Article Identification Number:
0065-731X/80/M030-0010/$03.00/0.

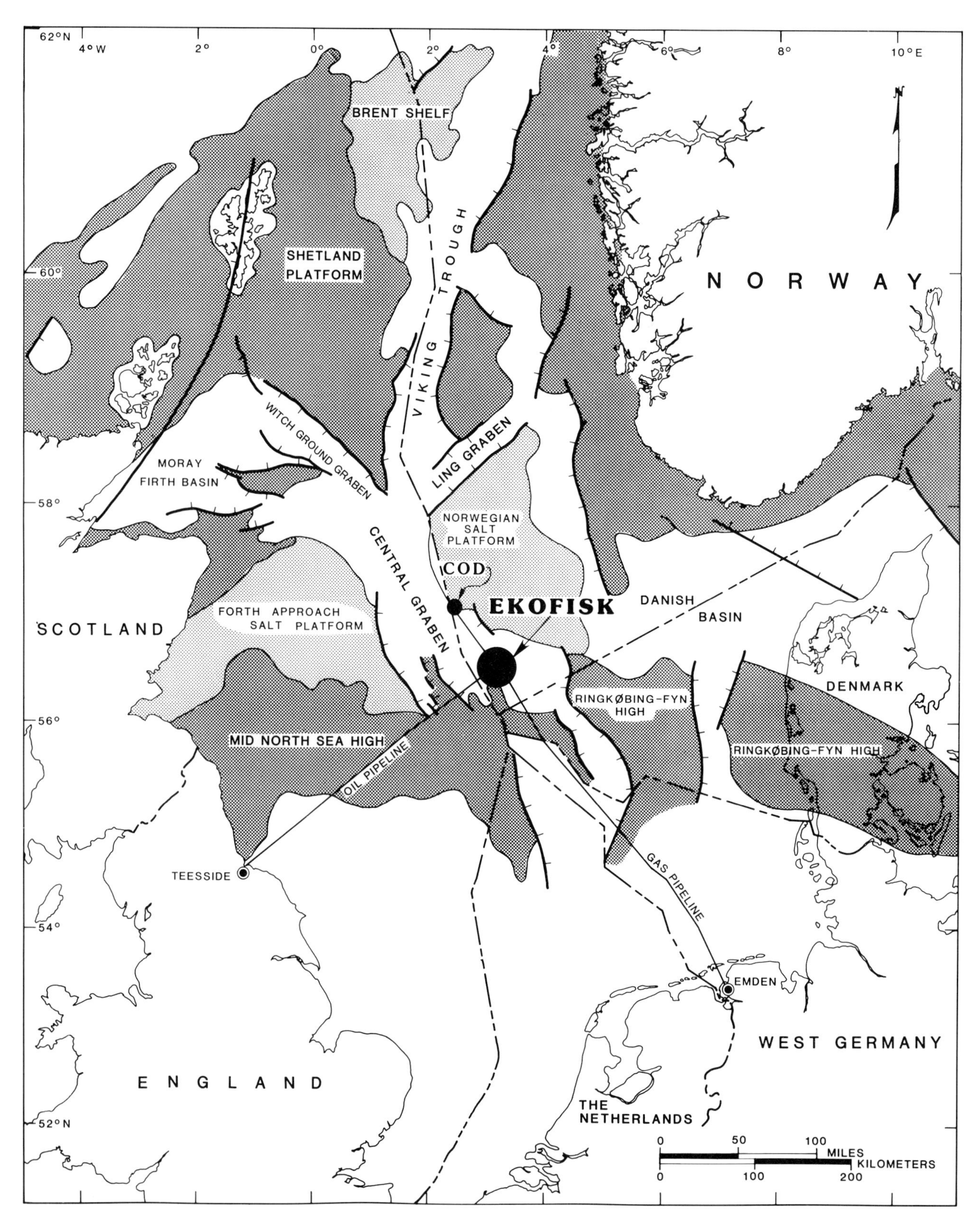

FIG. 1—Ekofisk field in the Central Graben, southern part of the Norwegian sector of the North Sea.

of Ekofisk. Over 200 exploratory wells were drilled in the North Sea, including 32 in Norwegian waters (seven of which were drilled by Phillips Petroleum Company). Although some of these wells discovered gas in the United Kingdom sector, none found commercial oil. Some companies were discouraged and abandoned the search while others considered diverting exploration funds to other more promising and less hostile environments.

Many considered it impractical or even foolhardy to drill in water more than 200 ft (61 m) deep at a location over 160 mi (256 km) from the nearest land, in a sea where gale-force winds are common, and storms with winds of hurricane force and 70-ft (21 m) waves are not uncommon. Those who persisted were looked on with deep misgivings by many; as has often happened in the history of the petroleum industry. A great discovery was made under discouraging circumstances. Even after initial reports of the Ekofisk discovery, the first reaction was undisguised cynicism because the reservoir was in chalky limestone. However, after four wells were placed on production within 18 months of the discovery and consistently produced at a rate of 40,000 b/d, even the most skeptical became believers. These initial wells eventually produced 28 million bbl of oil before being abandoned in favor of wells from permanent platforms.

Although Ekofisk field now is assured prominence in the history of petroleum exploration and development, except for early treatment in papers on the North Sea basin (Dunn, 1975; Byrd, 1975) the geology and development of the field itself has not been fully presented in the literature.

DISCOVERY

The Ekofisk structure was mapped by the common reflection point seismic system. A reflector from just above the top of the Danian showed about 800 ft (244 m) of closure over an area of 12,070 acres (49 sq km). However, a graben or downfaulted block appeared to cover a large part of the crest of the structure. Seismic sections showed two crests with reflectors turning sharply downward into this central "graben." The first well the Phillips 2/4-1X, was drilled as high as possible on the structure without penetrating this apparent "collapsed area." The well encountered an oil show in Miocene carbonate at about 5,500 ft (1,676 m) but began to kick.

Consequently a decision was made to plug and abandon the hole and spud again 3,300 ft (1,006 m) farther south-southeast. This second hole, designated the Phillips 2/4-A 1X, drilled uneventfully through the Miocene and at a depth of 9,831 ft (2,860 m) encountered a strong show of oil in Danian chalk. Oil shows persisted through 600 ft (183 m) of section. A drill-stem test over the interval from 10,364 to 10,464 ft (3,159 to 3,190 m) flowed 1,071 b/d, 37.2 API gravity oil through a 34/64 in. (1.3 cm) choke. This first test was limited by the capacity of the equipment, however, and later tests sustained flow rates exceeding 10,000 b/d. This was in December 1969. Extremely rough seas prevailed for some time and because of this and other reasons further testing and drilling were suspended until early the following year. In 1970 a second well (the 2X) was drilled about 2 mi (3.2 km) to the east and penetrated 540 ft (165 m) of pay in the chalk which tested, 3,850 b/d. A third evaluation well, drilled 1.5 mi (2.4 km) south of the discovery well, and a fourth test drilled 2.5 mi (4 km) north of it penetrated 200 ft (61 m) and 470 ft (143 m) of net pay and tested, 3,788 b/d and 3,230 b/d, respectively.

After permanent platforms were installed in the Ekofisk field a well was drilled in the center of the "collapsed area." It found 1,033 ft (315 m) of pay and proved the absence of the hypothetical crestal graven or "collapsed area." At that time, the full potential and huge dimensions of the field were confirmed.

GEOLOGICAL SETTING

The Ekofisk field is in the Central Graben in the southern part of the Norwegian sector of the North Sea (Fig. 1). Although several periods of tectonism have affected this area it has remained an intercratonic basin since Devonian time. The main elements of the tectonic fabric were established during the Caledonian and Hercynian orogenies and later remained as controlling features for facies and sediment distribution.

The post-Hercynian tectonic related sedimentary units are represented in the Ekofisk area by thick Permian Zechstein Salt overlain by nonmarine Triassic red beds and shallow-water marine shale. Uplift and erosion in Middle Jurassic time resulted in the absence of Lower and Middle Jurassic rocks in the Ekofisk area.

The next period of tectonism occurred during the Cimmerian orogeny when substantial relief developed following tensional faulting. The Central Graben was clearly established at this time and received a thick accumulation of marine shale during Late Jurassic time. This rapid influx of sediment initiated halokinetic movements in the form of salt swells or pillows which eventually created the important hydrocarbon traps in the Ekofisk area (Dunn, 1975).

Subsidence continued in the Central Graben area throughout Cretaceous time, with a gradual transition from deposition of shale in shallow water to deep water chalk by the end of the Early Cretaceous. By Late Cretaceous Maastrichtian time, chalk deposition was widespread in the North Sea area with depocenters located in the Ekofisk area where over 3,000 ft (915 m) of chalk accumulated by the end of Danian time.

Chalk deposition ended in the Ekofisk area when Laramide movements in northwestern Europe reactivated existing faults and caused new regional relief that generated a rapid influx of clastic sediments. Sediment source areas then retreated from the Central Graben as subsidence continued throughout the North Sea area and more than 10,000 ft (3,048 m) of Tertiary and Quaternary clastic sediments were deposited at Ekofisk. A combination of this great thickness of overburden and the high heat flow associated with the graben development produced salt piercements in the basin during the late Tertiary.

It was these later movements which fractured the chalk cover over the earlier salt pillow structures and created avenues for hydrocarbon migration.

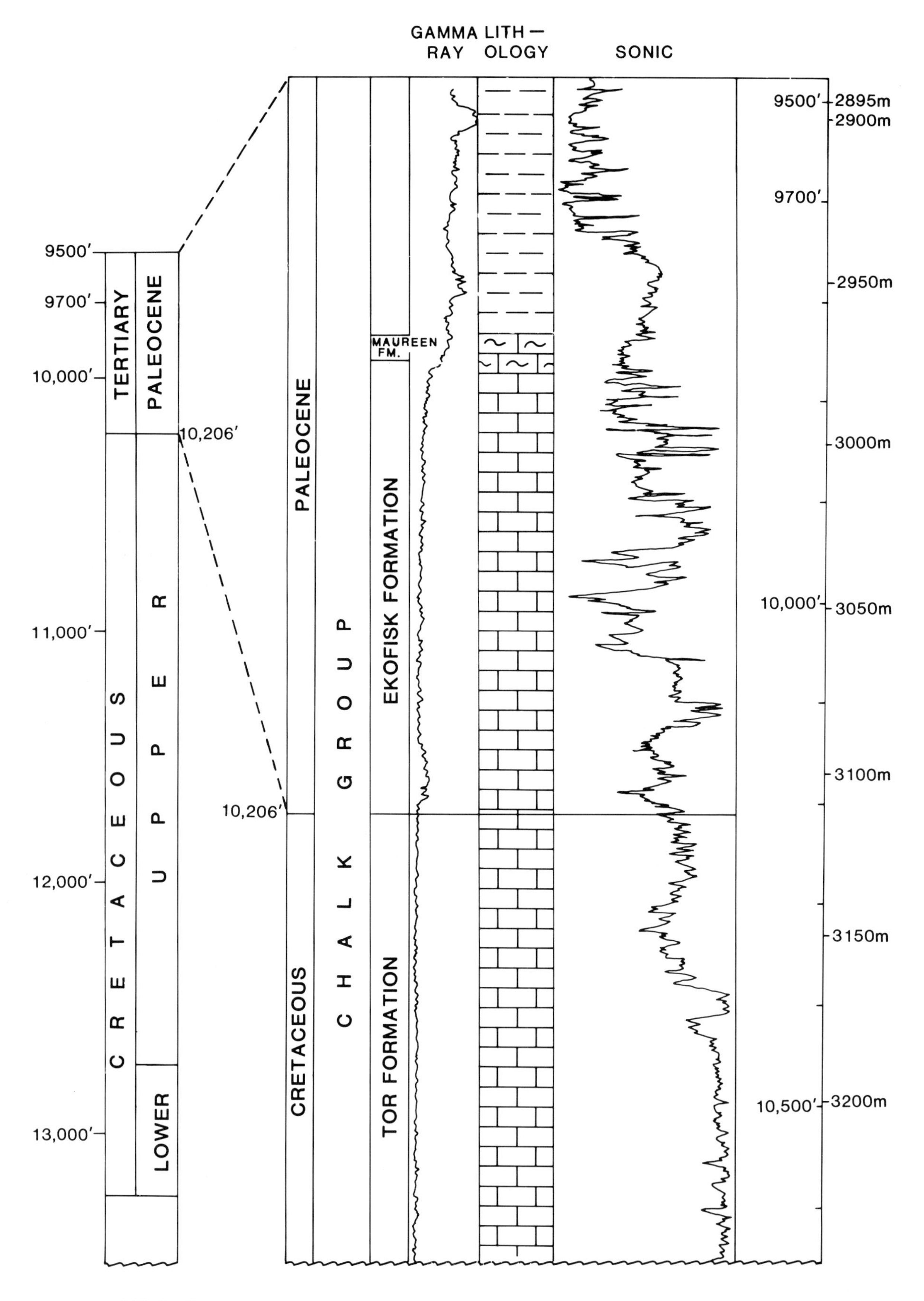

FIG. 2—Cretaceous and Lower Tertiary lithostratigraphic nomenclature Ekofisk area, North Sea.

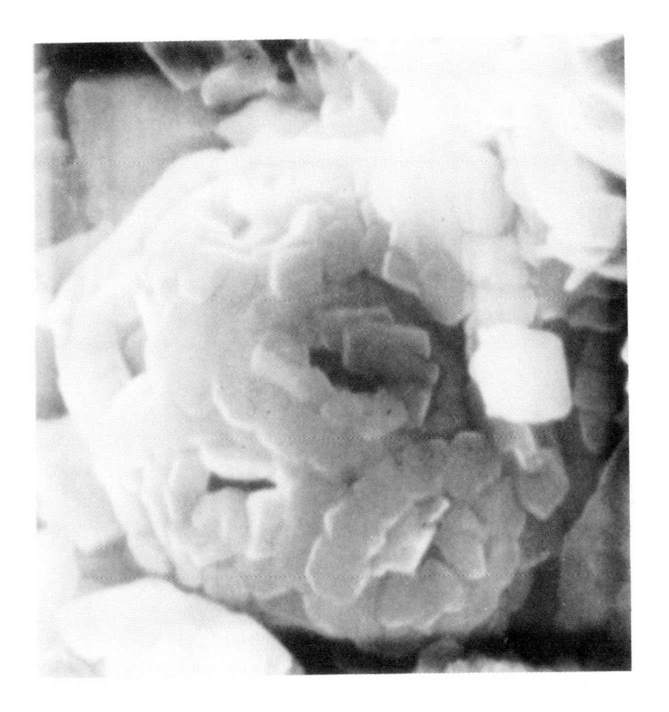

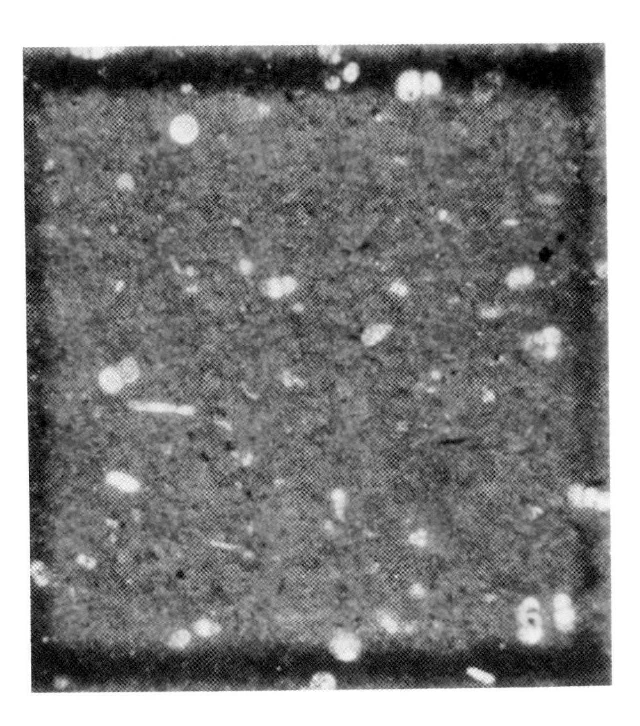

Fig. 3—**A.** Coccosphere, composed of coccoliths. SEM at 15,000 X. **B.** Numerous coccoliths, Ekofisk Formation. SEM at 5,000 X. **C.** Coccolith and small platelets resulting from the disaggregation of coccoliths. SEM is 5,000 X. **D.** Thin-section photomicrograph of Ekofisk Formation chalk, showing foraminiferal tests in a lime mud matrix. Plane light at 40 X.

FIG. 4—**A.** SEM photograph of Ekofisk Formation chalk that has been etched with dilute hydrochloric acid. Solution-resistant constituents, such as sponge spicules and radiolarians (upper right) stand up in relief. 200 X. **B.** Slab of Ekofisk core, showing numerous clay seams.

STRATIGRAPHY

The Cretaceous and Lower Tertiary lithostratigraphic terms used in the Ekofisk field are shown in Figure 2. These are based on the recommendations of the Norwegian and United Kingdom Subcommittees to the Joint United Kingdom-Norway Main Committee established by the Institute of Geological Sciences and Norwegian Petroleum Directorate for the purpose of determining common lithostratigraphic nomenclature. The productive zone of the Chalk Group at Ekofisk is confined to the Ekofisk and Tor Formations which are Danian and Maastrichtian in age, respectively.

Petrography of the Chalk Group

Primary Constituents—The reservoir rock of the Ekofisk field is a true chalk—a fine-grained limestone composed mainly of the skeletal remains of pelagic unicellular golden-brown algae or coccolithophores. The whole skeleton, a coccosphere, is occasionally found intact (Fig. 3a) but usually is disaggregated into the distinctive, button-shaped grains or coccoliths (Fig. 3b). Further disaggregation produces the plate or lath-shaped crystals termed coccolith platelets (Fig. 3c). Floating within this matrix of coccoliths and platelets are varied amounts of globular foraminiferal tests, mostly planktonic in origin (Fig. 3d). The combination of primary grain types within the Ekofisk Chalk gives it a polymodal grain size distribution with the diameter of the platelets from 0.2 to 1.0 μm, the coccoliths from 1 to 10 μm, and silt-size foraminiferal tests.

In comparison to coccolithophore debris and foraminiferal tests, other primary constitutents of the chalk are usually of minor importance, although some can locally reach significant concentrations. Minor primary constituents include siliceous sponge spicules, radiolarians (Fig. 4a), pelecypod shell fragments *(Inoceramus),* echinoderm debris, bryozoans, and bone fragments. The content and distribution of sedimentary clay in the Ekofisk Chalk is still not fully understood. In core samples, zones of apparent high clay content are obvious (Fig. 4b) but nonauthigenic clay is usually difficult to find in scanning electron microscope examinations of the chalk.

Secondary Constituents—Calcite overgrowths on coccoliths and their disaggregated crystal elements are major secondary features of the Ekofisk Chalk (Fig. 5a). Large, blocky calcite crystals also are present in some parts of the chalk (Fig. 5b) and most foraminifer chambers are filled with secondary calcite, quartz, pyrite, or clay (Fig. 6a). Secondary chert is present in some zones and occasionally forms nodules and beds (Fg. 6b). Minor amounts of secondary dolomite rhombs also have been observed (Fig. 7a).

Porosity—The matrix porosity in the chalk group consists of original, intergranular pore spaces between coccolith grains and platelets (Fig. 7b). The pores are extremely small (1 to 5 μm) but abundant with porosity measured as high as 40% or more.

Fractures are common in the Ekofisk Chalk and contribute to the effective porosity and permeability. Some fractures are large and are probably related to the later Tertiary tectonic history of the Ekofisk structure. Others are small and appear to be caused at least in some cases by microtectonic adjustments associated with stylolitization (Fig. 7c).

Environment of Deposition

The producing chalks of the Ekofisk field were deposited below wave base and below the photic zone, in clear waters to

FIG. 5—**A.** Coccolith with calcite overgrowths, particularly in the central part. SEM at 10,000 X. **B.** Large, blocky, secondary calcite crystal in Ekofisk Formation chalk. SEM at 3,000 X.

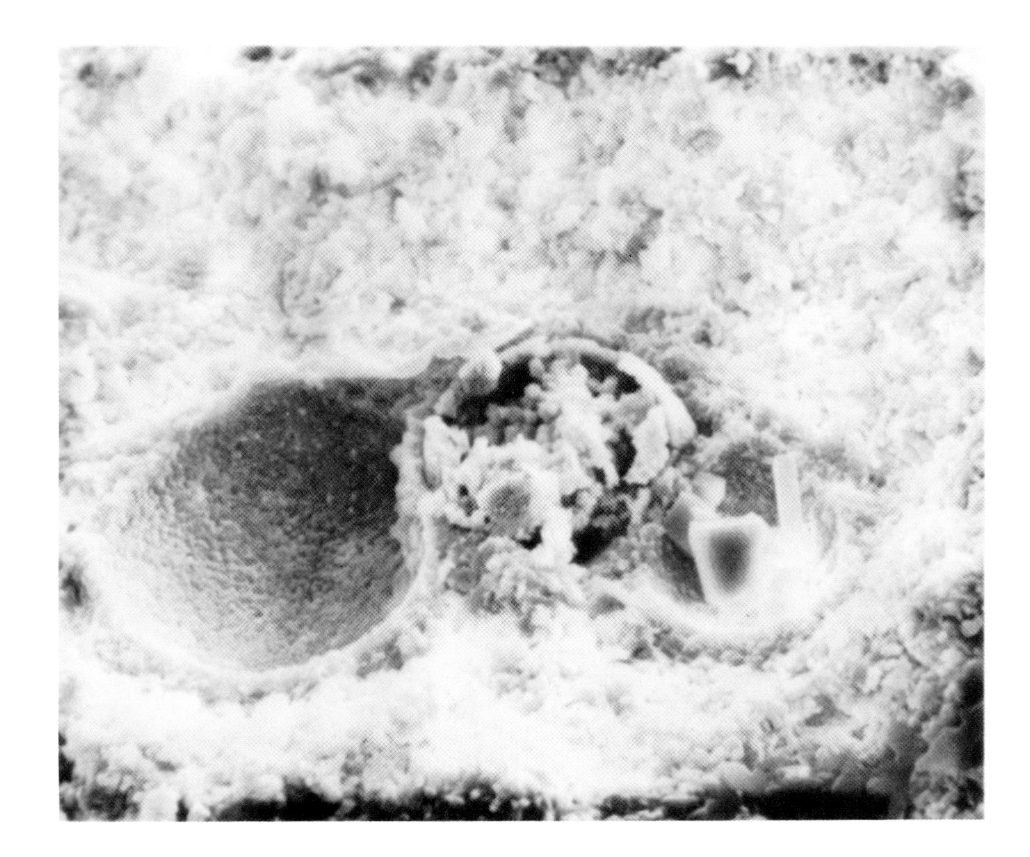

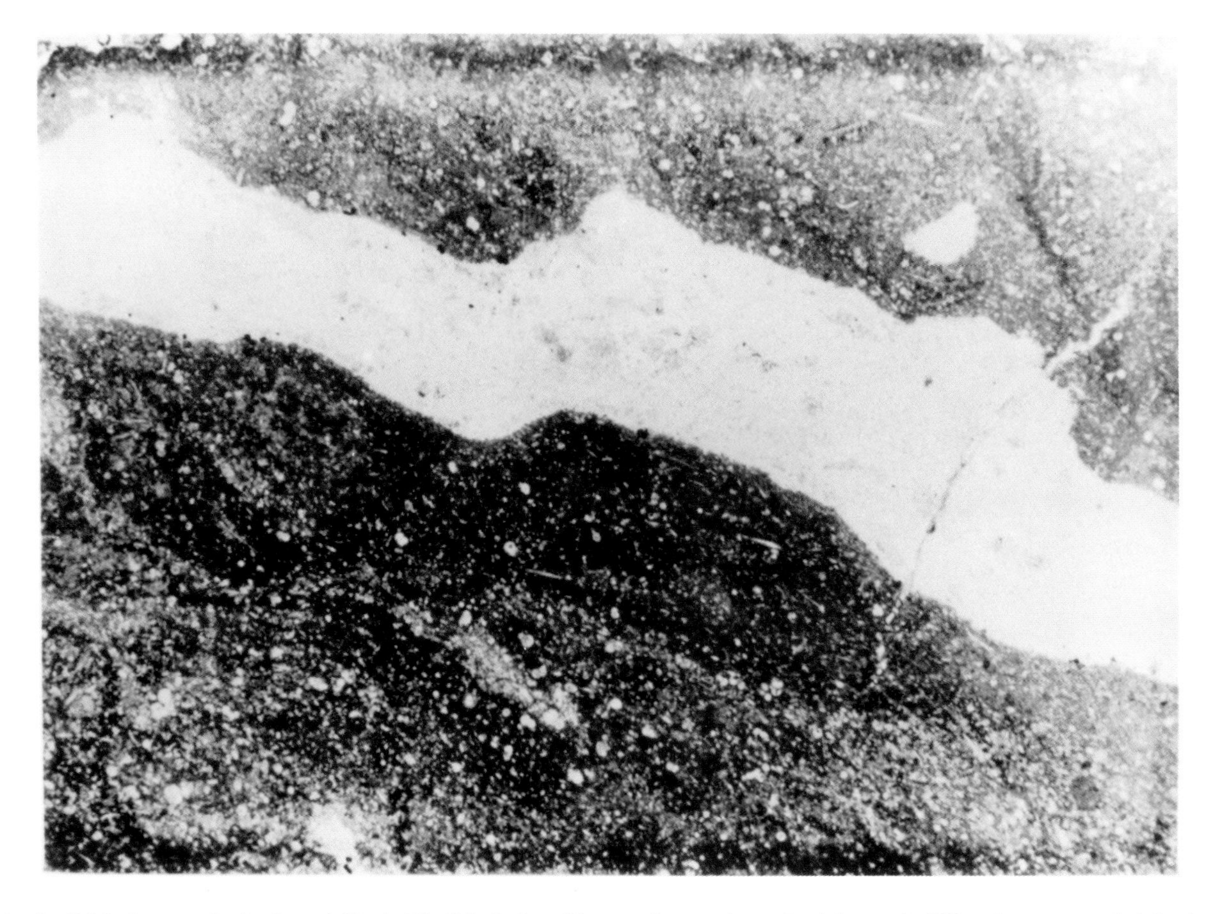

FIG. 6—**A.** SEM photograph of a foraminifer in Ekofisk chalk, with secondary pyrite and calcite partly filling the center and right chambers, respectively. 300 X. **B.** Thin-section photomicrograph of chert lens or layer (light gray) in Ekofisk chalk. 10 X.

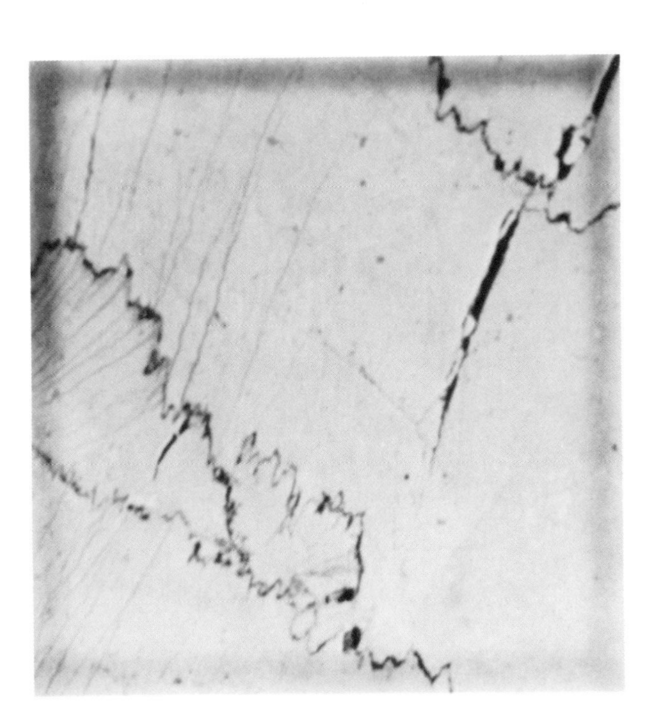

FIG. 7—**A.** SEM photograph of etched Ekofisk chalk, showing rhombic, solution-resistant dolomite. 5,000 X. **B.** SEM photograph of Ekofisk Formation chalk, showing excellent original intergranular porosity, measured at 37.9% for this sample. 5,000 X. **C.** Photograph of Ekofisk chalk core slab, showing stylolites and two generations of fracture. The very fine fractures (left) are related to the stylolitization process. 2X. **D.** Ekofisk Formation core showing numerous burrows. 0.75 X.

FIG. 8—**A.** Ekofisk Formation core showing possible cross-bedding in a chalk calcarenite. 1 X. **B.** Pebble conglomerate as seen in Ekofisk Formation core slab. 0.55 X.

depths of 600 to 1,500 ft (183 to 457 m) or deeper. The bottom waters were oxidizing, normally saline, and supported an abundant benthic fauna. Burrowing organisms thrived and produced intensely bioturbated sediments. The primary burrower is thalassinidean crustaceans resembling the modern burrowing shrimp *Callianassa* with lesser occurances of annelid worms and an organism that produced burrows similar to *Nereites* and *Zoophycos* (Fig. 7d).

Sedimentary structures are rarely seen in the Ekofisk Chalk due partly to masking by the homogeneity of the chalk and partly by bioturbation which destroyed the structures that were present. Evidence of sediment distribution or redistribution is confined mainly to calcarenites (Fig. 8a) and pebble conglomerates (Fig. 8b). The calcarenites consist of sand-sized biogeneic grains (mainly foraminiferal tests) that may have been formed by turbidity currents. The conglomerate pebbles consist of chalks of slightly different color than the enclosing sediment and overlie an undulating, possibly erosional surface.

Diagenetic History

From a petroleum geologists's point of view, everything (except fracturing) that happens to a chalk after deposition is detrimental because it reduces porosity and permeability. Therefore, considerable effort has been devoted toward understanding deposition and diagenesis of the Ekofisk Chalk.

All true chalks have a high (70 to 80%) initial porosity that is reduced by early dewatering to approximately 50 to 60% (Scholle, 1977). Subsequent porosity reduction is accomplished mainly through mechanical and chemical compaction under increasing overburden stress. Mechanical compaction evidence includes breakage, reorientation or repacking of the grains, deformed burrows, and wispy organic layers or "horsetails." This compaction then results in a grain-supported framework and from this point onward chemical compaction predominates.

The most important effect of chemical compaction is the reduction of porosity and permeability while the most obvious agent of chemical compaction is solution transfer. In this process, dissolution occurs preferentially at grain contacts where overburden stress is the greatest. The dissolved calcium carbonate is precipitated on nearby pores or as overgrowths on coccoliths or platelets (Fig. 5a). With increasing overburden stress and solution transfer, large, blocky calcite crystals are formed (Fig. 5b) and a tightly interlocking mozaic of calcite crystals results in limestone with little effective porosity (Fig. 9).

Tight Zone—Formation of the "tight zone"—an interval approximately 50 ft (15 m) thick with extremely low porosity located at or near the Cretaceous and Tertiary boundary was a significant diagenetic event in the development of the Ekofisk Chalk Group. This zone of low porosity is present in all Ekofisk wells and is both overlain and underlain by highly porous chalk.

The high degree of lithification that characterizes the tight zone is similar to the lithification of "diagenetic hardgrounds" that are exposed in European onshore outcrops. These "hardgrounds" are interpreted as surfaces of exposure, either subaerial or at the sediment-water interface. They are characterized by encrusting epifauna, reworked hardened pebbles, and borings that transect grain boundaries. Because the Ekofisk tight zone does not exhibit these latter characteristics, it is not considered to be a hardground.

It seems that the Ekofisk tight zone is a product of subma-

FIG. 9—SEM photograph of Ekofisk Formation chalk showing little effective porosity. 3,500 X.

rine lithification slightly below the sediment-water interface during a period of slow sedimentation. A low rate of sedimentation could result from a low rate of productivity of sediment, a high rate of dissolution, or both, either in the water column or on the seafloor. Because evidence of dissolution is not apparent within the tight zone, a low rate of nannofossil productivity probably is the likely explanation for the Ekofisk tight zone. Declines in productivity of lesser magnitude could be responsible for less dramatic decreases in the porosity of various intervals within the Ekofisk Chalk Group.

Preservation of Porosity

If the process of solution and reprecipitation described above simply increased with depth of burial, then the Ekofisk field would not exist. This process suggests that a chalk at the depth of the Ekofisk pay zone should have a porosity of less than 10% (Scholle, 1977). Therefore, it is important to understand why the diagenesis of the Ekofisk and Tor Formations have retained a porosity average of 30% and range to over 40%. There are several possible mechanisms for retaining porosity. At this time, it is appropriate to state that the anomalously high porosity in Ekofisk field is probably due to a combination of: (1) overpressuring of the reservoir; (2) magnesium-rich pore fluids; and (3) early introduction of hydrocarbons.

Overpressuring may have helped preserve the porosity by reducing the differential stresses at the grain contacts. As the hydrostatic pressure approaches the lithostatic pressure, such a reduction in grain contact stress would be expected. The early influence of magnesium-rich pore fluids is another factor. Neugebauer (1974) suggested that such pore fluids may retard solution transfer. The scattered secondary dolomite rhombs (Fig. 7a) observed in the Ekofisk Chalk may indicate the presence of magnesium-rich pore fluids.

The possibility of early introduction of hydrocarbons into the reservoir may have excluded or reduced pore waters, and therefore prevented solution transfer of calcite. Recent studies of chalk core samples directly above and below the base of hydrocarbons in a well in the nearby Tor field have lent some support to this concept. Samples appear to have had a high original porosity, but only those that are hydrocarbon-saturated retained much of this original porosity. Samples from below the base of the hydrocarbon zone showed extensive solution transfer effects and concomitant destruction of almost all porosity.

Another idea toward understanding chalk porosity at Ekofisk is based on the fact that in the modern ocean, calcareous nannofossils are dissolved as they settle to the deep seafloor and are preserved in the sediments only where the seafloor topography is above the carbonate compensation depth. Therefore, if the salt-cored structures in the Central Graben of the North Sea were active at the time of chalk deposition as believed, then these might provide the necessary seafloor relief to account for the thicker, better preserved chalk on the structural crests as compared to the flanks. In fact some correlation can be made between nannofossil preservation and porosity in the

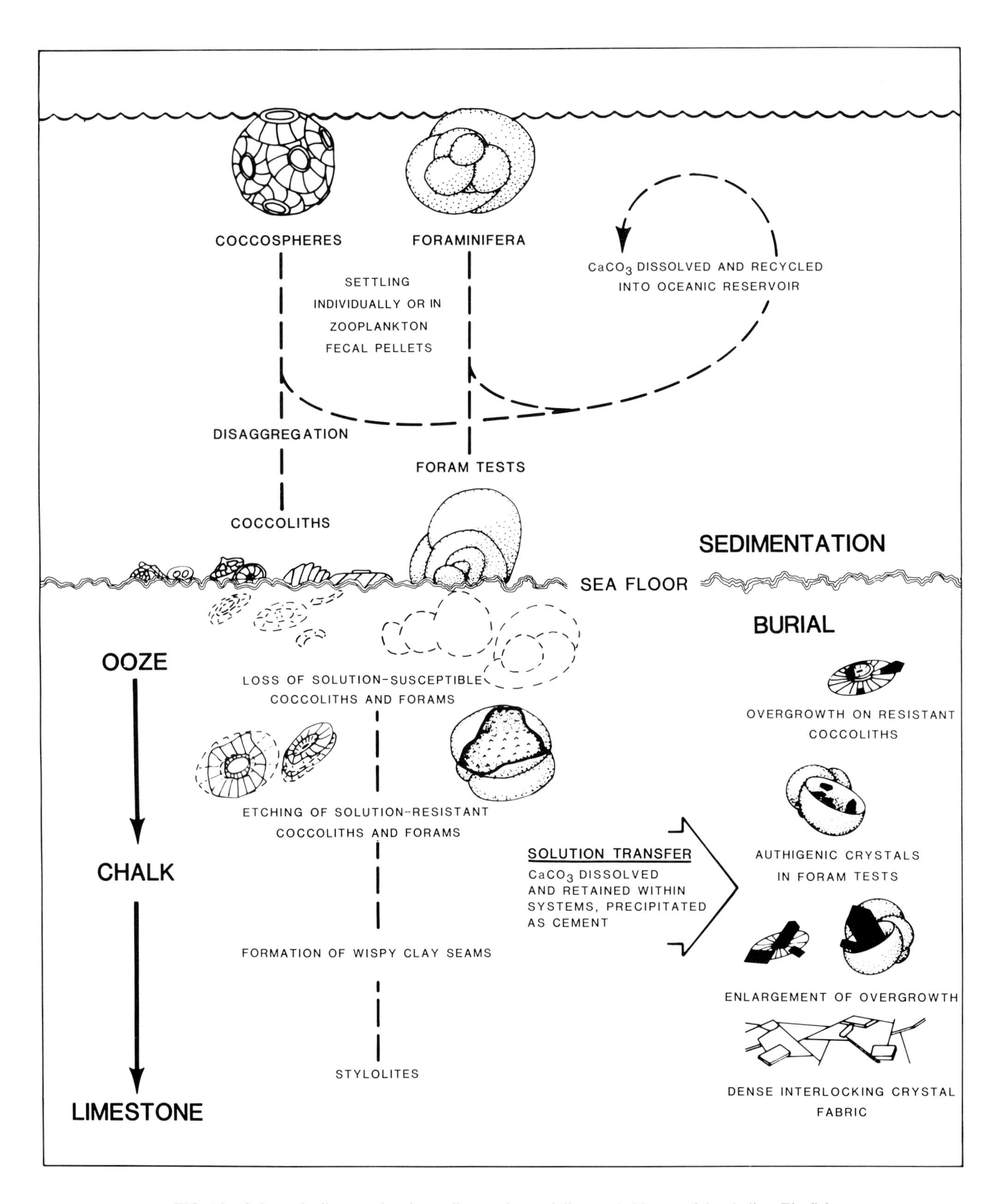

FIG. 10—Schematic diagram showing sedimentation and diagenetic history of the chalk at Ekofisk.

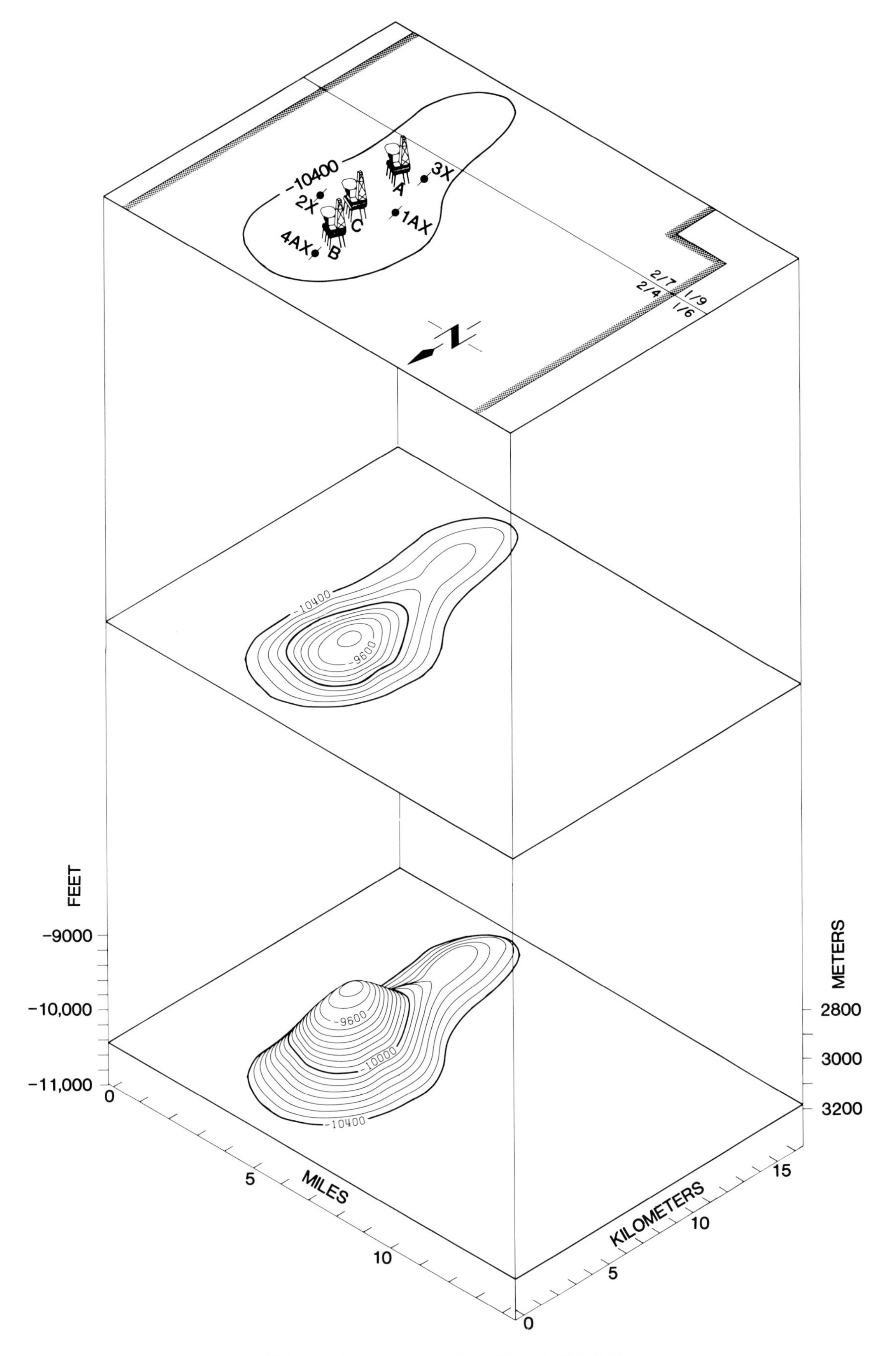

FIG. 11—Isometric projection of the Ekofisk field.

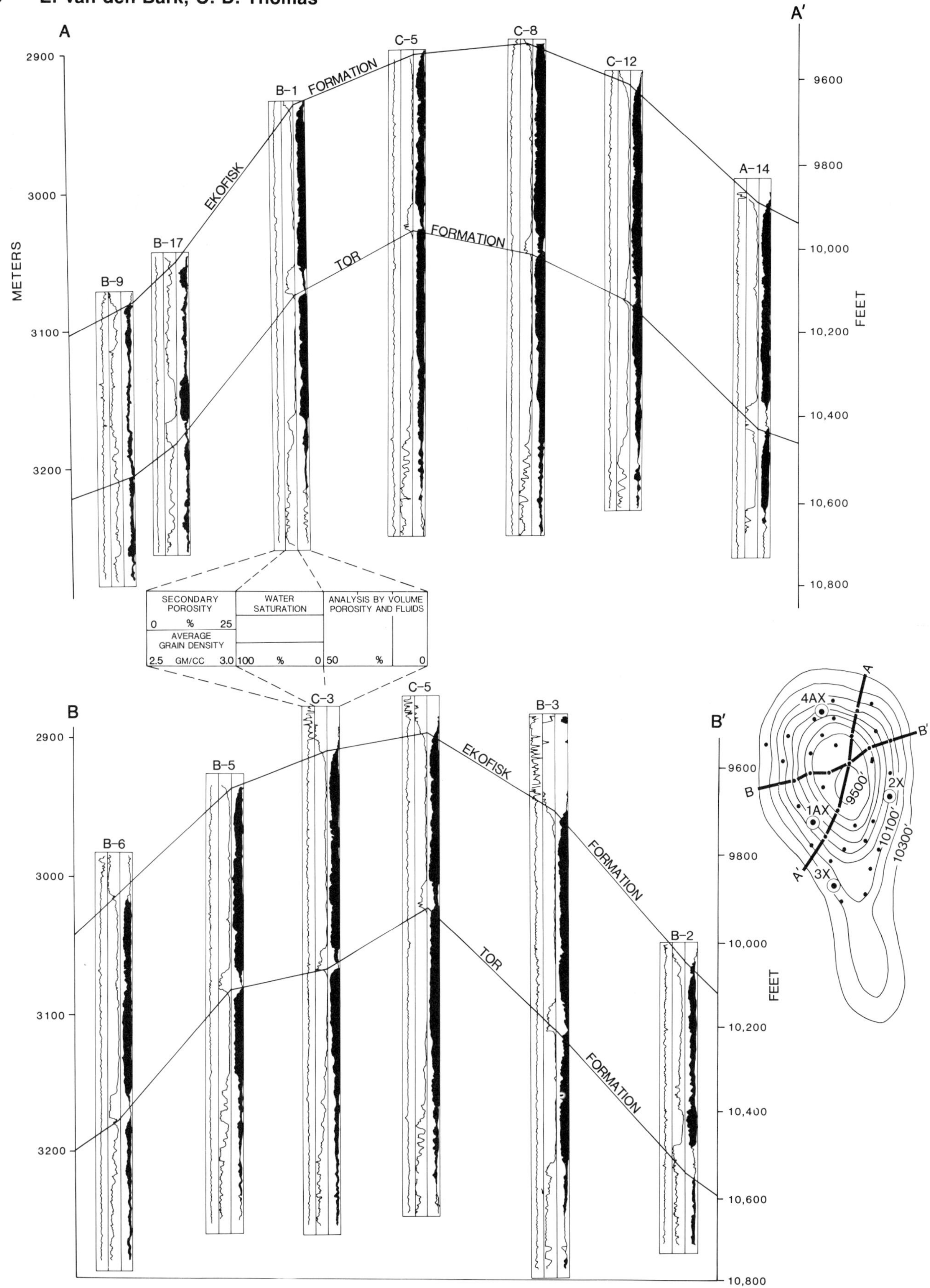

FIG. 12—Cross sections across the Ekofisk field show only minor changes in reservoir thickness.

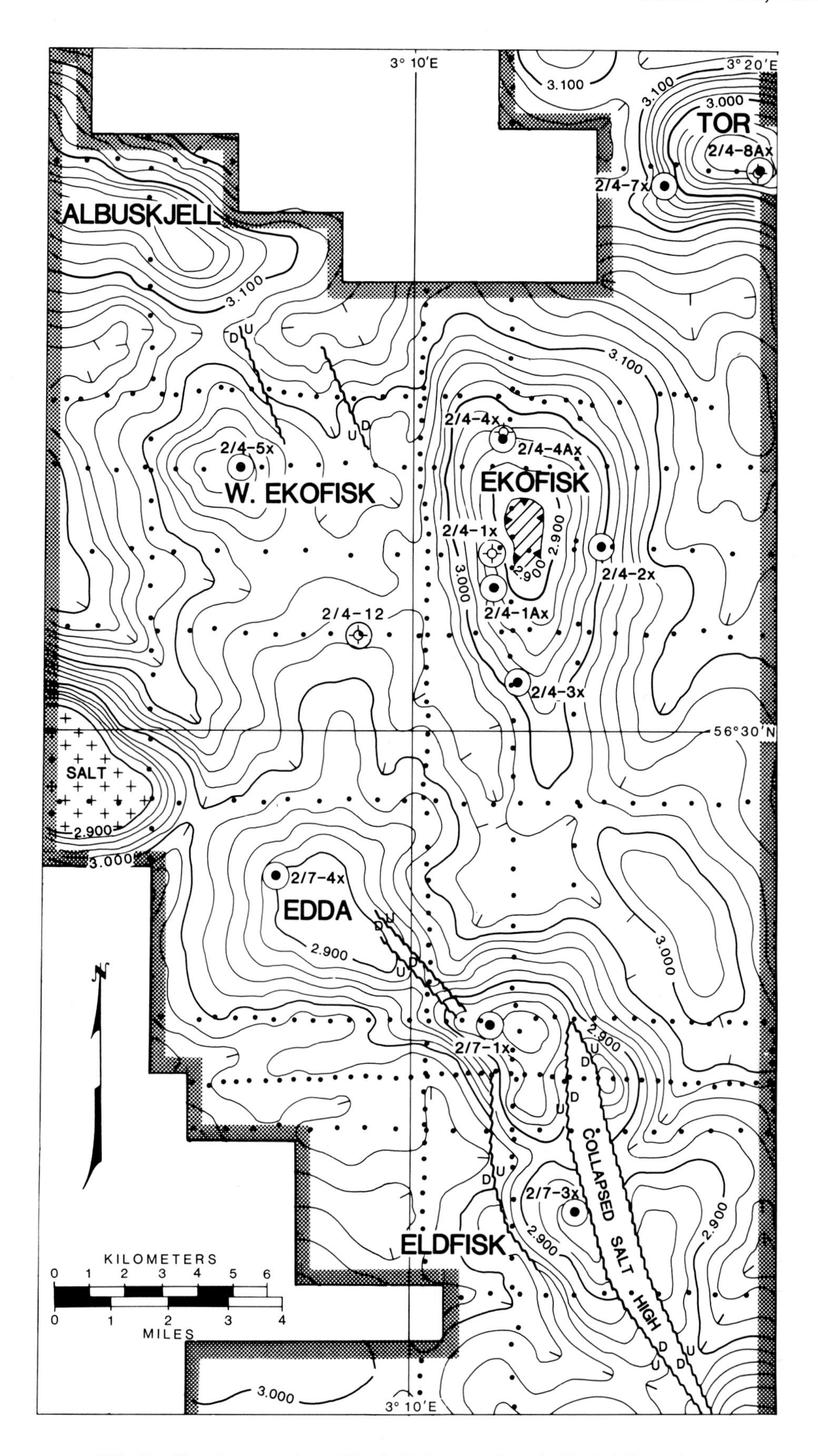

FIG. 13—Time interpretation within the Paleocene above the Ekofisk Formation (made in 1968). Subsequent exploraton wells have been plotted on the prospects. Contours in two-way time.

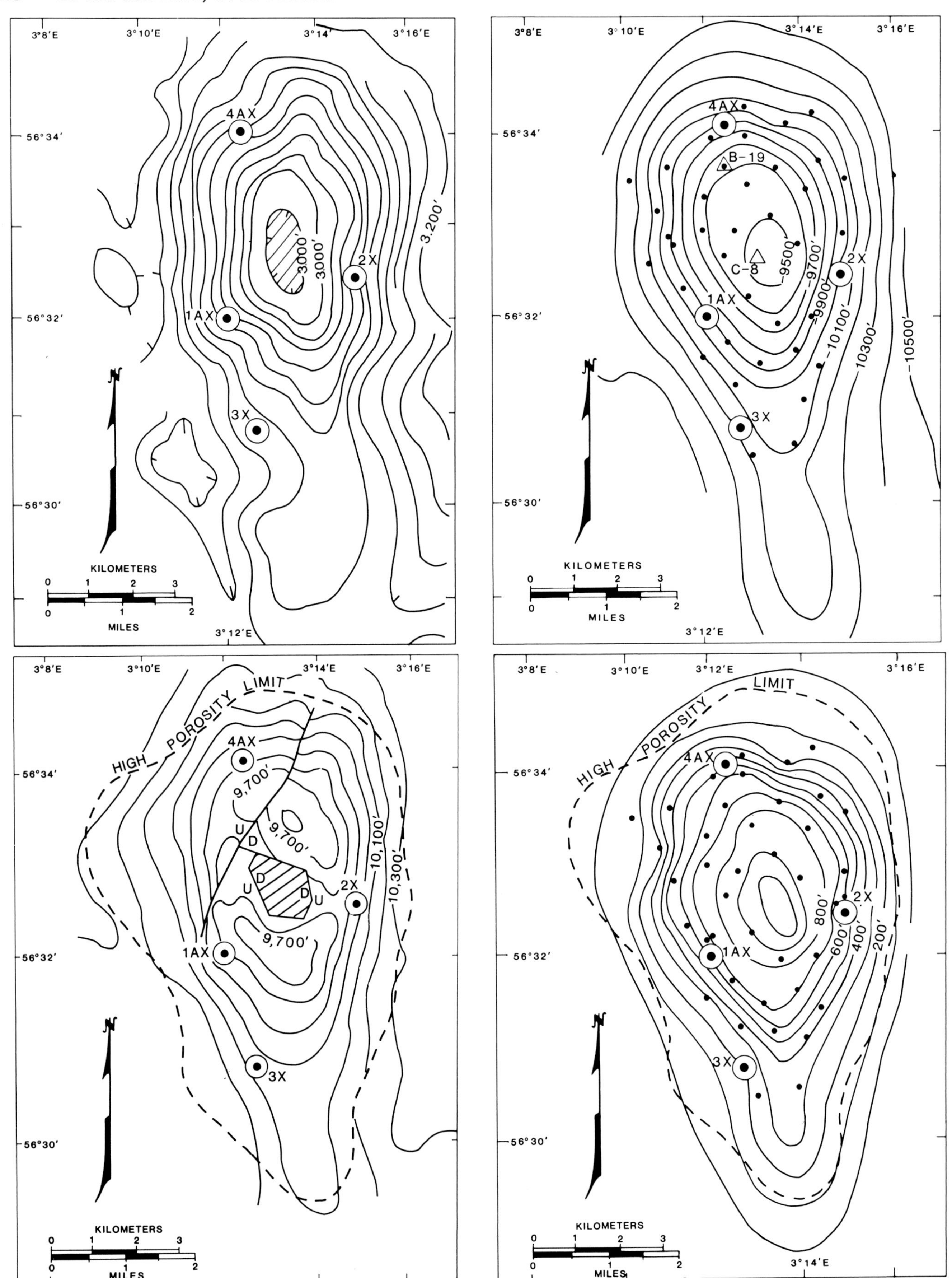

FIG. 14—**A.** Time interpretation mapped (1969) at the top of the Ekofisk Formation (contours in two-way time). **B.** Map of the Ekofisk Formation structure adjusted by drilling data (in 1979). Contours in feet below sea level. **C.** Structure map of the top of the Ekofisk Formation showing the hypothetical collapsed zone (1971 interpretation). Contours in feet below sea level. **D.** The porosity limit of the Ekofisk Formation compares very closely to the outline of the isopach map of the net pay based on development drilling results.

chalk and maybe due to the dissolution of the skeletal carbonates as they settled through the water column (Figure 10). Although the exact mechanisms controlling this relationship between porosity and nannofossil preservation are not well understood, they may involve a decrease in sorting caused by fragmentation of the skeletal particles resulting in lower porosities.

STRUCTURE

Ekofisk field is on a relatively simple north to south oriented anticlinal feature. It measures 12,070 acres (49 sq km) in areal closure and has 800 ft (244 m) of vertical closure at the top of the Ekofisk Formation (Fig. 11). The overpressured Paleocene shale above the chalk reinforces the anticlinal trap causing a long hydrocarbon column (1,000 ft or 305 m) which extends below the spill point of the reservoir.

The earliest salt movement is difficult to determine at Ekofisk, but the existence of structures of Late Jurassic age in the Greater Ekofisk area suggests that halokinetic movements were initiated by a combination of sufficient overburden and normal faulting during that time. Ekofisk was probably no more than a very low relief feature during chalk deposition since the gross reservoir units correlate across the structure with only minor changes in thickness (Fig. 12). Otherwise, contemporaneous uplift and deposition would have caused slumping of the coccolith ooze resulting in thinning on the structural crest. Salt movement continued into the late Tertiary and formed the anticlinal trap as well as fracturing the then-lithified chalk at Ekofisk (Byrd, 1975). Large-relief piercement structures are evidence of continued halokinetic movements, but to date these have not proved productive.

The degree of fracturing within the chalk increases with depth. This general trend also is reflected in the effective porosity. The upper part of the Ekofisk Formation is slightly less porous (25% to 30%) than the lower part (30% to 40%). The latter is separated from a porous zone in the Tor Formation (35% to 40%) by the "tight zone" with a porosity range from 0 to 20%. Porosity also varies with structural position. However, effective porosity increases in all zones toward the structural crest (Fig. 12) due largely to increased fracturing.

GEOPHYSICAL EXPLORATION

The six Greater Ekofisk fields now being developed were all located during the 1960s by reconnaissance seismic work. A comparison of the time interpretation mapped within the Paleocene above the top of Ekofisk Formation in 1968 (Fig. 13) and the Ekofisk portion from a 1969 map on the top of the Ekofisk Formation (Fig. 14a), with a more recent map of the Ekofisk structure corrected by drilling data (Fig. 14b) show little change in field outline. Nevertheless, considerable seismic shooting has been necessary for structural and stratigraphic refinement prior to further wildcat and development drilling. In addition, a combination of borehole data and reflection character (or wave shape) changes led to a clearer understanding of the seismic response. These studies also permitted the mapping of porosity variations over the Ekofisk structure.

Velocity Effects

Reflection quality in the area is excellent so interpretations usually are reliable. However, the middle Tertiary rocks which overlie the Ekofisk structure have extremely low seismic velocity because of abnormally high pressure. This creates a distortion of the structural interpretation and led to the early theory that the center of the structure was collapsed. (Fig. 14).

The east to west seismic cross section across the Ekofisk structure in Figure 15 shows the apparent collapsed zone which later proved to be only a "collapsed velocity zone" rather than an actual collapsed structure. This interpretation was determined by studying the seismic derived velocities which indicated the existence of the low velocity condition near the crest of the structure. However, these velocities precluded making an accurate depth map on the chalk. For example, (Fig. 16) the sonic log from the 2/4-C8 well drilled through the crest of the seismic anomaly in 1974 shows an extremely-low-velocity section from a depth of 5,800 ft (1,768 m) to 7,190 ft (2,192 m). Although the sonic log moves off scale on the low side, by using check-shot calibration the velocity in this zone is measured to be 4,950 ft (1,509 m) per second. This velocity is lower than any encountered at this depth in this part of the North Sea and remains an anomaly unique to the Ekofisk area.

Seismic Porosity Mapping

A north to south seismic cross-section (Fig. 17) integrates borehole and seismic data to show the high porosity limits of the field. Porosity over 16% creates an increase in amplitude. When mapped in the Ekofisk Formation this indicates a porosity limit which compares very closely to the isopach map of the net pay which was recently prepared from data generated during development drilling (Fig. 14d).

Additional seismic data have continued to be recorded since the discovery of Ekofisk, along with the application of new processing methods. These have helped in developing structural and stratigraphic refinements needed to assist in the orderly and efficient development of Ekofisk and the Greater Ekofisk Area. For example, seismic sections (Fig. 18) across Ekofisk show several amplitude anomalies and "bright spots" within the Paleocene shale which results from the presence of gas. It is possible that this gas leaked from the reservoir into the already overpressured shales and enforced the low velocity effect over the structure.

SOURCE ROCK ANALYSIS

The source of the hydrocarbons trapped in Ekofisk field has been the object of considerable interest and analyses since discovery. The large volume of hydrocarbons in the Ekofisk area requires a large volume of source rocks. This limits the potential source rock to either the Paleocene Lista Formation or the Upper Jurassic Kimmeridgian Clay.

The Upper Jurassic Kimmeridgian Clay was deposited in an open-marine environment under relatively anoxic conditions.

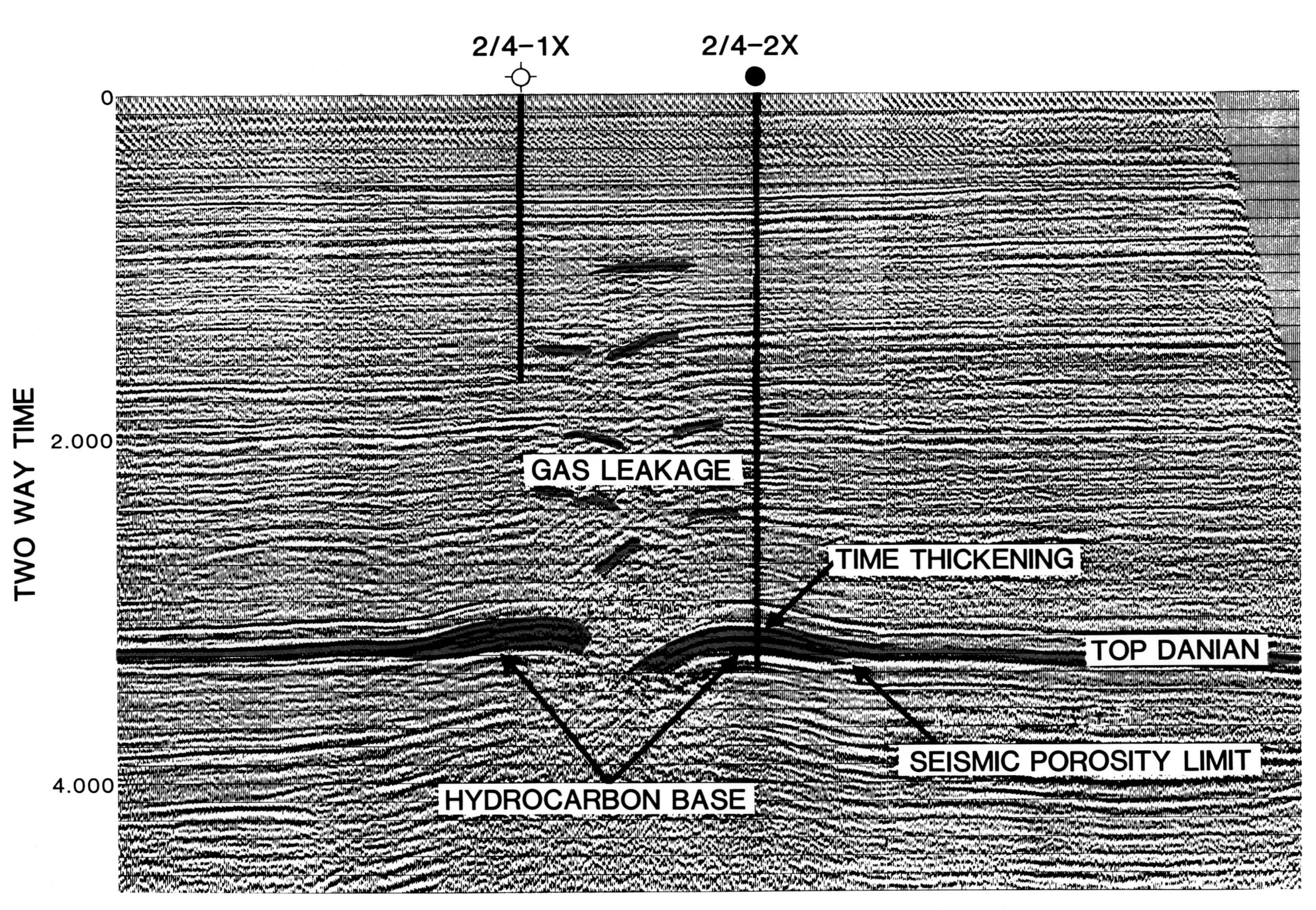

FIG. 15—East to west seismic cross section the Ekofisk structure showing the apparent collapsed zone resulting from low velocity conditions near the crest.

Table 1. Geochemical Data, Ekofisk Field, North Sea.

DEPTH	PERCENT ORGANIC CARBON	RATIO SOLUBLE/ TOTAL CARBON	RATIO PRISTANE/PHYTANE	$\overline{OEP}$
EKOFISK WELL 2/4-12			PALEOCENE	
9930'-9990'	1.8%	0.014	1.46	1.12
10000'-10060'	2.0%	0.012	1.46	0.98
EKOFISK WELL 2/4-19B			UPPER JURASSIC	
13400'-13490'	1.4%	0.059	1.27	1.01
14250'-14350'	2.3%	0.057	1.45	1.01
14550'-14650'	2.5%	0.047	1.42	1.08
14690'-14790'	2.6%	0.038	1.37	1.03

Reducing marine conditions are indicated by: (1) the limited number of terrestrially derived spores and pollen; (2) the abundance of amorphous organic matter; and (3) the low pristane to phytane ratios (Didyk et al, 1978; Tables 1, 2). Also unlike the organic faction of terrestrial matter, a relatively small amount of normal alkanes occur in the plant wax region of the saturates (Fig. 19). The Paleocene Lista Formation also appears to have been deposited in an open-marine environment.

In the initial geochemical analyses, the Paleocene shale samples from crestal Ekofisk wells were classified as source rocks because oil in the rocks showed the same general characterisics as the oil produced from the Tor and Ekofisk Formations. However, this is now considered to be oil migrating through the section from the underlying reservoir and is not indigenous. Recent vitrinite reflectance studies show that the Paleocene shales are at the lower threshold of thermal maturity (R_0 = 0.59% to 0.62%), and therefore cannot qualify as the source rocks for the large volumes of hydrocarbons in the Ekofisk field. Moreover, gas chromatography/mass spectral data of the saturate fraction of oils extracted from the Paleocene shales off structure show both rearranged and unrearranged steranes—a relationship characteristic of relatively immature organic matter (Seifert and Moldowan, 1978).

The Jurassic Kimmeridgian Clay samples from Ekofisk wells have vitrinite reflectance values ranging from R_0 = 0.93% to R_0 = 1.16%, indicating that they are in the peak range of thermal maturity for hydrocarbon generation (Fig. 20). Moreover, the maturity of the clays is confirmed by the presence of only rearranged steranes. Therefore, although both the Paleocene and Jurassic shales have sufficient volume and organic content to qualify as source rocks, only Jurassic beds have experienced a thermal history of high enough temperature to have generated the hydrocarbons at Ekofisk. How these hydrocarbons migrated from the Jurassic up into the uppermost Cretaceous/Paleocene chalk at Ekofisk is not yet clear, although faulting appears to have served as important conduits in other parts of the Greater Ekofisk area.

Further and even more definitive evidence that the Kimmeridgian Clay is the source of the Ekofisk oil is based on techniques which distinguish unique similarities and differences between rocks and oils. Conventional methods for correlation such as the comparison of normal alkane distribution (Fig. 21 with Figs. 19, 22) and carbon isotope data have been inconclusive. High resolution gas chromatography coupled with sensitive detectors (such as element selective photometric detectors for sulfur compounds, or mass spectrometer data systems for chemical structure identification) provides a highly definitive analytical tool for differentiation of rocks and correlation of rocks and oils. These "chemical fossils" found in rock extracts and crude oils show that compounds unique to the Paleocene shales are not present in the Ekofisk oil whereas compounds relatively abundant in the Upper Jurassic shales are relatively abundant in Ekofisk oil (Table 2).

Altogether, based on relative maturity and "chemical fossils" the Upper Jurassic shale in the vicinity of Ekofisk appears to be the source for oil in the Ekofisk structure.

GEOPRESSURES

The oil in the Ekofisk reservoir is a single accumulation in an abnormally pressured environment. The pattern of abnormal pressure distribution was first interpreted from the four exploratory wells and later confirmed during development drilling.

Oil Accumulation Continuity

Oil was produced from reservoirs of Danian age in all four exploratory wells from a total of thirteen tests. Oil was recovered in three additional tests from rocks of Maastrichtian age in the 2X and 4AX exploratory wells. Because these tests were conducted over separate intervals, it was difficult to determine whether or not these oil zones represented one or more accu-

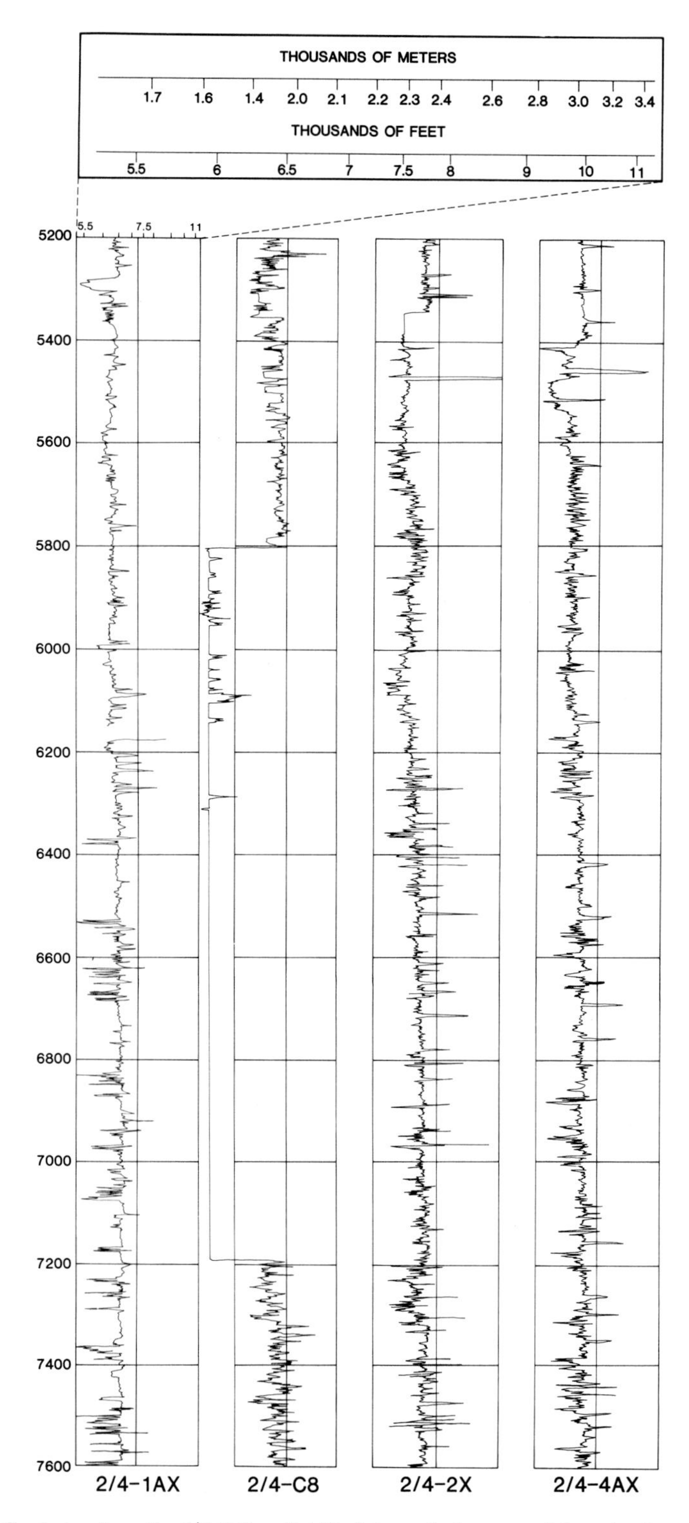

FIG. 16—Sonic log from the 2/4-C-8 well drilled through the crest of the seismic anomaly in 1974, compared with the discovery wells. Note that the C-8 well shows an extremely low velocity section from 5,800 to 7,190 ft (1,768 to 2,192 m).

mulations. However, because any oil or gas accumulation in its original state underground is static, the initial pressures plotted at datum will fall on or within the limits of pressure data accuracy of the subsurface fluid pressure gradient for that accumulation. Therefore, pressure-elevation correlations were based on initial reservoir pressure and showed that all of the oil zones tested in rocks of Danian and Maastrichtian age are in the same accumulation (Figure 23).

Table 2. Comparison of Rock Extract and Crude Oil Chemical Fossils.

Chemical Fossil Type / Sample	Unique Chemical Fossils: 30 Carbon Pentacyclic Triterpane and Methyl Hopanes	Branched and Cyclic Hydrocarbons Smaller than 25 Carbons	Ratio of Cholestane to Ergostane	Ratio of Rearranged Steranes to Natural Steranes	4 Carbon Substituted Benzothiophene	Dibenzothiophene 1 and 2 Carbon Substituted Dibenzothiophenes
	Analysis by High Resolution Capillary Gas Chromatography/ Mass Spectrometry/Data System				Analysis by High Resolution Capillary Gas Chromatography/Flame Photometric Detector	
Ekofisk Crude Oil	Absent	Minor	>1	>1	Minor Constituent of Sulfur Compounds in the Aromatic Fraction	Most Abundant of Sulfur Compounds in the Aromatic Fraction
Paleocene Shale Extracts 2/4-12	Present	Abundant	< 1	< 1	Major Constituent of Sulfur Compounds in the Aromatic Fraction	Major Constituent of Sulfur Compounds in the Aromatic Fraction
Upper Jurassic Shale Extracts 2/4-19B	Absent	Minor	> 1	>1	Minor Constituent of Sulfur Compounds in the Aromatic Fraction	Most Abundant of Sulfur Compounds in the Aromatic Fraction

Apparently, oil has migrated into the Ekofisk structure forming a vertically and laterally large and continuous accumulation without regard to geologic rock units and reservoir heterogeneities.

Pressure Conditions Influencing Entrapment

Abnormal pressures start at a depth of about 3,400 ft (1,306 m) below the rotary Kelly bushing (RKB) in the Tertiary shales and continue into the Upper Cretaceous rocks. A pressure regression is the most significant feature in the shales above the oil reservoir. This suggests a downward pressure drop from the shales which have an estimated pressure gradient of about 0.75 psi/ft (16.97 kPa/m) into the underlying oil reservoir which has a measured pressure gradient of 0.6882 psi/ft (15.57 kPa/m). Such a downward pressure drop serves to strengthen the shale caprock and increase its ability to trap oil. A formation pressure profile for a typical Ekofisk well is shown in Figure 24. It is based on pressures estimated from drilling data and sonic-log shale travel time data while the pressure profile through the reservoir is based on measured pressures.

Base of Hydrocarbons

Exploratory and development drilling have shown that a well defined oil-water contact does not exist in the Ekofisk field. In fact, the only clear evidence of an actual oil-water contact is in the east flank C-15 well (2X replacement well) where it occurs at a depth of 10,786 ft (3,288 m) subsea. Preliminary studies based only on the 2X and 3X wells suggested the possibility of a tilt in the oil-water contact. Tests in the 2X well recovered oil to a subsea depth of 10,711 ft (3,265 m) while tests in the 3X well recovered water beginning at a subsea depth of 10,651 ft (3,246 m). Also pressure-elevation data suggested that an oil-water contact occurred at 10,585 ft (3,226 m) subsea in the 3X well and at 10,745 ft (3,275 m) subsea in the 2X well. Moreover, the water pressure in the 3X well was slightly higher than that pressure in the 2X well. Therefore, if the reservoir was characterized by homogenous porosity, lateral flow through the water phase would be from the 3X well toward the 2X well. Such a direction and magnitude could be capable of tilting the oil-water contact from the 3X well toward the 2X well at about 60 ft per mile (12 m/km). However, it was recognized during early development drilling that flow was reduced significantly with less then 15% porosity and greater than 50% water saturation. Consequently, the base of hydrocarbons occurs at different elevations in each well and the base of hydrocarbons throughout the field is a function of reservoir quality. A map of the base of hydrocarbons shows approximately 350 ft (107 m) of relief from the crestal part of the structure to the lowest eastern flank well. A net pay isopach of the total Ekofisk reservoir from the top of the Ekofisk Formation to the base of the hydrocarbons is shown in Figure 14d.

The dome shape of the base of hydrocarbons can be explained by early migration of the hydrocarbons into a low relief anticlinal trap, which arrested secondary cementation within the oil column. Porosity and permeability below the oil column were greatly reduced by secondary cementation. Subsequent Late Tertiary salt movement beneath the chalk created structural relief at all levels including the base of hydrocarbons.

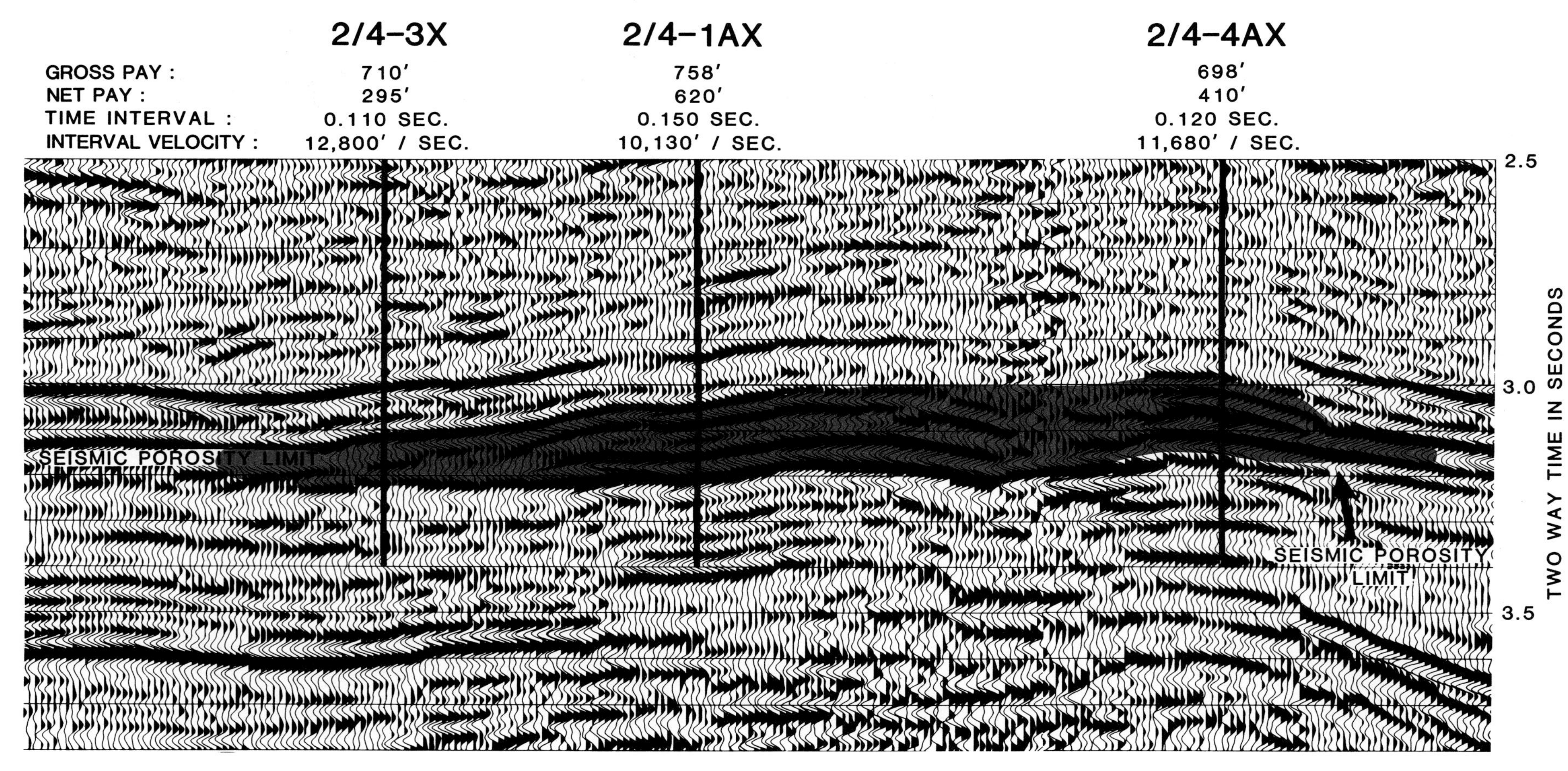

FIG. 17—North to south seismic cross section integrates borehole and seismic data to show the high porosity limits of the field.

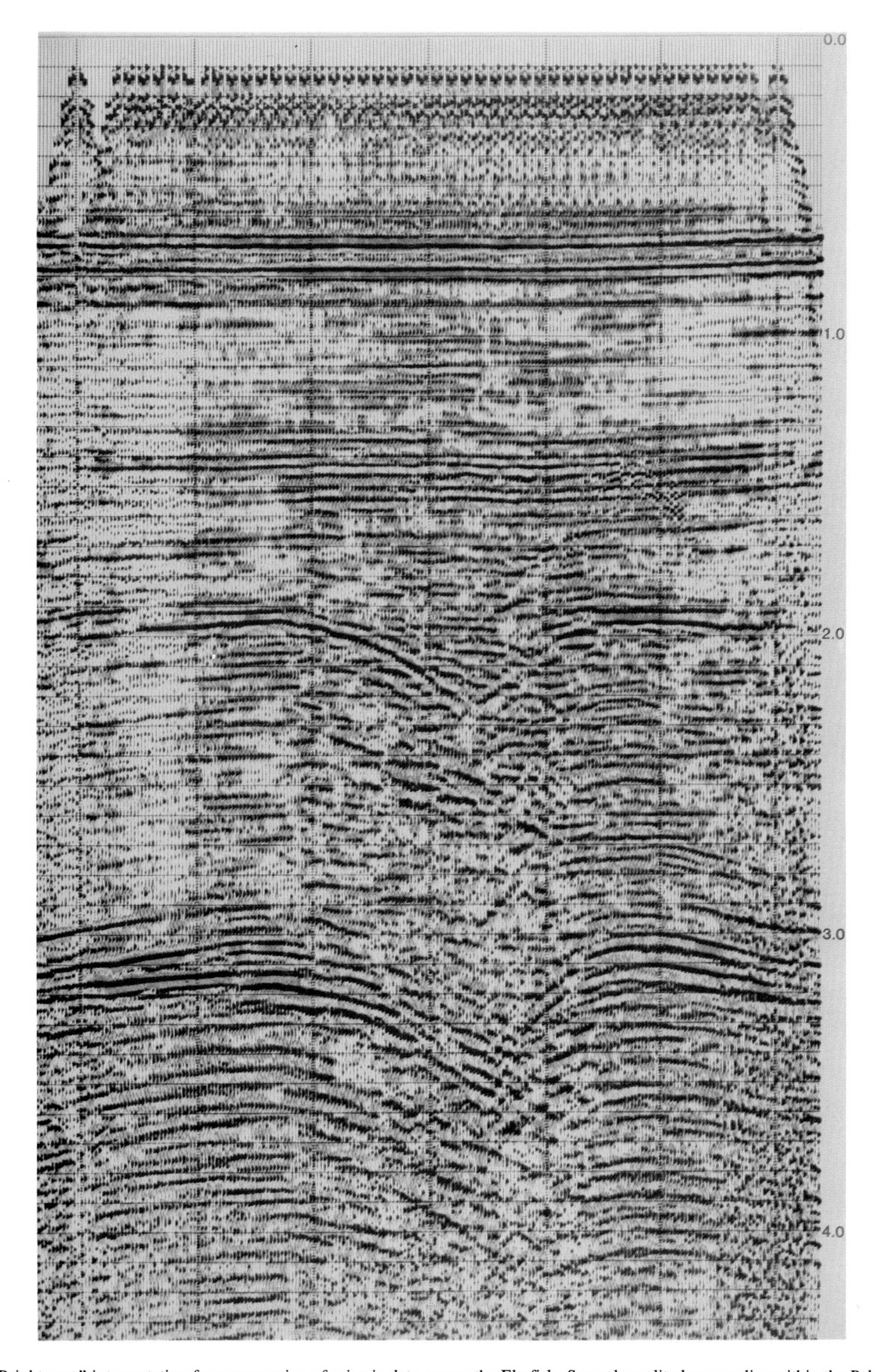

FIG. 18—"Bright spot" interpretation from processing of seismic data across the Ekofisk. Several amplitude anomalies within the Paleocene shale result from the presence of gas.

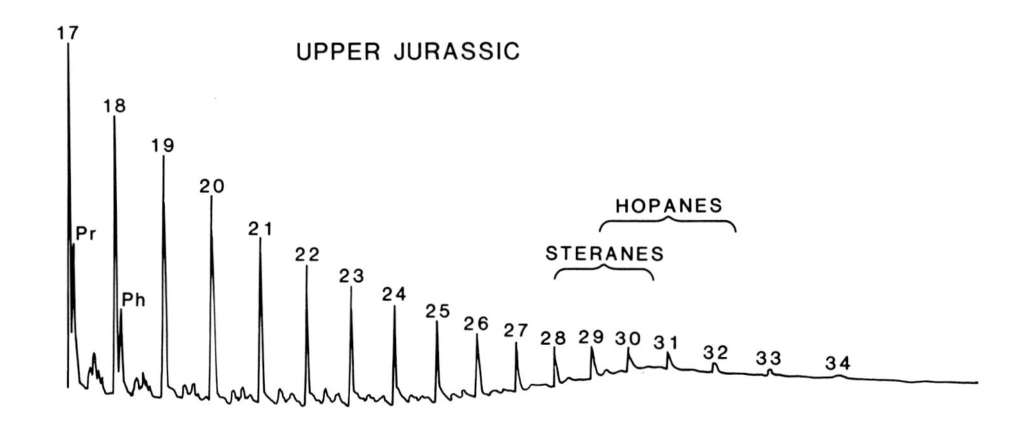

FIG. 19—Chromatograph showing the relatively small amount of n-alkanes which occur in the plant wax region greater than n-C_{25} of the saturates extracted from Upper Jurassic rocks.

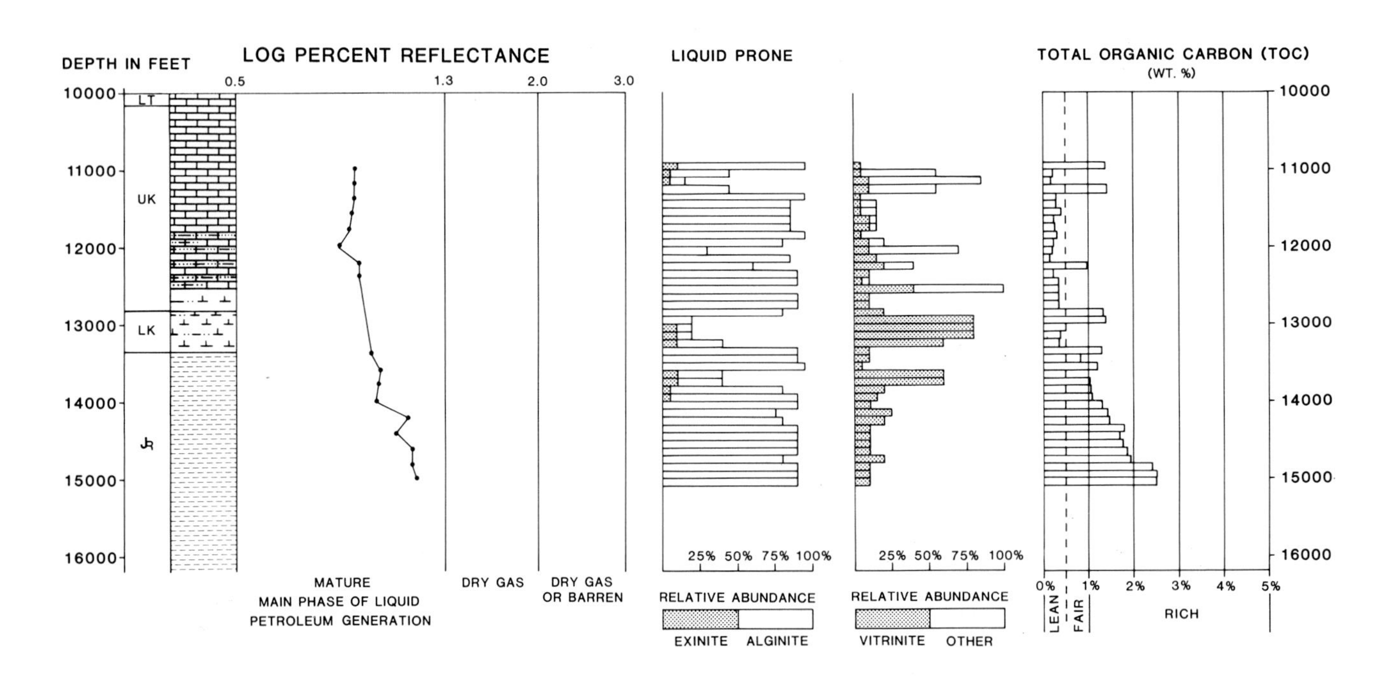

FIG. 20—Thermal alteration and source rock potential of 2/4 B-19 well.

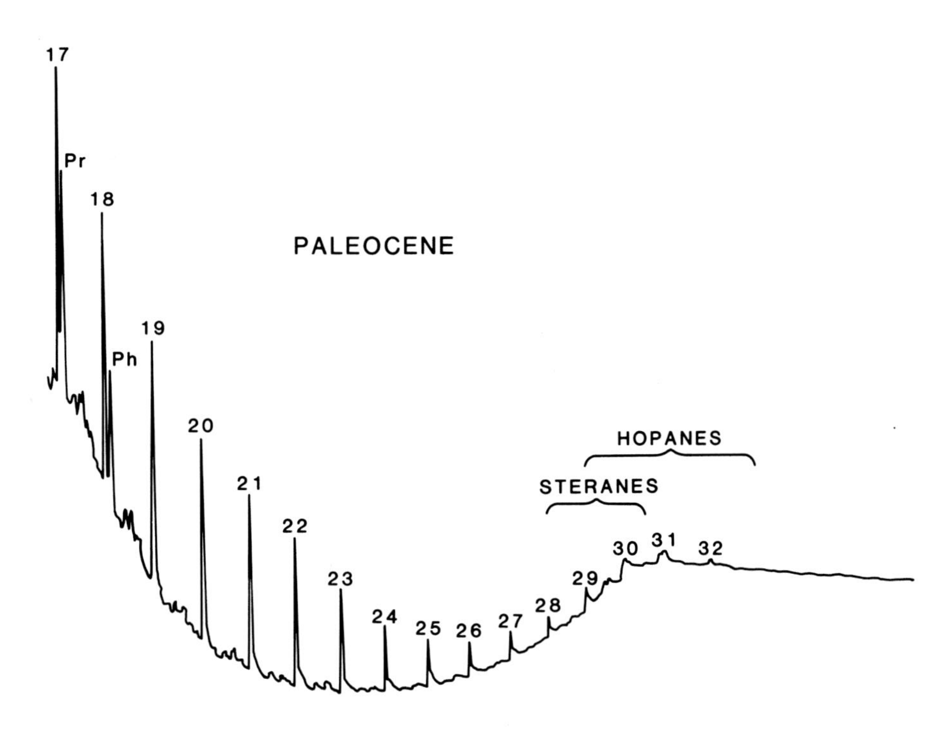

FIG. 21—Chromatograph of n-alkane distribution in Ekofisk crude.

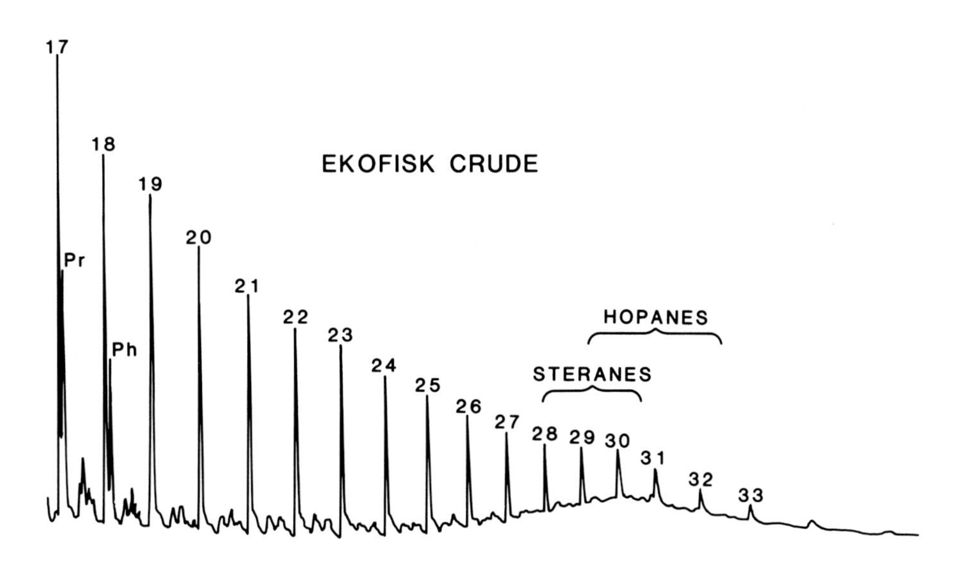

FIG. 22—Chromatograph of oil extracted from the Paleocene Lista Formation. Distribution of n-alkanes is characteristic of an open marine environment.

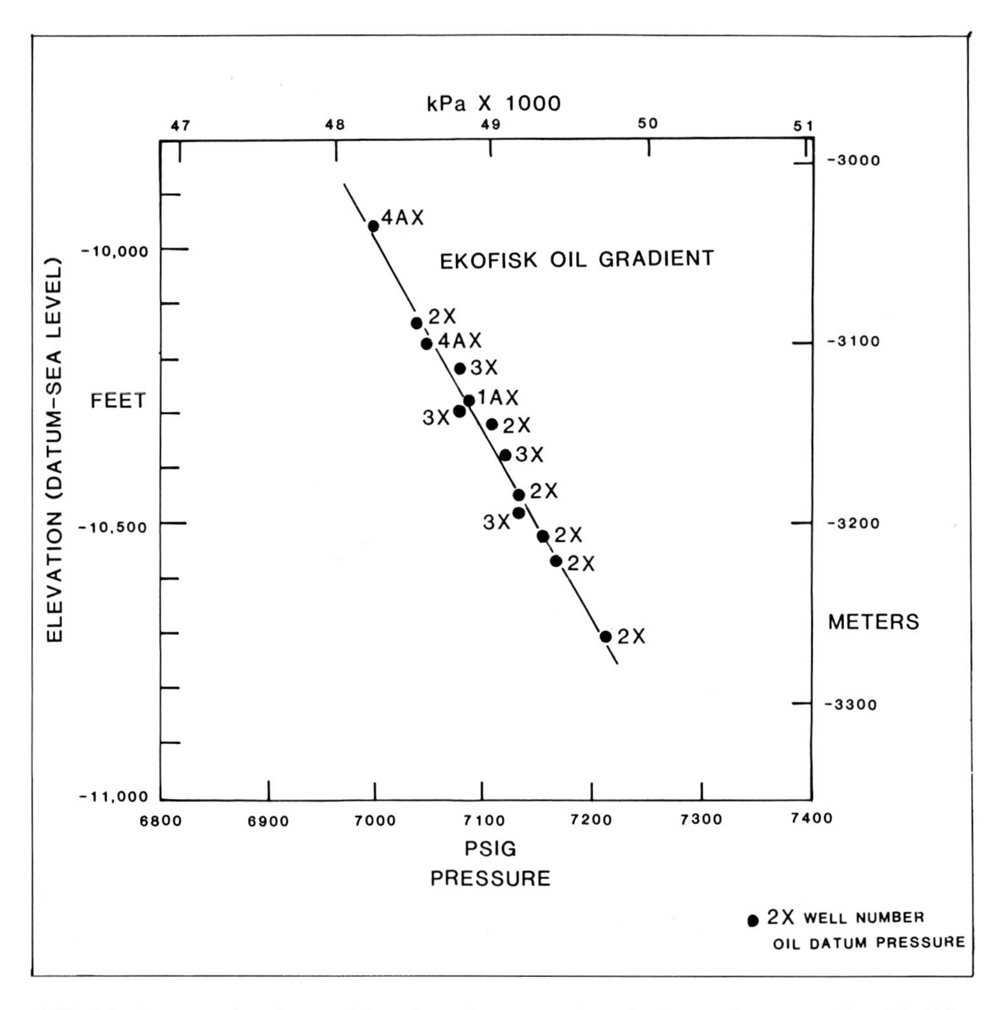

FIG. 23—Pressure elevation graph based on oil pressures from the four exploratory wells at Ekofisk.

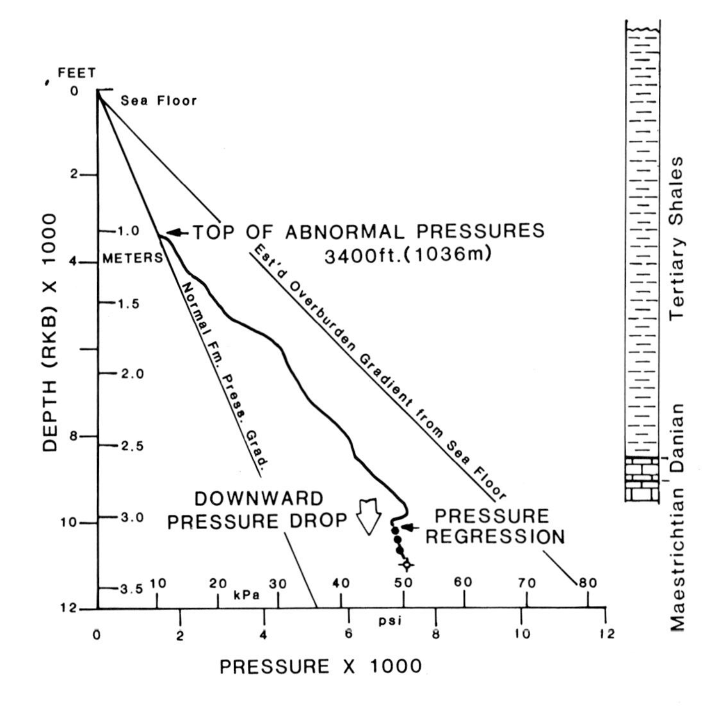

FIG. 24—Pressure-depth plot of a typical well at Ekofisk.

Table 3. Reservoir Parameters, Ekofisk Field, North Sea.

LITHOLOGY	CHALK
DEPTH	-10,400 ft (-3170 m) reservoir midpoint
AREA	12,071 acres (49 Sq Km)
NET PAY	590 ft (180 km) with Ø >15%, Sw< 50%
GROSS PAY	832 ft (254 km) avg., 998 ft (304m) max.
NET/GROSS RATIO	0.709
AVERAGE POROSITY	31.7%
PERMEABILITY RANGE	1-100 MD (matrix >1 to 10)
RESERVOIR PRESSURE	7,135 psi at 10,400 ft. (49,194 kPa at 3170m)
PRESSURE GRADIENT	0.685 psi/ft (15.4951 kPa/m)
RESERVOIR TEMPERATURE	268% F at 10,400ft (131%C at 3170m)
TEMPERATURE GRADIENT	2.5% /100 ft. through the reservoir (0.046% C/m)
INTERSTITIAL WATER	23.6%
GOR	1547
FVF	1.78
OIL GRAVITY	36% API (Density 0.8443 Kg/Lit 15% C)
AFTER ACID SKIN	-4 to -6
PRODUCING MECHANISM	Solution gas drive

RESERVOIR CONDITIONS

Although the gross hydrocarbon column in the Ekofisk reservoir is over 1,000 ft (304 m) the average net pay is 590 ft (180 m) so the net/gross ratio is 0.709. Perforations are restricted to the three most porous zones which are: the upper Ekofisk Formation, the lower Ekofisk Formation, and the upper Tor Formation. The Ekofisk crude is an undersaturated volatile oil.

A list of average parameters that we derived from a three dimensional phase computer simulated model for the Ekofisk field is given in Table 3.

The model incorporated provisions for gas injection above the original bubble point. The initial model was based on the four exploration wells and the reservoir was divided into four layers: the upper Ekofisk Formation, the Lower Ekofisk Formation, the "tight zone", and the upper Tor Formation. After 39 development wells were drilled, a 10-layer model was constructed using each of the wells as control for reservoir properties distribution. The resulting model had 1,053 cells described as net with 3,259 total cells and indicates 5.4 billion stock tank bbl of oil in place. The ultimate recoverable reserves with depletion and gas injection are estimated to be 1,220 million standard bbl., 4,410 billion standard cubic feet, and 110 million bbl of natural gas liquids.

GREATER EKOFISK DEVELOPMENT PROGRAM

The development of the Ekofisk field was tied into the Greater Ekofisk Area. At the present time this area consists of seven fields, namely: Ekofisk, West Ekofisk, Edda, Tor, Eldfisk, Albuskjell, and Cod fields. Because all these fields were connected to use joint facilities, the development of the Ekofisk field should be discussed as part of the total development of the Greater Ekofisk Area (Fig. 25).

Preliminary studies showed that the development of the Ekofisk field would require substantial investment. Therefore, before further financial commitment, it was necessary to acquire more reservoir and production information on the Ekofisk and Tor Chalk Formations in order to confirm reserve estimates and proper design for the equipment. Moreover, it was desirable to obtain income from the field as quicky as possible. To attain these objectives it was decided to develop Ekofisk in three phases.

Phase I

The first phase was designed as a testing and data gathering program while generating income. To meet these objectives the four original exploratory wells were brought on production as subsea completions some 18 months after discovery. Undersea flow and control lines, a temporary production platform, separation equipment, and an open-sea tanker loading system, were installed (Rickards, 1974).

A jack-up drilling platform, was modified by removing the drilling equipment and installing production equipment necessary for processing the production from the four subsea wells. Two undersea loading lines were connected to two single-point mooring bouys (SBM) to load the oil production into tankers for delivery to refineries. Two tankers were modified for

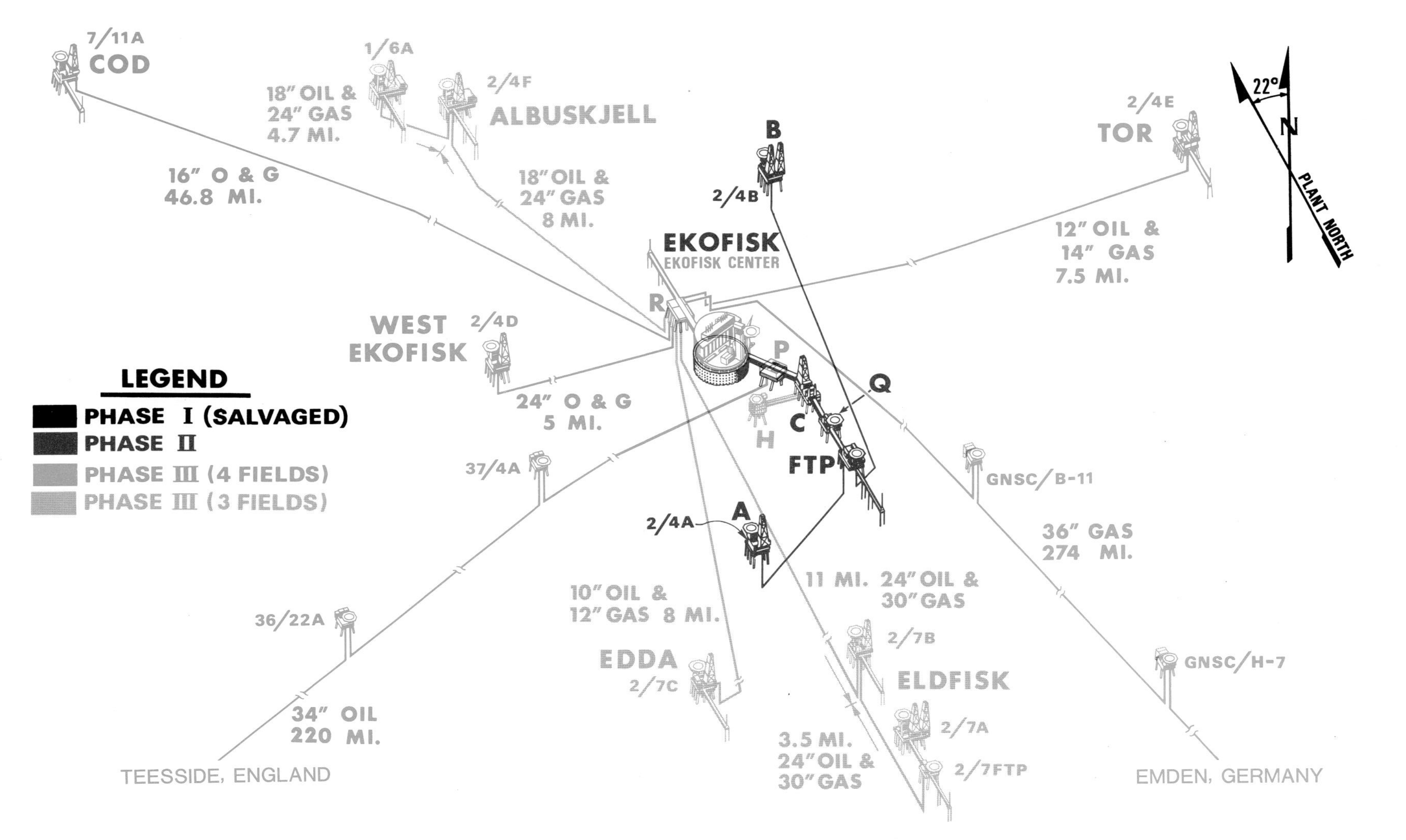

FIG. 25—Greater Ekofisk area showing relationship of facilities in total development of satellite fields.

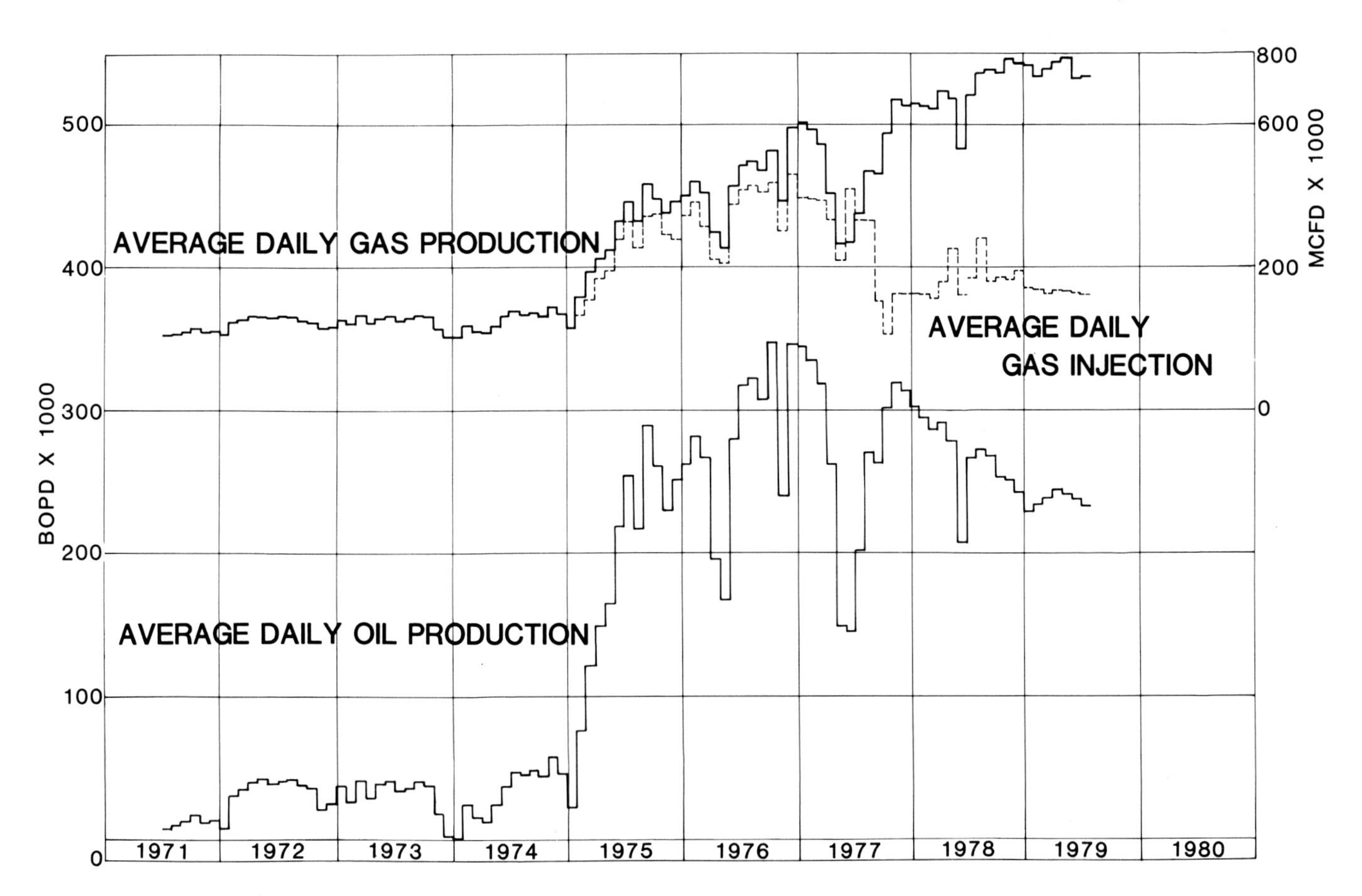

FIG. 26—Production profile of Ekofisk field, North Sea.

mooring and bow-loading of crude oil from the four exploratory wells and intial production began July 7, 1971. Phase I lasted until May 1974. During that time about 28 million bbl of oil were produced from the Ekofisk field at a rate of approximately 42,500 b/d when all four wells were on production. Gas production of about 42 mcf/day was flared during the period.

Phase II

Phase II covered continued exploratory drilling in the Greater Ekofisk Area including 40 wells in the Ekofisk field. This phase also included the installation of permanent risers, quarters, and drilling-production platforms with production and gas-injection equipment. Also included was the installation of a one-million-barrel capacity concrete storage tank to provide a surge to allow the wells to be operated when sea conditions required SBM loading to cease.

This second phase was planned and constructed while the first phase was on stream. It was designed to bring production to approximately 300,000 b/d from the Ekofisk field while returning to the reservoir all produced natural gas (approximately 440 million cf/day) which was not utilized for fuel. This gas reinjection program necessitated the drilling of 8 injection wells and the installation of compressors to raise the pressure to ovecome the high formation pressure and friction. Three drilling and production platforms (A, B, and C), one field terminal platform (FTP), a quarters platform (O), and the million barrel crude oil storage tank were constructed. The SBM system for loading oil tankers offshore was used in handling Ekofisk field production until October 15, 1975 when the crude oil pipeline to Teeside began operation.

Phase III

The third phase was designed to develop, produce, transport, and process the crude oil and natural gas from the six other fields (Cod, Tor, West Ekofisk, Edda, Eldfisk, and Albuskjell) in addition to Ekofisk. The additional steps required in Phase III, most of which are in the final stages of completion, were:

1. Complete the development drilling in all seven fields—163 wells.
2. Gather the well fluids from all seven fields into the Ekofisk Center (on top of the storage tank).
3. Construct the Ekofisk Center to process the well fluid from all fields to produce a residue gas suitable for transportation to shore; and a 100 psi vapor pressure liquid gas—crude oil mixture also for transportation to shore.
4. Construct the necessary pipeline facilities: (a) Oil pipeline 34 in. diameter (86.36 cm), 1 million b/d capacity, 220 mi (372 km) in length to Teeside, England; and (b) Gas pipeline 36 in. diameters (91.44 cm), 2.2 billion cf/day capacity 275, mi (466 km) in length to Emden, Germany.
5. Construct the necessary facilities to separate the liquid

gas (NGL) from the crude oil at Teeside and produce ethane, propane, iso-butane, normal butane, and 5-pound crude, and load into tankers.

6. Construct facilities at Emden to treat the gas for pipeline delivery.

7. Provide a communication system to link all operating units together for information gathering and dissemination, and control of the built-in safety features.

With the completion of the above steps, offshore loading was discontinued. However, the system remains so that offshore loading can be resumed if needed. The majority of the gas is transported to Emden for sale, but the injection system continues to operate to inject some as and serve as a "flywheel" on gas deliveries. Gas was started down the pipeline to Emden, Germany on September 17, 1977, and has reached a level of 1.6 billion cf/day. Ekofisk itself contributes some 600 million cf/day. Oil began flowing in the pipeline to Teeside, England, on October 15, 1977, and now is averaging 500,000 b/d, of which Ekofisk produces 235,000 b/d (Fig. 26).

CONCLUSION

North Sea oil has already led to dramatic changes in he economies of the host countries. Nevertheless, it is all too easy to forget the enormous costs, difficulty, and risks that have been involved. However, the tough and optimistic-persistence characteristic of the "oil finders" has never been needed more. Much of the future lies further offshore and into deeper waters and hostile environments. We are convinced that there are many giant fields remaining to be found for those who meet the challenge.

REFERENCES CITED

Byrd, W. D., 1975, Geology of the Ekofisk field, offshore Norway, *in* A. W. Woodward, ed., Petroleum and the continental shelf of north-west Europe: New York, John Wiley and Sons, v. 1, p. 439-445.

Didyk, B. M., et al, 1978, Organic geochemical indicators of paleoenvironmental conditions of sedimention: Nature, v. 272, p. 216-222.

Dow, W. G., 1977, Kerogen studies and geological interpretation: Jour. Geochem. Exploration, v. 7, p. 79-99.

Dunn, W. E., 1975, North Sea basinal area, Europe—an important oil and gas province: Norges Geologiske Undersoklse, Universitetsforlaget, p. 69-97.

——— S. Eha, and H. H. Heikkila, 1973, North Sea is a tough theater for the oil-hungry industry to explore: Oil and Gas Jour., v. 71, no. 2, p. 122.

Hill, G. A., W. A. Colburn, and J. W. Knight, 1961, Reducing oil-finding costs by use of hydrodynamic evaluations, *in* Economics of petroleum exploration, development, and property evaluation: Southwestern Legal Foundation, Int. Oil and Gas Educational Ctr. (Prentice-Hall, Inc.), p. 38-39.

Hubbert, M. K., 1953, Entrapment of petroleum under hydrodynamic conditions: AAPG Bull., v. 37, p. 1954-2026.

Neugebauer, J., 1974, Some aspects of cementation in chalk; K. J. Soo and H. C. Jenkins, ed. Pelagic Sediments: On Land and Under Sea: International Assoc. of Sedimentologists, Special Publication No. 1, p. 149-176.

Rickards, L. M., 1974, The Ekofisk area—discovery to development: Stavanger, Norway, Offshore North Sea Tech. Conf., Proc.

Scholle, P. A., 1977, Chalk diagenesis and its relation to the petroleum exploration—oil from chalks, a modern miracle?: AAPG Bull., v. 61, p. 982-1009.

Seifert, W. K., and J. M. Moldowan, 1978, Applications of steranes, terpanes, and monoaromatics to the maturation, migration, and source of crude oils: Geochim. et Cosmochim, Acta, v. 42, p. 77-95.

Tissot, B. P., and D. H. Welte, 1978, Petroleum formation and occurrence: New York, Springer-Verlag, 538 p.

Geology and Petroleum Fields in Proterozoic and Lower Cambrian Strata, Lena-Tunguska Petroleum Province, Eastern Siberia, USSR[1]

A.A. Meyerhoff[2]

Abstract A minimum of nine commercial and 13 noncommercial discoveries has been made since 1962 in the Lena-Tunguska petroleum province of Eastern Siberia. However, the petroleum potential of this province has scarcely been tapped because the prospective area is greater than 1,737,000 sq km. Discovered reserves are in Proterozoic marine terrigenous clastic reservoirs and Early Cambrian fractured carbonates interbedded with evaporites. The Cambrian reserves are small and are associated with salt swells and salt pillows, whereas the Proterozoic discoveries are large and are in both stratigraphic and structural traps. Most of the Proterozoic accumulations were found as a result of drilling through overlying Lower Cambrian structural traps. Generally, the discovered hydrocarbons are gas and gas condensate, although some oil has been found.

The original discovery field for the basin, Markovo field found in 1962, has proved and probable reserves of 622 Bcf of gas, 16 million bbl of condensate, and 10 million bbl of oil. The second largest field to date is the giant Sredne-Botuobin gas-condensate field, found in 1970. Sredne-Botuobin is a structural trap in both Proterozoic and Lower Cambrian strata. Proved plus probable reserves are 5.95 Tcf of gas and 149 million bbl of condensate. In both fields, the major reserves are in the Proterozoic. The largest field, Verkhnevilyuy, was found in 1975, and has 10.5 Tcf of proved plus probable reserves of gas and about 260 million bbl of condensate.

The Yaraktin oil field was discovered during 1971 in Proterozoic strata. Proved, probable, and potential reserves are about 210 million bbl. The field is a very large stratigraphic trap. The reserve estimate is subject to upward and/or downward revision as stepout and infill wells are drilled.

The potential of the area is very great. Several hundred, perhaps several thousand, oil and gas fields remain to be found. The province appears to be gas prone, but the discovery of the Yaraktin oil field indicates that some areas will have oil production. In terms of ultimate recovery, this region has the potential for producing 100 billion bbl of oil and 200 Tcf of gas, together with condensate. The existence of major to giant fields in Proterozoic strata—containing hydrocarbsons known from paleontological and geological data to have originated in Proterozoic beds—emphasizes the fact that nowhere should the Precambrian be regarded as economic basement.

[1]Manuscript received August 9, 1979; accepted for publication, October 2, 1979.

[2]Meyerhoff and Cox, Inc., Tulsa, Oklahoma 74104.

This paper, in its original and more detailed form, was prepared for the Cambridge University Arctic Shelf Programme (CASP) for private distribution. The present version is a contribution to the goals of the International Geological Correlation Project (IGCP) 157.

The writer thanks W.B. Harland, J.H. Oehler, N.J.R. Wright, P. Miles, and K. Fancett for criticism and help in editing; Ernestine Voyles, Karin Fancett, and Donna Meyerhoff for typing; Marie Wells for assistance with references; and Kathryn L. Meyerhoff for drafting.

Article Identification Number
0065-731X/80/M030-0011/$03.00/0.

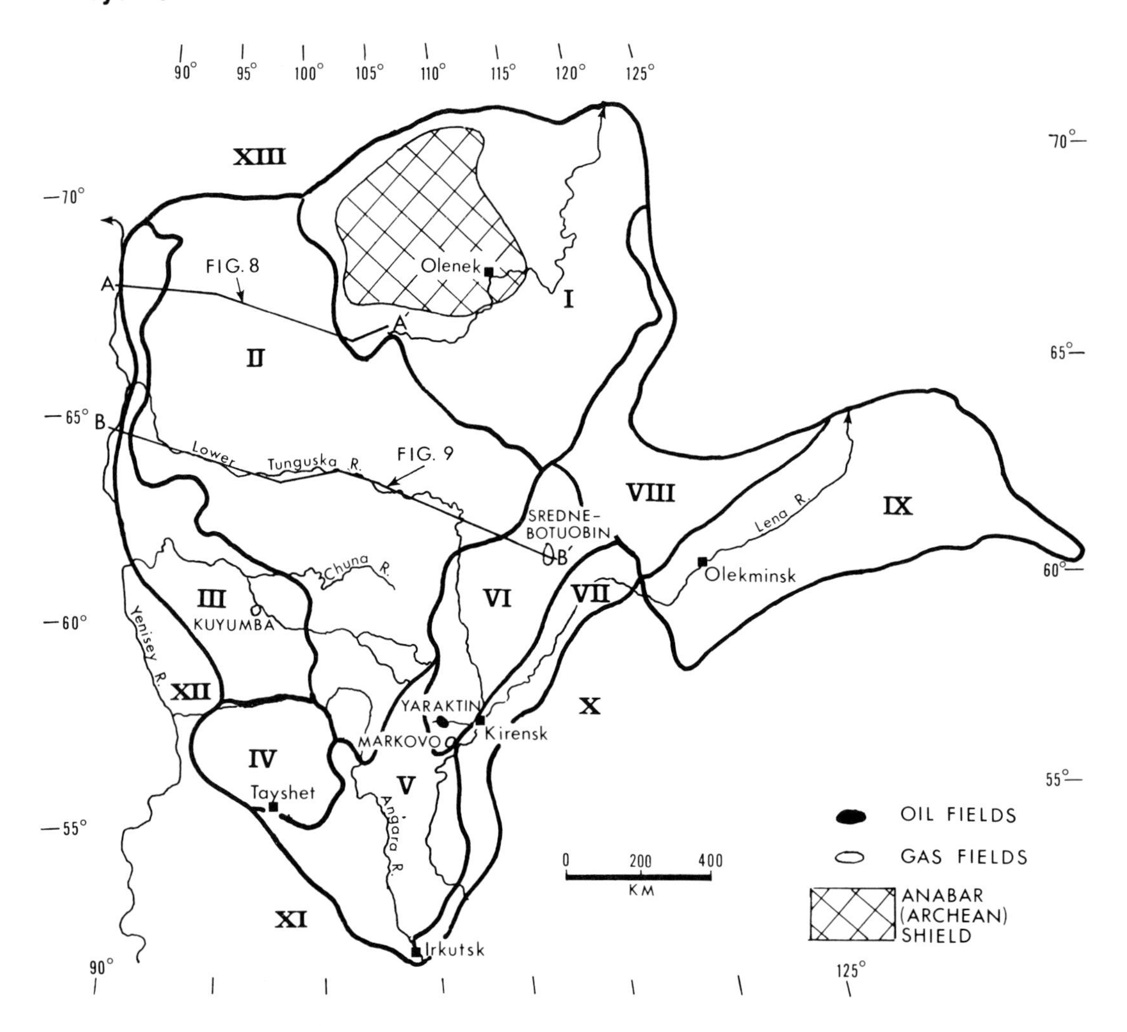

FIG. 1—Index map of the tectonic subdivisions of the Lena-Tunguska petroleum province (from Dikenshteyn et al, 1977). I - Anabar anticlise; II - Tunguska syneclise (basin); III - East Yenisey structural terrace; IV - Pre-Sayan-Yenisey syneclise (Kansk basin); V - Angara-Lena structural terrace; VI - Nep-Botuobin anticlise (V and VI together form the original Markovo-Angara arch of Trofimuk et al, 1964); VII - Lena trough (this includes the Baykal and Angara-Lena troughs of Provodnikov, 1965; compare with Fig. 3); VIII - Vilyuy syneclise (basin); IX - Aldan anticlise; X - Baykal-Patom fold region; XI - East Sayan anticlinorium; XII - Yenisey Ridge; XIII - Lena-Anabar trough. Locations of Figure 8 and 9 are shown. The fact should be stressed that this is but one of many conflicting subdivisions of this region. Data still are sparse in many areas and new tectonic schemes are still to be published.

INTRODUCTION

The Lena-Tunguska petroleum province of the Central Siberian platform is receiving increasing attention from USSR petroleum officials. The total area, as defined by Dikenshteyn et al (1977, p. 100-101, 112-113), is 2,827,000 sq km (Figs. 1, 2). However, this province includes the exposed Archean-middle Proterozoic rocks of the Anabar shield and the structurally complicated Aldan anticlise, where exploratory drilling has been discouraging. If the last two areas are excluded, the remaining prospective region still is 1,737,000 sq km.

Most exploration has been carried out in the south and east—the Tunguska syneclise, the pre-Sayan-Yenisey syneclise, the Angara-Lena structural terrace, and the Nep-Botuobin anticlise. These four zones, except for the northern half of the Tunguska syneclise, comprise what once was termed "the Irkutsk amphitheater," so-called because of the amphitheater-like shape on relief maps of the southern, curved part of the Lena-Anabar province (see Fig. 3). The area where most drilling and geophysical work has been done is shown on Figure 2. The prinicipal discovered fields, some noncommercial, are shown. Not all discoveries are shown because some appear to be of negligible importance.

The area is one of great geologic interest, because the reserves are the world's largest deposits of indigenous late Proterozoic gas, condensate, and oil. Most of the discoveries have been gas, and the largest are in Proterozoic reservoirs.

Although the region is considered to be gas prone on the basis of discoveries to date, detailed investigations now are under way to locate possible zones of oil genesis (Mitroshin and Gvizd', 1974; Balitov, 1977; Fuks et al, 1977; Babintseva et al, 1978; Vysotskiy et al, 1978; Fuks and Fuks, 1979). The 1971 discovery of the major Yaraktin oil field has given rise to

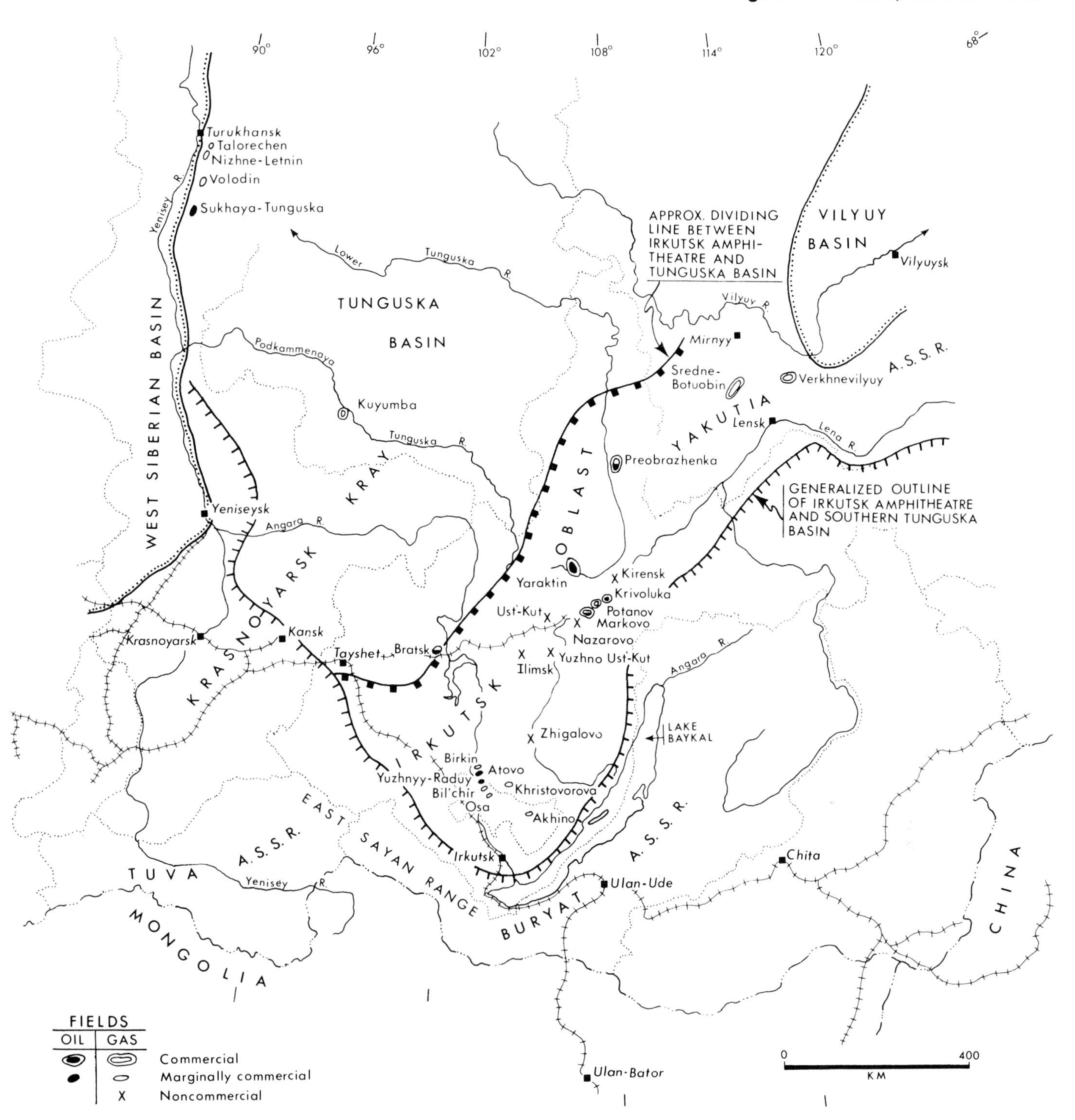

FIG. 2—Index map to localities, oil, gas, and condensate fields and to localities where flows of oil and gas have been found. The map is not complete except for the locations of commercial and important noncommercial fields.

some optimism about the future of the area as an oil province (Bazanov, 1973).

A major problem in the exploration of this region is the fact that most Proterozoic discoveries are in stratigraphic traps which have not been detectable with the reflection seismograph. Almost all drilling has been for the shallower, seismically detectable salt pillows in the Lower Cambrian carbonate-evaporite sequence, where the reserves are in fractured carbonates, mainly dolomite. Drilling through the Cambrian to the Vendian and Riphean (late Proterozoic; 570 to 925 m.y.) led to the serendipitous discovery of the much larger hydrocarbon accumulations in Proterozoic stratigraphic traps below the Lower Cambrian salt pillows. The Soviet seismic equipment currently used generally cannot "see" through the

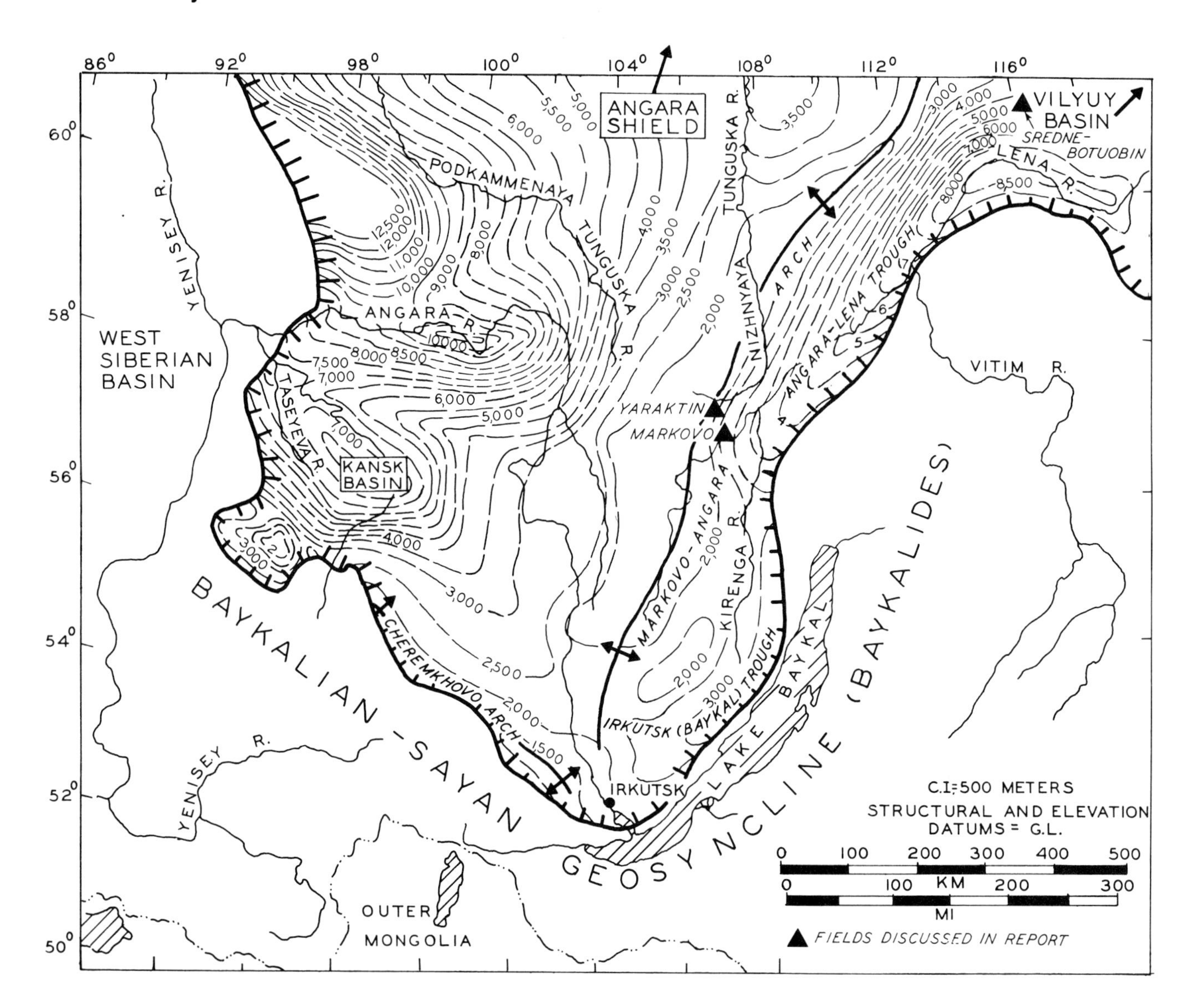

FIG. 3—Structural-contour map, of the southern part of the Lena-Tunguska petroleum province, with structural datum at the base of the unmetamorphosed section (top of crystalline basement). The area shown includes the Tunguska syneclise (II in Fig. 1), the Pre-Sayan-Yenisey syneclise (Kansk basin: IVof Fig. 2), the East Yenisey structural terrace (III of Fig. 1), the Angara-Lena structural terrace (Vof Fig. 1), the Nep-Botuobin anticlise (VI of Fig. 1), and the Lena trough (VII of Fig. 1). Contours, based mainly on geophysical data, are in kilometers. Compare with Figure 4. From many sources, expecially Provodnikov (1965).

Cambrian carbonate and evaporite section to the Proterozoic and, to complicate matters further, the Proterozoic traps commonly are unrelated to the Cambrian structural traps above. The fact that important Proterozoic reserves are being found beneath numerous Lower Cambrian salt pillows (and also beneath Permian-Triassic and Lower Jurassic flows and sills) suggests that many more Proterozoic accumulations remain to be found.

Geologically, the Lena-Tunguska region occupies one of the largest stable platform regions in the world. Since the last local incursion of Silurian marine waters into the Tunguska syneclise on the western side of the platform, the platform has remained largely stationary (except for the Permian-Jurassic marine incursion in the Vilyuy basin on the east). Even the outpouring of more than 1,000,000 sq km of Late Permian through Early Jurassic flood basalts and their tuffs did not result from, or produce, tectonic instability on the platform.

The Lena-Tunguska petroleum province—a term adopted here from Dikenshteyn et al (1977)—includes the entire Central Siberian platform (the East Siberian platform of some authors). It is bounded on the northeast and east by the Late Carboniferous through Jurassic Verkhoyansk foldbelt, a former geosyncline which was thrust toward the platform from the east; on the northwest by the Paleozoic Taymyr foldbelt, which was thrust southeastward against the platform; and on the southeast, south, and southwest by the horseshoe-shaped Baykalian foldbelt, last deformed in Middle Silurian to Middle Devonian time. This foldbelt, the "Baykalides" of geologic literature, extends uninterrupted from the Bering Sea at Anadyr Gulf to the Barents Sea, via the Timan Ridge. It may con-

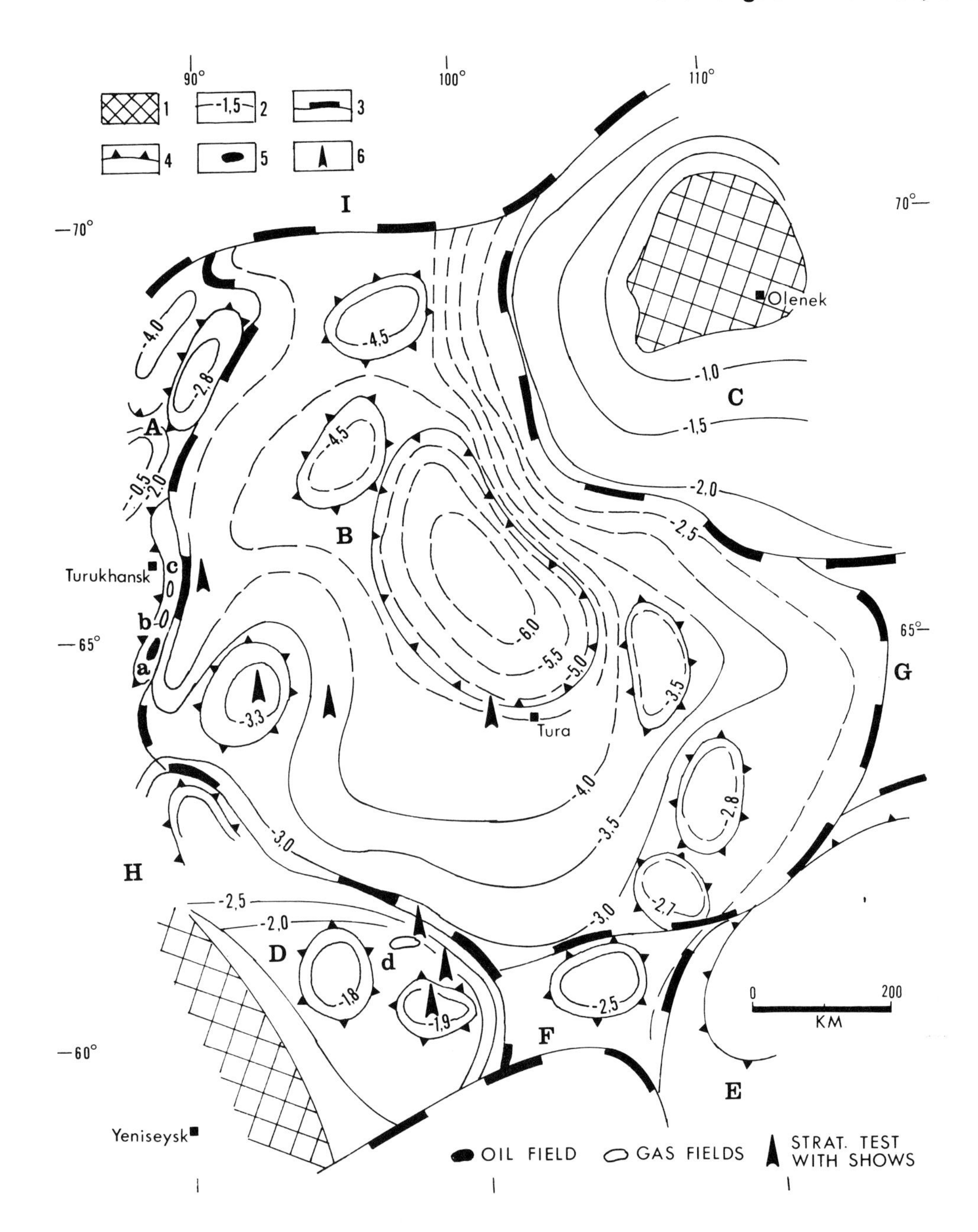

FIG. 4.—Structural-contour map of the Tunguska basin area, from Levchenko (1975) with structural datum at the base of the Cambrian, equal to the top of the Vendian. *Legend (top left):* 1 - Archean and early to middle Proterozoic crystalline and metamorphic rocks; 2 - contours in kilometers; 3 - boundary of tectonic unit; 4 - boundary of first-order structural element; 5 - oil or condensate field; 6 - stratigraphic test location. *Petroleum fields:* a - Sukhaya Tunguska; b - Volodin; c - Nizhne-Letnin; d - Kuyumba (see Fig. 2). *Tectonic subdivisions:* A - Turukhan-Noril'sk Ridge; B - Tunguska syneclise (basin); C - Anabar anticlise; D - Podkammenaya Tunguska structural terrace (East Yenisey structural terrace of Dikenshteyn et al , 1977); E - Anabar-Lena structural terrace; F - Katanga saddle; G - Botuobin (Botuoba) structural terrace; H - West Siberian basin; I - Yenisey-Khatanga trough.

tinue westward into the Caledonides of northwestern Europe. It is the world's longest geosynclinal system, more than 8,200 km long within the Soviet Union alone.

A branch of the Baykalides swings around the western flank of the Central Siberian platform, beneath the eastern side of the West Siberian basin. It is exposed on the platform margin at various places along and east of the Yenisey River. Folds produced during the last phase of Baykalian deformation in this region (Devonian) are thrust against the Central Siberian platform from the southeast, south, southwest, and west; the folds die out abruptly on the platform. Many of these folds were drilled in the early days of exploration of this region but were found to be "bald-headed." The folds show synsedimentary growth over a long period from the late Riphean onward (Mitroshin and Gvizd', 1974). In the deeper section—Cambrian and Proterozoic—the facies are different from those penetrated in platform wells and are unfavorable for petroleum generation and entrapment (Levchenko, 1975; Fuks et al, 1977).

Within the Lena-Tunguska petroleum province—and excluding the Archean and early to middle Proterozoic metamorphic and igneous complexes exposed in the Anabar shield on the north (Fig. 1)—the Central Siberian platform is overlain by an average thickness of 3,000 to 4,500 m of mainly marine, unmetamorphosed, flat-lying to gently dipping late Riphean, Vendian, Cambrian, and Ordovician strata. The nearly total absence of orogenic—even epeirogenic—movements across this vast area makes it one of the world's largest stable regions. In the Tunguska basin or syneclise, an exceptionally thick section is present which includes marine Silurian and continental Devonian-Early Carboniferous rocks. Although small flows of oil have been recorded in Early Silurian carbonates, the commercial fields being found in the Lena-Tunguska region are in Early Cambrian, Vendian, and late Riphean rocks. The Riphean through earliest Cambrian is a marine terrigenous-clastic sequence that grades upward into a thick Early Cambrian carbonate section interbedded with anhydrite, halite, and heavier salts. Figure 3, a thickness map of the southern part of the Central Siberian platform, is now more than 15 years out of date. It was prepared mainly by Provodnikov (1965) and illustrates the original terminology used for different subbasins and arches within the platform. A more recent paper by Levchenko (1975) shows some of the more recent estimates for thicknesses within the Tunguska syneclise (Fig. 4) and more nearly conforms with tectonic subdivisions of Dikenshteyn et al (1977), shown on Figure 1. The two maps, Figures 3 and 4, are presented together so that those familiar with the older and simpler terminology of the so-called Irkutsk amphitheater region can compare that terminology with the more recent terms.

From the central part of the Tunguska syneclise northward, and from the position of Yaraktin field northward (on the Nep-Botuobin anticlise), Late Permian, Triassic, and Early Jurassic basaltic trap-rocks and their tuffs cover at least 1,000,000 sq km of the platform. Continental deposits, including Permian and Triassic coal, are widespread in the same areas as the traps. Numerous rich metallic mineral deposits are associated with the traps, as well as with Mesozoic and older kimberlite and other ultramafic pipes that contain some of the largest and richest diamond deposits in the world. The presence of the traps, as noted by Levchenko (1975) and Dikenshteyn et al (1977), can affect adversely the petroleum potential of the region, so that care must be excercised in selecting drillsites through trap rocks. However, Yaraktin field (Proterozoic oil) is beneath 1,550 to 1,750 m of traps (Bazanov, 1973) and is the largest oil field found to date in this province (see Levchenko, 1975, for a good discussion of the relation between traps and oil occurrence).

Production of (or strong shows of) oil, condensate, and gas has been reported from 14 separate zones, 9 in the Lower Cambrian, and 5 in the Proterozoic. The determination of nine separate productive zones within the Lower Cambrian carbonates is not definitive, because correlations within this sequence still are poor and inexact. Some of the zones may prove to be equivalent.

Dikenshteyn et al (1977) reported that, through 1976, there were 22 separate localities where good, strong flows of oil, condensate, or gas had been recorded. At least 9 of these discoveries are considered to be commercial fields by Soviet standards, and some Soviet geologists believe that 13 of the discoveries are commercial. If these discoveries had been in the densely populated countries of Western Europe, Japan, or the United States, all 22 would be considered commercial.

The four largest fields are Sredne-Botuobin, discovered in 1970 (5.95 Tcf of gas and 149 million bbl of condensate in recoverable reserves to date); Yaraktin, discovered in 1971 (210 million bbl of recoverable oil); Markovo, discovered in 1962 (the first commercial discovery for the province, with 622 Bcf of gas, 16 million bbl of condensate, and 10 million bbl of oil); and Verkhnevilyuy, discovered in 1975, with 10.5 Tcf of gas and about 260 million bbl of condensate (Bakirov, 1979).

The hydrocarbons were generated in Proterozoic and Cambrian strata. Evidence for this comes from many sources: (1) There are no strata younger than Ordovician—and apparently never have been—in that part of the Lena-Tunguska province where most of the commercial hydrocarbons have been found (Nep-Botuobin anticlise). (2) Riphean and Vendian pools are in traps which formed during late Riphean and Vendian times. Some of these traps, such as the Parfenovo reservoir at Markovo, are "closed reservoirs," in the petroleum engineer's sense of this term. (3) Cambrian accumulations are in traps which were created during Cambrian time (Bakirov and Ryabukhin, 1969; Dubronin, 1976). (4) The halogenetic processes which produced the Cambrian traps did not disturb the Vendian and Riphean traps and had little effect on overlying beds. (5) The sporophytes found in the oils and condensates are, without exception, late Riphean, Vendian, and Early Cambrian forms. Even the spores found in oils and condensates in post-Early Cambrian rocks are of late Proterozoic age, thereby suggesting leakage upward through fractures. (6) Finally, the entire province has been stable tectonically since Middle Devonian time. Tectonism and erosion have not adversely affected the region for more than 400 m.y. Like the West Siberian basin on the west and the Vilyuy basin on the east, the Lena-Tunguska province has remained essentially unaltered by time.

The general Soviet petroleum geologist's attitude about Proterozoic petroleum is that, if the strata are unmetamorphosed and there was sufficient life at the time the beds were deposited, the age does not matter. Soviet geologists were pioneers in the paleontologic zonation of the Proterozoic—long before the Markovo field was found in 1962. Therefore, petroleum personnel in the USSR had few qualms about drilling Proterozoic rocks. Only those Soviet geologists who were influenced by Western thinking—namely, that there was no important life before the Cambrian—had any second thoughts about drilling rocks older than the Cambrian.

As far as ultimate production from this province is concerned, a conservative estimate (based on area, sedimentary volume, and projected density of structural and stratigraphic traps) is the Btu (calorific value) equivalent of 100 billion bbl of oil and 200 Tcf of gas. This estimate may seem far too high. However, if one takes into account the sedimentary-basin area of the conterminous 48 states of the US, approximately equivalent to the Lena-Tunguska petroleum province in prospective sedimentary volume, the estimate for the Lena-Vilyuy province is reasonable. The only major adverse factor may be the 1,000,000 sq km of basaltic traps, and the discovery of the large Yaraktin field beneath these traps may have answered this problem in part.

HISTORY OF EXPLORATION

General

Although geologists visited and studied large parts of the region long before the 1917 Revolution, a systematic regional reconnaissance study was not begun until 1932. Since 1932, surface and subsurface mapping has been carried out slowly, although the most intensive efforts were not begun until after World War II. From 1941 until 1960, stratigraphic tests and some wildcat wells were drilled at Osa, Bokhansk, Bel'ya, Atovo, and other places. The type sections of the Osa (basal Cambrian) and Parfenovo (Vendian) productive formations were established during 1941 and 1942. Subsurface work and surface mapping led to the delineation of the huge north to south Nep-Botuobin anticlise in 1952 and 1953 (Markovo, or Markovo-Angara Arch of Fig. 3). This arch is the largest of the structural features found by subsurface and surface mapping—800 km long and 25 to 40 km wide at its crest. The overall strike of the arch is north-northeast to south-southwest (Figs. 1, 3). Surface flank dips generally are ½° or less.

Discovery of the arch (and other, smaller structures) led to more detailed studies of selected areas. Core drilling and seismic-reflection surveys were carried out to define local structural anomalies. Many shallow boreholes were drilled, although not all were for structural mapping. The region already was known for the large salt deposits along the western side of the basin, and the first exploration drilling for potash salts was nearly as intensive as the later exploration for petroleum. At least 160 tests, many of them shallow, were drilled from 1948 through 1967 (Vasil'yev, 1968a). In 1954 a small, noncommercial discovery was made at a depth of 1,560 m in the Cambrian "Osa beds" near Osa, 130 km N 10° W from Irkutsk (Fig. 2) on the southwestern plunge of the Nep-Botuobin anticlise. This led to more intensive drilling for petroleum, and on March 18, 1962, the Markovo 1 well on the Lena River (Fig. 2) tested 2,040 b/d of 42° API oil on a 25-mm choke in the Osa Horizon ("Osa beds") at the base of the Early Cambrian Usol'ye Suite from a depth of 2,156 to 2,164 m (Weaver, 1962; Trofimuk et al, 1964; Vasil'yev, 1968a, 1968b; see Fig. 5 for Markovo stratigraphic column).

The Markovo 2 well (Fig. 2) was noncommercial, but it was not dry. The productive carbonate rocks of the Osa zone were "tight," and potentially productive strata below the Osa did not look promising. It was not until the pre-Osa zones (the Parfenovo, Markovo, and Bezymyannyy Horizons) were tested in adjacent wells that Soviet management decided that more intensive drilling of the Markovo area should be carried out.

A well then was drilled at Krivoluka (Fig. 2), 60 km northeast of Markovo down the Lena River. Production (271 b/d) was found in the Vendian Parfenovo Horizon. Apparently, only one productive offset well has been drilled to this discovery (Drobot and Isayev, 1966). Nevertheless, Krivoluka is listed officially as a potentially commercial field (Vasil'yev, 1968a).

Most drilling from 1962 through 1965 was concentrated at Markovo. Although several dry holes and noncommercial wells were drilled, a sufficient number of commercial wells was completed to make feasible the ultimate exploitation of the field. During the development of Markovo, a noncommercial Vendian and Early Cambrian discovery was made at Nazarovo, 16 km south of Markovo along the Lena River (Fig. 2).

By 1976, 1,490,000 m (about 600 wells) had been drilled in wildcat exploration through the Lena-Tunguska province. This was in addition to 159,000 m of stratigraphic tests (about 110 wells) and 19 more wells drilled to test minerals other than hydrocarbons. Almost all of these were drilled in the areas of the Angara-Lena structural terrace (Fig. 1), the Nep-Botuobin anticlise, the pre-Sayan-Yenisey syneclise (Kansk basin), and the Tunguska syneclise (or basin; Dikenshteyn et al, 1977). Drilling has intensified since the discoveries of Sredne-Botuobin, Yaraktin, Kuyumba, and Verkhnevilyuy from 1970 through 1975. Exploration using two-fold CDP seismic-reflection data began in 1972 and has continued to the present.

Problems in Development

The problems of development in this petroleum province have been touched on briefly. The principal problem relates to the ease of finding the first fields in the West Siberian basin (discovered in the 1953-1964 period) versus the difficulty in finding fields within the Lena-Tunguska province. The entire matter reduces to simple economics: (1) the West Siberian basin is closer to industrial and civilian markets than the Lena-Tunguska province; (2) Soviet seismic equipment can handle shallow structural features in the sandstone-shale sequences of the West Siberian basin, but cannot "see" through the dense carbonate and evaporite sequence of the Lower Cambrian to the stratigraphic traps of the late Proterozoic; and (3) discoveries, once begun in the West Siberian basin, came in rapid succession, whereas they were difficult to make in the Lena-Tunguska area. Therefore, exploratory efforts were concentrated in the West Siberian basin, and the Lena-Tunguska province was left with few funds and little equipment.

Because of the paucity of drilling, the nature of the stratigraphy of the Riphean and Vendian sections was not understood at first. It still is difficult, without adequate well control, to predict the sandstone-porosity trends. The Lower Cambrian porosity, associated as it is with salt pillows, is more predictable, but knowledge of Cambrian trends provides no clues to the trends in the deeper Proterozoic reservoirs which are both isolated and interconnected sandstone channels of deltas,

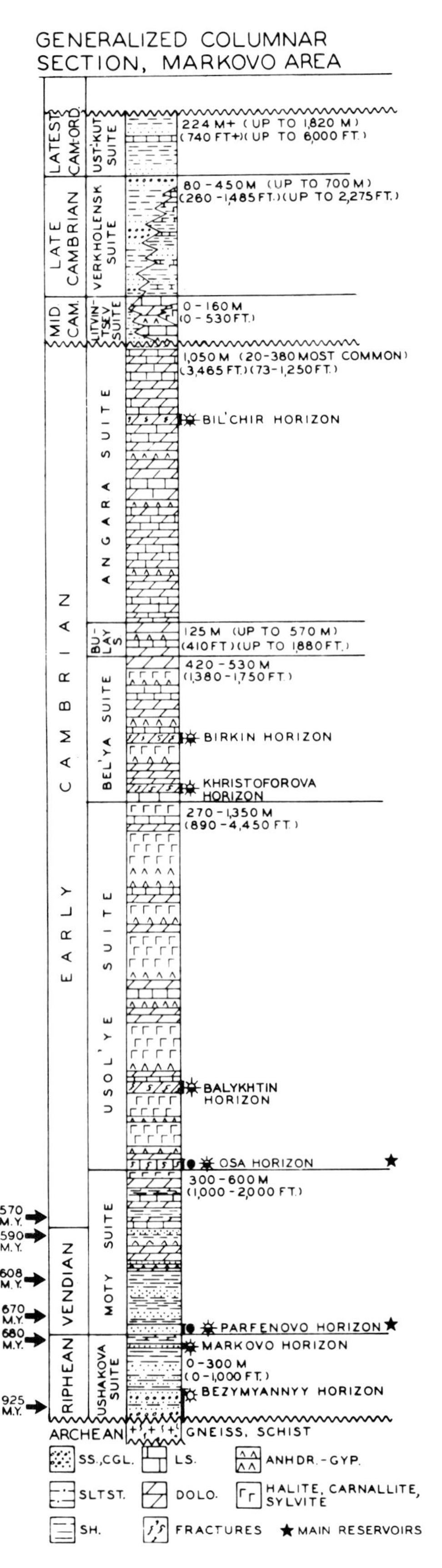

FIG. 5—Lithologic column, Markovo field, Irkutsk oblast', eastern Siberia. Thickness in meters. Data from many sources, expecially Vasil'yev (1968a, 1968b), Vassoyevich et al (1970), and Fuks et al (1977).

beach bars, and offshore shelf bars. However, once the sandstone porosity patterns of the late Riphean and Vendian reservoirs at Markovo were understood (Zolotov et al, 1968), it became possible to formulate a pattern for development drilling.

Intensive structural studies also were begun. It soon was discovered that few, if any, of the major basement fractures extend into the sedimentary cover (Gladkov et al, 1970). Study of cores showed that the Early Cambrian productive carbonates have a well-developed system of interconnected fractures, but that the sandstone beds of the underlying Riphean and Vendian deposits are not fractured; in fact, the sandstones are almost undisturbed. Structural-contour maps of the upper surfaces of the Osa and Parfenovo Horizons (see Fig. 5 for the relative positions of the two units) showed that the upper, or Osa Horizon generally exhibits anticlinal closure, whereas the lower, or Parfenovo Horizon exhibits only a gentle, regional dip. The reason for the different structural configurations of the two zones was not difficult to find—salt movements in the younger (Cambrian) rocks.

STRUCTURE

Regional Structure

The combined Angara-Lena structural terrace and Nep-Botuobin anticlise (which together comprise most of what was—and by many still is—called the Irkutsk amphitheater) form the southernmost prong of the Central Siberian platform. Their southern boundary is the amphitheater "wall" formed by the folded Baykalides along the southwestern, southern, and southeastern margins of the platform.

The basement of this area is broken by a series of large regional north to south fractures and by smaller east to west, northeast to southwest, and northwest to southeast fracture systems. The dominant basement structure is the Nep-Botuobin anticlise or arch which continues through the Angara-Lena structural terrace. Post-basement thicknesses of sedimentary rocks within this general area range from 2 km or even less at the crest of the Nep-Botuobin anticlise to more than 6 km in the adjacent basins (Levchenko, 1975; see Fig. 4) and perhaps more (Provodnikov, 1965; see Fig. 3).

Close to the margin of the amphitheater, adjacent to the Baykalides, a narrow zone of sharp folds separates the gently dipping strata of the platform from the fold and thrust belt of the Baykalides. Several of these folds have been drilled, but all have been found to be barren of petroleum. The sedimentary rocks within these folds commonly are a gradational facies between the platform strata of the amphitheater and the geosynclinal strata of the Baykalian geosyncline. Redbed molasse facies commonly are present in the youngest beds. In the pre-Sayan-Yenisey syneclise, thick salt deposits form the cores of many folds, although salt generally is thicker in the synclines than in the anticlines (Yanshin, 1962).

Tectonic History

A strongly negative geosynclinal basin occupied the present western, southern, and eastern zones of the Baykalides through much of Proterozoic and early Paleozoic times. Strong orogenic pulses began to deform the geosyncline during middle Proterozoic time and continued to deform it repeatedly through Middle Devonian time.

The tectonic history of the orogenic belt is reflected accurately in the platform sequence. Middle Riphean and older rocks, which are metamorphosed, folded, and faulted, are overlain unconformably by late Riphean strata.

As orogeny continued, the margins of the Irkutsk amphitheater were depressed increasingly. Vendian sediments were deposited across a much greater area than those of the late Riphean, which are restricted mainly to the edges of the platform (except in the Tunguska syneclise), and the Vendian successively oversteps older Precambrian units northward. Early Cambrian deposition was even more widespread. Continued folding, followed by uplift, led to a cessation of deposition in the Irkutsk amphitheater by Ordovician time, and in the Tunguska syneclise by Middle Devonian time. Outside of the Tunguska syneclise, Devonian marine beds are limited to a few small basins within the Baykalides. After Middle Devonian time, the entire region stabilized.

During the entire period of folding of the Baykalian geosyncline, the geosyncline was thrust toward the platform, with the result that, locally, parts of the foldbelt override small areas of the platform.

In general, this tectonic history took place in five gross stages:

1. Consolidation (intrusion and metamorphism) of the Archean-middle Proterozoic basement (Central Siberian platform) took place.
2. A late Riphean-Vendian open-marine transgression took place with gradual overstepping of the metamorphic and igneous basement. Sediments are terrigenous clastic.
3. There was a gradual shift to restricted-marine or lagoonal conditions during earliest Cambrian time. The terrigenous clastics grade upward into lagoonal carbonates and evaporites. The Proterozoic and Early Cambrian strata form the thickest section on the platform.
4. An unconformity is present at the top of the Lower Cambrian sequence, and a thin Middle Cambrian and younger, partly lagoonal carbonate-evaporite section overlies the older strata unconformably. The sea withdrew gradually from the area during Late Cambrian time, leaving behind smaller and smaller marine embayments which essentially disappeared by Middle Devonian time.
5. The Baykalides were uplifted during Silurian-Middle Devonian time, and a continental regime took over during Devonian-Carboniferous time.

Local Structure

Salt flow was extremely important in the development of the Lower Cambrian fractured dolomite reservoirs that surround the salt pillows and small-scale salt walls. Figure 6, from Bakirov and Ryabukhin (1969), shows the interpretation of the salt pillowing which took place at Markovo field and produced the fractured reservoir in the Osa Horizon. It is clear from Figure 6 that the salt flow took place before a large thickness of sediments was deposited above the Osa Horizon. Isopach studies of these structures demonstrate that they are synsedimentary (Yanshin, 1962). Figure 7, from Dubronin (1976), shows several forms of salt structures in the southern part of the Lena-Tunguska province—in the pre-Sayan-Yenisey syneclise, the Angara-Lena structural terrace, and the Nep-Botuobin anticlise.

Another factor which has played a role in the localization of oil and gas fields in the province includes rejuvenated basement faults, which strongly influenced oil and gas distribution at various fields (e.g., Yaraktin, the structure of which is discussed in a subsequent section).

A surprising phenomenon was discovered in 1968—cratonic-basin, basement-drape anticlines of the Saudi Arabian and West Siberian types. Before 1968, all important Proterozoic petroleum accumulations in the Lena-Tunguska province were found in stratigraphic pinchouts on the flanks of regionally dipping arches. Seismic work after 1968 in the Botuoba River area, just west of the Vilyuy syneclise (basin), identified several large basement arches in the Proterozoic and Cambrian sequences, arches which had been rejuvenated through Middle Cambrian time. The first of these to be delineated seismically and drilled became the Sredne-Botuobin field; a second was the Verkhnevilyuy structure (Belen'kiy and Kunin, 1978). Thus, for the first time (in 1968), more conventional structural traps were found in the late Riphean and Vendian sequences of the Lena-Tunguska province, and the discovery of stratigraphic traps by serendipity (because of poor seismic equipment) no longer was the sole means for exploring Proterozoic objectives in this huge area (some basement movements were involved in structure zone I of Figure 1, but this basement relief is not visible on seismic sections and is unproved).

The presence of synsedimentary structures around the rim of the Central Siberian platform where it is adjacent to the Baykalides has been mentioned. However, the importance of synsedimentary structures on the platform itself, as at Markovo, Bratsk, and Atovo, has not been emphasized (Figs. 6, 7). The fact has been mentioned that the Baykalian structures around the southeastern, southern, southwestern, and western rims of the Central Siberian platform also are synsedimentary and were rejuvenated repeatedly from late Riphean through Devonian times (Yanshin, 1962; Lokhmatov, 1966; Mitroshin and Gvizd', 1974). Thus synsedimentary structure is as important on the platform as it is in the deeper basins surrounding it.

STRATIGRAPHY

Introduction

The stratigraphy of this huge region is only partly known. Although more stratigraphic detail is given in some of the shorter articles listed in the bibliography, the following books will give the interested researcher considerable insight to the stratigraphic history of the Lena-Tunguska petroleum province: Brod and Vysotskiy (1965), Vasil'yev (1968a, 1968b), Bakirov and Ryabukhin (1969), Karasev et al (1971), Vasil'yev and Zhabrev (1975), and Dikenshteyn et al (1977).

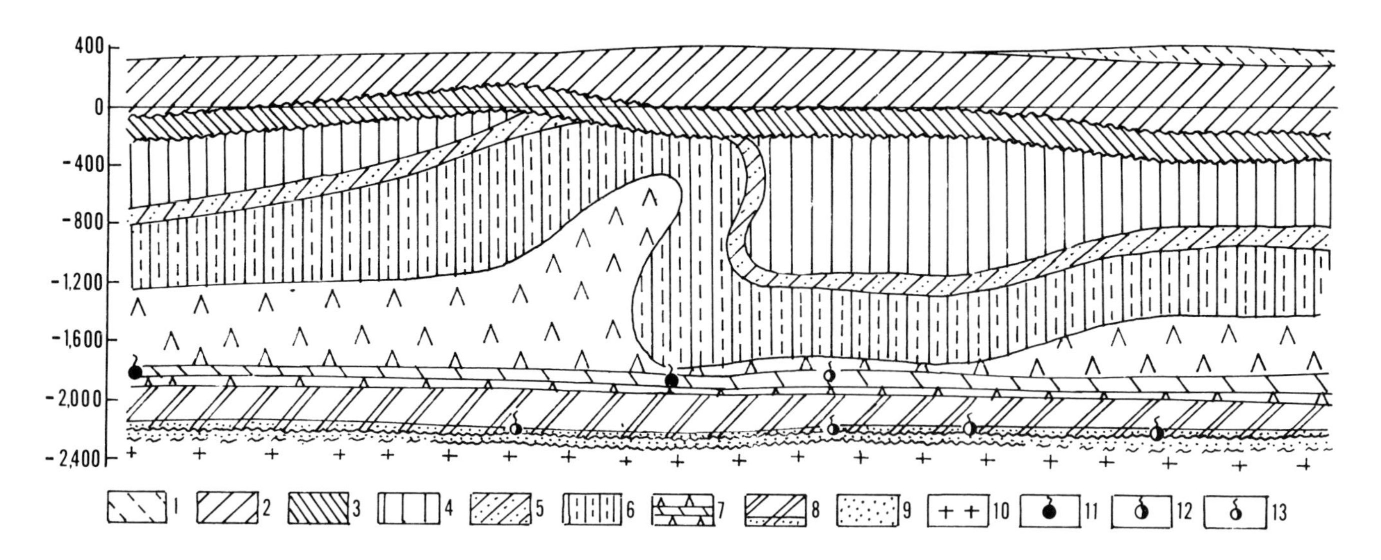

FIG. 6— Geologic cross section of Markovo gas-condensate field, Irkutsk oblast', From Bakirov and Ryabukhin (1969). *Legend:* 1 - Ordovician, Ust'-Kut Suite; 2 - Upper Cambrian, Verkholensk Suite; 3 - Middle Cambrian, Litvintsev Suite; 4 - Angara Suite; 5 - Bulay Suite; 6 - Bel'ya Suite; 7 - Usol'ye Suite; 8 - Moty Suite; 9 - Ushakova Suite; 10 - lower Proterozoic-Archean crystalline basement; 11 - commercial oil and gas flows; 12 - commercial gas and condensate flows; 13 - noncommercial shows of gas and condensate. Note synsedimentary nature of the structure. Vertical section in meters.

Relations between Baykalides and Central Siberian Platform

The stratigraphic section of the Irkutsk amphitheater is very different from that of equivalent strata in the Baykalian geosyncline. The facies changes between the two regimes are abrupt, and take place in the narrow foldbelt at the rim of the Central Siberian platform. For more than 30 years it was widely believed that the Precambrian rocks of the geosyncline had no equivalents on the platform. The discovery of hydrocarbons on the platform enhanced this belief, because some Soviet geologists thought that hydrocarbons could not have formed in Precambrian rocks. Then, as glauconites from the Irkutsk amphitheater section were dated by the potassium/argon (K/Ar) method and as Precambrian stratigraphic zonation by microfossils ("microphytoliths") evolved, the equivalence of much of the platform section with that of the geosyncline became apparent.

The principal characteristics of the geosynclinal sequence are great thickness and a greater "sameness" of lithologic types. The geosynclinal section adjacent to the edge of the platform is 2 to 4 times thicker than the equivalent platform section. Thus, 3,000 to 10,000 m of platform sediments is represented in the geosyncline by 12,000 to 20,000 m of strata.

Evaporites, which characterize the Cambro-Ordovician section, are not present in the geosyncline. Instead, polymict sandstones, flysch with graded beds, clastic and argillaceous carbonate rocks, and redbeds comprise the geosynclinal sequence. Precise correlations with the platform have not been established, but potassium-argon dating of glauconite and careful studies of microphytoliths are leading to more accurate correlations. More detailed discussions of the relations and facies changes between the platform and the geosyncline can be found in numerous works. Among the more accessible papers are those by Yanshin (1962), Sulimov (1964), Komar and Semikhatov (1965), Zharkova (1965), Anatol'yeva et al (1966), Korolyuk and Sidorov (1969), and Karasev et al (1971).

Stratigraphic Section, Markovo area, Nep-Botuobin Anticlise

The stratigraphy of the Markovo area is selected as a "type," because the first petroleum discoveries were made here. Therefore, Figure 5 (a composite section from the Markovo gas and oil field) should be referred to during the following discussion. The suite and horizon names are those used in standard Soviet literature on the area, although there are some differences from place to place. An abbreviated summary of the differences for each area was given by Fuks et al (1977, p. 14).

The complete section at Markovo (late Riphean through Ordovician) consists mainly of Lower Cambrian rocks (Fig. 5). The thickness of the remaining section, both above and below the Lower Cambrian, is only a small percentage of the total. Whether this also is true of the Tunguska syneclise still is not known.

Basement

The middle Proterozoic and older basement is deeply eroded in many parts of the amphitheater (Zhuravleva et al, 1966), and a substantial basement relief is present in some areas (Karasev et al, 1971). Ancient soil profiles have been found both in outcrop and in wells. In places, especially on the Nep-Botuobin syneclise, 4 to 25 m of residuum is present. The residuum consists of deeply weathered and jointed, commonly reworked basement, with greenish-gray shale and chloritized sediments.

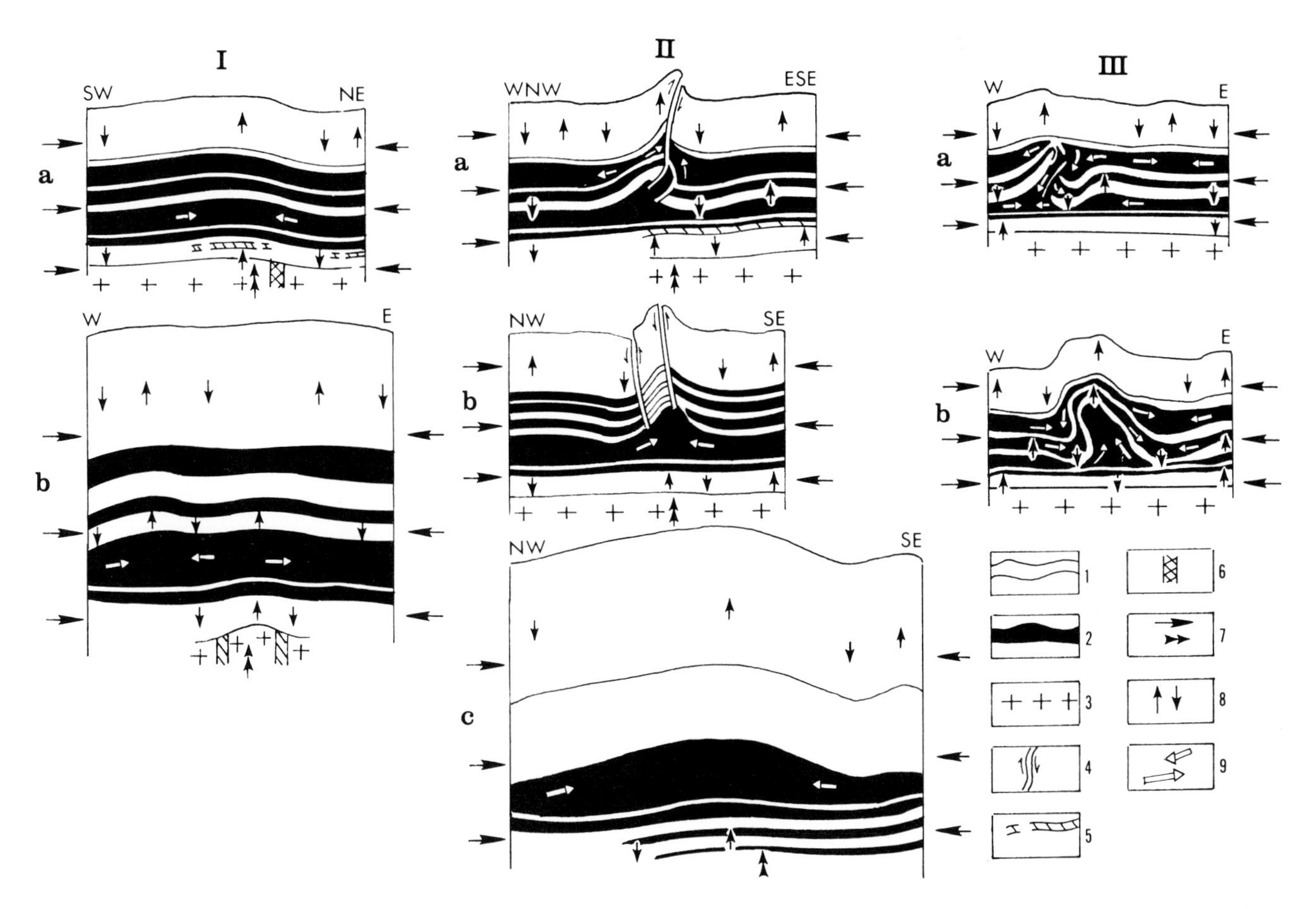

FIG. 7— This figure from Dubronin (1976) illustrates the types of synsedimentary (and some post-sedimentary) salt deformation which have taken place at different fields. Column I - structures in the Cambrian salt that involve slight amounts of basement flexuring: e.g., Bratsk, Atovo, and Birkin; Column II - salt movements accompanied by faulting, as at Litvintsev, Zhugalov, and Akhino; and Column III - intrabed deformation (almost synsedimentary) of the Markovo and Kirensk types. *Legend:* 1 - bedding planes; 2 - salt; 3 - crystalline basement; 4 - faults with sense of movement; 5 - traps; 6 - deep basement fault; 7 - compression directions; 8 - vertical sense of movement of basement, sediments, etc.; 9 - sense of salt flow.

Ushakova Suite

This is the oldest unmetamorphosed sedimentary unit known in the Irkutsk amphitheater. It ranges in thickness from a feather edge on the north to approximately 300 m along the platform rim, adjacent to the geosyncline (Vasil'yev, 1968a, 1968b). Between the East Sayan Range and Lake Baykal, the Ushakova consists of polymict conglomerate, gritstone, and sandstone, much of it red. Evaporites appear in the sequence along the eastern side of the East Sayan Range. Potassium-argon (K/Ar) dates from glauconites in this unit range in age from 680 m.y. near the top to 925 m.y. near the base (Anatol'yeva et al, 1966).

In the Markovo area and the upper Lena River region in general, the Ushakova Suite consists of 63 to 130 m of sedimentary rocks (Zhuravleva et al, 1966). At the base is the Bezymyannyy Horizon, 5 to 24 m thick (Fuks et al, 1977). It consists of poorly sorted, angular-grained, polymict sandstone and conglomerate with kaolinitic cement or matrix. The Bezymyannyy (also called the Bokhansk) is a noncommercial gas reservoir. Its exact age is unknown, but it is older than 680 m.y., and may be as old as 925 m.y.

A unit of greenish-gray to dark-gray mudstone separates the Bezymyannyy Horizon from the next productive zone, the Markovo Horizon. The Markovo is the second oldest commercial hydrocarbon-bearing reservoir discovered in the world (Table 5), after the Bitter Springs Formation of central Australia (850 to 1,050 m.y.). It consists of 0 to 30 m of gray-green to gray, medium-grained sandstone, which ranges from quartzose to feldspathic. Generally, the Markovo is better sorted than the Bezymyannyy. Some conglomeratic layers and mudstone partings are present. Locally, the unit is calcareous and "tight."

The uppermost part of the Ushakova Suite consists of about 40 m of dolomitic mudstone with thin gray partings of pyritized, micaceous mudstone. The Ushakova is overlain directly by the Parfenovo Horizon, the basal unit of the Vendian Moty Suite.

Moty Suite

The Moty Suite is divided by most workers into three informal subsuites, the upper, middle, and lower. The upper is of Early Cambrian age; the middle and lower are of Vendian age (Vassoyevich et al, 1970), although some workers still argue for a Cambrian age for the middle unit. The Vendian age is proved by the presence of glauconite potassium-argon dates of 609 to 670 m.y. (Postnikova and Kotel'nikov, 1969). The *youngest* glauconite age found in the middle Moty is 590 m.y. Therefore, whether the middle Moty is Cambrian or Vendian depends entirely on the precise age of the base of the Cambrian as established by radiometric studies around the world.

Around the amphitheater margins adjacent to the geosyncline, the lower and middle Moty consists of red conglomerate, sandstone, siltstone, and shale (Korolyuk and Sidorov, 1969; Karasev et al, 1971). Some anhydrite is present (Anatol'yeva et al, 1966). The entire unit in the geosynclinal margin area is about 600 m thick (Vasil'yev, 1968a, 1968b). Northward the thickness diminishes gradually to 70 to 100 m (Postnikov and Postnikova, 1964) and consists of gray to greenish-gray marine sandstone, siltstone, and mudstone. "Shoestring" sandstones are common in the basal Parfenovo Horizon (Vasil'yev, 1968b), and studies of these by Zolotov et al (1968) have shown them to be beach and offshore shelf bar sandstones.

At the base of the Moty Suite is the widespread productive Parfenovo unit (Fig. 2). This is a moderately well-sorted quartz sandstone, in places feldspathic. It is the most important reservoir unit in the Irkutsk amphitheater. The producing sandstone ranges in thickness from 15 to 90 m.

Two other sandstones have been found in the Moty Suite (see Table 3 and Fuks et al, 1977, for a summary). One is the oil-productive sandstone at Yaraktin field, northwest of Markovo (Fig. 2). This has been called the "Yaraktin Horizon," although it may be equivalent to all or part of the Parfenovo. It possibly is a somewhat younger zone, because it is a transgressive sandstone updip from Markovo. However, definitive age determinations have yet to be made. The Yaraktin sandstone was discussed in more detail by Bazanov (1973). The thickness is 0 to 30 m.

The other sandstone within the Moty Suite in which oil and gas shows have been found is the Verkhnetarsk, named after a well drilled in the upper Tarsk River valley. Because of its distance from correlated wells, its age also is in doubt. It could be equivalent to the Parfenovo (see Table 3). It is 2 to 14 m thick.

The upper subsuite of the Moty, although assigned to the Early Cambrian, does not contain characteristic Cambrian faunas. Instead, the fossil forms, mainly microphytoliths and problematica, suggest a transition between the Vendian and Cambrian forms. The assignment to the Early Cambrian is based on glauconite potassium-argon dates of 570 m.y. and less.

This subsuite consists of 265 to 300 m of dolomite, plus a 12 to 27-m-thick sandstone near the base (Postnikov and Postnikova, 1964). The dolomite grades into clastic and argillaceous limestone and dolomite toward the margins of the platform adjacent to the geosyncline (Korolyuk and Sidorov, 1969). Since 1970, the upper Moty has been found to contain two potentially productive zones which may be equivalent to one another (see Table 3). One of these is the main producing zone at the commercial Preobrazhenka field (Fig. 2). For lack of good correlations, it has been called the Preobrazhenka Horizon, 17 m thick at its type locality. The main productive zone in the small Ust'-Kut field (Fig. 2) is called the Ust'-Kut Horizon, but it could correlate with the Preobrazhenka. It ranges in thickness from 0 to 90 m.

Usol'ye Suite

The Usol'ye Suite overlies the Moty and ranges in thickness from 270 to 1,500 m. It consists of thick, massive units of dolomite, limestone, anhydrite, and halite (Vasil'yev, 1968a, 1968b; Karasev et al, 1971). Halite predominates. Two productive zones are present—the Osa Horizon along the central and southern parts of the Markovo-Angara arch, and the Balykhtin Horizon, restricted mainly to the southern part of the arch (Fuks et al, 1977). Both are fractured and cavernous limestone and dolomite zones (Sulimov, 1964). The Osa is the second most important of the productive zones in the Irkutsk amphitheater after the Parfenovo. It ranges in thickness from 15 to 100 m; the Balykhtin ranges in thickness from 9 to 10 m, where present. Both are productive on closed anticlinal and domal structure produced by salt movements. Oolitic, pseudo-oolitic, and pelletal fabrics show that the Usol'ye carbonates were deposited in shallow, agitated water. Many individual beds within the Usol'ye persist across great distances, a fact which demonstrates widespread uniformity of depositional conditions and tectonic stability across great areas (Lokhmatov, 1966). Some of the carbonate beds in the Usol'ye and succeeding suites are believed by numerous Soviet geologists to be primary precipitates.

Yanshin (1962) first predicted the presence of substantial thicknesses of potassium salts in the Usol'ye. Sulimov (1964) reported thicknesses of the Usol'ye in the order of hundreds of meters in the Sayan region on the west, and up to 1,500 m in the Lena and Taseyeva River basins farther east (Fig. 2). Most of this thickness is halite, with subordinate dolomite and anhydrite. Subsequently, Zharkova (1965) proved the existence of carnallite and sylvite in substantial quantities within the Usol'ye across the entire amphitheater. The evaporites grade abruptly into terrigenous and carbonate sediments at the edges of the geosyncline.

More recent drilling around all parts of the Siberian platform has demonstrated the presence of at least 558,000 cu km of halite and heavier salts (Zharkov, 1969). Zharkov wrote that the true volume of chloride salts is closer to 800,000 cu km.

Bel'ya Suite

The Bel'ya Suite ranges in thicknesses from 420 to 530 m (Vasil'yev, 1968a, 1968b). It consists mainly of limestone, dolomite, and anhydrite (Postnikov and Postnikova, 1964; Karasev et al, 1971). Some gypsum, rock salt, calcareous shale, and marl are present (Vasil'yev, 1968b). The suite is marked by the lowest occurrence of the *Bulaiaspis* trilobite zone.

Two fractured and cavernous dolomite zones in the Bel'ya Suite have hydrocarbon shows, and one of these is commercially productive. The Khristovorova Horizon (50 to 64 m thick) has some gas and condensate, but the quantities are only marginally commerical in the fields where the zone has been tested. Higher in the section, the Atovo Horizon (60 to 90 m) has commercial accumulations of gas in the southern part of the Markovo-Angara arch (Fig. 3).

In the Yenisey Hills of the northwestern part of the amphitheater (Turukhansk area), Yanshin (1962) made a study of thickness changes in the Bel'ya Suite. He found that at the crests of anticlines, the Bel'ya is much thinner than in the adjacent synclines. Moreover, the amount of halite, carnallite, and sylvite increases toward the synclinal axes. In one structure, Yanshin showed the Bel'ya to be 130 m thick on the crest (14 m, or 11% evaporites), 177 m thick on the western flank (39 m, or 22% evaporites), and 237 m thick on the eastern flank (135 m, or 55% evaporites).

Bulay Suite

The Bulay Suite, which overlies the Bel'ya Suite, is Early Cambrian in age. The average thickness is 125 m, and the thickness remains fairly constant across long distances. However, in some areas of the Kansk and other marginal basins, thicknesses up to 570 m have been found. Anomalous thicknesses in the Bulay and Bel'ya Suites may be caused by the presence of reefs (Postnikov and Postnikova, 1964). The Bulay consists mainly of dolomite with streaks of limestone, gypsum, and anhydrite.

The Birkin Horizon (90 to 100 m thick) is the only potentially productive zone within the Bulay Suite (Fuks et al, 1977). It is a fractured and cavernous carbonate zone similar to those described in preceding pages and is found only in the Angara-Lena structural terrace area.

Angara Suite

The Angara Suite, the highest unit of the Early Cambrian in the region, in most places is 20 to 380 m thick, although thicknesses up to 1,050 m are known (Vasil'yev, 1968b). In the Markovo area, the thickness ranges from 492 to 847 m (Postnikov and Postnikova, 1964). The suite consists of dolomite, with thin layers of dolomitic limestone.

There are two fractured dolomite and limestone reservoirs in the Angara. The Bil'chir, below, is 60 to 120 m thick; the Kelorskiy, above, is 16 to 30 m thick. These productive units are the youngest known in the entire Lena-Tunguska oil and gas province, and to date they have been found only in the Angara-Lena structural terrace.

Litvintsev Suite

After deposition of the Angara Suite, the region was tilted and arched gently. The Middle Cambrian Litvintsev Suite, the Middle to Late Cambrian Verkholensk Suite, and latest Cambrian-Early Ordovician Ust'-Kut Suite progressively overlap the Angara with gentle angular unconformity (Karasev et al, 1971).

The Litvintsev Suite consists of dolomite, limestone, anhydrite, and halite in the interior part of the amphitheater. Although no oil or gas fields have been developed in these rocks, Dikenshteyn et al (1977) reported one Middle Cambrian discovery at Talorechen, near Turukhansk (Fig. 2). These lagoonal strata grade into limestone and dolomite toward the amphitheater margins and, close to the margins, into redbeds. The redbeds commonly have been interpreted to be the Late Cambrian Verkholensk Suite, but they now are recognized to be a lateral facies equivalent of the Litvintsev (Karpyshev, 1965). Similar gradations are found in both the eastern and western margins of the amphitheater (Zharkov et al, 1963). The thickness of the Litvintsev ranges from 0 to 160 m.

Kolosov et al (1968) discovered numerous potash and heavier salt deposits in the pre-Sayan-Yenisey syneclise (Fig. 1). Among the precipitated minerals found are rinneite and erythrosiderite interbedded with carnallite, sylvite, halite, and anhydrite.

Verkholensk Suite

The Litvintsev Suite grades upward and laterally into the redbed molasse-like facies of the Verkholensk Suite; the redbed facies is developed adjacent to and within the Baykalian foldbelt. Thus, the Verkholensk ranges in age downward into the Middle Cambrian on the south, east, and west. The upper part is Late Cambrian. In the interior of the amphitheater the Verkholensk is marine marl, shale, and sandstone. Total thickness of the Verkholensk is 80 to 1,400 m, the average being about 500 m (Karasev et al, 1971).

Latest Cambrian and Ordovician

The uppermost Cambrian and Ordovician in the area also are marine. The basal unit, the Ust'-Kut Suite, exhibits a return from continental molasse-type conditions to more normal marine conditions. The average thickness of the latest Cambrian-Ordovician is 224 m.

Ordovician and Silurian of Western Lena-Tunguska Province

In the western part of the Angara-Lena structural terrace, the pre-Sayan-Yenisey syneclise, and the Tunguska basin, up to 1,820 m of Ordovician and 1,000 m of Silurian are present. The Ordovician consists of varying combinations of sandstone, limestone, and dolomite with some marl, siltstone, and conglomerate. Shale is abundant in some areas. The Upper Ordovician of the pre-Sayan-Yenisey syncline is entirely marine sandstone, siltstone, and shale.

The Silurian consists of sandstone, siltstone, shale, limestone, and dolomite. It includes a well-developed Wenlockian and Llandovirian fauna. Dark-gray fractured limestones in the Tunguska syneclise have yielded small amounts of oil, bearing a late Proterozoic and Early Cambrian microflora. Conglomeratic beds are more common toward the geosyncline on the west and southwest. As the sea withdrew, the depositional environment became lagoonal and beds of anhydrite were deposited in the upper part of the sequence (Karasev et al, 1971; Akul'cheva and Fayzulina, 1976).

PALEONTOLOGY

General

Only the paleontological data which establish the ages of the late Riphean, Vendian, and earliest Cambrian strata are considered here (i.e., all strata below the *Bulaiaspis* trilobite zone). The remaining faunal zones are standard boreal assemblages which have counterparts in and similarities with northwestern North America, the Urals area, and the Baltic syneclise on the Baltic Sea. The paleontological and other bases for determining the ages of the pre-Early Cambrian sequences are particularly important, because Soviet scientists have used micro- and mega-paleontology for many decades in their studies of Proterozoic sections. The great abundance of fossils in Proterozoic sequences seems to be poorly appreciated by the petroleum industry. As a result, the importance of Proterozoic source materials for petroleum is not recognized, and the Proterozoic is not generally regarded as a reasonable petroleum objective by explorationists outside the USSR and China. Only recently have Western geochemists begun to appreciate the source-rock and petroleum potential of Proterozoic and other Precambrian strata (e.g., Meinshein et al, 1964; Barghoorn et al, 1965; McKirdy, 1974).

The age of the late Riphean, Vendian, and earliest Cambrian beds was first established on the basis of detailed potassium-argon dating of glauconites, a method first perfected in the Soviet Union after it had been dismissed as equivocal in Western countries.

K/Ar dating of glauconites has been carried on in the USSR for nearly 30 years and is a routine correlation tool.

While glauconite dating was being refined, USSR paleontologists were attempting to establish a paleontological basis for zoning Proterozoic strata. Sporophytes and problematica have been collected in profusion from many Proterozoic sections of Russia and Siberia, and reasonably reliable zonation has been achieved and used successfully from European Russia to the Pacific Ocean. Similar techniques have been used successfully in China also.

Late Riphean

Popova and Glazunova (1965) and Zhuraleva et al (1966) reported the following forms from the late Riphean Ushakova Suite, Markovo Horizon:

Sporophytes

Archaeosacculina atava, Archaeosiscina atava, Bavlinella sp. (type species is in the Riphean of the Moscow basin), *Leiominuscula rugosa, Lopholigotriletum semiinvolutum, Lophominuscula prima, L. rugosa, Margominuscula antigua, M. prisca, M. rugosa, M. tennela, M. tremata, M. verrucosa, Minutissima prima, Polyedrosphaeridium septiforme, Protoleiosphaeridium colliculosum, P. compactum, P. muna, Protonucellosphaeridium* cf. *patelliforme, Protozonosphaeridium agaense, Stictosphaeridium sinapticuliferum, Symplassosphaeridium biglume, S.* cf. *subcoalitum.*

Problematica

Asterosphaeroides stellatus, Radiosus cristosus, R. elongatus, R. limpidus, R. striptus.

Vendian

Vendian forms reported by Korolyuk and Sidorov (1969), and by Postnikova and Kotel'nikov (1969) include:

Sporophytes

Asperatopsophosphaera partialis, Asterosphaeroides radiatus, Brochopsophosphaera simplex, Spumosata partialis.

In addition, some of the Riphean genera and species are found in the Vendian.

Problematica

Jurusania judomica (small) (a stromatolith), *Medularites lineolatus, Vermiculites concretus, V. irregularis, V. lobatus, V. tortuosus, Vesicularites lobatus.*

Earliest Cambrian

These forms are characteristic of the lower and middle subsuites of the Moty Suite. In the earliest Cambrian part of the Moty Suite, the following forms are characteristic:

Boxonia allachjunica, B. divertata, Collenia singularis, Collumnaefacta schancharia, Jurusania judomica (large), *J. sibirica, Linella sinica, Uricatella urica,* and assorted renalcids.

At the base of the Usol'ye Suite, the first trilobites appear in the *Bulaiaspis* zone. These characteristically are 200 to 300 m above the renalcids which form a zone that overlies the Vendian *Vermiculites* zone. Although many species (both of sporophytes and problematica) are present through a broad stratigraphic range from late Riphean through Early Cambrian, careful sampling usually results in the collection of a sufficiently large assemblage to permit dating.

The Bel'ya Suite also has a characteristic sporophyte assemblage which, although similar to some older assemblages, is sufficiently distinctive that the Bel'ya can be identified correctly on the basis of its sporophyte content. Popova and Glazunova (1965) have identified the following typical Bel'ya forms: *Polyedrosphaeridium semiconnatum, Protoleiosphaeridium conglutinatum, Protolophosphaeridium* sp., *Prototrachysphaeridium vavosum, P. samsonvielui, Stictosphaeridium tortulosum, Symplassosphaeridium vesiculiferum.*

PETROLEUM GEOLOGY

General

Dikenshteyn et al (1977) stated that, as of mid-1976, 22 different structures had been drilled with strong flows of oil, gas, or gas-condensate. Of these, from 9 to 13 are commercial, even in this remote part of Siberia. In a densely populated area such as Western Eruope, all of the discoveries would have been

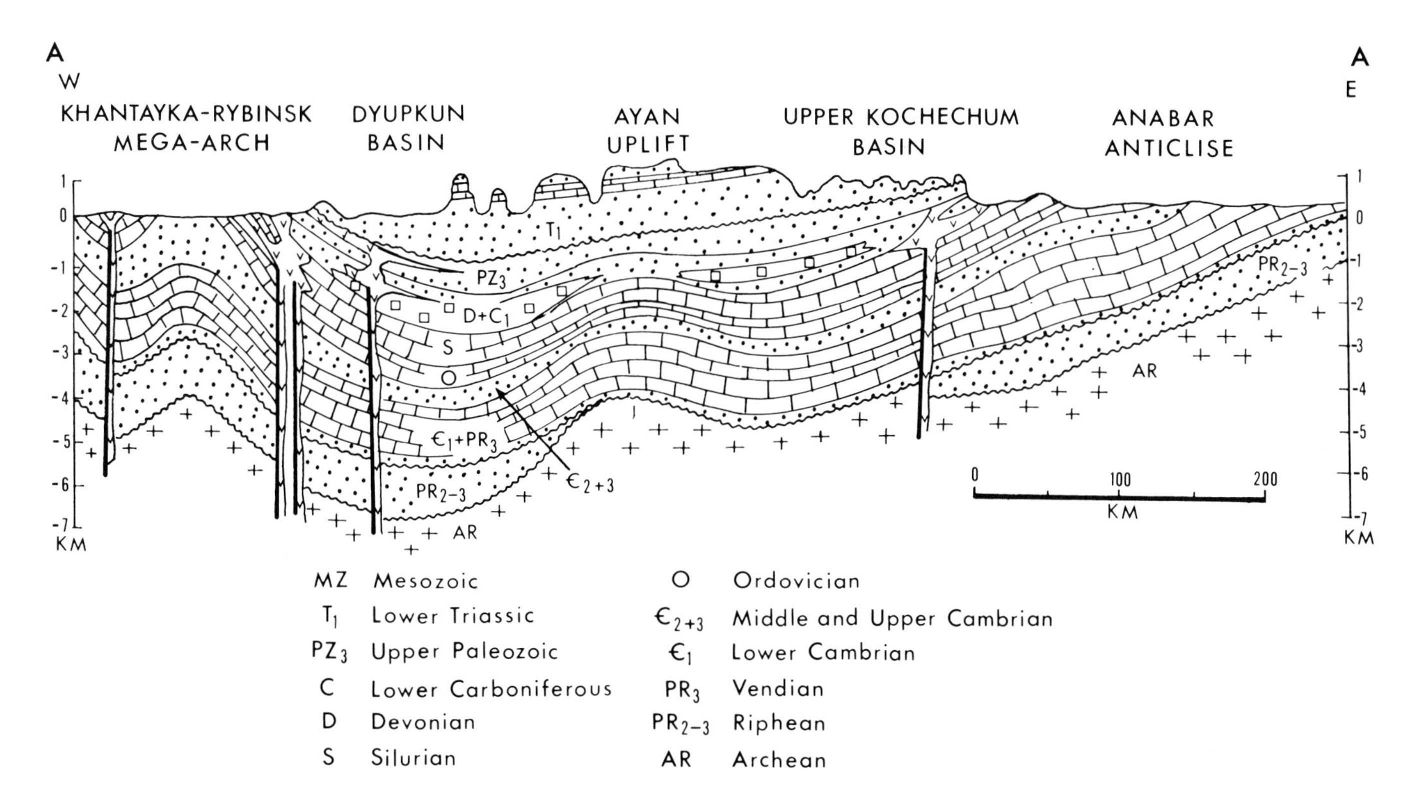

FIG. 8— West to east structural cross section of northern part of Tunguska syneclise. Location is on Figure 1. From Dikenshteyn et al (1977).

considered "commercial."

In addition, numerous wells have been drilled that have oil and gas shows, some of them with small flows. To date, all but one of the 22 strong flows have come from rocks of Vendian and Early Cambrian ages. There are multiple pays at many fields. Small noncommercial flows have come from Riphean rocks; one discovery was made in Middle Cambrian dolomite in the Tunguska basin (the youngest potentially commercial well to date in this area); and small shows have been found in rocks as young as the Silurian. However, as in the case of the Silurian, the oil carried only late Proterozoic and earliest Cambrian sporophytes, suggesting that it came from below.

Of the various tectonic provinces outlined by Dikenshteyn et al (1977; see Fig. 1), the Tunguska syneclise, with 775,000 sq km, has the thickest section and the greatest promise for the future. However, because of its remoteness, it is among the least explored areas, except where boats and winter-ice roads can be used along such streams as the Yenisey, Lower Tunguska, Podkammenaya Tunguska, and the Angara. The distribution of commercial and noncommercial discoveries shown on Figure 2 reflects these transport conditions.

The second most important area—and the most intensely explored because of the proximity to population centers, railheads, roads, and rivers (especially the Lena)—is the Irkutsk amphitheater, consisting of the Angara-Lena structural terrace area of 520,000 sq km. Of the 22 discoveries and strong flows discussed by Dikenshteyn et al, 17 come from this area.

The third area is the pre-Sayan-Yenisey subprovince, with an area of 160,000 sq km. Although close to population centers and railheads, it is structurally and stratigraphically complex, with not as much promise as the other subprovinces.

As of 1979, at least one more discovery of importance had been made in the Nep-Botuobin anticlise, and drilling for new fields continue, as well as development of those already discovered (particularly the big ones).

Many data from the fields, both commercial and noncommercial, have been plotted on tables. Tables 1, 2, and 3, taken from many sources, show the physical properties of various reservoirs, the flow rates from many wells, and other pertinent data that relate to the physical properties of the reservoirs themselves. Table 4 shows gas compositions from several fields, and Tables 5 and 6 give information on other parameters, including oil compositions. Complete data are not available.

Tunguska Syneclise

In the Tunguska basin on the northwest, four structures may be commercial; opinions differ considerably. The four structures are (Fig. 2): (1) Sukhaya-Tunguska, which produced 214 b/d of oil from a Lower Cambrian dolomite; (2) Volodin, which produced 450 Mcf/d of gas, also from Lower Cambrian dolomite; (3) Nizhne-Letnin, which produced 770 Mcf/d of gas from a Lower Cambrian dolomite; and (4) Talorechnen, which produced 7,000 Mcf/d of gas from a Middle Cambrian dolomite. This is the youngest producing bed found to date in the Lena-Tunguska province. Talorechnen is close to Nizhne-Letin and the town of Turukhan.

The porosity of these four reservoir beds ranges from 6 to

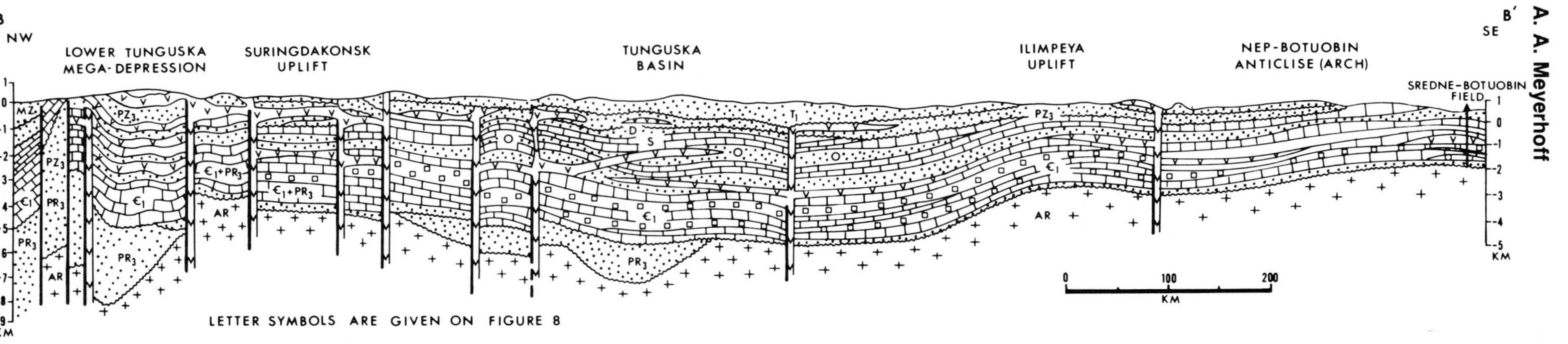

FIG. 9—West-northwest to east-southeast structural cross section of the central Tunguska syneclise. Note that the Sredne-Botuobin field location is shown on the right. Location of the figure is on Figure 1. From Dikenshteyn et al (1977).

Table 1. Average Properties of Productive Reservoirs, Lena-Tunguska Province.

Age	Suite	Horizon	Lithology	Thickness in meters	Average effective column in meters	Porosity in %	Permeability in md	Production type	Average DST	Field where productive
Early Cambrian	Angara	Bil'chir	fractured dolomite, fractured limestone	20	5	Up to 12	—	gas	2,625 to 3,500 Mcf/d (Bil'chir); 7,000 Mcf/d (Khristovorova)	Bil'chir; Khristovorova
	Bulay	Birkin	fractured dolomite	8 to 15	6	—	—	gas	15,750 Mcf/d	Birkin
	Bel'ya	Khristoforova	fractured dolomite	—	—	—	—	gas	—	Khristoforova
	Usol'ye	Osa	fractured dolomite, fractured limestone	8 to 40	20	0.13 to 11.2	up to 2,000 (av. = 100 to 150)	oil	11-6,700 b/d	Markova, Atovo, Yuzhno-Raduy, Sredne-Botuobin
Vendian	Moty	Parfenovo	sandstone	Up to 60-80 (aver. 16-37 m)	7	10-20	9 to 780 (av. 0.5 – 30)	oil gas cond.	270 b/d 14-10,600 Mcf/d; av. = 8,400 Mcf/d 369 b/d	Markovo, Krivoluka, Atovo Markovo, Sredne-Botuobin, Yaraktin, Bratsk Markovo
Late Riphean	Ushakova	Markovo	sandstone	0 to 29	5	8.9 to 16.7	1	gas cond.	6,475-8,625 Mcf/d 548 b/d	Markovo Markovo
		Bokhansk (Bezymyannyy)	sandstone	5 to 15	3	1 to 15	Up to 113	gas	up to 49 Mcf/d	Markovo

14% but is augmented by fractures. Permeability is low, rarely greater than 60 md, with some zones having values up to 215 md. In the opinion of Dikenshteyn et al (1977), the entire late Riphean through Silurian section is prospective. One well near Turukhan, close to Nizhne-Letin, produced a few barrels a day from Early Silurian fractured carbonates. However, the oil carried a late Riphean-earliest Cambrian sporophyte assemblage.

In the central part of the Tunguska syneclise, the Kuyumba field (Table 5) was discovered in 1973. The discovery was in earliest Cambrian dolomite at a depth of 2,170 m. Flow rates on different chokes ranged from 2,520 Mcf to 14,000 Mcf per day. This field is commercial and is being developed at present.

The entire Tunguska syneclise, with more than 775,000 sq km of area, is underexplored, has the thickest sections of the Lena-Tunguska petroleum province, and holds the greatest promise for the future. At the present time, its potential has scarcely been tapped (Levchenko, 1975).

Irkutsk Amphitheater (Angara-Lena Structural Terrace and Nep-Botuobin Anticlise)

In the Irkutsk amphitheater, the following structures are or appear to be commercial: Verkhnevilyuy, Sredne-Botuobin, Preobrazhenka, Krivoluka (this name also appears in the literature as Krivaya Luka), Potanov, Markovo, Yaraktin, and Bratsk. The following are considered by Soviet geologists and engineers to be noncommercial: Kirensk, Severo-Markovo, Nazarovo, Ust'-Kut, Olimsk, Zhigalova, Birkin, Atovo, Yuzhnyy Raduy, Bil'chir, Osa, Akhino, and others. Most are shown on Figure 2. Not shown are two noncommercial finds called Tokmin and Shamanov (Kazanskiy et al, 1978) and some localities noted on the tables. The data which pertain to these fields and to some of the noncommercial finds are given in Tables 1-6.

Traps

The types of traps have been discussed. The important points are: (1) The Lower Cambrian traps are salt pillows and syndepositional; the porosity is mainly fracture porosity in dolomite; salts are the seals. (2) Most Proterozoic traps are stratigraphic; porosity is intergranular porosity in sandstones; the seals are shale. Some traps (e.g., Sredne-Botuobin) are basement arches which have been rejuvenated. (3) Faults play a role in entrapment (e.g., Yaraktin), but are not major causes of traps in the fields found thus far. Kazanskiy et al (1978) discussed some of the porosity problems in the area, because the porosity values generally are low. Reservoir-stimulation techniques need to be used on a greater scale.

Oil and Condensate Compositions (Tables 4-5)

Drobot and Isayev (1966) made a thorough study of differences in liquid compositions in the different productive zones of the Anabar-Lena structural terrace and Nep-Botuobin anticlise. They found that crude oils and condensates of the region belong to the "methane class" (i.e., they are paraffinic). On the basis of physico-chemical properties, the hydrocarbons are divided readily into two groups: (1) condensate and light crude oil from the older, terrigenous part of the section; and (2) relatively heavier oil and condensate in the younger, evaporite-carbonate part of the section.

Condensates in the Proterozoic sandstone reservoirs of the Markovo and Parfenovo Horizons have low specific gravity (0.707 to 0.786), low kinematic viscosity (0.78 to 3.10 centistokes), low molecular weight (100 to 150), and low optical activity (0.11 to 0.28°/mm). The liquids are pale greenish-yellow to colorless. They have low tar, sulfur, and ash contents, and high alkane contents (up to 90% in the 300° C distillation fraction).

Light oils from the Parfenovo Horizon have specific gravity values of 0.810 to 0.816, viscosities of 6.5 to 10.5 centistokes, molecular weights of 240 to 255, and optical activity levels of about 0.39°/mm. Their colors are yellow, yellow green, pinkish green, and cherry brown. The total contents of sulfur, tar, and ash are higher than in condensates from the same horizon.

Light oils and condensates from the higher Late Cambrian carbonate reservoirs have a specific gravity range of 0.792 to 0.855, viscosities of 2.7 to 15.6 centistokes, high molecular weight values (255 plus), optical activity values of 0.87°/mm, and low alkane contents. Colors are medium to dark yellow green and cherry brown. Tar content is much higher—12% vs. 3 to 4% in the Parfenovo; sulfur also is higher (0.43 to 1.12% vs. 0.007 to 0.26); and ash is higher (up to 0.11%). Sulfur comes from the evaporites.

Properties common to the Proterozoic sandstone and Early Cambrian carbonate hydrocarbons are: (1) none contains asphaltenes; (2) both are of the methane class; (3) aromatics are absent in the higher boiling fractions; and (4) porphyrins are absent.

Catalysts are present in the late Riphean and Vendian reservoirs but not in the younger Early Cambrian carbonate reservoirs. As a result, the oils and condensates of the older zones show much more "metamorphism" or maturation. The older oils and condensates contain more nickel, chromium, and manganese, but less vanadium, iron, and copper.

Drobot and Isayev (1966) concluded that, "The observed differences in the chemical composition, optical activity, and microelement distribution in oils from the Osa and Parfenovo Horizons indicate probable differences in the original organic matter and oil-generation conditions; i.e., they suggest different cycles of oil generation." Unquestionably, the differences in catalyst contents and in composition of the older and younger petroleums account for many of the differences in the crudes. However, Drobot and Isayev presented a strong case for hydrocarbon generation, migration, and entrapment during the deposition of both the terrigenous strata of the Riphean-Vendian and the carbonate-evaporite strata of the Early Cambrian. Their data suggest that generation, migration, and accumulation were nearly continuous processes during the more than 120 m.y. involved in the deposition of the Ushakova, Moty, and Early Cambrian strata.

Additional and related geochemical studies bearing on the origin and time of oil and gas accumulation, especially at

Table 2. Flow rates and permeability values, Angara-Lena structural terrace and Nep-Botuobin anteclise, USSR.

Field (place) and well no.	Horizon (lithology)	Flow rate			Permeability (md)		
		Gas (Mcf/d)	Condensate (b/d)	Oil (b/d)	Core Measurement	Water Displacement Method	Log Measurement
Yaraktin	Yaraktin (Sandstone)						
8		—	—	1,080	—	—	780
10		—	—	18	5.88	—	—
11		—	—	123	28.23	31.0	9.6
13		—	—	—	89.0	0.871	4.48
14		—	—	140	189	5.94	4.46
15		—	1,299	—	176	106	140
16		—	—	1,144	477	39	237
18		—	—	—	595	855	950
19		—	2,837	—	382	88.0	43.9
20		—	265	—	8.3	—	—
21		—	796	—	160	135	173
22		—	—	453	46.4	5.94	20.4
(Tokmin)							
105		—	1,420	—	17.0	10.0	36.0
Sredne-Botuobin	Parfenovo (Sandstone)						
3		5,950	—	—	—	120.0	265.8
8		5,890	—	—	—	336	120.6
9		19,600	—	—	609.6	105	184
13		3,542	—	—	155	8.8	6.6
18		4,354	—	—	—	4	33
25		14,651	—	—	525.4	190	36.6
8	Osa (Carbonate)	282	—	—	—	—	—
1		5,880	—	—	—	—	—
3		252	—	—	—	—	—
18		1,096	—	—	0.124	—	—
25		945	—	—	—	3.7	—
37		9,940	—	—	—	14.0	—
Bratsk							
8	Parfenovo (Sandstone)	6,374	—	—	0.31	1.8	1.0
13		—	—	—	0.09	0.1	2.7
(Shamanov)							
12		3,500	—	—	7.3	4.51	7.3
Preobrazhenka	Preobrazhenka (Carbonate)						
135		175	—	—	—	—	—
137		175	—	—	—	—	—
135	Ust'-Kut (Carbonate)	70	—	—	—	—	—
106	Preobrazhenka (Carbonate)	—	—	414	—	6.5	20.8
Ust'-Kut							
4	Ust'-Kut (Carbonate)	2,870	—	—	—	—	—

From Fuks et al (1977, p. 15)

Markovo field, were published by Samsonov and Tyshchenko (1970) and Gavrilov et al (1971).

Additional Evidence Bearing on Times of Oil Generation and Accumulation

1. Field mapping and lithofacies studies indicate that no strata younger than Middle Devonian ever were deposited in the Lena-Tunguska area (except the Vilyuy syneclise) until continental flows, sills, and fluviatile-lacustrine sediments accumulated locally in late Paleozoic, Triassic, and younger times. Therefore, the hydrocarbons in the Lena-Tunguska area had to originate in sediments of late Riphean through Devonian ages.

2. The Riphean and Vendian pools are in traps which formed at the time of deposition—i.e., they are syndepositional stratigraphic traps. These traps today lie on a very gentle basement slope (less than 0.5°) which must approximate the late Riphean-Vendian paleoslope.

3. Some of the reservoirs, such as that of the Parfenovo, are "closed"; that is, they are sealed on all sides by shale. Therefore the sources of the petroleum had to be the beds surrounding and close to the reservoirs.

4. Traps in the Lower Cambrian certainly began to form shortly after or during the deposition of the latest Early Cambrian or Middle Cambrian beds (Figs. 6, 7). Early trap formation (i.e., "timeliness") would make it possible for very early generation, migration, and entrapment (Samsonov and Tyshchenko, 1970; Gavrilov et al, 1971). The fact that halogenetic processes did not disturb the underlying Vendian and Riphean strata supports this conclusion—that the Early Cambrian hydrocarbons came from Early Cambrian or older rocks and that the Proterozoic hydrocarbons came from Proterozoic rocks.

5. Sporophyte contents of the hydrocarbons show that the

Table 3. Physical parameters of Lower Cambrian and Vendian, Angara-Lena structural terrace and Nep-Botuobin anteclise, USSR.

Age	Suite	Productive Horizon (Lithology)	Thickness (m)	Porosity (%)	Permeability (md)
Early Cambrian	Angara	Kelorskiy (Carbonate)	16 to 30	1.4 to 7.0	0 to 185
		Bil'chir (Carbonate)	60 to 120	0.5 to 11.0	0 to 24
	Bulay	Birkin (Carbonate)	90 to 100	0.06 to 1.0	0 to 253
	Bel'ya	Atovo (Carbonate)	60 to 90	0.3 to 16.0	0 to 24
		Khristovorova (Carbonate)	50 to 64	0.5 to 1.5	0 to 100
	Usol'ye	Balykhtin (Carbonate)	9 to 10	0.6 to 8.0	Fractures up to 10
		Osa (Carbonate)	15 to 100	0.1 to 25.0	Up to 163
	Moty	Ust'-kut (Carbonate)	Up to 90	0.7 to 12.0	0 to 36
Early Cambrian to Vendian		Preobrazhenka (Carbonate)	17	0.3 to 12.0	0 to 8
		Verkhnetarsk (Sandstone)	2 to 14	3.0 to 19.0	0.5 to 77
Vendian (Proterozoic)		Parfenovo (Sandstone)	15 to 90	8 to 23	Up to 4,300
		Yaraktin (Sandstone)	Up to 30	0.2 to 17.5	Up to 4,000
		Markovo (Sandstone)	Up to 30	2 to 13	Up to 20
		Bezymyannyy (Sandstone and conglomerate)	Up to 24	4 to 6	0.8 to 3.0

From Fuks et al (1977, p. 14). Fuks et al also gave values for the Kansk, and central and northwestern Tunguska basins, but the above figures are representative for the whole region. Those who wish to see the additional data may refer to page 14 of the Fuks et al (1979) article.

original *reservoirs* were Proterozoic and/or Lower Cambrian, but do not prove the age of the oil, condensate, and gas. Nevertheless, the fact that all oils, condensates, and gases, whether in Middle Cambrian, Ordovician, or Silurian strata, have sporophytes only of late Riphean, Vendian, or Early Cambrian ages (Timofeyev and Bagdasaryan, 1964) is strongly suggestive of the times of origin of the hydrocarbons.

Markovo Field

Markovo 1 was completed on March 18, 1962, and tested 2,040 b/d of oil in the Early Cambrian Osa Horizon. The oil is 42° oil; completion was from a depth of 2,156 to 2,164 m (Weaver, 1962; Trofimuk et al, 1964; Vasil'yev, 1968a, 1968b). The second well was noncommercial, but eventually 23 oil and gas-condensate wells were completed in the Lower Cambrian Osa Horizon and the Vendian Parfenovo Horizon. Figure 10 is a structural-contour map, with structural datum at the top of the Early Ordovician Ust'-Kut Suite (hence the above-sea-level values on the map). Figure 5 is the columnar section for the Markovo area (this section also applies to Yaraktin, except that basalt traps are present at Yaraktin). Figure 11 shows structure at the Parfenovo level. The bar-sandstone nature of the reservoir can be seen clearly. Figure 6 is a structural cross section at the Osa level and shows the salt deformation. Figure 12, a structural cross section at the Parfenovo level, shows a gentle dip toward the southeast and—if contrasted with Figure 6—emphasizes the complete structural discordance between the Cambrian and late Proterozoic. Figure 13 is an electric and gamma-neutron log through the Vendian and Riphean parts of the section where the Riphean overlies basement.

Thus the shallow trap is structural and is small. In contrast, the deep trap is stratigraphic and much larger than the Osa trap.

The field is not a giant but is of respectable size. Proved reserves are 622 Bcf of gas and 16 million bbl of condensate in the Parfenovo, and 10 million bbl of oil in the Osa. An oil ring is present at the Parfenovo level (Fig. 12).

Many similar structures have been drilled along this part of the Lena River, and several commercial and noncommercial

Table 4. Gas Analyses from Lena-Tunguska Fields (from Vasil'yev, 1968b; Vassoyevich et al, 1970; Vasil'yev and Zhabrev, 1975).

				Composition (% vol.)								
Field	Age	Unit	Type of Gas	C_1	C_2	C_3	C_4	C_{5+}	N_2	CO_2	H_2	Depth (m)
Markovo	Early Cambrian	Osa	Associated	71.4	13.2	7.9	—	—	—	—	—	—
	Early Cambrian	Balykhtin	Associated	72.4	9.3	7.8	4.9	3.96	1.55	1.0	—	—
	Vendian	Parfenovo	Associated	73.7	7.6	2.7	—	—	—	—	—	—
	Vendian	Parvenovo	Nonassociated	82.5	7.0	2.9	0.8	2.9	3.3	0.6	—	—
	Vendian	Parvenovo	Nonassociated	78.9	7.4	3.65	1.07	5.25	3.13	0.64	—	—
	Riphean	Markovo	Nonassociated	72.5	11.0	6.5	4.8	—	—	—	—	—
	Riphean	Bezymyannyy	Nonassociated	81.8	10.7	3.8	1.6	—	—	—	—	—
Bil'chir	Early Cambrian	Bil'chir	Nonassociated	78.4	12.8	0.56	0.56	0.12	7.5	—	—	—
Birkin	Early Cambrian	Birkin	Nonassociated	92.0	5.6	1.5	0.5	0.35	—	—	0.6	—
Preobrazhenka	Early Cambrian	Preobrazhenka	Nonassociated	81.07	5.54	2.26	1.03	0.46	8.71	—	—	1,680-1,669
Sredne-Botuobin	Early Cambrian	Osa (?)	Nonassociated	87.7	4.0	1.2	0.1	0.0	6.1	0.9	—	1,487-1,509
	Early Cambrian	Osa (?)	Nonassociated	87.3	4.5	1.5	0.2	tr.	6.2	—	—	1,489-1,493
	Vendian	Parfenovo	Nonassociated	85.4	4.8	1.4	0.1	0.0	7.2	1.1	—	1,876-1,886
	Vendian	Parvenovo	Nonassociated	83.14	6.15	2.25	0.55	0.22	7.7	0.45	—	1,845-1,910

discoveries have been made. To map and find new fields in an area such as this, highly refined seismic equipment is needed to find zones of pinchout. Well control is needed to map favorable facies. Only in the last decade has sufficient control become available even to attempt detailed paleogeographic studies. One of the best attempts in this area was by Samsonov et al (1977).

Sredne-Botuobin Field

The Sredne-Botuobin gas field was discovered in 1970 on a huge, gently dipping (less than 1°) basement-arch closure—about 100 m of closure (Fig. 15). The field was the first found in this area on a Riphean-Vendian closed structure. Production was found both in the Osa (?) at a depth of 1,500 m and in the Parfenovo at a depth of 1,880 m. Basement is shallower than 2,000 m at the crest of the structure, which has an area of about 1,700 sq km. The Osa production is in Osa carbonates, and is rather poor. The amount of fracturing that took place was not sufficient to make an extensive reservoir. However, intergranular porosity (the carbonate is a dolomite) keeps the porosity in some parts of the field at the 3 to 13% level. Properties of the field are given on Tables 2-5.

Figure 14 includes a structural-contour map (contours are subsea values), a natural-scale cross section (with the Osa and Parfenovo productive zones shown), and a columnar section of the Osa and Parfenovo lithologies. The similarities with the Markovo section (Fig. 5) are clear.

Porosity in the Parfenovo averages about 18% in a medium- to fine-grained, light-gray feldspathic quartz sandstone. Permeability values range from 5 to 525 md, and average about 30 to 100. The unit thickens southward from 12 to 28 m. Its distribution is erratic, so that it is not everywhere present. The reservoir pressure is 40 atm lower than normal and the temperatures are actually cold (3 to 7 ° C). Fuks and Fuks (1979) published an analysis of the pressure and temperature to demonstrate how much change in both had to take place to explain (1) present temperature and pressure values, and (2) the presence of hydrocarbons under such conditions.

Other large arches are in the area. One is Verkhnevilyuy (Fig. 16) which, together with Sredne-Botuobin, has been studied by Belen'kiy and Kunin (1978). Reserves at Verkhnevilyuy have been announced to be 10.5 Tcf of gas and about 260 million bbl of condensate; See Bakirov, 1979.

Sredne-Botuobin is a giant gas field with 5.95 Tcf of proved reserves and 149 million bbl of condensate. Sixty percent of these reserves (3.75 Tcf and 100 million bbl) is in the Proterozoic Parfenovo Horizon. In all probability, development drilling will prove additional reserves. This field is the first Precambrian giant petroleum field ever recorded.

Yaraktin Field

Yaraktin was discovered in 1971 beneath 1,550 to 1,750 m of Permian, Triassic, and Lower Jurassic basalt traps (Bazanov, 1973; Bazanov et al, 1977). It is a bar sandstone in the Vendian (Yaraktin Horizon, which is equivalent to the Parfenovo). The field fault (Fig. 17) partly controls the oil distribution. The field is discussed here because (1) it is a large oil field, with 210 million bbl of proved, probable, and potential reserves to date; and (2) it is an *oil* field. Most fields in the region have been gas. It is still under development with 180 sq km of oil found to date. The pay sandstone ranges in thickness from 0 (where it wedges out updip) to 26 m at the oil/water contact. Porosity is 9 to 10%, and production per well (during drill-stem tests) has been as great as 876 b/d.

The field was discovered largely by luck, because no seismic penetration was obtained. However, subsurface mapping

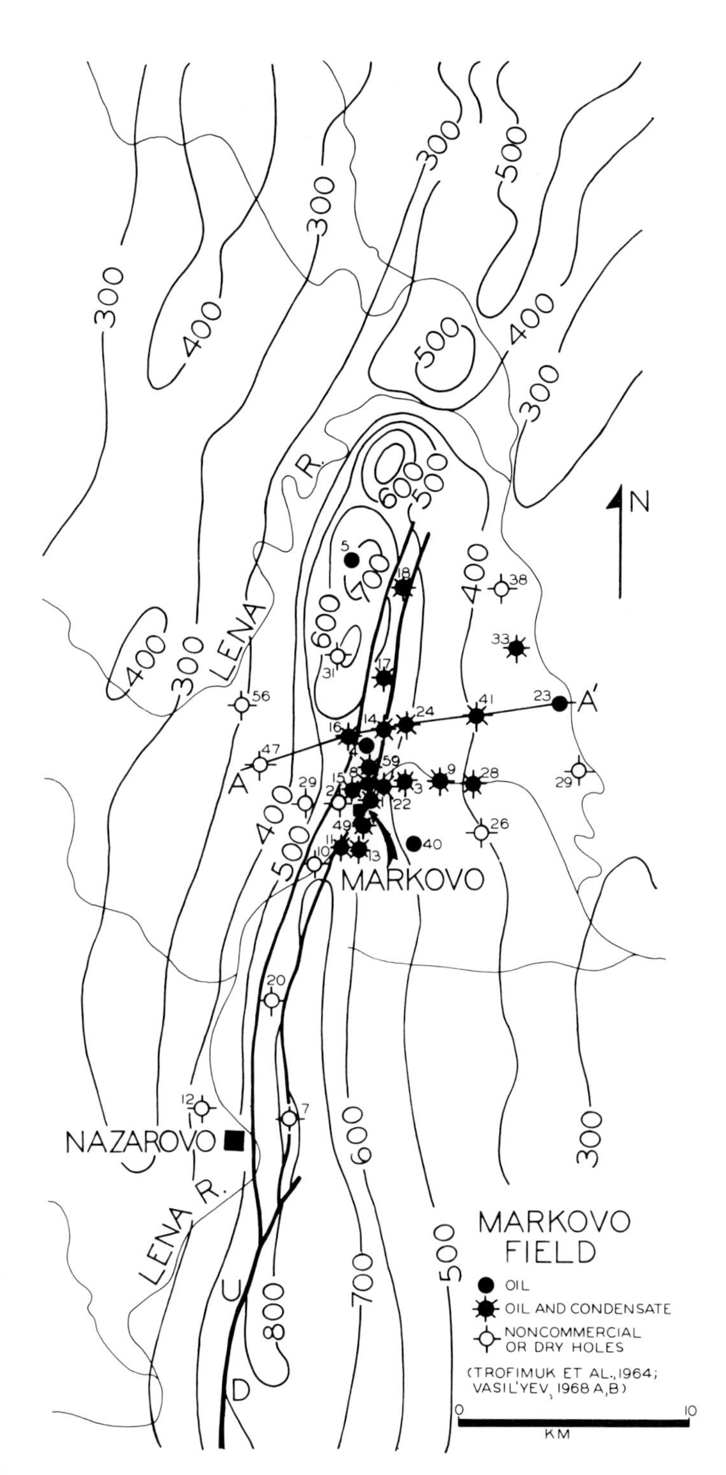

FIG. 10—Structural-contour map, Markovo field. Structural datum is the top of the Ust'-Kut Suite (Ordovician). Depths are in meters. From Trofimuk et al (1964). The line of cross section is Figure 12.

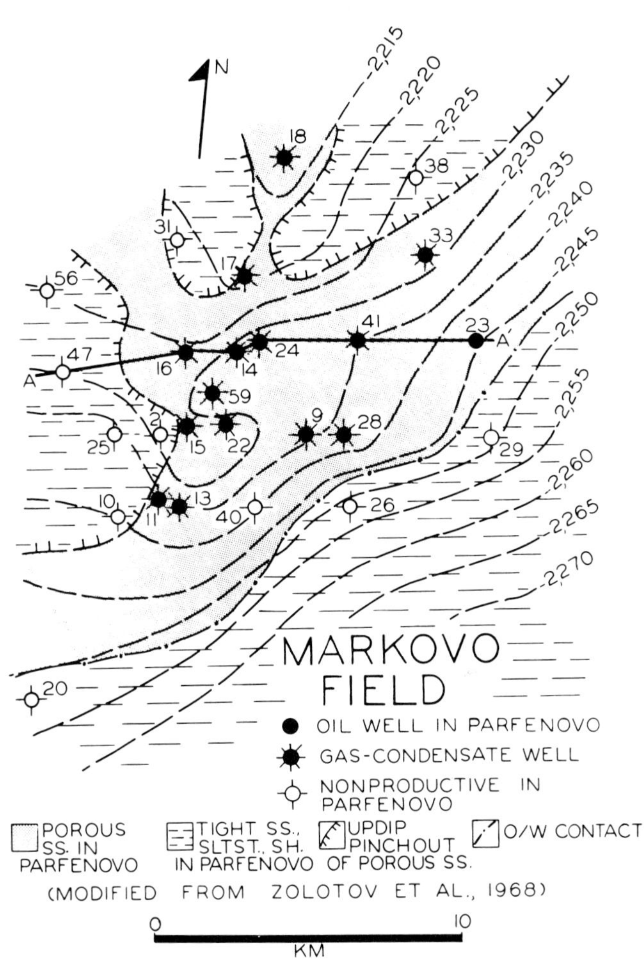

FIG. 11—Lithofacies map, Parfenovo Horizon, Markovo field. The trap is a submarine bar sandstone. Contour datum is the top of the Parfenovo. From Zolotov et al (1968). The line of cross section is Figure 12.

suggested that a series of marine bar sandstones farther east strike toward the Yaraktin area. This, coupled with the fact that the field is near the crest of the giant Nep-Botuobin arch, led to the drilling and discovery of the field.

Proterozoic Production or Production Potential in Other Areas

Production—small but real—has been known from Proterozoic rocks of the Amadeus basin of Australia for many years. It was the first Proterozoic production ever found. Murray (1965) wrote a paper on the Proterozoic in which he advised petroleum geologists to look at all unmetamorphosed rocks, whatever their age. Trofimuk wrote much the same thing after the Markovo discovery (Trofimuk et al, 1964). More recently oil production from Proterozoic rocks has been found near Tientsin, China, but this oil very likely originated in younger strata. A small (58 b/d) Proterozoic well has been tested in the Perm region of the European USSR, in sandstones of the Bavly beds, and exploration there continues (Kutukov et al, 1977). Other likely areas for Proterozoic petroleum were listed by Vassoyevich et al (1970). Large Proterozoic gas reserves (probably indigenous to the Proterozoic) have been announced in the Sichuan basin of south-central China.

Table 5. Selected Production Data from Angara-Lena structural terrace, Nep-Botuobin anteclise, and Tunguska syneclise fields.

Field	Year Dis-covered	Well no.	Depth (m)	Reservoir	Choke (mm)	Oil/d (bbl)	Cond./d (bbl)	Gas/d (Mcf)	Cond./gas ratio	Gas/oil ratio	S.G. oil	S.G. con-densate	% S	Formation Pressure (kg/sq cm)	Temper-ature °C
Markovo	1962	1	2,156-2,164	Osa	75	6,700	—	17,500	—	350 m^3/m^3	0.815	—	0.8	268	33
		1	2,156-2,164	Osa	25	2,040	—	?	—	—	0.815	—	0.8	268	33
		1	2,156-2,164	Osa	16	1,441	—	?	—	—	0.815	—	0.8	268	33
		8	2,172	Osa	16	318	—	—	—	215 m^3/m^3	0.818	—	0.74	271	33
		11	—	Parfenovo	75	—	—	8,445	560 cm^3/m^3	—	—	0.713-0.764	0.02	—	34
		15	2,574-2,605	Markovo	25	—	549	8,655	—	—	—	—	—	—	—
		15	2,900	Bezymyannyy	75	—	—	35	—	—	—	—	—	—	—
Bil'chir	1963	1	270-275	Bil'chir	—	—	—	2,625-3,500	—	—	—	—	—	28	—
Birkin	1964	1	1,236	Birkin	—	—	—	15,750	—	—	—	—	—	120	—
Krivoluka	1965	3	2,527-2,560	Parfenovo	16	271	—	—	—	400 m^3/m^3	0.814	—	0.07	284	—
Khristoforova	1968	1	912	Bil'chir	22	—	—	7,000	—	—	—	—	—	—	—
Sredne-Botuobin	1970	—	1,500	Osa (?)	—	—	—	7,700	—	—	—	—	—	142	—
		—	1,880	Parfenovo	—	—	—	9,275 14,350	—	—	—	—	—	147	—
Preobrazhenka	1972	135	1,690	Middle Moty (Preobrazhenka)	—	—	—	2,170	—	—	—	—	—	153	—
Kuyumba	1973	—	2,170	Middle Moty (dolomite)	11.3	—	—	2,520	70 cm^3/m^3	—	—	—	—	—	—
		—	2,170		various	—	?	7,000-14,000	?	—	—	—	—	—	—

Table 6. Miscellaneous data from fields of Irkutsk amphitheater.

Field	Trap	Size of Trap	Dips	Productive Zone	Thickness (m)	Age	Porosity (%)	Permeability (md)	Depth (m)	Formation pressure	Estimated minimum reserve
Bil'chir	salt anticline	length: 10-11 km. width: 7-8 km	1 to 2°	Bil'chir (fractured carbonate)	5	Early Cambrian	3 to 6	120 to 365	270 to 275	28 kg/cm^2	20 Bcf
Birkin	salt anticline	length: 7 km width: 4.5 km	1°	Birkin	10	Early Cambrian	9 to 11	12 to 15	1,236	120 kg/cm^2	15 Bcf
Krivoluka	strati-graphic	unknown	—	Parfenovo (sandstone)	—	Vendian	16 to 24	Up to 140	2,527 to 2,560	284 kg/cm^2	—
Khristoforova	salt anticline	length: 15 km width: 4 km	—	Bil'chir (fractured carbonate)	—	Early Cambrian	—	—	912	—	20 Bcf
Preobrazhenka	salt anticline	—	—	Middle Moty	12.5	Early Cambrian	7 to 12	low	1,690	153 kg/cm^2	500 Bcf

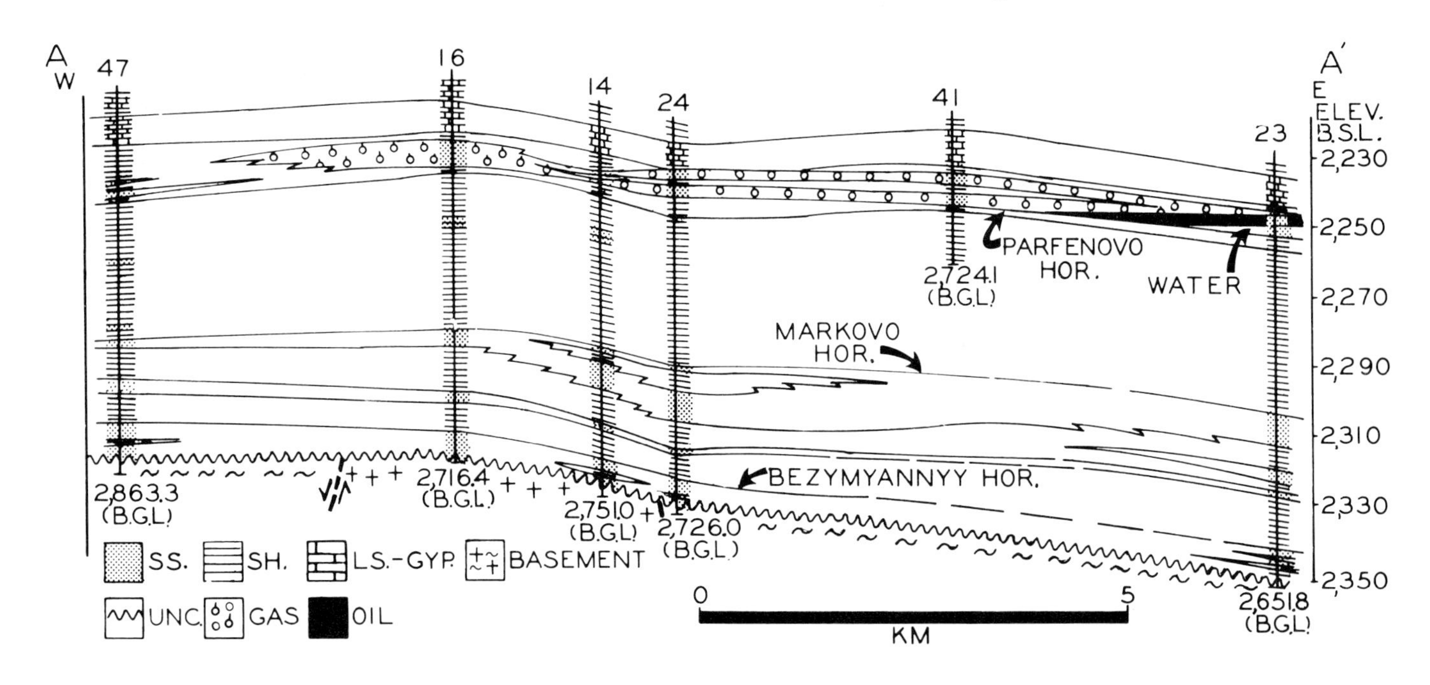

FIG. 12—West to east cross section, Parfenovo zone, Markovo field. Note the oil ring. Depths in meters. From Zolotov et al (1968). Location is on Figs. 10 and 11.

In the Canadian Arctic, in the graben zone between Norman Wells and Coppermine, and on Victoria Island, there is a great thickness of unmetamorphosed marine Proterozoic. During 1977 a gas discovery was made in the Cambrian of this basin, beneath a salt bed. As a result, the area became known as the "subsalt" play of the Northwest Territories. The similarity between the Canadian section and the section in Siberia suggests that Proterozoic objectives ultimately may be very important in this region as well. Even in the United States there are large areas of Proterozoic potential in Alaska, in the Belt Supergroup (northern Rocky Mountains), in the Great Basin (Misch and Hazzard, 1962), in West Texas and southeastern New Mexico, in the Appalachians, and possibly in the Great Lakes region.

CONCLUSION

What, then, are the lessons to be learned from this growing Siberian petroleum province? The most obvious is that commercial hydrocarbons have no respect for age. Becker and Patton (1968, p. 224) wrote: "Marine sedimentary rocks of Precambrian, Cambrian, and Ordovician ages constitute a major frontier for petroleum exploration. In regions where appreciable thicknesses of such rocks exist, test wells range from sparse in most Ordovician sections to almost nonexistent in Precambrian rocks. The prospects within these strata appear to improve with decreasing age, but the fact that the environment favorable for shelf sedimentation expanded progressively through the same span of time suggests that time is not the overriding factor and that no region of marine sedimentary rocks should be discounted merely on the basis of age."

Becker and Patton went on to observe (p. 237) that "Marine pre-Silurian rocks in untested or inadequately tested platform and basin regions should produce significant amounts of petroleum." They added "Cambrian and Ordovician reserves will remain difficult to discover because (1) much of the trapping will be stratigraphic and caused by change of facies; (2) the rocks are generally well lithified and difficult to drill; (3) in many localities they lie beneath unconformities and even multiple unconformities . . . ; and, most important, (4) the present density of control in most undeveloped regions is inadequate to permit effective use of new control . . .

"Stratigraphic age, within these ancient rocks as elsewhere in the sequence, is a less important factor than reservoir conditions, structure, unconformity, and facies.

"A few years ago a student in one of the writer's (sic) beginning geology classes defined the basement complex as '. . . a fear of being in mines and other deep places.' To some extent the petroleum industry has suffered from the basement complex in just this sense—a fear of being in rocks that are too old and too deep."

Halbouty et al (1970) and Holmgren et al (1975) noted, in addition, that the amounts of postdepositional deformation, uplift, and erosion are critical. They observed that, of all of the giant oil and gas fields in the world, 82% had no major unconformity in the section. The tectonic and erosional history, then, is a critical factor deciding whether oil and gas in commercial quantities are likely to be present. The amount and types of life determine whether or not hydrocarbon generation can take place. Thus, tectonic stability and organic remains are the keys to generation, accumulation, and preservation. Age is a less important factor.

It seems appropriate to end this report with a quotation from Vassoyevich et al (1970, p. 78): "The available materials on

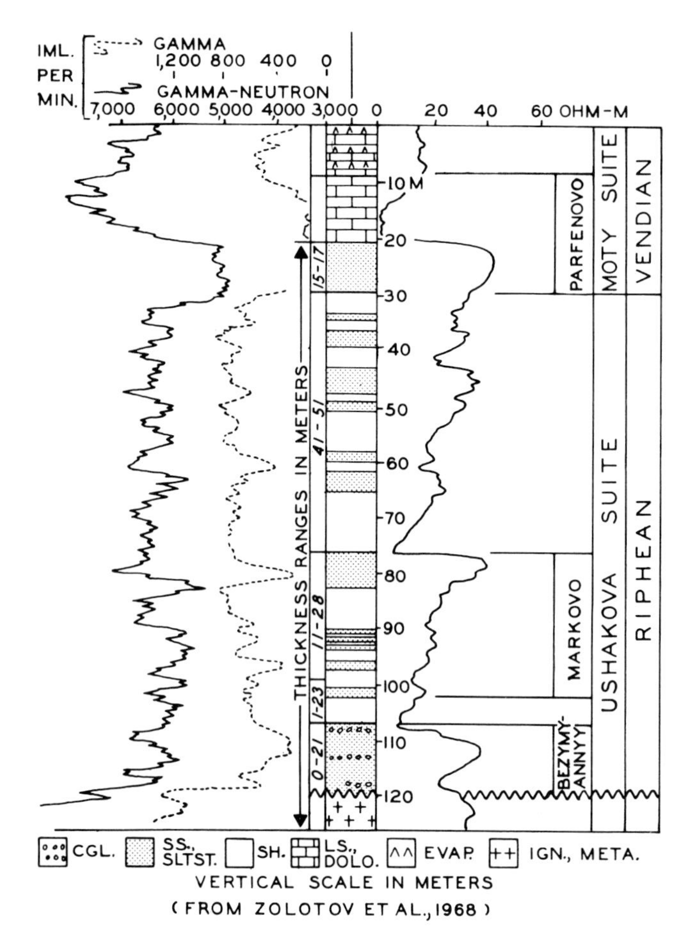

FIG. 13—Gamma-neutron and electric log, Parfenovo Horizon, Markovo field. From Zolotov et al (1968).

the distribution of Precambrian nonmetamorphosed sedimentary formations and the naphthide occurrences associated with them show that the oil and gas potential of the late Proterozoic deposits can and should be recognized as an important and potentially profitable matter of great scientific and economic significance. The scientific study of this problem will enrich our knowledge of one of the most interesting periods of the earth's development—the late Precambrian. The practical solution of the problem will expand the oil and gas reserves of the sedimentary mantle, both in the known oil and gas basins of the USSR and in new basins located in the areas of ancient platforms. The latest estimate of the oil and gas reserves in the Soviet Union quite properly took into account the oil and gas contents of the upper Precambrian deposits in the basins we have considered . . ."

REFERENCES CITED

Akul'cheva, Z.A., and Z.Kh. Fayzulina, 1976, O sopostavlenii nizhnepaleozoyskikh otlozheniy Prilenskogo i Botuobinskogo rayonov Nepskogo svoda: Geol. i Geofiz., No. 7, p. 10-17.

Anatol'yeva, A.I., M.A. Kharkov, and Yu.K. Sovetov, 1966, O korrelyatsii krasnotsvetnykh tolshch venda i nesov nizhnego kembriya yugozapadnoy okrainy Sibirskoy platformy: Akad. Nauk SSSR Doklady, v. 166, No. 2, p. 413-416 (Eng. trans. by Am. Geol. Inst., 1966, Acad. Sci. USSR Doklady, v. 166, p. 24-26).

Babintseva, T.N., et al, 1978, Gidrogeokhimicheskiye usloviya neftegazonosnosti Tungusskogo basseyna: Geol. Nefti Gaza, No. 3, p. 30-36.

Bakirov, A.A., ed., 1979, Neftegazonosnyye provintsii i oblasti SSSR: Moscow, Izd. Nedra, 456 p.

——— and G.Ye. Ryabukhin, 1969, Neftegazonosnyye provintsii i oblasti SSSR: Moscow, Izd. Nedra, 477 p.

Balitov, N.B., 1977, O genezise sernistykh neftey i serovodoroda v gazakh Osinskogo gorizonta Irkutskogo amfiteatra: Geol. i Geofiz., No. 9, p. 47-55.

Barghoorn, E.S., W.G. Meinschein, and J.W. Schopf, 1965, Paleobiology of a Precambrian shale: Science, v. 148, no. 3669, p. 461-472.

Bazanov, E.A., 1973, Geologicheskoye stroyeniye Yaraktinskogo neftyanogo mestorozhdeniya Irkutskoy oblasti: Geol. Nefti Gaza, No. 7, p. 15-18.

Bazanov, E.A., I.A. Vereshchako and B.M. Frolov, 1977, Sovershenstvovniye metodiki i puti povysheniya effektivnosti geologorazvedochnykh rabot na neft' i gaz v Irkutskom amfiteatre: Geol. Nefti Gaza, No. 2, p. 9-13.

Becker, L.E., and J.B. Patton, 1968, World occurence of petroleum in pre-Silurian rocks, AAPG Bull., v. 52, no. 2, p. 224-245.

Belen'kiy, V.Ya., and N.Ya. Kunin, 1978, Puti povysheniya effektivnosti seysmorazvedki pri podgotovke struktur v Zapadnoy Yakutii: Geol. Nefti Gaza, No. 5, p. 22-30.

Brod, I.O., and I.V. Vysotskiy, eds., 1965, Neftegazonosnyye basseyny zemnogo shara: Moscow, Izd. Nedra, 598 p.

Dikenshteyn G.Kh., et al, 1977, Neftegazonosnyye provintsii SSSR: Moscow, Izd. Nedra, 328 p.

Drobot, D.I., and V.P. Isayev, 1966, Novyye dannyye o sostave i svoystvakh nizhnekembriyskikh neftey Prilenskogo rayona Irkutskogo neftegazonosnogo basseyna: Akad. Nauk SSSR, Sibir. Otdel., Geol. i Geofiz., No. 10, p. 32-41. (Eng. trans. by Am. Geol. Inst., 1967, Internatl. Geol. Review, v. 9, No. 8, p. 1028-1035).

Dubronin, M.A., 1976, Usloviya obrazovaniya solyanykh struktur Angaro-Lenskogo progiba: Geol. i Geofiz., No. 6, p. 60-67.

Fuks, A.B., and B.A. Fuks, 1979, Genezis neftyanykh otorochek zalezhey Nepsko-Butuobinskoy anteklizy: Geol. Nefti Gaza, No. 2, p. 47-50.

Fuks, B.A., et al, 1977, Vskrytiye i ispytaniye produktivnykh plastov nizhnego kembriya Vostochnoy Sibiri: Geol. Nefti Gaza, No. 2, p. 13-18.

Gavrilov, Ye.Ya., I.B. Kulibakina, and G.I. Teplinskiy, 1971, O formirovanii zalezhey uglevodorodov Markovskogo mestorozhdeniya: Geol. Nefti Gaza, No. 2, p. 30-31.

Gladkov, V.G., V.P. Nikitin, and P.M. Khrenov, 1970, K voprosu o kinematike galozheniya razryvov v skladchatom obramlenii Yuga Sibirskoy platformy: Akad. Nauk SSSR Doklady, v. 190, no. 2, p. 405-408 (Eng. trans. by Am. Geol. Inst., 1970, Acad. Sci. USSR Doklady, v. 190, p. 42-45).

Halbouty, M.T., et al 1970, World's giant oil and gas fields, geologic factors affecting their formation, and basin classification: AAPG Mem. 14, p. 502-555.

Holmgren, D.A., J.D. Moody, and H.H. Emmerich, 1975, The structural settings for giant oil and gas fields: 9th World Petroleum Congr. (Tokyo), Proc. v. 2, p. 45-54.

Karasev, O.I., et al, 1971, Usloviya neftegazoobrazovaniya i neftegazonaklopleniya v vendskikh i kembriyskikh otlozheniyakh Yuga Sibirskoy platformy: Irkutsk, Izd. Vostochno-Sibirskoy Knizhnoye, 206 p.

Karpyshev, V.S., 1965, O vzaimootnoshenii mezhdu galogenno-karbonatnoy i krasnotsvetnoy formatsiyami kembriya v zapadnoy chasti Irkutskogo amfiteatra: Akad. Nauk SSSR Doklady, v. 160, no. 2, p. 425-448.

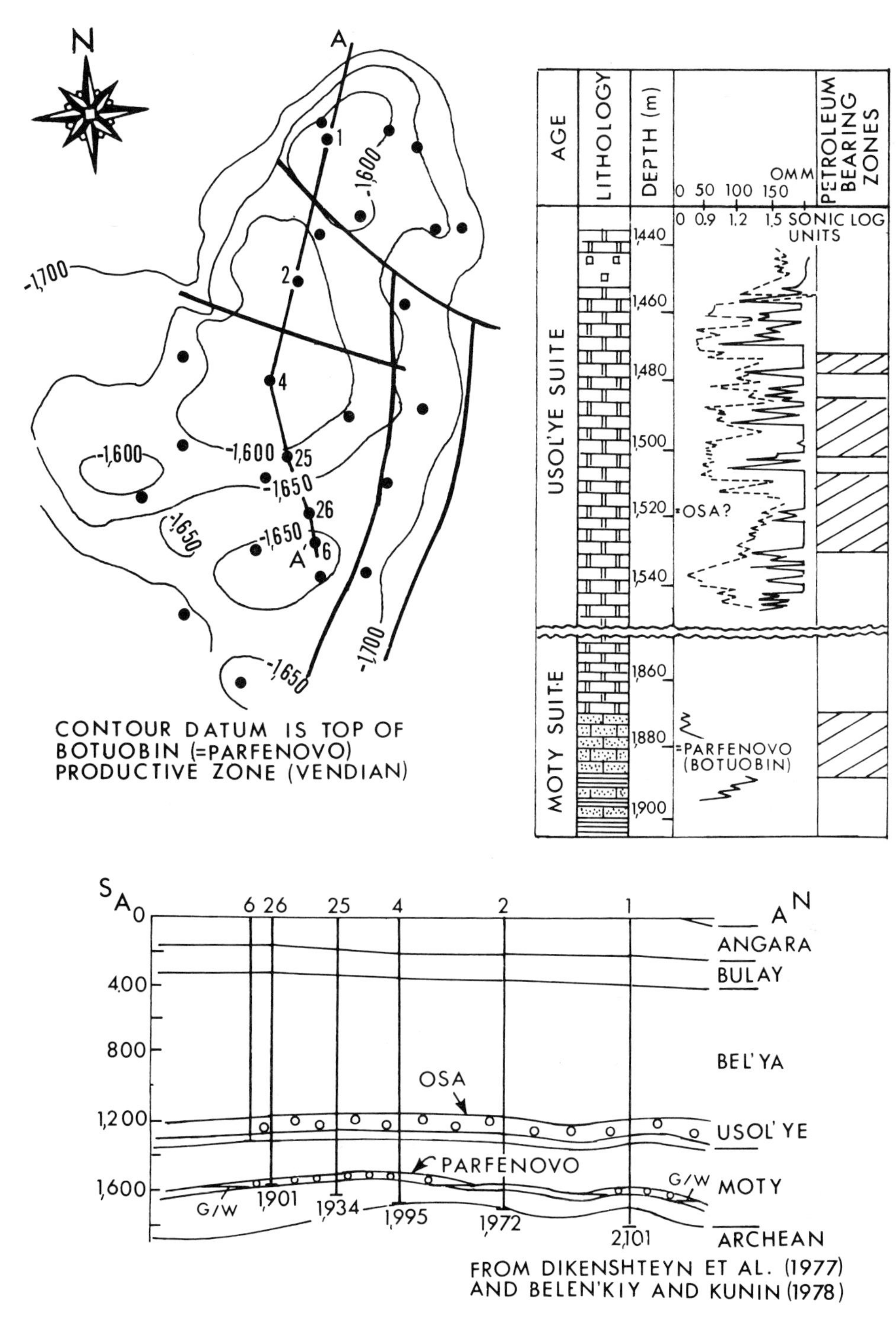

FIG. 14—Sredne-Botuobin field. Structural-contour map (in subsea meters) is at the top of the Parfenovo Horizon of the Proterozoic. Line of cross section is shown. Cross section strikes south to north and shows both Osa and Parfenovo producing zones. Columnar section shows lithologic details. From Dikenshteyn et al (1977) and Belen'kiy and Kunin (1978).

Kazanskiy V.V. et al, 1978, Metody vozdeystviya pri ispytanii nizkopronitsayemykh plastovkollektorov Vostochnoy Sibiri: Geol. Nefti Gaza, no. 4, p. 60-64.

Kolosov, A.S., A.M. Pustyl'nikov, and T.M. Kharkova, 1968, Slozhnyye khloridy zheleza i margantsa v kembriyskikh solyanykh otlozheniyakh Kansko-Taseyevskoy vpadiny: Akad. Nauk SSSR Doklady, v. 181, no. 6, p. 213-216.

Komar, V.A., and M.A. Semikhatov, 1965, K geologicheskoy istorii Sibirskoy platformy v pozdnem dokembrii: Akad. Nauk SSSR Doklady, v. 161, no. 2, 421-424 (Eng. trans. by Am. Geol. Inst., 1965, Acad. Sci. USSR Doklady, v. 161, p. 42-45).

Kontorovich, A.A., et al, 1977, Osnovvyye etapy i rezul'taty poiskovykh rabot v Zapadnoy-Sibirskoy neftegazonosnoy provintsii: Geol. Nefti Gaza, No. 11, p. 21-25.

Korolyuk, I.K., and A.D. Sidorov, 1969, Stromatolity motskoy svity Yuzhnogo Pribaykal'ya i Yugo-Vostochnogo Prisayan'ya: Akad.

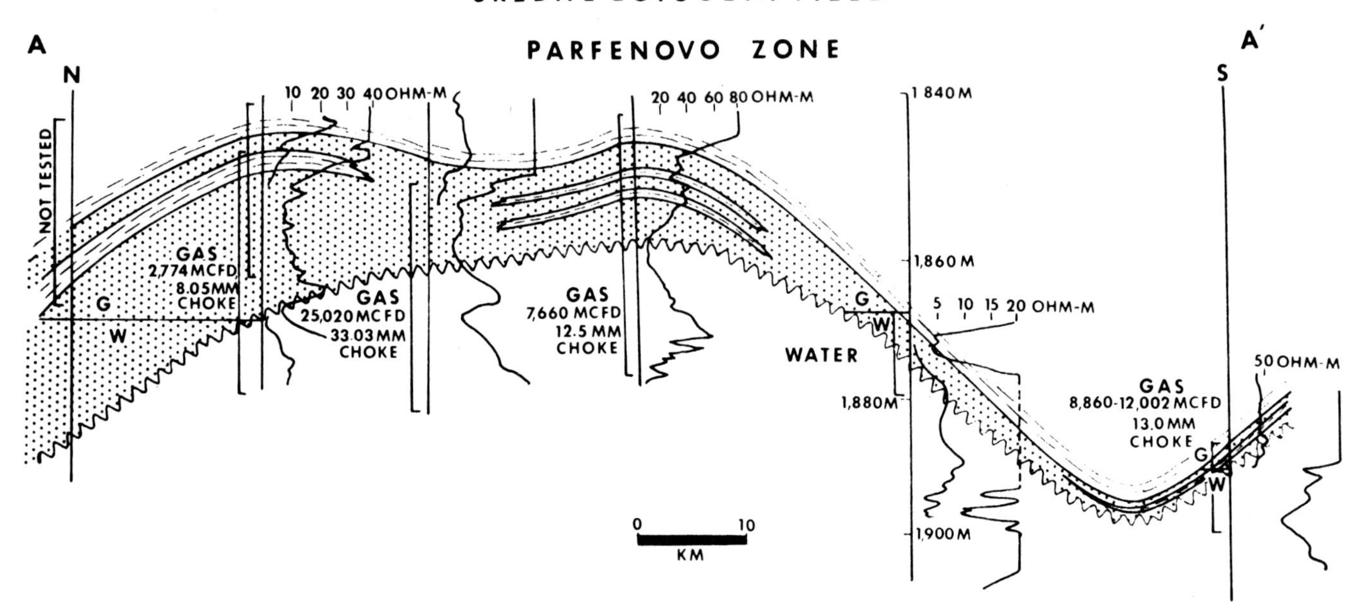

FIG. 15—West to east electric-log cross section of Sredne-Botuobin field. Gas flow rates are shown.

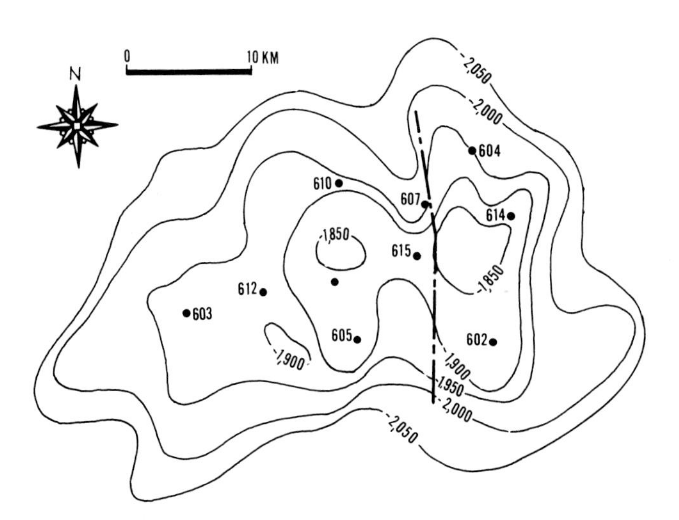

FIG. 16—Structural-contour map of Verkhnevilyuy field, Angara anticlise. Structural datum is the top of the Kharystan Horizon near the top of the Vendian. From Belen'kiy and Kunin (1978).

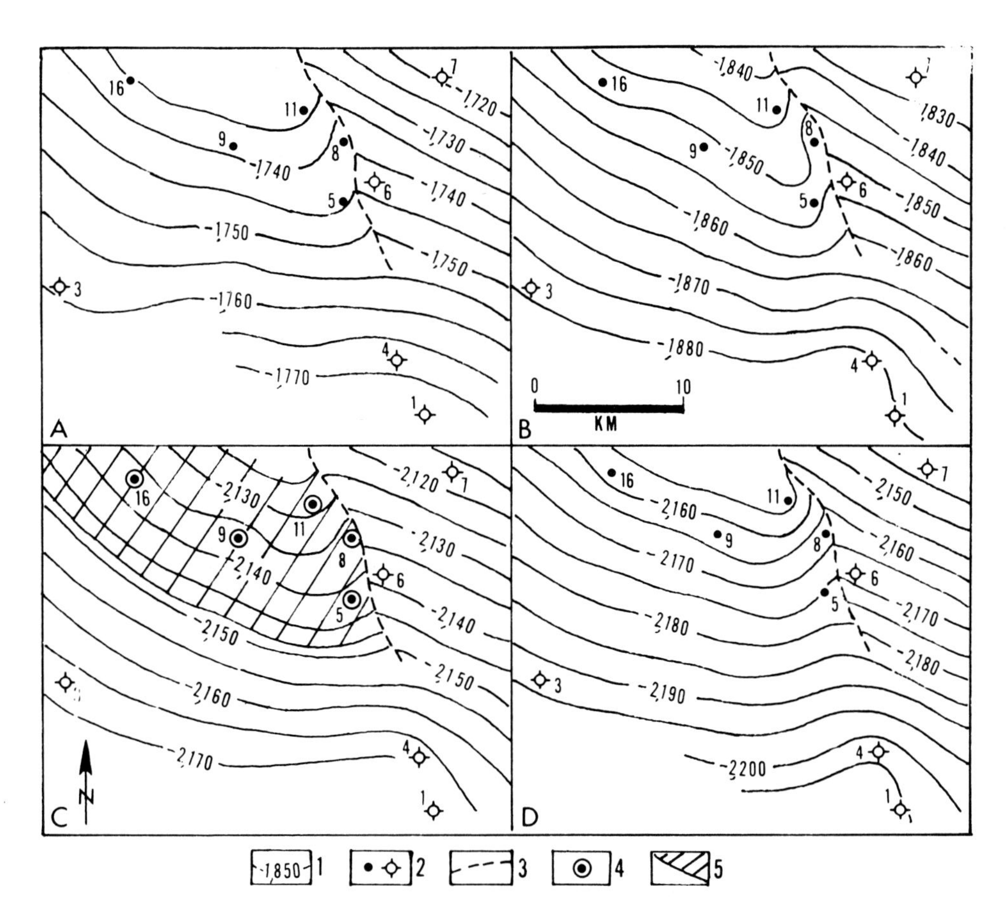

FIG. 17—Structural-contour maps, Yaraktin field, Angara anticlise. A - top of Osa; B - top of Moty; C - base of Parfenovo; D - top of basement. *Legend:* 1 - contours (meters); 2 - well control; 3 - fault; 4 - commercial oil well in Parfenovo; 5 - area of oil production with downdip oil/water contact. From Bazanov (1973).

Nauk. SSSR Doklady, v. 184, no. 3, p. 669-671 (Eng. trans. by Am. Geol. Inst., 1969, Acad. Sci. USSR Doklady, v. 184, p. 53-56).

Kutukov, A.V., S.A. Vinnikovskiy, and K.S. Shershnev, 1977, Perspektivy neftegazonosnosti vendskikh otlozheniy Permskogo Prikam'ya: Geol. Nefti Gaza, No. 11, p. 37-43.

Levchenko, I.G., 1975, Perspektivy neftegazonosnosti kembriya Tungusskoy sineklizy i yeye obramleniya: Geol. Nefti Gaza, No. 1, p. 1-9.

Lokhmatov, G.I., 1966, Izmeniye sostava nizhnekembriyskikh karbonatnykh otlozheniy pod vliyaniyem konsedimentatsionnogo formirovaniya geologicheskikh struktur (Yug Sibirskoy platformy): Akad. Sci. SSSR Doklady, v. 170, no. 3, p. 661-664 (Eng. trans. by Am. Geol. Inst., 1967, Acad. Sci. USSR Doklady, v. 170, p. 88-90).

McKirdy, D.M., 1974, Organic geochemistry in Precambrian research: Precambrian Research, v. 1, p. 75-137.

Meinschein, W.G., E.S. Barghoorn, and J.W. Schopf, 1964, Biological remnants in a Precambrian sediment: Science, v. 145, no. 3629, p. 262-263.

Misch, P., and J.C. Hazzard, 1962, Stratigraphy and metamorphism of late Precambrian rocks in central northeastern Nevada and adjacent Utah: AAPG Bull., v. 46, no. 3, p. 289-343.

Mitroshin, M.I., and D.I. Gvizd', 1974, Perspektivy neftegazonosnosti severo-zapadnoy chasti Tungusskoy sineklyzy: Sovet. Geol., No. 4, p. 45-51.

Murray, G.E., 1965, Indigenous Precambrian petroleum?: AAPG Bull., v. 49, no. 1, p. 3-21.

Popova, Zh.P, and N.N. Glazunova, 1965, Organicheskiye ostatki v nefti Markovskogo mestorozhdeniya: Akad. Nauk SSSR Doklady, v. 161, no. 3, p. 673-675 (Eng. trans. by Am. Geol. Inst., 1965, Acad. Sci. USSR Doklady, v. 161, p. 67-69.

Postnikov, V.G., and I.Ye. Postnikova, 1964, O vozmozhnosti rifoobrazovaniya v nizhnekembriyskikh otlozheniyakh na Markovskoy razvedochnoy ploshchadi (Irkustskaya oblast'): Akad. Nauk SSSR Doklady, v. 158, no. 3, p. 605-608 (Eng. trans. by Am. Geol. Inst., 1965, Acad. Sci. USSR Doklady, v. 158, p. 57-59).

Postnikova, I.Ye., and D.D. Kotel'nikov, 1969, Novyye dannyye o proyavlenii vulkanicheskoy deyatel'nosti v vendskikh otlozheniyakh Irkutskogo amfiteatra: Akad. Nauk SSSR Doklady, v. 186, no. 5, p. 1146-1149 (Eng. trans. by Am. Geol. Inst., 1970, Acad. Sci. USSR Doklady, v. 186, No. 5, p. 83-86).

Provodnikov, L.Ya., 1965, Rel'yef fundamenta Sibirskoy platformy: Akad. Nauk SSSR Doklady, v. 165, no. 8, p. 1379-1382 (Eng. trans. by Am. Geol. Inst., 1966, Acad. Sci. USSR Doklady, v. 165, p. 99-102).

Samsonov, V.V., and L.F. Tyshchenko, 1970, O geneticheskoy svyazi gazov zakrytykh por i gazov produktivnykh plastov: Geol. Nefti Gaza, No. 8, p. 33-36.

Samsonov, V.V., et al, 1977, Epigeneticheskaya tsementatsiya terrigennykh kollektorov i osobennosti razprostraneniya bitumonidov na yuzhnom sklone Nepskogo svoda: Geol Nefti Gaza, No. 2, p. 18-25.

Sulimov, I.N., 1964, Ob analogakh Usol'skoy svity i fatsial'nostrukturnykh zonakh nizhnego kembriya v predgor'yakh

Vostochnykh Sayan: Akad. Nauk SSSR Doklady, v. 156, no. 4, p. 838-840 (Eng. trans. by Am. Geol. Inst., 1965, Acad. Sci. USSR Doklady, v. 156, p. 73-75).
Timofeyev, B.V., and L.L. Bagdasaryan, 1964, O resul'tatakh mikropaleofitologicheskogo issledovaniya neftey Vostochnoy Sibiri: Akad. Nauk SSSR Doklady, v. 154, No. 1, p. 102-103 (Eng. trans. by Am. Geol. Inst., 1964, Acad. Sci. USSR Doklady, v. 154, p. 22-24).
Trofimuk, A.A., et al, 1964, Main problems of prospecting the Markovo oil field in eastern Siberia: Geol. Nefti Gaza (Eng. trans. in "Petroleum Geology," 1969, McLean, Va., v. 8, No. 1, p. 13-18).
Vasil'yev, V.G., ed., 1968a, Geologiya nefti, Tom 2, Kniga 1, Neftyanye mestorozhdeniya SSSR: Moscow, Izd. Nedra, p. 593-600.
Vasil'yev, V.G., ed., 1968b, Gazovye mestorozhdeniya SSSR: Moscow, Izd. Nedra, p. 377-382.
Vasil'yev, V.G., and I.P. Zhabrev, 1975, Gazovyye i gasokondensatnyye mestorozhdeniya, spravochnik: Moscow, Izd. Nedra, 527 p.
Vassoyevich, N.B., et al, 1970, K probleme neftegazonosnosti pozdnedokembriyskikh otlozheniy: Sovet. Geol., No. 4, p. 66-79 (Eng. trans. by Am. Geol. Inst., 1971, Internatl. Geol. Review, v. 13, no. 3, p. 407-418).
Vysotskiy, I.V., Ye.P. Larchenkov, and B.A. Sokolov, 1978, Perspektivy gazonosnosti yugo-vostochnoy chasti Leno-Vilyuyskoyo basseyna: Geol. Nefti Gaza, No. 5, p. 31-35.
Weaver, P., 1962, Challenge to Cambrian prospecting: AAPG Bull., v. 46, no. 10, p. 1941-1943.
Yanshin, A.L., 1962, Perspektivy otkrytiya mestorozhdeniy kaliynykh soley na territorii Sibiri: Akad. Nauk SSSR, Sibir. Otdel., Geol. i Geofiz, No. 10, p. 3-22 (Eng. trans. by Am. Geol. Inst., 1964, Internatl. Geol. Review, v. 6, no. 12, p. 2132-2147).
Zharkov, M.A., 1969, Ob ob'yemakh solenakopleniya v kembriyskuyu epokhu: Akad. Nauk SSSR Doklady, v. 184, no. 4, p. 913-914 (Eng. trans. by Am. Geol. Inst., 1969, Acad. Sci. USSR Doklady, v. 184, p. 72-74)
Zharkov, M.A., E.I. Chechel', and I.M. Knyazev, 1963, Kembriyskiye otlozheniya srednego i nizhnego techeniya reki Kirengi: Akad. Nauk SSSR Doklady, v. 149, no. 4, p. 922-924 (Eng. trans. by Am. Geol. Inst., 1965, Acad. Sci. USSR Doklady, v. 149, p. 65-67).
Zharkova, T.M., 1965, Karnallit v kamennoy soli kembriyskikh otlozheniy Sibirskoy platformy: Akad. Sci. SSSR Doklady, v. 164, no. 1, p. 177-178 (Eng. trans. by Am. Geol. Inst., 1966, Acad. Sci. USSR Doklady, v. 164, p. 144-145).
Zhuravleva, Z.A., et al, 1966, K stratigrafii Ushakovskoy svity Irkutskogo amfiteatra: Akad. Nauk SSSR Doklady, v. 166, no. 3, p. 678-680 (Eng. trans. by Am. Geol. Inst., 1966, Acad. Sci. USSR Doklady, v. 166, p. 53-55).
Zolotov, A.N., et al, 1968, Stroyeniye gazokondensatnoy zalezhi Parfenovskogo gorizonta Markovskogo mestorozhdeniya: Geol. Nefti Gaza, No. 6, p. 26-30.

Eugene Island Block 330 Field, Offshore Louisiana[1]

By D.S. Holland, W.E. Nunan, D.R. Lammlein, and R.L. Woodhams[2]

Abstract The Eugene Island Block 330 field is currently the largest oil-producing field in federally owned waters of the U.S. outer continental shelf. Located about 170 mi (272 km) southwest of New Orleans, the field was discovered by the Pennzoil 1 OCS G-2115 well in March 1971, after leasing on December 15, 1970. The field includes parts of blocks 313, 314, 330, 331, 332, 337, and 338, Eugene Island Area, South Addition, offshore Lousiana.

The field is an anticlinal structure on the downthrown side of a large northwest-trending growth fault. Production is from more than 25 Pliocene-Pleistocene delta-front sandstone reservoirs ranging from *Lenticulina* to *Trimosina "A"* zones and located at depths of 4,300 to 12,000 ft (1,290 to 3,600 m). The reservoir energy results from a combination water-drive and gas-expansion system. Recoverable reserves are estimated to be greater than 225 million bbl of liquid hydrocarbons and 950 Bcf of gas.

Considerable subsurface data provided by 220 exploration and development wells and several seismic grids form the basis for interpretation of the geology and geophysics of the Block 330 field and its producing zones.

INTRODUCTION

The Eugene Island Block 330 field is currently the largest oil producing field in Federal outer continental shelf waters. The field is located in the Gulf of Mexico offshore Louisiana about 170 mi (272 km) southwest of New Orleans (Fig. 1). It includes seven 5,000-acre (2,024 ha.) blocks (Fig. 2) in the west-central part of the Eugene Island Area, South Addition. Average daily production in December 1978 was 61,000 bbl of oil, 9,000 bbl of condensate (Fig. 3), and 400 MMcf of gas (Fig. 4). Cummulative production through September 1978 was 150 million bbl of oil and condensate and 570 Bcf of gas. Proven recoverable reserves are estimated to be over 225 million bbls of oil and condensate and over 950 Bcf of gas.

The field geology is typical of many Gulf of Mexico oil fields. The structure is a rollover anticline formed on the downthrown side of a large, northwest-trending, salt-diapir related growth fault. This growth fault and its associated faults are developed in the delta-front facies of delta lobes deposited on the shelf during the Pliocene and Pleistocene. Production of hydrocarbons is from more than 25 sandstone reservoirs located at depths from 4,300 ft (1,311 m) to 12,000 ft (3,658 m) subsea.

There are several outstanding features of this field in addition to the presence of large reserves. Production rates for several of the reservoirs are high with maximum single well rates of 2,915 b/d and 21,552 Mcf/day. Thick hydrocarbon columns are present in many of the reservoirs with some columns in ex-

[1]Manuscript received, September 10, 1979; accepted for publication, October 31, 1979.

[2]Pennzoil Exploration and Production Company, Houston, Texas 77001.

The writers express their appreciation to all the exploration and development personnel at Pennzoil Company who have contributed to the current understanding of this field. Those contributors are too numerous to mention here, but a few who have contributed to this paper include R.L. Lewis, P.A. Smith, S.E. Tripp, and T.W. Borawski. Steve Brown and Gary Ruth of Geochem Laboratories assisted in the geochemical analysis of the shale and hydrocarbons.

Article Identification Number:
0065-731X/80/M030-0012/$03.00/0.

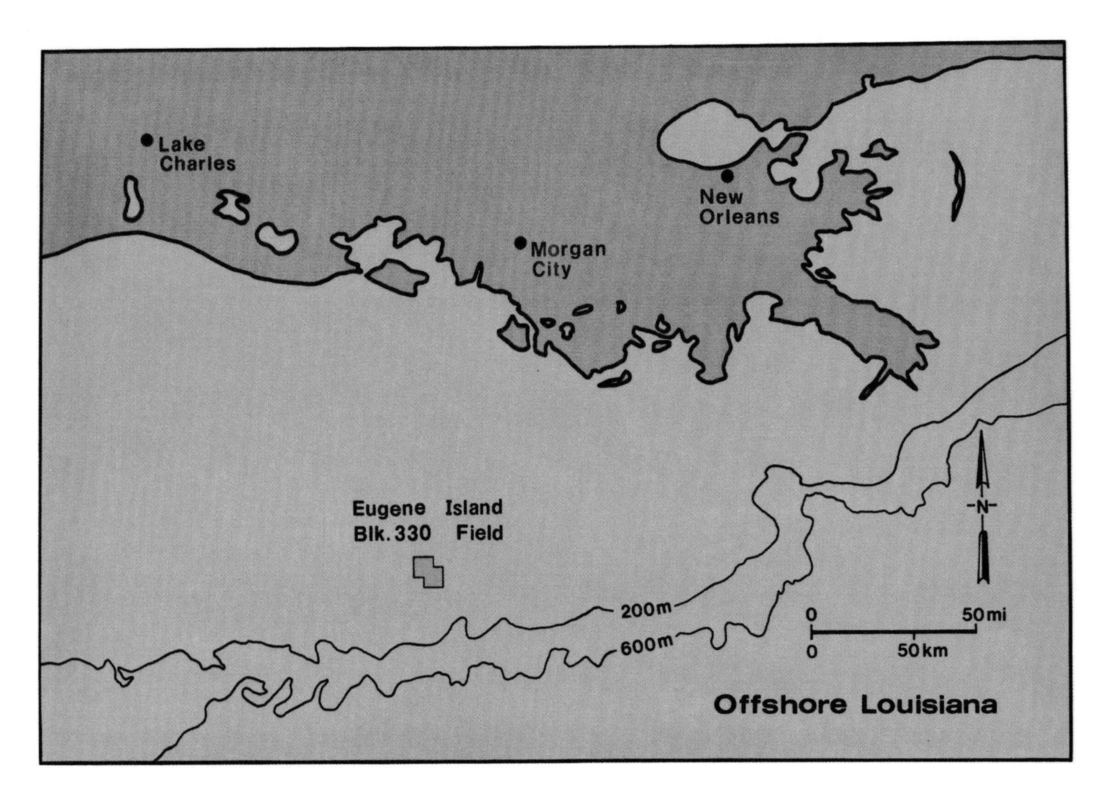

FIG. 1—Map showing the location of the Eugene Island Block 330 field.

cess of 1,000 ft (305 m). The amount of productive acreage from several of the sandstones is large with that of the JD sandstone in excess of 6,450 acres (2,611 ha.) Most of the producing zones, and over 90% of the reserves occur within an abnormal pressure regime.

Another interesting aspect of this field is the very young age of the reservoir rocks. All reservoirs are less than 2.5 m.y. and all but one range between 1.8 and 0.5 m.y. old. The mechanism by which this large volume of hydrocarbons accumulated in the very young reservoir rocks still is largely undetermined. This study describes the geologic setting under which the reserves exist, initiates an explanation for their existence, and discusses geological, petrophysical, and geophysical parameters of the field.

EXPLORATION AND DEVELOPMENT HISTORY

Evaluation of the study area began in the late 1960s as part of the industry's preparation for the 1970 federal lease sale. Prior to that time a few wells drilled north of the 330 field area had established shallow gas production in Pleistocene sandstones.

Reconnaissance seismic data consisting of 6-, 12-, and 24-fold CDP data, acquired and processed during the period 1966 to 1970, delineated growth faults and associated anticlines in the field area. These structures were similar to producing structures elsewhere in the Gulf of Mexico, and are well-defined in the 24-fold CDP seismic line shown in Figure 5. Reconnaissance maps, such as shown in Figure 6, were interpreted from these seismic data and delineate the major structural features. Several anticlinal culminations, separated by shallow synclines, occur downthrown to the major growth faulting and dominate the mini-depositional basin in the 330 field area.

In addition, the lithofacies and isopachous trends defined by well data (Norwood and Holland, 1974) indicated that large delta systems were present in the vicinity. The probability of thick sand accumulations and a structural closure on a rollover anticline on the downthrown side of a large growth fault highlighted the hydrocarbon potental of the area.

The seven blocks that comprise the field were purchased in three Federal OCS lease sales. Blocks 314, 330, 331, and 338 were acquired by various companies (Fig. 2) in 1970. Blocks 313 and 332 were purchased in 1974, and block 337 was leased in 1976.

Exploratory drilling began with the simultaneous drilling of the Pennzoil 1 well in block 330 and the Shell 1 well in block 331 (Fig. 6). The Pennzoil well reached a total depth of 6,300 ft (1,920 m) and logged 55 ft (16.8 m) of net oil and 55 ft (16.8 m) of net gas. The Shell well discovered 92 ft (28 m) of net oil and 106 ft (32.3 m) of net gas. These two wells established a cummulative hydrocarbon column of more than 1,500 ft (458 m) and indicated the discovery of a major oil and gas field.

As of January 1, 1979, nine production platforms have been set in the field and 192 development wells drilled. With the addition of the 26 exploratory wells, the total number of wells drilled to date is 218 (Fig. 7).

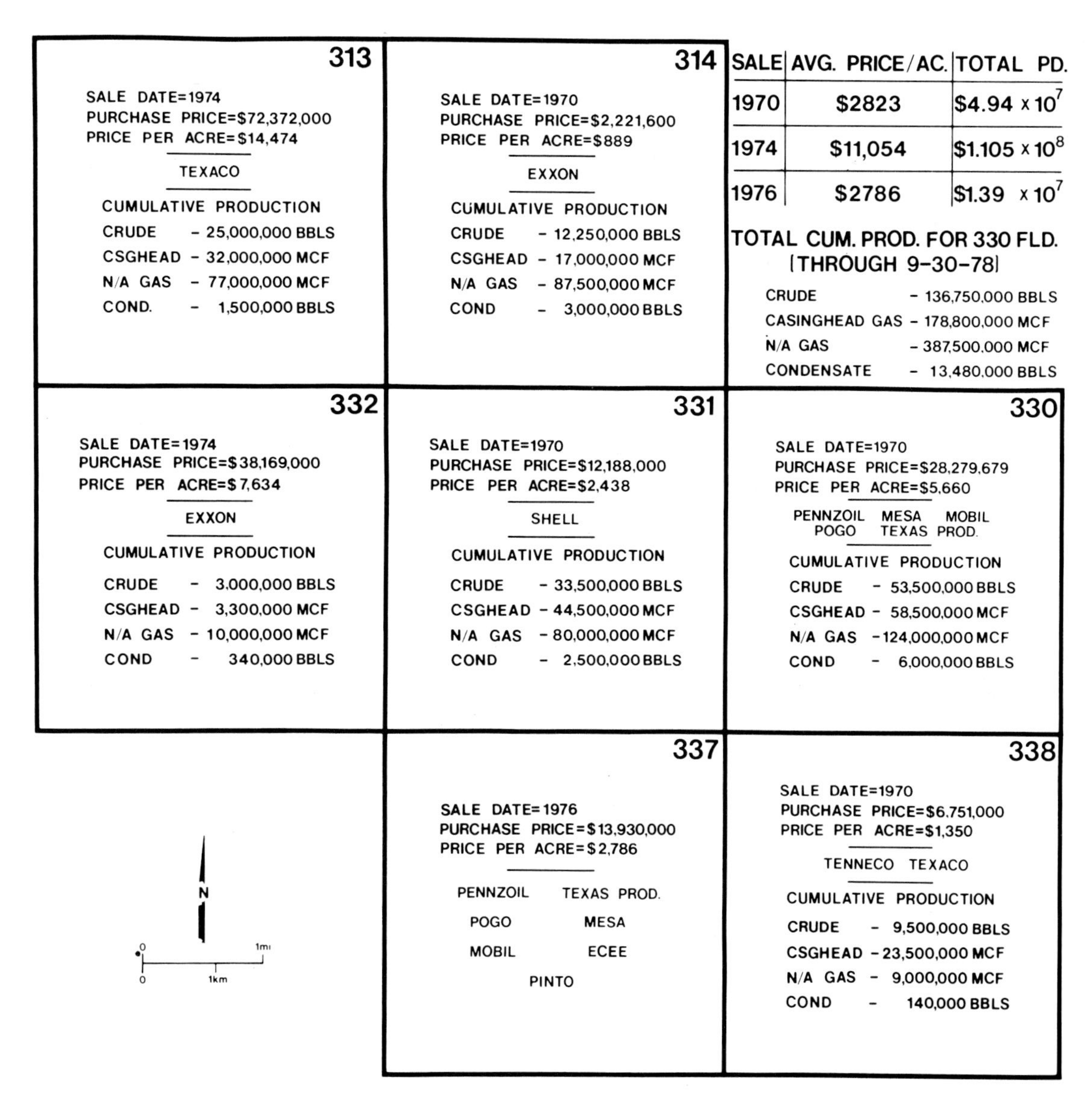

FIG. 2—Acquisition and production data from the Eugene Island Block 330 field.

GEOLOGIC SETTING

The scope of this paper does not permit a detailed treatment of the geology of the area. Those interested in details of the geology should refer to recent papers by Uchupi (1975), Martin and Case (1975), Martin and Bouma (1978), and Martin (1978).

Physiography

The study area is in the eastern part of the Texas-Louisiana Shelf (Fig. 8) of the Continental Shelf physiographic province of the Gulf of Mexico as defined by Martin and Bouma (1978). This part of the shelf is the submerged, seaward extension of the Gulf Coastal Plain and extends southwestward from the present shoreline to the approximate location of the 200-m isobath. In an east-west direction the shelf extends from the Mississippi River delta to the Rio Grande River. The shelf width ranges from 62 mi (100 km) to 125 mi (200 km).

The shelf has a gentle seaward slope of less than 1° over most of its extent. The surface generally is smooth with low relief irregularities resulting from the presence of relict Pleistocene stream and shoreline deposits, fault scarps formed by active growth faults, and sub-circular mounds produced by active salt and shale diapirism (Martin and Bouma, 1978).

The 100-m to 200-m bathymetric contours on the shelf show a prominent seaward deflection on the Louisiana Shelf. This pattern reflects the presence of major upper Tertiary depocenters on that part of the continental shelf (Woodbury, et al, 1973), the most recent of which formed during the Pliocene and Pleistocene in response to the sea-level fluctuations that accompanied glaciation.

The Eugene Island Block 330 field is close to the seaward edge of the Louisiana Shelf. Detailed bathymetric maps of the field area show water depths of 210 to 266 ft (64 to 81 m) and

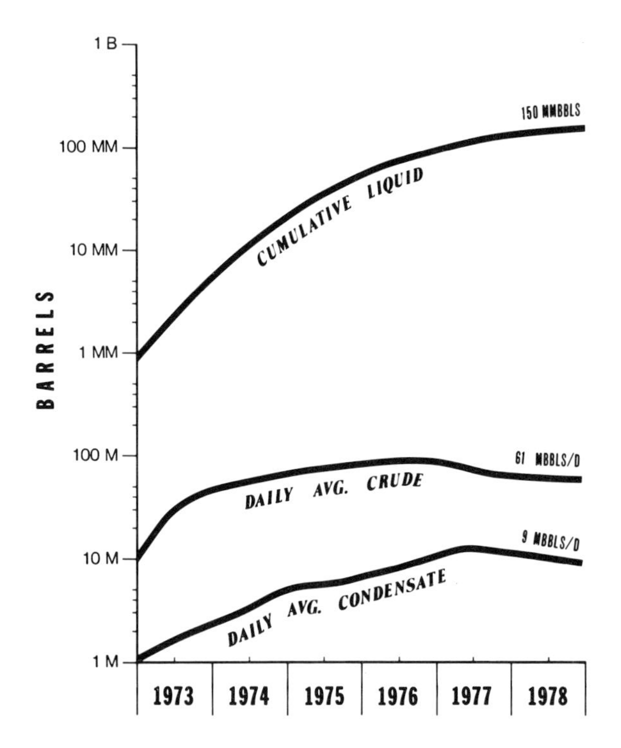

FIG. 3—Liquid hydrocarbon production history, Eugene Island Block 330 field.

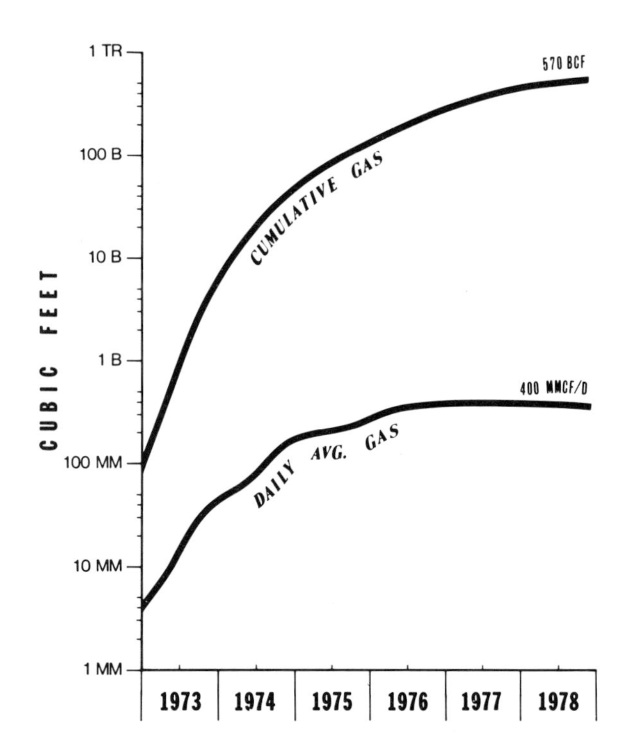

FIG. 4—Gas production history, Eugene Island Block 330 field.

the shelf's surface to be gently sloping to the south and smooth over most of the field area. A single mound-shaped feature rises to about a 138 ft (42 m) depth along the boundary between blocks 313 and 332, approximately 0.5 mi (0.8 km) west of the blocks' eastern borders.

Tectonics

The field is located in the Gulf Coast basin (Fig. 8). This basin extends from the updip pinchout of Coastal Plain sediments onshore United States southward to the Sigsbee Escarpment. The eastern boundary of the basin is approximately coincident with the axis of the Florida Peninsula and the western boundary of the basin is the East Mexico Shelf in northern Mexico. The basin is divided into an eastern carbonate province and a western siliciclastic province, the boundary between the two is approximately coincident with a line extending from the Alabama-Mississippi border onshore to the Florida Escarpment offshore west Florida.

The Gulf Coast basin is rimmed by exposed areas of the Ouachita and Appalachian orogenic belts. These systems extend for an unknown distance beneath the onlapping basin sediments (Fig. 9).

Graben systems and normal faults occur in a semicircular zone that extends from the Rio Grande River through Texas, south Arkansas, and central Mississippi to Mobile Bay in southwest Alabama (Fig. 8). Another system of high-angle faults parallels the present coastline from south Texas to southeast Louisiana.

The basin axis lies a few miles offshore from the present shoreline (Fig. 8). Several small basins are present in the province and are separated by broad uplifts such as the Sabine and Monroe uplifts.

Salt tectonism is the most important structural process active in the siliciclastic part of the Gulf Coast basin. Deformation related to salt movement has been in progress since the Cretaceous (Martin, 1978) and continues today (Humphris, 1978). Salt occurs in relatively small basins in east Texas, north Louisiana, south Mississippi, and in a large area including the southern Gulf Coastal Plain, most of the Texas-Louisiana Shelf, and the Texas-Louisiana Slope (Fig. 8). Woodbury et al (1973) subdivided the large area into three structural provinces based on the horizontal cross-sectional area of salt bodies at a depth of 12,000 ft (3,658 m) subsea (Fig. 10). These are a northern province characterized by small isolated salt diapirs with small areal extents, a central province containing large but discrete salt domes with cross-sectional areas of 25 to 30 sq mi (40 to 48 sq km) at −12,000 ft (−3,658 m), and a southern province of semicontinuous diapiric uplifts with cross-sectional areas of over 100 sq mi (161 sq km) at −12,000 ft (−3,658 m). The Eugene Island Block 330 field is on the northern edge of the southern province (Figs. 9, 10).

Salt tectonism in the northern Gulf of Mexico area has been closely related to deposition since the Cretaceous. Deposition of post-salt sediments has played a prominent role in the mobilization of the salt, growth of salt diapirs, and the development of associated faults. Salt withdrawal areas have become locals of sediment transport and the development of mini-basins, and many diapir-associated faults have developed into growth faults with thick sediment piles accumulating on their

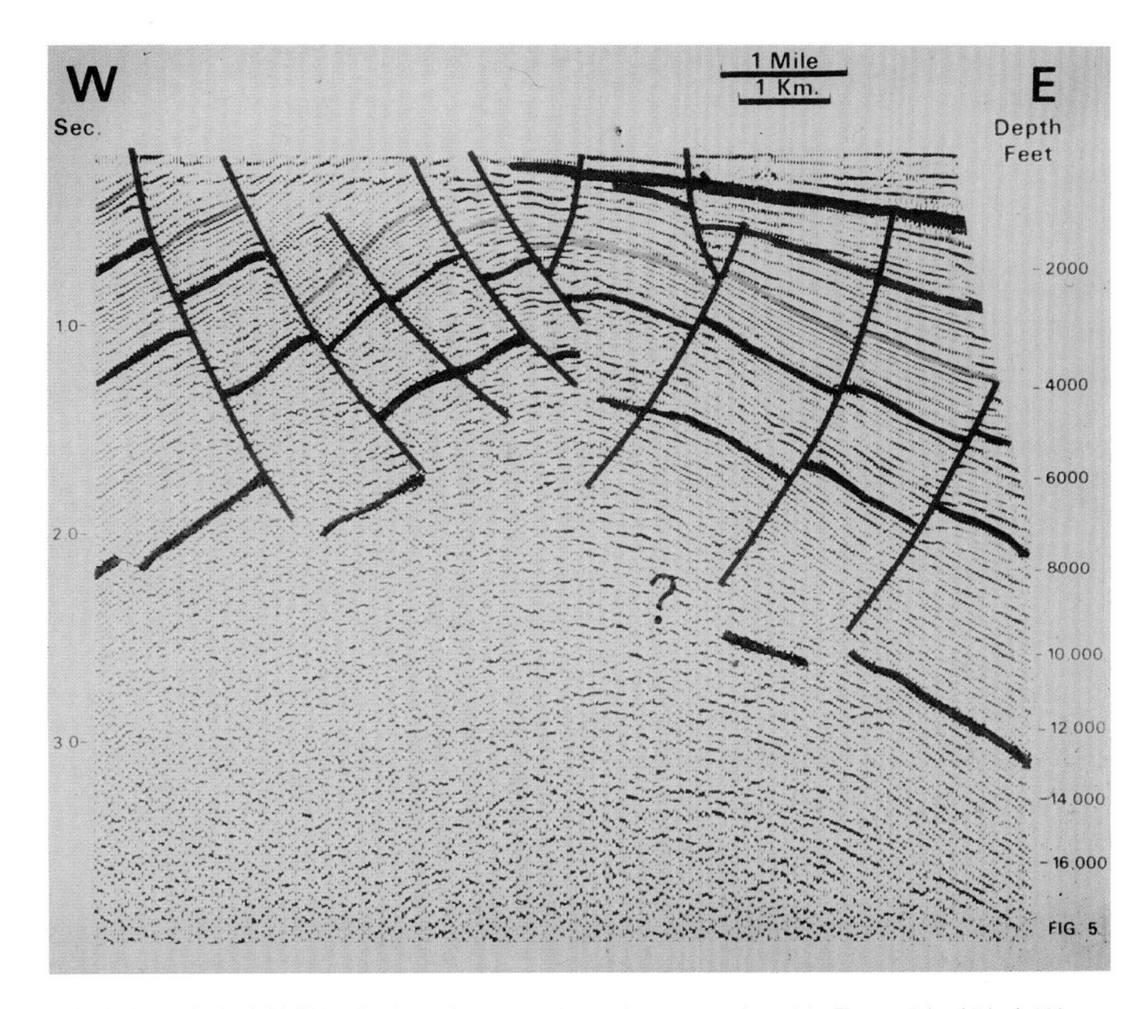

FIG. 5—Part of a 24-fold CDP seismic section across the southeastern portion of the Eugene Island Block 330 field. The location of the seismic section is shown in Figure 6. Bold numbers separated by vertical lines immediately above the section are block numbers. The interpreted horizons are the major oil and gas reservoir sandstone units.

downthrown sides (Woodbury et al, 1973).

Stratigraphy

Rocks deposited in the Gulf Coast basin are entirely Mesozoic and Cenozoic in age. These rocks occur in offlapping wedges that become younger towards the south. In the Gulf Coast basin sedimentation began with the emplacement of Triassic red beds and diabases in graben systems of restricted areal extent. This was followed by deposition of thick Triassic-Jurassic salt throughout most of the basin. Middle and Upper Jurassic rocks include red beds, platform carbonates (Smackover), evaporites, and deltaic siliciclastics. Lower Cretaceous carbonates developed in the western areas of the basin while siliclastic sediments accumlated in the east, and deep water conditions prevailed in the central Gulf (Garrison and Martin, 1973). Upper Cretaceous rocks include shale and deepwater chalks.

Tertiary rocks are predominantly siliciclastic, being deposited in deltaic, strand plain, and deepwater environments. The thickest accumulations of lower Tertiary rocks developed in the Rio Grande Embayment. Miocene deposits are thickest under the Louisiana coast, with Pliocene depocenters being concentrated under the central shelf and Pleistocene depocenters under the outer shelf and upper slope (Martin, 1978).

It has long been recognized that Cenozoic clastic sediments in the Gulf Coast area occur in belts consisting of a landward fluvial-deltaic facies, a delta-front neritic facies, and prodelta-bathyal facies. The delta-front facies are sand-rich with sand percentages ranging from 15 to 30% (Martin, 1978). Norwood and Holland (1974) recognized this alternating sand-shale facies in Pleistocene rocks in the Gulf of Mexico and drew attention to the high percentage of Pleistocene reservoired oil in these rocks. The Eugene Island Block 330 field is within the delta-front facies of a Pleistocene delta complex in the major Pleistocene depocenter.

FIELD GEOLOGY

Stratigraphy and Sandstone Distribution

Prodution is from ten sandstone units that range in age from late Pliocene (*Lenticulina* 1) to late Pleistocene (*Trimosina A*). Most of the production is from the eight Pleistocene sandstones shown in Figure 11. Each of these sandstone units is sub-

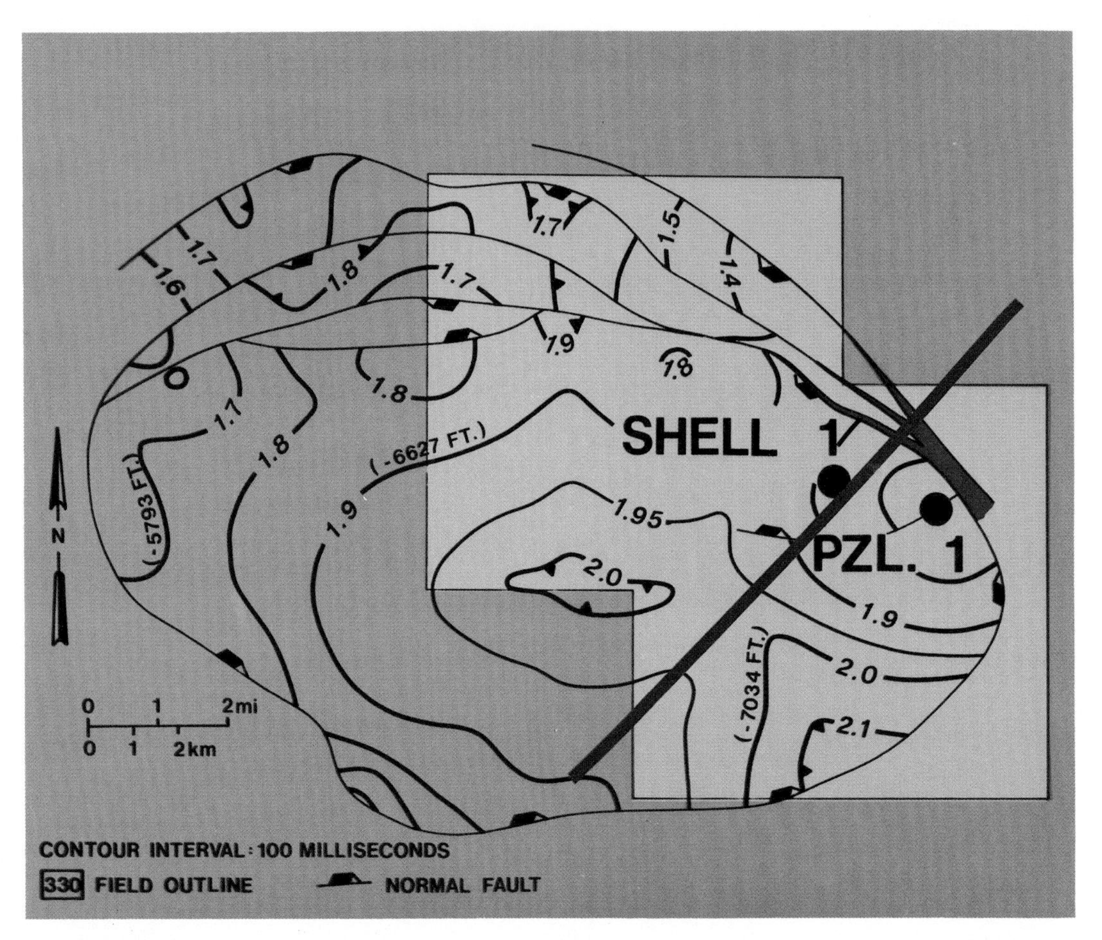

FIG. 6—Seismic structure map on the top of the "JD" sandstone, Eugene Island Block 330 field and vicinity, with the Pennzoil #1 and Shell #1 exploratory wells located. Heavy black line is seismic line 129 shown in Figure 5.

divided into two or more individual beds. For example, the GA sandstone unit consists of the GA-1, 2, 3, 4, and 5 sandstones, with production coming from the GA-1, 2, and 5.

The 25 producing zones occur in a 6,500-ft (1,981-m) thick interval of which about 20% is sandstone.

Several lines of evidence indicate that the Pleistocene pay zones are mostly delta-front sandstones. Regional lithofacies mapping (Norwood and Holland, 1974) demonstrates that the Eugene Island 330 field area is within the delta-front facies of late Pliocene and Pleistocene seaward deltaic progradations.

Seismic facies analysis also indicates a delta-front depositional mode. For example, a stratigraphic interpretation of the seismic line in Figure 5 is shown in Figure 12 with more than a dozen of the major seismic sequence boundaries indicated. Additional sequences may be identified but have been omitted for purposes of illustration.

The predominant seismic facies patterns present are the complex sigmoid-oblique and the shingled configurations as defined by Mitchum et al (1977). Both reflection configurations are interpreted as strata in which significant deposition is due to lateral outbuilding or prograding, such as occurs in the delta-front depositional environment.

The complex sigmoid-oblique facies pattern is characterized by both sigmoid and oblique progradational reflection configurations. The complex alternation of nearly horizontal sigmoid topset reflections and segments of oblique reflections implies strata deposited in a depositional regime of alternating upbuilding and depositional bypass in the topset segment.

When viewed parallel with the depositional dip as is the case for many of the sequences shown in Figure 12, the prominent sigmoid and oblique reflection configurations are readily apparent. Depositional dips range from 5 to 15°. The oblique reflection segments terminate from toplap at or near the upper surface and by downlap at the base. Although not shown, reflections from small channels also occur within these sequences and are best observed on depositional strike sections.

The topset reflections are interpreted as corresponding to a delta-plain environment. Usually, the seismic cycle width and amplitude are greatest in the upper foreset zone and decrease toward the bottomset zone interpreted as corresponding to a transition from delta-front sands to prodelta muds.

The shingled progradational reflection configuration are thin units with gently-dipping (5 ° or less) parallel, oblique reflectors that terminate by apparent toplap and downlap against par-

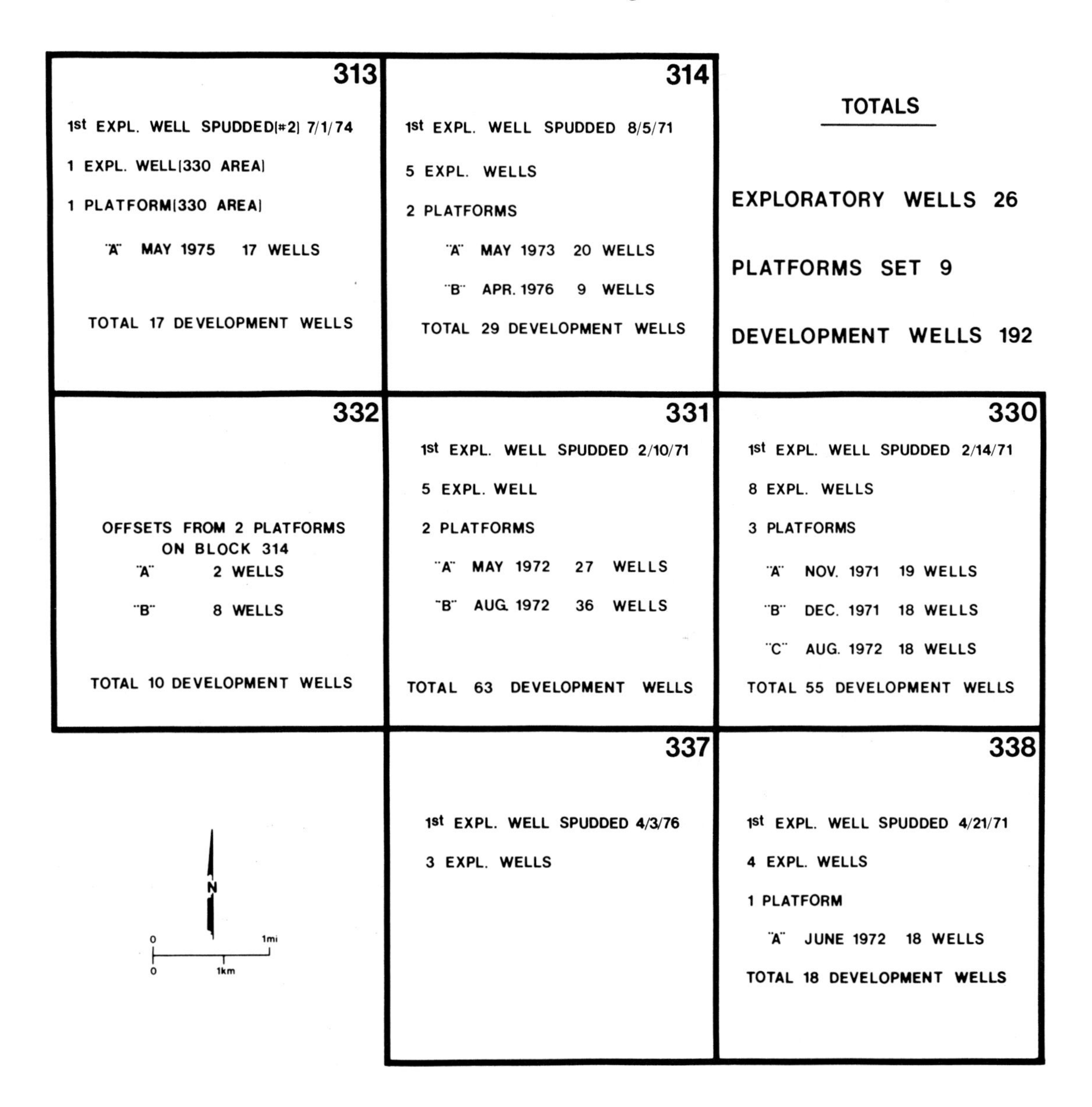

FIG. 7—Exploration and development drilling history of the Eugene Island Block 330 field.

allel upper and lower boundaries, respectively. Within each unit, the successive oblique reflectors show little and occasionally no overlap with each other. For display purposes many of the boundaries of the thin shingled sequences have not been interpreted in Figure 12.

The shingled reflectors (Fig. 12) appear as a series of semicontinuous or slightly offset events whose obliquity is barely resolved by the seismic data because the thickness of individual units and the separations between units are commonly about 50 ft (15 m) or less. This effect becomes more pronounced as reflection time or depth increases because of the general increase of seismic wavelength with depth.

The gently-sloping shingled seismic configurations are interpreted as representing delta-front depositional surfaces that prograded into shallow water. Individual shingled reflectors defined by seismic (Figs. 12, 26) and well control data are crescent or fan shaped (Fig. 28). Dipmeter data (Fig. 27) indicate current dip patterns that are characteristic of foreset deposition (Gilreath and Maricelli, 1964), with this same geometry. The fine grain size of the 330 field sandstones is also consistent with deposition on a subaqueous delta plain (Coleman, 1976).

Seismic data from the Eugene Island Block 330 area reveal that prominent hydrocarbon indicators are present and are associated with the field's oil and gas zones. For example, a relative amplitude processed version of the seismic line in Figure 5 is shown in Figure 13. These data are typical in that seismic hydrocarbon indicators are associated with each of the oil and gas reservoirs.

The most prominent hydrocarbon indicators are the relatively high amplitude reflections (bright spots) and the horizontal reflections from fluid interfaces (flat spots). These are shown in Figure 13 for each of the field's major oil and gas producing reservoirs from the HB through the OI. The prominent high-amplitude flat spot at a time of about 1.5 sec is a re-

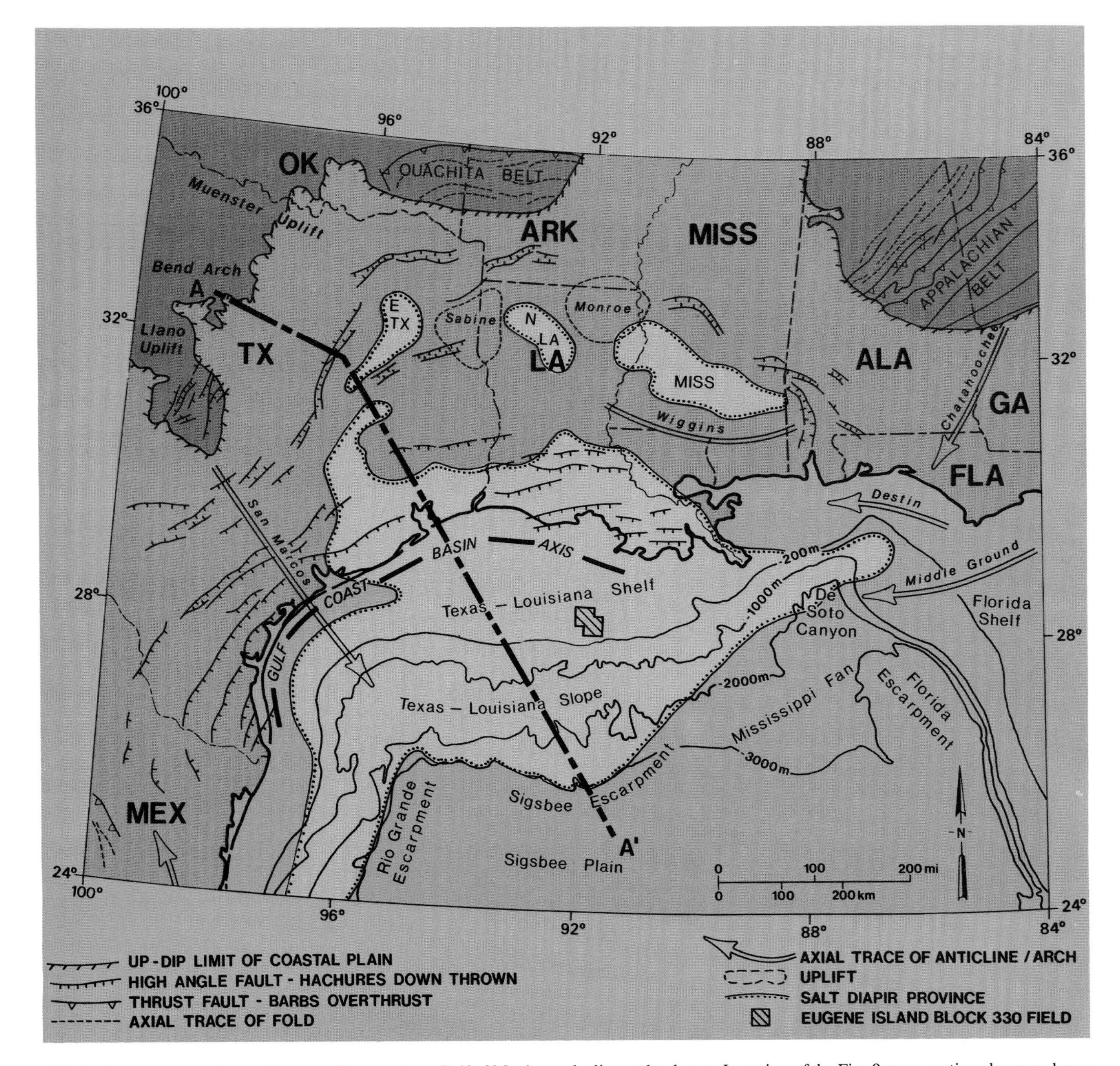

FIG. 8—Structural-physiographic map of the northern Gulf of Mexico and adjacent land area. Location of the Fig. 9 cross section shown as heavy dashed line between A and A′. Modified after Martin (1978).

flection from the oil-water contact within the HB reservoir. Other hydrocarbon indicators observed include reflection time sag due to lateral velocity variations, apparent frequency broadening and shadow zones beneath bright spots. Hydrocarbon indicators continue to be an important tool for identifying and delineating the oil and gas reservoirs in the 330 field.

Comparison of the seismic stratigraphic (Fig. 5, 12, 27), seismic amplitude (Fig. 13) and well (Fig. 28) data reveals that the seismic hydrocarbon indicators and the oil and gas reserves are associated with the delta-front sandstones defined by the shingled and sigmoid seismic reflections discussed previously. The bright spots and other hydrocarbon indicators present in the seismic data can be explained on the basis of the low velocity and density (acoustic impedence) values (Domenico, 1974, 1976) of the hydrocarbon sandstones relative to those of water sandstones and shales as observed in well data for the field.

The sandstones are thickest on the downthrown side of the major growth fault (Figs. 5, 12) where their accumulation was controlled by concomitant structural growth. Several of the sandstones are absent (Fig. 14) or are thin (Fig. 15) over the

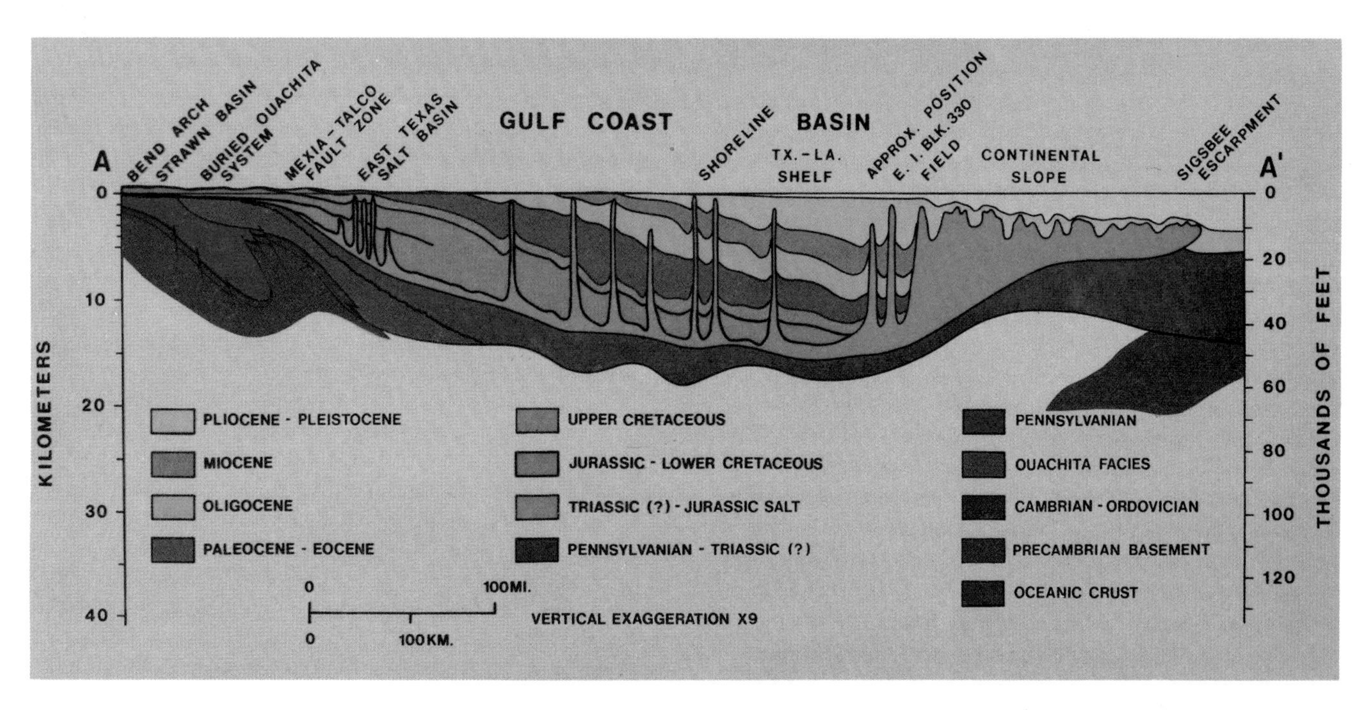

FIG. 9—Diagrammatic cross section of northern Gulf of Mexico margin. Location shown in Figure 8. Modified after Martin (1978).

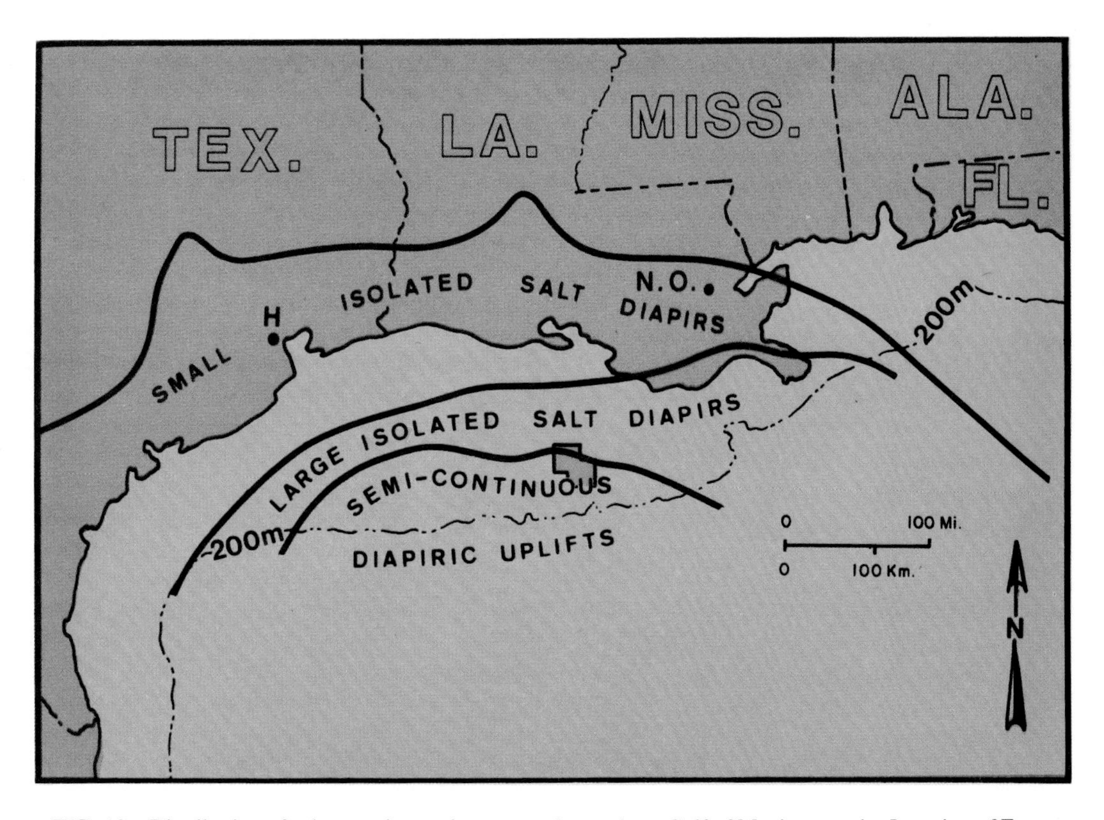

FIG. 10—Distribution of salt tectonic provinces near the northern Gulf of Mexico margin. Location of Eugene Island Block 330 field is indicated by yellow block.

crest of the rollover structure, but thicken away from the crest both into the growth fault and into the mini-basin (Figs. 5, 14, 15).

Most sandstones are replaced by shales basinward of the rollover structures. For example, only thin sandstone layers have been encountered in the three wells drilled in block 337. The nearly parallel, low-amplitude, seismic reflectors present basinward of the relatively high-amplitude, progradational units probably result from a basinward increase in mud content in laterally extensive beds.

Petrography

The producing sandstones are all either quartzose wackes, arkosic wackes, quartzose arenites, or arkosic arenites (Table 1). They consist of coarse silt to medium sand-sized clasts that are angular to well rounded. Most of the sandstones fall into the very fine to fine arenite range and are poorly to moderately sorted, but a few are well sorted (JD sandstone).

Framework grains comprise between 35 and 80% of the rocks, the remainder being divided between intergranular bonding material and pores. Of the framework, quartz accounts for 25 to 50% with feldspars comprising between 10 and 35%. Orthoclase is the most common feldspar (3 to 20%) followed by plagioclase (2 to 20%), microcline (1 to 7%), and perthite (Fig. 16). Lithic fragments amount to about 6% of the framework and include schist, quartzite, basic and acidic intrusive and extrusive igneous rocks, and trace amounts of sedimentary rock. Framework mica is mostly muscovite although some detrital chlorite is present. Zircon is the only common heavy mineral. Fragments of bivalves (Fig. 17), gastropods, and foraminifers occur in all the sandstones and account for up to 2% of the framework. Organic debris (kerogen) is present in many of the rocks and occurs in significant amounts (30%) in the JD sandstone (Fig. 18).

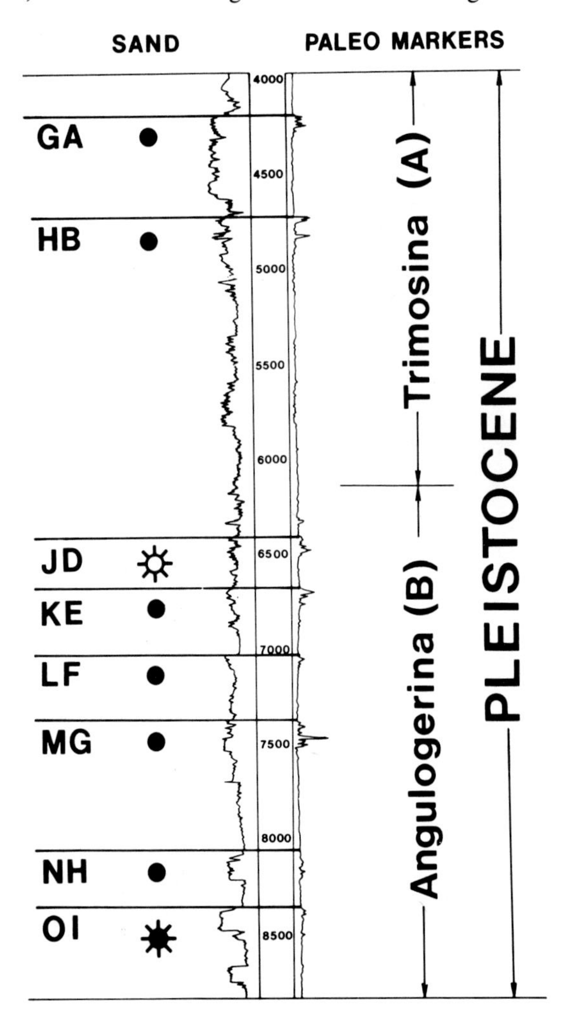

FIG. 11—Composite type log, Eugene Island Block 330 field. Solid circles indicate oil producers and open spoked circles indicate gas producers. Vertical scale in feet. Spontaneous potential curve on the left and resistivity curve on the right.

Framework grains show a variety of diagenetic phenomena. Relatively unaltered and intensely weathered feldspars (Fig. 19) are present. Some quartz grains have syntaxial overgrowths (Fig. 20) but pressure solution phenomena are not observed. Fragile grains such as micas (Fig. 21) and foram tests are seldom deformed.

Intergranular bonding material comprises 20 to 40% of the sandstones and consist of detrital mud matrix (Fig. 16), micrite (Fig. 22), syntaxial quartz cement (Fig. 20), and authigenic chlorite cement (Fig. 23).

Porosity is almost exclusively primary intergranular (Fig. 24) or reduced primary intergranular. Some secondary moldic porosity has developed where feldspars have been partially dissolved during diagenesis.

Sedimentary structures observed in thin section include fine lamination of very fine sand and medium-grained organic debris (Fig. 16), arenite-shale laminae, bioturbated laminated arenite and shale (Fig. 25). Graded bedding and cross-bedding were not observed. although they may be present at scales not discernable in a sidewall core thin section.

Structure

The field is part of a structural and depositional basin located on the downthrown side of a major growth fault system (Figs. 5, 6) that developed in the Pliocene-Pleistocene delta front facies. Salt diapirism is present at depth and was an active agent in localization and development of the growth-fault system. Salt domes occur at several places around the periphery of the fault-bounded area shown in Figure 6. Some of the western and southern parts of the boundary fault system are faults associated with the diapiric intrusion of those domes.

Anticlines are located in the northern part of the field area. These structures were formed by rollover of the deltaic sediments into the growth fault. The growth-fault system is complex and contains several down-to-the-south normal faults and a few down-to-the-north antithetic normal faults (Figs. 5, 6, 34). These faults effectively break the reservoir facies of the producing sandstones into separate reservoirs contained within separate faults. Fault growth was active throughout late Pliocene and Pleistocene time. Growth of the two anticlines included in this study has not been parallel. Each structure has its individual history, although both have followed the same mode

Table 1. Composition of major sandstones in Eugene Island Block 330.

Sandstone	GA-2	JD	LF	OI
Rock component				
Framework (% of whole rock)	75-80	35-70	35-80	60-75
Quartz (% of framework)	50	40-55	22-45	36-46
Feldspar (total) (% of framework)	30-35	15-35	10-24	9-20
Plagioclase	5-15	15-20	2-6	2-8
Microcline	2-7	2	2	2-3
Orthoclase	12-20	20-30	3-10	5-11
Lithic (total) (% of framework)	2-6	2-4	2-4	2-4
Metamorphic	1-3	1-3	1-2	1-4
Intrusive	<1	1-2	<1	1-2
Extrusive	2-4	1-2	TR-4	TR-3
Sedimentary	—	—	TR-1	TR-2
Heavy Minerals (% of framework)	1-2	TR-1	1-2	TR-3
Micas (% of framework)	TR-1	TR	1-2	TR-2
Fossils (% of framework)	3-6	2-3	3-6	2-4
Grain size	F-M Sd	C St-M Sd	C St-M Sd	VF-Md Sd
Roundness	A-SR	A-SA	A-SR	A-WR
Bonding matter (% of whole rock)	20-25	30-65	20-65	25-40
Mud Matrix (% of bonding matter)	16-23	26-59	15-26	23-35
Authigenic cement (% of bonding matter)	2-4	4-6	7-14	2-10

F: fine, M: medium, C: coarse, V: very, Sd: sand, St: silt, A: angular, R: rounded, S: sub, W: well.

of development. The eastern structure centered in block 330 had its most intense growth during *Angulogerina "B"* time with growth diminishing during *Trimosina "A"* and post *Trimosina* time. The northwestern structure apparently experienced some of its strongest growth late in its history.

Reservoirs

Eight of the 10 reservoir sandstone units are predominantly oil producing. The JD sandstone produces gas, and the OI sandstone is both a major oil and gas producer. The productive area ranges from a maximum of about 6,500 acres (2,632 ha.) in the JD sandstone to less than 200 acres (81 ha.) in some of the minor producing sandstones. Most of the reservoir sandstones are cut by faults (Figs. 5, 6, 31, 32) which break them into several distinct reservoirs that are identified by their respective fault blocks, such as the OI fault block A reservoir (Fig. 34).

Reservoir parameters of seven of the major oil and gas zones productive in Pennzoil's block 330 are listed in Table 2. Gross sandstone units range in thickness from 55 ft to 400 ft (16.8 to 122 m) and average net pay intervals range from 27 to 56 ft (8.2 to 17.1 m). Mean porosity values cluster around 30% and range between 25 and 35%. Permeabilities are quite variable ranging from 10 millidarcies to over 6 darcies, however more than 80% of the pay intervals have permeabilities over 100 millidarcies, and about 20% have permeabilities in excess of 1 darcy. Water saturation ranges from about 20% to over 40%, with average values mostly in the 30 to 40% bracket. Recovery

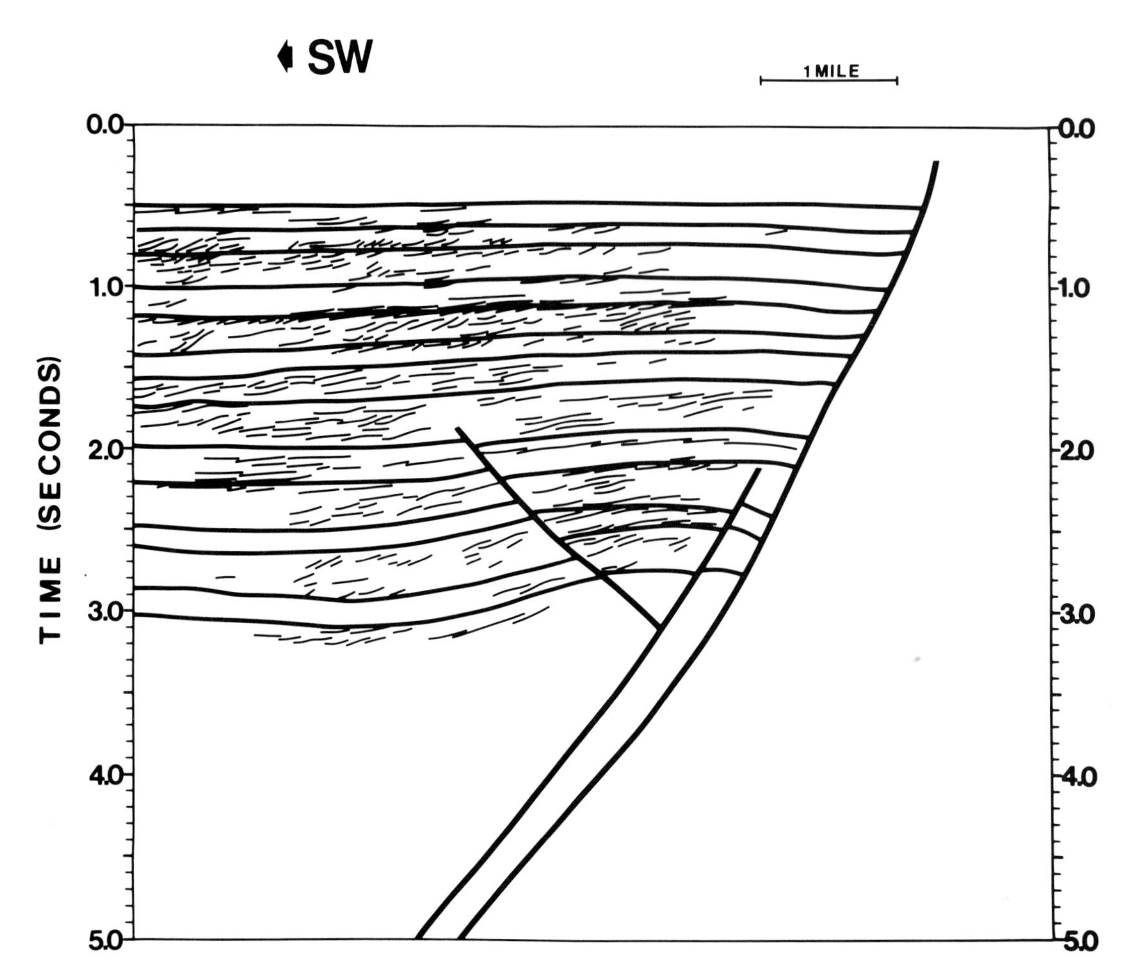

FIG. 12—Stratigraphic interpretation of the seismic line shown in Figure 5. The seismic sequence boundaries are defined by the truncation of oblique shingled and sigmoid reflections and seismic character variations. Additional sequence boundaries identified in Figure 5 have been omitted for the purpose of illustration. Along the left margin the GA top is at 1.5 sec, JD at 2 sec, LF at 2.2 sec, and the OI at 2.5 sec.

factors are fairly high throughout the field. In the oil sandstones, they range from 35 to 48% of the original oil in place, and in the gas sandstones recovery rates are expected to be as high as 75%.

Reservoir energy is provided by both water drive and gas expansion, and the particular drive mechanism operative in any one sand varies from fault block to fault block. Because both normal and abnormal pressure conditions exist, there are normal and overpressured water-drive systems present.

Abnormal pressure conditions exist within sandstone units occurring below the interval between the HB and IC units. Since the normally pressured GA and HB reservoirs contain less than 10% of the total recoverable reserves in the field, over 90% of the reserves occur in an overpressured regime.

GA Sandstone—The GA sandstone is the youngest and shallowest producing sandstone in the field with an average production depth of −4,300 ft (−1,311 m). Three sandstone beds, the GA-1, 2, and 5, are productive within this unit. The structure at the GA level (Fig. 26) is a simple rollover anticline.

The locations of hydrocarbons in the GA sandstone, and in most of the other reservoir sandstones in the 330 field, are determined by both stratigraphic and structural controls. The structural control over the areal distribution of hydrocarbons is shown by the entrapment of oil at the crest of the anticline (Fig. 26). However, detailed mapping of seismic and well data indicates that stratigraphic variations control the reservoir thickness.

For example, the seismic reflections from the GA sandstone shown in Figure 27 are the shingled progradational reflection configuration (Fig. 12) discussed earlier. The steeply inclined beds indicated by the dipmeter, as shown in Figure 27, correspond to the shingled GA seismic reflections. The seismic and dipmeter data (Fig. 28) indicate that the GA and other major reservoir sandstones have dip patterns which are the result of foreset current bedding.

Dip magnitudes range from several to about 7° as shown in Figure 28. With a dip magnitude spread of less than 10° the foreset deposits are more likely to be fan or crescent shaped as opposed to a more elongated shape (Gilreath and Maricelli, 1964). The detailed net oil isopach for the GA sandstone shown in Figure 29 indicates that the individual GA reservoir sandstone units are crescent or lobate shaped. Net oil isopach values from development and the location of the higher ampli-

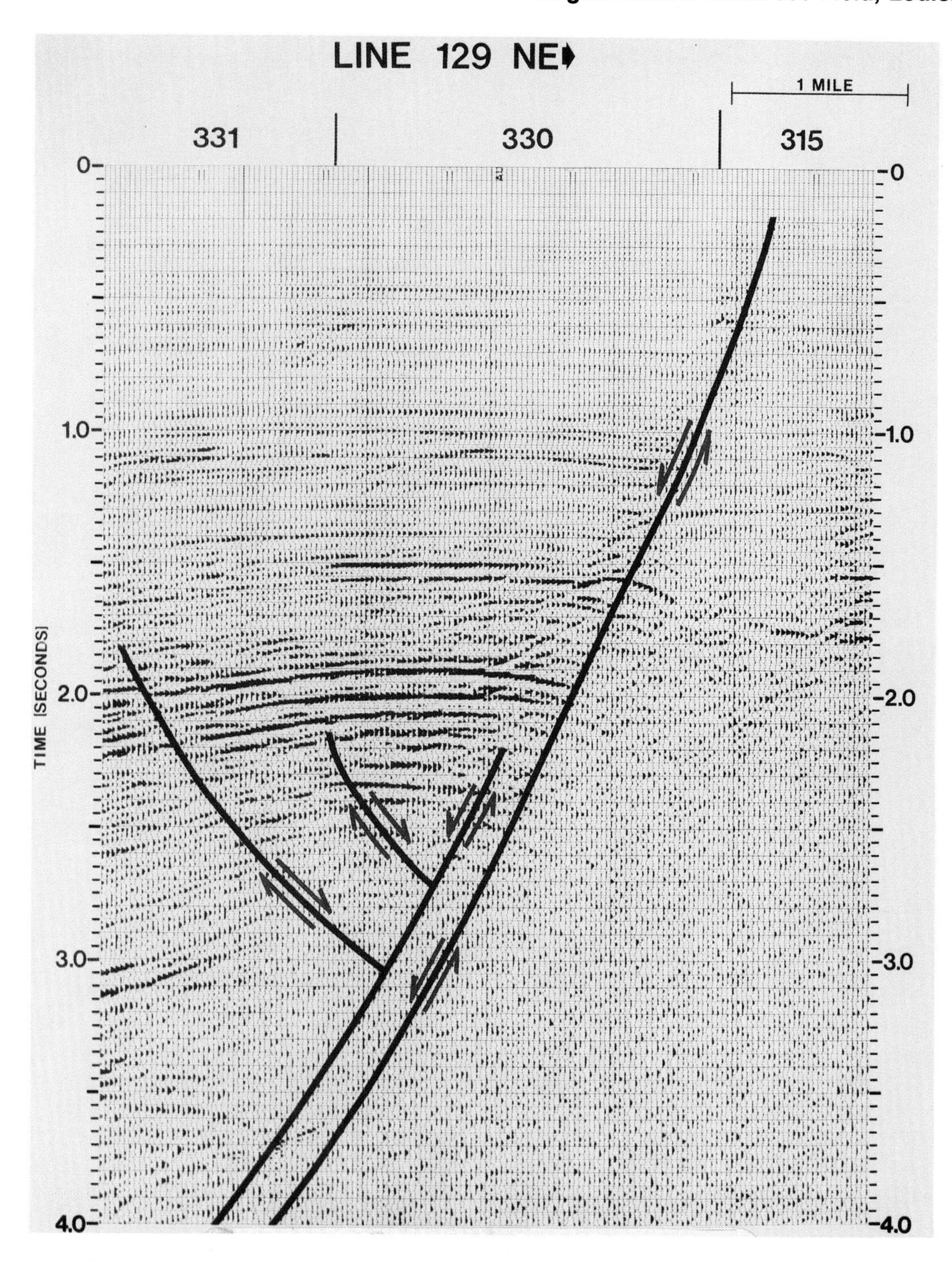

FIG. 13—Relative amplitude processed version of the seismic line shown in Figure 5 (location in Figure 6). Prominent seismic hydrocarbon indicators are associated with the oil and gas reservoirs that are traversed by the seismic line. At the crest of the anticline, the GA top is at about 1.35 sec, HB sandstone at 1.5 sec, JD sandstone at 1.85 sec, LF sandstone at 2.05 sec and the OI sandstone at 2.3 sec.

tude GA reflections were used to map the individual foreset lobes shown in Figure 29.

Development wells in the GA have a common water level and pressure history. It appears that the individual lobes have coalesced to form a common reservoir whose net lateral oil thickness variations are stratigraphically controlled.

Reservoir parameters for the GA sandstone are listed in Table 2. The original oil in place is estimated to be 25 million bbl of which 7.5 million bbl were produced by January 1, 1979. The production rate on that date was 3,161 b/d (BOPD) and the most prolific well completion in the GA sandstone, the Pennzoil C-2 well, was flowing at the rate of 743 b/d.

The GA sandstone is a normally pressured water-drive reservoir. The reservoir pressure (Fig. 30) decreased less than 8%

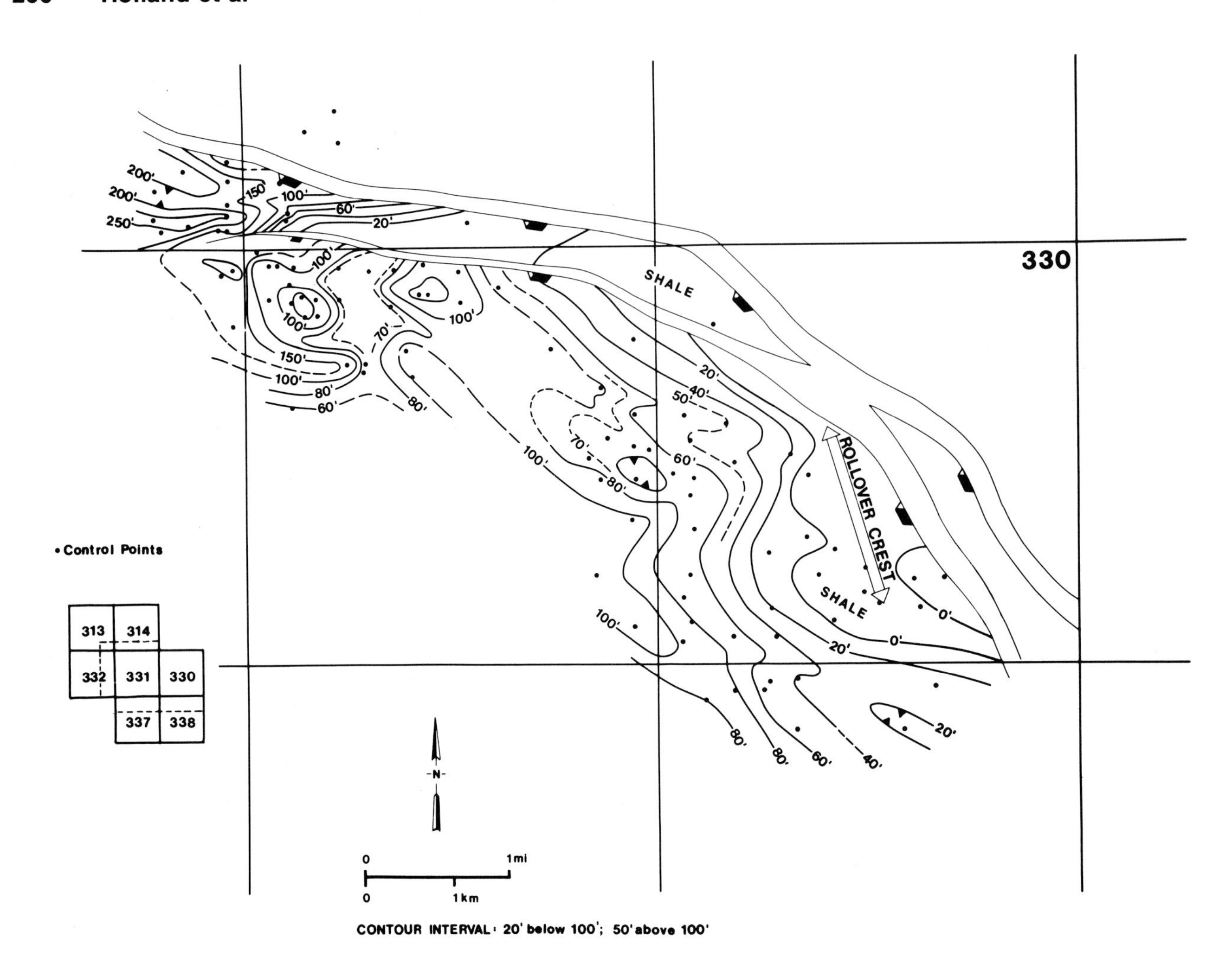

FIG. 14—Sand isolith map of the LF sandstone.

after the production of 71% of the estimated recoverable reserves. Ultimate recovery is expected to exceed 37% of the original oil in place.

JD Reservoir—The JD sandstone is the only reservoir sandstone in the field that is dominantly gas productive. Structure (Fig. 31) at the JD level consists of two anticlines traversed by minor growth faults. The southeastern anticline is also cut by two antithetic faults. Producing depths average −6,500 ft (−1,981 m).

The maximum hydrocarbon column present in the JD is 970 ft (296 m), of which 840 ft (256 m) is gas. The narrow oil rim present in the eastern structure has not been observed on the northwestern anticline. A satisfactory explanation for this phenomenon has not been found.

The JD sandstone has a productive area of over 6,450 acres (2,611 ha.). This is almost double that of any other productive zone in the field.

Average reservoir parameters for the JD sandstone in Pennzoil's block 330 are listed in Table 2. The average gross reservoir thickness is 150 ft (46 m) with an average net pay of 36 ft (11 m). Porosity averages 29%, and water saturation (Sw) averages 35%. Permeabilities range from 20 to 4,100 millidarcies and average 558. Ten percent of JD pay in block 330 has an excess of 1,000 millidarcies permeability.

The JD sandstone may be a combination delta-front/distributary-mouth bar sand. Some sidewall core samples are very well sorted, laminated, fine grained, quartzose arenite (Fig. 18) and probably were deposited on a distributary-mouth bar. Other JD samples consist of quartzose wacke and were deposited as delta-front sands.

The original gas in place estimate for the JD sandstone throughout the entire field is 700 Bcf of which 321 Bcf had been produced by January 1, 1979. The recovery factor is expected to be 75%. This gas contains approximately 30 bbl of condensate per MMcf. On January 1, 1979 production from the JD was averaging 259 MMcf of gas, 3,239 bbl of condensate (BOC), and 645 bbl of oil. The most productive well completion in the JD at that time was the Texaco A-11 well on block 313 which was producing 21.5 MMcf/day and 7 BOC/day.

LF Sandstone—The largest oil reservoir sandstone in the field is the LF sandstone. Oil productive area exceeds 3,500 acres (1,417 ha.) and a gas cap covers 600 acres (243 ha.). The maximum hydrocarbon column includes 658 ft (201 m) of oil and 364 ft (111 m) of gas (Fig. 31). Producing depths average about 7,100 ft (2,164 m) subsea.

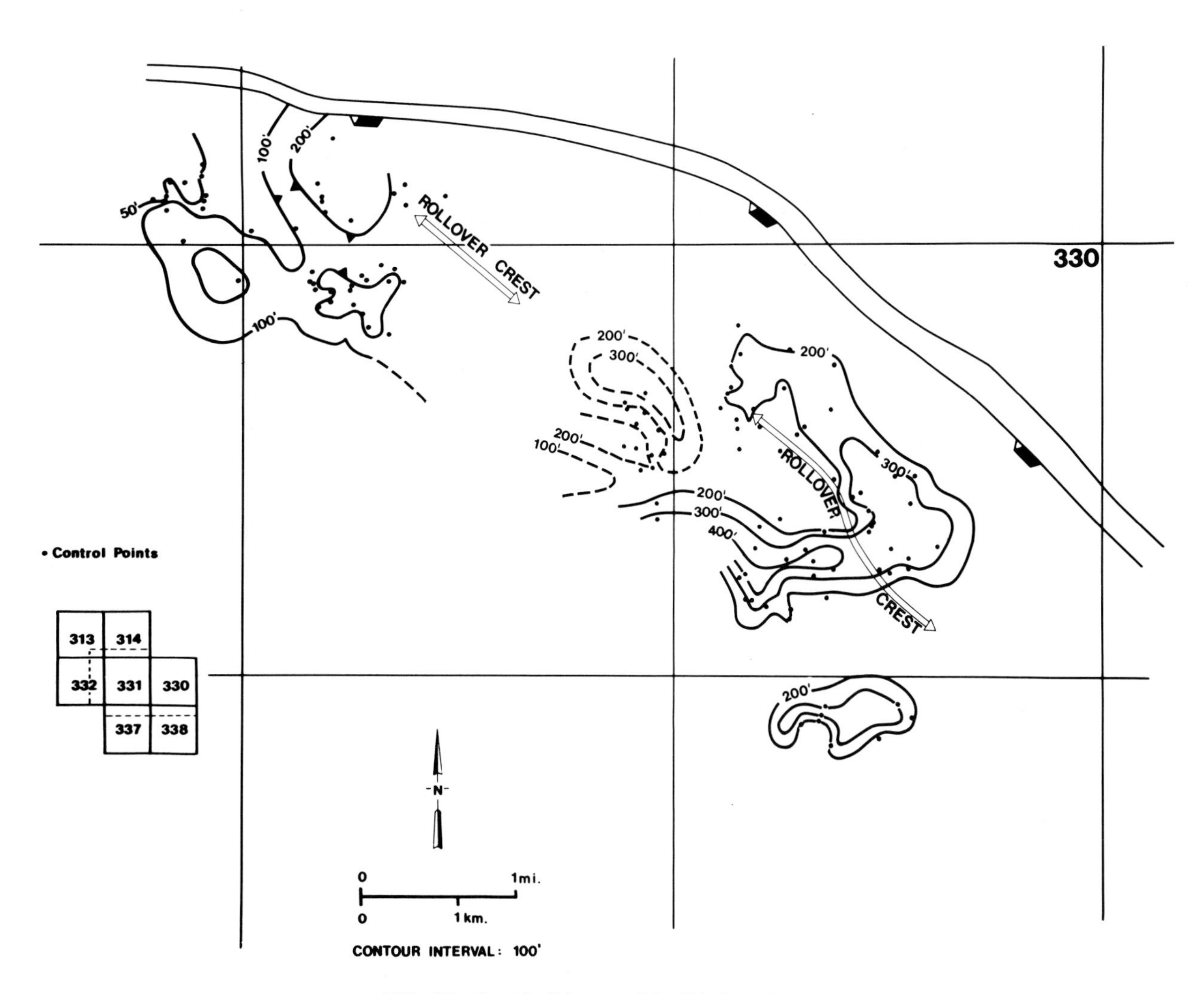

FIG. 15—Sand isolith map of the GA-2 sandstone.

A structure map of the LF sandstone (Fig. 32) shows it to be more complexly faulted than is the shallower JD sandstone (Fig. 31) but less so than the deeper OI sandstone (Fig. 34).

The LF sandstone unit probably is composed of several prograded wedges of delta-front sands. The LF sandstone isolith map (Fig. 14) shows several northwest-trending thick zones alternating with thin trends, and an updip shale interval. Oblique to parallel progradational reflection configurations (Mitchum et al, 1977) appear on many of the seismic lines (Figs. 5, 12) and those patterns support a delta-front interpretation for the depositional environment of the unit. The updip shale interval could be an area where the drill bit simply missed the sandy parts of prograding units.

Reservoir parameters for the block 330 part of the LF sandstone are shown in Table 2. Of special interest is the high water saturation (averaging 40%) found in most LF penetrations. This high value produces a low response on resistivity logs (Fig. 33). The high water saturation is believed to be due to high surface tension present between water and the very fine grains of the sandstone.

The LF sandstone had an estimated 180 million bbl of original oil in place, of which 58 million bbl had been produced by January 1, 1979. The well with the highest LF production as of that date was the Exxon A-23 well in block 314 with a 2,915 b/d flow. Approximately 47% of the original oil in place should be recovered.

The LF sandstone reservoirs have weak water-drive systems. Figure 30 shows a pressure history curve for the LF fault block B reservoir. During early competitive production the pressure dropped 34% after only 8 million bbl of oil had been produced. As a result, a voluntary production plan was designed to allow prodution to approximate the calculated water influx. This resulted in the production of the next 7 million bbl with only an additional 7% pressure drop. Recovery is now expected to exceed 40% of the original oil in place.

OI Sandstone—The OI sandstone contains the thickest hydrocarbon column in the field and is the deepest major reservoir sandstone. The only producing zone that is deeper is the *Lenticulina* sandstone which is a minor oil producing sandstone. Both oil and gas are produced from four zones within the OI unit. Producing depths range from 7,000 to 9,000 ft (2,134 to 2,743 m) subsea.

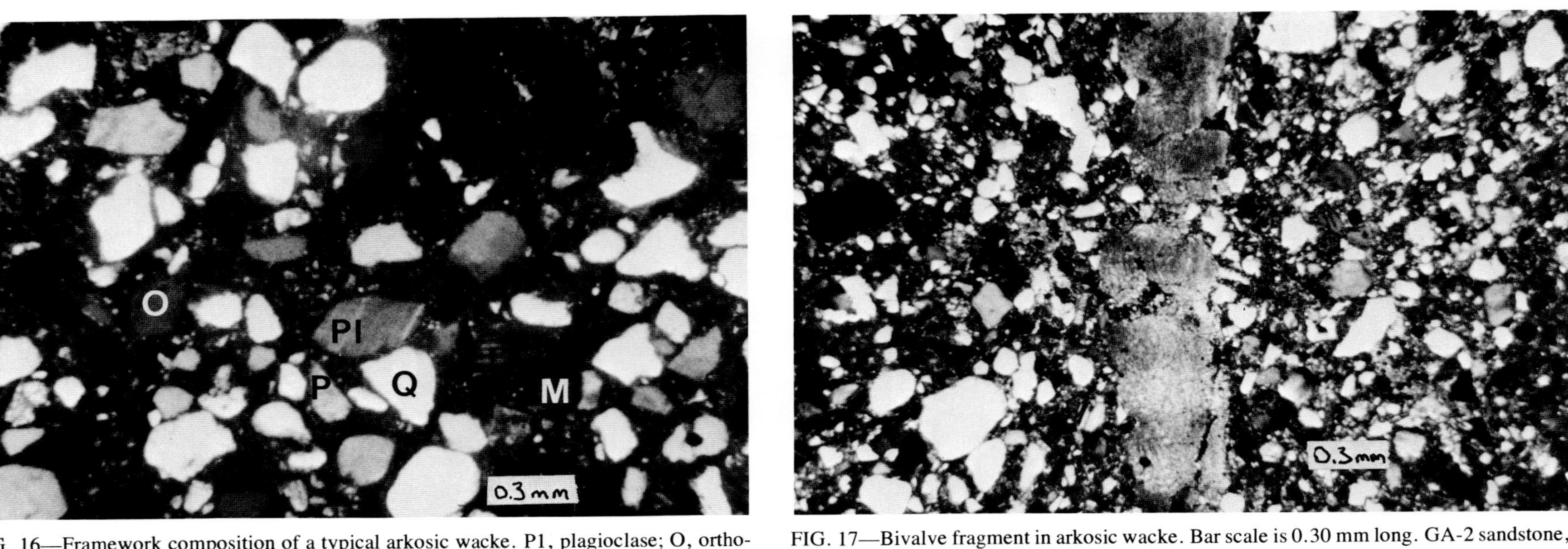

FIG. 16—Framework composition of a typical arkosic wacke. P1, plagioclase; O, orthoclase; M, microcline; P, perthite; Q, quartz. Bar scale is 0.30 mm long. OI sandstone, Pennzoil well B-13, −7,535 ft (2,297 m). Cross nicols.

FIG. 17—Bivalve fragment in arkosic wacke. Bar scale is 0.30 mm long. GA-2 sandstone, Pennzoil well A-13, 5, 253 ft (1,601 m). Crossed nicols.

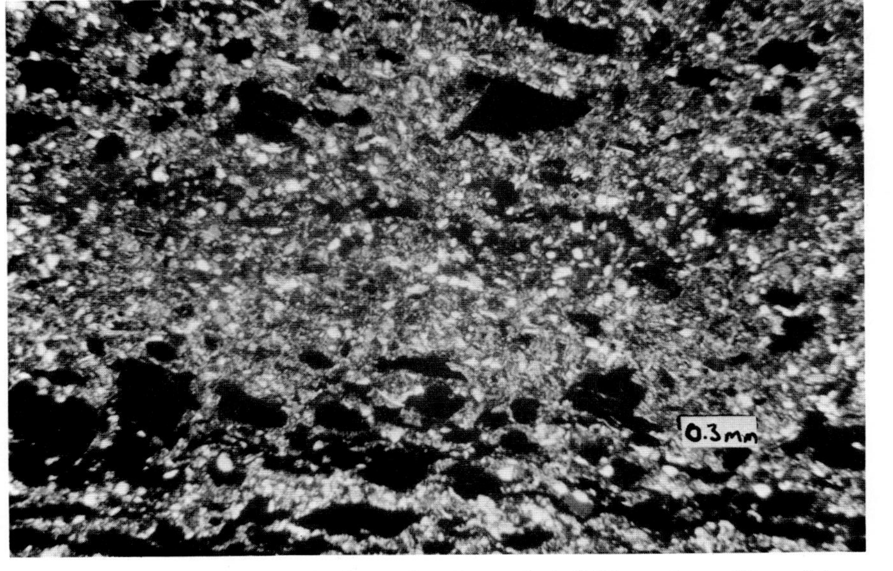

FIG. 18—Organic debris in arkosic arenite. Bar scale is 0.30 mm long. JD sandstone, Pennzoil well B-14, −7,465 ft (2,275 m). Crossed nicols.

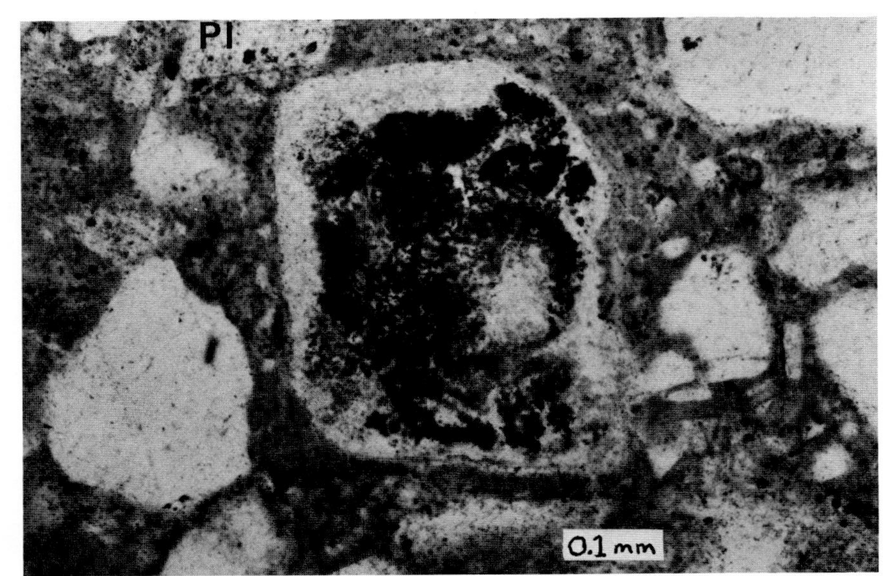

FIG. 19—Intensely weathered feldspar in arkosic wacke. A small, relatively unweathered plagioclase grain occurs above the upper left corner of the large weathered grain in the center. Bar is 0.1 mm long. OI sandstone, Pennzoil well B-13, −7,535 ft (2,297 m). Plain light.

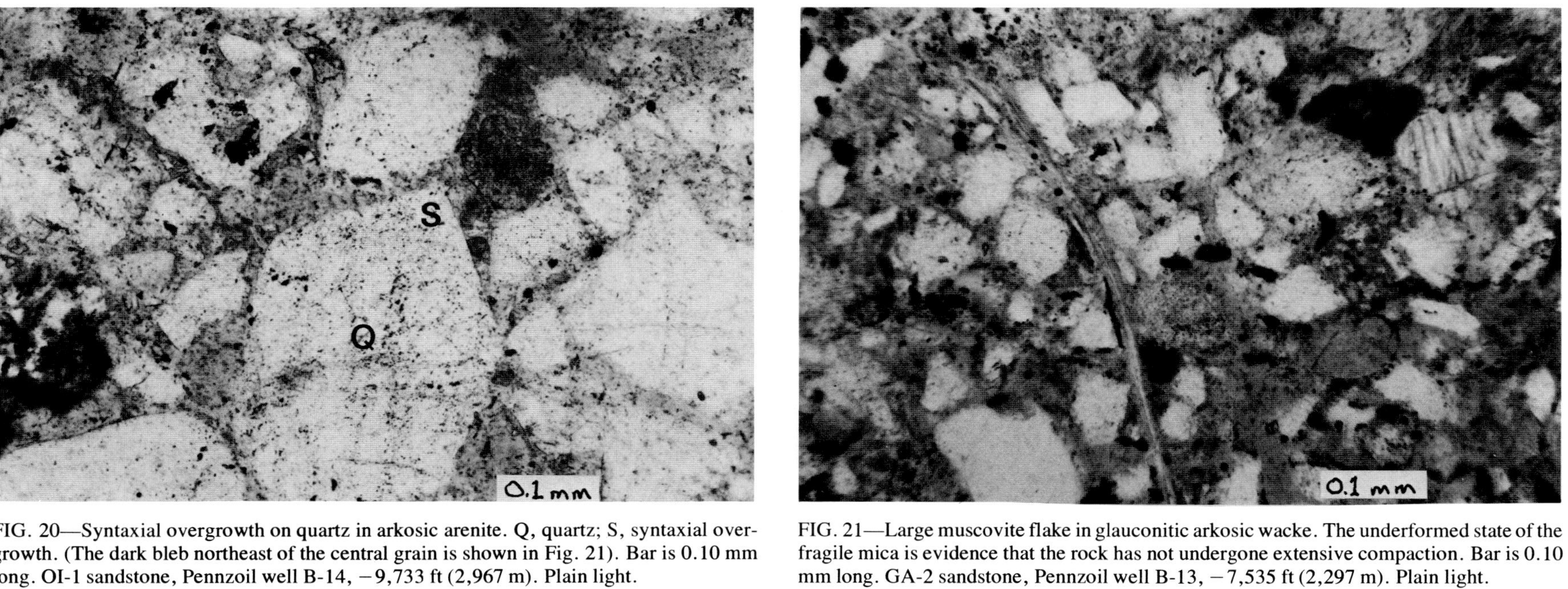

FIG. 20—Syntaxial overgrowth on quartz in arkosic arenite. Q, quartz; S, syntaxial overgrowth. (The dark bleb northeast of the central grain is shown in Fig. 21). Bar is 0.10 mm long. OI-1 sandstone, Pennzoil well B-14, −9,733 ft (2,967 m). Plain light.

FIG. 21—Large muscovite flake in glauconitic arkosic wacke. The underformed state of the fragile mica is evidence that the rock has not undergone extensive compaction. Bar is 0.10 mm long. GA-2 sandstone, Pennzoil well B-13, −7,535 ft (2,297 m). Plain light.

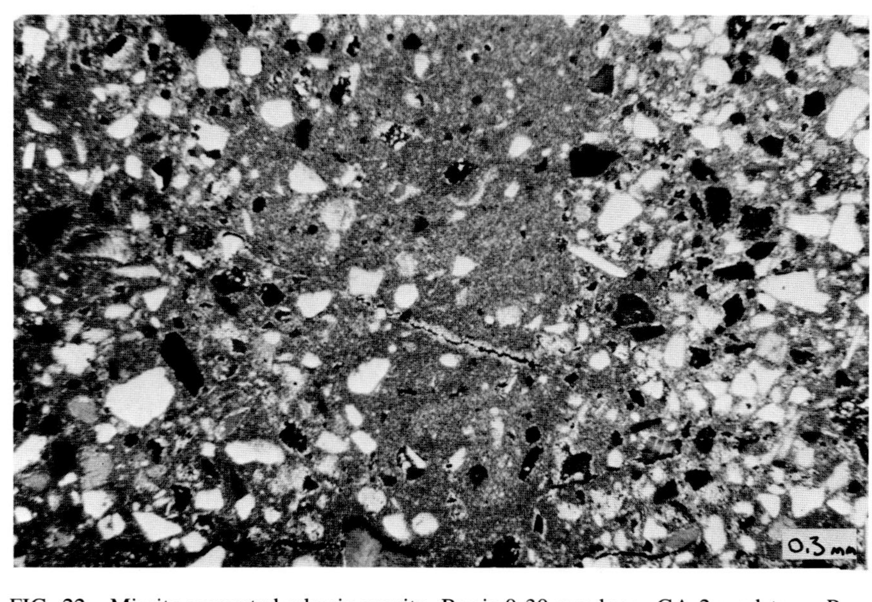

FIG. 22—Micrite cemented arkosic arenite. Bar is 0.30 mm long. GA-2 sandstone, Pannzoil well A-5, −4,490 ft (1,369 m). Crossed nicols.

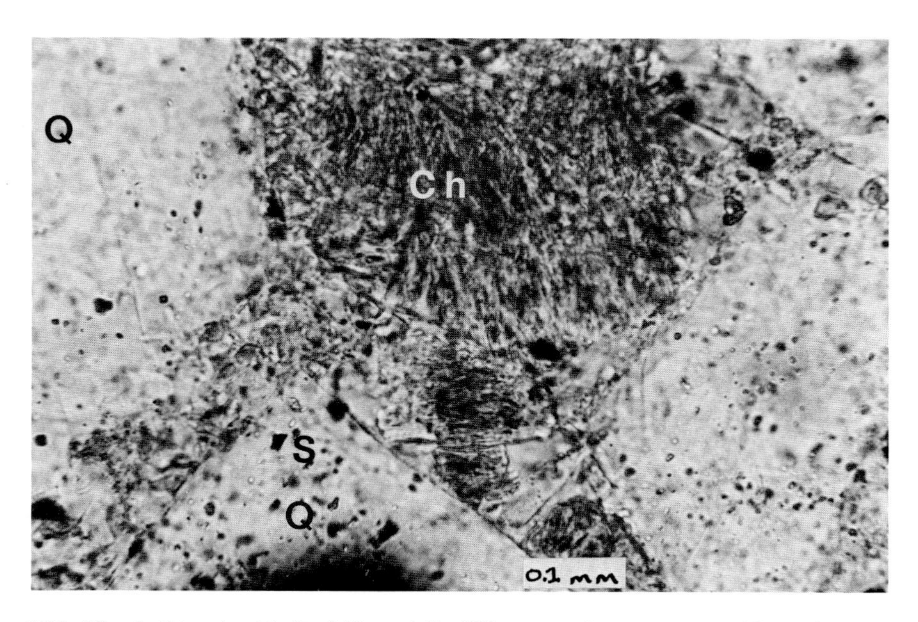

FIG. 23—Authigenic chlorite (ch) partially filling pores between quartz (Q) grains with syntaxial (S) overgrowths. Bar is 0.01 mm long. OI-1 sandstone, Pennzoil well B-14, −9,733 ft (2,967 m). Plain light.

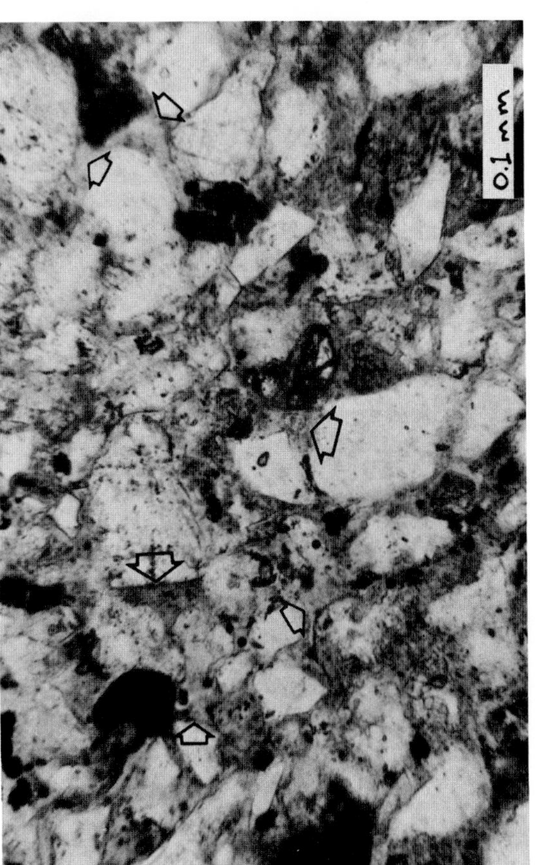

FIG. 24—Reduced primary intergranular porosity (arrows) in arkosic arenite. Bar is 0.10 mm long. LF sandstone, Pennzoil well A-7, −8,597 ft (2,620 m). Plain light.

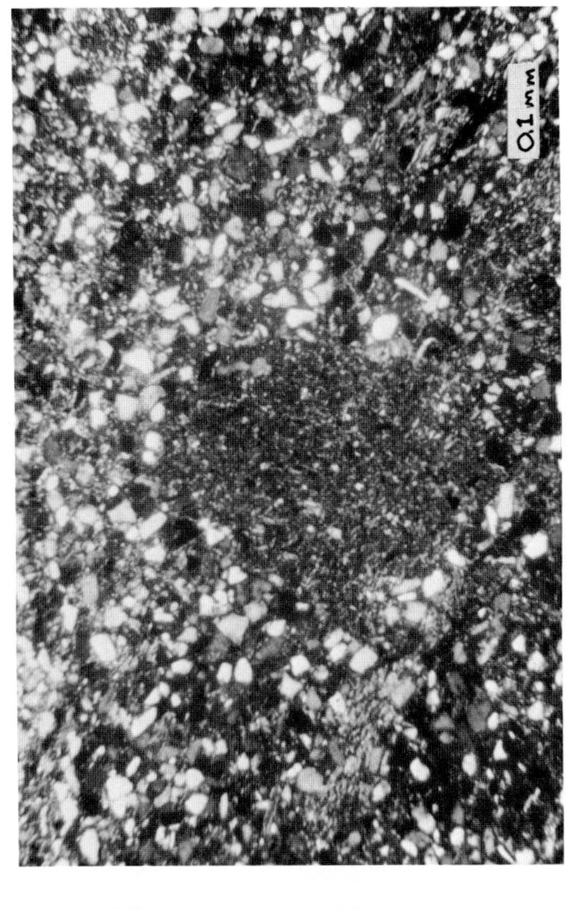

FIG. 25—Bioturbated quartzose wacke and mudstone. Bar is 0.1 mm long. LF sandstone, Pennzoil well A-7, −8,570 ft (2,612 m). Crossed nicols.

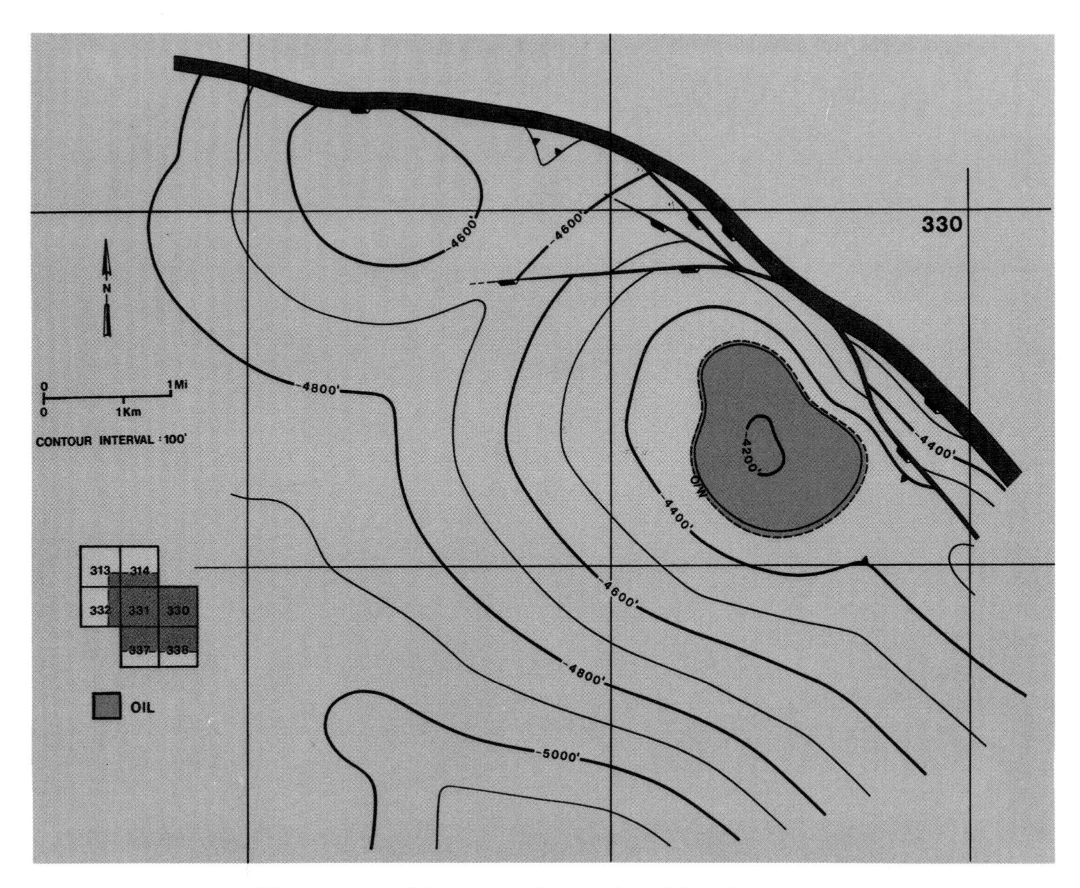

FIG. 26—Map of structure on the top of the GA sandstone.

The OI sandstone is the most complexly faulted major producer in the field (Fig. 34). Both anticlines are cut by several faults which have contributed to a complex hydrocarbon distribution pattern within the OI between the several fault blocks. In fault block A, there is a 520 ft (158 m) oil column with a 118 ft (36 m) gas cap. The oil column occurs between 6,905 and 7,425 ft (2,105 to 2,263 m) and is juxtaposed against an 1,800 ft (500 m) thick gas column between 6,950 and 8,750 ft (2,118 to 2,667 m) in fault block B. Fault block C includes parts of both anticlines and the intervening syncline. The OI contains water in the syncline which separates a thick gas column with a thin oil leg in the east closure from an oil accumulation to the northwest.

Reservoir parameters for oil and gas reservoirs in Pennzoil's block 330 are summarized in Table 2. Note that the average depth of the gas sandstones is deeper than that for the oil sandstones, a condition produced by faulting in the field. Oil production as of January 1, 1979 was 22 million bbl with 28 Bcf of solution gas, and 49 Bcf of gas plus 3 million bbl condensate.

The most productive OI oil well on January 1, 1979, was the Texaco A-5 well on block 313 which was producing at the rate of 1,625 b/d. The Pennzoil A-3 well was the largest producing gas well on that date averaging 10.5 MMcf/day with 435 BOC/day.

The OI sandstone reservoir in fault block A has a classic gravity-segregation drive mechanism. A sizable gas cap developed after a 16% decrease in pressure occurred (Fig. 30) and it became apparent that gas injection should begin as soon as possible. An injection program began in 1979.

Geochemistry

Crude Oils—Crude oils from the field range from API gravities in the middle 30s in the deeper reservoirs to 23 ° in the shallow GA sandstone (Table 2). Gas/oil ratios are also higher in the deeper reservoirs than in the shallower ones.

Gas chromatographic analysis of oils from several reservoirs (Fig. 35) show a distinct pattern related to reservoir depth. Chromatograms of C_{15}+ hydrocarbons from deeper reservoirs (Fig. 35 left) have patterns characteristic of mature oils with CPI_{A^3} values around 1.07 and CPI_{B^3} values around 1.00. Shallow reservoirs have oils that give C_{15}+ chromatograms with very low paraffin peaks (Fig. 35 right). There is a gradual change from deeply reservoired oils with normal paraffin peaks to shallowly reservoired oils with very small paraffin peaks.

Spectroflurescence analyses of whole oils and their aromatic fractions (Fig. 36 top) indicate that all of the reservoired oils belong to the same oil family, regardless of their reservoired depths. The general shapes of the napthene curves on the

Table 2. Reservoir Characteristics of Major Producing Zones, Eugene Island Block 330.

	GA	HB	KE	LF	MG	OI	JD	OI
Average depth	4,300 ft 1,311 m	4,850 ft 1,479 m	6,850 ft 2,079 m	7,100 ft 2,165 m	7,500 ft 2,287 m	7,200 ft 2,195 m	6,400 ft 1,969 m	7,600 ft 2,317 m
Producing acreage	776	1,029	745	633	230	327	2,244	485
Gross reservoir thickness	400 ft 122 m	150 ft 46 m	120 ft 37 m	55 ft 17 m	165 ft 50 m	350 ft 109 m	150 ft 46 m	120 ft 37 m
Average net pay	27 ft 8 m	32 ft 10 m	44 ft 13 m	49 ft 15 m	50 ft 15 m	44 ft 13 m	36 ft 21 m	56 ft 17 m
Average porosity	31%	28%	29%	31%	30%	30%	29%	29%
Average permeability	1,328	1,270	792	925	332	1,292	720	1,290
% < 100 md	19	16	17	21	15	19*	22	
% < 1,000 md	25	23	20	7	15	34*	10	
Highest permeability (md)	4,100	5,000	4,500	3,500	4,200	6,250*	4,100	
Water saturation	36%	33%	39%	40%	34%	20%	35%	25%
Original oil/gas in place**	25.2	36.5	31.6	30.4	18.2	64.4	155	148
Cumulative production to 1-1-79**	7.5	7.5	13.9	11.9	6.2	11.5	70.3	60.2
Remaining recoverable crude**	2.6	6.0	1.3	2.3	2.0	11.3	45.7	35.8
Prod rate b/d; MMcf/d	3,161	2,834	2,530	3,915	887	6,738	43	20
Recovery factor %	40	37	48	47	45	35	75	65
Formation pressure at depth (psig)	2,095 @ 4,300 ft	2,400 @ 4,900 ft	3,764 @ 6,900 ft	4,027 @ 7,200 ft	4,327 @ 7,400 ft	5,652 @ 7,400 ft	3,784 @ 6,400 ft	5,312 @ 7,600 ft
API	23°	25°	35°	34°	36°	32°		
BTU (d-dry, w-wet)							1,163 d 1,142 w	1,172 d 1,151 w
Gas/oil ratio (Mcf/bbls)	514	999	1,021	1,222	2,010	1,094		
Gas/condensate ratio (bbls/Mcf)							32	43
Bubble point pressure (psig)	2,147	2,420	3,796	4,036	4,355	5,692		
Sulfur (mol %)							0.06	None Rec.
Nitrogen (mol %)						0.15	0.06	
CO_2 (mol %)							0.08	0.09

*This data for oil and gas reservoirs combined.

**MMBbls/d : BCF/d.

NOTE: This table is intended to demonstrate representative reservoir characteristics for the field. The data presented is for just the major producing zones in Pennzoil's Block 330 only.

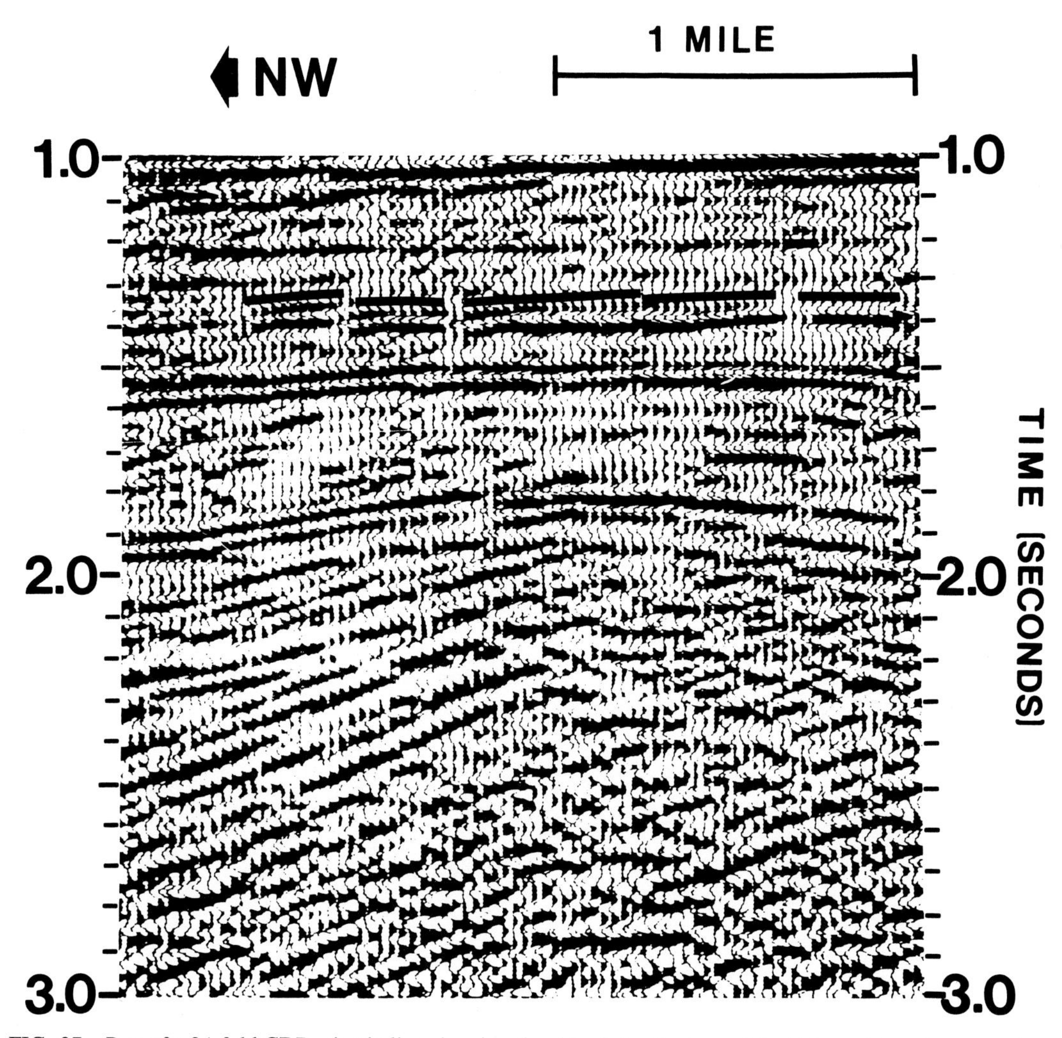

FIG. 27—Part of a 24-fold CDP seismic line showing the shingled reflectons of the GA sandstone between about 1.3 and 1.4 sec. The heavy bar above each shingled reflector indicates the extent of the unit used in constructing the net oil isopach map shown in Figure 29. See Figure 29 for the location of the seismic line.

gas-chromatograms (Fig. 35) also support this interpretation.

The trends of heavier oil and reduction of the paraffins with shallowing oil reservoir depths are interpreted to be due to bacterial degradation, and possibly water washing, of the shallow oils. Barker (1979) summarized the geologic conditions under which bacterial degradation occurs and the usual effects of such alteration, and Deroo et al (1974) described biodegradation of oils in the Western Canada basin that is similar to that present in this case. Alteration is enhanced by formation temperatures below 140 °F (60 °C) and oxygenated water that is in contact with the crude oil. In the Eugene Island Block 330 field growth faulting could have provided a hydraulic connection between oxygenated surface water and migrating crude oils. The present temperature regime in the field is such that reservoirs buried below 6,400 ft (1,951 m) have formation temperatures in excess of 140 °F (60 °C). The most altered oils (GA and HB oils) occur above that depth while the less altered (LF and MG) and unaltered oils (OI and NH) are in reservoirs below 6,400 ft (1,951 m) This situation suggests that bacterial degradation is a major agent responsible for crude oil alteration in the field.

Pleistocene Shales—Shale samples recovered in sidewall cores from the Pennzoil block 330 wells were studied to determine their source character and the state of maturation of their contained organic matter. Total organic content (TOC) of the shales form 0.30% to 0.71% of the rocks, with 70% of them containing over 0.50% TOC. Analysis of the kerogen from these shales indicates that most of it was derived from herbaceous and woody organisms (Fig. 37) of terrestrial origin and falls into Tissot and Welte's (1978) Type III kerogen. This type of kerogen is not a good source for oil (Tissot and Welte, 1978; Barker, 1979).

The results of virtinite-reflectance and thermal-alteration analyses of the kerogen is summarized in Figure 37. Two to three virinite populations are present in all of the shales studied. In each case the dominant population present has the lowest reflectance (%Ro) values. Average % Ro for the dominant populations are 0.31 to 0.47% Ro, all within the kerogen diagenesis stage of Tissot and Welte (1978), and thus below the level of significant generation of hydrocarbons.

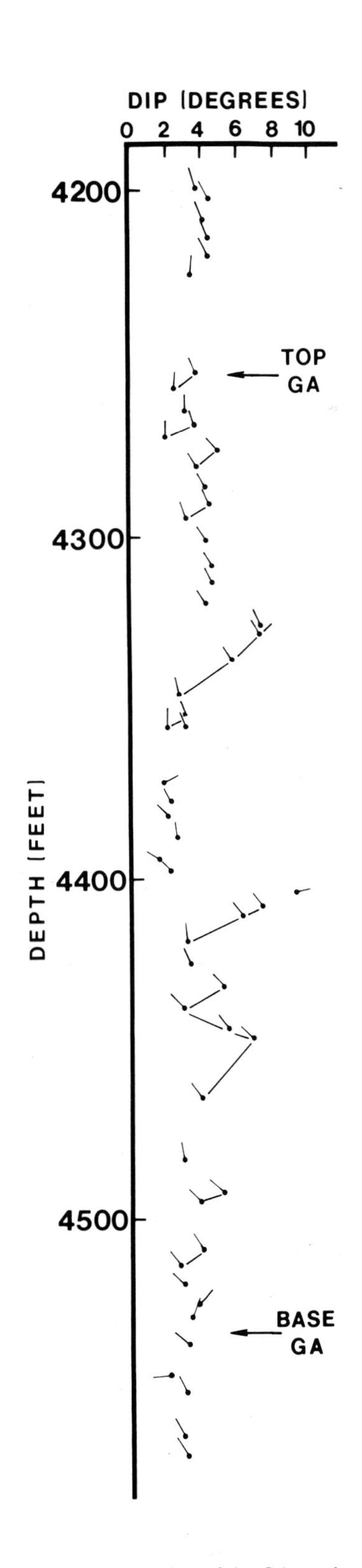

FIG. 28—Dip meter interpretation of the GA sandstone interval in the Pennzoil B-13 well, Block 330. The inclined GA sandstone beds indicated by the dipmeter correspond to the shingled seismic reflections from the GA shown on the seismic line in Figure 27.

The least dominant vitrinite population has % Ro values that cluster between 0.80 and 1.20% Ro. These values fall in the very mature alteration levels set by GeoChem Laboratories and into the catagenesis stage of kerogen degradation of Tissot and Welte, 1978. This population accounts for about 13% of all kerogen present and is interpreted to be a recycled population.

Thermal alteration analysis shows the kerogen present to fall mostly in GeoChem Laboratories 1+ to 2− range, or slightly to moderately altered, and well within their immature alteration field (Fig. 39).

Gas chromatographic analysis of the C_{15}^{+} extract from the shales produces curves characteristic of immature source rocks (Fig. 38). CPI_A values for these extracts average 1.37 and demonstrate the strong odd/even predominance of the higher *n*-alkanes present, a condition known to be associated with immature oils (Tissot and Welte, 1978).

Oil-Shale Relationship—In the previous two sections the oils in the field have been demonstrated to be relatively mature and the kerogen from the Pleistocene shales to be immature. The type of kerogen present in the Pleistocene shales plotted against their maturation level (Fig. 37) indicates that no appreciable amount of hydrocarbons has been generated in these rocks. This leads to the conclusion that the Pleistocene shales are not the source beds for the oil in the Pleistocene reservoirs, and therefore that the oils have probably migrated into their reservoirs from some deeply buried source rock.

Oil and Gas Distribution

The spatial distribution of oil and gas reservoirs in the field is complex. The OI sandstone (Fig. 11) is both a major oil and gas producer while the shallower NH, MG, LF, and KE sandstones are predominantly oil reservoirs with small gas caps. The JD is predominantly a gas reservoir with a thin oil column. Above the JD occurs the HB sandstone which is an oil pay with a small gas cap. The shallowest pay zone in the field is the GA oil sandstone that has no gas cap.

In addition to the vertical complexities, lateral complications exist. Faults with minor displacements have broken many of the stratigraphic units into several distinct reservoirs (Figs. 31, 32, 34). In the case of the OI sandstone, this has resulted in a situation where oil trapped in fault block A is in fault contact with gas trapped in fault block B. In fault block C, gas (with a thin oil leg) is in a reservoir on the east anticline while oil is present at similar depths on the west structure.

The timing of hydrocarbon migration into the field's reservoirs is not clearly understood, although it is probable that all migration occurred after *Trimosina A* time, or within the past 500,000 years. This interpretation is based on the distribution of the biodegraded oils present. As previously shown, these oils occur primarily at depths with modern conditions favorable for their biodegradation, whereas unaltered oils are found in reservoirs located in modern temperature regimes that preclude their bacterial alteration. The geologic setting, depositional environment, and tectonic processes in the field were all essentially constant throughout the Pliocene and Pleistocene. If migration had been occurring throughout the growth of the

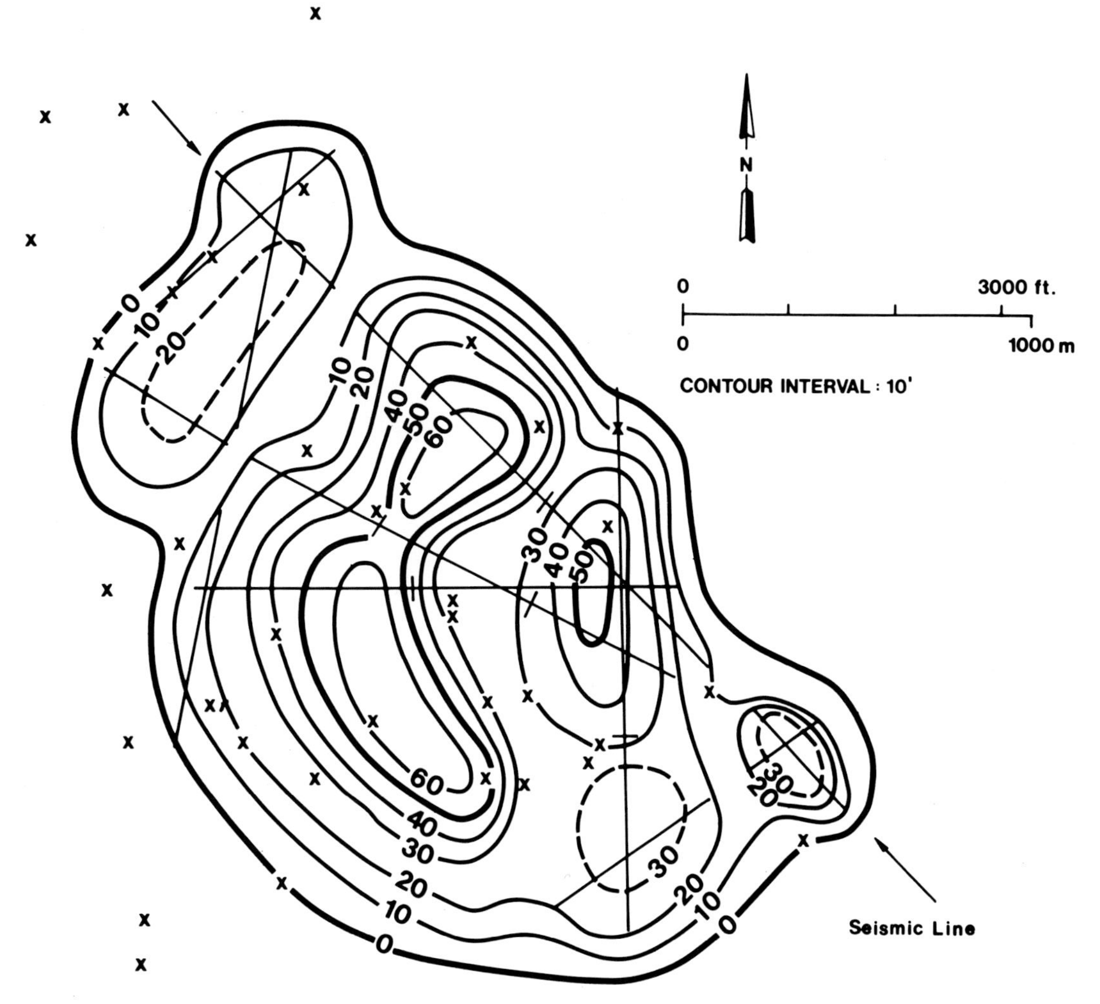

FIG. 29—Net oil isopach map of the GA sandstone. Net oil isopach values are from development wells(x). The locations of the shingled GA reflections (solid lines), as shown in Figure 28, were used to map the areal extent of the individual reservoir units.

structure, then oils migrated into shallower zones during structural growth (for instance, into the LF sandstone during KE deposition) might be expected to show the effects of biodegradation. On the other hand, if all migration occurred after the GA sandstone was deposited and was accomplished by fracturing of a deeply buried reservoir as a result of growth faulting and given the fact of crude oil biodegradation, the pattern of alteration present in the field today would be expected.

The causes of the complicated patterns of hydrocarbon distribution present in the field are not clearly understood. Migration is believed to have been primarily vertical along faults and fracture planes, and then lateral into the reservoirs. Invasion of different types of hydrocarbons into various sandstones may be due to their different capillary pressure characteristics.

The distribution pattern in the OI sandstone in fault block C might be explained by applying Gussow's (1954) theory of differential entrapment of oil and gas. Fluids could have migrated into the reservoir through the faults in the eastern part of the structure. Gas separation from the oil would have accumulated in the crest of that closure, forcing any spillage into the western fold to be oil.

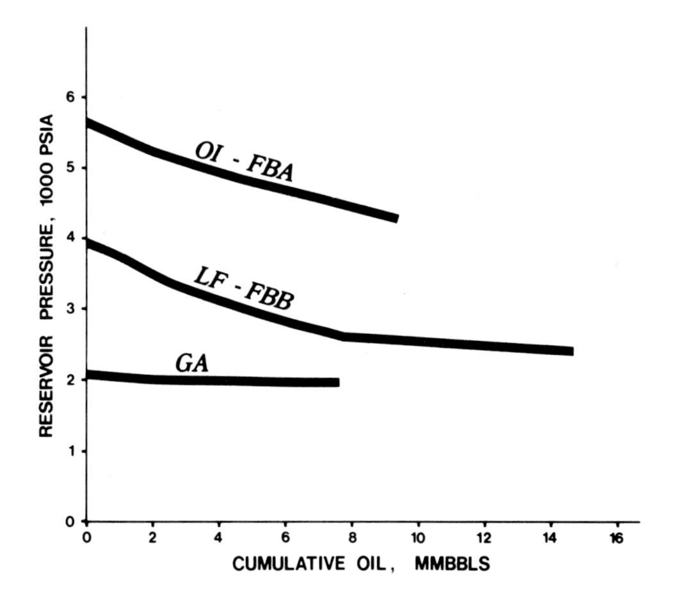

FIG. 30—Pressure history of three reservoirs in the Eugene Island Block 330 field.

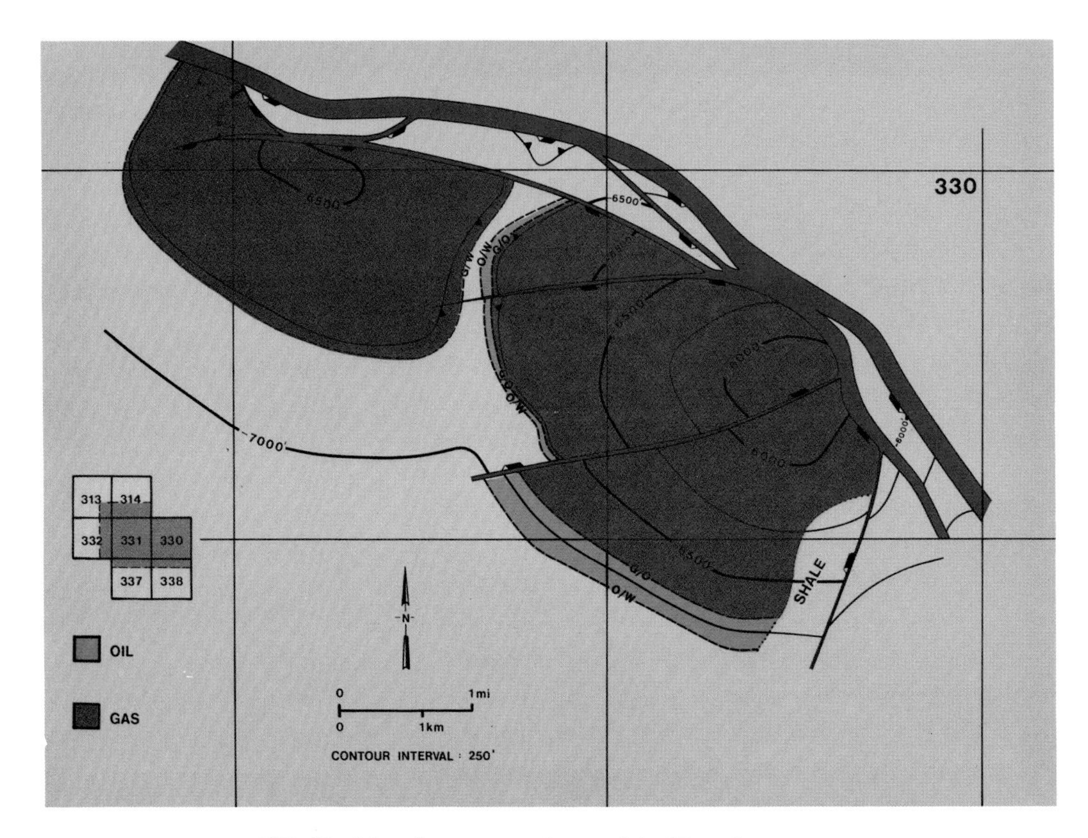

FIG. 31—Map of structure on the top of the JD sandstone.

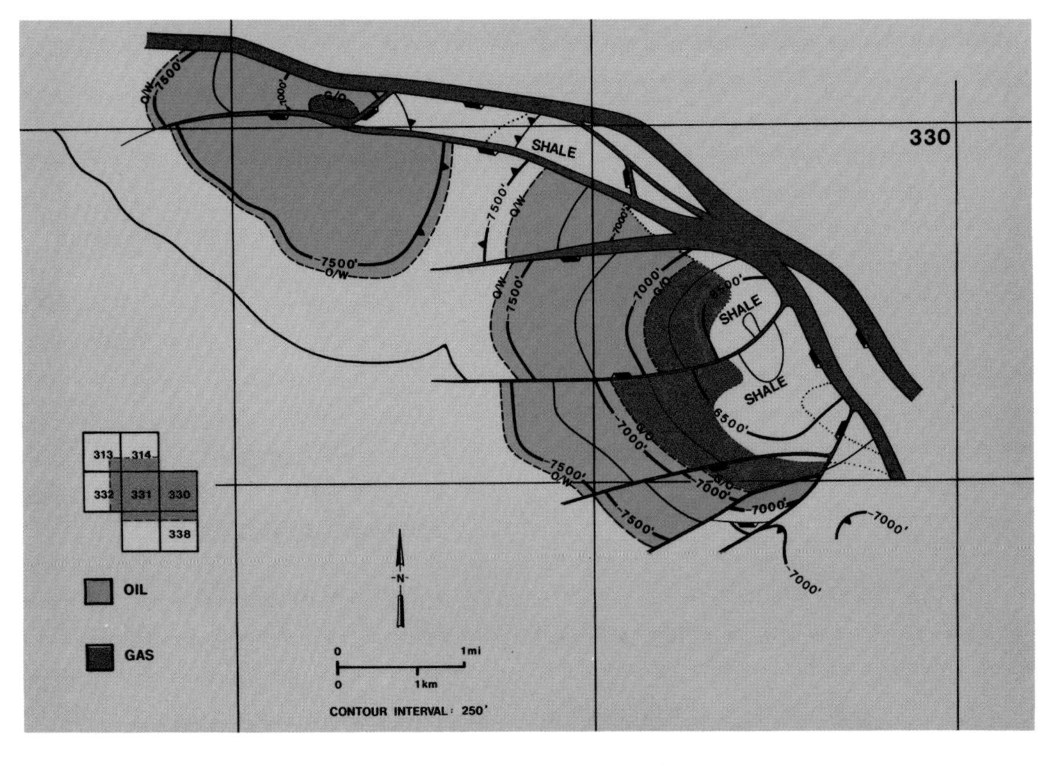

FIG. 32—Map of structure on the top of the LF sandstone.

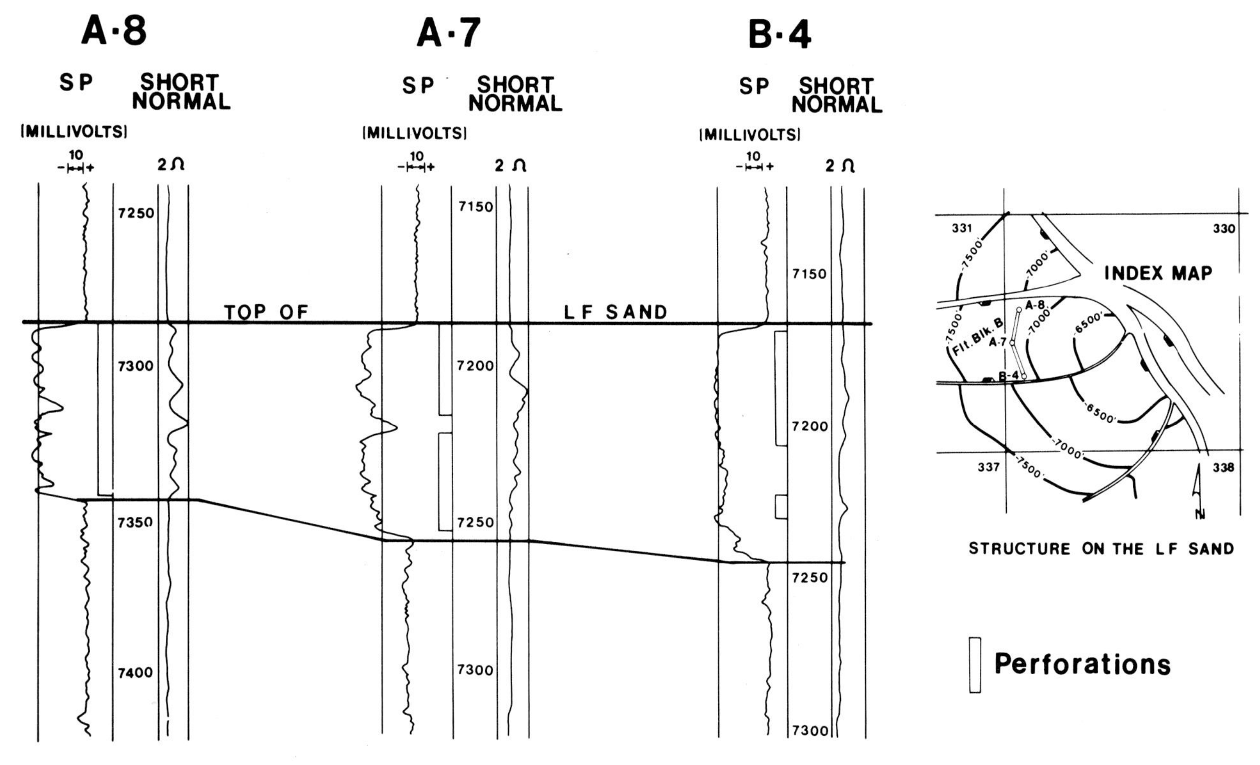

CUM. OIL TO 1-1-79: 2.25 MMBbls
PRESENT RATE: 1243 BOPD
AVG. RESISTIVITY 1.2 Ω

CUM. OIL TO 1-1-79: 1.49 MMBbls
PRESENT RATE: 277 BOPD
AVG. RESISTIVITY 1.0 Ω

CUM. OIL TO 1-1-79: 2.49 MMBbls
PRESENT RATE: 766 BOPD
AVG. RESISTIVITY 1.3 Ω

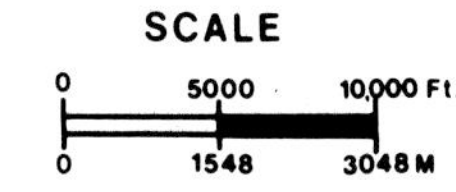

FIG. 33—Stratigraphic cross section showing the relationship between pay zones in the LF sandstone and the very low response of the short normal log.

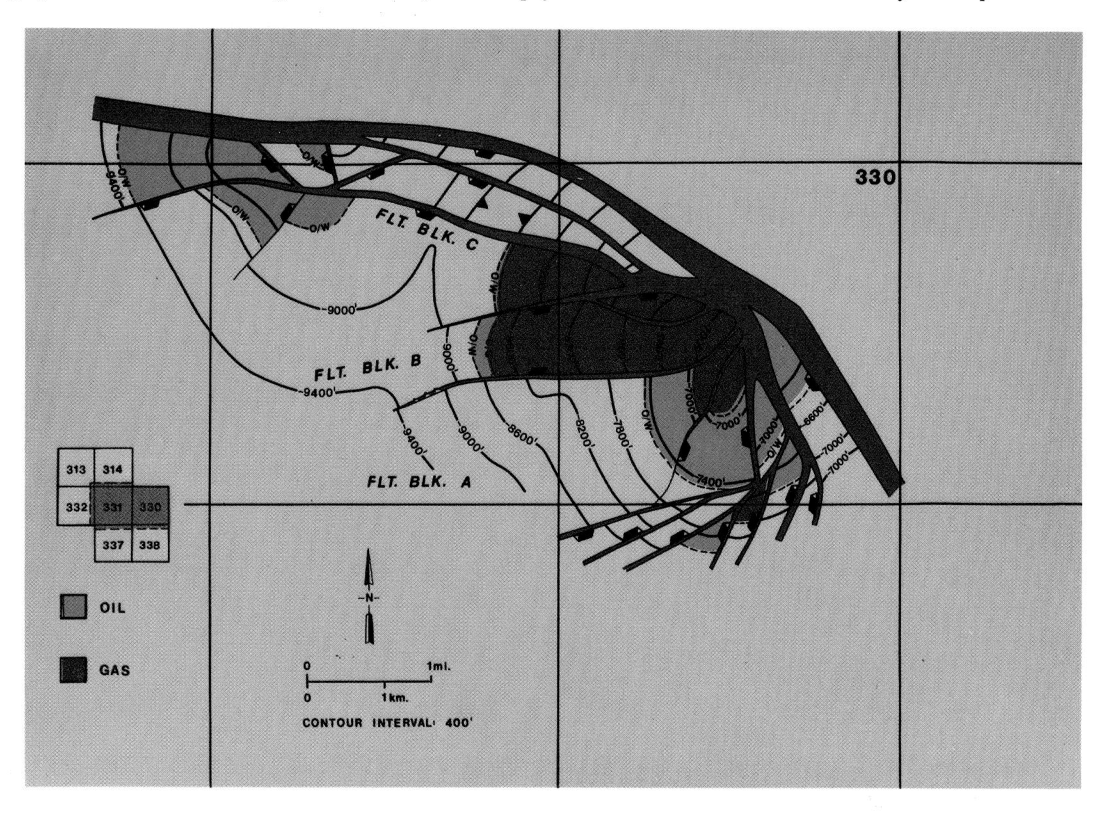

FIG. 34—Map of structure on the top of the OI sandstone.

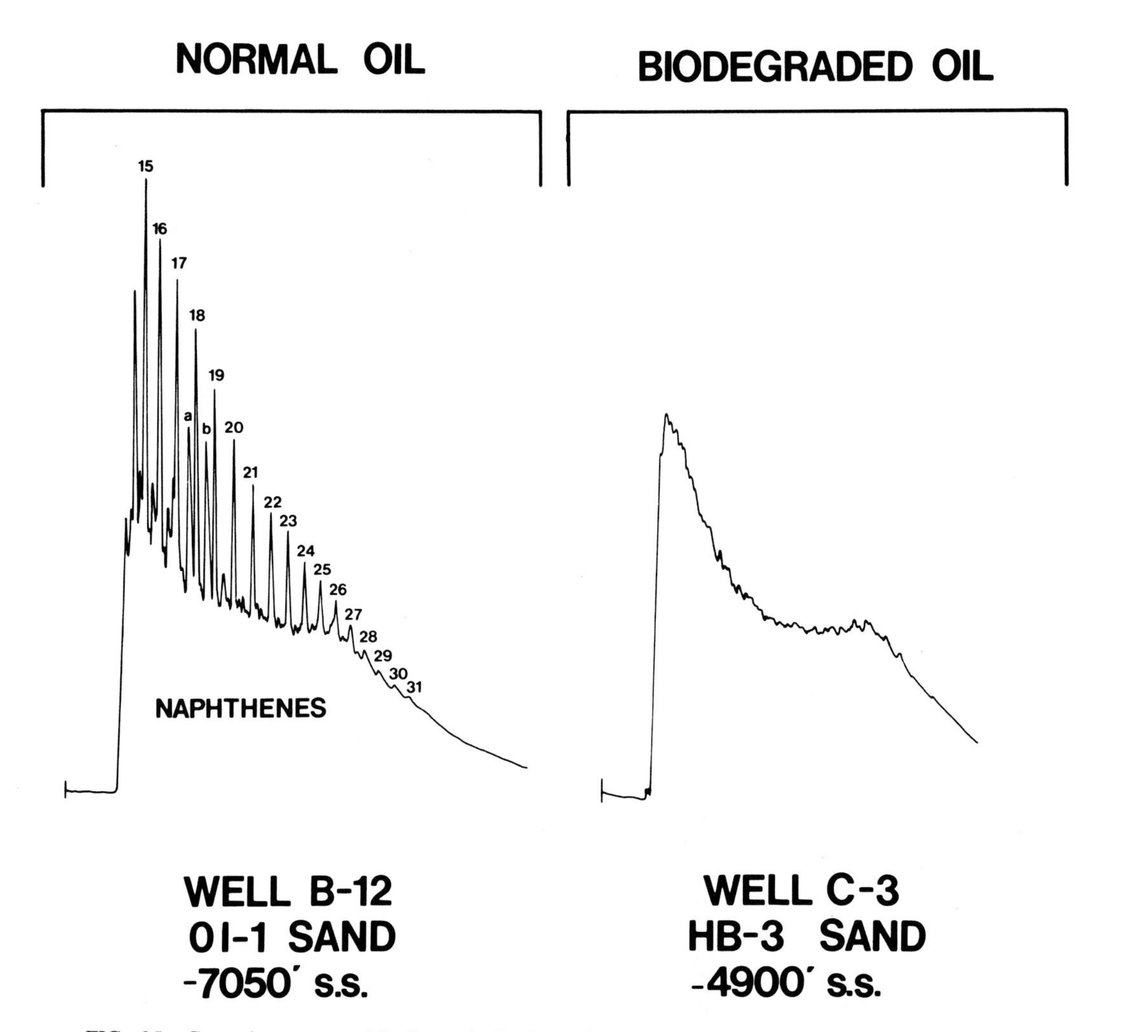

FIG. 35—C_{15}^{+} chromatographic data of oils from the OI-1 and HB-3 sandstones showing the relatively unaltered nature of the deeply reservoired oil, and the strongly altered nature of the shallower oil.

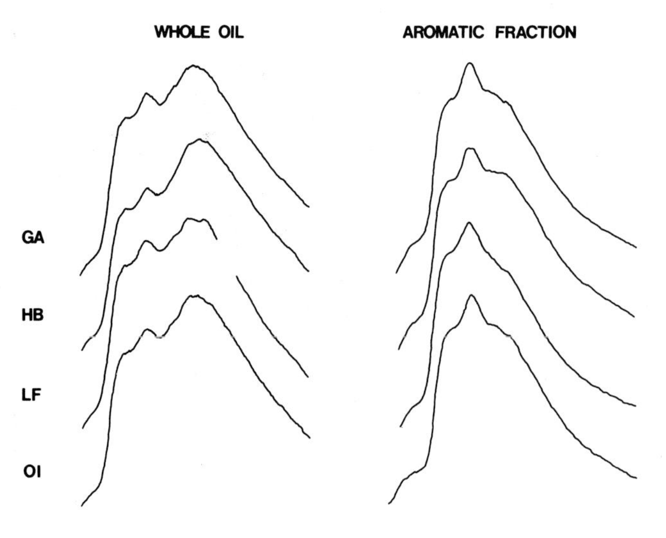

FIG. 36—Spectrofluorescence curves from several reservoir oils showing an apparent presence in the field of one oil family.

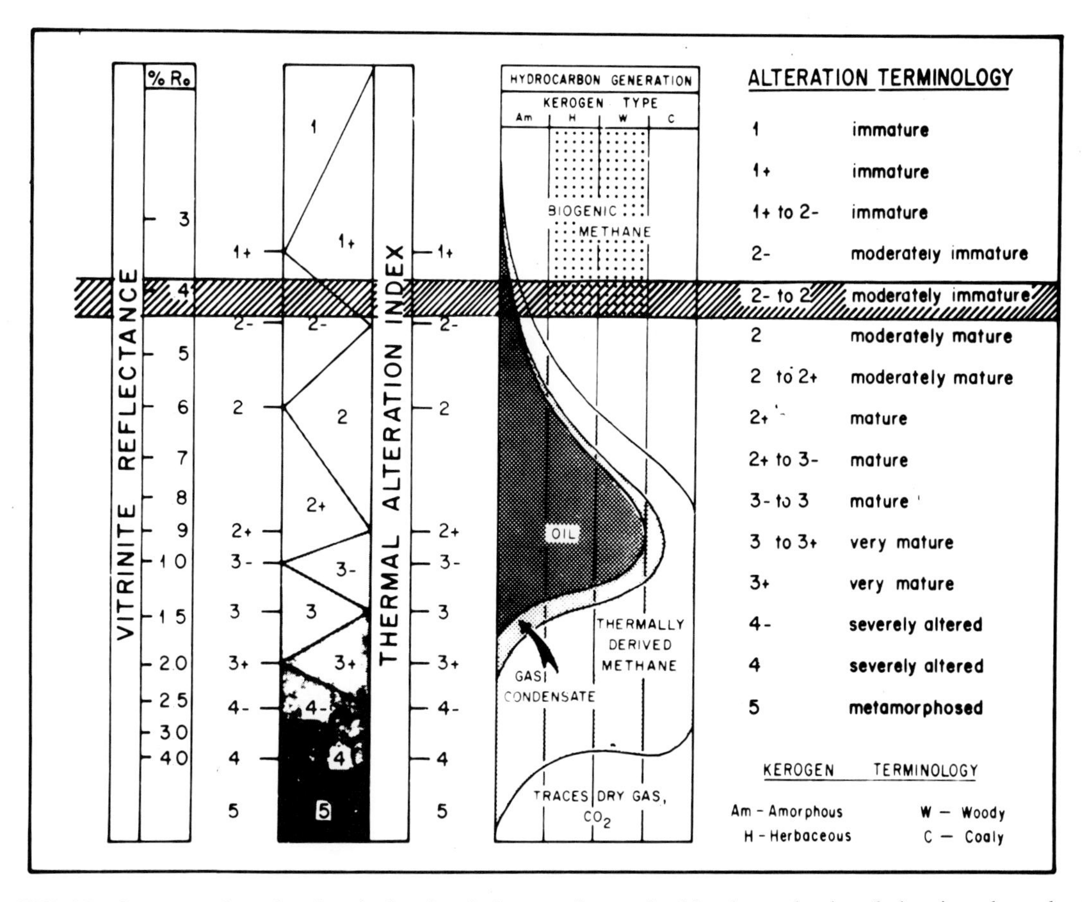

FIG. 37—Summary chart showing the low level of maturation attained by the predominantly humic and woody kerogen present in the Pleistocene shales in the field.

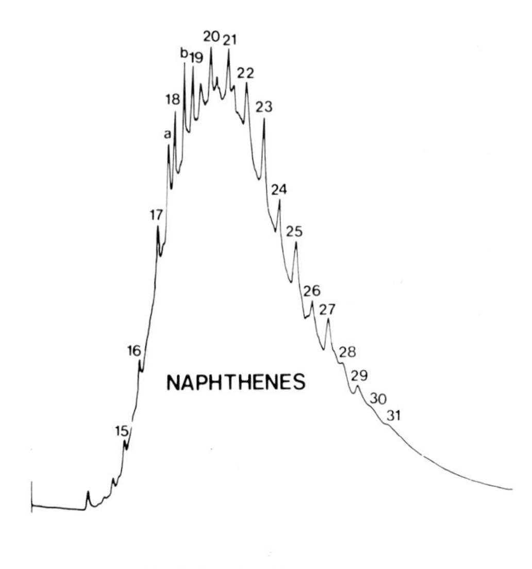

FIG. 38—C_{15}^+ chromatographic data of extractable matter from a Pleistocene shale.

SUMMARY

The Eugene Island Block 330 field is a major oil and gas field in which hydrocarbons have accumulated in more than 25 Pliocene and Pleistocene deltaic sandstone reservoirs. Trapping mechanisms are combinations of structural and stratigraphic varieties including 4-way dip closure, fault closure, and facies changes. Hydrocarbons were generated in pre-Pleistocene sediments and migrated into the reservoir beds via fault and fracture planes.

The field is in the Gulf Coast basin and is part of a Pliocene-Pleistocene deltaic complex that prograded seaward over Jurassic through Miocene deep marine and shelf sediments. Structures in the area are largely produced by buoyant salt tectonics involving the Jurassic Louann Salt, and growth faulting in the thicker parts of deltaic complexes. These two processes are often closely related with each other and with their associated depositional packages.

The field is formed on the downthrown side of a major salt-diapir associated growth fault system that encircles a small depocenter. Migrating lobes of Pliocene and Pleistocene deltas produced cyclic depositional packages that were deposited in this basin. The ideal composition of each package from the base upward is prodelta muds, steeply inclined (20°) lower delta-front sediments, gently inclined (2 to 5°) upper delta-front sediments, and nearly horizontal delta-fringe and delta-plain sediments. Seismic data indicate that each of these packages is almost complete. Periodically during this deposition, the thickness of the accumulated sediments produced unstable lithostatic conditions which resulted in growth faulting. Sometime after *Trimosina A* time, late in the development of this delta complex, the growth faults penetrated to a depth where hydrocarbons existed and migration into the field's reservoirs occurred. Since migration, the crude oils that are in reservoirs at shallow depths (less than 6,400 ft [1,951 m]) have suffered bacterial alteration.

The outstanding features of the Eugene Island Block 330 field are (1) the large reserves, (2) the high production rates, (3) the thick hydrocarbon columns present in several of the reservoirs, (4) the large amount of productive acreage, (5) the occurence of over 90% of the reserves in abnormally pressured reservoirs, and (6) the youth (less than 2.5 m.y. old) of the pay sandstones.

Our knowledge of the geology of this field is still in its infancy in many respects, but it is our hope that the data and interpretations presented herein can serve as an exploration model for other fields in this and similar geologic provinces.

REFERENCES CITED

Barker, C., 1979, Organic geochemistry in petroleum exploration: AAPG Course Note 10, 159 p.

Coleman, J.M., 1976, Deltas: processes of deposition and models for exploration: Champaign, Il., Continuing Education Publication Company, Inc., 102 p.

Domenico, S.N., 1974, Effect of water saturation on seismic reflectivity of sand reservoirs encased in shale: Geophysics, v. 39, p. 759-769.

——— 1976, Velocity and lithology: AAPG-SEG, Stratigraphic Interpretation of Seismic Data School (Monterrey, California).

Deroo, G., et al, 1974, Geochemistry of the heavy oils of Alberta *in*, Oil sands fuel of the future: Canadian Soc. Petroleum Geologist, Mem. 3, p. 148-167, 184-189.

Garrison, L.E., and R.G. Martin., 1973, Geologic structures in the Gulf of Mexico basin: U.S. Geol. Survey Prof. Paper 773, 85 p.

Gilreath, J.A., and J.J. Maricelli, 1964, Detailed stratigraphic control through dip computations: AAPG Bull., v. 48, p. 1902-1910.

Gussow, W.C., 1954, Differential entrapment of oil and gas: a fundamental principle: AAPG Bull., v. 38, p. 816-853.

Humphris, C.C., Jr., 1978, Salt movements on continental slope, northern Gulf of Mexico *in*, A.H. Bouma, et al, eds., Framework, facies, and oil trapping characteristics of the upper continental margin: AAPG Studies in Geology 7, p. 69-85.

Martin, R.G., 1978, Northern and eastern Gulf of Mexico continental margin: stratigraphic and structural framework *in*, A.H. Bouma, et al, eds., Framework, facies and oil trapping characteristics of the upper continental margin: AAPG Studies in Geology 7, p. 21-42.

——— and J.E. Case, 1975, Geophysical Studies in the Gulf of Mexico, *in* A.E.M. Nairn and F.G. Stehli, eds., Ocean basin and Margins; The Gulf of Mexico and the Caribbean: New York, Plenum Press, v. 3, p. 65-106.

——— and A.H. Bouma 1978, Physiography of Gulf of Mexico *in*, A.H. Bouma et al eds., Framework, facies, and oil trapping characteristics of the upper continental margin: AAPG Studies in Geology 7, p. 3-19.

Mitchum, R.M., Jr., P.R. Vail, and J.B. Sangree, 1977, Seismic stratigraphy and global changes of sea level, part 6: stratigraphic interpretation of seismic reflection patterns in depositional sequences *in* C.E. Payton, ed., Seismic stratigraphy—applications to hydrocarbon exploration: AAPG Mem. 26, p. 117-133.

Norwood, E.M. Jr., and D.S. Holland, 1974, Lithofacies mapping, a descriptive tool for ancient delta systems of the Lousiana outer continental shelf: Gulf Coast Assoc. Geol. Soc. Trans., v. 24, p. 175-188.

Tissot, B.P., and D.H. Welte, 1978, Petroleum formation and occurrence—a new approach to oil and gas exploration: New York, Springer-Verlag, 538 p.

Uchupi, E., 1975, Physiography of the Gulf of Mexico and Caribbean Sea, *in* A.E.M. Nairn and F.G. Stehli, eds., Ocean basins and margins; The Gulf of Mexico and the Caribbean: New York, Plenum Press, v. 3, p. 1-64.

Woodbury, H.O., et al, 1973, Pliocene and Pleistocene depocenters, outer continental shelf of Louisiana and Texas: AAPG Bull., v. 57, p. 2428-2439.

Painter Reservoir Field—Giant in the Wyoming Thrust Belt[1]

By Charles F. Lamb[2]

Abstract Painter Reservoir field is the largest of several recent Nugget Sandstone hydrocarbon discoveries in the Wyoming Thrust Belt province. The field is located in Uinta County, Wyoming, 5 mi (8 km) northeast of Evanston and on trend with the Clear Creek and Ryckman Creek fields, 5 and 10 mi (8 and 16 km), respectively, northeast, which are also productive from the Nugget.

The field discovery well, Chevron-Federal 22-6A, was drilled in mid-1977 on a seismically mapped anticlinal structure. The Nugget Sandstone was entered at 9,728 ft (2,918 m), and 1,355 ft (407 m) was penetrated to the total depth of 11,083 ft (3,325 m). After extensive testing, potential of the well on October 22, 1977, was 410 bbl of oil per day and 859 Mcf of gas per day on 15/64-in. choke, flowing tubing pressure 1,275 psi. Flow rates as high as 1,500 bbl of oil per day were recorded on larger chokes. Gravity of the oil is 48.4° API. Active development began immediately and is still in progress.

Field limits and structural configuration are not yet fully known, but seismic and drilling data indicate an overturned fold associated with the hanging wall of the Absaroka thrust. Present drilling has established an oil and gas column of over 1,000 ft (300 m). The producing Nugget is a cross-bedded, quartz sandstone over 850 ft (255 m) thick with an average porosity of 14.1% and permeability ranging from 0 to 1,000 md. Analysis of the oil suggests a Cretaceous source.

INTRODUCTION

Painter Reservoir field is currently undergoing development; therefore, a comprehensive report of the field is premature. This is a brief description of the regional setting and exploratory history, the structure and stratigraphy of the field area, and an idea of the size and shape of the reservoir based on present information.

Painter Reservoir field produces from the Nugget Sandstone of Triassic-Jurassic age and is located in the Thrust Belt province of western North America (Fig. 1). The fold and thrust belt is about 100 mi (161 km) wide and extends for nearly 3,000 mi (4,830 km) from Alaska through Canada to Arizona and probably beyond.

In southwestern Wyoming and adjacent states, the thrust belt contains a series of well-defined, northerly trending thrust sheets (Fig. 2). Royse et al (1975) described the period of compressional deformation as beginning in Late Jurassic and continuing through early Eocene. East to west shortening was achieved through motion on low-angle thrust faults and associated concentric folding. After years of discouraging exploration in the United States part of the thrust belt, only minor production was established, and then only in the leading or eastern edge. In 1975, however, new seismic processing techniques and the discovery at Pineview field started a new surge of drilling which has resulted in a series of important discoveries. Production is now established at fields which span a north to

[1]Read before the Association in Houston, Texas, April 4, 1979. Manuscript received, July 19, 1979; accepted for publication, October 2, 1979.

[2]Chevron U.S.A., Inc., Denver, Colorado, 80201.

Despite the early stage of development of the Painter Reservoir field, Chevron has been generous in granting permission to publish this paper. The writer thanks Chevron, and especially thanks his many colleagues for their contributions, assistance, and help.

Article Identification Number:
0065-731X/80/M030-0013/$03.00/0.

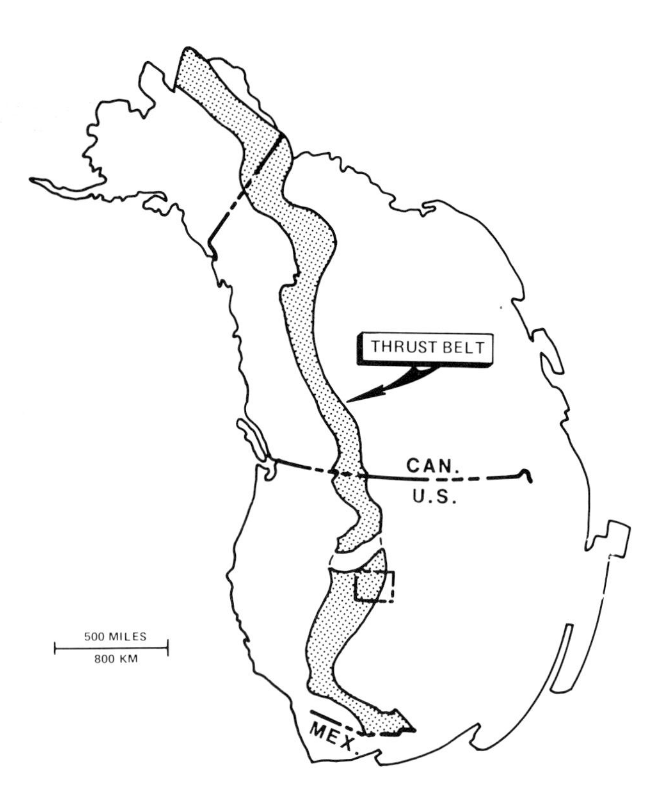

FIG. 1—North American continent showing Thrust Belt province (modified after Bally et al, 1966). Outlined area is Wyoming.

south distance of 70 mi (112 km) in two separate thrust sheets and 10 different formations from Cretaceous to Ordovician.

The hydrocarbon accumulation at Painter Reservoir field is found in a thrust fold involving the Nugget Sandstone in the hanging wall of the Absaroka thrust plate (Fig. 3). The Nugget has been thrust into position over the Cretaceous. Chromatographic analysis indicates that the hydrocarbons trapped in the Nugget originated from source rocks in the underlying, younger Cretaceous.

At the time the discovery well was drilled in mid-1977 by Chevron, only two fields were known to produce from the Nugget: Pineview and Ryckman Creek. The discovery well was drilled on a seismically defined structure on trend and 10 mi (16 km) south of Ryckman Creek. The original prospect map was based on seismic data that suggested a symmetrical anticline trending northeast to southwest. Earlier drilling on trend, however, indicated that the seismically interpreted gentle dipping eastern flank was probably caused by diffractions from a steep or overturned eastern limb. Particular attention was paid to locating the exploratory well on the better defined western flank. The well actually crossed the axis of the fold and entered the Nugget near the crest of the structure.

The discovery well, the Federal 22-6A, was drilled to a total depth of 11,083 ft (3,225 m) in structurally overthickened Nugget Sandstone which was penetrated at 9,728 ft (2,918 m). The well has 709 ft (216 m) of porous, productive Nugget, of which 319 ft (97 m) is in the oil column and 390 ft (119 m) is in the gas cap. The well was officially completed from the oil zone in October 1977, for 410 bbl of oil per day through 28 ft (8.5 m) of perforations from depths of 10,290 to 10,318 ft (3,136 to 3,145 m). During extensive testing, however, flow rates of 1,500 bbl of oil per day were recorded and the gas cap tested over 3 MMcf of gas per day with 335 bbl of condensate. Active development began immediately and is still in progress. In December 1977, a federal exploratory unit was formed with Chevron as operator. Amoco Production Co. and Champlin Petroleum Co. have leases within the productive area and are partners in the federal unit.

STRATIGRAPHY

Undifferentiated Green River–Wasatch of the Tertiary-Eocene is present at the surface at Painter Reservoir field. Figure 4 is a schematic columnar section of the field area. The Green River–Wasatch, together with the Paleocene-Cretaceous Evanston Formation, account for about 4,400 ft (1,341 m) of sedimentary rocks which angularly overlie the Lower Cretaceous Gannett Formation. The Gannett is a sequence 2,300-ft (701 m) thick of nonmarine, red and green siltstones, shales, and sandstones. Below the Gannett lies 950 ft (289 m) of dominantly red to green shales and siltstones of the Jurassic Stump and Preuss formations. Salt, up to 100 ft (30 m) thick, is present in the lower part of the Preuss.

Below the Preuss is the Jurassic Twin Creek Formation, which consists of 1,200 ft (366 m) of gray limestones and shales with some glauconitic sandstones in the upper members. The Twin Creek is nonproductive at Painter Reservoir; however, gas shows have been found during drilling. The anhy-

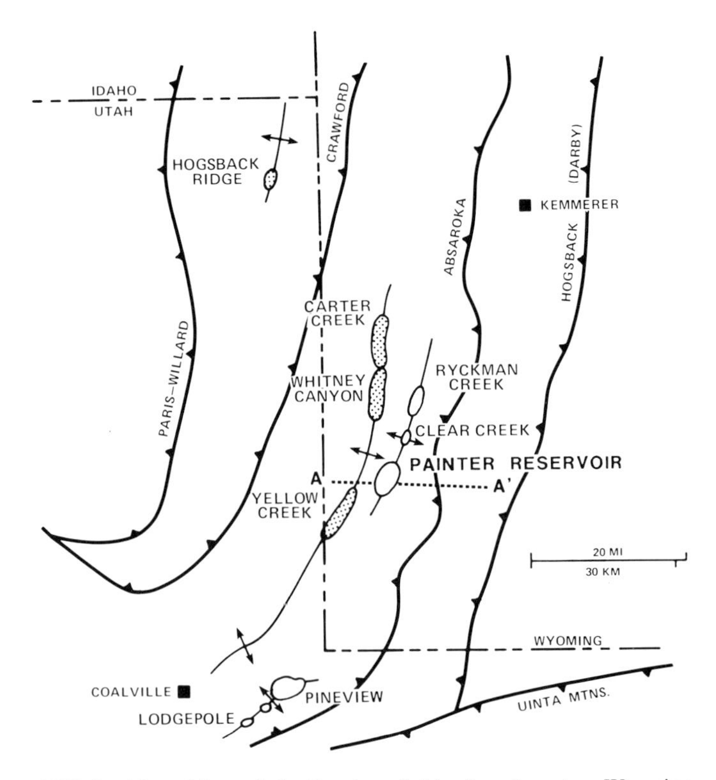

FIG. 2—Map of thrust-belt oil and gas fields of southwestern Wyoming and adjacent states, showing northerly-trending major thrust faults (modified after Royse et al, 1975). A-A′ is the line of section shown in Figure 3.

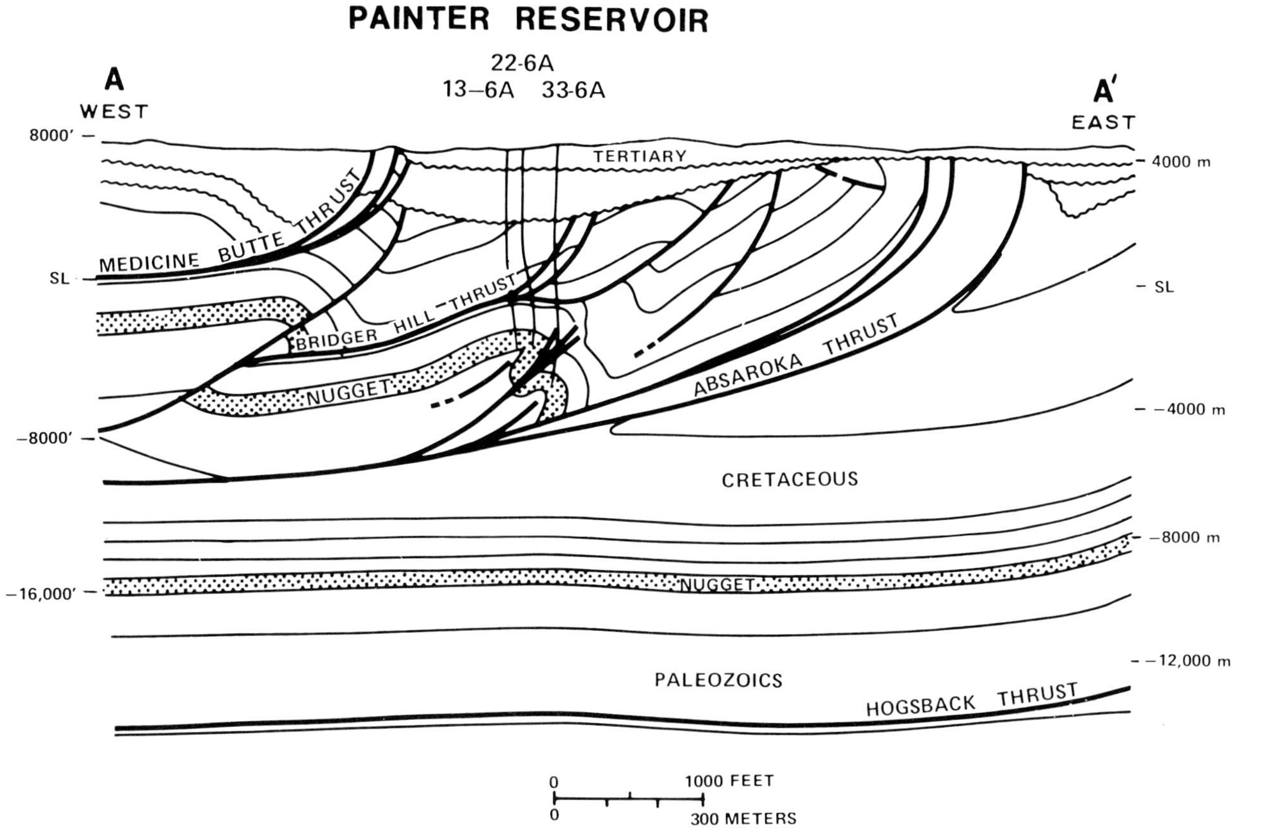

FIG. 3—West to east structure section through Painter Reservoir field. Depths on left are in feet. Location of section is shown in Figure 2.

dritic section of the lowermost Gypsum Spring Member of the Twin Creek forms the cap rock for the oil and gas accumulation in the underlying Nugget Sandstone.

The Nugget has a normal thickness of 880 ft (268 m). It is believed to be eolian in origin, and is typically white to pink, cross-laminated, massive with an average porosity of 14.1% but ranging up to 18%. Permeability averages 22.8 md but locally exceeds 1,000 md. Sand grains are mostly very fine to medium, well sorted, and subrounded. X-ray diffraction analysis shows the mineral composition to be 90 to 95% quartz, 5% feldspar, with minor amounts of calcite, dolomite, and illite clay in the form of authigenic overgrowths.

Figure 5 shows scanning electron micrographs of the Nugget Sandstone in the discovery well core from 9,734 to 9,735 ft (2,967 to 2,967.3 m). Sand grains are about 0.5 mm in size. On the lower magnification, the sand grains appear relatively clean, the pore throats open. The higher magnification, however, shows extensive secondary quartz overgrowth on the grain surface, and illite clay beginning to fill the pore spaces.

STRUCTURE

The production at Painter Reservoir is at the crest of an overturned anticline (Figs. 6, 7). The relatively minor Bridger Hill thrust glides in the lower Preuss salt and produces a fold and fault complex which is completely detached from the underlying productive structure. Seismic and sub-surface data show a doubly plunging, northeast-trending fold at the Nugget level with a well-defined western flank dipping about 20°. The southeastern flank is overturned, complexly faulted, and consequently much less well defined seismically.

The Federal 33-6A (the easternmost well on the cross section) is a dry hole drilled as the southeastern offset to the discovery. This well was drilled on the steep southeastern flank and found the Nugget in a separate fault block, 2,300 ft (701 m) low, water-productive, and dipping 15° southeast. It encountered an overthickened section of Twin Creek resulting from high, nearly vertical dips and at least two reverse faults. A second dry hole about 1 mi (1.6 km) southwest was drilled in a similar structural setting. These two wells have defined the eastern field limits. The discovery well (middle well on the cross section) drilled an overthickened Nugget of over 1,300 ft (396 m) and was still in Nugget at total depth. Dipmeter data show a southeasterly dip of 20° in the Twin Creek and upper Nugget. Overturning is indicated by the northwesterly formational dip in the bottom 100 ft (30 m) of the hole.

The wells on the more gentle northwest flank of the structure have essentially normal Twin Creek and Nugget sections. The 33-31B (0.5 mi or 0.8 km northeast of the discovery well) had 1,217 ft (371 m) of Nugget and was still in Nugget at total depth, suggesting a structural position similar to the discovery.

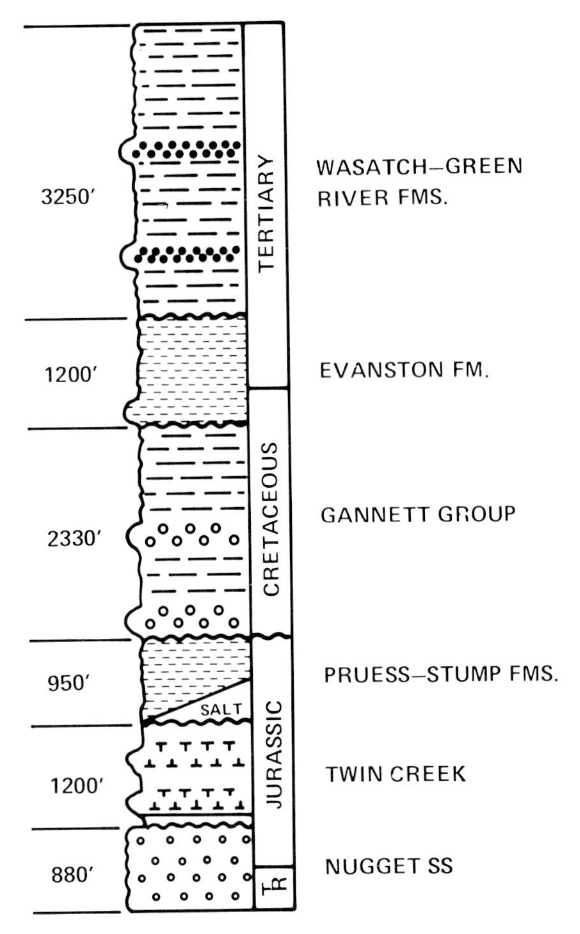

FIG. 4—Schematic columnar section of Painter Reservoir field. Thicknesses are in feet.

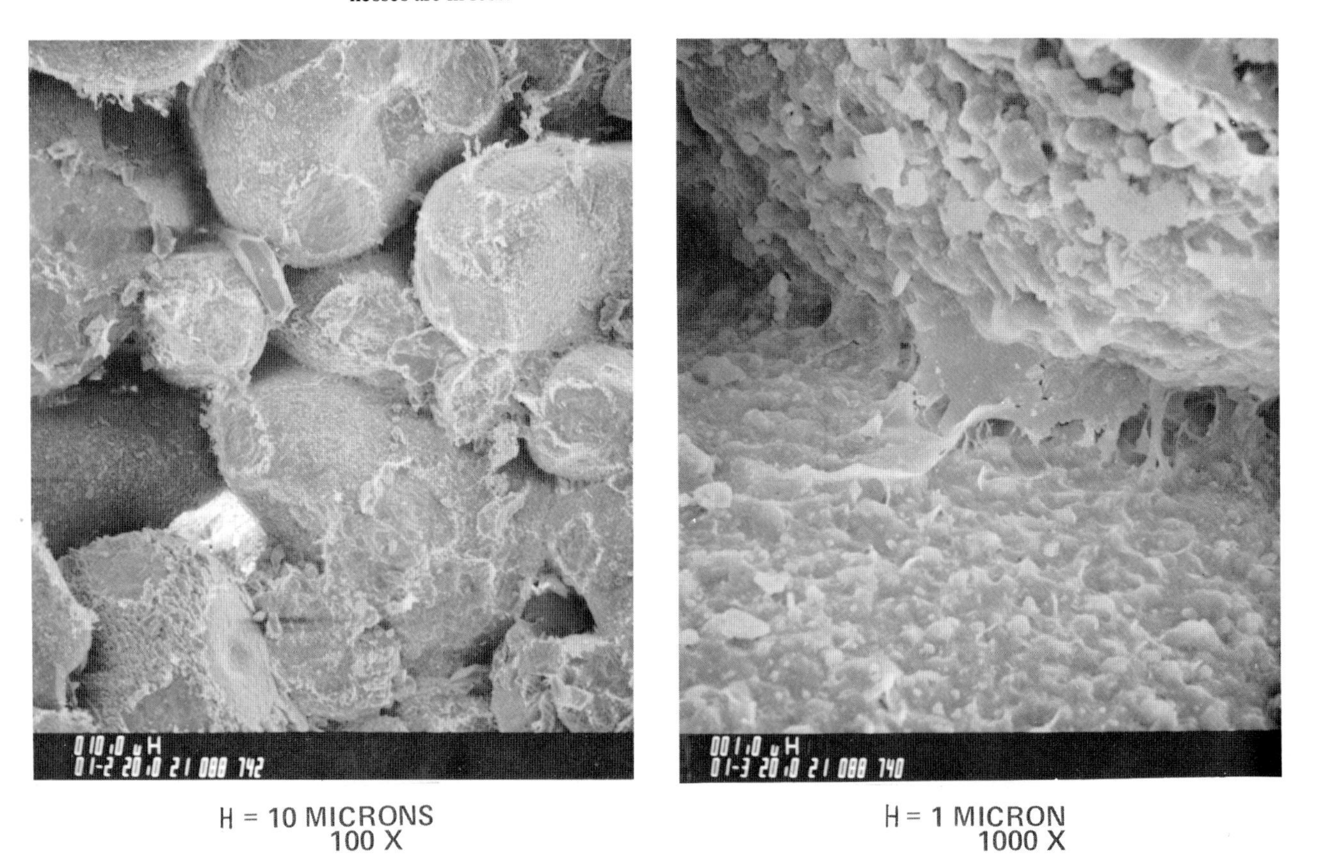

FIG. 5—Scanning electron micrographs of Nugget Sandstone at Painter Reservoir field. Sample is from core of discovery well from the depth interval 9,734 to 9,735 ft (about 2,967 m). In left photo, horizontal bar (scale) is 10 μ; on right, horizontal bar (scale) is 1 μ.

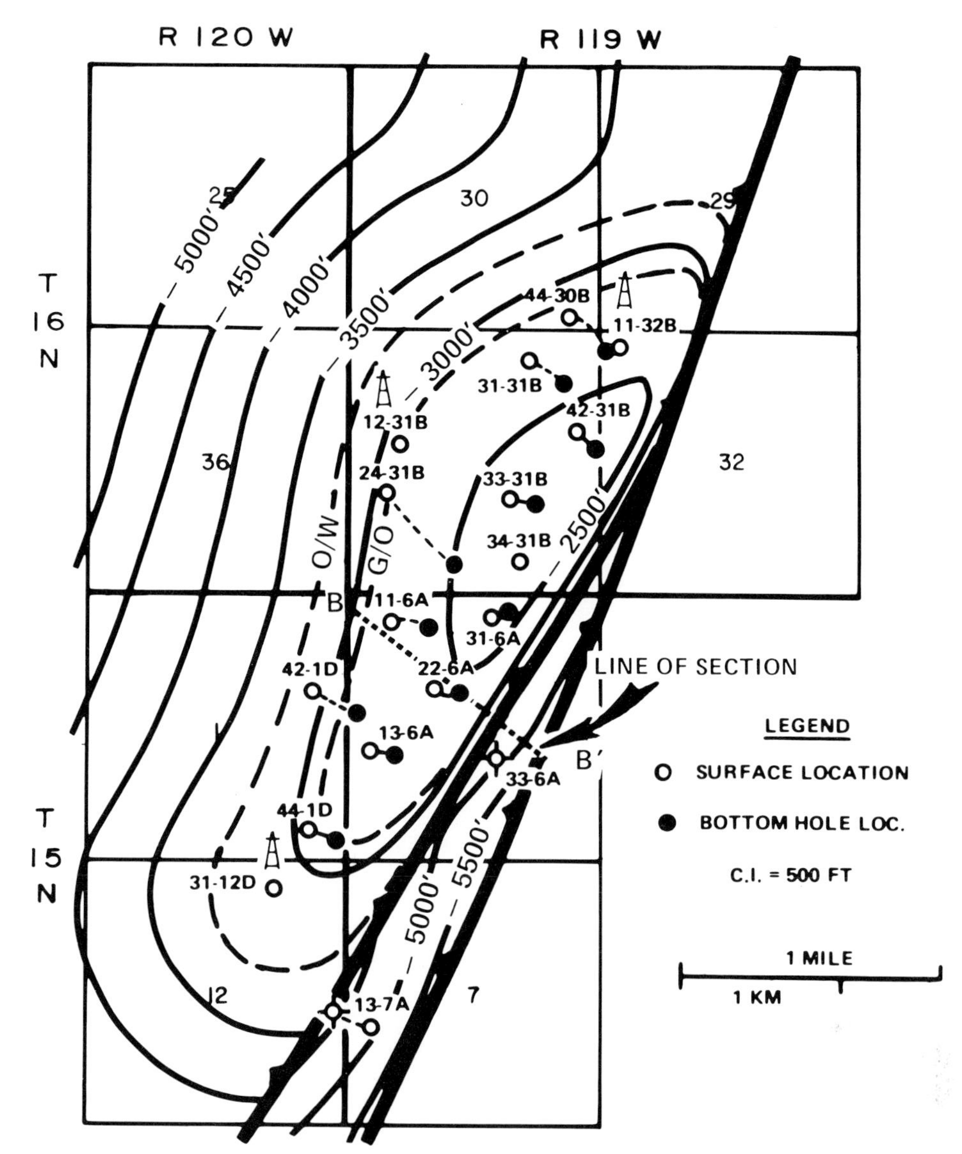

FIG. 6—Structural contours of the top of Nugget Sandstone at Painter Reservoir field. B-B′ is the line of section for Figure 7.

This well, incidentally, encountered the thickest pay section to date, over 1,025 ft (312 m) of total hydrocarbon column. Seismic data indicate that the structure has about 1,100 ft (335 m) of closure, with northeast plunge against the Ryckman Creek-Clear Creek structural trend being critical.

FIELD DEVELOPMENT AND RESERVOIR CHARACTER

Development drilling is currently in progress with three rigs working in the field. As of March 1979, 11 wells were completed in the oil zone, flowing a total of 4,000 bbl of oil per day. Two dry holes have been drilled. One well is waiting on completion. The cumulative production to February 1, 1979 is 526,000 bbl. Average perforations are 56 ft (19 m). Wells are currently drilled on 80-acre (32 ha.) spacing to a maximum depth of 11,000 ft (3,353 m). Completed wells cost about $2,300,000.

Figure 8 shows part of the neutron-density log of the discovery well. Fluid contacts are not apparent from wire-line logs, but have been partly defined by production testing. Reservoir characteristics are summarized in Table 1. An oil column of 319 ft (97 m) and a maximum gas column of 750 ft (228 m) are indicated. The gas-oil contact appears to be at a sea-level datum of −2,933 ft (−894 m) and the oil-water contact at −3,252 ft (−991 m).

Preliminary studies indicate that the reservoir energy is derived from an expanding gas cap and an active water drive. Currently, the field is producing from the oil zone, and produced gas is reinjected into the gas cap to maintain reservoir pressure and conserve gas. Chevron is also evaluating the prospect of producing both the oil and gas and injecting nitrogen into the gas cap to maintain reservoir pressure. The oil is sweet, 48.4° API gravity, with a solution gas-oil ratio of 2,133 cu. ft. of gas/bbl oil. Gas is 1,276 Btu. It is expected that modeling studies now in progress will indicate a spacing of 40

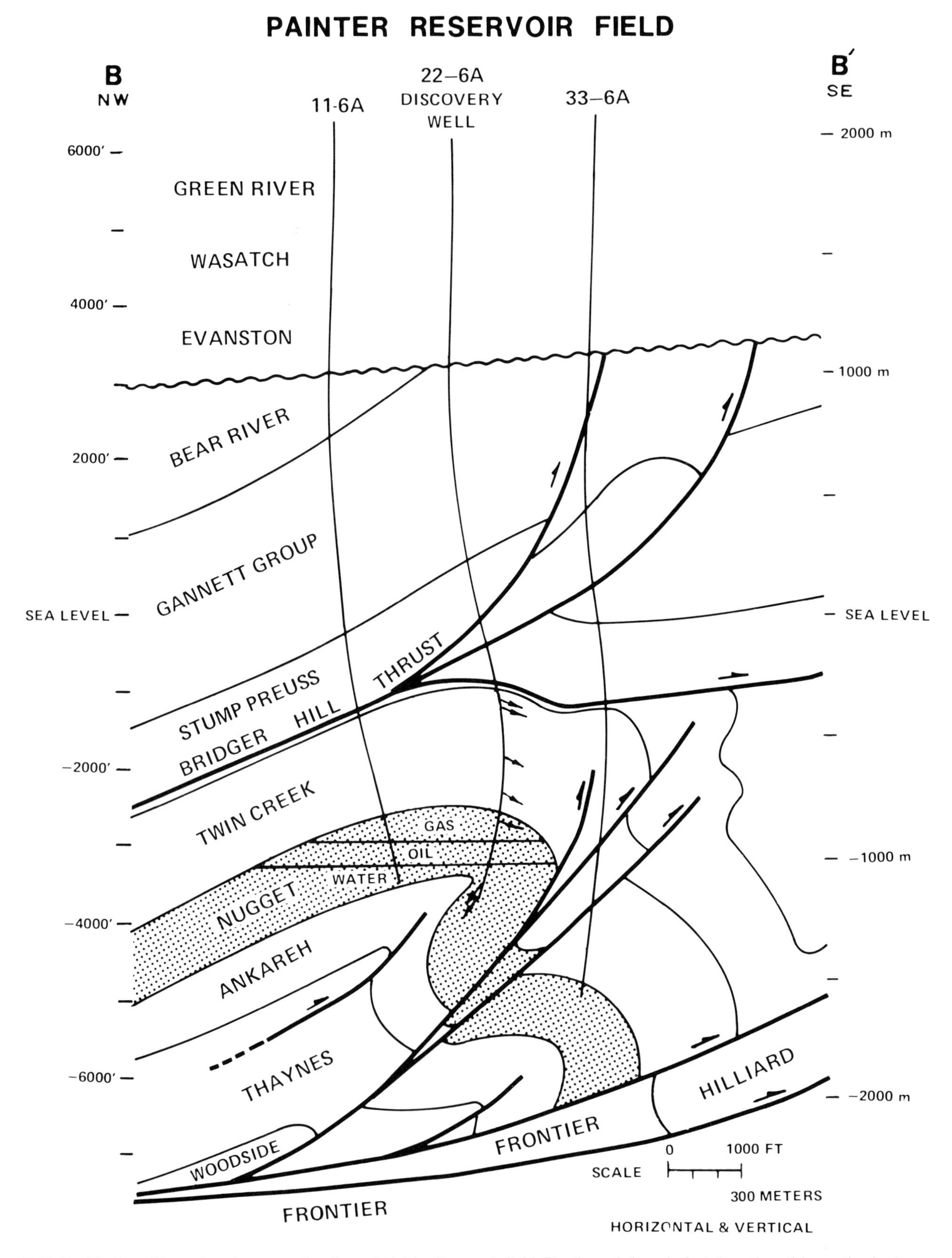

FIG. 7—Northwest to southeast cross section through Painter Reservoir field. Depths on left are in feet. Location of the section is shown on Figure 6.

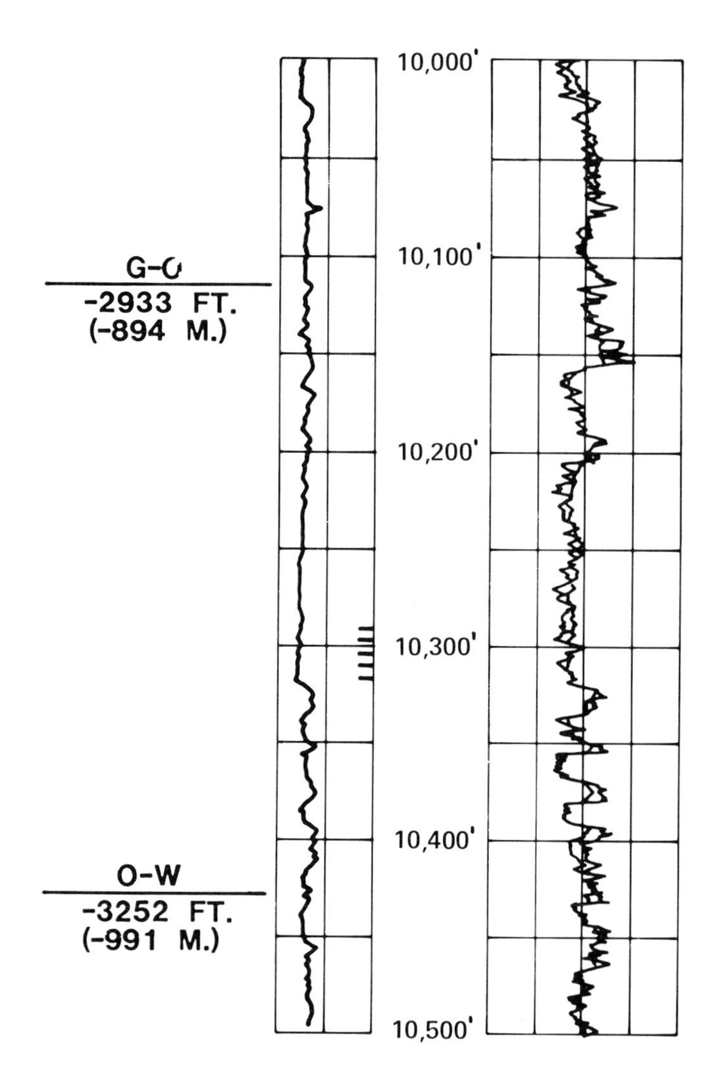

FIG. 8—Part of neutron-density/gamma ray log of Chevron's Federal 22-6A discovery well at Painter Reservoir field, showing lack of "gas effect" in gas cap. Depths are in feet. Initial potential on October 22, 1977, was 410 bbl of oil per day plus 859 Mcf of gas per day from perforations in the depth range of 10,290 to 10,318 ft (3,136 to 3,145 m).

Table 1. Reservoir Characteristics of Painter Reservoir Field, February 1979

Characteristic	Amount/Type
Average porosity	14.1%
Average permeability	22.8 md
Gas-oil contact	-2,933 ft (-894 m)
Oil-water contact	-3,252 ft (-991 m)
Maximum gas column	750 ft (228 m)
Oil column	319 ft (97 m)
Reservoir energy	Expanding gas cap and water drive
Solution gas/oil ratio	2,133 cf/bbl
Gas heat value	1,276 BTU
Bubble point	4,035 PSI @ BHT 164° F (73°C)
Original BHP	4,150 PSI
Type of crude	48.4° API, paraffin base, sweet, pour point 35°F

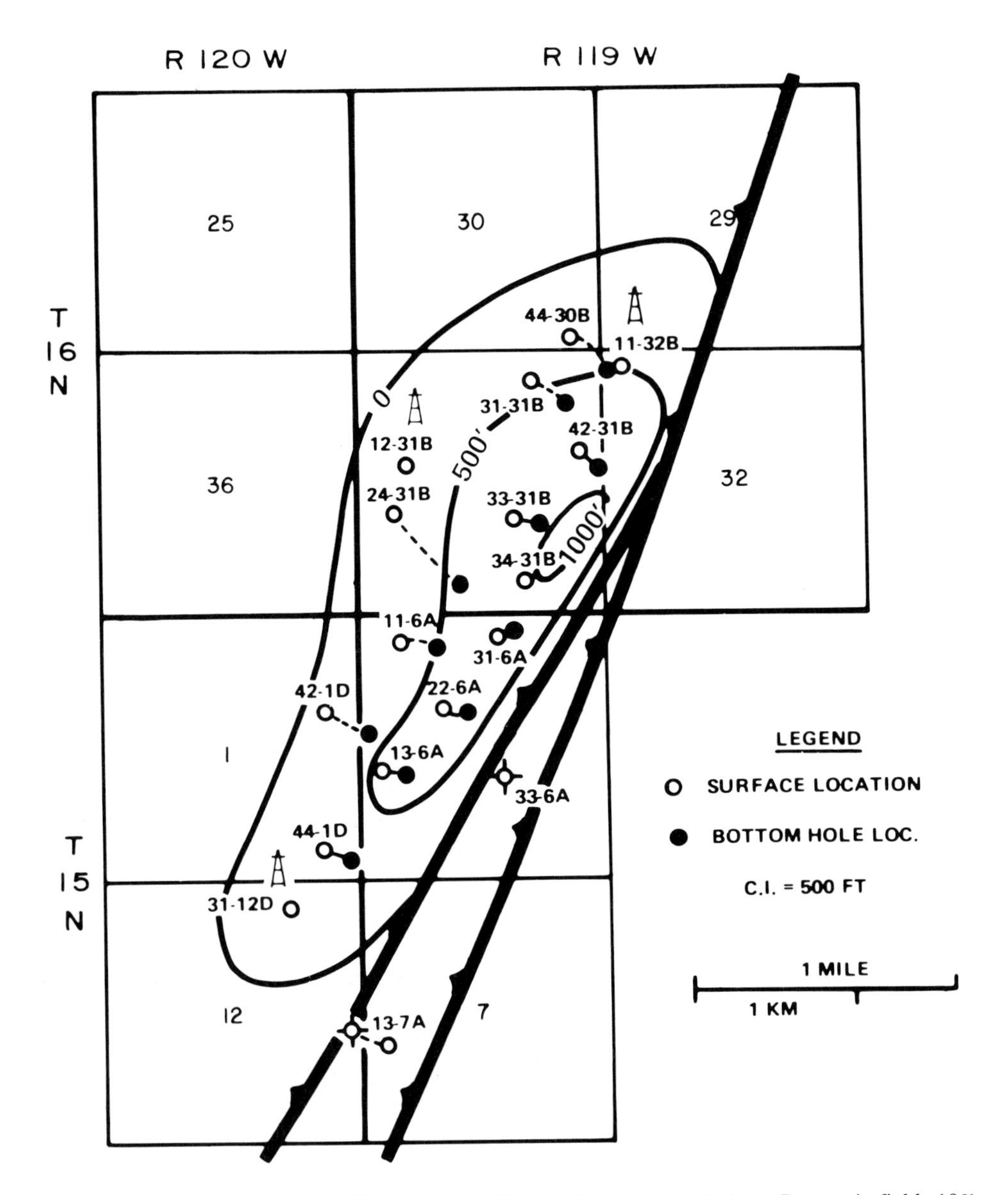

FIG. 9—Isopach map of net effective pay in Nugget Sandstone at Painter Reservoir field; 10% minimum porosity is used to define effective sandstone.

acres (16 ha.) per well for most efficient reservoir depletion.

Figure 9 is a map of the net effective pay in the hydrocarbon column. To define effective sand, 10% minimum porosity is currently used. The similarity to the structure map emphasizes that a high percentage of the gross sandstone is effective in the Nugget.

Although field limits are not well defined, it is estimated that something over 1,600 acres (647 ha.) will prove productive. Recoverable reserves are currently estimated to be over 100 million bbl of oil and oil-equivalent gas; 85 million bbl are estimated in the oil column and 380 Bcf in the gas cap.

SUMMARY

Although still in the development stage, Painter Reservoir field apparently will qualify as a giant discovery. It is located in the southwestern Wyoming part of the Thrust Belt province. Production is from the Nugget Sandstone of Jurassic-Triassic age. The field has an oil column of 319 ft (97 m) with a gas cap of 750 ft (228 m) in an overturned fold with over 1,100 ft (335 m) of closure; the fold trends northeast to southwest in the hanging wall of the Absaroka thrust plate. It is expected that the field will cover approximately 1,600 acres (647 ha.) when fully developed and have ultimate recoverable reserves of over 100 million bbl, including the gas cap reserves.

REFERENCES CITED

Bally, A. W., P. L. Gordy, and G. A. Stewart, 1966, Structure, seismic data and orogenic evolution of southern Canadian Rocky Mountains: Bull. Canadian Petroleum Geology, v. 14, p. 337-381.

Royse, F., Jr., M. A. Warner, and D. L. Reese, 1975, Thrust belt structure geometry and related stratigraphic problems, Wyoming-Idaho-Utah, *in* Deep drilling frontiers of the central Rocky Mountains: Rocky Mtn. Assoc. Geologists, p. 41-54.

Prudhoe Bay—A 10-Year Perspective[1]

By H. C. Jamison[2], L. D. Brockett[2], and R. A. McIntosh[3]

Abstract The Prudhoe Bay field is recognized as the largest oil field in the United States. The Permian-Triassic reservoirs, estimated to contain reserves of 9.6 billion bbl of oil and 26 Tcf of gas, have overshadowed other known substantial accumulations of hydrocarbons in formations ranging in age from Mississipian to Cretaceous in the general area of Prudhoe Bay. This study is a summary of the geology of the Lisburne carbonate rocks, as well as the Kuparuk River sandstone reservoirs and their potential. The regional structure and stratigraphic relationships of other less significant Permian-Triassic and Cretaceous accumulations are also included.

Perhaps unrecognized, except in retrospect, is the significance of the planned sequential availablility of both Federal and State lands on the North Slope beginning in 1958. An 11-year period of land availability followed a 14-year moratorium. The history of exploration that led to the discovery in 1968 is presented from that viewpoint. This period culminated with the (September, 1969) State of Alaska "Billion Dollar Sale."

The post-discovery sequence of exploration, development, and production in the area has been characterized by environmental, social, legal, political, and economic complexity and controversy. Comparison of the status of petroleum exploration today on the North Slope of Alaska with the history of the 1950s through the early 1970s is an object lesson for explorationists.

INTRODUCTION

More than a decade has passed since the announcement of the discovery of the Prudhoe Bay field and the completion of the Prudhoe Bay State 1 well on April 15, 1968. The well was drilled by Atlantic Richfield Company as operator for itself and Humble Oil and Refining Company (now Exxon). Since that time at least ten more North Slope hydrocarbon accumulations have been discovered by these and other companies. Industry exploration activity and interest is now at the highest level since the post-discovery 1969-1970 peak period because of the December, 1979, Federal-State Beaufort Sea Lease Sale.

Petroleum exploration in northern Alaska has focused on the part of the Arctic basin between the Brooks Range and the Beaufort Sea. This region, which contains the north slope of the Brooks Range drainage system, the Arctic Foothills and the Arctic Coastal Plain, is referred to as the North Slope. Its geology is characterized by complex stratigrahic and structural relationships. The sedimentary section ranges in age from pre-Devonian through Tertiary and contains numerous zones with reservoir potential. Regional tectonics have resulted in combinations of structural elements and stratigraphic variations favorable for the entrapment of hydrocarbons. This is

[1]Manuscript received, April 10, 1979; accepted for publication, July 30, 1979. Read before the Association at Houston, April 4, 1979.

[2]ARCO Oil and Gas Co., Dallas, Texas.

[3]Teck Corporation, Calgary, Alberta.

The writers thank the ARCO Oil and Gas Company for permission to publish this paper. Our appreciation is extended to company geologists of the Alaska North District Office in Anchorage, and especially to P. A. Barker, J. K. Lawrence, O. P. Majewski, J. C. Merritt, G. F. Player, and R. W. Tucker for contributions to the content and preparation of the paper. We also acknowledge the expert assistance of Charles Steiglitz for preparation of the illustrations.

Article Identification Number:
0065-731X/80/M030-0014/$03.00/0.

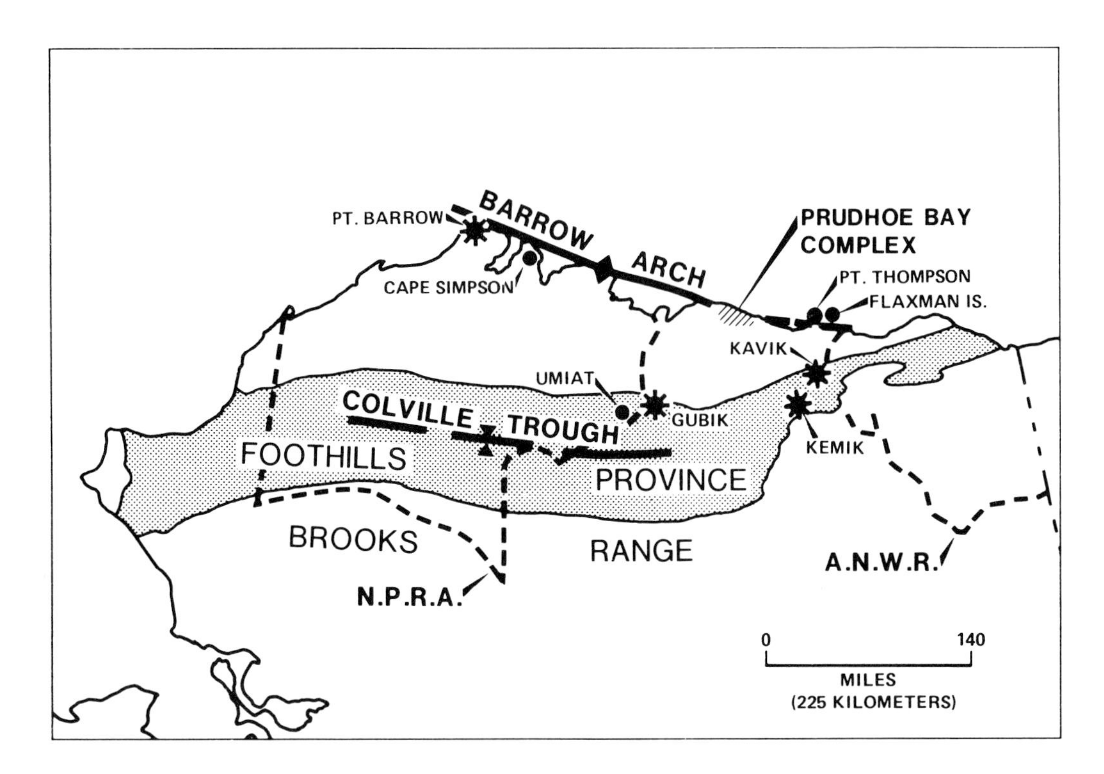

FIG. 1—North Slope index map showing major structural elements.

particularly true of the "Prudhoe Bay Complex," defined as that part of the Barrow arch between the Colville and Sagavanirktok Rivers and including associated Mississippian through Tertiary age reservoirs on its crest and flanks.

The Prudhoe Bay Complex is in the northeastern part of the 65,000 sq mi (168,300 sq km) North Slope region. This northeast area lies between the National Petroleum Reserve-Alaska (NPR-A) and the Arctic National Wildlife Range (ANWR), both withdrawn from entry by industry (Fig. 1). Onshore hydrocarbon accumulations discovered by industry have been limited to this northeastern area. A recent opinion by the Solicitor of the Department of Interior would restrict near-term offshore federal-state lease availability to a seaward projection of this same northeastern area.

All physiographic and climatologic obstacles to industry operations that exist within the region have been met successfully in the past, either on the North Slope or in the adjacent Mackenzie River delta region of Canada. Although environmental regulations cause time delays, operational problems and added costs, these difficulties are being surmounted. Successful exploration in the future will hinge on land availability and economic viability based upon known and reasonably predicatable federal and state policies.

NORTH SLOPE EXPLORATION HISTORY

Several published geological papers briefly documented events which preceded the Prudhoe Bay discovery in 1968, and more particularly contributed to the growing body of geologic knowledge through 1975. Among these is the summary of exploration history by Gryc (1970), and comments by Rickwood (1970), Morgridge and Smith (1972), and Jones and Speers (1976). A more recent chronology by Jamison (1978) amplifies a specific thesis—that planned sequential land availability was a dominant factor in the evolution of exploration activities and eventual success. This thesis is one that becomes more apparent with the passage of time and the awareness of restrictions placed on the normal activities of petroleum explorationists in our nation today.

Early Phase—Federal Government Programs

Oil and gas seepages were known to the Eskimos of the North Slope long before the advent of the white man. The first formal descriptions were recorded by Leffingwell of the U.S. Geological Survey in 1919. By 1921 prospecting permits were filed under the mining laws, and the expeditions of the geologists of the Geological Survey and their classic reports identified petroleum possibilities in northern Alaska. In 1923, during President Harding's administration, Naval Petroleumm Reserve No. 4 (now National Petroleum Reserve-Alaska) was established by Executive Order. The Geological Survey conducted reconnaissance mapping from 1923 through 1926 at the request of the Navy and published the results in 1930. In 1943 during World War II, the entire North Slope was withdrawn from public entry under Public Land Order 82, and the Navy, again with the U.S. Geological Survey, initiated a major exploration program.

This program was to last almost a decade and was to lay a solid foundation of both scientific and practical operational knowledge for the future. Reed (1958) thoroughly recorded the history and significance of this period of exploration. The eval-

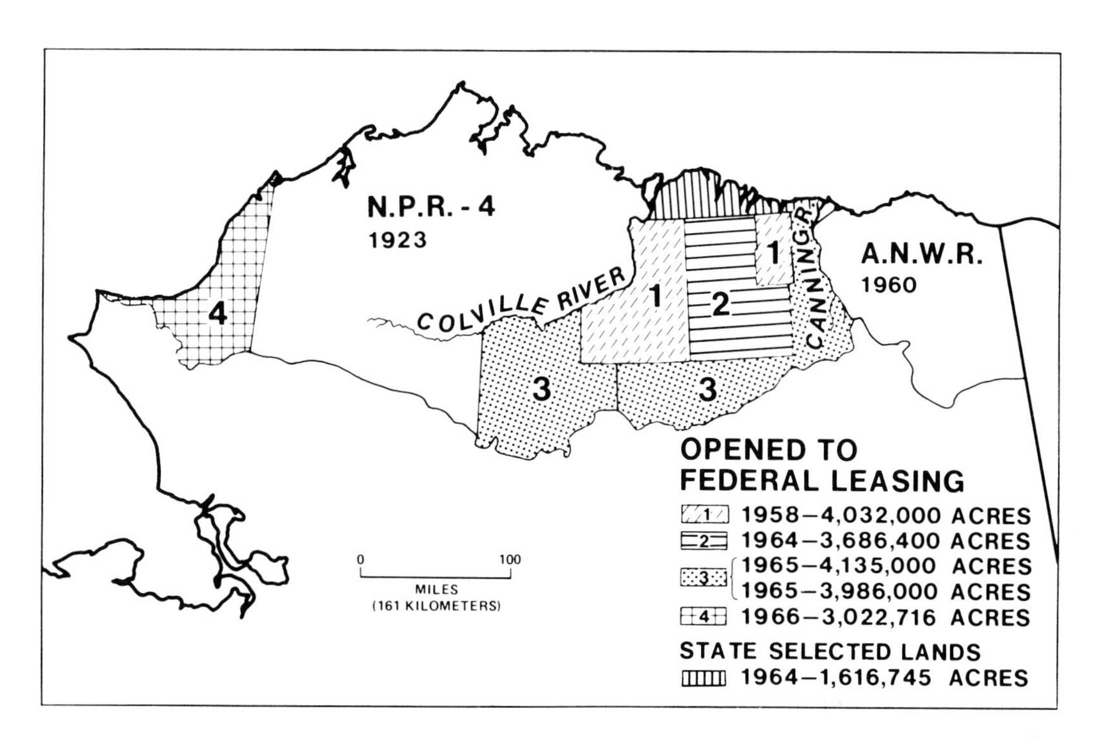

FIG. 2—North Slope land availability from 1958 through 1969.

uation of a 23-million acre (9.3 million ha.) area lying generally west and north of the Colville River, extending from the Beaufort Sea on the north to the Northern Foothills of the Brooks Range on the south, called for a monumental effort. Extensive geologic surface mapping, seismic, gravity, and magnetic surveys comprised the basic framework, and 45 shallow core holes aided in local and regional interpretation. When the program terminated in 1953, the Navy had drilled 37 test wells and found 3 oil accumulations at Umiat, Cape Simpson, and Fish Creek, as well as 6 gas accumulations at Gubik, South Barrow, Meade, Square Lake, Titaluk, and Wolf Creek. Of these, only 2 could be classified as sizable, namely, Umiat with potential reserves estimated at 30 to 100 million bbl of oil, and Gubik with potential reserves of 370 to 900 Bcf of gas. Even these fields were, and are, uneconomic to produce. However, the minor accumulation at South Barrow has been extended and still supplies gas to the native village (Fig. 1).

By 1953, the federal government effectively ceased 30 years of exploration activity in NPR-4 having spent an estimated $50 to $60 million in an unsuccessful effort to find expoitable petroleum resources. The succeeding 5-year period was quiescent, but two significant events were to occur which would bring a resurgence of exploration to the North Slope and eventually result in a new productive petroleum province of worldwide importance. First was the discovery of the Swanson River field, and second was the end of the federal moratorium on North Slope land availability.

Prediscovery Phase—Initial Industry Programs

Of primary importance was the initial commercial oil discovery in the Alaska Territory, the Richfield Oil Corporation Swanson River Unit 1 well on the Kenai Peninsula in August, 1957. This discovery contributed significantly to Alaskan statehood in 1959. It also signified the beginning of intensive industry exploration efforts in all the sedimentary basins of Alaska. One of the focal areas for industry interest was the North Slope. The potential for major petroleum reserves was well known from the pioneering work of the U.S. Geological Survey. The incentive was greater because of the Swanson River field discovery, the imminent probability of statehood, and the announced intention of the Bureau of Land Management (BLM) to make lands available for lease acquisition. The first lands became available in early 1958 when 16,000 acres (6,475 ha.) were put up on a competitive bidding basis in the Gubik Gas field area, classified as a Known Geological Structure. Later in the same year, approximately 4 million acres (1.6 million ha.), lying generally east and southeast of NPR-4 were offered by the BLM for simultaneous filing and subsequent drawing (Fig. 2).

This knowledge of substantial avails of federal lands for exploration and drilling and possible development under the same basic conditions that had been established throughout the western United States was the key factor in assuring serious interest on the part of the petroleum industry in exploring the North Slope.

During the summer of 1958, Sinclair Oil Corporation initiated surface geologic work on the North Slope with a field party based at Umiat for the 3-month field season. In September, after a 60-day simultaneous filing period, the BLM held the drawing from the 7,500 offers to lease the 4 million acres (1.6 million ha.) divided among 1,300 "blocks" (a lease comprised of 4 contiguous sections). This was the first sizable area to become available from lands withdrawn under PL0 82 in

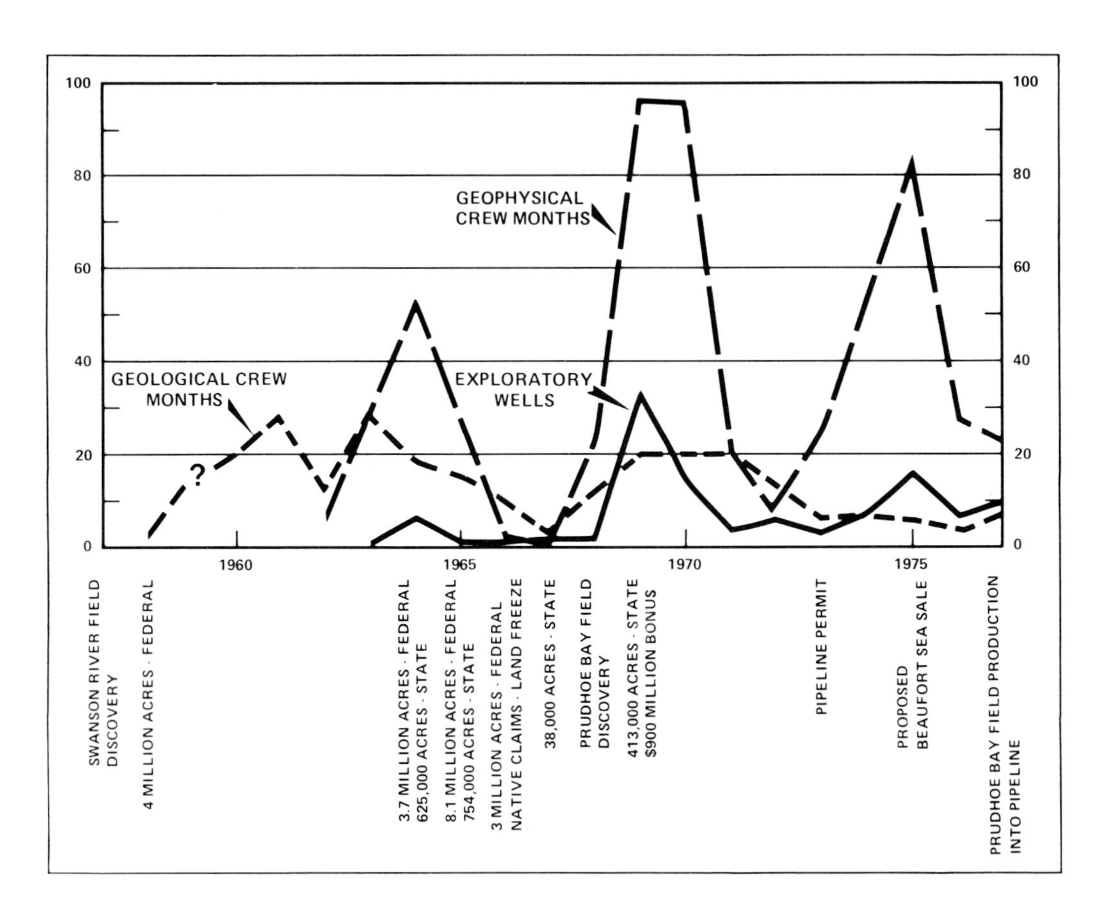

FIG. 3—North Slope exploration activity from 1958 through 1977 (excluding National Petroleum Reserve–Alaska).

1943. In the summer seasons of 1959, 1960, and 1961, 5 to 7 companies operated geologic field parties and major companies began to acquire substantial acreage positions. Sinclair and British Petroleum Company signed a joint exploration agreement in 1961 that initiated a commitment of major efforts and substantial funds to exploration on the North Slope.

In December of 1960 the federal government established the Arctic National Wildlife Range (ANWR) covering almost 9 million acres (3.6 million ha.) in an area extending from the Canning River on the west to the Canadian border on the east; and including the Brooks Range on the south to the Beaufort Sea on the north (Fig. 2). The creation of the Wildlife Range and the continued existence of NPR-4 thus limited industry entry to an area lying between the Colville and Canning Rivers. Additional areas west of NPR-4 and south to the Arctic Foothills would later become available for leasing, but two major potential areas comprising 32 million acres (12.9 million ha.) were eliminated from leasing and most industry exploration activities. They remain so to the present.

During 1962, 1963, and 1964 industry exploration programs expanded rapidly on the North Slope. Surface geologic parties representing a maximum of 10 companies accounted for 60 crew-months in the 3 field seasons. Seismic-crew activity was begun by Sinclair and British Petroleum (BP) in 1962. This was the first seismic work conducted since the termination of the Navy program in 1953, and marked the first such industry operation on the North Slope. Seismic crew-months advanced from 6.5 in 1962 to 29.25 in 1963 and culminated with 53.5 in 1964.

In the same period BP/Sinclair drilled a total of 7 exploratory wells located by surface geology and the recent seismic surveys. All of the wells penetrated a Cretaceous clastic section and all were unsuccessful in finding economic reserves.

In 1964 Richfield and Humble signed a joint exploration agreement with Richfield acting as operator. Richfield conducted surface geologic work during 1959, 1960, and 1963, and began seismic work near the confluence of the Sagavanirktok and Ivishak Rivers in late 1963. The reconaissance geologic program, although encompassing virtually all the North Slope, was focused on evaluation of the Northern Foothills and the coastal plain between the Colville and Canning Rivers where BLM lands were to become available to acquisition by simultaneous filing. The reconnaissance seismic coverage was designed as a grid which would broadly define the structural trends and potential closures on folds in the specific area of land avails. In 1964 seismic coverage was extended north to the coast and it was at this time the two companies obtained first definition of the Prudhoe Bay structure.

Department of Interior (DOI) policy allowed the formation of Development Contracts in remote areas of Alaska which removed lands leased by the operating companies in these contract areas from federal chargeability limits. The conditions of

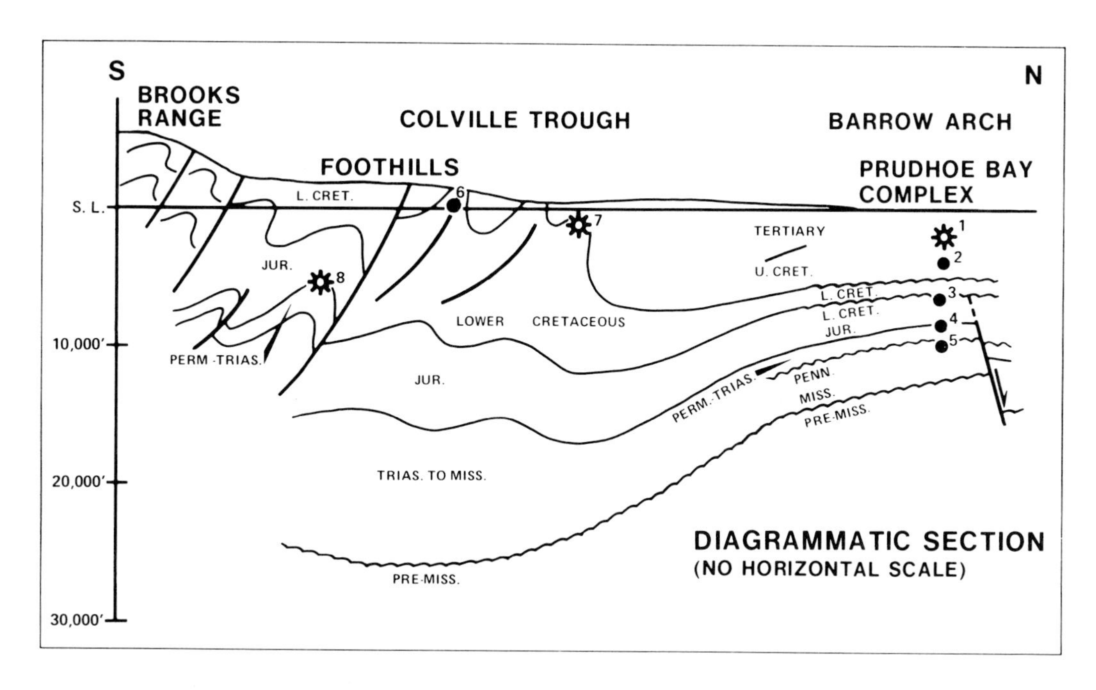

FIG. 4—North to south diagrammatic cross section from Brooks Range to the Prudhoe Bay Complex, Stratigraphic position of accumulations in several geographic locations indicated by symbols. *Prudhoe Bay Complex:* (1) Tertiary gas, (2) West Sak Sands, (3) Kuparuk River Formation, (4) Permo-Triassic, (5) Lisburne Group. *Umiat:* (6) Lower Cretaceous. *Gubik:* (7) Upper/Lower Cretaceous. *Kavik and Kemik:* Permo-Triassic.

the DOI approval required an exploration program and minimum annual expenditures. Seven such contracts were approved in the early 1960s and were a substantial incentive for companies willing to invest the capital necessary for thorough exploration, including drilling exploratory wells in a high-risk environment.

In mid-1964 the Bureau of Land Management held the second major simultaneous filing and drawing on 3.68 million acres (1.49 million ha.) in the "Corridor" area (between the east and west segments of the 1958 filing) of the Sagavanirktok and Kuparuk Rivers. Under the Statehood Act, the State of Alaska selected some 80 townships across the northern tier of lands between the Colville and Canning Rivers and received tentative approvals on the 1.6 million acres (0.65 million ha.) from the federal government in October. In December the state held the 13th State Competitive Sale (the first on the North Slope) of leases covering 625,000 acres (253,000 ha.) in the area east of the Colville River delta (Fig. 2). Individual company exploration programs had been planned to acquire geologic and geophysical data on a regional and detailed basis so as to be competitive and to participate intelligently in land acquisition. Knowledge of the schedule of land availability and legalities of operating under both federal and state laws and regulations in this remote, high-cost, and high-risk region played a major role in company internal decision-making.

Prediscovery Phase—Industry Transition Programs

By 1965 the emphasis of the industry's exploration activity began to shift from surface geology to seismic mapping, from the general vicinity of NPR-4 and the Foothills anticlines to the extreme Northern Foothills and Coastal Plain, and from Lower Cretaceous objectives to Upper Cretaceous or Tertiary, as well as pre-Cretaceous, potential reservoirs. The shift was caused by discouraging results of the earlier exploratory drilling to the east and south of Umiat and by the availability of lands for exploration and leasing farther to the east and north.

In 1965 federal simultaneous filings and subsequent drawings were held covering 3,150 blocks, or approximately 8 million acres (3.2 million ha.; Fig. 2). The area was to the east, south, and west of the first two federal offerings (1958 and 1964), thus being mostly in the Canning River drainage near the Sadlerochit and Shublik Mountains and in the Foothills of the Brooks Range. Lease offers totaled 428, mostly in the Canning River and eastern Foothills areas.

In July of 1965 the State held the 14th Competitive Lease Sale, which was the first opportunity to obtain leases on the anticlinal structure mapped by seismic methods in the vicinity of Prudhoe Bay. Richfield-Humble acquired 28 blocks in the crestal area. BP, bidding alone after Sinclair withdrew, acquired 32 blocks flanking the Richfield-Humble position.

Continued shifting of exploration emphasis was evident in the decrease of geologic crew-months to 15.5, and geophysical crew months to 26.75 in 1965. Sinclair began drilling the Colville 1 well as a joint venture with BP on the extensive acreage position obtained in the 13th State Sale east of the Colville River delta. The Colville 1, and the Union-BP Kookpuk 1 well drilled the following year in the same area, were the first industry tests to penetrate pre-Cretaceous objectives. Both were abandoned (in 1966 and 1967, repsectively) after encountering prospective reservoirs in the Permo-Triassic section and Mississippian-Pennsylvanian sections. However, the data on

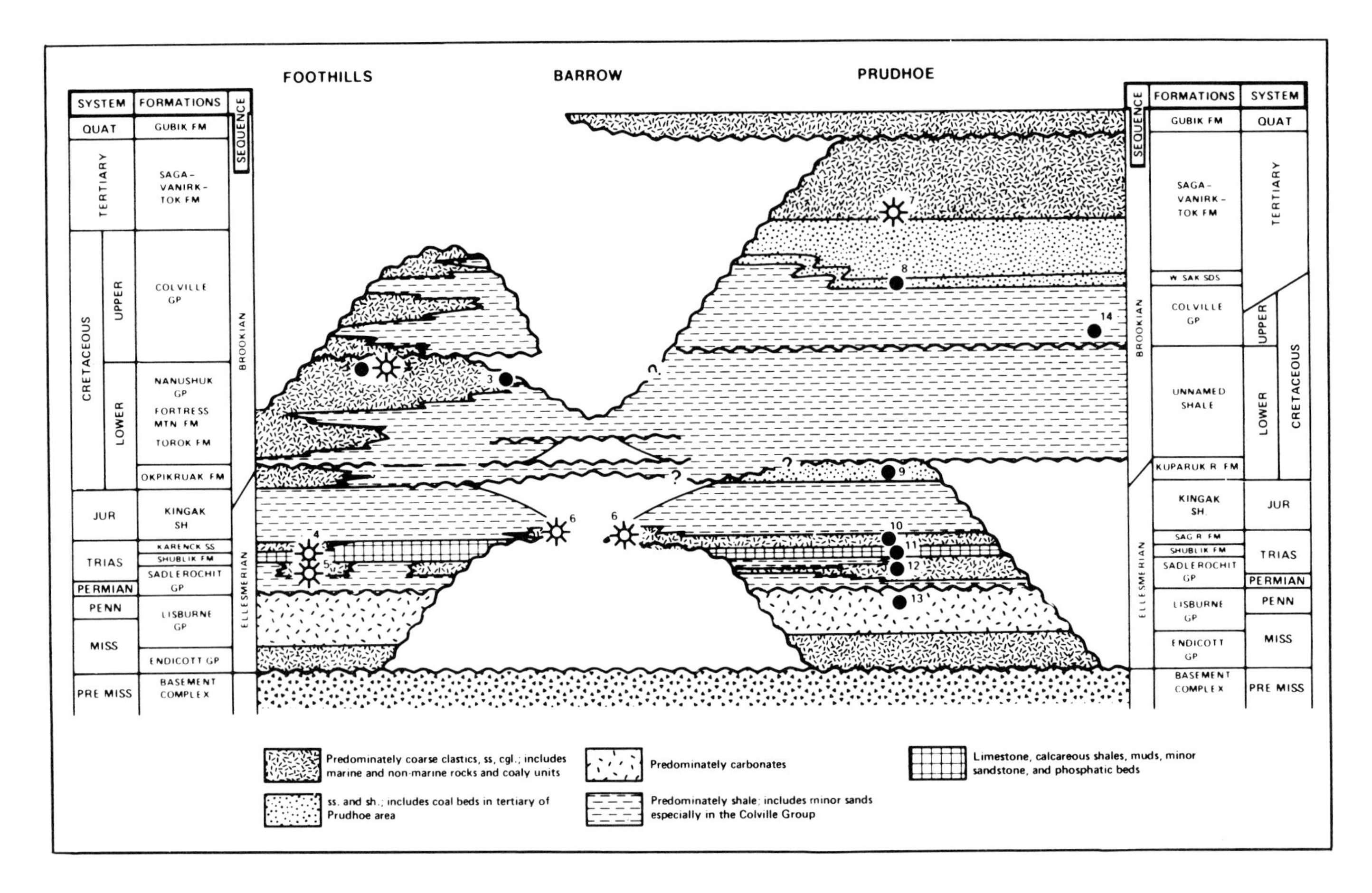

FIG. 5—Generalized stratigraphic diagram showing significant oil and gas occurrences, North Slope, Alaska. (1) Umiat, (2) Gubik, (3) Simpson, (4) Kemik, (5) Kavik, (6) Point Barrow, (7, 8, 9, 10, 11, 12, 13) Prudhoe Bay Complex, (14) Flaxman Island.

these wells were held in confidential status and thus were unavailable to other operators.

Early in 1966 Atlantic Richfield Company air-lifted a drilling rig from Fairbanks to the ARCO-Humble Susie Unit 1, the first well in the Sagavan Development Contract area and in the federal "Corridor" acreage filed on in 1964. This location was on a seismically mapped anticline with the primary objectives in the Upper Cretaceous, a section not tested by previous North Slope drilling.

Geologic and seismic crew-months continued declining to 10 and 2.75 respectively.

Later in 1966 the Bureau of Land Management opened 3 million acres (1.2 million ha.) west of the NPR-4. A total of 209 blocks received only 765 offers and no leases were issued because of the indecision resulting from the Native Land Claims (Fig. 2). In January 1967, in the State's 18th Competitive Sale, ARCO-Humble acquired seven critical offshore Prudhoe Bay tracts which covered the remainder of the crestal area of the structure. The 90,000-acre (36,000 ha.) lease block was now intact and enabled the companies to proceed with drilling.

Discovery Phase

After the Susie Unit 1 well was abandoned as a dry hole in January 1967, the rig was moved over the frozen tundra 60 mi (96 km) north to the location of the Prudhoe Bay State 1 well which commenced drilling in April. Operations were shut down in May because of ice breakup and resumed after freeze in late fall. ARCO-Humble announced the discovery in January 1968. In June, after drilling the confirmation well, the Sag River State 1 located 7 mi (11.3 km) southeast of the discovery, the companies release the DeGolyer and McNaughton estimate that Prudhoe Bay ". . . could develop into a field with recoverable reserves of some five to ten billion barrels of oil, which would rate it as one of the largest petroleum accumulations known to the world today."

The discovery came at a time when all other North Slope exploration had come to a halt. In 1967 only three crew-months of surface geology were recorded and no seismic programs were conducted by the industry. All drilling other than at Prudhoe Bay had ceased. Thus, the discovery focused industry attention on State of Alaska lands in the unleased part of the 1.6 million acres (0.65 million ha.) selected in 1965, both because of proximity, and because of the 1966 federal "land freeze" (BLM moratorium on leasing) resulting from the Native Land Claims. The state announced plans to accept nominations for a late 1969 sale and the industry responded with a surge of activity. By late 1968, five wildcats were drilling and

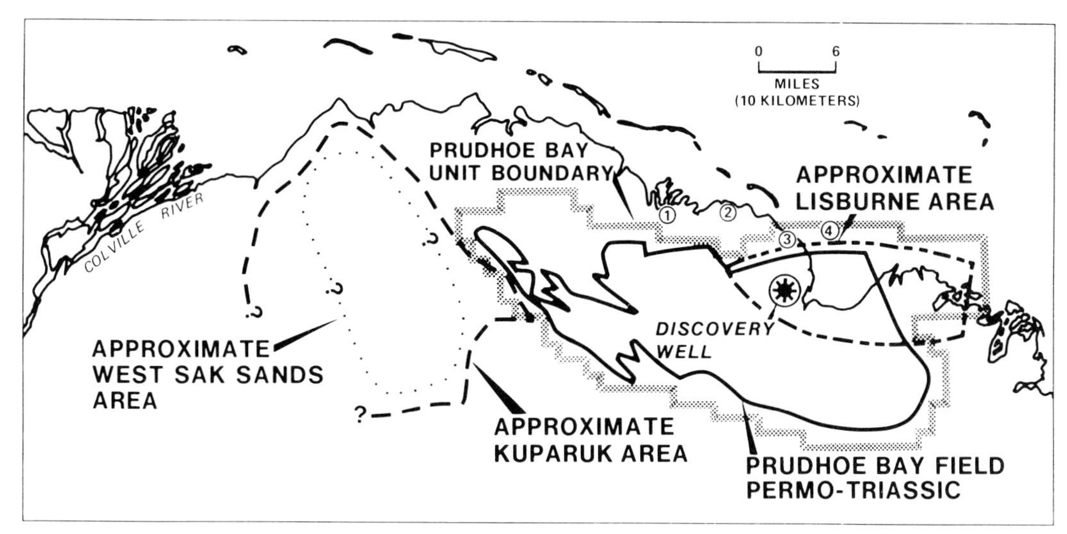

FIG. 6—Index map of known accumulations, Prudhoe Bay Complex. In addition to those labeled on the map, accumulations occur at: (1) Gwydyr Bay State 1, (2) Point Storkersen 1 and Kuparuk Delta 51-2, (3) North Prudhoe State 1, and (4) Gull Island State 1.

six locations were being prepared. Geologic surface work increased to 12 crew-months and seismic crew months went from 0 to 24.

Plans were underway for a pipeline from the North Slope to an ice-free port in south Alaska. Eventually Valdez was selected as the southern terminus and formal announcements of the plans for the 800-mi (1,280 km) Trans-Alaska Pipeline System were announced in February 1969. The cost was estimated at $900 million. Eight major companies later formed Alyeska Pipeline Service Company to design, construct, and operate the pipeline. (Alternates to pipelining were explored and Humble proved the possibility of use of tankers with the 1969 voyage of the *SS Manhattan* from the U.S. East Coast via the Northwest Passage to Prudhoe Bay.)

As expected, exploratory drilling during 1969 reached a dramatic new high of 33 completed wells (following two in 1968) and geologic crew-months reached 20 with seismic crew-months increasing to 97. All data from wells drilled in the area were kept in strict confidential status pending the 23rd State Competitive Lease Sale in September. The state put up almost 413,000 acres (167,000 ha.) along the Arctic Coast between the Colville and Canning Rivers. The successful bonus bids on 164 tracts totaled $900,041,605 making it the highest on record. Total bids submitted exceeded $1.68 billion. This was the last sale held on the North Slope.

Post-discovery Phase

The course of North Slope exploration following the discovery and the resultant enthusiasm of the industry indicated in the so-called "Billion Dollar Sale" can best be described by quoting Lian (1971): "The most significant development in Alaska in 1970 was the beginning of a statewide exploration decline brought about by the combination of the Federal land freeze and the continued delay in obtaining permission to construct the trans-Alaska pipeline from the Arctic Slope to the Gulf of Alaska." He further stated, "Contrary to all predictions, exploratory drilling in Alaska in 1970 showed an alarming decrease of 38% compared with 1969. Only 28 exploratory wells were drilled, with 3 discoveries, all in the Prudhoe Bay area of the Arctic Slope." Adams' (1972) summary expressed the situation in this way: "Petroleum exploration and development drilling activity in Alaska during 1971 was virtually at a standstill. Three years after the discovery of Prudhoe Bay field, the largest oil field ever found on the North American continent with recoverable reserves estimated at 10 billion bbl of oil and 26 Tcf of gas, statewide exploratory drilling declined almost fivefold and development drilling was down 49.2% from 1970. Only six exploratory wells and twenty-nine development wells were completed in 1971."

Environmental organizations had succeeded through the courts in delaying the issuance of federal permits to begin pipeline construction. The continuing federal "land freeze" was at least partly remedied by the passage of the Alaska Native Claims Settlement Act in December 1971. Nevertheless, it was 6 years from the discovery to the beginning of construction of the pipeline haul road north of the Yukon River in April 1974. Construction was finally authorized in November 1973, following Congressional and Presidential action. Native Regional Corporation land selections were completed by December 1975 which afforded opportunities for companies to explore these areas under concession-type contracts with the various native associations.

From 1970 through 1974 only 35 exploratory wells were drilled on the North Slope, barely exceeding the 1969 single-year total of 33. Prudhoe Bay field development drilling continued at a moderate level pending pipeline approvals and actual construction. In late 1969 unitization efforts were initiated and continued throughout this period, although at a very slow pace after 1970 when the duration of the delays became more apparent. Geophysical crew-months again reached a high level of 96 in 1970 and then declined to a low of 8 in 1972,

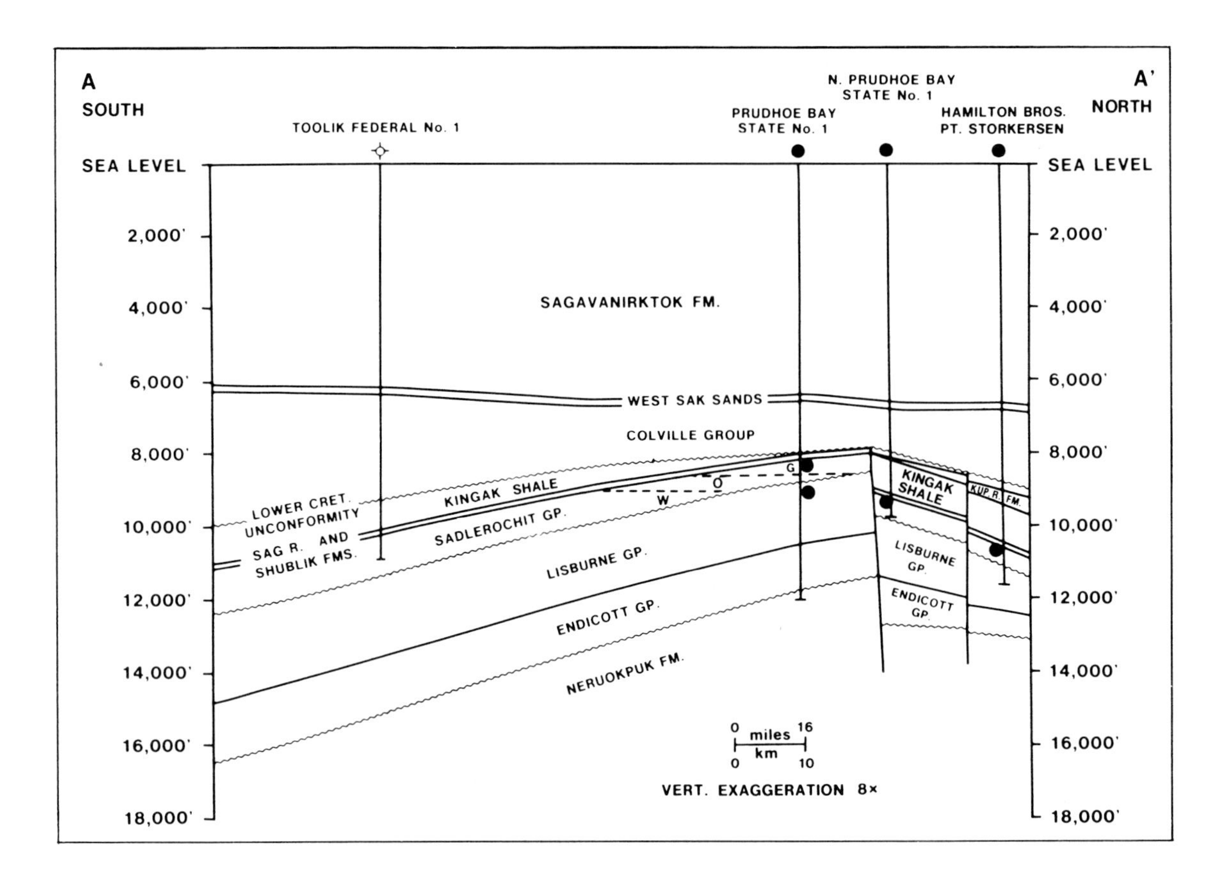

FIG. 7—North to south cross section, Prudhoe Bay Complex. For location, see structure maps of Ivishak (Fig. 11)., Lisburne (Fig. 14), Kuparuk (Fig. 20), and West Sak (Fig. 21).

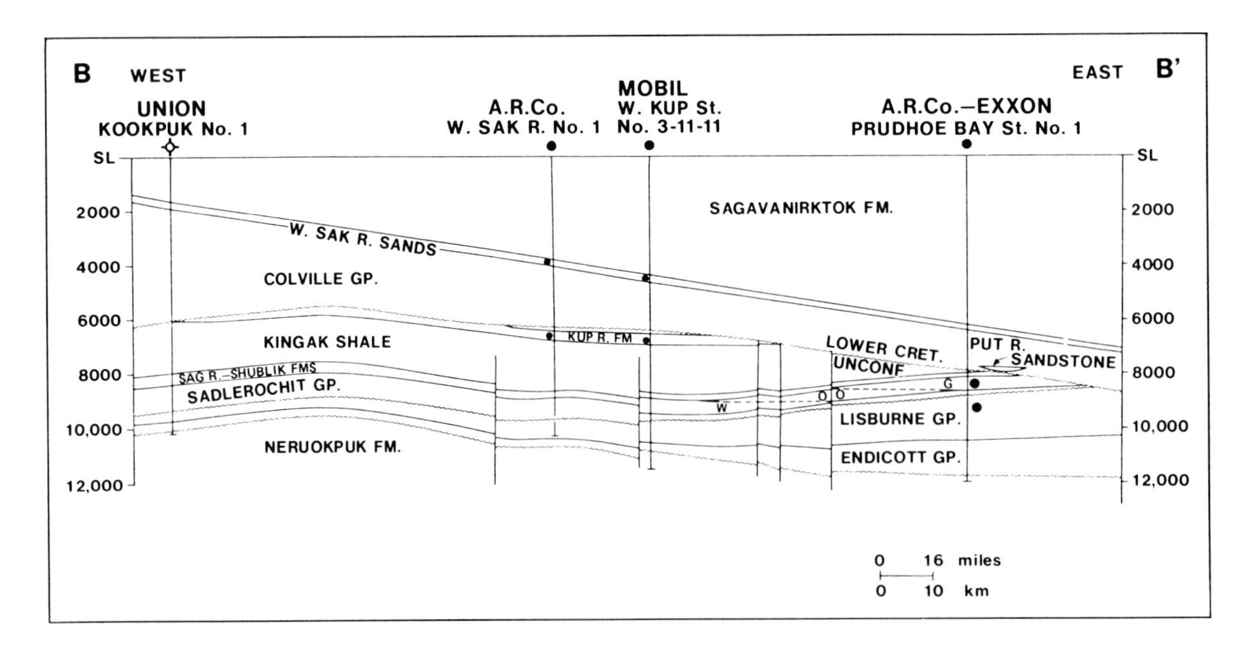

FIG. 8—East to west cross section, Prudhoe Bay Complex. For location, see structure map of Ivishak (Fig. 11), Lisburne (Fig. 14), Kuparuk (Fig. 20), and West Sak Sands (Fig. 21).

gradually increasing to 54 in 1974. Geologic field work held steady at 20 crew months from 1969 through 1971 and then decreased to about 6 in 1974.

Beginning in 1974, and continuing to the present, the U.S. Navy and the Department of Interior have been conducting a full-scale exploration program, mandated and funded by Congress, to evaluate the petroleum potential of the National Petroleum Reserve-Alaska (formerly NPR-4). From 1974 through 1977, 56.5 seismic crew-months of reconnaissance coverage defined or helped define locations for 19 wells. Through 1978 and into 1979, exploratory drilling has continued. Total cost of the program is estimated to be about $625 million (Petroleum Information, 1979).

The years 1975, 1976, and 1977 witnessed a modest resurgence of industry activity (exclusive of the NPR-A program) with 33 exploratory wells drilled and 16 crew-months of surface geology. Seismic crew months totaled 76. Encouraged by a possible state lease sale in the Beaufort Sea in 1975, the industry continued to evaluate the coastal and near-offshore areas with both marine and ice shooting. Unfortunately, the sale was delayed and apparently resolved only by the joint Federal-State Sale scheduled for December, 1979.

The graphic display of annual geologic and geophysical crew months and exploratory wells emphasizes the positive response of exploration activity to land availability, actual or implied by governmental announcement (Fig. 3). The prime indicators are geophysical crew-months and exploratory wells because of high investment costs. It becomes equally apparent that negative response is as rapid when governmental actions tend to disrupt an already tenuous exploration investment situation.

With passage of the Alaska Native Claims Settlement Act in 1971, a provision was included in Section 17 (d) (2) for federal classification of lands primarily devoted to wilderness uses. The Act authorized up to 80 million acres (32 million ha.) for inclusion in National Parks, Forest, Wildlife Refuges, and Wild and Scenic River systems. Various schemes to include much larger areas were devised, but the December 1978 deadline passed without Congressional resolution. As an "emergency" action in late 1978, Secretary of Interior Andrus used the 1976 Federal Land Policy and Management Act to withdraw 110 million acres (45 million ha.) in Alaska to entry or development for a 3-year period. President Carter immediately followed by invoking the Historic Sites and Antiquities Act to classify permanently 56 million acres (22.7 million ha.) of the withdrawal as National Monuments. These withdrawals cover large areas in the southern part of the North Slope.

Despite this long period of exploration doldrums which took its toll in expiring leases, removal of rigs from the North Slope, closing of exploration offices in Alaska, and lack of major or independent company risk capital, a few companies were still exploring. A number of discoveries resulted which extended the geographic and stratigraphic ranges of petroleum and natural gas accumulations in the region. Although none of these have been developed and produced on an economic basis, drilling and evaluation of production projects is continuing. Discovery wells are listed in Table 1.

The only established economic production on the North Slope is still confined to the Permo-Triassic section in the Prudhoe Bay field. After permits were issued to build the Trans-Alaska Pipeline System, development drilling and the systems for the surface production facilities proceeded at a full pace. Unitization negotiations among the working interest owners, including equity determinations, were concluded April 1, 1977 with final State of Alaska approval June 2, 1977. Eighteen days later oil began to flow through the pipeline and reached Valdez on July 28. Construction cost of the pipeline on completion was announced as $7.7 billion, over eight times the 1968 estimate of $900 million. Current production rate approximates design capacity of 1.2 million b/d. Capacity is to be increased to 1.35-1.4 million b/d by the end of 1979.

Discussion of the regional geology and descriptions of the more significant accumulations in the region follow.

REGIONAL GEOLOGIC SETTING

The Prudhoe Bay field and nearby known hydrocarbon accumulations are on the North Slope adjacent to the Beaufort Sea coastline, about 200 mi (320 km) east of Point Barrow, the most northerly point in the continental United States. Principal structural features of the North Slope are the Barrow arch which parallels the coastline eastward from Point Barrow, and the Colville trough to the south (Fig. 1). Sediments thicken from approximately 2,500 ft (760 m) on the arch at Point Barrow to more than 30,000 ft (9,100 m) in the deepest parts of the Colville trough (Fig. 4).

Figure 5 summarizes the stratigraphy of the areas of principal interest and indicates where known significant accumulations occur in the stratigraphic column. Details of North Slope stratigraphy have been reported by numerous authors (Gryc, 1970; Rickwood, 1970; Morgridge and Smith, 1972; Jones and Speers, 1976; Detterman et al, 1975; and Carter et al, 1977) for those areas of known hydrocarbon accumulations.

Pre-Mississippian rocks are considered to be economic basement, although there may be units within this basement complex capable of acting as petroleum reservoirs. Rocks above the pre-Mississippian can be divided into a "lower sequence" of Mississippian through lowermost Cretaceous (Neocomian) named the Ellesmerian sequence, and an "upper sequence" of Lower Cretaceous through Tertiary age named the Brookian sequence. This nomenclature was first used by Lerand (1973) to describe sediment source areas throughout the Beaufort Sea area and onshore, and was later adapted by Carter et al (1977).

Carbonate and clastic rocks of the Ellesmerian sequence lie unconformably on the pre-Mississipian section. In the areas of principal interest, the sequence from oldest to youngest consists of the Endicott, Lisburne, and Sadlerochit Groups, the Shublik Formation, the Sag River Formation, the Kingak Shale, and the Kuparuk River Formation. All of these units are believed to have had a northerly source of sediment supply from highlands near the present Barrow arch or perhaps from a source area now completely detached from north Alaska.

Although it may have local sediment sources from the south and east, the Kuparuk River Formation has been included in

Table 1. North Slope discovery wells drilled by industry through 1977.

NORTH SLOPE DISCOVERY WELLS

Completion Date	Well Name	Productive Zones	Depth	Gas/Oil	Rate
3/8/66	Sinclair Colville #1	Shublik	7,872-7,922	Gas	5 MMCFG/D
4/15/68	ARCo-Humble Prudhoe Bay St. #1	Ayiyak	6,876-6,998	Oil	280 BOPD
		Sag River	8,130	Gas	1.5 MMCF/D
		Sadlerochit	8,208-8,578	Gas	25.6 MMCF/D
		Sadlerochit	8,578-8,750	Oil	2025 BOPD
		Wahoo	8,750-8,883	Gas	22 MMCF/D
		Wahoo	9,200-9,410	Oil	434 BOPD
		Alapah	9,505-9,825	Gas/Oil	1.3 MMCF/D 1152 BOPD
5/9/69	Sinclair Ugnu #1	Kuparuk	6,160-6,182	Oil	1056 BOPD
8/8/69	Mobil Phillips W. Kuparuk St. #3-11-11	Kuparuk	6,574-6,585	Oil	2220 BOPD
10/24/69	Socal Kavearak Pt. 32-25	Kuparuk	6,898-6,947	Oil	1100 BOPD
11/8/69	Pan American Kavik #1	Sag River - Shublik	4,234-4,524	Gas	3.9 MMCFG/D
		Sadlerochit	4,524-5,125	Gas	10.5 MMCFG/D
11/25/69	Hamilton Bros. Point Storkersen #1	Sadlerochit	10,552-10,922	Gas/Oil	5.5 MMCFG/D 381 BOPD
1/19/70	ARCo Beechey Pt. St. #1	Kuparuk	6,690-6,715	Gas	2.5 MMCFG/D
4/9/70	ARCo-Humble North Prudhoe St. #1	Sag River - Shublik	9,124-9,176	Gas/Oil	3.6 MMCFG/D 132 BOPD
		Sadlerochit	9,240-9,256	Oil/Gas	1.1 MMDFG/D 2727 BOPD
8/13/70	Hamilton Bros. Kup Delta 51-2	Kuparuk	9,614-9,618	Oil	660 BOPD
		Shublik	11,562-11,638	Oil	520 BOPD
		Sadlerochit	11,638-12,040	Oil	695 BOPD
4/26/71	ARCo West Sak R. St. #1	West Sak Sands	3,745-4,000	Oil	112 BOPD
6/17/72	Forest Kemik Unit #1	Shublik	8,538-8,788	Gas	2 MMCFG/D
4/6/75	Mobil-Socal Gwydyr Bay State #1	Sadlerochit	10,076-10,132	Oil	2263 BOPD
9/21/75	Exxon Alaska St. A-1 (Flaxman Island)		12,565-12,635	Oil	2507 BOPD
4/1/76	ARCo-Exxon Gull Island St. #1	Sadlerochit	12,528-12,537	Oil	1152 BOPD
4/28/76	Sohio-BP Sag Delta #1	Alapah	9,276-10,114	Oil	7875 BOPD
12/8/77	Exxon Pt. Thomson #1		12,963-13,050	Oil	2300 BOPD

the Ellesmerian sequence. In the Prudhoe Bay area the formation conformably overlies the Kingak Shale and is unconformably overlain by Lower Cretaceous shales. This relationship makes it convenient to place the Kuparuk River Formation in the Ellesmerian sequence.

The Ellesmerian sequence is, in most part, unconformably overlain by the Brookian sequence. This unconformity, termed the Lower Cretaceous unconformity by Jones and Speers (1976), is most pronounced in the Prudhoe Bay area where it is a major trapping element of the field and to the east along the Barrow arch where it truncates the entire Ellesmerian sequence. Whether the unconformity exists or can be recognized toward the center of the Colville trough is problematical.

The Brookian sequence in the Foothills of the Brooks Range consists of the Lower Cretaceous Okpikruak Formation, Torok Formation, and Nanushuk Group; the Upper Cretaceous Colville Group; and the Tertiary Sagavanirktok Formation. All of these sediments were derived from southerly uplifted areas in what is now the Brooks Range. They form thick clastic wedges in the Colville trough and thin northward over the Barrow arch. The Upper Cretaceous and Tertiary units progressively thicken to the northeast across the eastern North Slope, due to

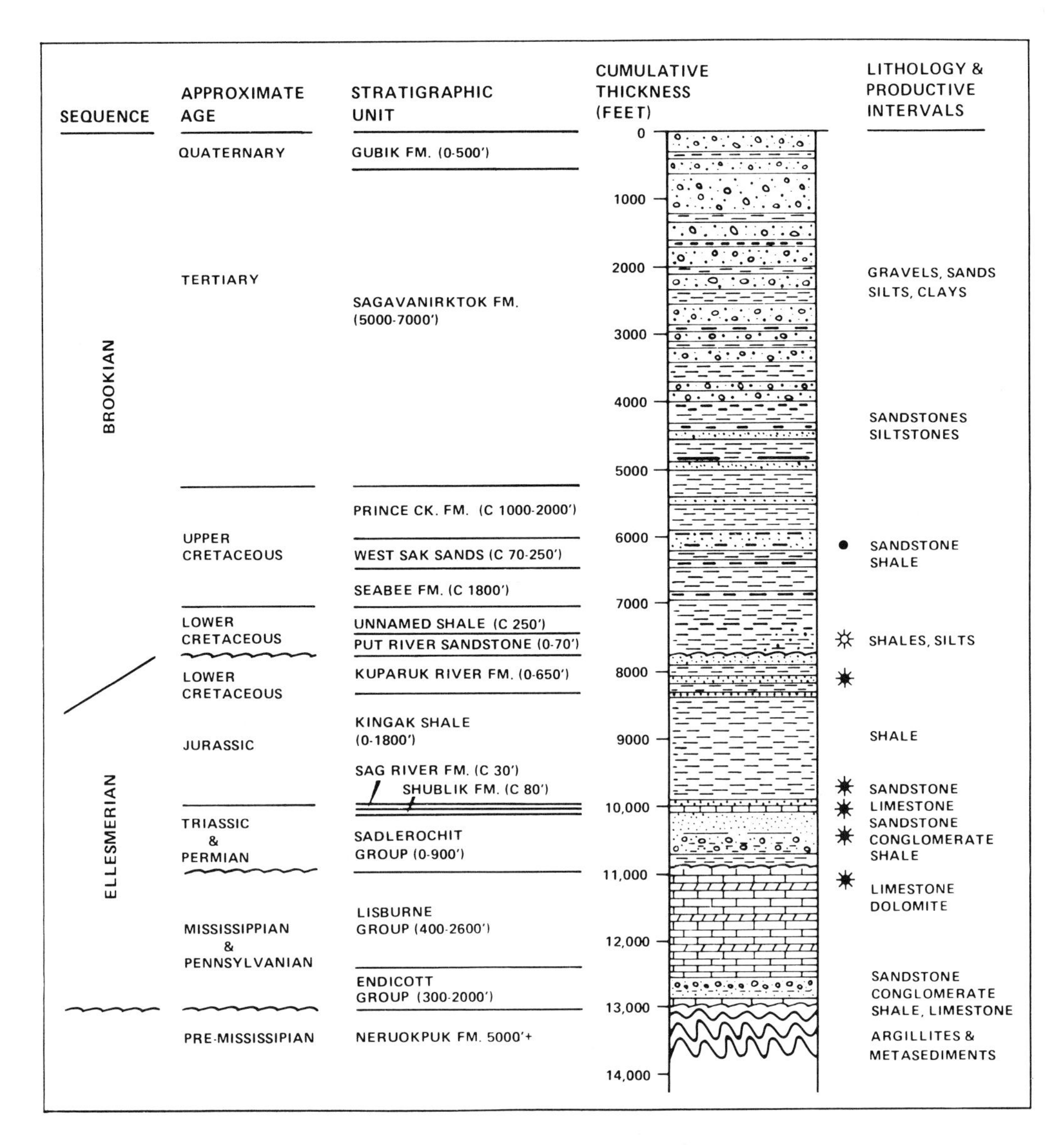

FIG. 9— Generalized stratigraphic column of Prudhoe Bay Complex. Oil and gas symbols indicate known accumulations.

northeastward migration of the depocenters.

The Prudhoe Bay field is located on the easterly extension of the Barrow arch and at the eastern end of the Prudhoe Bay Complex. Accumulations in the Ellesmerian sequence are trapped on this regional high by combinations of folding, faulting, and truncation by the Lower Cretaceous unconformity. Brookian sequence hydrocarbon accumulations in the Prudhoe Bay area appear to be due primarily to stratigraphic changes. Permafrost is believed to be a trapping mechanism for gas accumulations in the younger Brookian sequence. Proximity to the axis of the regional arch appears to influence trapping although the structural effects of the arch cannot always be demonstrated in the younger beds.

PRUDHOE BAY COMPLEX

Hydrocarbon accumulations of the Prudhoe Bay Complex generally are located on or near its crest; however, accumulations have been discovered in downfaulted blocks on the north flank (Fig. 6). Recent discoveries tend to indicate that the complex may logically be extended easterly beyond the Sagavanirktok River. Two cross sections depict the general structural and stratigraphic aspects of the complex and illustrate the trapping effect of the Lower Cretaceous unconformity, especially where it is combined with favorable structural position (Figs. 7, 8).

Subsequent sections of this paper will describe those major reservoir zones of the Prudhoe Bay complex which have exhibited capability of hydrocarbon productivity (Fig. 9). Permian-Triassic reservoirs of the Prudhoe Bay field will be discussed first because they contain the bulk of the hydrocarbons in place as well as the only reserves currently being produced. Other reservoirs, including Mississippian-Pennsylvanian carbonates of the Lisburne Group, Lower Cretaceous sandstones of the Kuparuk River Formation, and the Upper Cretaceous West Sak Sands, will be

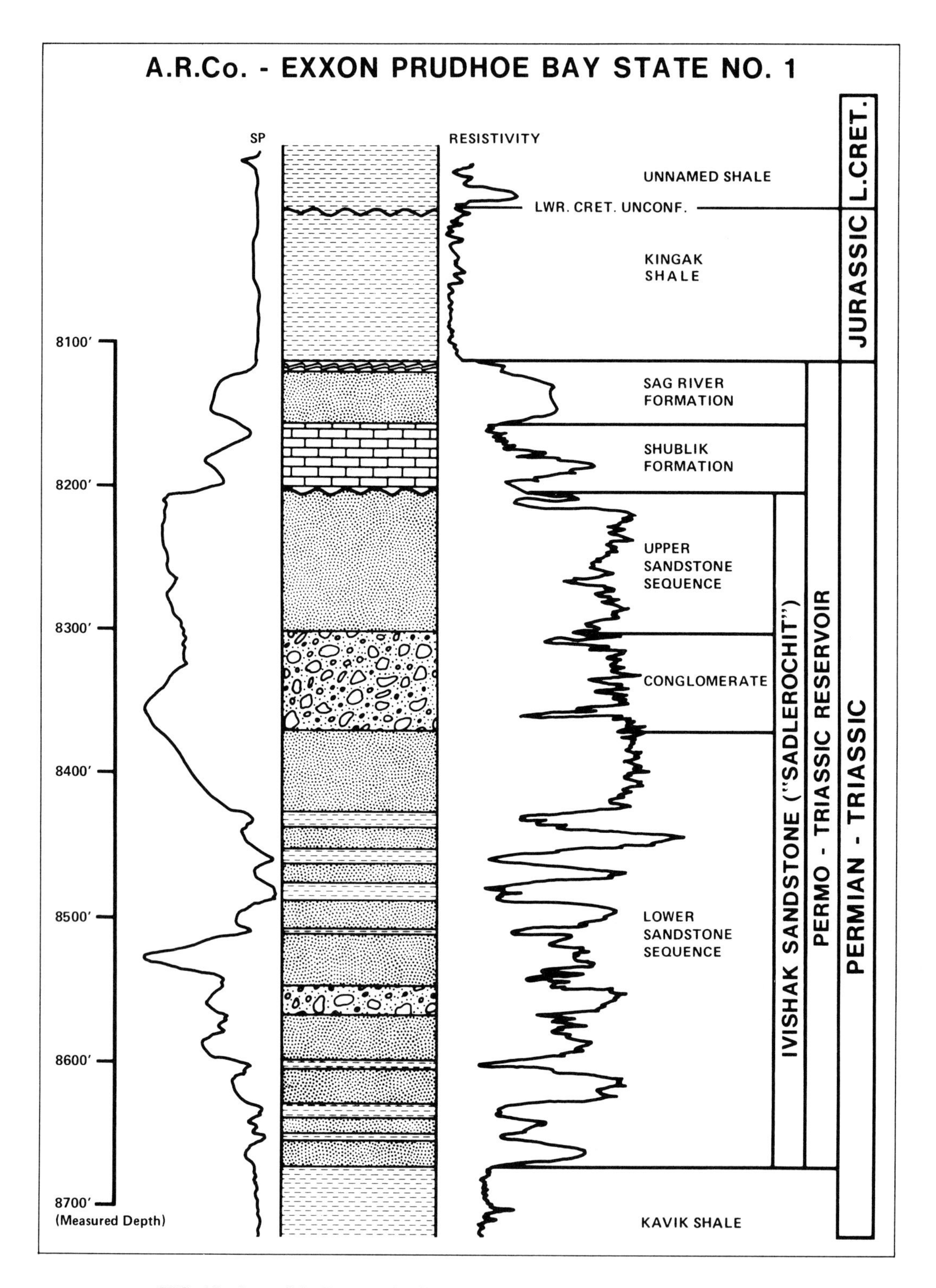

FIG. 10—Log of the Permo-triassic reservoir encountered in the discovery well.

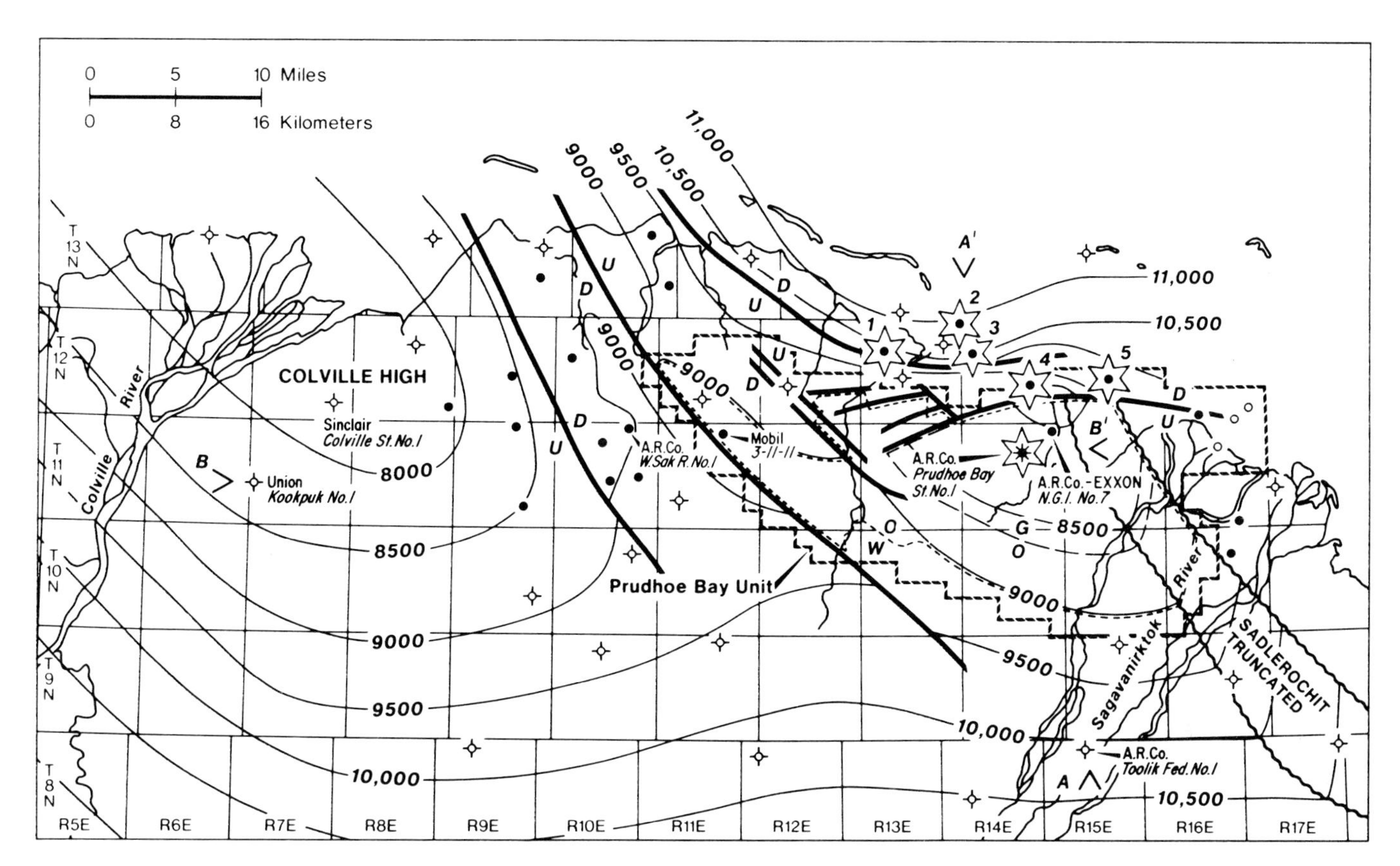

FIG. 11—Structure map of top of the Ivishak Sandstone ("Sadlerochit"). Star symbols indicate wells known to have discovered oil in the Permo-Triassic reservoir section: (1) Mobil, Gwydyr Bay State 1, (2) Hamilton Brothers, Kuparuk Delta 51-2, (3) Hamilton Brothers, Point Storkensen 1, (4) ARCO, North Prudhoe Bay State 1, (5) ARCO, Gull Island State 1. Faulting is simplified here. Subsea contour intervals in feet.

reviewed in order of oldest to youngest age.

The various reservoirs of the Prudhoe Bay Complex are estimated to contain at least 23.5 billion barrels of oil in place. Approximately 20 billion bbl of oil in place occur in the Permian-Triassic of the Prudhoe Bay feld (H. K. van Poolen and Associates, Inc., and the State of Alaska Division of Oil and Gas, 1974). An additional 3.5 billion bbl of oil in place are attributed to the Kuparuk River Formation (van Poolen and Alaska, 1978). Inadequate data exist to make reliable estimates of the volume of oil in place for other accumulations. Commercial production of hydrocarbons trapped in accumulations other than the Permian-Triassic of the Prudhoe Bay field will depend primarily on economics, and thus will be extremely sensitive to production rate, as well as reserves.

Permo-Triassic Reservoirs

The most prolific reservoirs of the Prudhoe Bay Complex occur within the Permian-Triassic section. They are the Sadlerochit Group, the Shublik Formation, and the Sag River Formation (Fig. 10). The Sadlerochit Group has been defined to include, in ascending order, the Echooka Formation, Kavik Shale, and Ivishak Sandstone (Jones and Speers, 1976). The Put River Sandstone of Lower Cretaceous age lies unconformably on the Sag River in a limited area of the Prudhoe Bay field, but is considered part of the field "Permo-Triassic Reservoir." A number of publications, most notably those of Detterman (1970), Detterman et al (1975), Morgridge and Smith (1972) and Jones and Speers (1976) have described various aspects of the Permo-Triassic sequence and, consequently, this paper will be limited to brief discussions of the productive reservoir zones.

Structure on top of the Ivishak Sandstone of the Sadlerochit Group in the Prudhoe Bay Complex is expressed as an east to southeast anticlinal trend containing two prominent highs, one at Prudhoe Bay, and one in the Colville River delta area, referred to, respectively, as the "Prudhoe high" and the "Colville high" (Fig. 11). The Complex is interrupted by a series of northwest to southeast trending down-to-the-west normal faults between the Colville and Prudhoe highs. These faults have displacements of up to 200 ft (61 m). A complicated series of down-to-the-north normal faults with throws of up to 1,000 ft (305 m) define the northeastern flank of the complex. The southern flank dips regionally into the Colville trough.

Although test results in the discovery well of 25.6 million cf/d of gas and 2,025 b/d of oil from the Sadlerochit indicated significant production rates, they do not compare to current field well rates. Some typical high rates reported in November 1978 ranged from 15,792 b/d to 19,909 b/d of oil from the Permo-Triassic.

Ivishak Sandstone—The principal productive unit of the Prudhoe Bay field is the deltaic sequence of the Ivishak Sand-

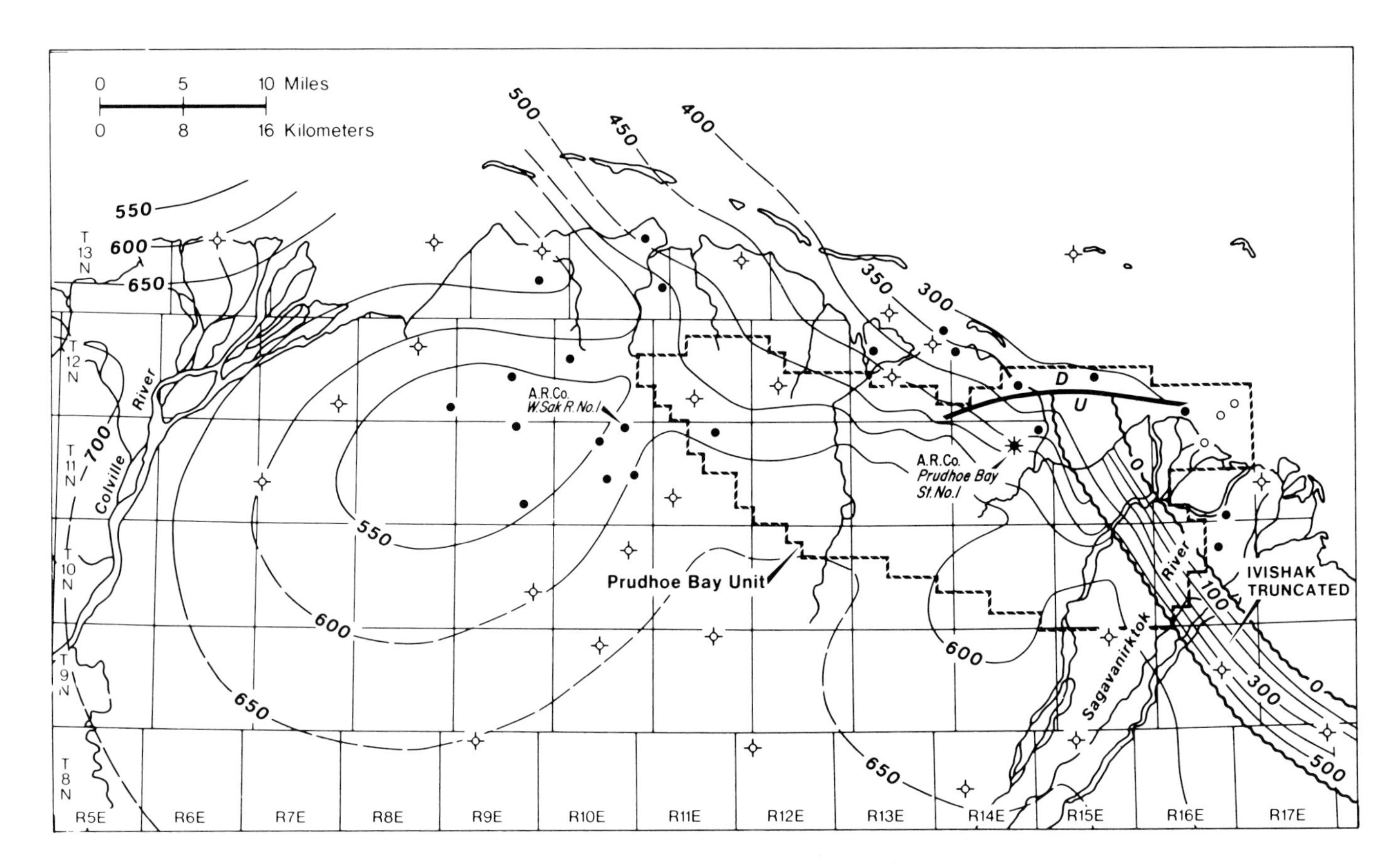

FIG. 12—Isopach map of Ivishak Sandstone ("Sadlerochit"). Contours in feet.

stone of the Sadlerochit Group (Jones and Speers, 1976), informally referred to as the "Sadlerochit." This reservoir was first encountered in the Atlantic Richfield-Humble (Exxon) Prudhoe Bay State 1 discovery well (Fig. 10) where it occurs from 8,206 to 8,673 ft (2,501 to 2,644 m) measured depth. Jones and Speers defined the type section for the Ivishak Sandstone as the strata occurring within the interval 8,935 to 9,513 ft (2,723 to 2,900 m), measured electric-log depths, in the British Petroleum 19-10-15 well.

Thickness decreases from more than 650 ft (198 m) in the south and southwestern part of the Prudhoe Bay Complex to less than 350 ft (107 m) in the northeast (Fig. 12). Jones and Speers (1976) attributed much of the northward thinning in the field area to pre-Shublik erosion. In the western part of the complex, the Ivishak Sandstone thins over the Colville high, whereas east of the Prudhoe high it has been completely removed by erosion.

In the Prudhoe Bay field the Ivishak Sandstone consists primarily of two fine- to medium-grained pebbly sandstone sequences separated by an interval dominated by massive conglomerates. The contact between the Kavik Shale and the Ivishak Sandstone appears gradational on mechanical logs and is arbitrarily placed at the lowest porous sandstone. The top of the Ivishak Sandstone is commonly placed at the base of a thin, radioactive, phosphatic conglomerate that is overlain by calcareous mudstones of the Shublik Formation.

Sandstones of the lower sequence are separated by major shale interbeds that were deposited in bays between the main distributary channels of a delta. The sandstones are clean, massive, occasionally conglomeratic, and grade downward into finer grained sandstones interbedded with siltstone and shale that overlie the Kavik Shale. This sequence is about 300 ft (91 m) thick. Porosities and permeabilities in the very fine- to fine-grained sandstones average about 20% and 75 md, while the fine- to medium-grained sandstones have porosities ranging from 25 to 30% with permeabilities ranging from 250 to more than 3,000 md.

The lower sandstone sequence is overlain by massive, nonmarine sandy conglomerates that mark the maximum southward advance of the Sadlerochit delta. The conglomerates are more than 140 ft (43 m) thick in the northeastern part of Prudhoe Bay field and thin to less than 40 ft (12 m) to the east, west, and south. Porosities within the conglomerates usually range from 10 to 20% with permeabilities ranging from less than 50 to more than 1,000 md.

Overlying the conglomerates are homogeneous fine- to medium-grained sandstones of the upper sandstone sequence. The basal part was probably deposited in braided streams, while the uppermost rocks indicate a nearshore marginal-marine environment. The upper sandstone sequence is more than 200 ft (61 m) thick in the southwest and thins northeastward. Porosities generally range from 25 to 30%,

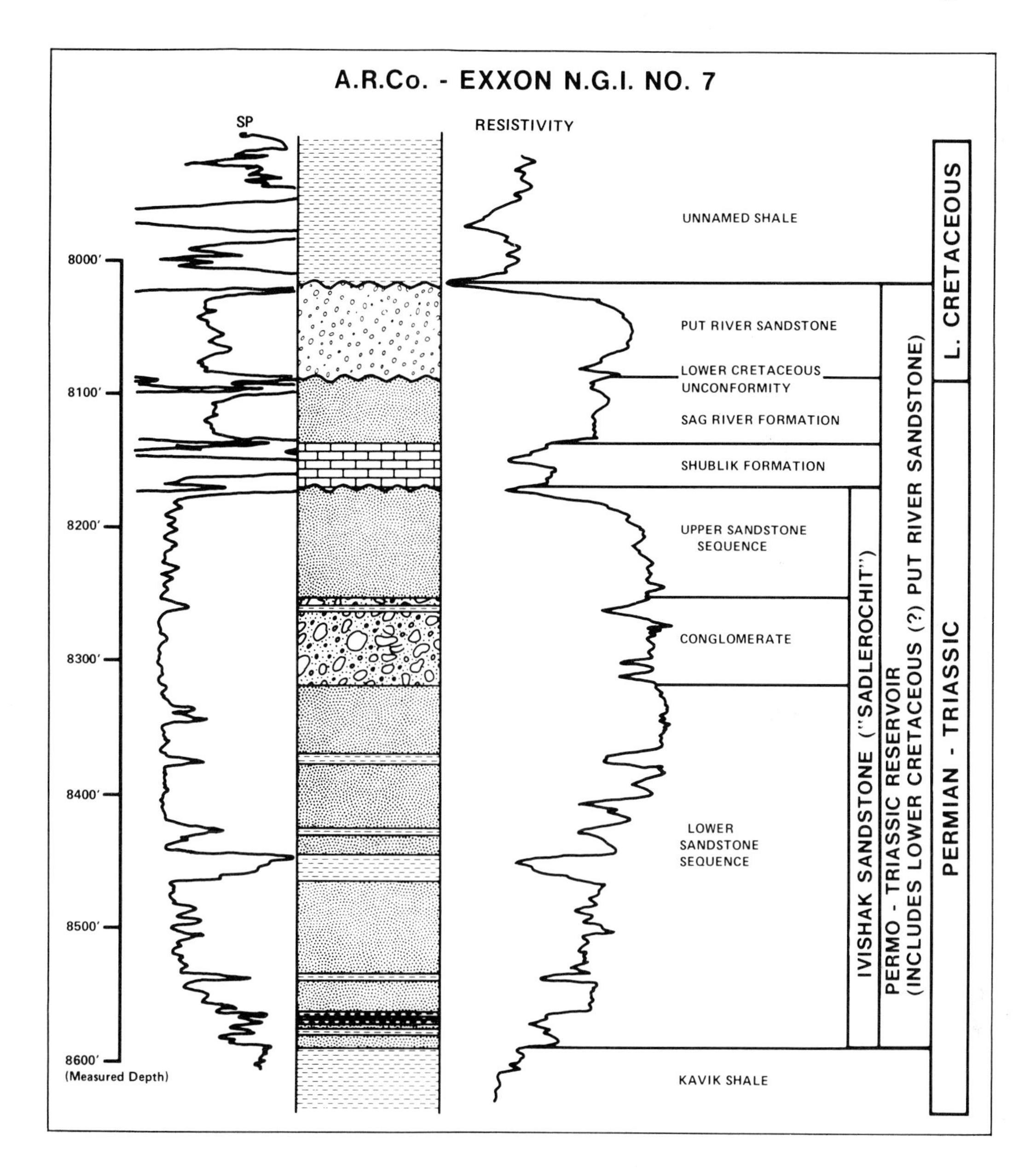

FIG. 13—Well log from ARCO-Exxon, North Gull Island 7, showing position of Put River Sandstone within the Permo-Triassic reservoir at Prudhoe Bay field.

with permeabilities ranging from less than 500 to more than 4,000 md.

The Ivishak Sandstone is barren of megafauna or microfauna, however palynomorphs in interbedded shales indicate a Late Permian to Early Triassic age (Jones and Speers, 1976).

Shublik Formation—The Shublik Formation was defined by Leffingwell (1919) at Shublik Island in the Canning River. A reference section of the Shublik occurs in the Prudhoe Bay State 1 well between 8,158 and 8,206 ft (2,487 to 2,501 m), measured electric-log depth (Fig. 10). It consists of argillaceous bioclastic limestones, calcareous shales, and phosphate rock, probably deposited in a low energy marine shelf environment with moderate water depths between 200 ft (61 m) and 1,000 ft (305 m), (Detterman, 1970).

Thickness of the Shublik Formation ranges from less than 80 ft (24 m) in the field area to more than 200 ft (61 m) west of the field, with the most pronounced thickening occuring in a basal sandy section (Jones and Speers, 1976). Porous and permeable reservoirs of limited and unpredictable extent are found in sandy and pelletal limestones in the main field area. The basal sandy part forms a reservoir in the western area of the field. Porosities range from 5 to 15%, but permeabilities are low.

Sag River Formation—The "Sag River Sandstone" was defined as "the sandstone interval lying between the Jurassic Kingak Shale above and the shales and limestones of the Triassic Shublik Formation below, in wells drilled in the Prudhoe Bay field, North Slope, Alaska" (Alaska Geological Society, 1970-1971). A reference section of the Sag River Sandstone occurs in the Prudhoe Bay State 1 well between 8,117 and

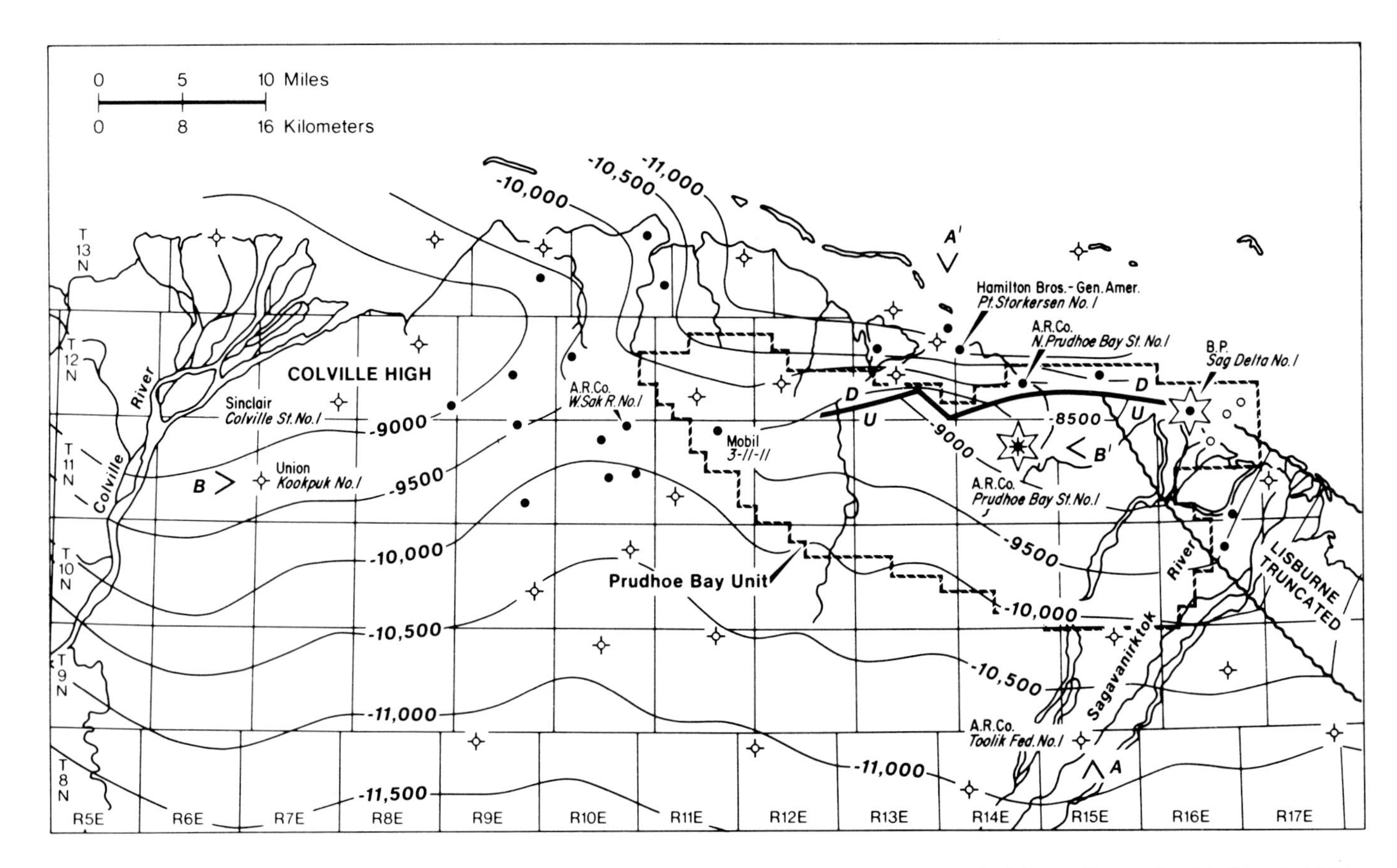

FIG. 14—Structure map of top of the Lisburne Group. Star symbol indicates wells with announced Lisburne discoveries. Faulting is highly simplified. Subsea contour intervals in feet.

8,163 ft (2,474 to 2,488 m), measured electric-log depth (Fig. 10). Jones and Speers (1976) redefined the Sag River, adding an overlying shale interval to the sandstone sequence and defining the combined sandstone and shale as the Sag River Formation.

The "sandstone member" is a uniform, well-sorted, fine-grained bioturbate sandstone to coarse siltstone consisting primarily of quarz with minor amounts of glauconite and chert. It forms a thin but continuous reservoir across the Prudhoe Bay field except where it is truncated by erosion to the east. The unit ranges from less than 20 ft (6 m) thick in the south to about 70 ft (21 m) near the coast of Prudhoe Bay. Porosity and permeability reach 25% and 270 md in the northern part of Prudhoe Bay field.

Put River Sandstone—Although the Lower Cretaceous unconformity surface is generally planar to moderately undulating, local channels have been eroded into the underlying rocks. Reservoir sandstones of Early Cretaceous (?) age occupy the basal sections of these channels near the center of the field and along the western shoreline of Prudhoe Bay. These rocks are herein defined as the Put River Sandstone.

The type section of Put River Sandstone occurs in ARCO-Exxon NGI-7 well in the interval from 8,027 to 8,091 ft (2,447 to 2,466 m), measured electric-log depth (Fig. 13). The rocks unconformably overlie the Sag River Sandstone and in turn are overlain by unnamed Lower Cretaceous shales.

Wherever the Put River Sandstone has been encountered in wells at Prudhoe Bay field, it occurs directly above the Lower Cretaceous unconformity. This unconformity truncates progressively older rocks toward the northeast edge of the field, and the Put River Sandstone consequently lies on rocks of differing age and lithology. The Put River Sandstone has not been tested for hydrocarbons, but as a result of the direct contact between the Sag River Sandstone and the Put River Sandstone it is considered part of the Permo-Triassic reservoir of the Prudhoe Bay field.

The Put River Sandstone consists of gray-brown, conglomeratic, quartzose chert sandstone with lesser amounts of glauconite. Many of the sand-sized chert grains are angular and may be fragments of broken chert pebbles, whereas the quartz grains typically have silica overgrowths. Gravel-sized fractions contain clasts ranging in size from granules to pebbles 3 in. (7.6 cm) in diameter. Pebbles are white quartzite, gray chert, tan mudstone and light gray, glauconitic, fine-grained sandstone probably eroded from the underlying Sag River Sandstone. Porosities in the interval from 8,055 to 8,085 ft (2,455 to 2,464 m) average about 12%. Permeabilities range from less than 10 to 404 md, but are typically less than 100 md.

Sparse micropaleontologic and palynologic data from well cuttings and a few sidewall core samples suggest that the Put River Sandstone is Early Cretaceous in age.

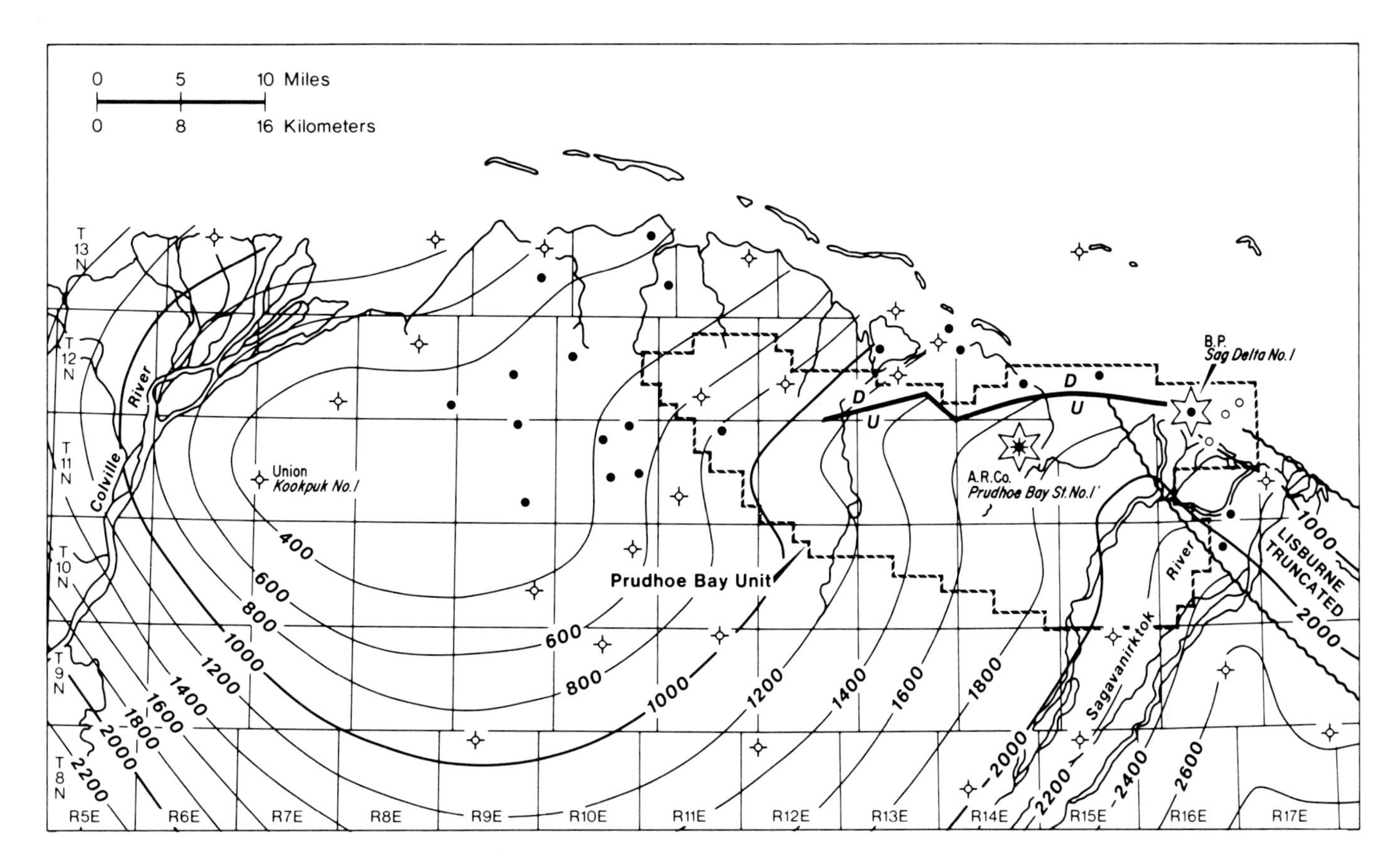

FIG. 15—Lisburne Group isopach map. Contours in feet.

Lisburne Group Reservoirs

The Lisburne Group is a thick, shallow-marine sequence of Mississippian and Pennsylvanian carbonate rocks with minor interbeds of terrigenous clastics and chert and occasional thin beds of evaporite.

As the most widespread potential reservoir in the subsurface of the North Slope, the Lisburne has long been a major exploration objective. From its type section near Cape Lisburne on the northwest coast of Alaska, it is exposed in thrust sheets of the Brooks Range eastward into the Yukon Territory of Canada. It underlies the Prudhoe Bay Complex and is now known to occur in more than 60 wells in the area between the Colville and Canning Rivers. Oil and gas was discovered in the Lisburne in the Prudhoe Bay State 1 well, and test rates ranged from 434 to 1,152 b/d of oil and 1.3 to 22 million cf/d of gas. The recent BP-Sag Delta 1 well also tested Lisbune hydrocarbons and is apparently a significant easterly extension to the discovery. Press reports indicated a 9,380 b/d rate of oil decreased to 6,700 b/d with increasing gas rate during a 16.5-hour test in which the flow was not stabilized (Table 1).

Structure and Stratigraphy—Structure of Lisburne conforms generally to that of the Sadlerochit (Fig. 14). From structural highs of the Prudhoe Bay field (Prudhoe high) and over the Colville River delta (Colville high), the Lisburne dips northward toward the Beaufort Sea and southward toward the axis of the Colville trough.

The Lisburne is the carbonate phase of a Carboniferous carbonate-clastic sequence which transgressed eroded Devonian and older terrain. In the central Brooks Range the Lisburne consists of two formations, the Alapah and the Wachsmuth (Bowsher and Dutro, 1957), overlying dark marine shales of the Kayak Shale assigned to the Endicott Group. In the eastern Brooks Range, Brosge et al (1962) identified the Wachsmuth and Alapah and named a younger carbonate unit the Wahoo Formation.

Northward in the subsurface of the North Slope only the Wahoo and the Alapah are recognized. Here the Alapah lies on gray shales of the Kayak or on red shales, sandstone, and limestones of its lateral equivalent, the Itkilyariak Formation.

Within the mapped area, Lisburne isopachs reflect the outline of an ancient positive area east of the Colville River delta (Fig. 15). The carbonates thin rapidly northward and westward from a maximum in excess of 4,000 ft to less than 400 ft (1,219 to 122 m). This can be accounted for by depositional thinning and convergence of lithic units onto the ancient high by truncation of the Wahoo by pre-Permian erosion. Much of the thickening southeastward is due to the addition of carbonate beds within the Alapah at a rate greater than it is truncated by the Lower Cretaceous unconformity.

Age—The Lisburne has been zoned and dated by Armstrong and Mamet (1970, 1974), and Armstrong, Mamet, and Dutro (1970,1971) using worldwide microfossil assemblage zones of benthonic Foraminifera and algae. In the mapped area the basal Alapah has been dated as late Visean (early Chester) and

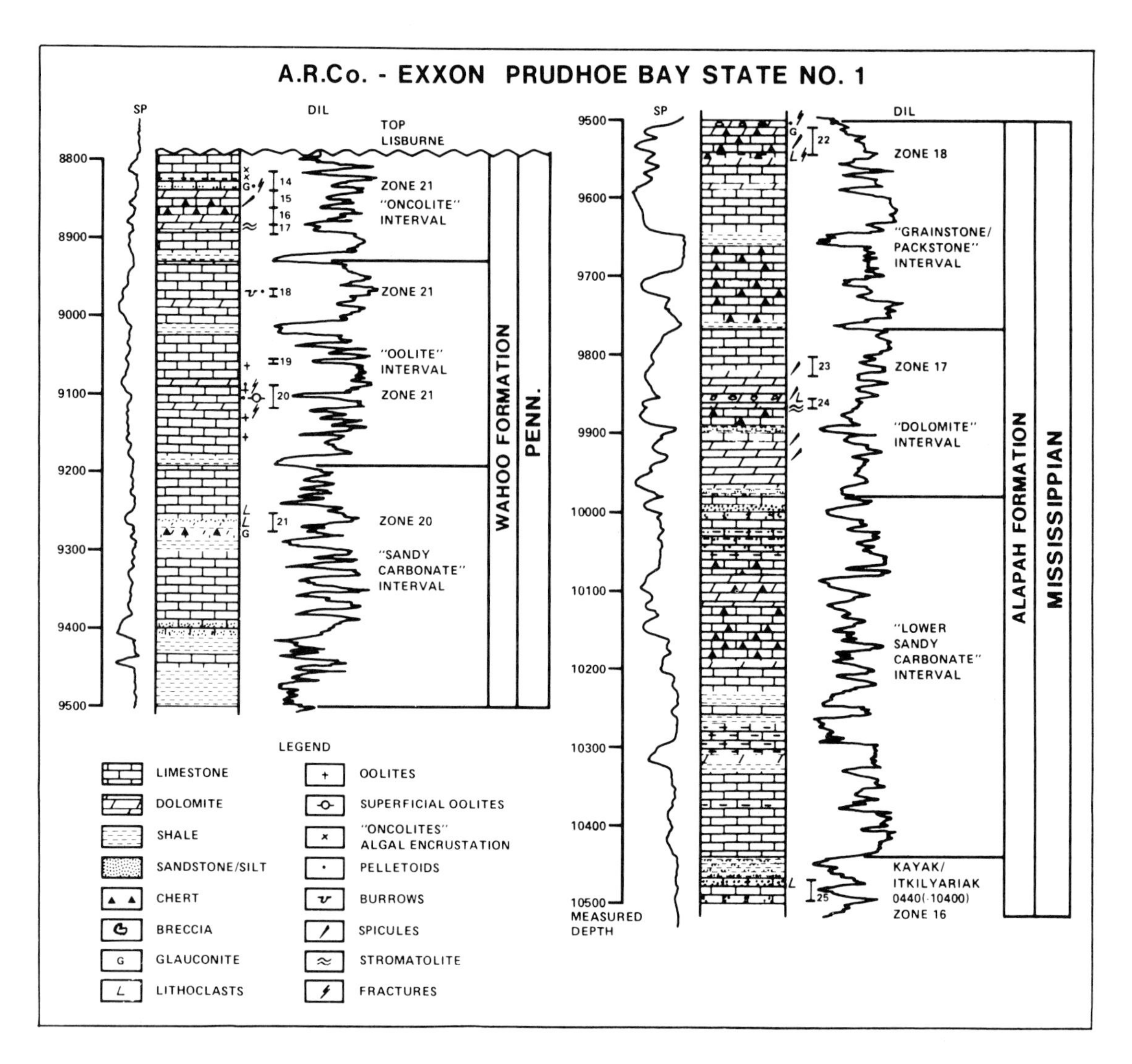

FIG. 16—Log of the Lisburne section encountered in the discovery well.

its top as middle Namurian (late Chester). The Wahoo carbonates in the subsurface range from late Namurian to early Westphalian (Morrow and Atoka).

Within the Prudhoe Bay Complex lithic units and formation boundaries appear to be synchronous. The top of the Lisburne in the subsurface is unconformably overlain by Permian, Triassic, or Cretaceous clastic rocks. The nature of the carbonate rock types suggests that the original sediments were frequently subaerially exposed, yet no major unconformities or significant hiatuses have been recognized within the Lisburne Group.

Lithology—Lisburne carbonate rocks are grain-supported (grainstones and packstones), high-energy, well-winnowed, well-sorted, spar-cemented calcareous sandstones composed predominantly of finely particulate echinoderm and bryozoan skeletal debris with minor amounts of forams and algae. Superfical oolitic coatings are common but true oolites with multiple rinds are concentrated only in the Wahoo.

Mud-supported rocks (wackestones and mudstones) representing protected quiet water or tidal-flat environments are now largely represented by early diagenetic finely crystalline calcitic dolomites. The dolomites are the main reservoirs within the Lisburne. Intercrystalline porosity is the most common type with values as high as 20% but with associated average permeability of only 1 md. Vuggy porosity is sometimes associated with intercrystalline porosity in the most dolomitic rocks. The vugs commonly are small (approximately 0.1 mm), "pin-point" voids caused by solution-removal of sponge spicules and other fragments of indeterminate fossils. Open fractures ranging from a fraction of a millimeter to several centimeters provide added porosity and enhanced permeability.

Chert is common in thin beds or in small nodules commonly associated with concentrations of sponge spicules. Evaporites (anhydrite or gypsum) are very local in extent and found only in thin beds in the lower Alapah. Terrigeneous clastic deposits are represented throughout by green, brown, and red shales, silts and occasional thin beds of fine-grained quartz sands, which may contribute a small amount to reservoir capacity.

Lithostratigraphic Framework—The Lisburne is subdivided and correlated using wireline logs and sample data. Within the field area, six lithic units are recognized which have value in

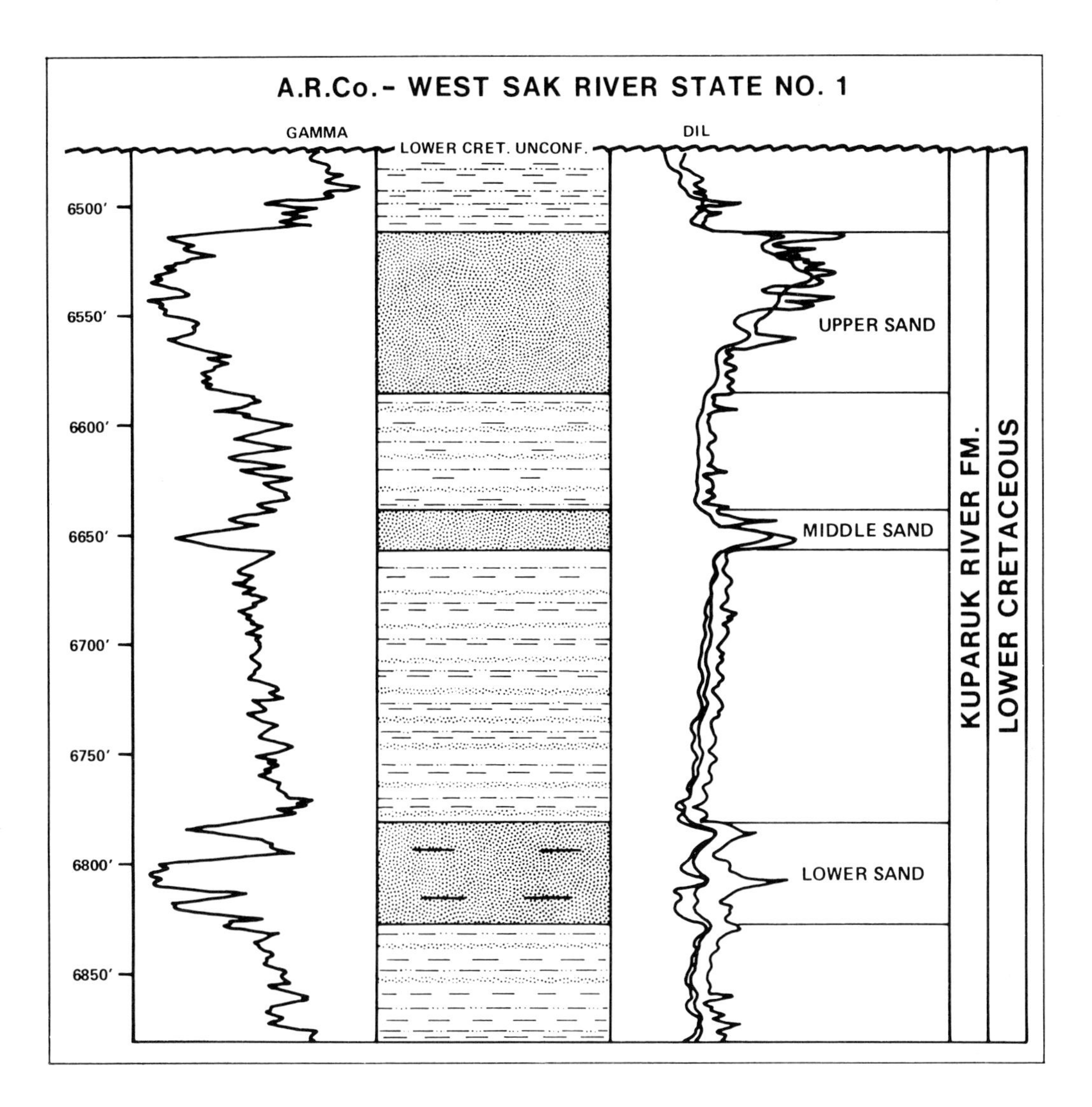

FIG. 17—Log of the Kuparuk River Formation in the ARCO West Sak River State 1 well.

correlation. These are referred to as "Intervals" on the Lisburne type log and identified by conspicuous grain or rock type (Fig. 16). The Wahoo and Alapah each contain three lithic units. In the Prudhoe Bay State 1 well, in ascending order, they are:

Alapah Formation:

(1) Lower sandy carbonate interval—9,980 to 10,440 ft (3,042 to 3,185 m)—The lower Alapah is characterized by a decrease upward in quartz-bearing and lithoclast-bearing carbonates. The base of the Alapah is gradational with the Kayak and/or the Itkilyariak Formation which is identified as Zone 16s in core no. 25 by Armstrong and Mamet (1974).

(2) Dolomite Interval—9,765 ft to 9,980 ft (2,976 to 3,042 m)—The middle part of the Alapah is characterized by relatively persistent beds of dolomite with good porosities (to 2061%) and permeabilities (approx. 3 61 md). These values are the result of complete dolomitization and subsequent leaching of sponge spicules and other fossil fragments. The interval is identified as Zone 17 in core no. 23.

(3) Grainstone/Packstone Interval—9,500 to 9,765 ft (2,896 to 2,976)—The upper part of the Alapah in the field area is primarily a crinoid-bryozoan grainstone and packstone with minor amounts of calcitic dolomite, some stromatolitic. The unit lacks other distinctive features and is identified as Zone 18 in core no. 22.

Wahoo Formation:

(1) Sandy Carbonate Interval—9,190 to 9,500 ft (2,801 to 2,896 m)—Like the lower Alapah the lower part of the Wahoo is characterized by a decrease upward in the amount of clastics. Sandy and lithoclast-bearing carbonates are common. The interval is identified as Zone 20 in core no. 21.

(2) Oolitic Interval—8,930 to 9,190 ft (2,557 to 2,801 m)—The middle part of the Wahoo is characterized by several different levels of oolites with multiple rinds

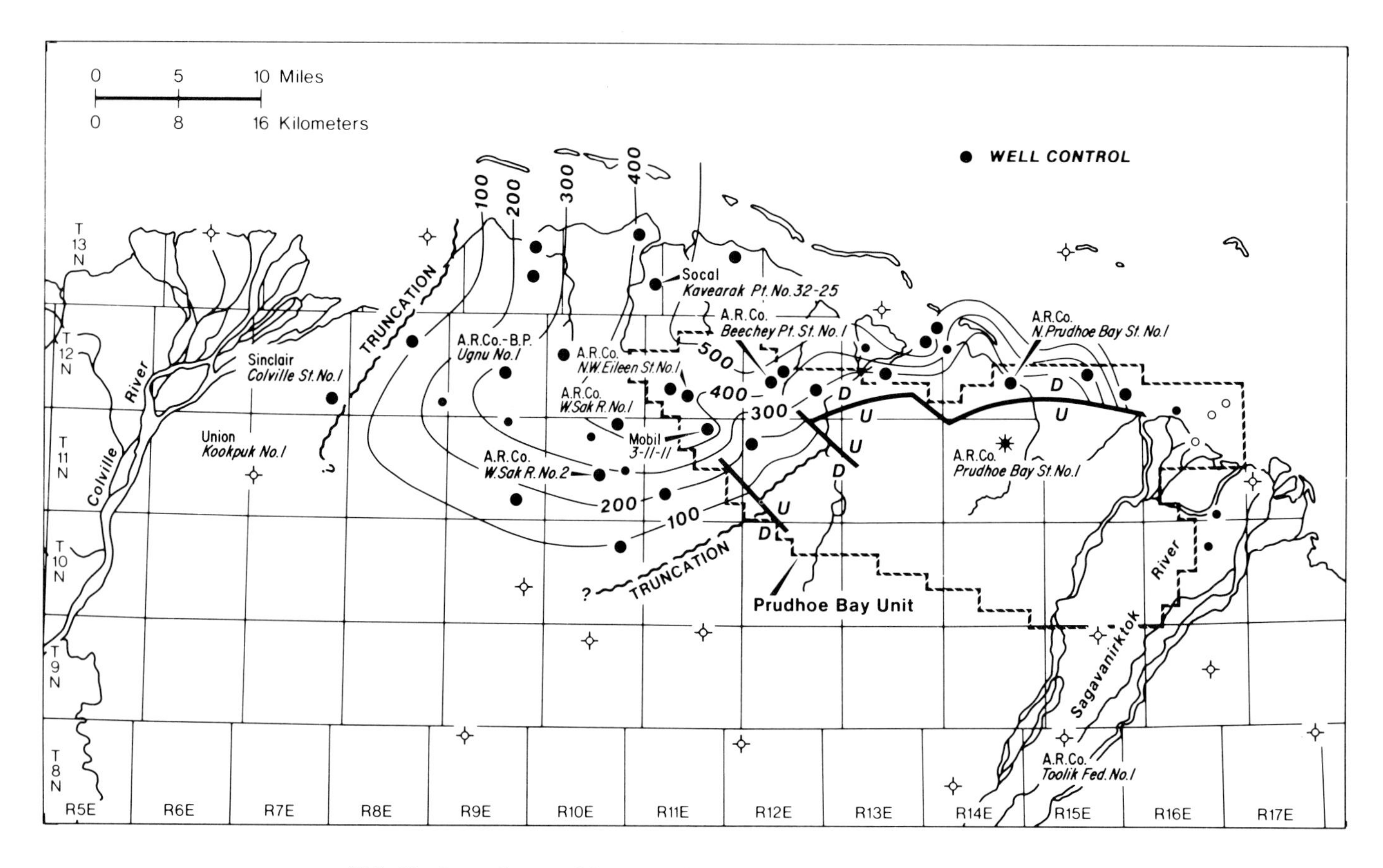

FIG. 18—Isopach map of the Kuparuk River Formation. Contours in feet.

and diameters of 1 mm or less (Zone 21).

(3) "Oncolite Interval"—8,790 to 8,930 ft (2,679 to 2,557 m)—The upper part of the Wahoo is characterized by an abundance of small (6fi1 mm) diameter, algal-encrusted grains with crenulate outer margins. Associated Foraminifera identify this interval as Zone 21.

Kuparuk River Formation Reservoirs

The Kuparuk River Formation consists of sandstones, siltstones, and shales of Lower Cretaceous age deposited in a shallow marine, tidal-influenced environment and derived, at least in part, from a southerly or easterly source. It is present over the western and southwestern part of the Prudhoe Bay Complex, as well as along the northern flank. It lies conformably on marine Kingak Shale. The North Slope Stratigraphic Committee of the Alaska Geological Society (1970-1971) informally named this interval the Kuparuk River Sands and defined the type section. Detterman et al (1975) used the term, "the Kuparuk River Sand," while Jones and Speers (1976) refer to it as the Kuparuk River Formation. The interval between 6,474 and 6,880 ft (1,973 to 2,097 m), in the ARCO West Sak River State 1 well, was selected as illustrative of the Kuparuk River Formation for this paper because of the well's location near the center of the Kuparuk depositional area and because much of the interval was cored (Fig. 17).

Along the shore of the Beaufort Sea the Kuparuk has been identified in wells in an area at least 18 mi (28.9 km) north to south and greater than 35 mi (56.3 km) east to west. Although the northern and southern limits are not defined, existing well control indicates that the formation thickens to more than 600 ft (183 m) northeastward beneath the Beaufort Sea and thins to less than 90 ft (27 m) in the west, south, and southeast (Fig. 18). This thinning can be, for the most part, attributed to truncation by the Lower Cretaceous unconformity. However, thinning may also be attributed to local erosional unconformities or depositional hiatuses within the formation (Fig. 19).

In the subsurface the top of the Kuparuk River Formation has been encountered at depths ranging from less than 5,800 ft (1,768 m) to deeper than 9,000 ft (2,743 m). It is contoured as an easterly plunging anticline truncated on the northwest and southeast by the Lower Cretaceous unconformity (Fig. 20). The southwestern flank of the structure is not defined. The northeastern flank is interrupted by normal faults that control hydrocarbon accumulations in this area. The Eileen fault trends northwestward from the ARCO Northwest Eileen 1 well and may separate accumulations in the West Sak area on the south from those on the north. The structure depicted north of the Eileen fault is generalized, and hydrocarbon trapping appears to be related to fault closures or stratigraphy or both. However, because of the lack of adequate well control, these accumulations are difficult to define.

Lithologies and Sedimentation—Three sand members, in-

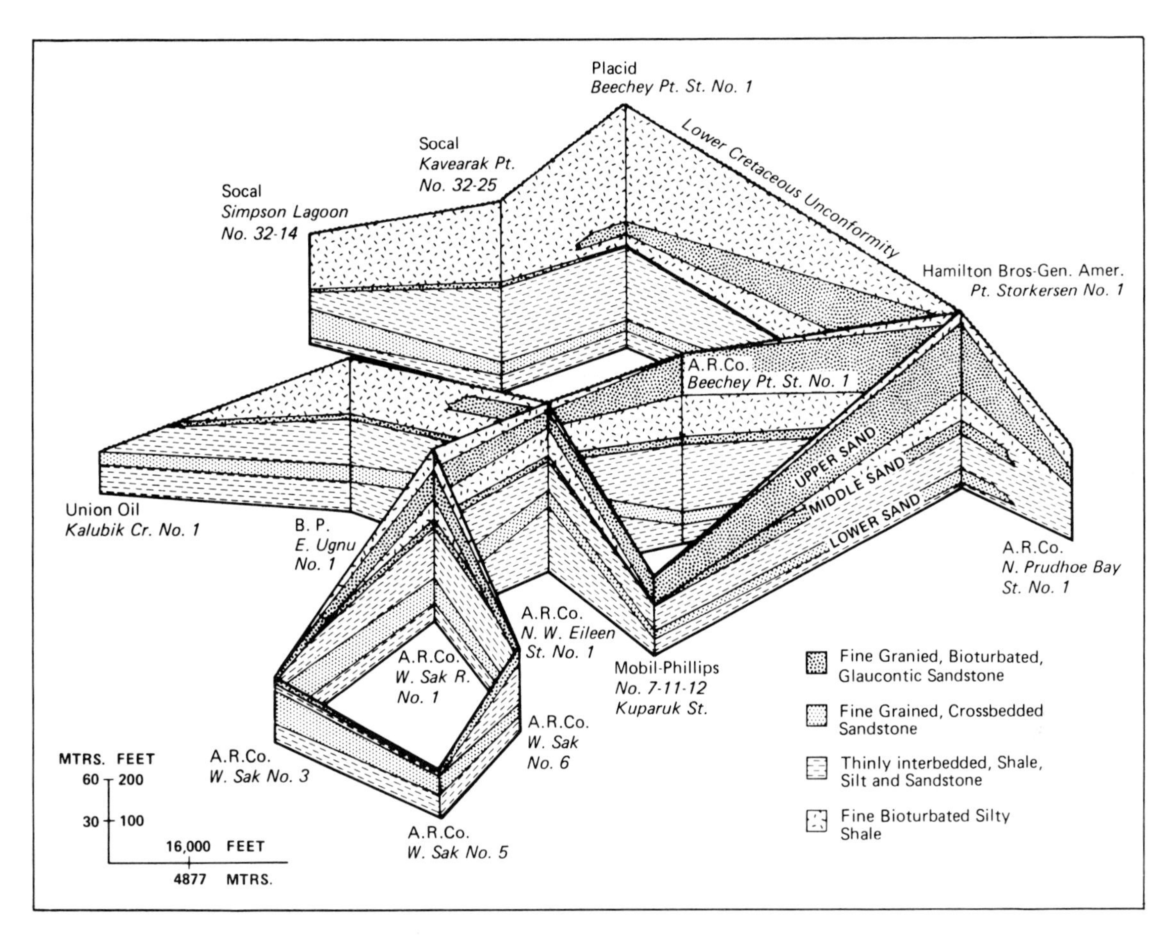

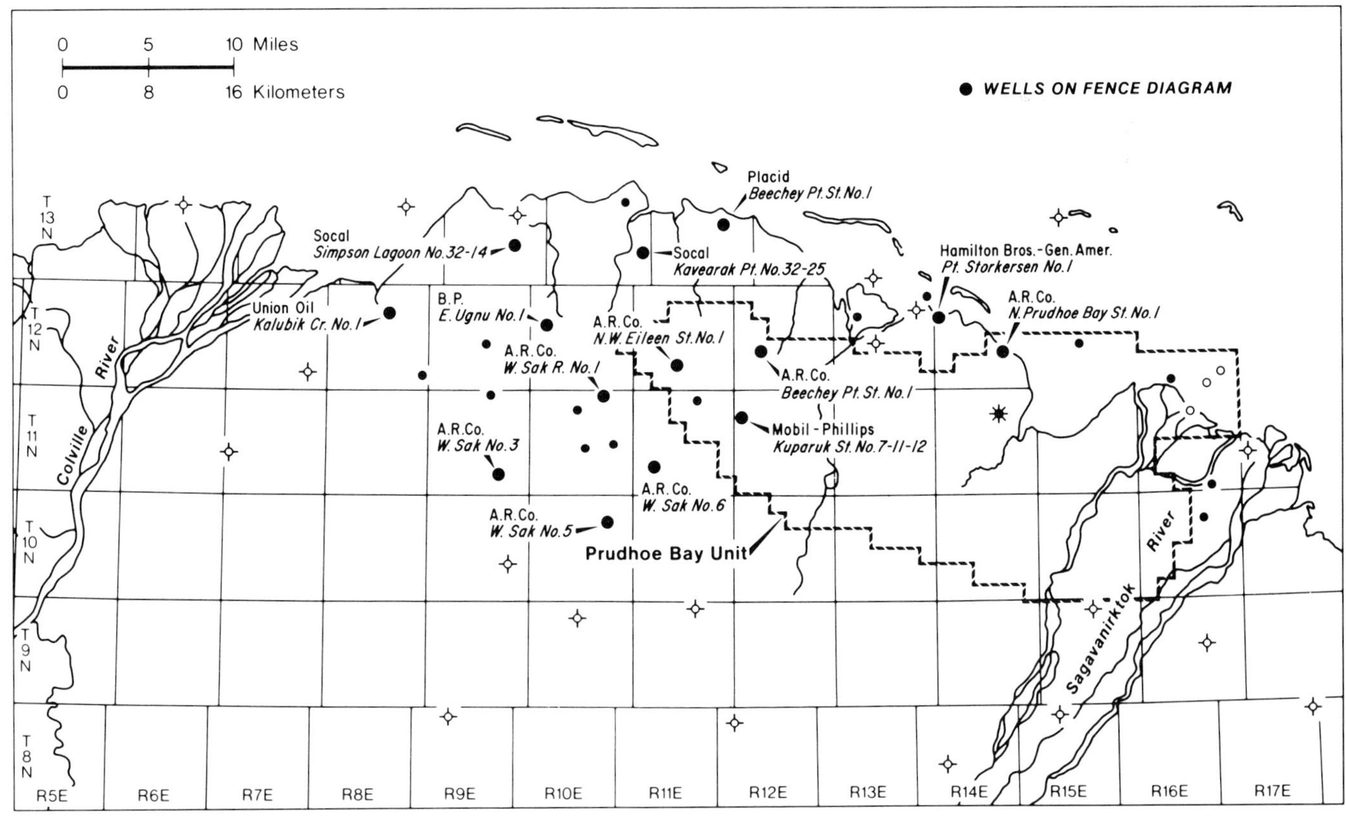

FIG. 19—**A.** Fence diagram, Kuparuk River Formation. Shows relationship of lower, middle, and upper Kuparuk sandstones. **B.** Fence diagram index.

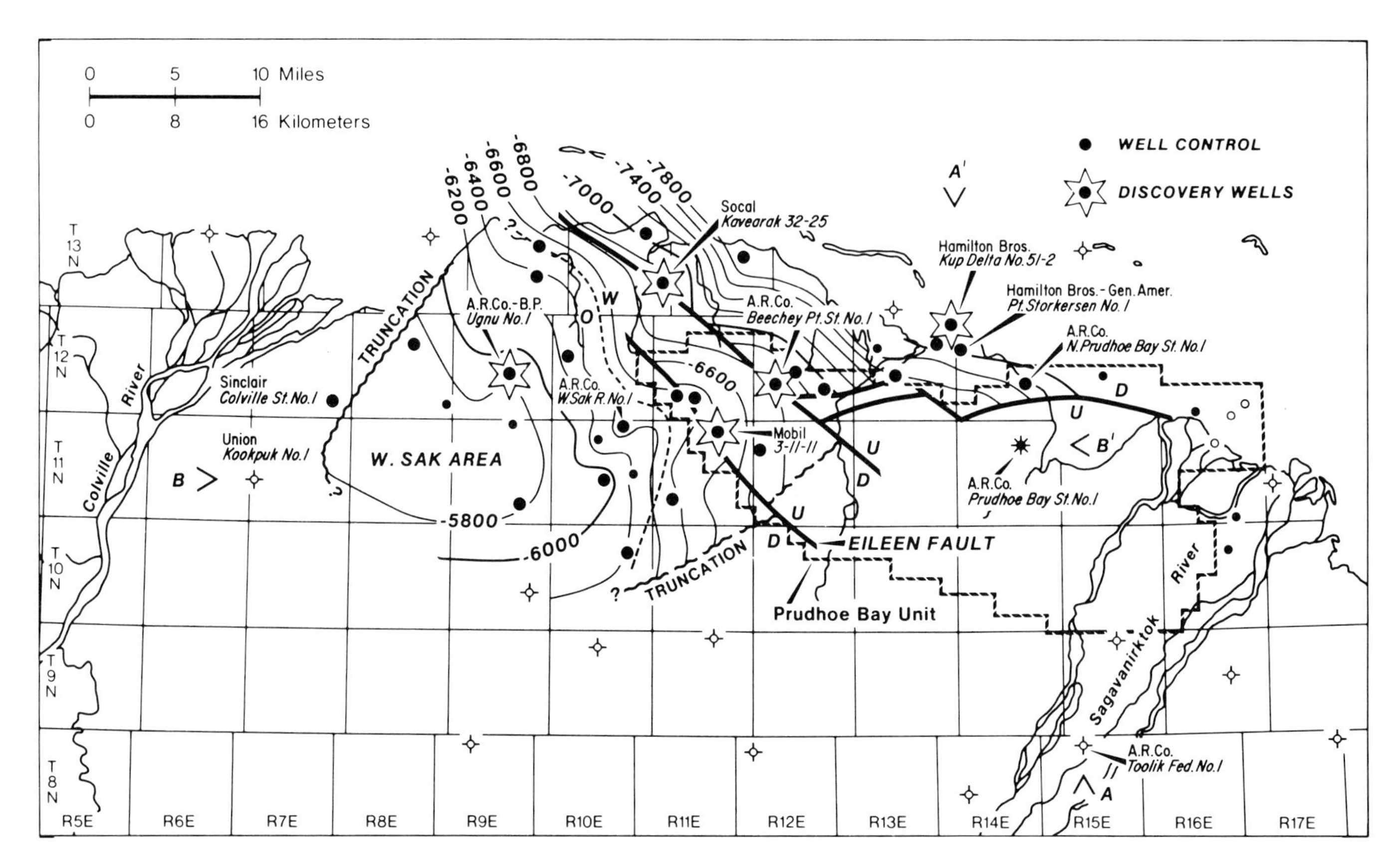

FIG. 20—Structure map of top of Kuparuk River Formation. Star symbols indicate wells that have discovered hydrocarbons. Subsea contours in feet.

formally termed the "Upper, Middle and Lower Sands," separated by shales and silty shales, are the primary Kuparuk reservoirs of the Prudhoe Bay Complex (Fig. 17). The formation is unconformably overlain by an unnamed Lower Cretaceous shale.

The "Lower Sand" is the most widespread of the sand members and can be correlated throughout most of the mapped area (Fig. 19). It ranges in thickness from about 35 to 75 ft (10.7 to 22.9 m), with average porosities ranging between 20 and 27% and permeabilities up to 500 md. The "Lower Sand" differs from the "Upper and Middle Sands" in that it exhibits extensive sedimentary structures characteristic of current and wave action. Low-angle crossbeds and planar or horizontal laminations are common, as well as ripples. Silt and clay layers indicate changing environmental energies. The "Lower Sand" occurs as either two or three electric-log benches probably formed by the interfingering of two or more sand bodies. Fine sand and silt are the predominant constituents with less than 20% discontinuous shale layers. The lens-shaped shale layers are commonly a few millimeters to several centimeters thick. Characteristics of the "Lower Sand," such as planar laminae and ripples, clay laminae, and variable development of the benches, indicate that it was deposited in a nearshore environment, possibly adjacent to an emergent area east of Prudhoe Bay. The "Lower Sand" overlies an interbedded siltstone and dark gray, pyritic shale member that becomes shalier toward the base with fewer and fewer interbeds of silt and very fine sand.

Overlying the "Lower Sand" is an interval of thinly interbedded cyclic sands, silts and shales probably deposited in a tidal environment. Shale layers 1 to 2 cm thick are draped over lenses and thin beds of silt and sand that are ripple and planar laminated. Normally, the lenses are 1 to 3 cm thick and 3 to 7 cm long. The sands are very fine to fine-grained and commonly contain thin laminae of clay. Worm burrows are common in the shales and mollusk burrows, although uncommon, are present. The sands and silts in this interval are oil saturated, and therefore, may contribute to a small degree to the productivity of the Kuparuk.

The "Middle and Upper Sands" differ from the "Lower Sand" in that they are glauconitic and lack sedimentary structures. Biologic activity or turbulence at the site of deposition could have prevented formation of or destroyed sedimentary structure. The "Middle and Upper Sands" are similar lithologically, but differ in thickness and distribution (Fig. 19). Both sands are very fine to medium grained, glauconitic, and frequently moderately to poorly sorted. Glauconite is considered an excellent indicator of marine conditions and frequently occurs as an alternation of fecal pellets. This appears to be the case in the Kuparuk sands and suggests that bioturbation is the cause of the lack of sedimentary structures.

Of the three sands, the "Middle Sand" is thinnest and ranges

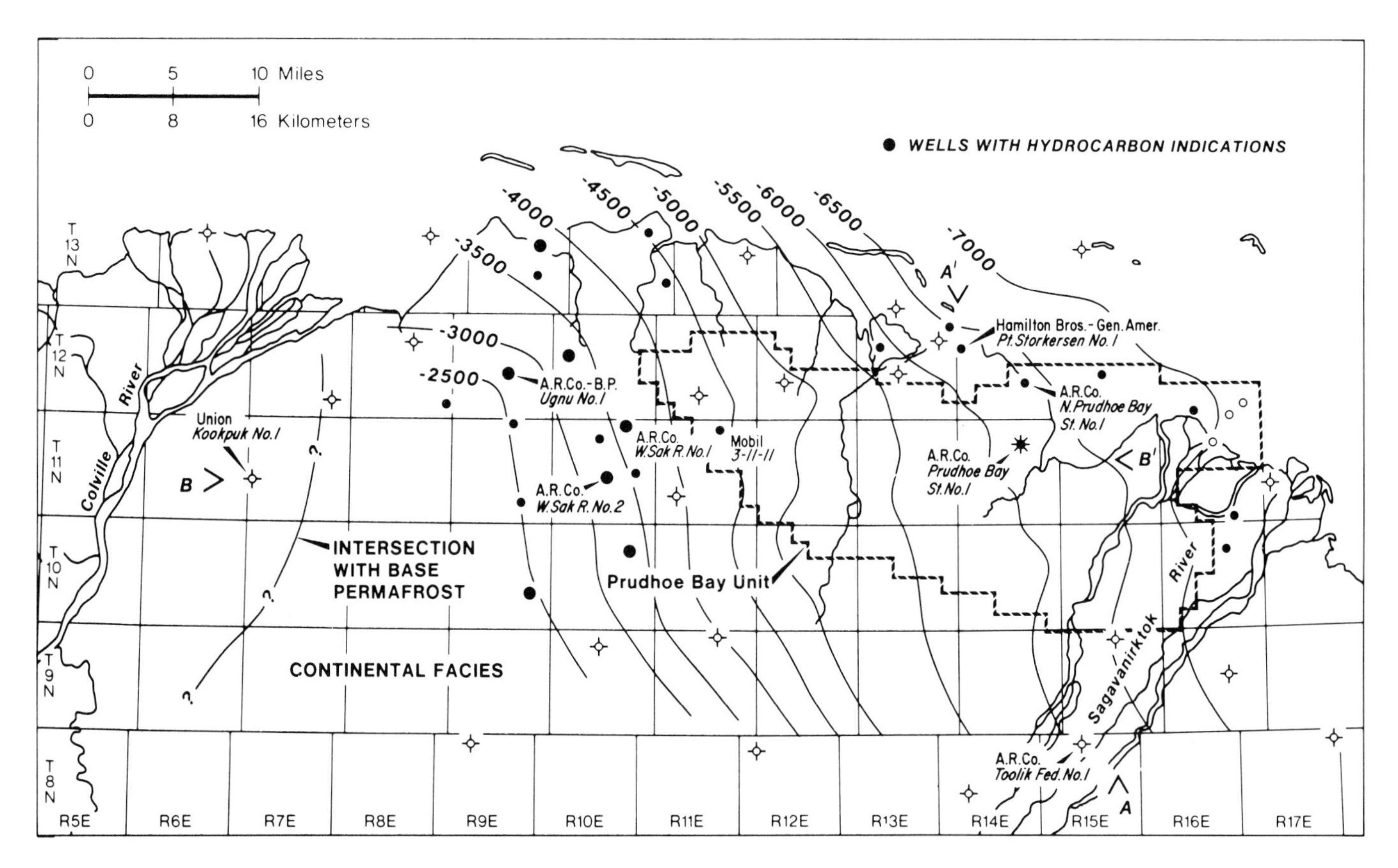

FIG. 21—Structure map of top of West Sak Sands. Subsea contours in feet.

from less than 10 ft (3 m) to almost 20 ft (6 m) thick. Average porosity ranges between 20 to 25% with permeabilities up to 525 md. At present, the "Middle Sand" is thought to have been deposited as a subtidal shoal. The fine grain size and bioturbation suggest that the "Middle Sand" was deposited in a relatively low energy environment.

The "Upper Sand" attains a thickness of more than 150 ft (46 m) in the Hamilton Brothers Point Storkersen well. Average porosity ranges from 20 to 22% with permeabilities up to 266 md. It has been truncated to the south and southeast and shales out to the northwest, suggesting a southerly or southeasterly source (Fig. 19).

Between the "Middle" and "Upper Sand" is a silty shale member which exhibits a transitional change from shale at its base to silts at the top. The silts, in turn, grade into the "Upper Sand" with no visible break in deposition. Bedding tends toward planar rather than lenticular. The lower shaly part is not extensive, occurring only in the ARCO West Sak River State 1 and Northwest Eileen 1 wells. The silt part is found wherever "Upper Sand" is present. An unconformity appears to be present between the silt and shale. Overlying the "Upper Sand" is bioturbated, silty, muddy shale.

Age—Tabbert and Bennett (1976) identified the Kuparuk River Formation in the ARCO West Sak River State 1 as being Neocomian in age. Thirty-four microplankton species were documented from conventional cores obtained from this well. The Neocomian assemblage can be correlated with sediments from southern Alaska and northern Canada.

However, Foraminifera present in the Kuparuk River Formation are of mixed Late Jurassic and Early Cretaceous ages and can be correlated with surface sections collected by the U.S. Geological Survey in northeastern Alaska. The fauna from these sections has been called a transitional fauna by Bergquist (Detterman et al, 1975), suggesting that the Kuparuk represents continuous deposition beginning Late Jurassic and extending into the Early Cretaceous. The reason that the faunas are mixed is not clear. Reworking is not considered likely because Jurassic species occur in great abundance and outnumber Cretaceous species 2 to 1; and, additionally, the mixed assemblage occurs in every Kuparuk well examined in the area. Possibly the Jurassic and Cretaceous species which occur in these sands have a longer range than previously thought.

Exploration History—In 1969 after the discovery at Prudhoe Bay in 1968, the industry drilled numerous wells in the region between the Colville and Canning Rivers to evaluate acreage to be leased in the September 1969 lease sale. During this spurt of activity, Sinclair discovered the productive Kuparuk "Lower Sand" in the Ugnu State 1 which tested 1,056 b/d of oil; ARCO's Northwest Eileen 1 recovered small amounts of oil from the " Middle and Upper Sands" and Mobil's West Kuparuk 3-11-11 tested 2,220 b/d from the "Upper Sand." Exploration activity remained high in the Kuparuk area into 1970 as various companies continued to evaluate "Sadlerochit" and/or Lisburne structural anomalies on their existing and

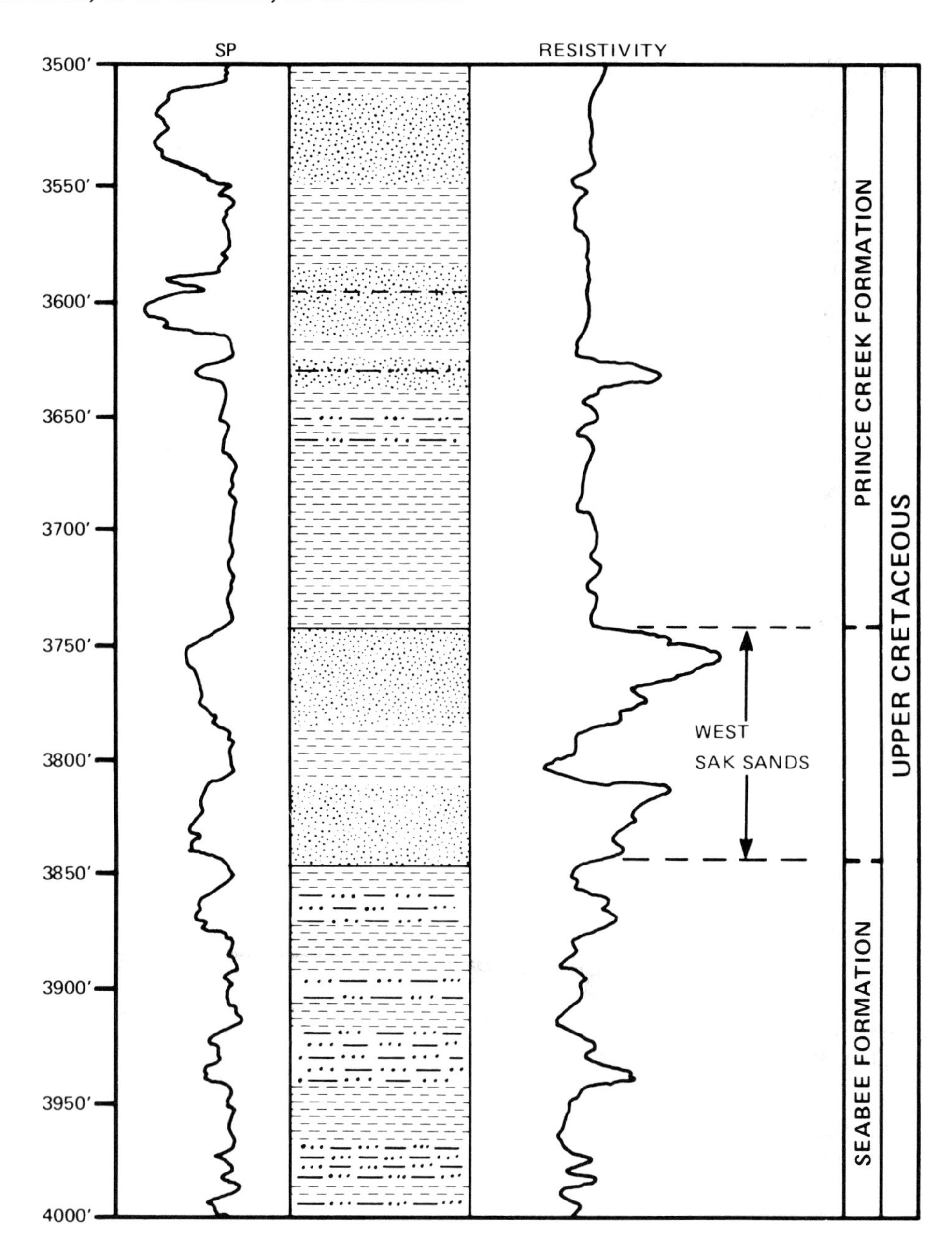

FIG. 22—Type log for West Sak Sandstones, ARCO West Sak River State 1 well.

newly acquired leases.

ARCO's West Sak River State 2, drilled in 1974, was the first well drilled expressly for Kuparuk evaluation. This marked the beginning of a series of exploratory and development wells in the West Sak area designed to define the extent and producibility of the Kuparuk reservoirs. In 1975, three wells were drilled, followed in 1978 by the drilling of six more. Two exploratory wells are being drilled during the 1979 winter season.

West Sak Sands Reservoirs

Oil was first recovered from the West Sak Sands in the ARCO West Sak 1 well about 24 mi (38.6 km) west of the Prudhoe Bay State 1 discovery well (Fig. 21). These sands are defined in the West Sak 1 well where they occur in the interval 3,742 to 3,842 ft (1,141 to 1,171 m), electric-log depths (Fig. 22).

The West Sak Sands are transitional from the underlying Upper Cretaceous marine shales of the Seabee Formation to the younger Upper Cretaceous nonmarine sandstones of the Prince Creek Formation. This transition also is supported by faunal evidence that indicates a nearshore marine environment. The microfauna consists of sparse arenaceous and calcareous Forminifera which are identical to forams collected from exposures of the Upper Cretaceous Sentinel Hill Member of the Schrader Bluff Formation and described by Tappan (1962).

In the ARCO West Sak 1 well average porosity is 29% and average permeability is about 500 md. The sands are fine grained, quartzose, well sorted, subangular, argillaceous, glauconitic, and very friable. The interval normally consists of two sand units, each about 40 ft (12 m) thick, separated by a 25-ft (7.5 m) mudstone. Above the upper sandstone, a 100-ft (30-m) mudstone sequence separates these sandstones from the overlying nonmarine sandstones. Occasionally, thin sandstones occur in this clay sequence. Below the West Sak Sands the lithology becomes progressively more fine grained and clay rich.

The West Sak Sands dip gently to the east-northeast and exhibit gentle nosing coincident with the anticlinal trend of the Prudhoe Bay Complex. The trapping mechanism for the accumulation is currently not defined, although both stratigraphic changes to the west and an updip seal provided by permafrost are possibilities.

Drill-stem test rates ranged from 112 to 192 b/d of oil. Oil recovered ranged in gravity from 17° to 23° API and in viscosity from 25 to 50 centipoise. Primary oil recovery is anticipated to be very low because of these unfavorable fluid properties. Owing to the sparse information related to this accumulation, the amount of oil in place cannot be determined with any accuracy.

CONCLUSIONS

In attempting to develop a meaningful perspective of the discovery of the Prudhoe Bay field as seen by the petroleum geologist 10 years later, we have reached these basic conclusions:

1. Successful oil-finding on the North Slope depends on continuity and persistence of the overall exploration effort. This effort was, and is, directly responsive to knowledge of a firm schedule of land availability.

2. A stable and predictable investment climate is of utmost importance in North Slope exploration and production operations.

3. The variety of accumulations found in the Prudhoe Bay Complex would indicate the probability of future sizable discoveries in other areas of the North Slope, if exploration in such areas is not unduly restricted by federal, state and local regulatory procedures.

Because the earth's surface and subsurface form the explorationist's laboratory, freedom of access to that laboratory is a prerequisite for future success. We believe the example of Prudhoe Bay, as the largest field ever discovered in the United States, serves as an object lesson for explorationists. By documenting what occurred in the 10 years prior to the discovery and comparing that record of success to the relatively minimal progress of the last decade, we have presented a case history that should be helpful in guiding future exploration policies in areas such as Alaska.

REFERENCES CITED

Adams, W. D., 1972, Developments in Alaska in 1971: AAPG Bull., v. 56, p. 1175-1187.

Alaska Geological Society, 1970-1971, West to east stratigraphic correlation section, Point Barrow to Ignek Valley, Arctic North Slope, Alaska, North Slope Stratigraphic Committee, Anchorage, Alaska.

Armstrong, A. K., and B. L. Mamet, 1970, Biostratigraphy and dolomite porosity trends of the Lisburne Group, *in* W. L. Adkinson, and M. M. Brosge, eds., Proceedings of the geological seminar on the North Slope of Alaska, Los Angeles: Pacific Section of AAPG, p. N1-N16.

——— ——— 1974, Carboniferous biostratigraphy, Prudhoe Bay State 1 to north-eastern Brooks Range Arctic Alaska: AAPG Bull. v. 58, p. 646-660.

——— ——— and J. T. Dutro, 1970, Foraminiferal zonation and carbonate facies of carboniferous (Mississippian and Pennsylvanian) Lisburne Group, Central and Eastern Books Range, Arctic Alaska: AAPG Bull, v. 54, p. 687-698.

——— ——— ——— 1971, Lisburne Group, Cape Lewis-Niak Creek, northwestern Alaska, *in* Geological Survey Research, 1971: U.S. Geol. Survey Prof. Paper 750-B p. B23-B34.

Bowsher, A. L., and J. T. Dutro, Jr., 1957, the Paleozoic section in the Shainin Lake area, Central Brooks Range, Alaska: U.S. Geol. Survey Prof. Paper 303a, 39 p.

Brosge, W. P., et al, 1962, Paleozoic sequence in Eastern Brooks Range, Alaska: AAPG Bull. v. 46, p. 2174-2198.

Carter, R. D. et al, 1977, The petroleum geology and hydrocarbon potential of Naval Petroleum Reserve No. 4, North Slope, Alaska: U.S. Geol. Survey Open File Rept., p. 77-475.

Detterman, R. L., 1970, Sedimentary history of the Sadlerochit and Shublik Formations in northeastern Alaska, *in* Proceedings of the geological seminar on the North Slope of Alaska: Pacific Section AAPG, p. 01-013.

——— et al, 1975, Post-Carboniferous stratigraphy, northeastern Alaska: U.S. Geol. Survey Paper 886, 46 p.

Gryc G., 1970, History of petroleum explortion in Northern Alaska, *in* Proceedings of the geological seminar on the North Slope of Alaska: Pacific Section AAPG, p. C1-C8.

Jamison, H. C., 1978, *in* Tailleur and others, Folio of maps with a section on exploration: U.S. Geol. Survey Misc. Field Stud. 928-A.

Jones, H. P., and R. G. Speers, 1976, Permo-Triassic reservoirs on Prudhoe Bay Field, North Slope, Alaska, *in* Jules Braustein, ed., North American Oil and Gas Fields: AAPG Memoir 24, p. 23-50.

Lefingwell, E. dek., 1919, The Canning River region, northern Alaska: USGS Prof. Paper 109, 251 p.

Lerand, Monti, 1973, Beaufort Sea, *in* R. G. McGrossan, ed., The future petroleum provinces of Canada—Their geology and potential: Canadian Soc. Petroleum Geologists, Memoir 1, p. 315-386.

Lian, E. B., 1971, Developments in Alaska in 1970: AAPG Bull., v. 55, p. 943-957.

Morgridge, D. L., and W. B. Smith, Jr., 1972, Geology and discovery of Prudhoe Bay field, Eastern Arctic Slope, Alaska, *in* Robert E. King, ed., Stratigraphic oil and gas fields—classification, exploration methods, and case histories: AAPG Memoir 16, p. 489-501.

Petroleum Information, 1979, The Alaska Report: February 14 issue, Sec. 1, p. 3.

Reed, J. C., 1958, Exploration of Naval Petroleum Reserve No. 4 and Adjacent areas, northern Alaska, 1944-53, Pt. 1, History of the exploration: U.S. Geol. Survey Prof. Paper 301, 192 p.

Rickwood, F. K., 1970, The Prudhoe Bay field, *in* Proceedings of the geological seminar on the North Slope of Alaska: Pacific Section AAPG, p. L1-L11.

Tabbert, R. L., and Bennett, J. E., 1976, (abstract), Lower Cretaceous microplankton from the subsurface of northern Alaska, Geoscience & Man, LSU Press, v. 15, p. 146.

Tappan, Helen, 1962, Foraminifera from the Arctic Slope of Alaska, pt. 3, Cretaceous Foraminifera: U.S. Geol. Survey Prof. Paper 236-C, p. 91-209.

Van Poolen, H. K., and Associates, Inc., and State of Alaska, 1974,

In-place volumetric determination of reservoir fluids, Sadlerochit Formation, Prudhoe Bay field: Department of Natural Resources Division of Oil and Gas.

——— 1978, In-place hydrocarbon determination, Kuparuk River Formation, Prudhoe Bay, Alaska: Report prepared for State of Alaska, 13 p.

Strachan and Ricinus West Gas Fields, Alberta, Canada[1]

By M.E. Hriskevich, J.M. Faber, and J.R. Langton[2]

Abstract Exploration for reef reservoirs in the "Deep Basin" of Alberta during the mid-1960s resulted in the discovery at Strachan and Ricinus West of 1.9 Tcf of sales gas, 50 million bbl of condensate, and 24.5 million long tons of sulfur in two reefs of Late Devonian age. The reefs were discovered in 1967 and 1969, respectively, by adapting the seismic common-depth point (CDP) techniques of data acquisition and processing that were being developed (particularly in the Rainbow area, in the shallower part of the Western Canada sedimentary basin).

The key well for these discoveries was the Gulf-Strachan well in lsd. 12-31-37-9 W5M, which was drilled in 1955. This well encountered a partial buildup of an Upper Devonian reef which yielded some gas and salt water at a depth of 13,900 ft (4,237 m). CDP seismic data were acquired and, after considerable experimentation in processing with orientation to the appropriate geologic model, showed that the key well was on the flank of what is now called the Strachan reef. In 1968, Banff and Aquitaine drilled a full reef buildup of 900 ft (274 m) in lsd. 10-31-37-9 W5M with a pay section of 536 ft (165 m). A separate pool, the Ricinus West reef, was discovered in 1969 by Banff and Aquitaine in lsd. 6-25-36-10 W5M. The well showed a reef buildup of 800 ft (245 m) and a maximum pay of 634 ft (193 m). Remaining reserves of marketable natural gas at Strachan and Ricinus West, after 6 years of production, are about 1 Tcf.

INTRODUCTION

Major reserves of sour gas and liquids are being produced from deeply buried reef reservoirs of Late Devonian "Leduc" age in the Strachan-Ricinus area of Alberta. This production is located in the deepest part of the Western Canadian Sedimentary basin at the eastern edge of the Rocky Mountain thrust belt, approximately 100 mi (62 km) northwest of Calgary (Fig. 1).

The significance of "Leduc" age reefs as reservoirs for hydrocarbon accumulation in western Canada is well documented. However, the discovery of the Strachan and Ricinus fields marked the first occurrence of commercial production from depths in excess of 14,000 ft (4,267 m) and in such a tectonic setting.

Earlier Leduc reef oil and gas fields, discovered during the late 1940s to early 1960s, accounted for some 3.9 billion bbl of proved oil reserves and 17.8 Tcf of gas[3]. These discoveries are attributed almost entirely to seismic subsurface investigation using single fold analog instrumentation. Much of the Western Canadian Sedimentary basin, including the Strachan-Ricinus part of the deep basin was evaluated in this manner. However, in this area the Leduc reefs went undetected. This paper presents the sequence of events that led to the ultimate discovery of these deeply buried reefs, a discovery that came in response to technological change in seismic processing—the adaptation of seismic acquisition and data processing techniques to a particular geologic setting. This study also attempts to out-

[1]Manuscript received, August 10, 1979; accepted for publication, October 29, 1979.

[2]Aquitaine Company of Canada, Ltd., Calgary, Alberta, Canada, T2P 3J6.

[3]Alberta Energy Resources Conservation Board Reserve Report.

Article Identification Number:
0065-731X/80/M030-0015/$03.00/0.

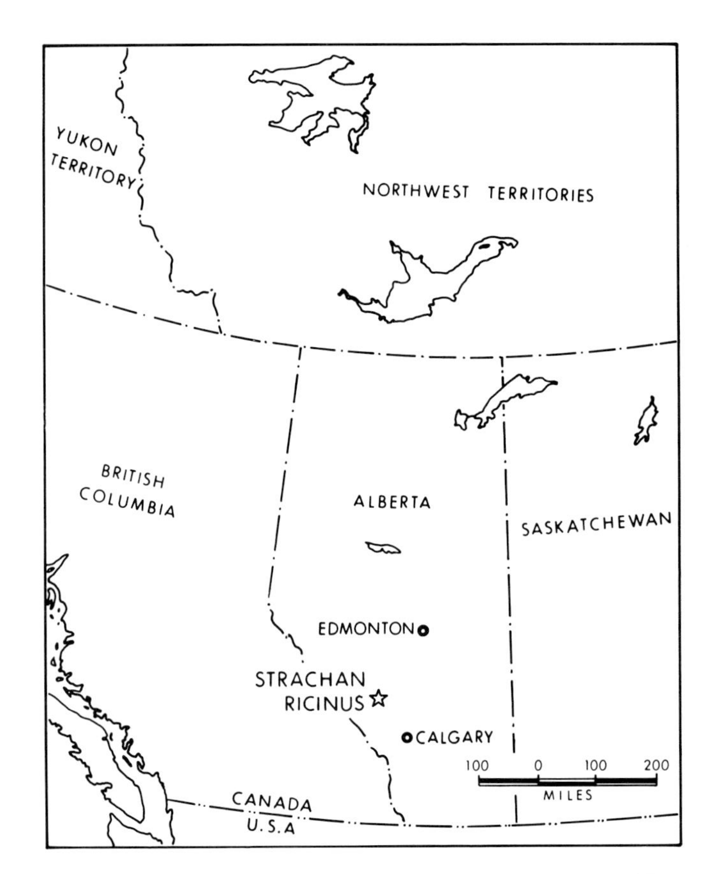

FIG. 1—Index map of Western Canada showing the location of Strachan-Ricinus West fields.

line the manner in which the Strachan-Ricinus reefs evolved and how facies distribution and variations in diagenetic history resulted in significant variations in reservoir quality within the pools and in ultimate individual well performance.

STRATIGRAPHY

Several episodes of reef growth occurred during the Middle and Late Devonian in western Canada. Figure 2 illustrates the stratigraphic succession and relative ages of these reef building episodes. Middle Devonian reefs (Rainbow Member) contain major oil reserves in northwestern Alberta and gas reserves in northeastern British Columbia. Reefs of earliest Late Devonian age (Swan Hills Member) contain major reserves of both oil and gas in the Swan Hills area of central Alberta. Most prolific in their distribution and in total reserves are the Upper Devonian reefs of the Leduc Formation which contain both oil and gas reserves throughout central and western Alberta. The discovery of oil at Leduc in 1947 represented the first significant production from a reef reservoir of any age in Canada and marked the beginning of Canada's present oil boom. The reefs of Strachan and Ricinus West are of the same age. Of less significance economically is the Grosmont Formation in which gas and oil accumulations essentially are restricted to areas of northeastern Alberta where the reef subcrops against impervious beds of Cretaceous age. The youngest reefs recorded in the Devonian of western Canada (Nisku Formation) were discovered as recently as 1977 and contain significant reserves of both oil and gas in the West Pembina area of central Alberta.

The Leduc Formation and its lateral "offreef" equivalents, the Cooking Lake, Duvernay, and Ireton Formations constitute the Woodbend Group. With commencement of Woodbend deposition the Alberta basin became differentiated into areas of scattered and relatively widespread carbonate shoal development surrounded by an extensive shale basin. In the southeast part of the basin, where conditions throughout this period were more stable, evaporitic subbasins were developed. The distribution of Woodbend facies in the Alberta basin is illustrated in Figure 3.

The carbonate shoal areas, represented by the Cooking Lake Formation, consist of fine-grained limestones developing locally into lime sands and biostromal facies, which at Strachan-Ricinus West attain a thickness of 120 ft (37 m). These shoals provided the platform on which the Leduc reefs developed. The Duvernay Formation consists of 180 ft (55 m) of finely fragmented, argillaceous limestones and bituminous shales interpreted to have been deposited in a stagnant basin. This sequence is overlain by 600 ft (183 m) of fine clastics, shale, and argillaceous limestone deposits of the Ireton Formation. Sediments of both the Duvernay (Newland, 1954) and the lower Ireton (McCrossan, 1959) are interpreted in central Alberta as being partly reef derived. In the Strachan-Ricinus area the upper Ireton becomes more calcareous, particularly near the reefs, where the sequence comprises typically calcareous shales and fine-grained argillaceous limestones. This contributes to the problems of reef identification on seismic data. The Leduc reefs of Strachan and Ricinus are biohermal developments that have grown to heights in excess of 900 ft (274 m) suggesting that their evolution was in a deep part of the Alberta basin that was subsiding throughout Woodbend time. In common with other Leduc reefs of Alberta, the Strachan-Ricinus reefs were a stromatoporoid-coral assemblage that was extensively dolomitized in all but a part of the Strachan reef.

Figure 4 shows the stratigraphic relationship between the reef and "offreef" equivalents at Strachan and also the structural drape of overlying beds near the 10-31 reef well. This relationship is due to differential compaction of the Ireton shale. The magnitude of drape in the beds directly overlying the reef has been calculated to be 150 to 200 ft (46 to 61 m). The amount of structural drape decreases upwards in the stratigraphic section indicating that compaction proceeded as the sedimentary load increased. The presence of drape in the Upper Cretaceous beds, which are separated from the Paleozoic by several major unconformities, clearly demonstrates that the process of differential compaction was operative throughout geologic time from Late Devonian to (at least) early Early Cretaceous time.

In the Alberta basin the average interval velocity of Ireton shale is typically lower than that of the reef. In the Leduc area of central Alberta, the locale of the first reef discovery, this interval velocity is approximately 13,000 ft/sec (3,962 m/sec) compared to that of 20,000 ft/sec (6,096 m/sec) for the reef. This velocity contrast results in the relatively easy identifica-

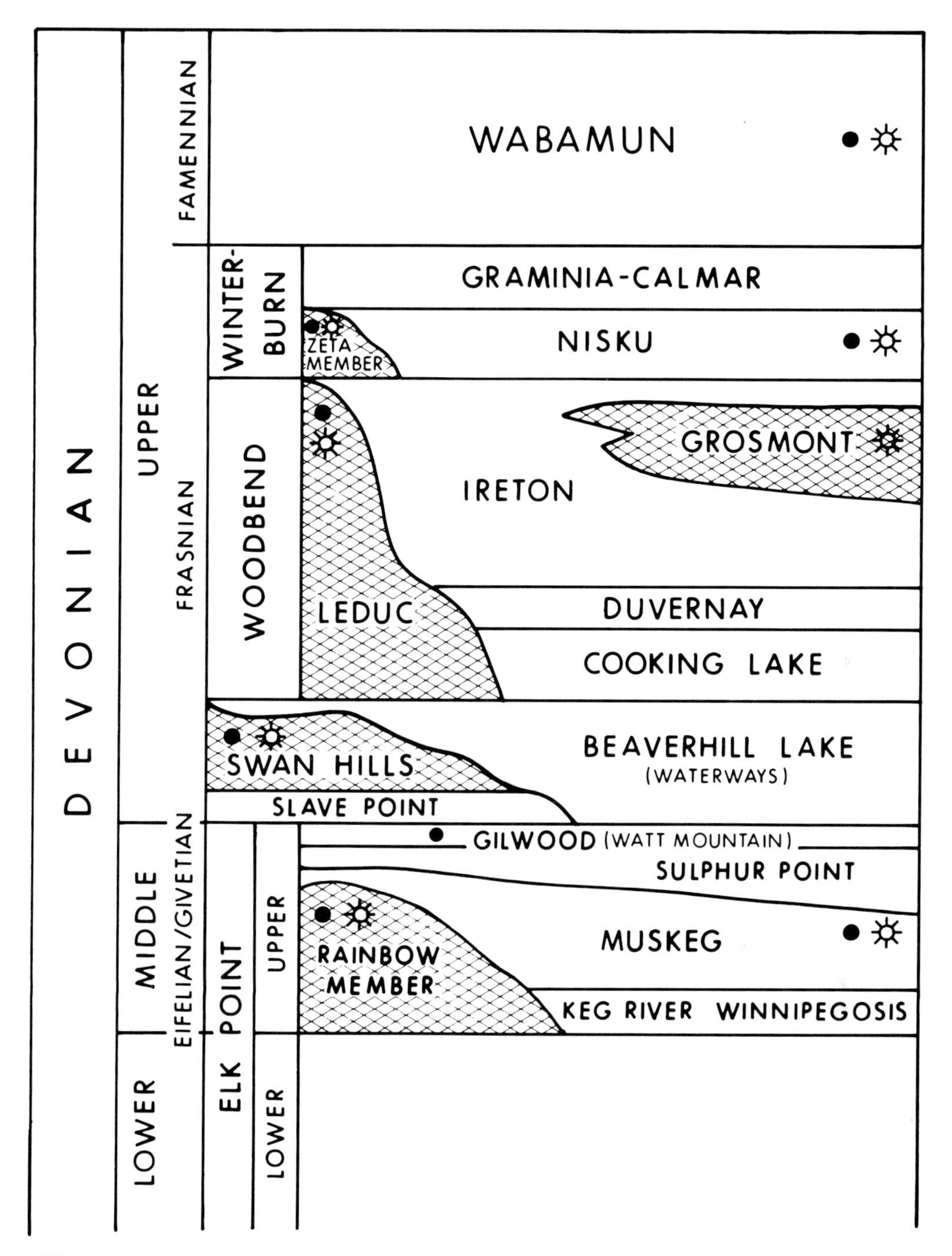

FIG. 2—Devonian nomenclature chart showing the reef horizons in the Strachan-Ricinus West field area.

tion of reefs on seismic data without the need for any refined processing techniques. In the Strachan-Ricinus area, the Ireton Formation is more deeply buried (14,000 ft [4,267 m] versus 5,000 ft [1,524 m] at Leduc) and is more calcareous resulting in a higher interval velocity (18,000 ft/sec or 5,486 m/sec), but still providing some contrast with the Leduc reef (20,000 ft/sec or 6,096 m/sec). In some parts of Strachan-Ricinus, however, the upper Ireton becomes very calcareous near the reef buildups and consequently the velocity contrast between it and the reef is too small for seismic resolution. It is the combination of structural drape, due to differential compaction, and the contrast in interval velocity that is responsible for the successful use of seismic technique in the isolating and mapping of these reefs.

HISTORY OF DISCOVERY

Figure 5 shows the location of wells at Strachan-Ricinus West that penetrated the Woodbend Group before the discovery of the Strachan Field in 1968. The first indication of the hydrocarbon potential of the area was in 1956 when the Gulf Strachan 12-31 well penetrated a partial Leduc reef buildup of 550 ft (168 m). On testing, the reef yielded traces of gas and a large

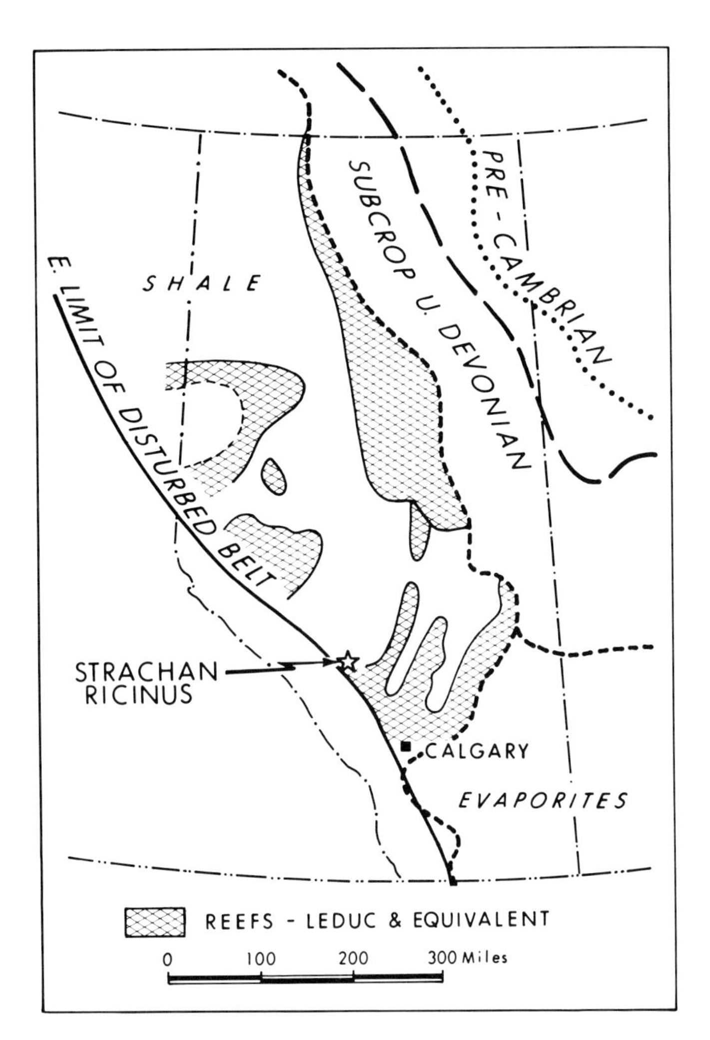

FIG. 3—Facies distribution in the Woodbend Group showing also the location of the Strachan-Ricinus West field area in relation to the Rocky Mountains disturbed belt.

quantity of formation water. The "discovery" was not considered commercial due to the quality of the reservoir and thin pay zone (79 ft or 24 m) and the well was abandoned. At that time the market for gas was limited and a wellhead price of only (approximately) 9 cents per Mcf was offered. Moreover the presence of hydrogen sulfide (11%) in the gas was another deterrent to completion of this well in an area remote from a sour gas processing facility.

During the period from 1956 to 1967, two further attempts were made to contact the Strachan reef at a higher elevation. The first attempt (in 1964) was by Chevron, who drilled the 10-25 well about 1 mi (1.6 km) southwest of the Gulf location. This well penetrated a 310-ft (94 m) sequence of interbedded basinal and forereef detrital sediments. The location of the well was based on seismic data which were 100% digital tape-and analog-processed. Figure 6 is a display of this type of data on a seismic line shot from an off-reef position west of the field, through the Gulf 12-31 well, to a location in section 31 that was subsequently drilled and found to have a complete reef buildup. Although the Lower Cretaceous Blairmore Formation is recognizable on the seismic line, the resolution is too poor to be able to map or even identify any deeper horizons.

Recognizing the potential of the Strachan-Ricinus area, and anticipating that improved seismic techniques could significantly improve the subsurface identification of Leduc reefs at drilling depths in excess of 14,000 ft (4,267 m), Banff and Aquitaine embarked on a program of land acquisition, initially with the purchase of two releases and two exploration reservations from the Crown on which the Gulf and Chevron abandonments were located. Having established a good land position the next step was to conduct a seismic survey that tied to the two key wells and to an off-reef well some 4 mi (6.4 km) to the north. This was the first attempt to use multi-fold subsurface seismic coverage in the deep basin of Alberta. In this survey 600% coverage was used with a gapped-spread design of 9,000 ft (2,743 m) in length which was selected for enhancement of the deep reflections. However, the geologic setting of the area presented the seismic interpreter with a variety of problems, the most significant of which was the lateral velocity variations in formations above the Upper Devonian reef. This resulted in part from the area's proximity to the Rocky Mountain disturbed belt (Fig. 3) and the fact that the overlying Tertiary and Mesozoic formations were involved in thrust faulting. It was necessary therefore to revise previous interpretation techniques that assumed constant lateral velocity in the section above the reef. Additional problems were caused by the Brazeau Thrust, a major thrust sheet that brought high velocity Paleozoic carbonate rocks to the surface within a few miles of the reefs. This created velocity anomalies and interference patterns over a broad area.

Other geological complexities that presented problems for the interpreter were: (1) the general absence of velocity contrast from reef to off-reef due to the presence of high velocity carbonates in the upper part of the Ireton; and (2) velocity changes within the reef itself due to variations in reef porosity, reef matrix velocity, and fluid content—changes which tend to be even more difficult to predict.

One of the criteria used early in Alberta for reef detection (Pallister, 1965) was identification of a velocity "pull-up" of the pre-reef horizon due to the presence of high velocity carbonates offset laterally by the lower velocity Ireton sequence. In some places at Strachan-Ricinus, where porosities are particularly high, the interval velocities in the reef are lower than that of the adjoining Ireton and a velocity delay or "pull-down" occurs.

The seismic program conducted by Banff and Aquitaine in 1967-1968 was shot using PT 100 instrumentation. This was analog data with 600% CDP coverage. The intial phase of the program required that the line trending east to west through the Gulf 12-31 location be shot twice using different acquisition parameters in an attempt to improve the signal-to-noise ratio. These data were digitally processed and reprocessed numerous times using a new deconvolution program which eventually identified a complete reef buildup within half a mile (0.8 km) of the original Gulf 12-31 abandonment. Also important in this processing was the space variant velocity analysis and application to compensate for the lateral velocity variations in the area.

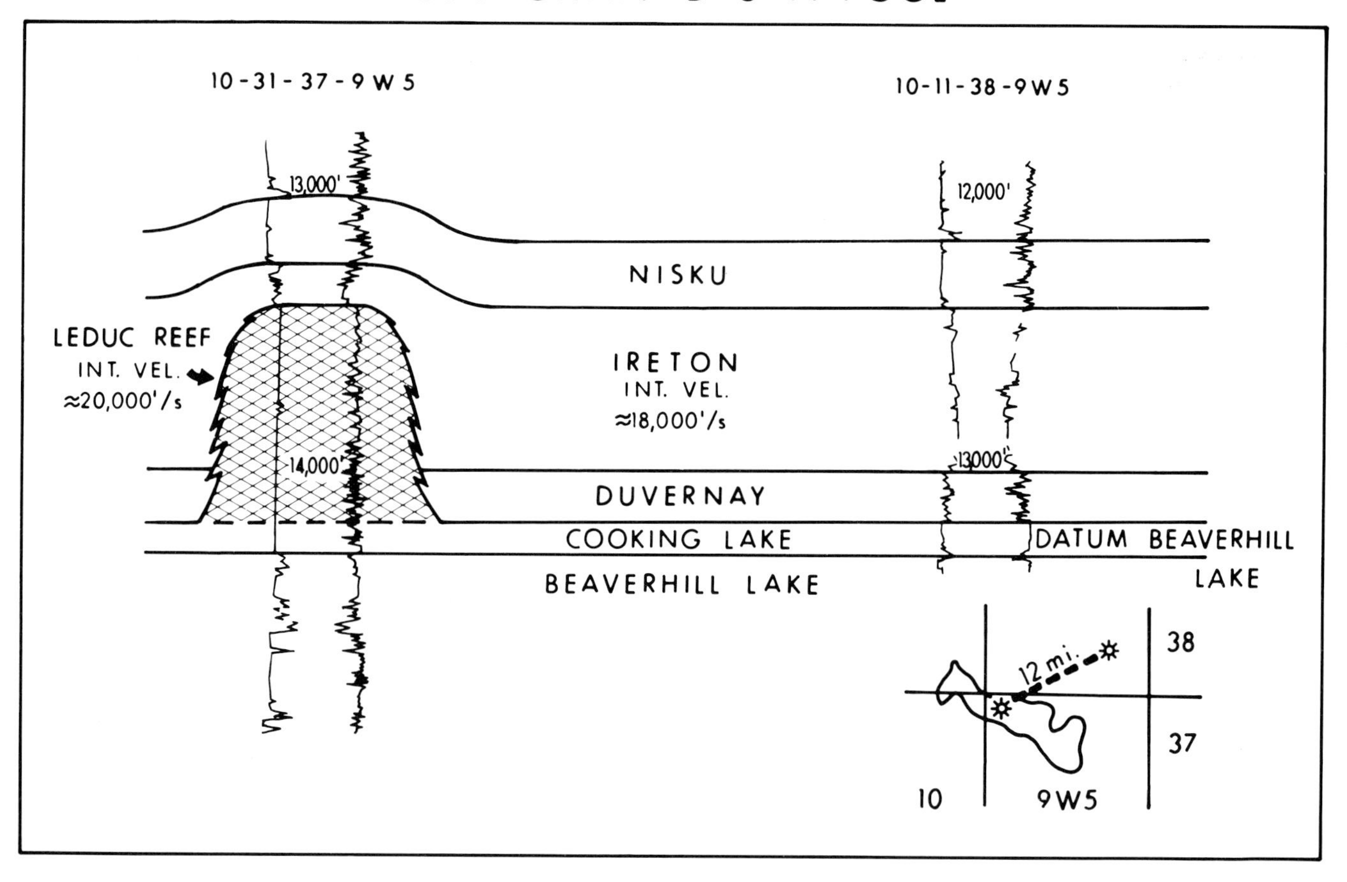

FIG. 4—Northeast to southwest cross section of Strachan field showing reef to off-reef relationship and interval velocities in the Leduc reef and Ireton Formation.

While Banff and Aquitaine were interpreting and reprocessing the new seismic data, a third well (drilled 2 mi [3.2 km] to the east by Stampede Oils on 100% seismic data) actually penetrated a partial Leduc reef buildup at an elevation 32 ft (10 m) higher than the Gulf abandonment. The thickness and quality of the pay zone at this location were sufficient to warrant completion as a commercial gas well, with 111 ft (34 m) of pay. It was not until the summer of 1968 however that a "complete" buildup was encountered. This was achieved by Banff and Aquitaine in their 10-31 well which was the first location based on new multi-fold seismic data. The 10-31 well penetrated a pay zone of 536 ft (163 m). The thickest pay zone in the Strachan field is in the 11-27 well which has 736 ft (224 m) of gas saturated reef section at a location 2 mi (3.2 km) regionally updip from the 10-31 well. Figure 7 shows the seismic line on which the location of the 10-31 discovery was based. The upper section shows a display of 600% CDP data without deconvolution. The lower section is the same seismic line with deconvolution. This section demonstrates a change of seismic character at the Leduc reef level as well as drape in the overlying beds in spite of the fact that the section was flattened on the top of the Lower Cretaceous Blairmore Formation. The absence of any velocity "pull-up" or "pull-down" in the pre-reef beds can be attributed to the moderate levels of porosity in this part of the field. All seismic interpretation to this point was based on analog processing.

The seismic techniques used and experience gained at Strachan were immediately applied to reconnaissance surveys in the general area. The presence of a reef of similar magnitude was soon identified 5 mi (8 km) to the south at Ricinus West. The discovery well in this field was drilled by Banff and Aquitaine in 1969 at a location in 6-25-36-10 W5M. This well penetrated a gas saturated reef section 634 ft (193 m) thick.

Figure 8 is a current well penetration map of the Strachan-Ricinus area showing the distribution of Leduc reef gas fields. Several other reefs have been discovered by drilling but were either water bearing or lacked porosity.

FACIES OF PRODUCING REEFS

The Upper Devonian carbonate reefs of Strachan and Ricinus developed as distinct biohermal complexes isolated from the major reef chain of the Cheddarville - Leduc trend. They developed their own distinct geologic facies, diagenetic history, and reservoir characteristics.

The geometry of the Strachan and Ricinus West reefs has been well delineated through seismic and development drilling which have outlined reef complexes of 7 by 2.5 mi (11 by 4

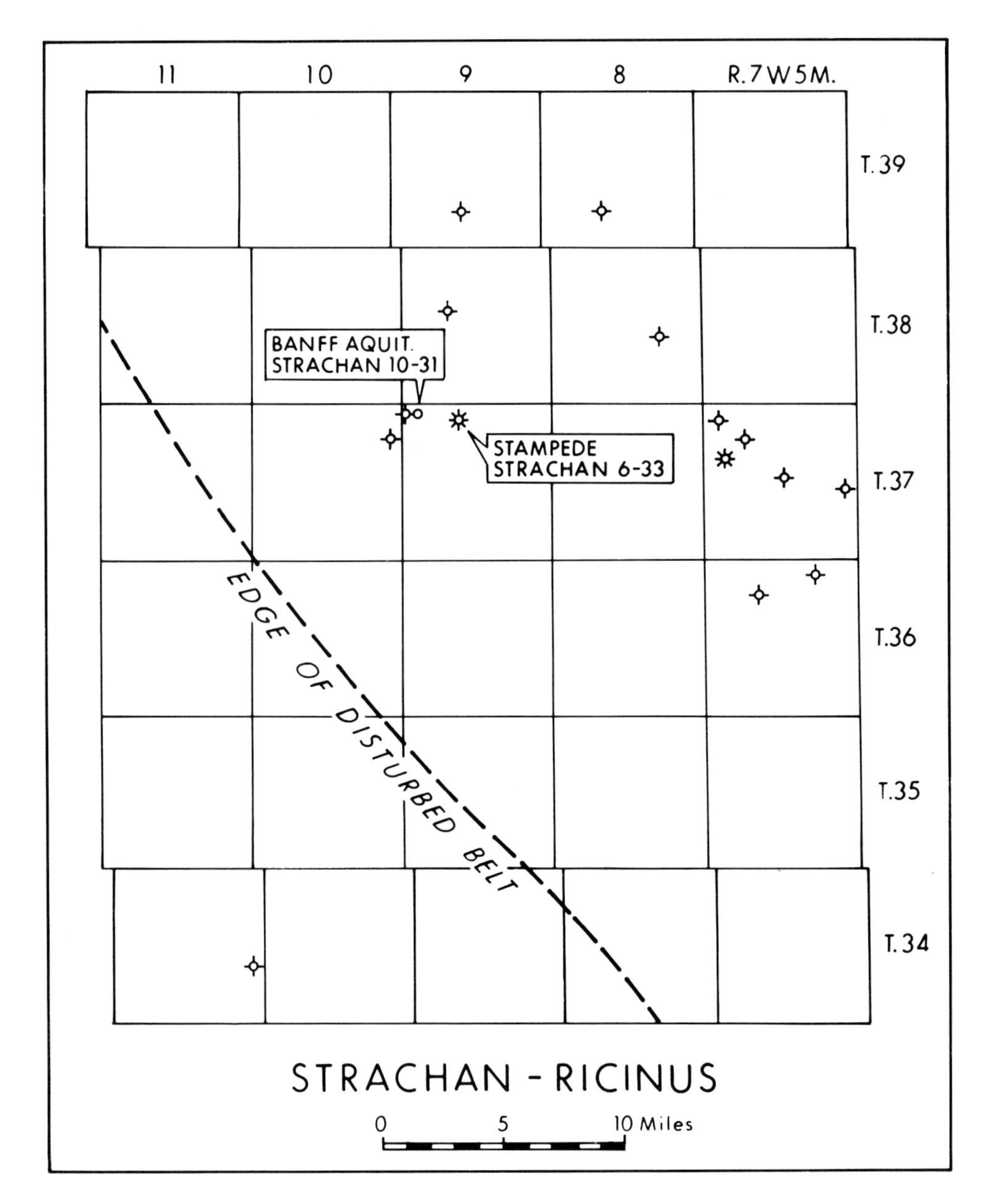

FIG. 5—Map showing all wells drilled into the Woodbend Group prior to the "complete reef discovery" of 1968.

km) with both complexes being oriented in a northwesterly direction. Both reefs attained heights in excess of 900 ft (274 m); the Strachan reef however, has a separate culmination at its north end. The composition of the gas and reservoir pressure in this culmination are different from those in the main Strachan pool.

Reconstruction of the manner in which these Upper Devonian reefs developed and the identification of distinct geologic facies has been severely impeded by the widespread dolomitization of the entire Ricinus reef and of the eastern half of the Strachan reef. Fortunately however, virtually all of the field wells penetrated to the reef platform and the majority of these were cored, thus affording some opportunity for facies interpretation and prediction. With these data it is possible to determine the distribution of geologic facies within the Strachan reef with some confidence. It is also possible to predict why the Strachan complex developed where it did. Although good exposures of the reef platform or Cooking Lake equivalent are limited, they do indicate a particularly high concentration of potential reef-building organisms including tabular and dendroid stromatoporoids and the dendroid coral, *Thamnopora*. Whereas these organisms are more indicative of deeper shoaling conditions, they are diagnostic of a relatively high energy environment and one in which organisms could proliferate. That these shoaling conditions permitted the establishment of a massive stromatoporoid facies is clear from an examination of undolomitized reef cores at Strachan. Figure 9 is a cross section through the narrower dimension of the complex in an area where much of the section remains undolomitized.

An organic reef facies is well developed and probably occurs

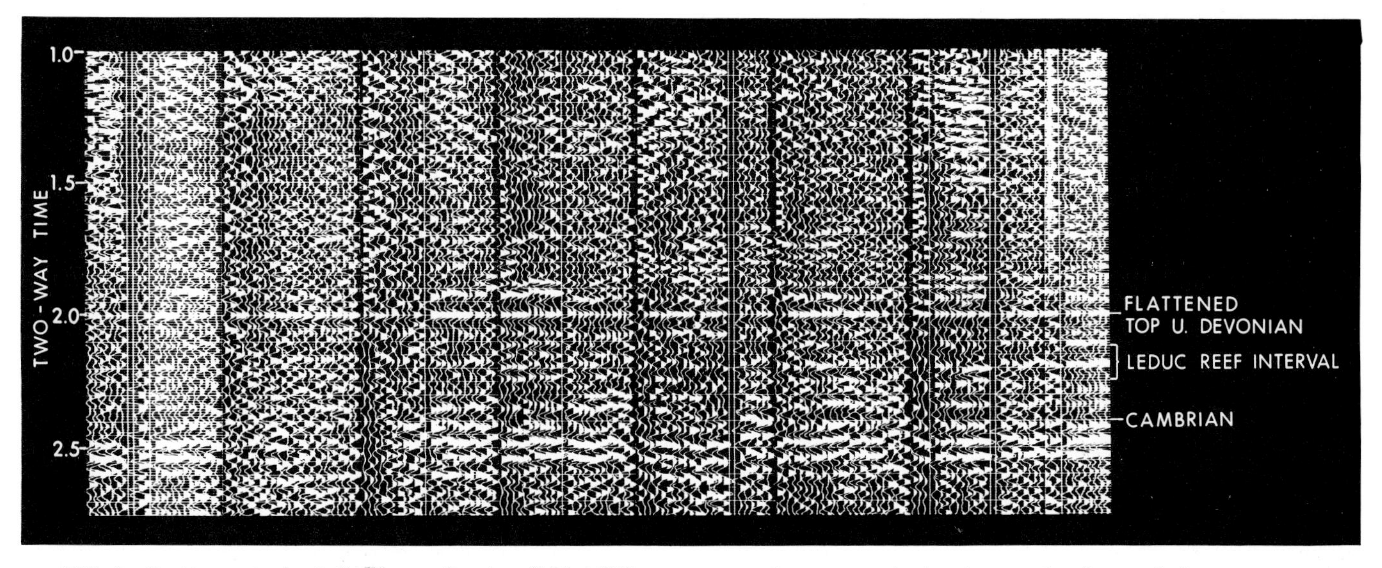

FIG. 6—East to west seismic line across Strachan field, 100% coverage, analog processed using time varying deconvolution program.

as a discontinuous barrier along the length of the complex. The framework of the organic reef was constructed of massive stromatoporoids and colonial corals *(Alveolites* and *Phillipsastrea)* enclosed typically in a biomicritic matrix, presumably derived from the destruction of the organic reef itself by wave action. The large stromatoporoid and coral heads appear to have provided shelter for the finer grained detritus to accumulate as growth continued into the surf zone above, an observation not inconsistent with the findings in similar geologic settings in other Upper Devonian reefs (Klovan, 1964). In the organic reef, massive stromatoporoids are the most important reef building organism and are generally found in growth position. Elsewhere on the reef complex they are observed in fragmented form having been transported either down the fore reef slope or behind the organic reef.

In the back reef or interior reef, assemblages of coarse massive corals and massive stromatoporoid fragments constituting rudites and arenites are found. Nodular or bulbous stromatoporoids and amphipora are also common. The interfingering of the reef and back reef facies suggests the frequent disruption in growth of the organic reef framework by periodic storm action and the dumping of this material in shallower waters behind the reef, an environment in which the energy of wave action gradually dissipated with distance from the surf zone. The interior of the reef complex appears to have been an area continually open to the circulation of marine waters. The periodic influxes of large volumes of organic material and seawater would have precluded the development of a restricted lagoonal environment in which chemically precipitated limestones could have been formed. The lime muds observed in this position are believed to be derived from the continued destruction of the organic reef with deposition in the quieter waters of the back reef.

Basinward from the organic reef a sequence of reef-derived sediments are present. Transported fragments of massive stromatoporoids and colonial corals occur immediately seaward of the organic reef to be replaced farther seaward in the less turbulent deeper waters of the fore reef by tabular stromatoporoids and colonial corals which generally are found in growth position. Farther basinward, the dendroid stromatoporoid *Stachyodes* predominates in association with open-marine fauna consisting of rugose corals, crinoids, brachiopods, and gastropods in a more argillaceous fine-grained matrix. Some distance from the organic reef in the lower fore reef position the facies becomes very fine grained and argillaceous and contains both indigenous skeletal material and fragments of reef derived skeletal material.

Cessation of reef growth at Strachan and Ricinus West occurred simultaneously as the result of a sudden influx of argillaceous material and decrease in the rate of subsidence which had permitted the continuous production of reef building organisms throughout the time of Leduc deposition.

DIAGENESIS AND RESERVOIR CHARACTERISTICS

Several diagenetic processes have exerted an influence on the Strachan and Ricinus reef complexes, the most significant of which was the complete dolomitization of the Ricinus reef and partial dolomitization of the Strachan reef. In its indiscriminate alteration of these carbonate masses and obliteration of much of the original limestone texture, dolomitization has significantly increased porosity within the matrix in both coarse and fine grained sediments. The most significant porosity in dolomites developed where the primary matrix was altered while fossil material remained calcitic. This was subsequently leached out leaving molds or vugs—hence the highest porosity in these reefs occurs in dolomitized sections of the organic reef and adjacent areas of the fore reef and back reef, which are facies containing the highest proportions of faunal material.

Despite the widespread dolomitization that occurred in these reefs it is still possible to predict the distribution of porosity within the carbonate mass from a knowledge of the faunal distribution.

A later stage diagenetic process which served to diminish effective porosity was the cementation by calcite of an original

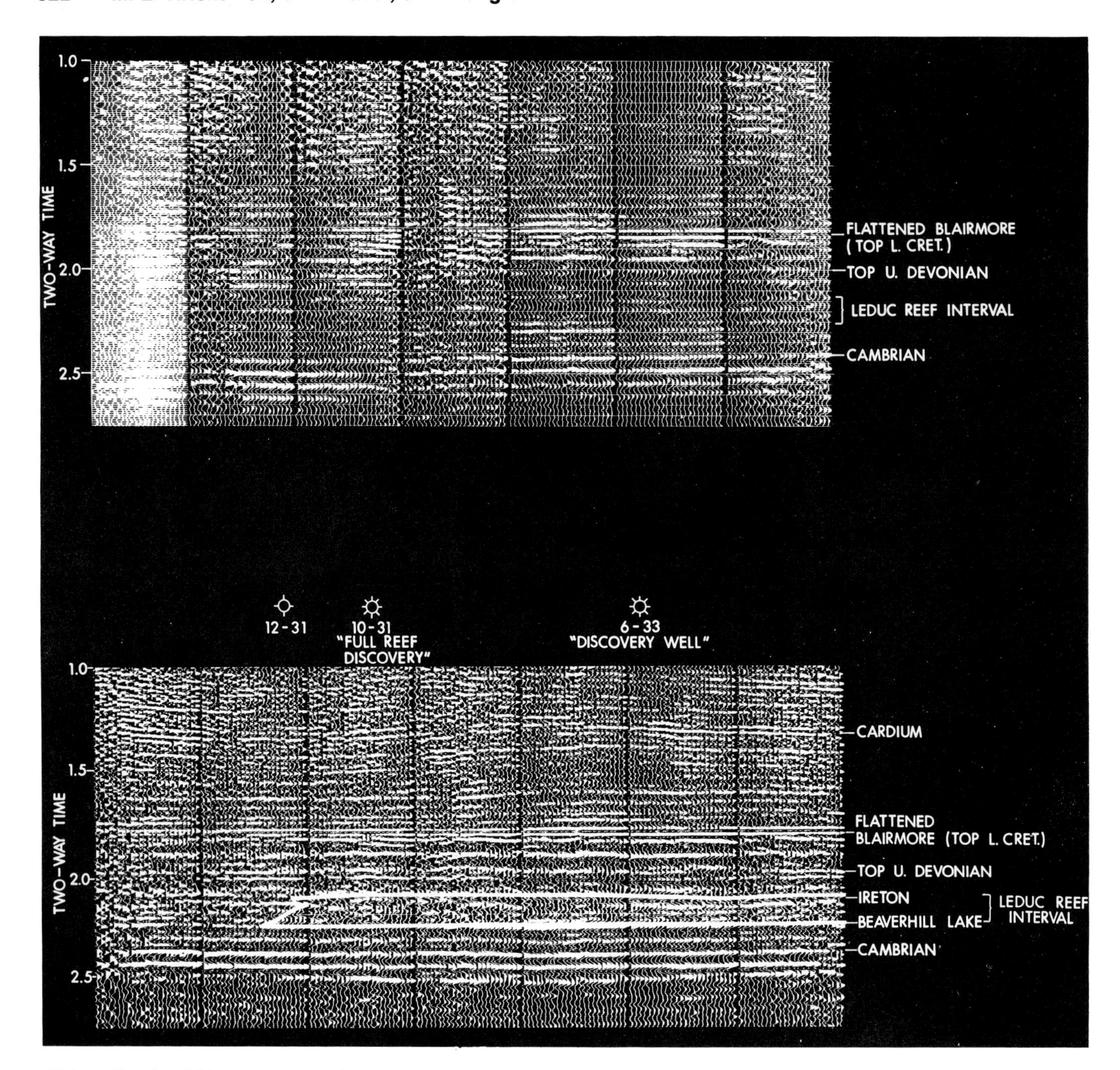

FIG. 7—Strachan field east to west seismic section. Upper section shows 600% coverage without deconvolution and lower section is the same line with deconvolution.

extensive pore system. This accounts for the low porosity in the wells in the western part of the Strachan reef. A further process occurring late in the diagenetic history of these reefs was the partial plugging of pores and fractures in both the limestone and dolomite reservoirs by bitumen. This process led to some deterioration of effective porosity. In some places in the limestone reservoir, where original porosity was already reduced by cementation, bitumen plugging has reduced porosity to essentially zero.

RESERVES AND PRODUCTION

Following the discovery of the Strachan and Ricinus reefs, seismic techniques were used to predict the size of reserves. By this time seismic data were able to map the geometry of the reef complex and delineate with some accuracy the productive limits of the reservoirs. Predictions of sour gas reserves based on these data were subsequently confirmed by development drilling and today there are eight commercial gas wells in each field.

Strachan

The two separate seismically defined structural culminations in the Strachan reef were confirmed by the 6-1 development well, which contacted a gas-water interface at an elevation 100

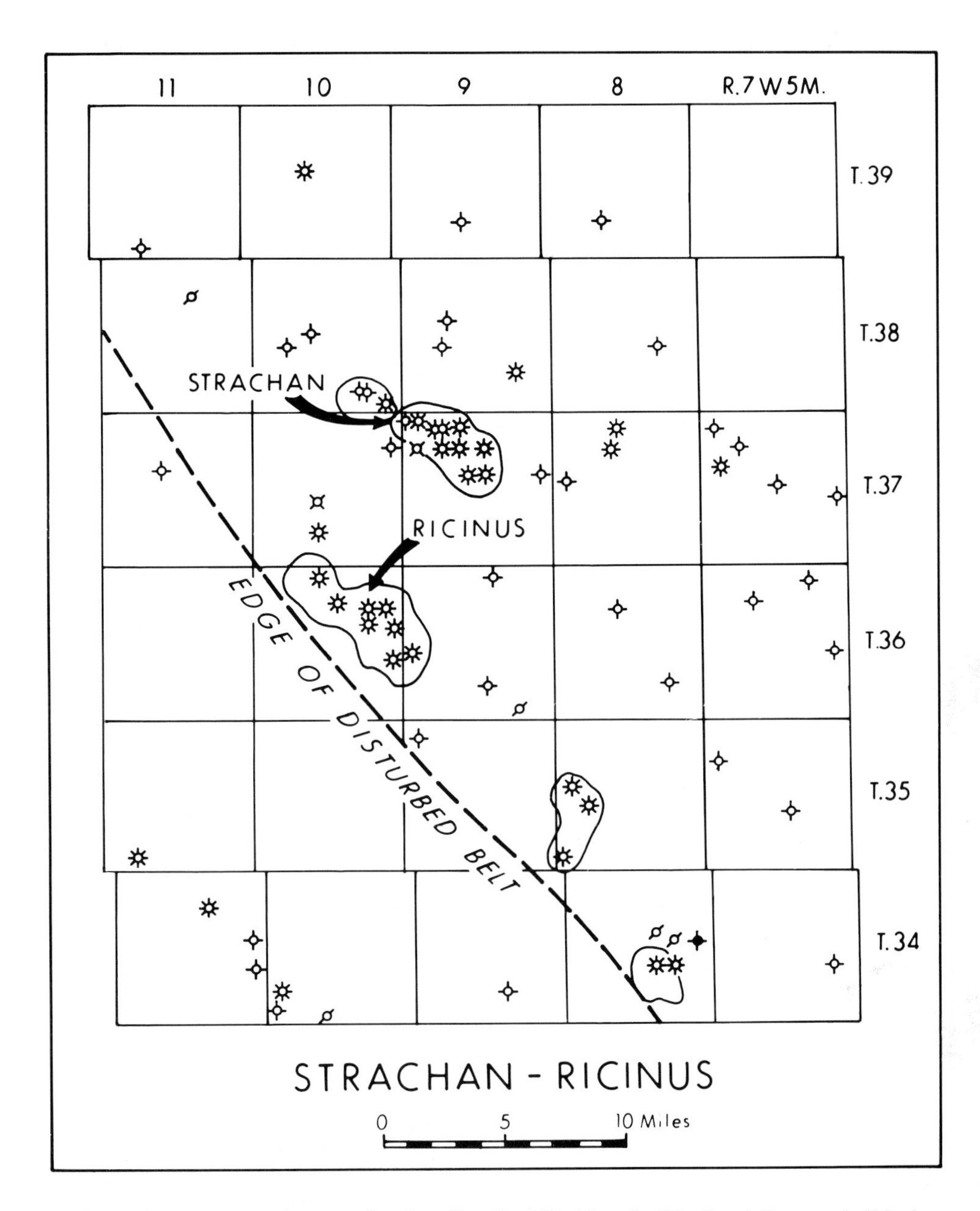

FIG. 8—Present penetration map showing all wells drilled into the Woodbend Group and all Leduc producing fields. Field outlines are those defined by the Alberta Energy Resources Conservation Board.

ft (30 m) higher than in the eastern part of the reef. Figure 10 is a northwest to southeast cross section through the field illustrating the pool separation and extent of dolomitization in the reef. The east and west pools, which were designated Strachan Leduc (D-3) "A" pool and Strachan Leduc (D-3) "B" pool respectively, had in fact many significant differences. The "A" pool gas composition is 9.3% H_2S, 2.1% CO_2 with 20 bbl of stabilized condensate and 3.5 long tons of sulfur per MMcf of raw gas. The "B" pool is 6.9% H_2S, 2% CO_2 with 2.5 bbl of condensate and 2.6 long tons of sulfur per MMcf of raw gas. Initial reservoir pressures varied by approximately 50 psig. Original reserves of gas in place were 1.49 Tcf for the "A" pool and 39 Bcf for the "B" pool. Reserves were calculated using the net pay isopach shown in Figure 11 and hydrocarbon pore volumes predicted from facies mapping. Porosities in the Strachan field range from 2 to 10% and permeabilities from 1.0 to 40 millidarcies. The most favorable reservoir characteristics are found in the organic reef and proximal areas of the fore reef and back reef facies. The poorest reservoir is found in the lower fore reef areas. The "B" pool is generally poorer than the "A" pool in overall reservoir quality due to the absence of dolomitization and existence of limestones that have undergone extensive calcite cementation. This is reflected in the average daily production for the "A" and "B" pools which during 1978 amounted to 191 MMcf and 2.8 MMcf, respectively. Production at Strachan commenced in 1971 and by the

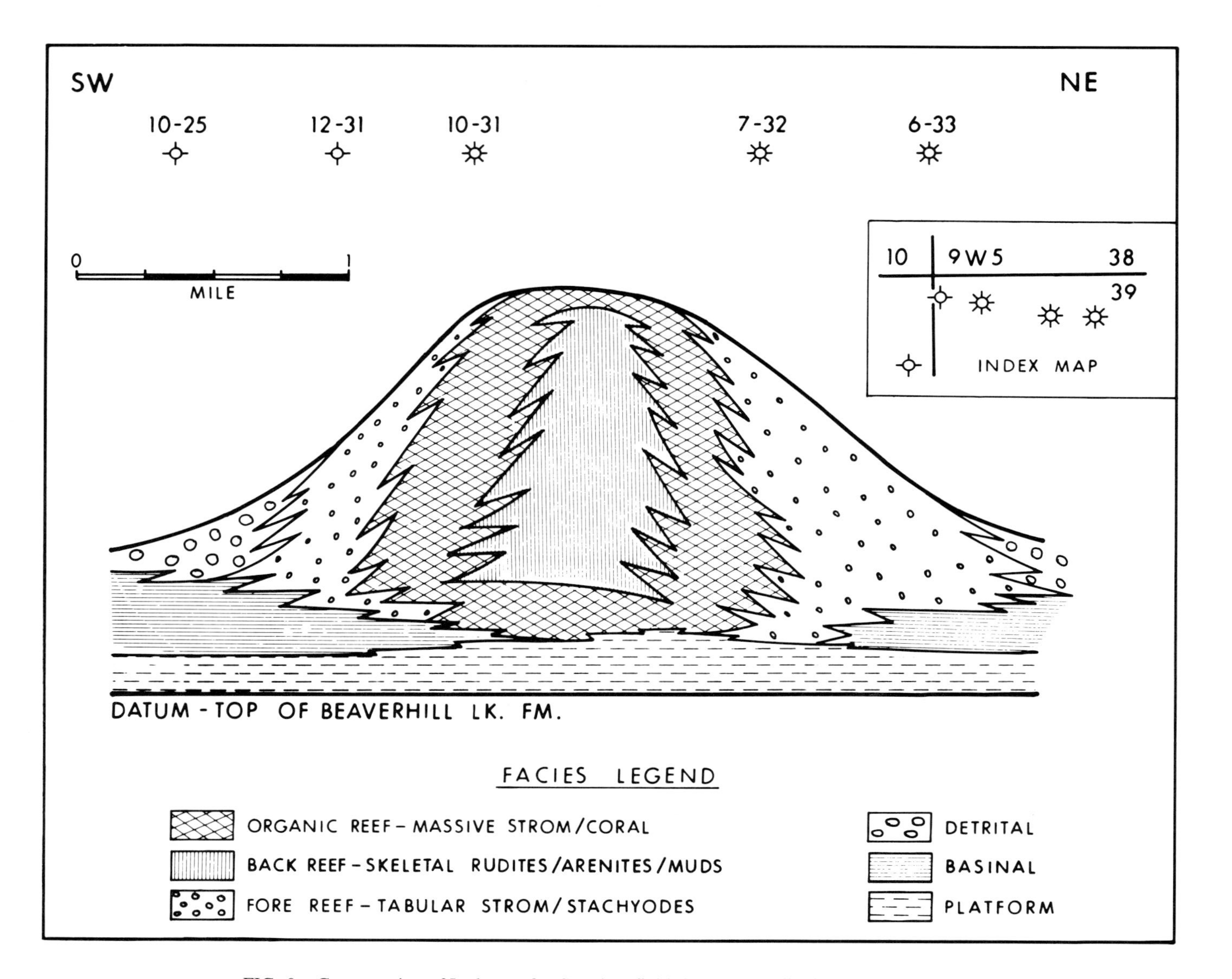

FIG. 9—Cross section of Leduc reef at Strachan field showing the distribution of facies.

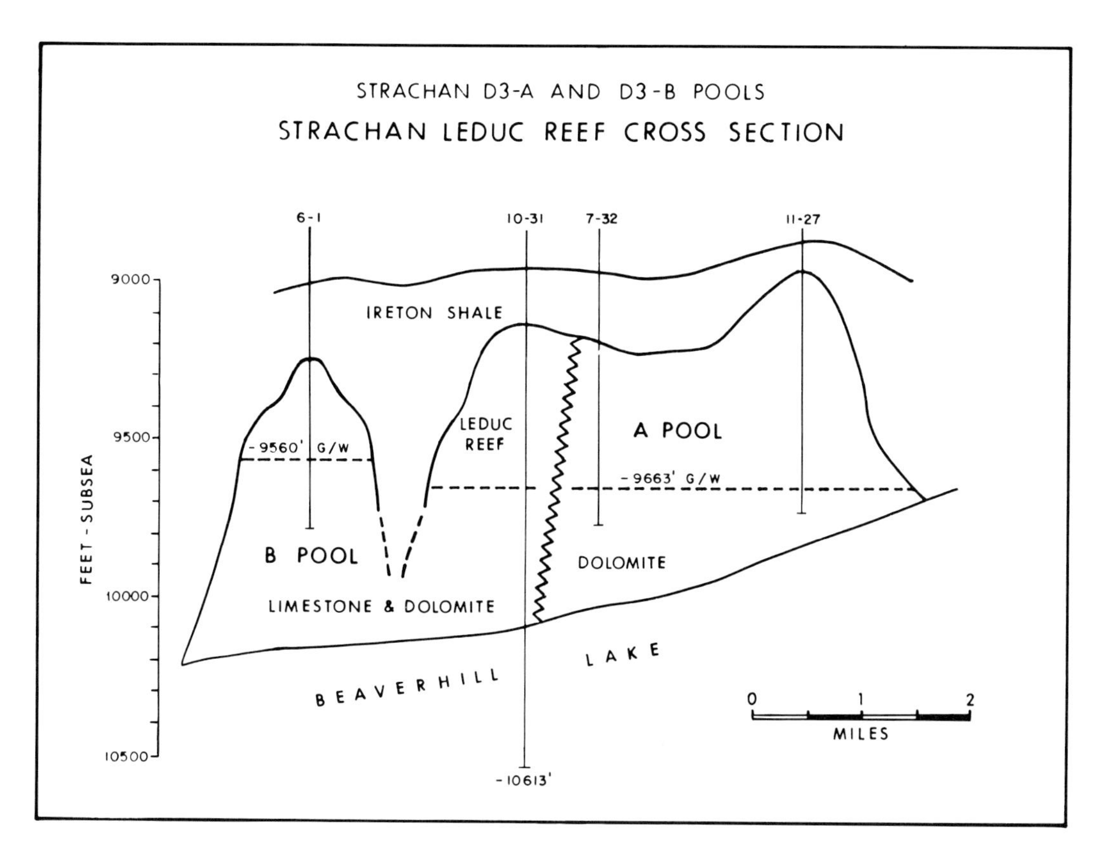

FIG. 10—Northwest to southeast cross section through the Strachan field showing the pool separation and extent of dolomitization in the reef.

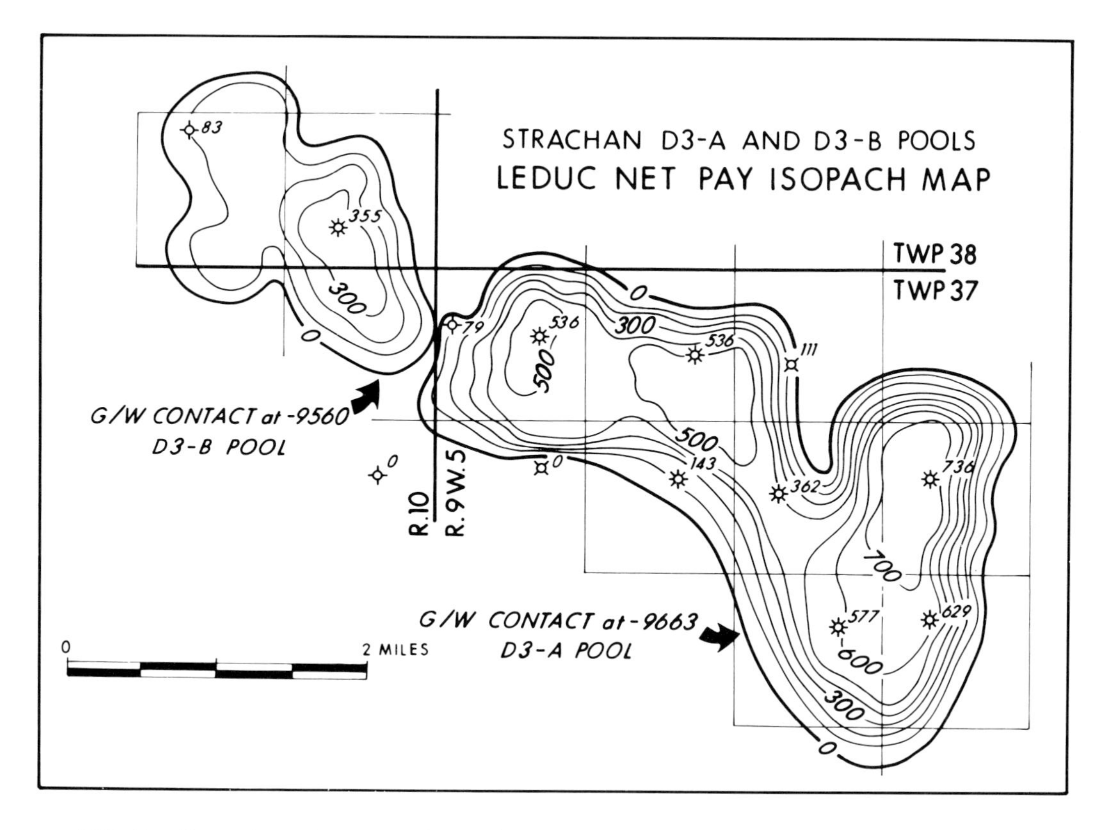

FIG. 11—Net pay isopach map showing the Strachan D3-A and D3-B pools, Strachan field, Alberta.

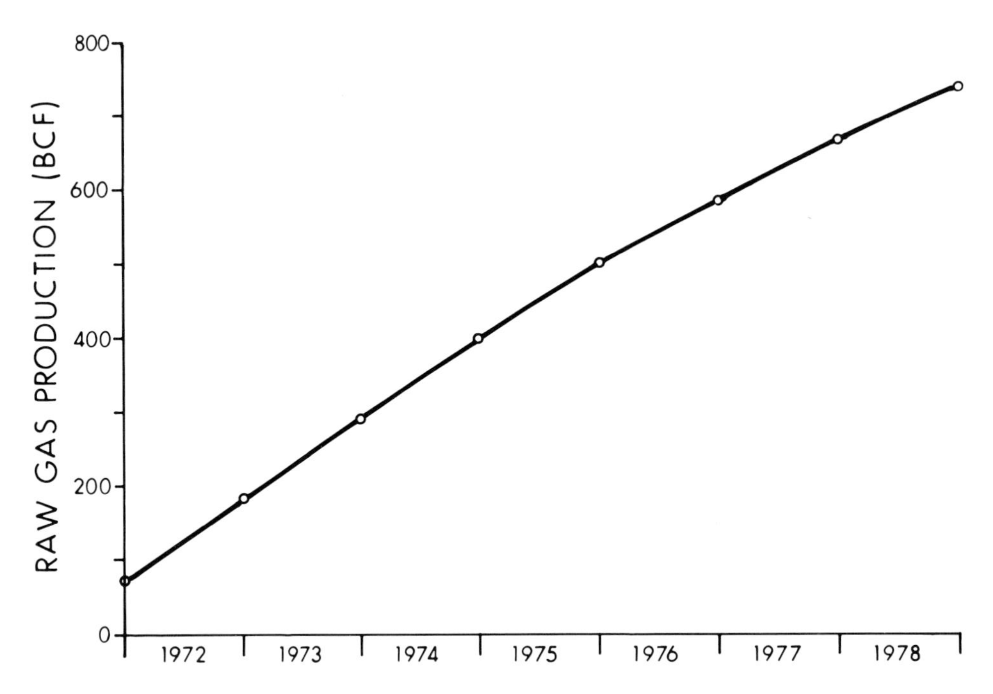

FIG. 12—Cumulative Strachan raw gas production from 1972 to 1978, Strachan field, Alberta. Production measured in billion cu ft (Bcf).

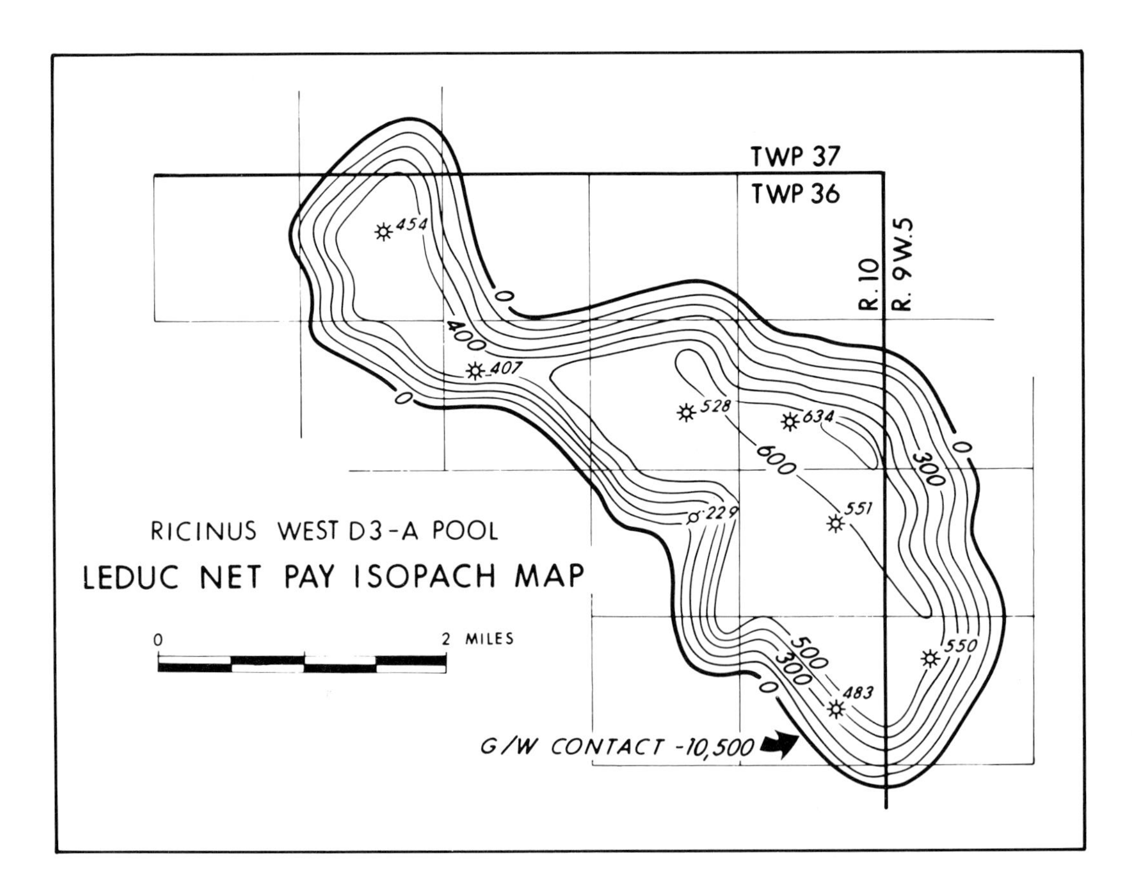

FIG. 13—Leduc net pay isopach map showing the Ricinus West D3-A pool, Ricinus West field, Alberta.

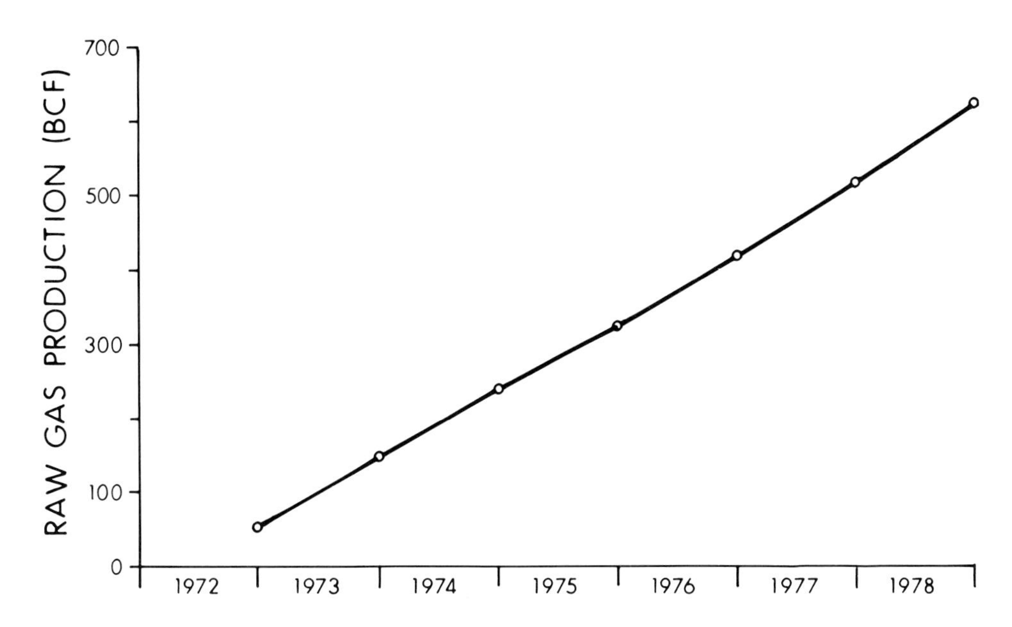

FIG. 14—Cumulative raw gas production in the Ricinus West field from 1972 to 1978. Production measured in billion cu ft (Bcf).

end of 1978 cumulative raw gas production was 740 Bcf (Fig. 12). By 1990 it is estimated that 80 to 84% of the reserves will have been produced.

Ricinus West

The Ricinus West field has a gas composition significantly different from the Strachan field. The gas is much more sour with an H_2S content of 33%. Condensate production is only 0.88 bbl per MMcf of raw gas. Initial reservoir pressure was 5,800 psig which is 1,400 lb (635 kg) lower than at Strachan. Original reserves of gas in place, based on the net pay map shown in Figure 13 were 1.7 Tcf. Porosities in the field range from 6 to 10% and the permeabilities from 14 to 25 millidarcies. As in the Strachan field the best reservoir is developed in the organic reef, and in adjacent areas of the fore reef and back reef. The complete dolomitization of the Ricinus reef has resulted in improved overall porosity and permeability and a higher average daily raw gas production rate, which was 316 MMcf during 1978. Sustained gas production for the field commenced in 1972 and by the end of 1978 cumulative production amounted to 624 Bcf (Fig. 14). By the year 1990, 85% of the original gas in place will have been produced.

SELECTED REFERENCES

Andrichuk, J.M., 1958, Stratigraphy and facies analysis of Upper Devonian reefs in Leduc, Stettler, and Redwater areas, Alberta: AAPG Bull., v. 42, p. 1-93.

Belyea, H.R., 1960, Distribution of some reefs and banks of the Upper Devonian Woodbend and Fairholme Groups in Alberta and Eastern British Columbia: Canada Geol. Survey Paper 59-15, 7 p.

Klovan, J.E., 1964, Facies analysis of the Redwater Reef complex, Alberta, Canada: Canadian Soc. Petroleum Geologists Bull., v. 12, p. 1-100.

McCrossan, R.G., 1959, Resistivity mapping of the subsurface Upper Devonian Inter-Reef Ireton Formation of Alberta: Alberta Soc. Petroleum Geologists Jour., v. 7, p. 121-130.

Newland, J.B., 1954, Interpretation of Alberta reefs based on experience in Texas and Alberta: Alberta Soc. Petroleum Geologists News Bull., v. 2, p. 1; v. 4, p. 3-6.

Pallister, A.E., 1965, The application of seismic techniques to reef traps: Univ. Alberta, Calgary, Division of Continuing Education.

The Namorado Oil Field: A Major Oil Discovery in the Campos Basin, Brazil[1]

By Giuseppe Bacoccoli, Roberto Gamarra Morales, and Odimar A. J. Campos[2]

Abstract The Campos sedimentary basin is in the Rio de Janeiro State continental shelf between 21° and 23°S lat. The first oil field, Garoupa, was discovered in 1974 when Petrobras was drilling its ninth offshore wildcat. Since then, 10 more significant accumulations have been discovered and are in the delimitation or early development phase. Geologic interpretations suggest that at least four fields may have a reserve (volume of proven recoverable oil) of about 100 million bbl of oil. However, at this time, the 20 sq km area of the Namorado oil field alone is calculated to contain reserves of 250 million bbl.

The Namorado field was discovered in 1975, when Petrobras was drilling the wildcat 1-RJS-19 (22°27′S lat., 40°25′W long.) in a water depth of 166 m. The location was selected based on seismic interpretation of a structural high at the top of the Macae Formation (Albian limestones). At a depth of 2,980 to 3,080 m the well penetrated thick oil-bearing sandstones near the previously mapped horizon. Production testing flowed 6,000 bbl from the highly porous (30%) and permeable (more than 1 darcy) reservoir.

According to sedimentologic studies of ditch samples and cores, the reservoirs are marine turbidite deposits related to the first important transgression over the Albian limestone shelf.

The field was entirely delimitated using five extension wells and seismic interpretation based mostly on acoustic impedance seismic logs. The reservoir was named the Namorado Sandstone and appears to have formed by the coalescence of channels and lobes deposited over an irregular depositional surface. At the time of deposition, the area of the present oil field was a relative low where the turbidites were trapped.

As a consequence of structural activity related to salt flow in the Late Cretaceous, there was an inversion of relief. The reservoir now is located in an elongate, dome shaped, partly faulted structural high. Differential compaction over the structure enhanced the positive relief. The Namorado field accumulation is controlled both by structure and stratigraphy.

INTRODUCTION

Campos basin (Fig. 1) occupies a 500 sq km onshore area of the recent delta of Paraiba do Sul River. The remainder of the basin is offshore on the Brazilian continental margin northeast of Rio de Janeiro State between 21° and 23°S lat. Oil exploration involves the total 30,000 sq km to the 200 m isobath.

On the continental shelf, the Campos basin is limited on the south by the Cabo Frio arch and on the north by the Vitoria arch. Both of these highs have very thin sedimentary cover (less than 1,000 m). In deep waters, on the present slope, there

[1]Manuscript received, April 27, 1979; accepted for publication, July 23, 1979.

[2]Petrobras, Rio de Janeiro, Brazil.

Because practically all previous works related to the Campos basin and the Namorado oil field are Petrobras internal reports, the writers do not present an extensive list of **References Cited.** However, they would like to acknowledge the contributions of all geologists and geophysicists who have worked in the Petrobras Campos basin exploration group: A. A. Arruda, A. B. Silva, A. Goncalves, A. L. R. Rosa, C. F. Lucchesi, E. Tessari, H. Schaller, G. S. de Aquino, C. Della Favera, J. M. L. Perrella. K. Tsubone, L. C. Toffoli, M. C. de Barros, M. Saito, M. V. Dauzacker, N. R. Camos, P. C. M. Castro, and S. Possato.

Article Identification Number:
0065-731X/80/M030-0016/$03.00/0

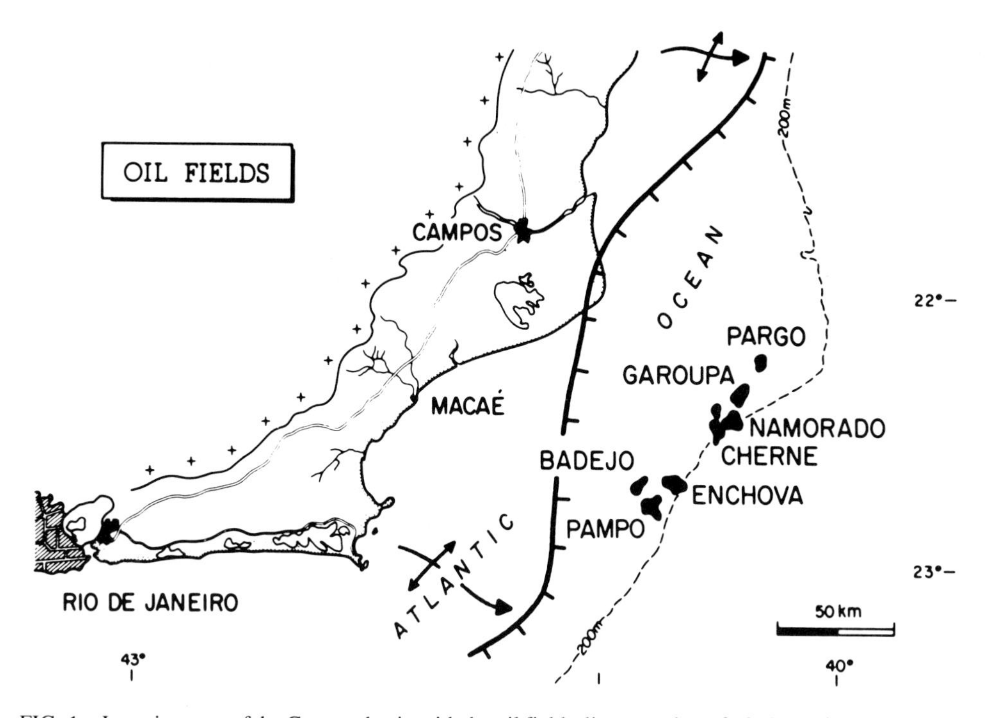

FIG. 1—Location map of the Campos basin with the oil fields discovered as of 1978. Arches shown in the figure are Vitoria in the north, and Cabo Frio in the south. Major normal fault is indicated by a hatchured line.

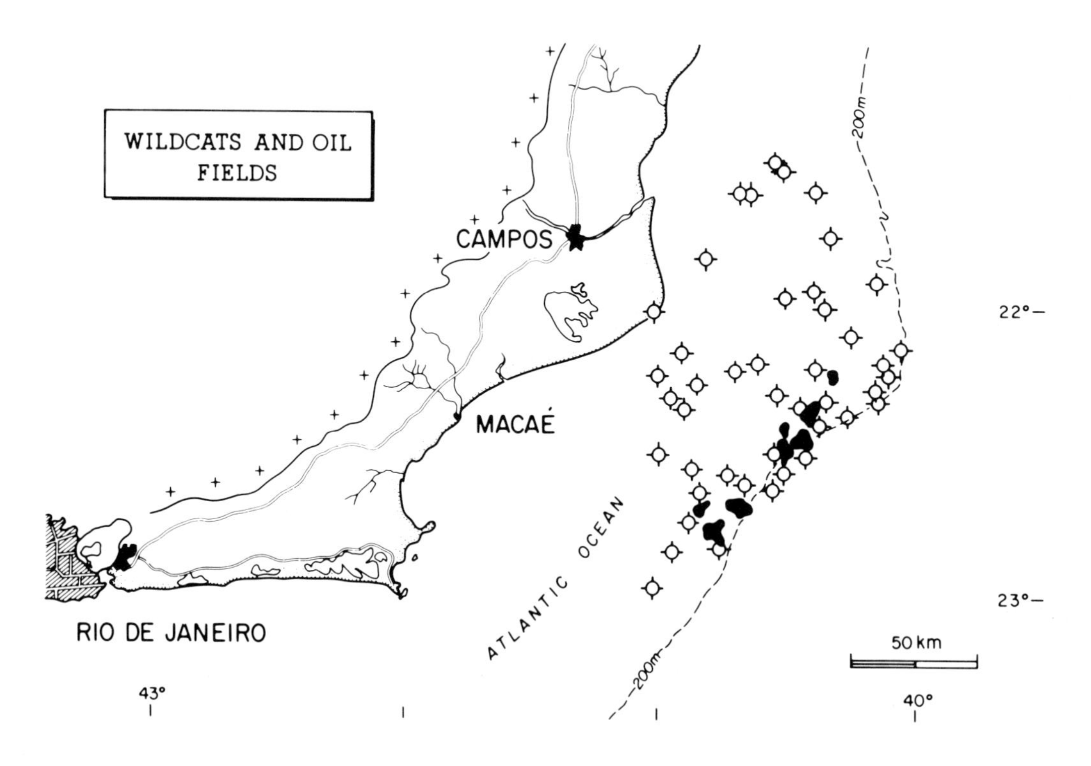

FIG. 2—Wildcat and oil field location map.

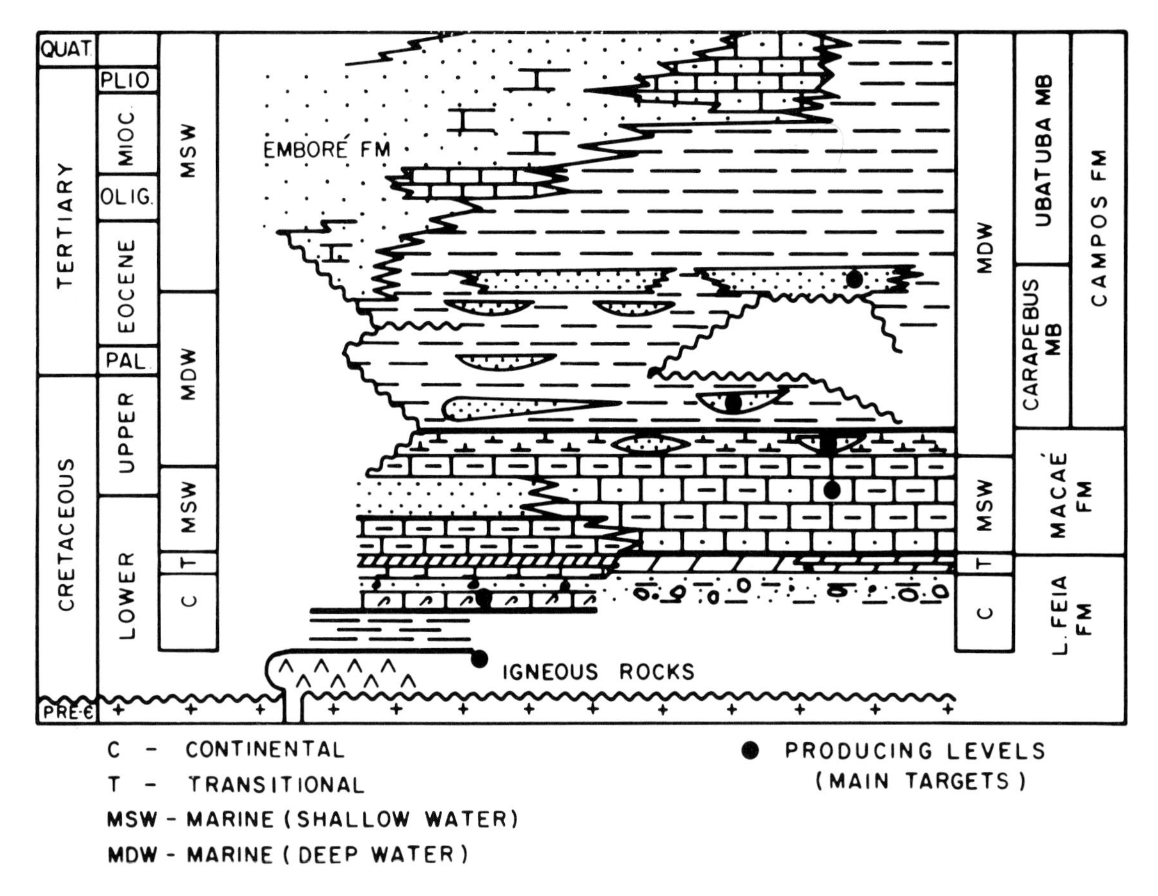

FIG. 3—Simplified lithostratigraphic column. The Namorado sandstone is represented as sand lenses on the top of the Macae Formation.

is evidence of continuous sedimentation from the Santos basin in the south, through the Campos basin to the Espirito Santo basin in the north.

One onshore stratigraphic test was drilled in 1959, an offshore gravity survey was performed in 1967, and modern marine seismic data was obtained from 1968 to the present. More than 20,000 km of seismic lines have been completed.

The Garoupa oil field was the first commercial discovery in the Campos basin. It was found by Petrobras in 1974 while drilling the ninth offshore exploration well. The producing zone is porous limestone of Albian-Cenomanian age. Other important discoveries since Garoupa include the Namorado, Enchova, Cherne, Pampo, and Badejo fields. Oil accumulations have been found in Aptian and pre-Aptian reservoirs of the continental rift phase in Albian-Cenomanian shallow-water marine carbonates and deepwater sandstones, and in turbidites of the open-marine deepwater phase of Late Cretaceous and early Tertiary ages.

Despite regional exploration efforts in the basin (more than 40 wildcats; see Fig. 2) oil accumulations tend to be concentrated along a northeast to southwest trend near the present shelf edge. This appears to be related mostly to the structural framework of the basin, as well as to the distribution of the source and reservoir rocks.

As several of the recently discovered accumulations are still in the delimitation stage and proved oil volumes are still low, it is estimated from geologic and geophysical interpretation that at least three oil fields, Cherne, Enchova, and Garoupa, will have an ultimate recoverable reserve of about 100 million bbl.

This study deals with the Namorado oil field with a recoverable oil reserve of about 250 million bbl. At Namorado, strong stratigraphic control causes serious problems in delimitation and development of the field. In addition, production problems due to water depth are numerous. Except for Badejo, all the discoveries in the Campos basin are in waters more than 100 m deep.

STRATIGRAPHIC FRAMEWORK

Figure 3 shows the Campos basin lithostratigraphic column, as used by Petrobras geologists. Precambrian basement has not been reached in the offshore part of the Campos basin. Basalt flows dated 120 m.y. are considered to be the economic basement.

A continental rift stage syntectonic sequence of fluviolacustrine sediments of the basal part of Lagoa Feia Formation overlies the basalt. A thick layer of organic-rich lacustrine shales informally called "green shales" normally occurs at the base of the sequence. Locally, the Green Shales are covered by lacustrine limestones which in some places have good secondary porosity, and by continental sandstones and conglomerates. The basal part of the Lagoa Feia Formation is syntectonic with

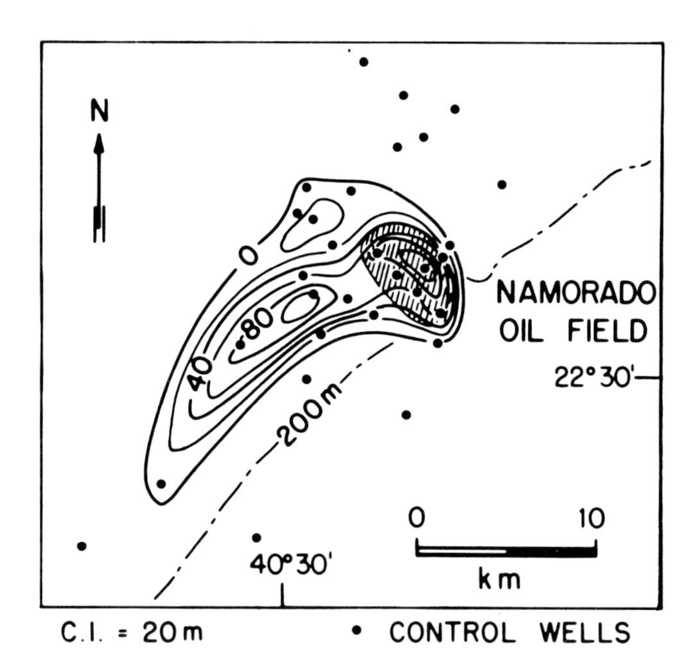

FIG. 4—Regional isopach map of the Namorado sandstone. The map is schematic due to strong thickness variations in short distances.

the rifting phase; and for this reason, it is characterized by facies changes and variations in thickness. In the upper part of Lagoa Feia Formation, there are transitional marine sediments with evaporites, limestones, and limestone altered dolomites. In some wells, several hundred meters of halite have been drilled, but generally, only a few meters of anhydrite are present. According to the tectonic-sedimentary model used in this study, much of the salt was lost due to salt flow that occurred mostly during the Late Cretaceous.

The Macae Formation is almost entirely composed of limestones deposited in a shallow-water marine environment. Dolomites are common near the base of the unit, and limestones are mostly algalic calcarenites interbedded with micrites. The Macae has an anomalously sparce faunal content that suggests some type of environmental restriction during deposition of the "Lower Macae" massive limestone. Important oil accumulations are present in the "Lower Macae" where adequate porosity is present.

The Namorado sandstone is an informal rock unit named after the discovery of the Namorado field. It is present in the upper Macae sequence just on top of the lower Macae massive limestones. The Namorado sandstone consists of several turbidite sandbodies interbedded with basinal shales, sometimes coalescing to form a thick (more than 100 m) sandbody. The isopach map (Fig. 4) shows regional distribution of the sandstone in the southeastern part of the basin near the present shelf edge. The sandstones attain maximum thickness near the Namorado. However, the map is very schematic because of the complex distribution of this unit even in a short distance. While the sandstones may exceed a thickness of 100 m within the Namorado field, they pinch out rapidly to the north, east, and south margins of the field, as indicated by the zero sandstone isolith.

In the area of the field, the Namorado sandstone is medium-grain sized, rich in fresh feldspars and locally conglomeratic with a minor amount of matrix. It is commonly very porous (20 to 30%) and highly permeable (1 darcy). Locally there are some thin zones where porosity is completely obliterated by carbonate cementation.

The turbidite interpretation of Namorado sandstones is based on their stratigraphic relationships, core study of sedimentary structures in cores, and log studies by Della Favera, Tessari and Arruda et al (unpublished Petrobras report).

The Carapebus Member of the Campos Formation was deposited during the Late Cretaceous and early Tertiary in a marine deepwater environment. The Carapebus Member is rich in turbidite sandstones very similar to those of the Namorado sandstone. An important hiatus is present in the Campos basin as Paleocene sediments are found only near the present slope; sediments of early Eocene age also may be absent. Deposition of the Carapebus Member was affected by active salt flow. A layer of widespread, almost blanket sandstone with good reservoir characteristics was deposited over the previously mentioned hiatus surface.

The Ubatuba Member of the Campos Formation and the Embore Formation together form a normal sequence of prograding slope and shelf deposits, respectively. The shelf-edge limestones of the Siri Member locally have algal reefs with exceptionally high porosity. In the southern part of the basin the reefs locally are saturated with immature oil.

STRUCTURAL FRAMEWORK

Normal faulting occurred in the basin during the deposition of the rift phase Lagoa Feia Formation. Only a few important basement faults seem to have been active after the deposition of the evaporites.

According to the tectonic-sedimentary model used in this study, deposition of the massive limestones of the Macae Formation was coincident with the halokinesis of the Lagoa Feia Formation salt, forming slump faults, small salt pillows, and domes. In the upper part of the Macae limestones, as in the Garoupa oil field, there is an obvious control of porosity by structure.

Salt flow was active during Late Cretaceous time when the salt escaped through the depositional surface leaving salt scars. Important inversions of relief, collapse structures, and symmetrical and assymetrical scars resulted from the salt escape (Fig. 5).

NAMORADO DISCOVERY

Discovery of the Garoupa field in 1974 in porous calcarenites encountered in the upper part of the massive lower Macae Formation limestones resulted in intensified exploration in the area. In the same area and south of Garoupa, a structural high associated with a seismic amplitude anomaly was mapped near the top of the Macae Formation (Fig. 6). The amplitude anomaly was interpreted as possibly indicating more porous limestones.

In 1975, the wildcat 1-RJS-19 (22°27′S lat., and 40°25′W

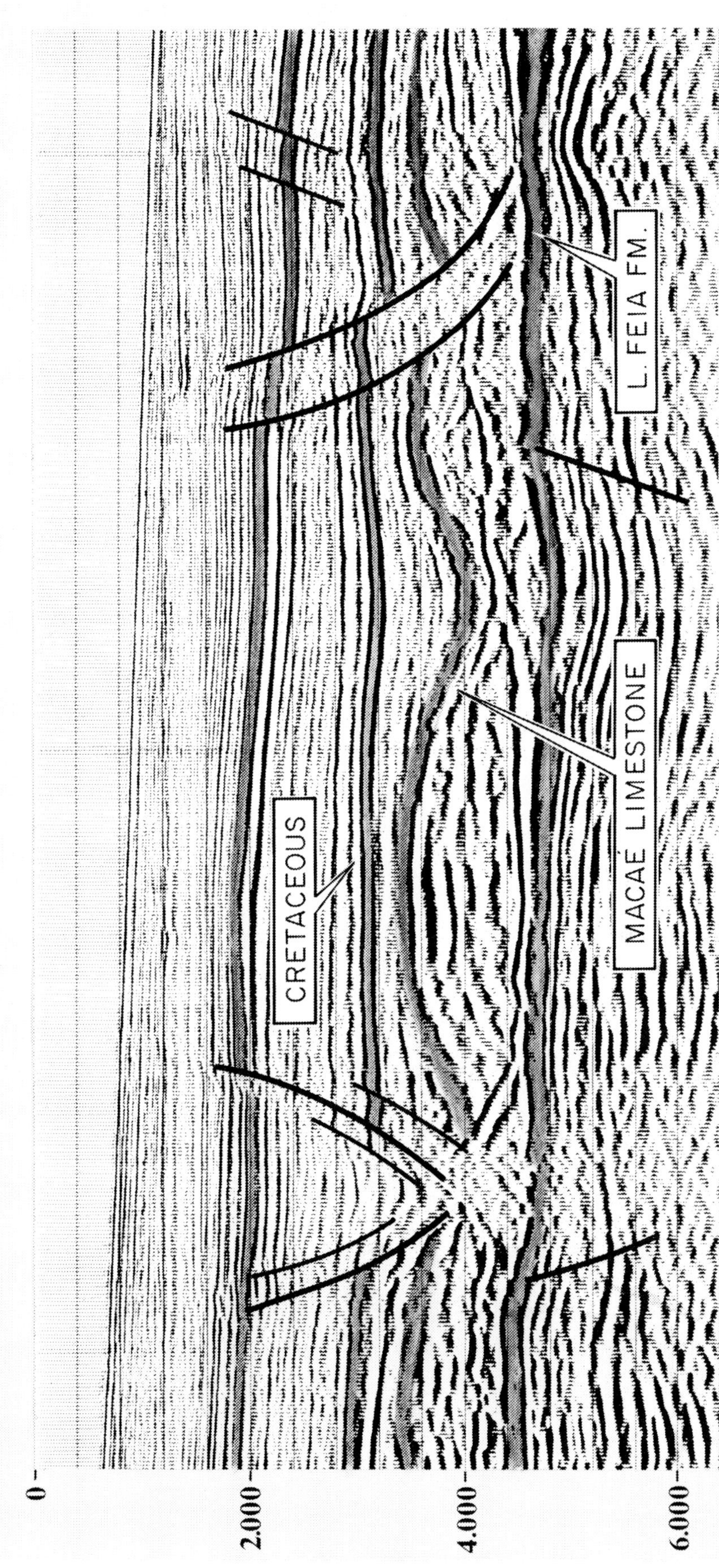

FIG. 5—Seismic dip section near the present shelf edge. Structural low on the left is interpreted as a collapse structure or symmetrical scar. On the right are slump faults.

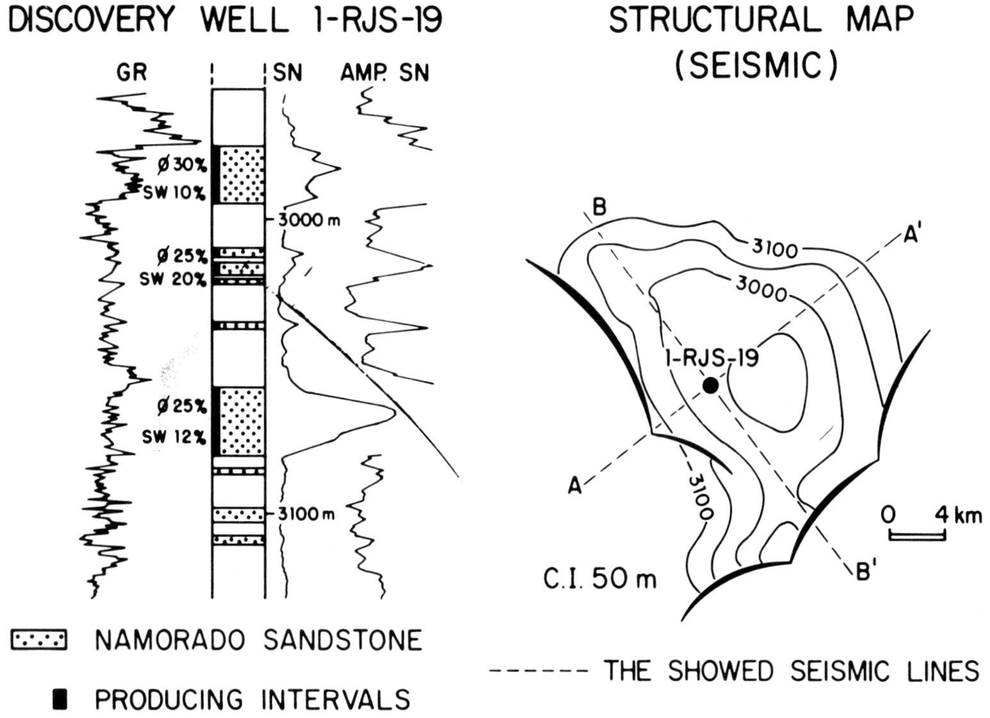

DISCOVERY WELL 1-RJS-19
GR
SN
AMP. SN
Ø 30%
SW 10%
3000 m
Ø 25%
SW 20%
Ø 25%
SW 12%
3100 m
STRUCTURAL MAP
(SEISMIC)
B
3100
A'
3000
1-RJS-19
A
3100
0 4 km
B'
C.I. 50 m
NAMORADO SANDSTONE
PRODUCING INTERVALS
THE SHOWED SEISMIC LINES

FIG. 6—Structural seismic map of the Namorado sandstone in the Namorado oil field. Seismic lines A-A′ and B-B′ are shown in Figures 7 and 8. Log abbreviations: GR = gamma ray; SN = short normal; and AMP. SN = amplified short normal.

NAMORADO OIL FIELD
4 km
A
A'
0.0
1.0
2.0
3.0
4.0
time in seconds
RESERVOIR

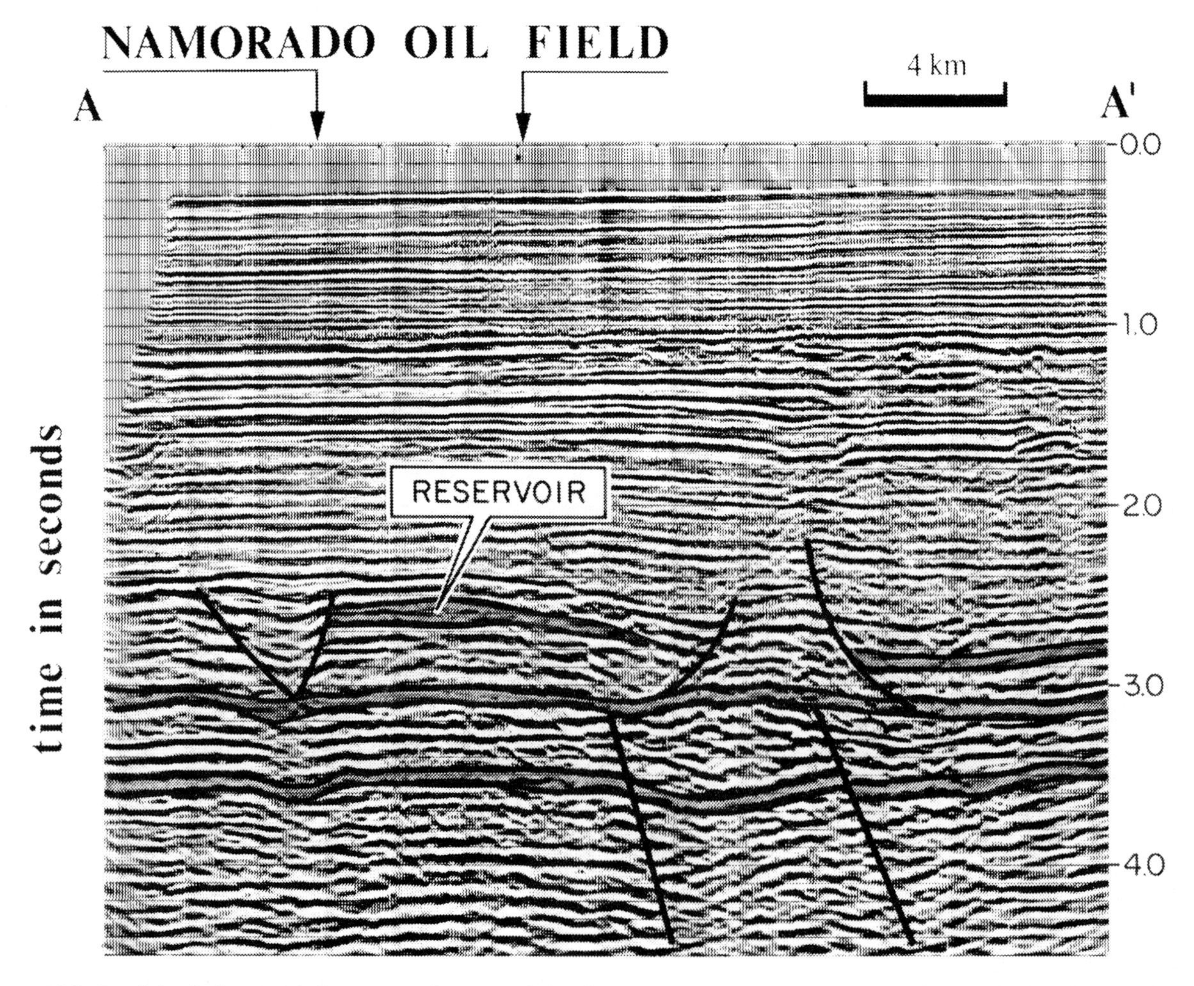

FIG. 7— Seismic line A-A′. Structure at the reservoir level seems to be due to salt flow. See Figure 6 for location. Vertical distance in two-way travel time.

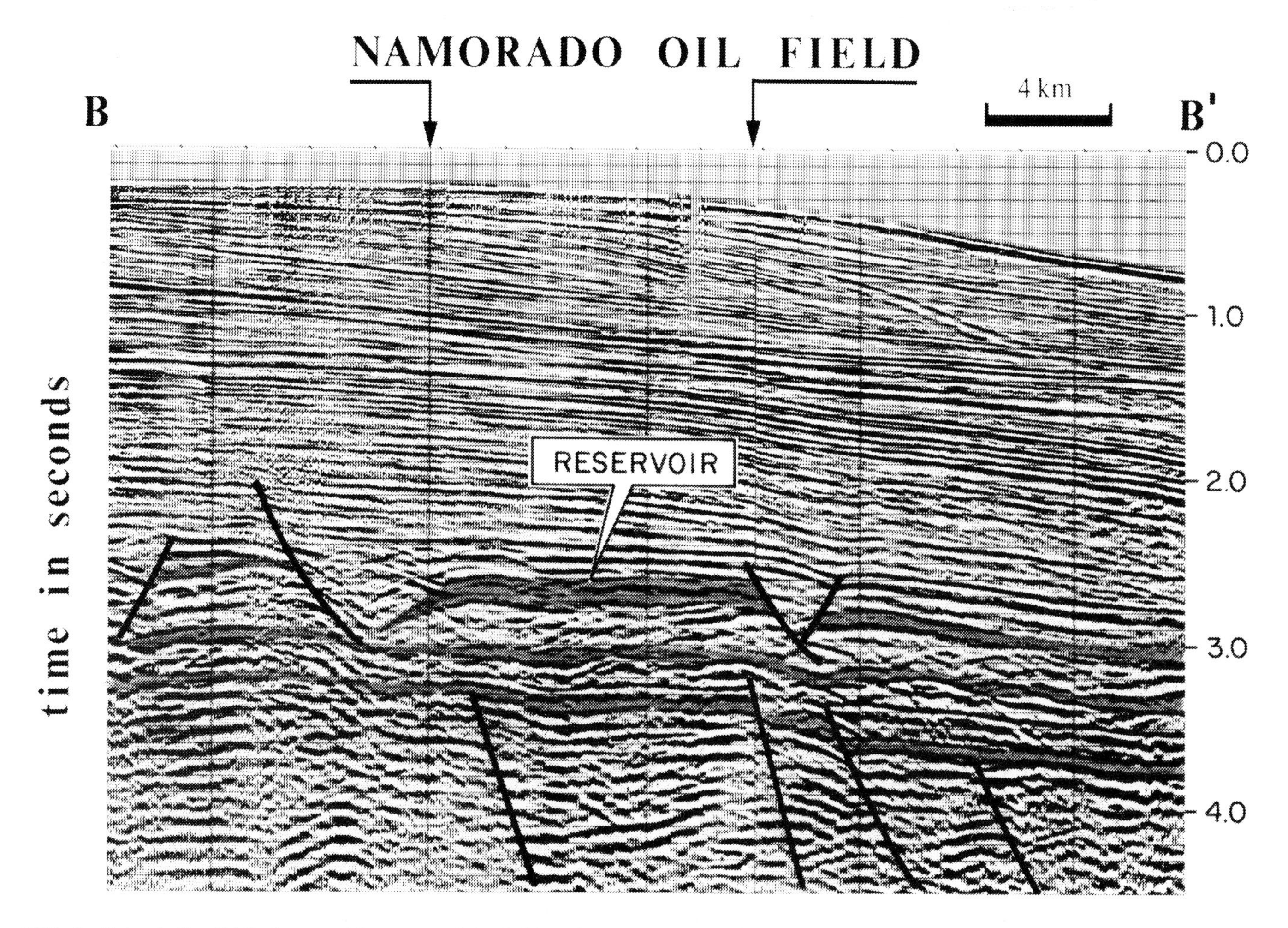

FIG. 8—Seismic line B-B'. Structural lows around the field are interpreted as collapse structures. See Figure 6 for location. Vertical distance in two-way travel time.

long.) was drilled at a water depth of 166 m. While drilling the upper Macae Formation, several bodies of oil-bearing sandstones with good porosity (25 to 30%) and permeability (1 darcy) were penetrated in the depth interval 2,980 to 3,080 m. Production testing resulted in flow rate calculations of about 6,000 b/d of 31°API oil.

Discovery of the Namorado field also meant the first important oil occurrence in the Namorado sandstone in the basin. Due to the turbiditic sandstone nature of the reservoir, the presence of stratigraphic control unrelated to structure was soon suspected.

DELIMITATION

The first two delimitation wells 3-NA-1-RJS and 3-NA-2-RJS had excellent results especially the 3-NA-1-RJS well that found more than 100 m of almost continuous oil-bearing sandstones. Extension well 3-NA-3-RJS was dry because the reservoir was absent. At that time, Petrobras geophysicists were using acoustic impedance seismic logs searching for porosity-enchanced zones in the Macae Formation limestone. The method was applied to the Namorado field.

Figure 9 shows the acoustic impedance seismic log in the area of the third extension well—a pinchout of the Namorado sandstone is apparent near NA-3. This information was used to orient drilling of the directional well NA-3D from the same location; the reservoir was found as predicted, which raised the confidence in the utilization of such seismic techniques for delimitation of the field.

Three more delimitation wells have been drilled; only the 3-NA-5-RJS is a dry well also because the reservoir is absent. The well was drilled in a structural high, but with strong indications of absence of the reservoir suggested by acoustic seismic logs.

The Namorado oil field has an area of about 20 sq km and now is completely delimitated (Fig. 10) both by extension wells and by seismic data; a volume of proven recoverable oil of 250 million bbl has been calculated.

GEOLOGIC CONSIDERATIONS

As shown by the seismic lines in Figures 7 and 8, basement faults active during the deposition of the Lagoa Feia Formation were not active during the deposition of the Macae Formation; structures in the upper Macae Formation and Carapebus Mem-

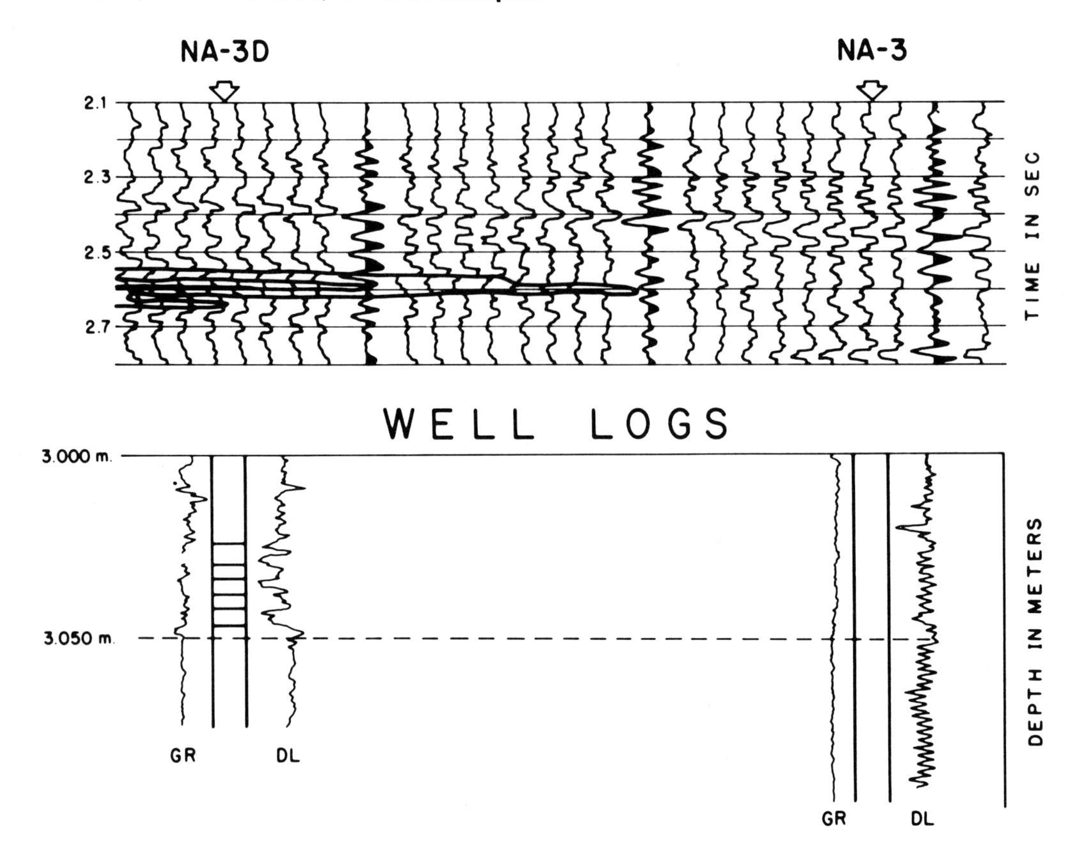

FIG. 9—Acoustic impedance seismic log showing the Namorado sandstone pinchout between the NA-3 and NA-3D wells. Log abbreviations: GR = gamma ray log; DL = dual induction log.

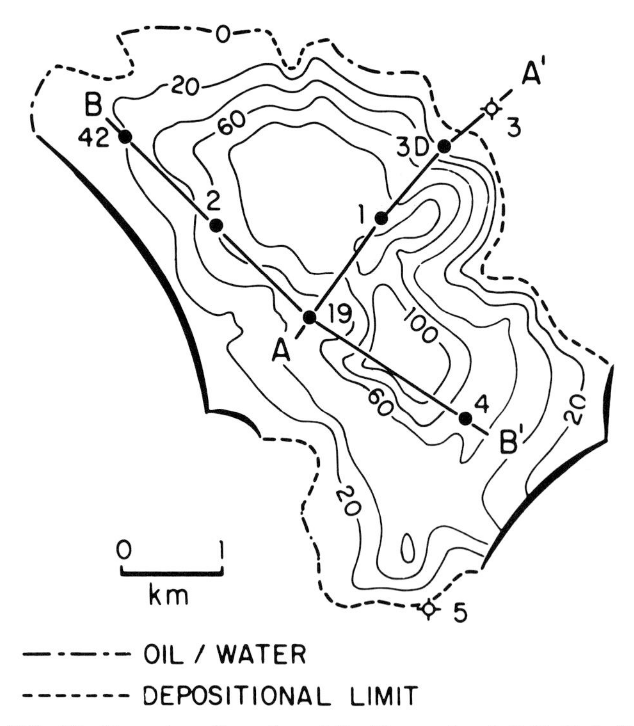

FIG. 10—Present configuration of the Namorado oil field. Note that despite the presence of a structural high the accumulation has strong stratigraphic control with pinchout of the reservoirs in the north, east, and south. Geologic sections A-A′ and B-B′ are shown in Figure 11 and 12. Contour interval is 20 m.

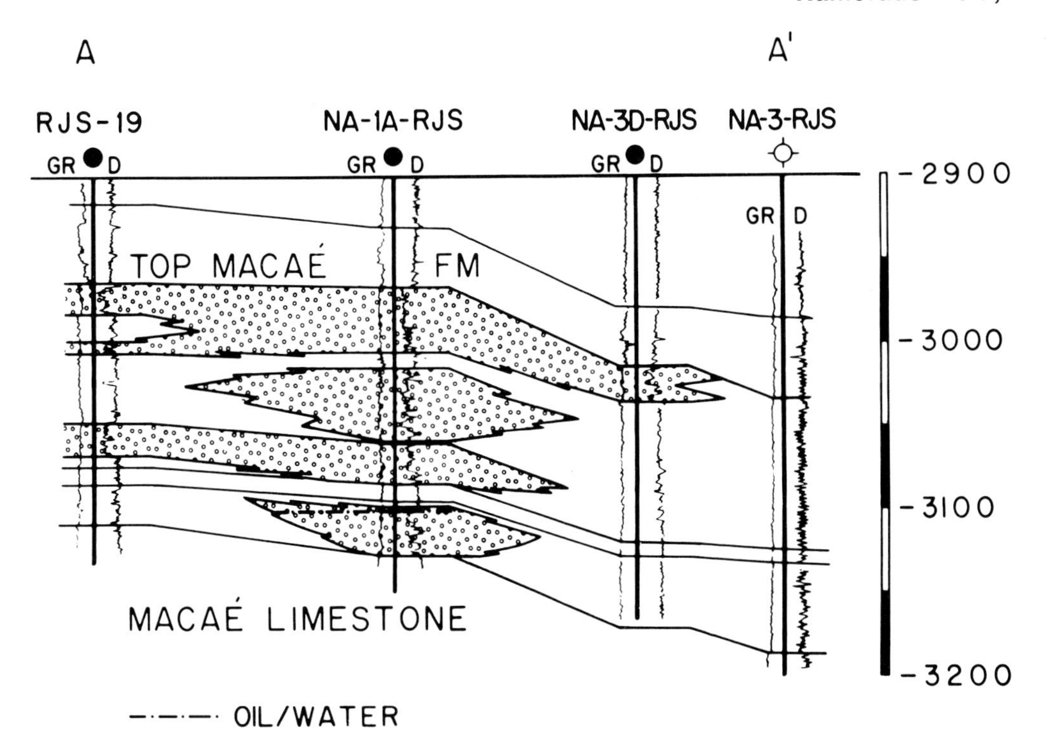

FIG. 11—Geologic schematic cross section A-A′ showing amalgamation of sand bodies in the NA-1A and the shale-out of reservoir between NA-3D and NA-3. Depth measured in meters. Location shown in Figure 10.

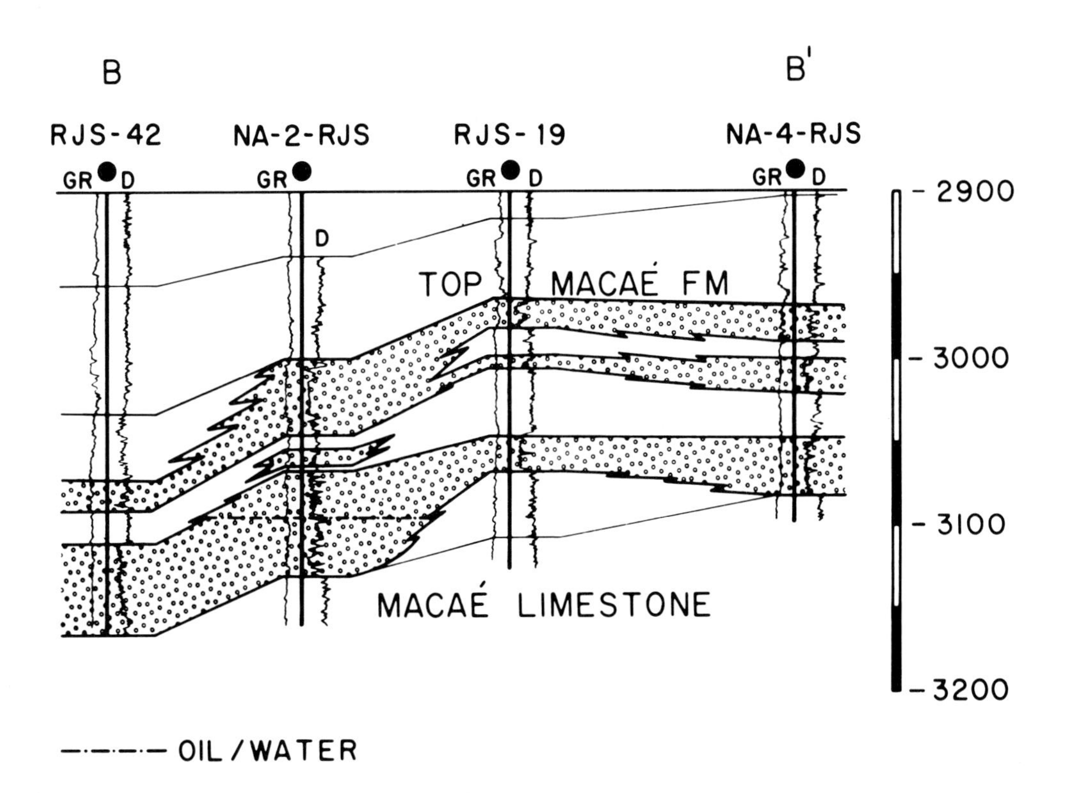

FIG. 12—Geologic schematic cross section B-B′. The well NA-2 is one of the few with a clearly defined oil-water contact in the reservoir. Depth in meters. Location shown in Figure 10.

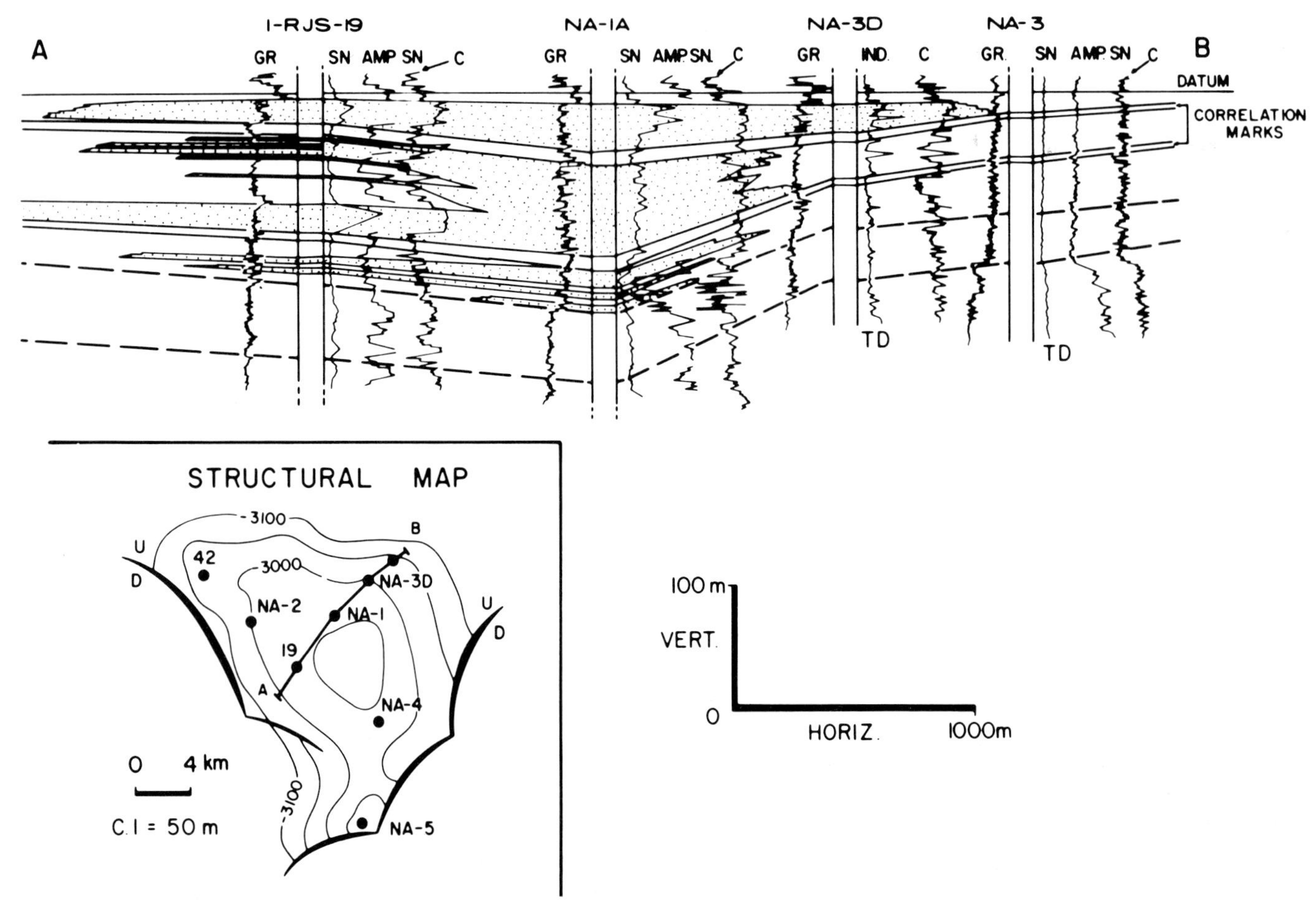

FIG. 13—Geologic cross section through Namorado oil field using a radioactive layer near the top of the Macae Formation as a datum. The Namorado sandstone accumulated in a contemporaneous structural or depositional low. Log abbreviation: C = conductivity log.

ber apparently resulted from salt flow. The Namorado high is a dome elongated in the northwest to southeast direction and faulted on the south and west.

Differential compaction over the Namorado sandstones enhances the structural configuration at the reservoir level because the highest structural point coincides with the maximum sandstones thickness. Core of the oil-bearing Namorado sandstones show it to be a medium-grain sized sandstone, normally massive, with few sedimentary structures apparent and possible amalgamation of layers. No typical coarsening and thickening upward, or thinning and fining upward trends, could be identified in cores and logs to distinguish channel and lobe facies.

Detail correlations prepared using a radioactive marker above the youngest sandbody, and also other good correlation markers (Fig. 13) indicate that the Namorado sandstone was actually deposited in contemporaneous lows; therefore, it seems that coarse sediments carried by turbidity currents were dumped in those lows thus the lack of typical turbidite facies characterization.

The depositional low can be tentatively interpreted as peripheric synclines surrounded by salt pillows or salt domes. The later escape of the salt to the surface during the Late Cretaceous could have produced structures and inversions of relief.

Further detailed studies of the Namorado sandstone in the field area are in progress, because although the reservoir is probably amalgamated, it was suspected some problems will rise in several points during production and injection.

SELECTED REFERENCES

Carozzi, A. V., et al, 1977, Microfacies and depositional-diagenetic evolution of the Macae carbonates (Albian-Cenomanian) of the Campos basin: Petrobras-Dexpro Internal Rept. 6013.

Paula Couto, C., and S. Mezzalira, 1971, Nova conceituacao geocronologica de Tremembe, Estado de Sao Paulo, Brasil—in Symposia: Acad. Brasileira Cienc. Anais., Bras. Paleontologia, v. 43, supp., p. 473-488.

Schaller, H., 1973, Estratigrafia da Bacia de Campos: 27th Cong. Brasileira Geologia, v. 3, p. 247-258.

Tessari, E., K. Tsubone, and N. Brisola, 1978, Petroleum bearing deposits in the Campos basin: Offshore Brazil 78 Conf., Proc., (Rio de Janeiro), 14 p.

Toffoli, L. C., and M. C. de Barros, 1978, Esforcos e resultados de pesquisas de trapas estratigraficas nas bacias sedimentares Brasileiras: 1st Cong. Brasileira de Petroleo (Rio de Janeiro), I.B.P., p. 51-62.

Giant Fields of the Southern Zone—Mexico[1]

Jose Santiago Acevedo[2]

Abstract Giant oil and gas fields in the southern part of Mexico and out through the Yucatan area are discussed. "South Zone" exploration (as defined by the exploration program of Petroleos Mexicanos) includes production areas in the Mexican states of Guerrero (southern part), Oaxaca (southern part), Chiapas, Quintana Roo, Yucatan, Campeche, Tabasco, and Veracruz (southern part), and includes some areas offshore Mexico in the Gulf of Mexico.

A history of exploration and production is highlighted since 1938 (nationalization), and stratigraphy, structure, and accumulation of petroleum of giant fields are described.

Fields in the Isthmus Saline basin which are included here are: Tonala-El Burro, El Plan, Cinco Presidentes, Magallanes, and Ogarrio. Fields in the Macuspana basin include: Jose Colomo, Chilapilla, and Hormiguero. Sections of the manuscript also discuss the generalities and stratigraphy of the Chiapas-Tabasco Mesozoic Area (including the Sitio Grande field, Cactus field, and the Antonio J. Bermudez complex), and the general stratigraphy and hydrocarbon accumulation of the Campeche Marine Platform.

INTRODUCTION

This work deals with the giant oil fields of Southern Mexico, including not only the fields discovered during the period from 1967 to 1977, but also several important fields discovered prior to that time. The classification used for a giant field is one having an estimated recovery of 100 millions bbl of oil or 1 Tcf of gas.

The south zone of Mexico, for PEMEX exploration purposes, includes the states of Guerrero (southern part), Oaxaca (southern part), Chiapas, Quintana Roo, Yucatan, Campeche, Tabasco, and Veracruz (southern part, Fig. 1).

Hydrocarbon exploration in South zone actually began in 1863 when Father Manuel Gil y Sainz discovered what he called "San Fernando Mine" near Tepetitan Town, Tabasco. He sent 10 bbl of oil to New York City for analysis.

Later on in 1883, Dr. Simon Sarlat, Governor of Tabasco, drilled a well to 27.4 m on the Sarlat anticline, and in 1886 a small production of light oil was obtained, but was not for commercial exploitation.

During the first years of the 20th Century, foreign companies working in Mexico found commercial production in the Capoacan (1905) and San Cristobal fields (1906). Both fields are located in the southern area of Veracruz and are associated with salt domes near Coatzacoalcos City.

Early oil production was discovered in southen Veracruz on the Ixhuatlan structure in 1911, Concepcion structure in 1929, on the Tonala-El Burro in 1928, and El Plan field in 1931. The production was from Miocene and Pliocene sandstones associated with salt domes. The two most important fields are Tonala-El Burro and El Plan.

After March, 1938, the Mexican national oil company, PEMEX, began to explore the South zone. This area is divided

[1]Manuscript received, June 11, 1979; accepted for publication, August 7, 1979.

[2]Petroleos Mexicanos, Coatzacoalcos, Veracuz, Mexico.

Article Identification Number:
0065-731X/80/M030-0017/$03.00/0.

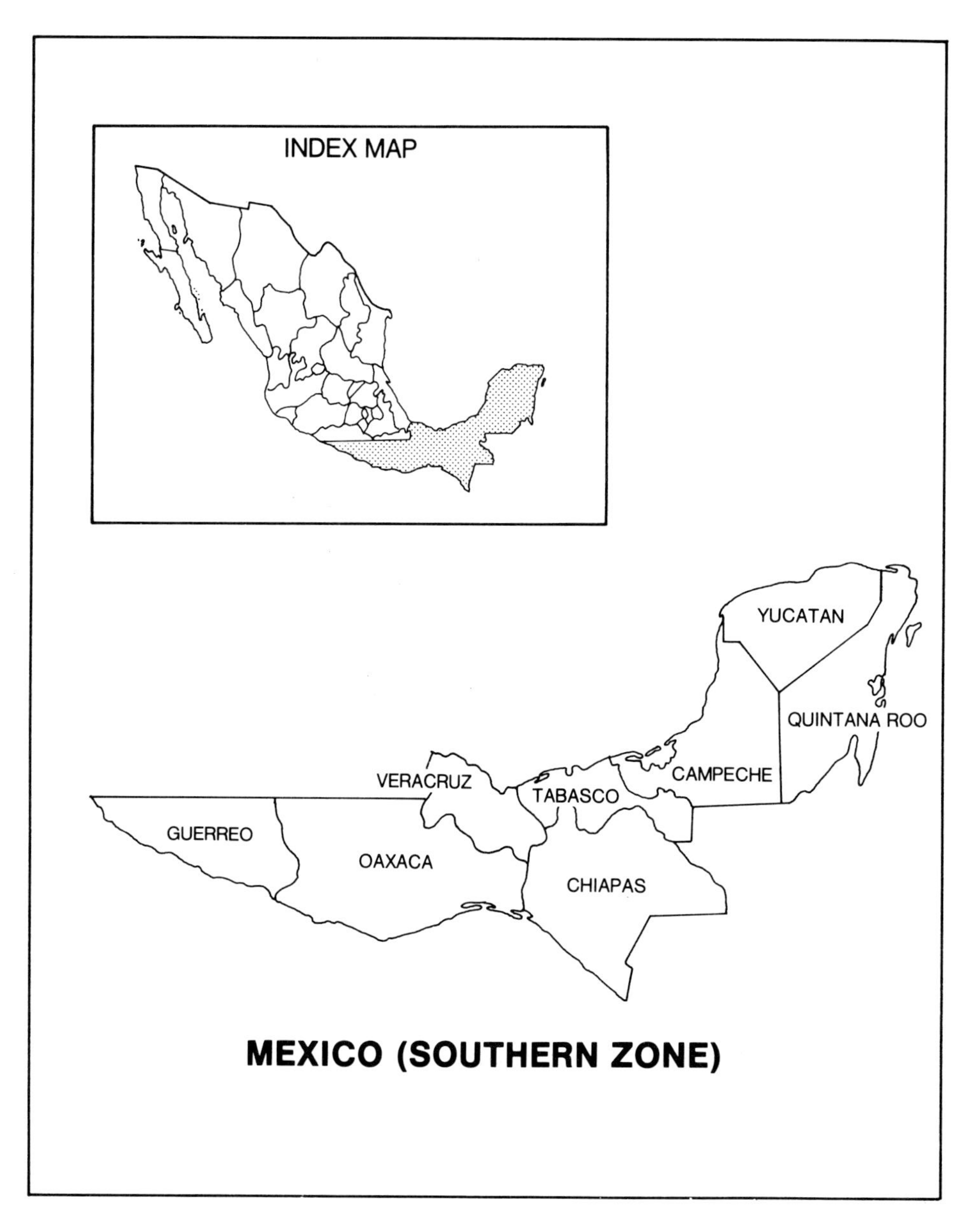

FIG. 1—Location map of southern zone, Mexico.

geologically into the Isthmus Saline basin, Macuspana basin, Comalcalco basin, and the new Chiapas-Tabasco Mesozoic area. This area was explored during the ensuing years, and in 1976, oil production from dolomitic breccias of early Paleocene age was discovered offshore in the Campeche Platform.

Giant field descriptions in this study are from west to east, finishing with the Chiapas-Tabasco and the offshore field (Fig. 2).

GIANT FIELDS OF ISTHMUS SALINE BASIN

Because of shallow salt deposits over the west and southwestern parts of this basin (Fig. 3) and the associated salt diapirs, it has been called the Saline basin. Until recently, the salt was believed to be present only in this area; however, new drilling data, surface geology, and geophysical data (mainly reflection seismology) have led to the conclusion that salt is present up to the Yucatan Platform in the east, the Chiapas Massif to the south, and offshore into the Gulf of Mexico to the north (Fig. 4). The total area of the Saline basin of southeastern Mexico is about 6,000 sq km. The age of the salt is Triassic-Jurassic (pre-Kimmeridgian). In different parts of the Saline basin, salt beds with thickness ranging from 1 to 1,200 m thick are emplaced within Lower Tertiary, Cretaceous, and Jurassic sediments. Most of these salt beds are the result of migration along faults and fractures; others are product of salt dissolution.

The largest oil fields produce from Tertiary sandstones associated with salt domes in the central part of the Macuspana basin. The most productive gas fields also are found in Tertiary

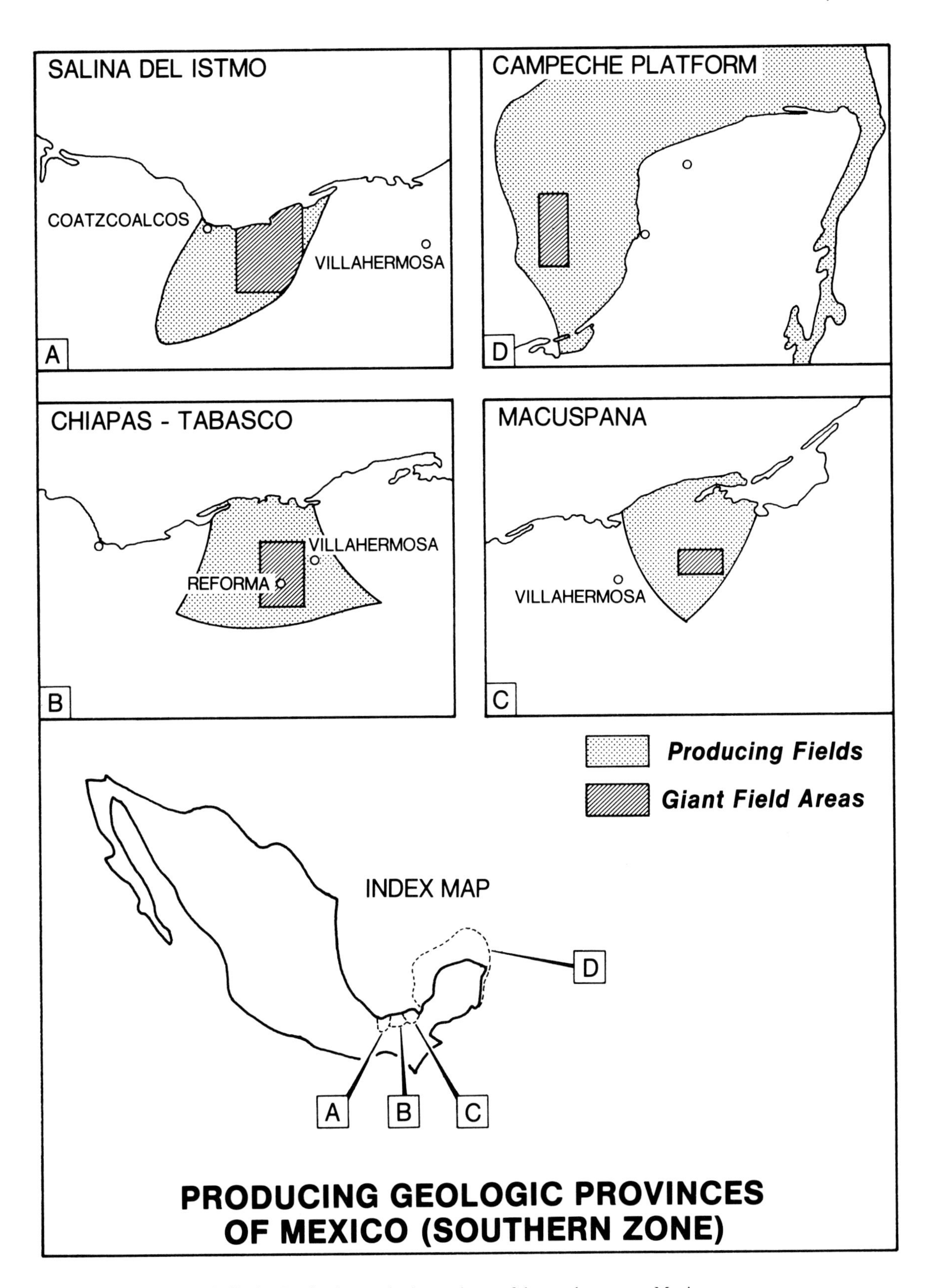

FIG. 2—Productive geologic provinces of the southern zone, Mexico.

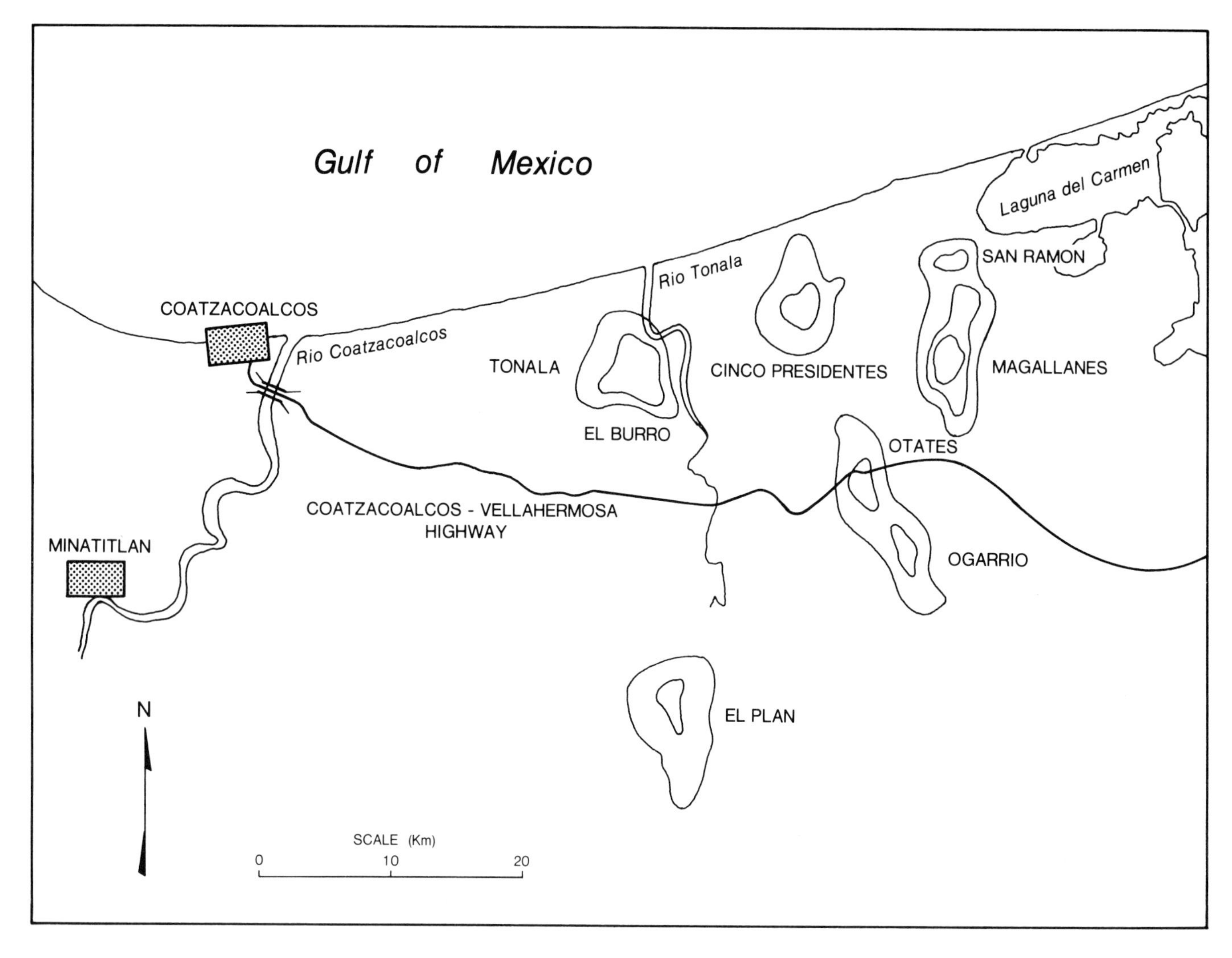

FIG. 3—Giant fields of the Saline basin of southeastern Mexico.

sandstones in the eastern part of the basin; however the gas is trapped in anticlines.

Tonala-El Burro Field

Tonala-El Burro (Fig. 5) is located in the Gulf Coastal Plain, and was discovered in 1928 on the eastern flank of a salt dome.

Stratigraphy

Wells in the field have penetrated Triassic-Jurassic (anhydrite-salt) section. Erosion has removed the Cretaceous and Paleocene. The Tertiary section thickens on the flanks of the dome and some wells have not penetrated the Eocene. The most common section encountered in the Tonala-El Burro field is recent, Miocene, and Oligocene deposits (Fig. 6).

Structural Geology

The structural geology of the field is an elliptical salt dome 8 km long by 7 km wide, and associated normal faults trend predominantly northeast to southwest (Figs. 7, 8). The fault blocks are present over the flanks of the dome and act as hydrocarbon traps. The most productive traps were discovered during the 1960s using a combination of geologic and seismic data.

Hydrocarbon Accumulation

Production is primarily from the Encanto, upper and lower Concepcion of Miocene age, and the Filisola and Paraje Solo of Pliocene age. There have been 358 wells drilled in the field, of which 247 were productive and 111 were dry. Spacing is 200 m in the oldest part of the field and 400 m over the flanks of the dome. Cumulative oil production at the end of 1978 was 93,366,000 bbl of oil, and expected recovery is 13,000,000 additional barrels. Oil density ranges from 0.857 at 37°C (33.5° API) and 0.877 at 37°C (29.8° API). The average porosity of the productive formations is 23% and average permeability is 175 md.

El Plan Field

Similar to Tonala-El Burro, El Plan (Fig. 9) is on the Gulf Coastal Plain in what is known as the Isthmus Saline basin.

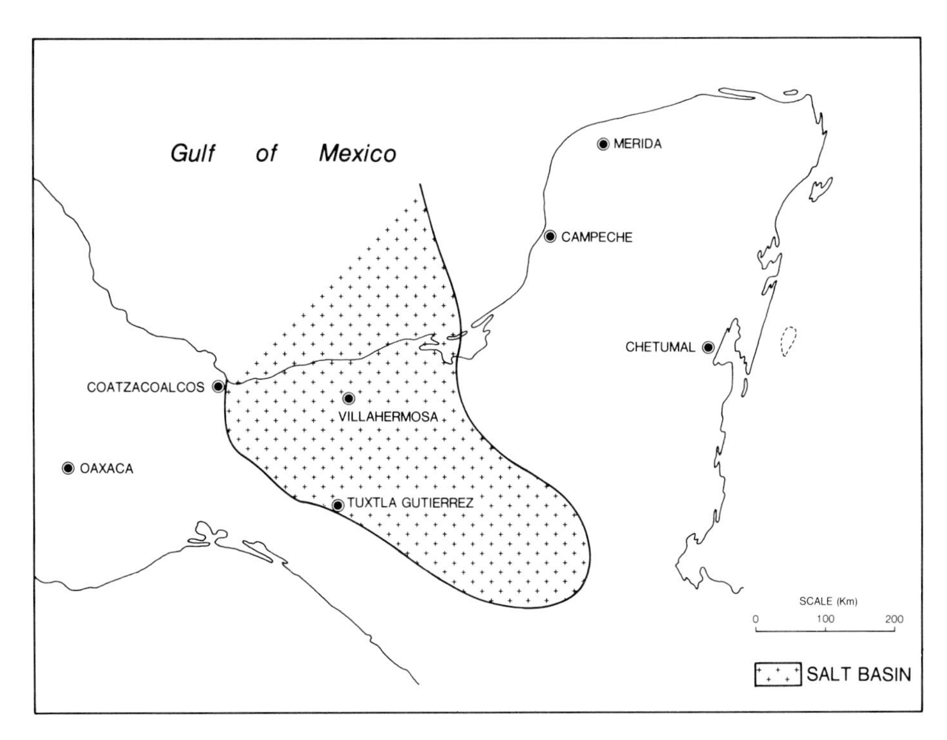

FIG. 4—Location map of salt in the Saline basin of southeastern Mexico.

Exploration began in 1929 on a recognized northwest to southeast trending anticlinal structure.

The Plan 1 was the discovery well drilled to 665.3 m with initial production of 179.5 cu m of oil and 19,138 cu m of gas from Pliocene sandstones of the Filisola Formation.

Stratigraphy

Overlying the Triassic-Jurassic salt is a sedimentary section consisting of Oligocene, Miocene, Pliocene, and recent rocks. Fractures in the lower Miocene rocks on the west flank of the salt dome have been filled with salt. Formation thicknesses are similar to those of the Tonala-El Burro field.

Structural Geology

The salt dome is 7 km long by 5 km wide, trending in a north to south direction (Figs. 10, 11). The maximum salt intrusion penetrates the Cedral Formation (Pliocene) causing numerous normal faults and separate fault blocks; each block has a different oil-water contact. One of the most productive blocks is on the southwestern flank. This block also is the area of deepest production, with the well reaching a producing depth of 2,500 m.

Hydrocarbon Accumulation

Oil production comes from Miocene sandstones associated with salt movements. Some wells produce as much as 80 cu m of oil/day, although the average is about 30 cu m/d.

Two types of oil are found in the field, waxy oil with a density of 0.860 at 20°C (28° API), and a no-wax-content oil with a density of 0.915 at 20°C (24° API).

Well spacing is 200 m in the older part of the field, and 400 m in the newer parts of the field (flanks). The accumulative production at the end of 1978 was 145 million bbl of oil. An additional recovery of 7 million bbl of primary oil is expected. The average porosity of the producing zones is 20 to 25% and permeability about 180 md.

Cinco Presidentes Field

The field is in the Isthmus Saline basin (Fig. 12). Because the field is in a swampy lowlands, development has been hampered because of poor access. The Cinco Presidentes complex was discovered in 1946 by Petroleos Mexicanos, but from 1953 to 1957 additional seismic data defined the Yucateco salt dome, and in 1960 the Yucateco 1 produced oil and gas from a sandstone in the Encanto Formation (lower Miocene) at a depth of 2,092 to 2,095 m. Total depth of the well was 2,286.3 m, and the salt was penetrated at 2,268 m.

Stratigraphy

A sequence of Oligocene, Miocene, and Pliocene-age sandstones unconformably overlie the Triassic-Jurassic anhydrite and salt. The most productive formation is the lower Miocene Encanto. Different formation thickness varies according to salt movement. The Encanto Formation is as much as 1,000 m thick on the flanks of the salt dome and is absent over the crest of the dome.

Structural Geology

The structure is an oblong dome 8 km long by 6 km wide, trending north to south (Figs. 13, 14). The piercement dome has formed several faults and fault blocks in which different oil-water levels are present.

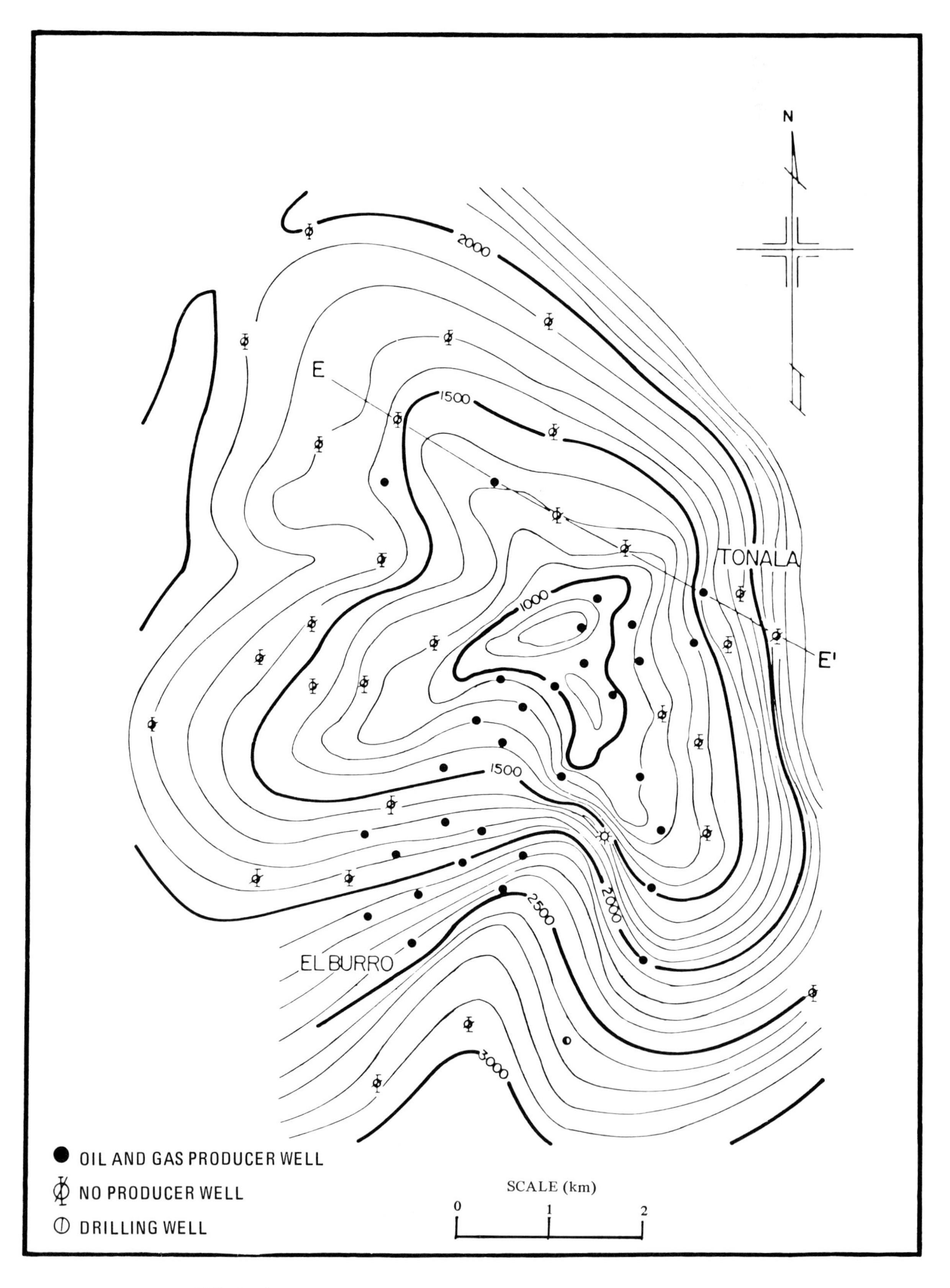

FIG. 5—Structure contour map on the top of salt at Tonala-El Burro field.

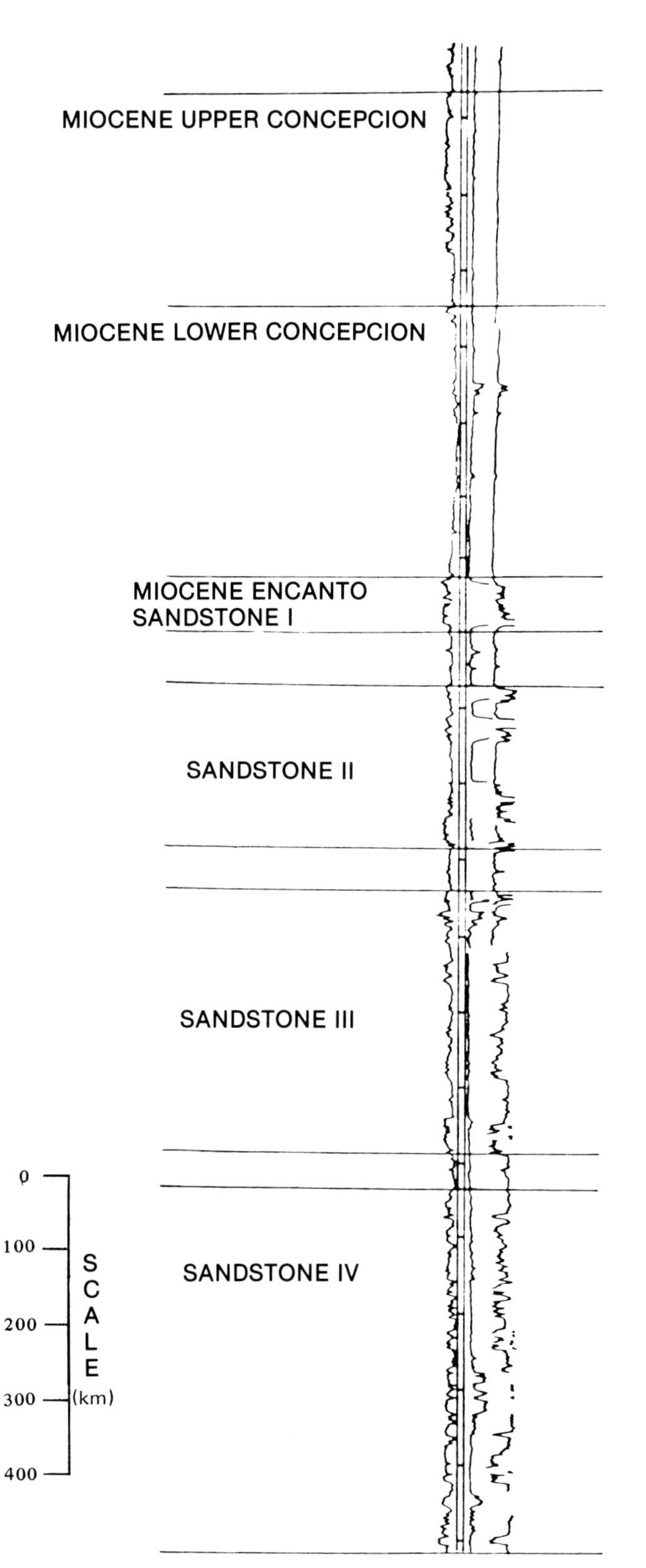

FIG. 6—Typical geologic column in the Saline basin, Mexico.

Hydrocarbon Accumulation

By the end of 1978, 331 wells had been drilled at 200 and 400 m spacing; 284 were productive. The cumulative production through 1978 was 217,857,000 bbl of oil, with 86,000,000 bbl of additional recoverable reserves (total 303,000,000 bbls). The average porosity of the producing formation is 20%, permeability is 176 md, and oil density is 0.853 at 37°C (34.2° API).

Magallanes Field

Magallanes field also is located in the Gulf Coastal Plain in the easternmost part of the Isthmus Saline basin (Fig. 15). The discovery well, Magallanes 3, was completed in 1957 to a depth of 1,600 m, producing 104 cu m of oil per day from three intervals (1,167 to 1,183; 1,248 to 1,251; and 1,292 to 1,294 m) in the lower Miocene Encanto Formation. This well was drilled at the top of the structure. Some wells in the field have reached depths of 4,200 m, exploring the lower Miocene Deposito Formation.

Stratigraphy

The geologic column is similar to previously described fields. Production is from the sandstones of the Encanto Formation. As with other described fields, formation thickness varies considerably with thicker sandstones over the flanks of the structure.

Structural Geology

This structure is the result of salt movement with several north to south trending closures (Figs. 16, 17). The largest production is obtained from the central and southeastern part of the dome. Over the northernmost part of the structure there is another closure where the San Ramon field is located. The major axis of the structure is 18 km (including San Ramon field) and the minor axis is 7 km. Salt flow has caused many normal faults. One of the major normal faults crosses in a north-northeast to south-southwest direction near the middle structure. The downthrown block is to the southeast and has a vertical displacement of 100 to 500 m. In general, the dominant fault system trends north-northeast to south-southwest.

Hydrocarbon Accumulation

Of the 620 wells drilled, 529 are productive with a cumulative production of 122,276,005 bbl of oil by the middle of 1978. An estimated 108,000,000 bbl of oil remains to be recovered (total 230,276,000 bbl). Well spacing is 200 m and 400 m. Average porosity of producing zones is 19%, average permeability is 180 md, and oil density is 0.864 at 37°C (32° API).

Ogarrio Field

Ogarrio is located in the Tabasco Gulf Coastal Plain (Fig. 18). Blasillo 1 (Ogarrio 1) well was completed as producer in

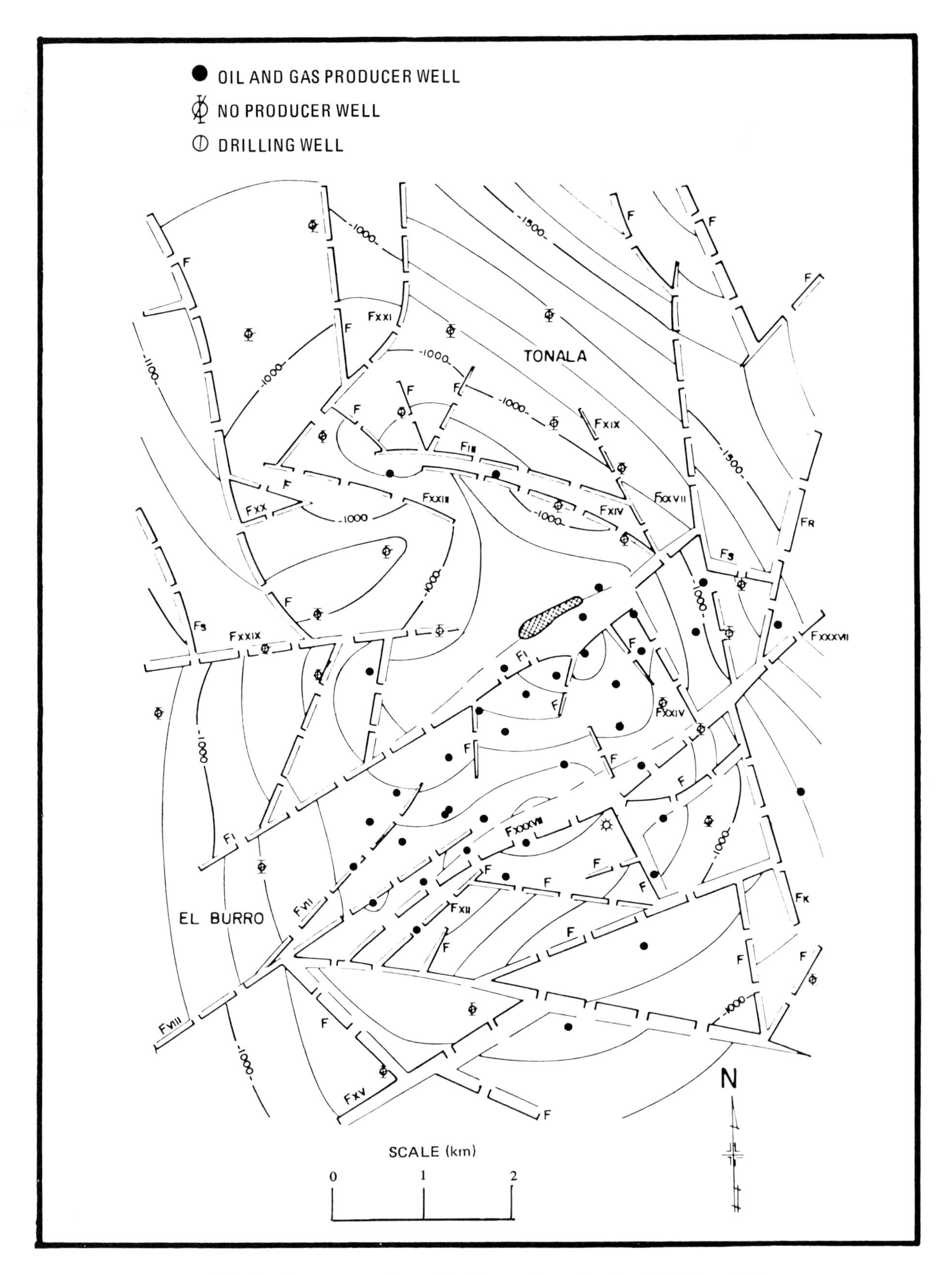

FIG. 7—Structure map of top of the Encanto Formation, Tonala-El Burro field.

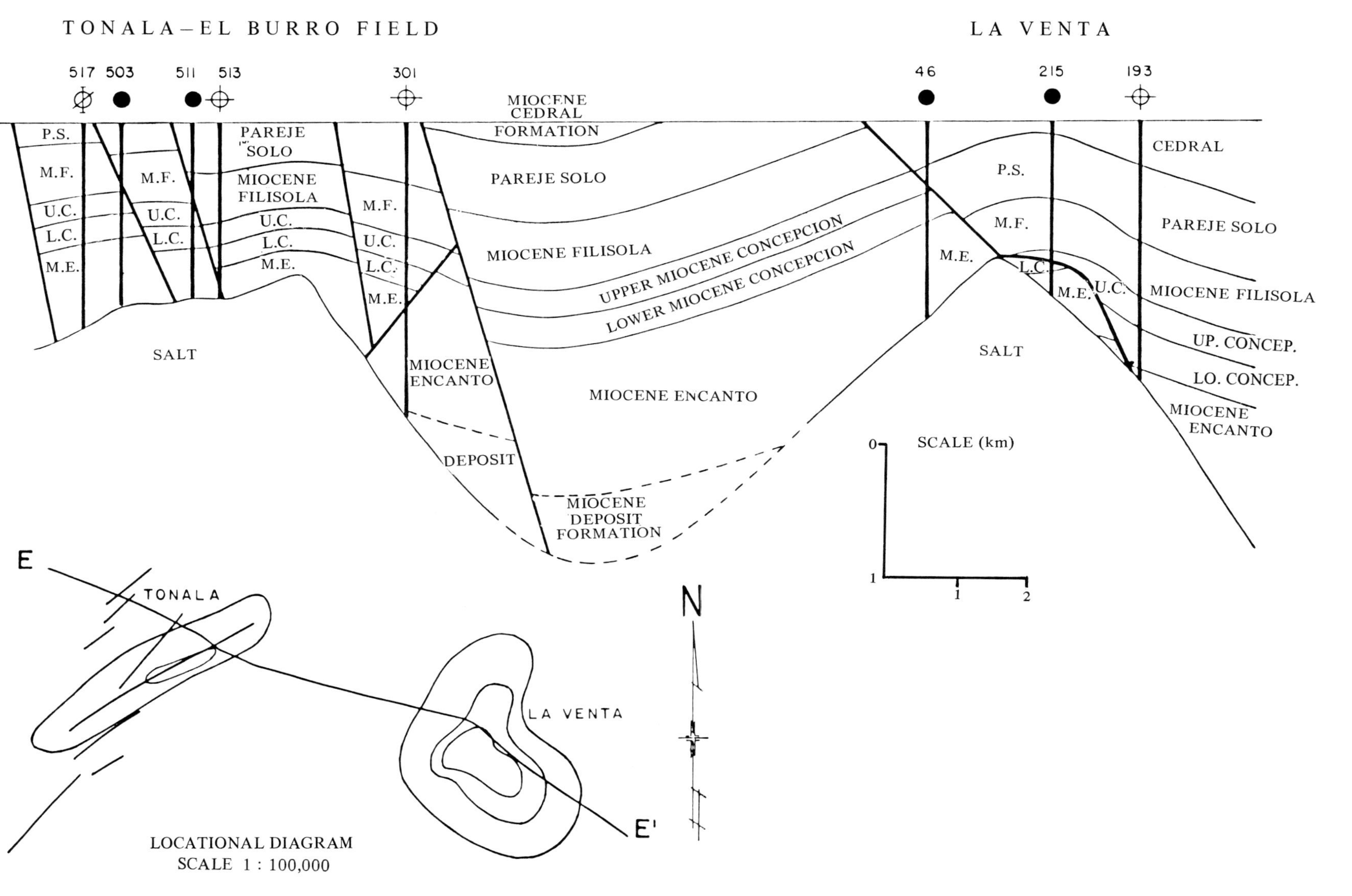

FIG. 8—Geologic cross section through the Tonala-El Burro and La Venta fields.

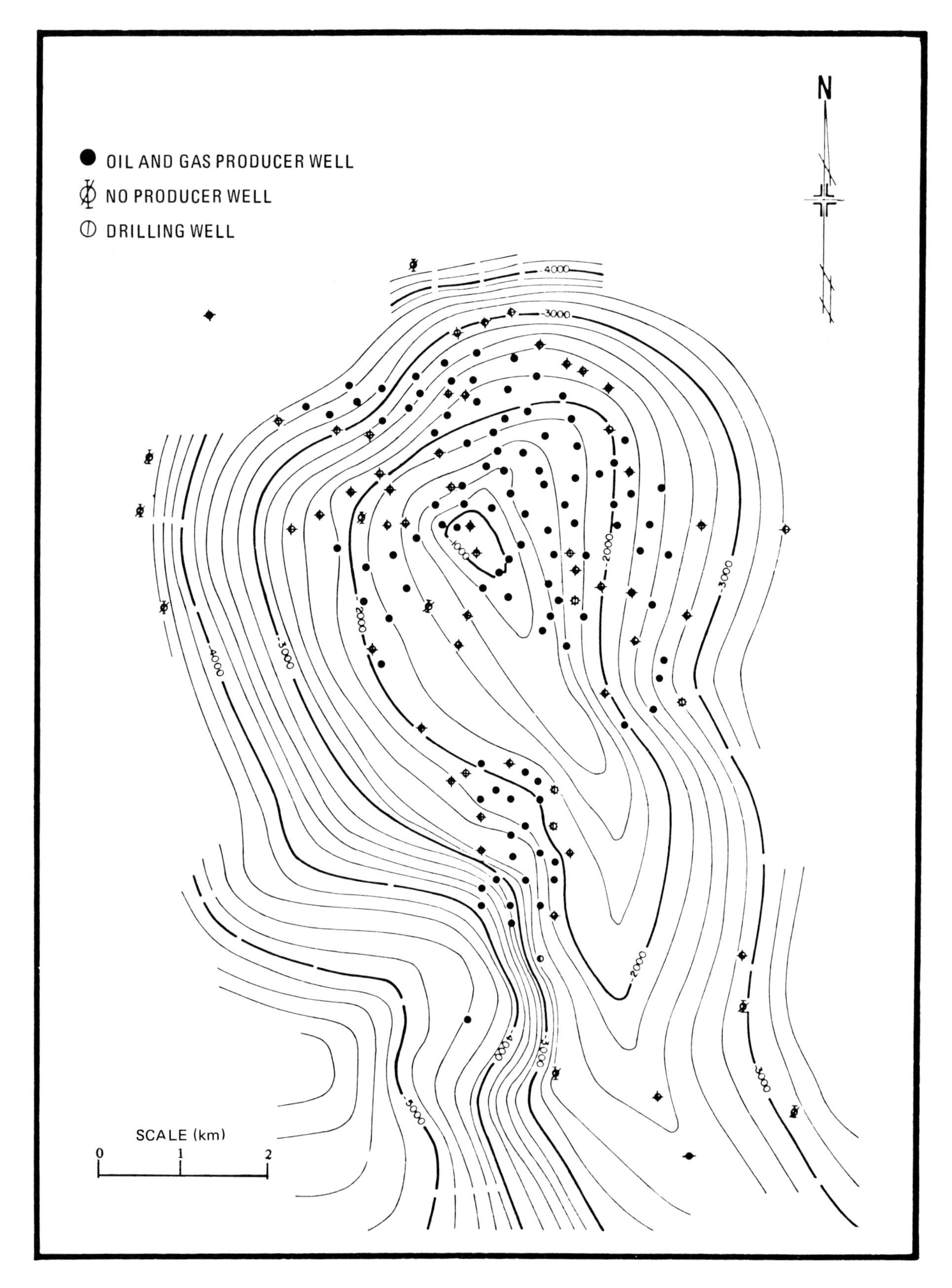

FIG. 9—Structure map on the top of the anhydrite-salt in the El Plan field.

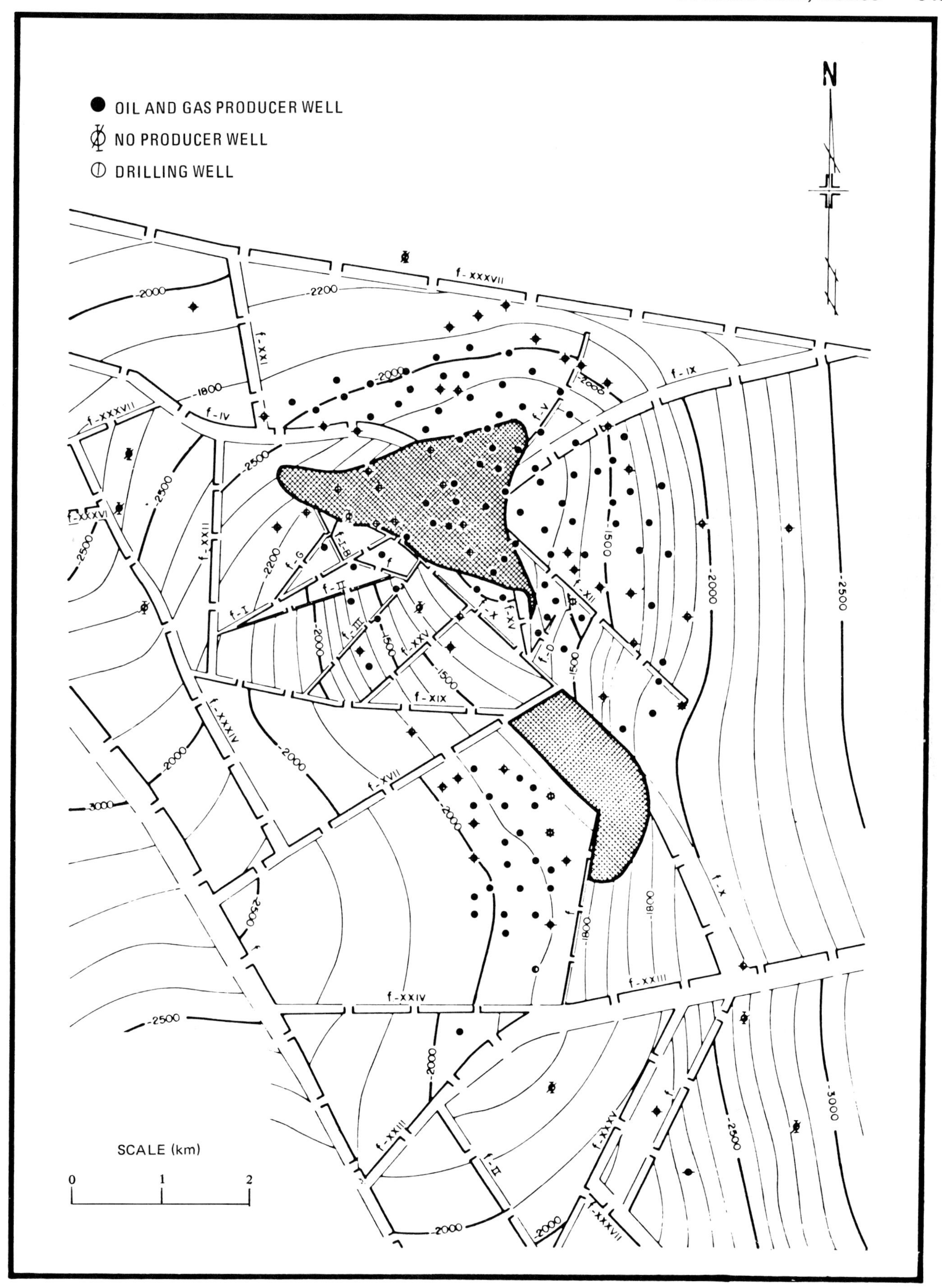

FIG. 10—Structure map of the top of the Miocene Encanto Formation, El Plan field.

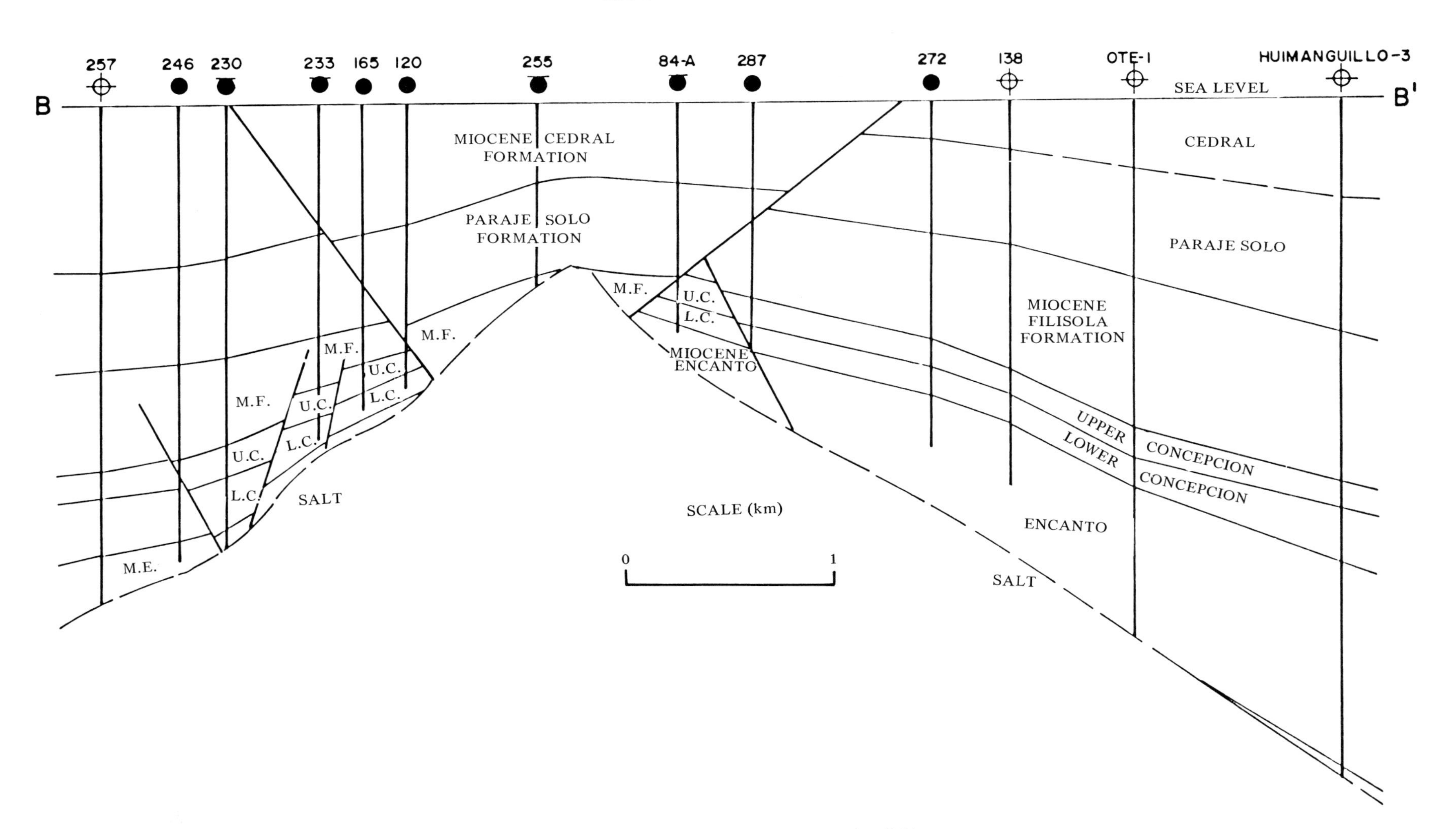

FIG. 11—Geologic cross section through the El Plan field.

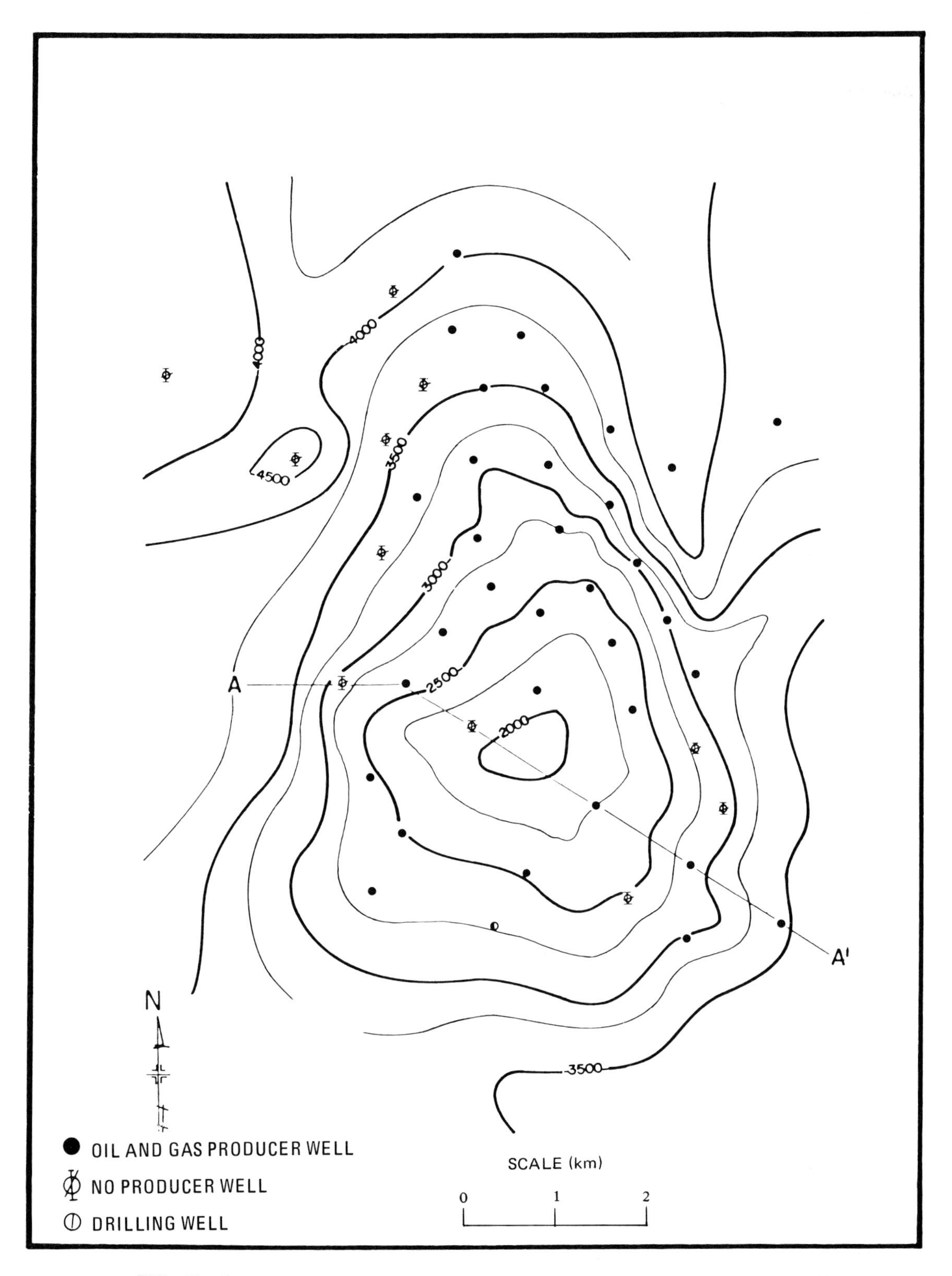

FIG. 12—Structure map of the top of the salt in Encanto Formation, Cinco Presidentes field.

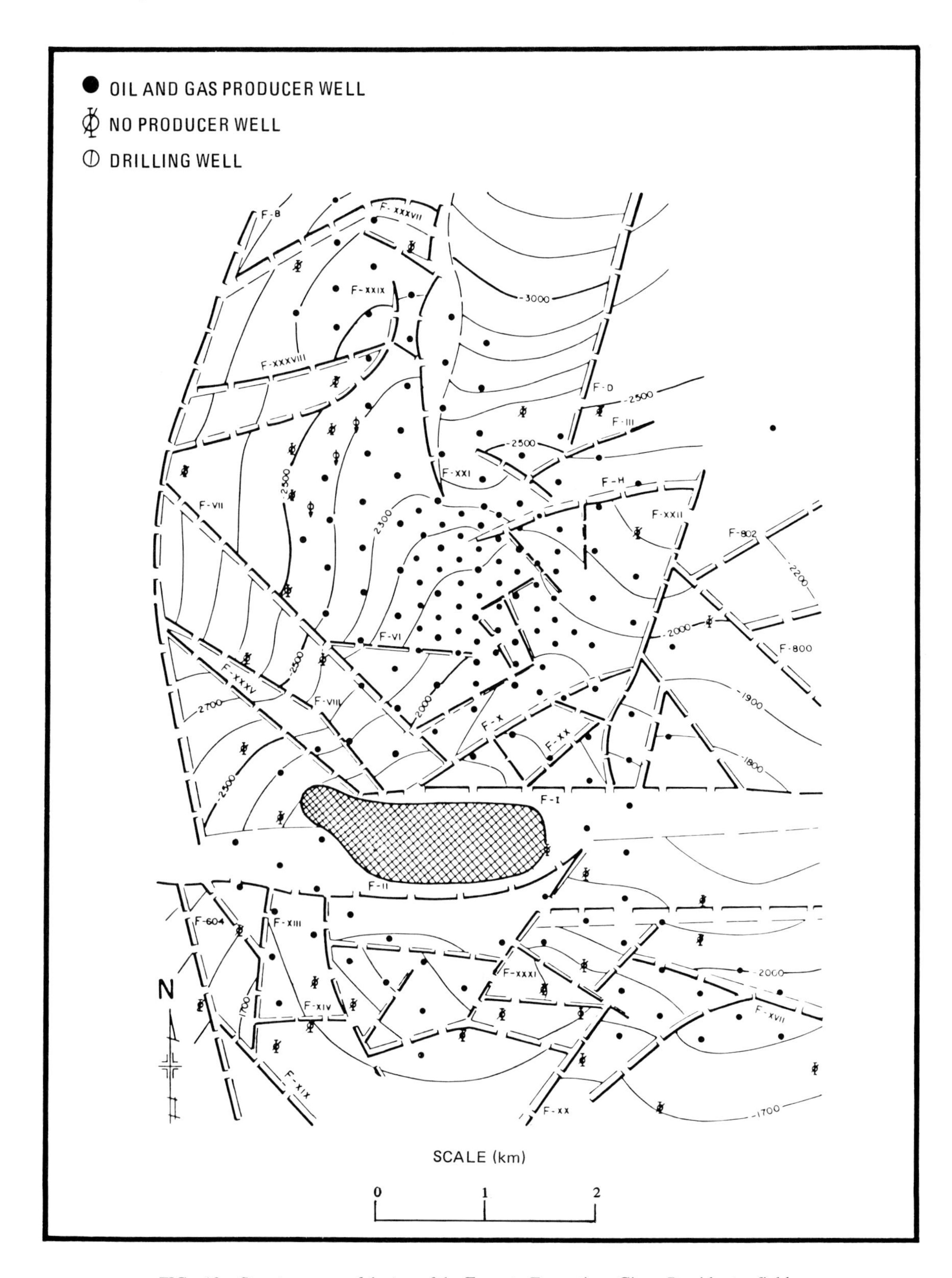

FIG. 13—Structure map of the top of the Encanto Formation, Cinco Presidentes field.

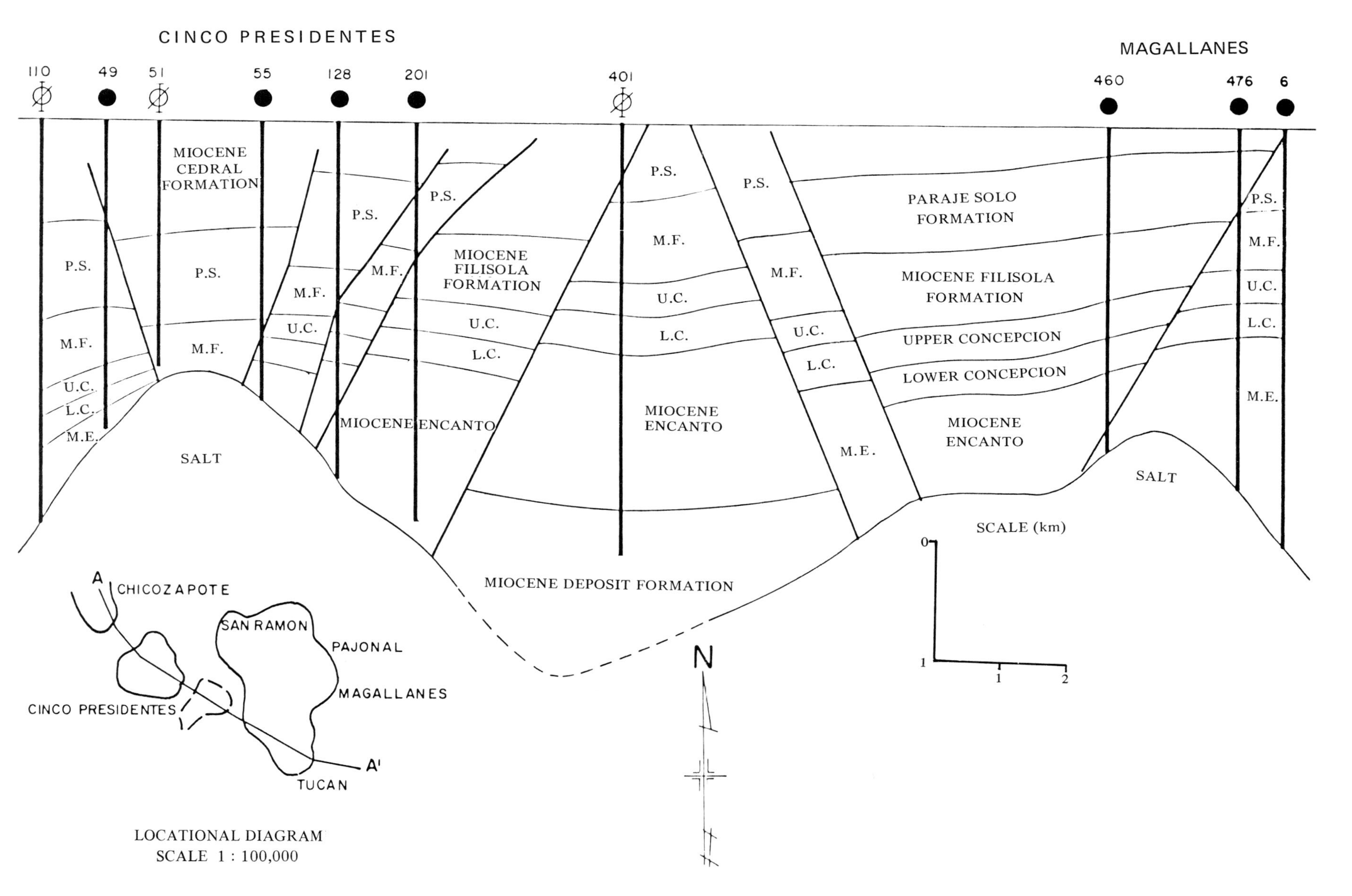

FIG. 14—Geologic cross section through the Cinco Presidentes and Magallanes fields.

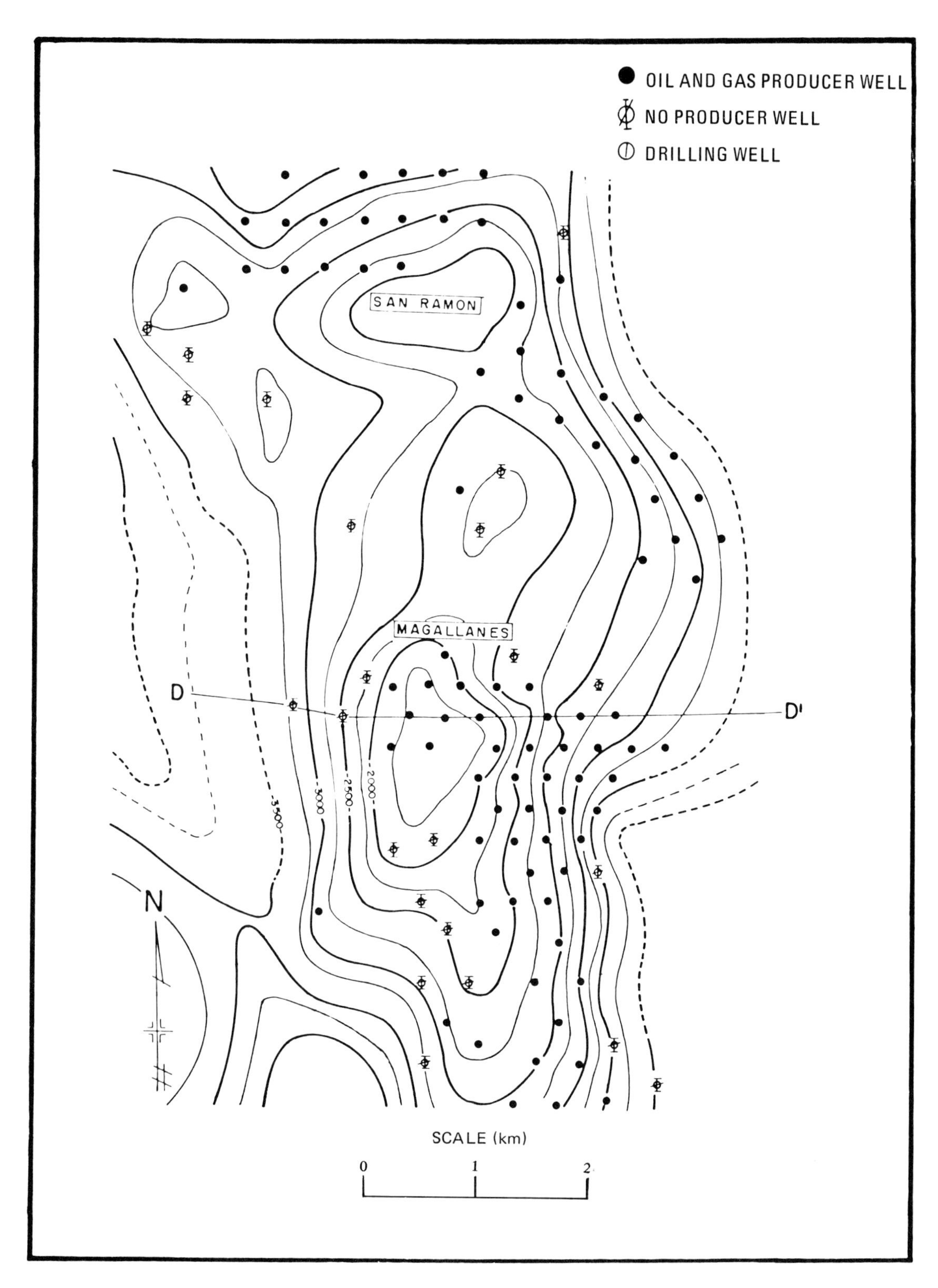

FIG. 15—Structure map of the top of the salt, Magallanes field.

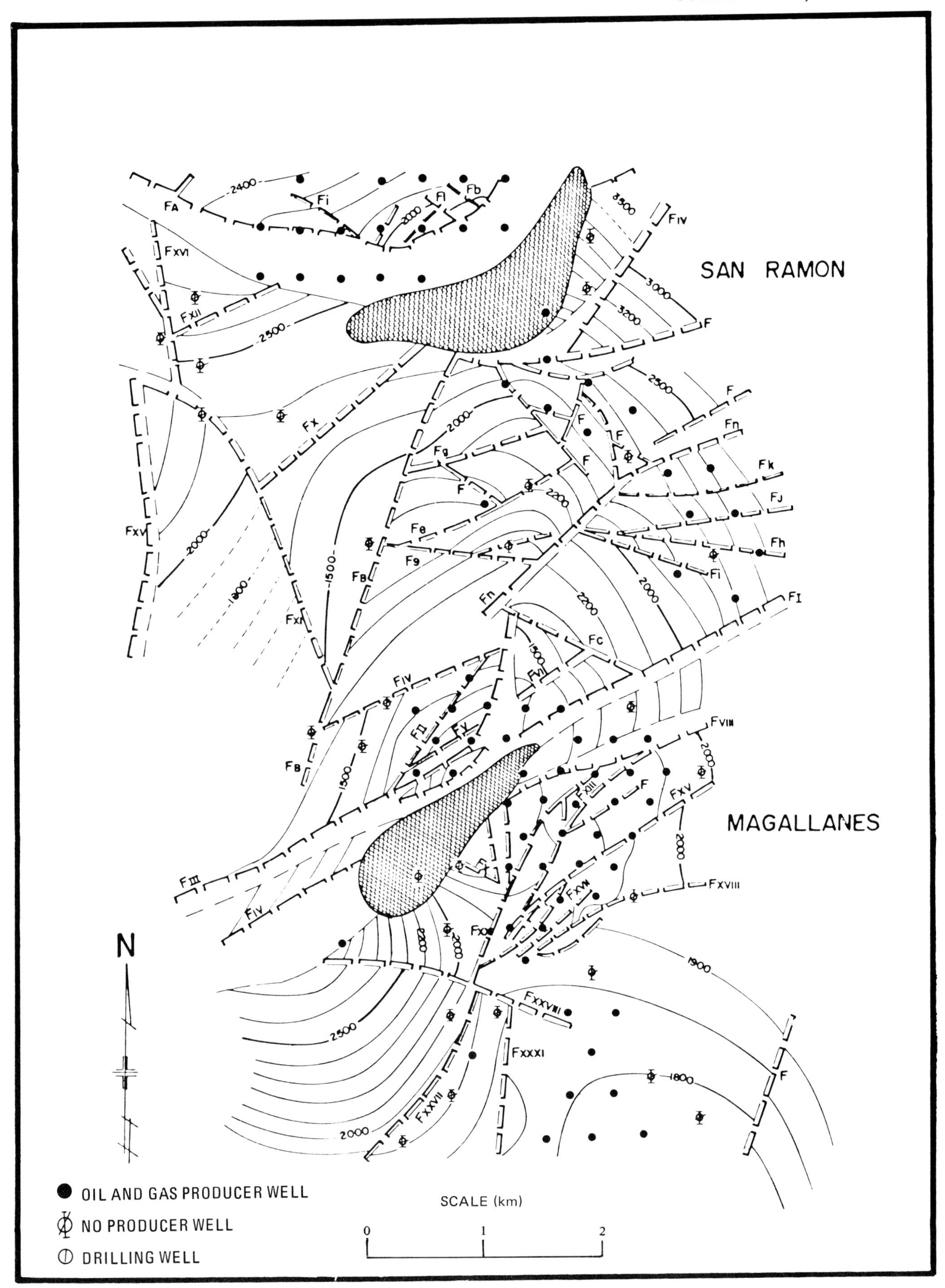

FIG. 16—Structure map of the top of the Encanto Formation, Magallanes field.

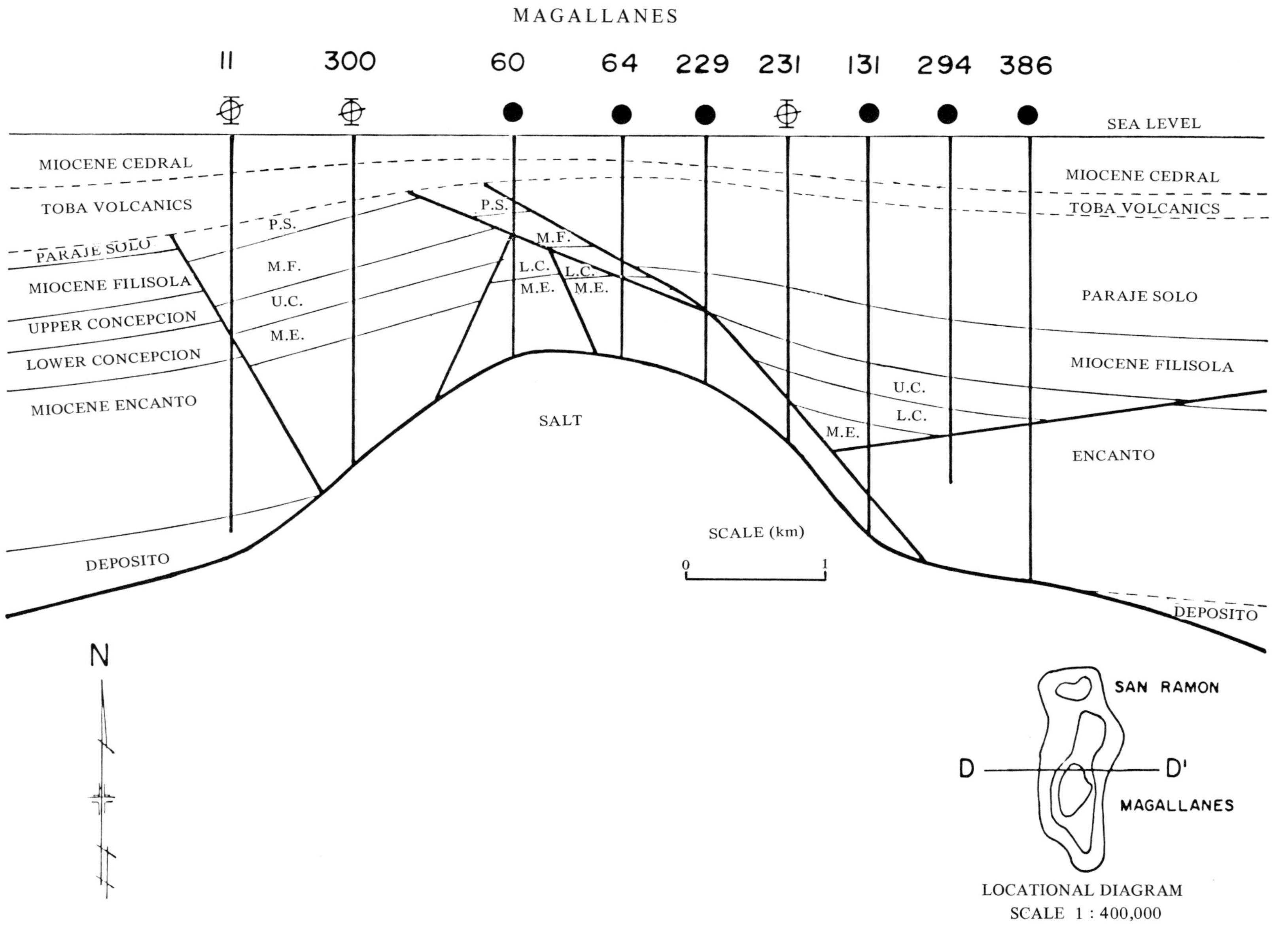

FIG. 17—Geological cross section, Magallanes field.

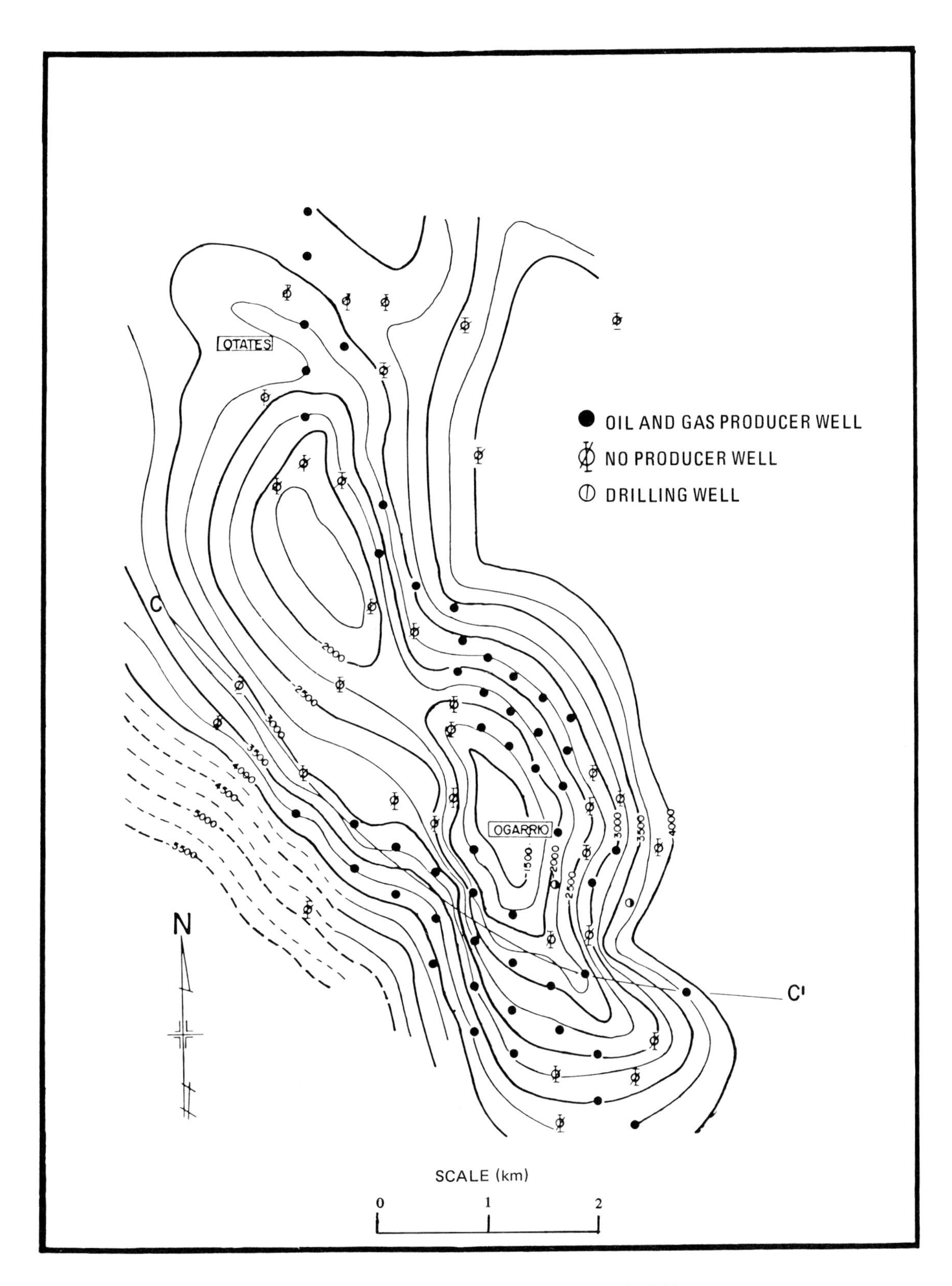

FIG. 18—Structure map of the top of the salt, Ogarrio field.

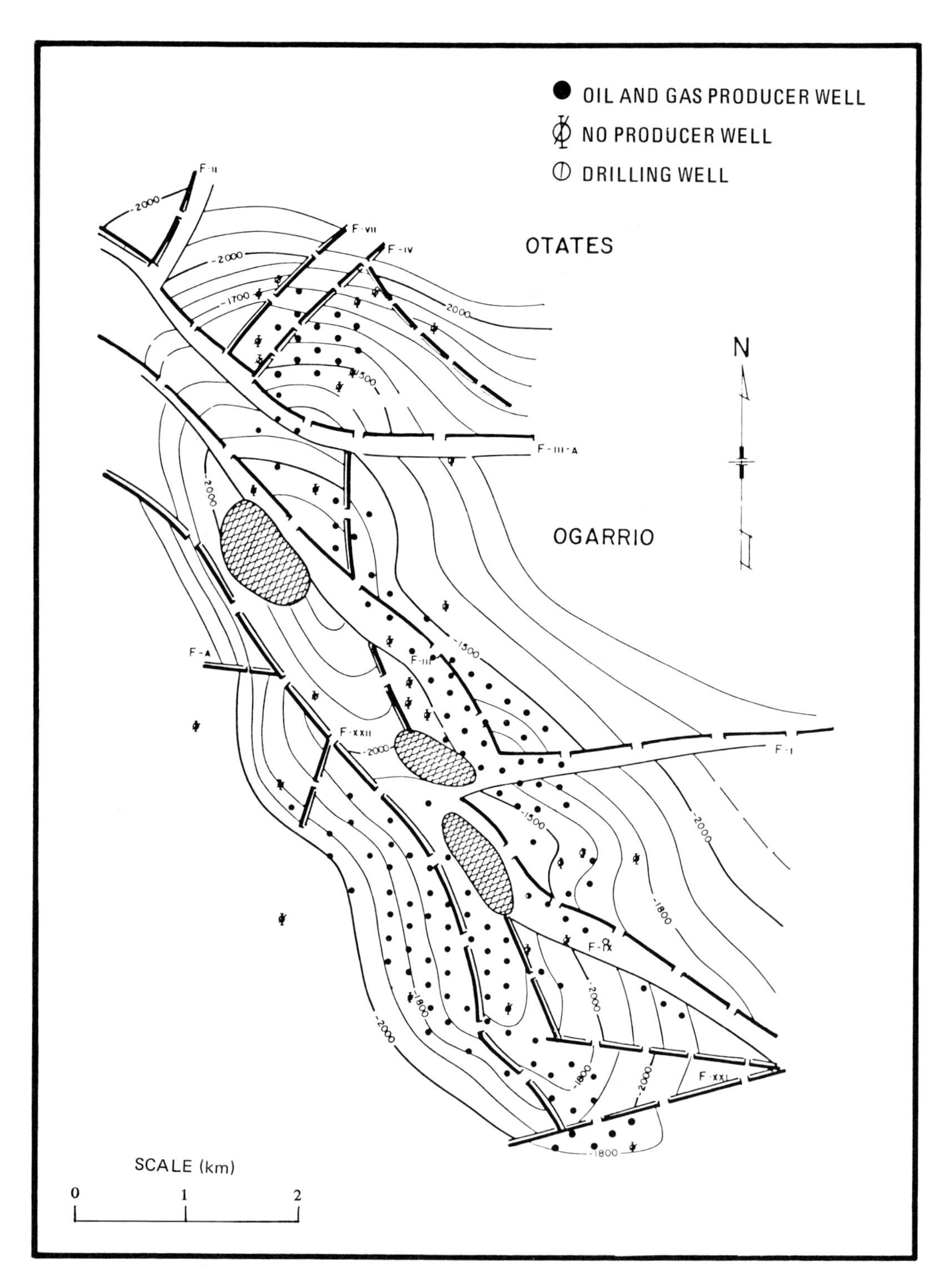

FIG. 19—Structure map of the top of the Encanto Formation, Ogarrio field.

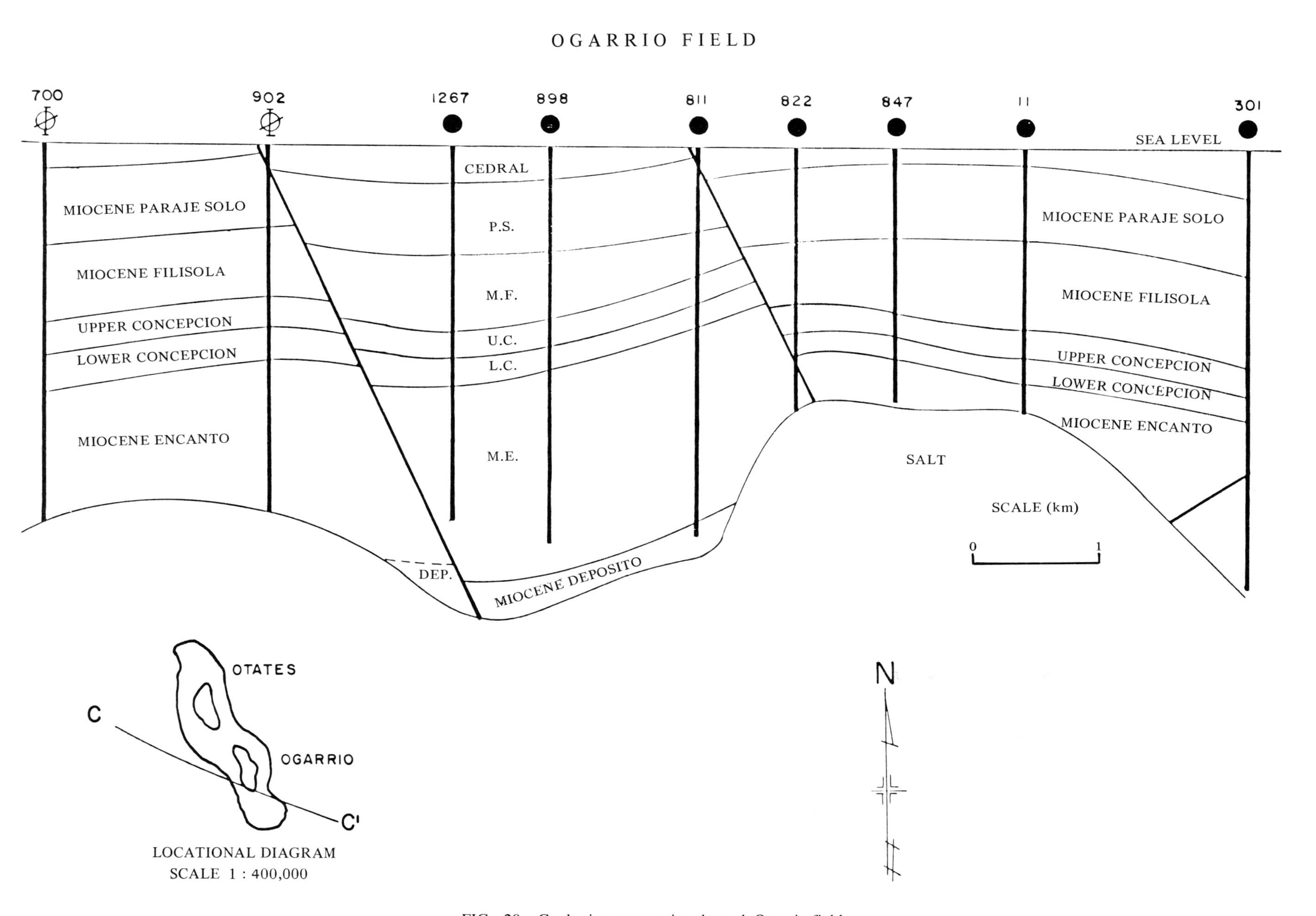

FIG. 20—Geologic cross section through Ogarrio field.

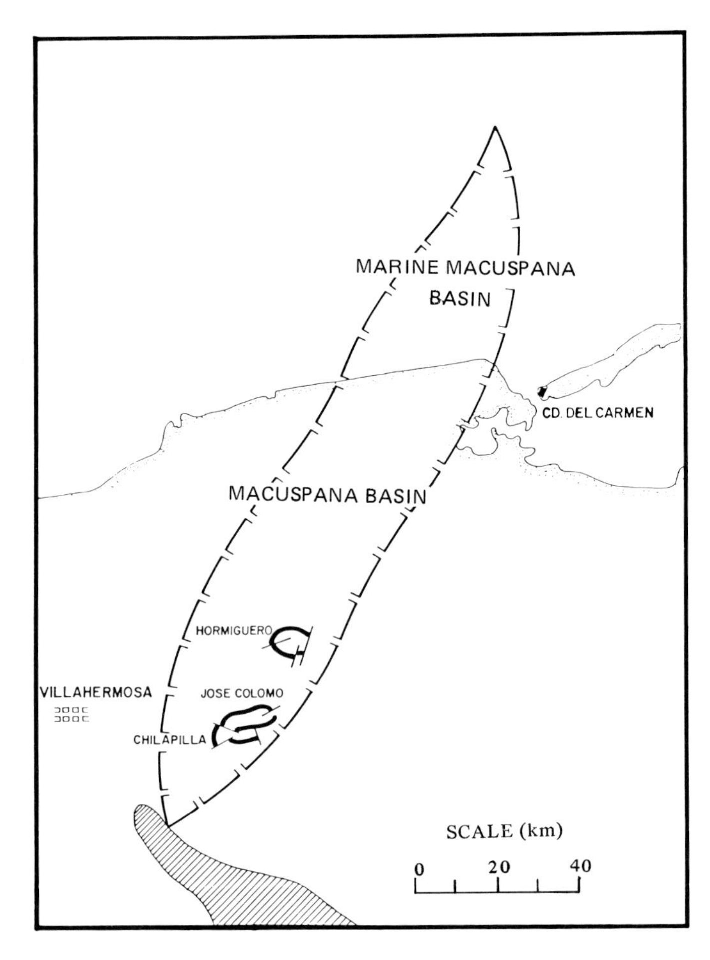

FIG. 21—Location map, Macuspana Basin, Mexico.

1957 from sandstones of the Encanto Formation (lower Miocene). The initial production was 39 cu m/d of oil from three intervals, 1,744 to 1,758 m; 1,766 to 1,769; and 1,787 to 1,791 m. The total depth is 1,825 m.

Stratigraphy

The geologic column is similar to that of the Magallanes and Cinco Presidentes fields. The Encanto Formation is the most productive section with a thickness of about 1,200 m.

Structural Geology

The structure (Figs. 19, 20) is a southeast to northwest trending salt dome 11 km long and 5 km wide. The northeastern flank dips more than the opposite flank. There is a secondary uplift over the northwest corner. The piercement salt caused faulting in the strata and formed two major fault systems: one parallel with the direction of the structure and the other transverse to it. One of the main faults dips to the south with an east to west strike and 400 m of vertical displacement, decreasing upward to 200 m.

Hydrocarbon Accumulation

To date, there have been 406 wells drilled in the field at 200 and 400 m spacing, and 331 have been productive. At the end of 1978, 113 million bbl of oil was produced and an estimated 233 million bbl is recoverable by primary means. The oil density is 0.848 at 37°C (35° API), average porosity of the producing formations is 19%, and permeability is 177 md.

MACUSPANA BASIN

The Macuspana basin (Fig. 21) is located between the Chiapas-Tabasco Mesozoic fields and the Yucatan Platform (Fig. 21). The basin extends northward and joins the Campeche Marine Platform. The southern limit of the basin is the Sierra de Chiapas. Production from this basin is primarily gas from anticline structures in the central and south parts of the basin.

Jose Colomo Field

Initial structures were defined (Fig. 22) using gravity data and subsequent seismic work indicated an east to west elongated anticline coinciding with minimum-gravity anomaly. The discovery well, Chilapa 1, was completed in 1951 and produced gas and condensed oil from a sandstone in the Amate Superior Formation in the upper Miocene. The total depth was 1,700 m. Initial production was 7,874 cu m/d of gas and 40.5 cu m/d of oil.

Stratigraphy

The section penetrated by wells in this field includes the Oligocene, middle Miocene (Amate Inferior), upper Miocene (Amate Superior), Pliocene (Encajonado, Zargazal, and Belem formations), and recent (Fig. 23). The two productive formations, Amate Superior and Amate Inferior, average 800 to 1,000 m in thickness. The Amate Superior formation consists of several thick sandstones such as the "D Sand" which is 300 m thick.

Structural Geology

The structure is an east to west oriented anticline 7 km long and 3 km wide (Fig. 24). A large north-dipping normal fault with 500 m of vertical displacement marks the southern limit of the structure. Most of the Jose Colomo field is located on the downthrown block of the fault. Another smaller fault affects the southeastern flank of the structure and seems to control water levels in both blocks. This fault dips southward and has a vertical displacement of about 40 m.

Hydrocarbon Accumulation

Well spacing throughout the field is 500 m. Of the 123 wells drilled, 106 are producers. The accumulative gas production for the field is 2.08 Tcf and 25 million bbl of condensate. Average porosity of the productive zones is 18 to 31%, permeability is 50 md, and the oil density is 0.761 at 20°C (55° API). Some (small) oil has been produced from the A and C sandstones of the Amate Superior in the central part of the field.

Chilapilla Field

Chilapilla field (Fig. 22) is on a westward continuation of

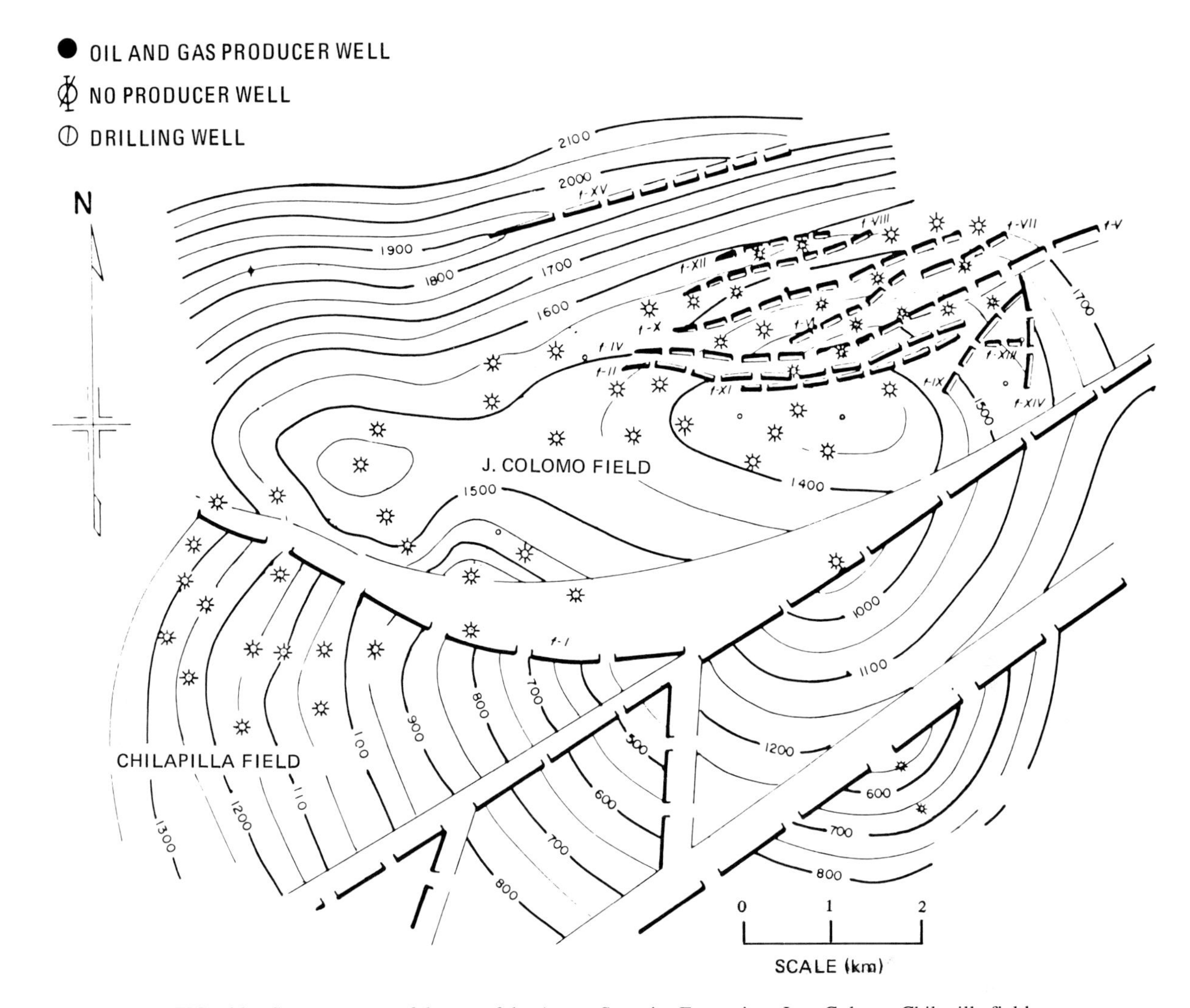

FIG. 22—Structure map of the top of the Amate Superior Formation, Jose Colomo-Chilapilla field

the structure of the Jose Colomo field but with slightly different structural and accumulation conditions. Chilapilla 1, the discovery well, was completed in 1956 to a total depth of 2,600 m. It produced gas and condensed oil from the "G Sand" of the Amate Inferior (lower Miocene).

The oil saturated sandstones of Jose Colomo field are the "A", "C", and "D", but at Chilapilla they are completely invaded by saltwater. However, the stratigraphically lower "G", "H", and "I" sandstones are the producing zones at Chilapilla. The main productive zone is about 220 m thick.

Stratigraphy

The geologic column at Chilapilla is the same as Jose Colomo. The main difference is that many of the sandstone units at Chilapilla have a greater shale content than at Jose Colomo Field.

Structural Geology

Chilapilla structure is an anticlinal nose cut by a large normal fault with a vertical displacement of 500 to 550 m.

Hydrocarbon Accumulation

Of the 59 wells drilled in the field, 53 are producers. The well spacing is 500 m and accumulative gas and condensate production to date is 757 Bcf of gas. It is estimated that an additional 1 Tcf is recoverable. The condensate oil density is 0.763 at 20°C (56° API). The average porosity of the producing zone is 18 to 31%, and the average permeability is 50 md.

Hormiguero Field

The Hormiguero structure (Fig. 25) is an east to west anticline 5 km long and 3 km wide. The discovery well, Hormiguero 1, was completed at a total depth of 3,000 m. The well produced from the Amate Superior (upper Miocene) sandstones and initial production was 42,586 cu m/d of gas and 14.4 cu m/d of condensed oil.

Stratigraphy

The typical geologic section at Hormiguero field is middle Miocene (Amate Inferior), upper Miocene (Amate Superior),

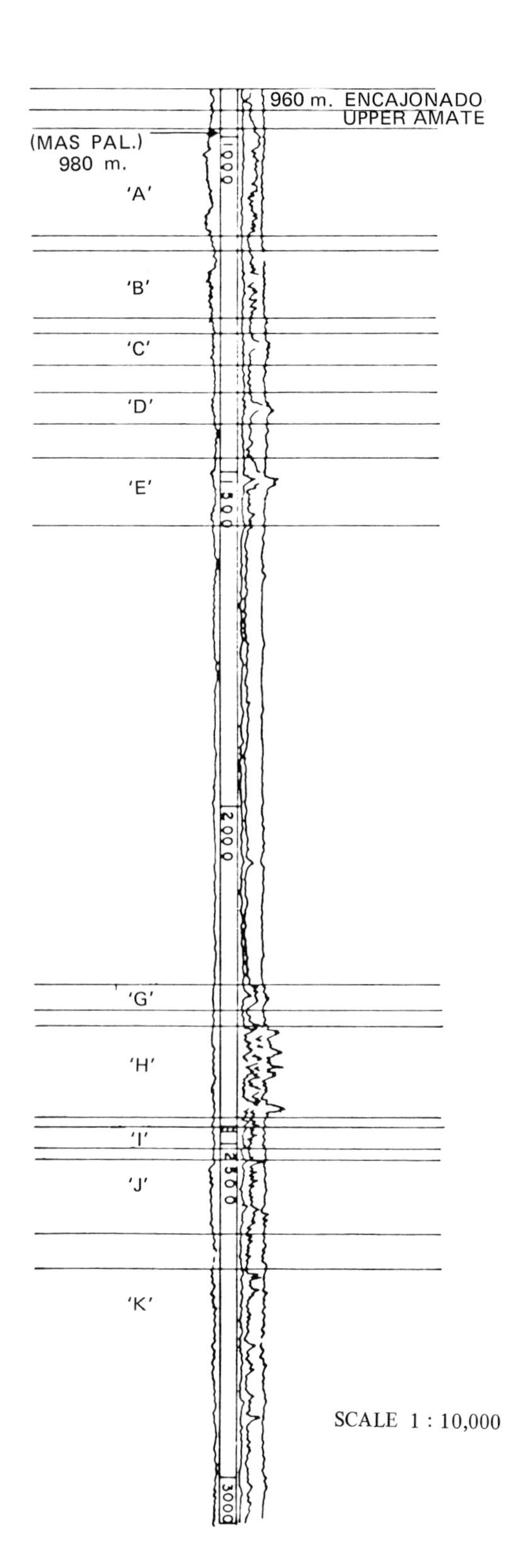

FIG. 23—Typical geologic column of the Macuspana basin.

Pliocene (Encajonado and Zargazal) and Recent. The most productive formation is the Amate Superior.

Structural Geology

The structure is an east to west anticline and an associated normal fault almost parallel with the major anticlinal axis (Fig. 26). The vertical displacement of the fault is 200 to 500 m. The field is on the upper block.

Hydrocarbon Accumulation

Of the 51 wells drilled, 46 were successful for a cumulative production of 450 Bcf of gas. An additional recoverable reserve of about 1 Tcf is estimated. The wells are set at 800-m and 400-m spacing in two blocks with 17 wells. The average porosity of the producing zones is 26% and average permeability is 46 to 56 md.

CHIAPAS-TABASCO MESOZOIC AREA

This area covers 7,500 sq km and is located between the Isthmus Saline basin and the Macuspana basin (Fig. 27). Its southern boundary is the Sierra de Chiapas and its northern limit is the Gulf Coast. In 1960, oil and gas was discovered in Cretaceous carbonate rocks in the Cerro Nanchital Structure, about 120 km southwest of the Antonio J. Bermudez Complex. Production is from dolomite of Early Cretaceous age; however, because the production in the 7 wells drilled was marginal and refinery localities were not nearby, the wells were temporarily plugged.

Because of the success at Cerro Nanchital, additional geologic and seismic work was done on the Gulf Coastal Plain. In 1969, the Jalupa 3 well was drilled 8 km north of the Antonio J. Bermudez Complex. This well penetrated Cretaceous sediments and created additional interest in the area. Shortly thereafter, based on seismic studies, the Cactus 1 and the Sitio Grande 1 wildcats were drilled in the Reforma area. Both wells were oil and gas producers from Upper Cretaceous sediments in the Cactus 1, and middle Cretaceous sediments in the Sitio Grande 1. These successes were the first in a series made in the Reforma area, making it the most important productive province in Mexico.

Mesozoic Area—General

Middle Cretaceous sediments of this area were deposited from south to north (Fig. 28, i.e., from the south to an imaginary line passing north of the Mundo Nuevo structure, south of Sitio Grande structure, and terminating at Agave field). Directly northward is a narrow belt of carbonates containing rudists of mid-Cretaceous age (platform border environment). North of this belt, open-water carbonate sediments (dolomitic limestones and dolomites) were deposited. The dolomites commonly contain clastic fragments derived from the northwest. Along a line from Cardenas on the northwest, Jalupa in the center, and Caparroso on the northeast, a facies change occurs abruptly, changing to shaly limestones without dolomite.

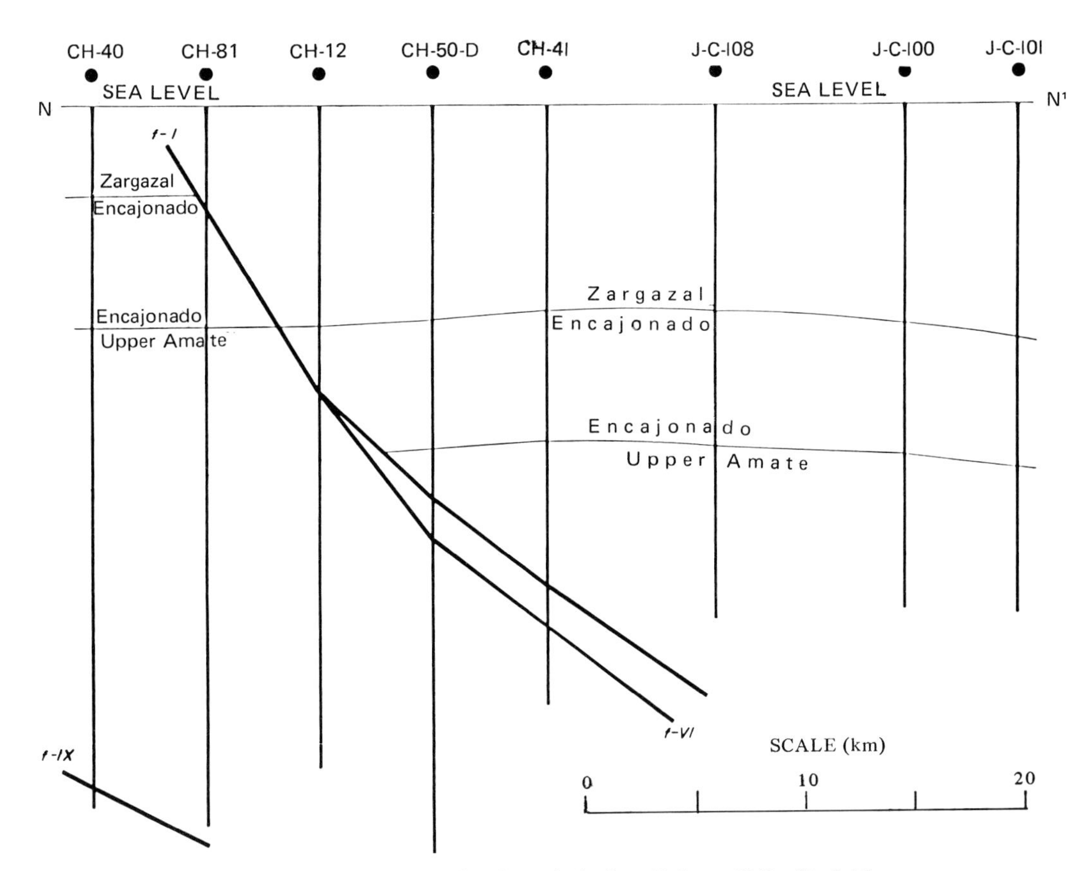

FIG. 24—Geological cross section through the Jose Colomo-Chilapilla fields.

Stratigraphy

Jurassic—The Jurassic (Fig. 29) ranges in thickness from 800 to 1,660 m. The lower part (± 600 m) represents inner platform deposits consisting of pellets, oolites, and bioclastic packestone which are dark gray and partly dolomitized. The packestones alternate with anhydrite and mudstone beds. Abundant ostracods, valves, and other shell fragments are present. The upper part of the Jurassic section (± 500 m) consists of dark brown to light gray argillaceous mudstone.

Lower Cretaceous—The thickness of the Lower Cretaceous ranges from 200 to 600 m. The Lower Cretaceous conformably overlies the Jurassic and consists of dolomitic limestones and dolomites with some bands of dark brown chert.

Middle Cretaceous—At the Sitio Grande field, the middle Cretaceous ranges in thickness from 70 to 300 m with the thickest section in the southern part of the field. The Central part is only 70 m thick because of emergence and erosion. The dominant sediments are dolomites and dolomitic limestones containing dissolution pores 2 cm in diameter. The dolomite and dolomitic limestones alternate with dolomitic breccias which also have solution porosity caves and fractures. Black chert interbeds are present throughout the section.

Upper Cretaceous—The description of Upper Cretaceous is based on Sitio Grande field data. The Upper Cretaceous consists of three lithologic units equivalent to Agua Nueva Formation (± 140 m), San Felipe Formation (± 50 to 200 m), and Mendez Formation (± 20 to 200 m) of the North Zone of Petroleos Mexicanos (Tampico-Poza-Rica). All three formations were deposited in an open-marine environment.

Agua Nueva Formation—This formation tends to pinch out to the south and locally disappear (close to Sitio Grande wells 112, 100, 91, and 93). The stratigraphic sequence consists of alternate white, dark brown, and black crystalline and microcrystalline limestones, with green interbedded bentonites and gray and black chert bands. Some dolomitic breccias within dolomitic limestones are present in the formation. These lithologic characteristics differ from those of the Agua Nueva Formation to the north, but because of stratigraphic position the Sitio Grande section is considered a dolomitic member of the lower part of the Agua Nueva.

San Felipe Formation—The San Felipe also tends to be thinner in the south. The deposition of the San Felipe was influenced by the paleomorphology of the Sitio Grande structure and is thicker on the flanks of the structure than on the top. On the southeastern edge of the field it lies directly on middle Cretaceous rocks. The San Felipe consists of argillaceous and bentonitic limestones of thin stratification, microcrystalline texture, light gray to dark gray color with thin interbeds of bentonite shale.

Mendez Formation—As with the other two formations the Mendez tends to be thinner in the south and also thins over the top of the Sitio Grande structure. The Mendez conformably overlies the San Felipe Formation; however, argillaceous and

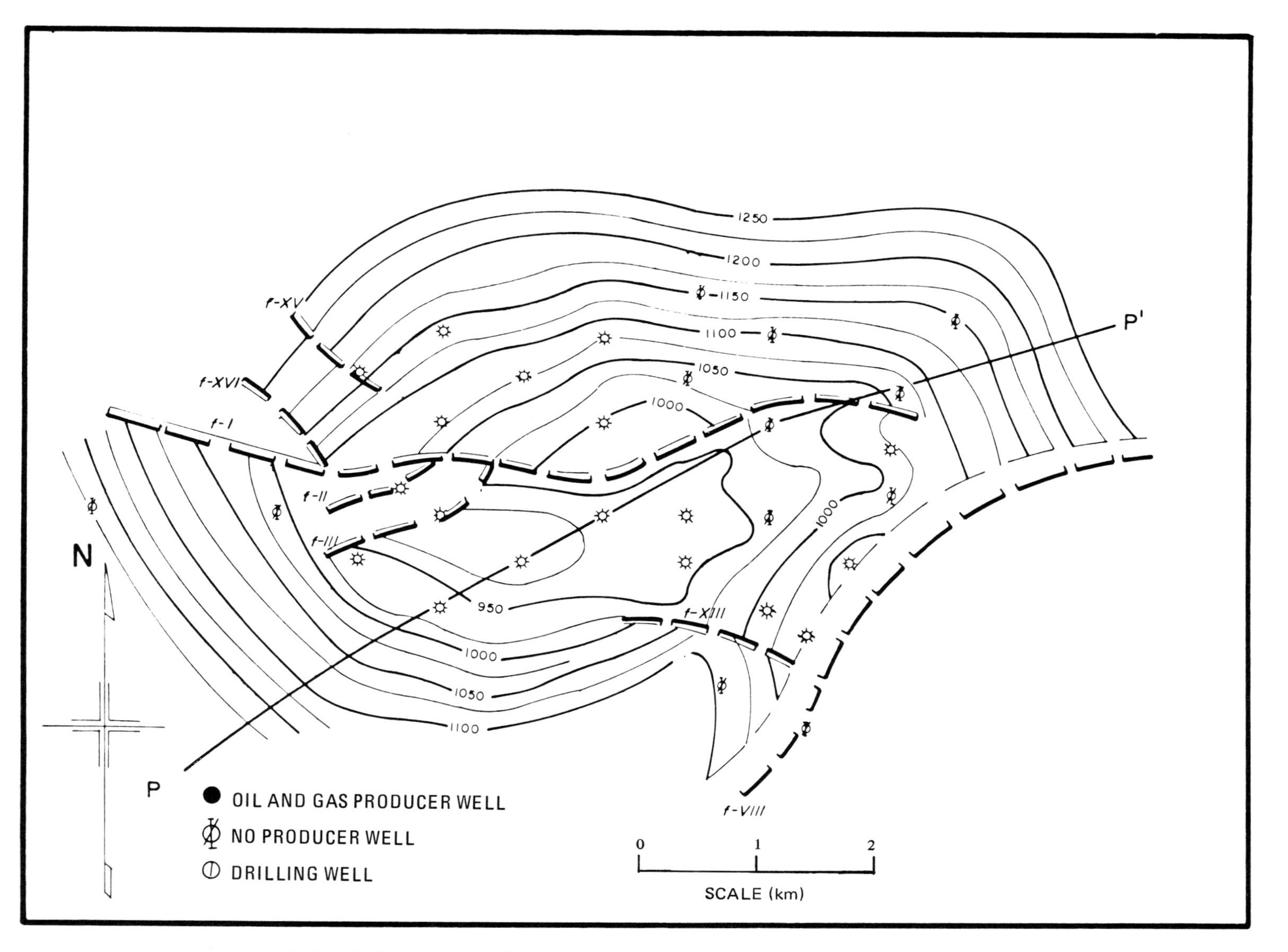

FIG. 25—Structure map of the top of the Amate Superior Formation, Hormiguero field.

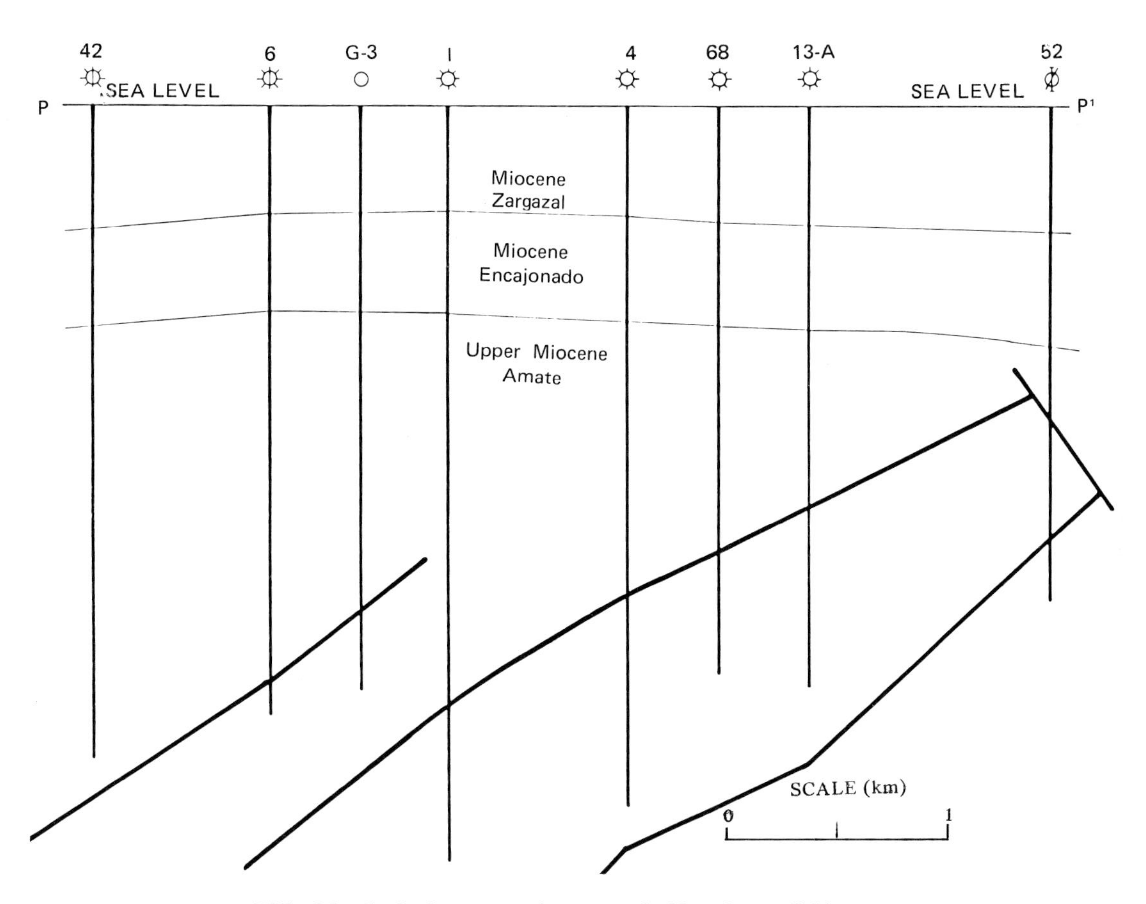

FIG. 26—Geologic cross section across the Hormiguero field.

calcareous sediments in the upper Mendez interfinger with the overlying Paleocene breccias. This is believed to be in response to Laramide tectonic movements. The Mendez is an alternate sequence of dark brown, light gray, and dark gray poorly stratified marls.

Paleocene—The Paleocene generally is uniform in distribution with thicknesses ranging from 400 m in the northern part of the Chiapas-Tabasco region to 300 m in the southern area. It lies in unconformity on the Mendez Formation and consists of limestone breccias in a marl matrix with abundant bioclastic fragments including rudists and microfauna of middle Cretaceous origin. The rest of the Paleocene consists of calcareous shale with a few sandstone lenses and some bodies of lenticular breccias as described before.

Eocene—The Eocene is a marine sequence of thin calcareous shale beds alternating with thin, light gray, fine-grained sand lenses. The average thickness of the Eocene is 800 m, reduced to 400 m over the southern flank of the region due to a regional unconformity.

Oligocene—The Oligocene is a marine section of hard, dark gray shales alternating with thin beds of fine to medium grain, light to dark gray sandstones. Maximum thickness is 450 m but an erosional unconformity has removed the upper part of the section.

Pliocene (Paraje Solo Formation)—The Pliocene is in angular unconformity with the Oligocene and/or Eocene deposits. The maximum thickness of the Pliocene is 3,200 m and consists of greenish-gray and greenish-blue shales, alternating with fine to coarse, light gray sandstones commonly changing to fine to middle grain gravel. There also is present thin, fine grain, calcarious cemented sandstone. These sediments are of inner neritic origin.

Sitio Grande Field

The Sitio Grande field is located in the Gulf Coastal Plain 25 km north of Sierra de Chiapas (Fig. 30). The discovery well, Sitio Grande 1, was completed in 1972 as an oil and gas producer from a middle Cretaceous carbonate section at a total depth of 4,197 m. Initial production from two intervals was 274 cu m/d of oil and 109,000 cu m/d of gas. Oil density is 0.851 at 20°C (35° API).

Stratigraphy

The stratigraphic column of the Sitio Grande field includes rocks from Jurassic (Callovian-Oxfordian) to recent age. Several unconformities are present within the Tertiary and Mesozoic section. Most notable is the one between Pliocene and mid-Cretaceous where argillaceous sandstones of Paraje Solo Formation lie directly over mid-Cretaceous dolomites. In some places, the unconformities are both angular and parallel.

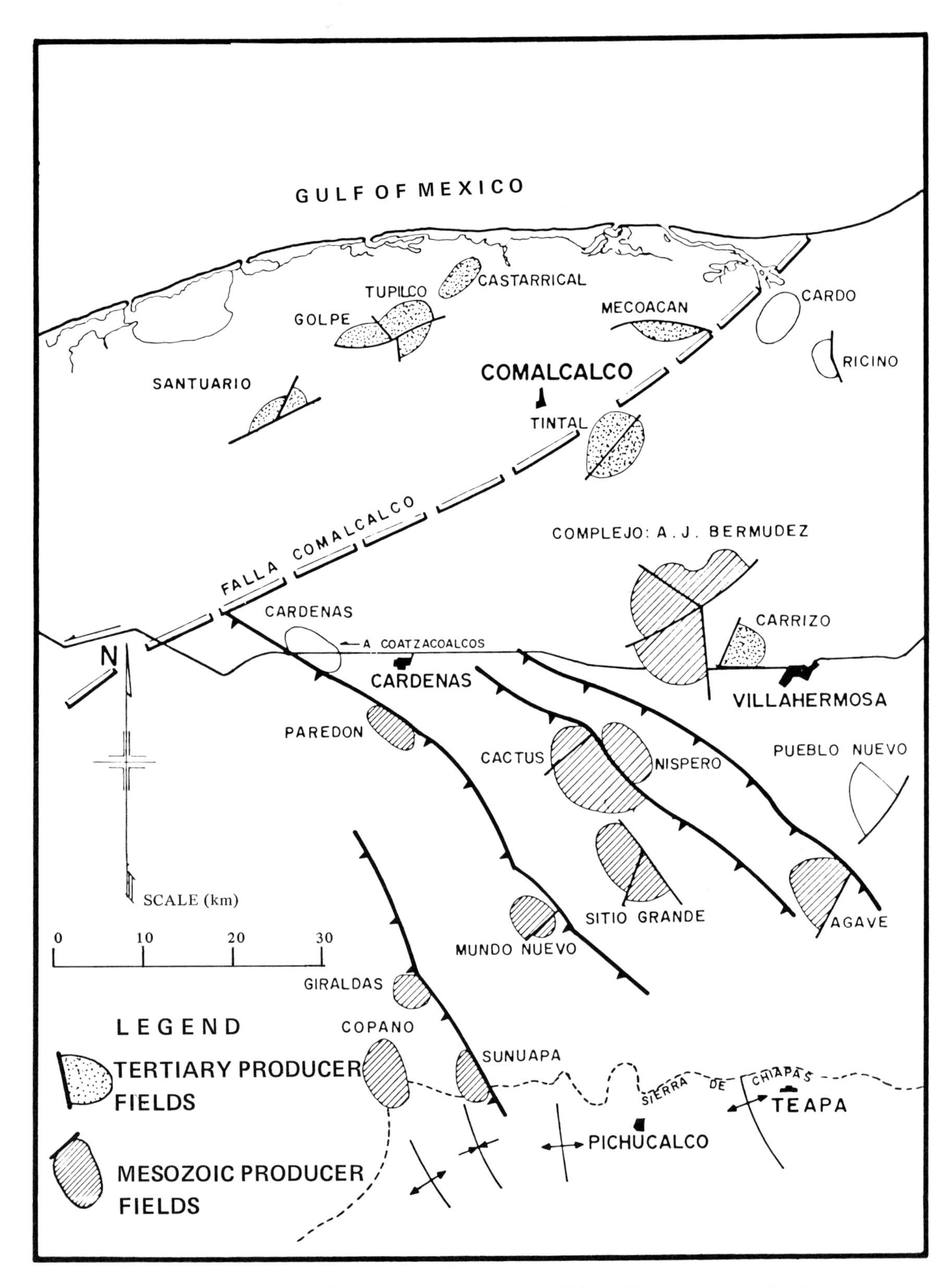

FIG. 27—Location map of major producing fields in the Chiapas-Tabasco Mesozoic Province.

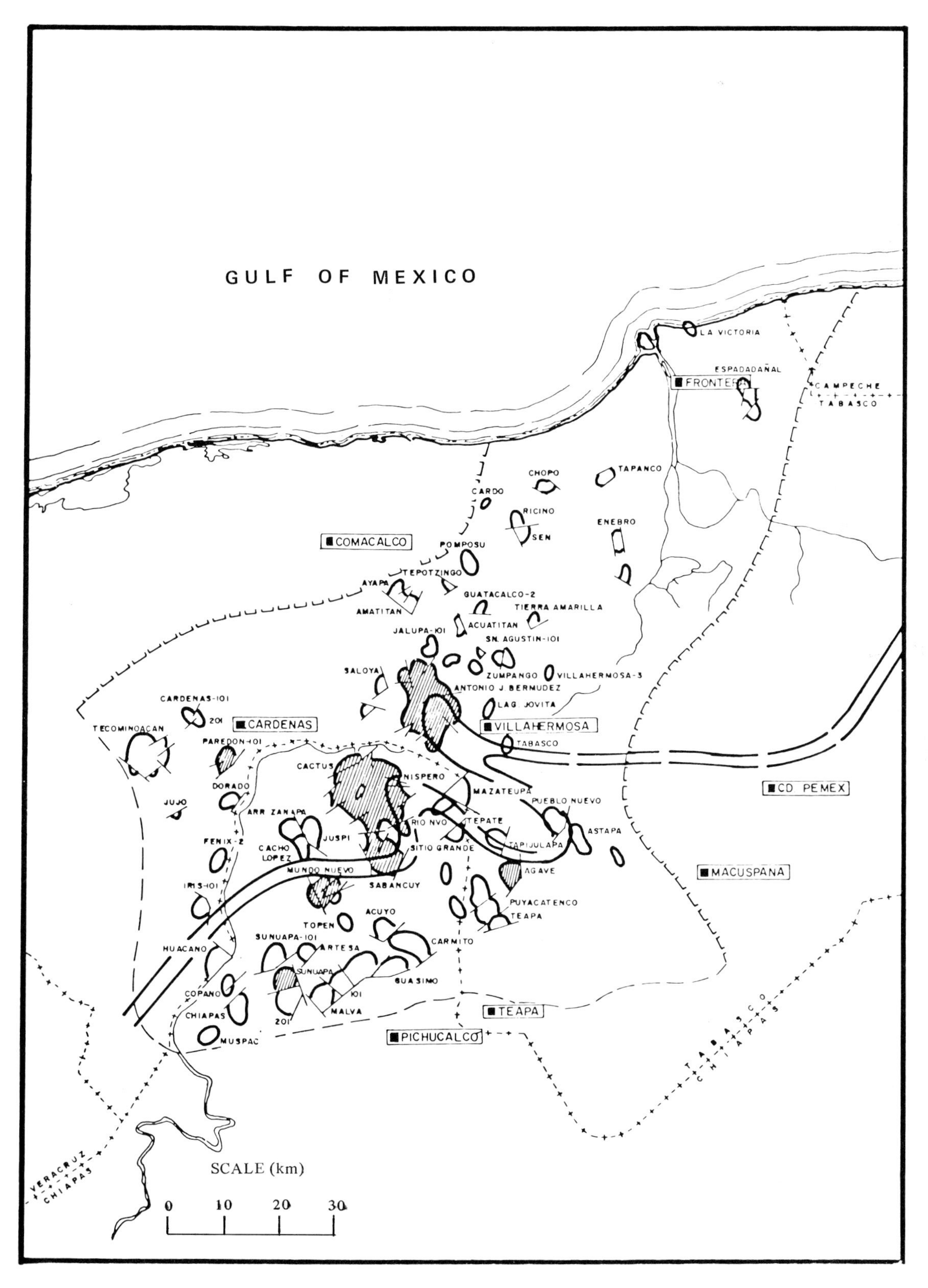

FIG. 28—Middle Cretaceous paleogeographic map of the Chiapas-Tabasco Mesozoic Province.

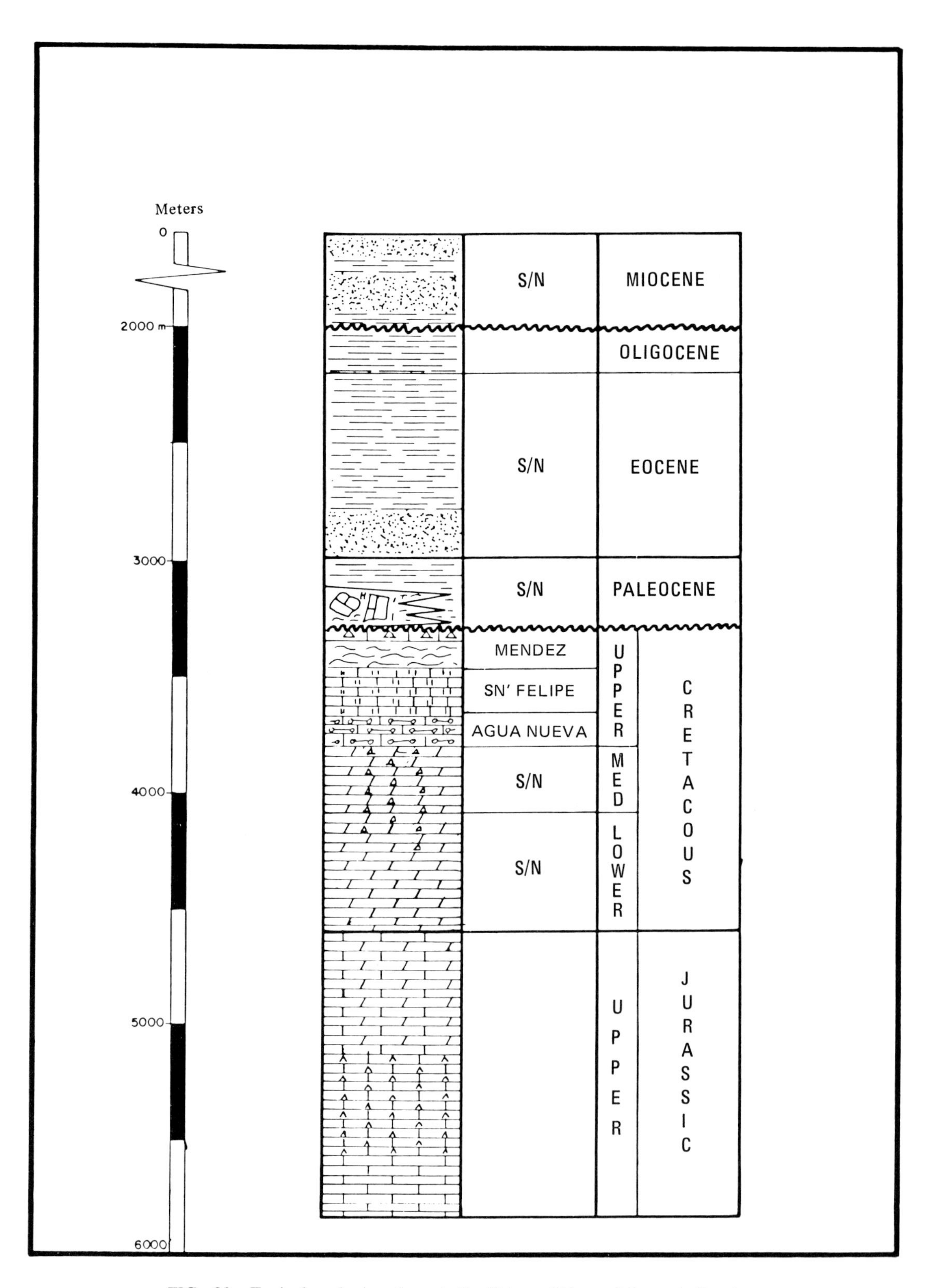

FIG. 29—Typical geologic column in the Chiapas-Tabasco Mesozoic Province.

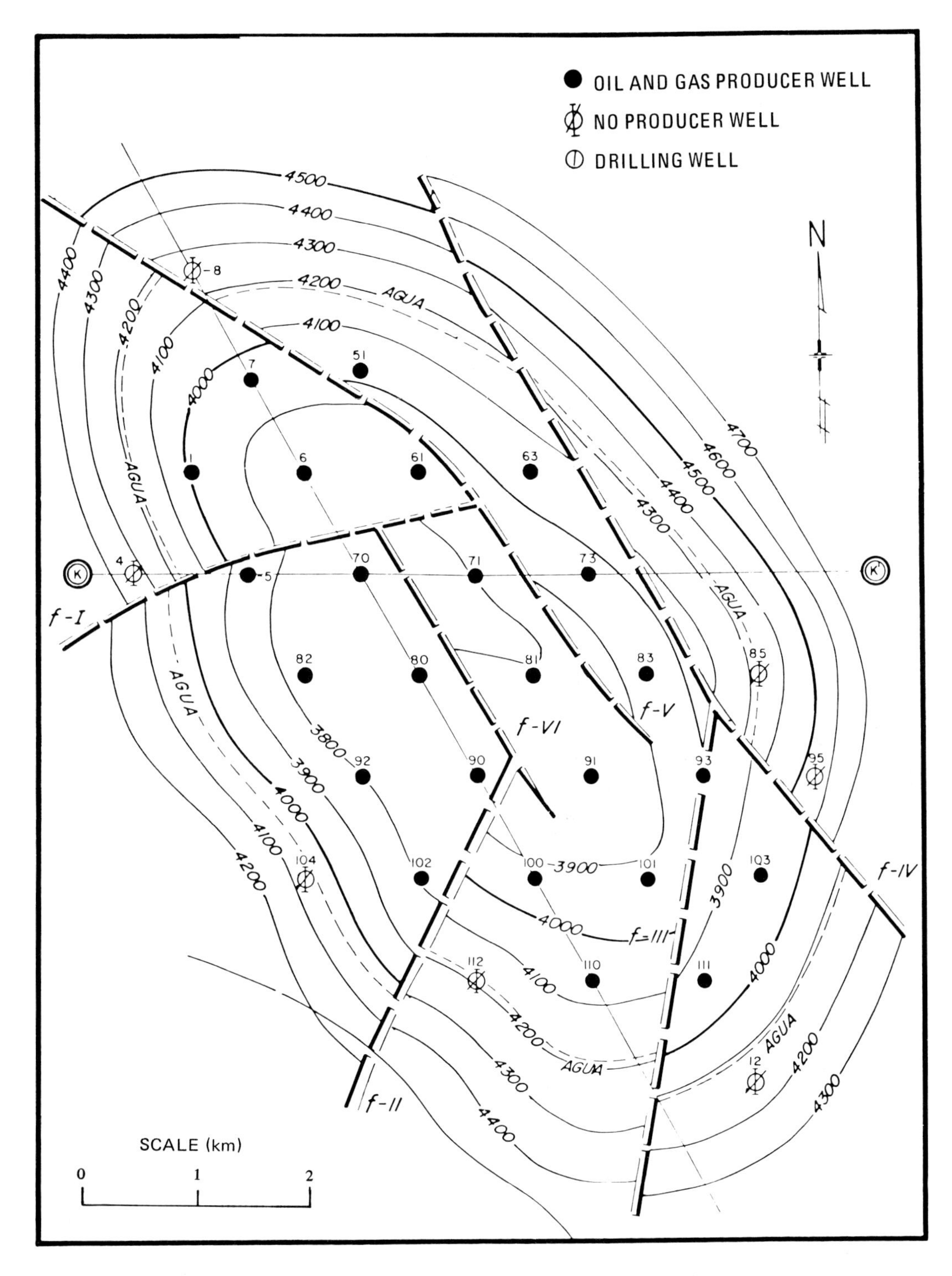

FIG. 30—Structure map of the top of Sitio Grande field.

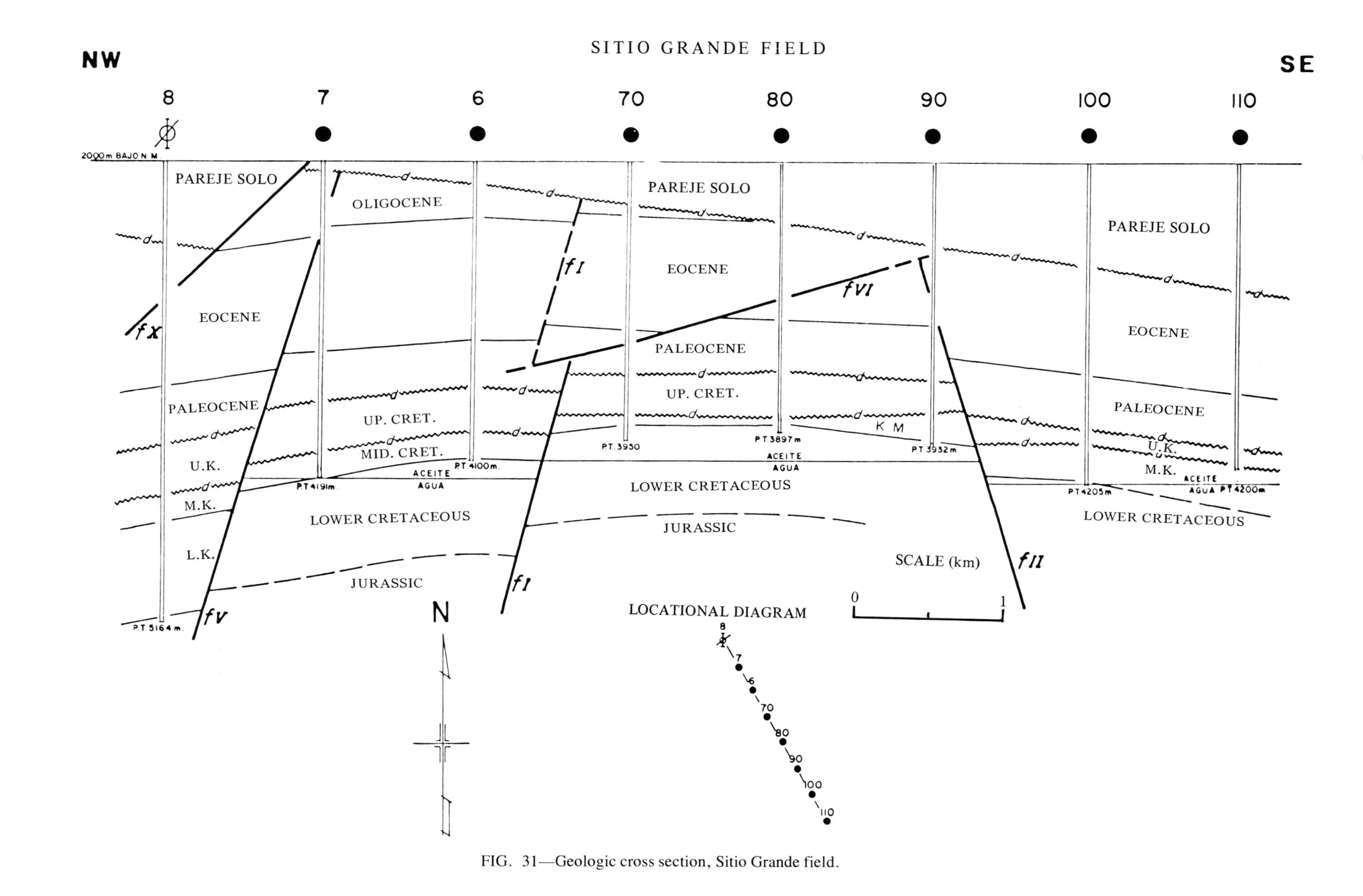

FIG. 31—Geologic cross section, Sitio Grande field.

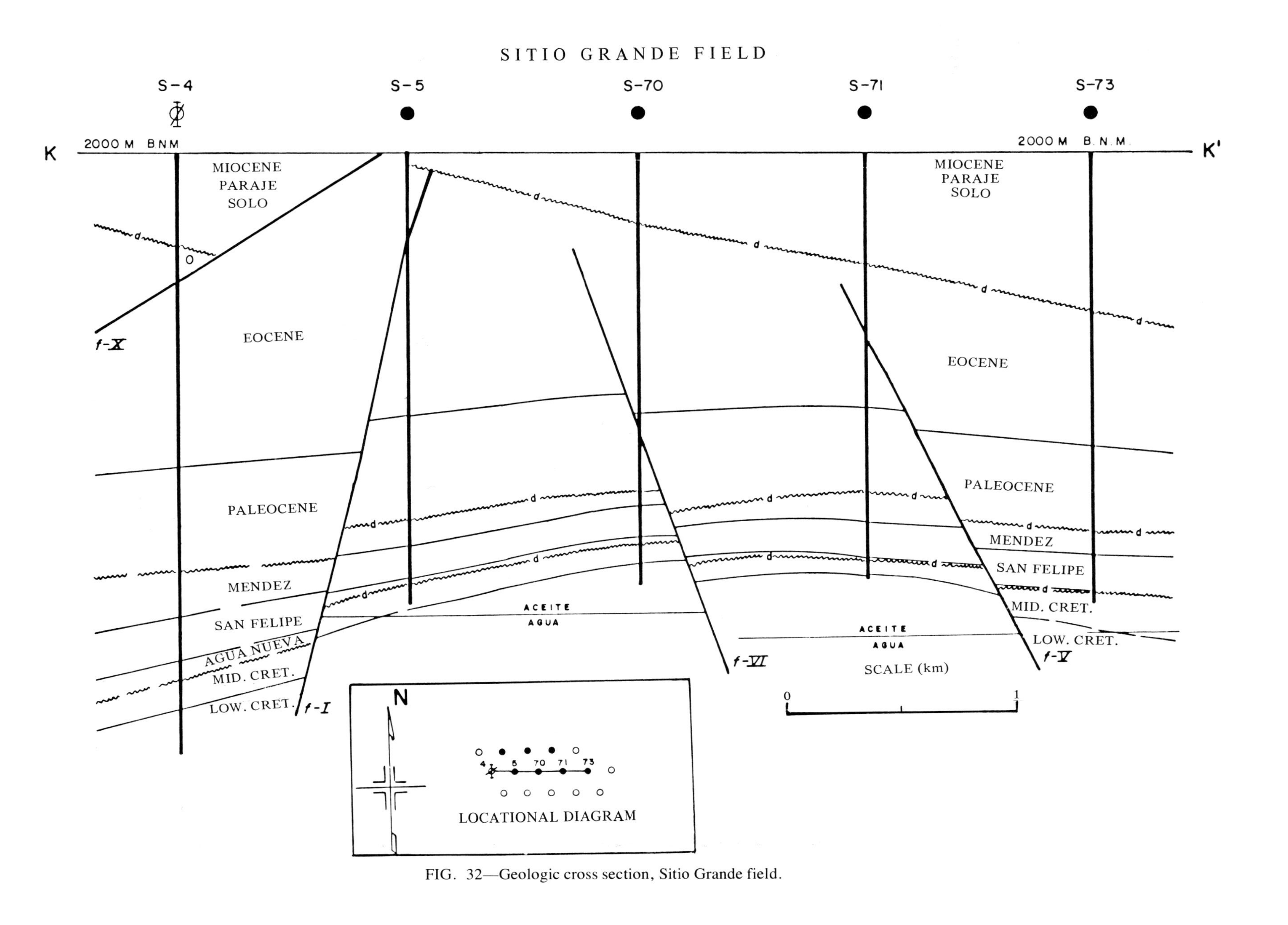

FIG. 32—Geologic cross section, Sitio Grande field.

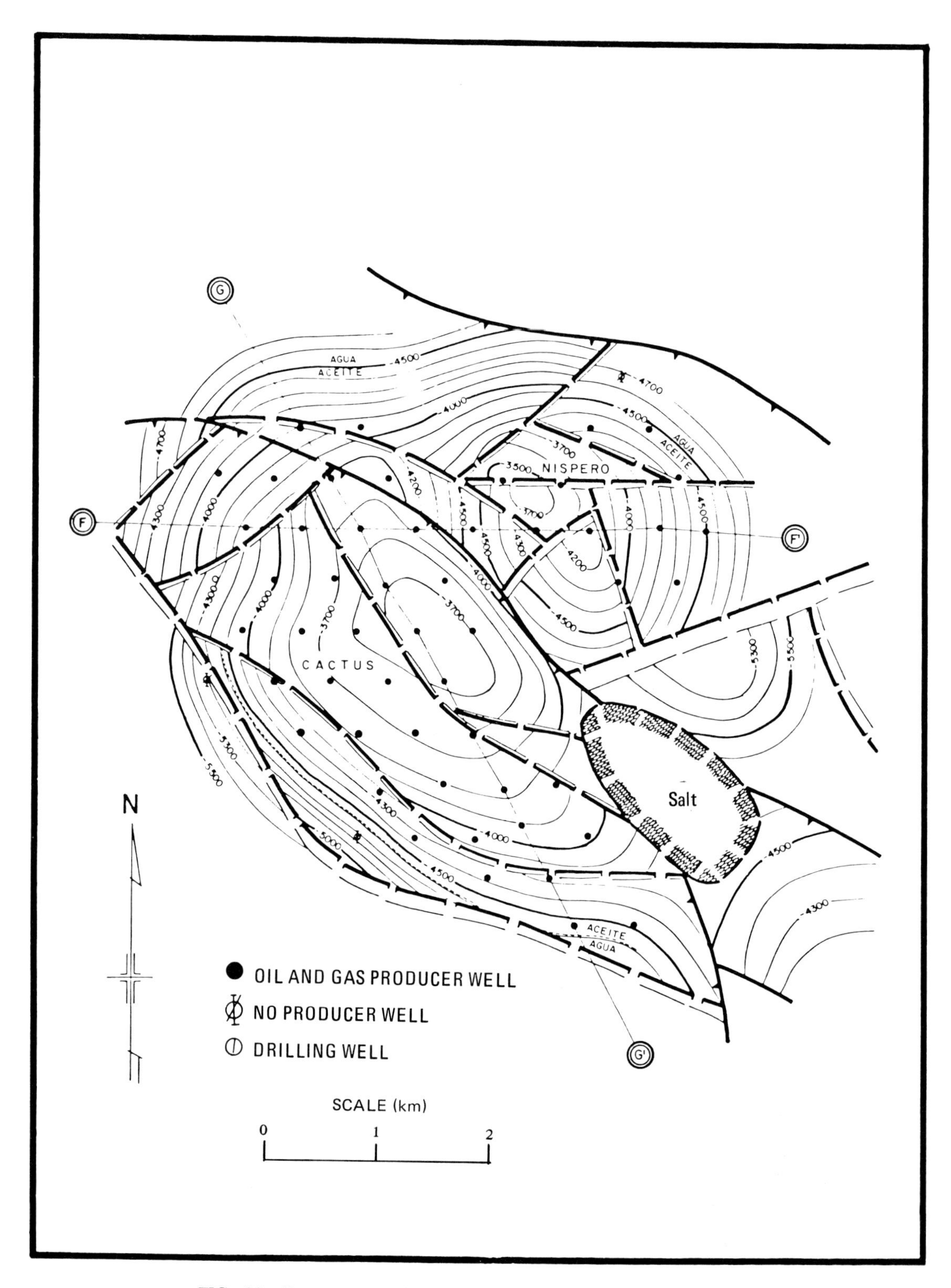

FIG. 33—Structure map of the middle Cretaceous, Cactus-Nispero field.

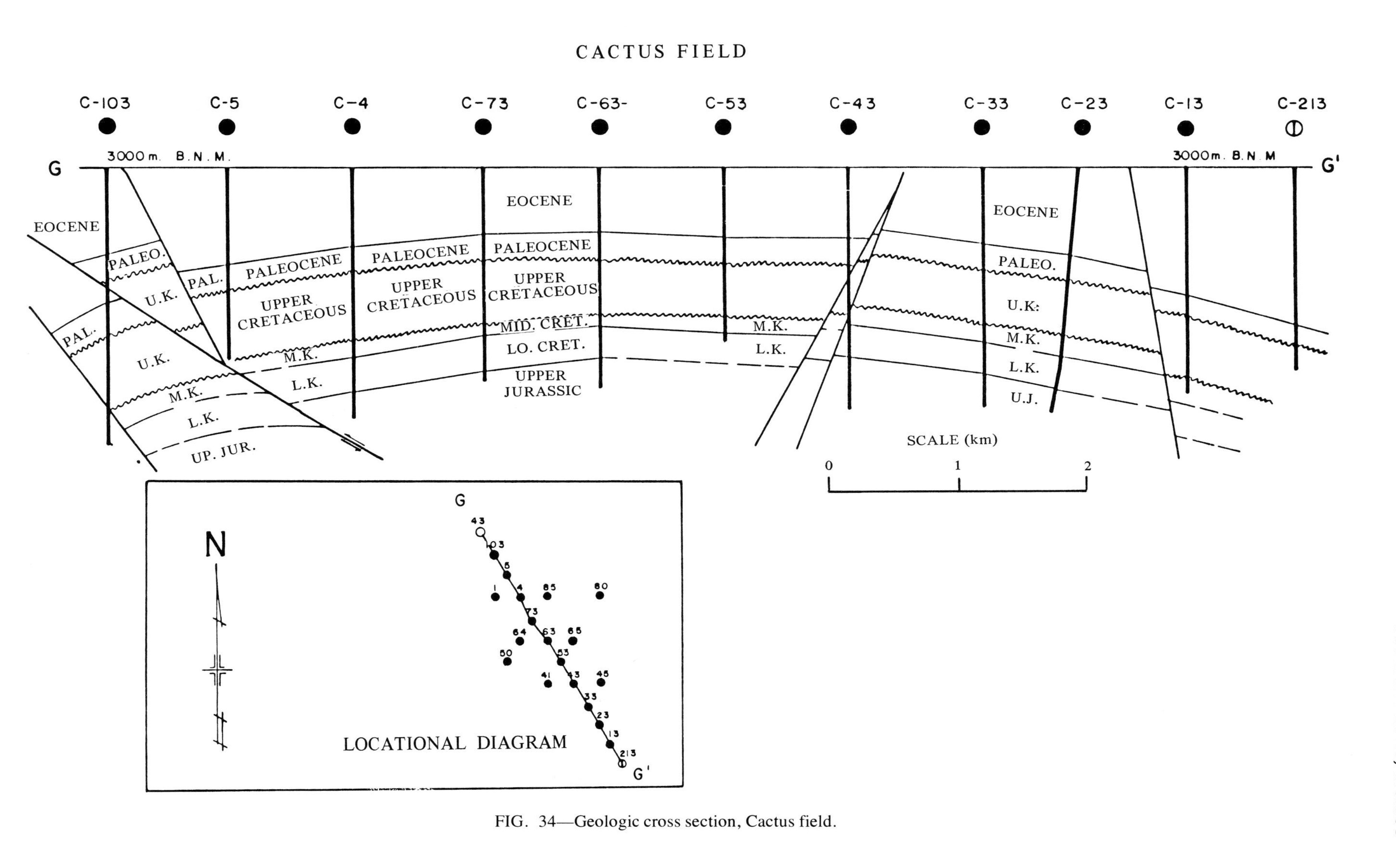

FIG. 34—Geologic cross section, Cactus field.

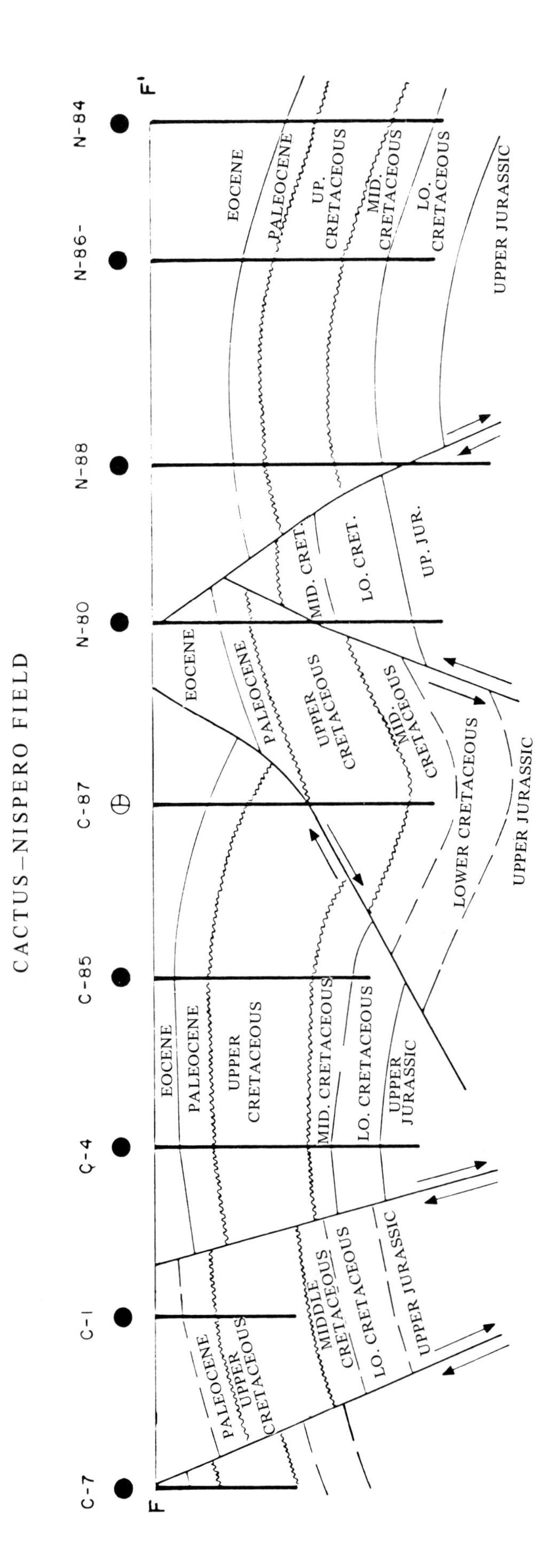

FIG. 35—Geologic cross section, Cactus-Nispero field.

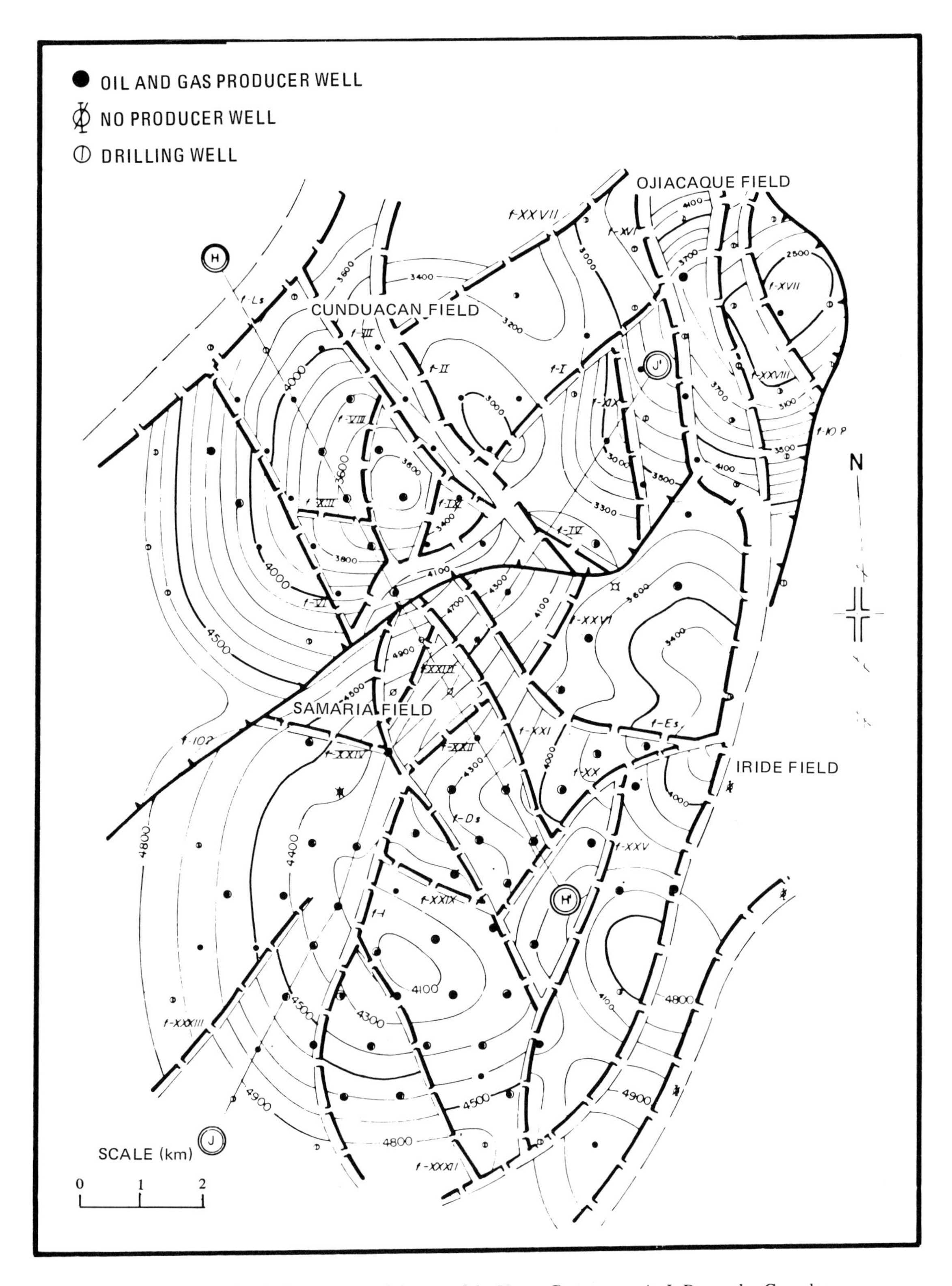

FIG. 36—Geologic structure of the top of the Upper Cretaceous, A. J. Bermudez Complex.

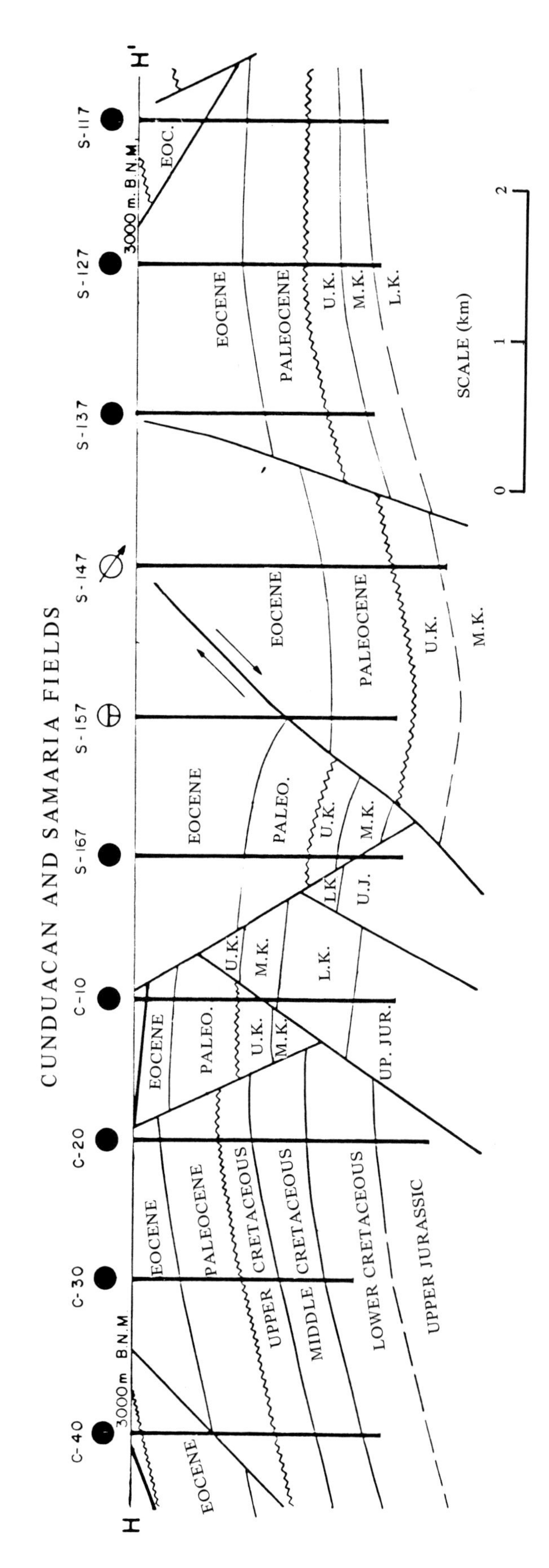

FIG. 37—Geologic cross section, A. J. Bermudez complex.

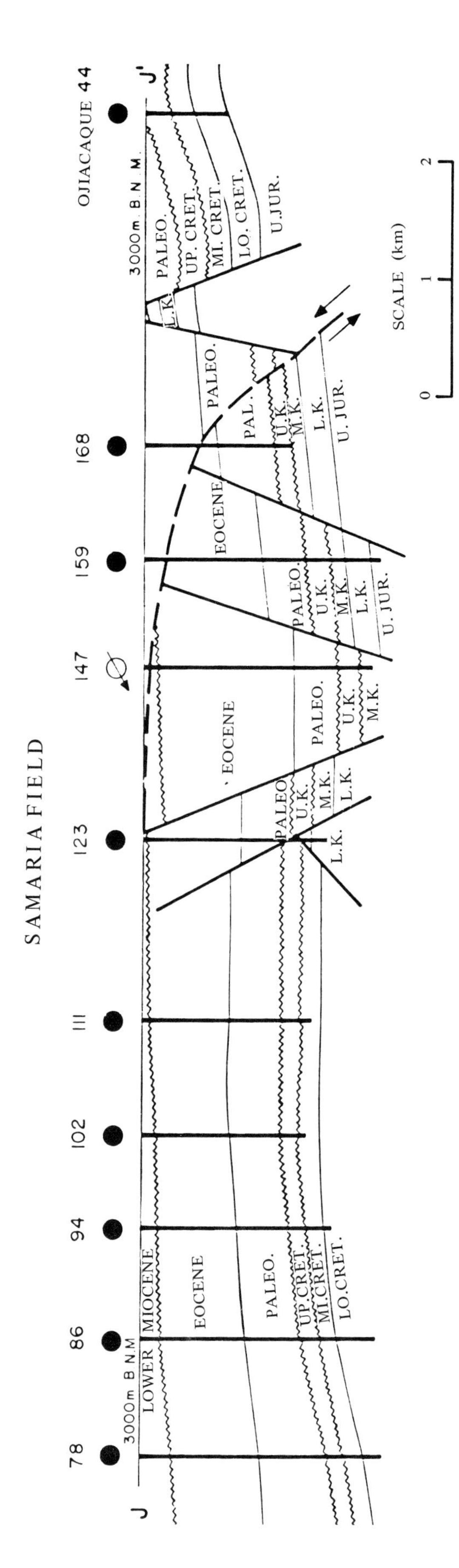

FIG. 38—Geologic cross section, A. J. Bermudez complex, Samaria field.

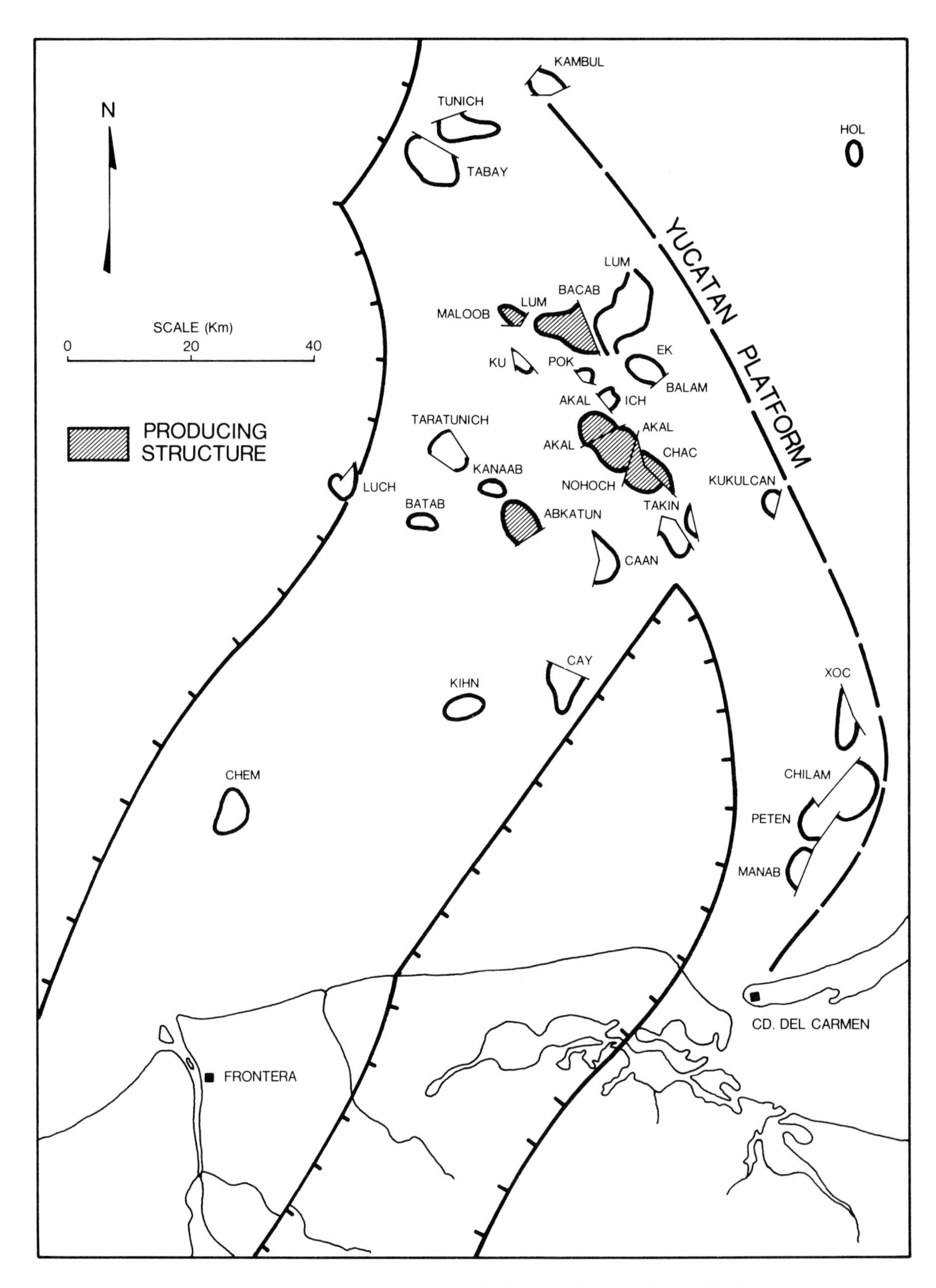

FIG. 39—Location of major producing fields in the Campeche Marine Platform.

Structural Geology

The Sitio Grande structure (Figs. 31, 32) is an elongate dome 10 km long by 6 km wide, trending northwest to southeast with a closure of 500 m. Two systems of normal faults are associated with the structure. The largest fault strikes east to west across the east flank of the dome and dips to the east with a maximum vertical displacement of 600 m, decreasing northward to 100 m.

Hydrocarbon Accumulation

Faulting has divided the structure into blocks with different oil-water levels. Oil-water contact in the middle Cretaceous section varies from 4,300 to 4,030 m below sea level. The pay zone of the dolomites of the Lower and mid-Cretaceous range in thickness from 100 to 150 m. There have been 34 wells drilled on 1,000 m spacing, and 26 are productive. The deepest wells are the No. 8 (5,166 m) and 82 (5,600 m).

Some wells on the northwestern flank of the dome have oil and gas accumulation in sandstone of the Pliocene Paraje Solo Formation. At the end of 1978, the cumulative production for Sitio Grande was 145 million bbl of oil with an additional 600 million bbl recoverable. Cumulative gas production was about 200 Bcf; this gas contains 3.52% sulfur. Oil density is 0.851 at 20°C (35° API), and the average porosity for producing zones is 8%; permeability is 3,080 md. Water injected by means of peripherial wells began two years ago for secondary recovery.

Cactus Field

The Cactus field (Fig. 33) discovery well, Cactus 1, produced from carbonate rocks of the Agua Nueva Formation (Upper Cretaceous) at a depth of 3,760 m. Initial production was 102 cu m/d of oil and 39.511 cu m/d of gas. This field is located about 13 km north of the Sitio Grande field.

Stratigraphy

The stratigraphic column of the field is similar to those described for the Chiapas-Tabasco Area relative to the Mesozoic section except for the lowest Upper Cretaceous. Production of the Cactus 1 well comes from calcarenite beds (lowest Upper Cretaceous). This same interval in Sitio Grande field is entirely dolomized and there are not calcarenites.

As in the Sitio Grande field, the deposits conform to the same sedimentologic pattern, but all units generally are thicker at Cactus field. Sediment thicknesses include a maximum for the Jurassic of about 400 m, 500 to 600 m for the Lower Cretaceous, and 50 to 200 m for mid-Cretaceous. In the Cactus field, middle Cretaceous sediments seem to have been eroded.

The Agua Nueva Formation (Upper Cretaceous) shows the same depositional characteristics of the mid-Cretaceous—that is it thins to only 60 m over the top, and measures about 200 m over the flanks of the anticline. There is an erosional unconformity between the Upper and middle Cretaceous represented by a breccia deposit. The San Felipe Formation of Upper Cretaceous is slightly thicker (about 250 m) at Cactus field. The Upper Cretaceous Mendez Formation is a more uniform deposit with a thickness of 250 to 300 m. Tertiary sediments have the same lithologic characteristics as those at Sitio Grande field.

Structural Geology

The structure is a dome 10 km long and 7 km wide oriented northwest to southeast (Figs. 34, 35). Two fault systems are well defined, one parallel with the main axis, and the other one transverse to it. Among the major normal faults there is one over the southwestern flank of the dome, parallel with the main axis of the dome and dipping northeast with a vertical displacement of approximately 400 m. Toward the eastern flank, a reverse fault is present, also parallel with the main axis of the structure.

Data from well 37 indicates salt migration along fault planes. Mesozoic and Oligocene sediments have been broken by salt intrusion.

Hydrocarbon Accumulation

Oil production in Cactus field is from Cretaceous strata. The pay zone is one of the biggest in the Chiapas-Tabasco Mesozoic area with 1,000 m of oil-saturated section. On top of the structure the oil-water contact is at 3,660 m below sea level, and over the southwestern flank is at 4,600 m below sea level. Several different blocks have different oil-water contact levels due to faulting. In well 84, on the upthrown block of a normal fault, the oil-water contact is at 4,500 m; in well 104, over the downthrown block of the fault, the oil-water level is at 4,700 m. Good structural control and reliable seismic interpretation are necessary for secondary recovery because water injection is based on the influence of the structure. By the end of 1978, 41 out of the 48 wells drilled had a cumulative production of 200 million bbl of oil. It is expected that an additional 1,500 million bbl (primary and secondary reserves) will be produced. The cumulative production of gas was 328,000 MMcf with 2.36% sulfur content. Oil density from the Upper Cretaceous calcarenite oils is 0.845 at 20°C (36° API), and 0.851 at 20°C (34.8° API) for the oil from the dolomites. Average porosity of the productive zones of the field is 7%, and permeability is 6,228 md.

Antonio J. Bermudez Complex

In 1958, in the Samaria area (Fig. 36) heavy oil was discovered in Pliocene sandstones. Due to high density of the oil, some studies were done to increase production either by heat or gas injection. After the Cactus and Sitio Grande discoveries in 1972, the Samaria 101 well was drilled in 1973 on the top of a Tertiary structure. Initial production of this well was 544 cu m/d of oil and 171,420 cu m/d of gas from the Upper Cretaceous. Later in 1974 the Cunduacan 1 well was drilled 8 km north of Samaria 101. Initial production in the well was from the middle Cretaceous, at a rate of 773 cu m/d of oil and 208,700 cu m/d of gas. Another productive well, Iride 2, was drilled in 1974 about 4 km east of Samaria 101. Initial production from a dolomitic section in the Lower Cretaceous was 203 cu m/d of oil, 54,710 cu m/d of gas. The Ojiacaque 1 well, completed in 1977 as an oil and gas producer from dolomitic sediments of Kimmeridgian age, is located 10 km northeast of

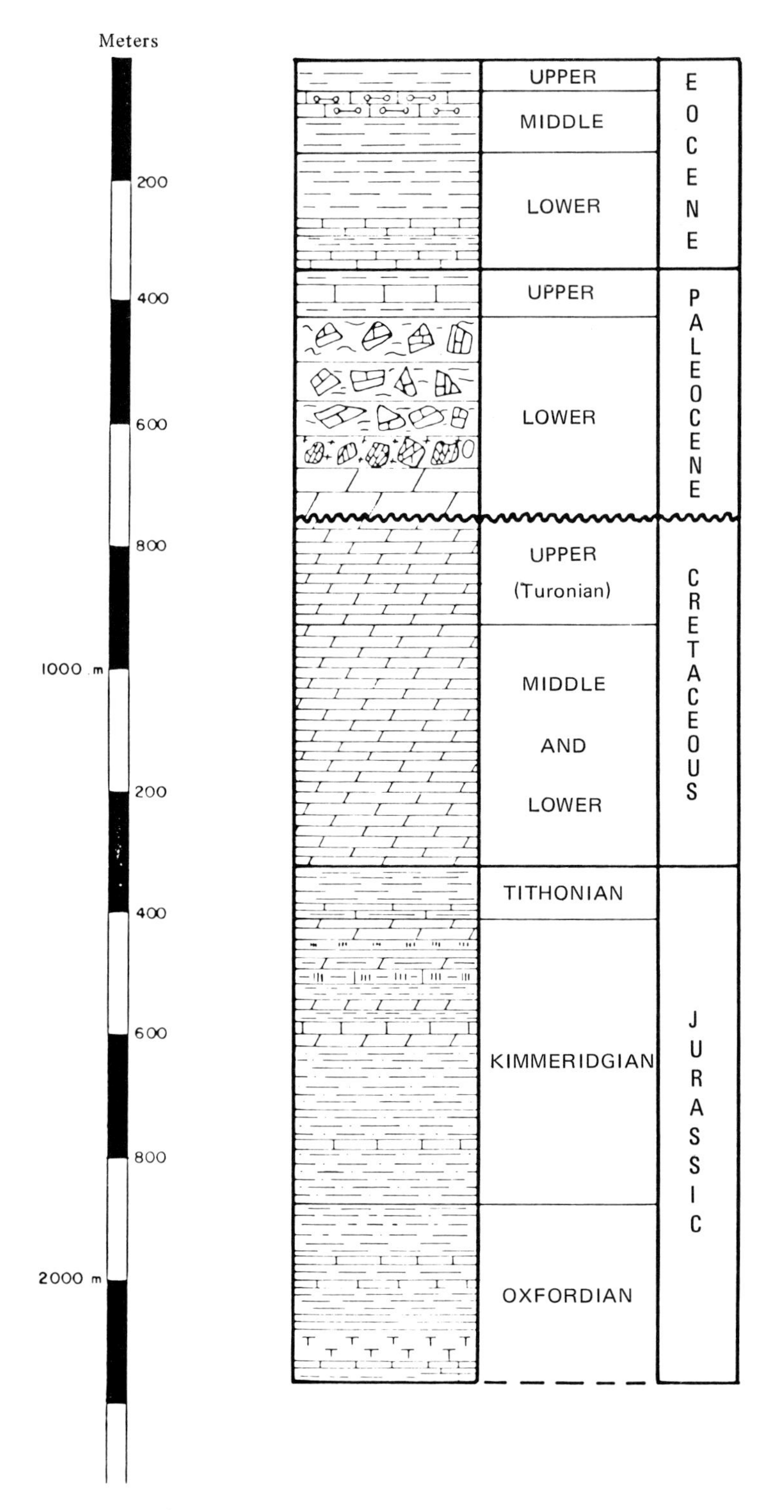

FIG. 40—Typical geologic column of Cantarell Complex.

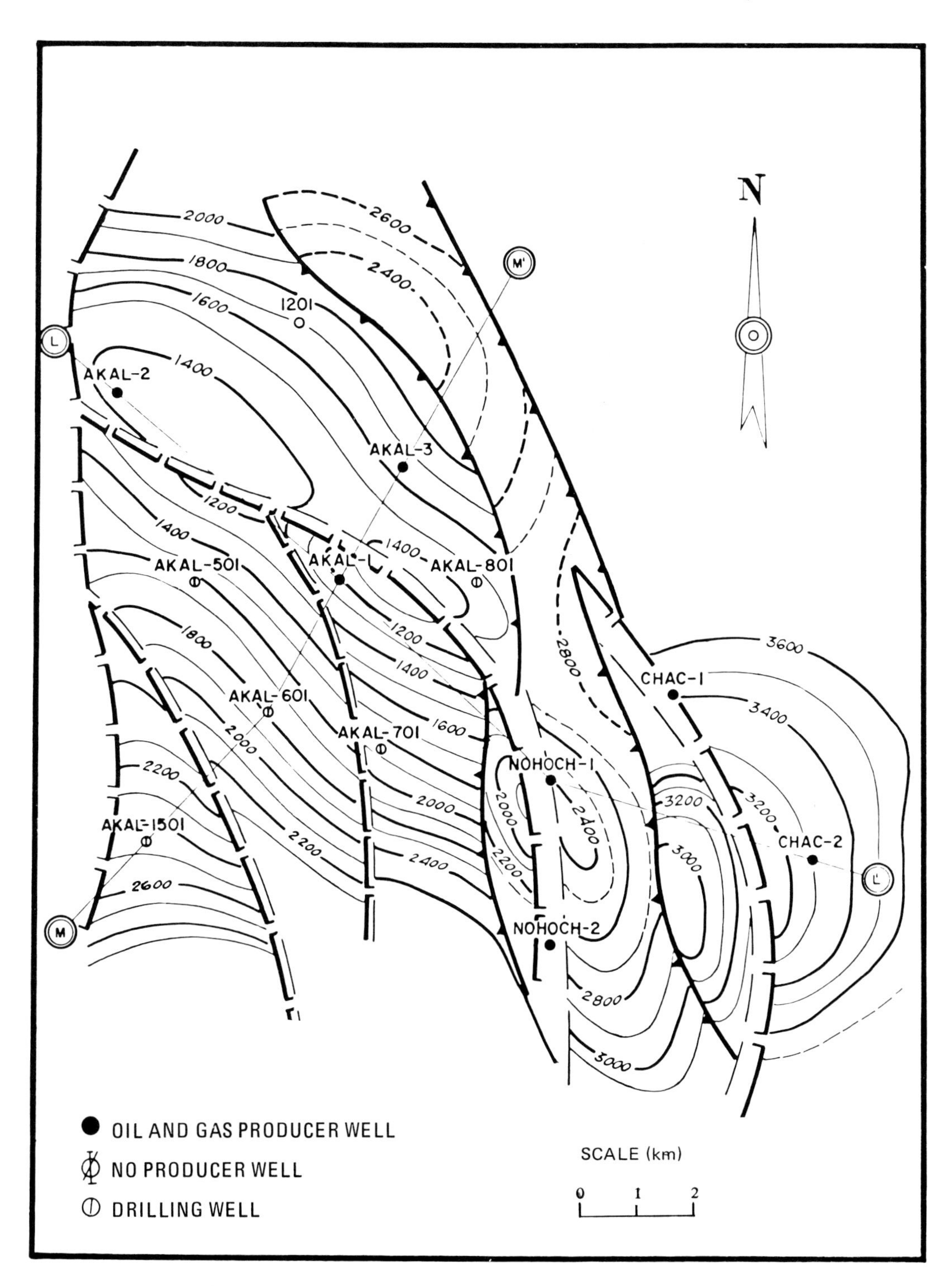

FIG. 41—Geologic structure map of the top of the Cantarell Complex.

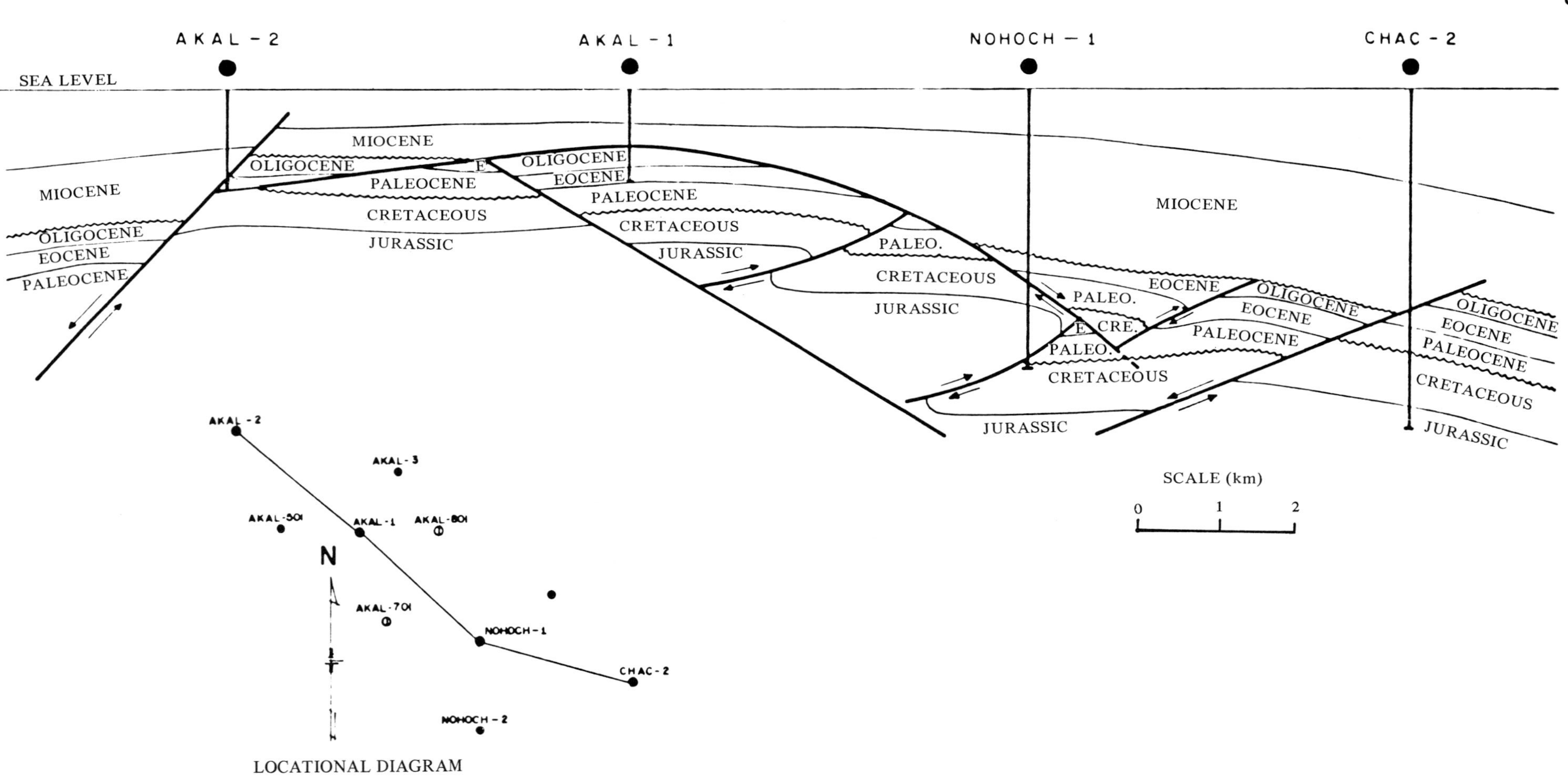

FIG. 42—Geologic cross section of the Cantarell Complex (Akal-Nohoch-Chac).

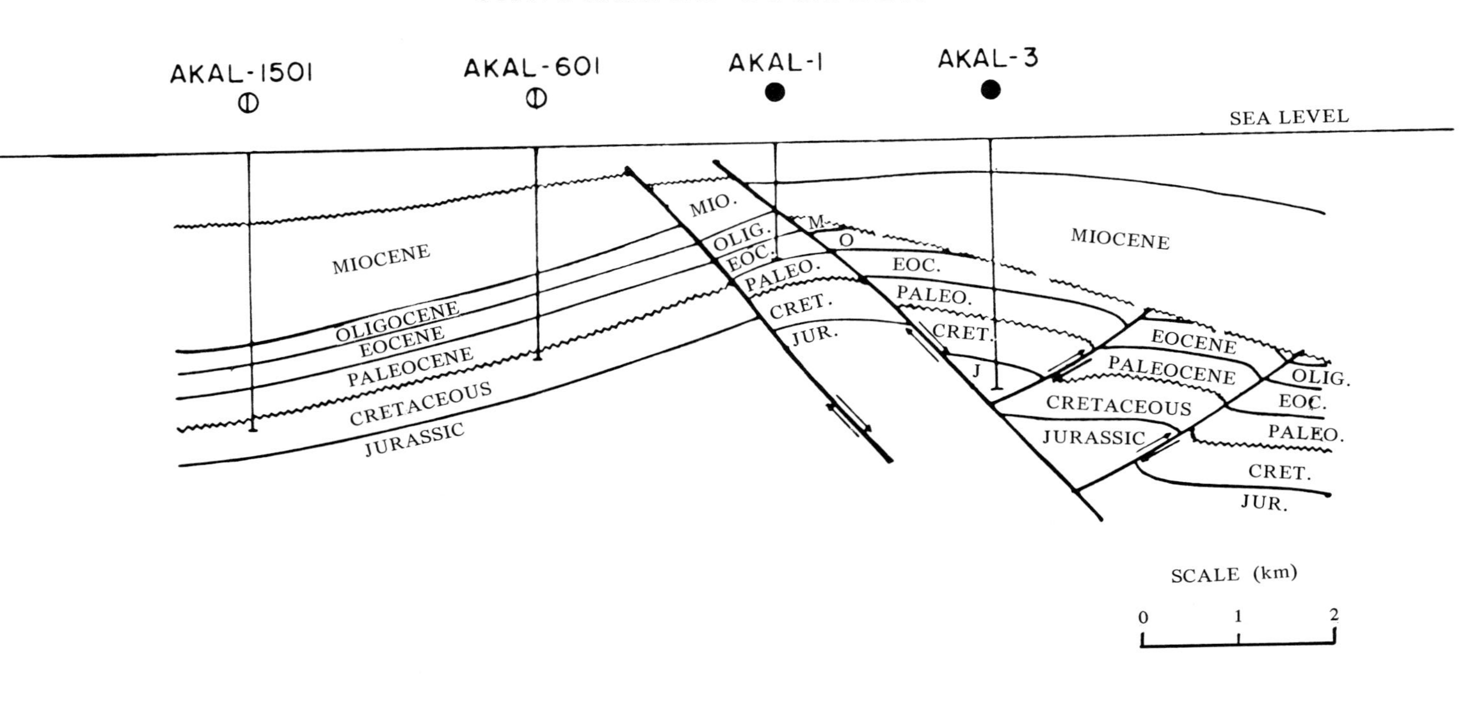

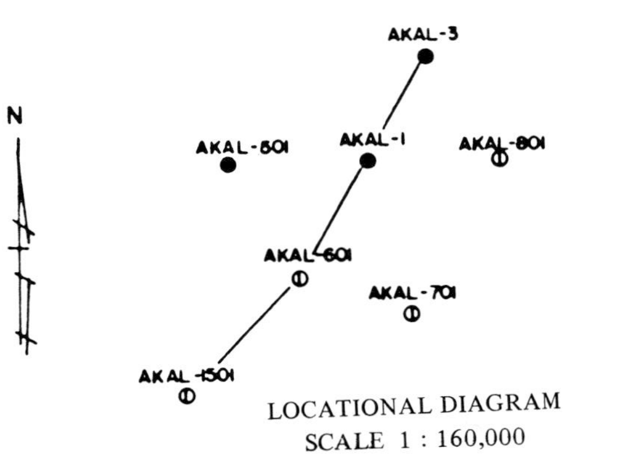

LOCATIONAL DIAGRAM
SCALE 1 : 160,000

FIG. 43—Geologic cross section, Cantarell Complex-Akal field.

Samaria 101. Initial production was 586 cu m/d of oil and 174,628 cu m/d of gas. Also in 1977 the Tres Pueblos 1A, another oil and gas producer in the A. J. Bermudez Complex, was completed 2,300 m east of the Cunduacan 1 well. Initial production was 338 cu m/d of oil and 46,644 cu m/d of gas. Production is from the Upper Jurassic, and the well touched bottom in the Middle Jurassic (Bathonian), the oldest formation penetrated in the area. These described discoveries collectively defined the fields which make up the Antonio J. Bermudez Complex.

Stratigraphy

Middle Jurassic—The section consists of gray to brown mudstones with thin beds of white anhydrite.

Upper Jurassic—Maximum thickness of the Upper Jurassic is 1,660 m consisting of dolomites, argillaceous mudstone, oolitic sandstones, and anhydrite beds. Cavities and fractures are present throughout the Upper Jurassic. Three depositional environments are defined in this section—inner platform, platform border, and open marine.

Lower Cretaceous—The Lower Cretaceous ranges in thickness from 500 m in the south of the complex to 200 m in the north. Sediments are marine, argillaceous, occasionally dolomitic, highly fractured mudstones with fractured interbedded dolomite and black chert beds.

Middle Cretaceous—Consist of dolomites with large cavities and fractures, except in the Tres Pueblos area, where the limestone is not dolomitized. The limestone is fractured with dissolution cavities and local white interbedded anhydrite.

Upper Cretaceous—Over the southern part of the complex (Samaria and Iride fields), the Upper Cretaceous consists of clastic biogenic mudstone and grainstone breccias within an argillaceous mudstone matrix. Caves and fractures are present in this section. Over the northern part (Cunduacan, Tres Pueblos, and Ojiacaque fields) of the complex three formations are distinguished (Agua Nueva, San Felipe, and Mendez) and are lithologically similar to the description under the section headed "Mesozoic Provinces—General."

The thickness of the Upper Cretaceous is reduced to about 200 to 300 m due to erosion as evidenced by breccias over the southern part of the complex.

Paleocene—The lower Paleocene consists of brown clastic mudstone breccias in an argillaceous matrix derived from middle to Upper Cretaceous deposits. Thickness of the lower Paleocene ranges from 60 to 200 m. The upper Paleocene is an argillaceous and calcareous section typical of the Paleocene throughout the Mesozoic Area.

Eocene—The Eocene is predominantly a sequence of greenish-gray shales interbedded with fine grained light gray sandstones. The upper part of the section is a clastic gray mudstone breccia with an argillaceous matrix. The average thickness of Eocene is about 800 m.

Oligocene—Absent by erosion.

Lower Miocene—The lower Miocene lies unconformably over the Eocene and consists of an argillaceous sandstone sequence.

Structural Geology

The Bermudez structure (Figs. 37, 38) is a wide anticline (16 by 14 km) trending northwest to southeast. Numerous normal faults are present in the complex. Over the central part of the Bermudez a reverse fault divides the complex and the northern block overrides the southern block (the Tres Pueblos, Cunduacan, and Ojiacaque fields override the Samaria and Iride Block).

Hydrocarbon Accumulation

Cumulative production for the 94 producing wells at the end of November 1978 was 600 million bbl of oil, with an expected additional 5.2 billion bbl recoverable. The cumulative gas production was 697,760 Bcf with about 3.5% sulfur content. Oil density ranges from 0.883 (28.7° API) to 0.879 (29.5° API) at 20°C, and the average porosity of the producing zones is 8%; permeability is 7,800 md.

Campeche Marine Platform

Several significant wells have been drilled recently on the Campeche Platform (Fig. 39) and are described. In 1976, the Chac 1 well was completed as an oil producer from an interval of lower Paleocene calcareous breccias, 3,545 to 3,567 m. The total depth of the well was 4,934 m and initial production was 952 bbl of oil per day. Oil density is 0.935 a 20°C. This discovery began an intense exploratory program to test the potential of several structural highs defined by marine seismic work. The prospective area on the platform covers about 8,000 sq km and some seismic work has been completed.

At the present time, 11 productive wells have been discovered including Chac 1. Three additional wells (Akal 601, Akal 701, and Akal 801) appear to be good prospects. Chac 1 well is on the southeastern flank of a large anticlinal structure, 8 to 12 km wide and 30 km long. This has been named the Cantarell Complex. Chac 1 is about 85 km north of Ciudad del Carmen, Campeche. In February 1979, the Abkatun 1A well was completed as an oil and gas producer from the lower Paleocene. Oil density is 0.860 at 26°C. This well is about 23 km southeast of Chac 1 and is on a structural trend oriented northwest to southeast, 30 km long and 7 km wide. Another well is being drilled to test the northwestern flank of the structure. The well is located about 15 km northwest of the Abkatun 1 well and is designated the Tarantunich 1. Another well, the Tunich 1, is located about 140 km northwest of Ciudad del Carmen, Campeche. It was completed but has been invaded by salt water. In order to test the hydrocarbon potential of this large area, Petroleos Mexicanos drilled some wells close to Yucatan Platform. The Chilam 1 well, 33 km northwest of Ciudad del Carmen, Campeche, was drilled to a total depth of 5,334 m; the well was "salt water invaded." Also on the Yucatan Platform, the Kukulcan 1 well, 76 km northeast of Ciudad del Carmen, Campeche, resulted in a salt water-invaded test at a total depth of 3,522 m.

The Luch 1 well, about 100 km northwest of Ciudad del Carmen, Campeche, is now at 3,280 m depth. There are some oil shows in Miocene sandstones. This well has two objectives;

the Tertiary and the Mesozoic sediments.

Two more wells, Ek 1 at a depth of 2,855 m, and the Hol 1, at a depth of 2,017 m, are scheduled for deeper drilling. At the end of February 1979, six more wells were being drilled (Akal 601, Akal 701, Akal 801, Akal 1501, Ixtoc 1, and Maloob 1). This complex is to be developed by means of 10 fixed platforms. One is already built, three more are in the installation process, and the rest are scheduled in the future.

Stratigraphy

Upper Jurassic (Kimmeridgian)—The Upper Jurassic (Fig. 40) is 500 m thick. The upper part is predominantly argillaceous with some oolitic limestones. Lower part is a sequence of shales, dolomites, and bentonites, with a few interbedded anhydrites.

Upper Jurassic (Tithonian)—The Tithonian is 85 m thick. The lower part consists of argillaceous limestones changing upward to shales and a few sandstone lenses.

Undefined Cretaceous—This section consists of 425 m of a crystaline dolomite with fractures and dissolution caves. Because of strong dolomitization, it is very difficult to make age determination in this section. However, it was possible to determine middle Cretaceous, Lower Cretaceous, and Upper Jurassic within the section in the Nohoch 2 well, and Lower Cretaceous and Upper Jurassic sediments in the Tunich 1 well.

Paleocene—The Paleocene is 220 m thick. The lower part consists of a sequence of dolomitic breccias with fractures and dissolution caves. The upper part is predominantly calcareous and above it there is a shale section changing again to calcareous breccias.

Eocene—The thickness of the Eocene ranges from 95 to 170 m. It consists of greenish gray bentonite shales. The lower part occasionally has a dolomite section.

Oligocene—The Oligocene is 12 to 165 m thick, characterized by a sequence of hard greenish gray shales. Probably the sediments of this age were not completely represented due to emergence of the Yucatan Platform during the Oligocene.

Miocene—The sediments of this age vary in thickness according to paleogeographic position. The Miocene over the Yucatan Platform (Cantarell Complex) represents some erosion and nondeposition section resulting thickness from 600 to 1,050 m. However, to the west across the Yucatan Peninsula (platform border) the Miocene is more than 2,600 m thick.

Structural Geology

The so-called Cantarell Complex is located over an area of very complicated structure in which unconformities and normal and reverse faults occur (Fig. 41). This causes varied structural and stratigraphic conditions in the wells drilled in the area. For example, the lower Pleocene calcareous section in Akal 1 well is 2,300 m higher than in the Chac 1 well, only 7 km away. This also is evident in the Akal 3 well and the Nonhoch 1 and 2 wells; the evidence includes repetitions of section. In the Nonhoch 1 well the Paleocene overlies poorly defined Lower Cretaceous and Jurassic rocks, whereas in the Nohoch 2 well the middle Cretaceous (Cenomanian) directly overlies Upper Jurassic (Kimmeridgian) rocks, and at depths below this contact rocks are younger, passing from Kimmeridgian to Late Cretaceous and Tertiary-Paleocene in age.

This complicated structural situation is caused mainly by the presence of both reverse and normal faults (Figs. 42, 43).

Hydrocarbon Accumulation

The wells drilled and completed as producers have been closed in. Thus only an estimate of production can be made. The most important well is the Akal 2 well, which produced 5,000 bbl of oil per day. Only "marine tests" have been conducted without stimulation and, therefore, bigger production is expected. Oil density ranges from 0.937 at 42°C to 0.893 at 35°C. The largest oil column found is 900 m, in the Nohoch 2. In this well, oil is present not only in the lower Paleocene breccias but also in the entire Cretaceous section and in the Jurassic. Total estimated Cantarell reserves are 8 billion bbl of oil.

Production test in Kimmeridgian sediments (interval 2,492 to 2,522 m) of the Akal 3 well suggest an oil production of 200 cu m/d and 9,446 cu m/d of gas. Oil density is 0.916 at 27°C and the oil column is 650 m without reaching oil-water contact.

Test of the Kimmeridgian dolomite breccias (3,334 to 3,377 m) in the Nohoch 2 produced oil and water of 91,000 parts per million (ppm) and a probable oil-water contact is estimated at 3,250 m. Another production test in Upper Cretaceous dolomites (2,316 to 2,325 m) gave an estimate of 464 cu m/d oil and 23,750 cu m/d of gas. Oil density was 0.910 at 26°C and the oil column is 650 m without reaching oil-water level. There is a good possibility for production from the Oxfordian Jurassic sandstone found in the Chac 1 well.

Geology and Development of the Teak Oil Field, Trinidad, West Indies

By S. C. Bane, and R. R. Chanpong

Abstract The Teak oil field is located 25 mi (40 km) off the southeastern coast of Trinidad in the eastern part of the Venezuela Tertiary basin. The Teak feature, discovered in 1969 by seismic survey, is a broad asymmetrical anticline along a compressional fold belt between the Caribbean and South American tectonic plates. It is broken by many transverse antithetic and synthetic normal faults which divide the producing reservoirs into many separate pools. Production is presently from a depth range of 4,000 to 14,000 ft (1,219 to 4,267 m) subsea in seventeen "producing" sandstones of late Pliocene age ranging in thickness from 10 to 600 ft. The effectiveness of the faults as barriers to communication between fault blocks is demonstrated by variations in edgewater conditions, reservoir pressures, and gas to oil ratios. Migration of oil into the Teak feature may be related to deep-seated fault conduits communicating with underlying Miocene or older shales.

Teak field production began in 1972, and the area is still being actively explored and developed. A total of 51 productive wells have been drilled from 5 platforms with an additional 4 wells recently drilled for water injection purposes. As of January 1, 1979, the field had produced 101 million bbl oil and 107 Bcf gas.

INTRODUCTION

The Teak field is located 25 mi (40 km) off the southeastern coast of Trinidad in the eastern part of the Venezuela Tertiary basin in a water depth of 185 ft (56 m; Fig. 1). It was the first major oil and gas field to be discovered and developed in the east coast marine area of Trinidad.

The Teak structural anomaly encompasses approximately 16 sq mi (41 sq km) with an approximate productive area ranging from 100 to 900 acres (40 to 364 ha.) Important requisites are present for petroleum generation and entrapment on a giant scale. The field lies along a regional negative gravity anomaly, in an area which underwent continuous subsidence throughout the Tertiary (Fig. 2). This subsidence was the consequence of compressional forces directed north to south between the South American plate to the south and the Caribbean plate to the north. Synchronous with this tectonism was the deposition of an abundant supply of fine-grained clastic sediment from the Guayana shield area to the south-southwest, and the development of regional, contemporaneous growth faults along the shoreward (eastward) side of this major depocenter. These events led to a petroleum accumulation at Teak which has a composite hydrocarbon column in excess of 3,100 ft (945 m) in thickness and consists of 32 stratigraphic zones.

Teak field is still in an active state of exploration and development. From initial production in mid-February, 1972, to the end of 1978, the field has produced 101 million bbl of oil and 107 Bcf of gas.

[1]Read before the Association April 4, 1979; accepted for publication, July 30, 1979. Also read before the Fourth Latin American Geological Congress, Port of Spain, Trinidad, July 9, 1979. Published with permission of Amoco International Oil Company.

The writers thank Amoco International for permission, and gratefully acknowledge suggestions and criticisms of P. K. Bettis, H. E. Christian, E. B. Eggertson, and J. W. Sides.

[2]Amoco International Oil Co., Houston, Texas 77001.

[3]Amoco Trinidad Oil Co., Port of Spain, Trinidad, West Indies.

Article Identification Number:
0065-731X/80/M030-0018/$03.00/0.

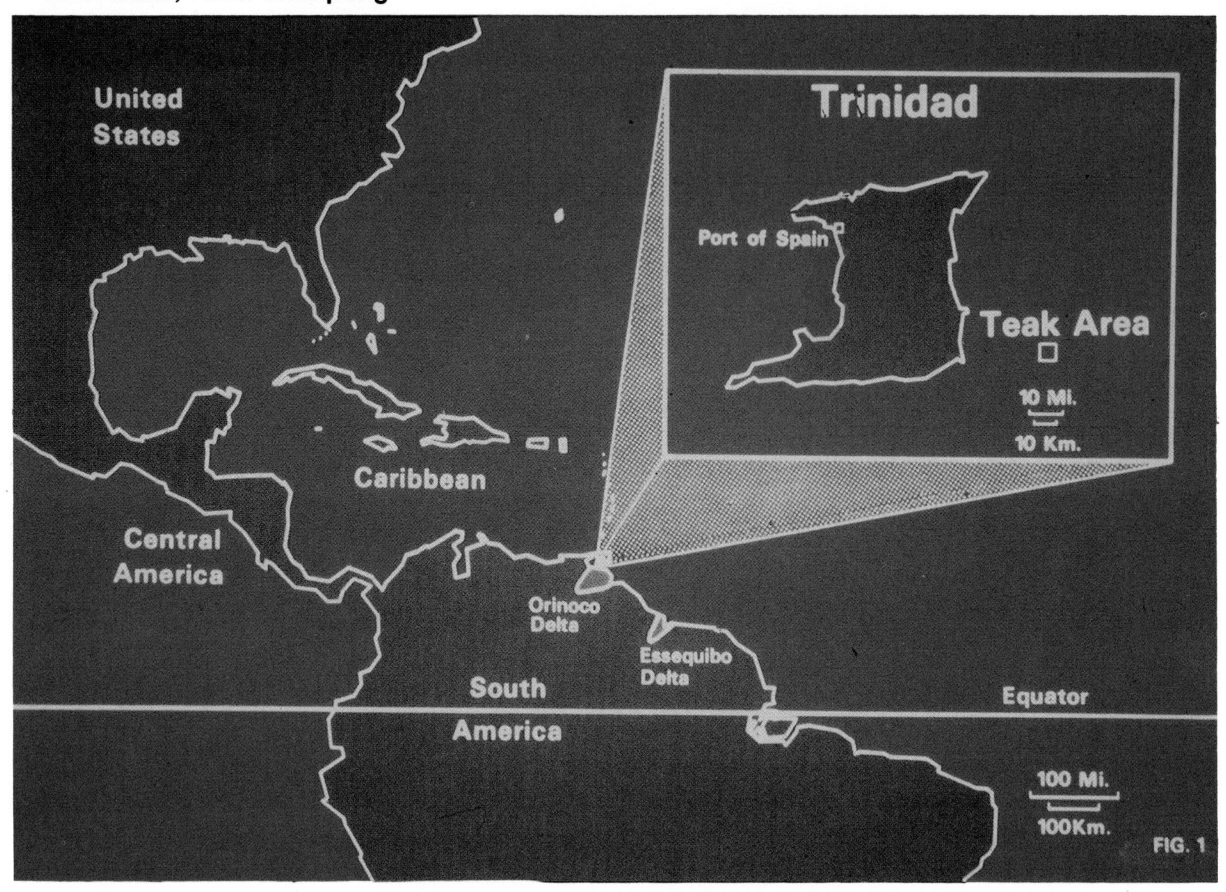

FIG. 1—Location map, Teak oil field area, Trinidad.

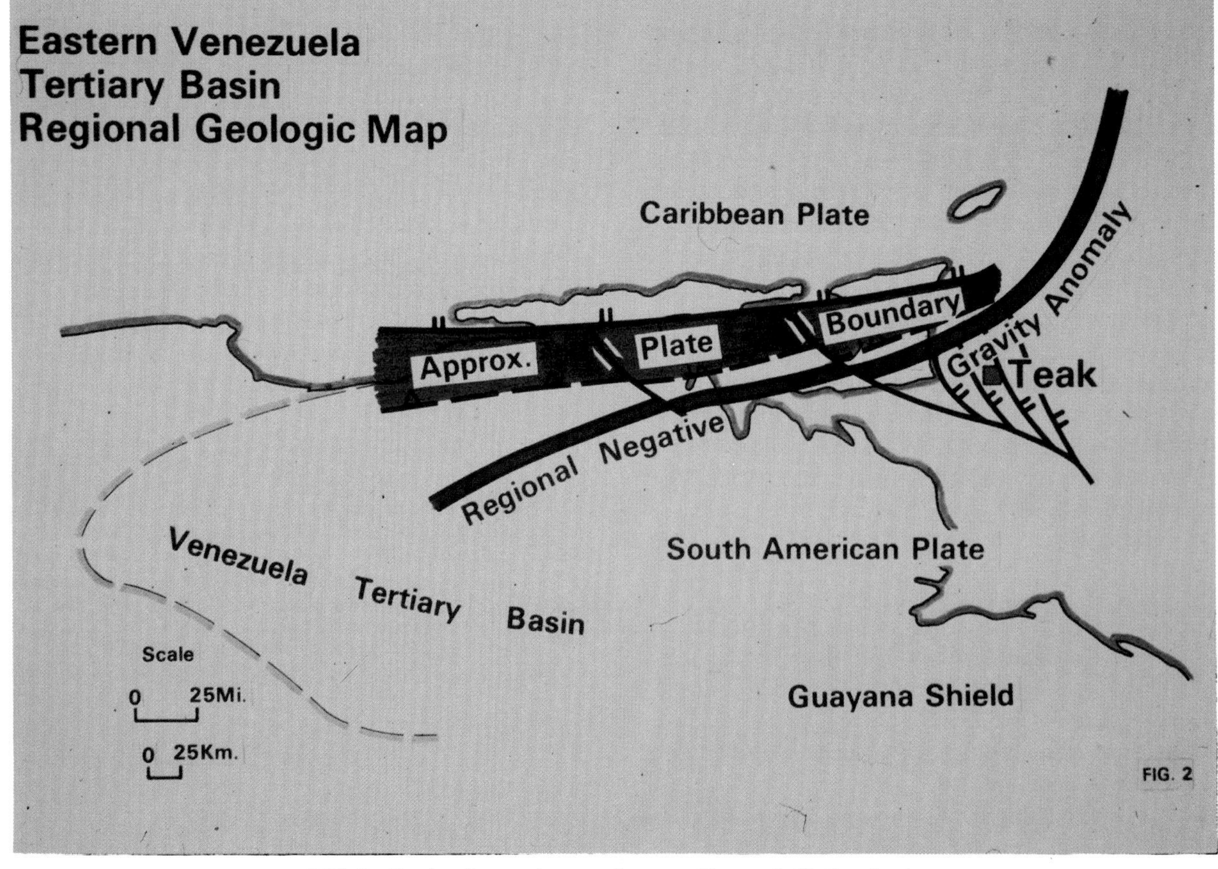

FIG. 2—Regional tectonic map of eastern Venezuela Tertiary basin.

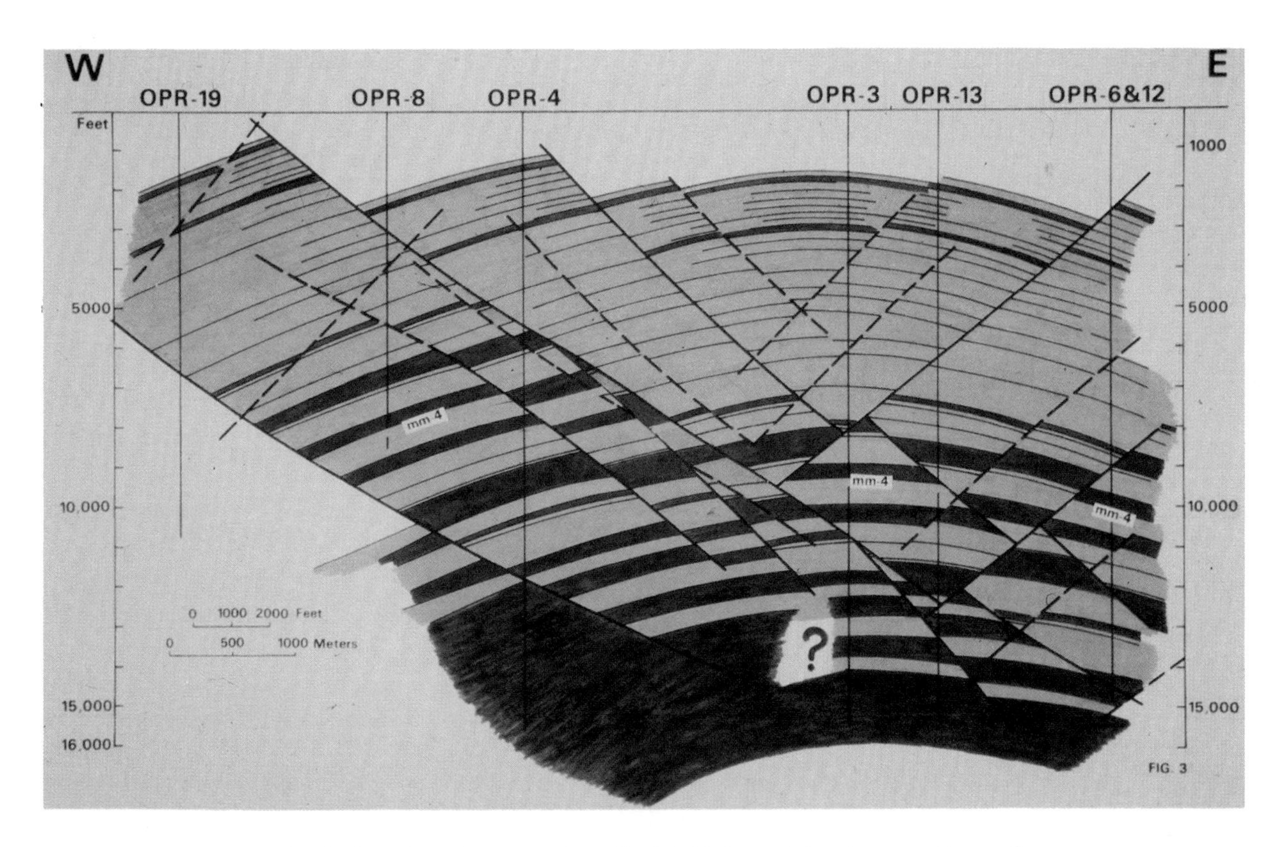

FIG. 3—West to east cross section of Teak field. For well locations, see Figure 10.

DEVELOPMENT HISTORY

The Teak structure is an intensely faulted anticline. Its numerous east- and west-dipping faults provide excellent trap boundaries, but complicate seismic interpretation and field development (Fig. 3). Regional seismic lines shot throughout Amoco's offshore Trinidad license area in 1968 and 1969 showed Teak to be a northeast- to southwest-trending asymmetrical anticline. Figure 4, which is a north to south line, shows the anticline with its pronounced south flank to the left, and the more gentle north flank to the right. Figure 5 is an east to west line which indicates the closure of the structure in those directions. On this line, the high-angle, east- and west-dipping normal faults can be traced down to about 2 seconds (approximately 7,000 ft or 2,134 m). Lack of seismic detail below this point was originally thought to indicate possible existence of a mobile shale mass at depth, similar to the diapiric shale in the nearby onshore Miocene Guayaguayare field (Fig. 6). Based on this interpretation, the Offshore Point Radix (OPR)-2 exploration well was spudded in September, 1968, on the northeast flank and drilled to a total depth of 15,708 ft (4,788 m; Fig. 7). It encountered 250 ft (76 m) of gas-bearing sandstone in the depth interval of 11,118 to 12,500 ft (3,389 to 3,810 m). No diapiric shale was penetrated at depth. Shortly thereafter, exploratory well OPR-3 was spudded two mi (3 km) south of OPR-2, and drilled to a depth of 14,650 ft (4,465 m) on the crest of the structural high (Fig. 8). The well encountered 400 ft (122 m) of gross oil pay updip from that discovered in OPR-2 and an additional 300 ft (91 m) of gross oil pay in the deeper section. The section was a sandstone and shale sequence with no evidence of mobile shale to total depth. Palynological evidence combined with a reinterpretation of seismic data suggested a major east-dipping, down-to-the-basin fault separating downthrown gas reservoirs from upthrown oil reservoirs. Subsequently, exploratory well OPR-4 was drilled on the west flank of the Teak anticline and penetrated oil-bearing sandstones upthrown to the major down-to-the-basin fault complex (subsequently designated the 'F' fault system; Fig. 9). OPR-4 penetrated over 500 ft (152 m) of gross oil pay and subtantiated almost 4,000 ft (1,219 m) of displacement on the 'F' fault system. Three platforms were set on the west flank of the anticline from 1971 to 1975, to develop the oil reservoirs in the upthrown block, and a fourth platform was set farther east to develop the gas reservoirs in the downthrown block. In 1976, a fifth platform was added for the dual purpose of development drilling and water injection into the oil reservoirs (Fig. 10). Through 1978, 4 water-injection wells and 51 oil and gas wells were completed in the development of Teak field.

REGIONAL GEOLOGY

The Eastern Offshore basin of Trinidad, in which the Teak feature is located, is the easternmost part of the Venezuela Tertiary basin (Fig. 11). Much discussion in the literature of the regional tectonic framework of Trinidad concerns its relationship to the Caribbean and South American plates (Kugler, 1959; Suter, 1960; Barr and Saunders, 1968; Potter 1965,

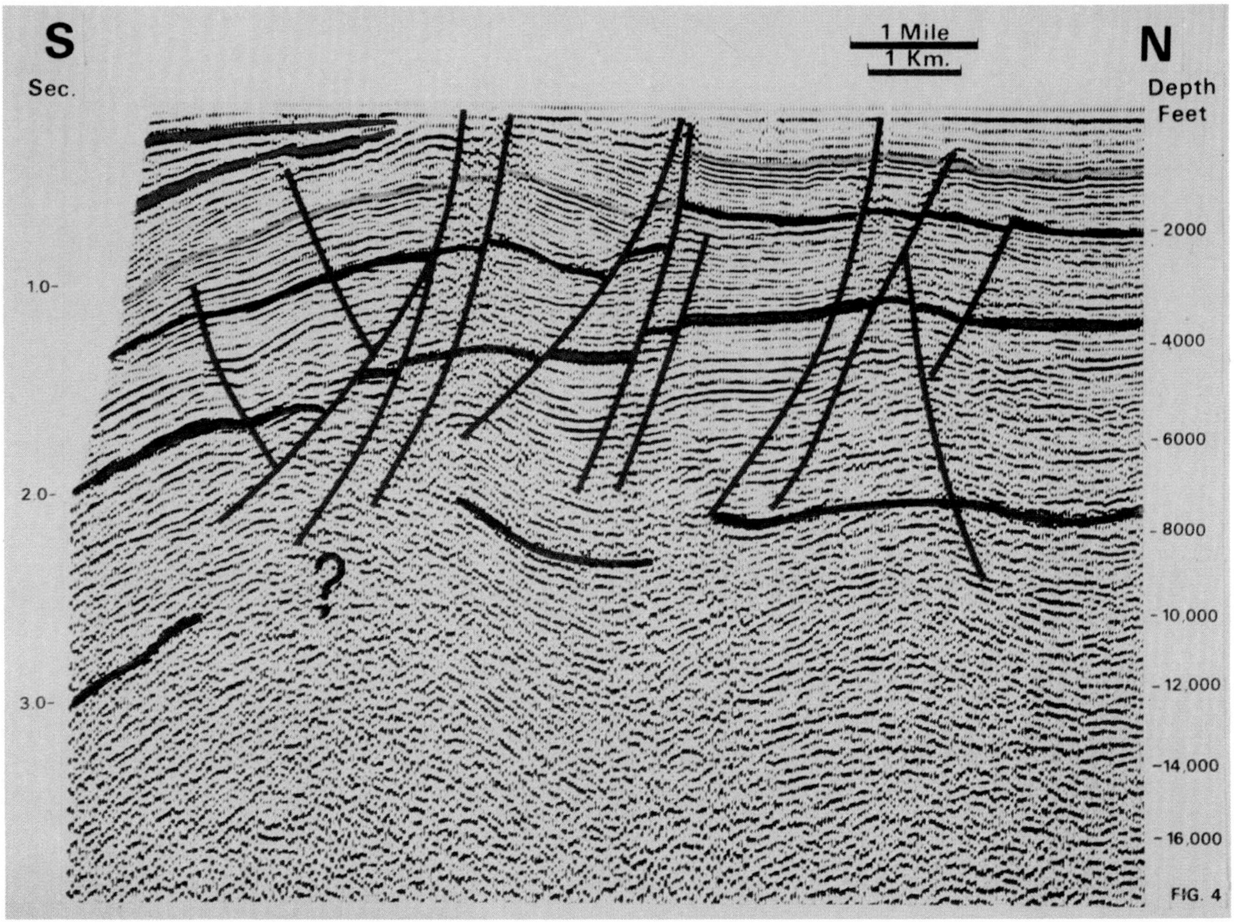

FIG. 4—South to north seismic cross section, Teak field. For location, see Figure 10.

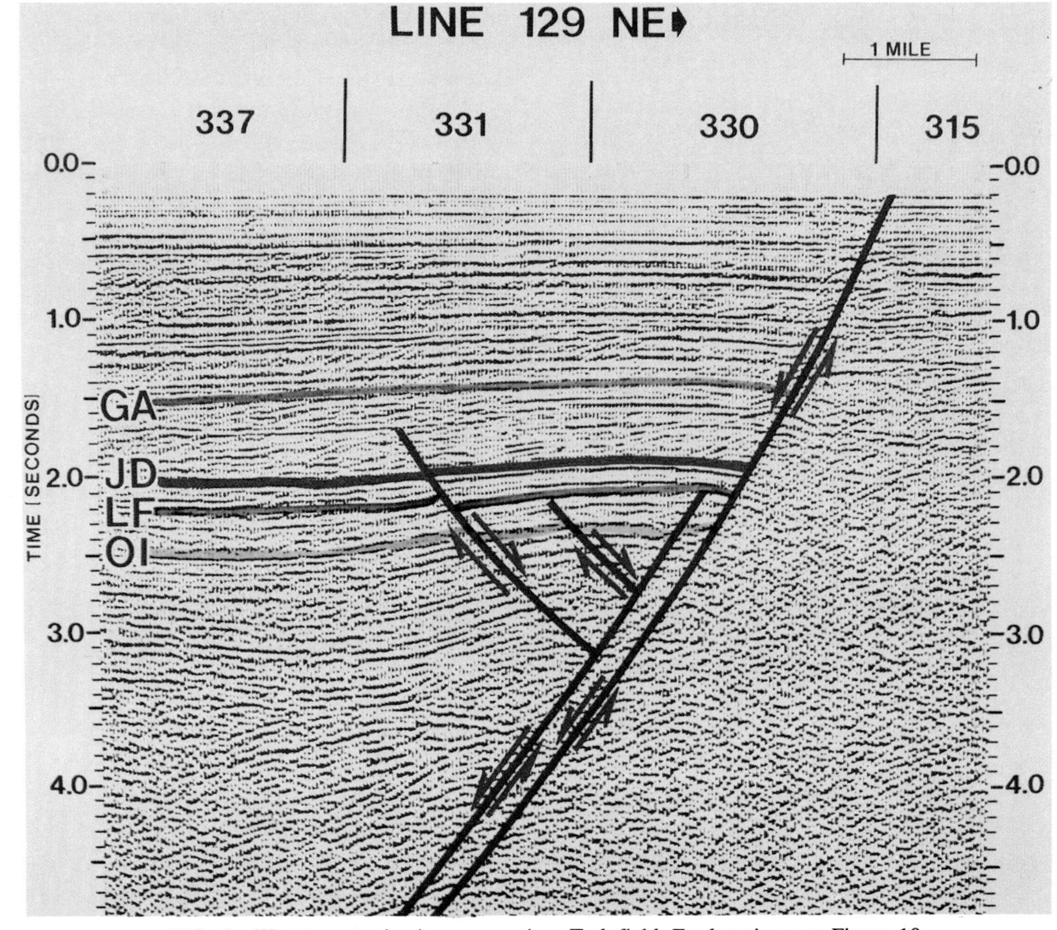

FIG. 5—West to east seismic cross section, Teak field. For location, see Figure 10.

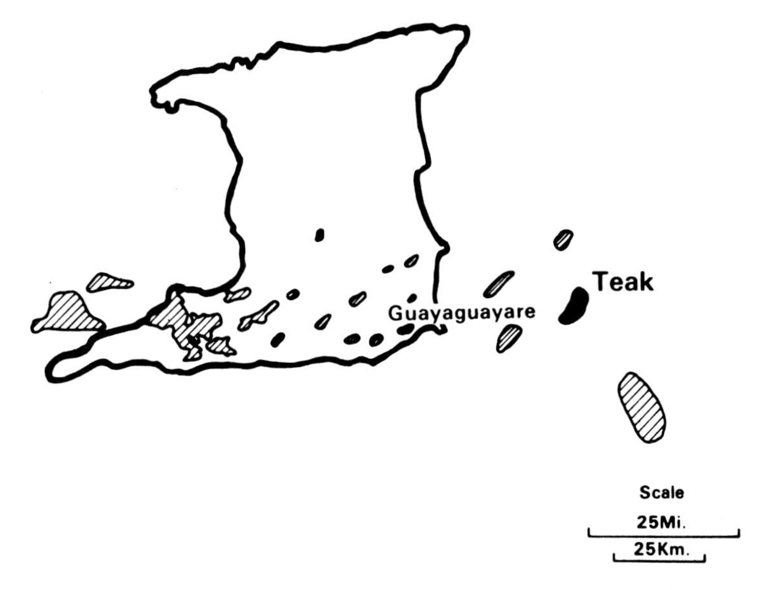

FIG. 6—Location map of oil and gas field in Trinidad.

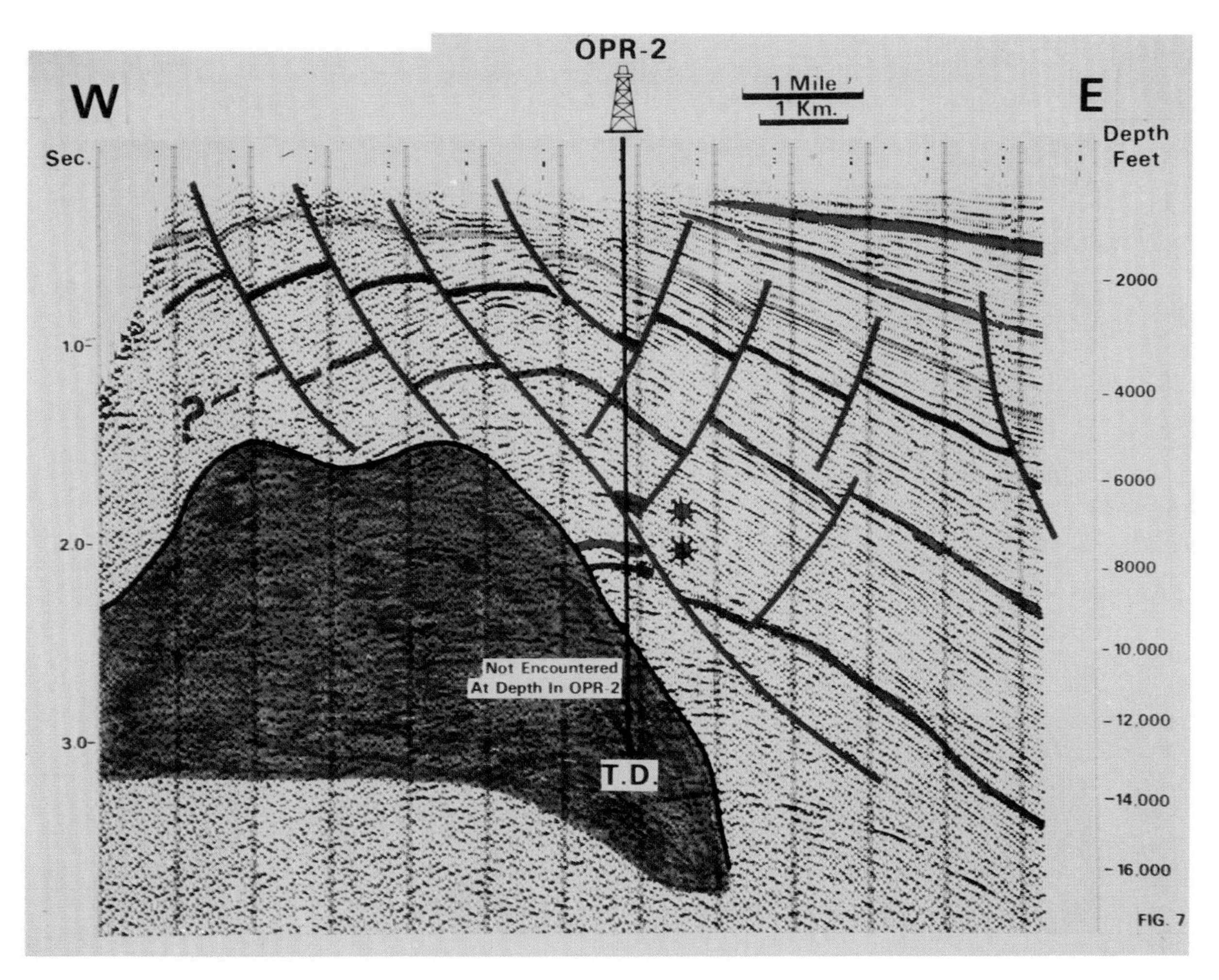

FIG. 7—West to east seismic cross section, Teak field. For location, see Figure 10.

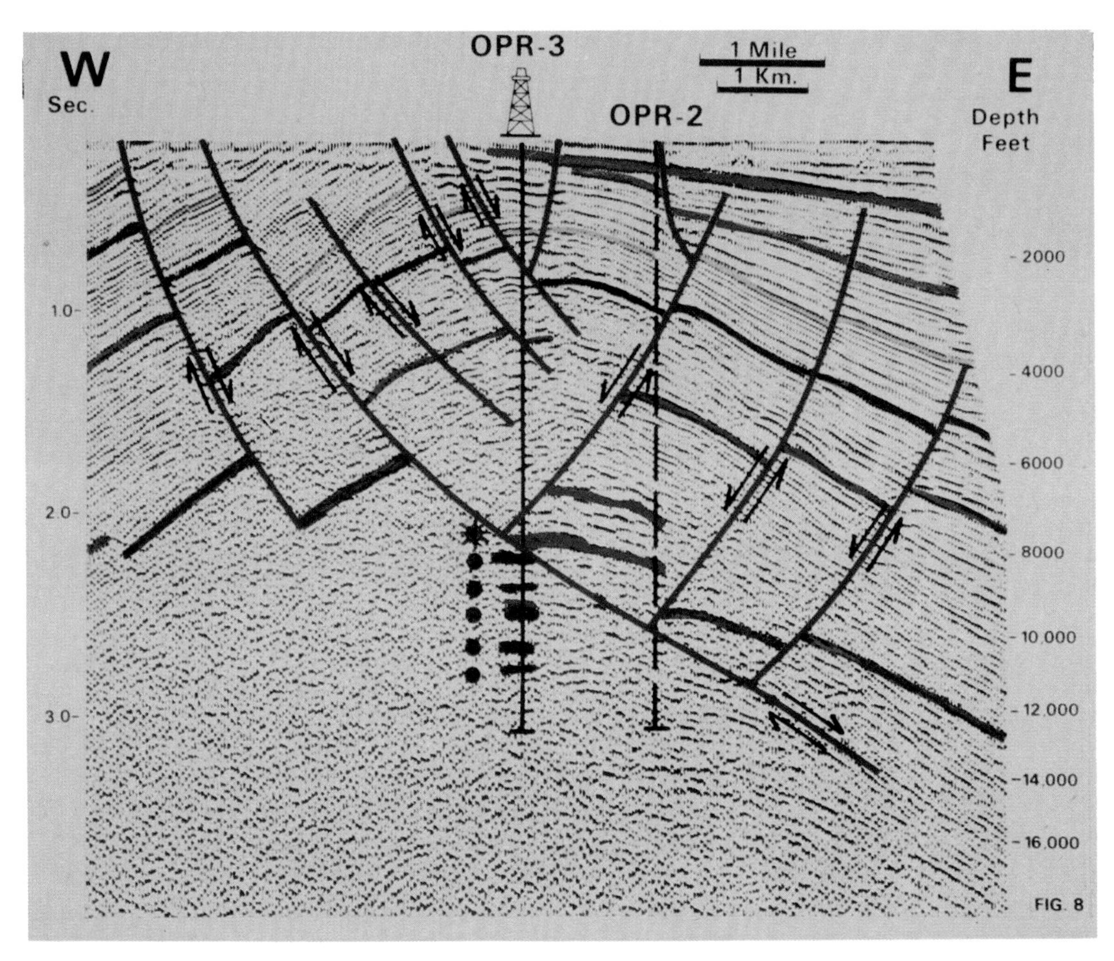

FIG. 8—West to east seismic cross section, Teak field. For location, see Figure 10.

1967; LePichon, 1968; Wilson, 1968; Molnar and Sykes, 1969; Weeks et al, 1971; Lau and Rajpaulsingh, 1965). The consensus is that in the Late Cretaceous and early Tertiary, the interaction of the Caribbean and South American plates produced a subsiding basinal area bounded by the coast ranges of Venezuela and the Northern Range of Trinidad on the north, and the Guayana shield area on the south. Concurrent with basin formation, the El Pilar-Oca fault system developed in the early Tertiary, and the Central Range thrust belt evolved in the Miocene and Pliocene. These two fault systems mark the shifting boundary of the interacting plates.

As the basin subsided, the basinal axis migrated southward, and by the Miocene the axis reached the position marked by a line between Maturin in Venezuela through the Erin-Siparia-Ortoire area of southern Trinidad and continuing offshore in an easterly direction. During the early and middle Miocene, the basin was tilted more to the east, and the Eastern Offshore basin received deep-water marine silts and clays (Stainforth, 1948, 1968).

By the late Miocene–early Pliocene another structural trend, the southeast-trending Los Bajos strike-slip fault, evolved. The basin sediments, reacting to continued plate movement, were compressed into en echelon folds, having east to west trending axes. The right lateral movement of the Los Bajos fault began a slow counterclockwise rotation of the area enclosed by itself and the plate boundary to the north, so that the grain of the en echelon fold system was slowly reoriented to a southwest to northeast direction by Pliocene-Pleistocene time.

The late Miocene and early Pliocene were marked by the onset of deltaic conditions in the Eastern Offshore basin. Palynological studies indicate that the sources of these sediments may have been the Proto-Essequibo drainage (Stainforth, 1978) as well as the Proto-Orinoco river, as other authors have concluded (Lamy, per. commun.).

By the early Pliocene, a prodeltaic environment existed in the Teak area of the Eastern Offshore basin where thick beds of fine-grained sediment were being deposited interbedded with thick marine shales.

In general, subsidence maintained pace with sedimentation creating a thick, uniform stratigraphic section. By the middle Pliocene the basinal axes shifted south of the Teak area and migrated even farther south toward the close of the Pliocene, thus slowing the rate of sedimenation in the Teak area. In the late Pliocene, the Proto-Orinoco delta prograded to a position in which the Teak area experienced a period of sand deposition in a delta-front environment. By the Pleistocene, the direct influ-

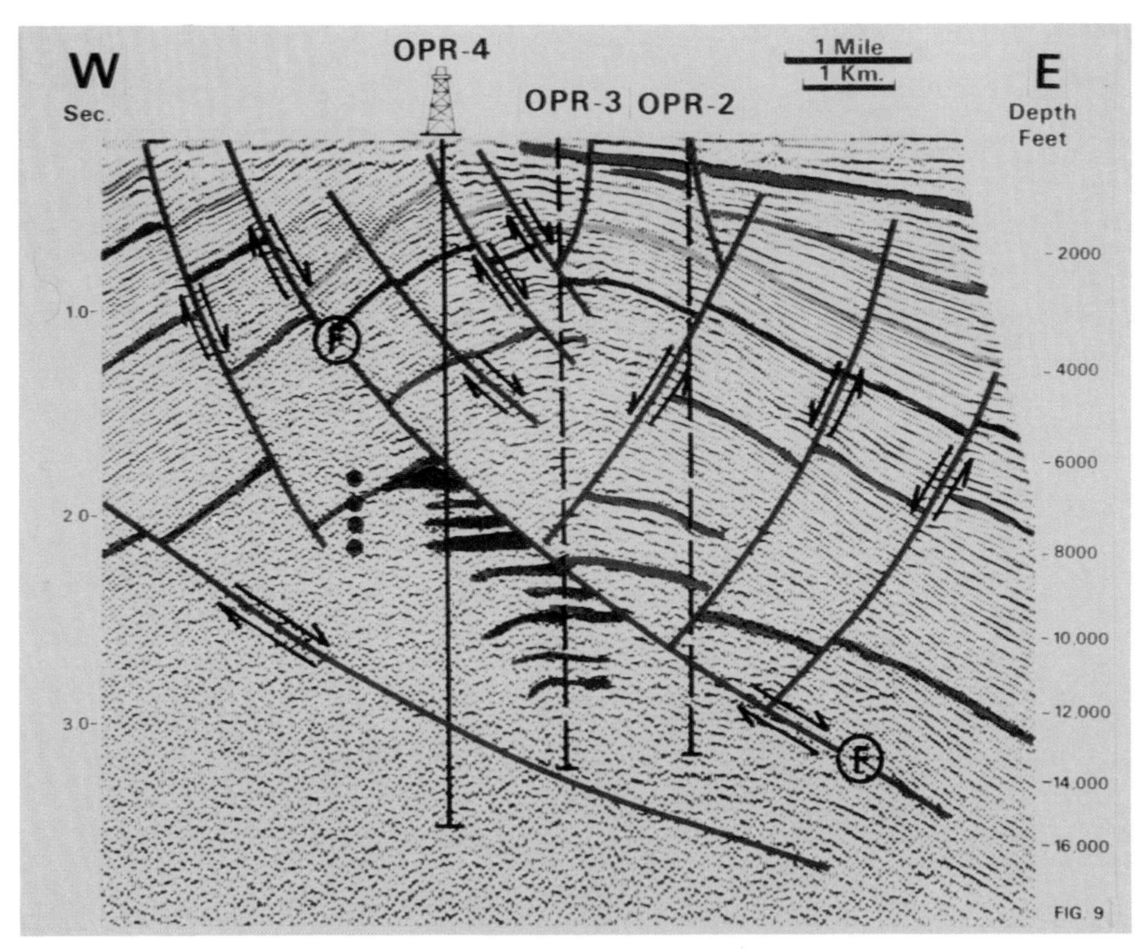

FIG. 9—West to east seismic cross section, Teak field. For location, see Figure 10.

ence of the river was lessened due to the rapid influx of sand from the newly emergent Southern Range of Trinidad. In the Pliocene and early Pleistocene, major growth faulting gave rise to a primary down-to-the-east and northeast system, and a resultant antithetic down-to-the-southwest system. These fault systems transected the northeast to southwest-trending en echelon folds. The primary fault system appears to be similar to the regional contemporaneous faults (or growth faults) of the United States Gulf Coast (Bruce, 1973). The primary faults extend laterally throughout the entire Eastern Offshore basin, joining in the southern end of the basin with the Los Bajos fault. The secondary fault system is antithetic to the primary system and the faults are generally of much smaller displacement.

Structurally, the Eastern Offshore basin is bounded on the north by the subsea extension of the Southern Range, on the west by major down-to-the-east and northeast faults, and on the south by the Los Bajos fault and Guayana shield. The oldest structural trend is the en echelon northeast to southwest fold system which is cut by younger primary and antithetic fault systems at right angles to the folding. This structural pattern is closely connected with the hydrocarbon accumulations in the Teak field.

STRATIGRAPHY AND PETROLOGY

In excess of 19,000 ft (5,791 m) of composite stratigraphic section of Pliocene to Holocene age sediments has been drilled in the Teak field (Fig. 12). A Miocene age section has been drilled nearby in the Eastern Offshore basin, but rocks older than Miocene were never penetrated. It is believed that the Miocene and older sediments lie unconformably on a Cretaceous-Jurassic sedimentary section which, in turn, overlies granitic rocks of the Guayana shield. The maximum thickness of the sediments in the Eastern Offshore basin is estimated to be in excess of 30,000 ft (9,144 m).

The drilled section consists of sandstones and shales ranging in thickness from 10 to 600 ft (3 to 183 m). The sedimentary units are predominately fine grained and shaly at depth and become less shaly toward the top of the section. The shales at depth are firm to hard and grade into softer shales, silty shales, claystones, and clays in the more shallow zones. The sandstones are silty at depth and become cleaner toward the surface. Generally, they are moderately consolidated at depth and become less consolidated in the more shallow zones. The sandstones are uniformly fine grained throughout the section except at or near the surface where an increase in the coarser grained

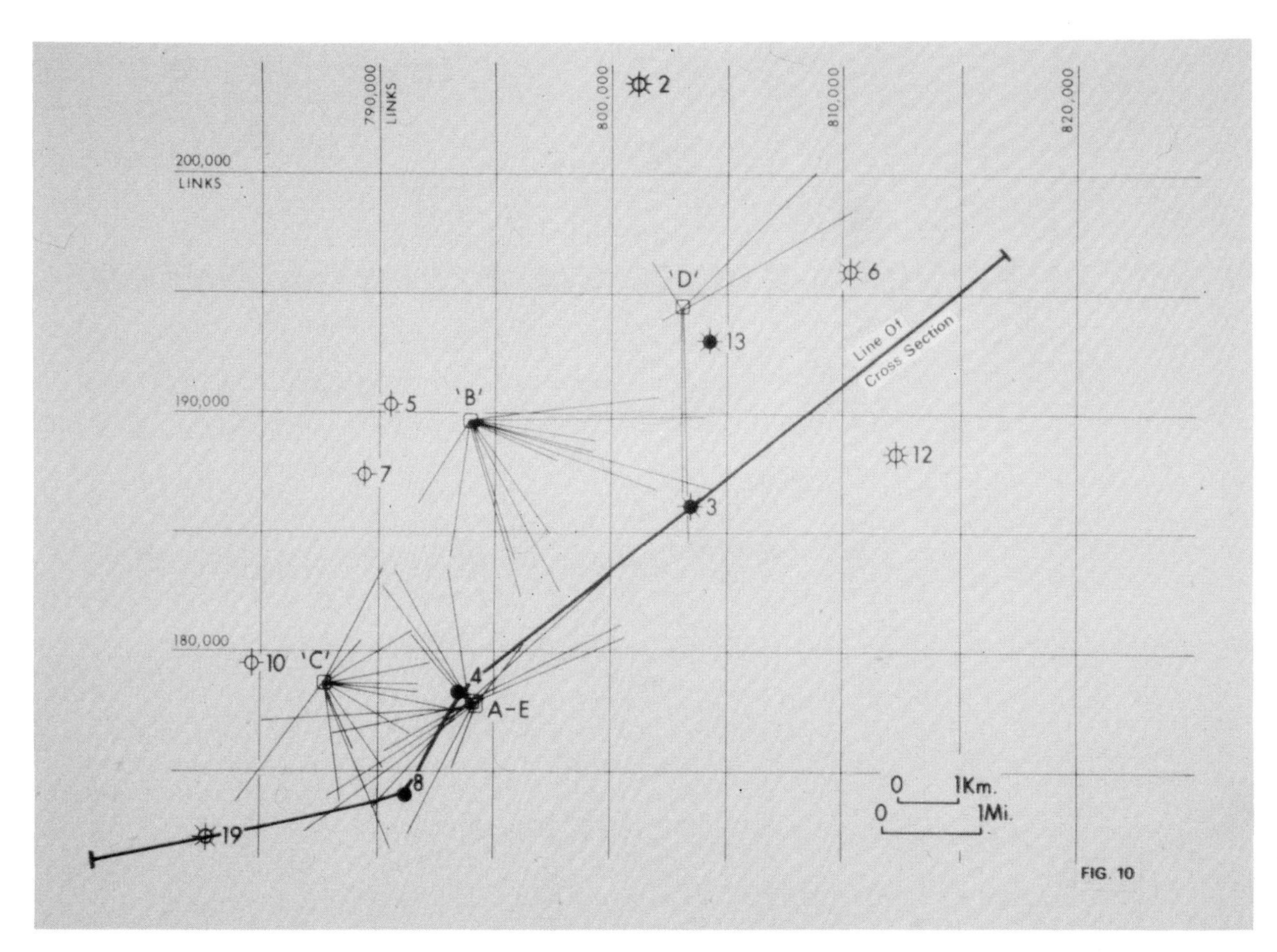

FIG. 10—Cross section and well location map, Teak field.

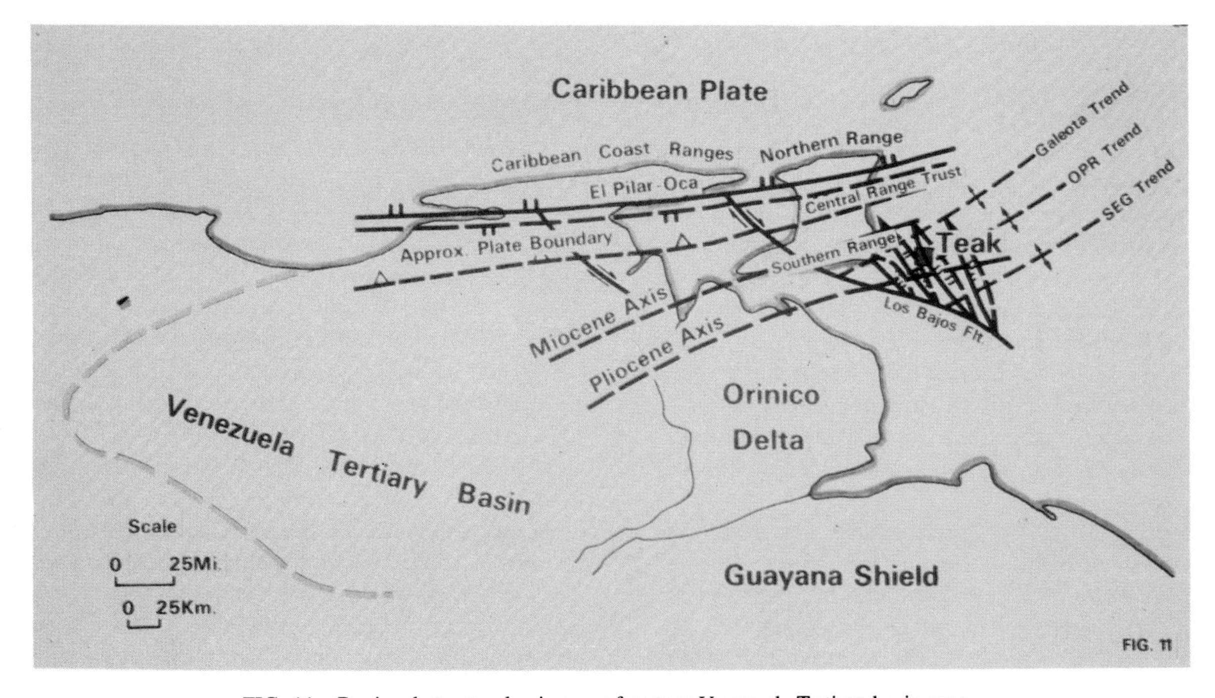

FIG. 11—Regional structural axis map of eastern Venezuela Tertiary basin area.

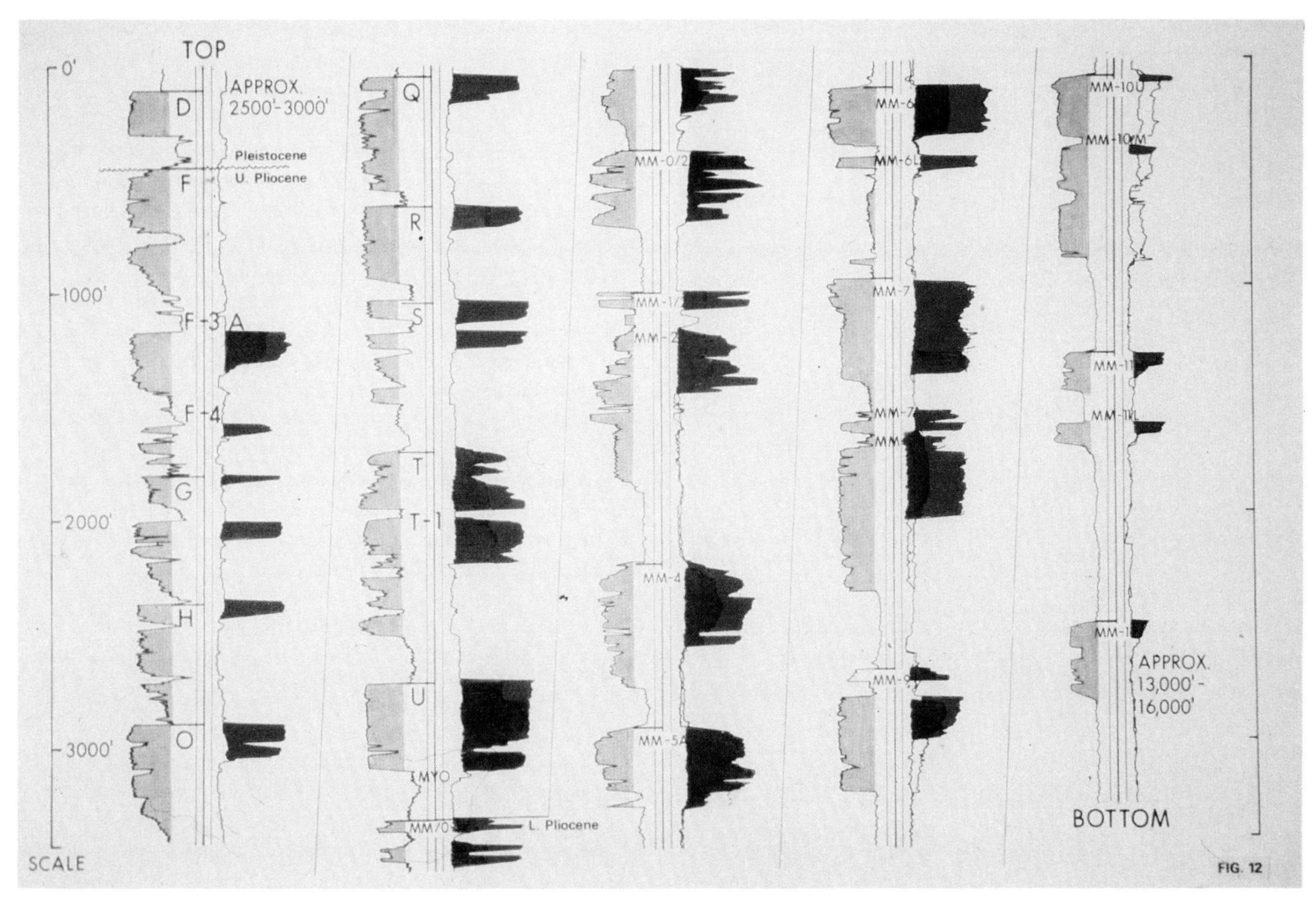

FIG. 12—Composite log section of Pliocene-Pleistocene, Teak field, Trinidad.

fraction apparently indicates a separate source area. A variety of accessory minerals are present in varied percentages throughout the sedimentary section. Lignite and other organic matter are ubiquitous and other minerals such as pyrite, muscovite, and chlorite occur, while kaolinite becomes abundant in the older sediments.

A formal stratigraphic nomenclature has not been established for the Teak area; however, onshore time-equivalent formations in eastern and western Trinidad can be used to differentiate the section. It must be realized that the section at Teak represents the basinward equivalents of these formations (Fig. 13).

PLIOCENE RESERVOIRS

The lower Pliocene is approximately 11,000 ft (3,353 m) thick in the Teak area and may be even thicker to the east. It consists of alternating sandstones and shales of different thicknesses. The deeper part of the section shows a general increase in the finer sediment fraction. The sandstones are generally well developed, ranging in thickness from 10 to 600 ft (3 to 183 m) and grading into thinner 10- to 200-ft thick (3 to 61 m) units toward the top.

The shales are laminated, medium to light brownish-gray, micaceous, silty, noncalcareous, firm to hard, and contain varying amounts of disseminated organic matter present as laminae and microlaminae. The sandstones are predominately very fine grained to silt sized, light gray, and consist of well to moderately well sorted, subangular, clear to frosted quartz grains. The sandstones may be iron stained, but are uniformly noncalcareous and moderately consolidated. Accessory minerals are predominately pyrite, calcite (rare), chlorite, glauconite, and kaolinite, the latter being most common deeper in the section.

Within the lower Pliocene, the Gros Morne Formation equivalent is marked by the presence of moderate to abundant amounts of kaolinite which appears as a mottled gray, white, soft, silty, amorphous substance. Usually the occurrence of a flood of kaolinite marks the top of the Gros Morne equivalent in the Teak field. This is indicated by a change in color of the shales from a uniform medium gray to a medium gray with a brownish and sometimes yellowish-brown hue. This color persists throughout the Gros Morne in the Teak area.

Above the Gros Morne equivalent, the Mayaro Formation equivalent is lithologically similar to the Gros Morne except the shales do not have a yellowish brown hue. The Mayaro Formation equivalent extends from the base of the MM-6 to within the MYO horizon. Their onshore equivalents are also

EPOCH		ROCK STRATIGRAPHIC UNIT	
HOLOCENE		MARINE SHELF SANDS (RELICT)	
PLEISTOCENE		UNNAMED	D sand member
PLIOCENE	UPPER PLIOCENE	UNNAMED	F sand member
			G-1 sand member
			Q sand member
			U sand member
			MYO sand member
	LOWER PLIOCENE	MAYORO FM	MM-01 sil. member
			MM-1 sand member
			MM-2 sand member
			MM-4 sand member
			MM-5A sand member
			MM-6 sand member
			MM-6L sand member
		GROS MORNE FM	MM-7 sand member
			MM-7A sand member (OPR-2)
			MM-9 sand member
			MM-9A sand member
			MM-10 sand member
			MM-11 sand member
			MM-12 sand member
	?		?

FIG. 13—Stratigraphic column for Eastern Offshore basin.

productive on both the eastern and western sides of southern Trinidad.

The upper Pliocene sediments are approximately 3,000 ft (914 m) thick and consist of alternating sandstone-shale sequences. The sandstones predominate and range from 10 to 350 ft (3 to 107 m) in thickness. Lithologically, the sandstones resemble those already described, except that they are not as consolidated and sometimes have a small amount of medium to course fraction. Lignite and shell fragments become more common with minor amounts of glauconite, micas, and siderite, but kaolinite is not present. The shales are laminated, medium gray, firm to soft, micaceous, silty, noncalcareous, and contain laminae of finely disseminated organic matter. Higher in the section, the shales grade into and are interbedded with soft to very firm claystone. Currently, the upper Pliocene, T, T-1, and U sand zones are producing.

Unconformities are present within the upper Pliocene section, but none represents a substantial hiatus. The F horizon, which marks the boundary between the upper Pliocene and Pleistocene, is unconformable contact.

PLEISTOCENE

The Pleistocene is represented by 3,000 ft (914 m) of interbedded unconsolidated to loosely consolidated sandstones, soft clays, and claystones in unequal amounts. The clays are gray, very soft, lignitic, micaceous, noncalcareous and silty. The sandstones are medium to very fine grained, white to gray, with poorly sorted, subangular, clear to opaque to frosted quartz grains and dark rock fragments. In general, the sandstones are loosely consolidated to very friable, with abundant foraminiferal tests, organic matter, lignite, wood, shell fragments, and black, fine-grained organic material present. Accessory minerals are siderite, micas, and occasional glauconite.

HOLOCENE

A veneer of loose quartz sands and soft, gray muds comprise the Holocene section in the Eastern Offshore basin. There are ubiquitious shell fragments, foraminiferal tests, organic matter, and plant material.

STRUCTURAL GEOLOGY

As development of the Teak field progressed, it was thought that any variance observed in stratigraphy between wells was caused by contemporaneous growth on the 'F' fault. However, recent detailed work in the field suggests that there is little stratigraphic variation across the field, and that absence of section or irregularities within individual sand lobes can be explained by a very complicated fault pattern (Figs. 14, 15). Within the field the 'F' fault is actually a family of five major east-dipping down-to-the-basin faults. At least six distinguishable west-dipping antithetic faults are present, and in some instances are interpreted to cross-cut the old pre-existing 'F' system. The combined throw on the east-dipping 'F' system is approximately 4,000 ft (1,219 m), whereas the combined throw on the west-dipping antithetics measures less than 2,000

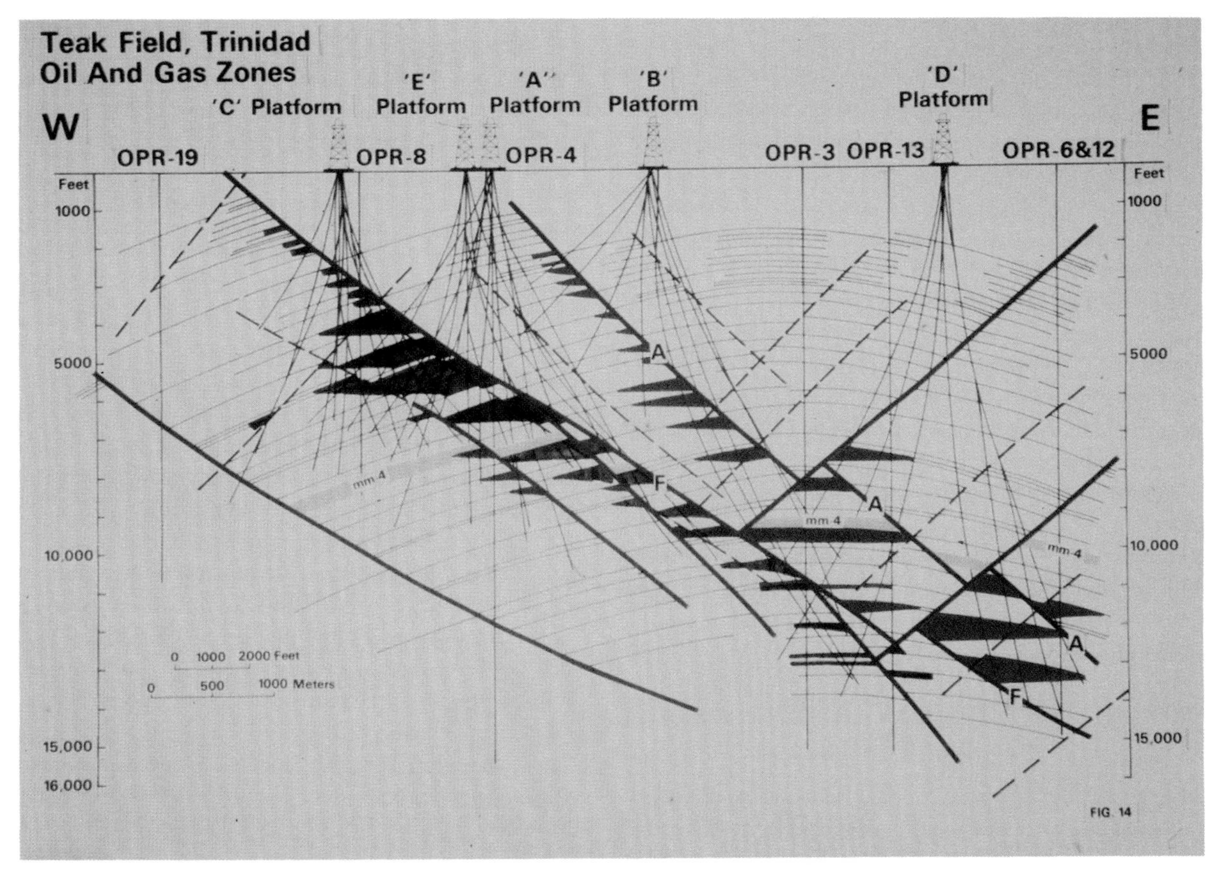

FIG. 14—Cross section, Teak field, showing oil and gas zones. For well locations see Figure 10.

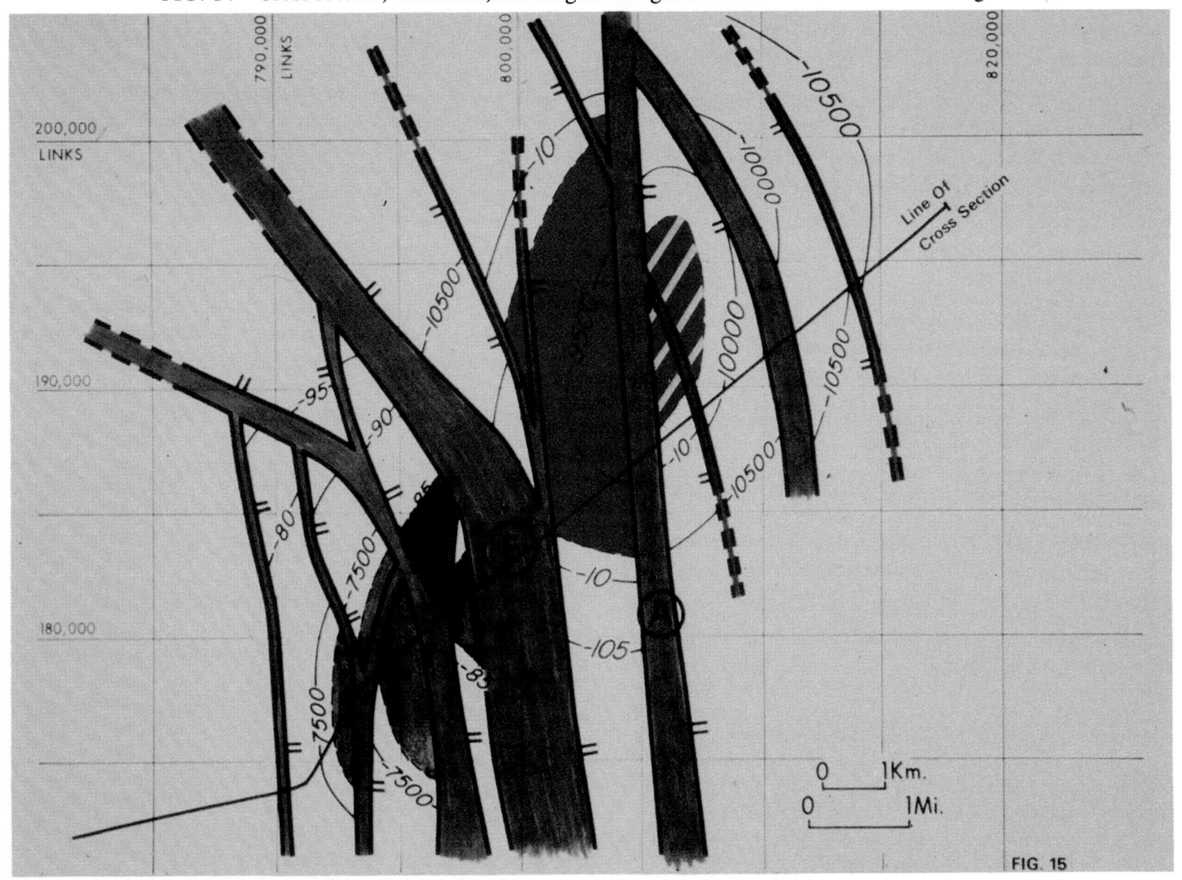

FIG. 15—Structure map of top of MM-4 sandstone, Teak field, Trinidad.

ft (609 m). This accounts for the steeper structural dip of the beds on the west flank.

The main 'F' fault and the 'A' fault are the two major east-dipping faults in the Teak field. Neither is a major growth fault in the Pliocene, but they appear to limit the distribution of hydrocarbons by providing an updip seal for the reservoirs. Almost all the oil in the field has been found in fault blocks upthrown to the main 'F' fault in 32 pay sandstones ranging in thickness from 10 to 600 ft (3 to 183 m) each. Similarly, gas, which is found downthrown to the 'F' fault, almost always occurs in the block upthrown to the 'A' fault in the same sandstones as the oil.

The distinct separation of the major oil and gas reservoirs in Teak has intrigued most geologists and engineers who work the field. It is generally agreed that primary migration of the oil and gas into the Teak structure was along the 'F' fault system from underlying Miocene or older shales before major structures occurred. The present reservoir configuration is the result of secondary migration of the oil and gas when major structural development occurred along the 'F' fault system and the antithetic down-to-the-southwest fault system.

RELATION OF HYDROCARBON ACCUMULATION TO STRUCTURE

The numerous east- and west-dipping faults which traverse Teak field form many separate fault blocks and subpools. Original reservoir pressures exhibited a normal hydrostatic gradient and were approximately the same in each oil pool upthrown to the main 'F' fault and in each gas pool upthrown to the 'A' fault. Original hydrocarbon-water contacts were similar for correlative zones in each of the pools; however, in a few places where faults have in excess of 200 ft (61 m) of throw, the hydrocarbon-water contacts are offset correspondingly. However, since production began, productive capacities of wells from correlative zones in different fault blocks have shown marked differences. Well productivity is related to structural position. As production from the oil and gas reservoirs progressed, it became evident that most of the faults are effective pressure seals, and subsurface pressures now differ across faults by as much as 1,000 psi. These pressure differences are the result of depletion which exceeded the available water drive. This results in different recovery rates dependent on the total surface area of the reservoir in communication with the water contact.

The subdivision of the Teak oil and gas reservoirs into many independent fault blocks with subpools has led to the development of the field on a fault-block-by-fault-block basis. Wells are completed to produce from one zone in a fault block to avoid pressure communication.

CHARACTER OF THE RESERVOIR FLUIDS

Generally, the character of the reservoir fluids changes little with depth in Teak field. The gravities of the oil range from 24.5 to 32°API with no wax or sulfur. Correspondingly the gas gravity ranges from 0.7 to 0.8%, with a solubility of 250 to 450 cu ft/bbl. The temperature and pressure gradients at Teak are approximately 1.25°F and 46 psi/100 ft (30 m) of depth.

The porosities range from 20 to 30% in the upper Pliocene reservoirs to 14 to 20% in the lower Pliocene reservoirs. Correspondingly, the permeabilities range from 300 to 500 and 150 to 200 md, respectively.

CONCLUSION

Structurally the Eastern Offshore basin is the result of compressional forces directed north-south between the South American and Caribbean plates. Concurrent with the tectonism was the deposition of an abundant supply of sediment throughout the Tertiary from the Guayana shield area to the south. These sediments were compressed into east to west trending en echelon folds. In the late Miocene–early Pliocene right-lateral movement of the Los Bajos fault produced a counterclockwise rotation of the en echelon folds to a northeast to southwest trend. In the Pliocene and Pleistocene major faulting gave rise to a primary down-to-the-east and northeast system and an antithetic down-to-the-southwest system which transected the northeast to southwest trending en echelon folds. This structural style resulted in an intensely faulted anticline at Teak with faults acting as trap boundaries for a composite hydrocarbon column in excess of 3,100 ft (945 m). Since the discovery of Teak this regional geologic model has been substantiated by the discovery of Samaan and Poui fields in the Eastern Offshore basin.

SELECTED REFERENCES

Barr, K. W., and J. B. Saunders, 1968, An outline of the geology of Trinidad: 4th Caribbean Geol. Conf., Trans. (1965), p. 1-10.

Bruce, C. H., 1973, Pressured shale and related sediment deformation: mechanism for development of regional contemporaneous faults: AAPG Bull., v. 57, p. 878-886.

Bassinger, B. G., R. N. Harbison, and L. A. Weeks, 1971, Marine geophysical study northeast of Trinidad-Tobago: AAPG Bull., v. 55, p. 1730-1740.

Kugler, H., 1959, Geological map of Trinidad: Onell Fussli Arts Giaphiques SA, Zurich, Switzerland.

Lau, W., and W. Rajpaulsingh, 1976, A structural review of Trinidad, West Indies in light of current plate-tectonics and wrench fault theory: 7th Caribbean Geol. Conf., Trans. p. 473-483.

LePichon, J., 1968, Sea floor spreading and continental drift: Jour. Geophys. Research, v. 73, p. 3661-3697.

Molnar, P. and L. Sykes, 1969, Tectonics of the Caribbean and American regions from focal mechanism: Geol. Soc. America Bull., v. 50, p. 1639-1684.

Stainforth, R. M., 1948, Description, correlation and paleoecology of Tertiary Cipero Marl Formation, Trinidad, B.W.I.: AAPG Bull., v. 32, p. 1292-1330.

———, 1968, Mid-Tertiary diastrophism in northern South America: Trans. 4th Caribbean Geol. Conf. Trans., p. 159-174.

———, 1978, Was it the Orinoco?: AAPG Bull., v. 62, p. 303-306.

Geology of the Handil Field (East Kalimantan—Indonesia)[1]

By A. C. Verdier[2], T. Oki[3], and Atik Suardy[4]

Abstract Handil field is in the swampy distributary area of the present Mahakam River delta in East Kalimantan (Indonesia), in the central part of the Kutei basin.

The anticline features (10.5 km long by 4.5 km wide) was mapped using seismic data in 1973, and the discovery well was drilled in April 1974. A east to west fault, perpendicular to the axis of the anticline, divides the field into two parts of roughly equivalent area.

The area of closure is 40 sq km and vertical closure increases with depth through the hydrocarbon-bearing section. Most of the 150 reservoir sandstones between the depths of 450 and 2,900 m are tidal to fluvial deltaic plain sediments of middle to late Miocene age; most of them are oil bearing with a gas cap. The depositional environments can be identified as channel fills, tidal bars, etc. A high pressure zone is present below 2,900 m where deeper prospects have not yet been drilled.

More than 70 significant lignitic or coaly marker beds are used to correlate the sandstones.

The field has been divided into 6 superimposed zones from the lower zone (2,800 to 2,450 m) through the very shallow zone (850 to 450 m) corresponding to changes in environment of deposition and/or oil characteristics. Isobath maps show a displacement to the southwest by approximately 3 km of the top of the anticline from deeper zones to the surface.

In the upper and shallow zones, hydrocarbons are found in the southern part only, which itself is divided into smaller blocks by a east to west fault pattern.

Daily production is 160,000 bbl of oil with cumulative production reaching 140 million bbl at the end of 1978.

INTRODUCTION

The Kutei basin, situated along the eastern margin of Kalimantan, ranks second only to the Central Sumatra basin in importance as a hydrocarbon producing area in Indonesia (Fig. 1). Drilling for hydrocarbons began in the late nineteenth century and several onshore oil fields were found: Sanga-sanga (1897), Balikpapan/Klandasan (1898), Semberah-Pegah (1906), and Samboja (1908). At the end of 1970 more than 300 million bbl of oil had been produced from these fields.

In 1966 Pertamina (National oil company of Indonesia) signed the first production sharing contracts with foreign oil companies and, as a result, the exploration activity increased and extended into offshore areas. In 1971 Total Indonesie, a

[1]Manuscript received, May 21, 1979; accepted for publication, July 23, 1979.

[2]Compagnie Francaise des Petroles (Total Indonesie), Jakarta, Java, Indonesia.

[3]Inpex Indonesia, Ltd., Japan.

[4]Pertamina, Indonesia.

This paper is the result of the efforts of a team of geologists, geophysicists, sedimentologists, and micropaleontologists that has worked on the Mahakam Delta and the Handil field for 5 years. The authors are indebted to all their colleages and friends for their work, and particularly to P. Lalouel who is largely responsible for the preparation of this paper.

The writers thank Pertamina, the Indonesian state company, for permission to publish the data included in this paper.

The writers also thank Total Indonesie (Compagnie Francaise des Petroles) and Inpex Indonesia, Ltd., who have kindly given permission to prepare and present this study.

Article Identification Number:
0065-731X/80/M030-0019/$03.00/0.

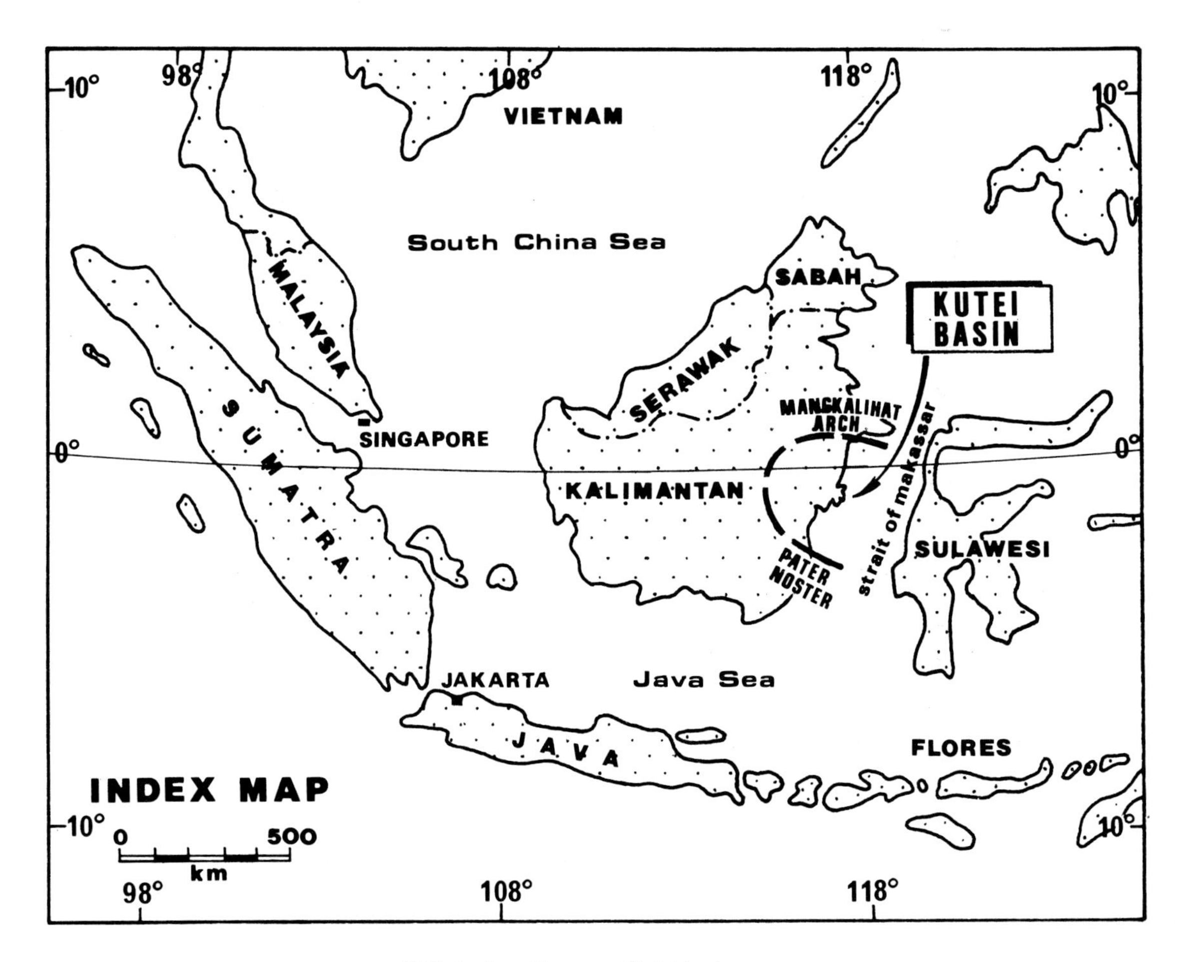

FIG. 1—Location map, Kutei basin.

subsidiary of Compagnie Francaise des Petroles, who had been active in Indonesia since 1967, bought into the Japex (now Inpex) Mahakam production sharing permit (Fig. 2). Since 1966 numerous hydrocarbon accumulations have been discovered in the basin, of which the most important are Attaka (1970), Bekapai (1972), Badak (1972), and Handil (1974).

The anticlinal structure of Handil, located in the Mahakam River delta, was first mapped in 1973 using seismic data, and the discovery well (Handil-1) was drilled in March-April 1974. Five delineation wells were drilled before the decision to develop the field was made.

An early stage of production (phase I) started in June 1975 and produced at an average rate of 40,000 b/d from a cluster of 4 wells. The oil was transported by shuttle tanker to a storage tanker which was being used at that time for Bekapai field production as well.

Phase II was completed in 1976. At the present, production of 160,000 b/d from 65 wells is transported by 20-in. (51 cm) pipe to Senipah Terminal (Fig. 3) and loaded into tankers via an offshore Single Buoy Mooring (SBM).

REGIONAL SETTING

Paleogeography

Throughout the Eocene and the early part of the Oligocene a predominance of marine shales were deposited in the Kutei basin.

In the late Oligocene an important phase of regression began and has continued through to the present. Clastic sediments transported by numerous rivers prograded from west to east over a depositional front of a few hundred kilometers between the Paternoster and Mangkalihat carbonate provinces (Fig. 1). In the Handil area during the middle and late Miocene, a major delta was formed by the principle river that drained the onshore area at that time. Production from Handil field comes from multiple-sandstone reservoirs deposited during this period.

The distribution of the sand during the middle and late Miocene indicates that the position of the Miocene delta coincides with the recent delta of the Mahakam River. Moreover the presence of the maximum thickness of the upper and middle Miocene fluvial facies in the axis of the present delta shows that the distributaries have not significantly changed position since that time.

Consequently, it was a "Proto"-Mahakam River located at the same place as the current Mahakam River which deposited the producing zones of Handil field.

General Stratigraphy and Lithology of Miocene Delta

A west to east section through Handil and Bekapai fields is useful to locate the Mahakam delta in its stratigraphic context.

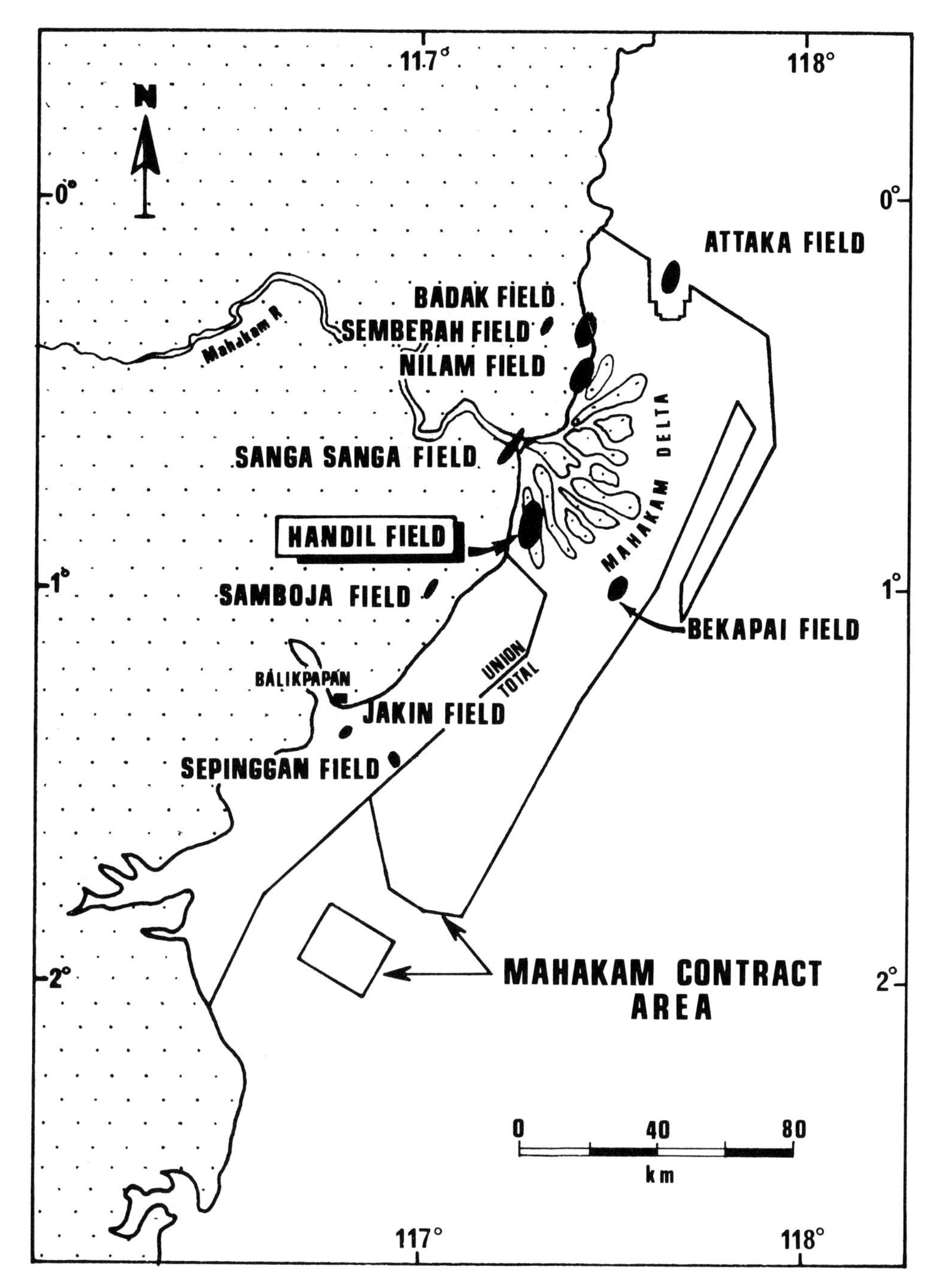

FIG. 2—Total/Inpex production sharing permit offshore East Kalimantan.

It clearly shows the progradation of the deltaic sediments from west to east. A synthetic lithologic log extrapolated from data on Handil field and neighboring wells show a regressive megasequence thickening and coarsening upwards, which reflects progradational deposition (Fig. 4).

Characteristic fauna aids in the identification of environments. From bottom to top of the megasequence is deep-marine fauna which are replaced progressively by fauna representative of shallower environments and which characterize, respectively, bathyal and shelf deposits. The appearance of sands containing shallow-marine fauna and arenaceous brackish-water organisms characterize delta-front and lower tidal delta-plain deposits; higher in the section the shallow-marine fauna disappear and only a few brackish-water arenaceous species remain, which corresponds to an upper tidal delta-plain environment. The upper sand-rich sequence is almost barren of fossils, but some radiolarians and diatoms are present and these sediments are interpreted as fluvial

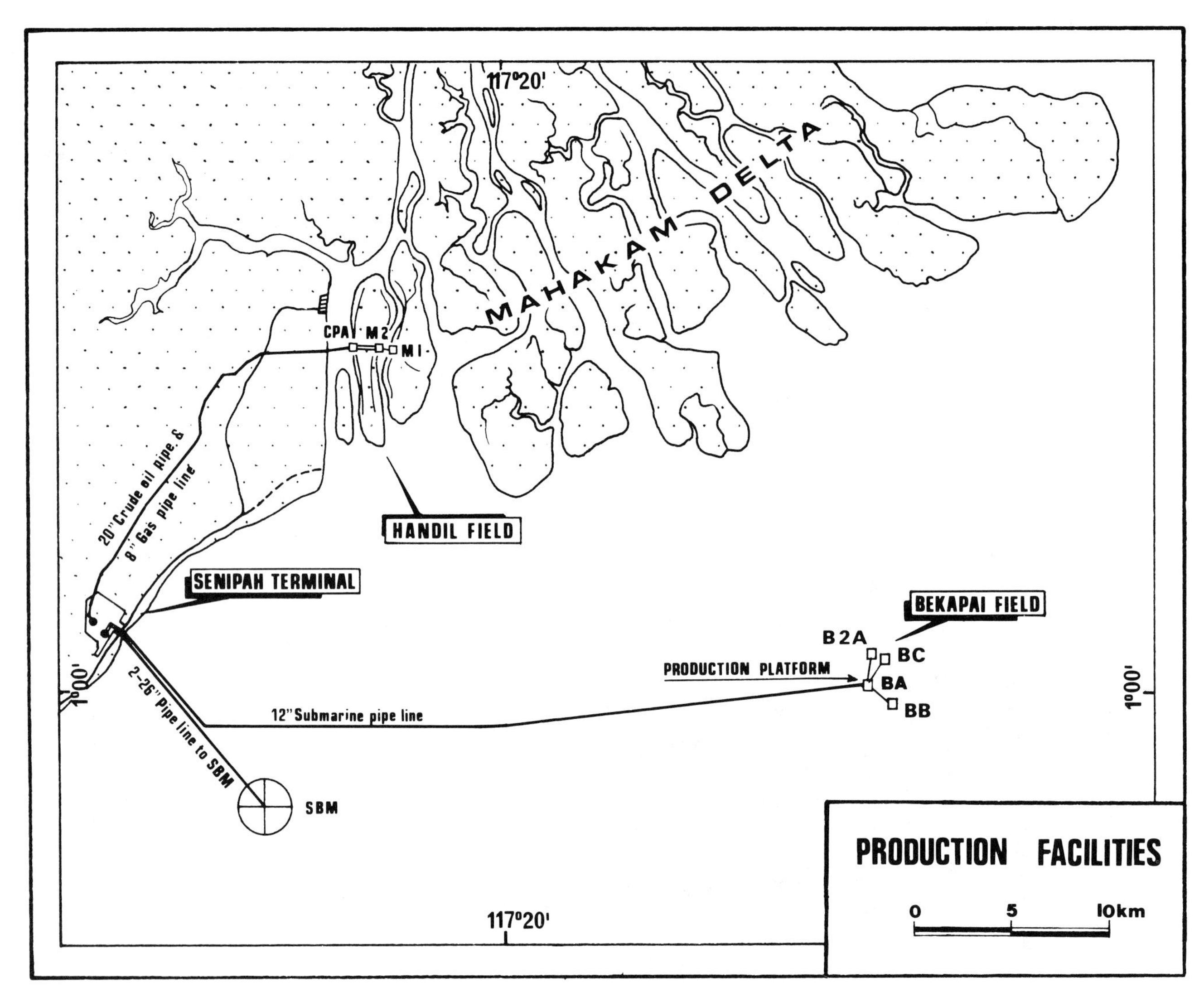

FIG. 3—Total Indonesie: location of production facilities.

delta-plain deposits.

Structural Setting

On the side-looking airborne radar (SLAR) imagery, a succession of long and narrow anticlines, oriented parallel with the East Kalimantan coastline can be seen in the Samarinda area (Fig. 5). In the delta and offshore the same north-northeast to south-southwest structural deformation is completely obscured by the rapid recent sedimentation, but is clearly seen on the interpretation of seismic data (Fig. 6).

It is thought that these trends are the result of the eastward sliding of the thick sedimentary cover which is made up mostly of shale. This folding was accentuated by shale diapirism in the center of the anticlines which has led in some places, to localized thrusting on the flanks of the structures. This motion toward the depocenter has abutted against a north to south oriented basement high in the Handil area resulting in crumpling, parallel with the high of the shallow more clastic-rich sediments.

Handil is the southern extension of the Badak-Nilam trend which is parallel with the Attaka-Bekapai trend (Fig. 6).

HANDIL STRUCTURE

Seismic Studies

Initial seismic surveys in this area used low-energy techniques and the results were not sufficient to detect the Handil anticline. Because unconsolidated material is thickest in the distributaries, it is likely that most of the downward directed energy was absorbed near the surface.

The Handil structure was first mapped after a seismic survey in 1973 (Fig. 7). This survey, rather than using the river courses to shoot seismic, avoided the water as much as possible and crossings of water areas were made perpendicular to the water whenever possible. Bridging was built above the high-tide level and holes were drilled for explosives. Data continuity was (in general) maintained across the rivers (Fig. 8).

The initial interpretation showed steep eastward dips with closure to the west defined by a saddle. Thinning over the top

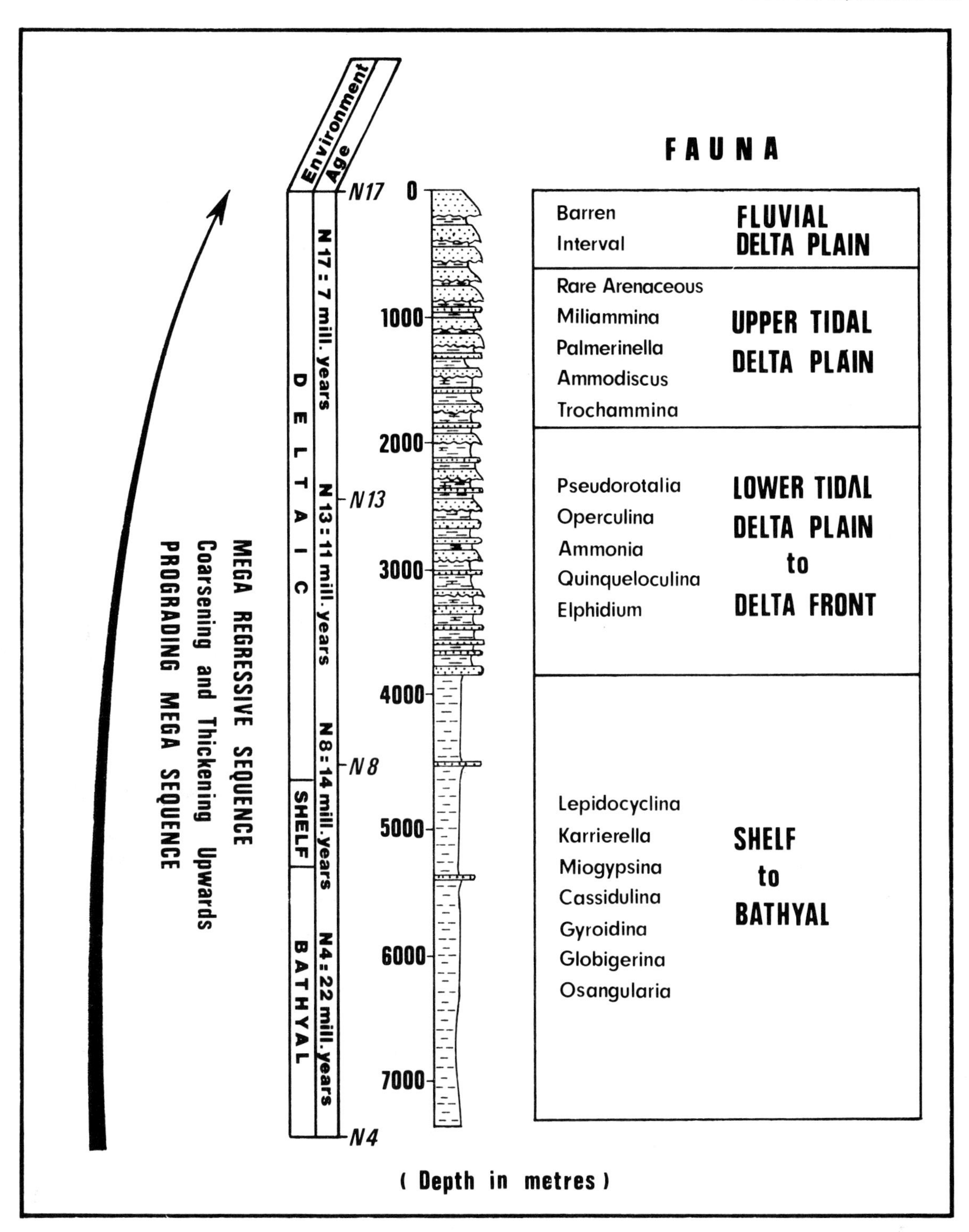

FIG. 4—Schematic Miocene stratigraphic column Handil field.

of the structure indicated early growth at the time of reservoir deposition. Vertical closure was defined at approximately 250 msec with a closed area of approximately 35 sq km.

A seismic program to delineate the field was undertaken in 1974 after the drilling of the Handil-1 discovery well. Again, essentially land techniques were used. The seismic correlations from north to south did not agree with the well correlations which suggested the presence of a fault.

The latest detailed seismic survey of Handil was conducted in 1976. The common-depth point (CDP) coverage was increased from 12- to 24-fold. It was hoped that this program would be of sufficient quality to permit definition of faults and even sand bodies interbedded with shales. Data quality was superior to the previous surveys and confirmed the presence of the east-southeast to west-northwest fault across the structure, but it was not sufficient quality to define the sand bodies. As

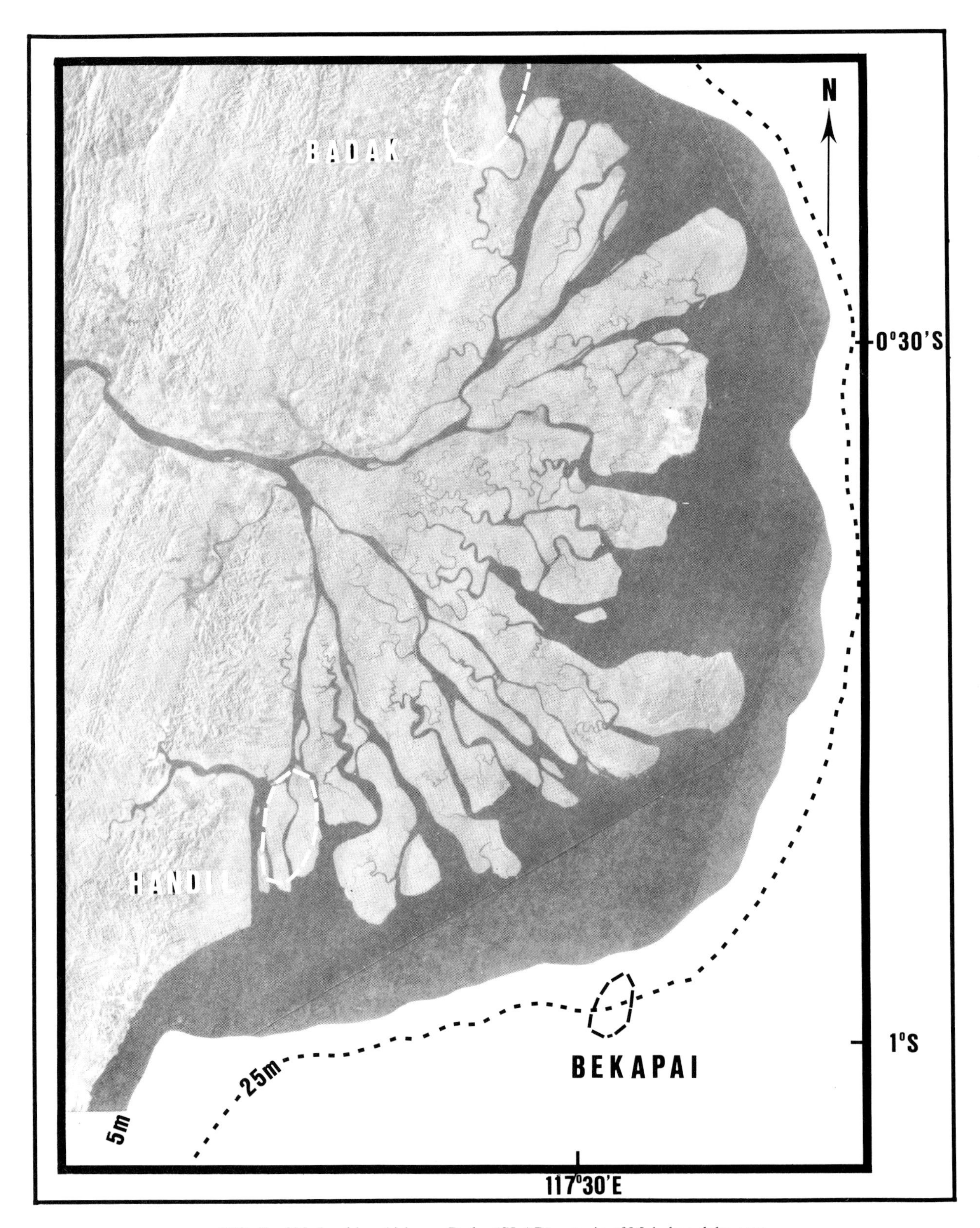

FIG. 5—Side Looking Airborne Radar (SLAR) mosaic of Mahakan delta area.

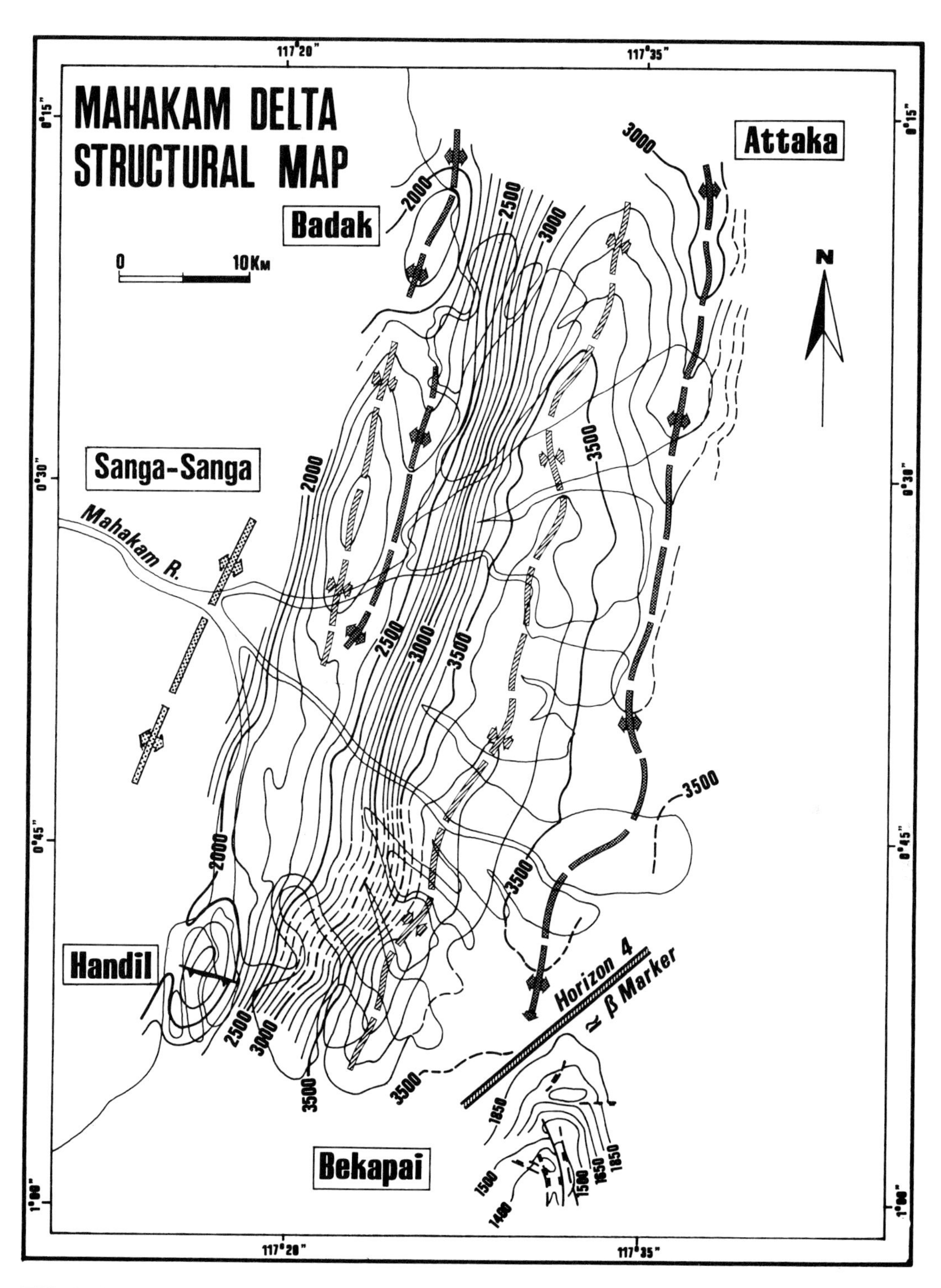

FIG. 6—Structural map of horizon 4 of the Mahakam delta area showing north-northeast–south-southwest trending axis. Contours in two-way time.

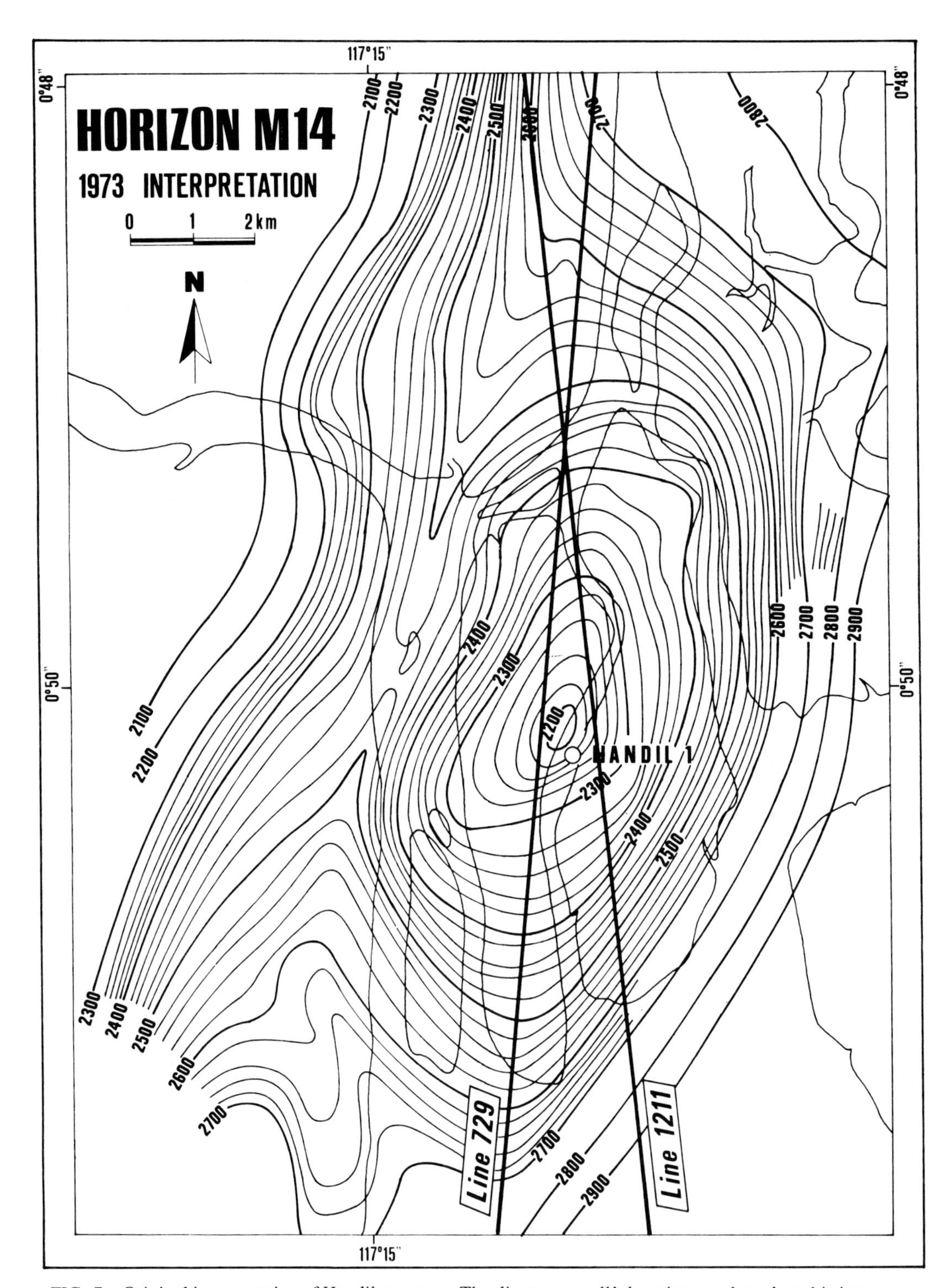

FIG. 7—Original interpretation of Handil structure. The discovery well's location was based on this interpretation.

can be seen on north to south line 1211 (Fig. 9), the fault decreases in throw with depth and does not seem to have penetrated the deeper zones.

The "M" horizons marked on line 1211 are seismic events that correspond as closely as possible to the "R" geologic markers. The interval between M-5 and M-20 corresponds to the main pay zone. The M-28 to M-30 interval has not yet been drilled. The M-30 horizon is the deepest event which can be mapped and shows the same structural picture as the shallower horizons although there is a suggestion of an unconformity. A deep well, H-9, is planned to test this zone.

To date, the seismic data have not been useful in defining small structural or stratigraphic features. Perhaps in the future, field techniques will provide data of higher frequency spec-

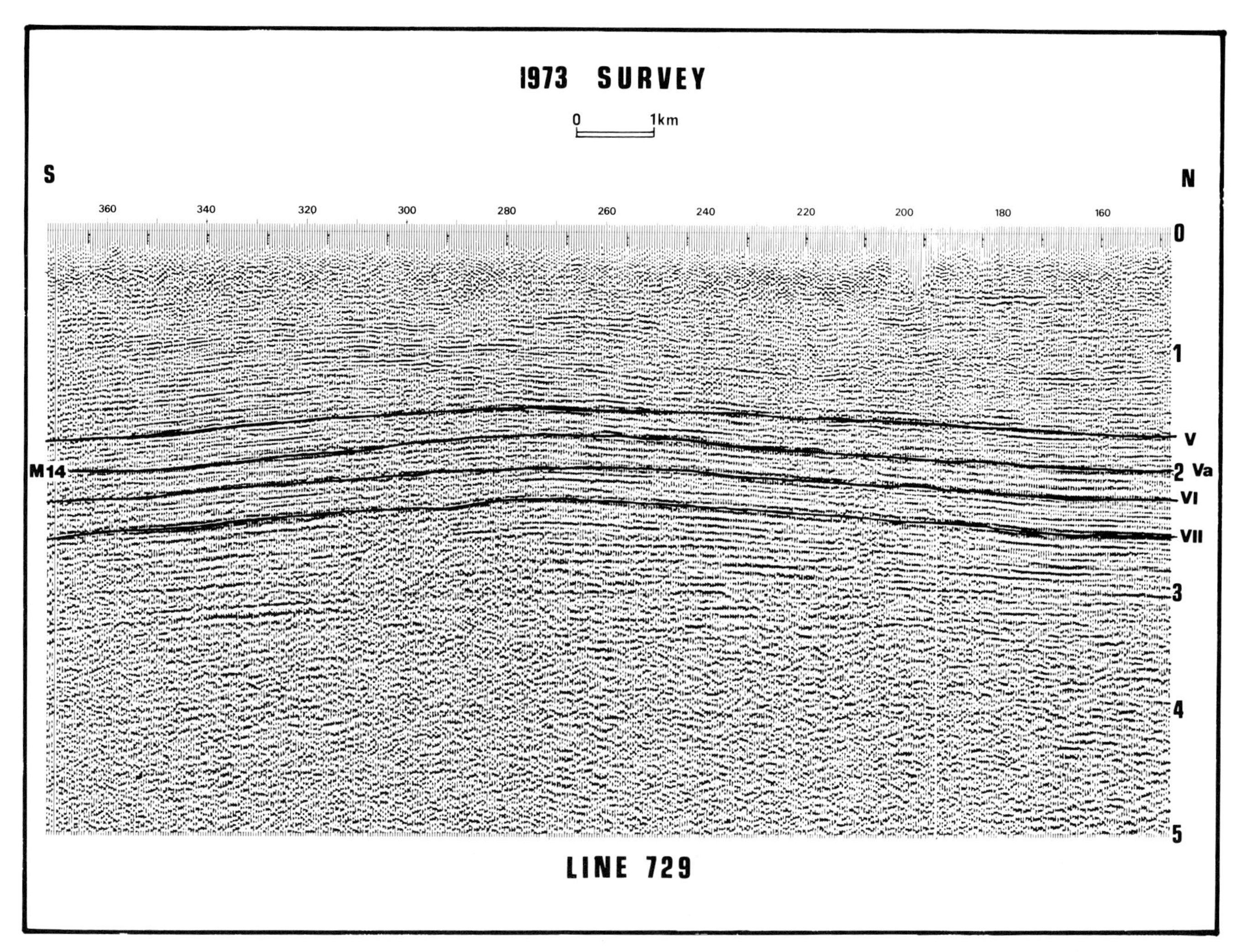

FIG. 8—Line 729 from 1973 survey. Definition of fault is difficult.

trum allowing more advanced interpretation techniques. This is particularly relevant to the drilling of additional wells on the flanks of the structure.

Structural Geology

Drilling data have confirmed the anticlinal structure defined by seismic data in 1973. The anticline has an average dip of 10° on the east flank and 8° on the west flank where there is a spill point toward the northwest (Fig. 10). Development wells have shown that dips along the structural axis are smaller than shown on the seismic data (4° instead of 7°) and the area of structural closure at the depth of the main hydrocarbon levels is 40 sq km. The vertical closure increases with depth, from 100 m at a depth of 850 m to more than 300 m below 2,200 m depth. The area of structural closure also varies partly in relation to the increasing vertical closure but also due to shifts in the top of the structure and changes in its length.

A major fault, which is difficult to detect on the lastest seismic data, cuts the field into two major parts. This normal fault is now defined in four wells and is oriented almost perpendicular to the axis (N105 against N20) with a dip of 70° southward and a throw of 80 m in the main reservoir interval. The displacement decreases with depth from 100 m at 1,200 m depth to 70 m at 2,400 m depth, and the fault appears not to cut the zones below 3,000 m but this has not been confirmed by drilling data. It is not known if the fault extends beyond the vicinity of the structure.

In the southern compartment, detailed well correlations have shown several other minor normal faults oriented generally parallel with the major fault and having throws of less than 40 m. However, these faults, which cannot be seen on the seismic lines, serve as impermeable barriers to hydrocarbons. There is no detailed correlation available in the upper part of the northern compartment because there are no hydrocarbon bearing reservoirs in this interval. The fault pattern of the whole field shows only four faults in the southern compartment and none in the northern (Fig. 11), but there are probably other small faults in the northern compartment as well as in the south.

Another major structural feature, which has a strong effect on the distribution of hydrocarbons within the field, is the displacement of the progressively higher stratigraphic horizons on the top of the anticline (some 3 km) southward along the axis of the anticline. The top, which is between the wells HH-1 and

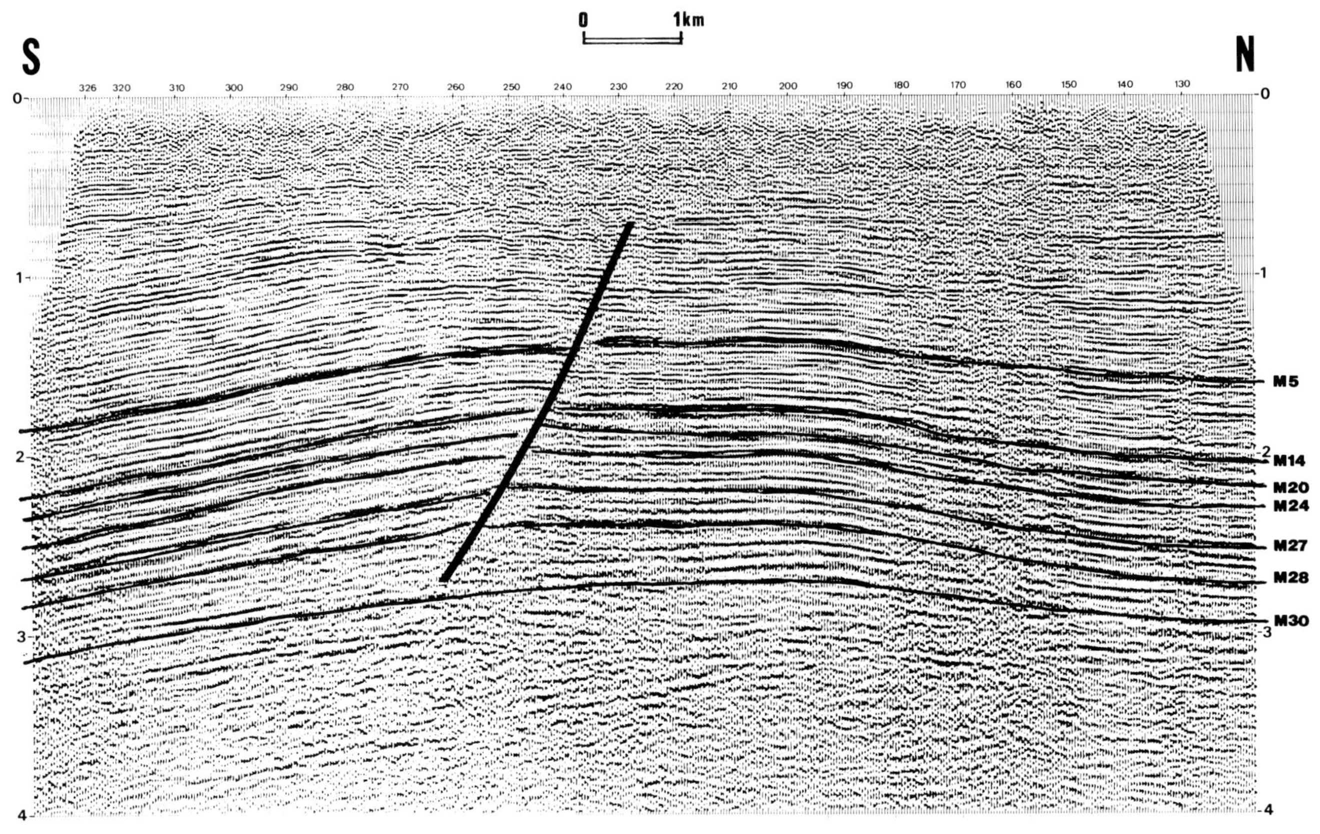

FIG. 9—Line 1211 from 1976 showing the fault position.

HR-1 at a depth of 2,800 m, moves into the southern compartment and, at a depth of 450 m, is near the HL well cluster (Figs. 10,11, 15). This displacement of the top may be due to the minor faults.

Figure 16 shows that, below 1,800 m, there are approximately as many hydrocarbon-bearing zones on both sides of the fault, but the southern compartment is richer in gas than the northern one. The contacts between fluids of any given reservoir are lower in the southern compartment than in the northern one. All the reservoirs are at hydrostatic pressure so an equilibrium was attained after faulting, and there is no noticeable diminution of the gas caps in the southern downfaulted compartment nor gas-cap extension in the north.

This fault was not an important factor in the accumulation of the hydrocarbons (at least for what is known as the main zone). It occurred after the initial migration of hydrocarbons and presently acts as a seal. Moreover, the displacement of the top of the anticline is not related to this major fault and controls the fluid distribution.

ENVIRONMENTS OF DEPOSITION

Modern Mahakam Delta and Sandstone Distribution

A modern Mahakan River delta has formed since the Holocene marine transgression (Fig. 12). The delta itself extends over about 5,000 sq km comprising subaerial deltaic plain, delta front (shallow platform, water depth $\leq$ 5 m) and prodelta slope. Wave action in the area is minor due to the narrowness of the Makassar Straits (150 km) and the tidal range is 1 m to 2 m in the delta area.

From the Holocene transgression to the present, sediments were deposited in these conditions. Thus the Mahakam delta has developed under the influence of both tidal and fluvial action. This has resulted in a specific sedimentary sequence.

Detrital sediments transported by the Mahakam River consist of fine well-sorted sands which are spread by funnel shaped distributary channels. Due to the lateral migration of the thalweg in these channels, sand accumulates in lateral accretion bars forming channel-fill sequences of clean massive

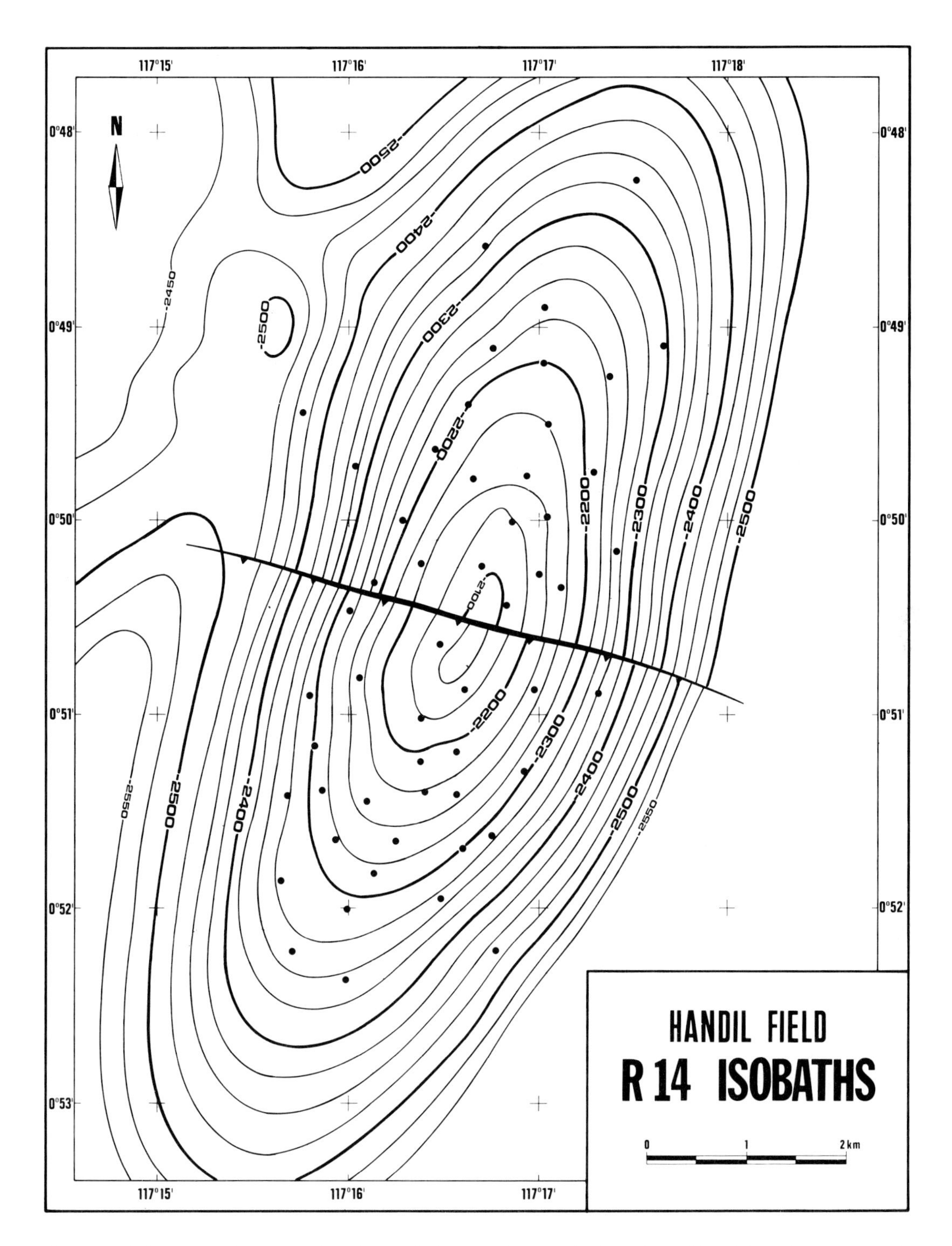

FIG. 10—Map of Handil structure based on well data. The R14 marker is in the middle of the main pay interval.

sands with errosive bases and shale content increasing upwards (Fig. 13). Such lateral accretion bars occur mainly in the upper deltaic plain where tidal action is reduced. They often form continuous sandstone units due to the coalescence of individual bars and result in elongated sandstone ribbons of several tens of kilometers long and up to a few kilometers wide.

In the lower deltaic plain, tidal currents influence the sediment distribution. This results in the widening of the channels which become estuarine in shape. Preferential current paths (downstream fluvial current and periodically upstream tidal current) favor the development of channel bars in the middle of distributary channels. The sedimentary sequence in these channel bars consists of coarsening-up and thickening-up vertical succession (Fig. 13). The clean sand formed at the top of each sedimentary sequence commonly is overlain by a deltaic-plain tidal-marsh sequence, composed of organic clays which become fixed by the advancing vegetation (nipah-nipah marsh). The lower tidal deltaic plain is also the preferential site of de-

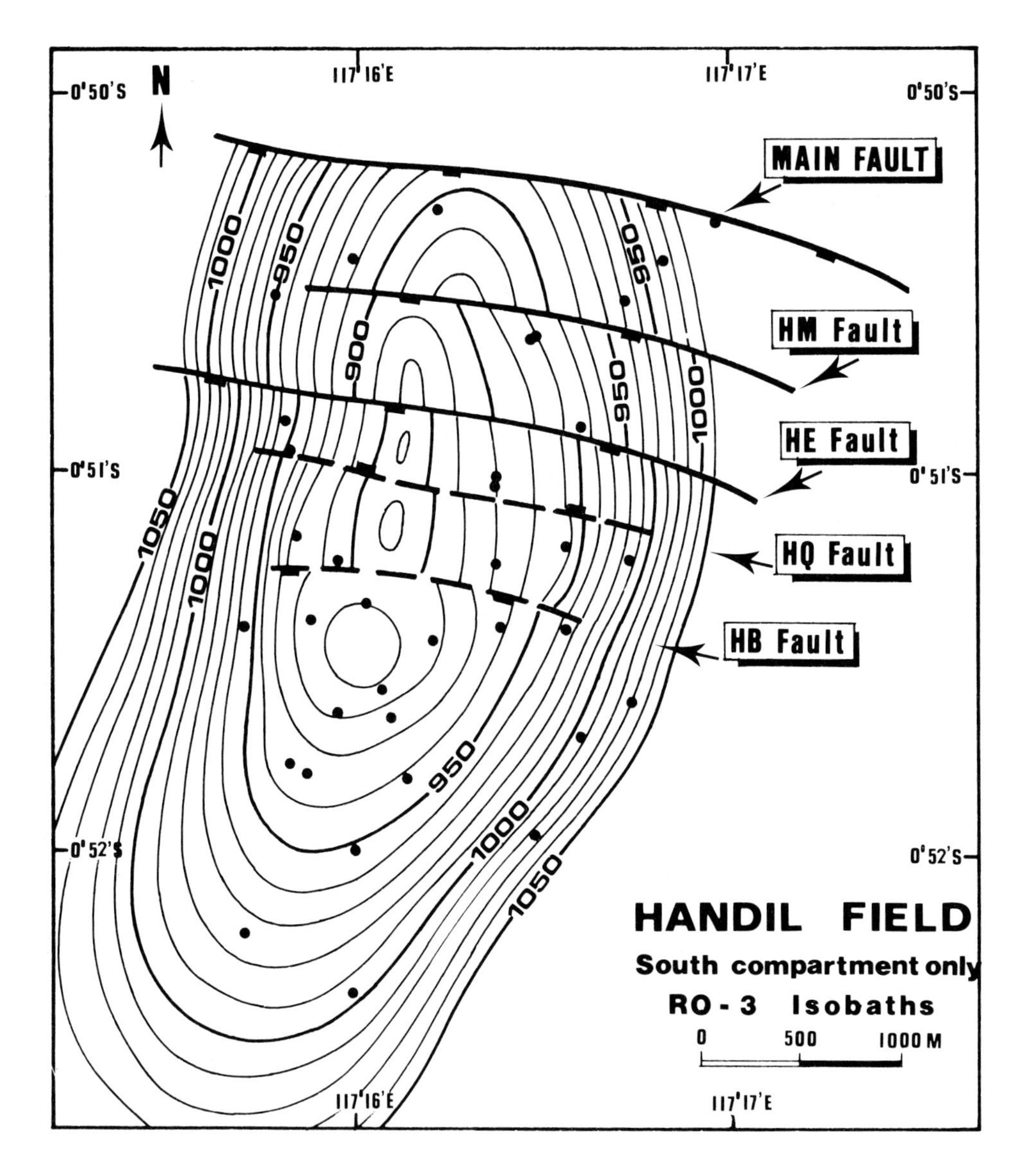

FIG. 11—Map of southern compartment of Handil field showing the fault pattern in the shallow zones.

velopment of tidal channels which are sinuous with subvertical sides and can be 20 m deep. These channels, which cut the lower delta-front sediments, are at present not receiving sediment.

In the delta front environment, the distributary channels discharge sediment on the shallow delta front platform (5 to 10 km wide) under tidal influence. Sand deposits consist of stream mouth bars, middle ground bars, and crescent bars depending on the strength of distributary currents (Fig. 14). The deposits consist of 5 to 10-m thick accumulations exhibiting a coarsening-up character. Clean sand occurs at the top of the units but sand/shale laminations form the main part of the sand bodies. The geometry of these bars are generally changed by tidal currents which reworks the sediment resulting in elongated or U-shaped sand ridges parallel with the tidal current direction. Clay-rich sediments are deposited between (mud flats) and sometimes above these sandy bars.

The prodelta slope represents around 30% of the total area of deposition of the delta but no sand accumulations have been recorded in the present prodelta slope environment.

Ancient Mahakam Delta

More than 200 discrete sandstone zones have been identified in Handil field between the depths of 450 and 2,900 m (Fig. 15). These are Miocene deltaic sediments and are interbedded with massive, silty or organic shales and coal beds. They range in thickness between 0.5 to 25 m, but coalescing sand zones may reach 50 m thickness.

These sands occur in two main types (channel and bar) recognizable on logs due to their distinct log character responses. The channel type, which has a strong erosional base, begins with massive clean sand with rare shale laminations, shaliness increases upwards, and the sequence is frequently topped by a coal or an organic-rich shale. The bar type is characterized by a transitional base with development of sand/shale alternations

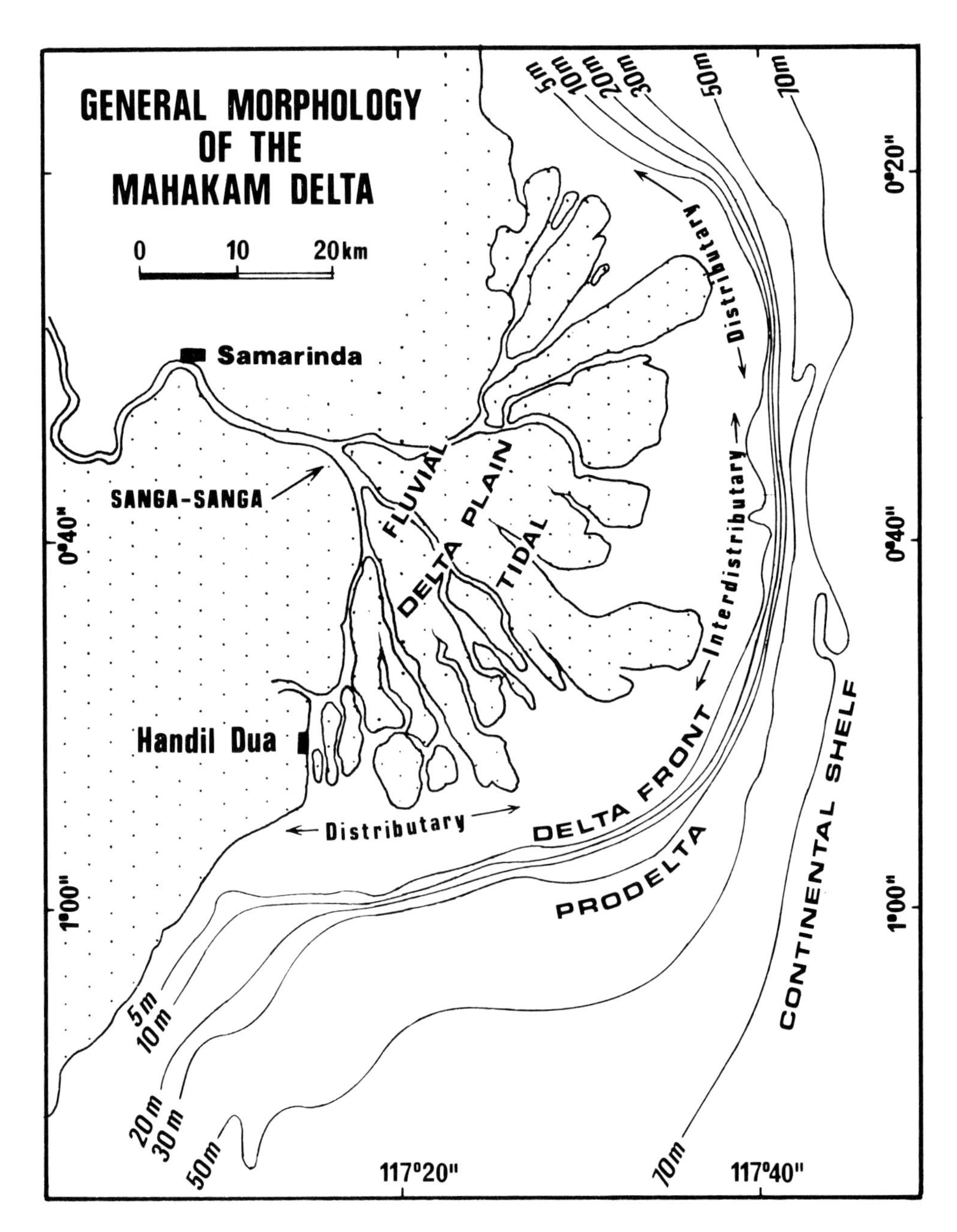

FIG. 12—General morphology of the present Mahakam delta.

with shaliness decreasing upwards. The upper part of the sequence is made up of clean sand with few shale laminations (Fig. 16). Such basic sequences are recognizable in nearly all the sandstones, but sometimes there are variations, either by development of shaly layers in bars of channel sands or by superposition of elementary sequences (Fig. 16).

Sedimentological studies, including core description, micropaleontology, geochemistry as well as electrofacies comparison, and field studies of the modern Mahakam delta have shown that the Miocene sediments are deltaic. They were deposited, as were the Holocene sediments, in a tide-dominated environment associated with a greater sediment discharge than at the present.

Channel sandstones were deposited mainly as distributary channel fill in a deltaic-plain environment, and commonly are associated with organic-rich shale and coal. Most of the channel sands represent several filling phases and result in "multi-story" sandstone reservoirs with thickness exceeding 20 m. Channel sandstones may also occur in the delta-front environment associated with sandstone bars indicating an active progradation of distributary systems.

Bar sand belongs mainly to a more open-marine environment and are the "classic" type of deposit found on very shallow platforms off distributaries (Allen et al, 1976). The intertidal delta-front platform is an environment where various types of bars are present.

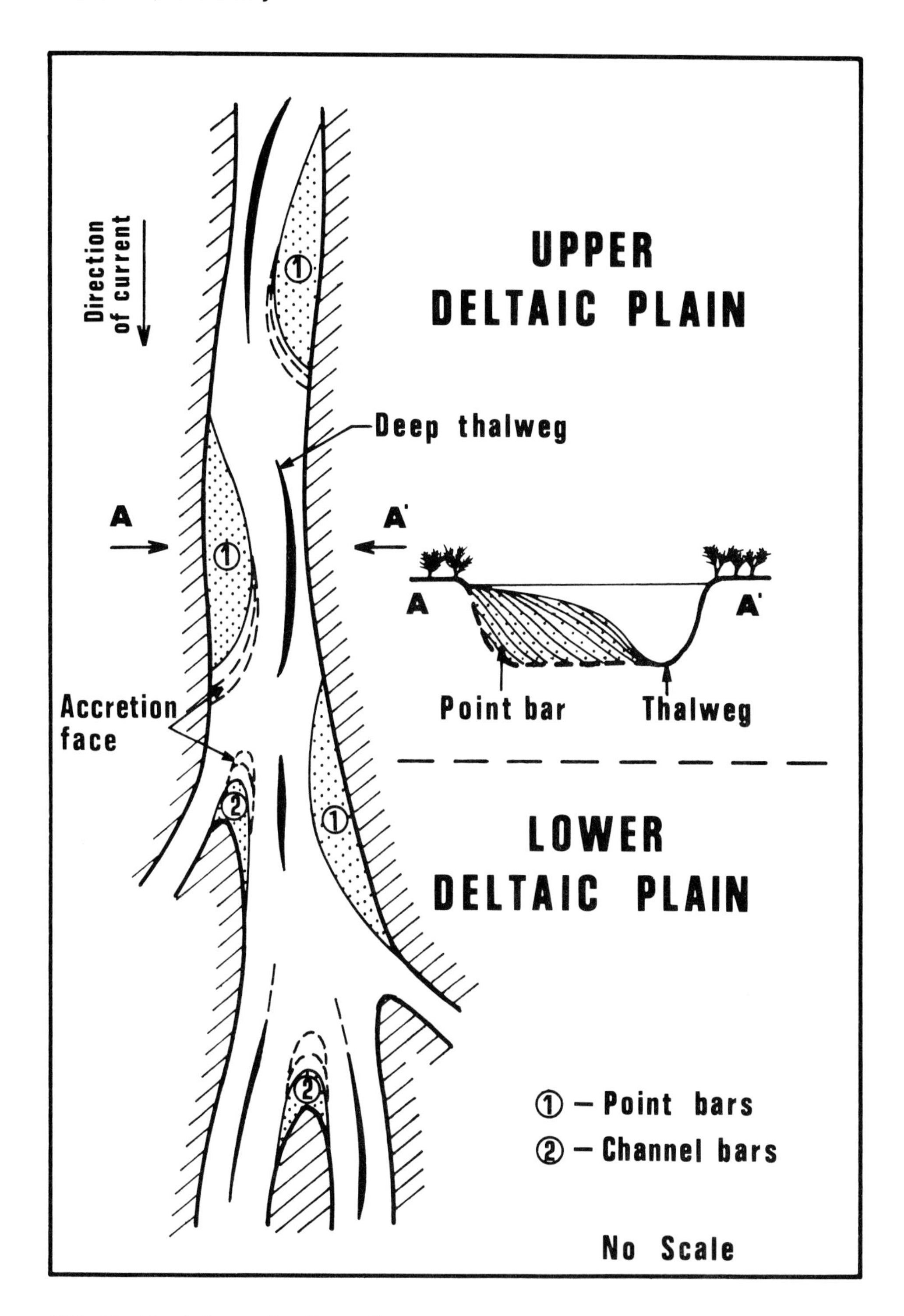

FIG. 13—Morphology and sediment distribution pattern of delta-plain distributary environment (modified from Allen, 1976).

Prograding channel sands which locally occur on the platform environment can greatly enhance the lateral and vertical reservoir continuity of these bar sandstones by providing connections between different bars. Small and elongated bars also sometimes occur in deltaic-plain environments in the middle of old distributary channels.

In this type of sedimentary environment, the progradation phases result in superposition of elementary sequences with bar dominant (delta front) or channel dominant (deltaic plain) reservoirs. Transgression phases are not marked by accumulation of detrital sediments and, as the previous delta construction may be partly destroyed by the advance of the sea, most of the sequences are incomplete. A complete progradational phase includes from base to top: prodelta clay, delta-front

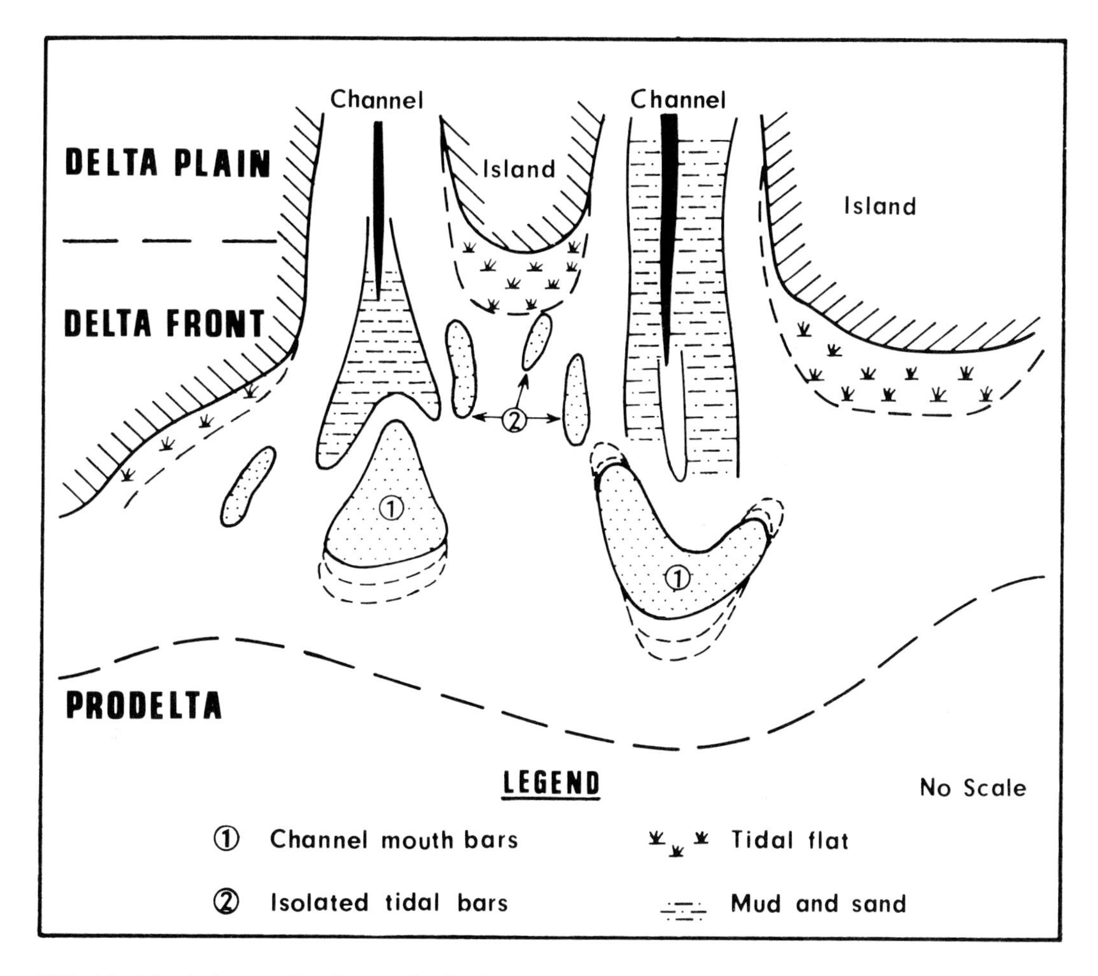

FIG. 14—Morphology and sediment distribution pattern of tidal delta-plain and delta-front environments (modified from Allen, 1976).

shale, alternations of bar sands and shales, alternations of bar and channel sands, channel sands and organic-rich clays, and finally coal beds (Fig. 16).

RESERVOIR DISTRIBUTION AND CHARACTERISTICS

Horizontal Reservoir Distribution

Well to well correlations are based on the continuity of the lignite and coal beds. More than 70 marker beds have been defined between the depths of 450 m and 2,900 m and most are lignite coal beds or organic-rich shale. They provide a helpful framework for reservoir correlations (Fig. 17). Generally, the interval between two markers corresponds to one progradational sequence and, as far as there are no fluid incompatibilities or discrepencies in the pressure measurements, all the sand bodies found within the same interval are considered as forming only one reservoir. Approximately 150 independent hydrocarbon bearing reservoirs, with their own gas-oil and oil-water contacts, have been defined.

The reservoirs producing hydrocarbons are often complex bodies made of bar and/or channel sandstones. As elementary units of deltaic sequences the reservoirs of channel or bar type have different reservoir characters. Channels create thick and good quality reservoirs of elongated and frequently sinuous form. Bars are thinner and more shaly, except at the top of the sequences, and although generally of wide areal extent, they have limits which are more related to the reservoir cut-off value rather than an identifiable sedimentary boundary. Such elementary units often coalesce vertically or laterally and this results in reservoirs of complex geometry and varied quality.

Vertical Reservoir Distribution

The vertical distribution of the reservoirs shows, from top to bottom, a succession of hydrocarbon-bearing and water-bearing sandstones as if several "pools" composed of different sandstones were superposed one over the other (Fig. 15). In each of these "pools," the uppermost reservoirs have the largest gas caps and the lowest sands of each unit are water bearing. Historically, these units have been defined one after the other during development drilling. These sandstones form individual pay zones which have different horizontal extension and relative importance (Table 1). The main and upper-main zones contain 85% of the reserves in place and produce 90% of the present 160,000 b/d.

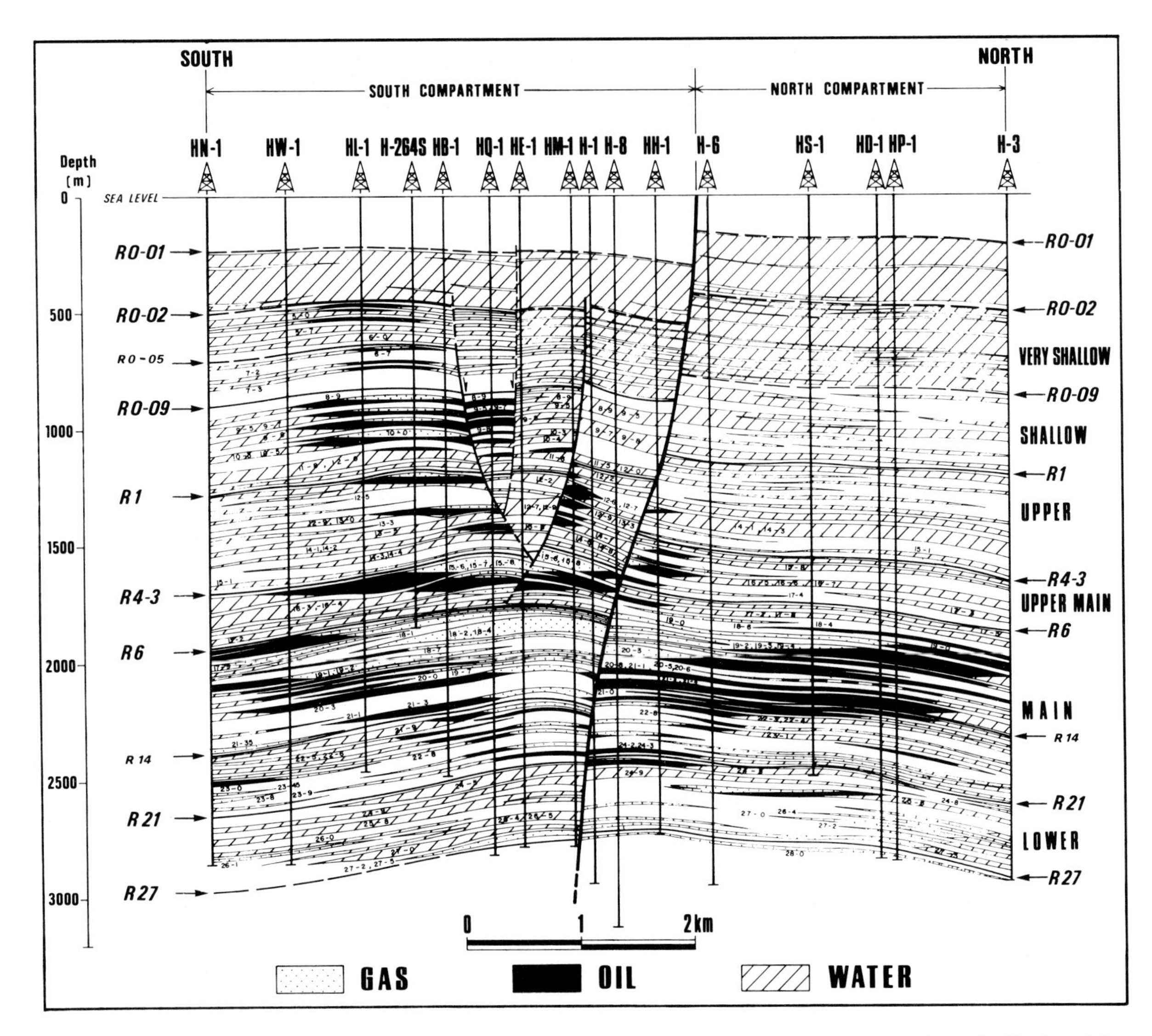

FIG. 15—Schematic north to south cross section of Handil field indicating the multiple reservoirs and the distribution of the hydrocarbons.

They are almost invariably topped by a particularly continuous and thick coal bed (R6, R21) or a thicker shaly sequence rather than the interbedded shales present between the reservoirs (R0-09, R1, R4-3). The termination of these "megasequences" must correspond to a major geologic event, marked by an important change of environment of deposition.

Reservoir 19-7

This reservoir unit illustrates the complexities common to Handil field, and clearly shows the relationship between channels and bars. The reservoir (called 19-7 because it was reached at a depth of 1,970 m in the reference well H-6) is located between one coal (R9) and one organic-rich shale (R10) which are good marker beds over the entire field. Most of the wells penetrated a number of typical bar sequences up to a maximum of 3 bars. Generally only one or two are present displaying variable facies grading from clean sand to silty shale. In a few wells, these levels are partly eroded and replaced by a thick massive channel sand (Fig. 18).

A detailed study of the vertical and horizontal distribution of these bars and channels in that interval clearly illustrates the complex organization of these sand bodies. Reservoir 19-7 is in fact composed of 5 different units (Fig. 19): a succession of three independent bars (lower, medium, and upper bar) which have poor north to south continuity, and two channels. The upper channel, which is located in the northeastern part of the field, erodes totally the upper bar and locally also the medium bar (Fig. 20). The medium channel located on the western flank erodes the medium bar and is thought to communicate with the lower bar, but this has not been verified by well control. Fluid content and pressure measurements have confirmed the good connection between the upper and middle bar in the northern compartment of the field where the erosion by the upper channel links the two sandstones.

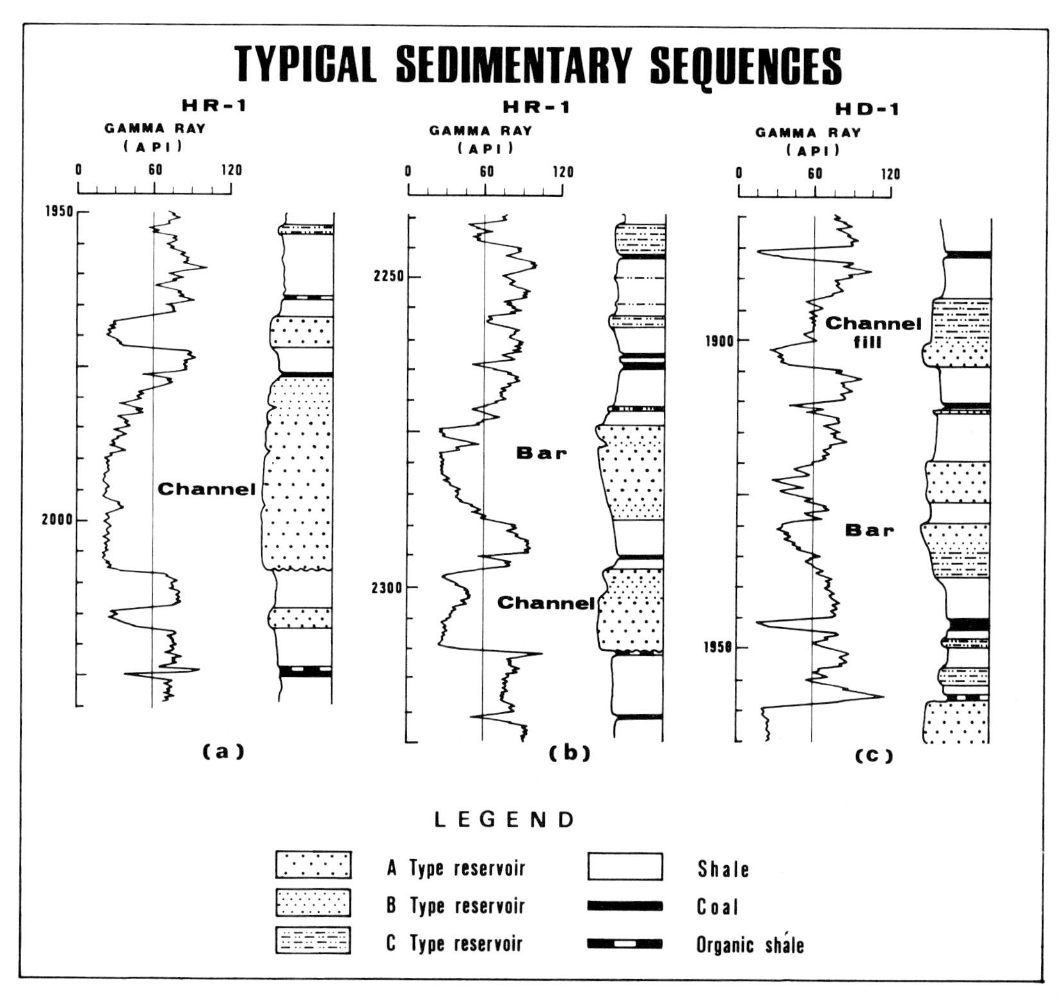

FIG. 16—Typical sedimentary sequences: (a) thick channel unit in a complete sequence, (b) bar sequence overlying a thinner channel unit, and (c) series of uncomplete sequences, no thick reservoir development.

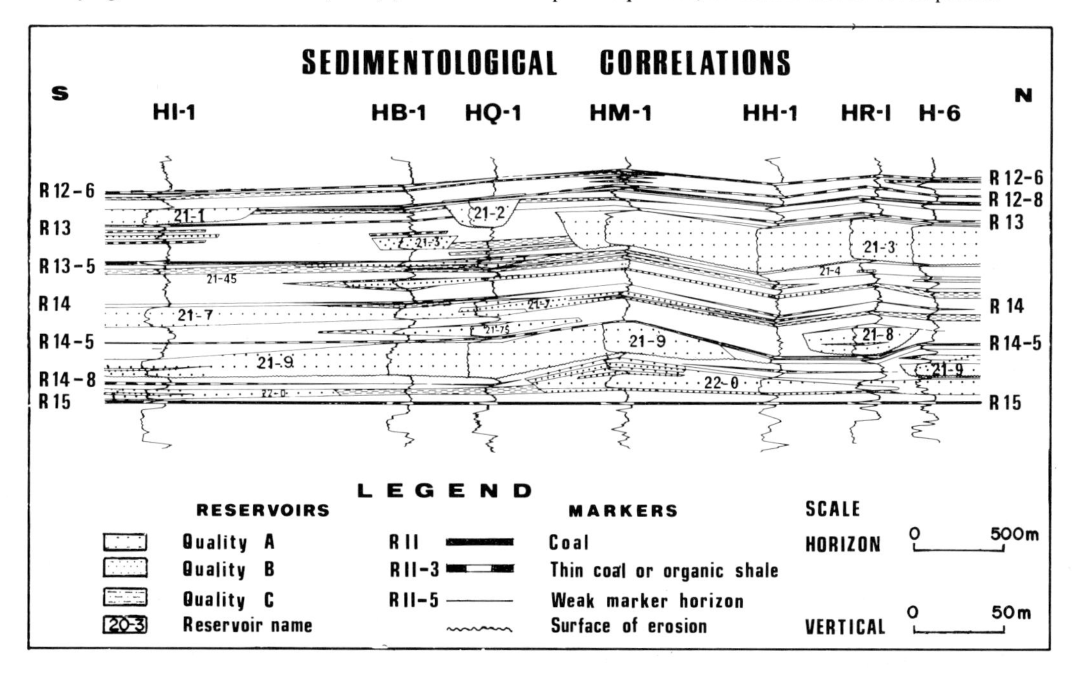

FIG. 17—Example of correlation within the Handil field based on the continuity of coal beds.

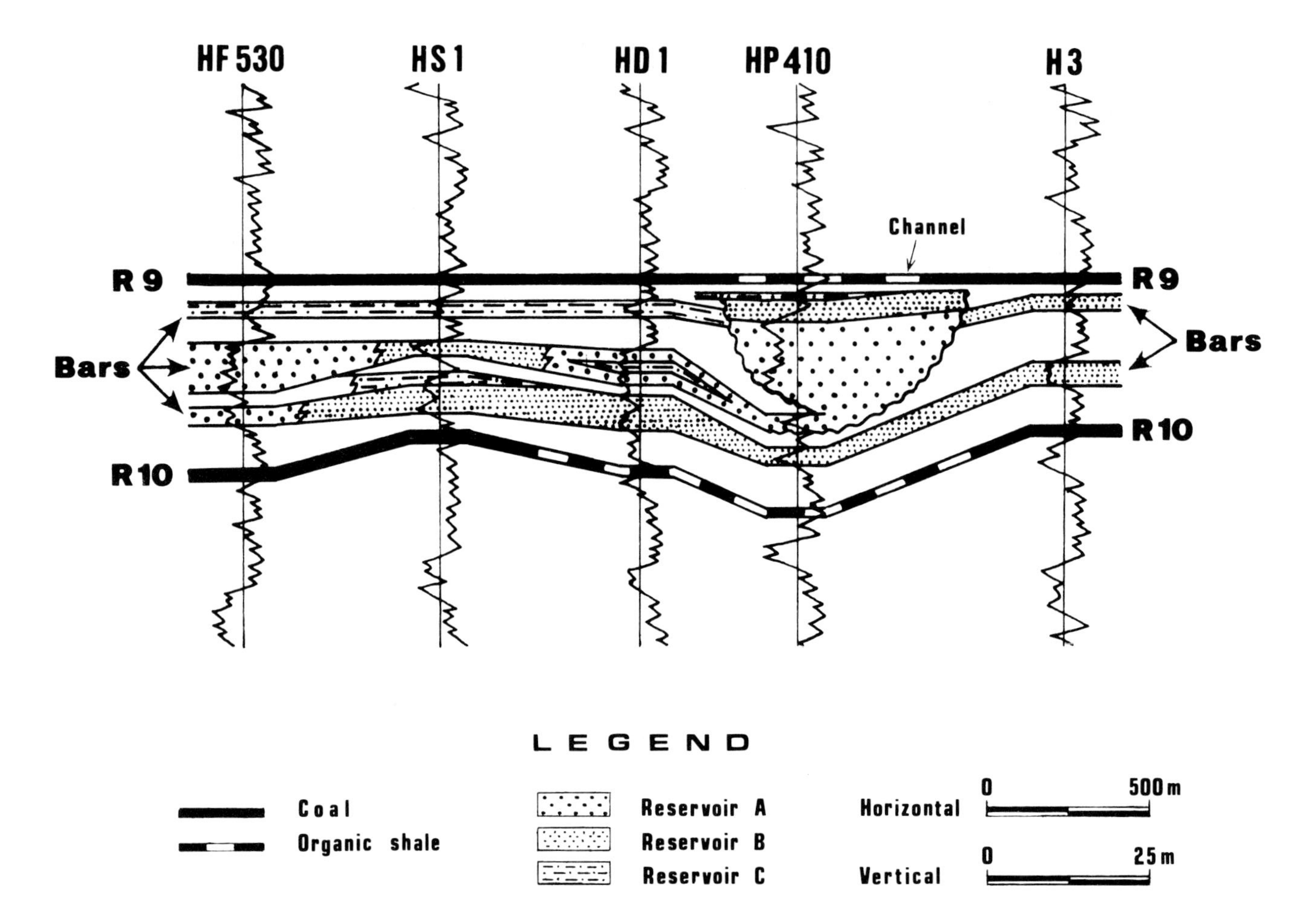

FIG. 18—Detailed correlation between R9 and R10 markers. The 19-7 reservoir is in this interval.

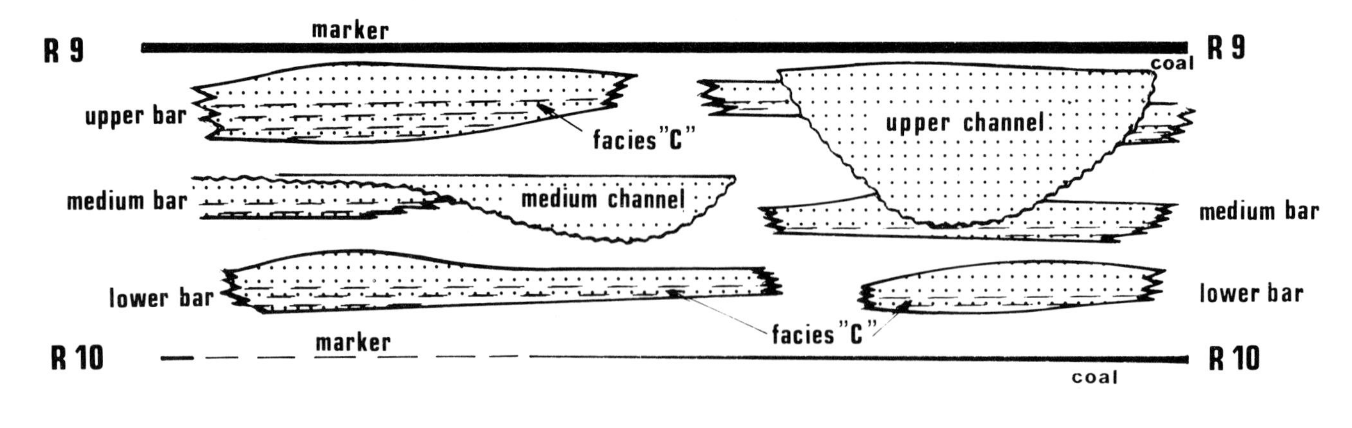

FIG. 19—Sketch representation of the 19-7 reservoir. Note that there are five discrete units, 3 bars and 2 channels with the channels serving as conduits between different bars.

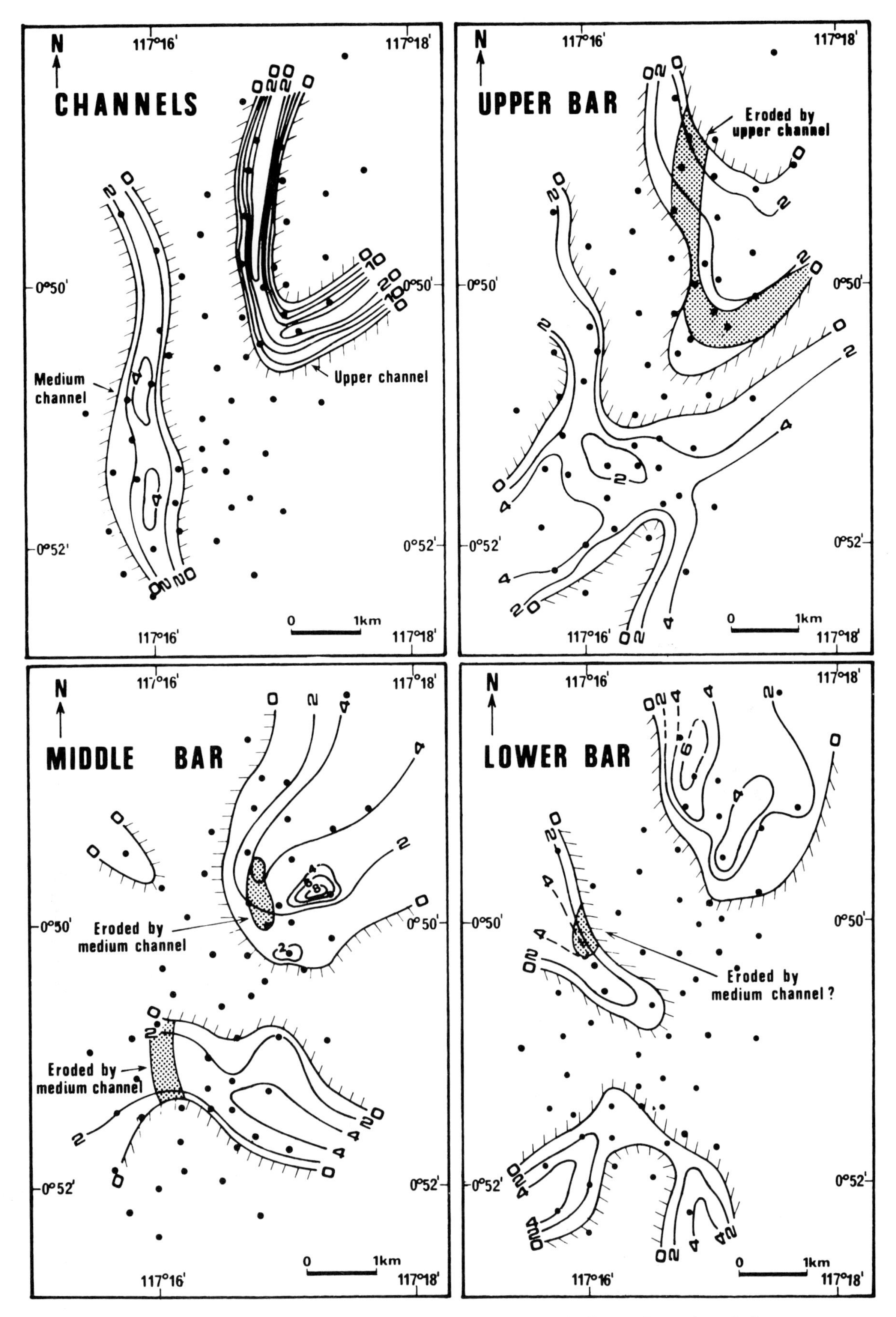

FIG. 20—Isopach maps of each unit within the 19-7 reservoir. Contour Interval equals 2 m.

Table 1. Definition of the different pay intervals and their relative importance.

ZONES	Marker	Indicative depths (m)	Discovery well and date	Number of oil bearing reservoirs	Maximum oil net pay (m)
VERY SHALLOW	RO – 02 (RO – 05) RO – 09	450 650 850	HL 370 Jan 78 H 264S Dec77	3 3	20 20
SHALLOW	R1	1200	HQ-1 Sept. 75	12	75
UPPER	R4 – 3	1550	HM-1 July 75	17	70
UPPER MAIN	R6	1800	H-1 March 74	14	205 (O + G)
MAIN	R21	2450		62	
LOWER	R27	2800		14	
DEEP ?			H-9		
			TOTAL	125	

The vertical superposition of such beds corresponds to a progradational sequence with a delta-front environment illustrated by the bars, overlain by a deltaic-plain environment represented by the channels. The end of the sequence is well marked by the covering coal layer which has been found in all wells in the field.

Such analyses are necessary to understand the complex geometry of the reservoir and these models are then used for the location of new development wells, reserve calculations, and as data base for computerized reservoir engineering studies.

Reservoir Characteristics

Three main types of reservoirs were described from initial core studies and log responses were calibrated on these cores.

The best quality reservoirs are composed of clean sand with shale confined to a small amount of dispersed clay and few shaly laminations. These sandstones have high porosities and good horizontal and vertical permeability. Although the vertical permeability is lower than the horizontal, they are of the same order of magnitude. Because there is a fairly good correlation between horizontal permeability and porosity, the permeability can be easily estimated from the porosity determined from the logs. This reservoir quality is found mainly in the lower and middle part of the channel fills and at the top of the bars when they are well developed.

Intermediate quality reservoirs are made of of shaly sandstones where the porosity is partly reduced by dispersed clay and many shale laminations. The horzontal permeability is still high but the vertical permeability is reduced by the laminations. This kind of facies appears near the top and the edges of the channel-fill sequences and forms the main part of the bars.

The poorest category of reservoirs are the very shaly sandstones with low porosities. There are in fact two types: (1) where shaliness is mainly due to dispersed clay mixed with the sand by bioturbation, and (2) where bulk shaliness seen on gamma-ray logs is due to thin intercalations of fine shaly sands and shale levels. Bioturbated levels are found especially at the top of the channel fills when the vegetation of the delta plain progrades over the channel; they have very low vertical and horizontal permeabilities. The thin intercalations were deposited in the base of the bars and sometimes as clay plugs of channels; they have extremely poor vertical permeabilities but do possess low horizontal permeabilities.

These different qualities of reservoirs are called A, B, and C type, and, for each of them, we have defined cut-offs on porosity and shale content so that the type could be deduced from logs. If the cut-off on the shale content is constant (Table 2), cut-off on porosity is related to depth. Due to the uncommon thickness of the pay zone (450 to 2,800 m) the effect of compaction on porosity cannot be neglected and, as core studies have shown that the maximum porosity decreases with depth, cut-off on porosity changes with depth (Table 3).

GEOCHEMISTRY

Nature Crude Oils

The oils found in the various reservoirs of Handil field have about the same composition and must have been generated from the same type of organic matter. They are paraffinic oils with up to 45% of alcanes and less than 25% of cyclanes (mainly di- and tricyclic). The amount of associated gas increases with depth and the size of the gas caps is related to the depth; the reservoirs of the very shallow zone (450 to 800 m) have no or very small gas caps, whereas in the lower zone, the gas caps are large and the oil annuli are thin.

The distribution of hydrocarbons in crude oils shows an evolution with depth (Fig. 21). In one well, the deepest oil has a predominance of the heavy molecular weight n-alcanes whereas the C_{15} to C_{20} range is more common at shallower depths.

This characteristic has been observed in other wells although not well marked in the diagram of API density (Fig. 22). However, there is a tendency toward lighter oil in the upper reservoirs. This gradient is more evident where pour point is plotted against depth (Fig. 22). The most fluid oils are located at shallow depths due to their higher mobility.

Type and Distribution of Organic Matter

The overall total organic content (TOC) of the shaly levels is good and commonly ranges between 1.5 and 4%, with an average of 2.2% (for 200-sample analyses). In the lignitic beds, normally from 10 cm to 3 m thick, the amount of organic carbon ranges from 45 to 70%. These organic-rich beds comprise about 10% of the thickness of the sedimentary column in the Handil field.

Because of the type of vegetation and its abundance, the organic matter found in both older and younger sediments is preponderantly vegetal and changes little throughout the drilled section (2,900 m). The petrographic composition of this lignitic material is humite-vitrinite 70 to 90%, exinite 10 to 20%, and inertite 2 to 20%.

Under microscopic examination, mainly vegetal debris and gelled wood can be seen. The amorphous organic matter seen in the kerogen, which generally has good oil potential, can be considered to come from the destruction of vegetal tissues. The sapropelic matter, which is considered to be mostly of marine planktonic origin, cannot be recognized due to the overabundant humic fraction associated with recognizable vegetal elements. As far as hydrocarbon potential is concerned, the exinite, always present and composed of exinite and resinite, is the most important.

Maturation

Thermal maturation plays an important part in the generation of hydrocarbons and it can be determined by measuring

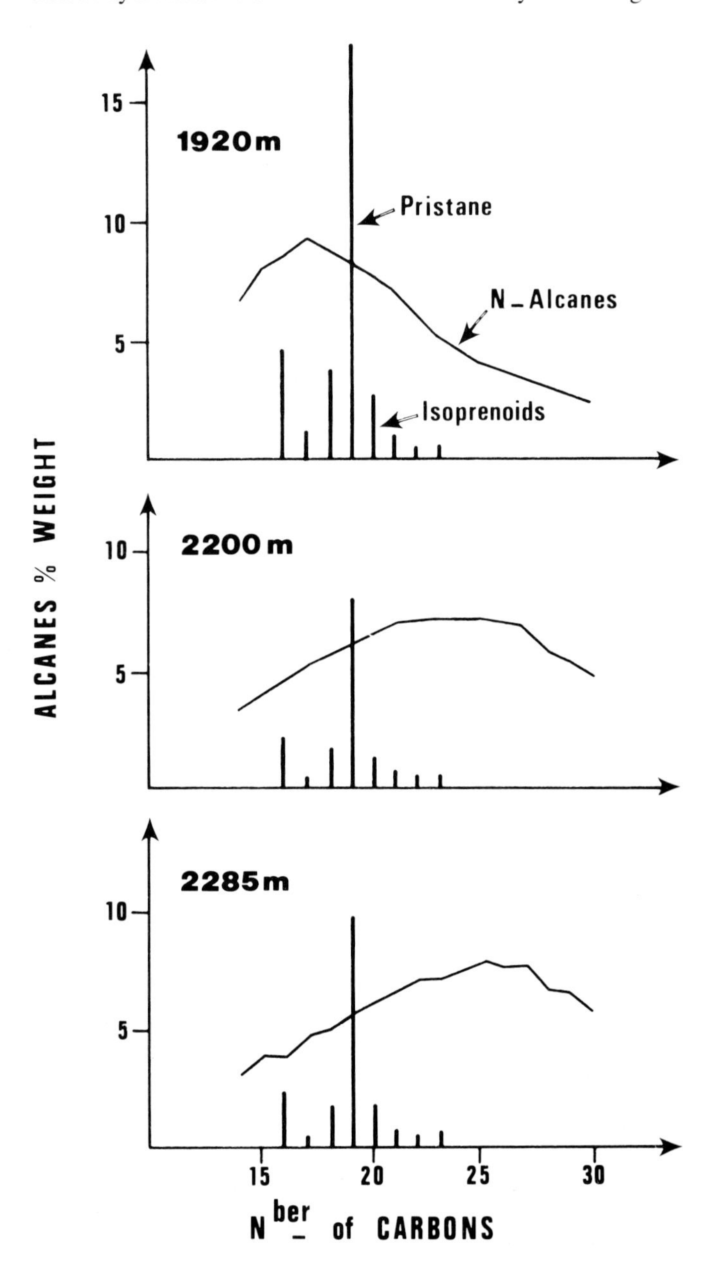

FIG. 21—Change in character of hydrocarbons with depth.

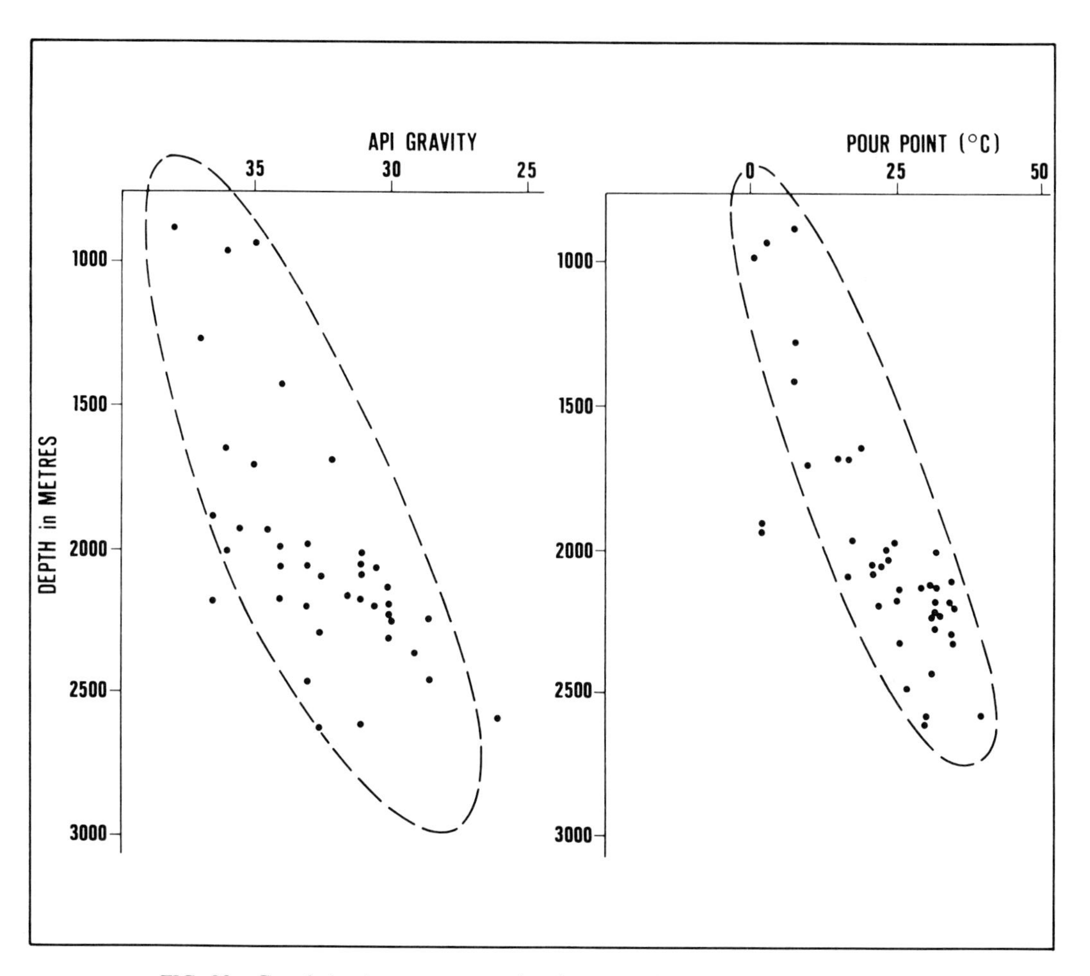

FIG. 22—Correlation between poor point changes and API gravity changes with depth.

the reflectance of the vitrinite (R_o). This reflectance increases with temperature and the boundary between diagenesis and categenesis, i.e. the beginning of oil generation (oil window), occurs at 0.5% to 0.7% reflectance.

The maximum temperature of the hydrocarbons formed during pyrolysis with temperature programming (T_{max}) also gives an evaluation of the degree of maturation. Because R_o and T_{max} correlate well together and because T_{max} is more readily obtainable, the geochemical studies carried out by Durand and Oudin (1979) are based on pyrolysis analysis.

They showed a large-scale correlation between isomaturation lines and isobaths (Fig. 23). It appears that for any one stratigraphic zone, the degree of maturation is the same whether structurally positioned relatively high or low.

As the main pay zones are located above or just at the beginning of the oil window, the hydrocarbons are above their zone of generation and therefore migration must have taken place.

Migration

A scheme of generation of hydrocarbons has been established and it shows generally that the greater the maturation, the lighter the hydrocarbons thus formed (Durand and Oudin, 1979). It also was shown that in the immature or just mature zone, all the stratigraphic levels with any significant porosity are invaded by hydrocarbons coming from the more mature zone.

These authors explain these phenomena by suggesting that the overpressured zone could play a key role in the first phase of migration (primary migration). The origin of the over pressure could be due to one or a combination of any of the following phenomena: (1) delayed compaction of the shales, (2) generation of large quantities of low molecular weight hydrocarbons, and (3) according to Barker and Bradley (1972) expansion of pore water due to increasing temperature. This pressure would rise continuously until fissuring occurs at the fracture pressure. The microfissures would let the more mobile fluids escape towards a lower pressure zone.

The hydrocarbons migrate to the reservoirs which lie mainly between the depths of 800 and 2,400 m. The lighter the hydrocarbons, the farther the migration, this is to be expected because they are the most mobile; whereas the heavier hydrocarbons have not moved far from their zone of formation. However, in a sequence containing abundant sandstones, it is difficult to separate primary and secondary migration (i.e. movements of hydrocarbons between reservoirs). The major fault which is a seal at the present may have been a path of migration in the past.

The oils with low API gravities and high pour points occur deeper. Migration with a preponderance of a lateral component is difficult to envisage and it is therefore recognized that there must be an important vertical component in the migration. The lighter the oils, the farther the migration.

The age of the primary migration can be estimated. The age of the deepest sediments is evaluated at 30 m.y. and the maturation level is reached after 22 to 23 m.y. with a period of optimum oil generation reached at 25 m.y. Migration is therefore considered to have taken place essentially during the last 5 m.y. Thus the migration of Handil oil is a recent phenomenon which is probably still taking place. The main phase of the formation of the structure dates back less than 5 m.y. Thus, the structure formation, the hydrocarbon generation, and migration are therefore contemporaneous.

SELECTED REFERENCES

Allen, G. P., D. Laurier, and J. P. Thouvenin, 1976, Sediment distribution patterns in the modern Mahakam delta: 5th Ann. Mtg., Indonesian Petroleum Assoc. Proc. (Jakarta), p. 159-178.

Barker, C., 1972, Aquathermal pressuring—role of temperature in development of abnormal pressure zones: AAPG Bull., v. 56, p. 2068-2071.

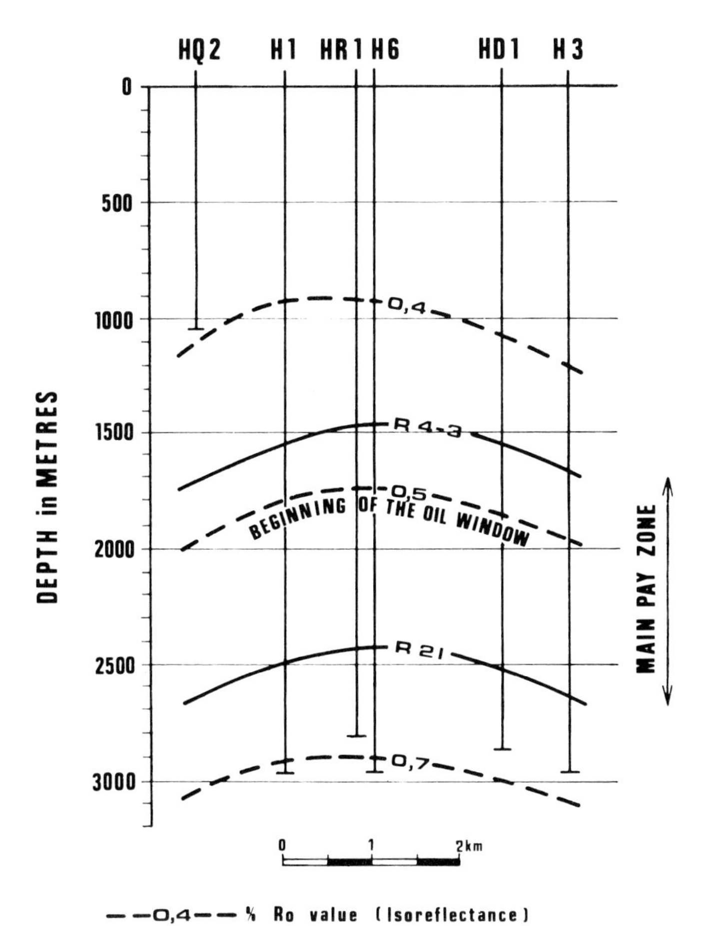

FIG. 23—Isoreflectance lines showing the top of the oil window in the middle of the main pay zone.

Table 2. Shaliness cut-off used for the different reservoir types.

Reservoir Type	A	B	C
Shaliness cut off (%)	≤ 25	≤ 40	≤ 55

Table 3. Porosity cut-off for the different reservoir types at different depths.

Interval depth (m)	Maximum porosity	Reservoir Type		
		A	B	C
Above- 800	38	24	20	14
800 - 1200	34	22	18	12
1200 - 1800	30	18	16	10
1800 - 2200	26	18	14	8
2200 - 2800	22	16	12	6

Billmann, H. G., and L. W. Kartaadipura, 1974, Late Tertiary biostratigraphic zonation, Kutei basin, offshore East Kalimantan, Indonesia: 3rd Ann. Mtg. Indonesian Petroleum Assoc. Proc. p. 301-310.

Combaz, A., and M. De Matharel, 1978, Organic sedimentation and genesis of petroleum in Mahakam delta, Borneo: AAPG Bull. v. 62, no. 9, p. 1684-1695.

Durand, B., and J. L. Oudin, 1979, Example of hydrocarbon migration in deltaic-type series: the Mahakam delta, Kalimantan, Indonesia: 10th World Petroleum Cong. (Bucharest, Romania), Sept. 1979.

——— et al, 1977, Etude geochimique d'une serie de charbons: Advances in organic geochemistry, Madrid, 1975, p. 601-631.

Gerard, J. and H. Oesterle, 1973, Facies study of the offshore Mahakam delta: 2nd Ann. Mtg. Indonesian Petroleum Assoc. Proc. (Jakarta), p. 187-194.

Lalouel, P., 1979, Logging interpretation in deltaic sequence: 8th Ann. Mtg., Indonesian Petroleum Assoc. (Jakarta), June 1979.

Magnier, Ph., and Ben Samsu, 1975, The Handil oil field in East Kalimantan: 4th Ann. Mtg., Indonesian Petroleum Assoc. Proc. (Jakarta), p. 41-61.

——— T. Oki, and L. Kartaadiputra, 1975, The Mahakam delta, Kalimantan, Indonesia: 9th World Petroleum Cong. Proc. (Tokyo), p. 239-250.

Oudin, J. L., 1970, Analyse geochimique de la matiere organique extraite des roches sedimentaires. Composes extractibles au chloroforme: Rev. Inst. Francais Petrole, v. 25, p. 3-15.

Samuel, L., and S. Muchsin, 1975, Stratigraphy and sedimentation in the Kutai Basin, Kalimantan: 4th Ann. Mtg. Indonesian Petroleum Assoc. Proc. (Jakarta), p. 27-39.

Tissot, B., et al, 1974, Influence of nature and diagenesis of organic matter in formation of petroleum: AAPG Bull., v. 58, p. 499-506.

Weeda, J., 1958, Oil basin in East Borneo, *in* Habitat of Oil: Tulsa, Oklahoma, AAPG symposium, p. 1337-1346.

Exploration in East Malaysia Over the Past Decade[1]

By F. C. Scherer[2]

Abstract Petroleum geology of the area offshore of the Malaysian states of Sarawak and Sabah in Northwest Borneo is discussed. Exploration activities have centered in this area for the past 20 years. Drilling activity over the past decade has led to the discovery of two large oil fields, six large gas fields, and several smaller oil fields.

The geological framework and evolution of the East Malaysian shelf is discussed, along with comment on basement rocks, the Northwest Borneo geosyncline (Cretaceous-Paleogene), middle Tertiary rifting, tectonics, and facies (Neogene) analysis.

Three giant oil and gas fields are examined: the Baronia field, Samarang field, and the Central Luconia field. Discussions include material about location, stratigraphy, structure, hydrocarbon reservoirs, and development to date.

A final section highlights petroleum generation and trapping mechanisms in the East Malaysia area.

INTRODUCTION

The article covers the shelf offshore from the Malaysian States of Sarawak and Sabah, in Northwest Borneo (Fig. 1). Shell has been exploring large parts of this offshore area for more than 20 years (since 1976 under the terms of a Production Sharing Agreement with *Petronas,* the Malaysian National Oil Company). Very active drilling over the past decade has led to the discovery of two large oil fields, six large gas fields, and a number of smaller oil fields. The gas will be liquified in a plant near Bintulu, as a joint venture by *Petronas,* Shell, and Mitsubishi. This plant construction is presently Malaysia's biggest single industrial project.

The exploration in East Malaysia is treated in four parts. The first is a brief account of the main exploration activities undertaken and the discoveries made offshore East Malaysia over the past decade. The second is the geologic evolution of the area and a summary of the regional geology, the tectonic framework, the stratigraphy, and the facies development, as far as it is considered important for the understanding of the particular geologic setting in which exploration took place. Part 3 is a more detailed examination of the two largest oil fields, Baronia and Samarang, and of the E.11 gas field. Finally a discussion highlights the specific generation and trapping conditions offshore Sarawak and Sabah and how they have affected exploration activities in this area.

Previous publications that refer to the offshore geology of East Malaysia include Doust (1978), Ho (1978), Whittle and Short (1977), Bell and Jessop (1974), two Woods Hole papers by Parke et al (1971) and Emery and Ben-Avraham (1972).

EXPLORATION ACTIVITIES OF THE PAST DECADE

Shell has been involved in hydrocarbon exploration in Sarawak and Sabah for many years, and achieved early success in

[1]Manuscript received, November 27, 1979; accepted for publication, January 7, 1980.

[2]Sarawak Shell Berhad, Sarawak, Malaysia.

The writer thanks PETRONAS (the Malaysian National Oil Company), Sarawak Shell Berhad, Sabah Shell Petroleum Company, Ltd., Pecten Malaysia Company, and Shell International Petroleum Maatschappij for permission to publish this article.

The writer particularly thanks his many colleagues, who are too numerous to be listed individually, but whose work provided the information and ideas presented in this paper. Special thanks go to Zainudin and his staff who drafted all the figures.

Article Identification Number:
0065-731X/80/M030-0020/$03.00/0.

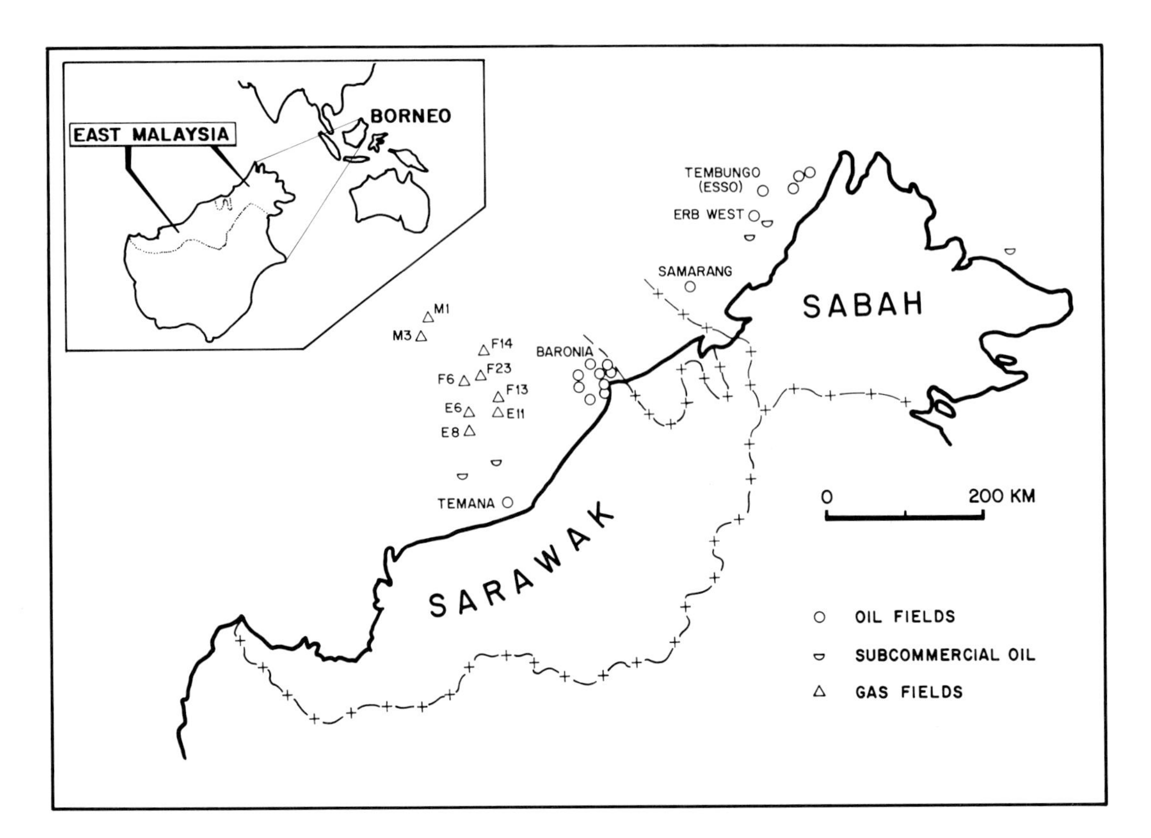

FIG. 1—Location of oil discoveries and major gas fields in East Malaysia.

1910 with the discovery of the onshore Miri field. Offshore drilling began in 1957 and accelerated around 1966. About two thirds of the total footage in Sarawak and Sabah have been drilled during the past 10 years. The sudden acceleration of exploration activities was due to three important technical advances:

1. The arrival in 1965 of a modern floating rig (*Sedco*-A), capable of drilling in deep, rough waters (where the earlier drilling barge encountered considerable problems) and in areas with soft sea bed conditions (which caused serious problems with a jack-up rig). The successful semisubmersible *Sedco*-A is still operating in East Malaysia together with several other mobile rigs.

2. An equally important event was the use of modern digital seismic techniques in the area in 1967. Although (by today's standards) data quality was crude, it was an enormous improvement to the previous analog records. Since then much effort has gone into continuously improving seismic data quality both at the acquisition and processing levels.

3. The successful launching of the first production platform and the start-up of oil production in 1968 from the West Lutong field discovered two years earlier.

About 1966, drilling concentrated on the Baram delta, where not only the best seismic records were obtained, but where earlier wells (1961-1963) already had encouraging oil shows. This resulted in the discovery of nine commercial oil fields, of which the largest is the Baronia field (discovered 1967, reserves 185 million bbl; Figs. 1, 9).

Extensive drilling began farther offshore, where seismic data located many large "buildup" features interpreted to be buried carbonate reefs. This interpretation was confirmed during 1968 to 1975 when 20 gas accumulations were found. At least 10 of them are of substantial size, and six contain reserves in excess of 1 Tcf each.

Exploration continued farther south, where oil shows were encountered in the 1960s. After drilling a series of wells over the complex Temana structure, the first commercial oil field was established in Sarawak outside the Baram delta.

While exploration of the Sarawak shelf was carried out exclusively by Shell, four other companies became involved in exploration in Sabah (Esso and Oceanic Oil in western Sabah and Aquitaine and Teiseki in eastern Sabah). After the first two oil discoveries in Sabah in 1971 (Esso's Tembungo and Shell's Erb West), exploration activities in that part of East Malaysia were increased considerably. Between 1971 and 1977 Shell discovered in five more oil fields in Sabah, among which was the large Samarang field (discovered in 1972; 230 million bbl reserves).

Since the acceleration of exploration activities (1966) 200 exploration wells have been drilled in offshore East Malaysia, including 30 wells drilled by Esso, Aquitaine, Teiseki, and Oceanic. These activities have led to the discovery of 16 oil fields, (two of them in the 200 million bbl reserves class), five subcommercial oil discoveries and ten gas fields (six with re-

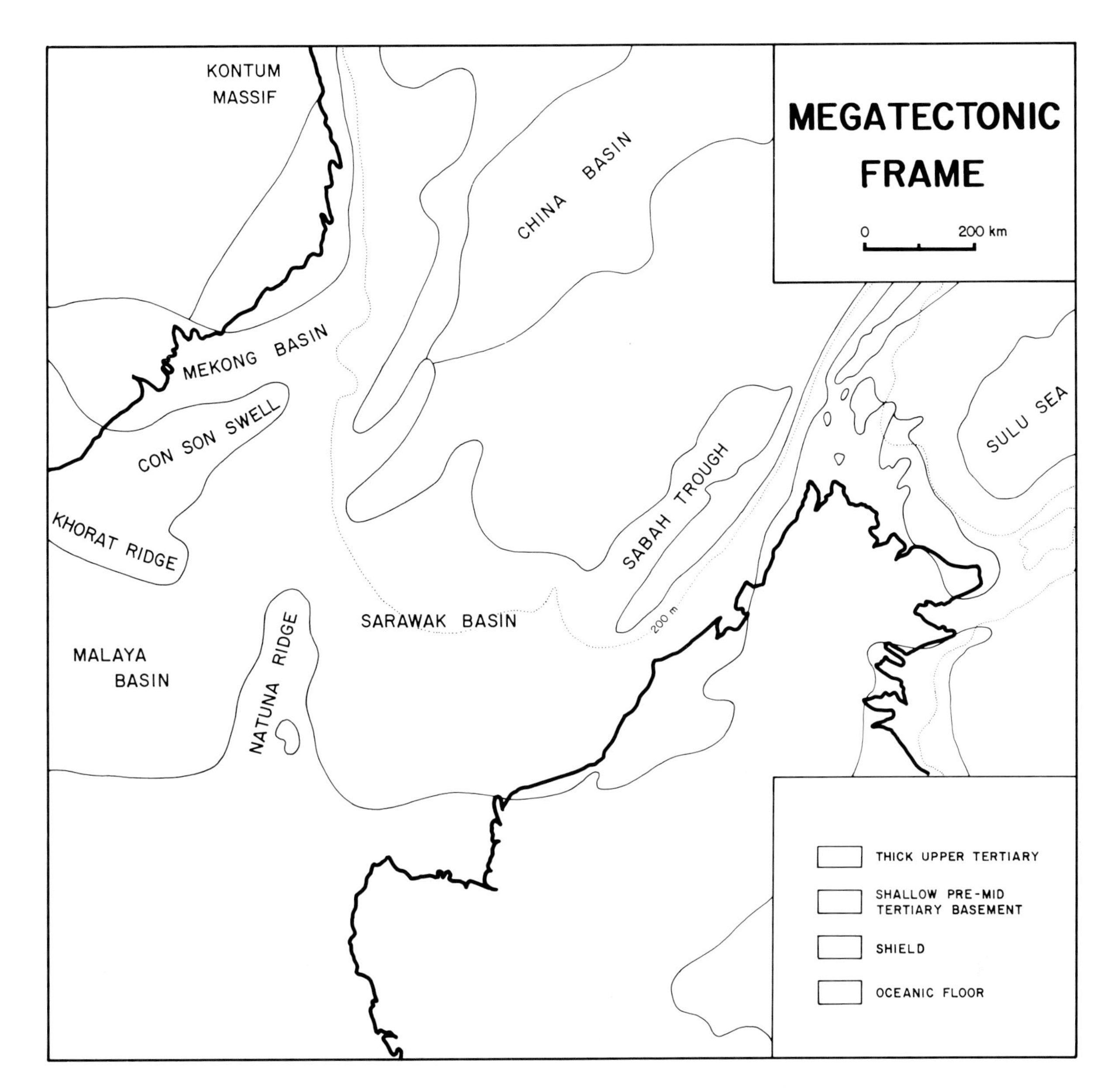

FIG. 2—Megatectonic framework showing thick upper Tertiary basins and areas of shallow pre-middle Tertiary basement. Oceanic crust of the abyssal plains of the South China Sea and Sulu Sea is probably of middle Tertiary age (South China Sea deepwater data partly from Emery and Ben-Avraham, 1972; and from Bowin et al, 1978).

serves in excess of 1 Tcf each).

GEOLOGICAL EVOLUTION OF EAST MALAYSIAN SHELF

Megatectonic Framework

The continental shelf offshore East Malaysia belongs to an extensive shallow-water area that connects Borneo with the Asian mainland (Fig. 2). Only the northern part of Borneo is separated from continental Asia by deep water areas of the South China Sea. Along central Sarawak the shelf is extremely broad, generally exceeding 300 km from shelf edge to coast. It becomes narrow toward northern Sabah, where it locally is less than 100 km wide.

Most of the shelf is underlain by a thick upper Tertiary sequence. Magnetic data, locally supported by seismic data, suggest the greatest sedimentary thicknesses are in central and northern Sarawak, close to the present coast (Figs. 3, 4). In Sabah, a zone of maximum thicknesses appears to occur 60 km offshore. The main source of these sediments was the orogenic belt that runs along the southern border of Sarawak northward into Sabah. These mountains, that were mainly uplifted in the Eocene, now form the landward boundary of the thick upper Tertiary basin.

In Sarawak, thick upper Tertiary sediments reach far beyond the shelf edge, covering large deepwater areas (Sarawak basin, Fig. 2). Farther north, in western Sabah, a deep, relatively narrow trough (Sabah trough) with mostly undisturbed, horizontal

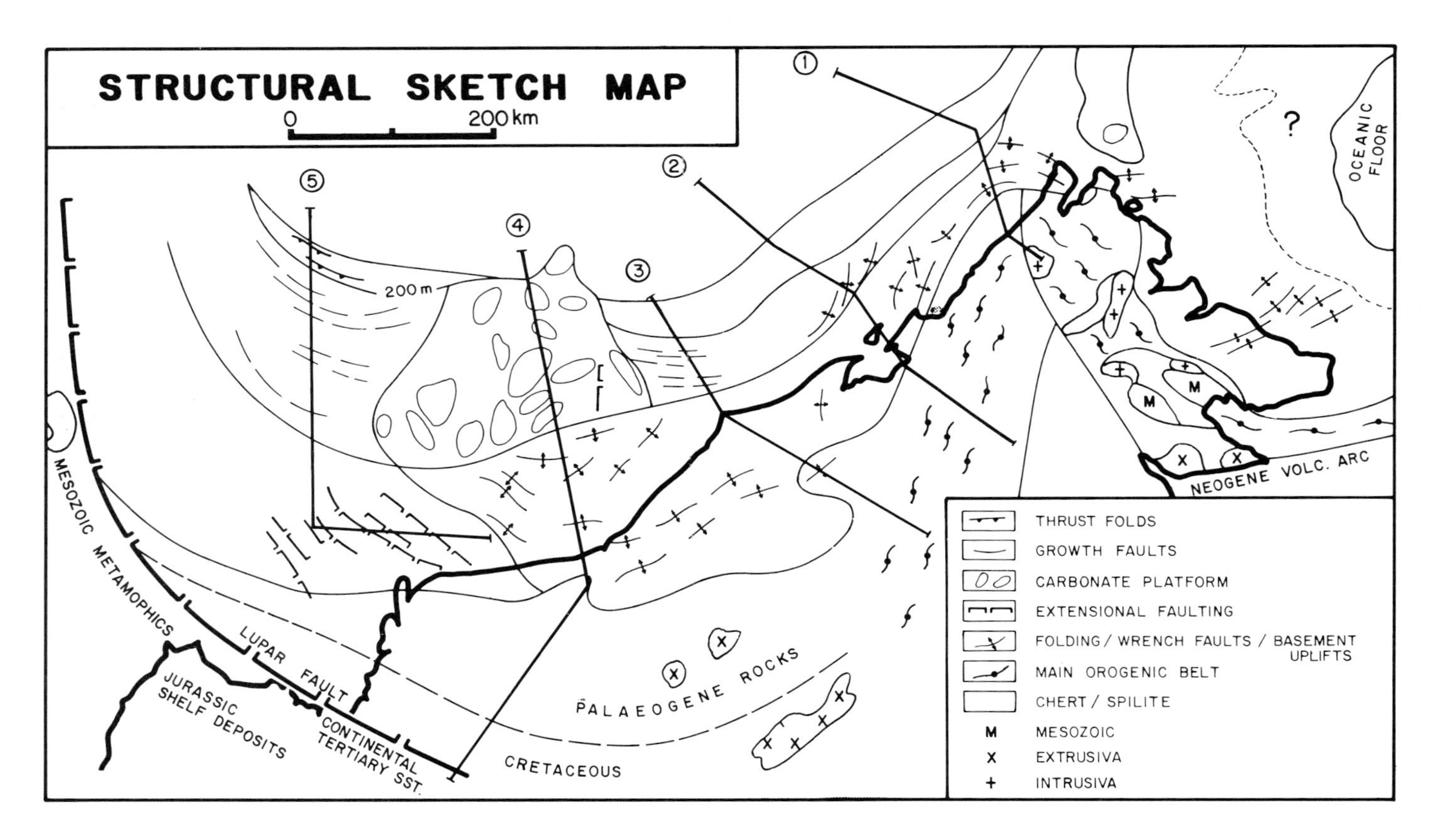

FIG. 3—Structural sketch map of northwest Borneo showing arcuate southwest to northeast oriented main orogenic belt of Cretaceous-Paleogene Northwest Borneo geosyncline, intersected by northeast to southwest oriented radiolarite/spillite-rich orogenic belt of northern and eastern Sabah. Map indicates main types of structure within prospective late and mainly post-geosynclinal sediments in front of the main orogene at a level 1 to 3 km below the seafloor or surface. Map also gives locations of cross sections shown in Figure 4 (data on eastern Sabah partly after Bell and Jessop, 1974).

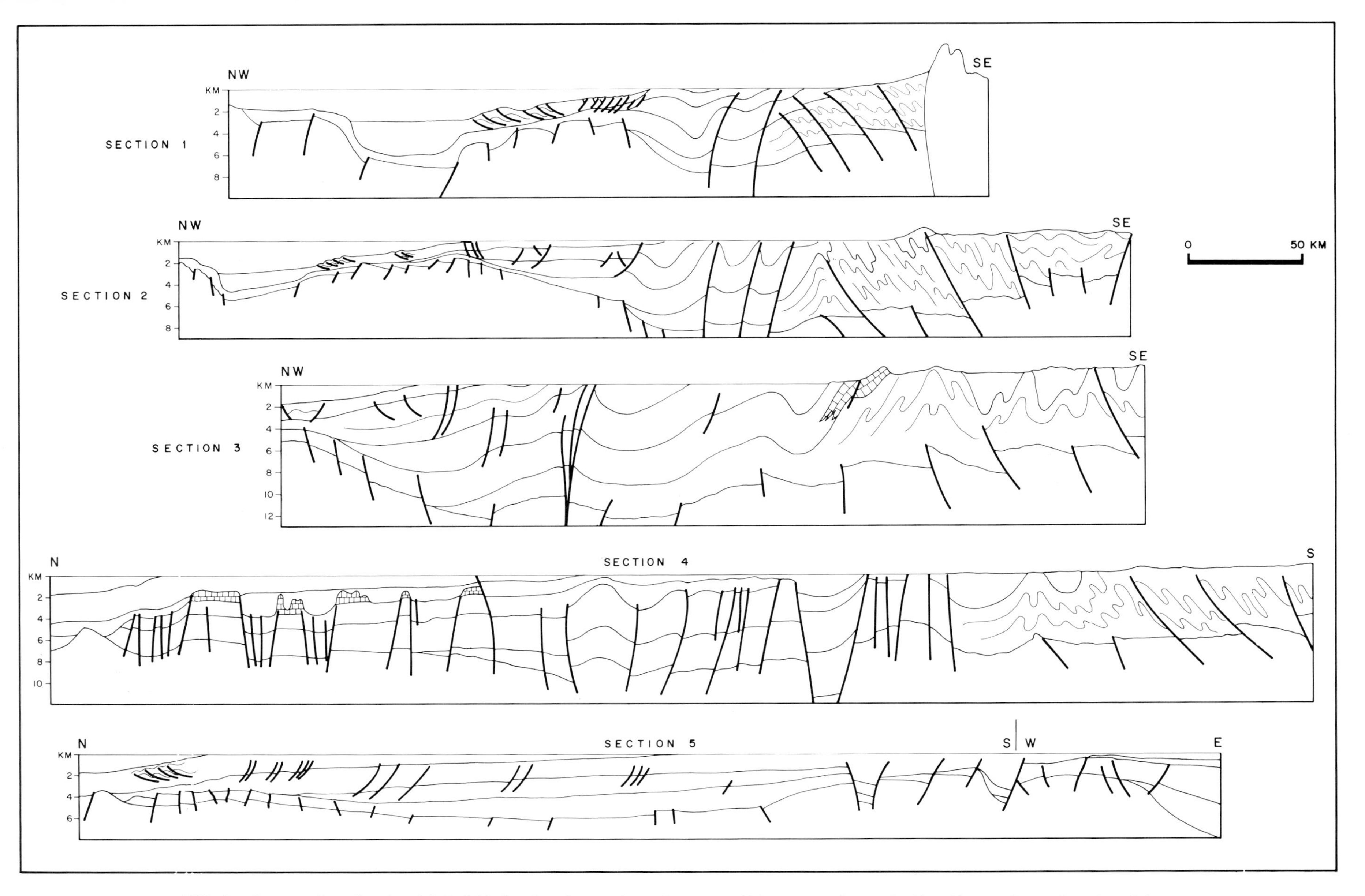

FIG. 4—Cross sections showing tightly folded and partly overthrust Cretaceous-Paleogene rocks (vertical hatching, to the southeast) wedging out towards the offshore (to the northwest). This contrasts with the general northward tilt and thickening of the overlying youngest Tertiary sediments. Basement uplifts are causing reverse faulting in the Neogene sediments in the central part of sections 1-4, synsedimentary faulting farther offshore (sections 1-3, 5), and thrust folds in Pliocene sediments in deep water on the slope. Vertical exaggeration is about 5x (deepwater parts of sections 1 and 2 are modified after Parke et al, 1971).

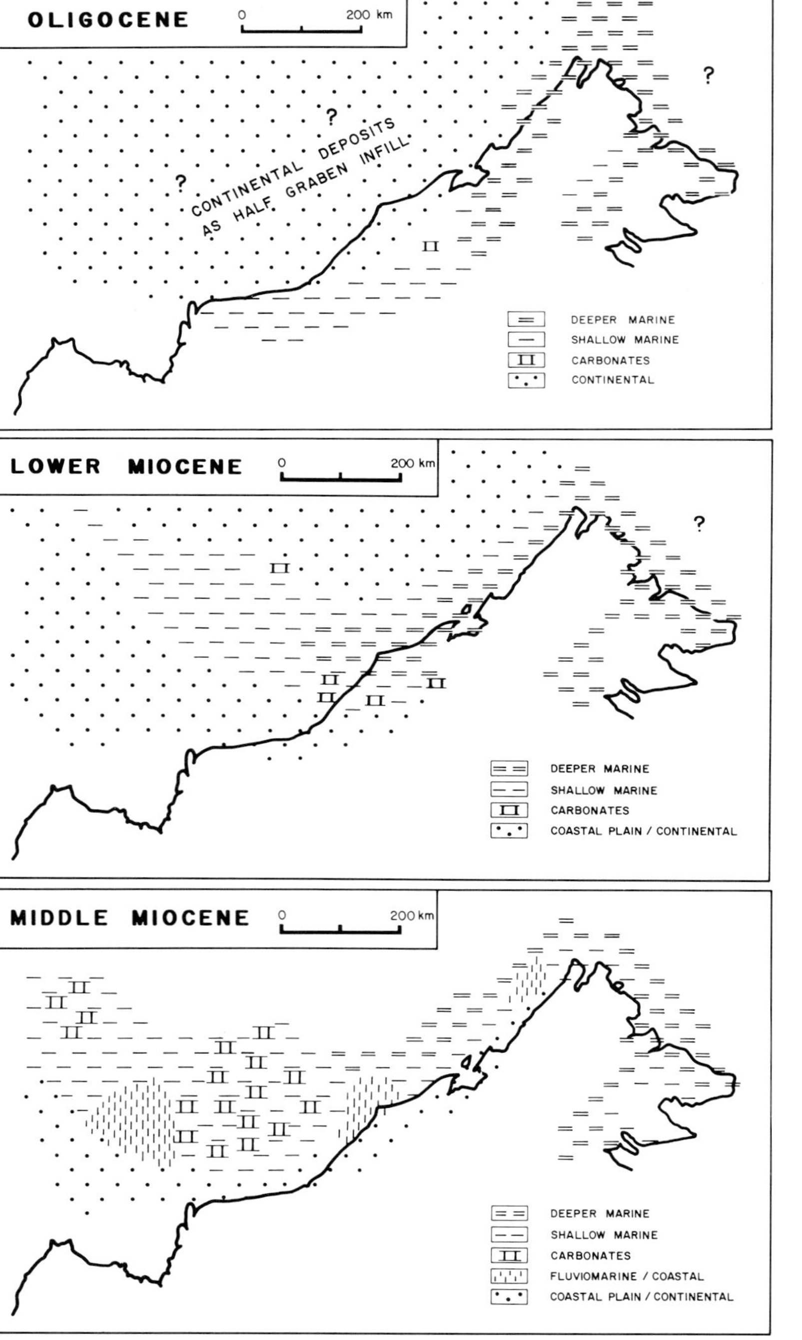

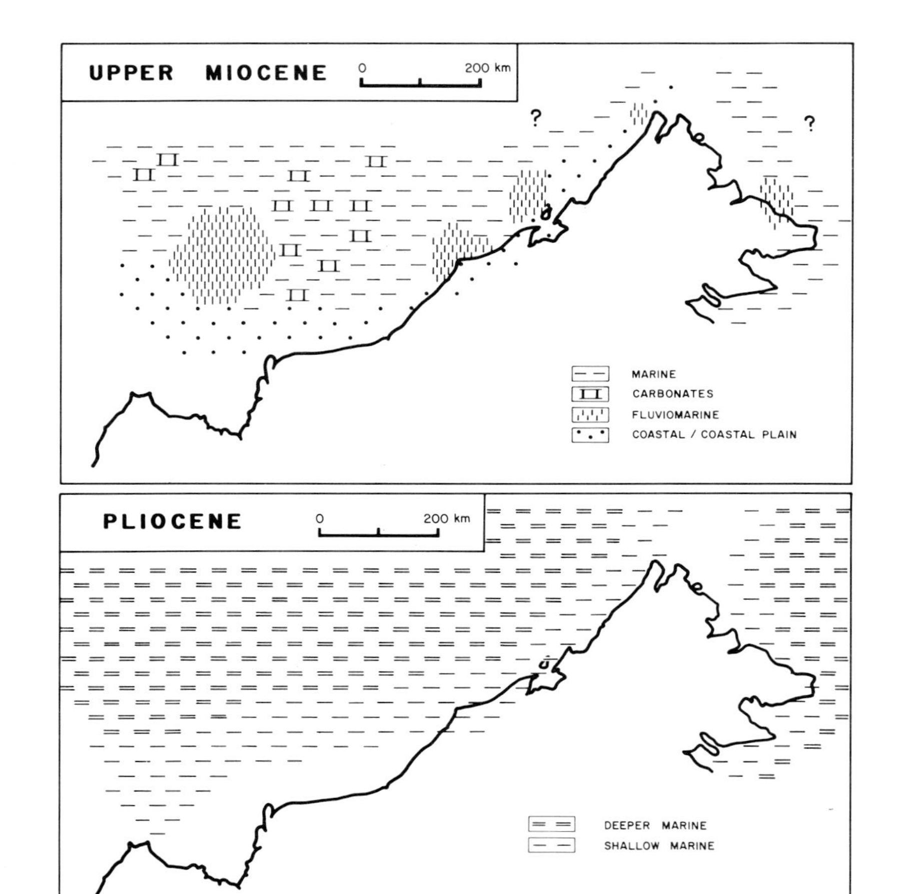

FIG. 5—Paleogeographic facies maps depicting a westward moving sea successively occupying the present shelf area. Middle Tertiary horst and graben tectonics (probably initiated through rifting of the China basin) and subsequent regional northward tilting made the area accessible to an advancing sea. Between early and middle Miocene time, deep depressions developed on either side of the relatively stable Central Luconia platform, where extensive carbonate platforms began to grow. A belt of coastal plain deposits developed along the entire southern margin of the basin while rapidly outbuilding deltas started to fill the depressions in western and northern Sarawak. An additional small delta began to form in northern Sabah (mainly during the late Miocene). The pliocene sea covered most of the shelf area and expanded over most of the present South China Sea and Sulu Sea.

SARAWAK

SABAH

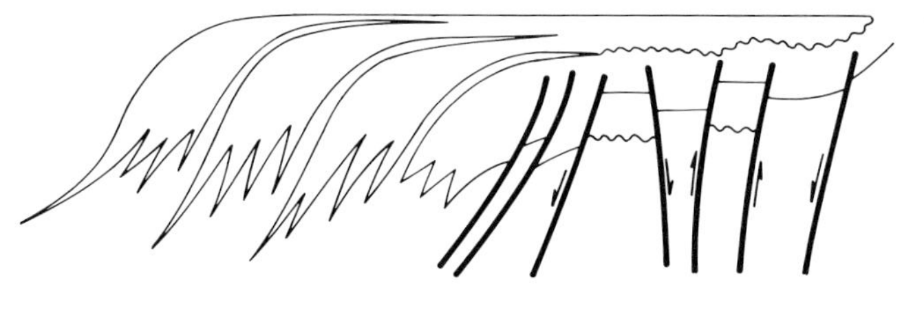

REGRESSIVE CYCLES

SEPARATED BY

SHORT TRANSGRESSIONS

POLLEN ZONATION

IN COASTAL PLAIN

SEQUENCE

UNCONFORMITIES

SUBDIVIDE

NEOGENE

SANDS IN REGRESSIVE INTERVALS

SANDS BOTH IN TRANSGRESSIVE AND REGRESSIVE INTERVALS

FIG. 6—Correlations within the prospective Neogene are hampered by the overall pronounced lithologic monotony. With the exception of local carbonate buildups the sequence consists of an extremely thick but rather uniform succession of interbedded sandstones, siltstones, and clays.

Where an overall regressive sequence has been interrupted by several short transgressions, such as in the Barnum delta and other parts of Sarawak, a number of stratigraphic cycles can be recognized (Ho, 1978). Each cycle starts with a transgressive base, followed by a regressive sequence which is, in turn, overlain by the basal transgression of the next cycle. Many of these transgressions have spread over a wide area within a very short time. Therefore, cycle boundaries generally provide an excellent means for rapid basinwide correlation. Dating (by paleontology/palynology) is only required at a few points in the cyclic sequence to assure the regional validity of these boundaries.

It is not possible to apply the cycle concept to the thick coastal-plain deposits of Balingian, where no transgressive intercalations exist. There, a subdivision based on pollen zones provides a means of reliable correlation. The cycle concept can neither be applied in Sabah where intense structural deformation strongly influenced the regional sedimentary pattern characterized by unconformities and large diachronous transgressions. Instead, stratigraphic subdivision is based on correlation of the various unconformities, which can be followed on seismic sections and detected in wells. They appear as abrupt and otherwise not explainable facies changes in the latter and are also confirmed by dipmeter data.

Note that in Sarawak the clastic reservoir objectives are always regressive sands, whereas in Sabah drilling targets also include sands deposited during a transgressive phase.

sediments of probably Pliocene age, separates the thick upper Tertiary sequence beneath the shelf from the much thinner Tertiary sequence which underlies the deep water farther offshore (Fig. 3, 4; sections 1, 2). A similarly deep, but shorter graben is found 250 km farther to the west-northwest (Fig. 2).

The abyssal plain of the China basin lies 350 km to the northwest of the Sabah trough, at a water depth of 4,000 m, and is underlain by oceanic basement with only a thin veneer of sediments. In this area, crustal extension led to the formation of oceanic basement, probably in middle Tertiary time, whereas in the south rifting never went beyond the initial graben formation.

Thick upper Tertiary sediments also underlie part of the shelf in eastern Sabah, extending landward across Dent Peninsula. However, in the deep waters to the northeast, oceanic basement appears to be at shallow depth beneath the Sulu Sea (Fig. 2).

Basement Rocks

Offshore East Malaysia, igneous or metamorphic basement has not been penetrated by any well. In most of the area of the continental shelf, basement occurs at depths that cannot be reached by the drill. In much of the shelf area basement cannot be mapped with reflection seismic equipment. In most areas seismic basement offshore corresponds to indurated Paleogene sediments of the Northwest Borneo geosyncline.

Based on projections from onshore western Sarawak and offshore well data from Peninsular Malaysia and Indonesia, basement is expected to consist of Mesozoic metamorphic and granitic rocks, and possibly at least partly of upper Paleozoic rocks similar to those exposed in Vietnam, Peninsular Malaysia, and western Sarawak. Mesozoic metamorphic rocks have been described from surface outcrops in eastern Sabah (Leong, 1974).

Cretaceous-Paleogene Northwest Borneo Geosyncline

While shelf conditions prevailed in western Sarawak, a deep trough developed in central Sarawak during Cretaceous-Paleogene time, extending northward over parts of Kalimantan and western Sabah. Several thousand meters of deepwater shales and turbidites accumulated in this trough, the axis of which appears to have been located 100 to 200 km inland from

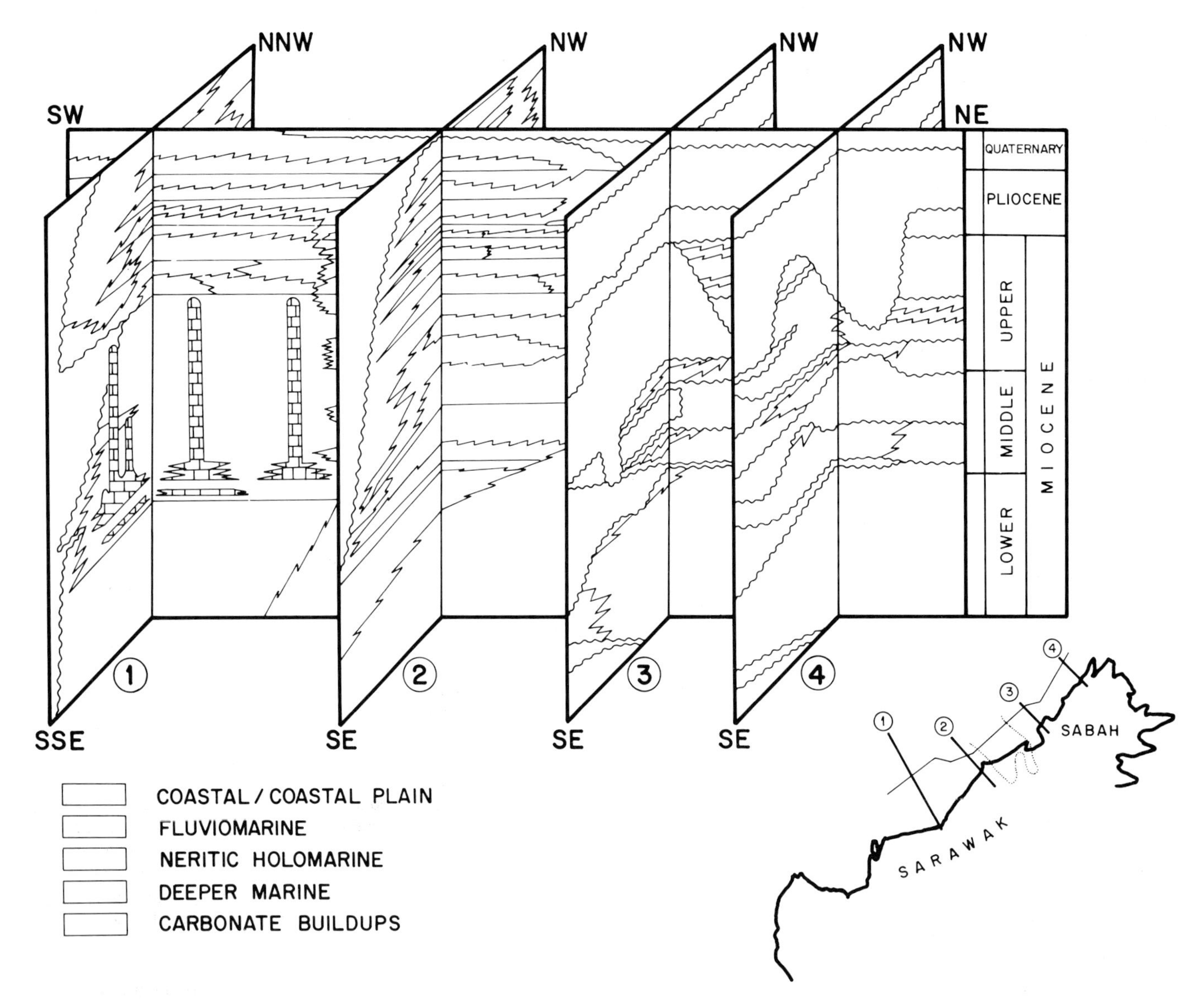

FIG. 7—Schematic fence diagram showing regional facies developments in Sarawak and Sabah during the Neogene. The various facies types are mainly defined on faunal and sedimentological criteria derived from the study of recent sediments of the Sarawak shelf. Note the lower Miocene dominance of coastal plain development in central Sarawak, of cyclic sedimentation in northern Sarawak, and of prominent unconformities in Sabah.

today's coast. Paleocene shallow-water limestones found in the subsurface of southwest Luconia indicate the presence of carbonate shoals along the western flank of the Paleogene depwater trough.

The main orogenic belt of the Northwest Borneo geosyncline was strongly folded and uplifted during Eocene time, thus becoming an important source for the younger Tertiary sediments.

Middle Tertiary Rifting, Late Geosynclinal Phase of Northwest Borneo Orogen and Subduction(?) in Sabah

Mid-Tertiary rifting in the China basin is thought to have exerted extensional stresses that led to the formation of a half graben and graben system in which mostly continental sediments were deposited (Figs. 2-5). At the same time a deep trough developed in front of the Eocene fold belt in Sabah and northern Sarawak. It rapidly filled with a thick shale and turbidite sequence (West Crocker and Temburong formations; Liechti et al, 1960), but carbonate shoals and reef buildups developed along the southwestern flank of the trough (Melinau Limestone; Liechti et al, 1960). In central Sarawak a shallower environment prevailed with a mainly argillaceous facies deposited (Kelabit formation, Setap shale, Penian marl; Liechti et al, 1960; partly Miri Zone, Hale, 1973).

Deep-marine, predominantly shaly sequences also were deposited in eastern Sabah, where they contain radiolarites and spilites. These have been interpreted as trench melanges indicative of a late Oligocene-early Miocene northwest to southeast oriented subduction zone (Hamilton, 1976; Beddoes, 1976). Although no blueschist metamorphism has been observed, this zone with its highly contorted shales and the frequent radiolarites and ophiolithes shows more indications of subduction than the southwest to northwest oriented trend of the main Northwest Borneo geosyncline, which lacks typical trench melanges.

Structurally, Sabah is the most complex area in northwestern Borneo, because of its megatectonic position between the island arc system of the western Pacific and the Asian mainland.

Neogene Facies and Tectonics

During the early Miocene the sea transgressed westward. Deeper marine deposits reached the present northern Sarawak shelf and a shallower marine wedge extended far into Indonesian waters (Fig. 5). Locally carbonate shoals and buildups fringed the basin (e.g. Subis Limestone, Melinau Limestone, Liechti et al 1960). Extensive coastal plain continental deposits formed along the basin margin, with a particularly thick development in the present area off central/western Sarawak. Northwest to southeast oriented horst and graben tectonics affected the area, but large parts of the area off western Sarawak have subsequently become fairly stable, elevated, and extensively eroded.

During the middle Miocene strong subsidence began off central Sarawak along a fault system of a general north-northwest to south-southwest orientation. The middle Miocene sea spread into the depressions that formed on either side of a relatively stable, elevated central area, where extensive carbonate buildups began to form (Central Luconia). At the same time gradually outbuilding deltas came into existence in western and northern Sarawak and in northern Sabah (Fig. 5).

During the late Miocene, much of the present area off central and southern Sabah underwent strong folding, initiated through basement uplifts and wrench faulting. Large parts of northern Sarawak, both onto and offshore, were also affected by this tectonic phase, though deformation generally has been weaker. Synsedimentary deformation took place in the thick sedimentary sequences that filled the deep depressions on either side of the Central Luconia carbonate platform. Deltaic outbuilding continued in western and central Sarawak and new deltas developed in southern and eastern Sabah (Fig. 5).

During the Pliocene, the sea rapidly expanded over the northward tilting shelf, depositing open-marine clays and sands (Fig. 5). On the shelf slope, thrust folds developed far offshore. Synsedimentary deformation continued in the deltaic areas, while another folding phase, probably again triggered by trough basement uplifts and wrench faulting, affected large parts of nearshore northern Sarawak and particularly northern Sabah (Figs. 3, 4).

"GIANT" OIL AND GAS FIELDS

Based on AAPG classification, two oil fields and six gas fields in East Malaysia qualify as "giant" discoveries (reserves in excess of 100 million bbl of oil or 1 Tcf of gas).

Baronia Field

Location — The Baronia field is in the Baram delta off northern Sarawak, 50 km northwest of the town of Miri. Average water depth is 75 m. The discovery well, Baronia-1, drilled in mid-1967, located at 04°44'17.48"N lat. and 113°44'29.27"E long., reached a total depth of 2,743.5 m.

Stratigraphy — The main prospective sequence consists of sandstone interbedded with siltstone and clays of late Miocene age (upper cycle V-lower cycle VI) and ranges in depth from 1,615 to 2,410 m.

The sandstones range in thickness from 3 to 75 m and the shales from 1.5 to 90 m. Porosities range from 13 to 25% in the poorly developed sandstones to 26 to 30% in the better sandstones. Most of the sandstones are fine grained, permeabilities range from 100 and 350 md.

Excellent log correlation across the field shows a considerable continuity of the sandstones bodies (Fig. 10). Micropaleontological and sedimentological data from cores and sidewall samples and comparison with the criteria derived from detailed studies of recent sediments of the Sarawak shelf indicate a coastal fluviomarine environment for the Baronia sandstones (Fig. 8). The fine grained sandstones have relatively low permeabilities and thus only fair reservoir quality when compared with those of the coastal plain channel sandstones elsewhere in Sarawak (> 3000 md). In the Baronia field, however, the lower reservoir qualities are more than offset by the great continuity of the sandstones.

Structure — The Baronia field is a simple domal structure between two major east to west trending growth faults (Fig. 9). Flank dips are about 2.5 to 3° in the south adjacent to the southern boundary fault. The most astonishing structural aspect of the field is the complete absence of faults over the entire central part of the field. Even today, after the drilling of 32 wells in the main field, no faults have been detected.

Hydrocarbons — The hydrocarbons in Baronia are found in stacked reservoirs. The discovery well, Baronia-1, logged a total of 81 m net oil sandstones and 162 m net gas sandstones in 13 different sandstone bodies.

The main reserves are distributed over ten sandstones with at least eight separate oil/water contacts. Five of these sandstones have gas caps, partly of considerable size (RM3, RN2, RN3, RR2 and RS2 sandstones). The remaining five lack gas caps (RP2, RT1, RT2, RU2/3 and RV2; Fig. 9).

Oil column heights range from 18 m (in the RT2 sandstones) to 55 m (in the RV2 sandstones) with an average of 27 m. Four sandstones, two at the top of the hydrocarbon-bearing sequence and two at the bottom, are completely gas-bearing. Only gas with minor quantities of 41°API crude were found in the R2 sandstone between depths of 2,956 and 2,972 m in an overpressured sequence (about 4,300 psi over hydrostatic, well Baronia 29, deepened to 3,185 m).

Accumulations in the RR and RS sandstones (possibly also in the RM3 sandstone) are partly controlled by the two major growth faults. At other levels, however, there is structural evidence that the trap fill is relatively poor (i.e. the available trap volume exceeds the volume of hydrocarbons trapped). The occurrence of thick water-bearing sandstones between the hydrocarbon sections is only in some cases explained by the absence of adequate sealing beds.

The oil produced from the Baronia field is light, waxy,

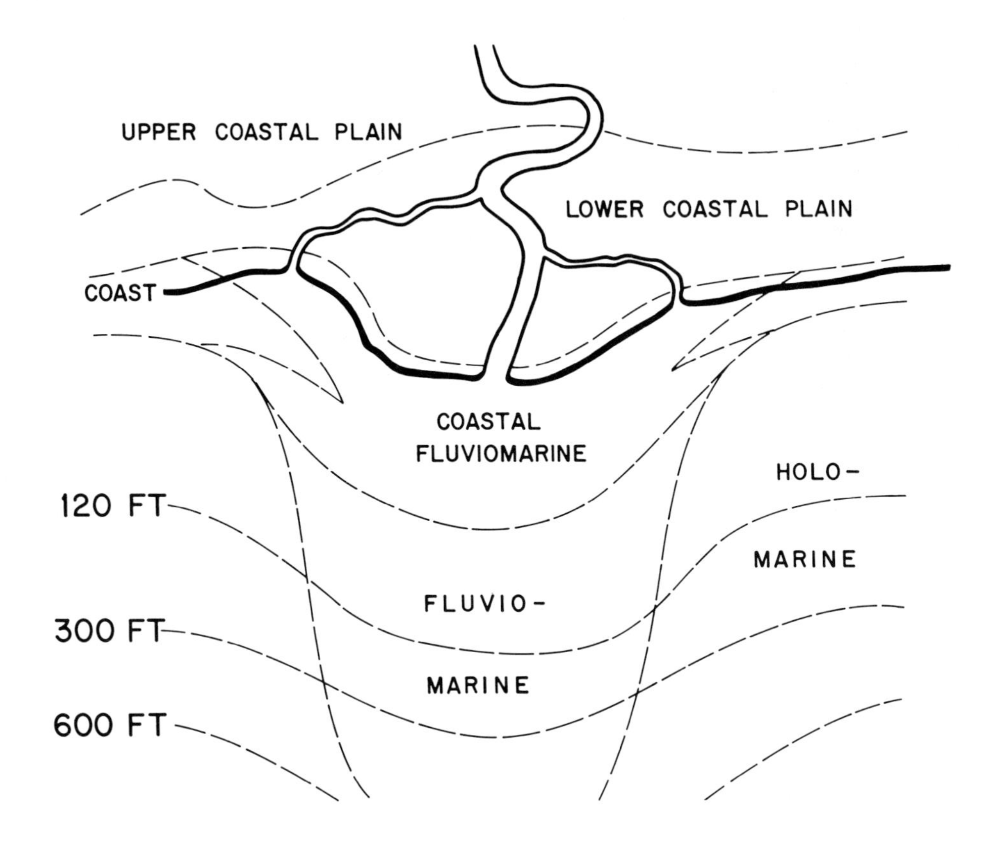
UPPER COASTAL PLAIN
LOWER COASTAL PLAIN
COAST
COASTAL
FLUVIOMARINE
HOLO-
MARINE
120 FT
FLUVIO-
MARINE
300 FT
600 FT
A

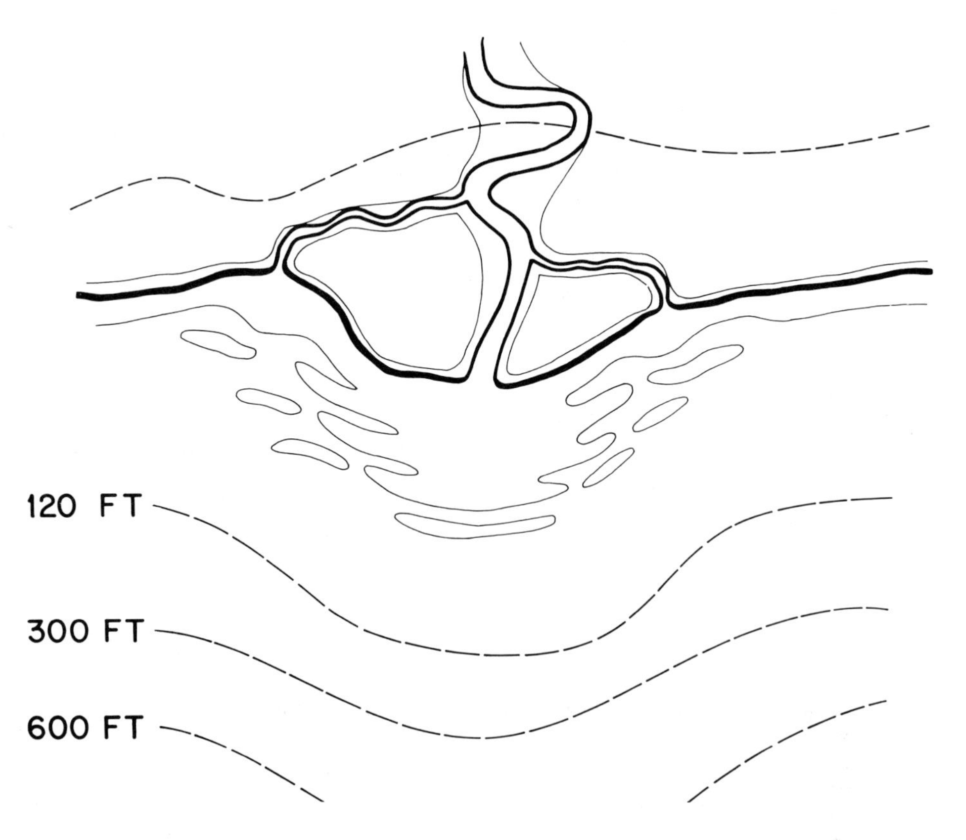
120 FT
300 FT
600 FT
B

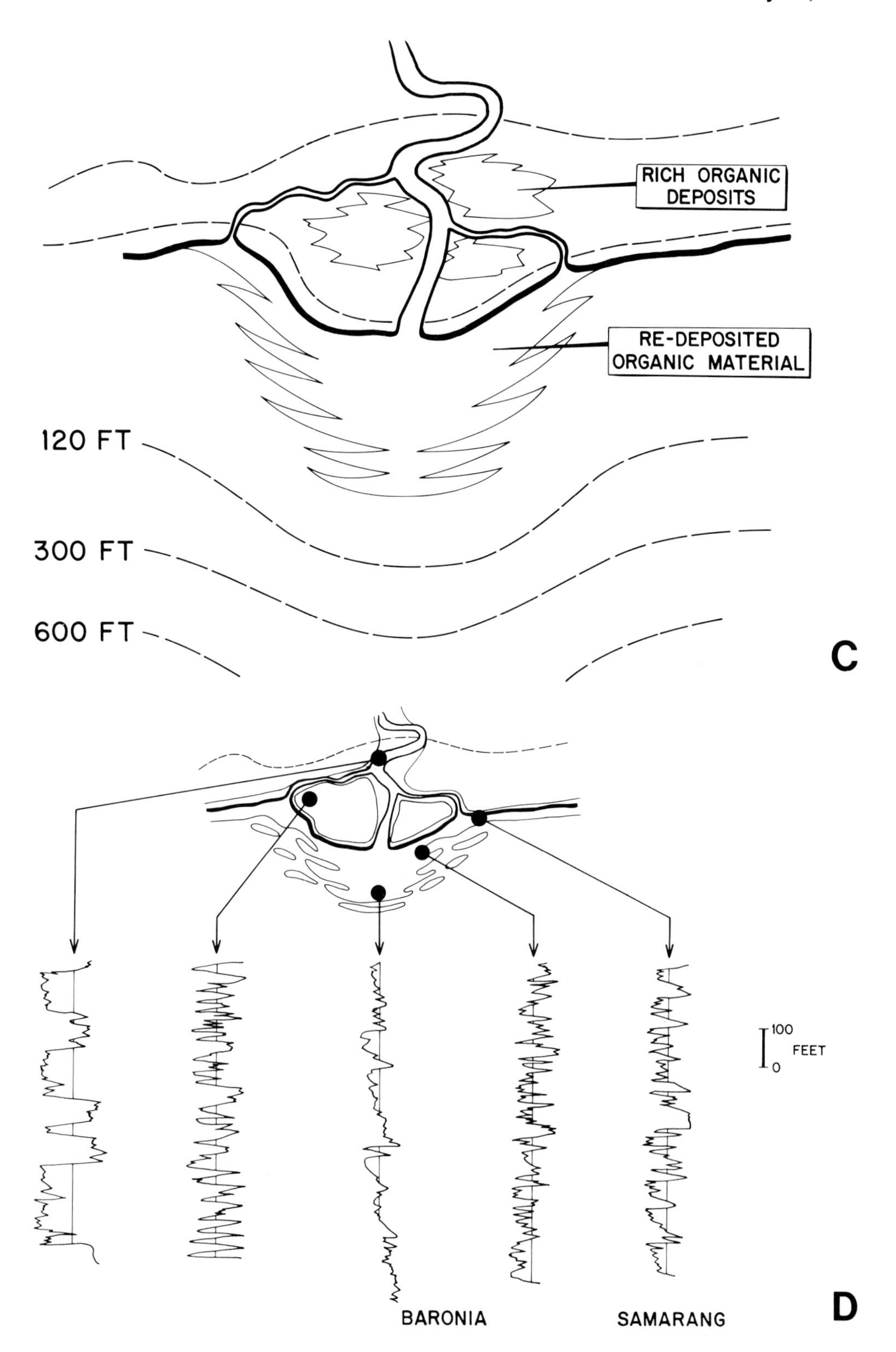

FIG. 8—Study of recent sediments of the Sarawak shelf. **8a**: environments of deposition; **8b, 8c**: models of main sand distribution. **8d** shows sections of gamma-ray logs from various wells offshore East Malaysia and an interpretation of the environment of deposition of the sandstones.

Most oils found in Northwest Borneo are of the light waxy type, with a low sulfur content, suggesting generation from land plant-derived source rocks deposited in a freshwater environment (Hedberg, 1968; Gransch and Postuma, 1973). Chromatographic oil analyses showing high pristine/phytane ratios support this interpretation of source material deposited in a peat swamp environment (Lijmbach, 1975). This is further supported by analyses of organic content in many East Malaysian well samples and from extensive sampling of recent sediments across the present Baram delta. Highest organic concentrations (waxes, resins, pollen cuticles) were found in coastal-plain deposits. In recent swamp and seabed samples, the organic content diminishes rapidly away from the large river mouths to the offshore areas, in line with observations from wells where fine grained marine sediments generally contain low concentrations of organic matter.

Table 1. Summary of Reservoir and Oil Data.

	Baronia	Samarang
Oil productive area (acres)	200 to 2,200	20 to 1,100
Oil column heights (feet)	60 to 180	20 to 240
Oil reserves, recoverable (million bbl)	185	230
Gas in place associated (Bcf)	1,658	680
nonassociated (Bcf)	422	0
Porosity (%)	13 to 30	17 to 32
Permeability (md)	100 to 350	100 to 1,000
Water saturation (%)	33 to 67	20 to 50
Initial reservoir pressure (psig)	2,400 to 3,400	850 to 3,500
Reservoir temperature (°F)	165 to 200	115 to 190
API gravity	38.5 to 41	17 to 40
Pour point (°F)	40 to 55	< 30 to 65
Viscosity (cp)	0.33 to 0.43	0.3 to 32
Sulfur content (% wt)	0.07	0.05 to 0.18
Wax content (% wt)	3 to 4	0.15 to 3.4
Formation volume factor	1.43 to 1.71	1.06 to 1.47
Gas/oil ratio (GOR) (cf/bbl)	650 to 1,100	100 to 925

land-plant derived oil with a low sulfur content (for details see Table 1). Recoverable reserves are 185.4 million bbl of oil and 1.26 Tcf of gas. Natural gas liquid reserves are estimated at 7.7 million bbl.

Development — Production development consists of two 12-slot platforms BNDP-A (12 producers) and BNDP-B (at present 5 producers) four 3-slot jackets plus two isolated underwater completions and an isolated WHPJ. The wells have multi-packer dual completions.

Samarang Oil Field

Location — The Samarang field is located in southern Sabah, 45 km northwest of the island of Labuan in water depths ranging from 9 to 45 m. The discovery well Samarang-1, drilled in late 1972, is at 5°37'41.6" N lat. 114°53'26.2" E long. It reached a total depth of 3,182 m.

Stratigraphy — The prospective sequence ranges in thickness from about 600 to 2,300 m and consists of alternating sandstones, siltstones and claystones of late Miocene age, overlying the regional shallow unconformity.

Sand thicknesses range from 1.5 to 25 m (modal thickness ± 10 m). Porosities range from 17 to 32% (modal value 23%) and permeabilities range from 100 md in the Q sands to more than 600 md in the M and K sands, reaching 1 Darcy in the shallow F, H, and I sands and locally also in the J sands.

Generally, log correlation across the field indicates a good lateral continuity of the reservoir units, although detailed sandstone development can only be accurately predicted within the main fault blocks. The sandstones are mostly medium to fine grained with generally better reservoir properties (higher permeabilities) than in the Baronia field. The main reservoir sequence was deposited in a coastal wave-dominated environment, in probably shallower water than the Baronia sandstones (Fig. 8). The sandstones progressively shale out off structure, towards the north, west, and southwest.

Structure – The Samarang field is structurally much more complicated than the Baronia field. It consists of a late Miocene/Pliocene growth fault structure, which was modified during late Pliocene time through further tectonic activities, probably related to wrench faulting in the basement (Figs. 11, 12). In some interpretations part of the structural deformation also is attributed to clay movements.

Samarang is a large (15 by 7 km) rollover anticline bounded on the east and southeast by several major growth faults of west hade. It shows a collapsed crestal area, dissected by a series of synthetic and antithetic faults. This contrasts with the relatively unfaulted western flank of the field, where the bulk of the commercial accumulations is found (Fig. 12b). The complicated, crestal, eastern part of the field is illustrated in Figure 12a. A series of minor faults of west and northwest hade further complicate both the northern and shallower southern ends of the fields. The fault system has been active throughout the entire depositional history of the field, growth being observed across several major anthithetic faults, at various stratigraphic levels. Pronounced crestal thinning of the reservoir units is found at shallow levels (above 1,770 m). Different thickness trends are observed at various stratigraphic levels, resulting in a structural culmination at deeper positions in the north part of the field and at shallower levels in the south.

Hydrocarbons — As in Baronia the hydrocarbons in Samarang are found in stacked reservoirs, but they are also distributed over different fault blocks. The discovery well, Samarang-1, logged a total of 98 m net oil sandstones and 91 m net gas sandstones in four separate intervals.

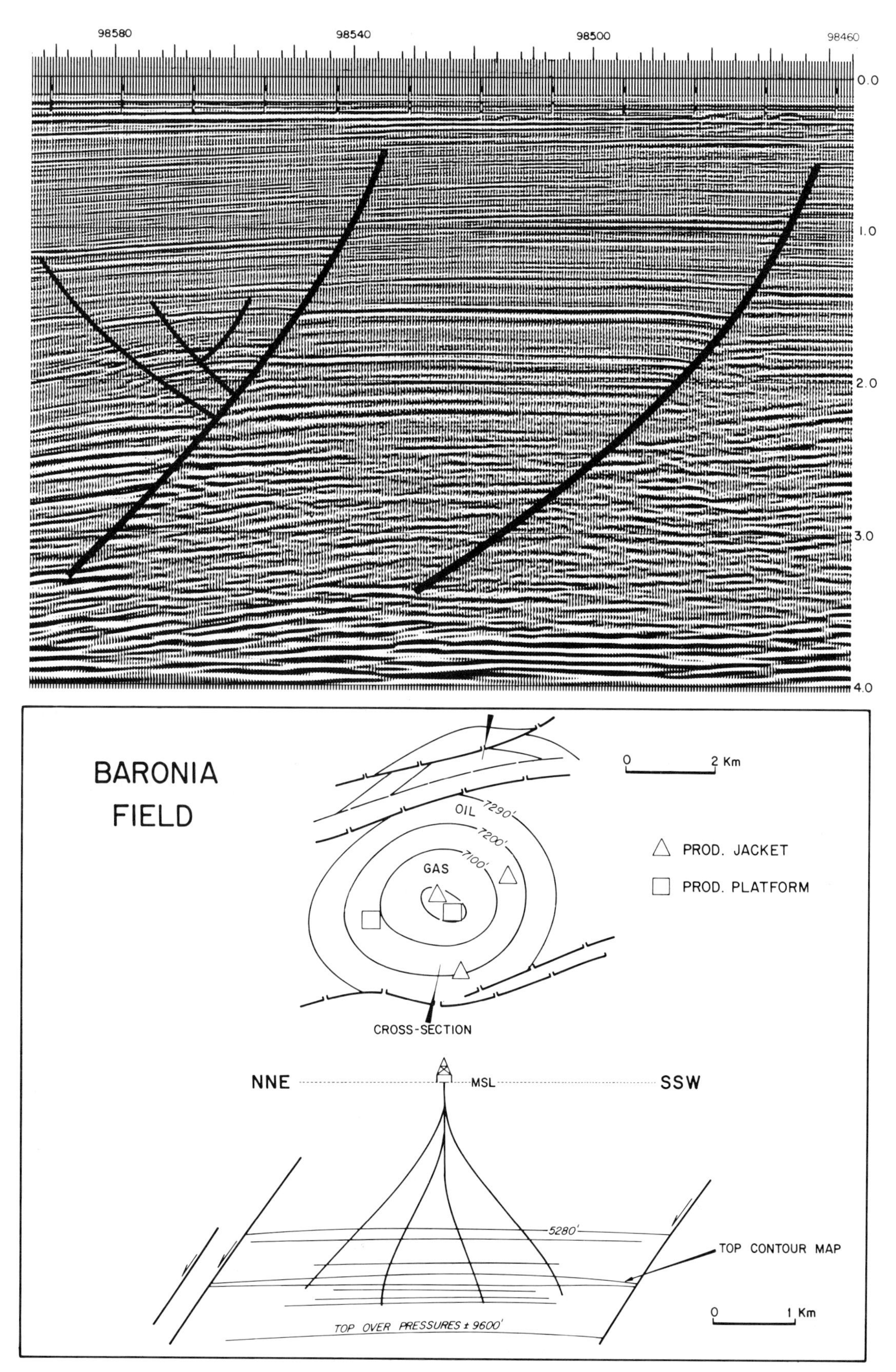

FIG. 9—Baronia field, seismic dip line, structural contours, and cross section. The field is a simple domal structure between two major growth faults. Note the complete absence of faults over the entire central part of the field and the large gas caps. (The writer is indebted to Raja Azhar, SSB Production for providing contour map and cross section).

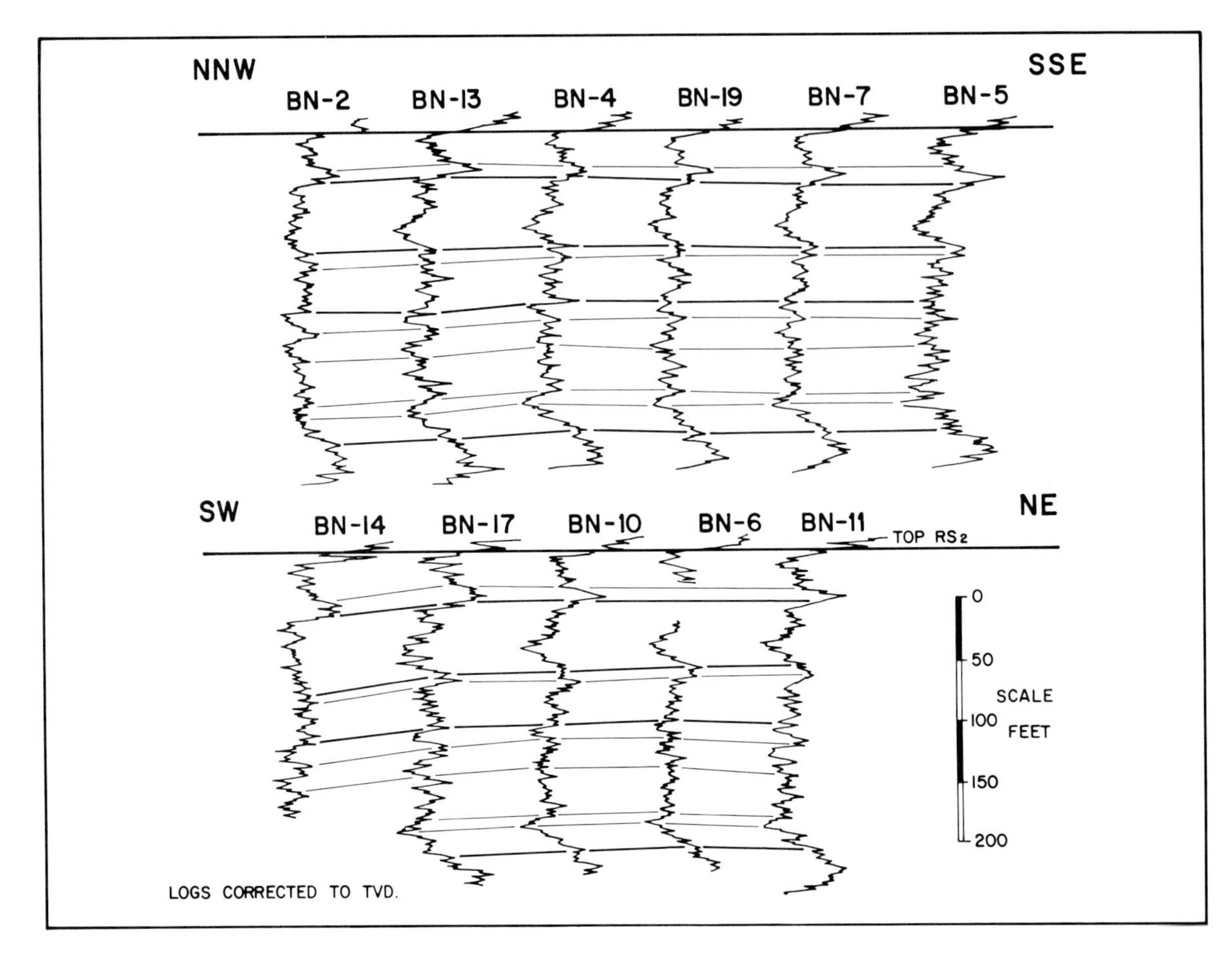

FIG. 10—Baronia field, RS 2 reservoir. Gamma-ray logs show excellent correlation across the field, demonstrating considerable continuity of the sandstone bodies which are believed to represent neritic sand sheets deposited in a fluviomarine coastal environment. (The writer thanks Raja Azhar, SSB Production for this figure).

The main reserves are found in four intervals between depths of 1,300 and 2,250 m in fault blocks 1 and 2 (reservoirs J1.0-J4.0, K5.0-K7.0, M1.0-M7.0, Q3.0-Q6.0). In the southwestern part of the field, hydrocarbons also are found at shallower levels, stacked behind antithetic faults.

Appraisal drilling has proved there are also hydrocarbons in the intensely faulted core of the collapsed crest, as well as farther east, where west-hading subsidiary faults trap steep east-dipping reservoirs.

Oil column heights range from 6 to 75 m. Most of the reservoirs have gas caps, but no significant accumulation of nonassociated gas is found. No hydrocarbons have been found in the overpressured deeper sequence (highest mud gradient 0.86 psi/ft).

Although locally some accumulations are fault-seal controlled and thus "filled to spill point," there is considerable evidence from detailed structural data and differential pressure measurements across faults that the overall trap fill in the field is relatively low and that the available trap volume considerably exceeds the volume of hydrocarbons trapped.

The bulk of the oil produced from the Samarang field is light, waxy, land-plant derived, with a low sulfur content (Table 1). The heavy, nonwaxy oil recovered from the shallow reservoirs has lost its wax content through bacterial degradation in the reservoir.

Recoverable oil reserves are 230 million bbl. Even though the Samarang oil reserves are larger than those of Baronia, Samarang contains only about one third of the amount of gas reserves of Baronia.

Development — Production development consists of a 19-slot and a 26-slot platform and three 6-slot jackets; additional jackets are planned. Whenever possible, the wells are completed as two string dual or multiple zone completions. The shallower sandstones are insufficiently consolidated and therefore are gravel-packed.

Central Luconia Gas Fields

Location — The Central Luconia gas fields are between 130 and 300 km offshore Central Sarawak in the area between 111° and 113°E long. and 3°30' and 5°30'N lat. (Fig. 1), in water depths between 60 to 100 m. In total, 20 accumulations were found from 1968 to 1975. Ten accumulations are of substantial size; six qualify as giant gas discoveries (Table 2; Fig. 1). The wells that encountered gas in Central Luconia range in total depths from 1,250 to 3,760 m.

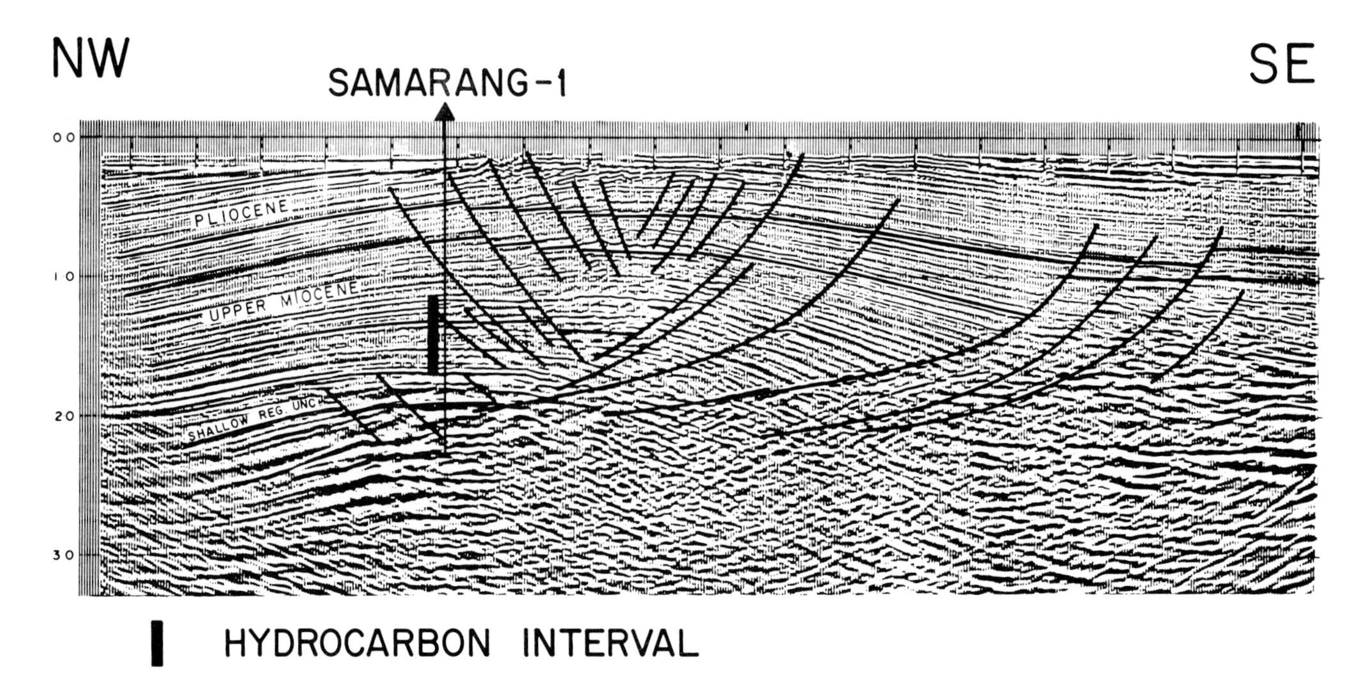

FIG. 11—Seismic dip line across Samarang field. The late Miocene/Pliocene growth fault structure was modified during the late Pliocene by tectonic movements which probably are related to wrench faulting in the basement. Note the collapsed crestal area with synthetic and antithetic faults. The bulk of the hydrocarbons is found along the relatively unfaulted western flank of the structure.

Stratigraphy — The main prospective sequence consists of cycle IVand V carbonates of middle to late Miocene age, penetrated at depths ranging from about 900 to over 3,000 m. In a few places gas was found in the cycle V clastic rocks overlying the carbonates.

The carbonate reservoirs are linked to carbonate buildups, the size, shape and structure of which was determined by the growth rate of corals and coralline algae, which in turn was mainly controlled by changes in sea level.

The reservoirs consist of limestones, dolomitic limestones, and dolomites. Reservoir quality varies according to environment of deposition and subsequent diagenetic processes, which eventually led to the development of a series of different rock types (Fig. 13).

Freshwater leaching of carbonates deposited in a reef (grain-supported) environment produced mouldic limestones, while in the muddy carbonate deposits in a protected zone behind the reef a chalkified limestone was found.

Dolomitization of carbonates deposited in a protected environment (e.g. lagoonal mudstones) was common and resulted in sucrosic dolomites. However, in most grain-supported sediments dolomitization was less complete and only the muddy matrix was dolomitized (over-dolomite).

The various environments of deposition, the sequence of diagenetic events, and the resulting rock types are found in almost all the buildups drilled. However, on a regional scale the relative importance of the various diagenetic effects in each buildup varies considerably. Thus, in the central part of the Central Luconia province, where the buildups had emerged for a long period of time, freshwater leaching was extensive, whereas in the southeast, where the buildups were covered with clastic deposits at an early stage, dolomitization prevails.

Structure — The structures which trapped all the giant gas accumulations in Central Luconia are reef carbonate buildups of cycle V sealed by cycle Vor, farther offshore, by cycle VI clastics. Two types of buildups are found: (1) platform buildups are large steep-flanked, very thick, flat-topped features which developed on highs, where the rate of carbonate sedimentation kept pace with the rate of subsidence; and (2) pinnacle buildups are conical-shaped, generally much smaller features, each of them covering only about one third or less of the area of the platform-type buildups. They are found in the transition zone between regional highs and the surrounding basins where the rate of subsidence exceeded the rate of carbonate sedimentation.

Hydrocarbons — The large gas accumulations in the Central Luconia carbonate buildups mostly occupy single reservoirs with a common gas-water contact. Vertical permeability barriers in the form of argillaceous transgressive layers dividing a buildup into separate reservoirs are only occassionally found, as in E.11 (Fig. 13).

In the main gas fields, gas occurs at depths ranging from 1,070 m (F.6) to 1,900 m (F.13), but significant quantities of gas have been found in Central Luconia as deep as 3,360 m.

Average gas column heights range from 65 to 175 m. No commercially producible quantities of oil have been found in Central Luconia, although in a few buildups small oil columns were logged underneath the gas. Water pressures in the carbonate reservoirs are about 80 to 300 psi above hydrostatic. The volume of hydrocarbons in almost all the buildups in Central Luconia is smaller than the available trap volume.

The bulk of the gas in the large Central Luconia fields consists of methane (Table 2). Total recoverable gas reserves from the largest six fields are estimated at 11.77 Tcf.

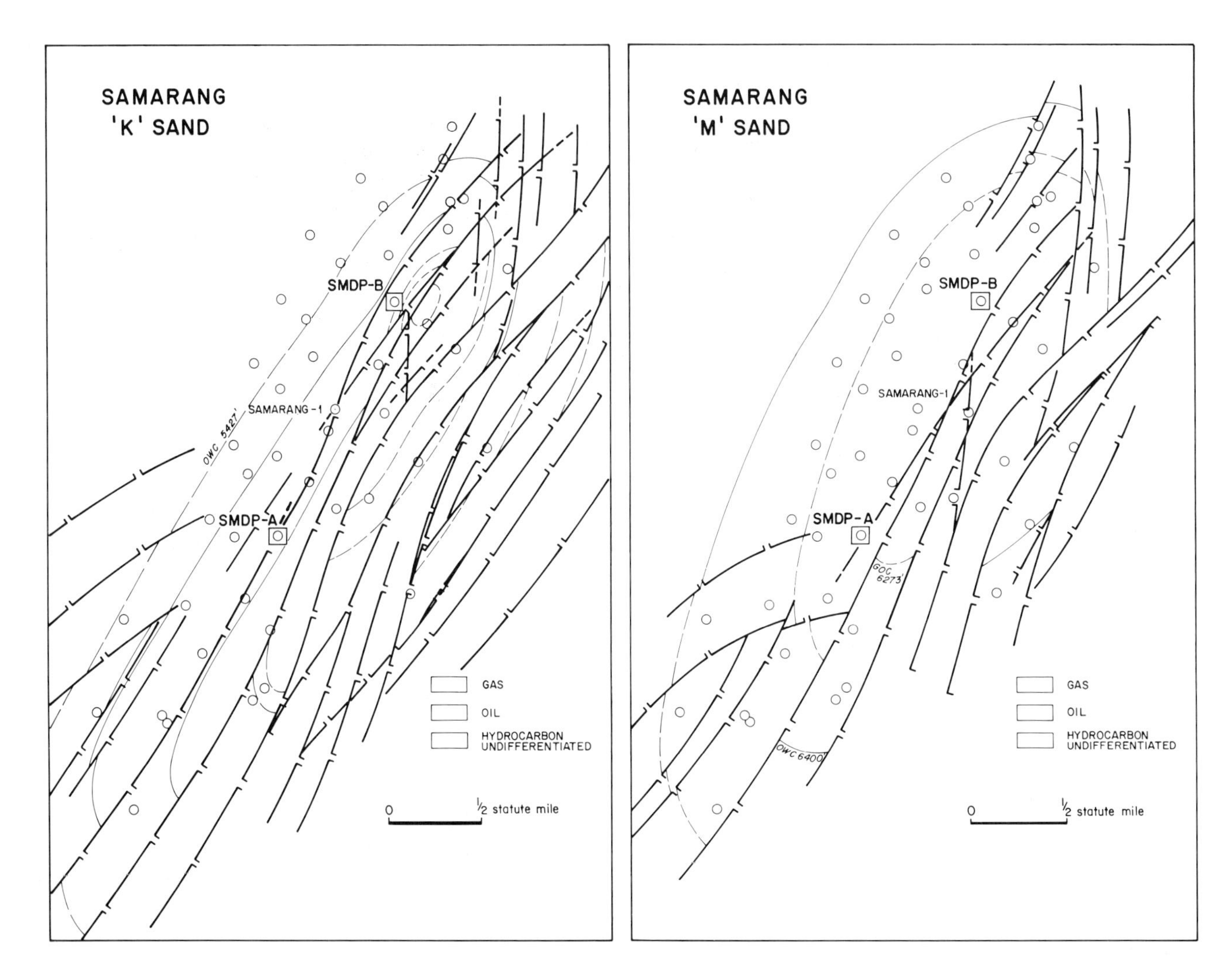

FIG. 12—Samarang field, contour maps of shallow K sands (**12a**) and deeper M sands (**12b**), illustrating the shallow crestal part of the field and little faulted deep western flank. (The writer is indebted to L.M. Kaye, SSB Production for the two contour maps).

Table 2. Summary of Reservoir and Hydrocarbon Data.

		F.6	F.13	E.11	E.8	F.23	M.3
Productive area	(acres)	16,500	12,000	5,500	5,000	9,000	4,000
Gas column heights	(avg., feet)	370	240	570	550	360	220
Reserves	(Tcf)	3.40	1.99	1.56	1.79	1.86	1.17
Porosity	(%)	20 to 32	21	15 to 28	15 to 22	16 to 29	24
Permeability	(md)	100 to 500	80 to 300	100 to 600	200 to 1,000	100 to 1,000	—
Water saturation	(%)	5 to 25	15 to 21	5 to 40	6 to 18	7 to 30	19
Initial reservoir pressure	(psig)	2,190	2,900	2,885	2,550	2,430	3,880
Reservoir temperature	(°F)	201	220	222	208	222	220
Gas composition methane	(%)	87.8	79.2	86.2	85.4	88.0	81.0
ethane	(%)	4.0	2.1	2.4	5.2	3.4	6.1
propane	(%)	2.9	1.1	1.3	3.5	2.5	4.5
C_4^+	(%)	3.4	1.8	1.6	4.1	3.5	4.4
Initial gas expansion factor		133	166	163	155	139	222

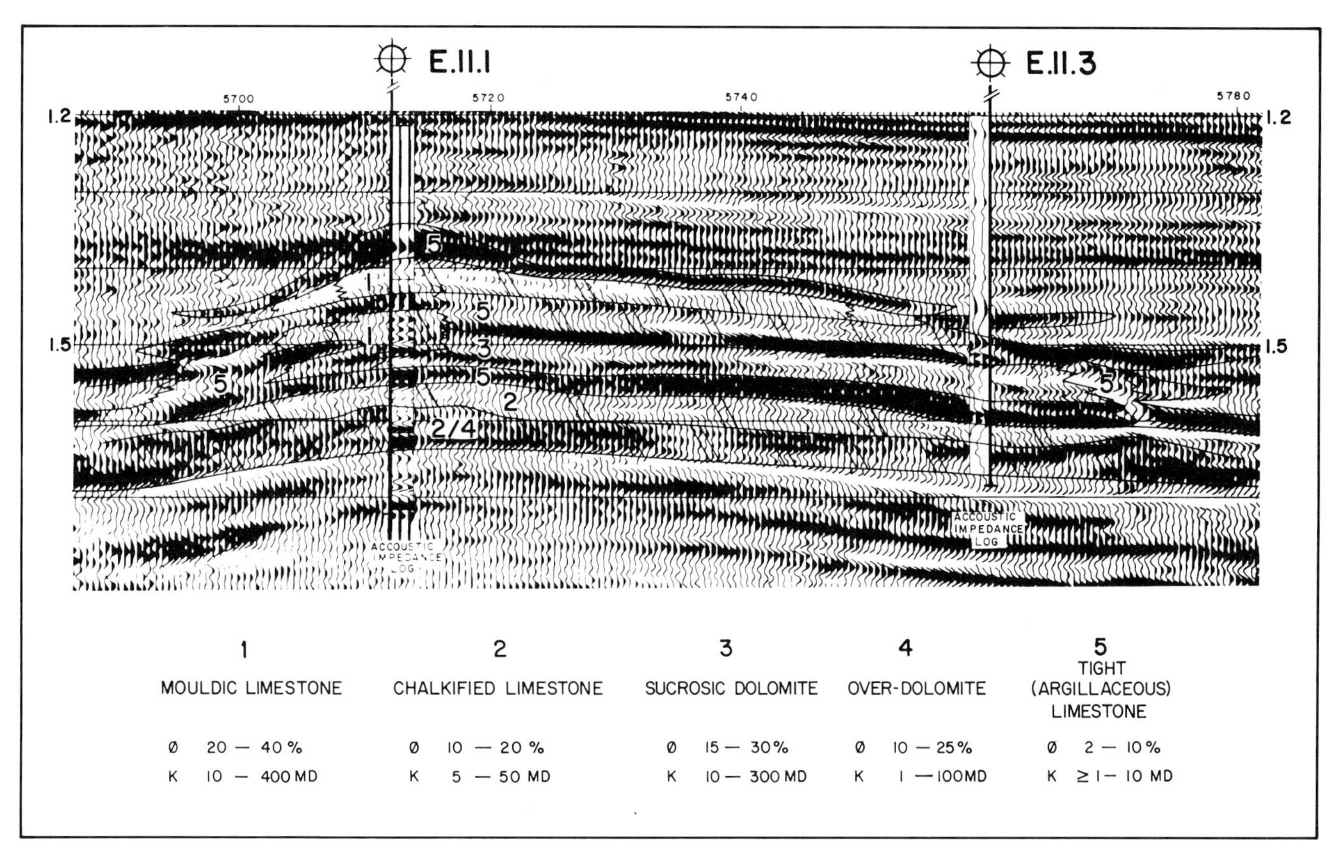

FIG. 13—E.11 gas field, seismic impedance section across carbonate buildup. The reservoirs are linked to the buildup, their quality varies according to the environment of deposition and subsequent diagenetic processes. Freshwater leaching in grain-supported reef produced mouldic limestones, whereas in the muddy carbonates behind the reef it produced chalkified limestones. Dolomitization of carbonates often produced chalkified limestones. Dolomitization of carbonates often produced sucrosic dolomites (in muddy sediments) or overdolomite (dolomitization of the muddy matrix only) in grain-supported carbonates.

Development — Development is planned to start in 1980 with the launching of a 20-slot drilling platform, a production platform, and a living quarter/control platform on the E.11 and F.23 buildups. Gas and condensate will be transported via a pipeline to the Malaysian Natural Gas liquefication plant in Bintulu.

GENERATION AND TRAPPING CONDITIONS IN EAST MALAYSIA

All producing East Malaysian fields and all discoveries made over the past decade are located offshore. The continental shelf is most favorable for hydrocarbon exploration in that it contains about 80% of all the prospective post-geosynclinal Neogene sediments present in East Malaysia. Moreover, structural deformation is generally less severe offshore than on land. However, offshore exploration had to be delayed to await the development of an efficient technology before the current intensive search for hydrocarbons along the continental shelf could begin.

Now commercial quantities of hydrocarbons are found in very different geographical areas off East Malaysia, from the north of Sabah to the south of Central Sarawak in all major stratigraphic intervals of the Miocene and in a variety of structural setting.

Hydrocarbon charge appears to have come from two principal sources: from source rocks that formed in the backswamps of the coastal plain and from land-plant derived material dispersed in the fluviomarine environment in front of the large river mouths. Generally these source rocks are relatively lean in organic material compared to oil source rocks and oil shales formed in a marine euxinic environment. The relative scarcity of source rock is offset by the source rocks being interbedded within the reservoir sequence which results in an efficient migration setting. On the other hand, the lateral continuity of many of the source beds appears to be irregular, especially within the lower coastal-plain sequence. This may explain the relatively low trap fill of several fields in this area, taking into account the lean source rock situation and the "competition" of available trap volume in the stacked reservoir sequences.

Where present, source rocks were probably buried sufficiently deep to reach thermal maturity in most synclines, as evident from the presence of the many commercial accumulations. Although the preferred presence of gas in the deepest reservoirs of some fields appears to be related to thermal maturity, there are also indications that the chance of finding increasing amounts of gas instead of oil rises in sequences changing from fluviomarine to marine.

Trapping conditions are mainly determined by the type of structural deformation and by the continuity of the reservoirs.

The most simple type of traps are the Central Luconia carbonate buildups. Although in several of them tight layers are interbedded within the main reservoir sequence (e.g. in E.11), the buildups generally consist of only one reservoir with a single gas-water, gas-oil and/or oil-water contact. This greatly contrasts with trapping in the clastic sequences, where hydrocarbon accumulations are in stacked reservoirs with several separate contacts. A relatively simple trap type is typified by the growth fault structure of the Baronia field, whereas the Samarang field is an example of a more complex trap, modified through deep-seated basement tectonics. It is obvious that the amount and quality of seismic data required to define a particular trap are closely related to the complexity of the structure.

The sealing capacity of faults is an important aspect of trapping in complex prospects depending on fault seals. Experience has shown most normal and synsedimentary faults to be sealing, sometimes even with throws of as little as 30 m.

For determining the size of a field the continuity of the reservoirs is at least as important as the dimensions of the structure. In this respect, the fluviomarine sequence is more favorable than the lower coastal-plain setting. However, individual reservoir sandstones are commonly considerably thicker in the lower coastal-plain facies, the reservoir quality better (much higher permeabilities), and the overall sequence less gas-prone than in the fluviomarine facies.

REFERENCES CITED

Beddoes L.R., 1976, The Balabac subbasin, southwestern Sulu Sea, Philippines, *in* Offshore South East Asia Conference: Southeast Asia Petroleum Expl. Soc. (Singapore), Paper 15.

Bell, R.M., and R.G.C. Jessop, 1974, Exploration and geology of the West Sulu basin, Philippines: Australian Petroleum Exploration Assoc. Jour., v. 14, p. 21-28.

Bowin, C., et al, 1978, Plate convergence and accretion in Taiwan-Luzon region: AAPG Bull., v. 62, p. 1645-1672.

Doust, H., in press, Geology and exploration history of offshore Central Sarawak, *in* M.T. Halbouty, ed., Geology and resources of the Pacific region: AAPG Studies in Geology Series.

Emery, K.O., and Z. Ben-Avraham, 1972, Structure and stratigraphy of China basin: AAPG Bull., v. 56, p. 839-859.

Gransch, J.A., and J. Posthyma, 1973, On the origin of sulfur in crudes: Cong. Int. Organic Geochemists, Program, p. 727-739.

Hale, N.S., 1973, The recognition of former subduction zones in Southeast Asia, *in* Implications of continental drift to the earth sciences: v. 2, p. 885-892.

Hamilton, W., 1976, Subduction in the Indonesian region (abs.): Southeast Asia Petroleum Expl. Soc., Proc. v. 2, p. 37-40.

Hedberg, H.D., 1968, Significance of high-wax oils with respect to genesis of petroleum: AAPG Bull., v. 52, p. 736-750.

Ho, K.F., 1978, Stratigraphic framework for oil exploration in Sarawak: Geol. Soc. Malaysia Bull., v. 10, p. 1-14.

Leong, K.M., 1974, The geology and mineral resources of the upper Segama valley and Darwel Bay area, Sabah, Malaysia: Malaysia Geol. Survey Memoir 4.

Liechti, P., F.W. Roe, and N.S. Hale, 1960, The geology of Sarawak, Brunei, and the western part of North Borneo: Borneo (British) Geol. Survey Dept., B. 3, v. 1, 360 p.

Lijmbach, G.W.M., 1975, On the origin of petroleum: 9th World Petroleum Cong. (Tokyo), Special Paper 1.

Parke, M.L., et al, 1971, Structural framework of continental margin in South China Sea: AAPG Bull., v. 55, p. 723-751.

Whittle, A.P., and G.A. Short, 1977, The petroleum geology of the Tembungo field, East Malaysia (abs.), *in* Seminar on petroleum geology of the Sunda Shelf: Geol. Survey Malaysia.

Discovery and Development of the Badak Field, East Kalimantan, Indonesia[1]

By Roy M. Huffington and H. M. Helmig[2]

Abstract The Badak field lies on the east coast of the island of Kalimantan (Borneo), Indonesia, about 22 mi (35 km) south of the equator. The Badak 1 discovery well was spudded on November 27, 1971, and completed on February 11, 1972. The well penetrated more than 1,000 ft (305 m) of net gas sandstone and about 120 ft (37 m) of oil sandstone. Drilling of Badak 1 was preceded by an exploration program, which started in December 1968 and included aerial photographic and magnetic surveys, geologic field work, and reflection and refraction seismic surveys.

Geologically, the Badak field is a part of the Mahakam delta, a wedge of upper Tertiary clastic sediments 20,000 ft (6,096 m) thick, deposited in the Kutai basin.

Badak reservoirs are coarse to very fine-grained quartz sandstones with an average porosity of 22% and average permeability of 200 md. The individual sand bodies are either channel, mouth, or finger bar sands, deposited in a deltaic environment.

Structurally, Badak is one of several culminations formed on a 50 mi long (80 km), north-trending structural axis, which also connects the Nilam and Handil fields to the south. The Badak culmination is a gentle anticlinal uplift with no known faults.

The Badak field has in-place reserves of more than 7 Tcf of gas, 130 million bbl of condensate and 60 million bbl of oil. It now produces 500 MMcf of gas, 15,000 bbl of condensate and 10,000 bbl of oil per day.

INTRODUCTION

The Badak field lies on the east coast of the island of Kalimantan (Borneo), Indonesia, approximately 22 mi (35 km) south of the equator (Fig. 1). In August 1968, Roy M. Huffington, Inc. (Huffco) and Virginia International Company obtained a Production Sharing Contract in Indonesia covering an area of approximately 1,990 sq mi (5,154 sq km) in South Sumatra and one of 4,870 sq mi (12,613 sq km) in East Kalimantan.

In November 1968, exploration operations began in the East Kalimantan area. As the area was previously explored by Shell (B.P.M.), a study of both published and unpublished data was the first phase of Huffco's activities in and around the area.

Base Map

In 1968, no consistent, suitably scaled, topographic map of the East Kalimantan contract area existed, so the next operational step was compilation of a set of preliminary base maps from a number of various topographic and geologic maps. To acquire a consistent base map of the entire area (on a scale of 1:50,000), an aerial photographic survey was undertaken using false-color infrared film. Persistent cloud cover in East Kalimantan forced the flight crew to remain on standby thoughout

[1]Manuscript received, April 4, 1979; accepted for publication, July 6, 1979.

[2]Roy M. Huffington, Inc., Houston, Texas 77002.

The writers thank Pertamina, Union Texas Petroleum Corp., Golden Eagle Refining Co., Superior Oil Co., Alaska Interstate Corp., and Universe Tankships, Inc., for permission to publish this paper. Daud Harahap and other members of Huffco's geological staff in Balikpapan supplied several of the illustrations and Robert LaRue provided assistance in drafting and final preparation of illustrations for the presentation of this paper.

Article Identification Number:
0065-731X/80/MO30-0021/$03.00/0.

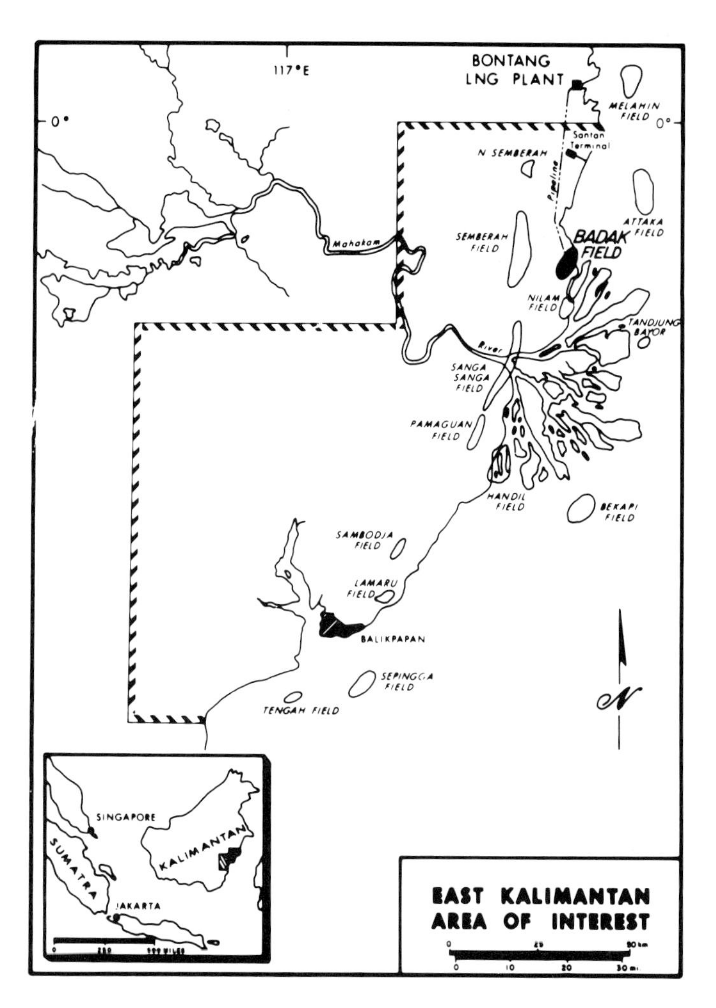

FIG. 1—Roy M. Huffington, Inc. Contract Area, East Kalimantan.

two dry seasons before 80% of the area was successfully surveyed.

Airborne Magnetometric Survey

Huffco's East Kalimantan contract area lies in the center of the Kutai sedimentary basin, which underlies that portion of East Kalimantan between the Mangkalihat Peninsula and the Meratus Mountains (Fig. 2). Because little was known about the gross configuration of the floor of the Kutai basin, a basinwide aeromagnetometric survey was taken in conjunction with Pertamina and Shell.

Surface Geological Surveys

Previous geologic field work was initiated with two field parties, primarily for obtaining firsthand information on the lithology and stratigraphy of the basin fill. Later geologic surveys mapped structural features in detail. Some geologic field work continues to take place, providing spot checks and additional details in problem areas.

Seismic Survey

Seismic surveying began in June 1970 and continued through June 1973. Another seismic survey took place from January 1975 to October 1976, and the most recent began in December 1978. The 1978 survey is designed to acquire higher resolution reflection data, enabling more subtle structural, as well as stratigraphic, traps to be detected.

Exploration Drilling

The first well drilled in Huffco's Indonesian operations, Badak 1, was spudded in November 1971, and was completed as an oil producer in February 1972. The well penetrated more than 1,000 ft (305 m) of gas and 120 (37 m) of oil pay in a section between depths of 4,500 ft (1,372 m) and 11,000 ft (3,353 m).

Development Drilling

Since the discovery well, 51 additional wells have been drilled in Badak field, of which 17 are completed as oil wells and 33 are gas wells.

REGIONAL GEOLOGY

Geologic history of the Kutai basin, in which the Badak field is located, has been the subject of several publications (the most recent by Rose and Hartono, 1978). Although differences of opinion exist among authors, there is consensus concerning the fact that the Kutai basin began its subsidence and sedimentary history during the late Eocene. The initial transgression reached its peak during the late Oligocene and earliest Miocene, as indicated by the predominantly fine-grained, widespread nature of the sediments of that age. Subsequently, from early Miocene to Holocene time, the rate of clastic deposition has outpaced that of subsidence, causing a general marine regression from west to east. This eastward regression was accompanied by a continuous, gradual shifting of the basin depocenter in the same direction, with increasing uplift and exposure to erosion of land areas to the west. In general, the Eocene and Oligocene rocks are thickest in the western part of the Kutai basin, while the Miocene-Pliocene reaches its maximum thickness around the present Mahakam delta.

The Badak field is one of the major fields producing oil and gas from sandstones deposited in the Mahakam delta. For more detailed descriptions of the Mahakam delta the reader is referred to Gerard and Oesterle (1973), Magnier et al (1976), and Allen et al (1977).

STRATIGRAPHY

The stratigraphic succession in the Badak field is a middle Miocene to Holocene shale-siltstone-sandstone sequence deposited in a deltaic environment, as depicted by the example of a stratigraphic column for Badak 6 (Fig. 3). Before the middle Miocene, predominantly deeper marine, prodeltaic shales of unknown thicknesses were deposited in the Mahakam delta region. As the delta steadily prograded to the east, terrestrial deltaic facies successively occupied the Badak field area. As can be expected, the eastward progradation of the Mahakam delta resulted from a series of minor oscillating movements (trans-

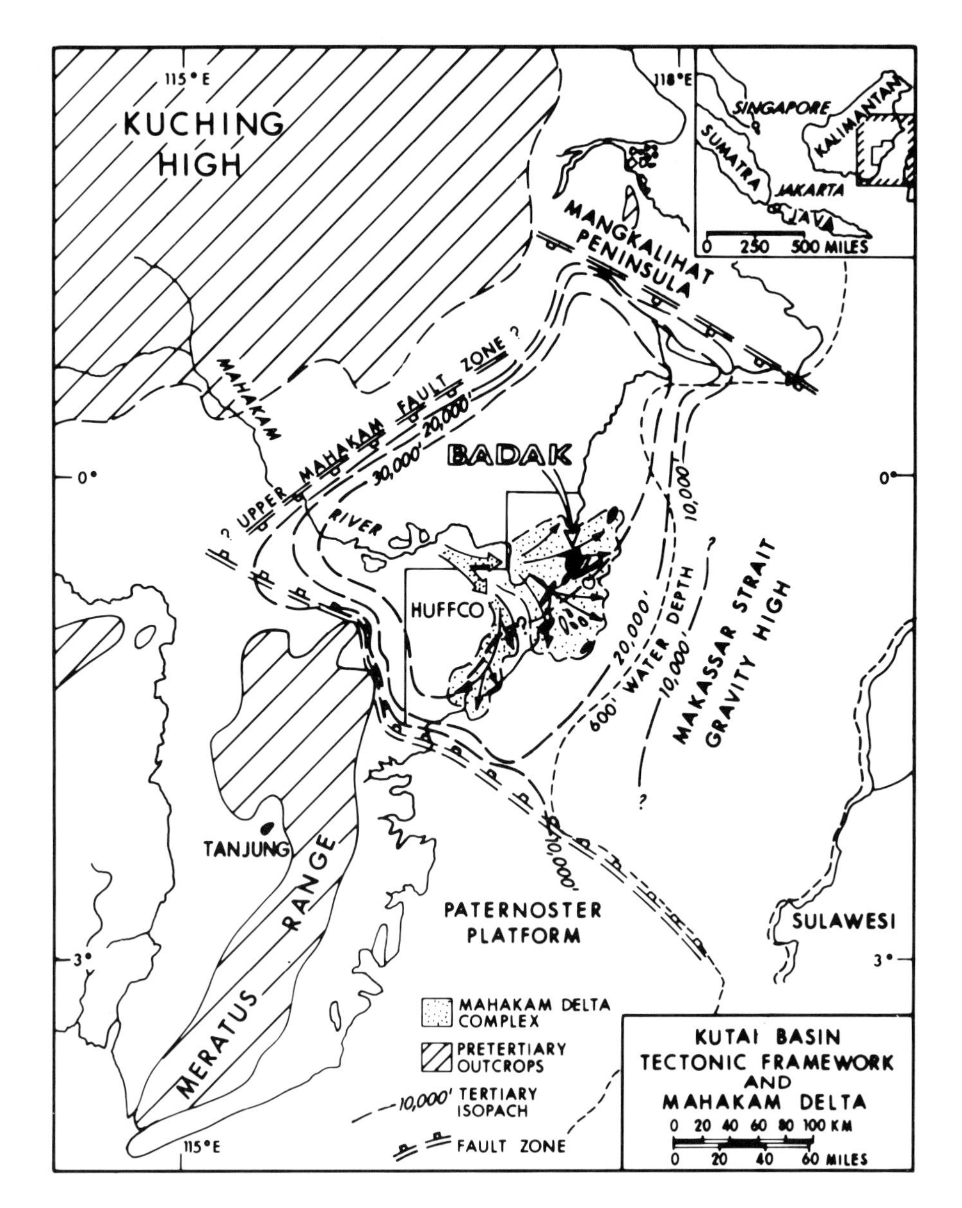

FIG. 2—Kutai basin tectonic framework.

gressions and regressions) of the deltaic zones through the Badak field area. This is to some extent demonstrated in the lithologic column of the Badak 6 (Fig. 3), which shows the oscillations in the, otherwise, generally upward-increasing sand percentage and upward-decreasing limestone content. This figure also presents a stratigraphic correlation of Badak reservoir zones to the Badak paleo-assemblages, depositional environments, and time stratigraphy.

For development purposes, the Badak E-log (electric log) stratigraphic succession was divided into a number of reservoir or E-log zones, alphabetically designated as zones "A" through "K", from top to bottom. Only zones "A" though "F" will be discussed, as our present knowledge of the deeper Badak zones is limited; however, zones "A" through "F" presently contain the majority of Badak hydrocarbon reserves. Within each Badak zone, the individual reservoir sandstones are numbered from top to bottom. In the correlation, lignite and limestone beds form good stratigraphic markers in the Badak field.

Each of the reservoir zones "A" through "F" will be described separately. Description will be primarily graphic, through use of a series of cross sections along the line of stratigraphic section shown in Figure 4. To give an impression of the sedimentological characteristics of Badak reservoirs, one typical example of a set of identifiable sand bodies out of each one of zones "B" through "F" is illustrated.

Zone "A" in Badak 1

Zone "A," the most shallow zone in the Badak field, consists predominantly of delta-plain deposits (distributary channel and braided stream sands and lagoonal and marsh deposits) represented by silts and clays with abundant lignites. Zone "A" has an average sand/shale ratio of approximately 45%. Lateral

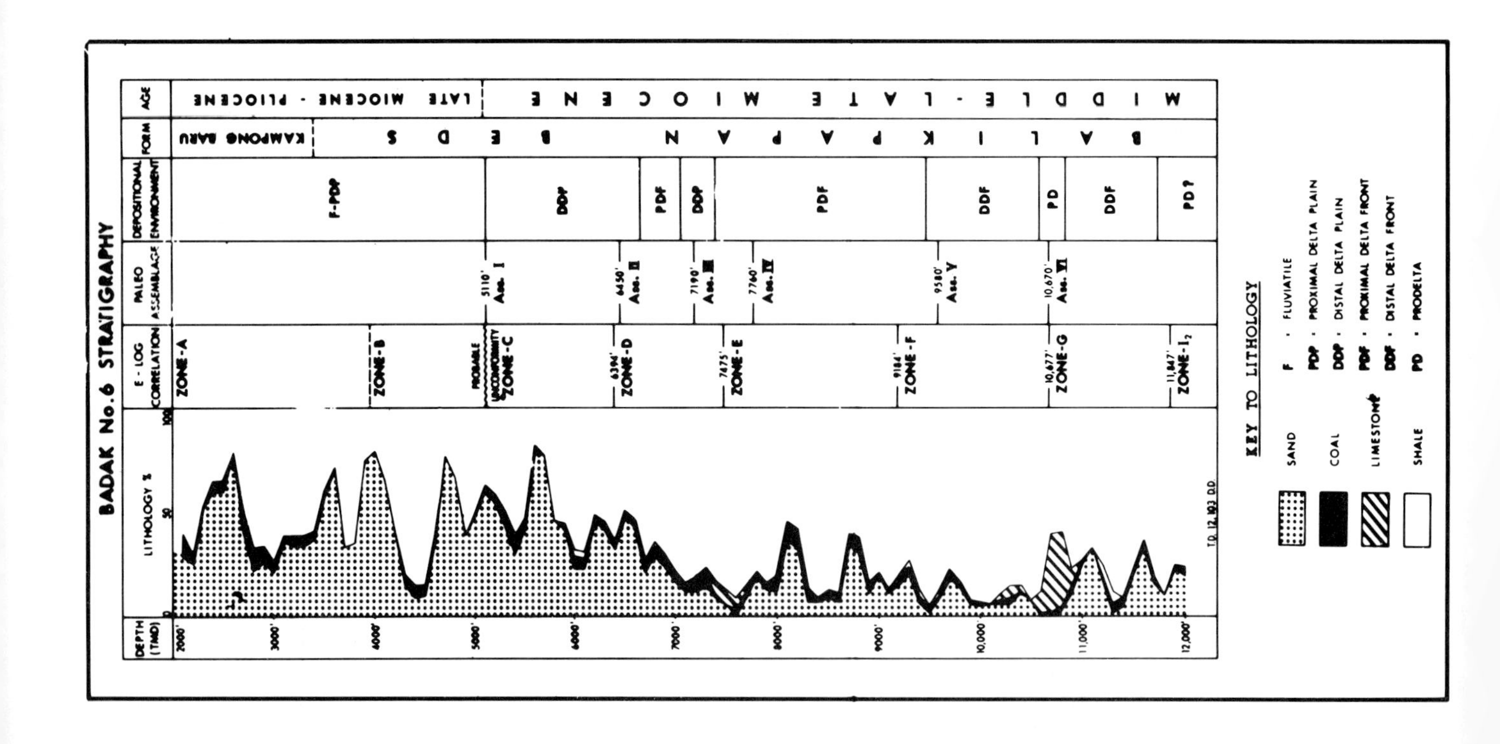

FIG. 3—Stratigraphy of the Badak 6.

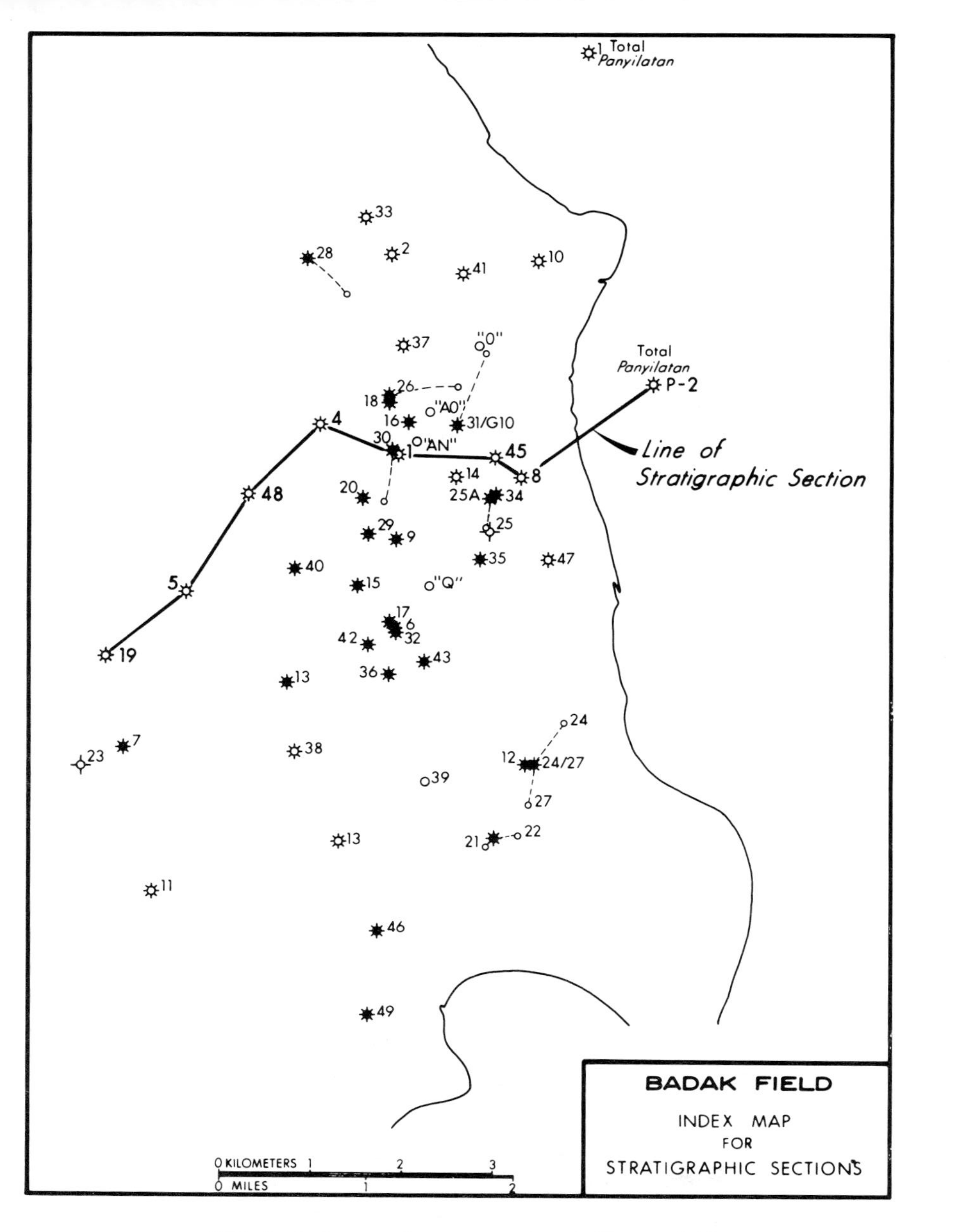

FIG. 4—Index map for cross-sections in Figures 5-10.

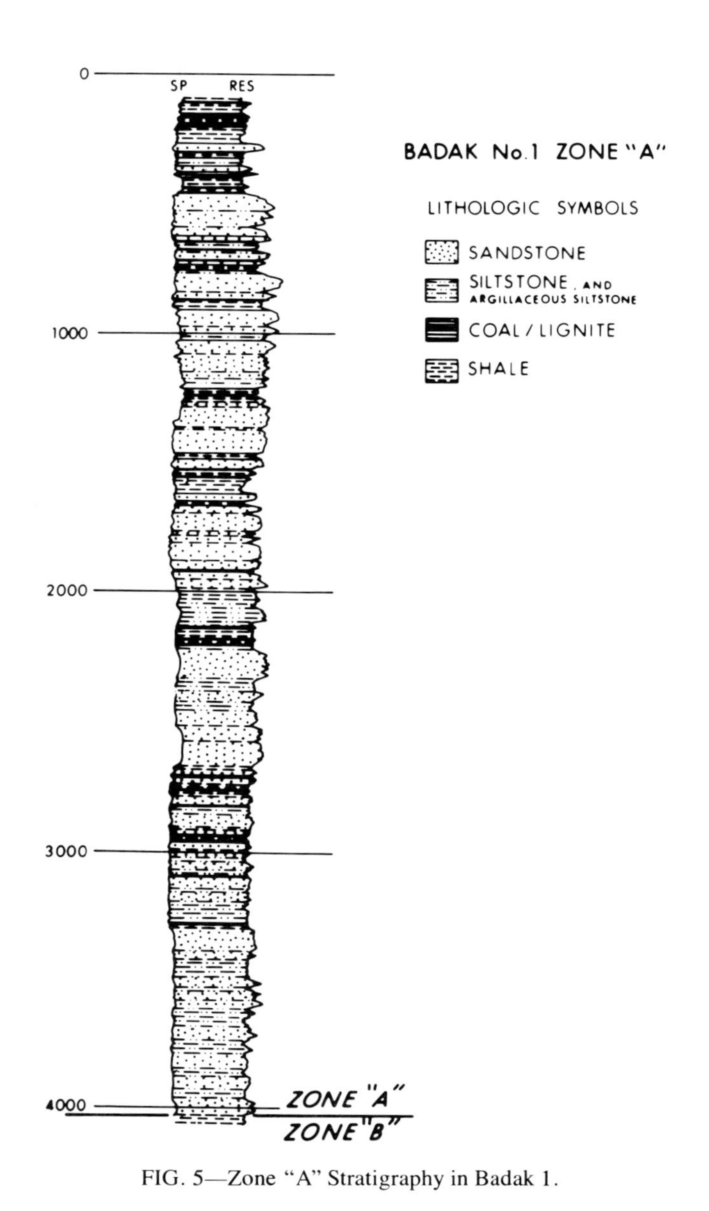

FIG. 5—Zone "A" Stratigraphy in Badak 1.

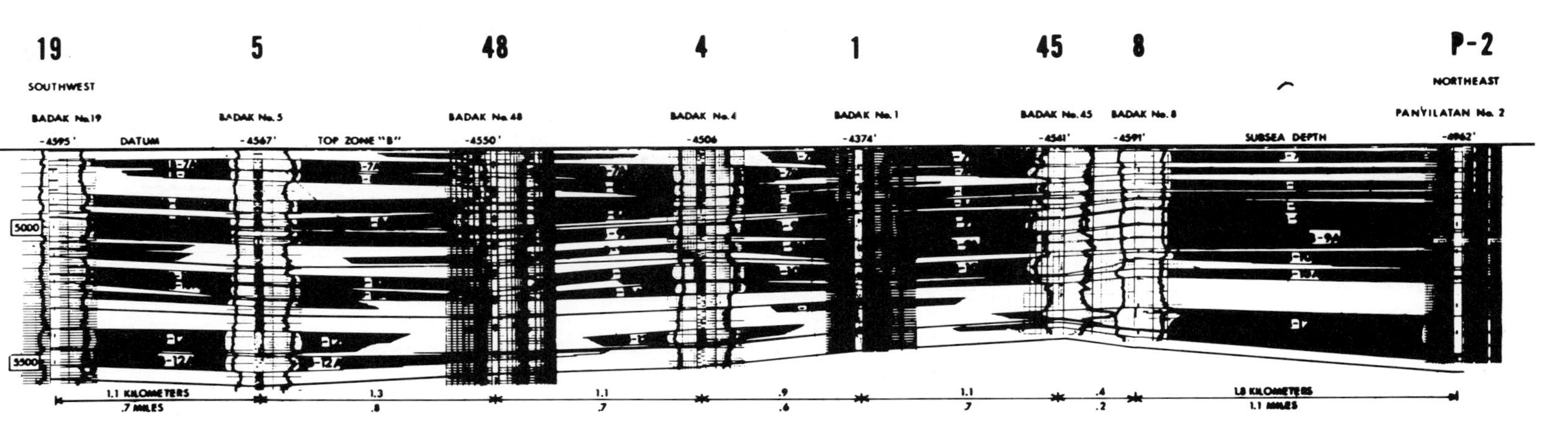

FIG. 6—Stratigraphic cross-section of Badak Zone "B."

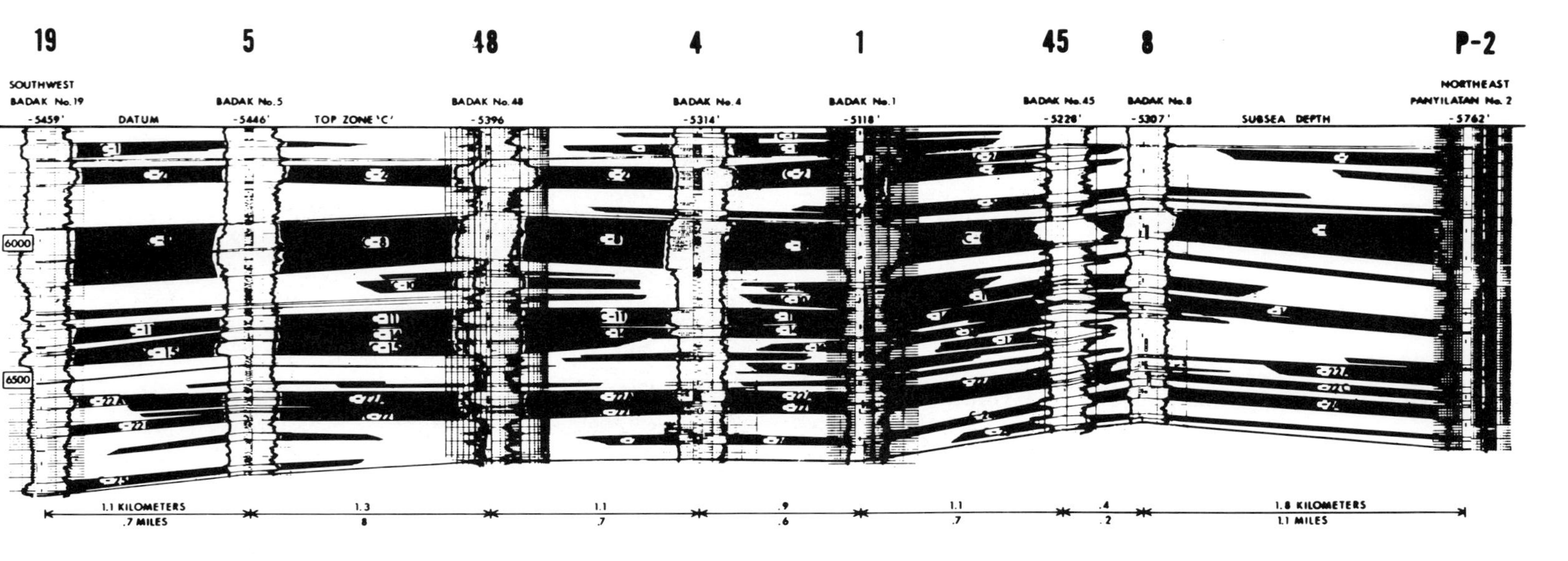

FIG. 7—Stratigraphic cross-section of Badak Zone "C."

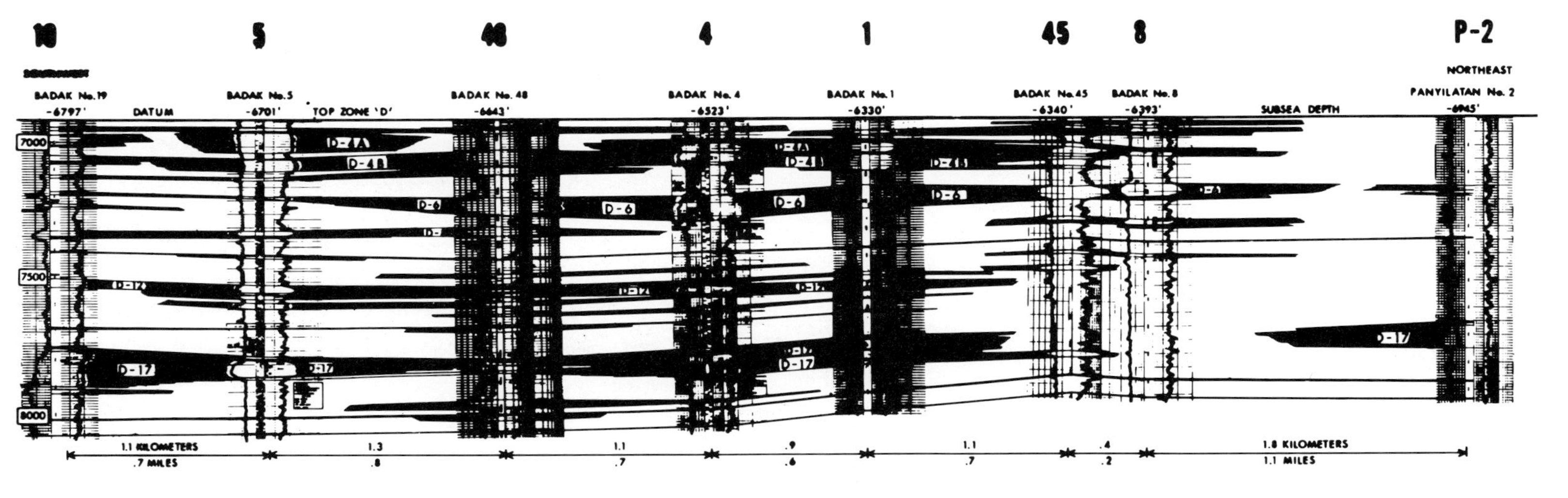

FIG. 8—Stratigraphic cross-section of Badak Zone "D."

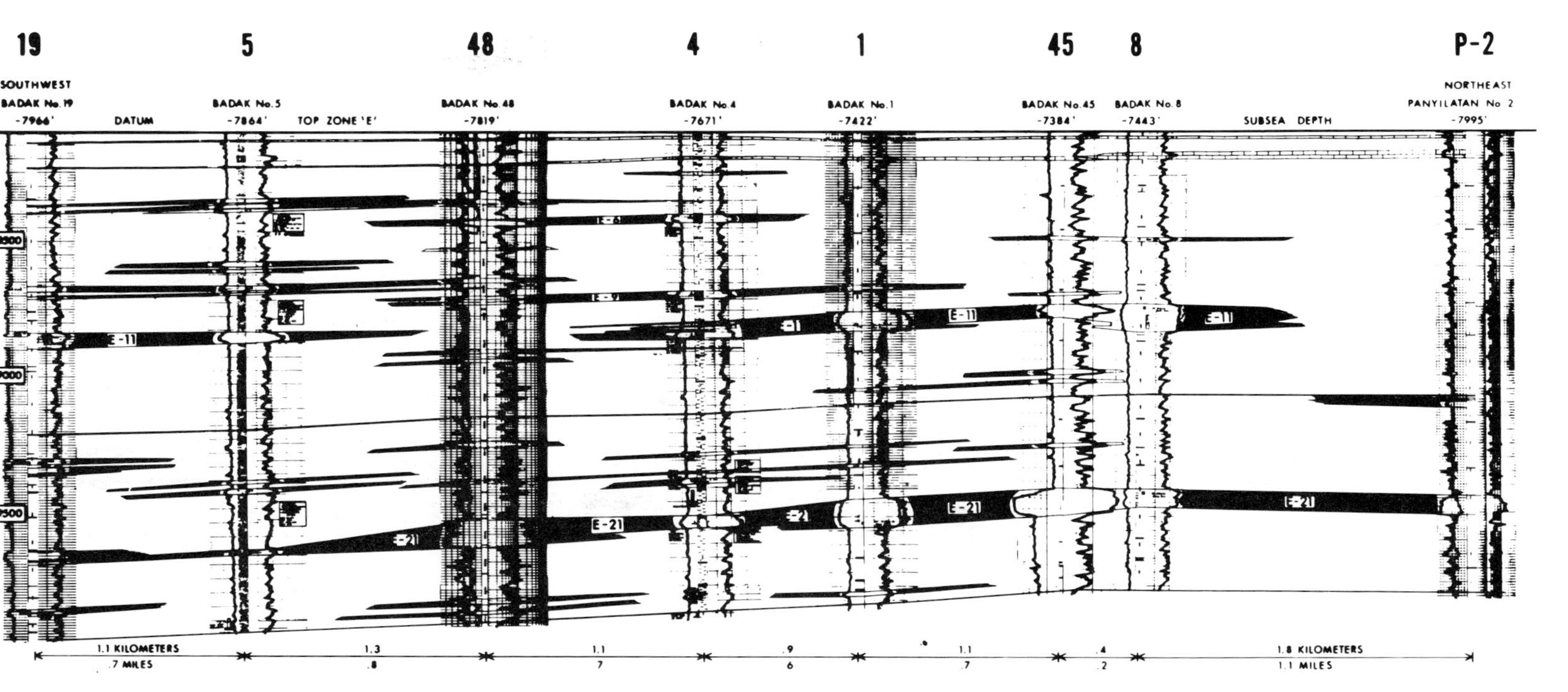

FIG. 9—Stratigraphic cross-section of Badak Zone "E."

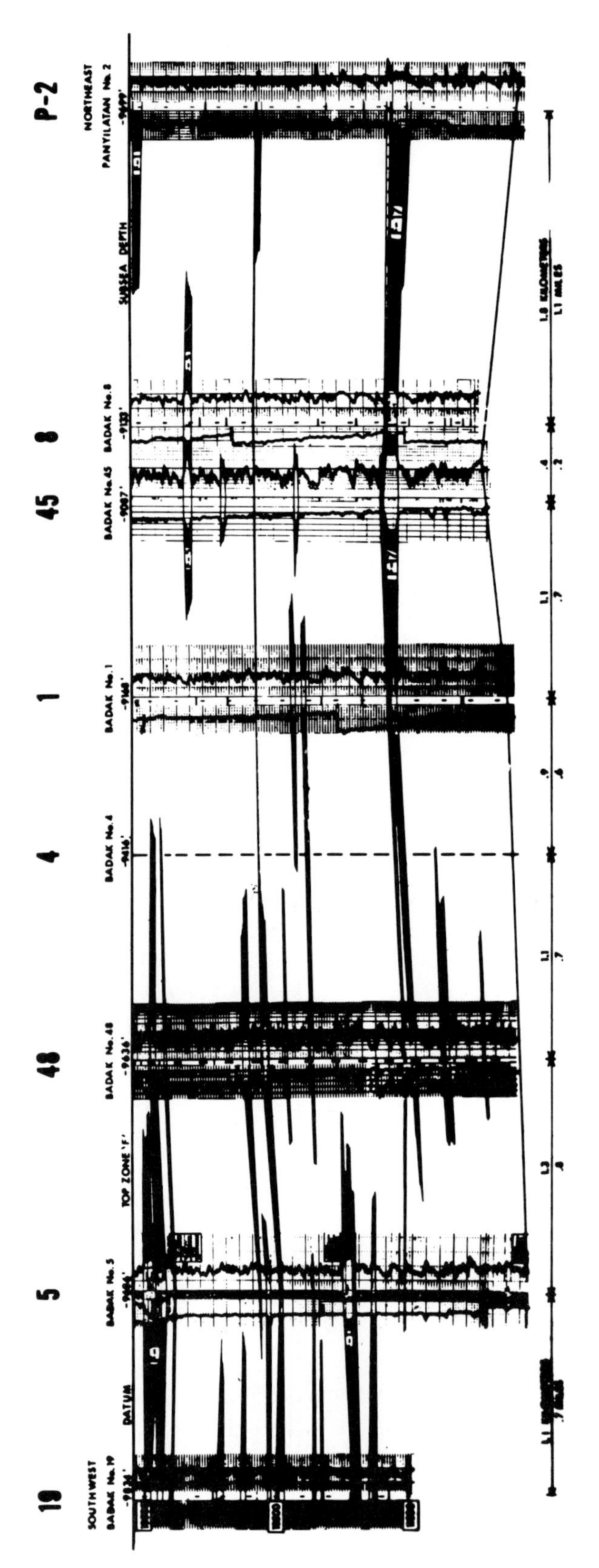

FIG. 10—Stratigraphic cross-section of Badak Zone "F."

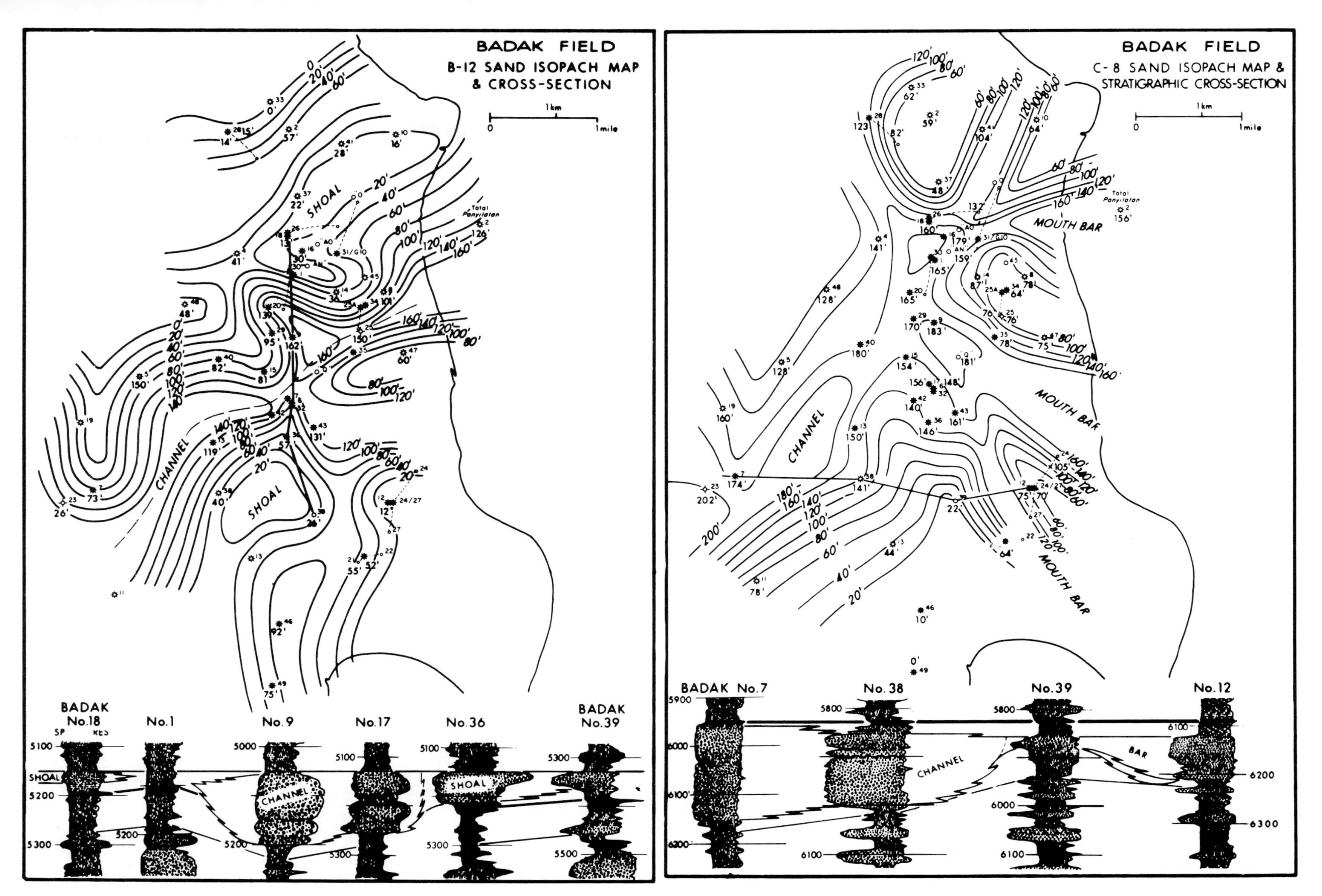

FIG. 11—"B-12 Sand" isopach map and cross-section.

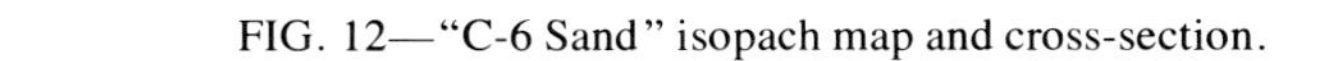

FIG. 12—"C-6 Sand" isopach map and cross-section.

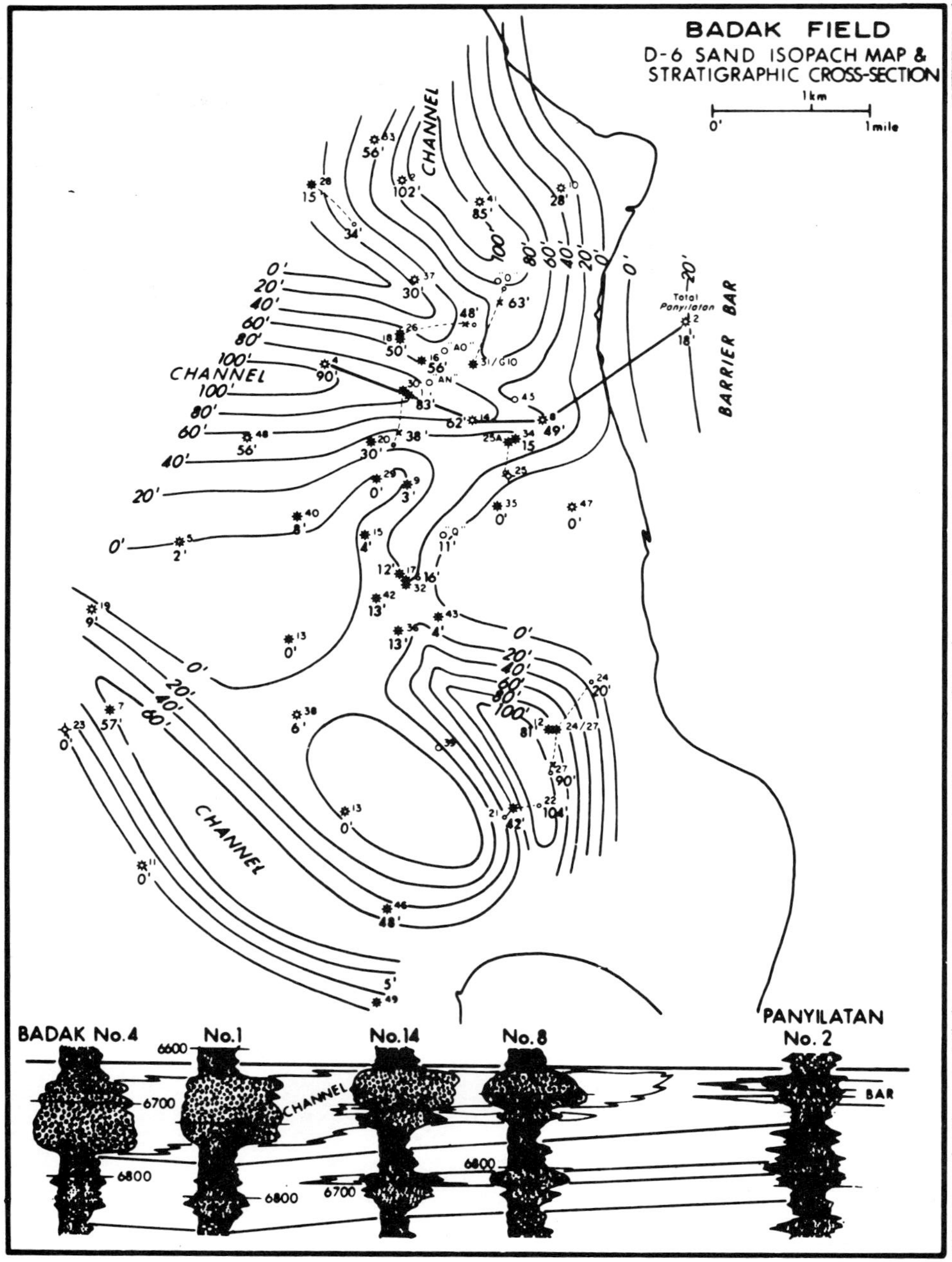

BADAK FIELD

D-6 SAND ISOPACH MAP & STRATIGRAPHIC CROSS-SECTION

FIG. 13—"D-6 Sand" isopach map and cross-section.

BADAK FIELD

E-11 SAND ISOPACH MAP & STRATIGRAPHIC CROSS-SECTION

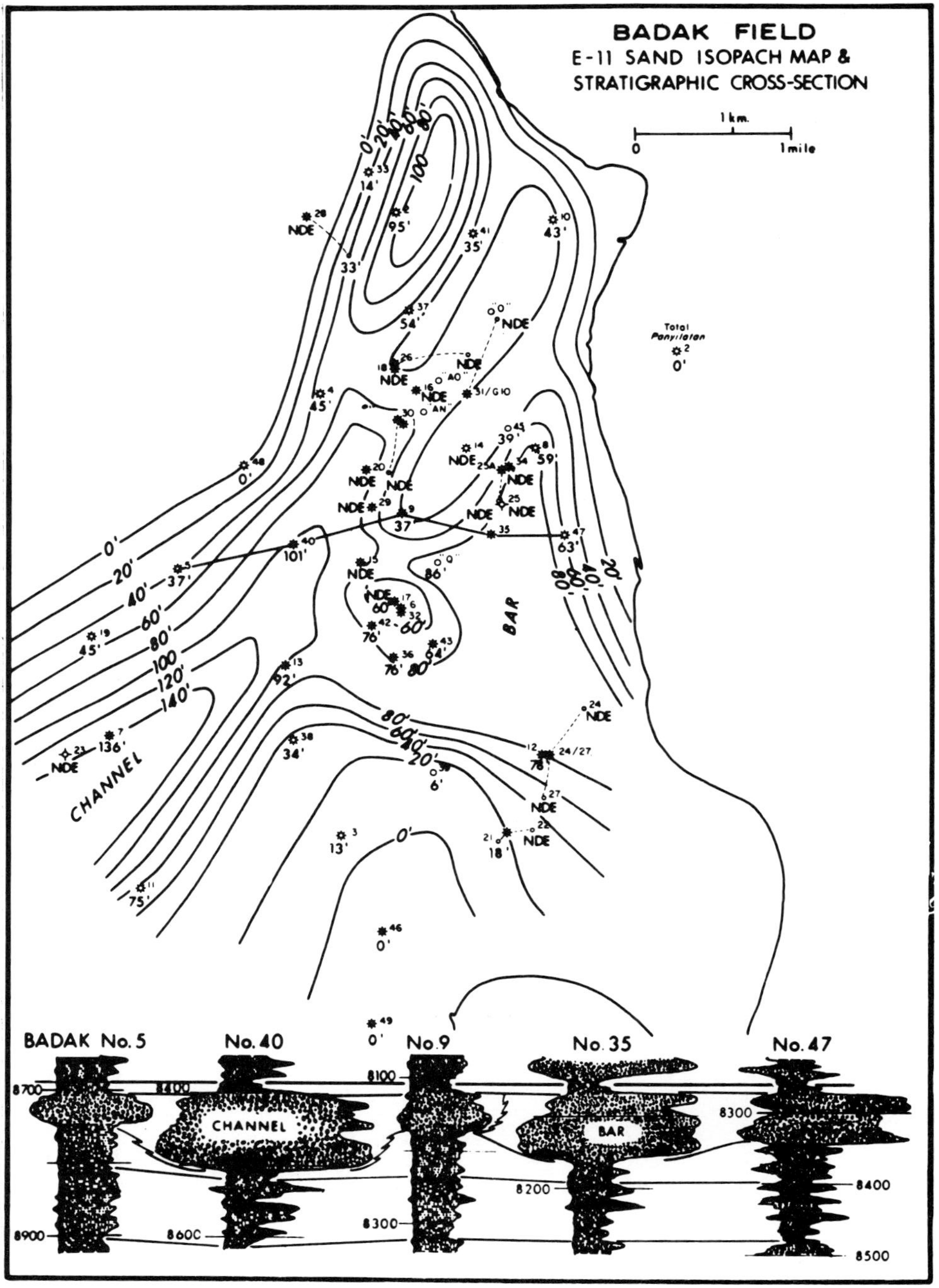

FIG. 14—"E-11 Sand" isopach map and cross-section.

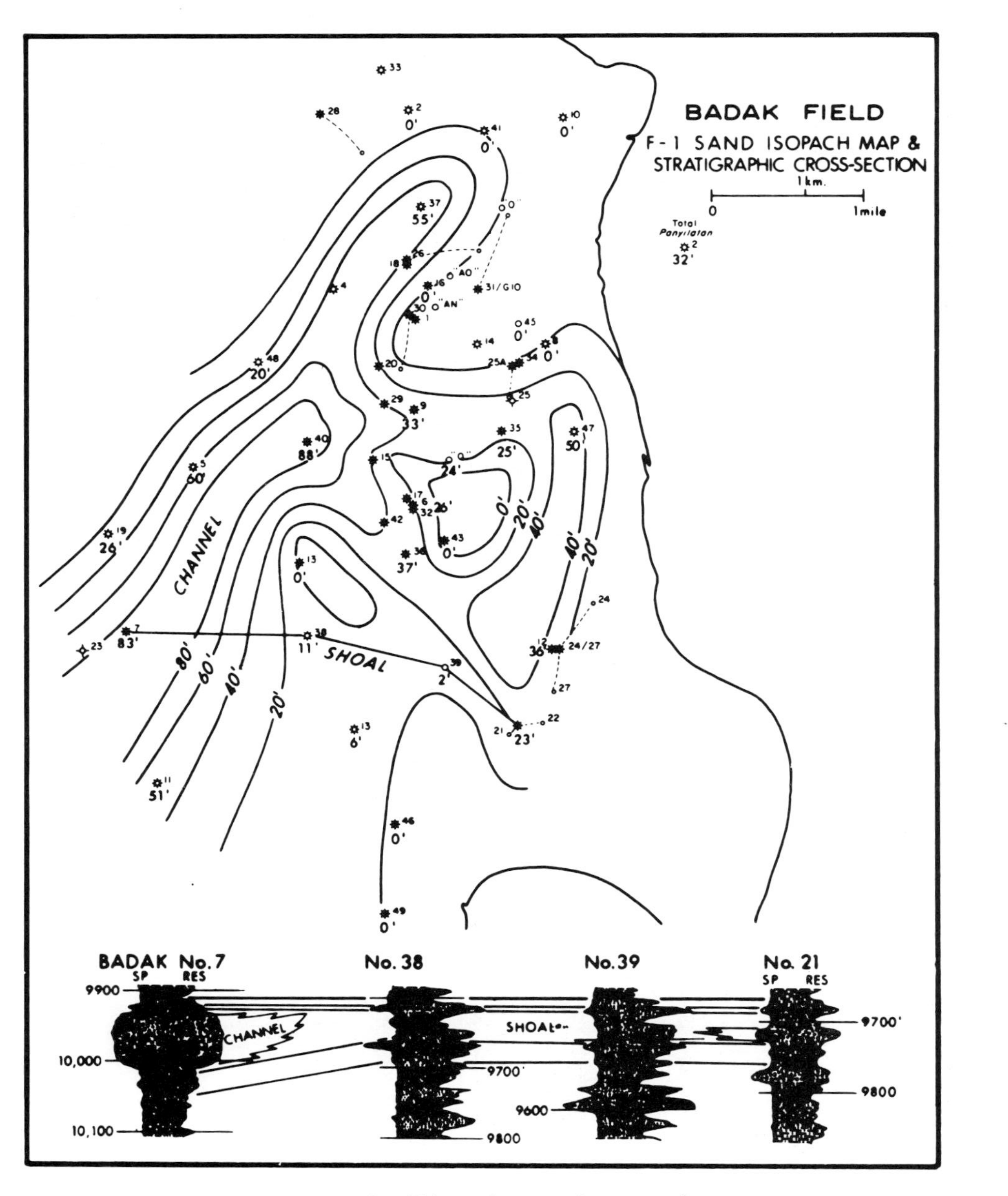

FIG. 15—"F-1 Sand" isopach map and cross-section.

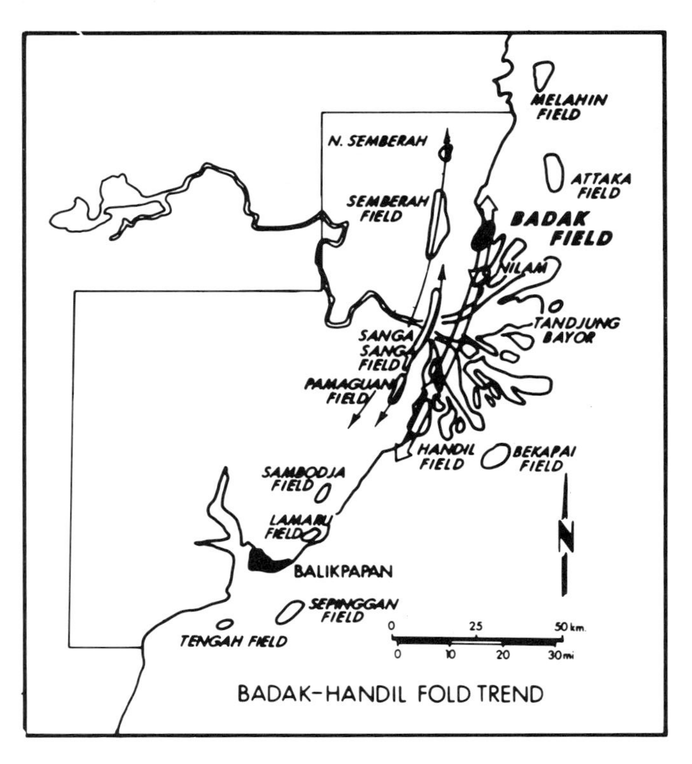

FIG. 16—Badak-Handil fold trend.

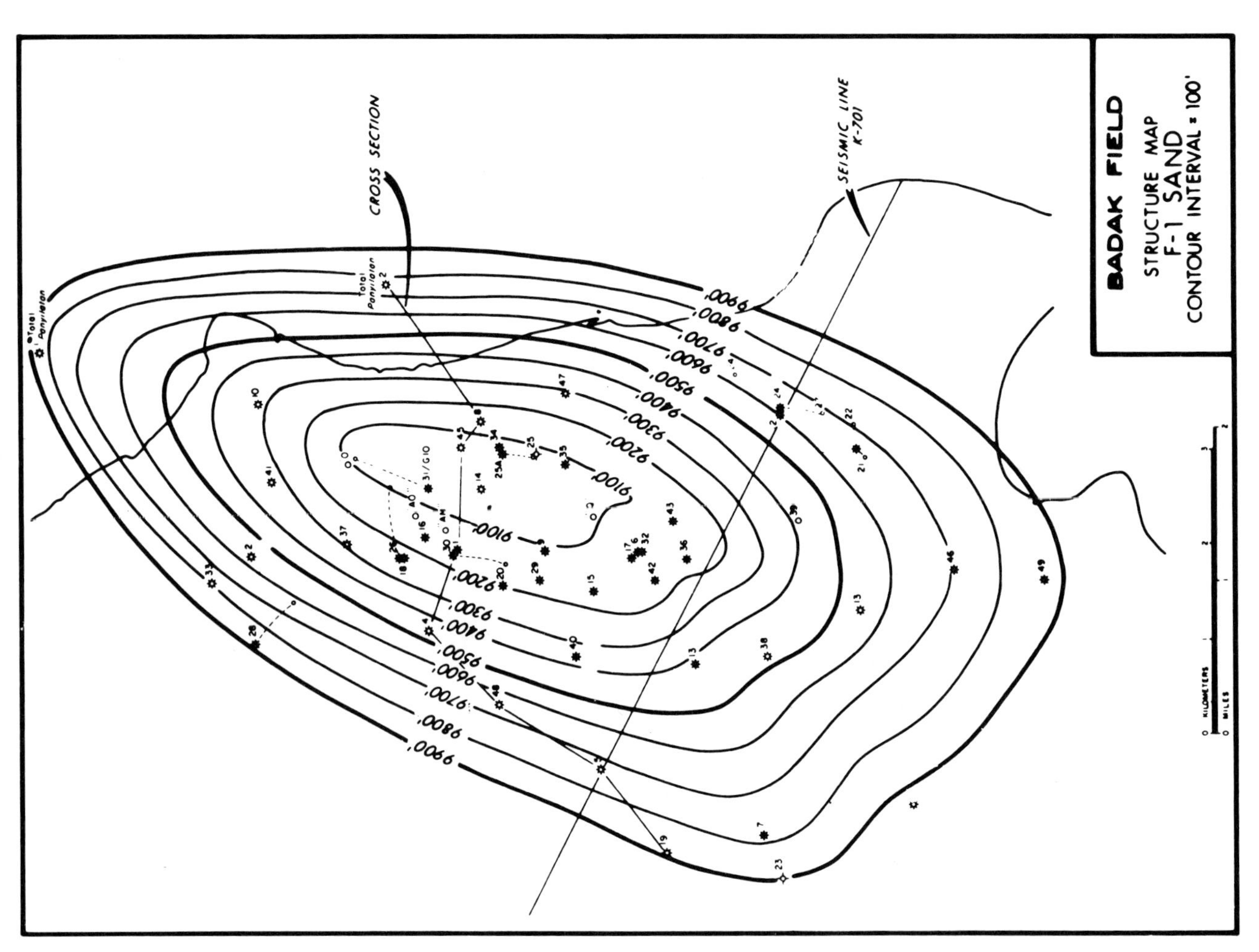

FIG. 17—Badak structure map on F-1 horizon.

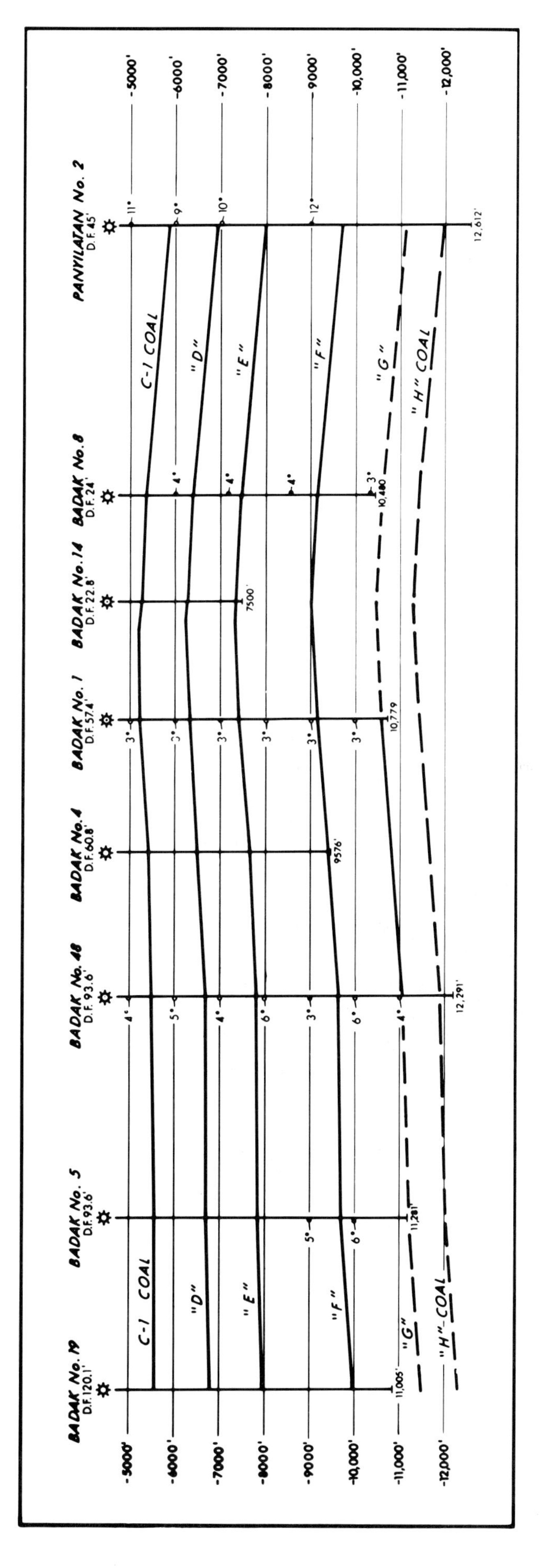

FIG. 18—Badak structural cross-section.

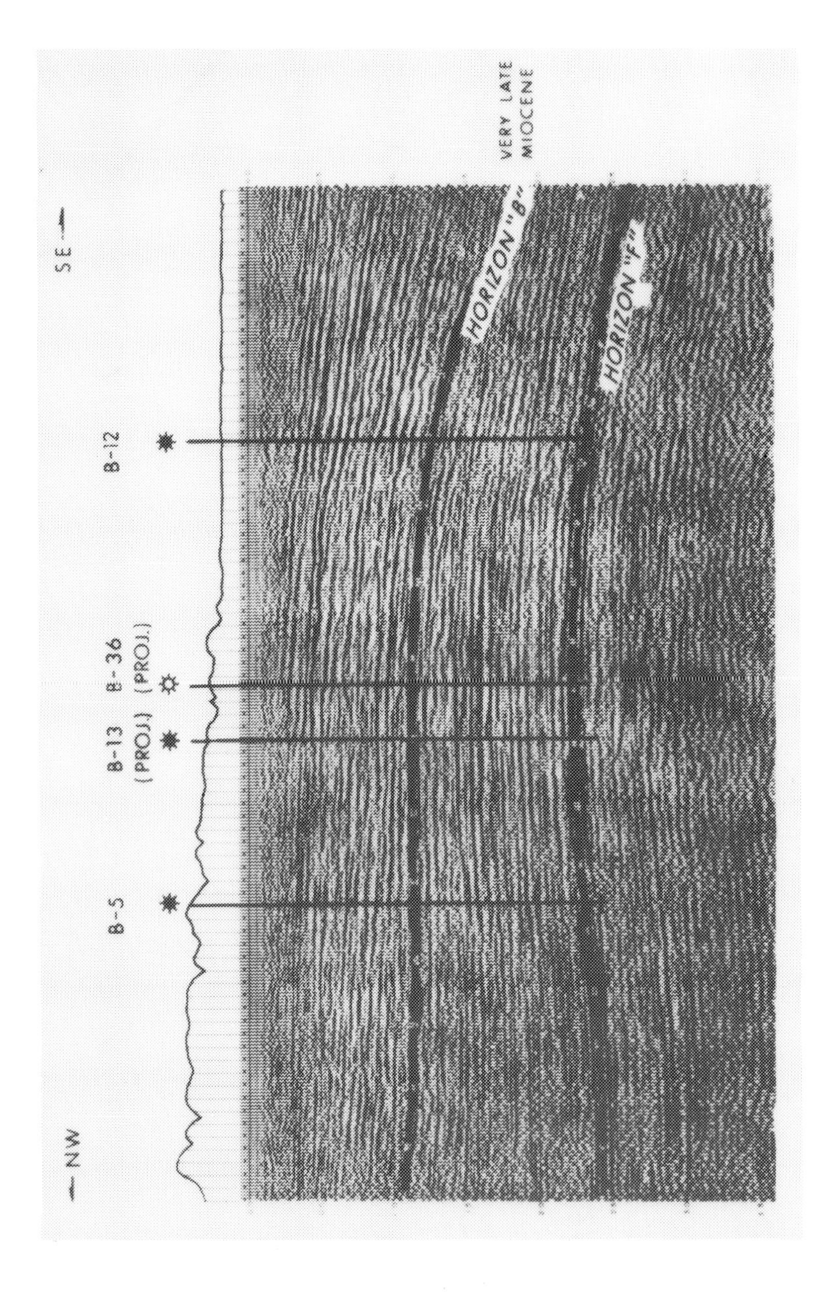

FIG. 19—Seismic line K-701.

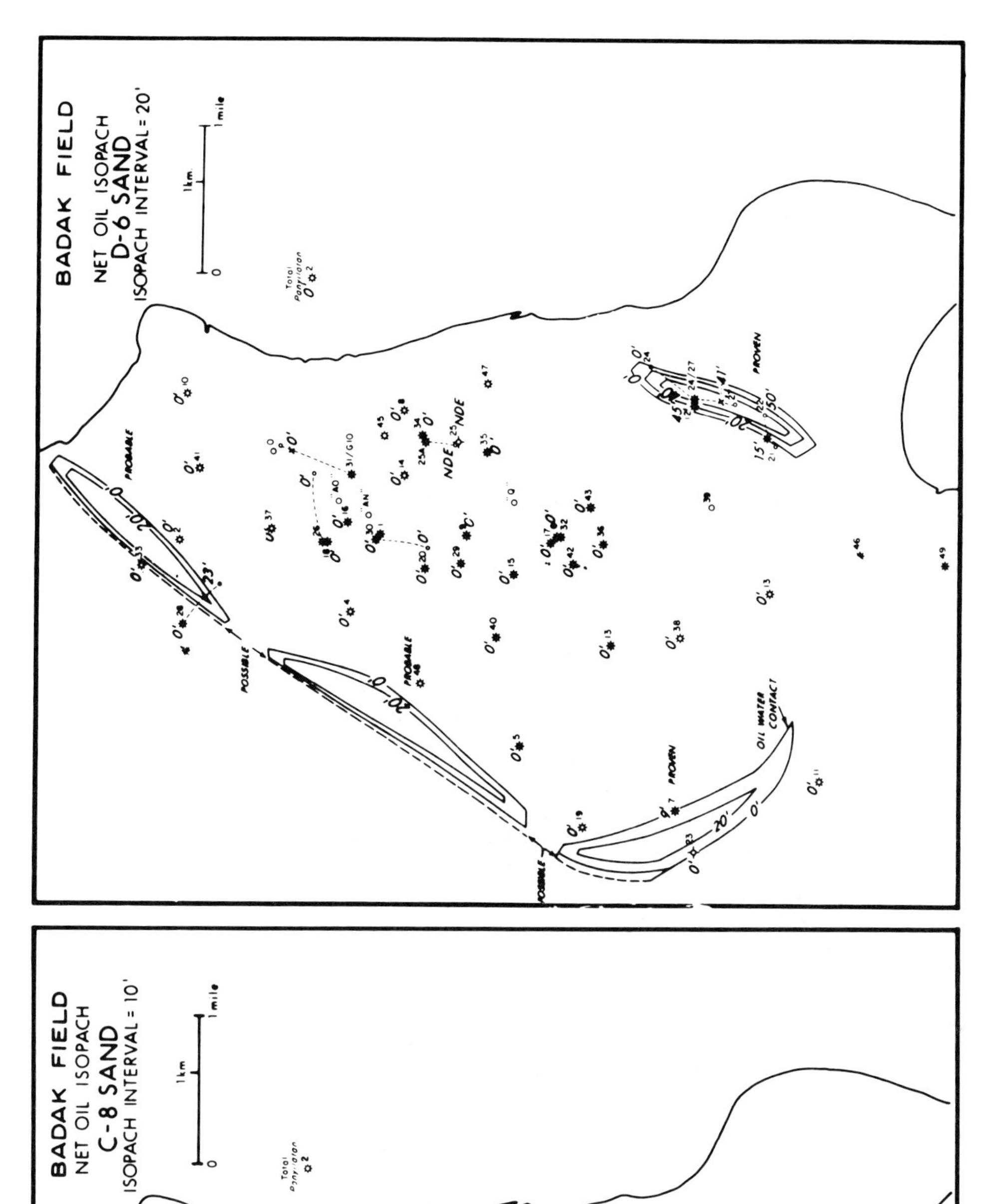

FIG. 20—"C-8" net oil sand.

FIG. 21—"D-6" net oil sand.

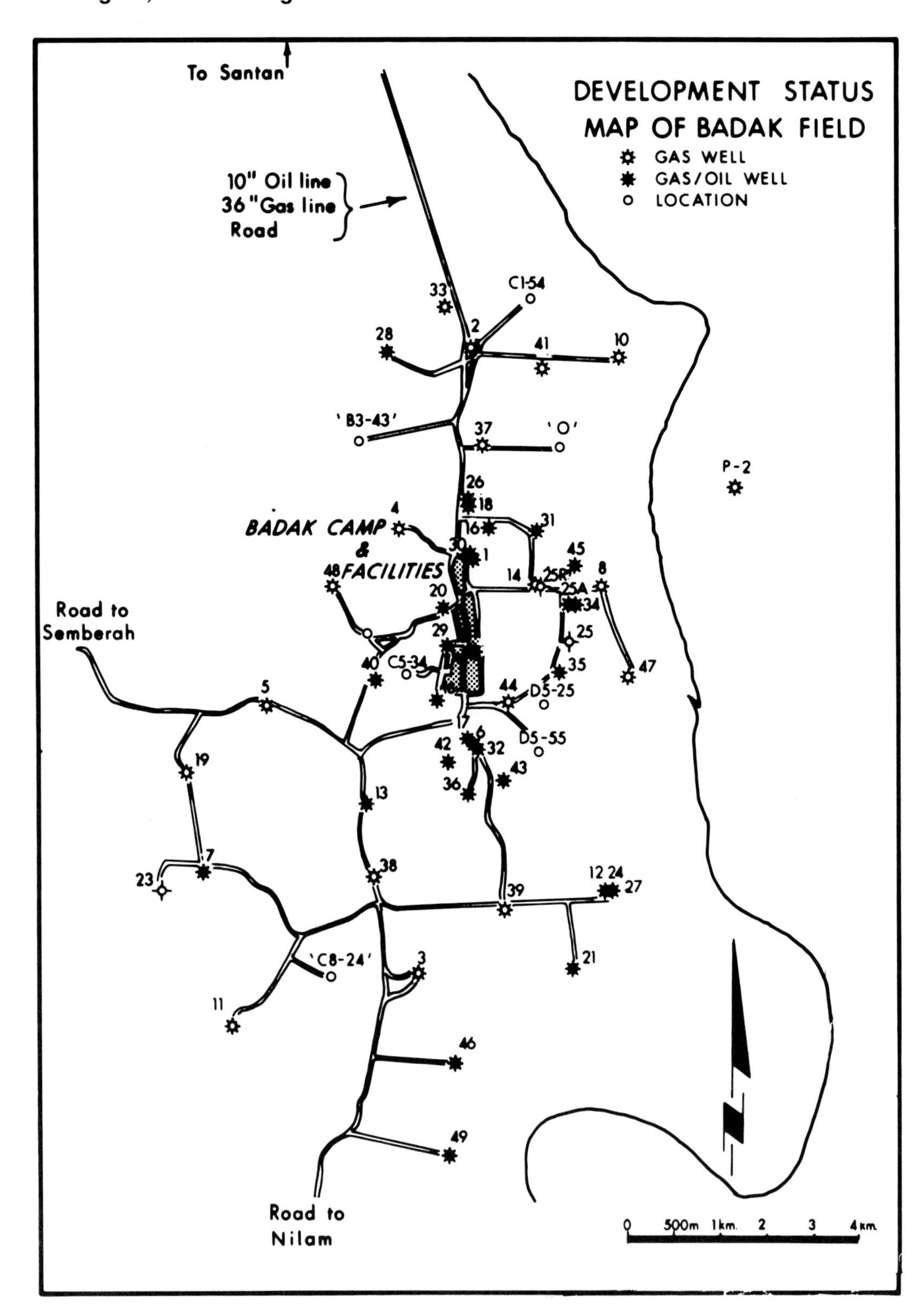

FIG. 22—Development map of Badak Field.

facies changes are common and abrupt, making stratigraphic correlation extremely difficult in this zone. As Zone "A" contains only minor amounts of hydrocarbons, only development in Badak 1 is shown (Fig. 5).

Zone "B"

The stratigraphy of this zone is similar to that of Zone "A," except that its slightly lower average sand/shale ratio and lignite content seems to indicate a more frontal deltaic position of the Badak area during deposition of this unit.

The base of a major freshwater sandstone in Badak 1 was chosen as the boundary between Zone "A" and Zone "B." Below this horizon, Zone "B" formation waters, in varying degrees, are predominantly saline. However, the top of Zone

"B" is usually difficult to pick on logs, as the previously mentioned freshwater sandstone is developed differently in different wells (Fig. 6).

Zone "C"

The top of Zone "C" in Badak 1 is defined at the base of a rare 350-ft (101 m) shaly interval and at the top of a thick sandstone bed, designated the "C-1 Sand." Zone "C" is very sandy in its upper 300 ft (91 m) and gradually becomes more shaly below. Its average sand/shale ratio is about 30%, again suggesting a more frontal deltaic position when compared to overlying Zone "B" (Fig. 7).

Zone "D"

Zone "D" is similar to Zone "C," and the repetition of the sedimentary pattern gives the impression of a certain cyclicity. However, the decline with age in the average sand/shale ratio is maintained in this zone, which has an average ratio of about 25% (Fig. 8).

Zone "E"

Zone "E" contains some of the major Badak gas reservoirs, such as the "E-11 Sand" and "E-21 Sand." Nevertheless, its average sand/shale ratio is only about 20%, confirming the tendency of this ratio to decrease with downward movement through the section (Fig. 9).

Zone "F"

The gradual downward trend toward more marine-like conditions reaches a maximum, or "deepest," marine condition at the base of Zone "F" (Fig. 10). At that point there is an increase in limestone content and the first appearance of prodeltaic shales. With the latter, the first indications of abnormal geopressures are encountered in the Badak field.

Examples of Deltaic Sandstone Deposition

Although a thorough discussion of Badak sedimentology and depositional paleoenvironments is beyond the scope of this paper, a brief explanation of the nature of Badak reservoir sandstone is appropriate. Most deltaic sandbody types, as described and classified by many authors, can be recognized in the Badak field. Figures 11 through 15 are self-explanatory illustrations of the nature of individual Badak reservoir sandstones selected from zones "B" through "F."

STRUCTURE

The Badak anticline is a pronounced culmination on a 50-mi (80 km) long, northerly trending line of folding which begins just south of the Handil field and ends just north of Badak (Fig. 16). The Badak structure is a simple anticline with gentle dipping flanks and no evidence of faulting (Fig. 17).

The cross section in Figure 18 and the seismic section in Figure 19 show that during deposition of the "F" to "B" zones, from the middle Miocene to Pliocene, sedimentary beds thinned eastward in the Badak area. Subsequently, during late Pliocene to Pleistocene time the direction of thinning was reversed, and sedimentary beds thickened in an eastward direction. These phenomena indicate that the Badak structure formed due to a sequence of events which began with relative subsidence on the west during the early stages, and was followed by subsidence on the east during a more recent period. The result is the structural reversal of the Badak anticline.

The absence of west dip in the more shallow beds of the Badak field is reflected in the virtual absence of hydrocarbons above a depth of approximately 4,000 ft (1,219 m), except for some minor stratigraphically trapped accumulations of gas.

HYDROCARBON CONTENT

All oil and gas found in the Badak field has accumulated in deltaic sandstone reservoirs. Net pay thicknesses in Badak 1 and average reservoir porosities are shown in Table 1.

Badak reservoirs are predominantly filled with gas of a condensate content averaging 12 bbl/MMcf of gas. The gas consists of hydrogen sulfide, nil; carbon dioxide, 2.94; nitrogen, 0.06; methane, 87.44; ethane, 4.51; propane, 2.84; butane, 1.29; pentanes, 0.43; hexanes, 0.17; and heptanes and heavier, 0.32. The heating value of the gas is 1,127 BTU/cf.

There are two types of oil occurrence at Badak. One occurs in the more shallow crestal sandstones with small associated gas caps, and is illustrated by the "C-8 Sand" (Fig. 20). The other occurs in some gas sandstones as oil "legs" downdip on the edge of the structure, and is illustrated by the "D-6 Sand" (Fig. 21). Because the fringe areas of the Badak structure have not yet been adequately explored, other types of oil occurances may possibly be found.

Badak oil has an average gravity of 39°API, is relatively rich in middle distillates, and contains less than 0.1% sulfur.

The original proved and probable reserves in-place in the Badak field, determined as of January 1, 1978, are 7 Tcf of gas, 130 million bbl of condensate, and 60 million bbl of oil.

DEVELOPMENT

The first three wells drilled after discovery of the Badak field served delineation purposes and demonstrated that the structure was filled primarily with gas and condensate. In addition, sizeable amounts of oil occurred in several of the Badak reservoirs. An additional ten wells were drilled in the Badak

Table 1. Net pay thicknesses in Badak 1 and average reservoir porosities.

Zone	Oil pay (ft)	Gas pay (ft)	Average porosity (%)
A	—	13	28
B	—	55	29
C	123	417	28
D	—	269	27
E	—	221	21
F	—	49	17

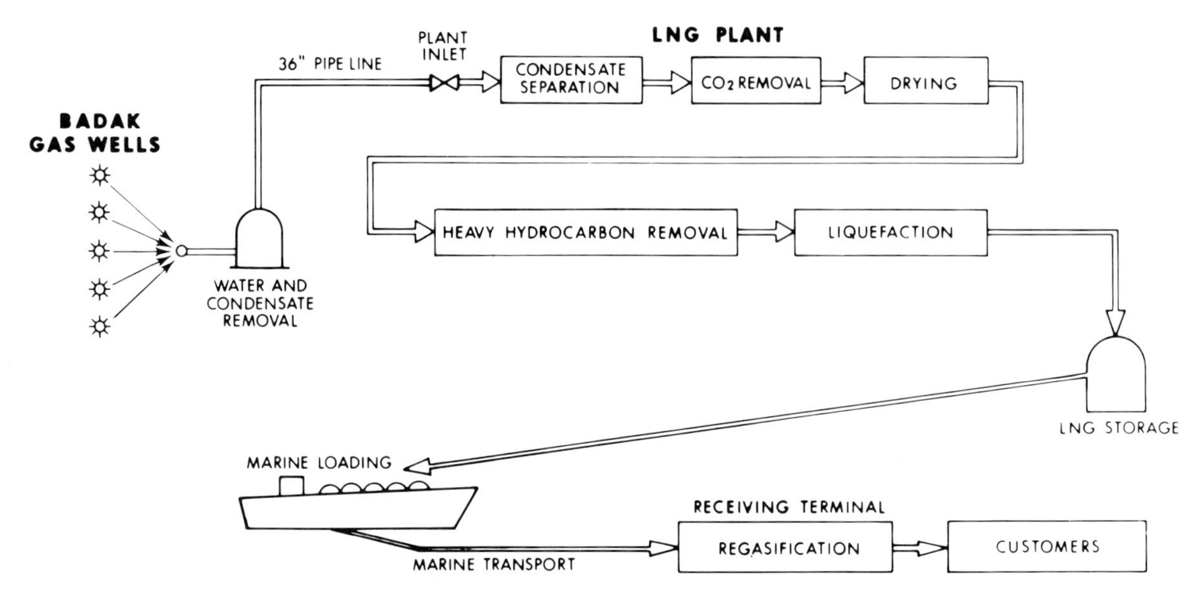

FIG. 23—Schematic diagram of Badak LNG system.

field for the purpose of proving sufficient gas reserves to support an LNG project. Figure 22 shows the present developmental status of the Badak field.

More than 100 reservoir sandstones are known to exist in the Badak field, all of which are more or less limited in area extent and may occur only in a few wells. Nevertheless, a typical Badak well may offer a choice of at least two possible oil completions and ten or more producible gas sandstones. Careful planning was clearly a necessity for the proper development and exploitation of the Badak field. A more detailed account of the development planning for required gas deliverability for the Badak LNG plan is given by Maggert and Anwar (1977).

Oil

The developmental plan for the Badak field provided for an initial oil development phase followed by a gas development phase. The oil development began in 1973 and has not concluded, as several potential oil locations remain to be drilled. The selection of Badak well locations required careful planning to optimize each location for maximum long-term deliverability of oil or gas. Total depths were selected so that Badak oil wells could be converted for use as gas wells after depletion of oil reservoirs. After completion of production facilities and construction of a 10-in (25.6 cm) pipeline to the Santan shipping terminal, 17 mi (27 km) north, oil production from Badak came on stream in October 1974. At January 1979, field production totalled 15.8 million bbl of oil. Production peaked at 15,000 b/d and by the end of 1978, daily production was approximately 10,000 b/d from 19 completions in 15 wells.

Gas

Badak gas production began in April 1977, coinciding with start-up operations of the LNG plant. In January 1979, the Badak field produced approximately 500 MMcf of gas per day from completions in 20 wells. The actual producing capacity of the field, however, was in excess of 700 MMcf of gas per day. In addition, some 15,000 b/d of condensate were being produced. The condensate has been commingled with and sold as crude oil through the Santan terminal.

ECONOMIC SIGNIFICANCE OF THE BADAK FIELD

At the time the Badak field was discovered, gas use in Indonesia was limited to relatively minor projects such as production of carbon black and fertilizers. The discovery of Badak spurred construction of a $700 million LGN plant at Bontang (Fig. 23). Together with exploration and production costs and expense of LNG tankers and receiving terminals in Japan, approximately $2.5 billion have been spent on the overall project.

REFERENCES CITED

Allen, J. P., D. Laurier, and J. M. Thouvenin, 1977, Sediment distribution patterns in the modern Mahakam delta: 5th Ann. Mtg., Indonesian Petroleum Association Proc., v. I, p. 159-178.

Gerard, J. and H. G. Oesterle, 1973, Facies study of the offshore Mahakam area: 3rd Ann. Mtg., Indonesian Petroleum Association Proc., p. 187-194.

Maggert, K. W., and A. S. M. Anwar, 1977, Use of a total field numerical simulator to plan Badak field gas development: 5th Ann. Mtg., Indonesian Petroleum Association Proc., v. I, p. 243-259.

Magnier, Ph., and B. S. Samsu, 1976, The Handil oil field in East Kalimantan: 4th Ann. Mtg., Indonesian Petroleum Association Proc., v. 2, p. 41-61.

——— T. K. Oki, and Luki Witoelar K., 1975, The Mahakam Delta, Kalimantan, Indonesia: 9th World Petroleum Congress Proc., v. 2, p. 239-250.

Rose R. and P. Hartono, Geological evolution of Tertiary Kutei-Melawi basin, Kalimantan, Indonesia: 7th Ann. Mtg., Indonesian Petroleum Association.

Geology of the Bekapai Field[1]

By M. DeMatharel, P. Lehmann, and T. Oki[2]

Abstract The Bekapai field was discovered in April, 1972, and is located approximately 15 km off the Mahakam delta, East Kalimantan in the central part of the Mahakam Contract Area. The exploration concession is held by Total Indonesie and Inpex Indonesia Ltd. under a production sharing contract with Pertamina.

The field structure is a large faulted anticline with an area of 210 sq km and a vertical closure of 400 m. The reservoir zones are at depths ranging from 1,300 to 1,600 m. The reservoirs are usually distributary front, channel, and bar deposits of a large eastward-prograding deltaic system.

Production started in July, 1974, from a single well as a pilot project. Production (slowly declining now) is at a rate of about 46,000 b/d from three platforms.

INTRODUCTION

This paper is a revised print of the paper published in the Proceedings of the Indonesian Petroleum Association Fifth Annual Convention (June 1976).

The Bekapai field is located 15 km offshore the Mahakam River delta, on the edge of the Kutai basin, along the east coast of Kalimantan (Fig. 1). Exploration activity began in the late nineteenth century and recently the most important discoveries have been Attaka (1970), Bekapai (1972), Badak (1972), and Handil (1974).

Bekapai field was discovered in March, 1972, when the Total-Japex Bekapai 1 (B-1) was drilled in a water depth of 35 m. The well discovered oil at a depth of 1,400 m and tested 1,930 b/d. This discovery was followed by several delineation wells that confirmed the presence of a commercial accumulation on the western part of the seismic anomaly.

Production began from a single well (B-2A) as a pilot project in July, 1974. The production from two sandstone zones at that time was 5,000 b/d. One year later the first multiwell production platform (BA platform) was completed and production from nine wells increased to 45,000 b/d. Oil was stored in a converted tanker, the *Wapiti* (50,000 ton capacity) which was linked to the production platform by a single buoy mooring system (Fig. 2).

Two more multi-well platforms (BB and BC) were completed to allow a higher recovery. Today, 26 wells produce 46,000 b/d which is shipped via a 40 km long 12 in. submarine pipeline to Senipah Terminal (Fig. 4).

GEOLOGIC SETTING

The Mahakam Contract Area is in the center of the Tertiary Kutai basin bounded on the north by the Mangkalihat arch and on the south by the Paternoster platform (Fig. 1). Since the Miocene, the Mahakam River has built a major deltaic complex into the basin.

From the wells drilled in the Bekapai area, two main pa-

[1]Manuscript received, September 7, 1979; accepted for publication, December 20, 1979.

[2]TOTAL Indonesie, Jakarta, Indonesia.

The contents of this paper result from the work of a team, to whom the writers express their sincere thanks. They are also grateful to Pertamina, Inpex, TOTAL Indonesie, and to the Indonesian Petroleum Association for permission to publish.

Article Identification Number:
0065-731X/80/MO30-0022/$03.00/0.

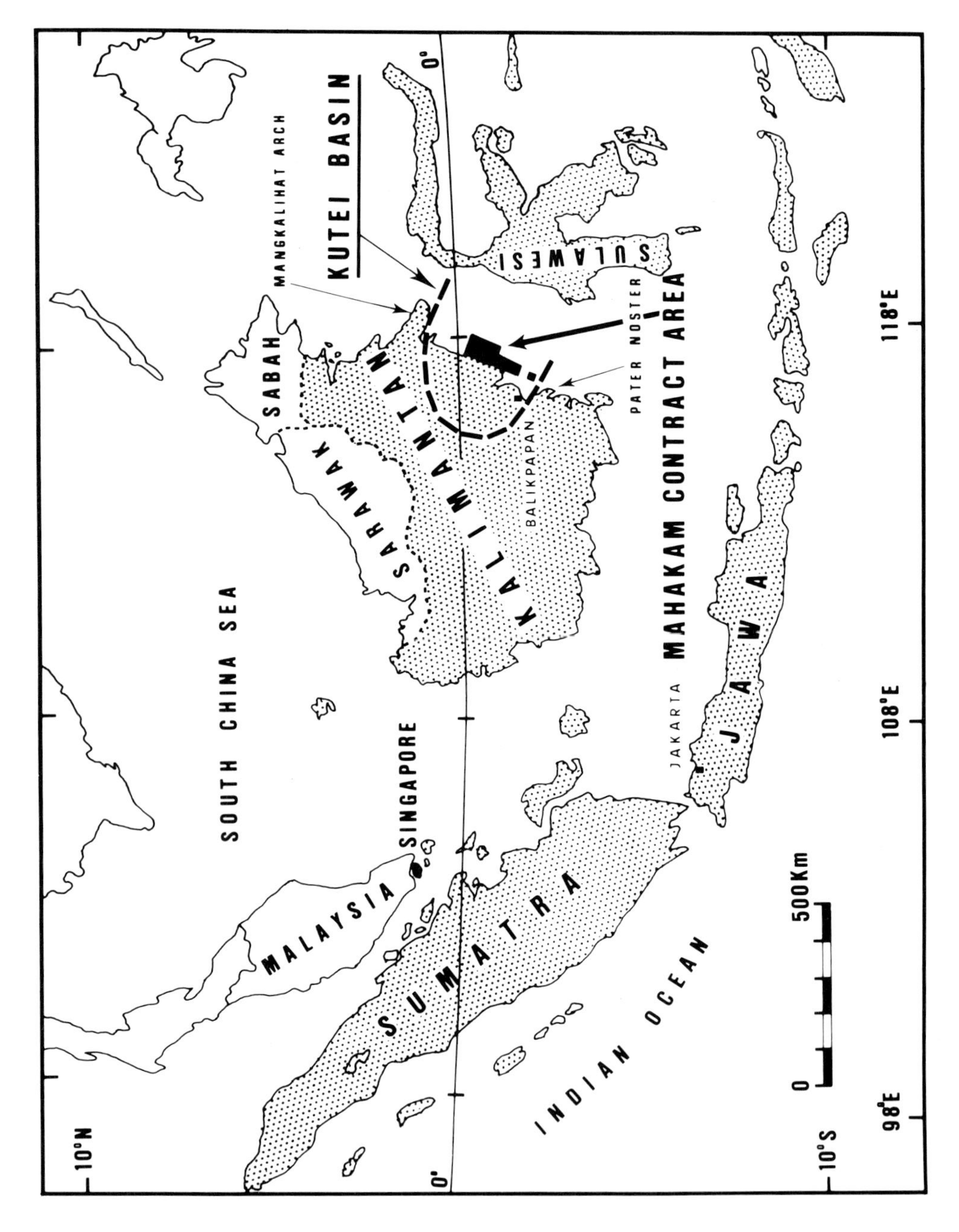

FIG. 1—Location of Bekapai field, offshore the Mahakam River delta, Kalimantan, on the edge of the Kutei basin.

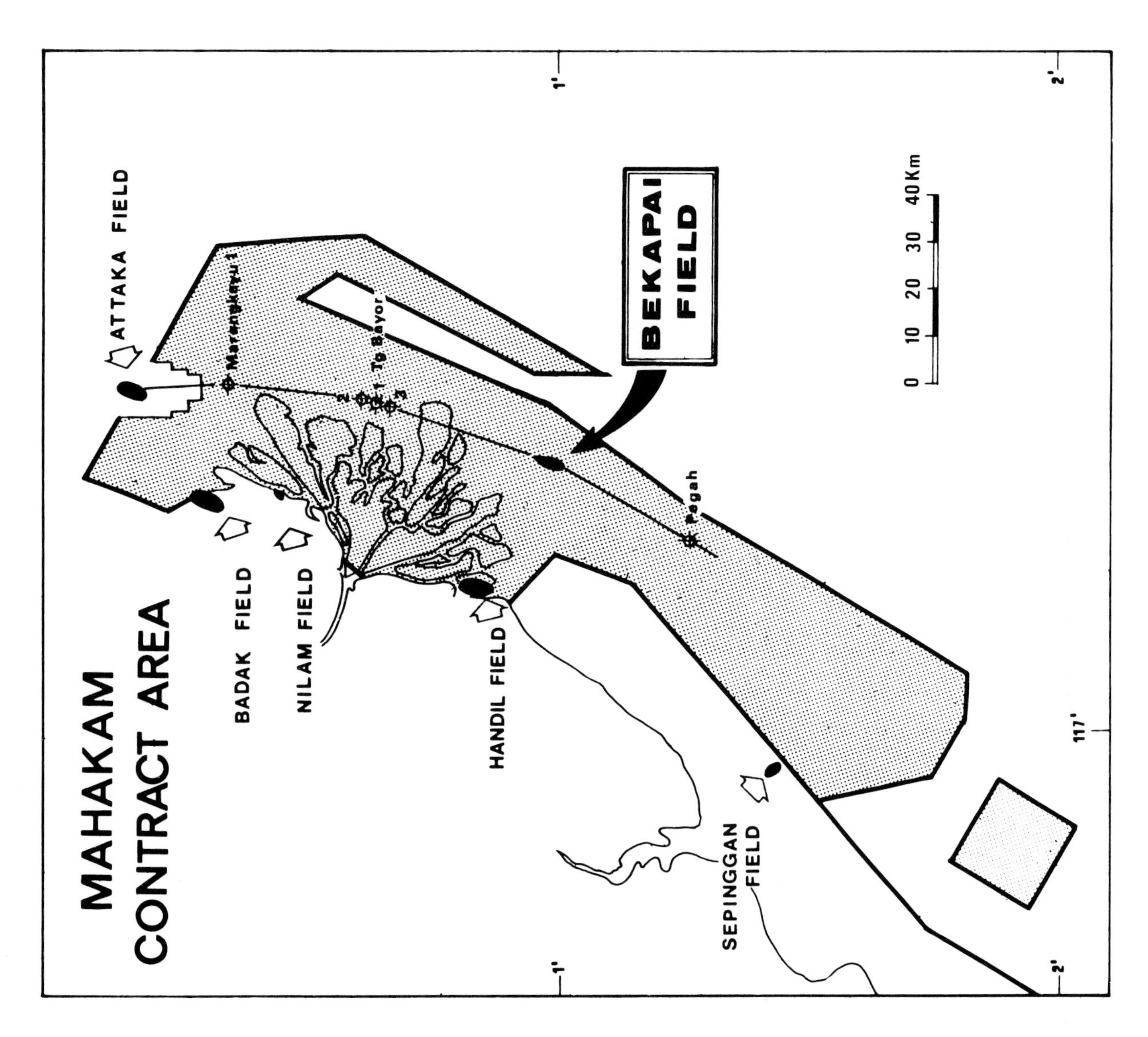

FIG. 2—Situation map of the Mahakam Contract Area, Bekapai field, Indonesia. For location, see Figure 1.

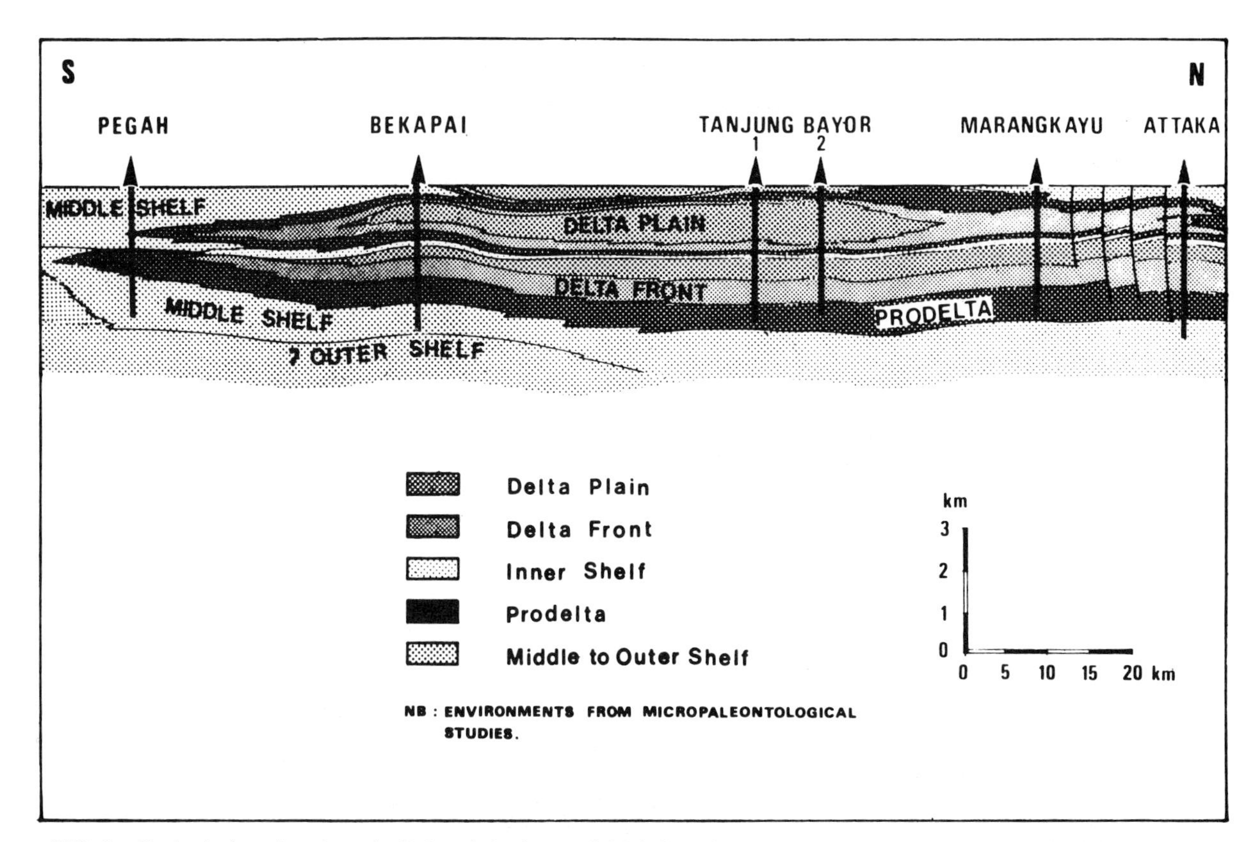

FIG. 3—Geological section along the Bekapai-Attaka trend, Mahakam Contract Area. Note the two main paleodeltas separated by a marine transgression.

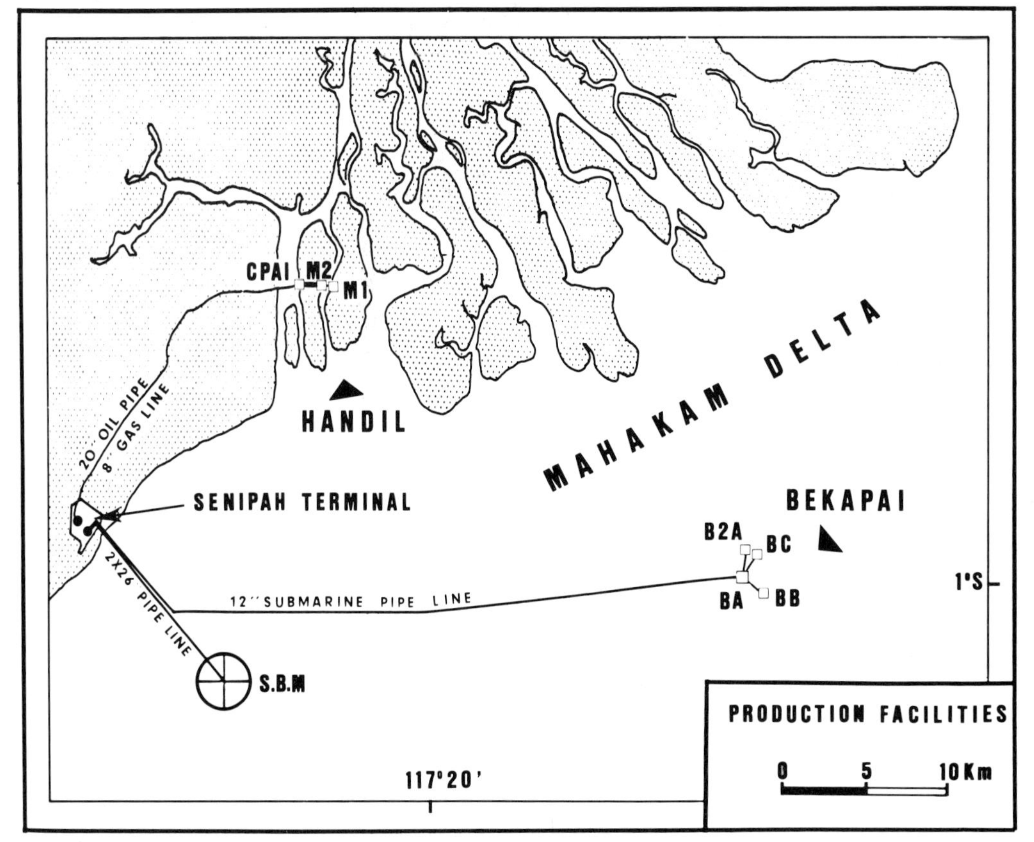

FIG. 4—Locations of production platforms in the Mahakam Delta, Bekapai field.

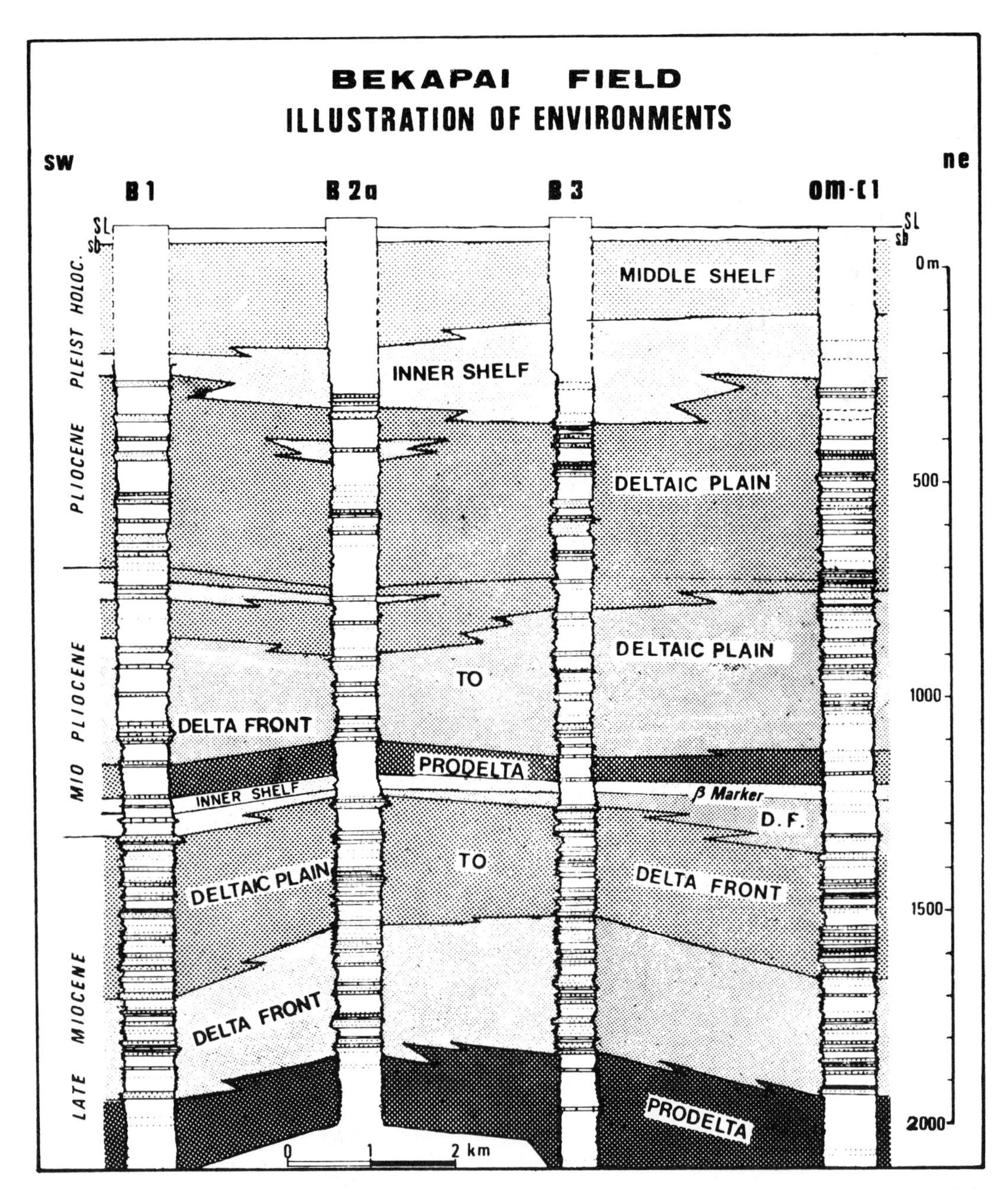

FIG. 5—Column illustration of depositional environments, Bekapai field, Mahakam delta. Ages on the left edge of illustration.

leodeltas of Miocene and Pliocene age can be recognized. They were separated in time by a marine transgression during which the delta retreated (Fig. 3).

A depositional environment study in the Bekapia field, based essentially on benthonic Foraminifera, shows the two deltaic sequences are separated by marine deposits. The lower sequence is subdivided (from bottom to top) into inner shelf, prodelta, delta front, and delta plain deposits (Fig. 5). On the top of this sequence a disconformity occurs at the base of the regional marine transgression. This major break is a consistent marker ("Beta Marker") and is detectable on electric logs as a sharp resistivity shift (Fig. 6). Above the Beta Marker the section can be divided into inner shelf, delta front, deltaic plain, delta front, inner shelf, and middle shelf environments. This complete geologic cycle preceded a regression sequence illustrated by the modern Mahakam River delta.

Hydrocarbons occur mainly in the delta-front sandstones of the lower delta. The same stratigraphic zones that produce at the Bekapai field also produce at Attaka field 75 km north. These sandstones are called "fresh water sands," as the water salinity is approximately 20,000 ppm which is low in comparison to sandstones below and above where the salinity ranges from 30,000 to 40,000 ppm.

STRUCTURAL GEOLOGY

Regionally, the Bekapai structure is part of an elongated north to south structural trend on which there are two major

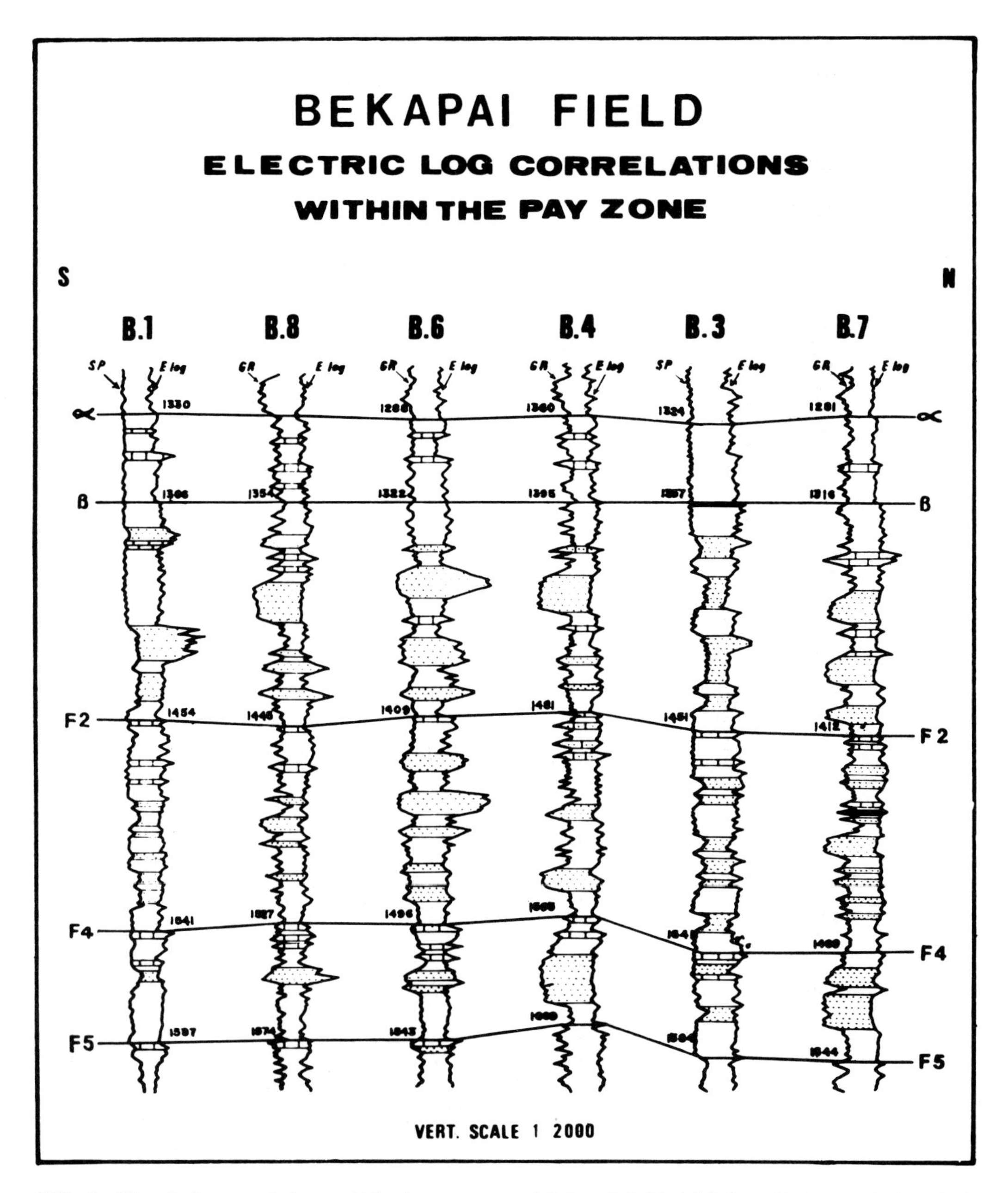

FIG. 6—Electric log correlations within the pay zone of Bekapai field, Mahakam Delta. Vertical scale is 1:2,000.

culminations: Attaka on the north and Bekapai on the south (Fig. 7). The structure is a large faulted anticline. Near the "Beta Marker" at the top of the pay zone the areal extent and vertical closures of the anticline are 210 sq km and 400 m, respectively. A rather complex north to south fault pattern transects the structure. The main period of movement of these faults appears to have been rather recent and resulted in the collapse of the central part of the anticline.

Well data (markers and reservoir correlations, dips, and fluid distributions) confirm two main faults (a western and eastern fault) separating a central collapsed part of the anticline from a western and eastern structurally higher segment of the anticline. The throw of these two main faults ranges from 100 to 150 m (Fig. 8). Only the western part of the structure, where oil is trapped against the western fault, has been developed.

FRESHWATER SANDSTONE

The lower part of the freshwater sandstone member is characterized by a predominance of shale interbedded with silts and few very fine to fine grained, argillaceous sandstones. Presence of bioturbated sediments and marine fossil debris support the interpretation for a delta-front depositional environment.

In the upper part of the freshwater sandstone member, the sand/shale ratio increases markedly. Whereas sands are usually

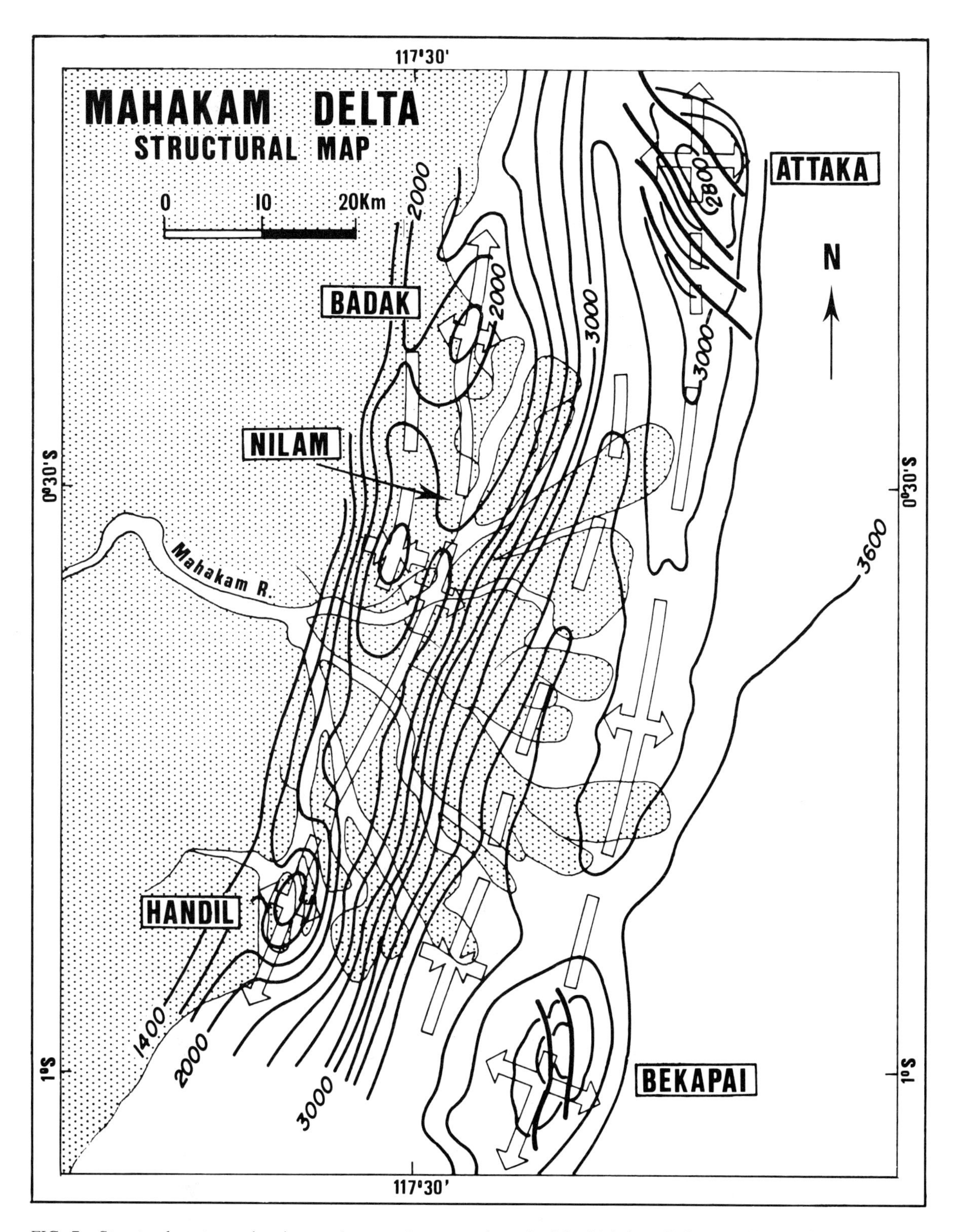

FIG. 7—Structural contours showing north-to-south structural trend of the Mahakam Delta area. Note Bakapai field at the southern end of the trend and Attaka field at the northern end.

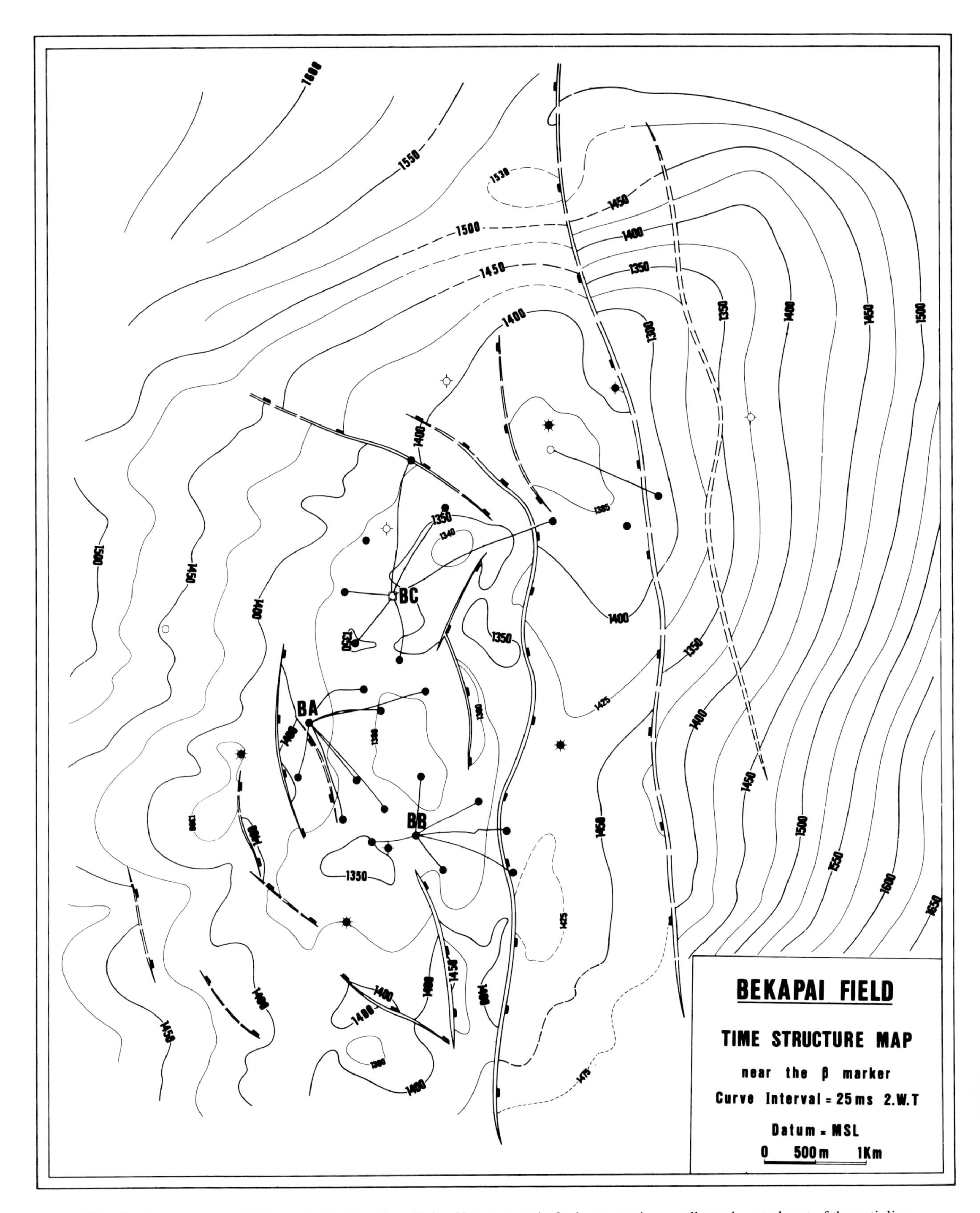

FIG. 8—Structure map of Bekapai field, Mahakam Delta. Note two main faults separating a collapsed central part of the anticline.

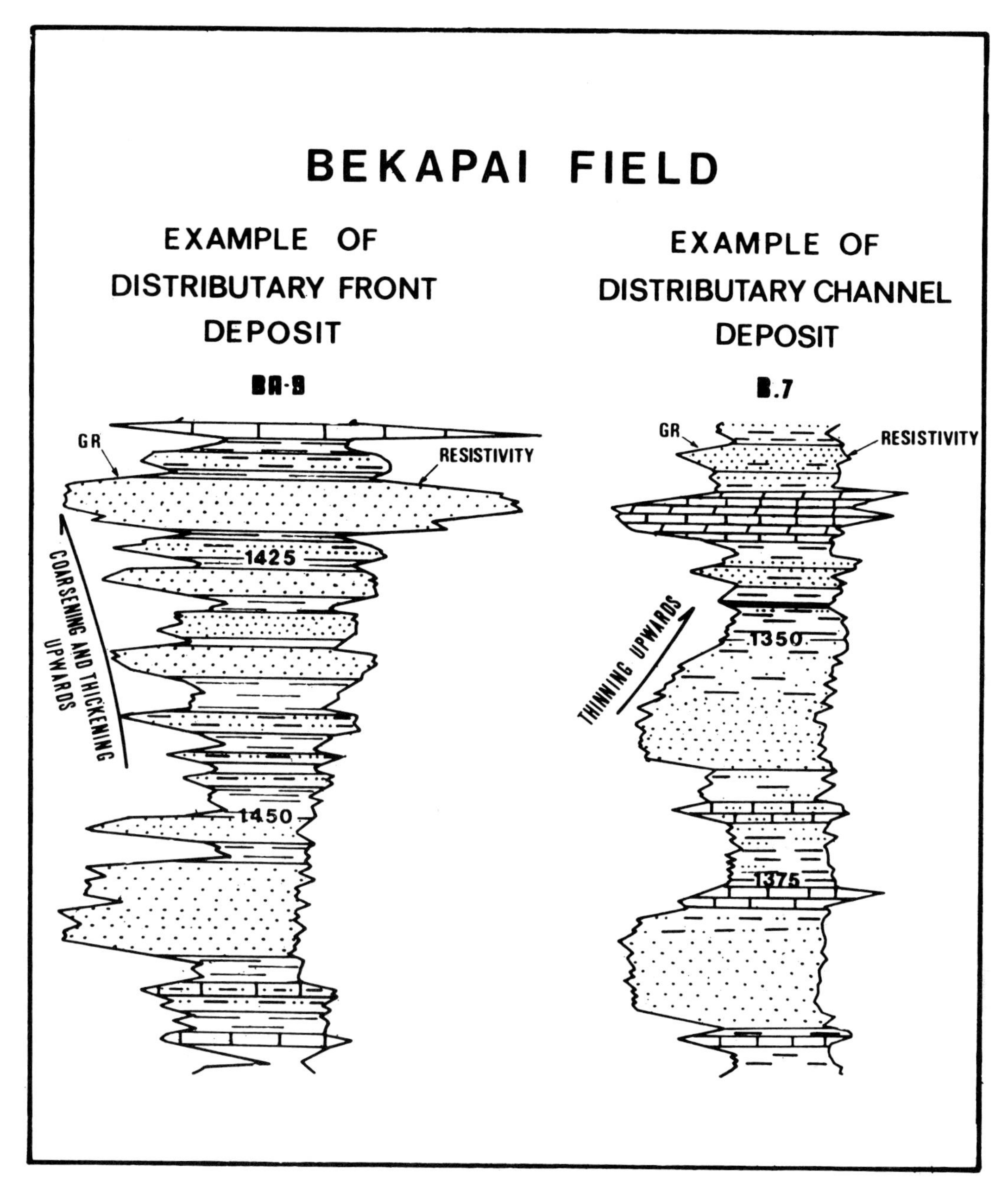

FIG. 9—Diagram showing examples of distributary front and distributary channel deposits, Bekapai field, Mahakam Delta.

fine grained, well sorted, unconsolidated, with chert and feldspars in minor quantities, the uppermost sandstones are generally coarser. The presence of aggradational sequences at several levels suggests a delta-plain rather than a delta-front environment. Within this part of the section, sandstones up to 20 m thick are interbedded with laminated shales and a few dolomitic limestone beds of varying thicknesses. Within the field boundaries these carbonate beds are reliable stratigraphic markers (F2, F4, F5).

At the "Beta Marker" level a few sandstones are characterized by the presence of volcanic rock debris and an abundance of feldspars. Zeolites and montmorillonite are commonly present, making the base of the transgression easy to identify. Within these series three types of sandbodies are present.

1. Distributary Front—The deposits are spread along the entire delta front by wave action (Fig. 9). They are characterized by a progradational pattern with an abrupt upper contact. These sandstones sometimes are capped by a 1- to 2-m thick limestone. According to published data, such sand bodies can be 3 to 4 km by 5 to 10 km in size with a thickness ranging from 5 to 30 m.

2. Distributary Channel Deposits—These deposits are incised along the delta front and oriented perpendicular to the shoreline. They are characterized by an aggradational pattern

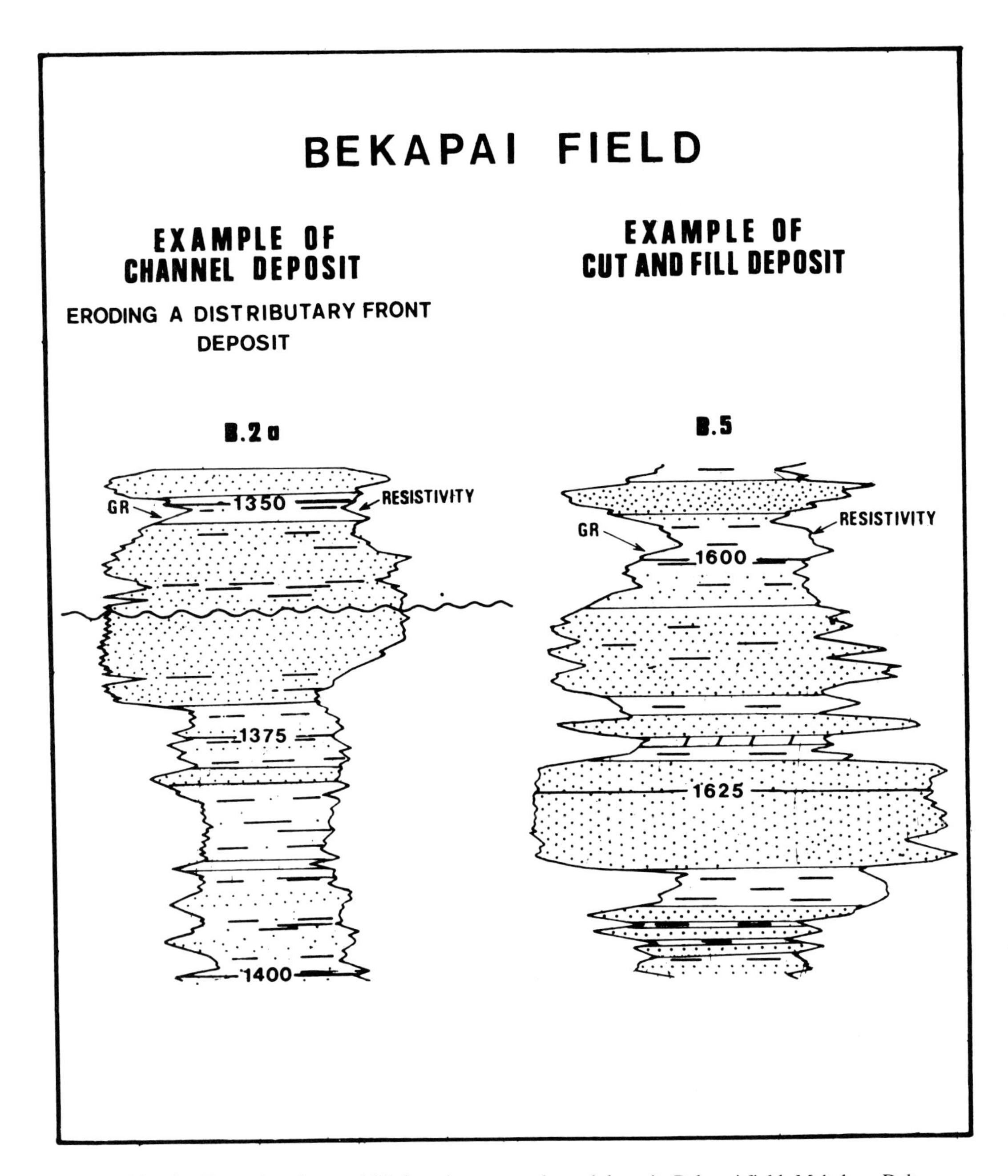

FIG. 10—Examples of cut and fill deposit versus a channel deposit, Bekapai field, Mahakam Delta.

with an abrupt lower contact. They can be 2 to 5 km wide and 5 to 20 m thick.

3. Cut and Fill Deposits—These deposits are generated by longshore current erosion creating submarine topographical irregularities parallel with the shoreline. They are filled by sediments transported in a direction perpendicular to the axis of the cut and fill structures (Fig. 10). Commonly located in the upper part of the freshwater sands, the reservoirs are a combination of two sand bodies in which distributary-front sands are overlain by channel sands.

RESERVOIR CHARACTERISTICS AND QUALITIES

Characteristics

Reservoirs in the Bekapai field are rather discontinuous, vary in thickness (0 to 25 m), and have a limited areal extent (Fig. 11). Most of the hydrocarbon-bearing sandstones are present between 1,300 and 1,600 m depth. The sandstones are unconsolidated and have excellent reservoir characteristics, porosity ranging from 25 to 35%, and permeability commonly more than one darcy. Water saturations calculated from the logs

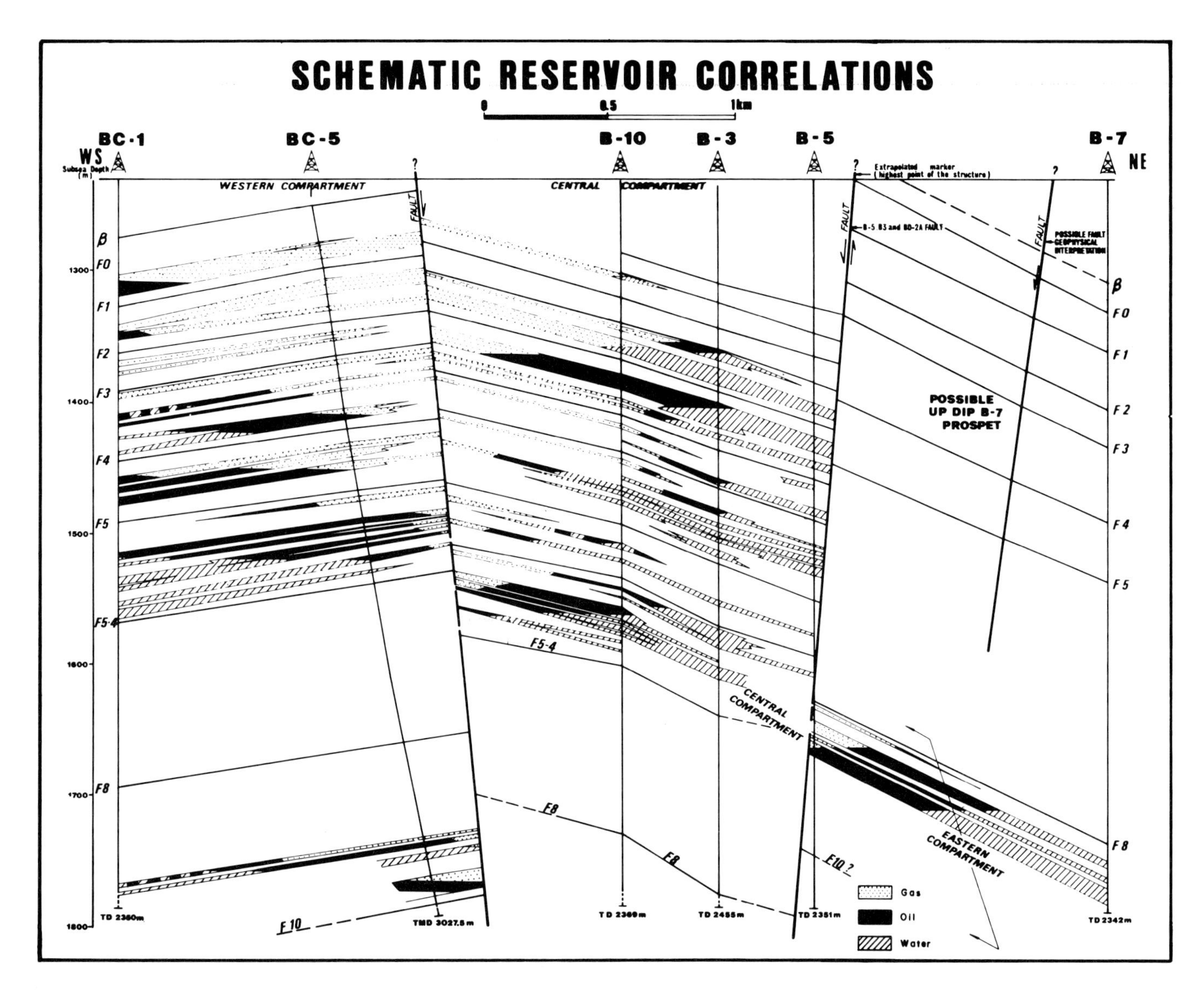

FIG. 11—Schematic reservoir correlations, Bekapai field, Mahakam Delta. Note that they are discontinuous, vary in thickness, and have limited areal extent.

are high, however, never less than 20%, and often reaching 50%. Productivity is high and some of the reservoirs are capable of producing 4,000 to 5,000 b/d.

Other reservoirs are located at greater depths (2,000 to 2,500 m) within the prodelta shale section. Sandstones in this stratigraphic section may be more continuous, but all reservoir characteristics are inferior to those of the stratigraphically higher sandstones. Drill-stem tests of these sandstones have yielded 1,000 to 2,000 b/d.

Oil Characteristics

The oil characteristics of the Bekapai field are as follows: Gravity, 41° API at 16°C; viscosity, 0.45 cp (140 kg at 82°C); pour point, −15°C; sulfur content, 0.08% wt; wax content, 1.66% wt; and asphalten content, 0.08%.

Geology Of Gudao Oil Field And Surrounding Areas[1]

By Chen Sizhong and Wang Ping[2]

Abstract The Gudao field is a large oil field formed in a small faulted basin. The source of hydrocarbon accumulation is in Miocene rocks draping over a graben and horst structure. The source of the oil is Oligocene continental deposits. Secondary oil and gas reservoirs in the Neogene section were formed when faulting breached the Paleogene reservoirs.

Several problems exist in the field including viscous oil, sand production, and poor water drive for reservoir recovery. These problems required well-planned water flooding and sand control in the early stages of development. Injection fluids and the injection pattern have been successful in lowering the viscosity of oil.

INTRODUCTION

The Gudao oil field is located near the mouth of the Huanghe River (Yellow River) in eastern China. It lies in the central part of Zhanhua basin (Fig. 1) and produces from Tertiary sandstones. It is one of the larger oil fields discovered in recent years in eastern China, with an area of 70 sq km and more than 100 million tons of oil in place.

Gravity, magnetic, and seismic reconnaissance surveys were completed over the entire basin during 1963-64. This information, combined with data from several preliminary exploratory wells, provided the base for interpretation of the tectonic framework of the basin. Based on the results of these data, the first well on the Gudao structure was spudded in 1968 and ultimately became the Gudao field discovery well. In 1969 ten more wells were drilled to delineate the field. After preparation for development drilling, oil production began in 1971.

ZHANHUA BASIN

The area of the Zhanhua basin is only 2,800 sq km and is one of the smallest faulted basins in the Bohai Bay oil and gas province. It was a part of the Northern China platform during the Paleozoic. Triassic red clastic deposits, Jurassic coal deposits, and intermediate to acidic volcanic rocks of Cretaceous were deposited in the basin. The major block-fault movement (Yanshan) became most active at the end of the Paleogene. As a result, the Zhanhua basin became a complicated geologic basin with faulted blocks formed in Mesozoic-Cenozoic sediments superimposed on the Paleozoic platform. The complexity is reflected in the intensive development of normal faults with predominant northwest and northeast trends. The northwest faults were developed by the tensional block-faulting in Mesozoic. The later development of northeast-trending faults proved the main factor controlling the sedimentation and the distribution of uplifts and depressions in the Paleogene section of the basin.

The two sets of faults resulted in a series of uplifts and depressions (horsts and grabens) that typify the structure of the basin (Fig. 1). Paleogene sediments were not deposited on most area highs except some thin accumulation overlapping the edge of horsts. Very thick continental sandstones and shales of Tertiary age were deposited in the grabens with maximum thickness of 6,000 m. The Kongdian Formation of the Eocene, composed of red sandstone and mudstones intercalculated with gray mudstones, is the oldest Paleogene deposit in the basin.

[1]Manuscript received, April 4, 1979; accepted for publication, July 23, 1979.

[2]China Oil and Gas Exploration and Development Corp., Peking, Peoples Republic of China.

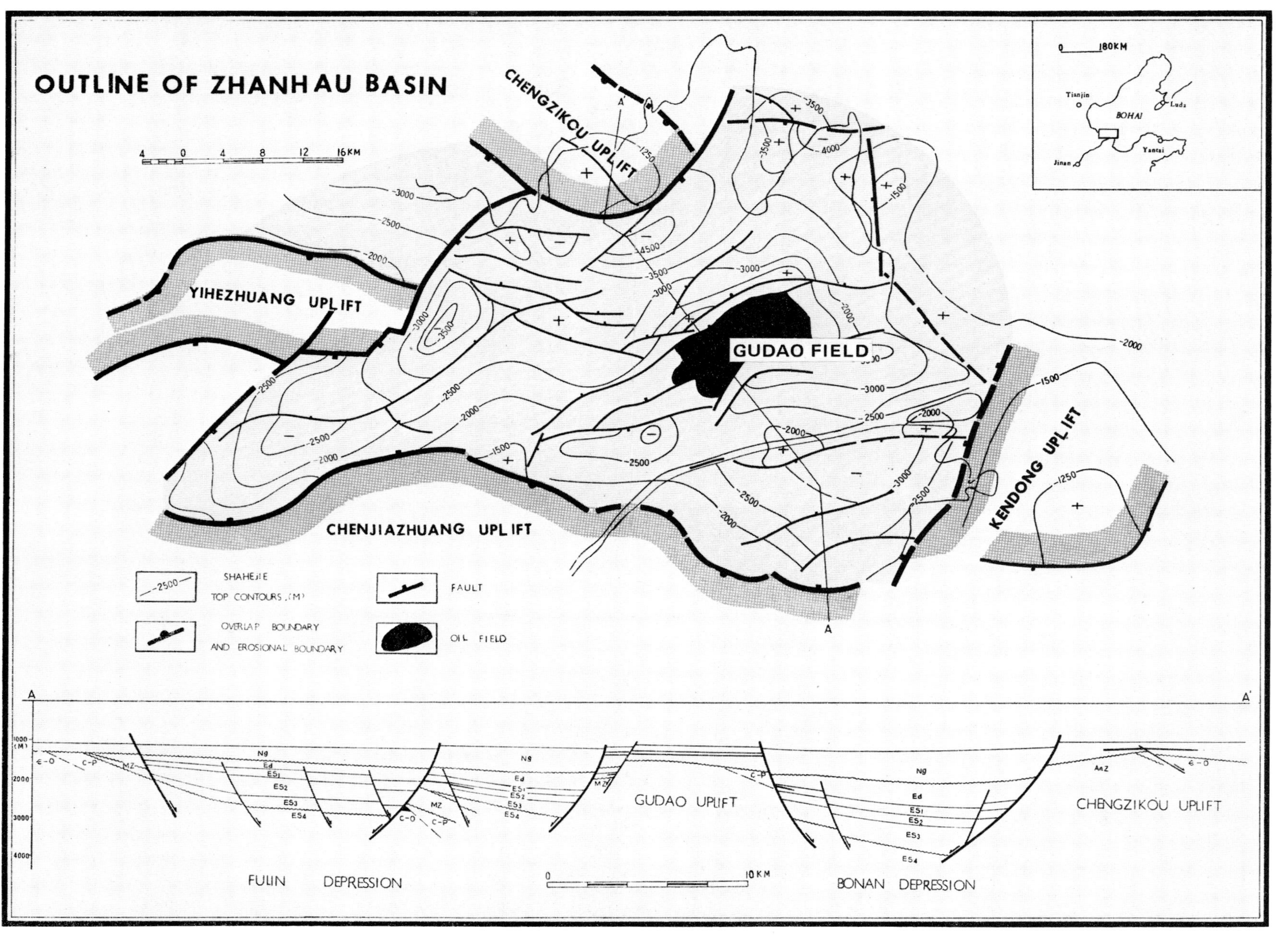

FIG. 1—Location of the Zhanhau basin. Structure contours of the top of the Shahejie Formation.

Table 1. Stratigraphy of the Zhanhau basin.

System	Series	Formation	Symbol	Thickness (meters)	Oil Occurrence	Lithology
Quaternary	—	—	Q	350 m	gas	gray silt, brownish-red clay
Neogene	Pliocene	Minghuazhen	Nm	910 m	gas	upper: gray siltstones dominant lower: brownish-red mudstone with gypsum
	Miocene	Guantao	Ng	730 m	main pay	thick bedded, variagated shales and gray siltstone
Paleogene	Oligocene	Dongying	Ed	570 m	pay	dark gray shale, gray siltstone
		Shahejie	Es_1	260 m	pay	dark gray shales, thin-bedded dolomites and biolimestones
			Es_2	70 m	pay	thin-bedded gray mudstone
			Es_3	560 m	pay	dark gray shale, oil shale
			Es_4	200 m	pay	thin-bedded limestone, dolomites, anhydrite, shale
	Eocene	Kongdian	Ek	400 m	—	brownish-red mudstone, conglomerate
Cretaceous	—	—	K	600 m	shows	andesite, tuffaceous sandstone, basalt
Jurassic	—	—	J	710 m	—	feldspathic graywacke, mudstone, and coral
Permian	—	—	P	220 m	shows	quartz arenite, gray shales intercalated with coal
Carboniferous	—	—	C	200 m	—	gray shale, sandstone, coal, with limestone and alumina in basal part
Ordovician	—	—	O	730 m	pay	thick-bedded limestone, dolomite
Cambrian	—	—	ϵ	650 m	shows	banded limestone, dolomite, red and green shale
Pre-Sinian	—	—	Anz	—	—	granite-gneiss

Table 2. Organic Matter compared to Transformation Conditions in the Gudao oil field area.

Horizon	Abundance of organic matter		Transformation Index		
	organic carbon (per cent)	amino acid (ppm)	chloroform asphalt "A" (per cent)	total hydrocarbon in the rock (ppm)	CPI
Es 1	2.20	700	0.24	530	1.19 1.33
Es 2	2.17	900	0.34	815	1.31

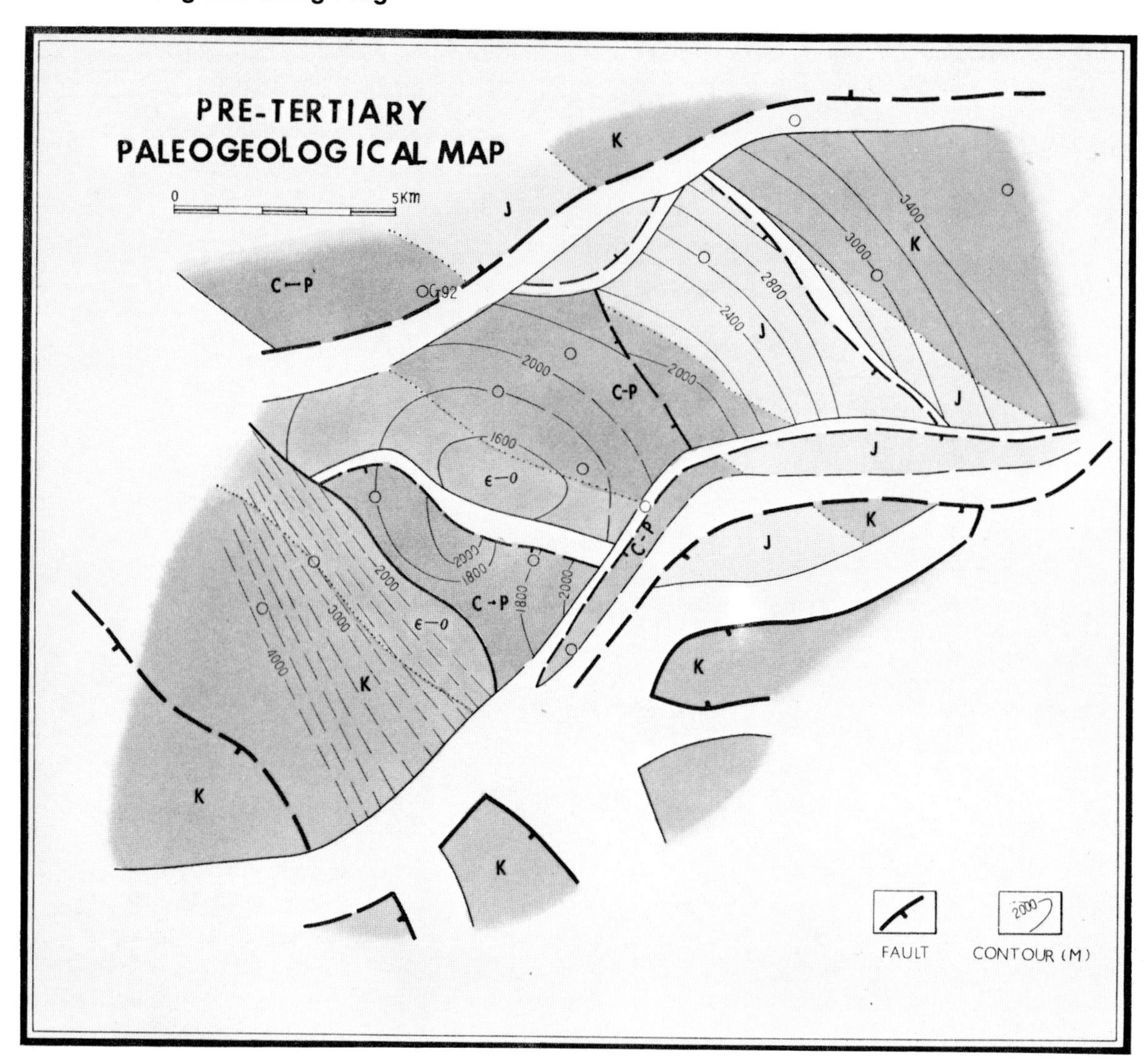

FIG. 2—Pre-Tertiary paleographic map. Contours of the top of the Ordovician. C-O, Cambrian-Ordovician; C-P, Carboniferous-Permian; J, Jurassic; K, Cretaceous.

The Shahejie and Dongying Formations of the Oligocene, consisting of two sedimentary cycles which began with the deposits of dark gray mudstones and oil shales of deep and semi-deep lacustrine origin, and ended with shallow lacustrine and lakeshore sandstones and mudstones, form a favorable source and reservoir combination. The latter two units constituted the main oil-bearing formations in the grabens. Lakeshore and fluviatile sediments of the Miocene Guantao Formation, consisting mainly of sandstones, together with 800 m of brownish-red mudstones of the Pliocene Minghuazhen Formation, form a favorable combination of reservoir and caprock and are the main oil-bearing formations on the basin highs.

Very thick sediments were deposited in the depression, with 4,500 m of Paleogene deposits alone and an average deposition rate of 0.127 mm per year. The differential block-faulted movement and subsidence created favorable conditions for deposition and preservation of organic matters within the depression. The two main source rocks of Oligocene totaling more than 1,000 m in thickness were very rich in fresh and brackish water organisms with an organic carbon content of more than 2%, chloroform asphalt "A" of 0.2 to 0.3%, total hydrocarbon in the rock of 530 to 810 ppm, and a CPI value of 1.1 to 1.3. This suggested not only abundant organic matter but also favorable conditions for hydrocarbon formation (Table 2), and proved the basic factors for the large Gudao oil field.

GUDAO FIELD STRUCTURE

Because the Zhanhua basin is small and has been subjected to intensive block-fault movement, no large folding is present in the basin. The field structure is a sediment drape over a buried horst. This uplifted block consists of Cambrian, Ordovician, Carboniferous, Permian, and Mesozoic rocks which dip to the northeast (Fig. 2).

The northeast faults dissected both the northern and southern flanks of the block during the Paleogene forming a northeast horst which took the shape of a dome. Relief on the dome-shaped uplift is approximately 500 m with a steep westward and gentle eastward dip. The Paleogene sediments

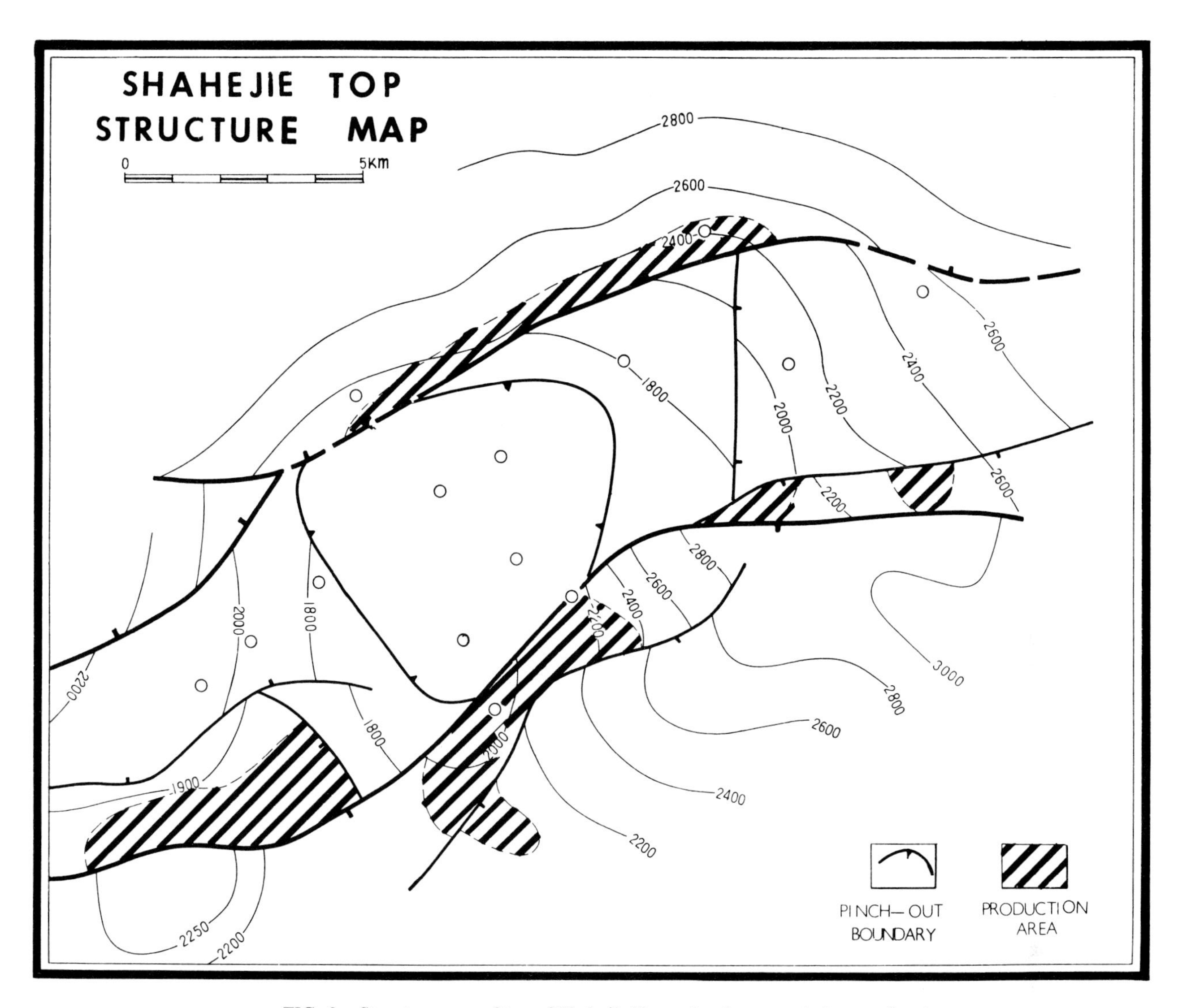

FIG. 3—Structure map of top of Shahejie Formation in meters below sea level.

successively overlap the buried high, being 400 to 500 m thick on the flanks and eroded from the top of the structure. More than 800 m of Paleogene section was deposited on a downthrown block near the horst (Fig. 3). During regional subsidence in the Neogene, the high was covered with about 800 m of sediments. Paleotopography and differential compaction of the Guantao Formation resulted in a drape structure with the formation being thin at the top and thick on the flanks of the high (Fig. 5). This dome-like anticline, 18 km long and 7 km wide, covers an area of more than 100 sq km, striking northeastward. Dip away from the dome is 0.5 to 1.5° and the structural closure is 120 m. All faults developing on the structure were small except at northern and southern boundaries of the anticline, where the faults continued to be active and have a displacement of 60 to 150 m (Figs. 4, 6, 7).

Continuous uplift in the central part of the basin made the Gudao structure favorable for the accumulation of oil and gas. Reservoir quality deposits were laid down in the Paleogene and Neogene, but 95% of oil in place was concentrated in the Miocene deposits. Faulting breached most of the Paleogene reservoirs, but the Miocene Guantao Formation, because of the alternation of sandstone and mudstone covered by over 800 m mudstone, formed an excellent trap. The reservoirs in the Guantao are controlled by structure with the oil-water contact at about 1,300 m and an oil column of about 110 m.

INTENSIVE REMIGRATION AND SECONDARY RESERVOIRS

The crude oils in various formations of Gudao oil field are derived from similar source rocks. All the reservoirs of Guantao and Minghuazhen Formations were deposited in an oxidizing environment, which is unfavorable for oil occurrence. Studies indicate the oil in Guantao Formation comes from Shahejie Formation which has been confirmed as the major source rock in the basin (Table 3). It is clear that some properties of the crude oil in Guantao are similar to those of the source rock and crude oil in Shahejie Es1 and Es3 units. On the other hand, all the algae, spores, and pollens found in the Guantas oil in Gudao field are common in the Paleogene of

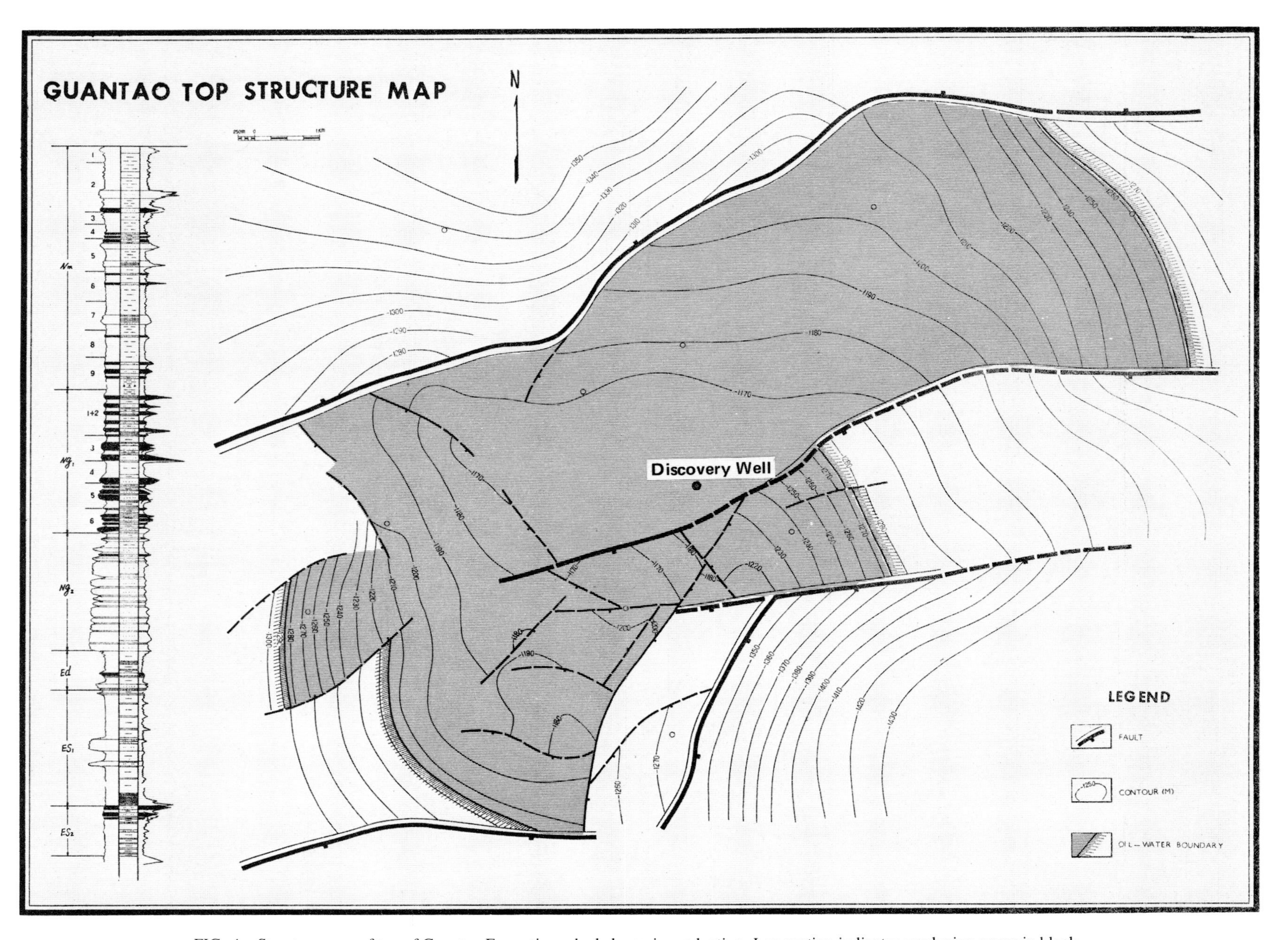

FIG. 4—Structure map of top of Guantao Formation, shaded area is productive. Log section indicates producing zones in black.

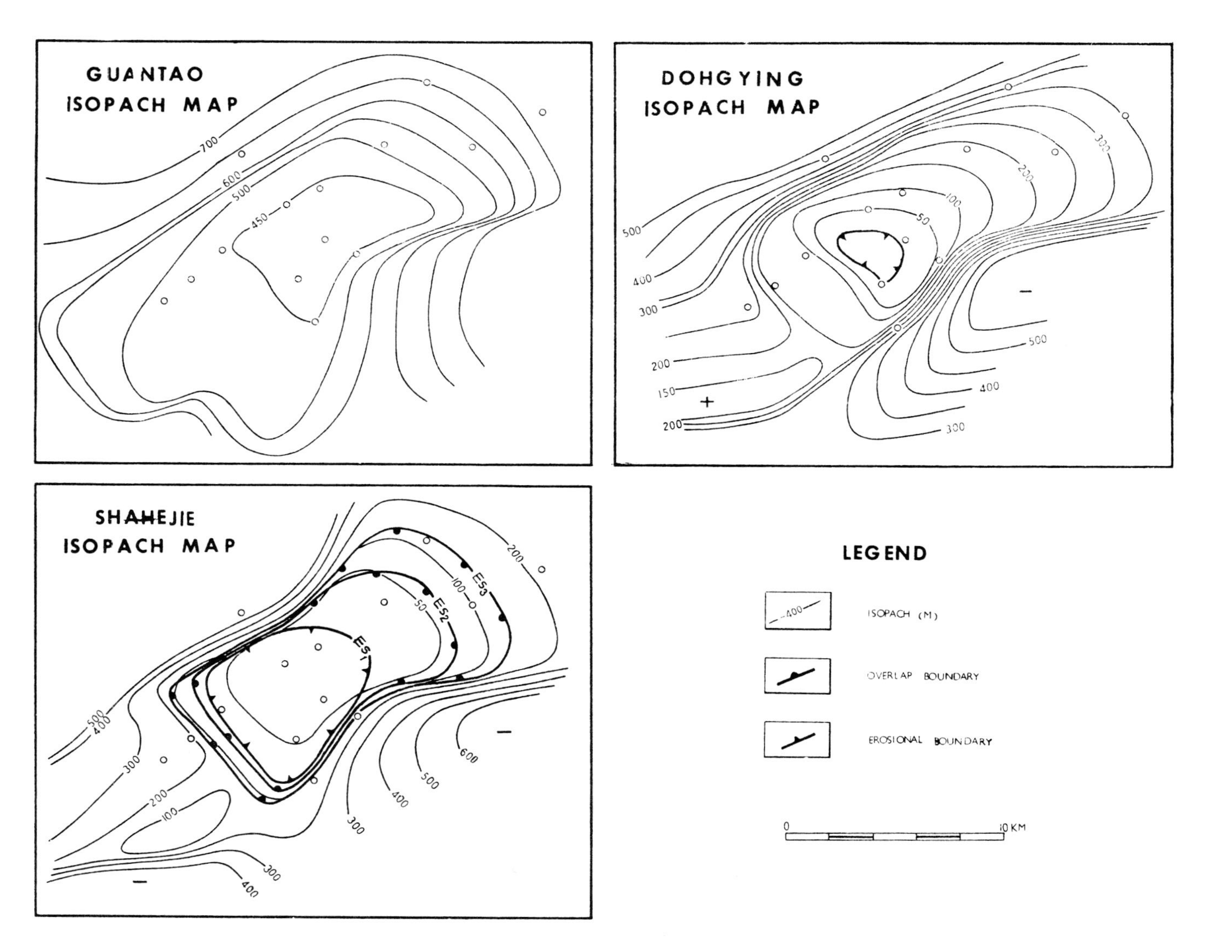

FIG. 5—Isopach maps of Tertiary formations on the Gudao uplift. C.I. = 50 m.

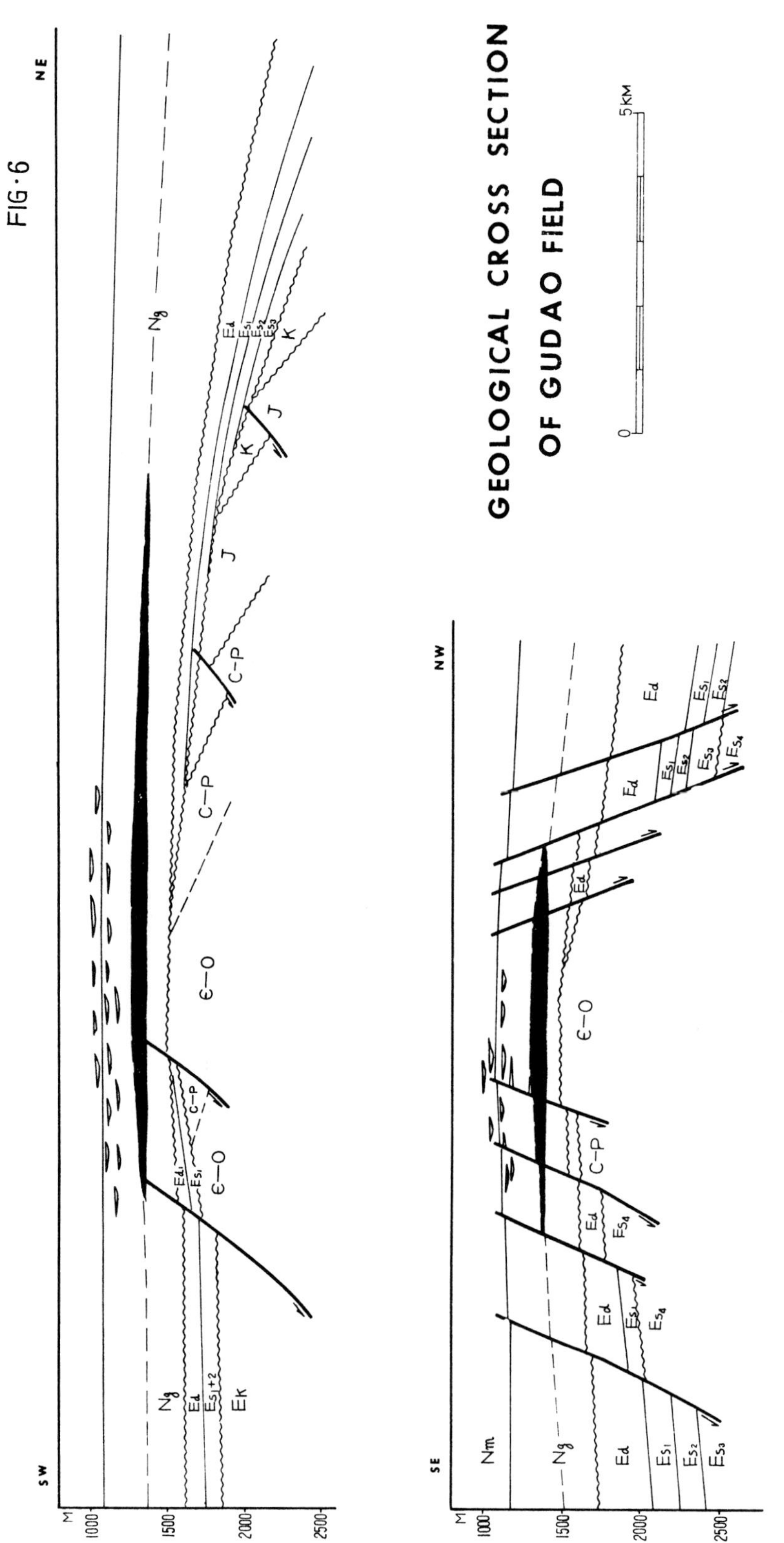

FIG. 6—Cross sections across Gudao field (see Table 1 for symbols).

Table 3. Comparison of properties of crude oil from the different producing zones of the Guantao Member of the Gudao oil field and the Shahejie Member in the depression.

Location	Zone	Sample	Chromatography of N-Paraffin			Pristane/ Phytane	δC13 (%)	Nickel/ Porphyrin
			main peak	CPI	$\frac{\Sigma C23^-}{\Sigma C24^+}$			
Depression	Es 1	source rock	C23	1.19	1.44	0.78	—	—
Depression	Es 3	source rock	C23	1.31	1.38	1.50	—	—
Depression	Es 4	source rock	C20 or C28	0.65 — 0.92	0.60 — 1.24	—	—	—
Depression	Es 1	crude oil	C23	1.14 — 1.29	1.13 — 1.67	0.39 — 0.64	(−26.2) — (−27.0)	45 — 106
Depression	Es 3	crude oil	C23	1.16 — 1.28	1.23 — 2.10	1.02 — 1.81	(−24.4) — (−25.5)	4.5 — 21
Depression	Es 4	crude oil	C20 or C28	0.73 — 1.01	0.73 — 1.71	0.26 — 1.22	(−24.4) — (−25.2)	—
Gudao oil field	Ng	crude oil	C23	1.15	—	0.62 — 0.67	(−24.3) — (−25.4)	0.0 — 30

Table 4. Comparison of physical properties of crude oil in the Guantano Member of the Gudao oil field with crude oil in the Shahejie Member in the depression.

Content	Shahejie oil in depression	Guantao oil of Gudao field
Specific gravity	0.854 to 0.879	0.933 to 1.0
Viscosity cp. 50°C	6 to 30	200 to 6,500
Pour point °C	28 to 40	(−4) to (−26)
Sulfur (%)	0.2 to 0.8%	1.7 to 3.0%
Oxygen (%)	0.08 to 0.9%	0.55 to 3.46%
Nitrogen (%)	0.14 to 0.62%	0.42 to 1.08%
Group analysis:		
paraffin (%)	46 to 52.3%	26.5 to 37.6%
aromatic (%)	12 to 16.9%	16.9 to 31.3%
nonhydrocarbon	25.6 to 38.4%	27 to 42.2%
asphalt (%)	1.1 to 8.7%	6 to 15.2%
Wax (%)	19.4%	4.1 to 7.2%

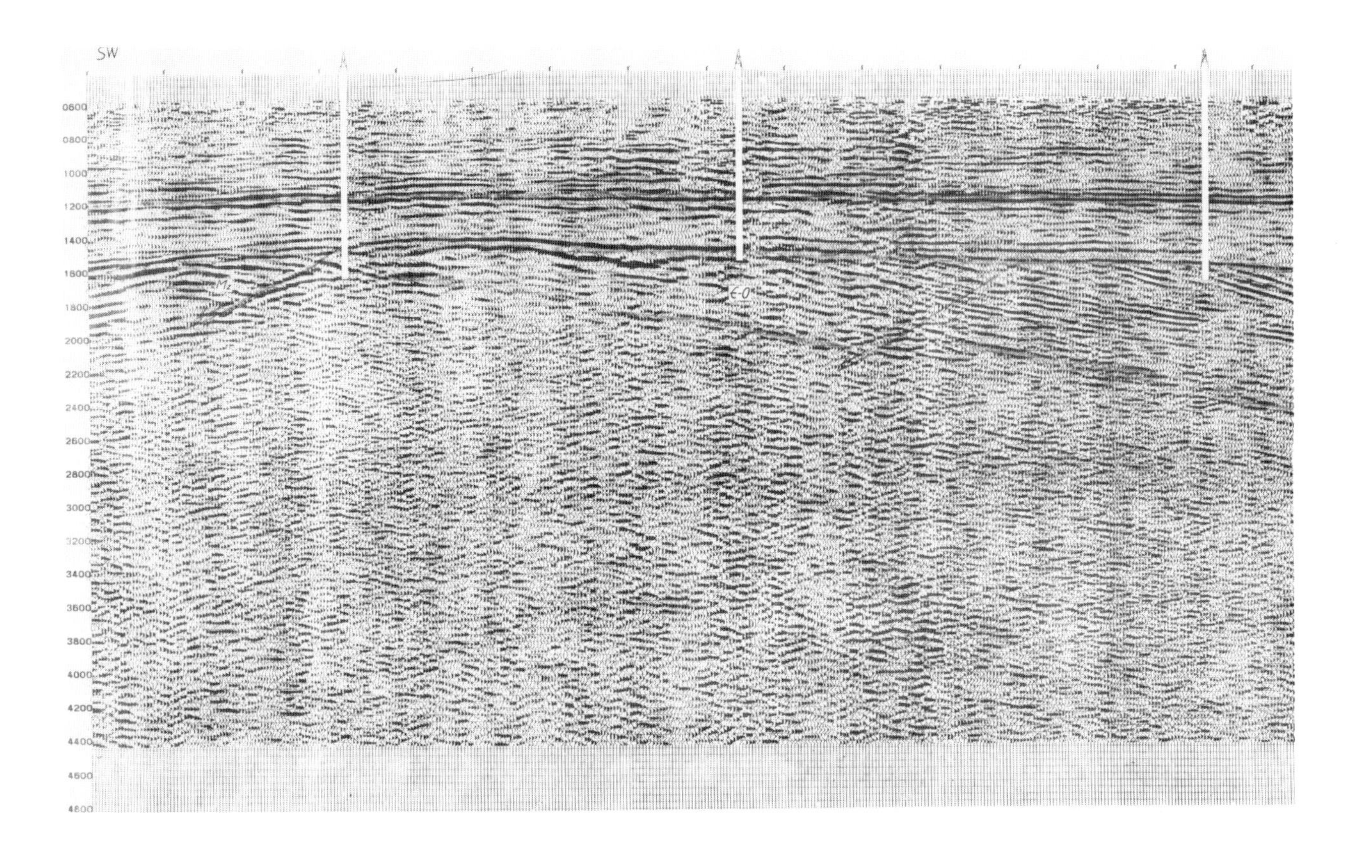

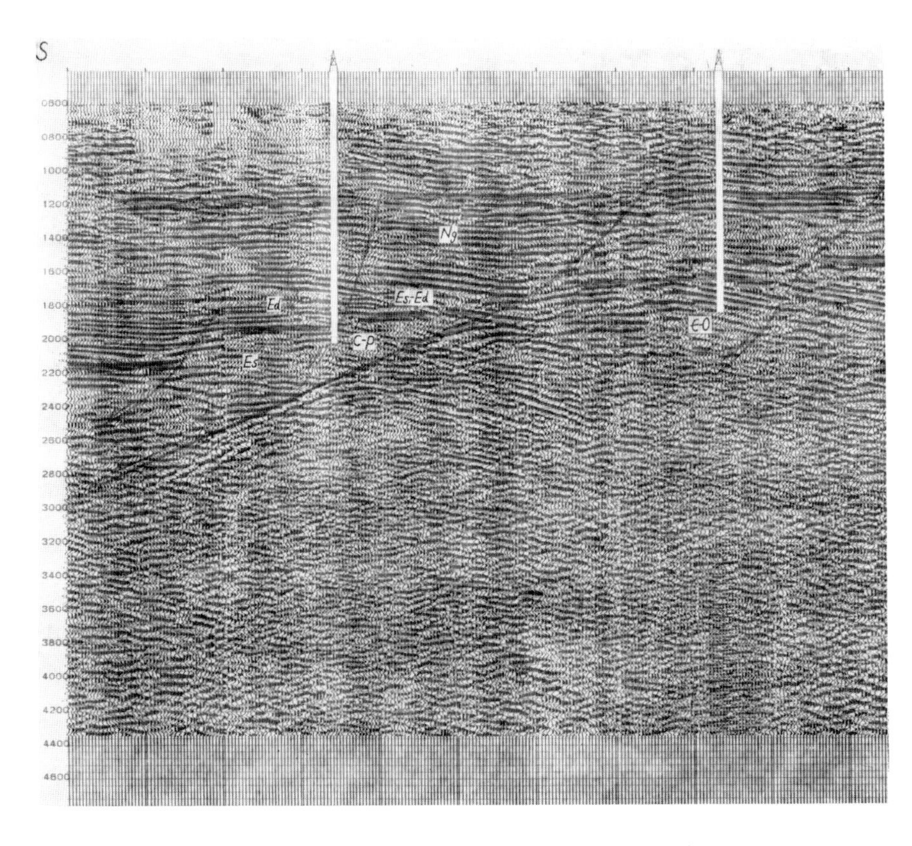

FIG. 7—Seismic profiles across Gudao field. Pre-Tertiary unconformity is obvious (see Table 1 for symbols).

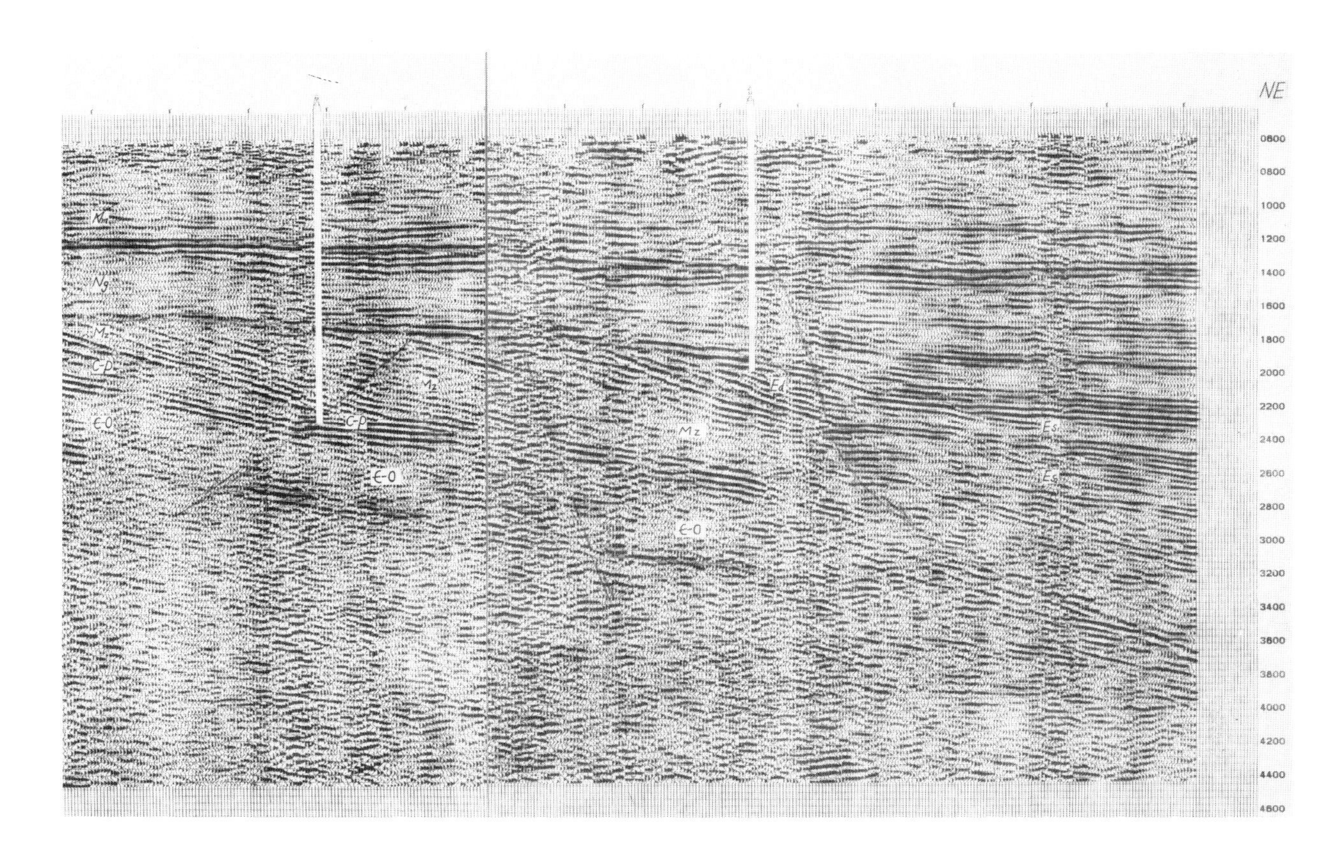
NE
Nm
Ng
Mz
C-P
Є-O
Mz
C-P
Є-O
Ed
Mz
Es
Es
Є-O
0600
0800
1000
1200
1400
1600
1800
2000
2200
2400
2600
2800
3000
3200
3400
3600
3800
4000
4200
4400
4600

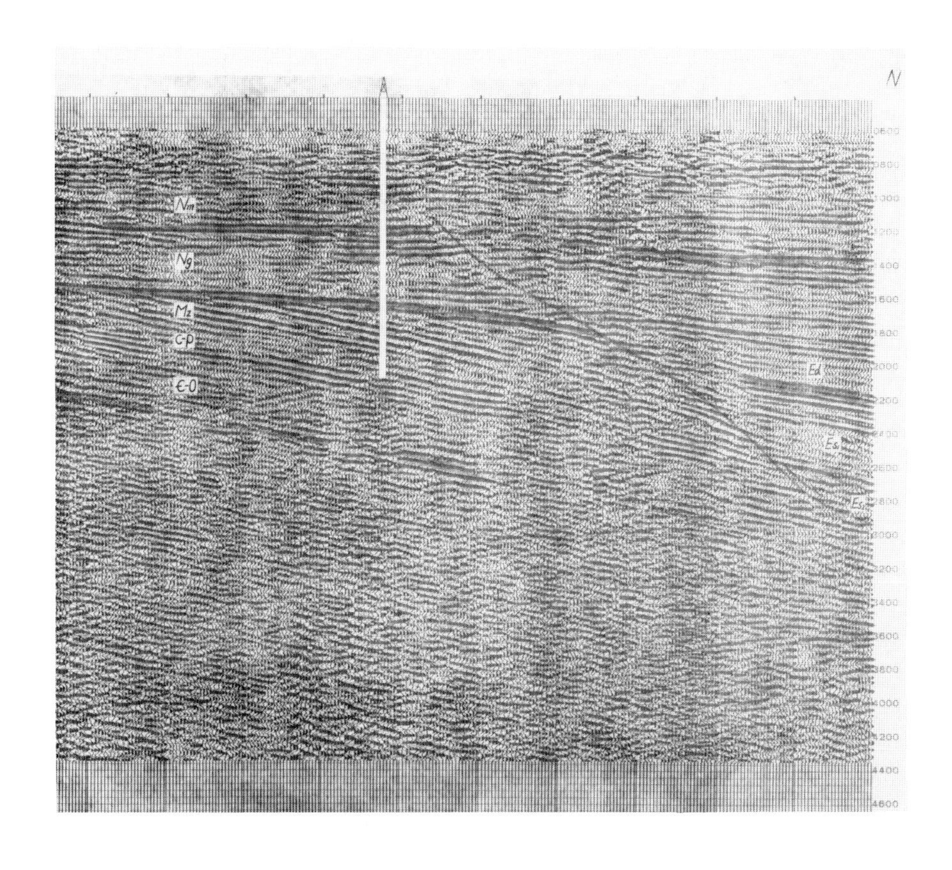
N
Nm
Ng
Mz
C-P
Є-O
Ed
Es
Es

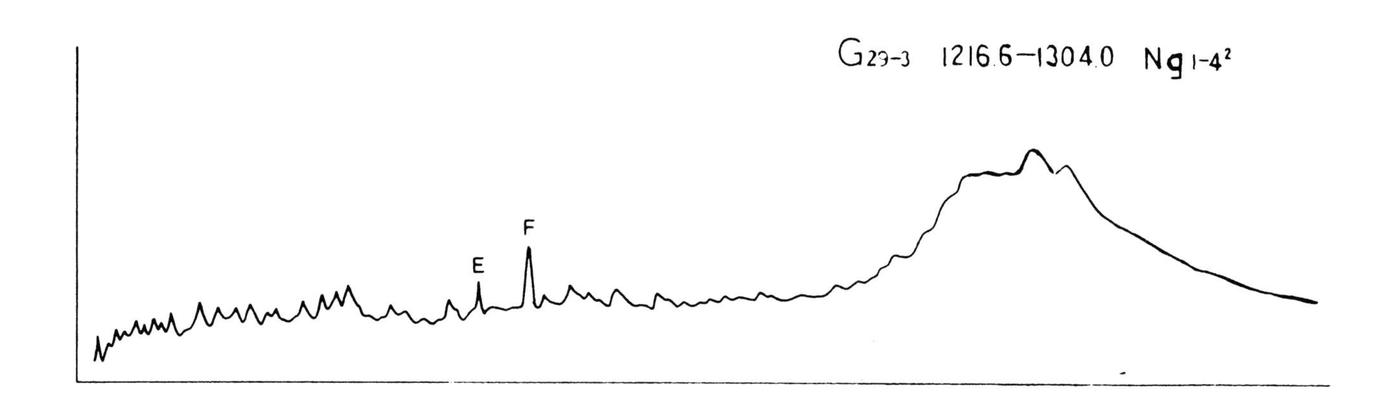

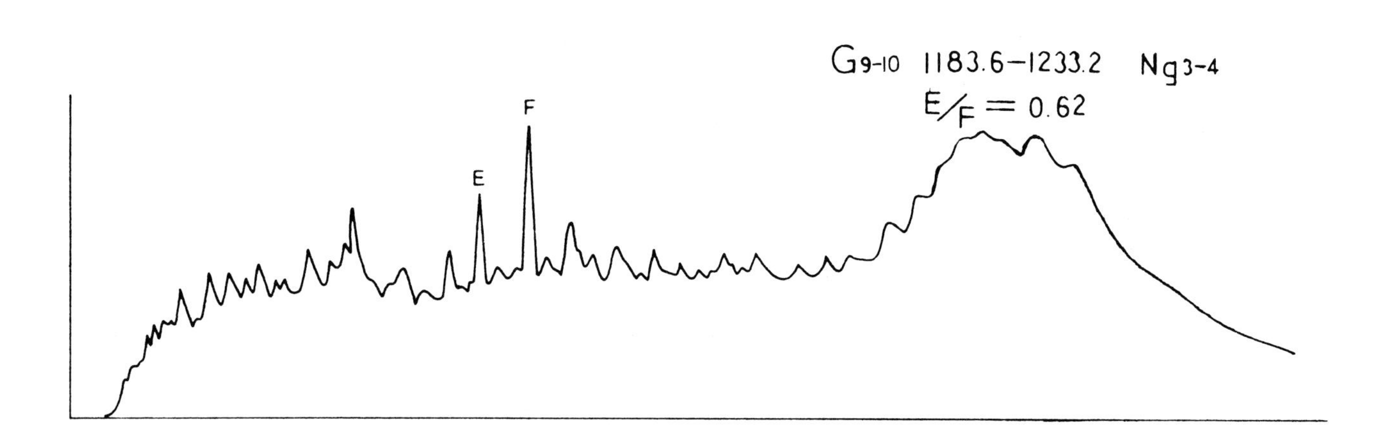

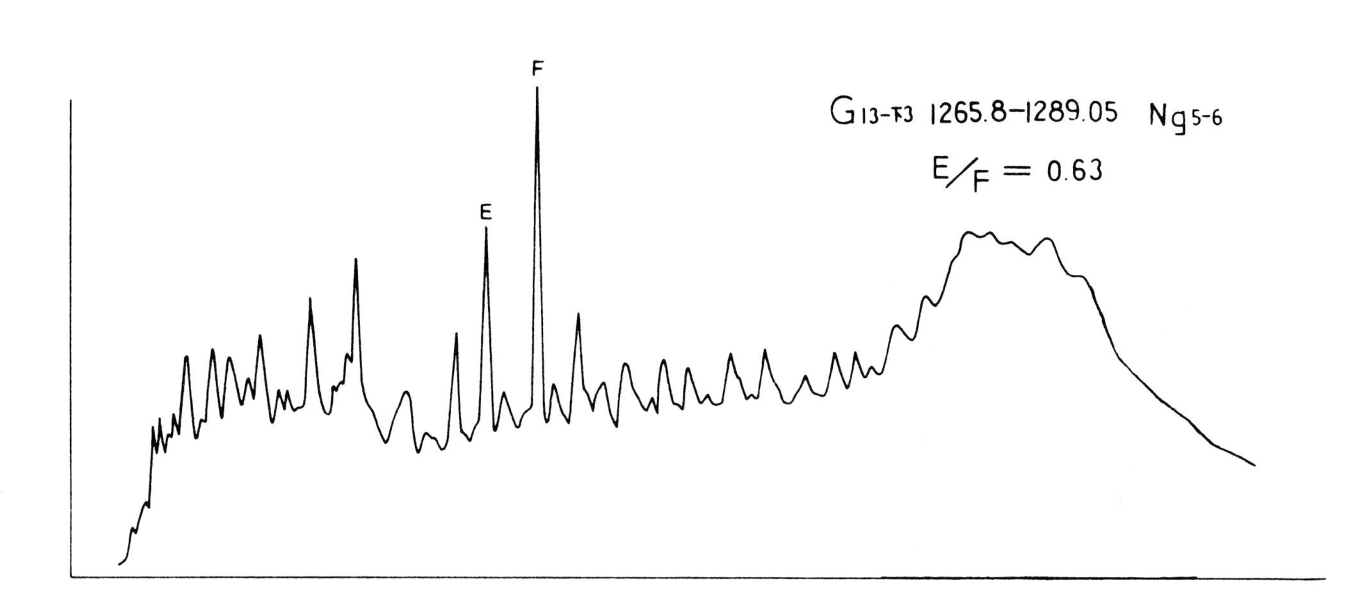

FIG. 8—Chromatographs of oils sampled in Gudao field. E/F (pristane/phytane) ratios are about the same for all oils.

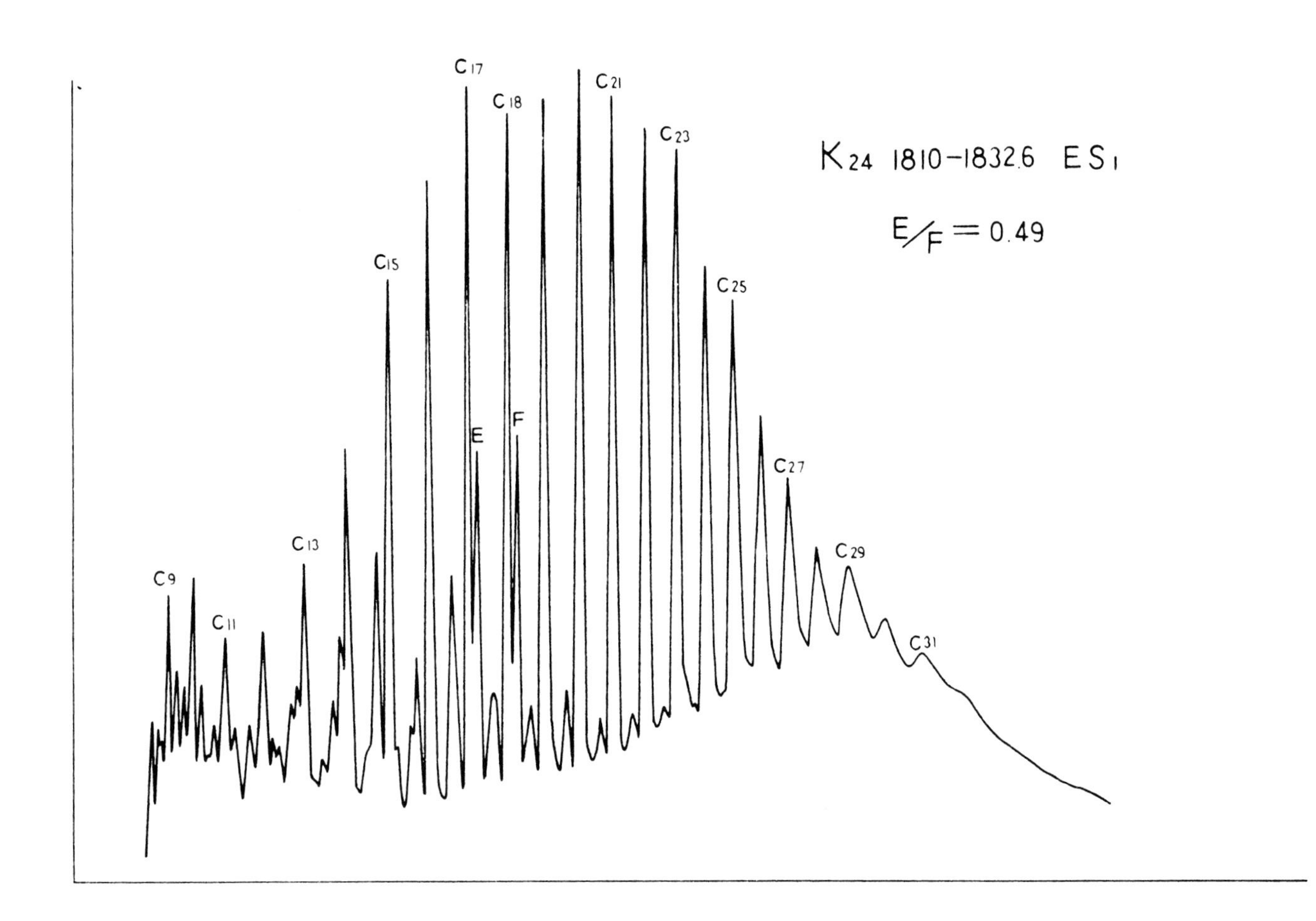
C17
C18
C21
C23
K24 1810–1832.6 ES1
E/F = 0.49
C15
C25
E
F
C27
C13
C29
C9
C11
C31

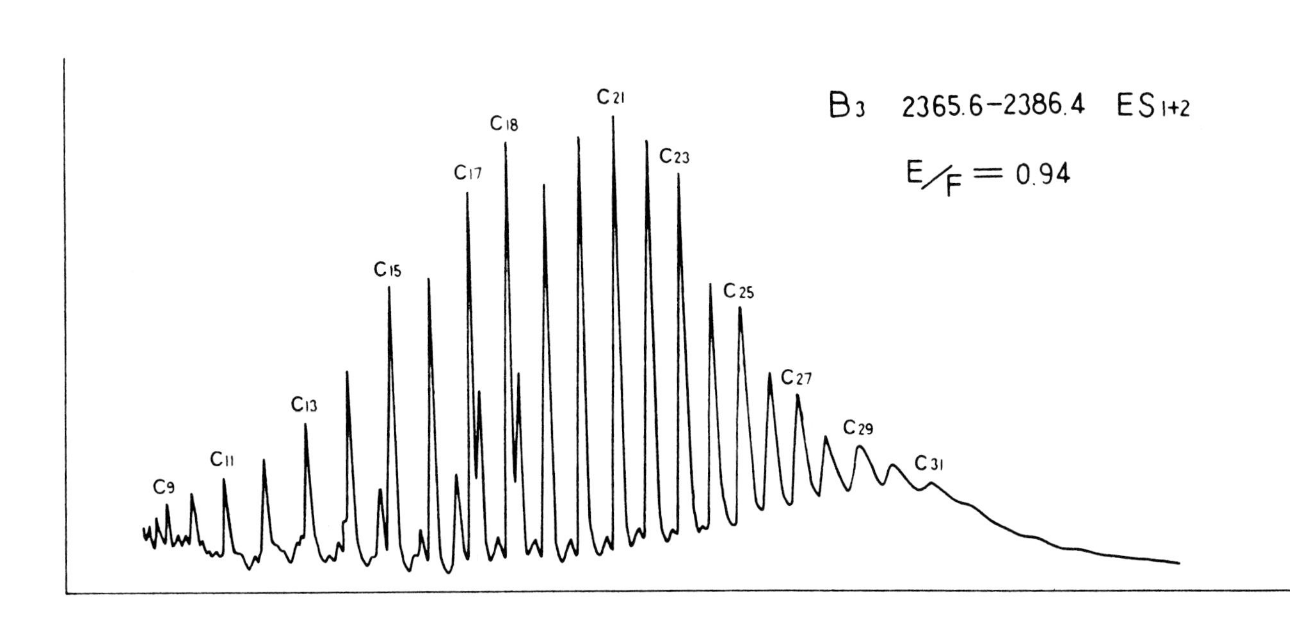
C21
B3 2365.6–2386.4 ES1+2
C18
C23
C17
E/F = 0.94
C15
C25
C27
C13
C29
C11
C31
C9

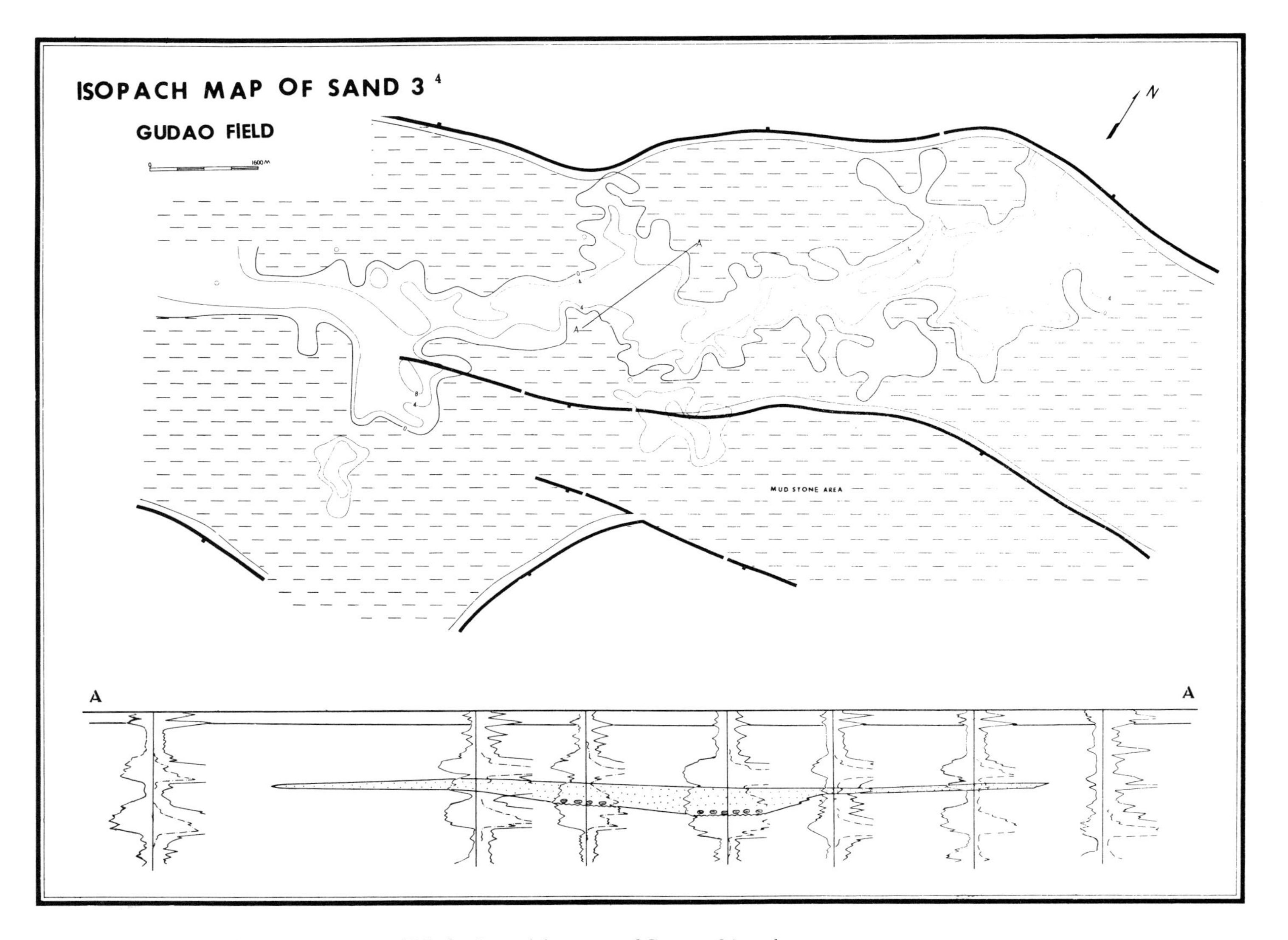

FIG. 9—Isopach in meters of Guantao 34 sand.

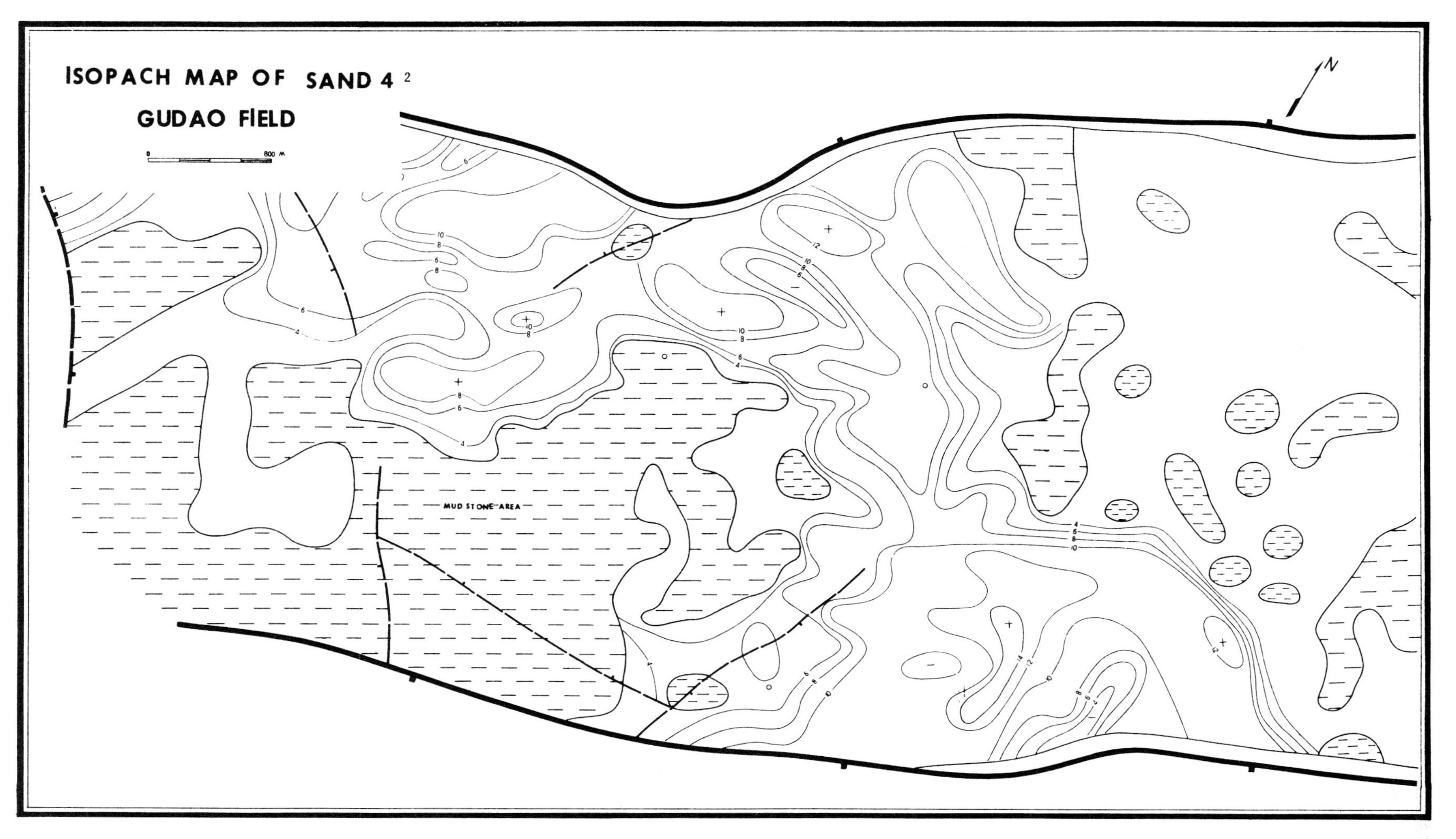

FIG. 10—Isopach of sand in meters of Guantao 42.

Zhanhua basin, especially *Bohaidina* which is found only in the Oligocene and is abundant in the lower part of Shahejie Formation. Evidence also indicates that the crude oil in the Guantao Formation comes from the "paleoreservoirs" in Shahejie Formation.

In addition to some small reservoirs, there are also widespread oil and gas shows in the Shahejie Formation in the Gudao field. Shahejie oil-bearing sandstones and bioclastic limestones ranging in thickness from 30 to 90 m have been encountered in many exploratory wells along the northern and western boundaries of the field. These formations mainly yield water with some viscous oil which seems to suggest that they are remnants of drained "paleoreservoirs." The present oil reservoirs in Guantao Formation and the gas reservoirs in Minghuazhen Formation are the result of oil and gas remigration and reaccumulation from the "paleoreservoir" during the process of continuous faulting and fault-block movement.

Faults were developed in the Zhanhua basin during the Mesozoic-Cenozoic, most of them being growth faults activated through a long period of time. Some faults acted as seals during the geologic history of the basin, and reservoirs that were formed were later destroyed through continuous faulting. Due to the intensive remigration and reaccumulation of oil and gas, secondary reservoirs of fairly large size were formed. Secondary changes are also apparently reflected in the properties of Guantao crude oil (Table 4).

In comparing the physical properties of the oil in the Guantao Formation with that in the Shahejie Formation it is noted that the specific gravity and viscosity increase, while the pour point decreases (Table 4). In comparing the chemical properties, the sulfur and oxygen contents increase, the wax and paraffin contents decrease, and the aromatics and asphalt contents also show considerable increase. This suggests that the crude oil in the Guantao Formation has undergone dewaxing under low temperature, biodegradation, and oxidation during its migration and accumulation (Fig. 8).

The Guantao consists of oxidized sediments. The content of total salts in the formation water, in general, is only 4,000 mg/1 and the water belongs to $NaHCO_3$ type. This implies that both the rock and the formation water possess strong oxidizing ability, therefore secondary changes in oil properties are the natural result of remigration and reaccumulation in the Guantao.

FLUVIATILE SEDIMENTS AND DEVELOPMENT WITH PROPORTIONAL WATERFLOODING

The sandstones of the Guantao Formation are the main reservoirs of Gudao oil field. They are fluviatile sediments consisting primarily of fine-grained moderately well-sorted sandstones and siltstones, commonly with gravels and clay ball at the base. The mineral contents of the rocks are rich in feldspar, generally 30 to 40%, while quartz is only 50 to 60% and other rock detritus 5 to 20%. The sandstones are classified as arkose and feldspathic graywacke with a high content of unstable heavy minerals. The sandstones generally contain 5 to 10% clay and little carbonate (less than 1%). The reservoir porosity ranges from 30 to 32%, permeability (air) is more than 1 darcy and effective permeability for oil 510 to 2,440 md. Due to rapid deposition and shallow burial, the sandstones are not well cemented and are poorly compacted. Cores taken from these beds are generally loose sands only.

The sandstones are distributed in a sinuous elongate geometry, e.g. the 3^4, 4^2 beds (Figs. 9, 10). The thickness of the sandstones varies from place to place, with the thickest part of the deposit in the main stream channel. The coarsest grain deposits are also in the main stream channel, easier for water to encroach during water flooding, and these deposits cause the severe sand problem in production of the oil wells.

The porosity, permeability, and crude oil properties differ considerably between different sandstone layers in the wells, as do their production capability. After detailed correlating, the Guantao Formation was divided into six sandstone groups according to sedimentary cycles. The sandstone group 3-6 which has been put into production is further divided into 20 oil sandstones. The distribution maps and interconnection diagrams for each oil sandstone were drawn as the basis for planning development of the field.

The oil viscosity of the Guantao crude oil is 20 to 130 cp. and has a mobility ratio to water of up to 50 to 300 cp. The oil is viscous and the field lacks enough energy to be produced by natural water drive. Although there are dispersed gas caps, they do not provide much driving energy. It has been calculated that with low initial gas/oil ratio, the primary oil recovery would be only 7 to 8% if produced with depletion drive, and the secondary recovery would be even more difficult and not economical. The method of pattern water flooding during an early stage of develoment has thus been adopted.

The proportional water flooding of each oil sandstone group was adopted to reduce the interference between different groups. The oil-bearing sandstones are divided into 3 to 4 groups for water injection. The quantity of injection water for the low permeable bed with heavy oil is increased and that for the high permeable bed with light oil is limited, so that all oil sandstones can be influenced by the water flooding and no water breakthrough will occur in the early stage.

Because the oil sands are loose and oil is viscous, another severe problem confronted in developing the oil field is production of sand in the wells. Wide application of sand control methods has solved the problem. The surface viscosity of the oil in about one-third of the wells is 1,000 to 4,000 cp. Measures to lower the viscosity have been adopted, and oil production in these wells has become normal.

Oil production of the field has increased continuously over the last eight years. By the end of 1978, the total oil produced reached 13.3% of the oil in place, the overall water cut was 32.0%. During 1978, the daily oil production remained the same as that of the early stage of develoment, the pressure of the reservoirs being stable. These factors show that good development results can be obtained in a viscous oil field like Gudao.

Petroleum Geology of Bombay High Field, India[1]

By R. P. Rao[2] and S. N. Talukdar[3]

INTRODUCTION

Bombay High field is a giant oil field discovered in 1974. The field is about 160 km off Bombay in the Arabian Sea at a water depth of about 75 m (Fig. 1). This field has almost half of India's known oil reserves. The discovery and development of the field has been a tremendous plus for the Indian economy. Initially, at least 20% of India's oil consumption (32 million bbl/year) is expected to be produced from this field.

The shelf area adjoining the Cambay Gulf was surveyed by conventional seismic reflection methods in 1964 by the Soviet ship *Akademic Arkhangelsky*. The Bombay High structure was discovered and mapped as a result of that survey. This structure is a north-northwest to south-southeast trending doubly plunging anticline with a faulted eastern limb. The anticline is about 65 km long and 23 km wide and covers an area of about 1,500 sq km. Before the first well was drilled, a detail survey was made of the structure by the French-owned company, CGG, in 1972.

To date, about 27 exploratory wells have been drilled on as many structures in the Bombay Offshore basin. Of these, nine have been found to be oil or gas bearing. Bombay High is the first and the largest field discovery.

OFFSHORE REGIONAL GEOLOGY

Bombay Offshore basin is on a broad shelf believed to be the largest continental terrace in the world (Shepard, 1973). The submerged forest at Bombay and the rill-like appearance of the Deccan lava coast suggest recent depression, and the littoral concrete present far behind the coast in the marshy areas of Bombay suggest emergence. The sagging due to the Deccan lava flows (60 m.y.) probably is now in isostatic equilibrium.

Except for the Deccan Trap outcrops and the intertrappeans (?) there is no indication of the expected geology in the Bombay Offshore basin. In the Saurashtra Penninsula (Kathiawar), a trap is present and is overlain by a consolidated Pleistocene wind-deposited sandstone called miliolites (also called the Porbunder sandstone; Wadia, 1966). East of Anklesvar oil field, Tertiary coarse continental clastic rocks crop out, interbedded with thin marine fossiliferous beds. A well drilled near the town of Surat penetrated an equivalent section of a shale and the facies are so different from those on outcrops that the anticipated geology of Bombay Offshore basin was at best guess work (Fig. 2).

SEDIMENTARY AREA OF THE VARIOUS OFFSHORE BASINS ADJOINING CAMBAY BASIN:

The sedimentary basin area offshore India, to the water depth of 200 m is 390,000 sq km. The Bombay Offshore basin

[1]Manuscript received May 21, 1979; accepted for publication, January 1, 1980.

[2]Basin Studies Division, Institute of Petroleum Exploration, Dehradun 248195, U.P., India.

[3]Institute of Petroleum Exploration, Dehradun 248195, U.P., India.

The views expressed in this paper are those of the writers, and not necessarily those of the organization for which they are working.

Thanks are due to S. Aditya, director (Geology), Oil and Natural Gas Commission, for the helpful discussions held during different times of the study. Thanks also are due to colleagues: P. C. Dhar, D. C. Srivastava, D. K. Pande, N. K. Lal, G. S. Misra, and many others.

Article Indentification Number:
0065-731X/80/M030-0024/$03.00/0.

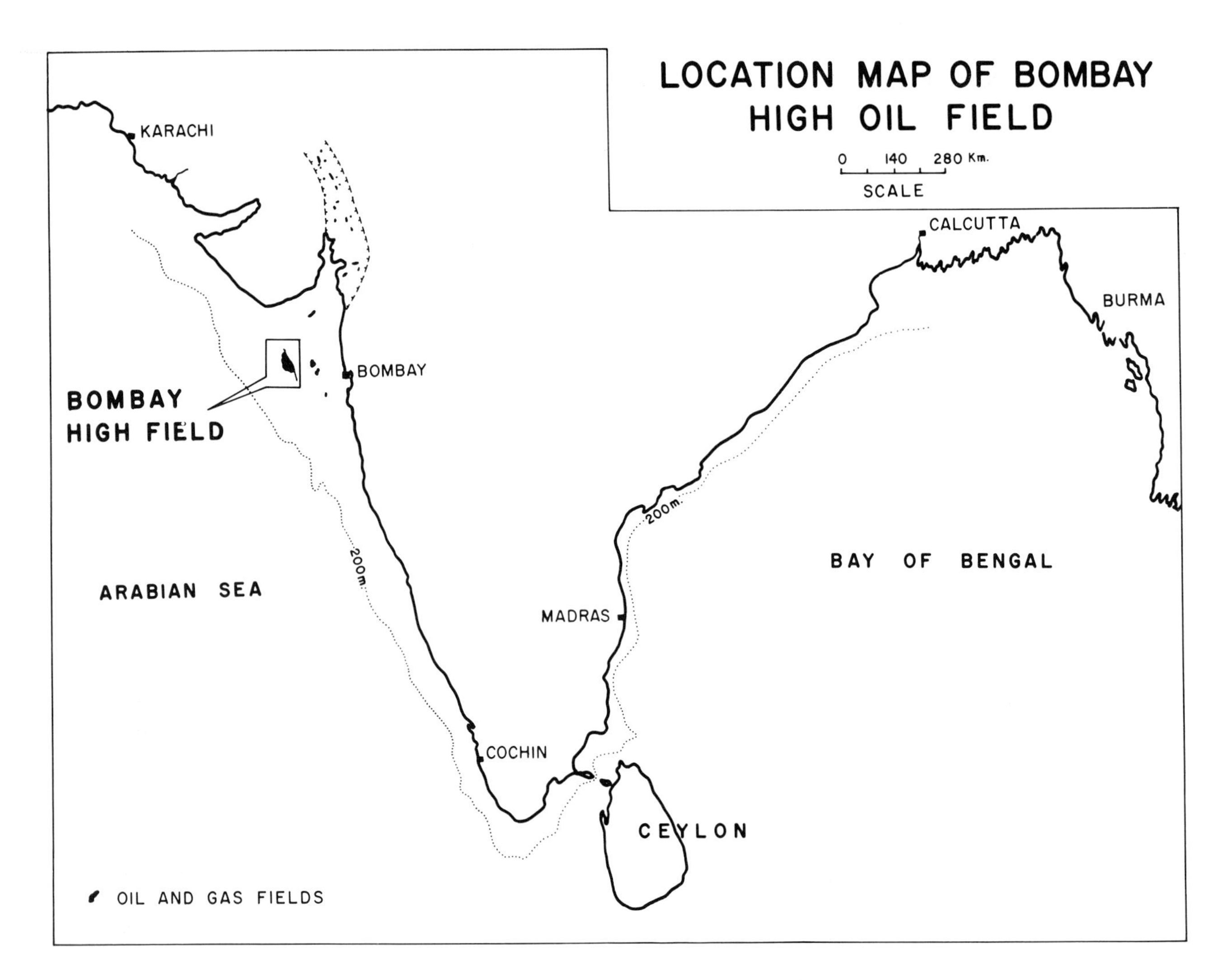

FIG. 1—Location map, Bombay High field, offshore India.

is about 120,000 sq km (nearly 30%) of the total offshore sedimentary area (Fig. 1). The west coast shelf has several sedimentary basins (Ramaswamy and Rao, 1978) from north to south: the Kutch basin, the Saurastra basin, the Bombay Offshore basin, the Konkan basin, and the Kerala basin. Only the Bombay Offshore basin has proved to be oil- and gas-bearing.

STRATIGRAPHY

Rock Stratigraphy

The sedimentary section drilled in the Bombay High field can be classified into three units: (1) the basal sands, lignite, and clays; (2) the middle limestone and alternate shale; and (3) the upper shale and claystone. Based on drilling data from the Bombay High field and elsewhere, lithostratigraphic classification has been proposed (Table 1; Rao, Dhar, and Pande, 1978). It appears that the entire Surat Depression consists of shale and claystone. The middle section in the Bombay Platform consists of alternate limestones and shales including the productive zones of the Bombay High field in the upper part. The shale in the sequence grades into a limestone west of Bombay High in the DCS ("Deep Continental Shelf") area and toward the south in the Ratnagiri block.

Seismic Stratigraphy

Bombay Offshore basin has four seismic reflection markers designated as H-1, H-2, H-3, and H-4. The lithologic, electric log, and paleontologic data obtained from drilling wells was fitted into this seismic framework. Seismic stratigraphic classification (Rao and Srivastava, 1979) suggests at least four seismic sequences, I, II, III, and IV, from bottom to top (Table 1). Each sequence is separated by an unconformity. Sequence I is reflection-free or chaotic with occasional diffractions and is identified with the basement which is either Archean or the Deccan Trap. Sequence II which has reflections with high amplitude, good continuity and high interval velocity, is a carbonate section. Sequence III, which has low amplitude, discontinuous reflections and low velocities, is a shale section. This sequence is seen to have an overlapping relation with the lower sequence II. The associated regression and transgression may be related to a geologic event at the end of Miocene time

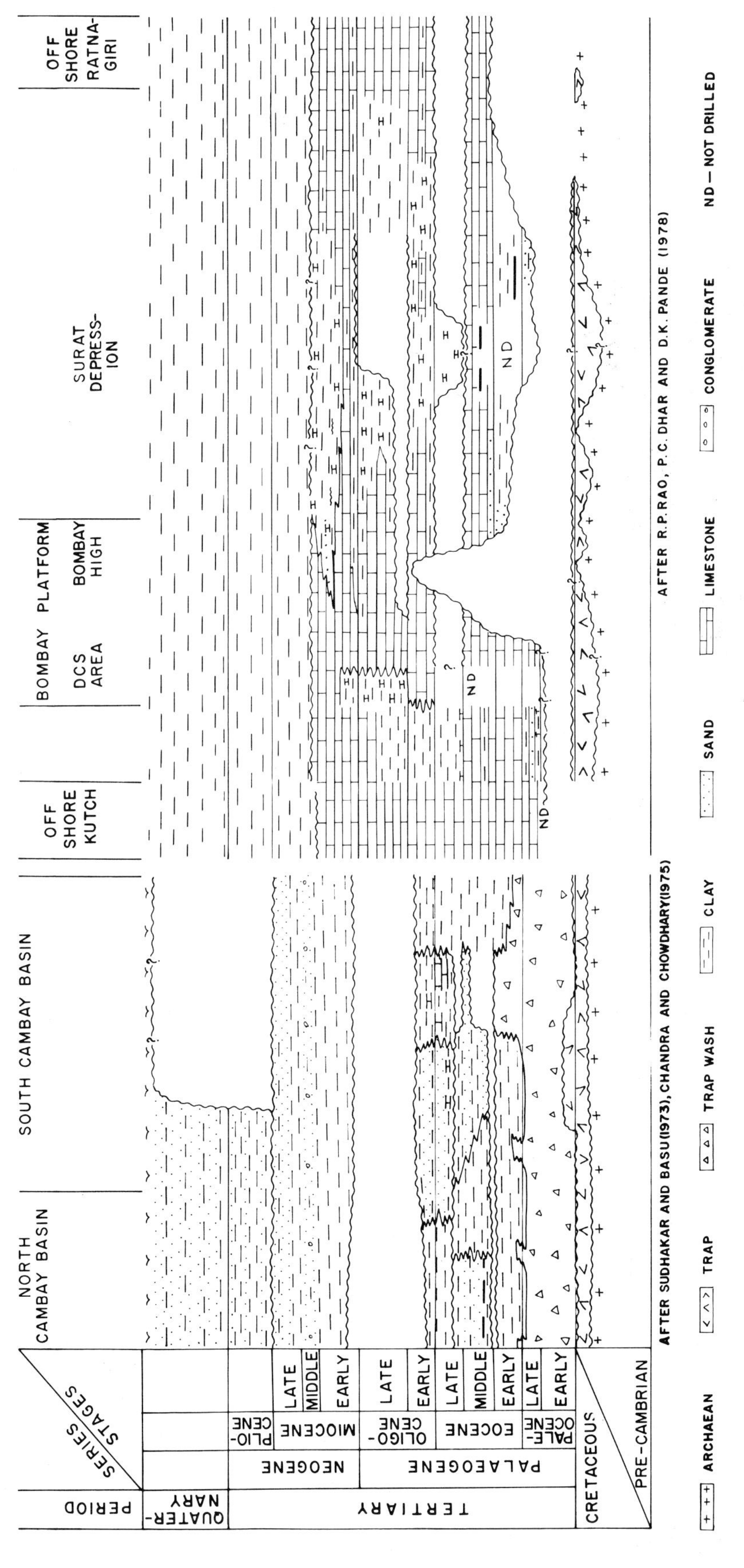

FIG. 2—Lithological correlation between Cambay basin and Bombay offshore basin.

Table 1. Biostratigraphy, Rock—Stratigraphy and Seismic Stratigraphy of the Bombay Offshore Basin.

BIOSTRATIGRAPHY*			ROCK STRATIGRAPHY※	SEISMIC STRATIGRAPHY✡		
			FORMATION ●	SEQUENCES	FACIES	INTERPRETATION
MIDDLE MIOCENE AND YOUNGER		NOT WORKED OUT	CHINCHANI FORMATION	SEISMIC SEQUENCE IV SEISMIC SEQUENCE III	LOW AMPLITUDE DISCONTINUOUS LOW AMPLITUDE DISONTINUOUS	OVERLAP OF CLAY STONE, POST GLACIAL SEALEVEL RISE 5. TRANSGRESSION REGRESSION
EARLY MIDDLE MIOCENE	EARLY LANGHIAN	Miogypsina antillea – M. cushmani Assemblage Zone	BANDRA FORMATION			OVERLAP OF CLAY STONE / SHALE SEALEVEL RISE 4. TRANSGRESSION
EARLY MIOCENE	BURDIGALIAN	Miogypsina excentrica Partial Range Zone.	RATNAGIRI LIMESTONE	SEISMIC SEQUENCE II	HIGH AMPLITUDE, CONTINUOUS, HIGH VELOCITY	REGRESSION (OSCILLATORY)
		Miogypsina globulina Range Zone	SALSETTE Fm.			
	AQUITANIAN	Miogypsina tani Partial Range Zone	BOMBAY Fm. DCS GROUP			CARBONATE SECTION WITH SHALE INTERCALATIONS 3. TRANSGRESSION
		Spiroclypeus Partial Range Zone.	ERRANGAL FORMATION			
LATE OLIGOCENE	CHATTIAN	Miogypsinoides complanatus Range Zone.				
EARLY OLIGOCENE	RUPELIAN LATTORFIAN	Nummulites fichteli Range Zone/ Globigerina tapuriensis Range Zone				REGRESSION
EOCENE — LATE	PRIABONIAN	Nummulites fabianii Zone Globorotalla centralis Globorotalia cerroazulensis Assemblage Zone	ALIBAG FORMATION			2. TRANSGRESSION
EOCENE — MIDDLE	LUTETIAN	Nummulites acutus – Assilina exponens – Fasciolites Assemblage Zone/Dictyoconus – Lituonella Fasciolites Assemblage Zone/ Rotalia trochidiformis – Fasciolites Assemblage Zone	BASSEIN LIMESTONE			REGRESSION
EOCENE — EARLY	YPRESIAN (PART)	Nummulites mamilla – Assilina spinosa Assemblage Zone.	VASAI FORMATION			
PALEOCENE	THANETIAN (PART)	Nummulites deserti – Miscellanea miscella Assemblage Zone.				1. TRANSGRESSION
PRE-TERTIARY				SEISMIC SEQUENCE I	CHAOTIC, REFLECTION FREE, DIFFRACTIONS	BASEMENT

* MADAN MOHAN et al 1978

※ R.P.RAO, P.C.DHAR & D.K.PANDE, 1978

● NOMENCLATURE INFORMAL ONLY.

✡ R.P.RAO & D.C.SRIVASTAVA 1979

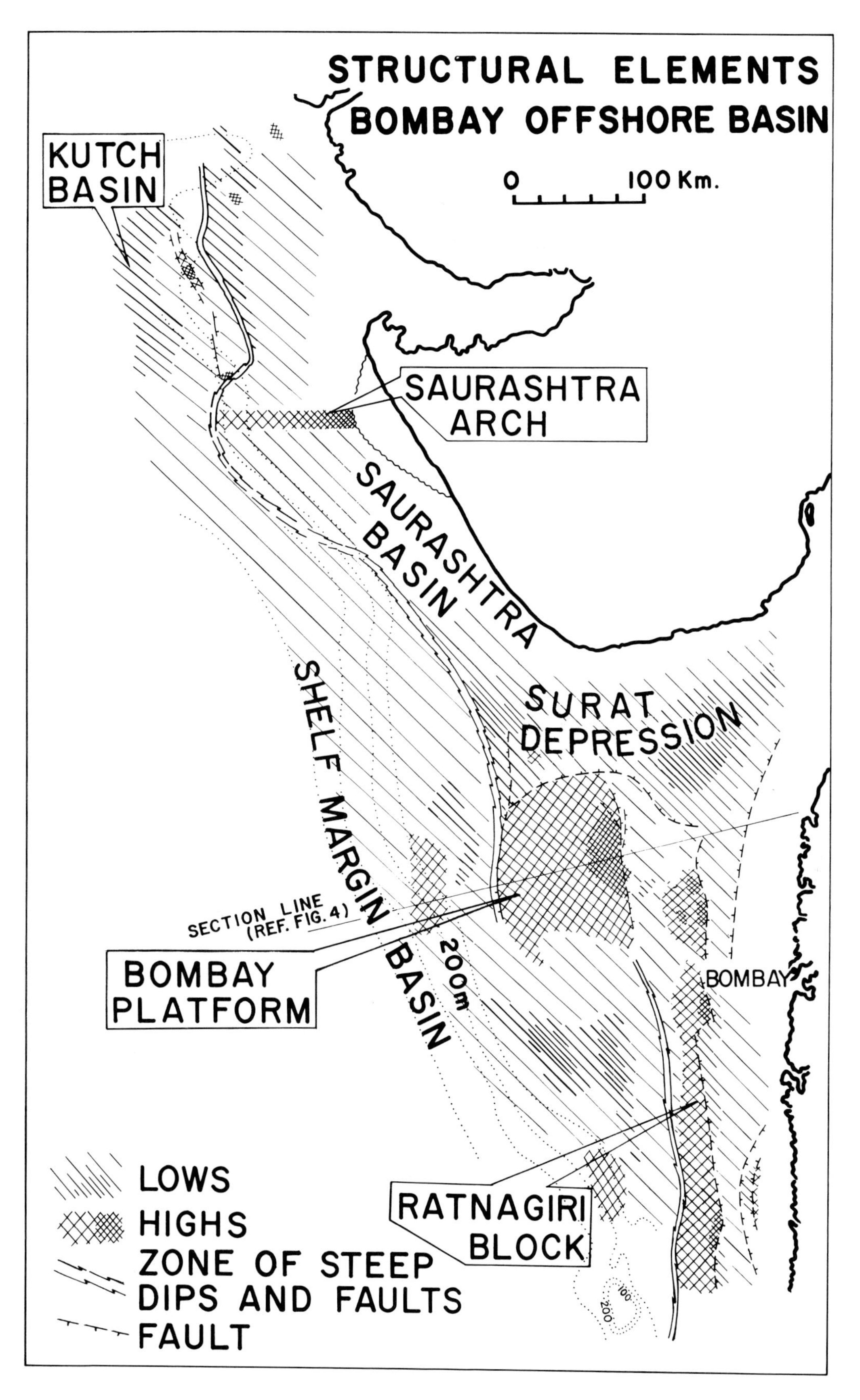

FIG. 3—Structural elements of Bombay offshore basin.

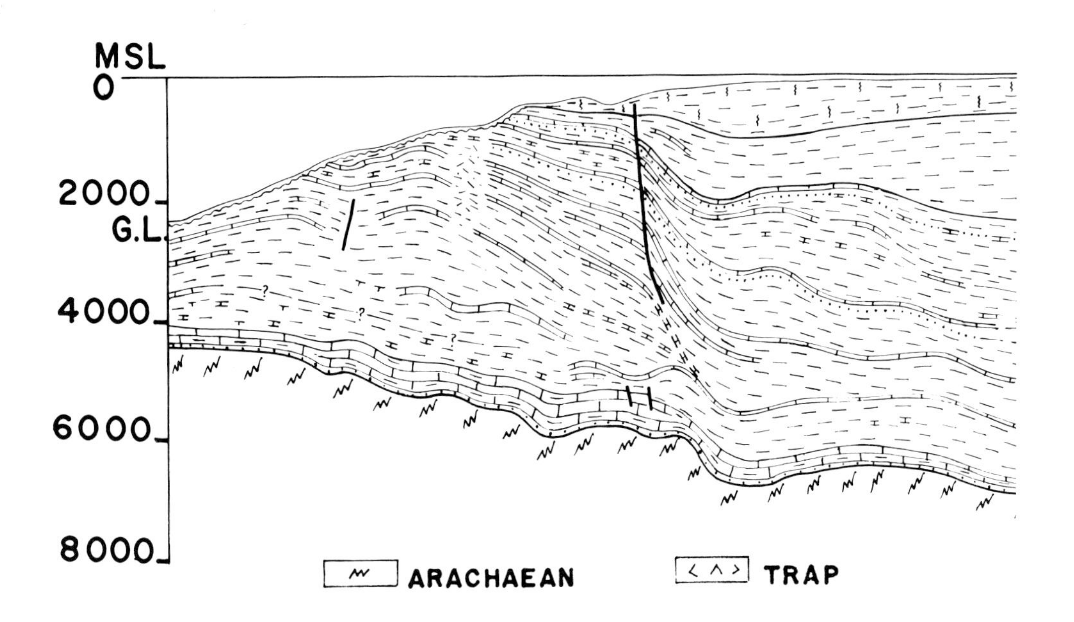

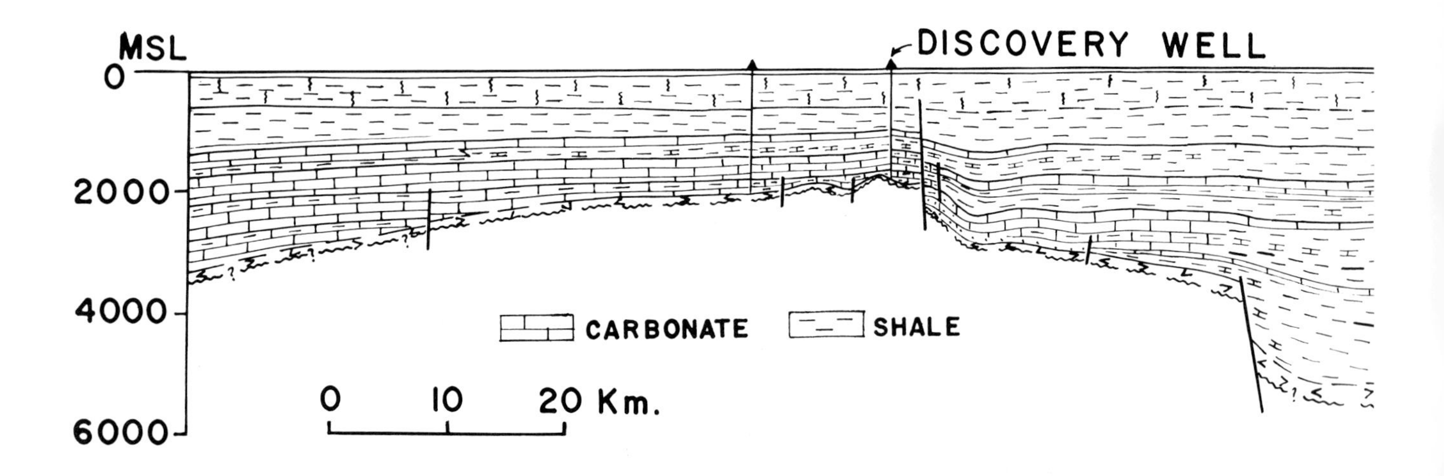

FIG. 4—Structural cross section (after seismic) across Surat Depression, Bombay Platform, and Shelf Margin basin.

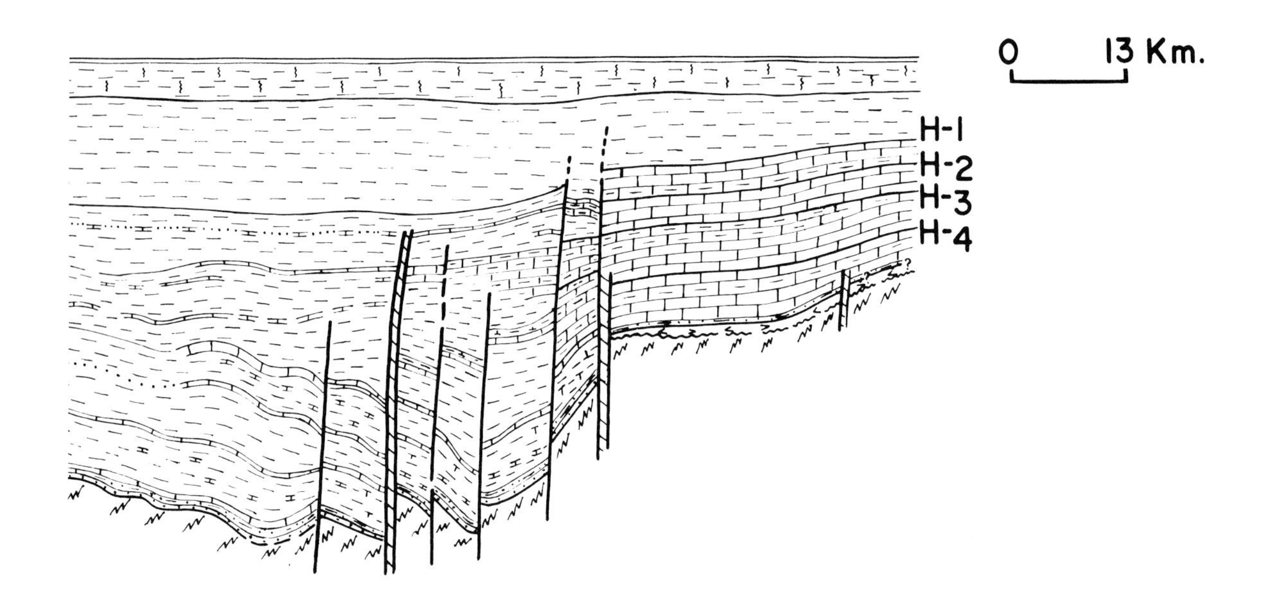
0 13 Km.
H-1
H-2
H-3
H-4

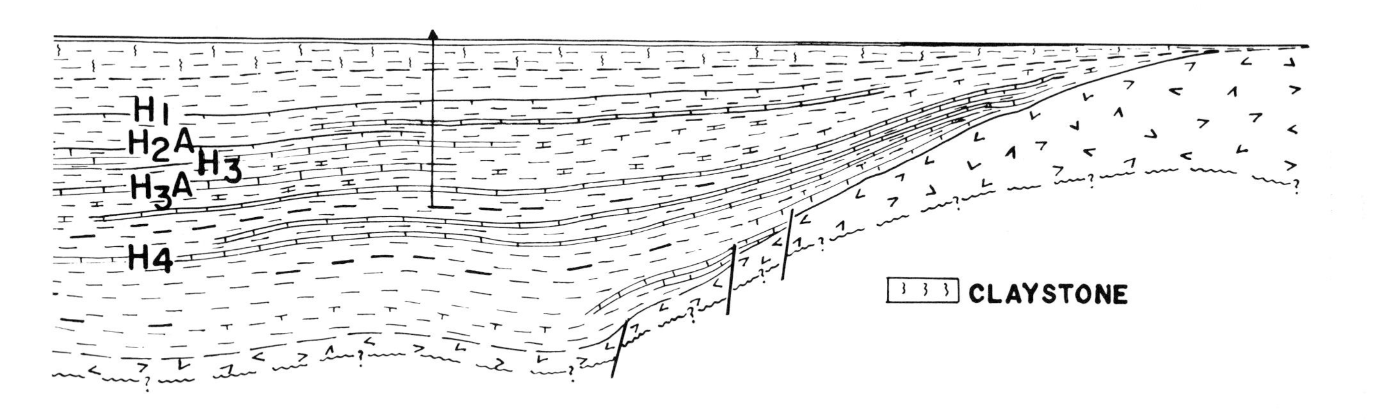
H1
H2A
H3
H3A
H4
CLAYSTONE

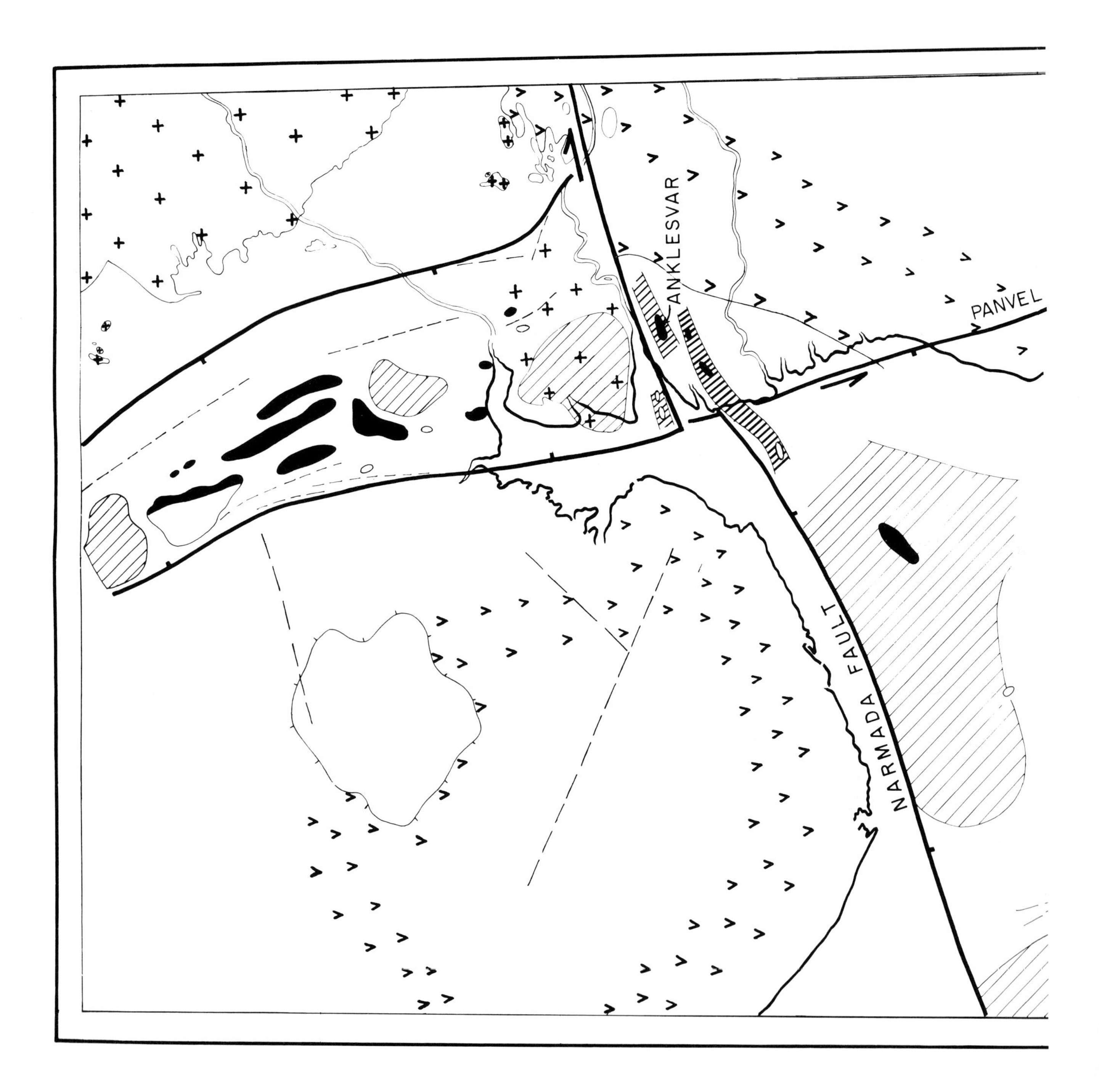
ANKLESVAR
PANVEL
NARMADA FAULT

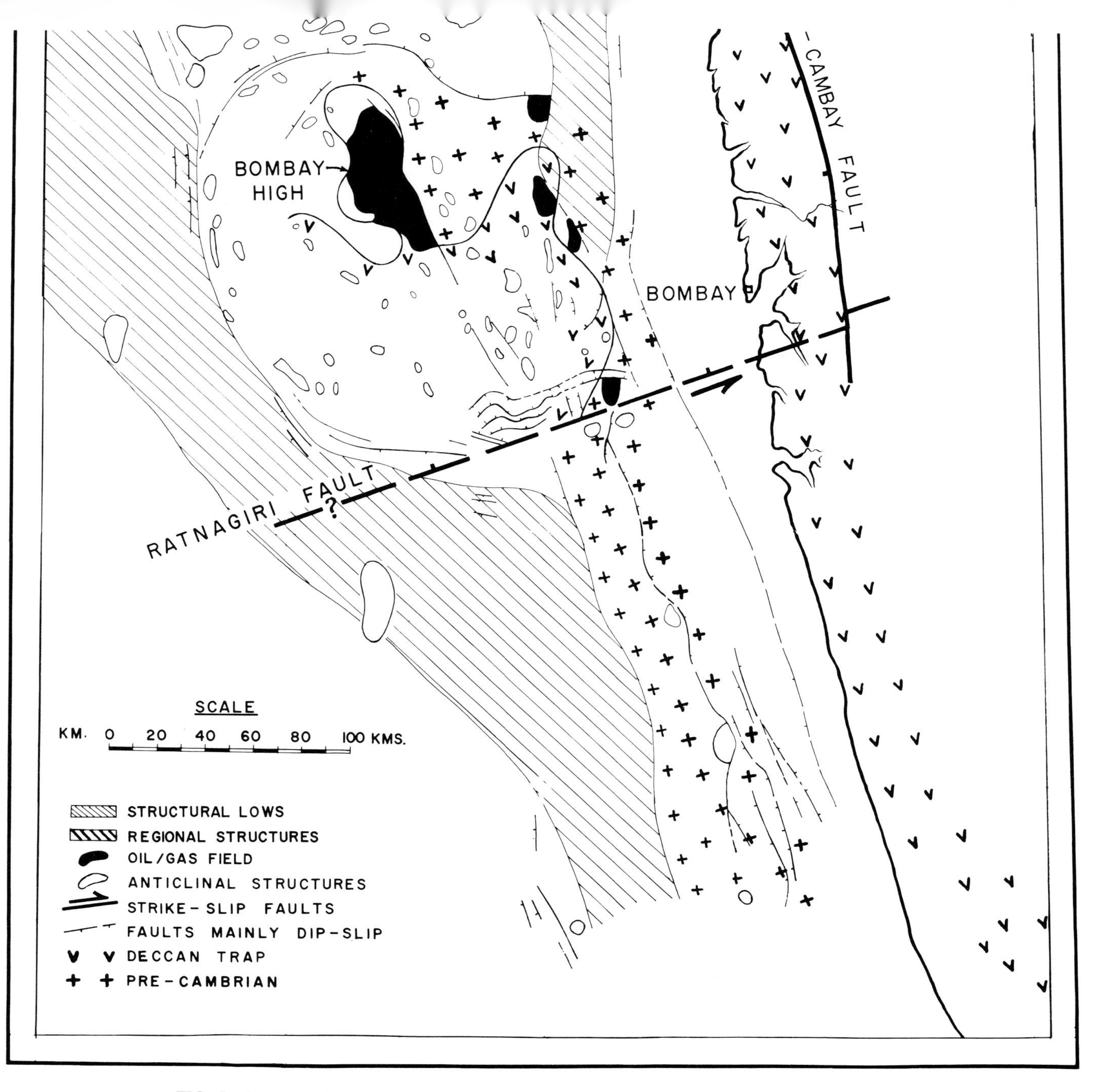

FIG. 5—Interpreted tectonic framework of the Cambay basin and Bombay offshore basin.

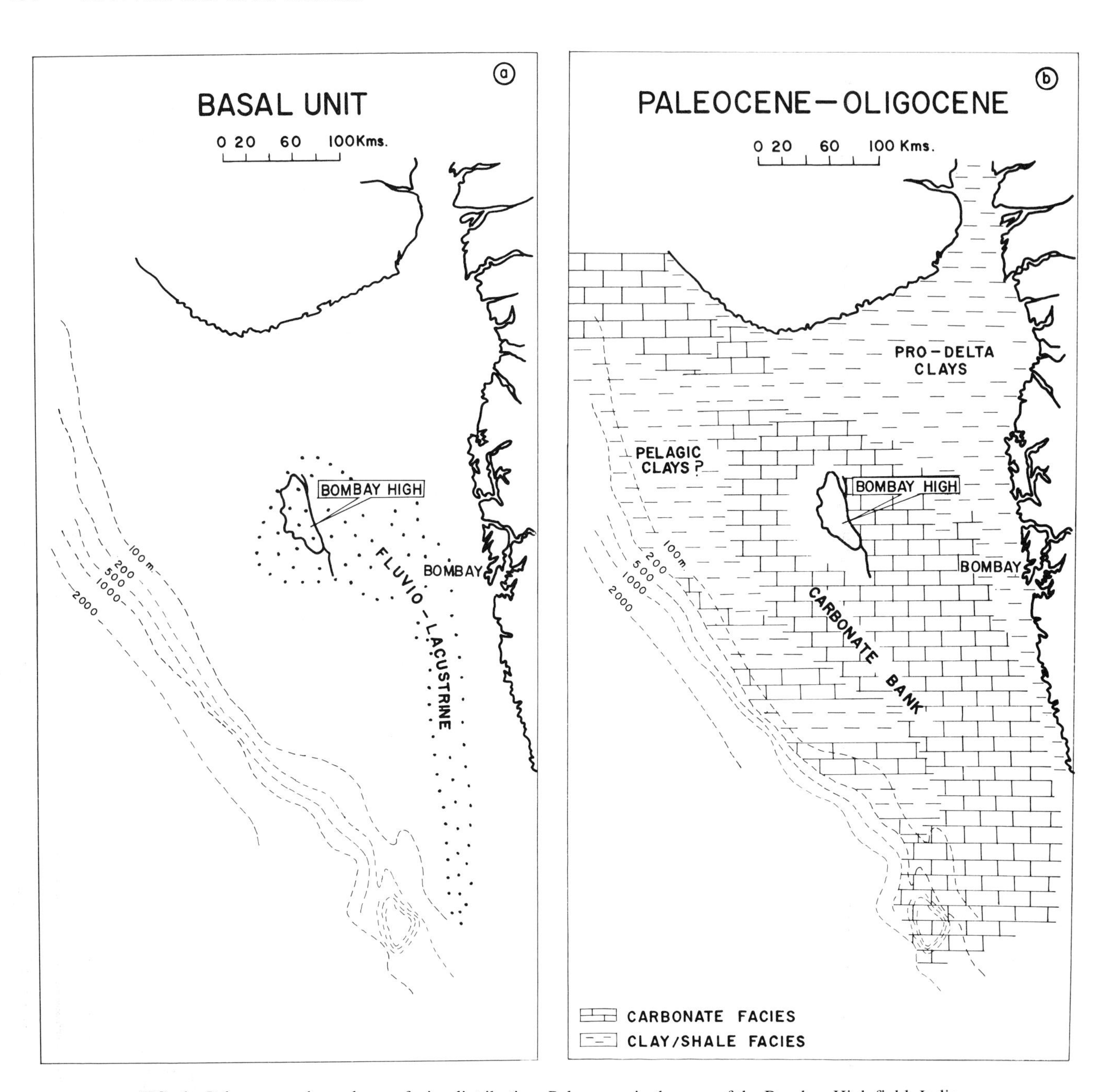

FIG. 6—Paleogeography and gross facies distribution, Paleocene, in the area of the Bombay High field, India.

(Berggren, 1978). Sequence IV is similar to sequence III and overlies sequence III unconformably. Here the regession associated with the unconformity and the transgression attributed to the overlap are probably related to the Pleistocene glaciation.

Biostratigraphy

Biostratigraphic classification (Table 1) is based on larger benthonic foraminifera (Madan Mohan et al, 1978) because planktonic forms are rare. Study was restricted to the carbonate and shale section below seismic horizon I, and showed that on the Bombay High the entire Paleocene and Eocene sections are missing; in the Shelf Margin basin an almost complete sequence from late Paleocene to early Oligocene is present. On the Bombay High, sedimentation started in late Oligocene and continued to the present. In the eastern part of the basin, south of Surat Depression, a thick sequence of middle Eocene carbonate rocks is directly overlain by early Oligocene rocks (Madan Mohan et al, 1978).

Presence of larger benthonic foraminifers in most of the Paleogene and Neogene section and the lack of planktonic foraminifers suggests a warm, shallow-water, marine (lagoonal?) environment in the Bombay High area as well as east of it. However, west of Bombay High, planktonic forms are present in Paleogene.

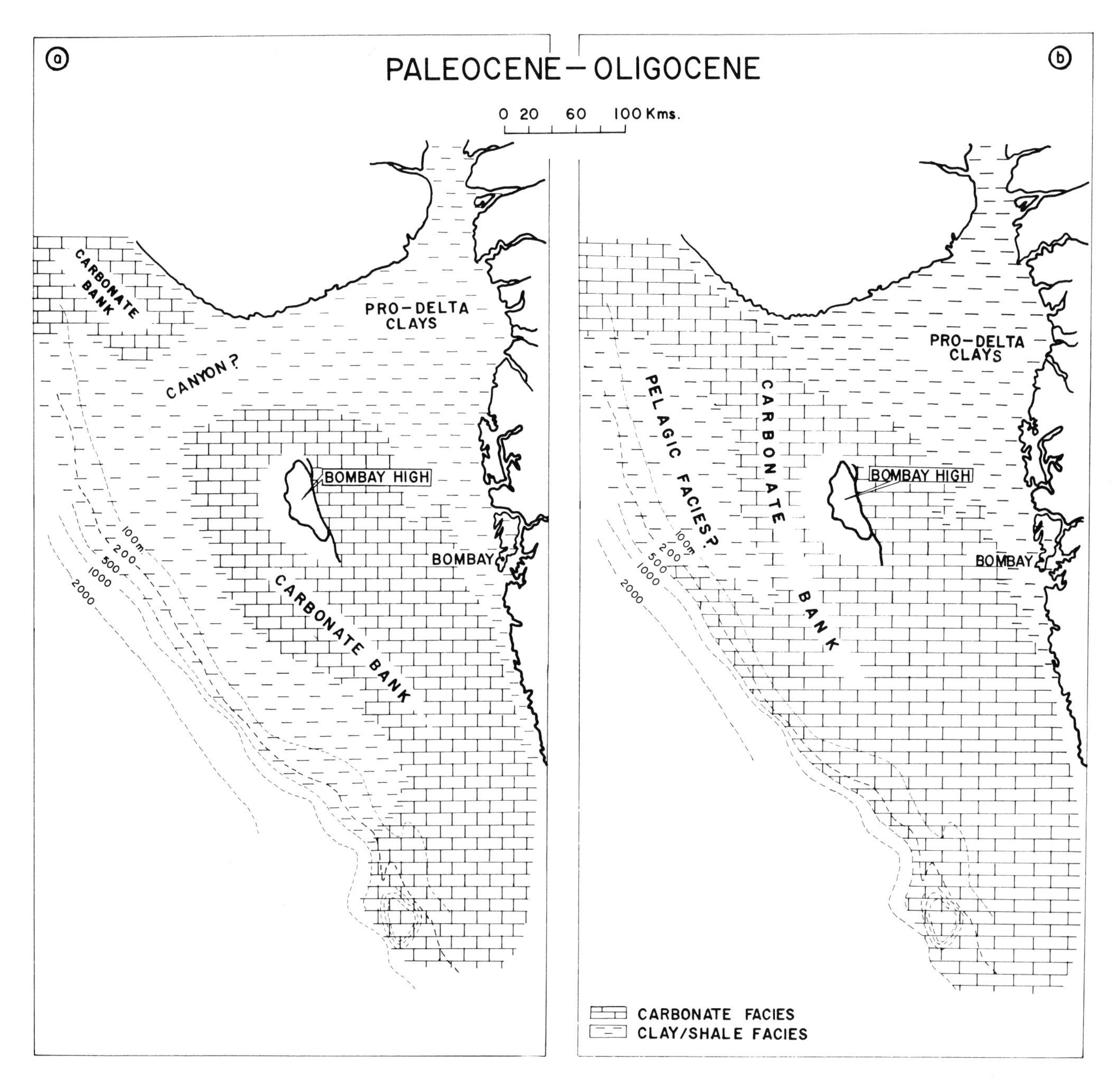

FIG. 7—Paleogeography and gross facies distribution, Paleocene to Oligocene, area of the Bombay High field, India.

STRUCTURE

The Bombay Offshore basin continues into the Saurashtra basin to a northern limit marked by the Saurashtra arch (Fig. 3). North of the Saurashtra arch is the Kutch basin. In the Bombay Offshore basin, the Bombay Platform separates the Surat depression on the east from the Shelf Margin basin on the west (Fig. 4). The present shelf as well as the sedimentary basin is narrow to the south and trends eastward into the Ratnagiri block. The Ratnagiri block and the Bombay Platform are separated by east-west faults. The drainage pattern onshore differs north and south of the Konkan coast. Also east-northeast to west-southwest trending fractures and dyke swarms are present onshore farther east. All the preceding support the idea of a major tectonic element called here "Ratnagiri Fault" (Fig. 5). This fault seems to have a sinistral strike-slip movement.

The Cambay basin appears to terminate abruptly south of Narmada River (Fig. 5), whereas the Bombay Offshore basin is cut off south of Saurashtra Peninsula. In the Cambay basin, north of Narmada River, all the structural elements have a north-northwest to south-southeast (Dharwarian grain?) trend. In the Bombay Offshore basin, the structural elements are essentially the same except in the northernmost part of the basin where they are east-northeast to west-southwest. The western marginal fault of the Cambay basin is down-thrown toward the east and apparently continues into the eastern margin of the

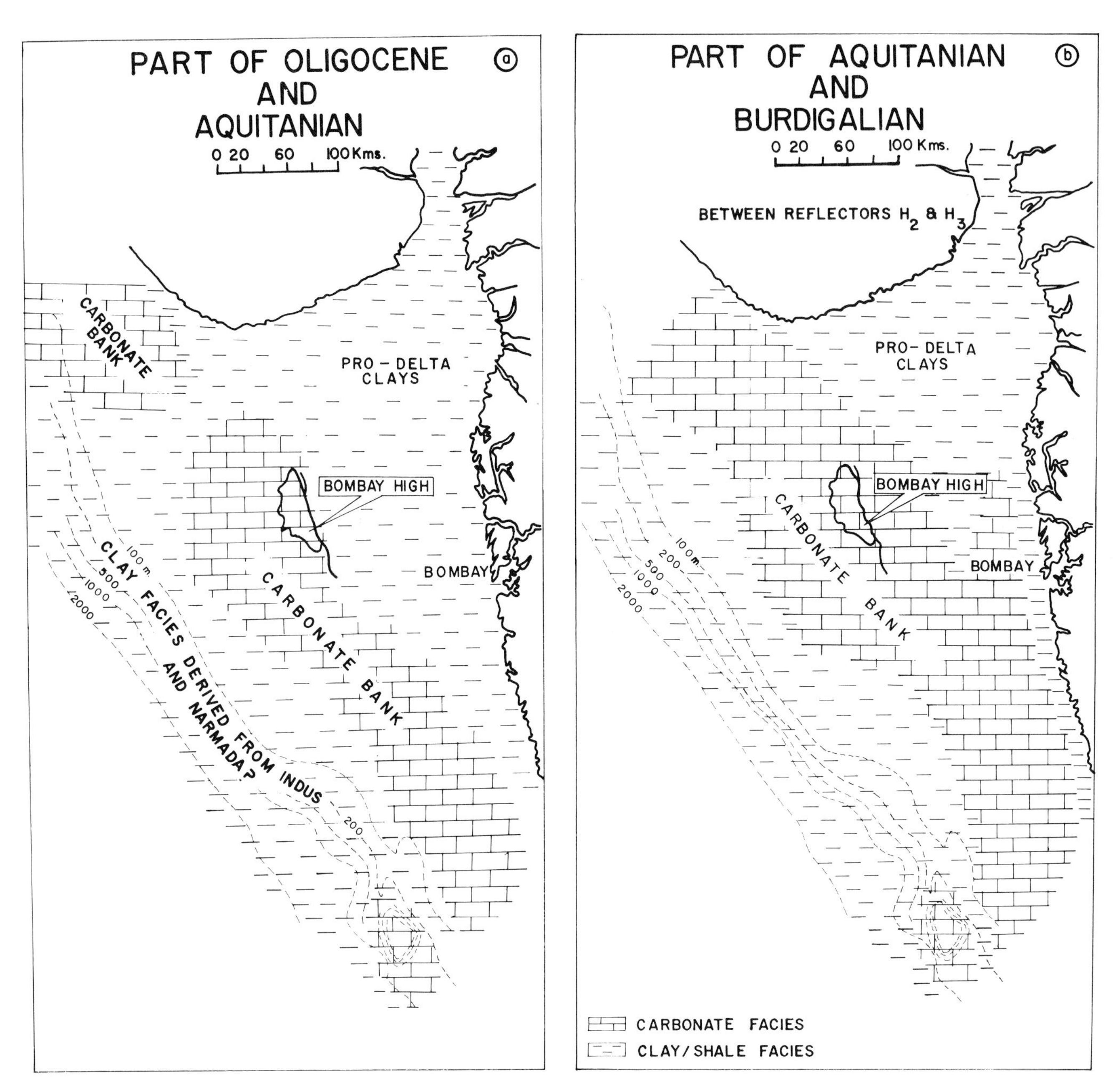

FIG. 8—**A.** Paleogeography and gross facies distribution, part of Oligocene and Aquitanian, Bombay High field area. **B.** Paleogeography and gross facies distribution, part of Aquitanian and Burdigalian, area of the Bombay High field, India.

Bombay Offshore basin. In the Bombay Offshore basin a number of wells penetrated Tertiary and entered Archean basement without encountering the Deccan Trap or the Mesozoic. A few wells bottomed in trap, but it appears that trap is thin in the Bombay basin. East of Bombay, the trap is as much as 2,500 m thick. North of Narmada and east of Cambay basin, the trap is very thin or absent. Thus, there are terranes which have thick trap juxtaposed with terranes of thin trap (Saurashtra Peninsula against Bombay Offshore basin, Sahyadris against Cambay basin and the terrane north of Narmada). This suggests a prominent tectonic element, the Narmada fault. A detailed discussion on this tectonic relation is beyond the scope of this article, but the suggestion of a dextral fault is interesting. In any case, the Bombay Platform appears to be bounded by the Narmada fault and Ratnagiri fault, with a probable westward movement of the Bombay Platform.

PALEOGEOGRAPHIC CONSIDERATIONS

Paleocene to Oligocene

Gross facies maps were prepared for intervals between the seismic reflectors H-1, H-2, H-3, and H-4. The Palaeontological Laboratory of the Institute of Petroleum Exploration (IPE) correlated these intervals to biostratigraphic boundaries (Ma-

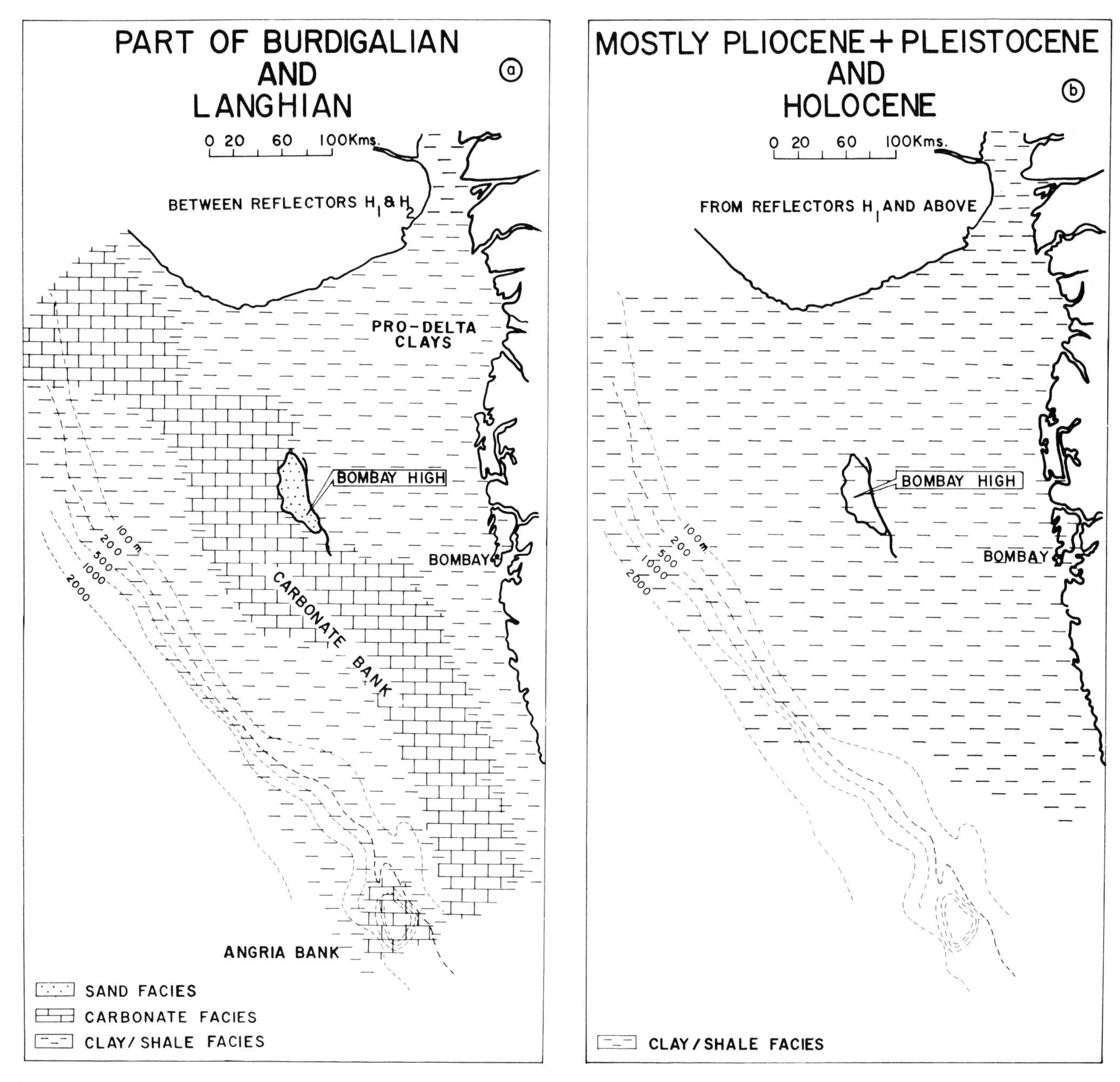

FIG. 9—**A.** Paleogeography and gross facies distribution, part of Burdigalian and Langhian, Bombay High field area, India. **B.** Paleogeography and gross facies distribution, mostly Pliocene-Pleistocene and Holocene, Bombay High field, India.

dan Mohan et al, 1978). The apparent absence of the Paleocene and Eocene sediments on the Bombay High and the presence of only part of the Oligocene sediments indicates that either each emerged during most of the early Tertiary or that the sedimentation was extremely limited. On Bombay High, sandstones are found directly above the basement. It is possible the "basal sands" of Bombay High represents the entire early Tertiary (Fig. 6a). Besides the general rise in the sea level due to rifting when the basin was formed, there were intermittent fluctuations of sea level. These fluctuations resulted in the Surat depression and the Shelf Margin basin being periodically connected across a carbonate bank (Fig. 6b). When the two were connected clay facies developed in the southern part of the Shelf Margin basin, and when the connection was cut off carbonate facies developed (Fig. 7a). It is speculated that the connection could be in the form of a submarine canyon resulting from an east to west fault (Narmada Fault?).

Part of Oligocene and Aquitanian

A further rise in the sea level broke the connection between the Bombay uplift/Ratnagiri carbonate bank and the Kutch-Saurashtra carbonate bank, and narrowed the Bombay-Ratnagiri carbonate bank (Fig. 8a). The Surat depression and the Shelf Margin basin were not affected.

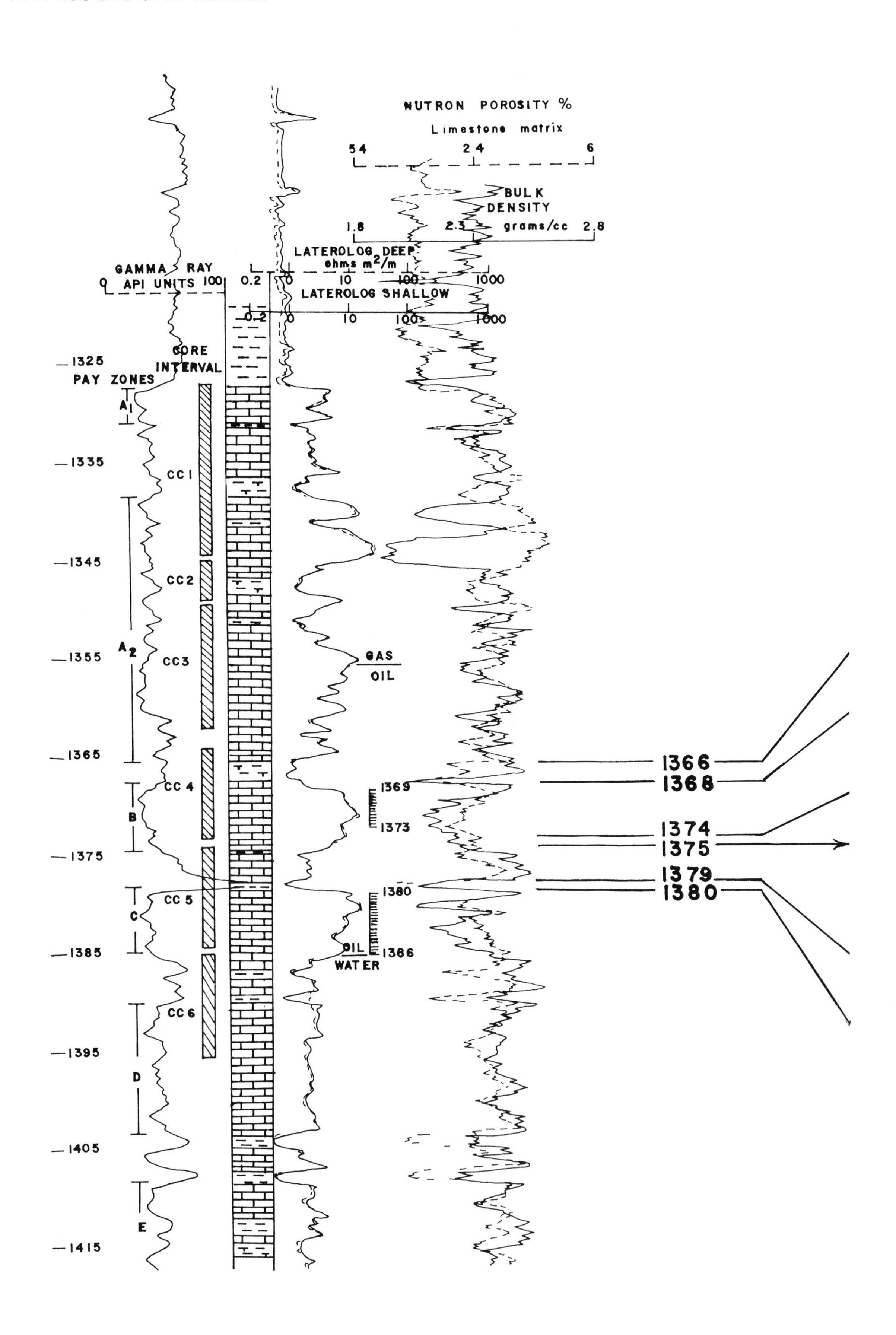
NUTRON POROSITY %
Limestone matrix
54
24
6
BULK
DENSITY
1.8
2.3
grams/cc 2.8
LATEROLOG DEEP
ohms m²/m
GAMMA RAY
API UNITS
0
100
0.2
0
10
100
1000
LATEROLOG SHALLOW
0.2
0
10
100
1000
CORE
INTERVAL
PAY ZONES
1325
1335
1345
1355
1365
1375
1385
1395
1405
1415
A1
A2
B
C
D
E
CC1
CC2
CC3
CC4
CC5
CC6
GAS
OIL
1369
1373
1380
1386
OIL
WATER
1366
1368
1374
1375
1379
1380

LAYER — B

	Shale	Showing high gamma ray count	Coastal Marshes
	Wackestone	With abundant to medium micrite consisting dominant larger foraminifera	Foramini-feral Mound
	Wackestone	With abundant micrite consisting codiacea forams and corals.	Algal Mound ?
	Wackestone	With abundant micrite dominant solitary corals subordinate forams ostracods and algae.	Lagoonal
	Shale	Dark grey showing high gamma ray count.	Coastal Marshes

FIG. 10—Layering in L-III main pay zone and microfacies in each layer.

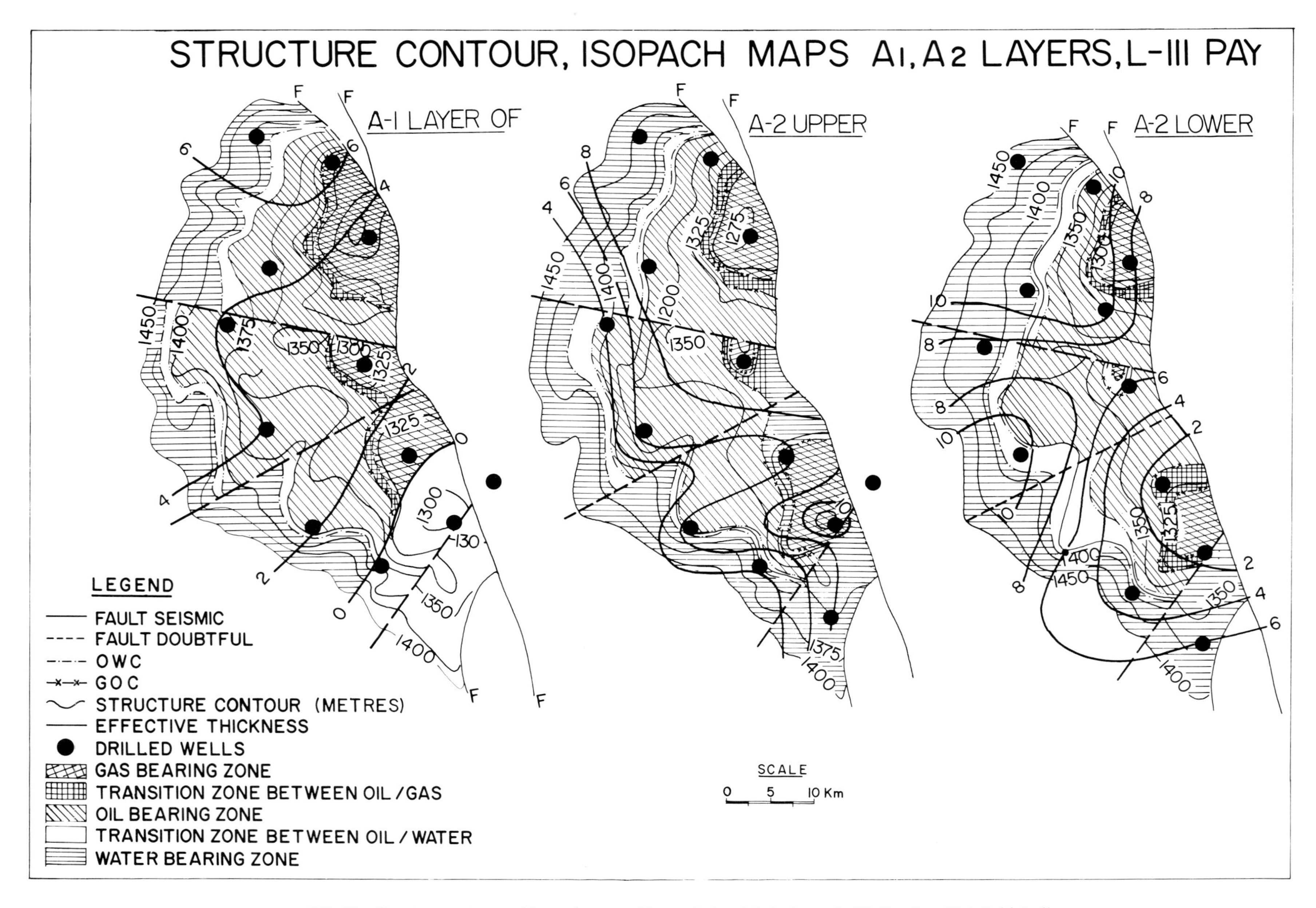

FIG. 11—Structure contour and isopach maps of layers A-1 and A-2 of zone L-III, Bombay High field, India.

Interval between reflectors H-2 and H-3: (Part of Aquitanian and Burdigalian)

With the fall in sea level, a carbonate bank developed from Ratnagiri to Kutch (Fig. 8a). The Shelf Margin basin started receiving sediments mainly from north (and west?). However, the Angria bank in the south (an extinct volcano?; personal communication, K. V. S. Murty, 1979) continued to be a site of carbonate sedimentation.

Interval between reflectors H-1 and H-2: (Part of Burdigalian and Langhian)

With the sea level remaining stationary or with a slight fall, an increase in the detrital supply from the east pushed the carbonate province farther west. Sands were deposited on the Bombay High, although it remained slightly above the wave base (Fig. 9A). These sands (S-1) are gas-bearing. The carbonate bank became narrow but remained continuous from Ratnagiri to Kutch. The Shelf Margin basin was receiving clastic sediments from the north and perhaps the west, but the Angria bank remained a carbonate mound.

Reflector H-1 and above: (Mostly Pliocene, Pleistocene and Holocene)

The destruction of the carbonate province during this time may be related to a major phase of the Himalayan uplift when both the Indus drainage system in the north and the Narmada drainage system in the east (more particularly the latter) discharged large volumes of clastic in highly turbid deposits (Fig. 9B). There appears to have been a rise in sea level which probably was associated with subsidence in the basin areas. The fall in sea level associated with the Pleistocene Glaciation did not result in carbonate deposition and the detrital supply continued unabated.

PETROLEUM GEOLOGY

The four oil- and gas-bearing zones identified in the Bombay High field are Miocene in age. These are designated from top: L-I, L-II (both limestones), S-I (sand), and L-III (limestone). Of these four, only two, L-II and L-III, are of economic significance, and L-III is the main pay zone.

The L-III unit has been tentatively divided into seven beds from top to bottom: A1, A2 upper, A2 lower, B, C, D, and E (Fig. 10). Production is from the A1, A2 upper, and A2 lower units.

Bombay High consists of three blocks: the northern, the central, and the southern, separated by east to west and east-northeast to west-southwest faults (Fig. 11). All the three blocks have different oil-water contacts for all the three productive zones of the L-III. A tilted oil-water contact is possible, but whether it is due to hydrodynamics or differences in the permeabilities is uncertain. In all the three layers and in all the three blocks, there is a gas cap. The gas-oil contact is the same in the middle and the southern blocks for all the three units, but A2 upper and A2 lower are separated by a saddle, thus increasing the oil column in the area (Fig. 11).

Lithology, Petrography, and Sedimentology of L-III

This zone is essentially wackestone with numerous corals, large forams, miliolids, algae, mollusks, and fossils. The unit contains shale seams, lignite lenses, traces of gypsum, small dolomitic patches, and ferrugenous encrustations. The limestone alternates with thin shales dividing L-III into a number of beds. The thin shales are dark, contain plant debris, and show high gamma radiation. Sometimes shales have associated calcareous nodules, some solution vugs, rare pinpoint porosity, and sparite-cemented fissures. A few shale beds are calcareous and do not contain plant debris (Rao, C.G., 1978a).

Petrographic and sedimentological studies of cores and cuttings show a cyclic nature to the L-III deposit and suggest four facies. In ascending order, the facies include the lagoonal, algal mound, the foraminiferal mound, and the coastal marsh facies. Each facies is identified based on matrix, biota, color, structure, clay material, and other megascopic and microscopic characteristics. The thin shales interspersed in the carbonate section represent a coastal marsh environment. These shales commonly are dark, rich in plant debris, pyritic, and contain some coal. These shales show a high gamma-ray count. The calcareous shales are different and represent a lagoonal environment. The algal mounds are believed to represent a deep-water facies compared to the foraminiferal mounds.

The interpretation of shallow-water, low-energy conditions in which L-III, as well as L-II and L-I, were deposited is supported by (1) the small size of montmorillonite, (2) conversion of aragonite to calcite but retaining the morphology of aragonite, (3) the presence of shallow-water benthonic foraminifers, (4) abundant plant debris, (5) ferrugenous encrustations, and (6) the presence of mud cracks.

Porosity and Permeability

Both primary and secondary porosities are present in the limestones but the primary intergranular porosity is the most important. The limestones are poorly compacted resulting in a wide network of solution vugs which interconnect with the intergranular pore space. This interconnection creates excellent permeability. Porosities and permeabilities improve upwards in the reservoir rocks of each cycle mentioned earlier. The shale at the top of each cycle suggests secondary porosities due to freshwater influx during periods of subaerial exposure. The humic acids resulting from the freshwater influx in the marsh shales percolated down to the carbonate section and caused leaching and solution vugs (Rao and Peters, 1978).

Interpretation of the Logs

In order to identify oil and gas bearing zones and porous zones in the exploratory and development wells, a variety of logs were run. These included the dual laterolog (DLL), micro-laterolog (MLL), induction log, compensated neutron log (CNL), borehole geometry tool (BGT), cement bond

Table 2. Reservoir L-III (Bombay Limestone)

Layer		: C
Interval	mts	: 1,380 to 1,386
Density	Kg/Liter	: 0.8286
Pour point	°C	: 30°
Wax content	%Wt.	: 10.7
Sulfur	%Wt.	: 0.19
Distillate up to 300°C	%Wt.	: 61.4
Light gasoline (ASTM Bureau of Mines)	%Wt.	: 12.3
Naphtha	%Wt.	: 22.74
Kerosine and gas oil	%Wt.	: 26.15
Lubricating oil	%Wt.	: 15.27
Residue	%Wt.	: 19.6
Correlation Index:		
Lower Fraction		: 30.3
Higher Fraction		: 32.0
Trace Metals:		
Nickel		: 3.0
Vanadium		: Nil

*Data from Ruby Kumar, Institute of Petroleum Exploration, Dehradun, India.

long—variable density log (CBL-VDL), and high resolution dip meter (HDT).

Research and Development Division and the Computer Services Division of the Institute of Petroleum Exploration, Dehradun, developed a computer program for interpreting the log data of the complex carbonate reservoirs.

Source Rocks

There are at least three views on the source rocks for the Bombay High field: (1) the pre-middle Miocene shales in the Surat depression, (2) the shales that alternate with the limestone reservoirs, and (3) the shales in the Shelf Margin basin which interfinger with the limestones of Bombay Platform, in addition to the previous two sources. The extensive post-middle Miocene shale that covers the entire basin is low in organic content, suggesting it would be poor source rock. Organic matter in all the shales mentioned under the three views have more than 0.5% organic content.

Caprock

The extensive post-middle Miocene shale acts as an excellent caprock. However, in the Ratnagiri block, the entire carbonate section is practically devoid of shales.

Development of the Field

In the three blocks present in the Bombay High, the northern block contains most of the oil reserves (54% of the total reserves), is comparatively less tectonically disturbed, and geologically is well understood. Thus, it was decided to develop the northern block first. Wells were to be drilled and completed separately for the beds A-1 and A-2 from one platform (Prasad et al, 1978), Initially, there were to be five platforms: A, B, C, D, and E. By May 1976, commercial production at the rate of 7,000 b/d began from Platform A and later increased to 15,000 b/d from the two wells on the same platform. A storage tanker *Jawahar Lal Nehru* was used to store the oil. However, a subsea pipeline was more economical, and a 30-in. oil and 26-in. gas pipeline was laid a distance of 220 km (137 mi) and completed in June 1978.

L-III reservoir is saturated and the saturation pressure is equal to the formation pressure which is 147.95 kg/sq cm (2,175 psi) at the gas-oil contact. The formation pressures are slightly higher than hydrostatic. The gas-oil ratio is 90 cu m/cu m (505 cu ft/bbl). The reservoir temperature is 115°C and the thermal gradient is 7°C/100 m. The thermal gradient in the adjoining Cambay basin is also high—6°C/100 m. All the calculations for the recoverable oil are based on the assumption of a depletion drive which is about 18.5%. There is no pressure decline and water cut. Production is uninterrupted.

Crude production by 1982-83 could be up to 12 million tons (91 million bbl), but it is likely to be kept down to 9 million tons (Offshore, Jan. 1979). A giant production platform BHN is to be installed at a cost of $60 million with all facilities, including accommodation for 74 persons. By 1980, plans are to install six more platforms in the northern block, and two more in the central block. In the northern block, each well would produce less oil but the total production would remain the same.

SUMMARY

1. Nearly half of India's oil reserves are in Bombay High.
2. This is the first oil discovery in a carbonate reservoir in India.
3. The reservoir rock is thin, shallow, and of wide extent.
4. Faults divide the field into three blocks—the southern, central, and northern blocks.
5. There are four hydrocarbon-bearing zones: L-I, L-II, S-I, and L-III; S-I is a gas-bearing sandstone and L-III is the main reservoir.
6. The main producting zone, L-III, consists of several beds separated by thin shales. These beds represent sedimentary cycles, each cycle represented by at least four different facies. The facies are in ascending order: the lagoonal, algal mound, foraminiferal mound, and coastal marsh.
7. Porosities are mainly intergranular but to some exent large solution vugs also contribute to the porosity. Interconnected solution vugs form a network of cavities with minor branching channels freely communicating with intergranular pores, and these give rise to high permeabilities. Porosities are the result of a low degree of compaction and freshwater leaching during intervals of emergence.
8. The post-middle Miocene shale acts as a regional caprock in the Bombay High basin.
9. The main source rock is believed to be the shale section of the Surat depression. This shale section is middle Miocene and older. The organic matter in these shales is more than 0.5%.

10. The reservoir is saturated and the formation pressures are above hydrostatic. The geothermal gradient is higher than normal.

11. The Bombay Offshore basin is characterized by the presence of an elongated carbonate bank separating a depression (Surat depression) in the northeast and Shelf Margin basin. The carbonate bank is believed to represent a low-energy environment. The thick clay facies of the Surat depression developed in a pro-deltaic regime. The carbonate bank grades into clay facies of the Shelf Margin basin. There is little paleontologic control in this basin but one well drilled indicated a pelagic facies. Seismic section indicates probable turbidite facies.

12. The carbonate bank was intermittently cut by an east to west trending submarine canyon probably related to faulting. This canyon might have carried clastic deposits from the east into the Shelf Margin basin. But even where the carbonate bank was continuous the Shelf Margin basin received clastic sediments. These could be from the west (micro-continent?) first, and later from the Indus cone on the north.

13. Bombay Platform appears to be a structural unit which experienced a westerly strike-slip movement along the Narmada and Ratnagiri faults since the Paleocene.

REFERENCES

Auden, J. B., 1949, A geological discussion of the Satpura hypothesis and Garo Rajmahal gap: India Natl. Inst. Sci. Proc., v. 15, p. 315-340.

Berggren, W. A., 1978, Terminal Miocene events: lecture delivered in the Inst. Pet. Expln., Dehra Dun, India.

Bhimasankaram, V. L. S. and P. C. Pal, 1973, Paleomagnetism and tectonism of the Narmada—Son lineament: ICSU Seminar on Geodynamics of the Himalayan Region, NGRI (Hyderabad), p. 195-196.

Chandra, P. K., and L. R. Chowdhary, 1969, Stratigraphy of Cambay basin, India: ONGC Bull., v. 6, no. 2, p. 37-50.

Chowbey, V. D., 1971, Narmada—Son lineament: Nature, v. 232, p. 39-40.

Crawford, A. R., 1978, Narmada—Son lineament of India traced into Madagascar: Geol. Soc. India Jour., v. 19, no. 4, p. 144-153.

Dighe, B. Y., 1978, ONGC Probing deep water possibilities: World Oil, v. 38, p. 7, p. 192.

Evans, P., 1964, The tectonic framework of Assam: Geol. Soc. India Jour., v. 5, p. 80-96.

Harbison, R. N., and B. G. Bassinger, 1970, Seismic reflection and magnetic study off Bombay, India: Geophysics, v. 35, no. 4, p. 603-612.

——— 1973, Marine geophysical studies off Western India: Jour. Geophys. Research, v. 78, no. 2.

ISSC Report 7, 1972, An international guide to stratigraphic classification, terminology and usage. Introduction and summary: Lethia, v. 5, no. 3, p. 283-323.

Kohli, G., and V. R. Rao, 1965, Status of offshore exploration in India: Proc. 3rd Symp. Dev. Pet. Resources, Asian Far East, ECAFE, Tokyo.

Kumar, R., and S. N. Bhattacharya, 1977, Geochemistry of Bombay High crude: Int. Cong. on Organic Geochem., Moscow.

Madan Modan, V. Narayanan, and P. Kumar, 1978, Paleogene and early Neogene biostratigraphy of Bombay offshore region: VII Colloquium in Micropaleontology and Stratigraphy, Madras, India (in press).

Mathur, S. P., 1977, Gravity anomalies and crustal structure, A review: Australian Soc. Expl. Geophys. Bull., v. 8, no. 4.

ONGC revises plans with program to boost production schedules: Offshore, Jan. 1979, p. 86-87.

Pal, P. C., and G. Srinivas, 1976, The Narmada—Son lineament: Jour. Geophys. Research, v. 14, no. 2.

Parrott, M., 1978, India: International report (description of the Bombay High oil field), Ocean Ind. (Houston), 13/4, p. 247-48.

Pitman, W. C., III, 1978, Relationship between eustacy and stratigraphic sequences of passive margins: Geol. Soc. America Bull., v. 89, p. 1389-1403.

Prasad, N. B., G. Ramaswamy, and K. L. N. Rao, 1978, Development continues in India's Bombay High field: World Oil, v. 186, no. 5, p. 81.

Qureshy, M. N., 1970, Relation of gravity to elevation, geology and tectonics in India: Proc. Second Symp. on Upper Mantle Project (Hyderabad), p. 1-24.

Raju, A.T.R., 1976, New global tectonics and petroleum exploration on continental margins—a review: July-Dec., Geonews, p. 24-40.

——— 1979, Basin analysis and petroleum exploration with some examples from Indian sedimentary basins: Geol. Soc. India Jour., v. 20, p. 42-60.

Ramaswamy, G., and K. L. N. Rao, 1978, Geology of continental shelf of the west coast of India in Facts and principles of world oil occurence: Canadian Assoc. Petroleum Geologists, Calgary, Canada.

Rao, C. G., 1977, Biogenic protodolomite from the limestone reservoir rocks of Bombay High oil field: Current Sci., v. 46, no. 3, p. 78-79.

——— 1978a, Carbonate clay mineral relation and the dolomite problem in the carbonate rocks of Bombay High oil field: 10th Int. Cong. Sedimentology (Jerusalem), Proc.

——— 1978b, Diagenesis of larger foraminifera (*Lepidocyclina* sp.) and its effect on porosity in the limestone reservoir rocks of Bombay Offshore area: Geol. Soc. India Jour., v. 13, no. 4, p. 165-168.

——— in press, An unusual solution deposition calcite fabric in an echinoidal skeletal fragment from Bombay offshore carbonate rocks: Bull. ONGC, India.

——— in press, Late diagenetic flaser structures and reservoir in the L-III carbonate reservoir rocks of Bombay High oil field: Geol. Soc. India Jour.

——— in press, Quantitative microfacies analysis and reservoir implications in the carbonate reservoir rocks of Bombay High oil field: Jour. Math. Geol.

——— and J. Peters, 1978, Application of Kotmogorov Semirnov 'D' statistic for comparison of two porosity samples distribution from Bombay offshore: Geol. Soc. India Jour.

Rao, R. P. and D. C. Srivastava, 1979, Seismic facies of parts of Bombay offshore basin between Bombay and Ratnagiri; AEG Symposium held at Roorkee, India (in press).

——— P. C. Dhar and D. K. Pande, 1978, Rock stratigraphy of Bombay offshore basin and paleogeographic and tectonic considerations, a discussion: 7th Colloquium Micropaleontology, Stratigraphy (Madras, India), in press.

Sastri, V. V., and L. L. Bhandari, 1973, Petroleum prospect of the Continental Shelves of India: Inst. Pet. Expln. ONGC, India (unpublished).

Sengupta, S. N., 1967, Structure of the Gulf of Cambay: Symposium on Upper Mantle Project (Hyderabad), Proc., p. 334-341.

——— 1969, Oil exploration in offshore areas in India: ECAFE symposium (Canberra), Proc.

——— 1972, Status of oil exploration in the offshore areas adjoining India: 4th Symposium on Development of Asia and Far East, Proc., p. 112-114.

Sharma, S. K., et al, in press, Laboratory studies for improved log evaluation in the Bombay High field: Symposium exploration geophysics in India (Calcutta), Proc.

Sheriff, R. E., 1976, Inferring stratigraphy from seismic data: AAPG Bull., v. 60, p. 528-542.

Shepard, F. P., 1973, Submarine geology: New York, Harper and

Row, New York, 3rd edition.
Sudhakar, R., and D. N. Basu, 1973, A reappraisal of the paleogene statigraphy of southern Cambay basin: ONGC Bull., v. 10, no. 1-2, p. 55-76.
Vail, P. R., et al, 1977, Seismic stratigraphy and global changes of sea level, *in* C. E. Payton, ed., Seismic stratigraphy—applications to hydrocarbon exploration: AAPG Memoir 26, p. 49-212.
Wadia, D. N., 1966, Geology of India: MacMilan, India.
West, W. D., 1962, The line of Narmada and Son valleys: Current Sci. v. 3, p. 143-144.
Whitmarsh, R. B., D. E. Waser, and D. A. Rose, 1974, Initial reports of the deep sea drilling project: Washington D.C., Govt. Printing Office, v. 23, p. 35-115.

Geology of a Stratigraphic Giant: Messla Oil Field, Libya[1]

By Harold J. Clifford, Roger Grund, and Hassan Musrati[2]

Abstract The Messla oil field is the most recent addition to the list of 20 giant fields discovered within the prolific Sirte basin of Libya. The field, discovered in 1971, lies in the southeastern part of the Sirte basin approximately 40 km north of the supergiant Sarir oil field. Although in an early stage of development the field is estimated to contain approximately 3 billion bbl of original oil-in-place.

The field is a seismically defined stratigraphic accumulation located on the east dipping flank of a broad Precambrian basement high. The reservoir is in the Lower Cretaceous fluvial Sarir Sandstone, which wedges out westward on the basement and is truncated by a basin-wide unconformity at the base of the capping Upper Cretaceous marine shales (considered to be the source rocks). The reservoir consists of two sandstones separated by a continuous shale bed. Porosity values average 17% and the permeability 500 md.

The oil column averages approximately 90 ft (27 m) and is productive from an average depth of 8,800 ft (2,682 m) over 200 sq km. Early 1978 production is in excess of 100,000 b/d of 40° API oil with a cumulative production of 45 million bbl.

INTRODUCTION

The Sirte basin of Libya has an onshore area of approximately 400,000 sq km (155,000 sq mi) and contains 20 giant oil and gas fields. The first giant field, the Nasser (Zelten) field, was discovered less than 20 years ago. The Messla field is Libya's latest addition to a worldwide list of 225 giant oil fields which have individual ultimate recoverable hydrocarbon volumes in excess of 500 million bbl. The field lies in the southeastern part of the Sirte basin (Fig. 1), approximately 500 km southeast of Benghazi, 115 km southeast of the giant Gialo field, and 40 km north of the supergiant Sarir oil field.

The Arabian Gulf Exploration Company (AGECO) is currently developing Messla with a multi-rig drilling program, and although in an early stage of development, the field is estimated to contain in excess of 3 billion bbl of oil-in-place. Messla differs from nearly all other Sirte basin giants in that it is neither a closed structure nor a carbonate "build up," but rather a clastic stratigraphic termination.

Three structurally controlled northwesterly aligned trends of accumulations are evident on Figure 1. The westernmost trend, corresponding to the Dahra-Hofra platform and extending from Bel Hedan on the southeast to Mabruk on the northwest, contains in excess of 3 billion bbl of ultimate recoverable hydrocarbons and includes the giant oil fields of Samah, Beda, Raguba, Dahra-Hofra, and Bahi. The central and more northerly aligned trend extending from Defa to Hateiba has an estimated ultimate production of 8 billion bbl. Giants on this trend, known as the Zelten platform, include Defa-Waha, Nasser, and Hateiba (gas). A less clearly defined eastern trend extends from Sarir on the south to Amal on the north. Giants on this trend include Sarir, Messla, Gialo, Buattifel, Intisar A, Intisar D, Nafoora-Augila, and Amal. Known recoverable oil in this

[1]Manuscript received April 4, 1979; accepted for publication, July 23, 1979.

[2]Exploration staff, Arabian Gulf Exploration Co., Benghazi, Libya.

The writers are indebted to the management of AGECO for permission to publish this work, and to the following exploration colleagues who contributed to the successful exploitation of Messla: Hamed Elhori, Jim Haebig, Salah Hashim, Muzaffar Husain, Carl Singleton, and James White.

Article Identification Number:
0065-731X/80/MO30-0025/$03.00/0.

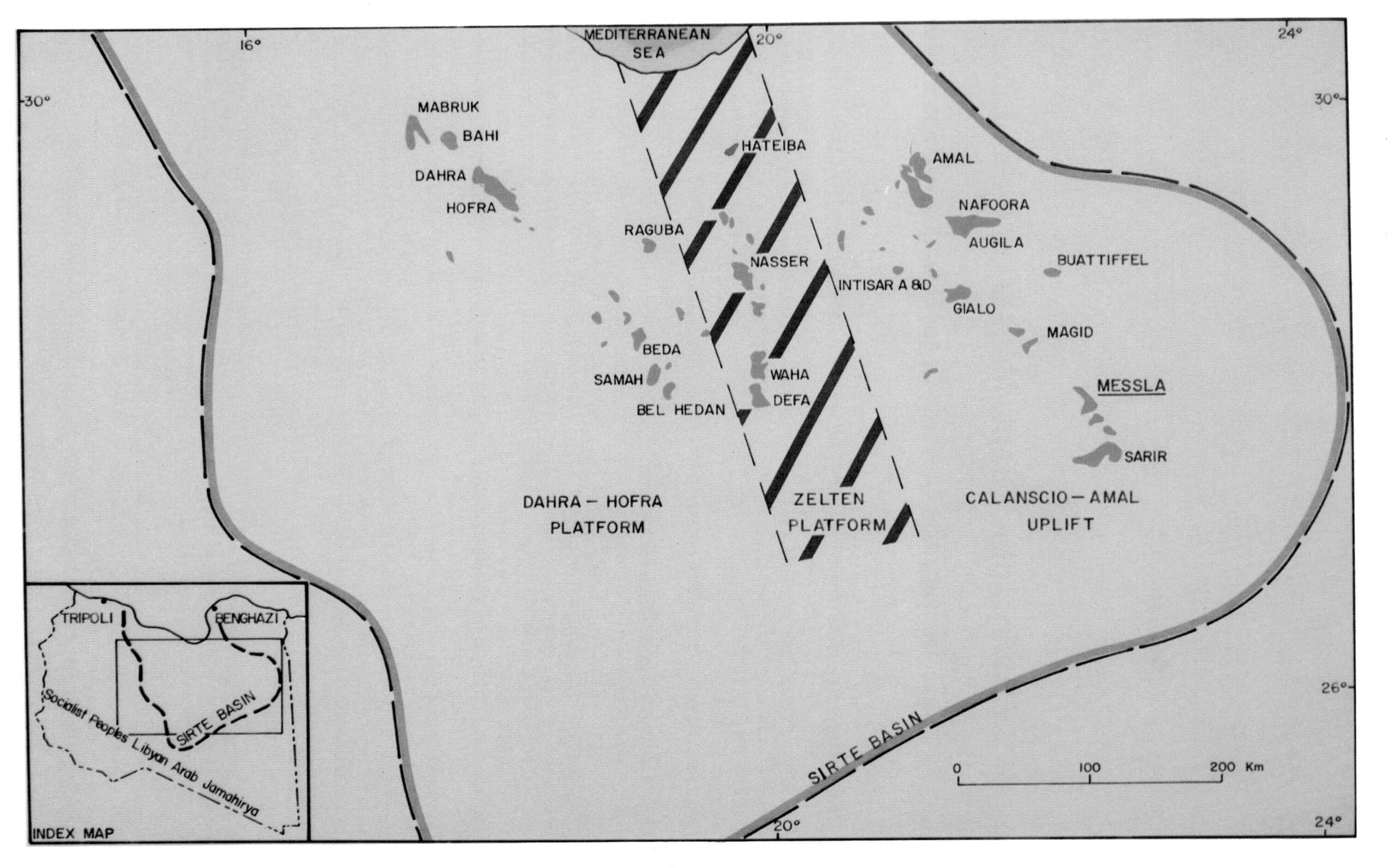

FIG. 1—Index map of Sirte basin oil fields.

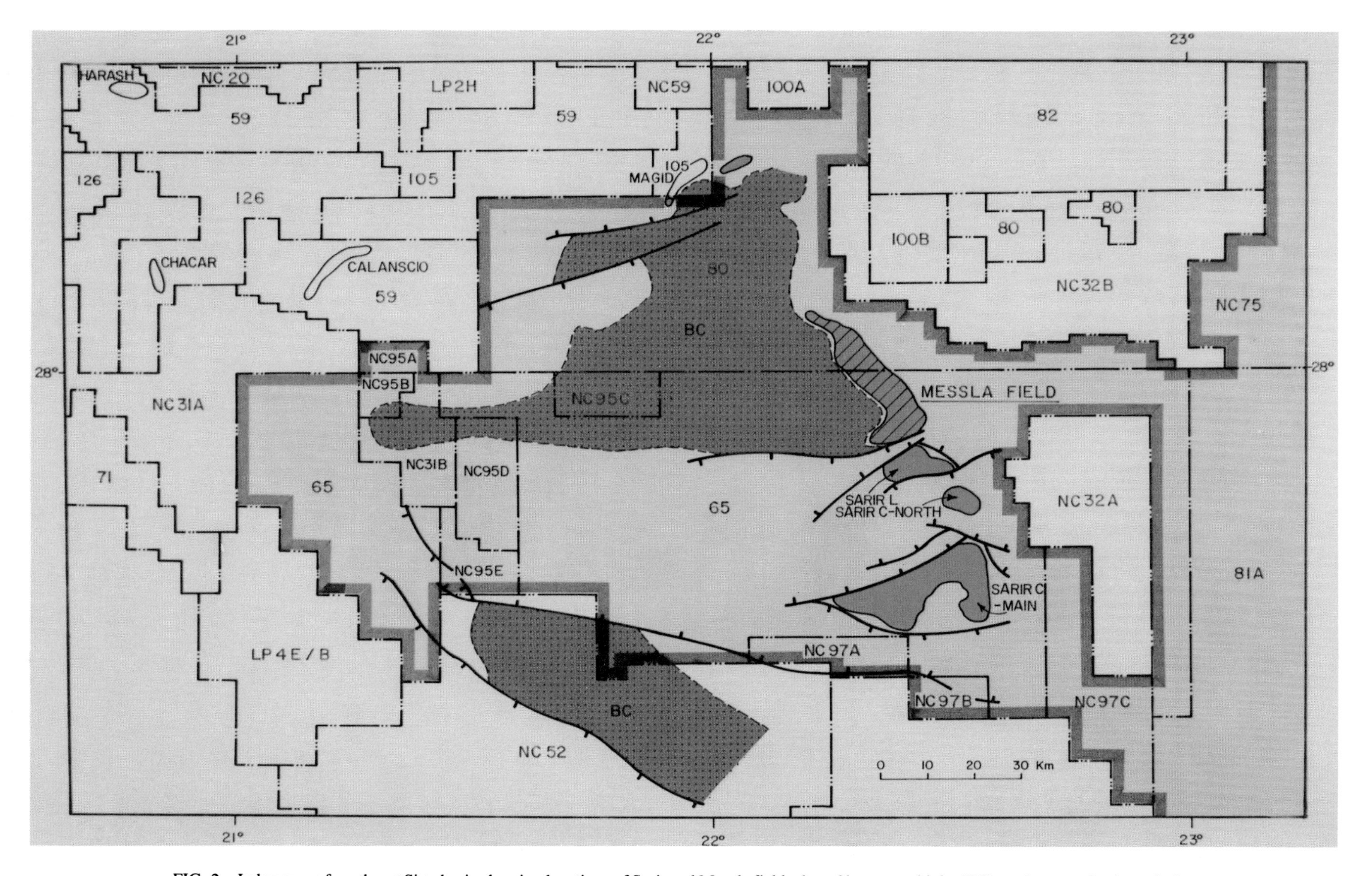

FIG. 2—Index map of southeast Sirte basin showing locations of Sarir and Messla fields, broad basement highs (BC), and concession boundaries.

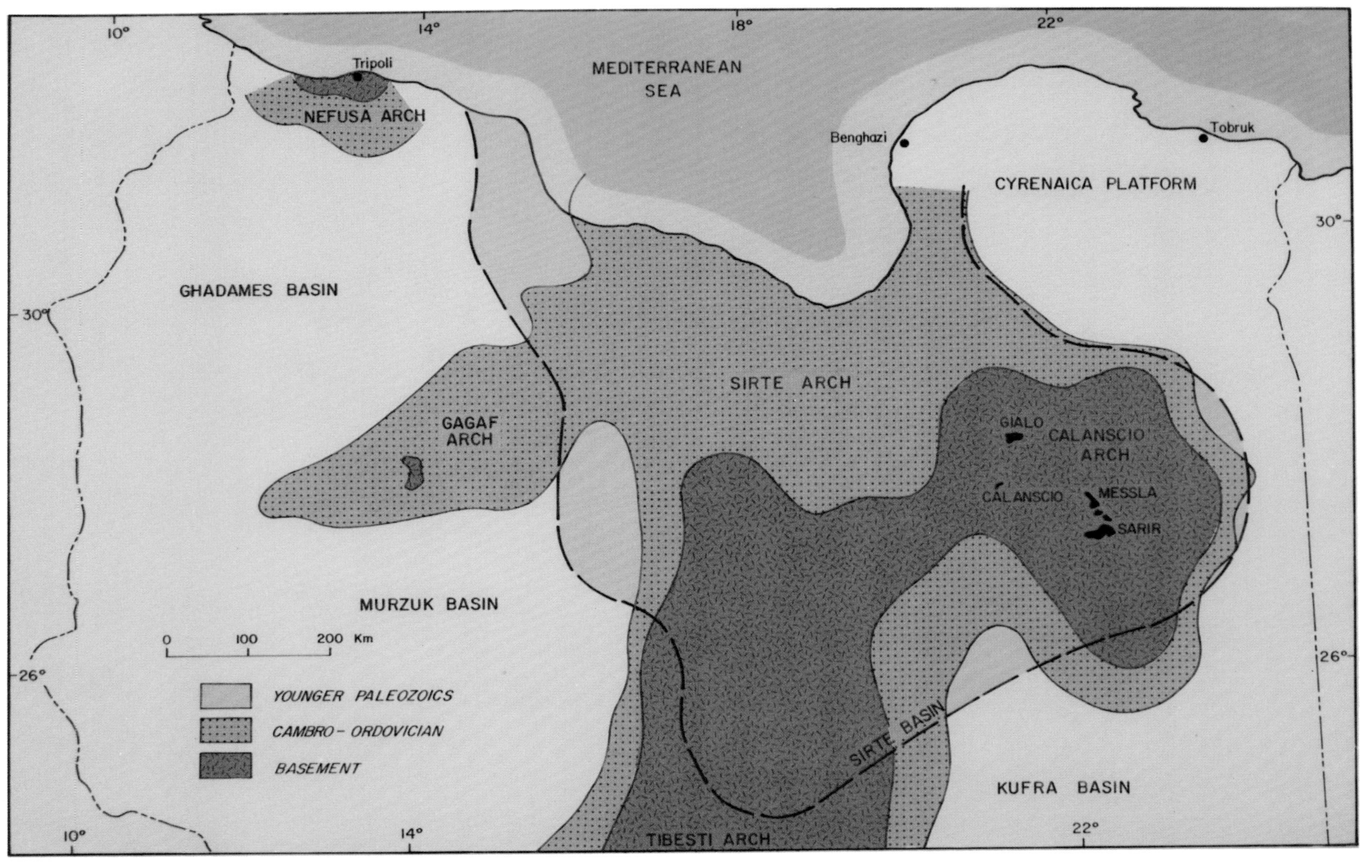

FIG. 3—Pre-Mesozoic subcrop in Libya.

trend is approximately 17 billion bbl. (These stated hydrocarbon volumes are an estimate of the distribution to these three producing trends from unofficial industry published cumulative and reserve volumes as of the beginning of 1977). In the western part of the Sirte basin, the bulk of the oil is in Upper Cretaceous and Tertiary carbonates, whereas in the eastern part the bulk of the oil is in pre-Upper Cretaceous sandstones with very substantial production also from younger reservoirs.

GEOLOGIC HISTORY - SIRTE BASIN

The onshore part of the Sirte basin encompasses much of the northeast quandrant of Libya (Fig. 3). The Sirte basin is a late Mesozoic-Tertiary cratonic rift, resulting from Suez-type crustal extension of older eroded basement and Paleozoic uplift. In excess of 25,000 ft (7,620 m) of late Mesozoic and Tertiary, predominantly marine, sediments have accumulated in the deeper segments of the basin. To the south is the Paleozoic Kufra basin, and to the west lie the Paleozoic basins of Murzuk and Ghadames. Thick Paleozoic sediments also are present to the northeast, but during the Mesozoic and Tertiary the area remained relatively stable and is called the Cyrenaican platform.

Although Paleozoic deposition was generally widespread across the northern part of the African continent, only a meager record of that era remains within the Sirte basin due to erosion associated with the Hercynian orogeny. Five major crustal upwarps, the Nefusa and Gargaf arches in western Libya, the Tibesti arch in southern Libya, and the Sirte and Calanscio arches of central and eastern Libya, were strong positive features by the end of the Hercynian orogeny.

During the early Lower Cretaceous the Calanscio arch, which consisted primarily of exposed metamorphic and igneous basement, was subjected to north-south extension resulting in its collapse and the initiation of the Sirte basin as an intracratonic rift depression. The collapse produced two major cross-basin ridges; an east to west trend through Calanscio and Messla, and a northwesterly trend through Gialo, Messla, and Sarir.

Initial sedimentation consisted of poorly sorted nonmarine sands and silts which partly infilled the irregular basement topography. This basal deposition was succeeded by a thick accumulation of relatively clean fluvial sand, the Sarir Sandstone, which is the major reservoir in the southeastern Sirte basin. The main source area is believed to have been a southern distal foreland with contributions coming from circumbasinal Paleozoic rocks. North of Messla, these massive sands were further overlain by a thick variegated tidal flat/brackish lagoonal shale, indicating partial access to the Tethys sea at that time. Volcanics also accumulated locally. At the end of the Lower Cretaceous a regional uplift resulted in an infilling of the depression with regressive fluviodeltaic and possibly eolian sandstones. This early Alpine-related (Austrian) period of uplift and block faulting resulted in substantial erosion, particularly of the high area just to the west of Messla. These Lower Cretaceous sediments, which form the Sarir Group, are over 5,500 ft (1,676 m) thick to the north of Messla, whereas to the southwest of this area, in the central part of Concession 65 (Fig. 2), over 3,000 ft (914 m) of Lower Cretaceous sediments are preserved.

During the Cenomanian the Sirte basin developed its present structural form. At that time, due to major northeast-southwest crustal extension, there was a collapse of the Sirte and Tibesti arches, and a renewal of the Calanscio downwarp allowing a major marine transgression of the Tethys sea into the eroded Lower Cretaceous surface in the southeast Sirte basin. The Tethys sea remained until the middle Miocene. The major Sirte basin faults were either initiated or strongly rejuvenated during this collapse, and a series of generally northwest oriented major horsts and grabens developed. The horsts have generally remained rigid and stable, forming loci for hydrocarbon accumulations, whereas the grabens have progressively subsided, providing restricted depocenters for marine shale in excess of 15,000 ft (4,572 m) thick and ideal for hydrocarbon generation. The largest of these grabens, the Sirte trough, extended southeasterly through western Concession 65 (Fig. 2), and south of Sarir C-Main field. A smaller trough paralleled the eastern side of Messla and Sarir fields. During subsequent structural movements in the late Upper Cretaceous and Tertiary, which were more pronounced in the Sirte trough, the high block containing the Calanscio, Magid, Messla, and Sarir areas was progressively tilted northwestward with respect to the top of the Lower Cretaceous Sarir Group, thus making the Sarir and Messla areas major focal points for hydrocarbon migration.

The initial Cenomanian sedimentation in the Sirte basin occurred in restricted grabens as evaporites which were subsequently buried during the late Upper Cretaceous by a thick shale and carbonate sequence. The transgressing Tethys sea did not cover the Messla area until the late Turonian. Paleocene deposition in the basin was similar to that of the Upper Cretaceous, providing mainly shallow-marine carbonates and local reefs on the highs, and deeper water shales and carbonates in the structurally low areas. A carbonate basin developed north of Messla and pinnacle reefs grew for the first time in the Sirte trough north of Calanscio. The Eocene was primarily a time of carbonate deposition, but near the end of this epoch there began a period of tectonic instability that peaked during the Oligocene and early Miocene. Regional uplift of the basin and a clastic infill of the Sirte trough occurred, and by the late middle Miocene the present onshore Sirte basin had emerged.

DISCOVERY HISTORY

Prior to the discovery of the Messla oil field British Petroleum, in partnership with Nelson Bunker-Hunt, discovered the multibillion barrel Sarir C-Main field in 1961. Production was extended northward with smaller but significant discoveries in 1963 of C-North, and in 1964 of the Sarir L field (Fig. 2). In 1971, HH1-65, the 38th wildcat in the then 8,200 sq km area of Concession 65, was drilled approximately 8 km north of the "L" field. HH1 was located on a seismic-mapped southeasterly plunging nose. The location was based on the geologic concept that the Lower Cretaceous Sarir Sandstone, productive in the "L" field, wedged out toward the west and northwest. In this

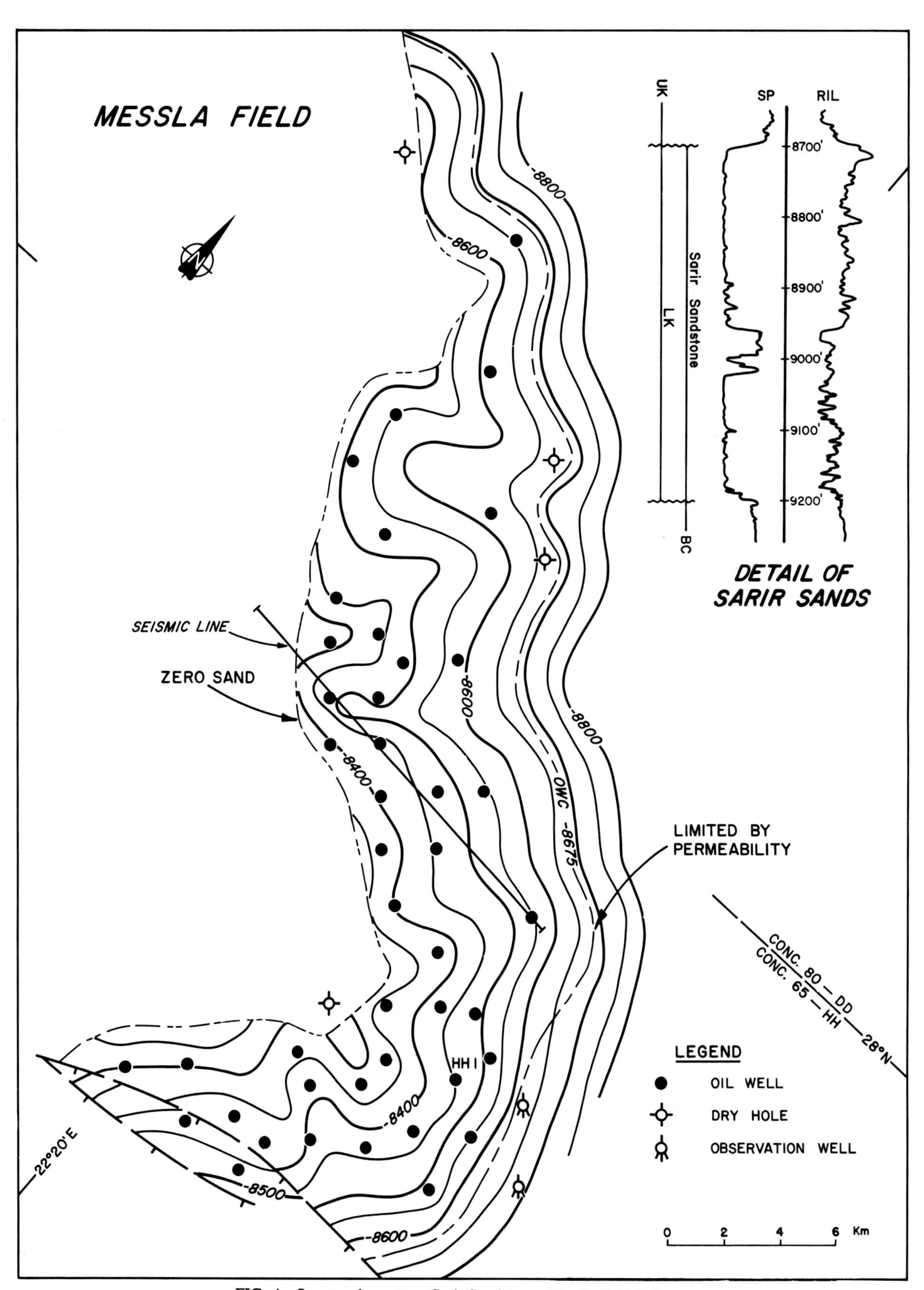

FIG. 4—Structural map, top Sarir Sandstone, Messla field, Libya.

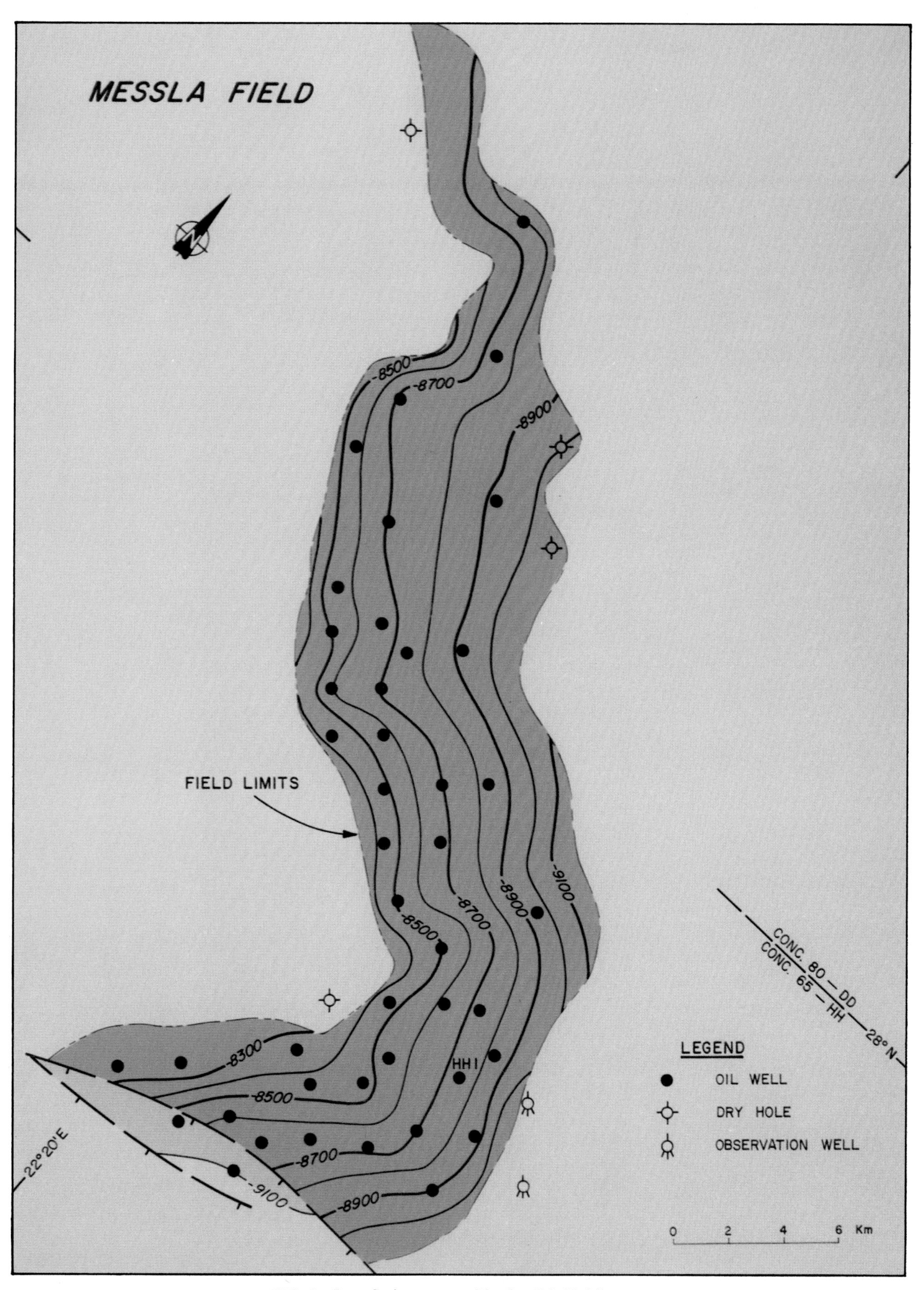

FIG. 5—Intra-Sarir structure, Messla oil field, Libya.

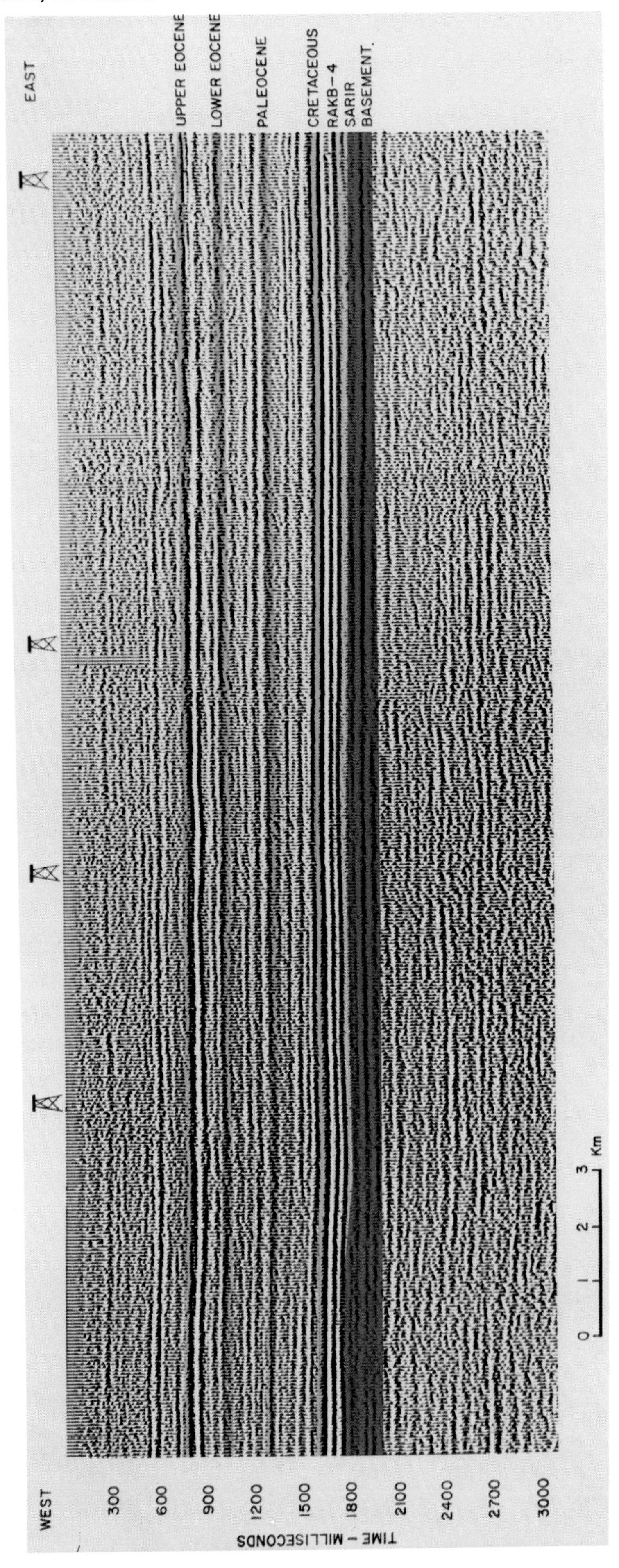

FIG. 6—Seismic line showing wedge edge of Sarir Sandstone and gentle eastbound dip of flank of basement high.

direction a prominent Sarir bald basement high had been encountered by earlier wildcats. The discovery well penetrated 110 ft (33 m) of net oil bearing Sarir Sandstone through the interval 8,768 to 8,896 ft (2,672 to 2,712 m) and was potential-tested at a rate of 10,900 b/d of 38° API oil.

Subsequent delineation drilling showed the accumulation to be stratigraphically trapped by a westerly pinchout of the Sarir Sandstone onto the basement, in conjunction with a regional unconformity at the top of the Sarir Sandstone.

STRUCTURE

The Messla field lies along a broad northeast dipping structural flexure that houses a weak eastward plunging anticlinal nose near its southern extremity (Fig. 4). Over a major part of the field the structure is a homocline, with a rather constant strike of N 40° W and eastward dip of ½ to 1°. The gentle dipping homocline terminates at a regional syncline 15 km northeast of Messla.

The variations in strike across the field, as indicated by contours on the unconformity surface (Fig. 4), are tentatively considered to be eastward flowing drainage patterns rather than structural depressions, as these strike variations are not evident on an intra-Sarir Sandstone zone (Fig. 5). As much as 100 ft (30 m) of possible erosion has occurred in these narrow "channels." A "channeling" interpretation is not, however, supported by all the data, and until additional drilling in these features is available, other causes, particularly faulting, are believed possible.

Faulting has not been recognized within the Messla field, which is surprising due to its frequency in the Sarir fields. Only at the southern boundary of the field is there a defined zone of westerly trending cross faults. These faults, downthrown on the south side, interrupt the southern flank of the nose and parallel the shallow syncline which separates Messla from the anticlinal and fault-controlled "L" field.

GEOPHYSICS

Various geophysical surveys were carried out in the years prior to the drilling of HH1-65. Airborne magnetometer and gravity surveys showed an eastward plunging basement nose in the general area of the field. Reconnaissance singlefold reflection and basement refraction seismic refined the interpretation of the nose but it was not until the 1969 completion of a 600% analog reflection survey that a comprehensive deep structural picture emerged. The additional information obtained from the initial discovery and early delineation drilling was used in a 1974 reinterpretation of the existing seismic data, which yielded a detailed structure map of a deep time event associated with the top of the Upper Cretaceous Rakb-4 Formation. This map forms the basis for the structural interpretation of the Messla field, and only a slight modification in its detail resulted from a 208 km, 1200% digital reflection survey in 1975. Subsequent drilling has confirmed the map's reliability.

The field area is characterized by good reflection results (Fig. 6). The seismic events tie to lithologic units, have reliable character, and can be correlated over relatively long distances. The Rakb-4 time event closely overlies the unconformity surface of the objective Sarir Sandstone and to a large extent conforms to the configuration of that surface. Due to multiples present just beneath the mapped horizon, early attempts to produce a basement map were unsuccessful.

Until sufficient well-velocity surveys had been conducted and a reliable velocity gradient established, considerable problems also were encountered along the downdip edge of the Messla field. The gradient is rather severe and meandering for such a small area and tends to cut across the structure rather than parallel it.

Although it is at the limits of seismic resolution, an approximate sandstone pinchout line was produced, based on certain subtle criteria. As the sandstone section thins to the west, small interpretational features such as phase changes, compression of pulse width beneath the mapping horizon, careful scanning of the lower returns, and sharp dip changes of the mapped horizon were used for determining the pinchout. When the Sarir Sandstone and basement are contained within a single low frequency pulse-width, the problem is beyond the limits of seismic resolution.

In an effort to overcome this problem in the northern area of the field where sandstone, basement, and the oil-water contact are convergent, it was necessary to use techniques and parameters designed to break down the low frequency pulse into its higher frequency components. In 1976 the vibroseismic system was successfully used. The parameters included an .002 ms sample rate, a short station interval, a nonoverlapping geophone spread, a directional off-end shooting, and a sweep frequency up to 80 hertz. Extreme care was used in choosing processing parameters, especially deconvolution operators and filters. Due to a combination of the above field and processing parameters it proved possible to break up the strong multiples below the Rakb-4 horizon and follow the true basement event. Limited experiments with cross-correlation before summation indicate that further resolution improvements can be anticipated.

Stratigraphy

The stratigraphic sequence in the Messla field is divided into two major depositional periods by a pronounced intra-Cretaceous unconformity (Fig. 7). The preunconformity sediments of the Lower Cretaceous age are predominantly nonmarine, and average 500 ft (152 m) in thickness. The post-unconformity Upper Cretaceous and Tertiary form a thick succession of deep-marine to near-shore deposits, averaging 8,800 ft (2,682 m) in thickness. Significant lateral facies variations within the field occur only in the Paleocene, which exhibits a strong north to south facies change.

Holocene to Oligocene

This interval consists of a 600-ft (193 m) thick upper sand, a medial 1,000-ft (305 m) thick dolomitic limestone and clay sequence, and a lower 1,000-ft (305 m) thick bed of unconsolidated sandstone. The sandstones are freshwater bearing and hold great promise for additional utilization in the vast agricul-

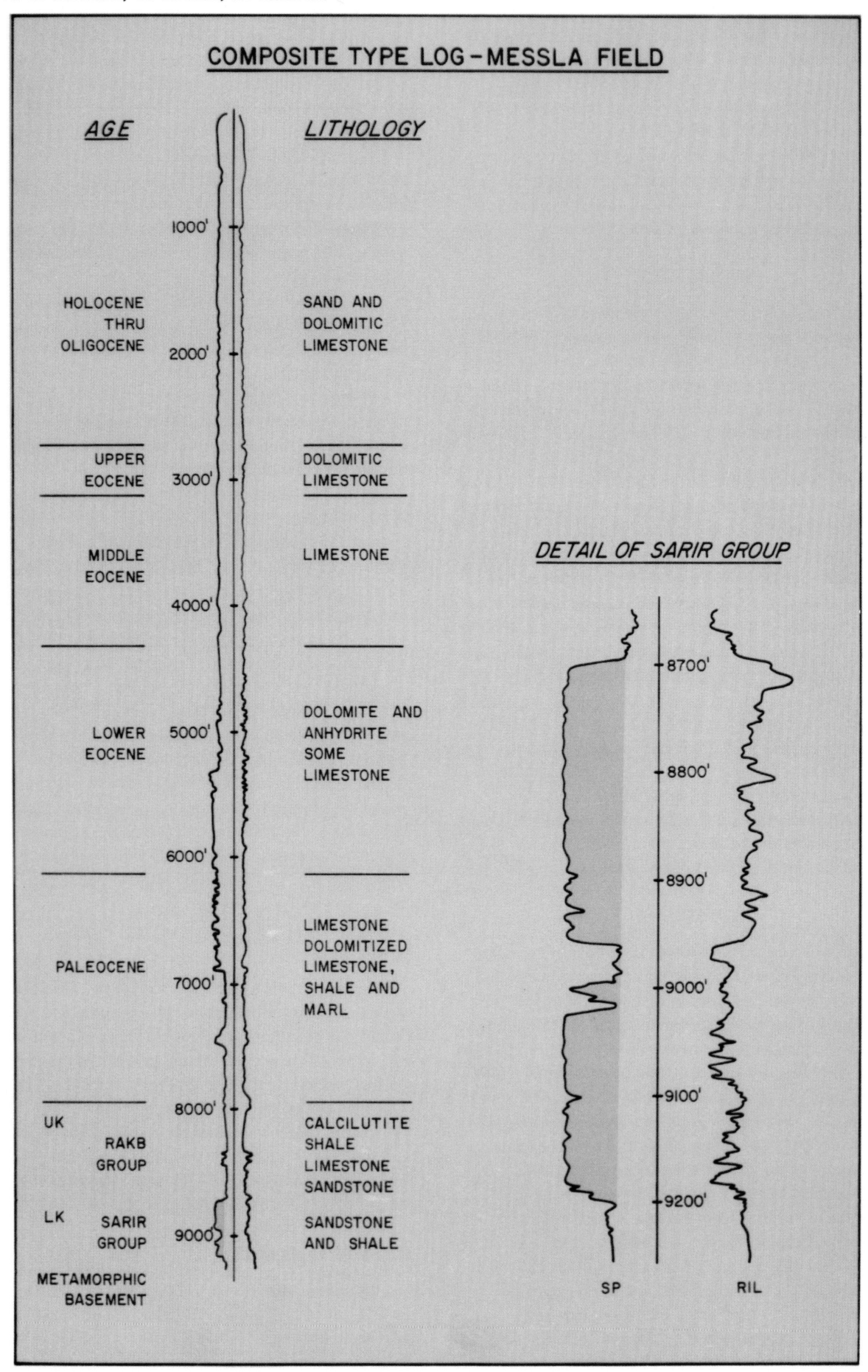

FIG. 7—Stratigraphic column at Messla field, Libya.

tural projects in the area. These rocks are poorly fossiliferous but, apart from the Holocene surface sands, are laterally equivalent northwards to marine beds of middle Miocene to Oligocene age.

Eocene

The upper Eocene consists of 250 to 300 ft (76 to 91 m) of sandy dolomitic limestone, sandstone, and claystone, representing a transitional facies with the over- and underlying units. A shallow-water, open-marine, nummulitic limestone with a thickness of 1,400 ft (427 m) is representative of the middle Eocene. The lower Eocene is composed of approximately 1,900 ft (579 m) of evaporitic interbeds of dolomite and anhydrite with occasional limestones, while the basal 100 ft (30 m) is a fossiliferous limestone which is depositionally related to the underlying Paleocene. The evaporite interval is poorly fossiliferous but it appears that its deposition was in a restricted, shallow evaporitic, marine environment.

Paleocene

The Paleocene sedimentation, generally about 1,900 ft (579 m) thick, is highly varied in facies. There is a facies change from shale to carbonate in a northerly direction which causes much of the horizontal velocity gradient difficulty encountered by geophysicists. Below a rather uniform 200-ft (61 m) thick nummulitic limestone, green-gray shale, and marl sequence, occurs the carbonate consisting mainly of fossiliferous limestone that is variably argillaceous and dolomitized. The dolomitized zones occasionally result in circulation losses during drilling, especially if the mud weight is not closely controlled. The carbonate becomes cleaner, and thickens from 600 to 1,350 ft (183 to 411 m) in a northerly direction. It is underlain by, and is a partial facies equivalent to, a gray calcareous shale which thickens in the opposite direction. It is 380 ft (116 m) thick in the northern part of the field and 1,100 ft (335 m) thick in the southern part. Both the carbonate and shale are richly fossiliferous. The shale was deposited in a middle to outer neritic open-marine environment, while the carbonate deposition was deposited mainly in a shallow open-marine environment.

Upper Cretaceous

The Upper Cretaceous, which is divided into five formations, shows a west to east down-flank thickening from about 750 to 870 ft (229 to 265 m) across the 6 km width of the field. The uppermost formation is about 50 ft (15 m) thick and is an argillaceous calcilutite in the southern part of the field, becoming chalky to the north. The remaining formations of the Upper Cretaceous form what AGECO calls the Rakb Group. The youngest formation of the group, the Rakb-1 Formation, is a highly fossiliferous, 330-ft (101 m) thick interval of middle to outer neritic dark-gray shale of Maastrichtian age. The Rakb-2 is generally argillaceous limestone, about 250 ft (76 m) thick, grading downward into shale. The limestone porosity varies, but generally is low and often provides hydrocarbon shows. Its fossils are Campanian in age from a partly restricted shallow-marine environment. The Rakb-3 is about 150 ft (46 m) thick and consists of dark-gray shale with some interbedded glauconitic sandstones. Its depositional environment is similar to Rakb-2 but more restricted. Rakb-3 is probably Coniacian to Santonian in age. The Rakb-4 is a transgressive formation exhibiting marked thickness variation, from 20 ft (6 m) in the south-central part of the field to 110 ft (34 m) at the northern limit, as well as an east and west thickening off the long axis of the field. It is a variegated sandy and anhydritic shale and its contact with the underlying Sarir Sandstone is usually marked by a 5 to 10-ft (1.5 to 3 m) thick anhydrite bed resulting from secondary anhydrite diagenesis. The Rakb-4, for which regional correlations suggest a late Turonian age, was deposited in an evaporitic, tidal dominated environment. The Upper Cretaceous "transgressive" sandstones which lie along the east flank of the Sarir "C" fields, and are described by Sanford (1970), do not occur at Messla. The whole of the Upper Cretaceous and the lower Paleocene forms a thick cap rock and a potential source rock in much of the southeast Sirte basin.

Lower Cretaceous

An unconformity separates the Upper Cretaceous shales from the underlying reservoir sandstones of the Lower Cretaceous. Regionally, the Lower Cretaceous rocks are assigned to three formations belonging to the Sarir Group, which is considered to be at least partly a Nubian equivalent (Pomeyrol, 1968). These formations comprise a lower massive sandstone (the Sarir Sandstone), a medial variegated shale, and an upper sandstone. Only the lower two of these formations are present at Messla.

The Sarir Sandstone was deposited primarily in a continental fluviatile environment with much oxidation and as a result is nearly barren of fossils. Meager flora suggests a mainly Neocomian age, with the basal part possibly ranging into the late Jurassic. The overlying formation was deposited in a tidal flat-lagoonal-coastal-sabkha environment, and consists mainly of variegated anhydritic shale and siltstone. However, in the region of Messla and Sarir the lower part of this formation changes facies to clayey sandstone, but these tight sandstones are confined to the lower flanks of Messla, below the oil-water contact. The age of this formation is Barremian.

Within the field area, the Sarir Sandstone is divided into three lithologic units; a massive upper sandstone which is designated the "Main Pay," a medial "Red Shale," and a restrictively distributed basal sandstone, the "Lower Pay" (Fig. 8). The "Main Pay" sandstone, which is in excess of 500 ft (152 m) in thickness along the southeastern edge of the field, consists of a coalescing series of dominantly braided but occasionally meandering stream-deposited sandstones, which appear and act as a single massive reservoir in both a lateral and vertical sense. Lithologically, these sandstones are white to light gray, fine to medium to occasionally coarse grained, cross-bedded quartz sandstones with minor amounts of feldspars and interstitial clays. Below the "Main Pay" is a field-wide unit of varied thickness which, because of its

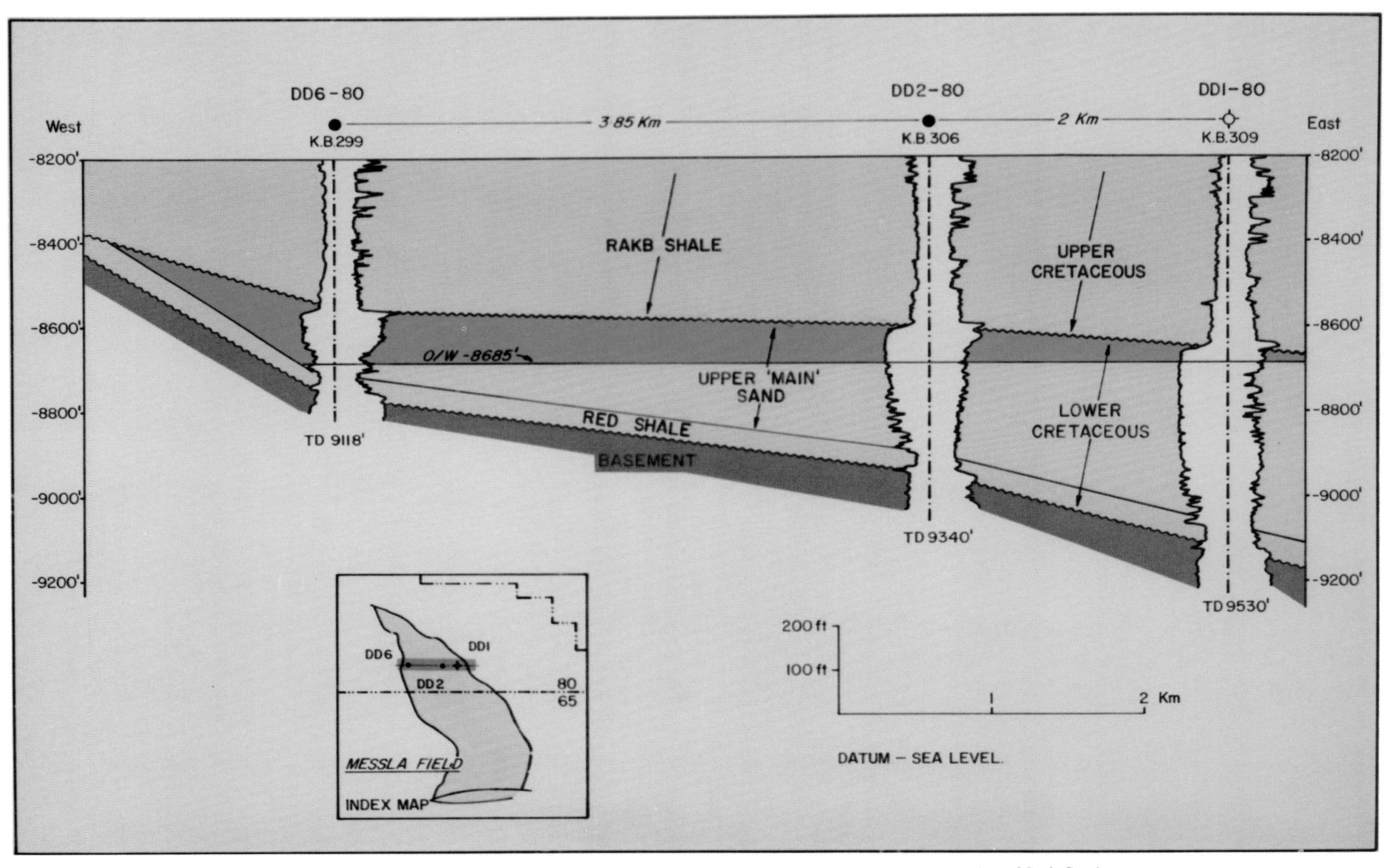

FIG. 8—West to east cross section of line A-A', Messla oil field, showing wedgeout and truncation of Sarir Sandstone.

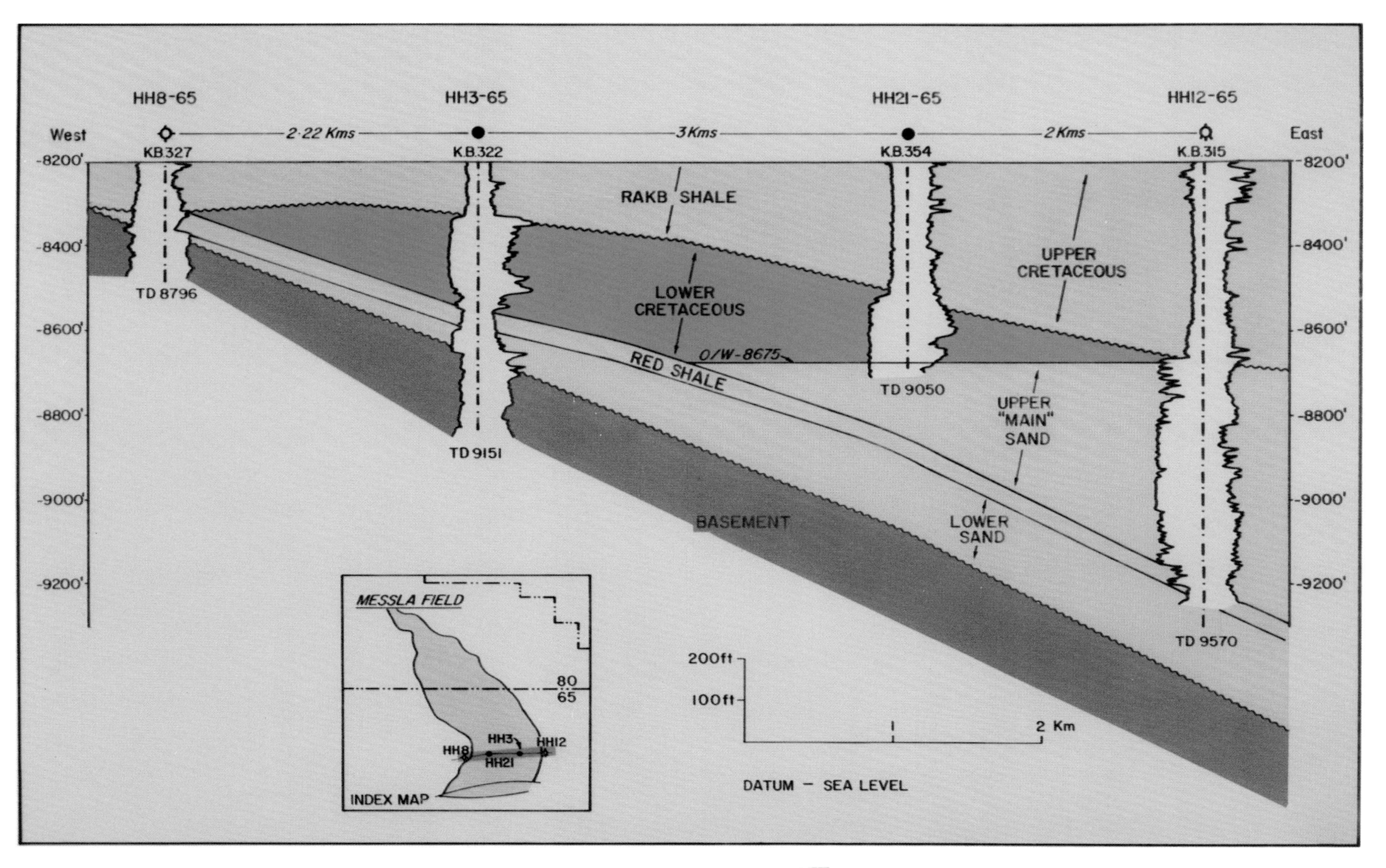

FIG. 9—West to east cross section of line B-B′, showing wedgeout and truncation of Sarir Sandstone.

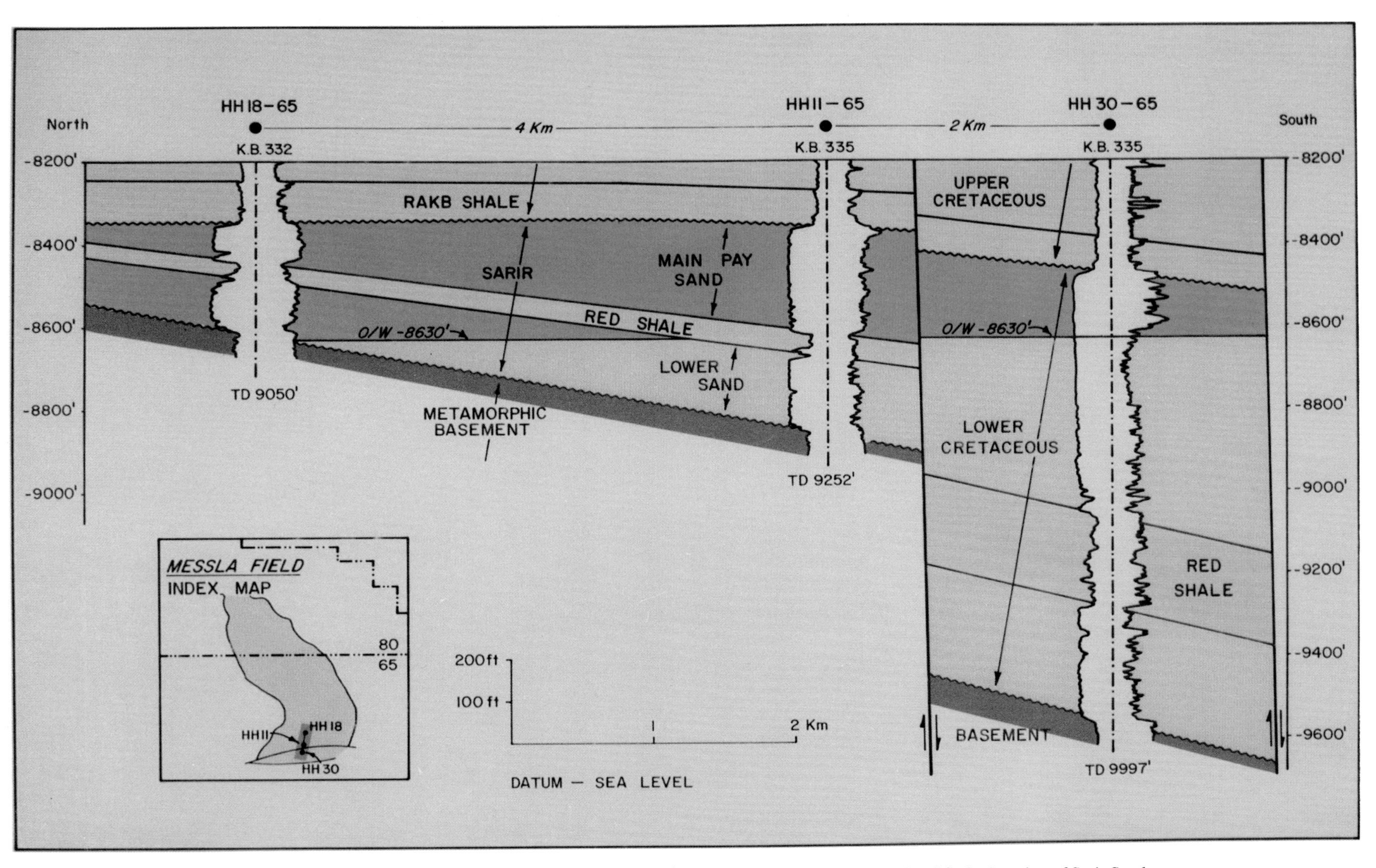

FIG. 10—North to south cross section of line C-C′, Messla oil field, showing a partly truncated and faulted section of Sarir Sandstone.

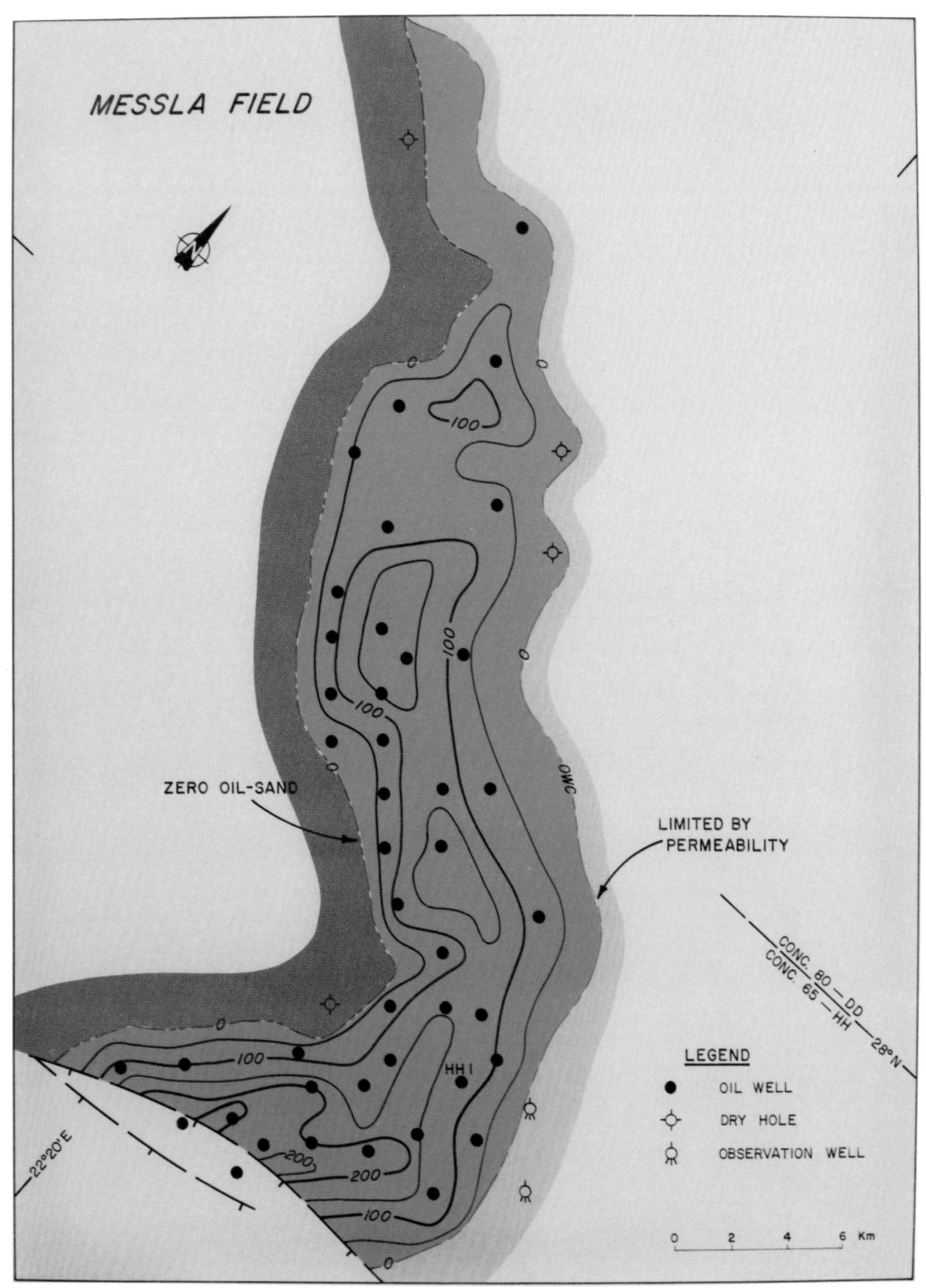

FIG. 11—Net pay isopach map, Sarir Sandstone, Messla field, Libya.

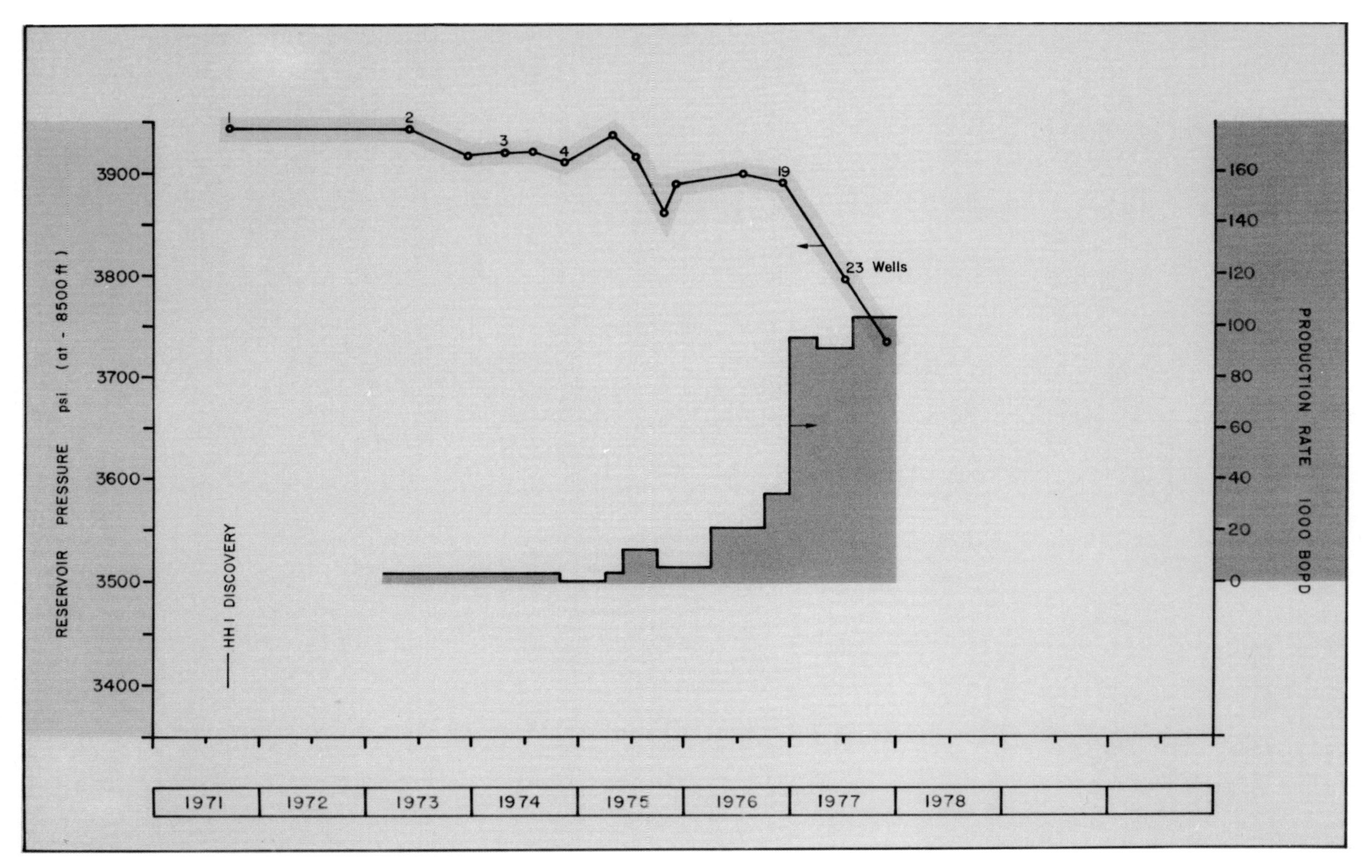

FIG. 12—Production graph of oil production in Messla field, Libya, beginning with the discovery of HH-1 in 1971.

diagnostic color, is named the "Red Shale" and is considered to be of a playa lake-inland flood basin origin. The "Lower Pay" sandstone is present in the southern part of the field and is lithologically similar to the "Main Pay." It is generally thin, except on the downthrown side of the south boundary faults where it attains a thickness of over 200 ft (61 m).

Basement

The basement consists of metamorphic schists and phyllites that have been intruded by granite. K/Ar radiometric datings of the granite indicate a late Precambrian to Early Cambrian age.

THE INTRA-CRETACEOUS UNCONFORMITY

The Intra-Cretaceous unconformity at the base of the marine Upper Cretaceous sequence is present throughout the Sirte basin, and is of profound significance to the entrapment of billions of barrels of petroleum in the underlying Sarir Group. The surface of the unconformity reflects cessation of the Lower Cretaceous nonmarine conditions, and the influx of a Late Cretaceous sea in Cenomanian-Turonian time. Fortunately, within the southeast Sirte basin, there was very little active winnowing of the reworked sandstones derived from the underlying Sarir Group by the transgressing sea, and consequently the overlying Upper Cretaceous shale is a regionally effective caprock and source rock.

In the Messla field, the unconformity is moreover the critical determinant of the accumulation in terms of the existence and spatial location of the trap. Although fields discovered many years prior to Messla desmonstrated the vast economic significance of the unconformity, the discovery of this giant trap justifies an accelerated exploration program along the truncated margins of the Sarir Sandstone. Sixteen oil fields in the southeast Sirte basin, including the giant fields of Sarir, Messla, and Buattifel, are productive from the Sarir Group sandstones directly below the Intra-Cretaceous unconformity. The known original oil-in-place of these sandstones is estimated to exceed 16 billion bbl.

Considering the extensive geologic and economic import of this unconformity, its insignificant structural discordance is surprising. In the Messla field, the dip discordance across this surface is approximately 1° and little strike divergence is recognized.

SOURCE ROCKS

Preliminary results of source rock potential and maturation parameters indicate that the crude oils in the Sarir and Messla fields were probably derived from both Upper and Lower Cretaceous shales occurring in the sub-basinal troughs of the southeast Sirte basin. The most probable areas of generation are considered to be located at least 80 km to the north and northwest of Messla. The Cretaceous shales contain organic carbon contents of up to 4% and are indicated to be mature at depths below about 11,000 ft (3,353 m). It is postulated that the Upper Cretaceous is the primary source for the low wax (8%) crude of the Messla field, whereas the high wax (15-25%) crude at the Sarir C-Main field may be derived from a mixture of both Upper and Lower Cretaceous sources.

TRAPPING MECHANISMS

The Messla field is defined on the west by the updip erosional limit of the Sarir Sandstone reservoir and to the east by the oil-water contact which occurs at an average depth of 8,675 ft (2,644 m; Fig. 7). The average width between these boundaries is 6 km and the long (northwest to southeast) dimension is 32 km (a total of 200 sq km).

The northwestern limit of the accumulation occurs where the structural top of the Sarir Sandstone, at the level of the oil-water contact (8,690 ft or 2,649 m subsea), intercepts the zero isopach of the sandstone (Fig. 9). Along the southern margin a zone of down-to-the-south faults has affected the oil accumulation, but the limit of the field in this direction has not been adequately established (Fig. 10). In this zone of faulting two semi-parallel branches are recognized and appear to be related to pre-Upper Cretaceous tectonics and have juxtapositioned 1,100 ft (335 m) of Sarir Sandstone in the south block against 500 ft (152 m) of sandstone in the main part of the field. The displacement at the basement level is 575 ft (175 m) while at the top of the "Red Shale" the displacement is 325 ft (99 m). Certain reservoir data, however, indicate an aquifer, if not an oil reservoir connection between the Messla and "L" fields.

The major trapping mechanism in the field is the westerly progression to total erosion of the Sarir Sandstone across a broad north to south structural flexure (Fig. 11). Approximately 600 ft (183 m) of the reservoir section is truncated by the Intra-Cretaceous unconformity across the 6 km average width of the field. The maximum vertical structural relief, which occurs in the south part of the field, is in excess of 400 ft (122 m).

RESERVOIR PARAMETERS AND PRODUCTION

The "Lower Pay" sandstone is relatively unimportant in terms of reserves, encompassing only 10 sq km of productive area in the southern part of the field. The "Main Pay" extends over the entire 200 sq km of closure and is thickest along the southern margin of the field where there is more than 250 ft (76 m) of net pay, in contrast to the average net pay of 90 ft (27 m) (Fig. 11). The oil-water contact in this sandstone intersects the "Red Shale" throughout the length of the field providing a substantial updip area of water-free sandstone of prime production potential. The "Main Pay" sandstone has an average porosity of 17%; the relatively low value is due to the generally fine- to medium-grained nature of the sand and to its interstitial clay content. Connate water saturations are about 35% with salinities of the formation water approximately 185,000 ppm sodium chloride. The reservoir volume of Messla is in the order of 5 million acre-feet, and the oil-in-place is estimated to be 3 billion stock tank bbl of 39.6° API, low-sulfur, paraffinic crude with a gas/oil ratio (GOR) of 380 cu ft/bbl. Given the expected active water drive, a recovery efficiency of 50% is anticipated,

Table 1. Messla field reservoir parameters (February 1978).

Rock:		Environments:	
Producing section (Sarir Sandstone)	(Lower Cretaceous)	Reservoir depth	8,250 ft (min.) 8,695 ft (max.)
Porosity	17% (avg.)	Oil-water contact depth	8,675 ft (avg.)
Permeability	500 md (avg.)	Vertical oil column	400 ft (max.)
Connate water saturation	35% (avg.)	Net pay	90 ft (avg.) 250 ft (max.)
Residual	32% (avg.)	Productive area	400 sq km
		Reservoir volume	5 million acre-feet
Fluids:		Datum	8,550 ft
Crude type	paraffin	Reservoir temperature	111°C at datum
Crude gravity	39.6° API	Original reservoir pressure	3,940 psig at datum
Wax content	8%		
Pour point	9°C	Recovery:	
Crude viscosity (centipoise)	0.79	Drive mechanism	natural water
Formation volume factor (ratio)	1.23	Recovery efficiency	50%
Shrinkage	18.7%	Recovery factor	330 bbl per acre feet
Formation water salinity (ppm, NaCl)	185,000 ± 10,000	Original oil-in-place	1.5 billion bbl
Bubble point pressure	1,290 psig	Ultimate recovery	20 to 25 years
Gas/oil ratio (ratio scf/bbl)	380	Projected economic life	2 sq km/well
		Well spacing	3 billion bbl

providing for an estimated ultimate recovery of 1.5 billion stock tank bbl of oil.

The field development is planned on a spacing pattern of 2 sq km/well, providing a total of about 100 wells. This spacing is planned to be less dense near the oil-water contact and somewhat denser updip in the water-free part of the field. The average well is forecast to produce 15 million bbl. At the beginning of 1978, with 40 producing wells, the cumulative oil production from Messla slightly exceeded 45 million bbl or approximately 1.5% of the original oil-in-place. The field production rate was just over 100,000 b/d (Fig 12), with an average water cut of 0.7% and a GOR of 380. Individual well flow rates range from 3,000 to 6,000 b/d. Projected production rates over a 20 to 25 year life are planned to average around 220,000 b/d and peak at about 300,000 b/d. Production today is being pipelined to Tobruk after being gathered at the Sarir C-Main oil field.

In September, 1971, the original static reservoir pressure was 3,940 psi at the datum of 8,550 ft (2,600 m) subsea (Fig. 12). Some evidence exists of a preproduction pressure drawdown in Messla due, perhaps, to the production in the adjacent "L" field. There appears to be an aquifer connection between these fields but it has yet to be demonstrated that a continuous oil column exists. In late 1977, the average pressure in the "Main Pay" had declined by 206 psi, to 3,734 psi at datum. The reservoir has closely followed the production changes and averages about 0.5 psi per 100,000 bbl of produced oil. The production history suggests that an active water-drive is sustaining the reservoir pressure.

CONCLUSION

Messla is a new giant oil field in which Lower Cretaceous Sarir Sandstone wedges out against the east and south flanks of a large basement high. The fluvially deposited sandstone is truncated by Upper Cretaceous marine shale which is both the seal and the source for the oil. These features are similar to those of the giant East Texas field. Both the structural and stratigraphic features responsible for entrapment are discernible from reflection seismic data.

The "Main Pay" sand, which is as much as 500 ft (152 m) thick, represents in large measure, a multistoried braided stream deposit with good lateral and vertical continuity of the reservoir.

Average porosity of the "Main Pay" sand is 17%, the average water saturation is 35%, and the sodium chloride content is approximately 185,000 ppm. Of the 3 billion bbl of 40° API low-sulfur crude in place, 1.5 billion bbl are expected to be recovered ultimately under anticipated water-drive conditions.

SELECTED REFERENCES

Barr, F. T., and A. A. Weeger, 1972, Stratigraphic nomenclature of the Sirte Basin, Libya: Petrol. Expl. Soc. Libya, 179 p.

Gillespie, I., and R. M. Sanford, 1967, The geology of the Sarir oil field, Sirte Basin, Libya: Elsevier Pub. Co., 7th World Petroleum Cong. Proc., Mexico, v. 2, p. 181-193.

King, R. E., ed., 1972, Stratigraphic oil and gas fields—classification, exploration methods, and case histories: AAPG Memoir 16, 687 p.

Klitzsch, E., et al, 1979, Major subdivisions and depositional environments of Nubia Strata, southwestern Egypt: AAPG Bull., v. 63, p. 967-974.

Parsons, M. G., A. M. Zagaar, and J. J. Curry, 1979, Hydrocarbon occurrences in the Sirte basin, Libya: *in* Miall, A., ed., Facts and principles of world petroleum occurence: Can. Soc. Pet. Geol. Memoir 6, 1000 p.

Pomeyrol, R., 1968, Nubian sandstone: AAPG Bull. v. 52, no. 4, p. 589-600.

Sanford, R. M., 1970, Sarir oil field, Libya—desert surprise, *in* M. T. Halbouty, ed., Geology of giant petroleum fields, AAPG Memoir 14, p. 449-476.

Malossa Field: A Deep Discovery in the Po Valley, Italy[1]

By G. Errico, G. Groppi, S. Savelli and G. C. Vaghi[2]

Abstract The Malossa condensate gas field was discovered in the Po basin of Italy, 15 mi (24 km) east of Milan. The basin is located in a depression between the Alps and Apennines, and opens eastward to the Adriatic Sea. Malossa production comes from a section ranging from Triassic to Quaternary in age; deepest production to date has been the Malossa 2 well, which reached a depth of 21,230 ft (6,471 m) and bottomed in the Middle Triassic Carnian dolomites of the Ensino Formation. The main pay zones of the Malossa field are in the "Dolomia Principale" of Noric age, and are part of the Liassic Zandobbio Formation.

Geological history is described along with the structural setting of the Po basin. Stratigraphy is described, citing Middle Triassic through Quaternary and mentioning fossil occurence as a criteria for identification.

The Malossa structure was detected with seismic survey, and velocity information is given. Reservoir characteristics are given for the Zandobbio and Haupt Dolomite formations, and reserves are discussed.

GEOGRAPHIC AND GEOLOGIC SETTING

The Malossa condensate gas field was discovered in 1973 and is located in northern Italy, 15 mi (24 km) east of Milan. This is the first Mesozoic pay zone discovered in the Po basin at any great depth (5,500 m; 18,000 ft; Fig. 1).

The field is in a plain area near the Adda River. The area is a continuation of Milan, and therefore densely populated, containing many towns, farms, and industrial sites. The producing section crops out about 16 mi (26 km) to the north in the folded belt of the Bergamo Alps (Fig. 2).

The Po basin, in which the Malossa field is located, is a depression between the Alps and the Apennines and opens eastward toward the Adriatic Sea. In this part of the southern Alps, which forms the northern boundary of the Po basin, the outcropping sedimentary section extends from Permian to Cretaceous in age. The typical section penetrated by the Malossa field wells extends from the Triassic to Quaternary. The deepest well, Malossa 2, bottomed at 6,471 m (21,230 ft) in the Middle Triassic Carnian dolomites of the "Esino" Formation.

The main pay zones are the "Dolomia Principale" of the Noric age and in the Liassic Zandobbio Formation.

GEOLOGIC HISTORY

In describing the Malossa structure and other structures in this part of the Po basin, it is appropriate to examine the geological evolution in the "Padana" area during the Alpine and Appennine tectonic phase.

There is little knowledge of the geologic history of the area before the Hercynian orogeny because only limited data are available. After the Hercynian orogeny, continental sandstones were deposited during the Permian. By Early Triassic time, however, marine conditions prevailed.

During the Triassic, the study area appears to have been a continental margin made up of Hercynian basement overlain

[1]Manuscript received, April 4, 1979; accepted for publication, July 23, 1979.

[2]AGIP, Milano, Italy.

Article Idenfication Number:
0065-731X/80/MO30-0026/$03.00/0.

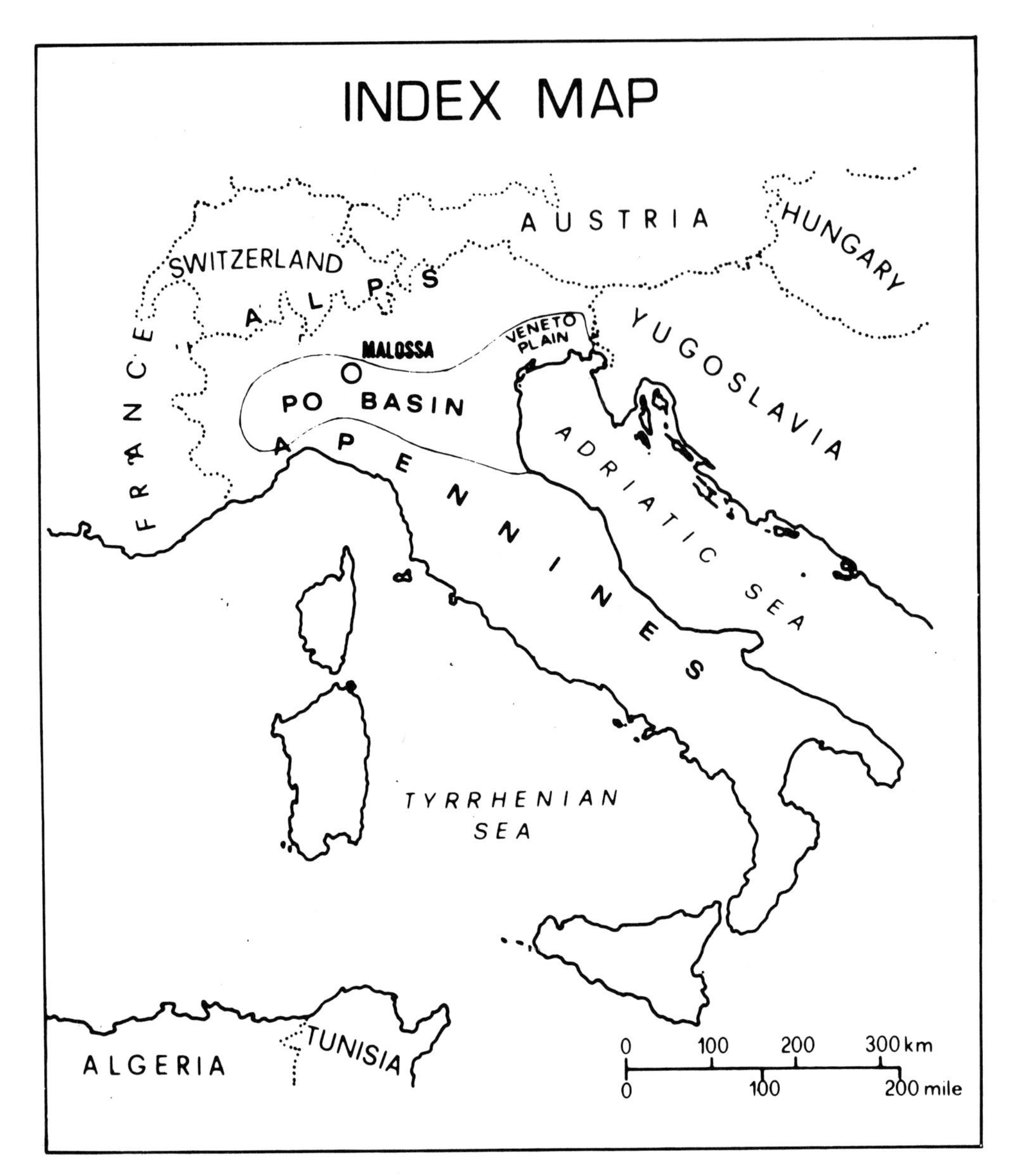

FIG. 1—Location map of Malossa field, Po basin, Italy.

by continental Permian clastics on which were deposited Triassic marine carbonate rocks.

This epicontinental environment existed over a large area of the Mediterranean region in the Late Triassic, as the "Dolomia Principale" Formation is widespread, and has a consistent lithologic character. In the Late Triassic and Early Jurassic, an extensional tectonic phase occurred and the adjacent continental plate began rifting and formed several blocks. The associated internal basins were areas with euxinic environments in which black argillitic-calcareous rocks of varied thickness were deposited.

With the continuation of the rifting, the sediments became more marine, with thickness variations due to continued active fault movement. The rifting phase reached the spreading phase in the Late Jurassic and an oceanic area divided the Paleoeuropean continent on the north from the southern plate which has been named Austro-Alpine, Apulia, or Insubric by different authors. During the Early Cretaceous, deep marine conditions were prevalent over the entire Po basin.

At the end of Early Cretaceous time the first Alpine compressional tectonic phase began. The first evidence of flysch sedimentation (Bergamo flysch) is present near the study area.

The compressional tectonism continued during Eocene time. During the Oligocene and early Miocene a dominant clastic depositional phase took place in the basin. The northern part took approximately the same shape as the present Po basin today except it was open on the south and southeast. A tremendous thickness of conglomerate sandstones and marls with interbedded sandstones were deposited southward into the basin.

Another strong compressional tectonic phase took place during the middle Miocene, and the Mesozoic and Oligo-Miocene section was folded and faulted. The subsurface configuration of the Po basin from Milano to the Garda Lake consists of numerous overthrusted folds which tend to arc from west to south to north. This tectonic phase was also responsible for the first development of the Apennine Mountains. During the late

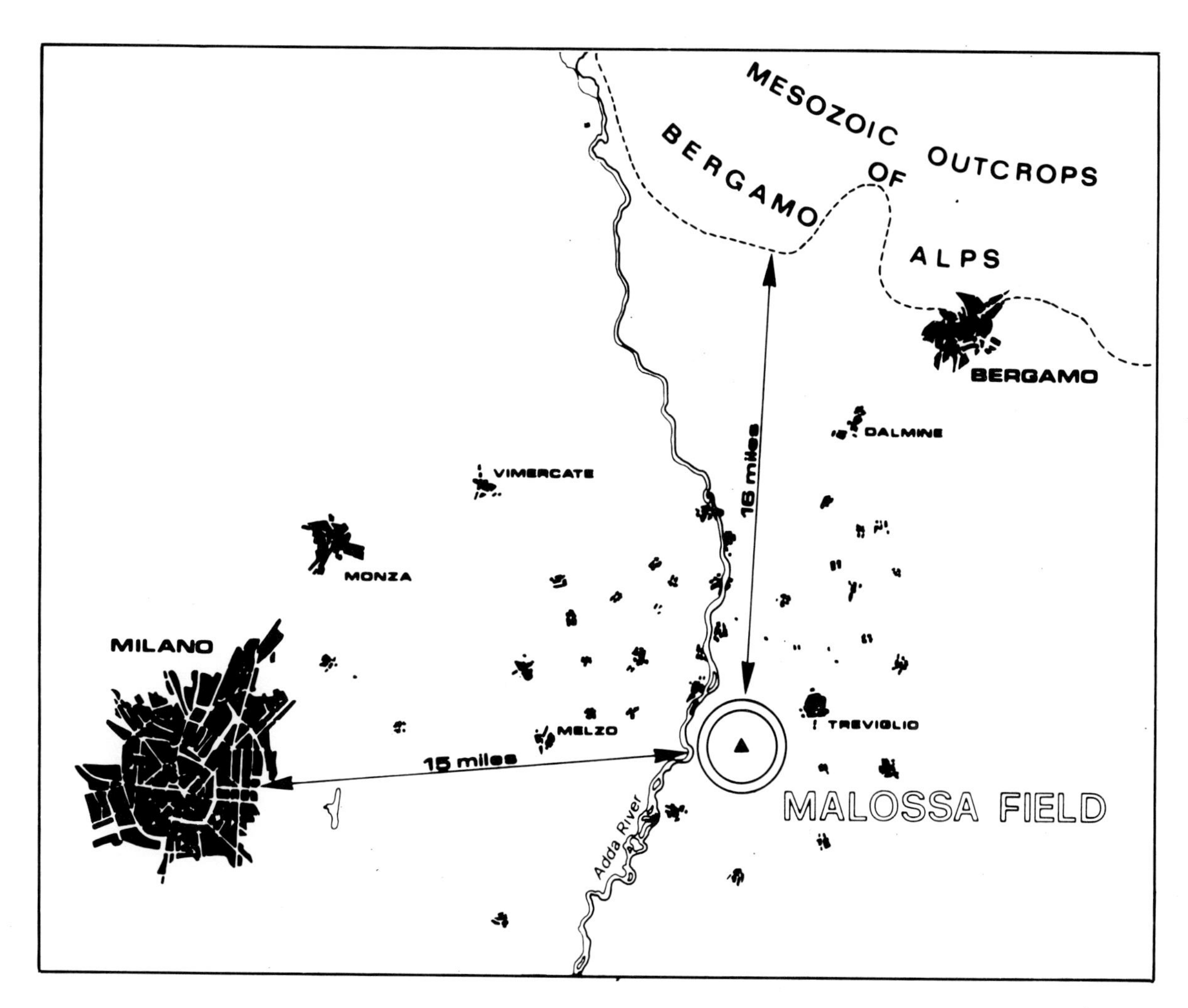

FIG. 2—Location of Malossa field.

Miocene, Pliocene, and Pleistocene, clastic deposition continued in the northern Po basin where no evidence of tectonic movement is present; however, in the southern Po basin, there is evidence of strong tectonics related to the Apennine folded belt.

STRUCTURAL OUTLINE

The Po basin, approximately 50,000 sq km area, is structurally complex in detail, but can generally be considered as a syntectonic post-tectonic basin, filled mainly with clastic sediments of Tertiary and Quaternary ages.

This region has maintained the characteristics of a syntectonic basin since the clastic Tertiary sedimentation replaced the Mesozoic carbonates as a result of an Eocene orogenetic phase.

From the structural point of view, the Po basin can be divided into three main areas (Fig. 3); (1) north of Po River, (2) south of Po River, and (3) Veneto plain.

The region north of the Po River and west to Garda Lake consists of anticlines and overthurst structures, generally with reverse faults on the southern flank. From Garda Lake eastward, on the contrary, there is a stable area gently sloping to the south (Figs. 4, 5).

Almost all the area south of Po River can be considered a folded belt. All the northwest and southeast trending folds are complicated by reverse faults and frequently overthrust toward the north and northeast. South of Po River it is possible to recognize three major elements consisting of a group of folds bent in the shape of an arch. These three arches are geographically located from west to east from Torino to Voghera, from Voghera to Cremona to Parma, and from Reggio E. to Ferrara to Ravenna (Fig. 3).

The Veneto plain is a stable area with thin Tertiary cover. In its easternmost part folds with Dinaric trend are present.

GAS AND OIL PRODUCTION

Hydrocarbon exploration in the Po basin began in the late 1940s. After World War II several substantial gas fields were discovered. Oil, mostly in the form of a condensate, was present in some zones of a few fields. Most of the gas fields discovered to date are within the Pliocene age range (Fig. 6).

During the 1950s the Mesozoic targets were tested in areas

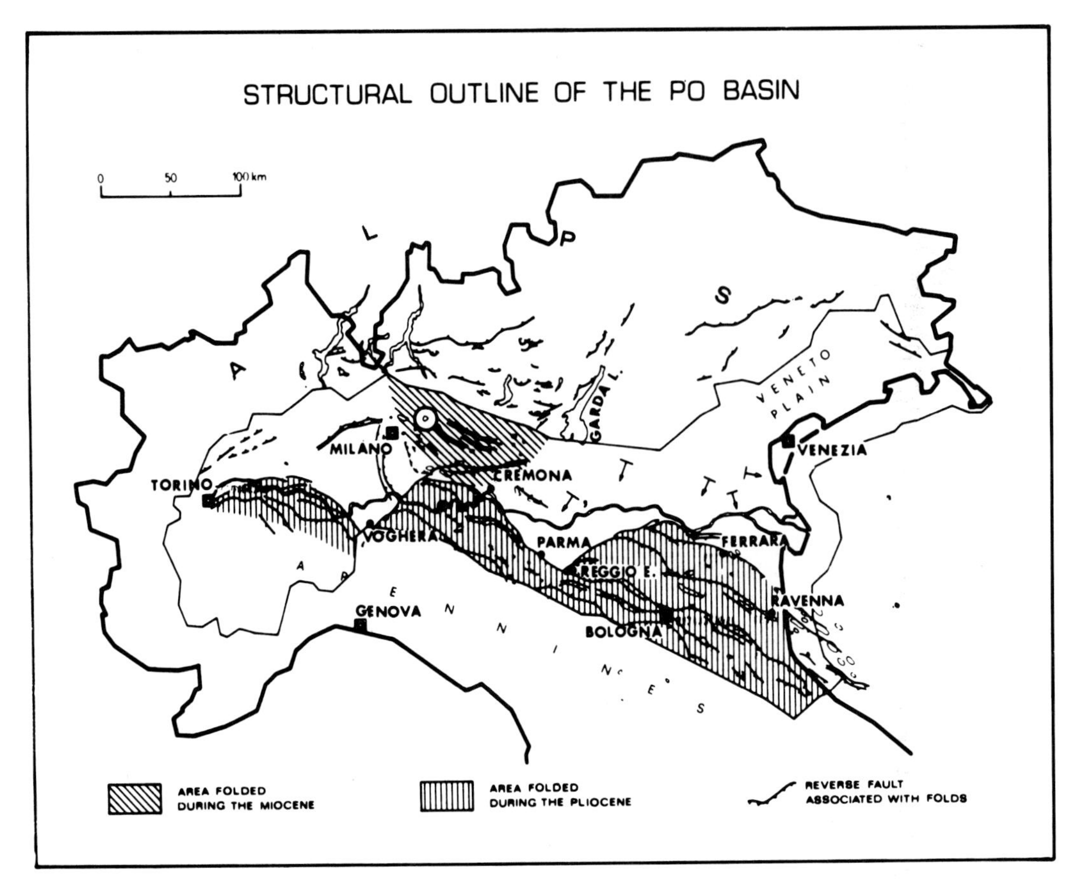

FIG. 3—Structural framework of the Po basin, Italy.

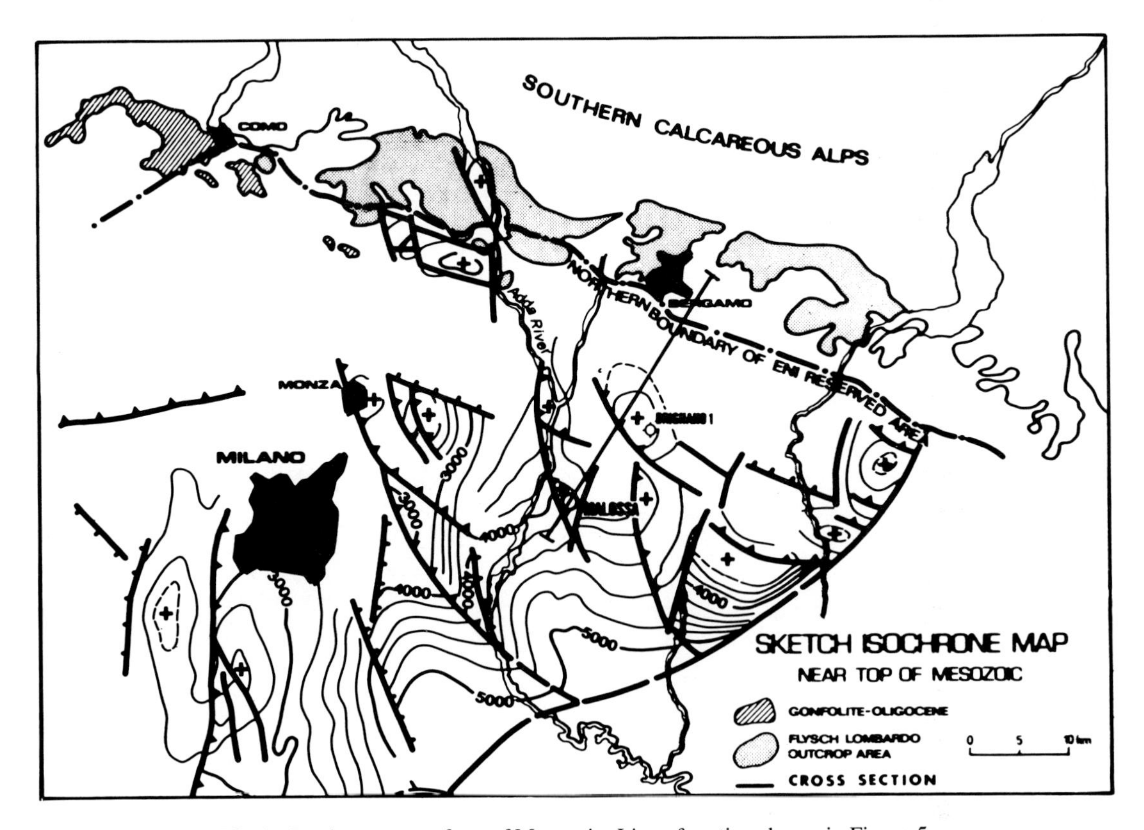

FIG. 4—Isochrone map of top of Mesozoic. Line of section shown in Figure 5.

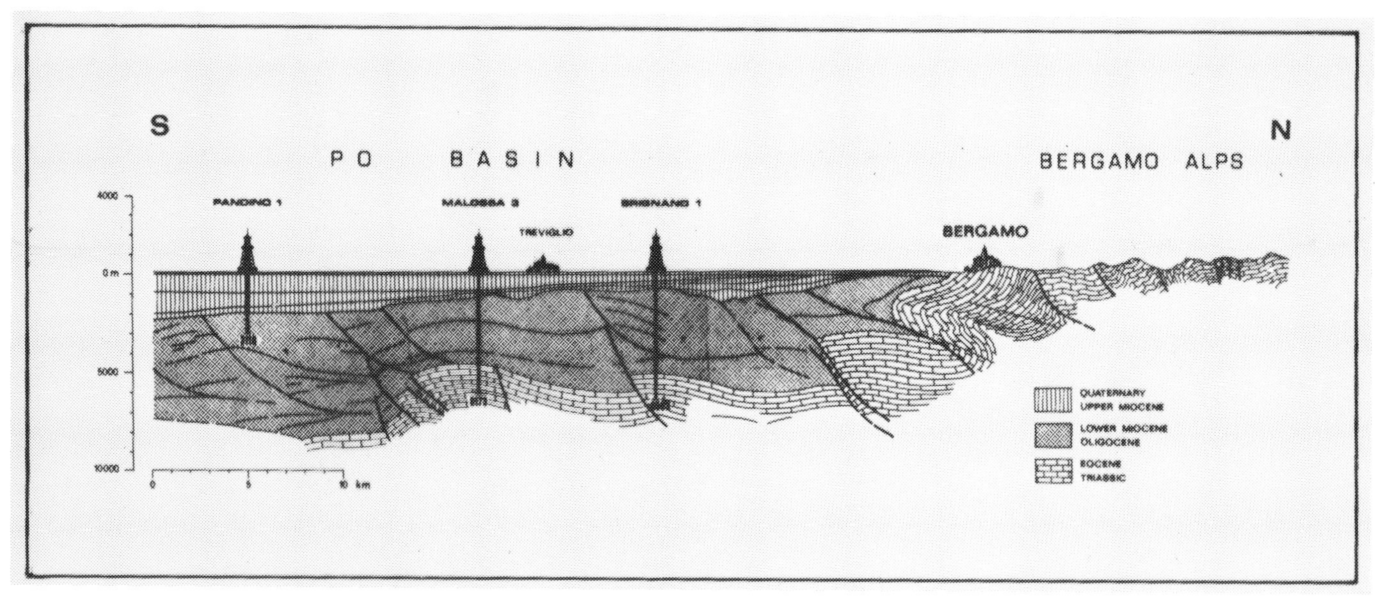

FIG. 5—South to north cross section Po basin. Location of section line shown in Figure 4.

south of Po River and in the Veneto plain, but failed to discover hydrocarbons. The wildcats S. Donato di Piave 1 in Veneto, and Casaglia 1 and Consandolo 1 in Emilia, although all dry holes, provided useful information about the lithology of the Mesozoic section.

The Malossa field was the first condensate gas discovery in Mesozoic rocks of the basins (1973).

DISCOVERY HISTORY

The region east of Milan and south of the Bergamo foothills, where the Malossa field is located, was explored during the late 1950s. The target at that time was a discontinuous gravel formation deposited on the erosional surface and covered by open-marine clay. An intensive seismic survey was carried out and provided information about the morphology of the erosional surface as well as some vague information about the gravel distribution. The seismic information proved accurate from the surface to the main unconformity (1,000 to 2,000 m depth) but below it data were discontinuous and scattered.

Based on this information numerous shallow wells were drilled resulting in the discovery of Sergnano, Brugherio, Orzivecchi, Orzinuovi, and other minor gas fields.

With the introduction of digital seismic techniques in geophysical prospecting during 1967-1968, the reflection seismic data now provided better information at greater depths with improved quality in the definition of the reflecting horizons. Seismic surveying was therefore carried out again over the entire Po basin.

It has been possible to reconstruct the structural layout of the Mesozoic formations at a considerable depth in a vast uniform area in the northern part of the Po basin. Malossa discovery is located in this area.

The first Malossa well was spudded in July 1972 and reached a total depth of 5,545 m (18,891 ft) in February 1973. During drilling, anomalous pressure was detected and the first drill stem test (DST) in the pay zone recorded a static pressure of 1,050.6 kg/sq cm (14,934 psi). The first production test in December 1973 gave the following results: gas, 509,136 cm m/d (17,980 MMcft/d); oil, 371 cu m/d (2,333.5 b/d), API°, 53.1; PTHP, 700 kg/sq cm (9,958 psi) and STHP 799 kg/sq cm (11,364 psi).

The second and third Malossa wells spudded in early 1974 and had two different targets. Well 2 was to define the southward extension of Malossa, while well 3 was to supply information on its pay zone thickness and lithologic characteristics.

A plant was installed at the Malossa 1 site to study the production conditions and all the parameters needed to design the final production facilities. Construction of the plant and drilling of 6 other development wells took place from 1974 to 1977. Production began in 1974 using the pilot plant and was switched through the final plant in October 1977.

Daily production from the 8 existing wells at the beginning of 1979 was 3.6 million cu m (127 MMcf gas, 2,892 cu m = 18,200 bbl oil, and 450 cu m = 2,830 bbl LPG).

STRATIGRAPHY

For a stratigraphic description of the Malossa field it is helpful to use the data from wells 2 and 3; well 2 because it is the deepest in the field, and 3 because the "Dolomia Principale," the main pay zone, is well represented. The sequence ranges from Triassic to Quaternary in age (Fig. 7).

Middle Triassic

The Middle Triassic Esino Formation is the oldest unit in the Malossa field. The nearest outcrops of the Esino are in the Southern "Calcareous" Alps.

The formation is a fossiliferous, tight, gray to light brown, very fine to medium grained dolomite with lithoclasts. The fossils are poorly preserved, consisting of Dasycladacean, So-

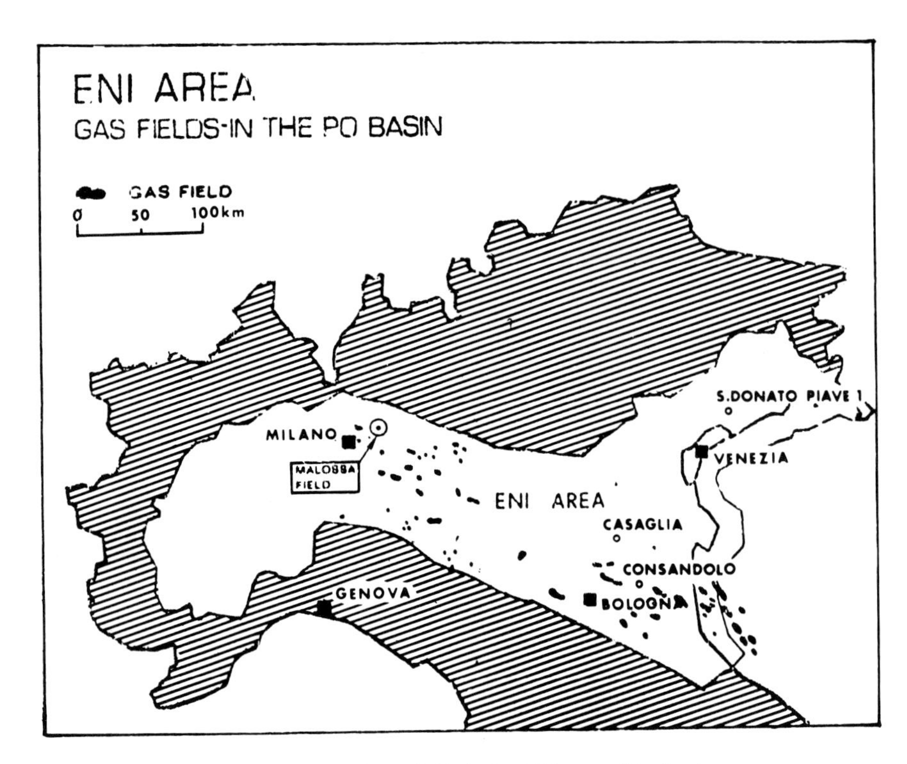

FIG. 6—Location of gas fields in the ENI area of Po basin.

lenoporacean algae, Molluska, and Ostracoda. Interbedded with the dolomites are thin layers of feldspathic sandstones and black silty shales, more frequent in the lower part of the section. Sedimentation is typical of the open shallow platform with littoral episodes evidenced by mud cracks.

Upper Triassic

The lowest part of the Upper Triassic early Carnian consists of light brown, fine to very fine grained shaly dolomites with chert nodules, and Radiolarian relic interbedded with dark gray dolomitized marls. Although these rocks have some peculiar characteristics, they are similar to the limestones and argillaceous limestones of the Gorno formation which crops out in the Bergamo Alps. This formation has been dated early Carnian because of its typically rich fauna. It was deposited in a deeper platform environment with anoxigenic episodes.

The late Carnian is represented by barren varicolored sandy shales interbedded with yellowish dolomitic marls and feldspathic sandstones. This formation, deposited in a lagoonal environment, is called the S. Giovanni Bianco Formation.

Overlying the S. Giovanni Bianco is the Norian which has a typically tidal flat facies and is widespread in outcrops all over the Southern Alps (Dolomia Principale). It consists of gray to light brown, fine to very fine grained dolomite with lithoclasts, and rare, badly-preserved fossils (algae, ostracods, and mollusks).

In the subsurface of the Malossa area this formation is thinner than on the outcrop. This is probably due to the erosion which occurred during an emersion phase of the Malossa area during Norian deposition. This interpretation is confirmed by the lack of the Rhaetian age units, which are always present in the Southern Alps as euxinic facies (Argilliti de Riva di Solto and Calcari di Zu).

The Norian sediments are overlain by fine grained, brown to reddish dolomites, sometimes brecciated, with some seams of gray-greenish shale. This interval has no fossils and thus it cannot be dated. However, it can be correlated on the basis of the lithologic similarity with the Zandobbio Formation cropping out at the Zandobbio quarry, west of the Iseo Lake. In this area some brachiopods and mollusks have been found and dated Rhaetian and, partly, early Liassic.

Liassic

The Jurassic sediments show lithologic characteristics and faunal contents suggesting a progressive lowering of the sea floor. The partly Liassic dolomites of the Zandobbio Formation are overlain by argillaceous, gray, sometimes silty limestones, with radiolarian and sponge spicules interbedded with dark gray marls and limestones with lithoclasts and small fossil fragments. The limestones are often silicified and dolomitized in the lower part.

This formation has been informally called "Medolo" and varies in thickness as a consequence of synsedimentary faulting in the Lombardian basin. These faults were particularly active during Medolo deposition.

The last Liassic sediments are represented by dark gray, argillaceous, cherty, limestones with radiolarians and pelagic pelecypods interbedded with greenish silicified marls and carbonaceous matter (Concesio Formation). Sedimentation occurred in a deeper platform environment with slope deposition episodes. The latter are represented by detrital sediments, which came from local highs in the Lombardian basin or from

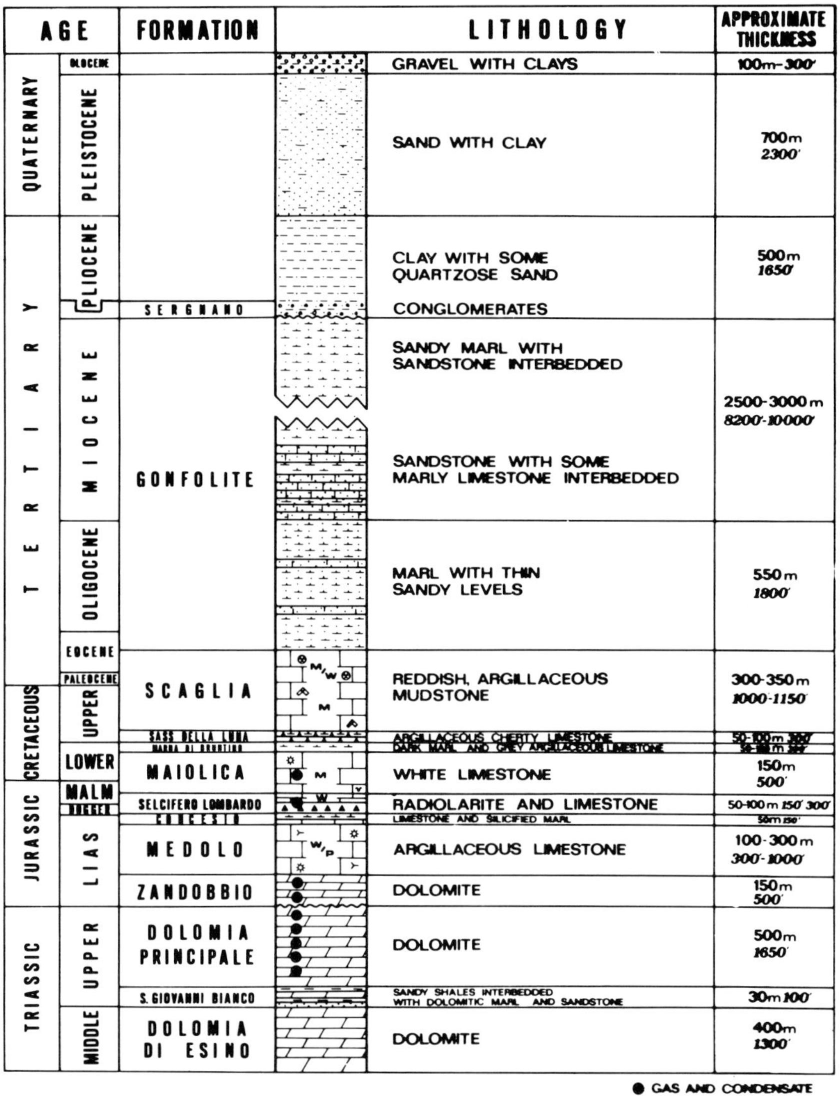

AGE		FORMATION	LITHOLOGY	APPROXIMATE THICKNESS
QUATERNARY	OLOCENE		GRAVEL WITH CLAYS	100m-300'
QUATERNARY	PLEISTOCENE		SAND WITH CLAY	700m 2300'
TERTIARY	PLIOCENE		CLAY WITH SOME QUARTZOSE SAND	500m 1650'
TERTIARY	PLIOCENE	SERGNANO	CONGLOMERATES	
TERTIARY	MIOCENE	GONFOLITE	SANDY MARL WITH SANDSTONE INTERBEDDED	2500-3000 m 8200'-10000'
TERTIARY	MIOCENE	GONFOLITE	SANDSTONE WITH SOME MARLY LIMESTONE INTERBEDDED	
TERTIARY	OLIGOCENE		MARL WITH THIN SANDY LEVELS	550m 1800'
TERTIARY	EOCENE			
TERTIARY	PALEOCENE	SCAGLIA	REDDISH, ARGILLACEOUS MUDSTONE	300-350m 1000'-1150'
CRETACEOUS	UPPER	SCAGLIA		
CRETACEOUS	UPPER	SASS DELLA LUNA	ARGILLACEOUS CHERTY LIMESTONE	50-100m [illegible]
CRETACEOUS	UPPER	MARNA DI BRUNTINO	DARK MARL AND GREY ARGILLACEOUS LIMESTONE	[illegible]
CRETACEOUS	LOWER	MAIOLICA	WHITE LIMESTONE	150m 500'
JURASSIC	MALM	SELCIFERO LOMBARDO	RADIOLARITE AND LIMESTONE	50-100m 150'-300'
JURASSIC	DOGGER	CONCESIO	LIMESTONE AND SILICIFIED MARL	50m 150'
JURASSIC	LIAS	MEDOLO	ARGILLACEOUS LIMESTONE	100-300m 300'-1000'
JURASSIC	LIAS	ZANDOBBIO	DOLOMITE	150m 500'
TRIASSIC	UPPER	DOLOMIA PRINCIPALE	DOLOMITE	500m 1650'
TRIASSIC	UPPER	S. GIOVANNI BIANCO	SANDY SHALES INTERBEDDED WITH DOLOMITIC MARL AND SANDSTONE	30m 100'
TRIASSIC	MIDDLE	DOLOMIA DI ESINO	DOLOMITE	400m 1300'

● GAS AND CONDENSATE

FIG. 7—Stratigraphic section Po basin.

the Venetian platform, still active east of the Garda Lake.

Dogger—Malm

Deposition of the Concesio Formation continued during the early Dogger and was followed by red and green radiolarite deposits which represent the lower unit of the Selcifero Lombardo Formation. This formation was deposited during the Middle to Late Jurassic in a deep marine environment, when the sea floor was at its maximum depth. Its upper part consists mainly of red marls and argillaceous limetones with Saccocoma, Aptychus and abundant siliceous radiolarians (Rosso and Aptici Unit).

Lower Cretaceous

From late Tithonian (*Calpionella alpina—C. elliptica* Zone) to Barremian, white, tight, and cherty limestones were deposited with thin black argillaceous seams. Radiolarians, sponge spicules, and *Tintinnidae* are found in these limestones (only in the lower part). The deposition environment of this formation (Maiolica) was deep marine.

During Early Cretaceous, lowering of sea level began and euxinic sediments of deeper platform were deposited (Bruntino Marls). This formation consists of gray to dark gray, slightly silty, poorly fossiliferous, marls interbedded with thin layers of gray argillaceous limestones with planktonic microfossils *(Ticinella* sp., *Hedbergella* sp., *Radiolaria).*

Upper Cretaceous

During Cenomanian time euxinic sediments were gradually replaced by argillaceous and cherty limestones, with planktonic Foraminifera (*Rotalipora appenninica, Praeglobotruncana* sp.), deposited in a deeper platform environment (Sass de la Luna Formation). Since Turonian time limestone deposition has prevailed.

Late Cretaceous sediments consist of reddish, argillaceous,

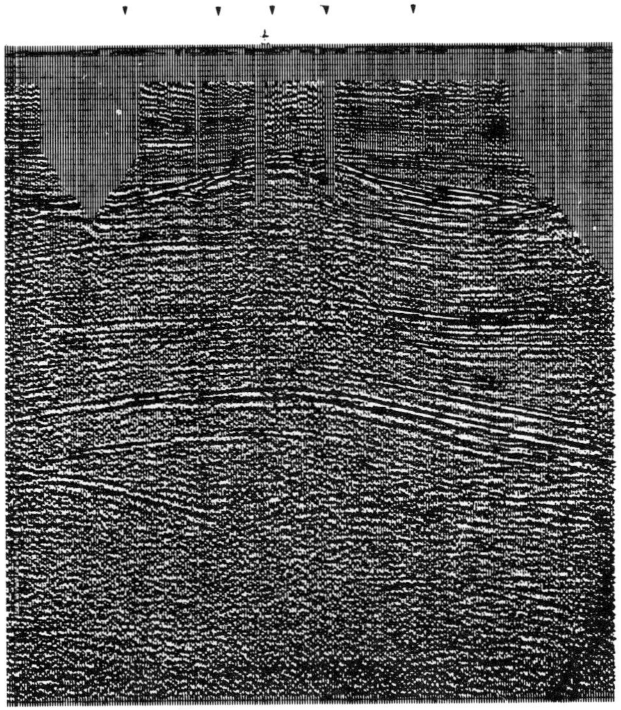

FIG. 8—Seismic line of Malossa structure shot in 1969 showing good Mesozoic reflections.

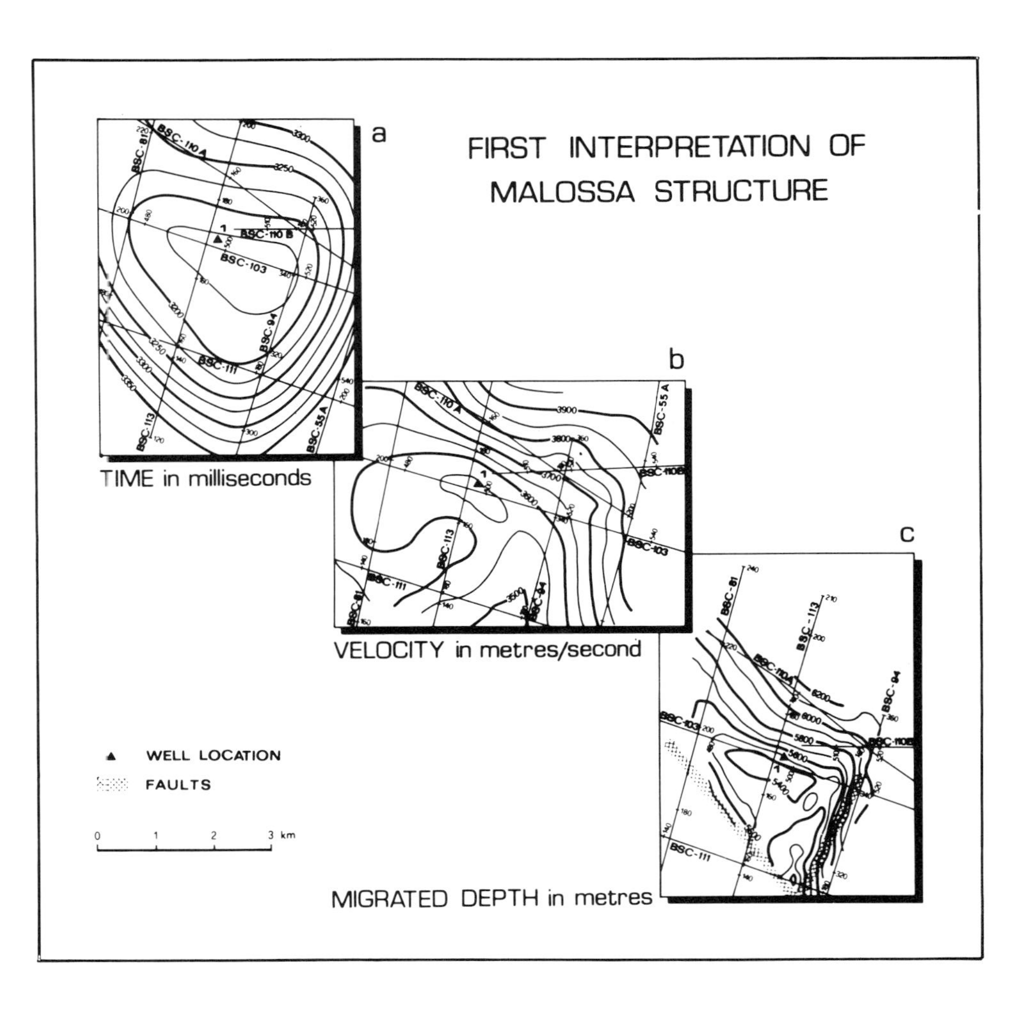

FIG. 9—First interpretation of Malossa field structure.

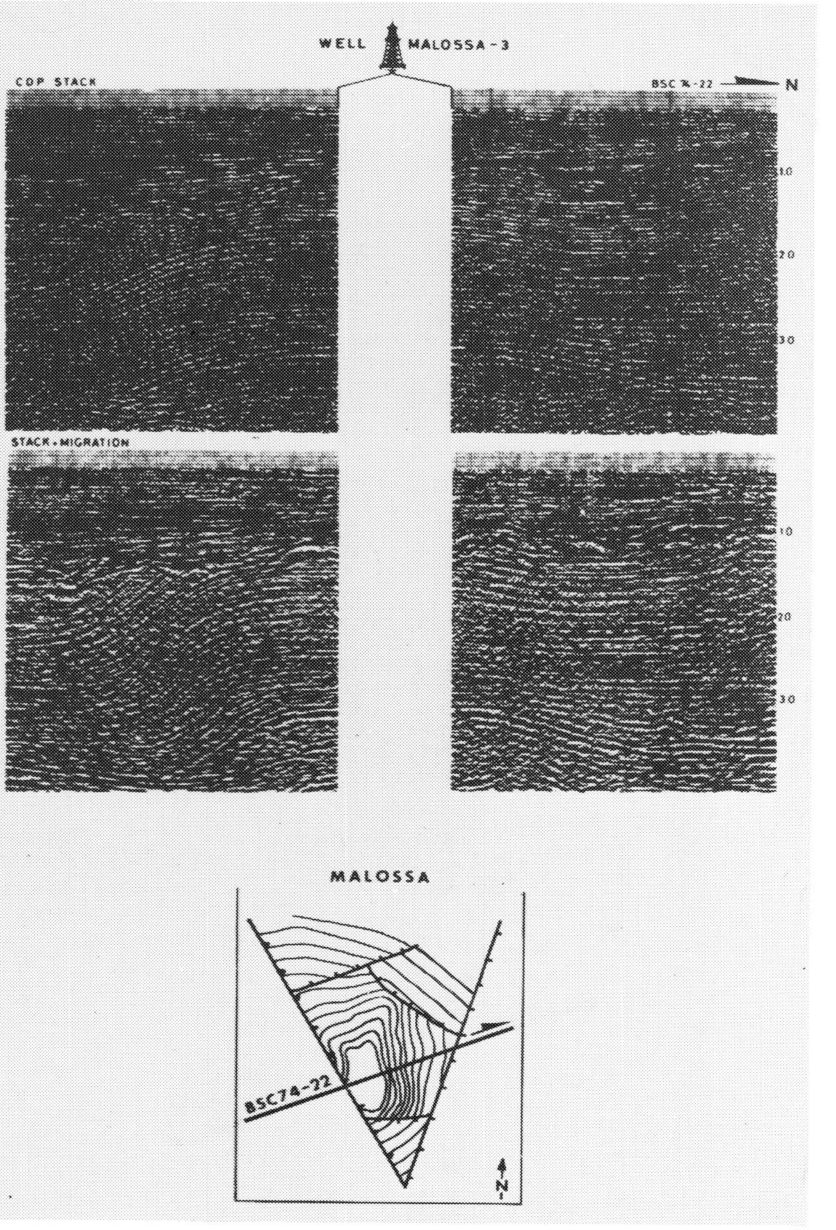

FIG. 10—Seismic data before and after migration processing.

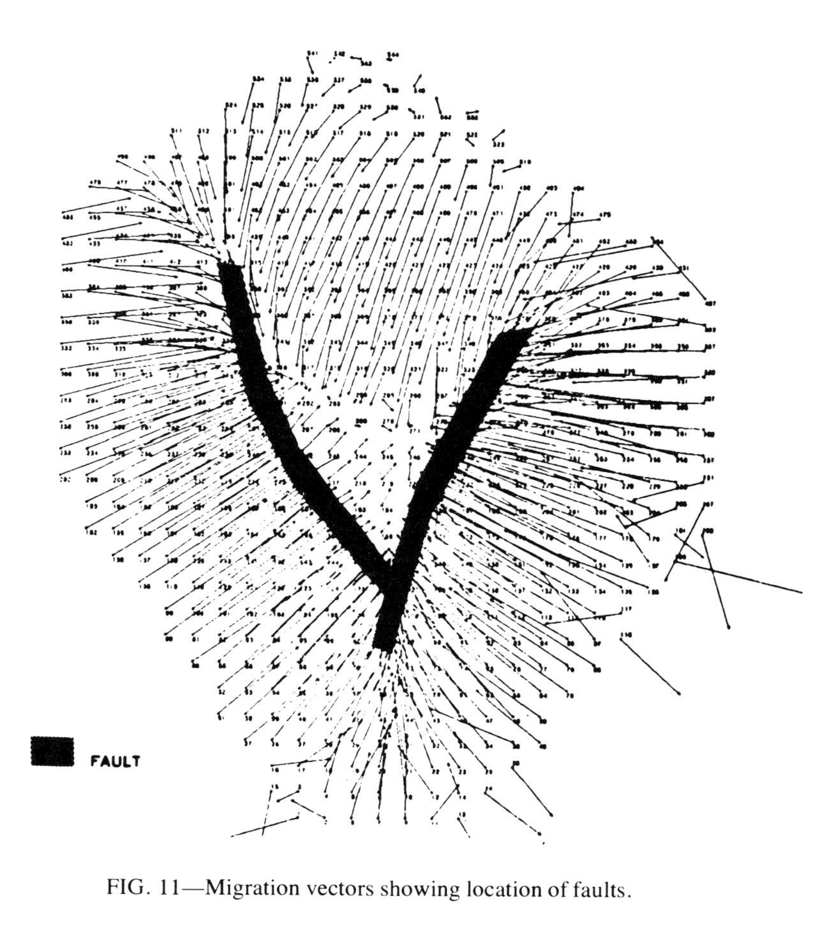

FIG. 11—Migration vectors showing location of faults.

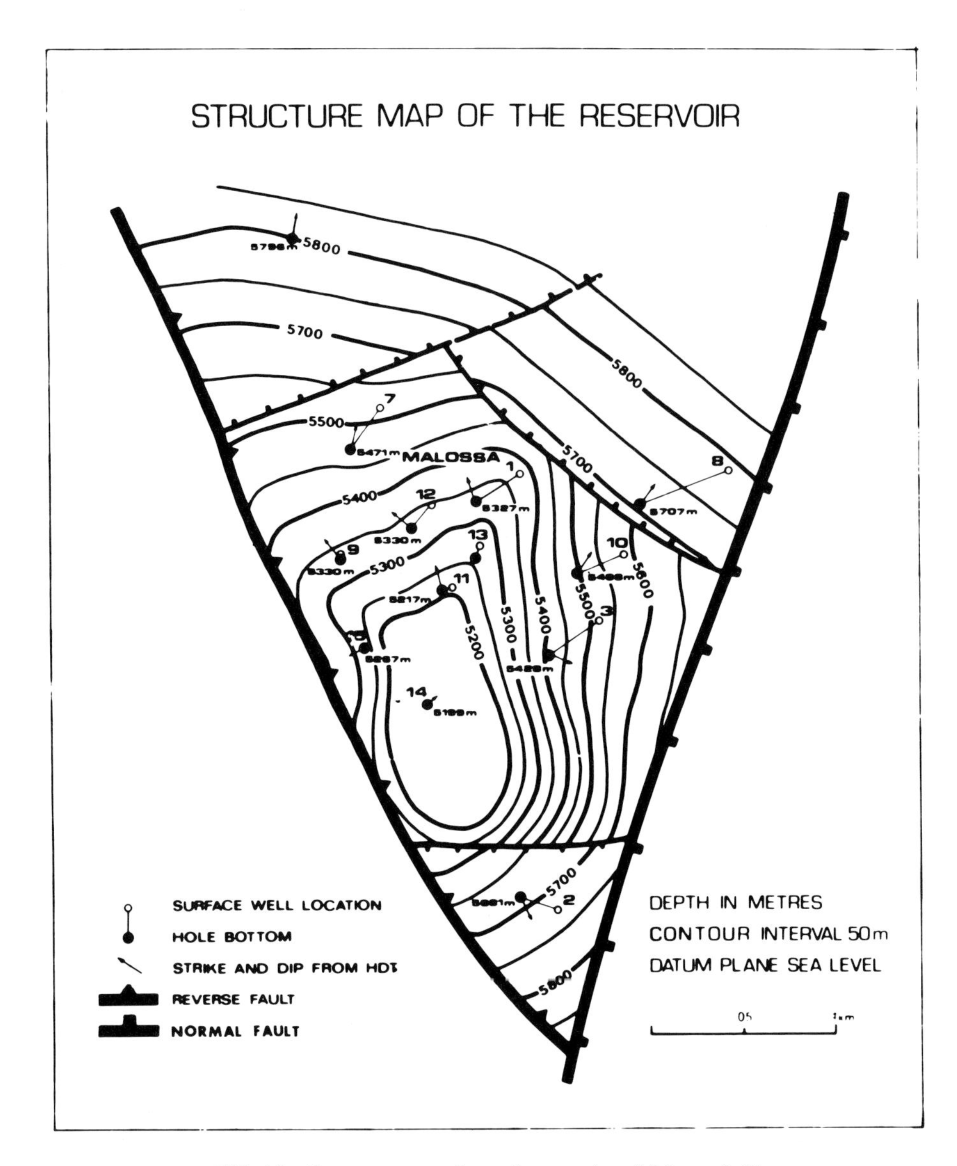

FIG. 12—Structure map of top of reservoir at Malossa field.

mudstones and wackestones with chert nodules and abundant planktonic microfossils. This formation represents the end of the carbonate cycle and is called "Scaglia." It maintained constant characteristics until the middle Eocene, when deposition of synorogenic sediments began.

Tertiary and Quaternary

The beginning of Tertiary is characterized by a hiatus during early and middle Paleocene. In late Paleocene time sedimentation began again with the deposition of the argillaceous pelagic limestones of the "Scaglia." Scaglia deposition continued until middle Eocene time and was followed by the deposition of a thick sequence of gray, sandy marls with planktonic Foraminifera (Marne di Gallare) interbedded with layers of lithic sandstones (Gonfolite). Sedimentation occurred in a deep marine environment with turbidite episodes. The youngest sediments of this sequence are Langhian.

Langhian sediments were followed by poligenic conglomerates of continental origin. This formation (Ghiaie di Sergnano) cannot be dated using paleontology, but has been ascribed to Messinian by means of correlation with the surrounding wells, and because of its stratigraphic position.

The Pliocene is represented by clays with some quartzose sand layers. Fauna is mainly planktonic, with deposition occurring in an outer shelf. The Pleistocene is sandy and deposited in a littoral environment representing the end of the marine cycle. It is overlain by Holocene alluvial deposits.

STRUCTURE DEVELOPMENT AND EVALUATION

The Malossa structure was detected by seismic survey, with the first line showing deep information shot during 1969. This line (Fig. 8) shows a good reflection at 3 seconds (two-way time) interpreted as the top of the Mesozoic. A grid of seismic

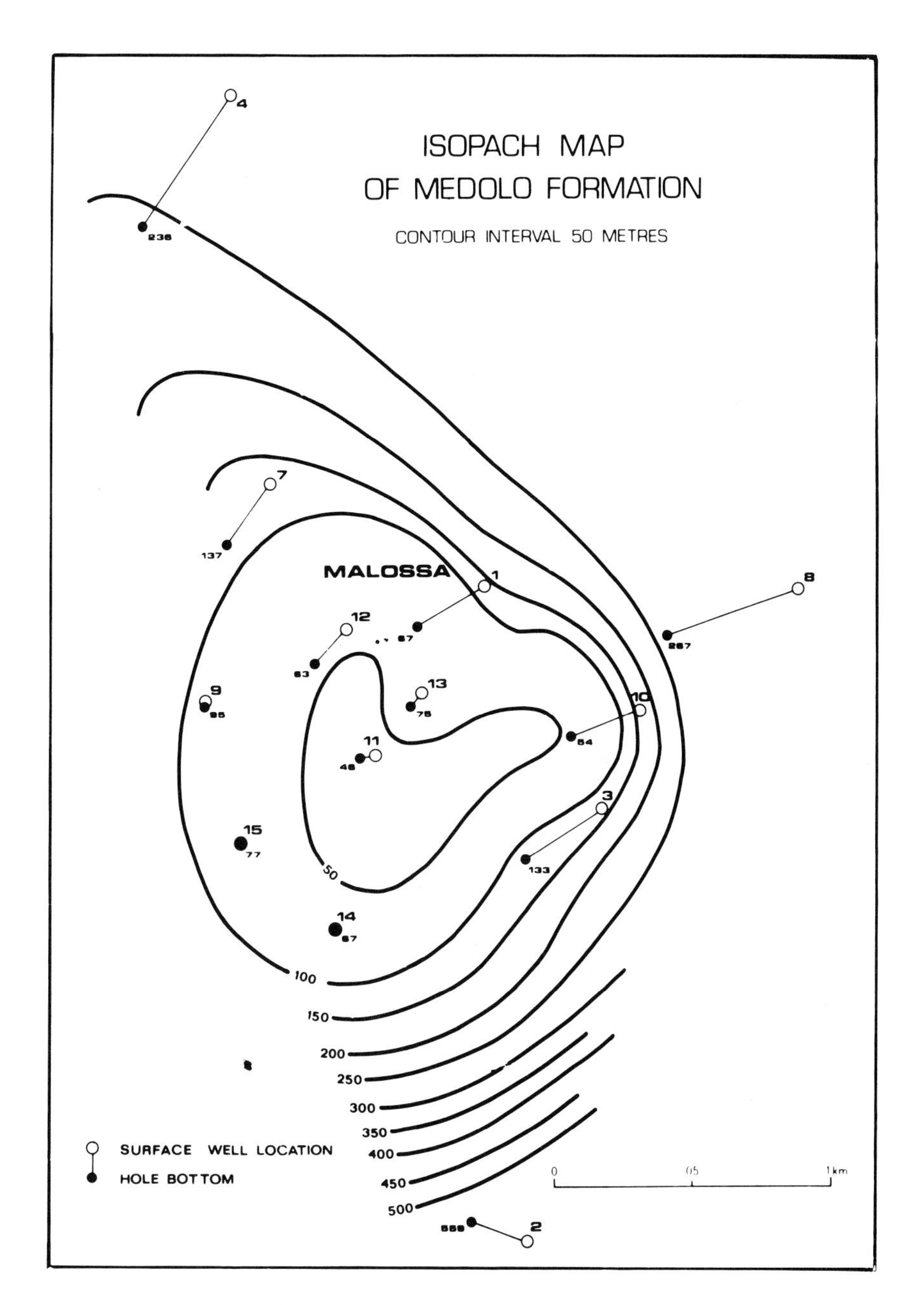

FIG. 13—Isopach map of Medolo Formation at Malossa field.

lines provided data for mapping of a regular, gently folded structure with a small elongation east to west (Fig. 9a). Well 1 was located using this preliminary isochrone contour map.

In late 1973, after the first well was successfully drilled, a preliminary attempt to obtain a reliable depth-migrated contour map of the structure was made by computer. Three tests were made using the following velocity assumptions: (1) well velocity data integrated by spatially smoothed velocity values derived from the seismic lines; (2) well velocity data laterally extrapolated on the basis of "formation velocity" assumption; and (3) well velocity data only. Figure 9b shows the velocity contour map of the assumption (1) and Figure 9c shows the derived depth-migrated map.

At the same time a synthetic seismogram was produced and compared with the seismic data. The comparison demonstrated that the strong reflection present on the seismic sections was related to the Maiolica Formation (Lower Cretaceous). The lack of seismic information below the Maiolica is explained by the small acoustic impedance contrast inside the carbonate sequence. In spite of this, the next two appraisal wells, Malossa 2 and 3, were located on the basis of the above-mentioned depth-migrated map of the Maiolica Formation supposing that the reservoir rock structure was not affected by lateral displacement. It is interesting to note that on the map the trend and location of the two main faults are convergent toward the south and are readily visible (Fig. 9c).

After the wells 2 and 3 were drilled in early 1975, a new revision of the structural interpretation took place. Meanwhile, new seismic lines were recorded in the area using more sophisticated acquisition techniques. The quality of the seismic cross

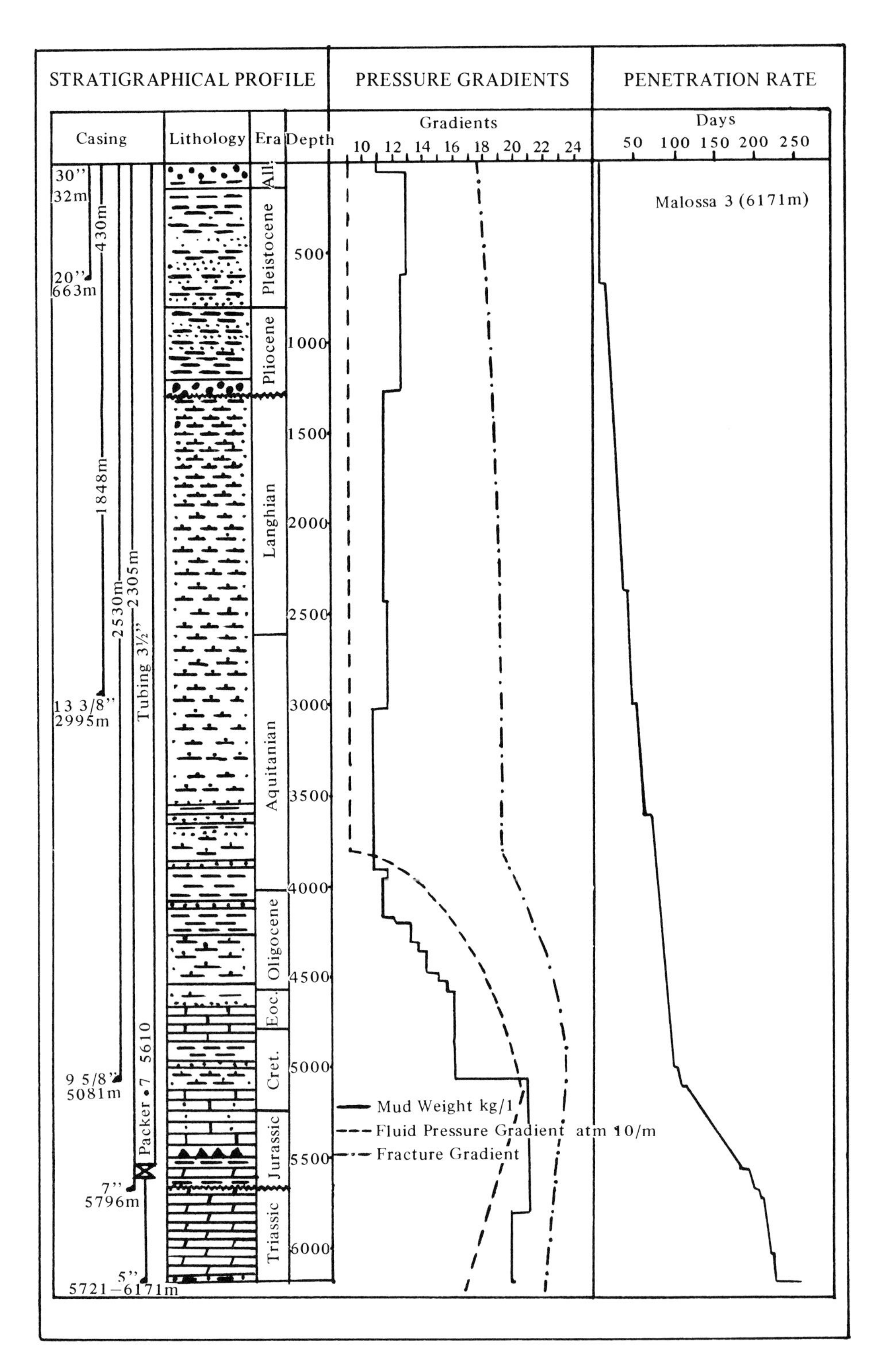

FIG. 14—Typical casing, pressure gradient, drilling time data for wells in Malossa field.

section was improved in such a way as to make possible an attempt to interpret the top of the reservoir rock. Figure 10 shows one of the new sections before and after migration together with information obtained from the Malossa 3 well. In this section the two main faults toward the east and west are very evident. The aim of the new interpretation and mapping was to obtain the actual shaping and location of the structure.

Production of a reliable isovelocity map was a very difficult problem to solve. Stacking velocities were considered ineffective because of the instability caused by the very irregular late Miocene unconformity. After several attempts new velocities were computed from the well velocity data and extrapolated (line by line) as formation velocities based on a stratigraphic model.

The interpretation was made on the time record sections, including diffraction hyperbolas and other seismic raypath distortion, to allow the spatial location of the two main faults with the best approximation. Figure 11 shows an intermediate stage of the 3D mapping with migrated vectors where the fault positions can be seen. Figure 12 shows the final depth-migrated contour map of the top reservoir rock. It indicates that the Malossa structure is a faulted and overthrusted anticline about 3 mi long and 2 mi wide (4.8 by 3.2 km) bordered on the southwest by a reverse fault. To the east, on the opposite side of the overthrust front, is a normal fault which is interpreted as a collapse fault caused by a release mechanism after compressional movements during Miocene time.

The lack of the Rhaetian deposition in this area and the thickness pattern of the Medolo Formation shown by well data indicate that during Triassic and Liassic time this area was a high. The favorable closure conditions present during the Jurassic were preserved through the time of the later tectonic movements (Fig. 13).

RESERVOIR CHARACTERISTICS

Condensate gas is found in the Maiolica (Lower Cretaceous-Upper Jurassic), Selcifero Lombardo (Middle to Upper Jurassic), Ammonitico Rosso (Lower Jurassic), Medolo (Lower Jurassic), Zandobbio (Lower Jurassic-Upper Triassic), and Dolomia Principale (Upper Triassic) formations.

The gas-bearing sequence at Molassa has a gross pay of 821 m (2,700 ft). The cap rock is assured by the Marne di Bruntino Formation (Lower Cretaceous) which consists of gray, sometimes reddish marls (50 to 100 m thick) with fossilferous mudstone-wackestone zones.

The main reservoir is the most porous and fractured reservoir and is capable of a fair production. It is represented by the Zandobbio and Dolomia Principale formations which are in vertical communication. Other reservoir formations are in vertical communication with the main reservoir through an irregular system of macro and micro fracture. The characteristics of the main reservoir are as follows:

Zandobbio Formation—a gray-greenish or reddish fine to medium grained dolomite and calcareous dolomite with dark gray and green shaly zones; highly fractured. The formation is early Liassic to late Raethian in age, with a deep platform (DP) to open shallow platform (OSP) sedimentary environment, and a thickness of 150 m.

Haupt Dolomite Formation—white, gray, fine to medium grained intraclastic, fossiliferous and sandy dolomite with gray marly dolomitic zones, common birdeyes and stromatolitic features; highly fractured.

The formation is Late Triassic (probably Raethian—Norian) in age, with a tidal flat complex (TFC) sedimentary environment, and a thickness of 450 m.

The water table is at the depth of 5,830 m (19,127 ft) on the basin of production tests performed in well 3.

As already noted, only the dolomitic complex, represented by the Zandobbio and Dolomia Principale Formations, is considered the main reservoir, and production wells have an open hole completion in the upper part of this complex particularly within the most porous section of the Zendobbio Formation.

FLUIDS

Reservoir fluid is in monophase flow with a pressure of 1,050 kg/sq cm (14,934 psi) and a temperature of 160 °C (320 °F). The composition is: C1 = 79.08%; C2 = 6.15%; C3 + C4 = 5.17%; C5 = 8.46%; N2 = 0.74%; and CO2 = 0.40%. Present normal flow conditions are 600,000 cu m/d (21 MMcf/d) per well.

With this flow rate the FTHP is 590 kg/sq cm (8,302 psi) and the FTHT is 110 °C (230 °F). Fluid at the well head is biphase, with only condensate water as liquid phase. The presence of free water in the system makes the CO_2 corrosive. An adustable choke is fitted on the well head reducing pressure to 150 kg/sq cm (2,133 psi), which is the working pressure of the gathering system.

Downstream of the choke, where pressure is 150 kg/sq cm and temperature has remained about 110 °C, the fluid is clearly biphase, since condensate hydrocarbons are added to the water. As a result, for every million standard cubic meters of dry gas there are 675 tons of liquid hydrocarbons. Separation will occur in the plant, as for every million standard cu m of gas, there will be 607 tons of oil having a density of 53° API and 68 tons of LPG consisting of 50% propane and 50% butane. Pressure-volume-temperature (PVT) tests have shown that at reservoir temperature the dew point is 398 atm.

ASSESSMENT OF RESERVES

A first assessment of field reserves based on the volumetric calculation gave an estimate of 50 billion cu m of gas and 40 million tons of oil. It has been impossible to check these data as the low cumulative production (1.1 billion cu m gas and 710,000 tons oil as of December 1979) and the likely presence of a water drive does not allow the material balance to obtain ultimate figures.

FIELD DEVELOPMENT

Due to production difficulties for this type of reservoir, and the lack of specific experience, even in the international field, it was decided to develop the reservoir in two production stages (an initial experimental period using a pilot plant connected with Well 1 immediately followed completion of the discovery well).

The first stage, with 10 producing wells at present, will make it possible to ascertain the productive capacity and maximum flow rate of each well. In this phase the maximum flow rate will depend more on completion problems than on reservoir geologic features; in fact, it will be necessary to control the sealing system, the reaction of piping to thermic variations and the corrosion inhibitor system.

Given the present depth of the water table, there should be no problems with the reservoir so long as the dew point limit is

Table 1. Casing program of Malossa wells.

Casing Diameter	Pressure Gradient	Drilling Data
20″	J55	106.5 lb/ft Buttres set at 600 m
13-3/8″	P110	68.0 lb/ft Buttres set at 3,000 m
9-5/8″	Q125	53.5 lb/ft Buttres set at 5,000 m
7″	Q125	38.0 lb/ft VAM set at 5,000 to 5,400 m

not exceeded.

At the end of this first stage, the optimum number of wells and the total field rate will be determined, and production will then enter the second and final stage.

The Malossa wells have been drilled to produce from open holes using the casing program shown in Table 1. The completion design consists of 03″ 1/2P 105 tubing, 12.9 lb/ft anchored to a hydraulic packer. A safety valve is set in a landing nipple at 1,300 m and tubing is hooked onto the well head with a hanger having metal-to-metal seals. The packer fluid is an oil phase mud with a density of 2 kg/l. The oil mud was designed to prevent tubing corrosion due to the rotting elements contained in a lignosulfonate mud. All the completion equipment, from the packers to the well head, are able to withstand a pressure of 15,000 psi.

A stainless steel seal block placed before the master valves ensures the capability of shutting the well in case of well head corrosion; for reasons of economy, a simple carbon steel well head was chosen. The well head assembly is remote-controlled with controls for the following: opening and closing controls on Wings 1 and 2; THP, THT, and Q signals; and alarms for high pressure on casings and metal seal area; low THP, closing Wing 1, fuse network; and the gearbox (low oil level, low oil pressure, power cut). All these are operated from the control room of the gathering and treatment center.

The fluid produced is corrosive due to the presence of CO_2. Though CO_2 level is limited to 0.4%, it becomes corrosive due to the high partial pressure. Corrosion control is assured by squeezing an anticorrosive mixed with the proper carrier into the well.

CONCLUSIONS

1. The discovery of the Malossa field has opened new targets at greater depths not only in this part of the Po basin but also in other areas where it is possible to reach the Mesozoic through a thick cover of Tertiary sequence.

2. The field area seems to have been a paleohigh area. If this characteristic is to be considered of primary importance in the history of hydrocarbon migration it is essential to conduct paleogeographic studies of the entire area.

3. The numerous anomalies in the pressure of the reservoir is a regional characteristic extending throughout this part of the Po basin. Overpressure could be due to tectonic deformation of the impervious sequence overlying carbonate formations.

4. The source rocks are believed to be shales of Rhaetian age. These shales of euxinic environment must have been widespread in the western part of the Mediterranean area.

5. The very difficult production conditions present due to anomalous pressure, high temperature, corrosion, and the need to impose strict safety controls has brought about the development of new techniques for well completion.

REFERENCES CITED

Blondi, G., R. Rocca, and S. Zanoletti, 1976, Automatic contouring of faulted subsurface: Geophysics, v. 41, p. 1377-1393.

——— ——— ——— 1977, Methods for Contouring irregulary spaced data: Geophys. Prospecting, v. 25.

Bogniorni D., V. Crisco, and D. Fenati, 1977, Geophysical and drilling problems encountered in the exploration of deeper structures in the Po basin: 10th World Energy Conference, Istanbul, Turkey.

Carissimo, L., and O. D'Agostino, 1960, Development of the new technique applied to the search for stratigraphic traps: Geophys. Prospecting, v. 3, p. 401-416.

Groppi, G., A. Muzzin, and G. Vaghi, 1976, Erdofeld Malossa Die erste Kohlenwasserstoffgewinnung aus dem tiefen Mesozoikum unter der Po-Ebene: 4 Gemeinschaftstagung Ogen/Dgmk - October 1976 (Salzberg).

Merlini, E., 1960a, A new device for seismic survey equipment. Geophys. Prospecting, v. 8, p. 4 - 11.

——— 1960b, The stratigraph—a new research device: World Petroleum, March, 1960.

Rocco, T., and O. D'Agostino, 1792, Sergnano gas field, Po basin, Italy; A typical stratigraphic trap, *in* R. E. King, ed., Stratigraphic oil and gas fields: AAPG Memoir 16, p. 271-285.

Petroleum Potential—Ouargla Region Triassic Basin, Algeria[1]

By Ait Hamouda[2]

THE OUARGLA REGION: OVERVIEW

The Ouargla Region is located in northeastern Algeria in the famous Algerian Triassic basin (see location, Fig. 1). This basin covers approximately 280,000 sq km and includes such giants as the Hassi R'Mel field. No giant individual fields have, to date, been discovered in the Ouargla Region; but collectively, numerous small fields have made the area a significant hydrocarbon province.

The Ouargla Region contains the necessary ingredients for hydrocarbon accumulations, in that there are source beds, primarily the Gothlandian which is a black graptolitic shale in excess of 150 m thick. The area also contains a prime reservoir of rock consisting of Triassic sandstones with local thicknesses as great as ± 70 m, overlain by a caprock of Triassic and Liassic salt in excess of 400 m in thickness.

Exploration

Exploration in the area has, until recently, concentrated on structural traps detected primarily by seismic methods. More than 10,640 km of seismic line have been shot to date with a more recent concentration on potential stratigraphic traps coming into play within the region.

Reservoir

The reservoir sequence is divided into two productive zones separated in most areas by an "eruptive sequence" of andesitic rock. Although the environment and deposition of the reservoir is still subject of additional interpretation, it appears to be a fluvial deposit with channels displaying an anastomosing pattern. The deposits represent a drainage system in an arid to semi-arid environment, and the area is apparently frequently inundated by marine transgressions. The andesitic sequences also provide an excellent source of seal rocks for traps in the Ouargla area.

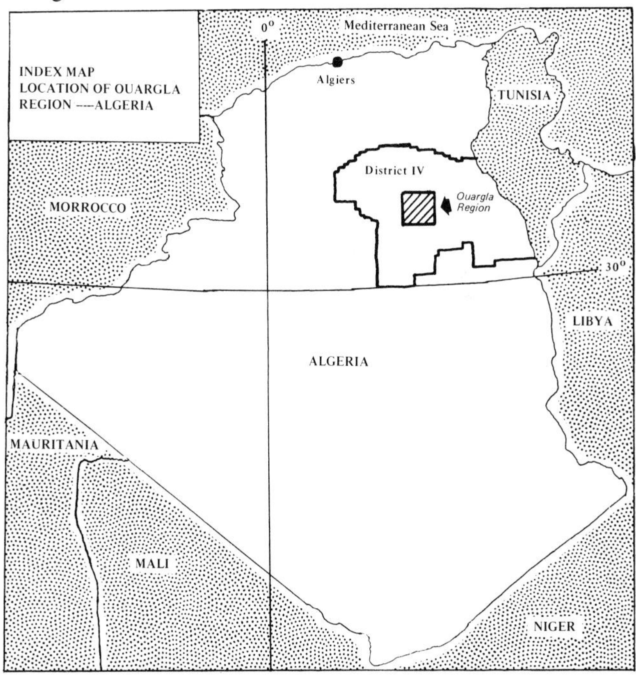

[1]Manuscript received, April 4, 1979; accepted for publication, January 17, 1980.

[2]Sonatrach, Place du Perou, Algeria. The writer thanks Mr. Garrett and Mr. A. El-Ouri for their assistance in translation from the original French text. In addition, he thanks Ali Lounes, B. Kermad, I. Donos, A. Senoussi, C. Garrett, and S. Gasior for their assistance in preparation of supplemental materials.

Article Identification Number:
0065-731X/80/M030-0027/$03.00/0.

Producing Fields

The two largest fields in the area are the combined fields of Haoud-Berkaolu and Ben-Kahla. These two accumulations are in the highest structural areas of the Ouargla Region and are owned by Sonatrach (51%) and TOTAL-Algeria (49%).The total estimated recoverable oil in this combined field is 383 million bbl. The accumulation is in Triassic sandstones directly overlying the Gothlandian and sealed by the andesitic flows mentioned previously. The second largest field in the area is Guellala which is owned solely by Sonatrach. The structure is an anticline and, again, the reservoir rocks directly overlie the Gothlandian and are sealed by the "eruptive sequence." The estimated recoverable oil from this particular field is 78 million bbls.

Other major fields and their potential recoverable oil are Draa et Temra, 21 million bbl; Bou Khezana, 17.7 million bbl; Guellala Nord-est, 11.7 million bbl; Kef el Argoub, 16.7 million bbl; N'Goussa, 39.6 million bbl; Takhoukht, 8.7 million bbl; Haniet-El-Beida, 39 million bbl; Ouargla, 2.4 million bbl; and Garet Ech Chouf, 16.7 million bbls. For field information, see Table 1.

Haoud-Berkaoui (1965) Ben-Kahla (1966)
Sonatrach 51% TOTAL-Algerie 49%

These two accumulations mark the (structurally) highest position of any producing features in the Ouargla region. These two are part of the same association, so will be considered together.

Oil column	
unit 2 (178 m max.)	3,000 to 3,178 m subsea
unit 1 (304 m max.)	3,025 to 3,329 m subsea
Oil/water contacts	
unit 2	3,321 m subsea (Berkaoui)
unit 1	3,329 m subsea (Ben Kahla)
Density of oil	0.82 (± 40° API)
Gas/oil ratio (GOR)	140 cu m/cu m (± 780 cu ft/bbl)
Original formation pressure	508 Kg/ sq cm (± 7,255 PSI)
Surface flowing pressure (9 mm choke)	160 Kg/sq cm (± 2,285 PSI)
Example of flow rate (OK-101 Berkaoui)	93 bbl/hour = 2,200 bbl/day
Total recoverable reserves	383 million bbl

Draa Et Temra (1971)
Sonatrach 100%

At the time of writing, two wells were drilled and a third was underway.

Oil column	
unit 2 (25 m)	3,350 to 3,375 m subsea (?)
unit 1 (25 m)	3,466 to 3,491 m subsea
Oil-water contacts	
unit 2	not established
unit 1	3,491 m subsea (?)
Density of oil	0.8 (± 45° API)
Gas/oil ratio (GOR)	112 cu m/cu m (± 625 cu ft/bbl)
Original field pressure	
unit 2	525 Kg/sq cm (± 7,530 PSI)
unit 1	538 Kg/sq cm (± 7,685 PSI)
Example of flow (DRT.1)	90 bbl/hour = 2,160 bbl/day (frequent plugging by salt)
Total recoverable reserves	21 million bbl

Guellala (1969)
Sonatrach 100%

At the time of writing, seven wells were completed and three were being drilled.

Oil column (56 m)	3,309 to 3,365 m subsea
Oil-water contact	3,365 m subsea
Density of oil	0.82 (± 41° API)
Gas/oil ratio (GOR)	110 cu m/cu m (± 615 cu ft/bbl)
Original formation pressure	529 Kg/sq cm (± 7,550 PSI)
Example of flow rate (GLA-2)	198 bbl/hour = 4,550 bbl/day
Total recoverable reserves	78 million bbl

Guellala Nord-Est (1973)
Sonatrach 100%

Seismic interpretation shows two separate areas, not fully defined by drilling. At the time of writing, three wells were completed and two were drilling.

Oil column (23 m)	3,512 to 3,535 m subsea
Oil-water contact	3,535 m subsea (GLNE-1)
Density of oil	0.81 (± 42° API)
Gas/oil ratio (GOR)	140 cu m/cu m (± 780 cu ft/bbl)
Original formation pressure	550 Kg/sq cm (± 7,857 PSI)
Flowing surface pressure	50 Kg/sq cm (± 710 PSI)
Example of flow (GLNE-1)	63 bbl/hour = 1,510 bbl/day
Total recoverable reserves	11.7 million bbl

Bou-Khezana (1972)

Sonatrach 100%

At the time of writing, three wells were drilled.

Oil column	
unit 1 (10 m)	3,772 to 3,782 m subsea
Gothlandian (25 m)	3,909 to 3,934 m subsea
Oil-water contact	
unit 1	3,782 m subsea
Gothlandian	not established at writing
Density of oil	0.83 (± 39° API)
Gas/oil ratio (GOR)	
unit 1	93 cu m/cu m (± 520 cu ft/bbl)
Gothlandian	150 cu m/cu m (± 840 cu ft/bbl)
Original field pressure	
unit 1	551 Kg/sq cm (± 7,880 PSI)
Gothlandian	600 Kg/sq cm (± 8,570 PSI)
Flowing surface pressure	
unit 1	45 Kg/sq cm (± 640 PSI)
Gothlandian	39 Kg/sq cm (± 555 PSI)
Example of flow (BKZ-1, unit 1)	92 bbl/hour = 2,215 bbl/day
Total recoverable reserves	17.7 million bbl

Kef El Argoub (1974)

Sonatrach 51% Hispanoil 49%

At the time of writing, four wells were completed.

Oil column	
unit 2 (25 m)	3,505 to 3,530 m subsea (?)
unit 1 (20 m)	3,603 to 3,623 m subsea
Oil-water contacts	
unit 2	not established
unit 1	3,623 m subsea
Density of oil	0.81 (± 42° API)
Gas/oil ratio (GOR)	
unit 2	140 cu m/cu m (± 780 cu ft/bbl)
unit 1	134 cu m/cu m (± 750 cu ft/bbl)
Original formation pressure	
unit 2	548 Kg/sq cm (± 7,635 PSI)
unit 1	540 Kg/sq cm (± 7,530 PSI)
Example of flow (KG-2)	65 bbl/hour = 1,550 bbl/day
Total recoverable reserves	16.7 million bbl

N'Goussa (1975)

Sonatrach 100%

At the time of writing, three wells were completed and one was drilling.

Oil column	
unit 2 (30 m)	3,404 to 3,434 m subsea
unit 1 (49 m)	3,539 to 3,588 m subsea
Gothlandian	3,711 to 3,735 m subsea (?)
Oil-water contacts	
unit 2	not established at writing
unit 1	3,588 m subsea
Gothlandian	not established at writing
Gas/oil ratio (GOR)	140 cu m/cu m (± 780 cu ft/bbl)
Density of the oil	0.82 (± 41° API)
Original formation pressure	
unit 2	552 Kg/sq cm (± 7,840 PSI)
unit 1	555 Kg/sq cm (± 7,880 PSI)
Gothlandian	582 Kg/sq cm (± 8,270 PSI)
Flowing surface pressure	120 Kg/sq cm (± 1,704 PSI)
Examples of flow (NGS.1)	
unit 2	75 bbl/hour = 1800 bbl/day
unit 1	160 bbl/hour = 3,840 bbl/day
Gothlandian	25 bbl/hour = 600 bbl/day
Total recoverable reserves	39.6 million bbl

Takhoukt (1976)

Sonatrach 100%

At the time of writing, only two wells were completed.

Oil column (9 m)	3,485 to 3,494 m subsea
Oil-water contact	3,494 m subsea
Gas/oil ratio (GOR)	80 cu m/cu m (± 450 cu ft/bbl)
Density of oil	0.81 (± 43° API)
Original field pressure	510 Kg/sq cm (± 7,240 PSI)
Flowing surface pressure	20 Kg/sq cm (± 285 PSI)
Example of flow (TKT-1)	40 bbl/hour = 1,150 bbl/day
Total recoverable reserves	8.7 million bbl

Haniet-El-Beida (1978)

Sonatrach 100% (formerly Hispanoil joint interest)

Discovery made in the permit of Kef-El-Argoub. At time of writing, one well was completed, another was drilling. The field was not defined.

Producing interval	3,430 to 3,448 m subsea
Oil-water contact	not established at writing
Density of oil	0.81 (± 43° API)
Gas/oil ratio (GOR)	75 cu m/cu m (± 420 cu ft/bbl)
Original field pressure	526 Kg/sq cm (± 7,460 PSI)
Flowing surface pressure	33 Kg/sq cm (± 470 PSI)
Example of flow (HEB-1)	92 bbl/hour = 2,200 bbl/day (open hole test)
Total recoverable reserves	39 million bbl

Ouargla (1978)

Sonatrach 100%

This field lies partly under the townsite, which will hinder development. At time of writing, one well was completed and another was staked. Directional drilling may be required for further development.

Producing interval	3,503 to 3,511 m subsea
Oil-water contact	not established at writing
Gas/oil ratio (GOR)	—
Original field pressure	515 Kg/sq cm (± 7,360 PSI)
Example of flow (ORG-1)	25 bbl/hour = 600 bbl/day
Total recoverable reserves	12.4 million bbl

Garet Ech Chouf (1978)

Sonatrach 100%

The first well here was drilled by Sopefal in 1969. Eight liters of oil were recovered on Formation Interval Test (FIT) of a sandstone zone within the andesite. The zone was not deemed commercial (5 m of shaley sandstone) and the well was abandoned.

The area was included in a joint interest permit with Sopetral. Another well (ZCR-1) was drilled to the south with no reservoir development and the permit was returned to Sonatrach.

GEC-2 was then drilled and confirmed the trapping potential of the sand pinchout.

Productive interval	3,500 to 3,515 m subsea
Oil-water contact	3,516 m subsea
Density of oil	0.81 (± 42° API)
Gas/oil ratio (GOR)	197 cu m/cu m (± 1,100 cu ft/bbl)
Original field pressure	550 Kg/sq cm (± 7,860 PSI)
Flowing surface pressure	40 Kg/sq cm (± 570 PSI)
Example of flow (open hole, GEC-2)	113 bbl/hour = 2,712 bbl/day
Total recoverable reserves	16.7 million bbl

Intisar 'D' Oil Field, Libya[1]

By T. J. Brady, N. D. J. Campbell,[2] and C. E. Maher[3]

Abstract The Intisar "D" reef oil field was discovered by Occidental in October 1967; the discovery well tested 75,000 b/d. The prospect was based on reflection seismic data, which indicated the presence of an isolated reef. Three such prospects were drilled previously with varying degrees of success.

The Paleocene of the Sirte basin is characterized by carbonate rocks and shales deposited in an epeiric sea. The Intisar reefs grew in a late Paleocene embayment bounded on three sides by carbonate banks. Three distinct stages of organic development are recognized.

The Intisar "D" reef is roughly circular in plan and approximately 5 km in diameter. Its maximum thickness is 1,262 ft (385 m). The reef is coral and algal with grain- and mud-supported biomicrites. Porosity averages 22% and is mostly solution and intergranular. Measured permeability is as high as 500 md and averages 87 md. The main reservoir is remarkably homogeneous without noticeable layering typical of other reefs in the area.

The reef was full to spill point with a maximum oil column of 995 ft (291 m). The 40° API gravity oil has a paraffinic base and is low in sulfur. The original gas/oil ratio (GOR) was 509 cu ft/bbl. Original stock tank oil in place is estimated at 1.8 billion bbl. The field currently produces 200,000 b/d oil from 13 wells; 11 water-injection and 7 gas-injection wells are used. Cumulative oil production as of September 30, 1978, totaled 777 million bbl. Ultimate recovery efficiency is expected to approach 75%.

No pressure support was expected. Supplemental recovery operations began early and include pressure maintenance by both water and gas injection. The reservoir pressure is now maintained at the 4,000-psi level, high enough for miscible gas displacement.

INTRODUCTION

The Intisar 'D' field lies in the Sirte basin of Libya (Fig. 1) in Concession 103, a land block granted to Occidental in 1966. The discovery well was drilled in 1967 and production was tested at rates up to 75,000 b/d (11,900 cu m/d). The main reservoir is an upper Paleocene pinnacle reef containing more than one billion bbl (160 million cu m) of recoverable oil.

This paper describes the geology and deposition history of this oil field, together with the exploration techniques and the production system used.

REGIONAL GEOLOGY

Sirte Basin

The Mesozoic-Tertiary Sirte basin formed in a tensional tectonic regime with large scale areas of uplift and subsidence and associated normal faulting, but a noticeable absence of com-

[1]Manuscript received, April 4, 1979; accepted for publication, August 6, 1979.

[2]Occidental of Libya, Inc., Tripoli, Libya.

[3]Occidental Petroleum Co., Aberdeen, Scotland.

The writers thank the management of Occidental of Libya, Inc., for allowing the time and facilities for the preparation of this paper. Special thanks go to the staff of Robertson Research International, Ltd., who carried out the petrographic and paleontological studies, and to the present staff of Occidental of Libya, Inc., who contributed by discussion and editorial assistance.

C. G. Fisher and C. L. DesBrisay of Occidental Exploration and Production Company made valuable contributions to the seismic and reservoir engineering topics respectively. The writers thank Mrs. Audrey Murrel who typed the manuscript.

Article Identification Number:
0065-731X/80/M030-0028/$03.00/0.

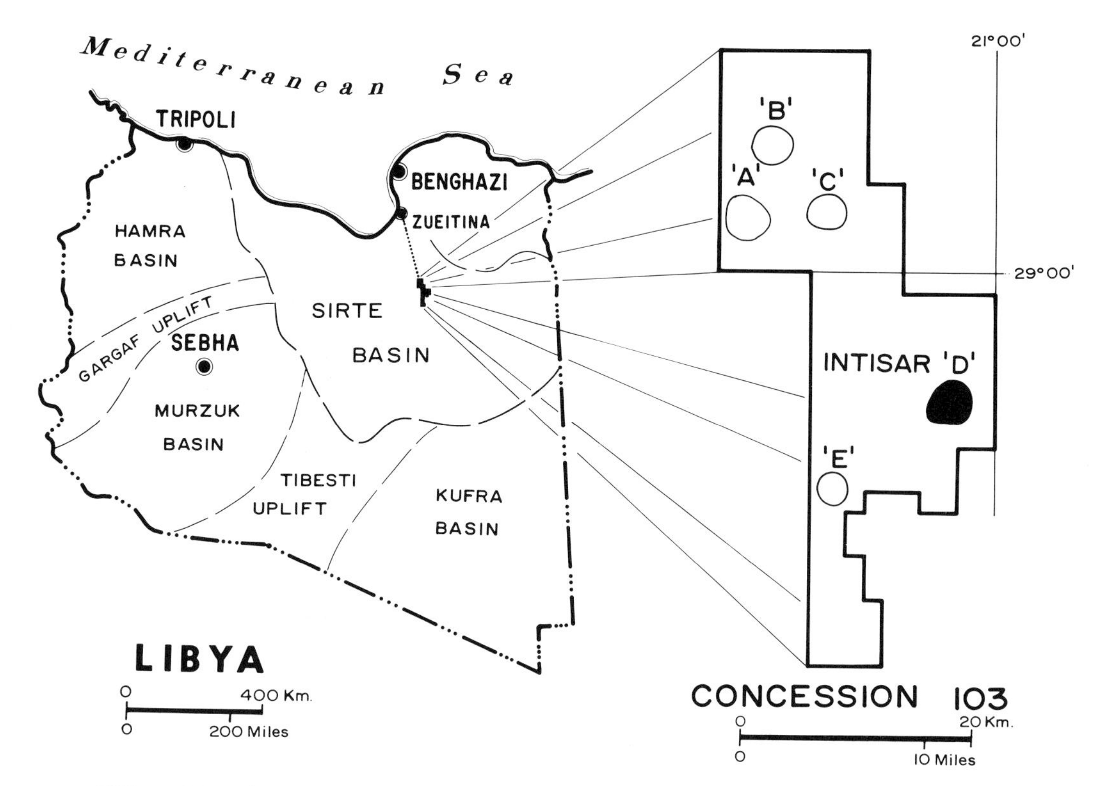

FIG. 1—Map of the Socialists People's Libyan Arab Jamahira showing location of Concession 103, and the Intisar 'D' field. Four other bioherms are also depicted.

pressional folds. The basin is bounded on the west by the Paleozoic Hamra and Murzuk basins, the south by the Tibesti uplift and the east by the Cyrenaica platform (Klitzsch, 1968). The major tectonic elements trend northwest to southeast (Fig. 2). The entire basin has been tilted down to the east with the stratigraphic section thicker in each trough as traced from west to east. The individual troughs are generally asymmetric with northeast dipping western flanks and fault-bounded eastern sides. These structural trends have been in existence since Late Cretaceous time and the deep eastern troughs contain thick salt sections that are loosely referred to as "pre-Upper Cretaceous," indicating an even older origin. (Conant and Goudarzi 1967).

The distribution of the Upper Cretaceous Rakb Group is well documented in the Sirte basin and an isopach of the group reflects the structure shown in Figure 2. The troughs have up to 5,000 ft (1,500 m) of sediments, though the section is noticeably thinner or absent on the highs. Limestones are common but arenaceous clastic sediments are absent except in the extreme southwest part of the basin. The absence of clastics, other than a thin basal unit, indicates that the highs, with their greatly reduced and partly attenuated sequences, were never above wave base, and that the relief across major faults was never great. One or two exceptions do exist, but they are localized and of relatively small magnitude. This group marks a major marine transgression.

By the end of the Late Cretaceous almost all the Sirte basin was a broad epicontinental sea. The open-marine Maastrichtian aged Kalash Formation can be traced throughout the area (Fig. 3), masking the previous tectonic and depositional history.

Marine sedimentation continued through the Paleocene and early to middle Eocene. This heterogeneous series of shales, carbonates, and evaporites shows considerable variations in formation thickness, which are related to the residual effect of the oil structures rather than renewed tectonic activity. The relatively rigid Precambrian and Paleozoic highs acted as stable areas in comparison to the synclinal regions, where differential compaction produced local, subtle differences in environment. Not until the Late Eocene to early Oligocene did rejuvenation of the major faults once more have a marked and direct effect on the sedimentation.

The stratigraphic section this study concerns is shown in Figure 4. The terminology is identical to that proposed by Barr and Weegar (1972). This is at variance with previously published papers which used terminology developed by individual oil companies. A previous paper on the Concession 103 area (Terry and Williams, 1969) used slightly different terminology from this presentation and the comparison is illustrated in Figure 4.

Concession 103 Area

The Concession 103 area is at the southern end of the Marsa Brega trough with the Amal/Gialo high to the east and the Zelten/Defa high somewhat farther to the west (Fig. 2).

After the blanketing by the open-marine, pelagic Kalash Formation, there was a differentiation into one area of continued carbonate deposition and another of nondeposition. The lower Sabil carbonate shelf (Fig. 5) attained a maximum thickness of 2,000 ft (600 m) before thinning abruptly to zero west

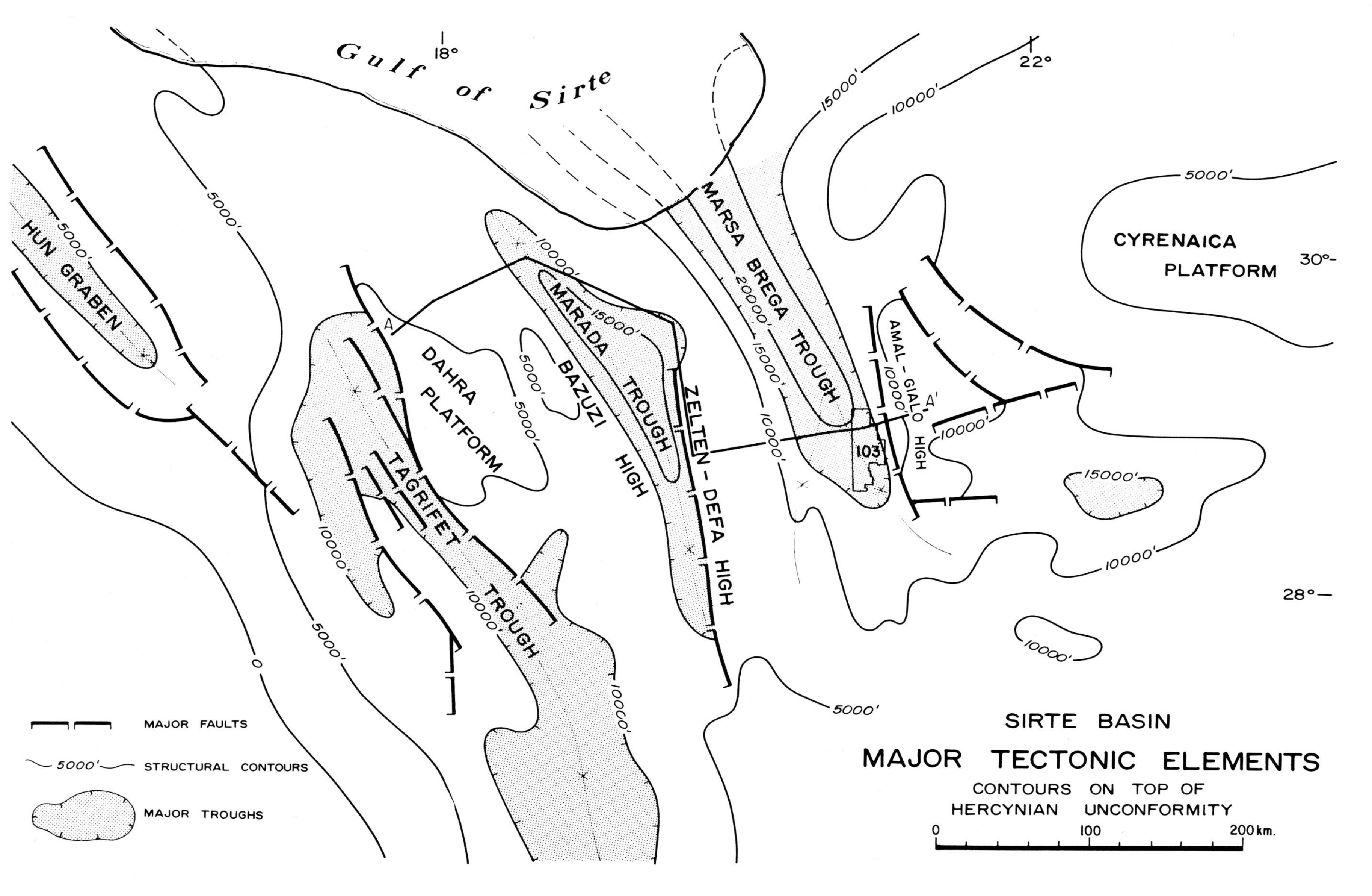

FIG. 2—Generalized tectonic map of the Sirte basin showing major depocenters. Structural contours on pre-Upper Cretaceous unconformity shown in Figure 3.

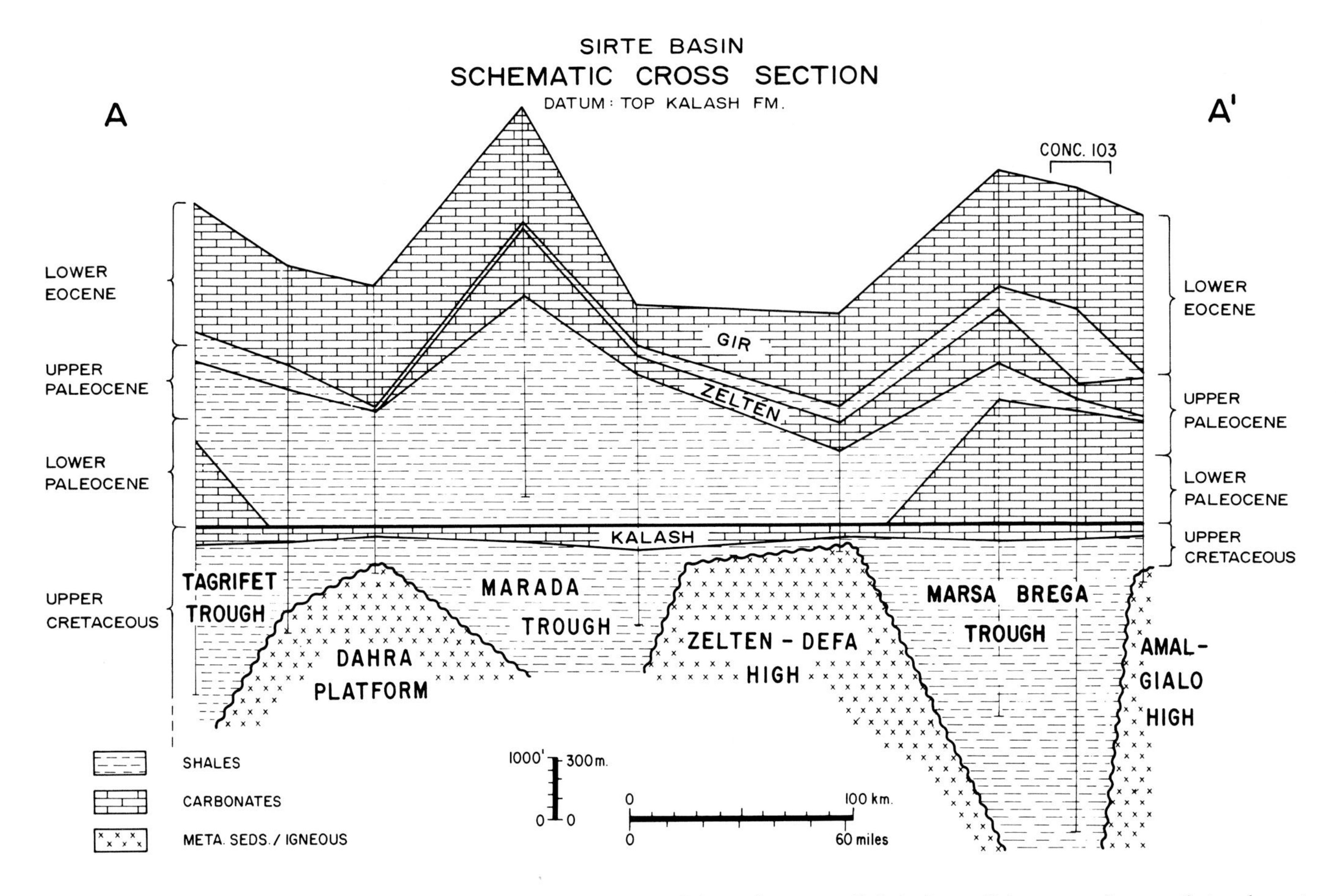

FIG. 3—Stratigraphic cross section, location on Figure 2, datum on top of Upper Cretaceous Kalash. Lower Paleocene carbonates that wedge out over troughs independent of location of earlier highs.

of Concession 103. The isopachs give little hint of the underlying structures. In the 103 subregion, the rate of subsidence never exceeded the ability of organisms flourishing at the time to manufacture sufficient material for production of a carbonate bank up to 2,000 ft (600 m) thick.

The close of lower Sabil deposition is marked by the influx of the Sheterat shale. Unlike previous authors (Terry and Williams, 1969), we see no conclusive evidence of a marine transgression or increased rate of subsidence, but prefer the interpretation that the appearance of the Sheterat shales marks the existence of a far distant provenance that suddenly supplied argillaceous material into a previously clear-water environment. This formation ranges in thickness from zero to over 2,500 ft (700 m) as a complement to the lower Sabil (Fig. 6).

A cessation of the supply of argillaceous material saw the return of widespread deposition of shelf carbonates. The upper Sabil carbonates have a much broader distribution than that of the lower Sabil (Fig. 7). The upper Sabil deposition continues to the west of Concession 103 without any noticeable change in thickness where it overlaps the lower Sabil shelf edge. In the Concession 103 area, however, the effect of the Marsa Brega trough is dramatic. While lower Sabil deposition was capable of compensating for any variable subsidence of the underlying strata due to compaction or slight fault movement, the upper Sabil was not. The re-entrant shown on Figure 7 depicts the inability of the biota to match the greater rate of subsidence in the trough. Without the effect of the Marsa Brega trough, the upper Sabil would doubtless be fully developed across the 103 area. As it is, isolated organic buildups of the same vertical magnitude as the carbonate shelf seem nature's best response to the inhospitable environment of the trough.

The reintroduction of shale to the region terminated upper Sabil deposition. This formation, the Kheir Marl, infilled the Concession 103 re-entrant and encased the organic buildups (Fig. 8). During the gradual infilling of the Marsa Brega trough two episodes of renewed biologic activity took place, and are discussed more fully later. The Kheir is conformably overlain by the limestones of the Gir Formation, which range up to 1,700 ft (520 m) thick, and which are in turn overlain conformably by the Gialo limestones.

The relationship of high stable blocks to deep slowly subsiding troughs, together with the facies variation of the Paleocene sediments, were known (although not fully understood) at the time Occidental acquired the concession. Figure 9 is a schematic cross section from west to east (location shown in Figs. 5, 7) using data available before Occidental began operations. It is clear that the central well has a section of upper Sabil and Kheir not in accordance with the other wells shown and can be

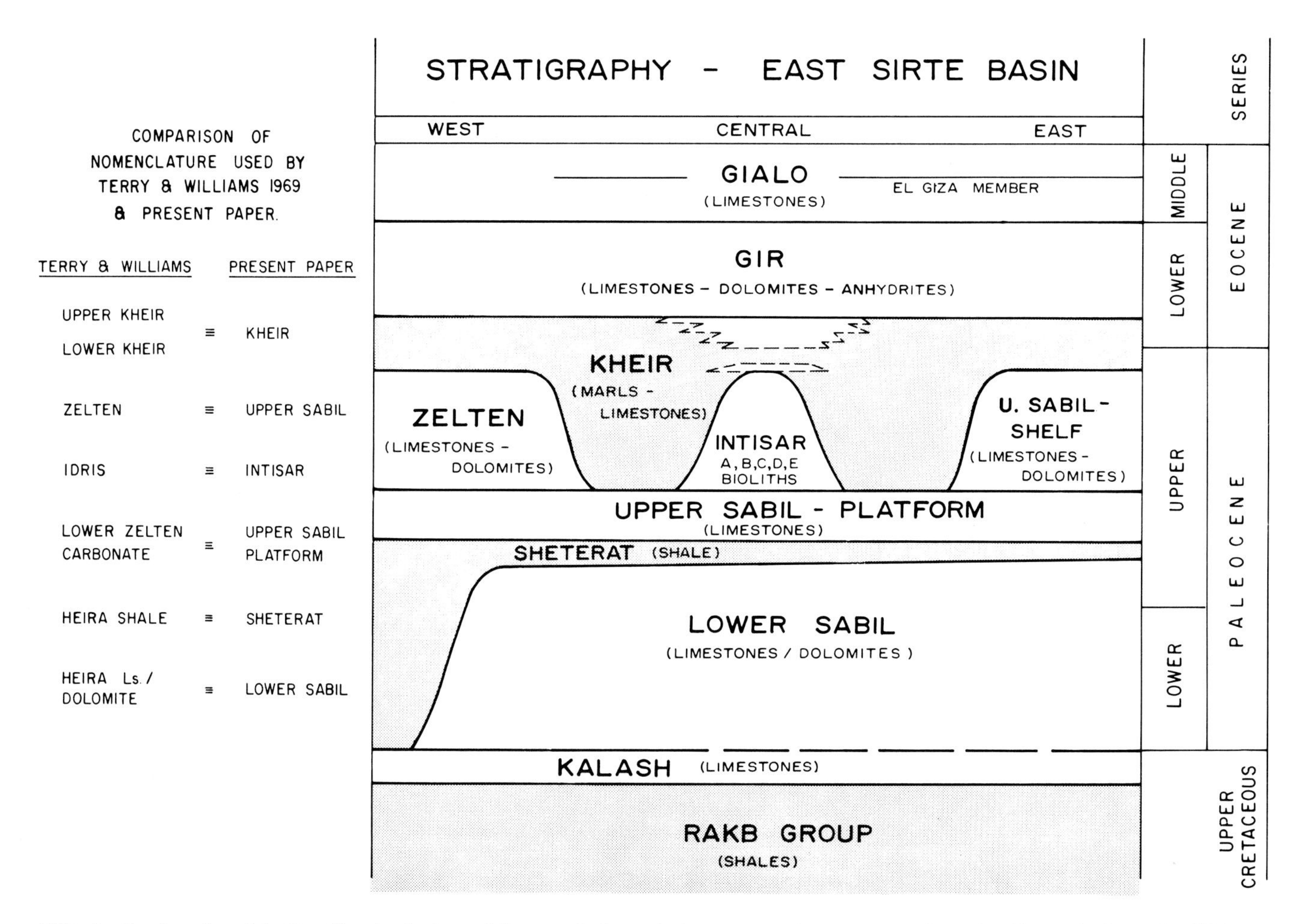

FIG. 4—Stratigraphy of the East Sirte basin. Dotted lines are facies variations. Heavy lines are generalized time lines. The lower Sabil/Sheterat interfacing in the west is more complex than depicted. Several limestone interbedded in the Sheterat have, by others, been equated with the lower Sabil, thereby making the two formations the same age. The interpretation in this paper is that the Sheterat post-dates the lower Sabil buildup.

interpreted as a classic reef/off-reef relationship.

SEISMIC EXPLORATION

Seismic field work in Concession 103 began approximately 9 months prior to spudding of the Intisar 'D' discovery well. The first two lines recorded crossed isolated reefs that eventually became the Intisar 'A' and 'B' oil fields. Seismic data allowed immediate identification of reefs, although the isolated pattern was not apparent until later. The 'A,' 'B,' and 'C' reefs were found and delineated by seismic data with 'A' and 'B' reefs being drilled before 'D' reef was detected on a seismic line; thus the type of prospect was well understood by the time 'D' reef was discovered.

The terrain is flat, gravel-covered desert ideal for seismic operations. Seismic data were digitally recorded, one of the early uses of that technique in Libya. A six-fold stack was used, after experimentation with 1200 and 2400% indicated that 600% multiplicity would be sufficient in most cases. An 18-hole pattern designed to eliminate two shot-generated, ground roll noise trains was used, as analog seismic surveys previously made of the area produced data quality insufficient for recognition of the reefs.

It was quickly determined that in most cases only the seismic lines crossing a reef near its crest would exhibit clear evidence of the reef. Figure 10 illustrates good seismic evidence for a reef: (1) convex shape at the reef top, (2) overlying drape, (3) break-up of data at reef edge, (4) little or no continuity through reef mass, (5) velocity sag under the reef which is due to interval velocity of the reef being lower than that of the surrounding rock.

Many seismic lines crossing well up-flank but still distant from the crest are insufficiently definitive to outline the reef mass. Figure 11 is a seismic section cutting the reef in a flank position, and is considerably less clear than the previous example. The exploration technique eventually used recorded lines radiating from an apparent center once reef evidence was observed on a line.

The four reefs, 'A' through 'D,' exhibited similar seismic characteristics, but drilling of the prospects so defined resulted in four oil fields of drastically differing reserves. While Intisar 'C' has less than 200 million bbl (30 million cu m) of oil in

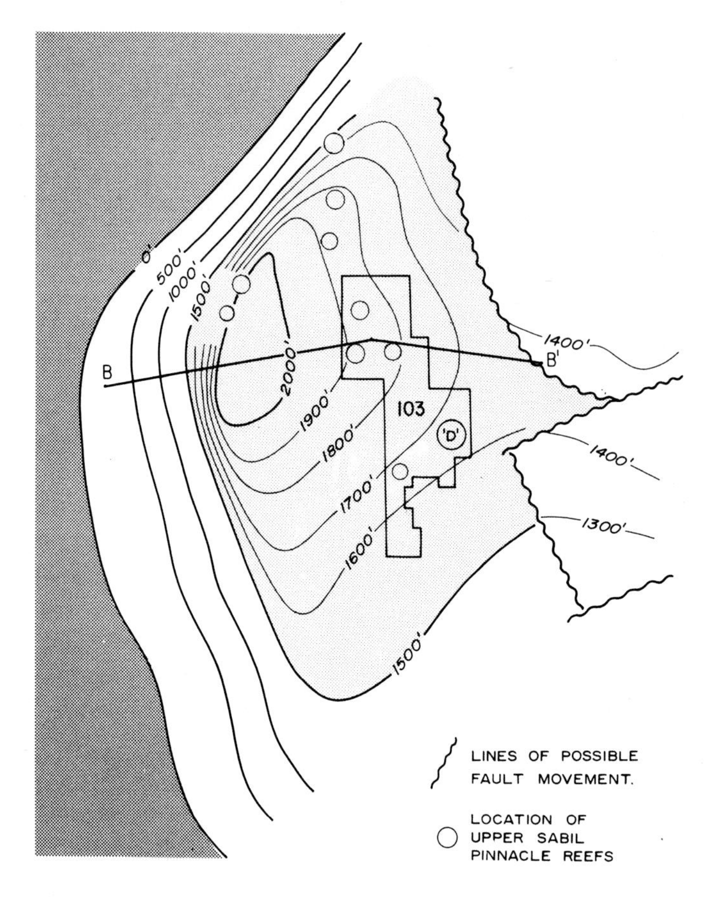

FIG. 5—Isopach map of the lower Sabil. Location of subsequent upper Sabil isolated bioherms is restricted to the area of full lower Sabil buildup.

place, Intisar 'D' has over 1.8 billion bbl (290 billion cu m) oil in place.

FIELD GEOLOGY

Reservoir Parameters

The D1-103 well was drilled to a total depth of 9,923 ft (3,024 m) and encountered oil-bearing carbonate rock from a depth of 8,946 ft (2,727 m) to 9,834 ft (2,997 m). A drill-stem test of the interval from 8,949 ft (2,728 m) to 9,000 ft (2,743 m) tested 40° API oil at a rate of 3,360 b/d (535 cu m/d) with a gas/oil ratio of 581 cu ft/bbl (103 cu m/cu m). Subsequent production tests yielded rates of up to 75,000 b/d (11,900 cu m/d). The productivity index was calculated at 423 b/d per psi drawdown (956 cu m/d per kgf/sq cm; DesBrisay and Gray 1975). Permeabilities in the field range from 4 md to over 500 md with an arithmetic average horizontal permeability of 87 md. However, individual well performances indicate that an average permeability of over 200 md is required to match the pressure buildups and productivities.

The induction-electrical, self-potential, caliper, and micrologs through the pay zone of the D1-103 are shown in Figure 12. Water saturation decreases from the oil-water contact (defined at water saturation of 50%) to less than 10% at the top of the reservoir. The uniform slope of the resistivity curve caused by capillary forces and the uniform microlog separation indicate the homogeneous nature of the reservoir. Average net porosity in the field is 22%, while actual values range as high as 35% and as low as 4%.

A total of 36 wells were drilled to, or through, the reservoir. An additional seven wells were drilled to develop a shallow pool in the El Giza Member of the Gialo Formation. Wells drilled in the central area of the field penetrated continuous porosity from the top of the pay zone to the oil-water contact and below. Wells drilled on the field periphery, particularly on the east flank, encountered one or more significant nonporous,

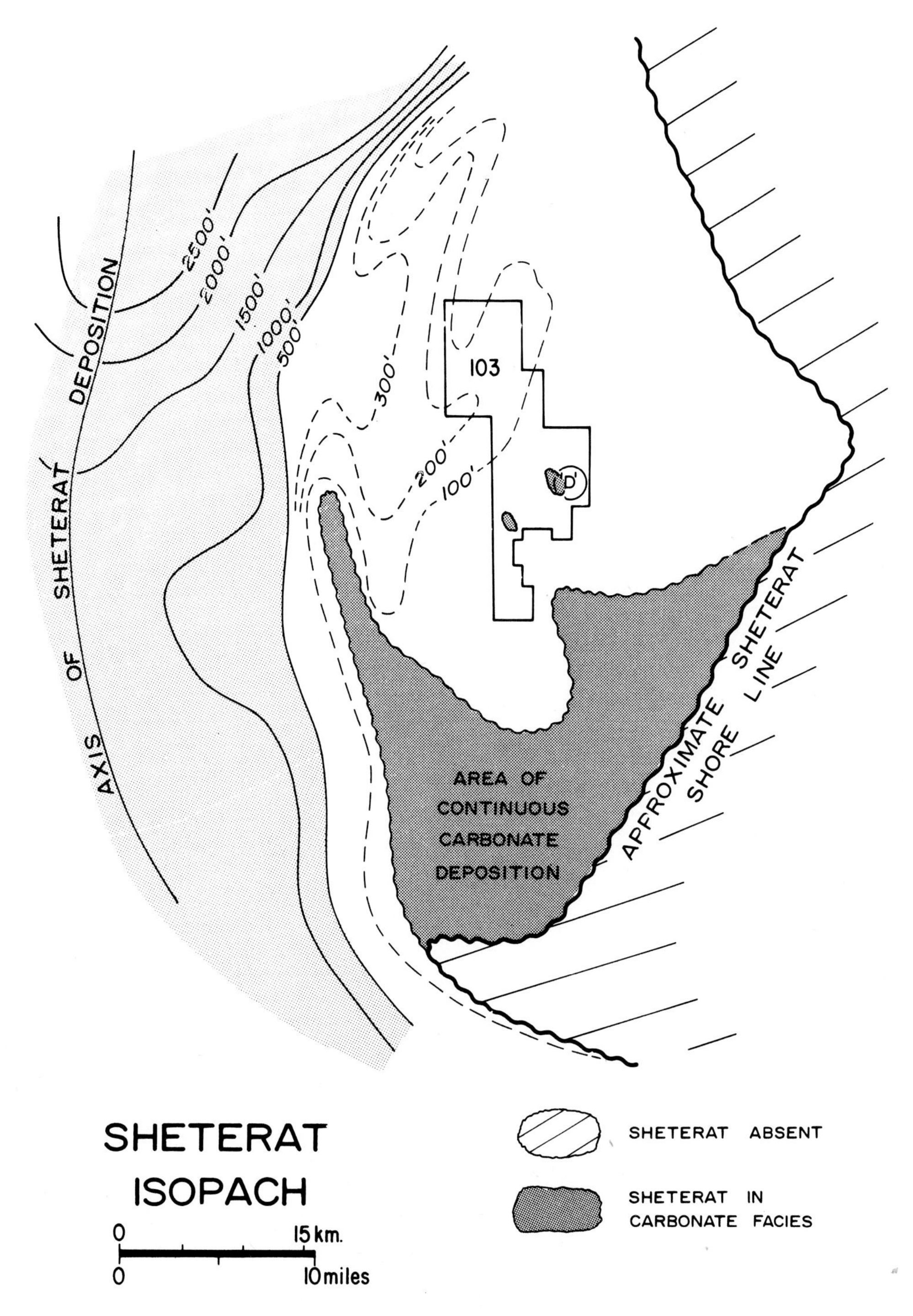

FIG. 6—The Sheterat shale thins abruptly onto the lower Sabil shelf and is developed in a carbonate facies south of Concession 103. The formation becomes progressively more difficult to trace eastward and is absent approximately where shown. This possible shoreline, together with the isopach and facies distribution indicates a marine regression rather than a transgression.

nonoil-bearing intervals within the reservoir. Figure 13 is a cross section showing this relationship and the division of the reservoir into three units.

The field, roughly circular in plan with a diameter of 3 mi (5 km), is shown in Figure 14. The structural contours top the first porous unit and portray the three elements; Intisar 'D' reef unit, Intisar unit 'A,' and Intisar unit 'B' (Fig. 13). These units will be discussed more fully later. The outer limit of the field was delineated by projecting the degree of slope of flank wells, correlating stratigraphically with wells outside the field area and intergrading these data with the seismic control. While the porosity of the lower part of the upper Sabil is too poorly developed to provide an active water drive, it is sufficient for fluid transmissibility and it is believed that the field was filled

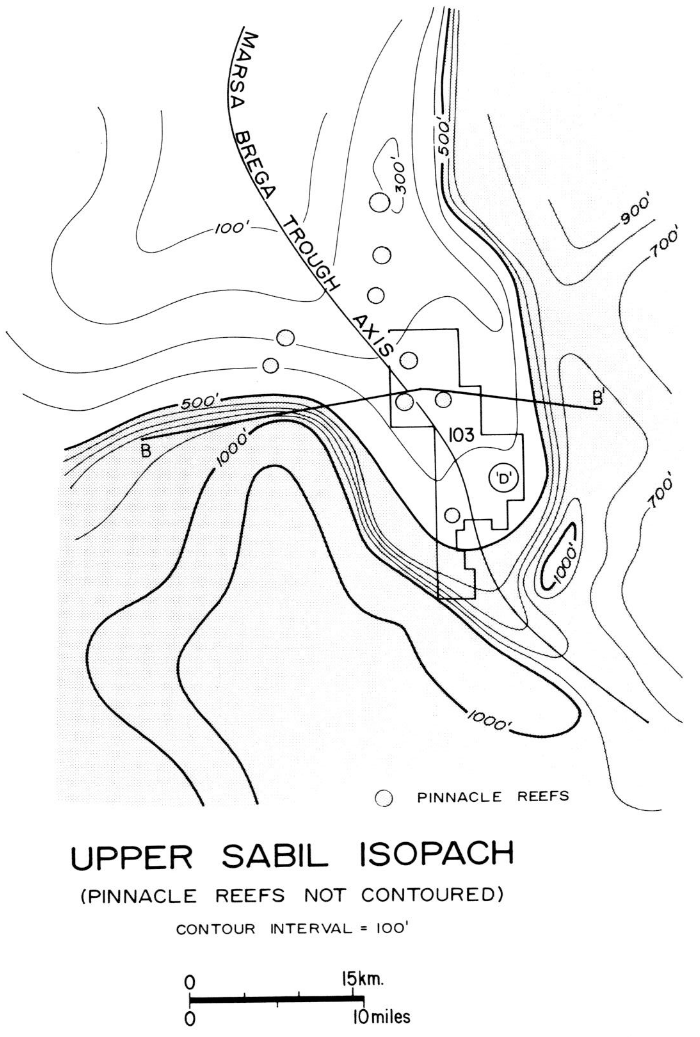

MARSA BREGA TROUGH AXIS

100'

300'

500'

900'

700'

1000'

103

'D'

B

B'

PINNACLE REEFS

UPPER SABIL ISOPACH

(PINNACLE REEFS NOT CONTOURED)

CONTOUR INTERVAL = 100'

0 15km.

0 10miles

FIG. 7—Upper Sabil isopach map showing location of upper Sabil pinnacle reefs in the Marsa Brega trough re-entrant. The 500 ft (152 m) contour is interpreted as being the approximate edge of an upper Sabil barrier reef.

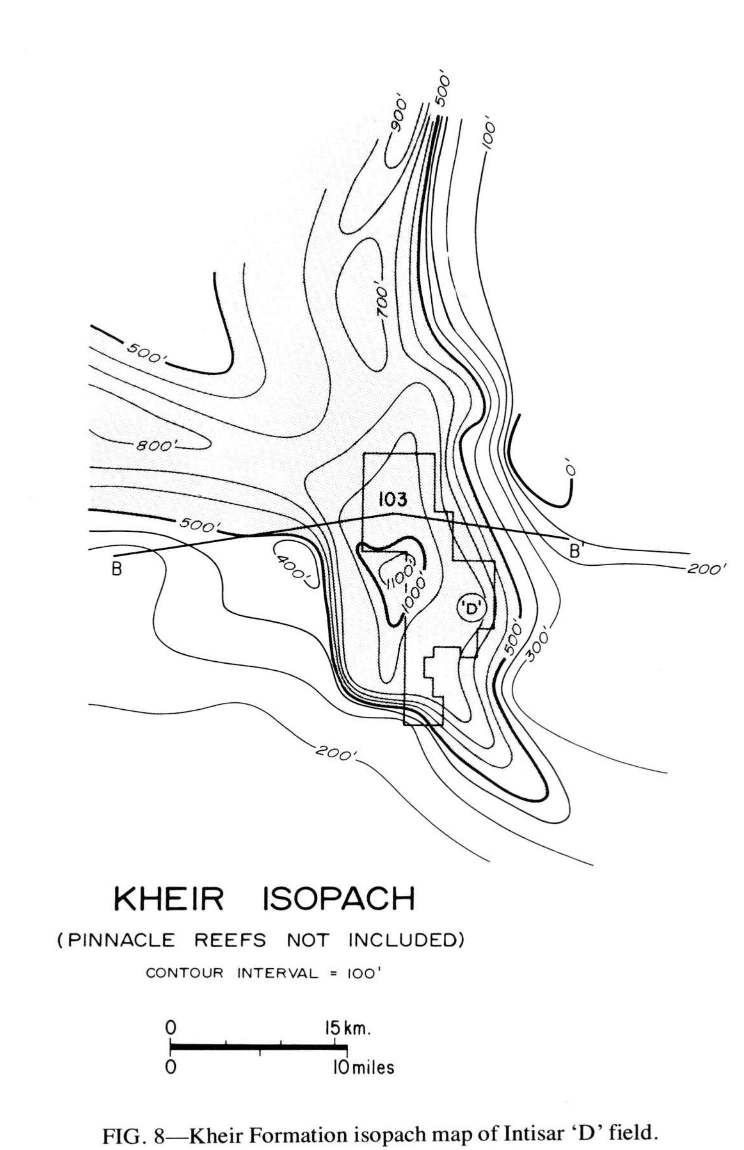

FIG. 8—Kheir Formation isopach map of Intisar 'D' field.

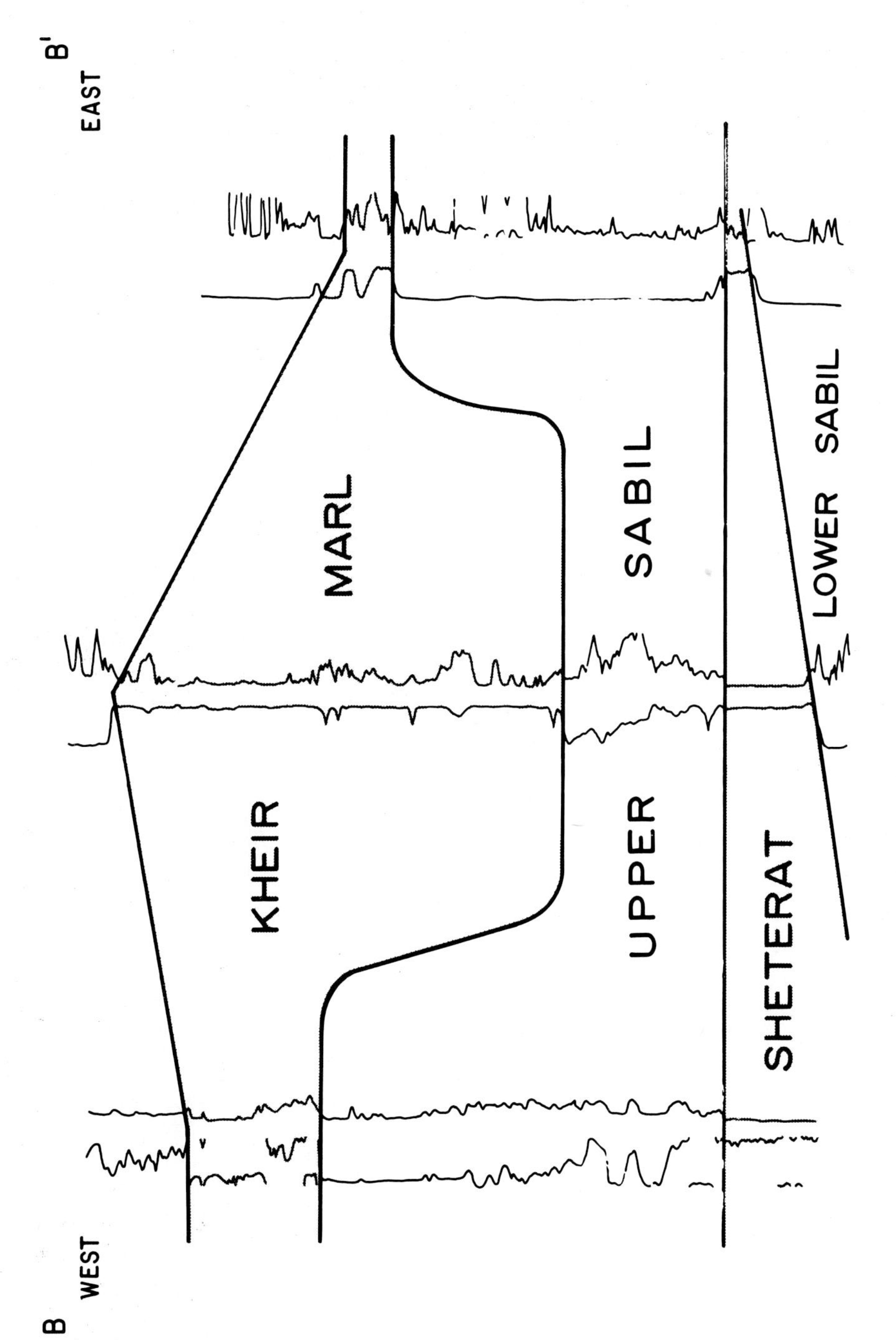

FIG. 9—Schematic cross sections, location on Figures 5 and 7, across Concession 103. Wells were drilled prior to 1966.

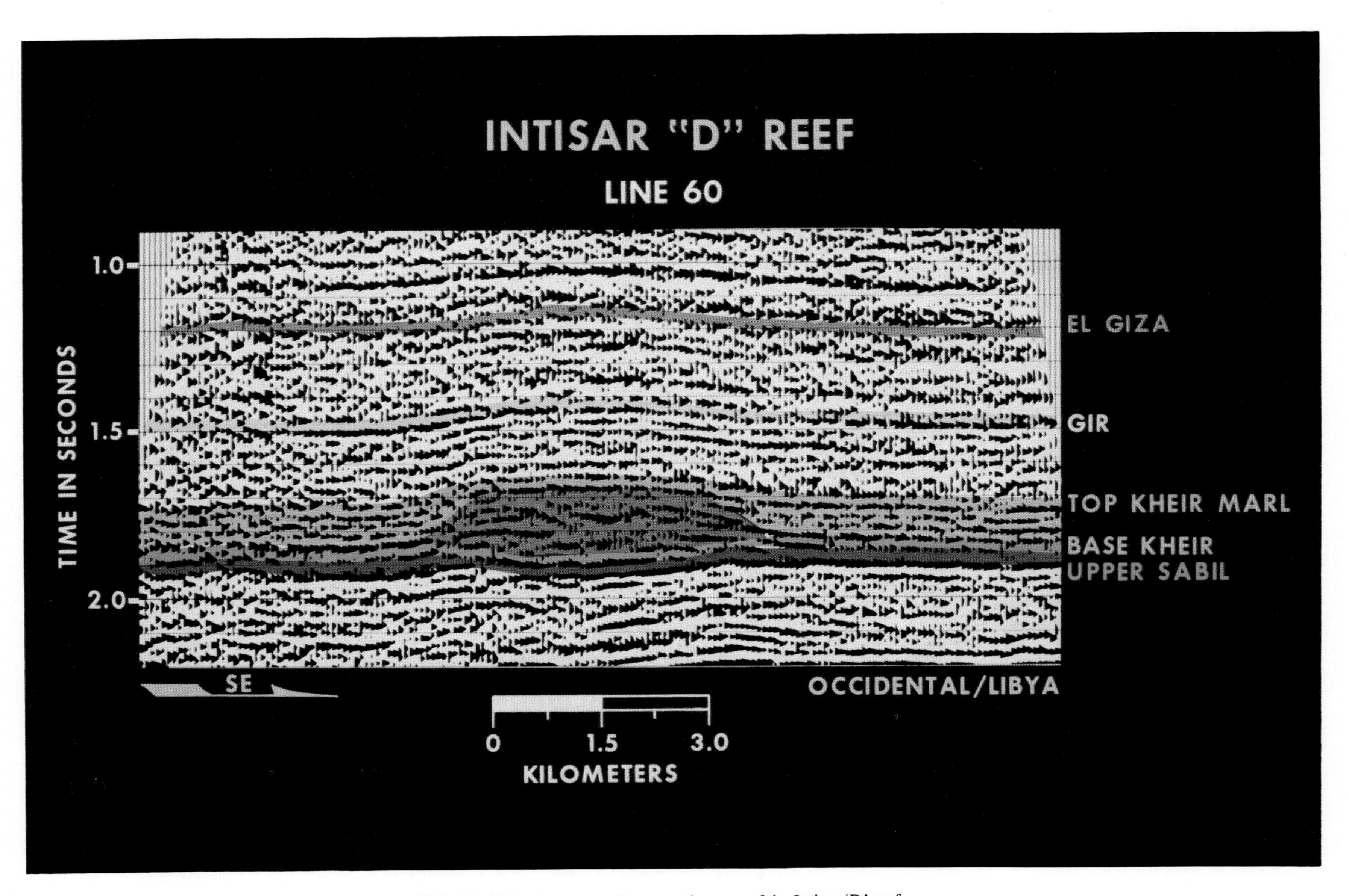

FIG. 10—Seismic cross section over the crest of the Intisar 'D' reef.

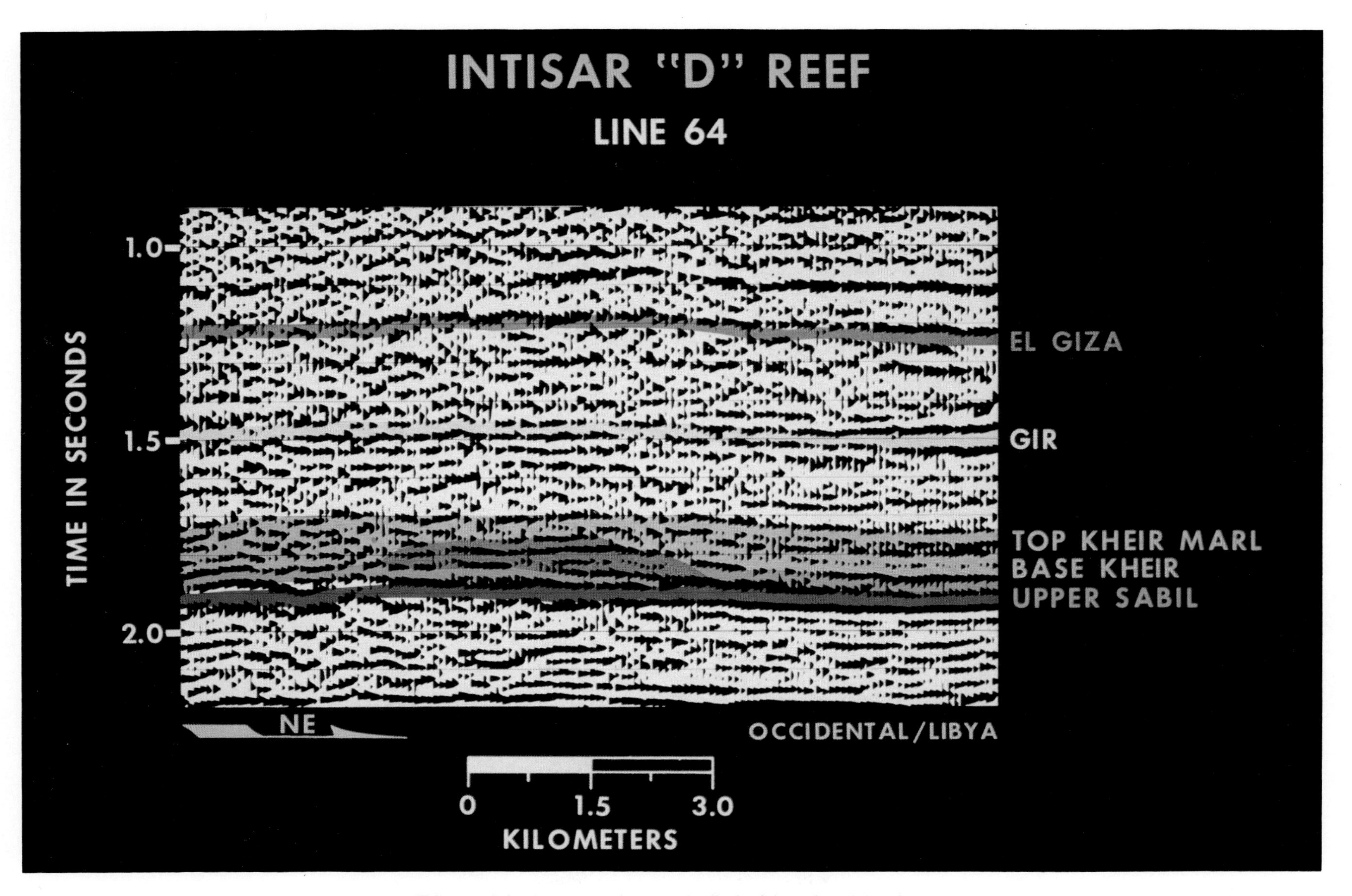

FIG. 11—Seismic cross section over the flank of the Intisar 'D' reef.

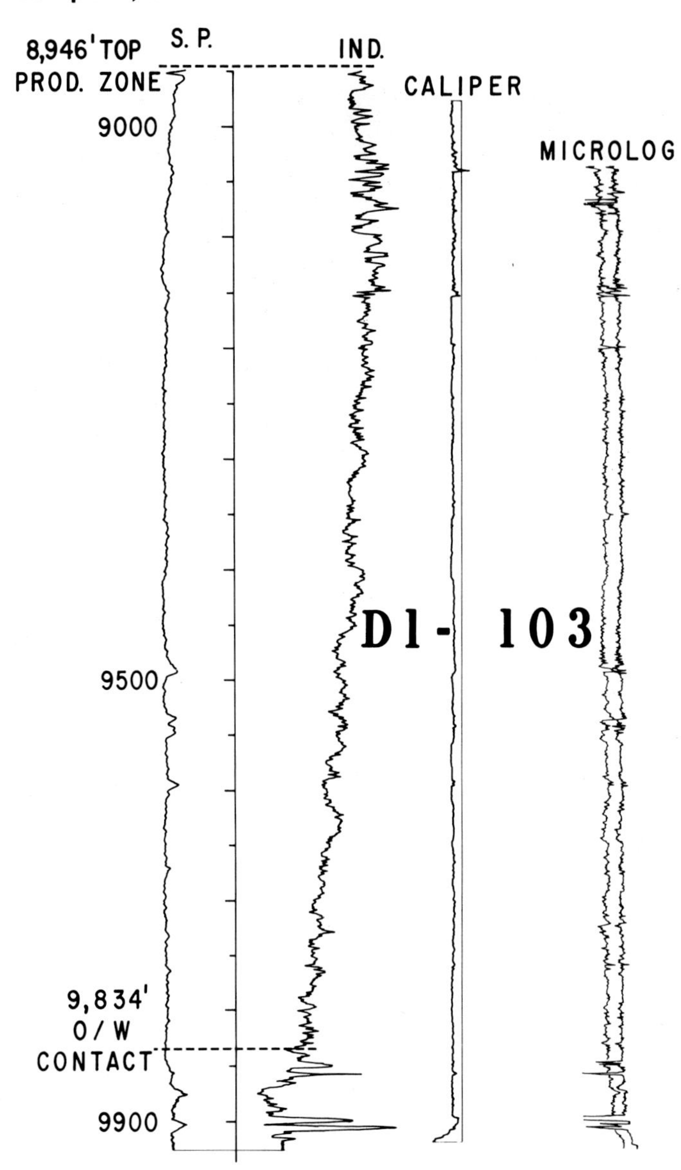

FIG. 12—Well D1-103 logs.

to spill-point by fluid migration through this zone.

Rock Types

The main lithofacies of the upper Sabil in the Intisar 'D' field are: (1) algal-foraminiferal biomicrite/biosparite, (2) biomicrites, and (3) coral biolithites. Lesser occurrences of pelbiosparites and microcrystalline dolomites are also recorded.

Algal-foraminiferal biomicrites—These are the dominant lithology. They are generally medium grained, mud supported, or grain supported, although the whole spectrum from fine to coarse grained is represented. Figure 15 illustrates a grain supported specimen. Calcareous red algae are the most abundant macrofossil. They, along with coral debris, echinoderm fragments, bryozoans, pelecypods, gastropods, and occasional crinoids (in varied proportions) make up the total assemblage. The microfossils consist of nummulites, discocyclinids, rotalids, miliolids, alveolinids, encrusting and calcareous benthonic forams. A few planktonic forams and ostracods are also reported. The algal-foraminiferal biomicrites that form the major rock type of the reservoir, exhibiting excellent solution and good intergranular porosities, are also the major lithofacies of the relatively tight basal section of the upper Sabil. The total quantity of allochems here is greatly reduced and the biota are mainly planktonic.

Biomicrites—Biomicrites differ from the algal-foraminiferal group only in relative fossil content with no one group dominating. Figure 16 is a fine grained, mud supported biomicrite and Figure 17 is a coarse grained, grain supported example.

Coral biolithites—Coral biolithites (Fig. 18) are composed of a framework of compound corals with interstitial biomicrite. Compound corals are the dominant component of the macrofauna with bryozoans, red algae, and mollusks common. The

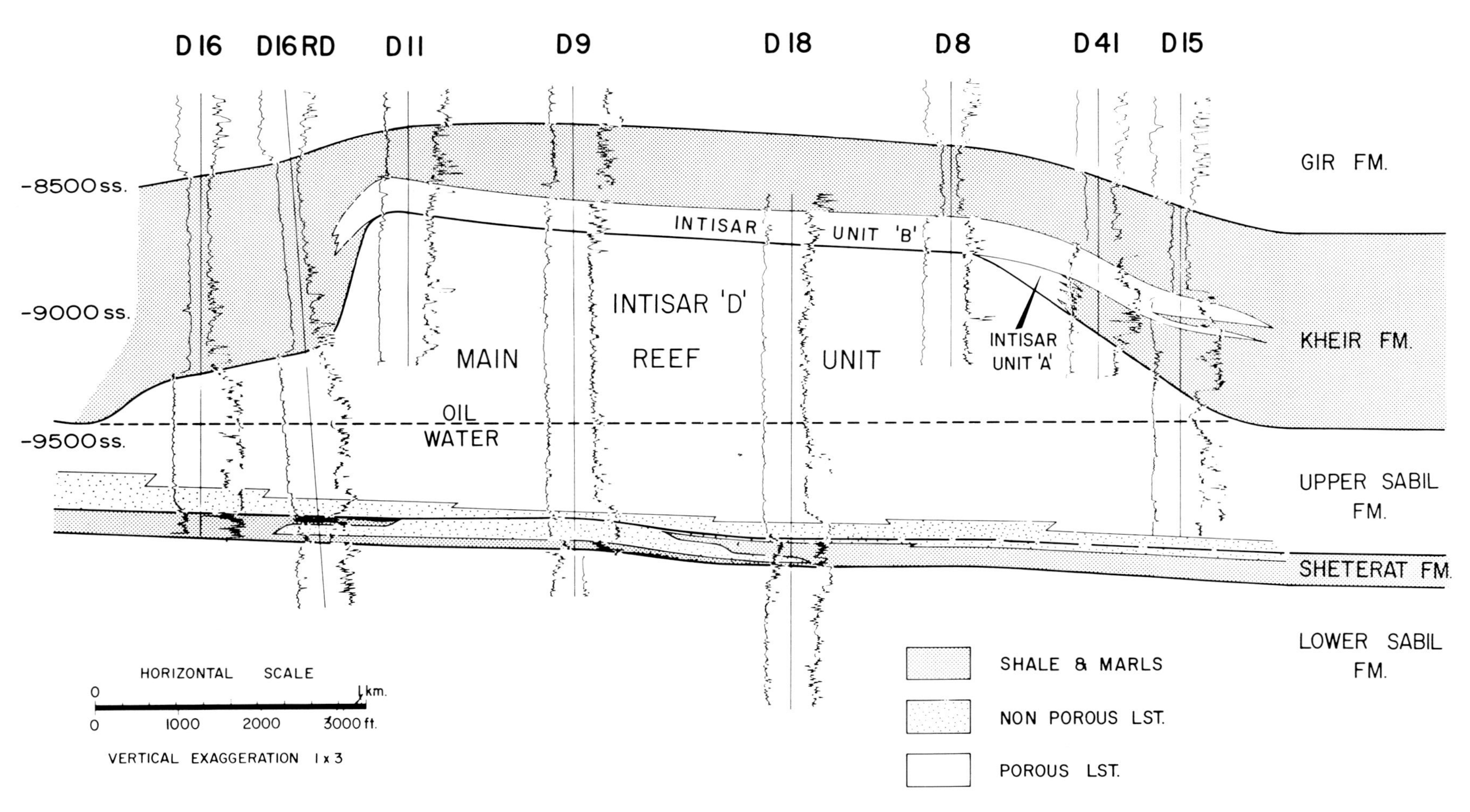

FIG. 13—Intisar 'D' reef cross section. Induction Electric logs (IES) logs are used down to top of reservoir and Gamma Ray/Formation Density Log (GR-FDC) from there to total depth. Wells D11 and D19 have GR-FDC only.

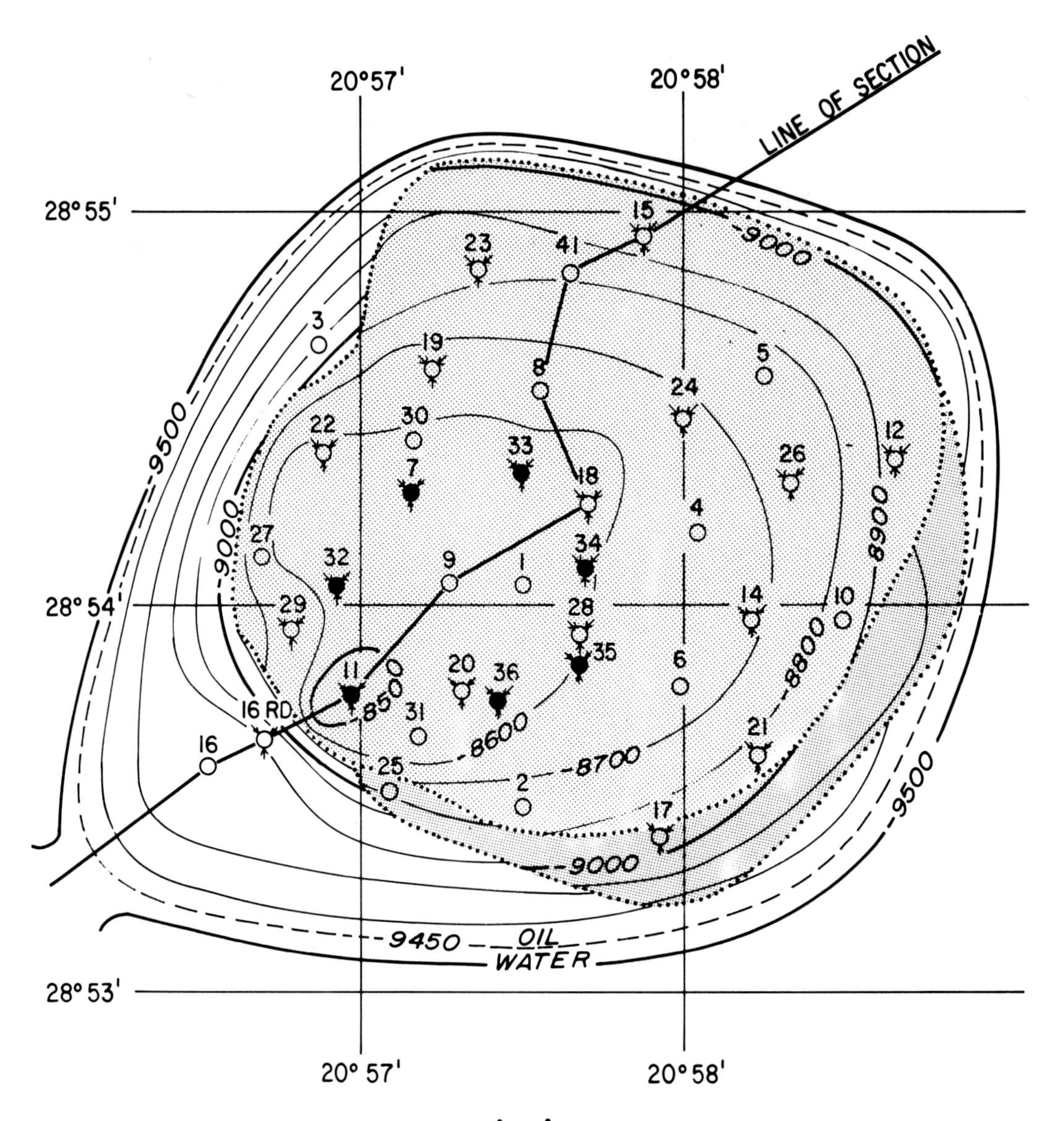

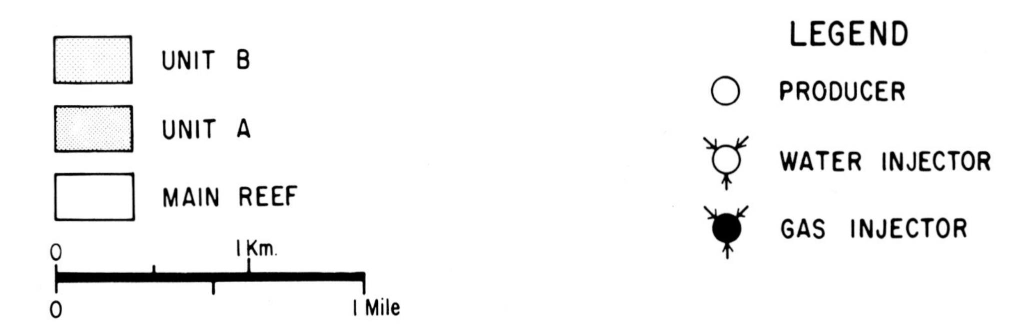

FIG. 14—Structural contour and well location map for Intisar 'D' field.

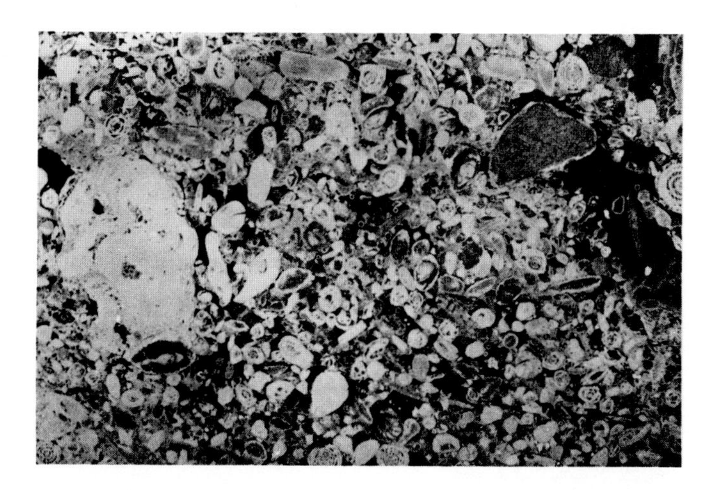

FIG. 15—Grain supported, algal-foraminiferal biomicrite from 8,987 ft (2,739 m) depth in well D1.

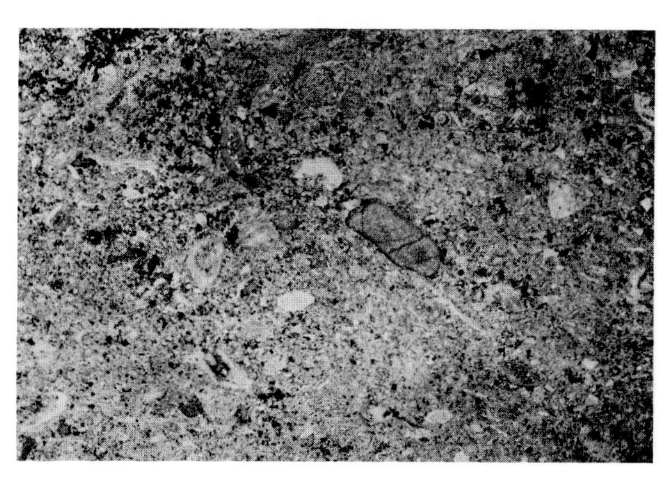

FIG. 16—Mud supported biomicrite from 9,832 ft (2,997 m) depth in well D9.

FIG. 17—Grain supported biomicrite from 9,794 ft (2,985 m) depth in well D9.

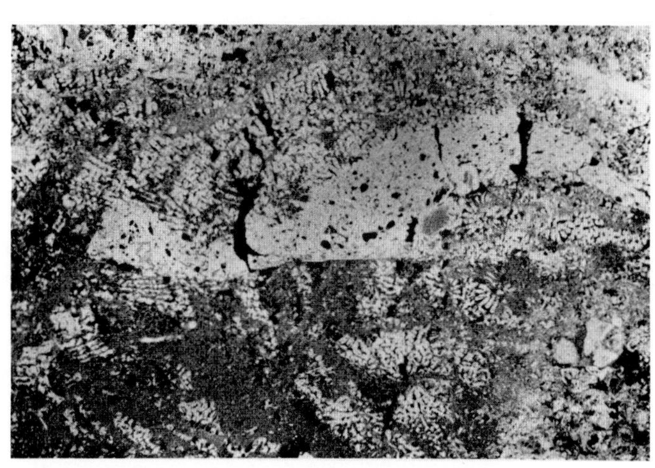

FIG. 18—Coral biolithite from 9,710 ft (2,960 m) depth in well D9.

microfauna commonly is represented by encrusting forams and both planktonic and calcareous benthonic forams occur. Porosity is mainly of solution origin but significant amounts of intergranular and intragranular porosity are present.

Pelbiosparites are scarce and restricted to centrally located wells. The macrofauna is dominantly crinoid debris and algae fragments, whereas alveolinids and miliolids comprise the microfauna. Porosity is moderate and apparently of solution origin. Secondary dolomites are scattered in 5 to 30 ft (1.5 to 9 m) beds apparently at random, in the section. These dolomites are microcrystalline, made up of equigranular aggregates of anhedral to subhedral crystals. Relic structures are generally well-preserved and the original rock types were grain or mud supported biomicrites. Porosity is intercrystalline and ranges from 4 to 12%, the higher values being enhanced by solution. The dolomites are more common in peripheral wells and when found in centrally located wells tend to be restricted to the lower part of the section. Their discontinuous nature is seen in the twin D13 and D18 wells. Although drilled within 525 ft (160 m) of each other, several of the dolomites present in the D13 cannot be traced to D18.

Unlike the Intisar'A' bioherm (Terry and Williams, 1969), the 'D' reef does not lend itself to division into horizontal layers by lithofacies. Figure 19 illustrates correlation within the reef, using gamma-ray markers, and the lithofacies as determined from cores and cuttings. The only consistent correlations are the gamma-ray markers. The lithofacies exhibit dramatic lateral variations and in the case of the algal-foraminiferal and nondesignated biomicrites, several correlations are possible.

DEPOSITIONAL HISTORY

As outlined previously, the lower Sabil of Concession 103 is a small part of a large carbonate shelf. The dolomitized biomicrites have an assemblage of echinoderms, gastropods, green algae, miliolids, rotalids, and arenaceous and calcereous benthonic foraminifers. Planktonic forams appear at the top of the

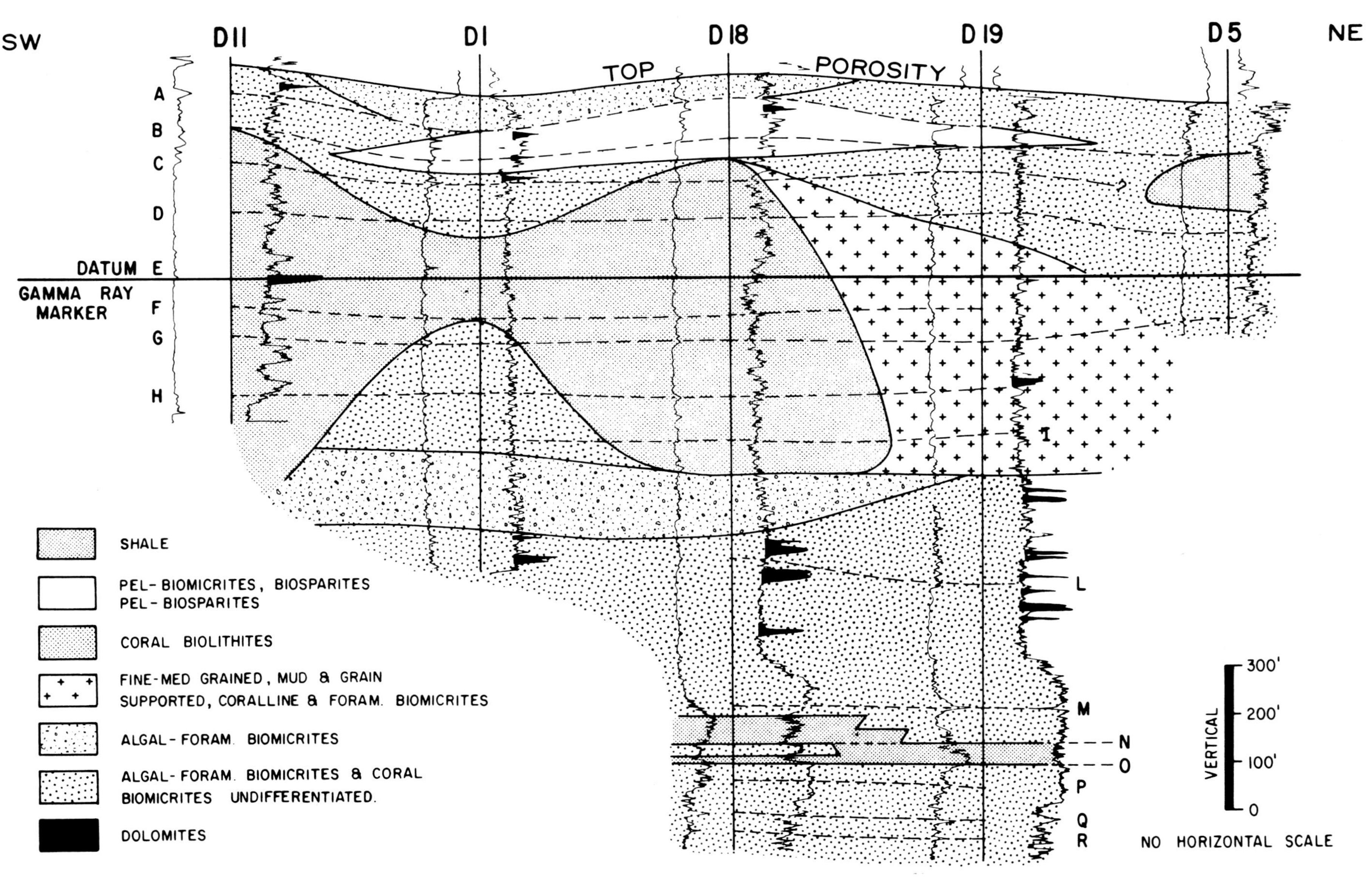

FIG. 19—Generalized lithofacies and gamma ray correlations. Logs displayed have gamma ray curve on the left and compensated formation density on the right.

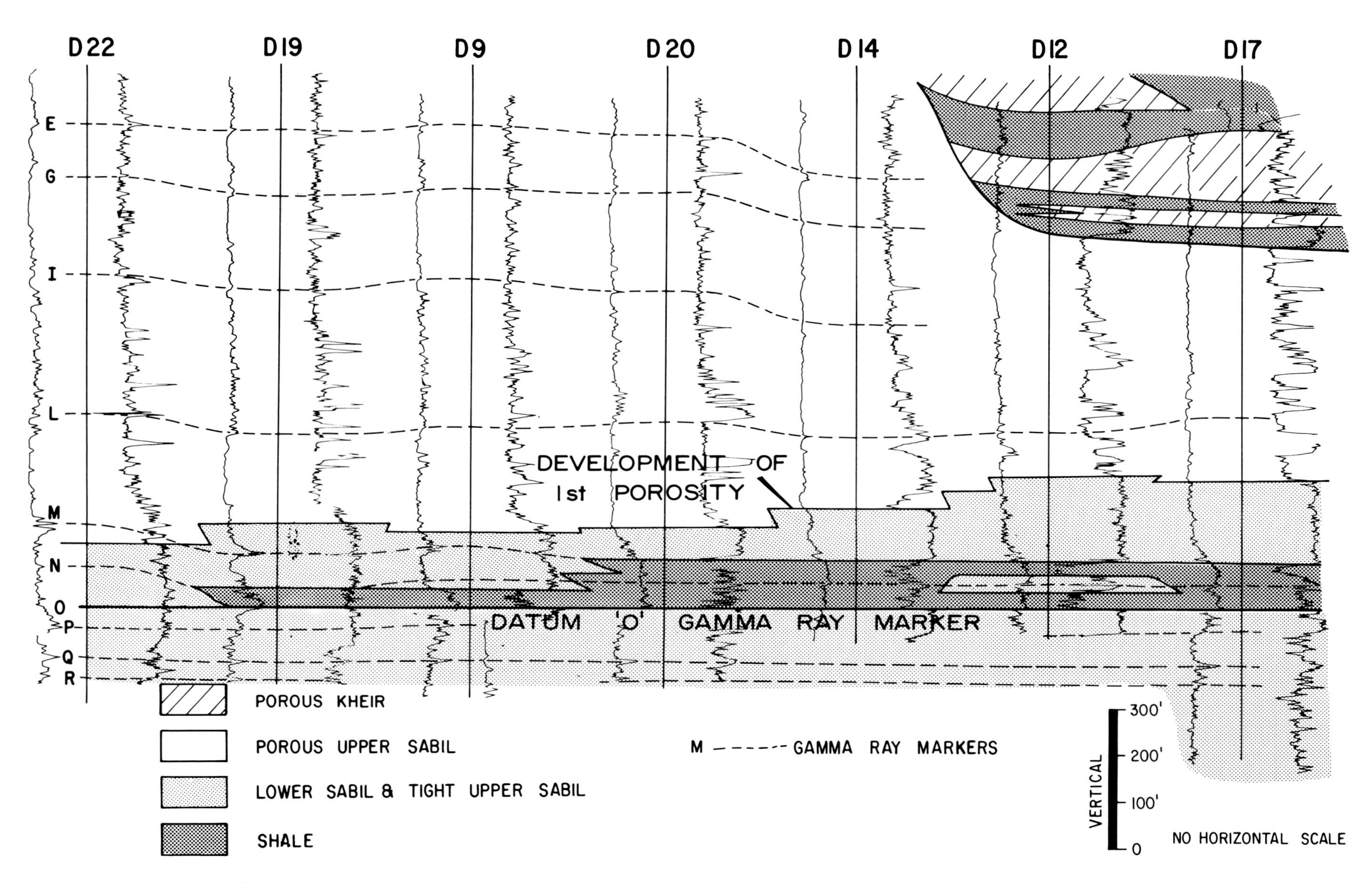

FIG. 20—Stratigraphic cross section showing Sheterat facies change and occurrence of first porosity with the upper Sabil. Gamma-ray curve on the left, compensated formation density on the right.

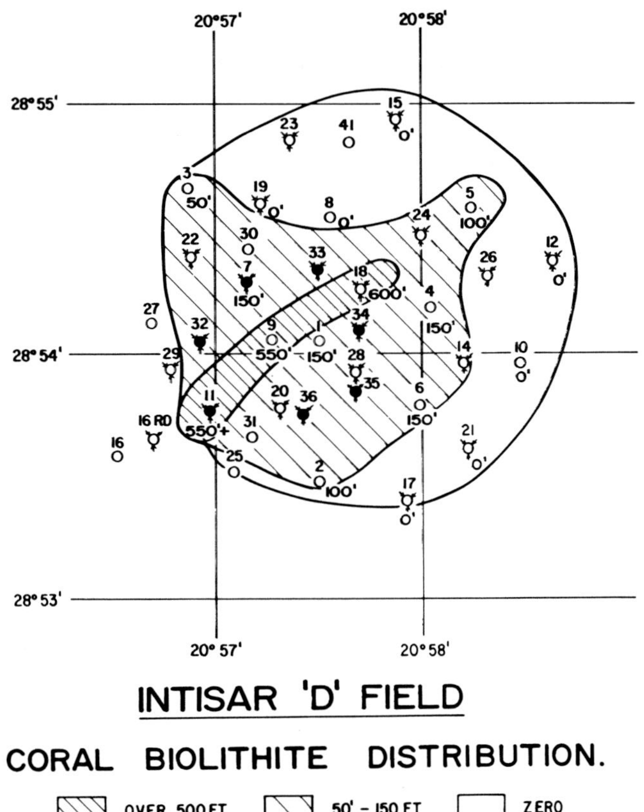

FIG. 21—Gross distribution of coral biolithite in Intisar 'D' field.

formation. The prevailing environment is interpreted as low-energy, shallow-water marine.

The influx of Sheterat shale proved fatal to the biota of the shelf edge and to most of the open-marine shelf. Occasional periods of clear-water conditions allowed a resurgence of biotic activity in the vicinity of the shelf edge and resulted in limestone/shale interfingering along that trend. Continuation of carbonate deposition took place but with a much reduced fauna. The shales have a fauna of planktonic forams, thin-shelled pelecypods, and scarce echinoderms. As illustrated earlier (Fig. 6), the Sheterat encroached into, but did not cover, the entire shelf area. In a narrow strip to the west of Concession 103, and in areas that were subsequently to become the Intisar 'D' and 'E' upper Sabil reefs, the Sheterat was developed as an argillaceous biomicrite. The D22 well, for example, exhibits a continuous carbonate section from early through late Sabil deposition. Gamma-ray markers, assumed to be time-equivalent, observed in all 12 wells that drilled to or through the Sheterat, allow recognition of the shale-biomicrite facies change and the increase in radial distribution of the biomicrites with time. These loci of biomicrite accumulation are interpreted as having stood somewhat higher than the surrounding areas in shallow water. Solitary corals, calcareous algae, and foraminifers that could tolerate the generally inhospitable conditions produced fine muds that were distributed locally, forming banks that with time coalesced, and accreted radially. The distribution of facies indicates that the shelf area occupied a low-energy environment with relatively shallow water depths compared to the main area of Sheterat deposition.

The return of clear-water conditions initiated the formation of widespread, open marine, mud supported biomicrites of the upper Sabil. These early members of the formation can be traced from the Amal-Gialo high across Concession 103 and farther west. Areas such as those at Intisar 'D' and 'E' were able to support a biota that included echinoids, red algae, and scarce corals in greater abundance than normally found in the Marsa Brega re-entrant. Conditions were sufficiently favorable to allow grain supported biomicrites to occur in the vicinity of the D22 well at the very onset of upper Sabil sedimentation. Figure 20 shows the occurrence of the first porosity of the upper Sabil, though at D22 this is below the "M" gamma-ray marker that elsewhere marks the top of the Sheterat shale and the base of the upper Sabil. The same figure shows the progressively higher and later occurrence of the first porosity as the wells are traced from west to east across the field. The D12 and D17 wells have tight, mud supported biomicrites for approximately 180 ft (55 m) above the "M" marker. From the onset of porosity development in any particular well, there is a transition zone of 20 to 40 ft (6 to 12 m) of increasing porosity. Above the transition zone the porosities are remarkably constant throughout the upper Sabil (with the exception of secondary dolomite streaks mentioned earlier). Higher energy conditions than those during Sheterat deposition are indicated by the well-winnowed, algal-foraminiferal biomicrites formed on the western edge of the 'D' area. The continued subsidence of the area allowed the expansion of the biomicrites, (envisaged as low relief shoals) over the whole field area.

Continued subsidence of the East Sirte basin and the slightly greater rate of subsidence in the Marsa Brega trough gave rise to two major differentiations in ecology and resultant sedimentation between Concession 103 and the surrounding areas. The broad, stable, shelf areas had enough biotic energy to allow sedimentation to keep pace with subsidence over the total area. The biota in Concession 103, although possessing the same potential as the neighboring region, could not survive the lowering of the seafloor in sufficient numbers to keep pace with subsidence. This resulted in thick, extensive carbonate shelf deposits of the upper Sabil over the stable areas, with an area (essentially) of nondeposition in Concession 103. In contrast, the shelf areas had only the linear front edges exposed to open-marine, high-energy conditions. Biomicrites, grain or mud supported, with well-developed porosities are more common toward the shelf edges, while micrites, dolomites, and anhydrites become predominant away from the edges. The greatest percentage of area covered by the full development of the upper Sabil lay in a semirestricted environment. Organisms that could sustain themselves in the Marsa Brega trough in any significant number, did so in colonies that were essentially cir-

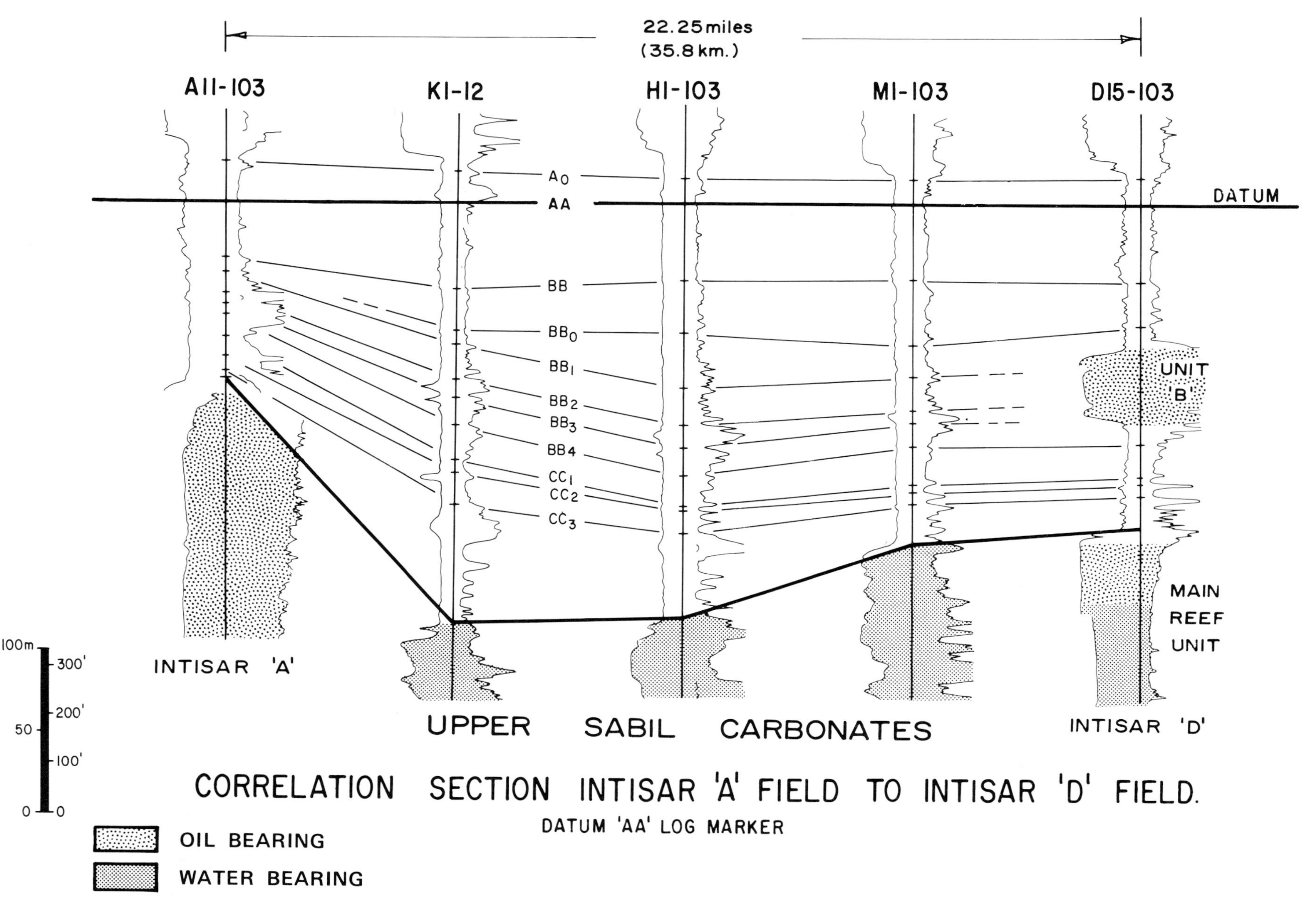

FIG. 22—Kheir marl correlations from the Intisar 'D' reef. Logs depicted are the spontaneous potential to the left and resistivity to the right of each well.

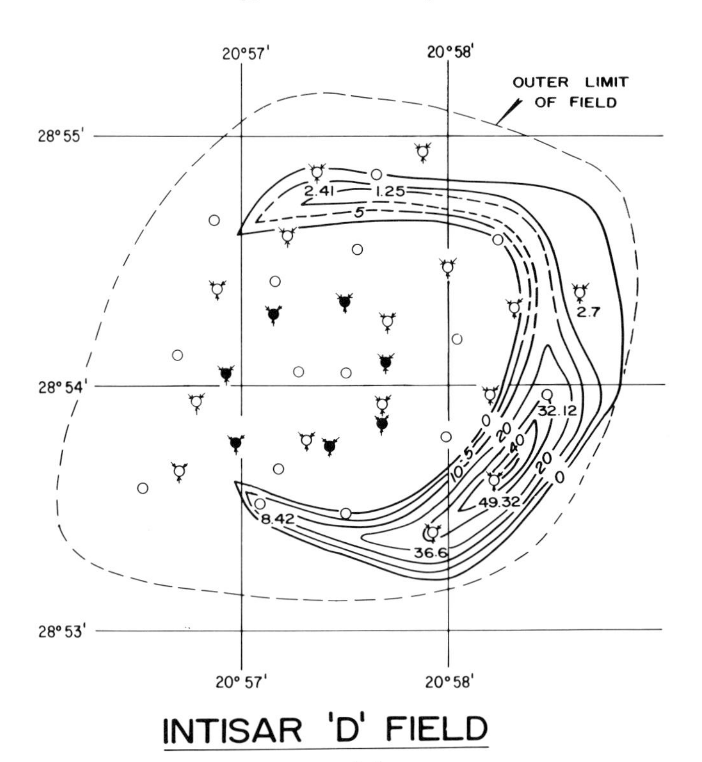

UNIT 'A'
DISTRIBUTION & POROSITY FEET

FIG. 23—Porosity feet of unit 'A.' This unit contained 66 million bbl of original oil in place (10.5 million cu m).

cular in plan. Several examples of bioherms that formed in the Marsa Brega trough show diameters that range from 2.5 to 3.75 mi (4 to 6 km). The angle of slope of the flanks is generally in the order of 18 to 22°. This limited size allowed for the maintenance of an overall energy level higher than that possible over the shelf area. Shaping was the result of similar energy and environmental conditions acting on all flanks of the colony. The slope of the flanks of the bioherms corresponds to the natural angle of repose of debris in the water and suggests that the energy levels dropped rapidly from the crestal to flank positions.

The Intisar 'D' area, an existing center of organic activity during early upper Sabil deposition, continued to be a favored venue for colonization.

Figures 19 and 20 show the correlations possible within the reef by use of the gamma curves. These markers (of approximately 20 API units maximum) appear in all the full reef wells and are independent of rock type. Correlative points such as the 'B' marker occur above a porous zone in some places and above a tight (dolomite) zone in others. Markers such as the 'E' and 'G' can be traced across the reef and occur in all three major rock types. (Interestingly enough, the four other bioherms in Concession 103 exhibit similar gamma-ray correlations, and the log of a well from one of the reefs can be inserted into a cross section drawn through any other reef and produce no noticeable anomalies). The writers attempt no explanation as to the cause of the concentration of radioactive minerals except to suggest that their widespread occurence and predictability indicates time equivalence.

Accurate mapping of the in-situ coral biolithite facies is hindered by lack of core data. Using cores and cuttings, every effort was made to record only those intervals where thanalocenosis represented the biocenosis of reef building organisms, as the writers have tried to remain objective as to the presence of what would be described as reef framework. Figure 21 exhibits the aerial extent of the coral biolithite. The gross isopach values show a thick core trending northeast to southwest through the D11, D9, and D18 wells in marked contrast to the surrounding wells, particularly to the D8 and D19 wells which, although endowed with a full carbonate section, have no coral biolithites. The north and east flanks of the field do show a paucity of this biofacies and an absence in the D8, D10, D15, D19, and D41 wells. The D5 stands alone with approximately 100 ft (30 m) of in-situ coral framework.

The origin of the dolomitization may have been by seepage reflux or evaporitic pumping. The lateral restriction and increased occurrence towards the reef edge certainly point to fluid movement along confined zones from (or toward) ephemeral pools formed on the reef top.

Peripheral wells present problems in correlation, particularly the north and east flanks, where there is a partial carbonate buildup. The gamma-ray markers are present in the lower part of the section but frequently cannot be traced with accuracy in the upper part. If the assumption that the markers are time equivalent and occur regardless of lithofacies is correct, then this departure from the norm must indicate that the time lines of flank wells are not flat lying as they appear in the full reef wells. Dipmeter logs run in edge wells indicate dips of from 5 to 40° with two patterns typical of small-scale crossbedding. Cores of the D2 well contain a great deal of intrabiomicrite and one 5 ft (1.5 m) interval exhibits slump structures in a series of finely bedded, fine grained, mud supported intrabiomicrites. Overall dips in this zone are about 16°. These observed data, together with the gross angle of slope of the reef flanks, are interpreted as being indicative of a zone of talus ringing the central area of sediment generation. The talus is not envisioned as consisting of large blocks of material broken from the reef by hydraulic action, but rather as underwater scree slopes of organic debris washed down from the reef top as individual grains, before lithofication, together with minor amounts of intraclasts.

The slope of the southwest side of the field, as seen between the D11 and D16 redrill, angles as high as 32°. This is somewhat high for a natural angle of repose in a fluid medium. The fact that the gamma-ray markers can be carried into wells such as the D11 and D22, which are located very close to the drop-off from full buildup, suggests that the talus zone is not well-developed on that side of the field. It is not thought likely that high energy generated by a prevailing southwest or northeast wind was responsible for inhibiting talus accumulation, as in that case the coral biolithites would be more liable to have a northwest to southeast alignment instead of being normal to it. Local, strong northwest to southeast currents that affected the west flank of the reef are only postulated.

Another anomaly, of which many examples exist elsewhere and have puzzled geologists in their study and understanding of reef growth, is the relationship of the porous units 'A' and 'B' that interfinger with tight, argillaceous biomicrites and shales. The question arises as to whether the reef grew at the same time as the deposition of surrounding marls of the Kheir Formation. The picture of a coral reef growing in a sea of muddy waters is not an easily supported one and is intuitively suspect. Comparison of all interfield wells allows widespread, accurate correlations to be made within the Kheir marls. These correlations have, in turn, been traced into the flanks of the 'A' and 'D' fields where excellent step-by-step correlations can be made from wells with thick Kheir/thin upper Sabil to crestal wells that have thin Kheir/thick upper Sabil. Figure 22 ties the 'A' field to the 'D' field passing through representative inter-

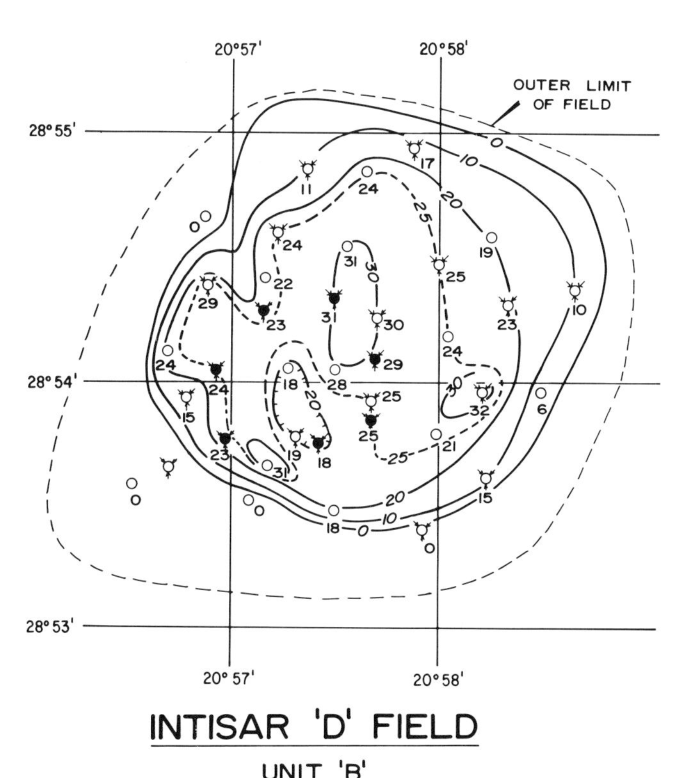

FIG. 24—Unit 'B' contained nearly 252 million bbl (40 million cu m) of the original oil in place.

reef wells. Several beds of the Kheir that post-date the 'A' bioherm deposition expand into the trough wells and thin again at the D15. The BB and CC_1 beds overlie the Intisar 'D' reef unit and are overlain by the Intisar unit 'B.' Similarly, the Intisar unit 'A,' best correlated in wells D12 and D17, can be equated with off structure beds of the Kheir.

Conclusions drawn from these data are:

1. The upper Sabil reef stood approximately 800 ft (240 m) above the surrounding seafloor when reef growth (sensu stricto) was terminated by the influx of argillaceous materials of the Kheir Formation.

2. Conditions at the 'A' bioherm during Kheir deposition never allowed an active resurgence of biotic activity.

3. The paleoecology of the 'D' area during Kheir deposition allowed continued development of algal-foraminiferal shoals with occasional periods conducive to compound coral growth. The Kheir Formation is extremely varied in lithology, ranging from shales through marls to limestones. The concentration of argillaceous material in the seas was probably less, and the rate of supply more varied, than was the case during the deposition of the Sheterat. The numerous thin limestones, traceable over wide areas, indicate periods of little shale introduction and a concentration of fines washed off the higher, uncapped regions of the shelf. Dasycladacean, blue-green algae are common in the interval between the BB_0 and CC_3 markers shown in Figure 22, and indicate periods of warm, clear water.

Figure 23 illustrates the occurrence of the Intisar unit 'A.' Its restriction to the north, east, and south flanks is interpreted to be caused by two factors: (1) The top of the reef was emergent and thus no sedimentation took place in the central area; and, (2) As the Kheir marl encased the reef, the morphology of the seafloor differed depending on the pre-existing attitude of the substrata. The north, east, and south flanks had an original slope of 18 to 22° and the Kheir seafloor had something less than this. The biota found their niche very precisely in these areas between two levels of water depth. On the west flank the original dip of the seafloor was about 30° and during unit 'A' deposition the slope was still too great and the water level too deep for the community to thrive.

Figure 24 outlines the Intisar unit 'B.' The infilling of the trough and the submergence of the reef allowed the 'B' unit to expand over a greater area than the 'A,' although the west flank still remained inhospitable. Recognition of the 'B' unit as a separate layer in centrally located wells was made early in the field development. Its definition was based on the porosity variability that occurs in the interval, but only with much later development drilling was the true relationship resolved.

The two shoal units contained 17.7% of the total volume of oil in place.

POST REEF COMPACTION

Present drape of the Kheir Formation over the reef is in the order of 450 ft (140 m). Part of this can be attributed to depositional thinning of the beds of the Kheir onto the reef, but the greater part is due to differential compaction. Two periods of compaction are observed. Shale units of the lower Kheir thin

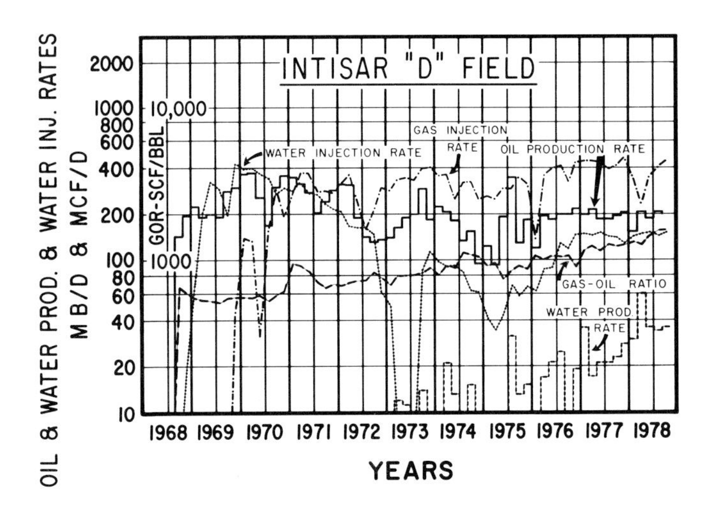

FIG. 25—Field performance curves for Intisar 'D' field.

onto the reef while the more calcareous units remain fairly constant. By the time of late Kheir deposition, the units show a constant thickness across the reef. The El Giza has an oil column of about 90 ft (27 m) with at least 170 ft (52 m) of closure. This closure falls directly over the reef. Units of the Gir and Gialo do not exhibit sufficient thinning to allow for attenuation over a pre-existing structure. This late compaction is attributed to pressure solution in the fine grained, less permeable material of the reef edges (Bathurst, 1975). Differential compaction continued through the Tertiary and shallow-water sandstones at 250 ft (76 m) subsea and shows about 30 ft (9 m) of closure.

OIL ENTRAPMENT

Source rocks are plentiful in the Sirte basin. The Sheterat is the most likely candidate for the origin of the oil in the upper Sabil in Concession 103, though the Kheir marl and Rakb shale may have contributed. Isolated bioherms, surrounded by and underlain by source rocks, are in an ideal position to receive migrating oil. They have a 360° sweep from which to accept fluids but only if structural configuration allows it. A large local rock mass, such as a reef, sitting on a thin carbonate platform that, in turn, overlies shales is likely to undergo subsidence due to compaction. This may form a shallow rim syncline in the surrounding and underlying rocks sufficient to alter the path of migrating fluids. Fortunately, this did not happen at the Intisar 'D' or 'A' fields but it may be the reason two other reefs have oil columns less than their stratigraphic and structural closure and another reef has no oil at all.

PRODUCTION HISTORY

The maximum oil column was 955 ft (219 m), and the reef was full to the apparent spill point. The oil is 40° API gravity, has a paraffinic base and is low in sulfur. The original solution gas/oil ratio was 509 cu ft/bbl (90 cu m/cu m), a highly undersaturated condition. Original stock tank oil in place is estimated at 1.8 billion bbl (290 million cu m). Because of the isolated nature of the reef, no natural pressure support was expected, and this together with the undersaturated state of the oil would have resulted in a low primary recovery efficiency. It was estimated that water injection alone would eventually yield a recovery of 40% of the oil but further studies indicated that, because of high permeability resulting from solution porosity and the massive nature of the reservoir, miscible gas injection combined with water injection would result in a recovery efficiency of between 75 and 80%. Bottom water injection (accomplished initially by dumpflooding and later supplemented by surface injection) began almost simultaneously with production, while crestal gas injection into unit 'B' began 15 months later. Reservoir pressure is now maintained at approximate 4,000 psi (280 kgf/cm^2), high enough for miscible gas displacement. In addition to enhanced oil recovery, the gas injection program will conserve nearly 3 Tcf (80 billion cu m) of associated gas from this and other fields when displacement of the reservoir is complete.

Field performance curves are presented in Figure 25. In January 1979 the field was producing 180,000 b/d (28,600 cu m/d) from 13 wells. In addition to these producers, 11 water-injection and 7 gas-injection wells are used for the input of 150,000 bbl of water (23,900 cu m) and 450 MMcf of gas (13 million cu m) daily. Cumulative oil production as of December 31, 1978, was 812 million bbl (129 million cu m). Cumulative water and gas injection volumes by the end of 1978 were approximately 600 million bbl (95 million cu m) and 1 Tcf (28 billion cu m), respectively.

REFERENCES CITED

Barr, F. T., and A. A. Weegar, 1972, Stratigraphic nomenclature of the Sirte basin, Libya: Grafiche Trevisan, Castle Franco, Italy.

Bathurst, R. G. C., 1976, Carbonate sediments and their diagenesis: New York, Elsevier, pp. 459-473, 532-534.

Braithwaite, C. V. R., 1973, Reefs: just a problem of semantics?: AAPG Bull., v. 57, p. 1100-1116.

DesBrisay, C. L., J. W. Gray, and A. Spivak, 1975, Miscible flood performance of the Intisar 'D' field, Libyan Arab Republic: Jour. Petroleum Tech., p. 935-943.

Folk, R. L., 1962, Spectral subdivision of limestone types: AAPG Mem. 1, p. 62-84.

Klitch, E., 1968, Outline of the geology of Libya in geology and archaeology of northern Cyrenaica Libya: Petroleum Exploration Society of Libya.

Terry, C. E., and J. J. Williams, 1969, The Idris 'A' bioherm and oilfield, Sirte basin, Libya, its commercial development, regional Paleocene geologic setting and stratigraphy; *in* exploration for petroleum in Europe and North Africa: London, Inst. Petroleum, p. 31-68.

Role of Reflection Seismic in Development of Nembe Creek Field, Nigeria[1]

By P. H. H. Nelson[2]

Abstract The Nembe Creek field, discovered in 1973 in the coastal swamp of the Niger delta, is developed in a middle Miocene deltaic sandstone-shale sequence. Oil accumulations, with some associated and unassociated gas, have been found in the depth range of 7,000 to 12,000 ft (2,134 to 3,658 m). The oils have gravities between 16° and 41° API, the heavier oils coming from the more shallow reservoirs. Drilling of 30 wells by the end of 1977 increased average recoverable oil reserves to 645 million bbl. Nearly one-third of these wells were proposed with direct support of reflection seismic data, used principally for structure-mapping and development of cross-sections. In addition, lateral predictions of lithology and sand pore fill for two appraisal wells came from comparison of true amplitude impedance profiles and synthetic seismograms generated from well data. It is concluded that the timely acquisition and interpretation of sufficient seismic data in an area of complex structure (but good reflection quality) permitted the successful drilling of long step-out appraisal wells. This approach led to an early delineation of the field's limits, a rapid growth of proved reserves, and a firm basis for proposal of an additional 70 development and appraisal wells (as yet undrilled).

INTRODUCTION

The goal of this paper is to define the uncertainties inherent in mapping a complexly faulted structure when the seismic line spacing is generally greater than fault spacing, and to demonstrate the improved solution obtainable after infill shooting halved the spacing of north-south dip lines.

SETTING

Operational Setting

The Nembe Creek field is about 9.5 mi (15.3 km) from the Nigerian coast in the coastal swamp of the Niger delta. It lies within Oil Mining Lease (OML) 29 of the Shell-BP Petroleum Development Company of Nigeria Ltd. The area of interest lies between Ekole Creek in the west and the smaller Surprise Creek on the east (Fig. 1). Aganatoku Island, between the east and west branches of the Brass River, is located over the central part of the oil field. Terrain between rivers and streams consists of mangrove swamp, in which ground level and sea level are virtually the same.

In shooting seismic lines a footpath was cut through the mangrove, and all recording and shothole drilling equipment was carried in by porters. Spread and geophone pattern geometries are necessarily simple in such confined pathways, but it is

[1]Read before the Association, April 4, 1979; accepted for publication August 6, 1979. Published by permission of Shell Internationale Petroleum Co., British Petroleum Co., and the Nigerian National Petroleum Corp. Modified from Shell-BP Petroleum Development Company of Nigeria, Ltd., Exploration Report 441 (July 1977).

[2]Shell-BP Petroleum Development Company of Nigeria, Ltd., Lagos, Nigeria. Presently with Pakistan Shell Petroleum Development, B. V. Karachi, Pakistan.

The writer acknowledges the work of numerous members of the Exploration and Petroleum Engineering departments of the above company, without whose contributions the rapid evaluation of this complex field would not have been possible. The text illustrations were prepared by the drawing office staff of Shell-BP Petroleum.

Article Identification Number:
0065-731X/80/M030-0029/$03.00/0.

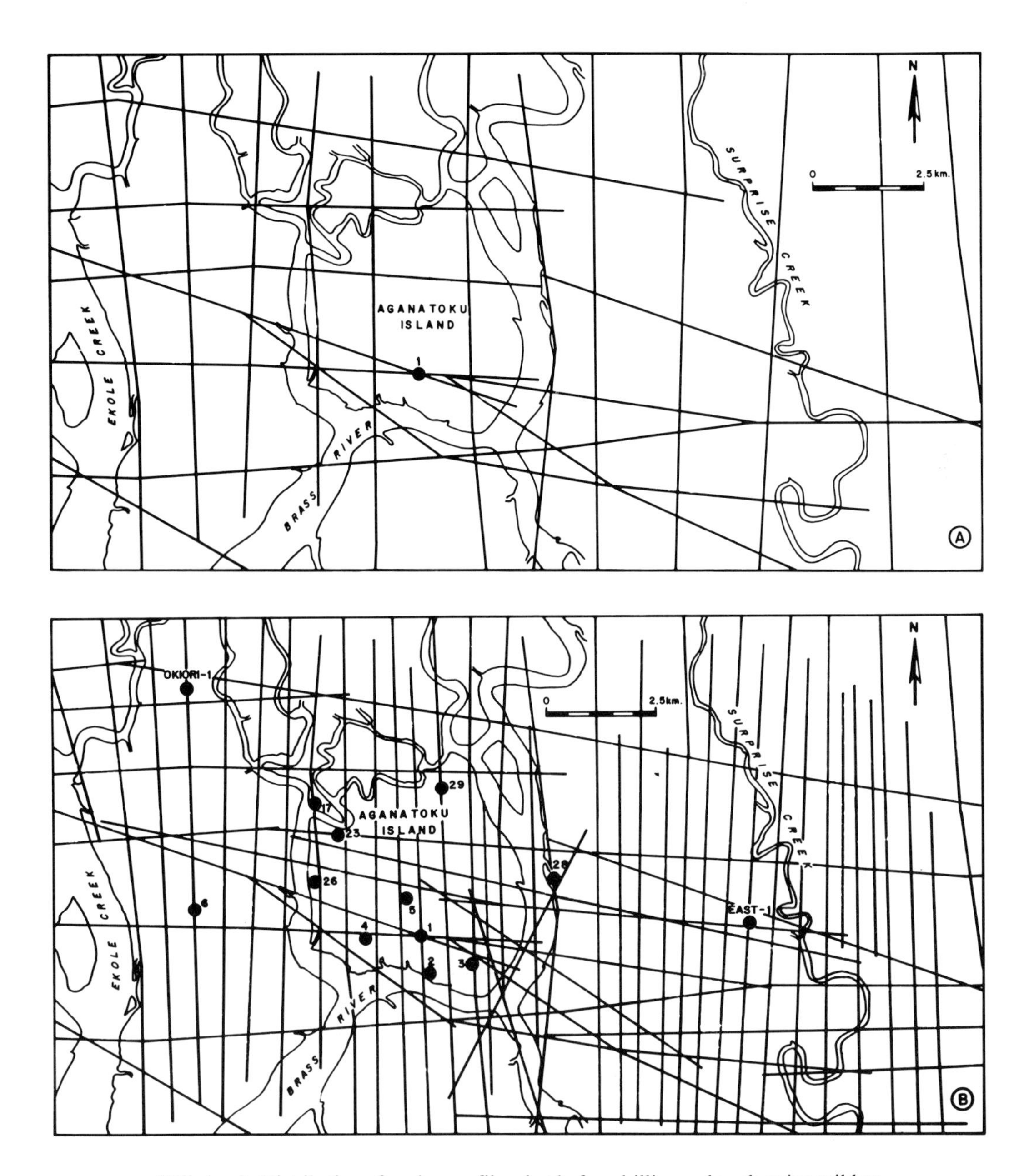

FIG. 1—A. Distribution of analog profiles shot before drilling and exploration wildcat.
B. Total coverage available now (Note the heavy emphasis on north to south infill profiles).

possible to use a 48-channel system with a station interval of 164 ft (50 m). Shots from both ends of the spread, updip and downdip for north to south profiles, are fired with 10-lb (4.5 kg) dynamite charges in shot holes about 130 ft (40 m) deep. River crossings are covered by under shooting when short enough, and with additional water shots when necessary. Recently, seismic coverage has been obtained beneath some of the more linear sections of the north to south trending rivers by broadside shooting from shot holes on one bank into geophone spreads on the other. Well locations are prepared for drilling from a swamp barge by dredging "slots" from the nearest navigable access point in a natural waterway. These "slots" can be little more than an artificial embayment, although in places they become lengthy canals.

Geologic Setting

The Nembe Creek structure is a complex anticline, about 9.5 mi (15.3 km) long and 3.75 mi (6 km) wide at its widest point (Fig. 3). Numerous subparallel, south-dipping growth faults dissect the structure, and these are interrelated with north-dipping, antithetic faults in both the central region and western plunge of the fold. Oblique cross faults, which apparently compensated lateral inequalities of displacement along the more extensive east to west faults, have been recognized locally. All the faults are normal, almost all are growth faults, steepening upwards, and have throws that diminish to zero.

The more shallow hydrocarbon-bearing sandstones in Nembe Creek are faulted such that they culminate at a more or

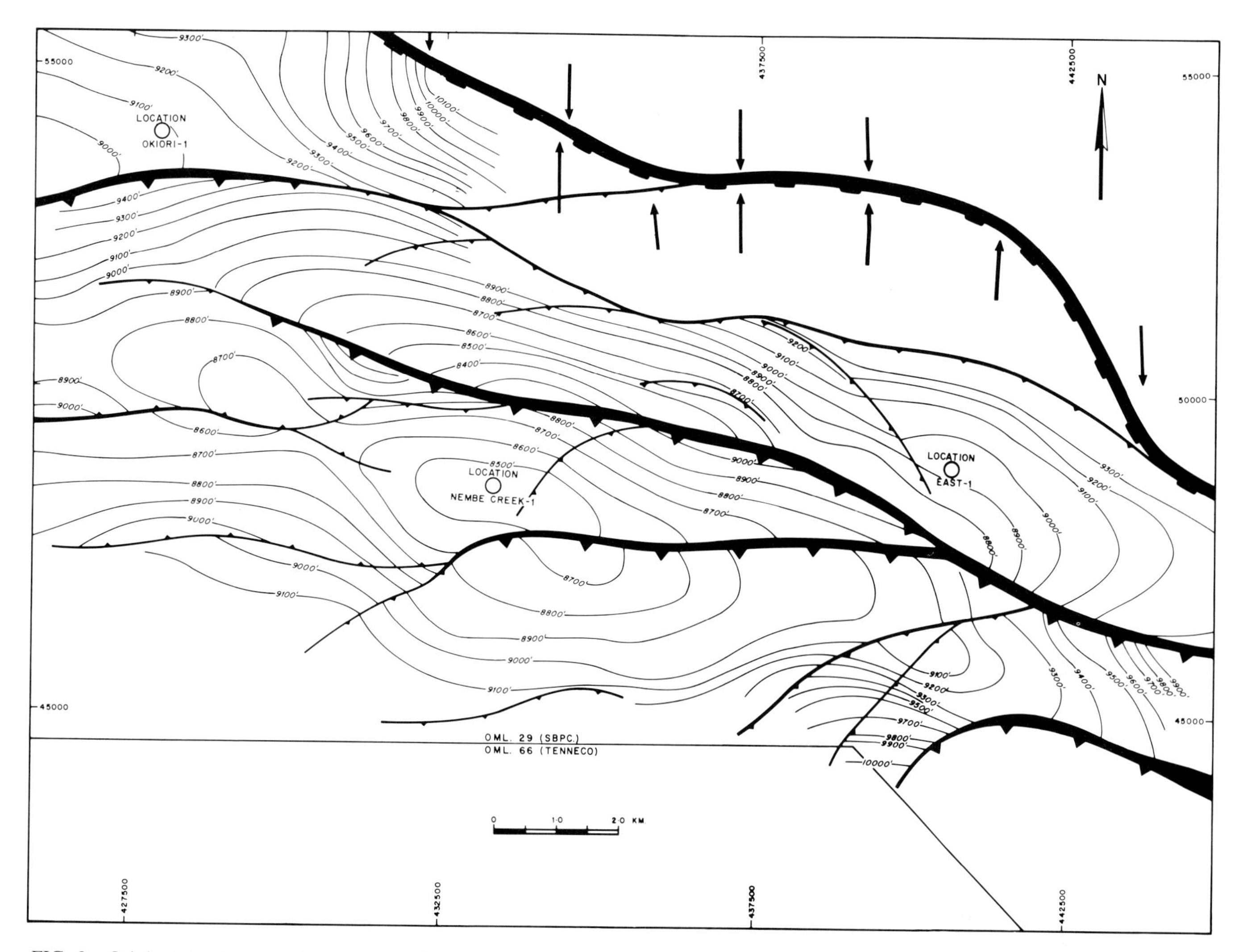

FIG. 2—Original structure map based on seismic coverage shown in Figure 1A illustrating the relative simplicity of the anticline and the varied azimuths of the faults.

less uniform structural level in all fault blocks (Fig. 4). It is believed that changes of fault throw along strike have generated spill points at remarkably uniform depths. For any one sandstone, this has led to a similarity of fluid contact depths, particularly oil and water, throughout the different fault blocks of the field.

This largely syndepositional structure was developed during the middle Miocene in a cyclothemic sequence of paralic sandstones and shales (sand/shale ratio approximately equal to one), overlain by a major transgressive shale that is the uppermost caprock of the field (Fig. 5). This seal, characterized by the presence of the foram *Uvigerina subperegrina* CUSHMAN, and locally known as the *Uvigerina* 8 shale, is 230 to 250 ft thick (70 to 76 m) and occurs in a depth range of 6,700 to 7,000 ft (2,042 to 2,134 m). Immediately overlying the *Uvigerina* 8 shale is 1,500 ft (457 m) of a more massive sandstone facies than is found in the paralic section. Salinity in this zone is distinctly lower, as "transitional" beds give way to freshwater-bearing continental alluvial sandstones at an approximate depth of 5,200 ft (1,585 m). From this depth to the surface, subordinate thin shales constitute no more than 10% of the section, and the relatively monotonous sandy lithology of this interval results in a sequence that by comparison is acoustically transparent. It is this transparency which contributes significantly to the success of the reflection seismic method in the area.

Beneath the *Uvigerina* 8 shale, velocity and density log evidence indicates that sandstones and shales possess contrasting acoustic impedances throughout the 5,000 ft (1,524 m) interval (known to be hydrocarbon-bearing). These contrasts produce positive and negative reflection coefficients large enough for generation of good reflections, even in those parts of the structure where the sandstones are brine filled.

The sandstone and shale cycles in the paralic succession become thicker with increasing depth. At a depth of 12,000 ft (3,658 m) shale thicknesses of approximately 300 ft (91 m) suggest that more open-marine conditions dominated the earliest stages of paralic deposition. Thickening of the cycles probably shares as much responsibility as do effects of seismic energy absorbtion for the increasingly low-frequency aspect of

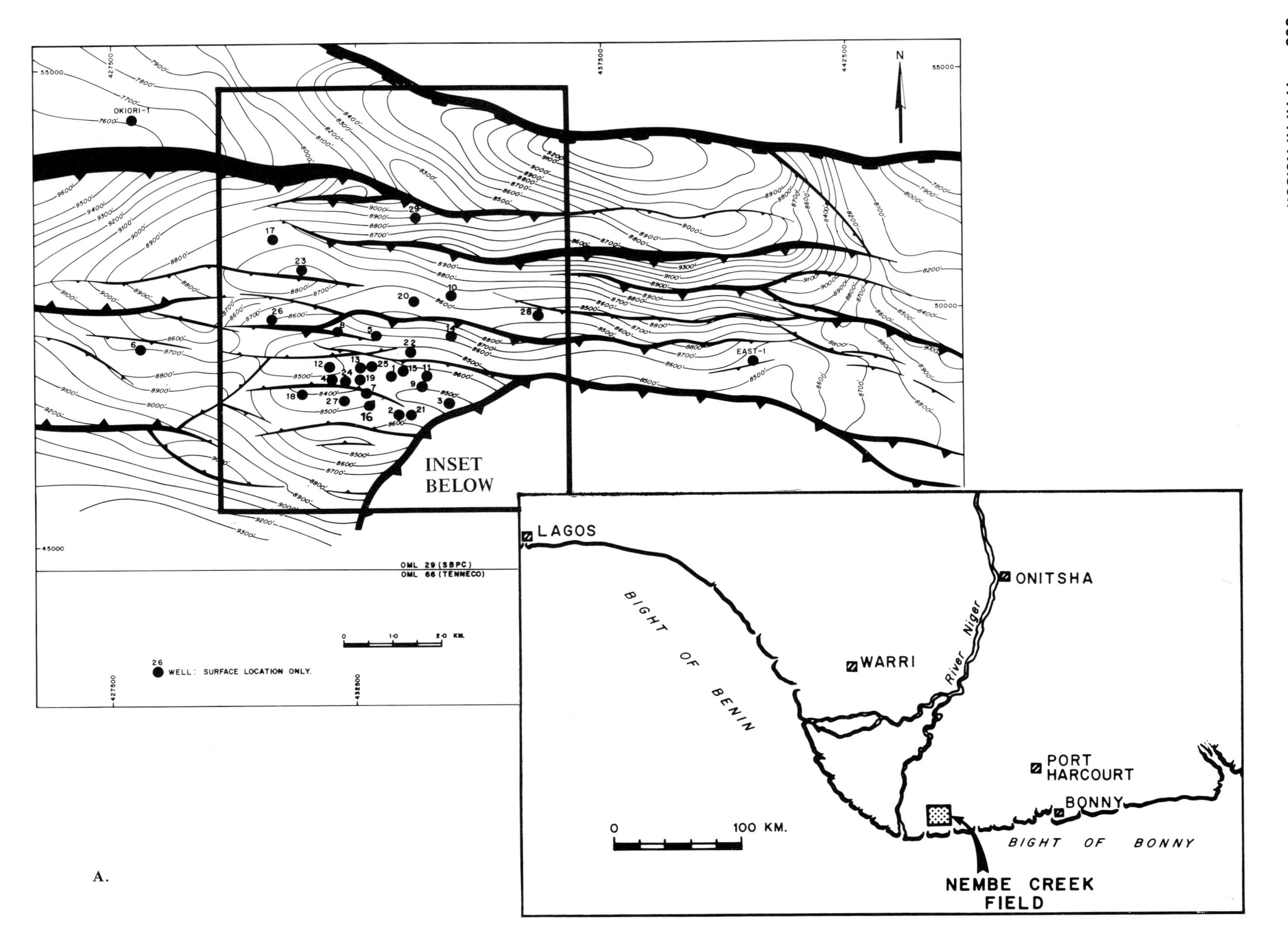
N
OKIORI-1
EAST-1
INSET BELOW
OML 29 (SBPC)
OML 66 (TENNECO)
0 1·0 2·0 KM.
WELL: SURFACE LOCATION ONLY.
LAGOS
ONITSHA
BIGHT OF BENIN
WARRI
River Niger
PORT HARCOURT
BONNY
BIGHT OF BONNY
NEMBE CREEK FIELD
0
100 KM.

A.

B.

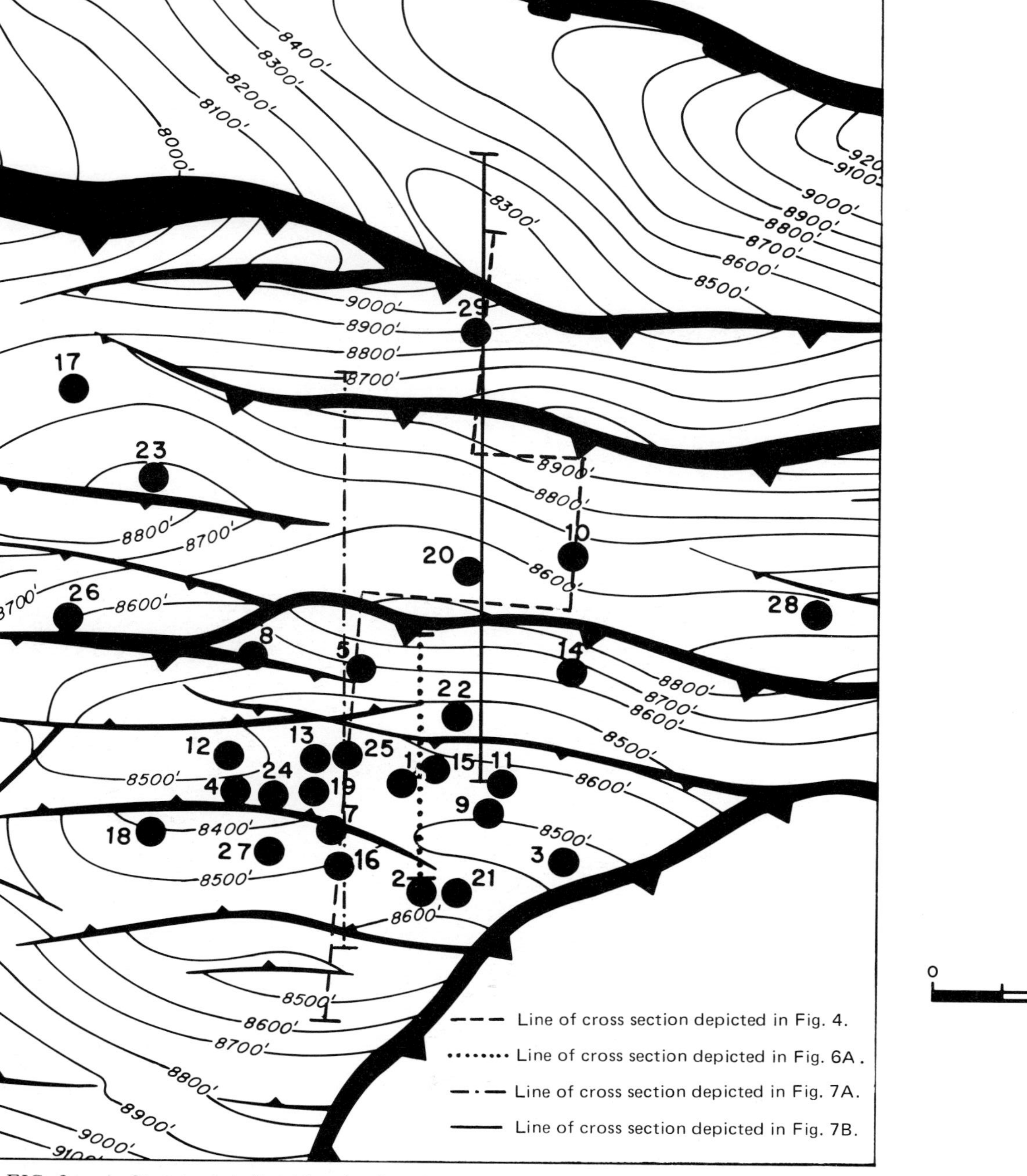

FIG. 3 -- **A.** Structure defined by the first 30 wells and the additional seismic coverage (Fig. 1B) showing the relative complexity and predominance of east-to-west faults. Bottom diagram depicts location of the field. **B.** Enlarged portion of Nembe Creek field showing the line of cross section depicted in Figures 4, 6A, 7A and 7B.

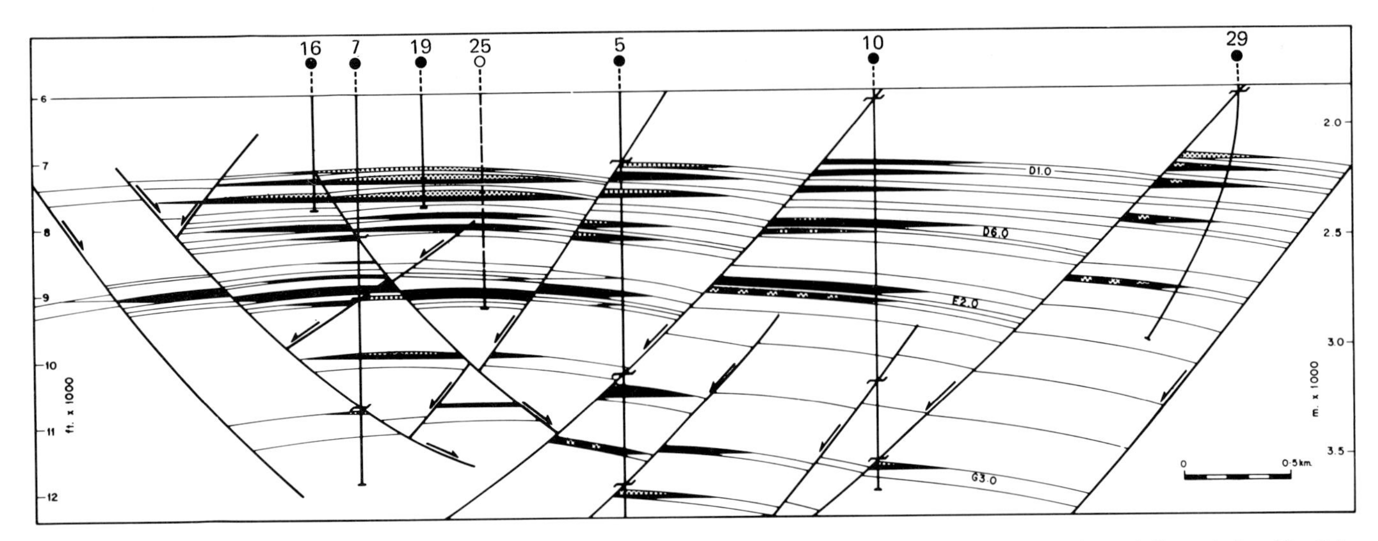

FIG. 4—Representative true-scale cross section. The latest seismic data (Fig. 7B) provide evidence of faults near the anticline axis (see Fig. 3 for location).

the profiles as reflection time increases.

Although the upper 2,500 ft (762 m) of the hydrocarbon-bearing sequence has been delineated with existing reflection seismic data, inferior resolution below this point has inhibited the mapping of deeper levels. Until it becomes possible to improve the seismic resolution of the deeper reservoirs, their structural configuration will be defined largely from deep well data as development and further appraisal drilling proceed.

In relation to other features along the regional east to west strike of this part of the Niger Delta, the crest of the Nembe Creek anticline is structurally one of the highest. The thick *Uvigerina* 8 shale is significantly more shallow in the area of the field than in wells to the east and west. Furthermore, this shale appears to be a regional seal since, with only one exception, all hydrocarbons in these and other wells on the same trend have been encountered in the underlying sequence.

Discovery

The wildcat, Nembe Creek 1, was spudded on the basis of an east to west seismic profile on August 19, 1973 (Figs. 1, 2). After reaching a total depth of 13,500 ft (4,115 m), a total of 477 ft (145 m) of net oil sandstone (NOS) in 11 intervals and 286 ft (87 m) of net gas sandstone (NGS) in 8 intervals was logged between depths of 7,092 and 11,536 ft subsea (2,162 to 3,516 m).

Follow-up (1974 to 1975)

Shortly after suspending the first well, five additional seismic lines were shot in the area. During the shooting, which included the first north to south profile across the location of Nembe Creek 1, digital recording techniques were used for the first time. The latest "Hi-Fi" data processing techniques were applied to the new profile in 1974, and a good match was obtained between an impedance section and a synthetic impedance trace from the logs of the well itself (Fig. 6A). An accurate lateral prediction of lithology and hydrocarbons was made prior to the drilling (in early 1975) of Nembe Creek 2, located 2,825 ft (861 m) south of the discovery well (Fig. 6B).

A third well, 4,265 ft (1,300 m) southeast of the discovery well, was drilled through two closely spaced south-dipping faults having a combined displacement of about 400 ft (122 m). No hydrocarbons were found in the downthrown blocks of these faults, which are believed to form the southern limit of the eastern half of the field (Fig. 3).

The drilling of well 4, located 4,150 ft (1,265 m) west of well 1, added sufficient log data to show that the deeper hydrocarbon zones in well 1 (the interval from 9,745 to 11,536 ft [2,970 to 3,516 m]) were in a separate upthrown fault block on the north. Based purely on subsurface geologic rationale, Nembe Creek 5 was proposed and successfully drilled later in 1975 at a location 2,855 ft (870 m) north-northwest of well 1.

Two important exploration wells were drilled in the western part of the complex in late 1975. The first, Okiori 1, tested a promising upthrown fault block at a seismically mapped culmination 4.75 mi (7.64 km) northwest of Nembe Creek 1. It was a curious disappointment that, although the uppermost three hydrocarbon-bearing sandstones of the field were encountered between 400 and 500 ft (122 to 152 m) higher than in the field, they contained only 10 ft (3 m) NGS, and 29 and 18 ft (8.8 and 5.49 m) NOS, respectively. Correlation shows the deeper sandstones (all water-bearing in Okiori 1) to be up to 1,500 ft (457 m) higher than in the field. Results at this well led to the current view that the northern limit of the field is defined by the large south-dipping major fault just south of Okiori 1 (Fig. 3).

The second exploration well, 3.25 mi (5.23 km) west of the discovery well, and along the same north to south seismic profile as Okiori 1, was drilled as Nembe Creek 6. Location was based on a seismic interpretation showing reflection times from the uppermost hydrocarbon-bearing sandstones at comparable reflection times to those near Nembe Creek 1. The importance of this well was due to oil/water contacts in three of the sandstones matching those of well 1.

Follow-up (1976)

Drilling in 1976 (wells 7 to 19) included appraisal well 10 (1.29 mi or 2.08 km, northeast of well 1) which was originally proposed as an exploration well (Figs. 5 and 7B). Another well, Nembe Creek 17 (2.35 mi or 3.78 km northwest of well 1), was originally proposed as exploration location Nembe Creek-Northwest. Both wells were located on interpretations of analog-recorded north to south seismic profiles.

Results from wells drilled near the center of the field spawned questions concerning the correlation of the different fault blocks. It soon became clear that neither the quality nor the quantity of available seismic profiles could provide adequate answers, and seven infill north to south profiles (dip lines) were shot across the field, bringing the average dip line spacing to about 1,970 ft (600 m). Six additional dip lines were shot in the area just east of the field as part of the same program (Fig. 1B).

East to west tie lines were deliberately omitted from this seismic program, since several profiles in a generally east to west azimuth had already been recorded. In most of the area, their only value is the control they provide for the relative time bases of adjacent dip lines and, with the digitally recorded data, as a means of comparing signal amplitude levels in adjacent profiles. They are often of little value for structural analysis when compared with north to south profiles which, because of the pronounced east to west strike of the fault blocks and the comparatively simple velocity function, yield effective 2-D migration stacks in the dip direction (Fig. 7A, B). Furthermore, the "character" of the reflection sequence from the upper 2,500 ft (762 m) of the prospective interval is very persistent across the field area, and jump correlation is straightforward from one dip line to the next. When the north to south profiles were spaced too widely across this field, however, with its closely spaced east to west faults, ambiguities occurred in the correlation of fault planes between profiles. For a more explicit structural interpretation it was found that profile spacing should be approximately equal to the shortest distance between faults.

By the end of 1976, new seismic data resulted in several significant changes to the isochron map of a reflection associated with the E2.0 sandstone. However, the long lead times required for location preparation of the first wells of the 1977 drilling round resulted in drilling programs and locations being prepared and constructed on the basis of an earlier reinterpretation.

Follow-up (1977)

The 1977 drilling, which began in January with well 20, generally supported the revised structural interpretation. Detailed refinements of structure were made according to the drilling results, but the arrangement of faults and fault blocks was generally confirmed. Shortly after the beginning of this drilling, and with the realization that the revised structure map was dependable, the emphasis was switched from development drilling near the center of the field to long step-out appraisal wells. Five wells were eventually drilled on the basis of the updated seismic map and the structural control it afforded; i.e. Nembe Creek 26, 28, 28 sidetrack, 29, and East 1.

Well 26—A location 1.7 mi (2.74 km) west-northwest of well 1 was prepared for a westward-deviated hole to appraise an apparently high block at the level of the E2.0 sandstone. However, the revised seismic interpretation showed the origi-

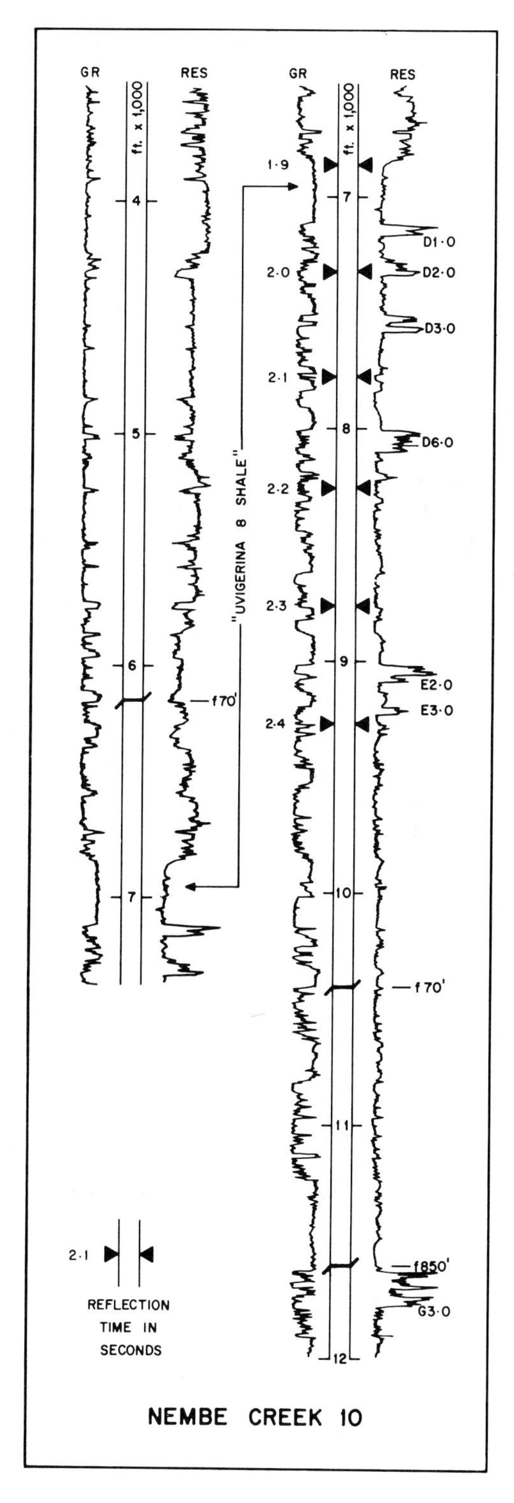

FIG. 5—Logs of well 10. The distribution of facies units should be compared with the seismic response (Fig. 7B).

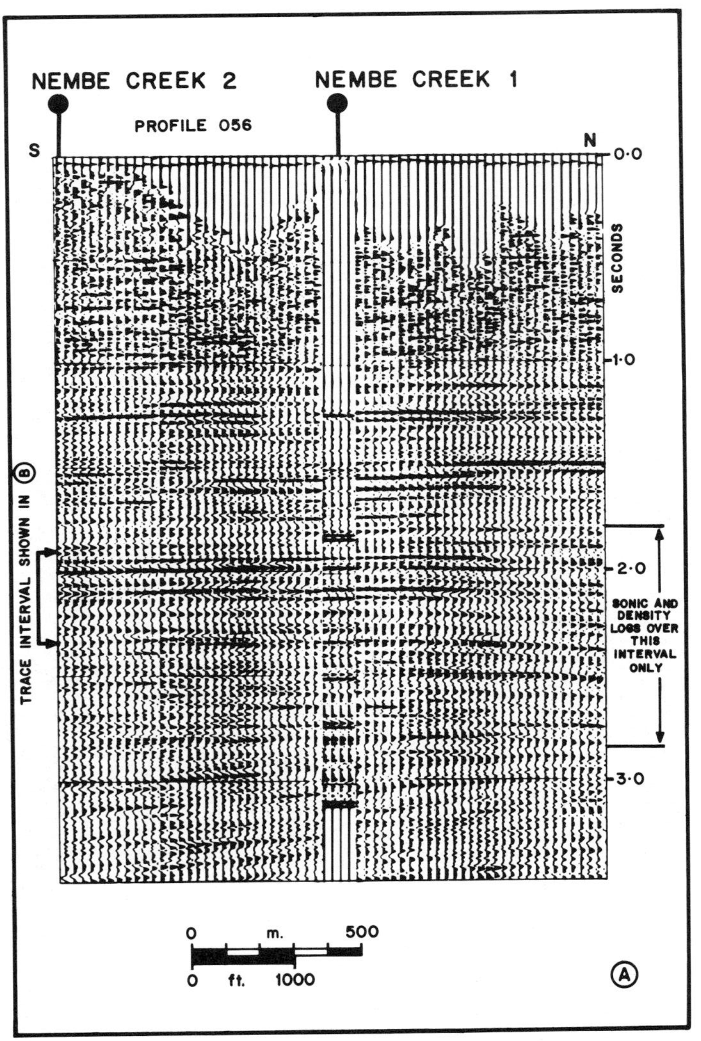

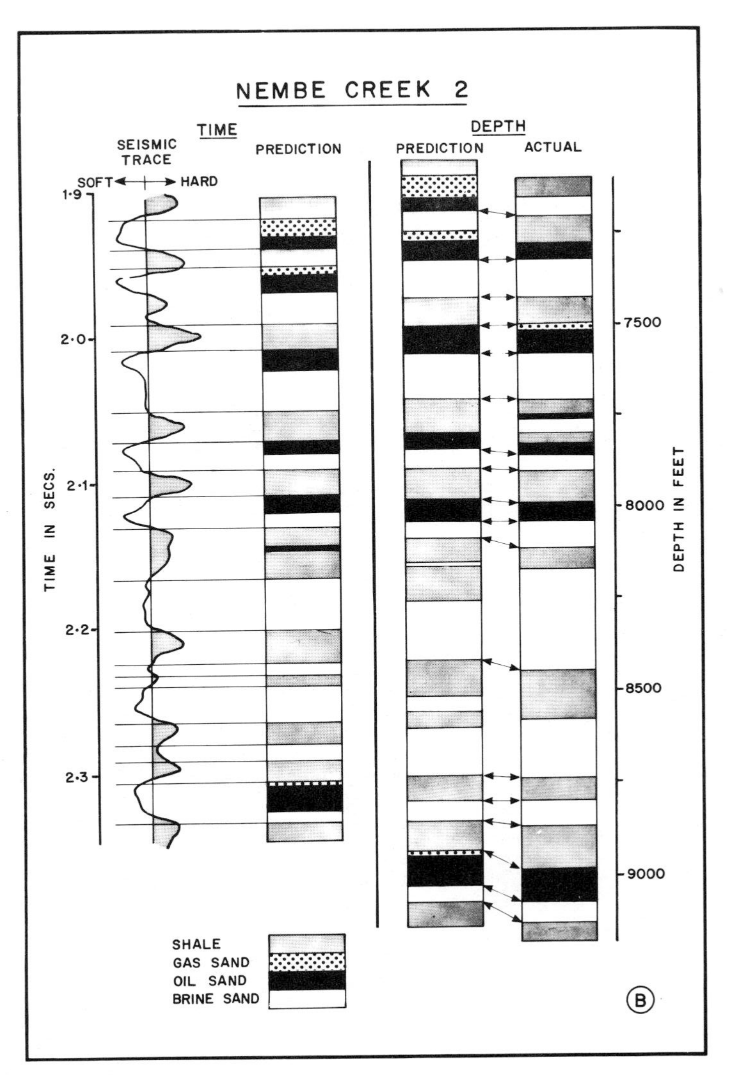

Fig. 6—A. Synthetic trace from well 1 (repeated four times for clarity) set in north-to-south impedence profile. The match is good over the interval with both sonic and density logs (see Fig. 3 for location).
B. The lithology and pore-fill prediction and result of well 2.

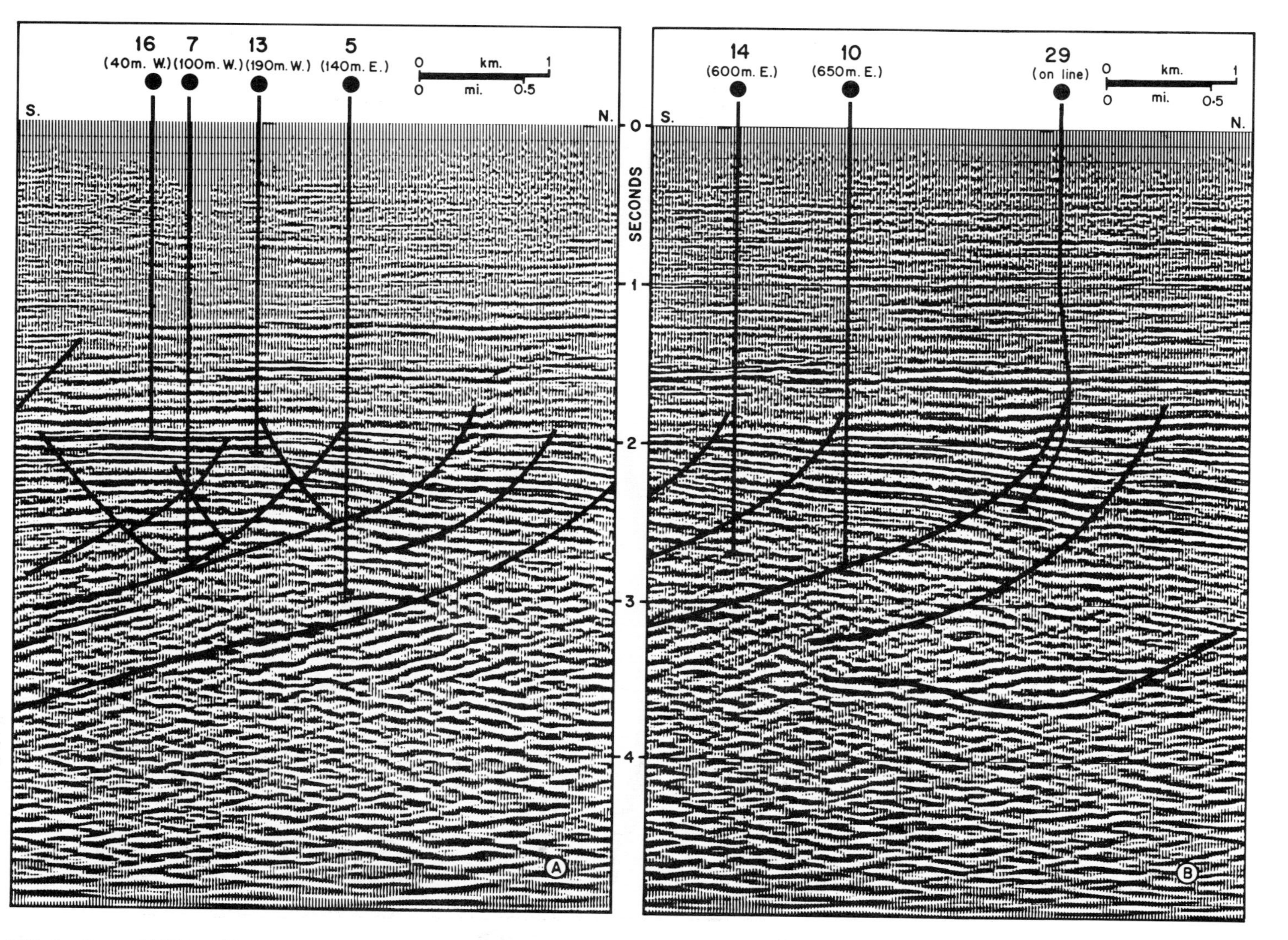

FIG. 7—Parts of 1976 profiles 119 (A) and 120 (B). Strong regular reflections between 2 and 3 seconds come from prospective sand/shale sequence (see Fig. 3 for location).

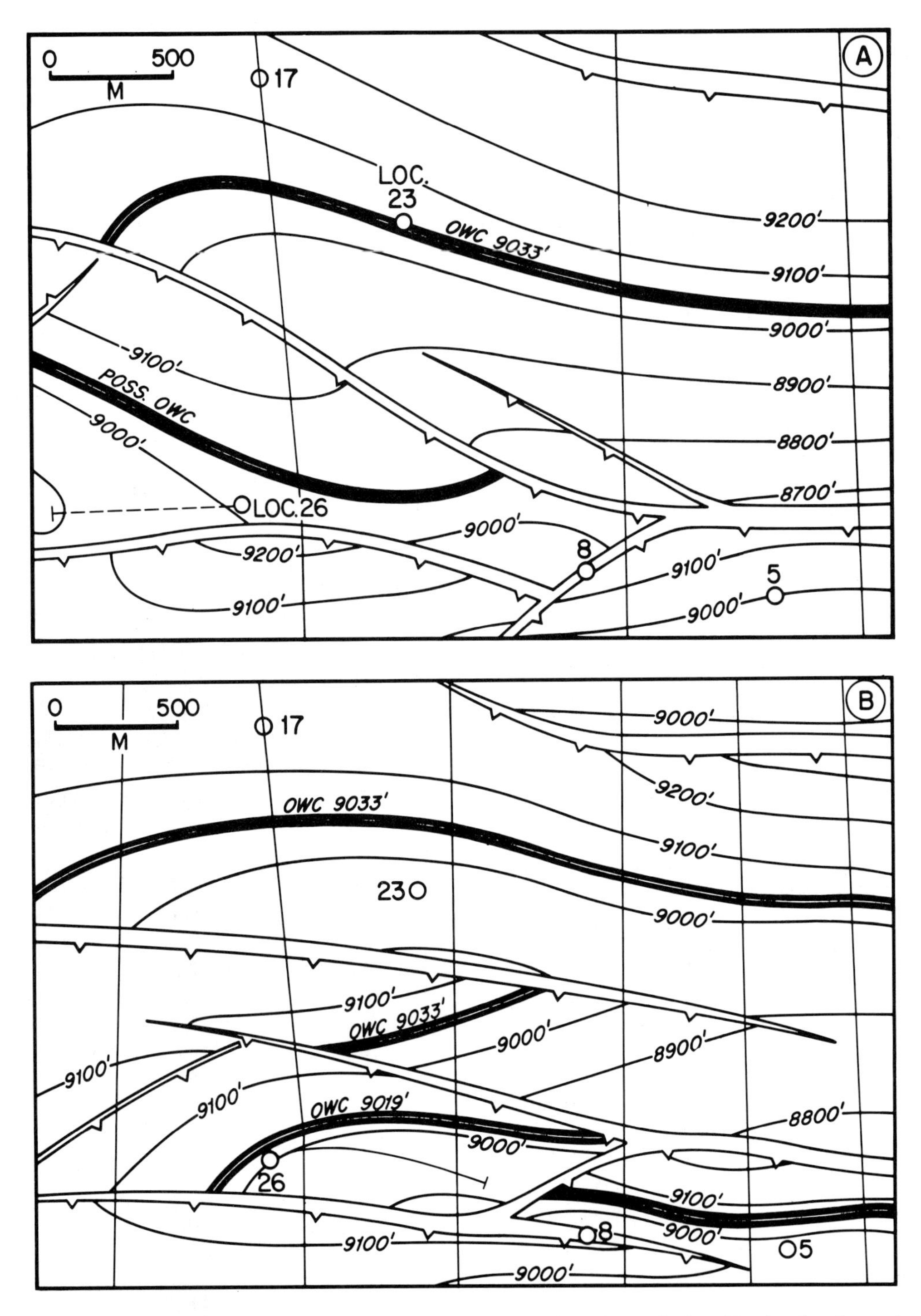

FIG. 8—Part of the structure map at E2.0 sandstone level before (A), and after (B), the shooting and interpretation of infill dip lines (first half of calendar 1977).

nal target likely to be below the probable oil-water contact (Fig. 8). To utilize the location, it was decided to drill an eastward-deviated hole to an E2.0 sandstone target on a new profile passing just east of well 23, a horizontal offset distance of 2,887 ft (880 m). The azimuth of the deviated hole was determined by comparison of well results and seismic data from Nembe Creek 23, on the premise that the oil-water contacts would be essentially uniform across the fault blocks.

A tentative suggestion that the well be drilled due east was rejected in favor of azimuth 97°, after due consideration of the reflection dip at E2.0 sandstone level and the uncertain location of the fault plane to the south. Ultimately, a dip of 11° north was logged in this part of the well. On the assumption of constant dip, had the well been deviated due east only 26 ft

(7.9 m) NOS would have been logged in the E2.0, instead of the total of 147 ft (45 m).

Well 29—Prior to the drilling of this well (located 2.06 mi or 3.32 km north of well 1) a lateral prediction was made of the lithologies, hydrocarbons, and fluid contacts to be encountered. Although the fault block to be evaluated was penetrated at a depth of 11,538 ft (3,517 m) by well 10 (1.12 mi or 1.8 km to the south-southeast; Fig. 4), the sequence immediately underlying the *Uvigerina* 8 shale had not previously been tested. Drilling results eventually showed that the lateral prediction was accurate with respect to horizontal hydrocarbon-water contacts, but that dipping shale-sandstone boundaries were estimated about 70 ft (21.3 m) too high, predicting 425 ft (130 m) of hydrocarbon-bearing sandstone instead of the 222 ft (68 m) discovered. Subsequent study showed that the overestimate resulted from the use of unmigrated seismic data. The use of migrated data (Fig. 7B), normally considered unreliable when preserving signal amplitudes, was later found to produce a more accurate prognosis.

Nembe Creek East 1—This well was drilled as an exploration test from a location 4.59 mi (7.38 km) east of the original discovery well. A straight hole from the preferred surface location would have risked drilling a fault cut-out of the E2.0 sandstone expected at almost the same depth as in the central part of the field (Fig. 3). The well was therefore deviated slightly southward to miss the fault in the E2.0 sandstone, then allowed to drop back to vertical toward its total along the hole depth of 13,015 ft (3,967 m).

An eastward thinning of the paralic section beneath the *Uvigerina* 8 shale was interpreted from seismic data. This was confirmed, as was the prediction that the E2.0 sandstone would share the same structural level it occupied farther west. A thin, but highly significant, 6 ft (1.83 m) NOS was logged in the E2.0 sandstone, with an oil-water contact differing only 4 ft (1.22 m) from that logged in well 5, a distance of 4.78 mi (7.7 km) to the west. In well 5 the sandstone was penetrated in what is probably the same fault block.

Deeper in Nembe Creek East 1, 56 ft (17.1 m) of NOS were logged in the F6.0 sandstone at a depth of 11,336 ft (3,455 m). Only a 47 ft (14.3 m) difference exists between the levels of the highest oil in this well and of the deepest gas in the same sandstone in well 5, suggesting possible communication within the deeper sandstones.

Follow-up (1978 and After)

With recoverable oil reserves currently estimated at 645 million bbl, and adequate seismic data to yield a reliable structural evaluation, it is possible to determine locations for an additional 70 wells for the efficient drainage of this complex field. Of these 70 locations, about seven considered critical for appraisal were drilled in early 1979. Ultimately, a total of about 100 wells should be capable of sustaining a production rate of 200,000 b/d, which will be handled by construction of a 50 mi (80 km) pipeline eastward to the trunk-line system leading to the oil terminal at Bonny (Fig. 3, inset).

TIME-DEPTH CONVERSION

Well velocity surveys through 1976 were made in four wells (3, 6, 8, and 17) using an air gun suspended in water surrounding the drilling barge. The results from wells 6, 8, and 17 showed a standard deviation of only ± 10 msec two-way time over the zone of interest. The data from the survey of well 3 fell a little outside this range, which was not surprising because the well drilled considerably deeper than the field average before passing from a barren downthrown fault block into the field area at a depth of 8,140 ft (2,481 m).

Confidence was shaken early in 1977, however, by the well velocity survey results from well 20, which produced a subparallel time-depth curve of about 70 msec two-way time slower than the curve previously in use. Observation also showed that an impedance trace match between a synthetic seismogram of well 10 and an appropriate trace of seismic profile 120 could only be accomplished by delaying the synthetic seismogram (timed according to the velocity surveys at wells 6, 8, and 17) by approximately 80 msec. Finer tuning and a loop-by-loop comparison of the well 10 synthetic seismogram with seismic data of profile 120 have resulted in a total two-way time delay of about 90 msec relative to the consistent time-depth curves from the surveys in wells 6, 8, and 17.

Although the source of the error in the first four velocity surveys has not been defined, unknown thicknesses and low velocities in the near surface layers can most likely be blamed. These parameters, therefore, cannot simply be copied from the values computed for the static correction of nearby seismic profiles and used for the correction of air gun well velocity surveys. In the future, reflection identification and resultant time-depth control probably will be achieved exclusively by trace matching (Fig. 6A).

The erroneous time-depth curve affected seismic interpretation by suggesting selection of all sandstones too early (shallow) on the VAR-sections. Thus, the obvious "brightness" of the signal associated with the dominantly gas-bearing D1.0 sandstone was overlooked, and this reflection was assigned to the D3.0 sandstone (compare Figs. 5, 7). Fortunately, the error was recognized prior to the design of well 26, whose very precise deviation azimuth of 97° was based on correct reflection identification. Had a reflection event been picked 90 msec earlier (more shallow), the hole azimuth would have been steered farther north to remain in the upthrown block of the fault, and the well would have been unsuccessful.

CONCLUSION

Nembe Creek field was discovered 18 years after the first commercial oil drilling in the Niger delta and is already the second largest field in Nigeria. Recent experience has shown that the timely acquisition and interpretation of sufficient seismic profiles in an area of complex structure, but good reflection quality, can result in an early delineation of the probable field limits. With a foreknowledge of the structure from seismic mapping, and with constant revision of time-depth relationships as new logs become available, it is possible to drill successful long step-out appraisal wells. This led to a rapid evaluation of the recoverable 645 million bbl reserves, after the drilling of 30 wells.

An accurate time-depth relationship was only achieved by comparison of well log synthetic and profile seismic traces. Such positive identification of relations is essential when subsurface targets for development drilling have to be specified with precision.

Geology of Grondin Field[1]

By Jacques Vidal[2]

Abstract During the formation of the Atlantic Ocean, which was preceded in Aptian time by the deposition of a thick layer of salt, the Gabon basin filled with two kinds of sedimentary deposits; mainly sandy continental and littoral deposits in the east and marine deposits in the west. The latter are generally shaly, but a few thick sandstones are present far to the west intercalated in marine shales. The Grondin field was discovered in one of these sandstones called the Batanga, which is of Maastrichtian age.

The sandstones are generally clean and have good porosity. They contain some shales and can reach a gross thickness of more than 200 m. The trap at Grondin is an anticlinal salt structure without noticeable piercing. A median fault divides the structure in two parts, with the upper area's productive formation thinner due to an internal unconformity. The oil field is a unique pool, with a unique oil-water contact, and a small gas cap.The source rocks are marine shales of post-salt age, particularly the Turonian.

The Grondin field, situated 40 km offshore, was discovered in 1971 by Elf Gabon and was developed rapidly. Initial reserves are estimated at 30 million tons (approximately 200 million bbl).

POST—SALT GABONESE SEDIMENTARY BASIN

The formation of the Gabonese sedimentary basin began in the Early Cretaceous at the time Africa separated from America. The basal stratigraphic series consists of thick continental deposits in the rift which preceded the openings of the Atlantic. The first transgression of the sea occurred in the Aptian and resulted in the deposition of a thick salt sequence. This was followed by the opening of the Atlantic, and the basin has since been part of the eastern edge of this ocean.

Three facies are effectively observed in the entire stratigraphic series from east to west (Fig. 1): (1) continental facies, (2) littoral facies, and (3) open sea.

The latter facies are generally shaly (Fig. 2). However, locally thick sand deposits are present. This also applies to the sandstones of the Batanga Formation (Maastrichtian), which constitute the chief reservoir of Grondin field.

Batanga Formation

Facies—The Batanga consists of a collection of sandstones, shaly silts, and silty shales in differing amounts, with local limestone and porcelanite seams.

Age—The Batanga Formation dates from the Maastrichtian in the upper part and in places is Campanian at the base.

Thickness—The Batanga varies in thickness regionally, ranging from 200 to over 400 m. At Grondin field (Fig. 3), it ranges from a maximum thickness of 250 m to a minimum of 50 m. This is probably due to the presence of two internal unconformities.

Extension—The Batanga Formation has an entirely shale/silt lateral equivalent and is difficult to define accurately within the Maastrichtian and Campanian shaly series. Thus, the definition of the formation is restricted to the zones in which it con-

[1]Manuscript received, April 4, 1979; accepted for publication, July 24, 1979.

[2]Elf Aquitaine Production, Paris la Defense, France.

Article Identification Number:
0065-731X/80/MO30-0030/$03.00/0.

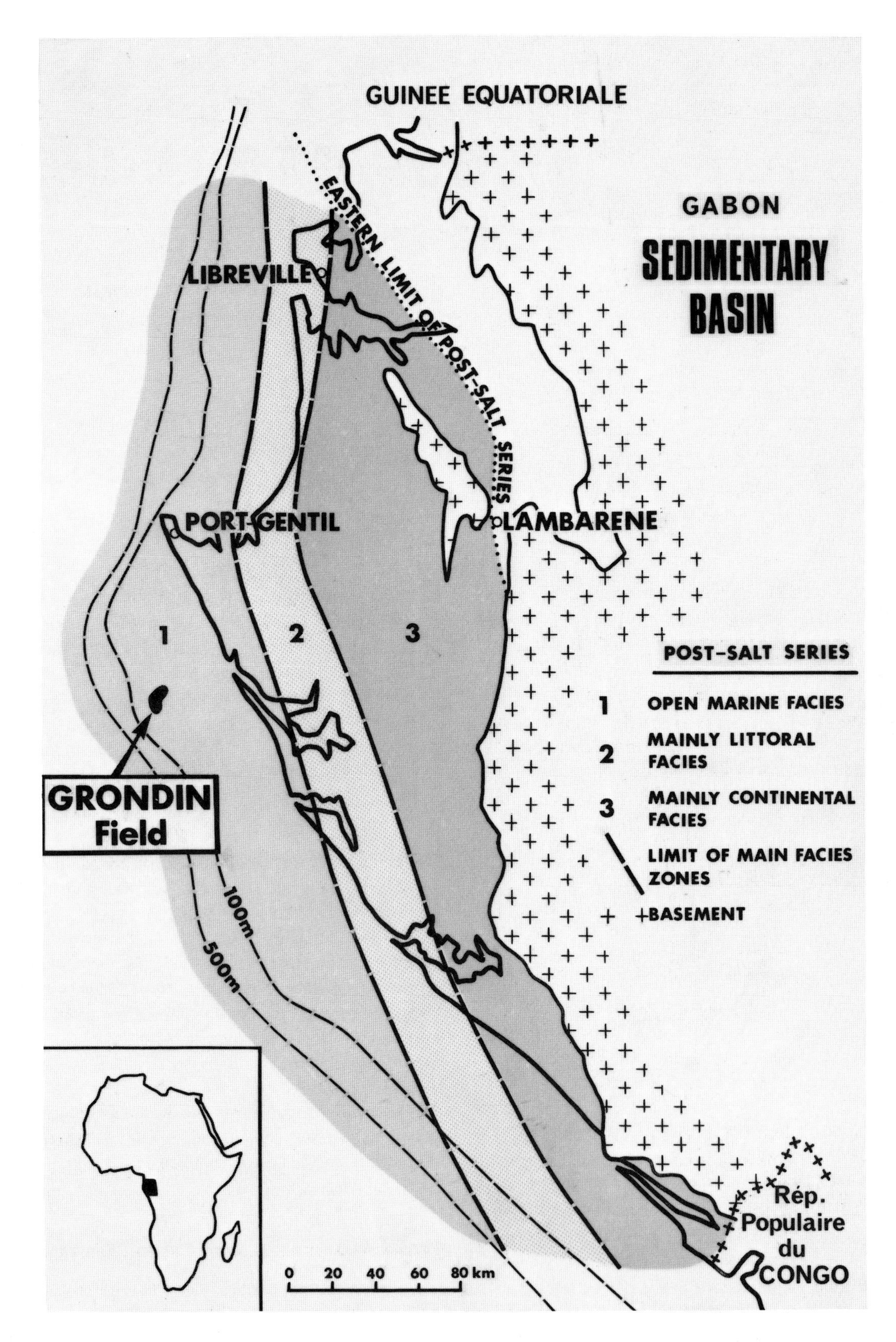

FIG. 1—Map showing the sedimentary basin of Gabon, with the three main zones of facies of the post-salt series.

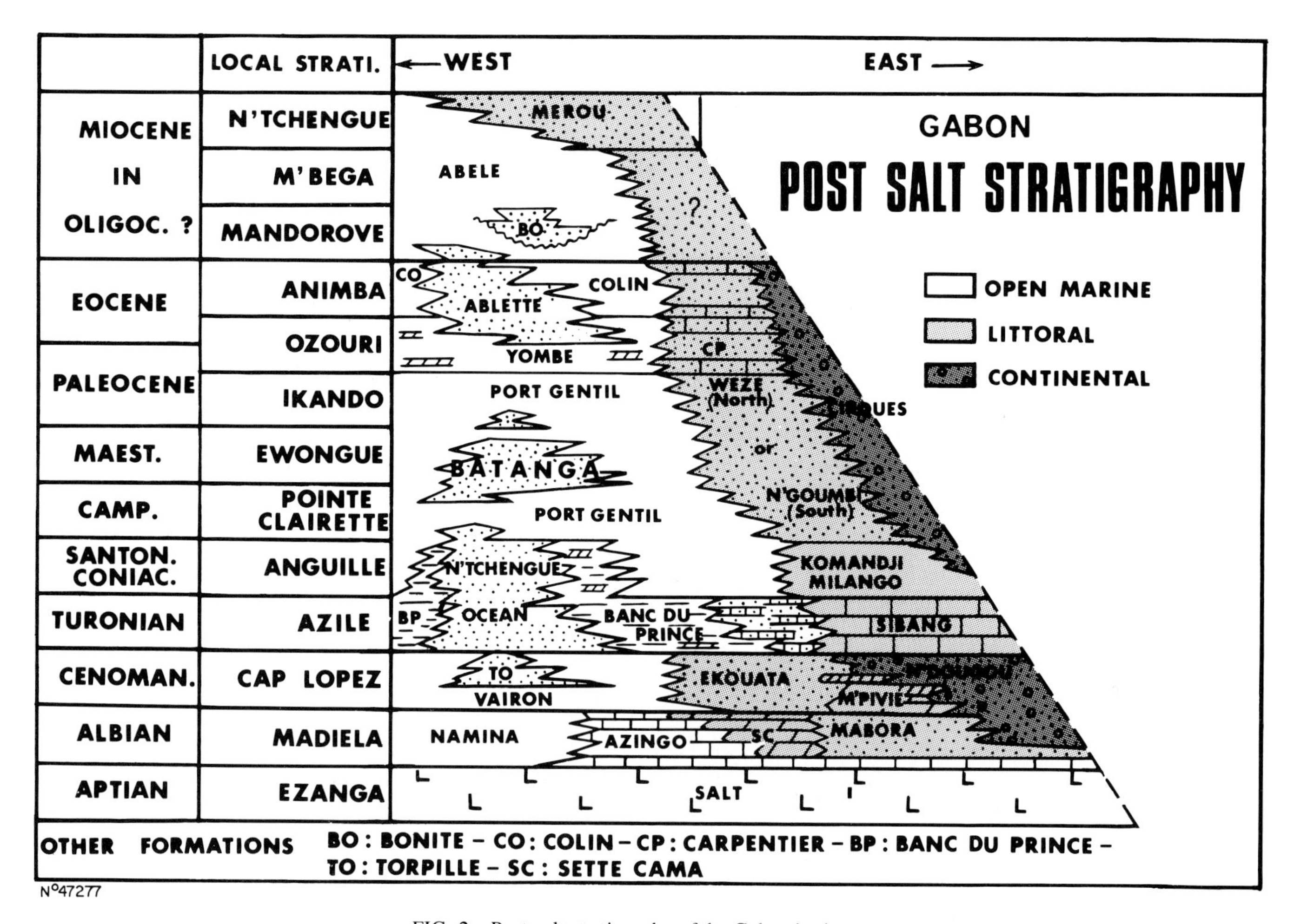

FIG. 2—Post-salt stratigraphy of the Gabon basin.

tains sandstone or strongly silty facies.

The formation's extension zone has been identified in Figure 4. At present (Fig. 5), the northern boundary of the Batanga is relatively clear; however, the southern boundary is still debatable because it is not known if the absence of the Batanga in certain wells is the result of purely local conditions.

Sedimentation Environment—The first assumptions concerning the sedimentation environment of the Batanga Formation considered it a fluviatile or deltaic deposit. This was based on outcrop data, where the deposits clearly exhibit a littoral to continental character.

Throughout the subsurface, however, micropaleontology and palynology clearly indicate an open marine environment. Two alternative hypotheses are available:

1. Assumption of a prograding gravity deposit—this hypothesis is based on the spatially limited character of the formation, and on sedimentary structures considered indicative of gravity deposits (i.e., turbidites and deep sea fans).

2. Assumption of a neritic shallow marine deposit—this hypothesis is supported by the writers. Because the Batanga Formation is perfectly synchronous over the 100 km region where it is identified, as well as the surrounding shales above and below it, this contradicts the progradating character. Furthermore, seismic data have never revealed any evidence of progradation (slopes and canyons) which should characterize deposits outlined by the first hypothesis. On the contrary, seismic data show far more agreement with deposition by upbuilding, whereby the deposits acquired their present position by progressive subsidence. Finally, an important argument is derived from the existence of large sediment gaps over many structures. In view of their characteristics, these gaps are considered as unconformities and probably represent emersion or subemersion phases.

For these reasons the Batanga sandstones appear to represent a shallow marine platform facies where the sediments are dispersed by wave and tidal actions. The disappearance of the Batanga sands northward is explained by a lowering of the seafloor below the wave base.

Petroleum Accumulations in Batanga Sandstones

At present, some ten fields have been discovered in the Batanga sandstones (Fig. 5). With the exception of Pageau field, located near the pinchout of the sandstones and probably a stratigraphic trap, all other fields are of structural origin. They are either on the limb of structures penetrated by the salt (Batanga, Breme, Barbier), or on top of salt anticlines or domes (Mandaros, Gonelle, Grondin).

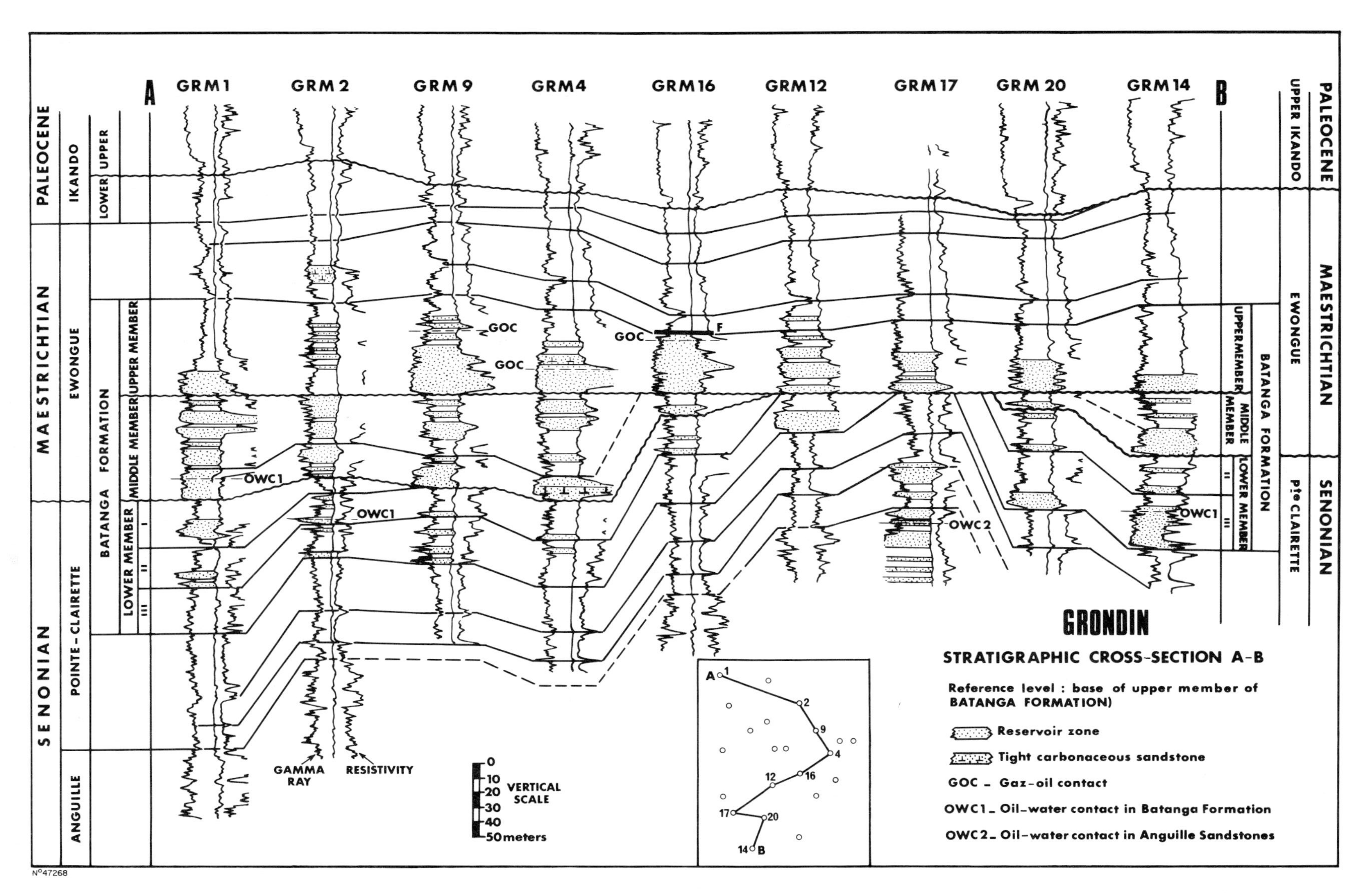

FIG. 3—Stratigraphic cross section across Grondin field, with the hypothesis of reduction of thickness of the Batanga Formations by internal erosion.

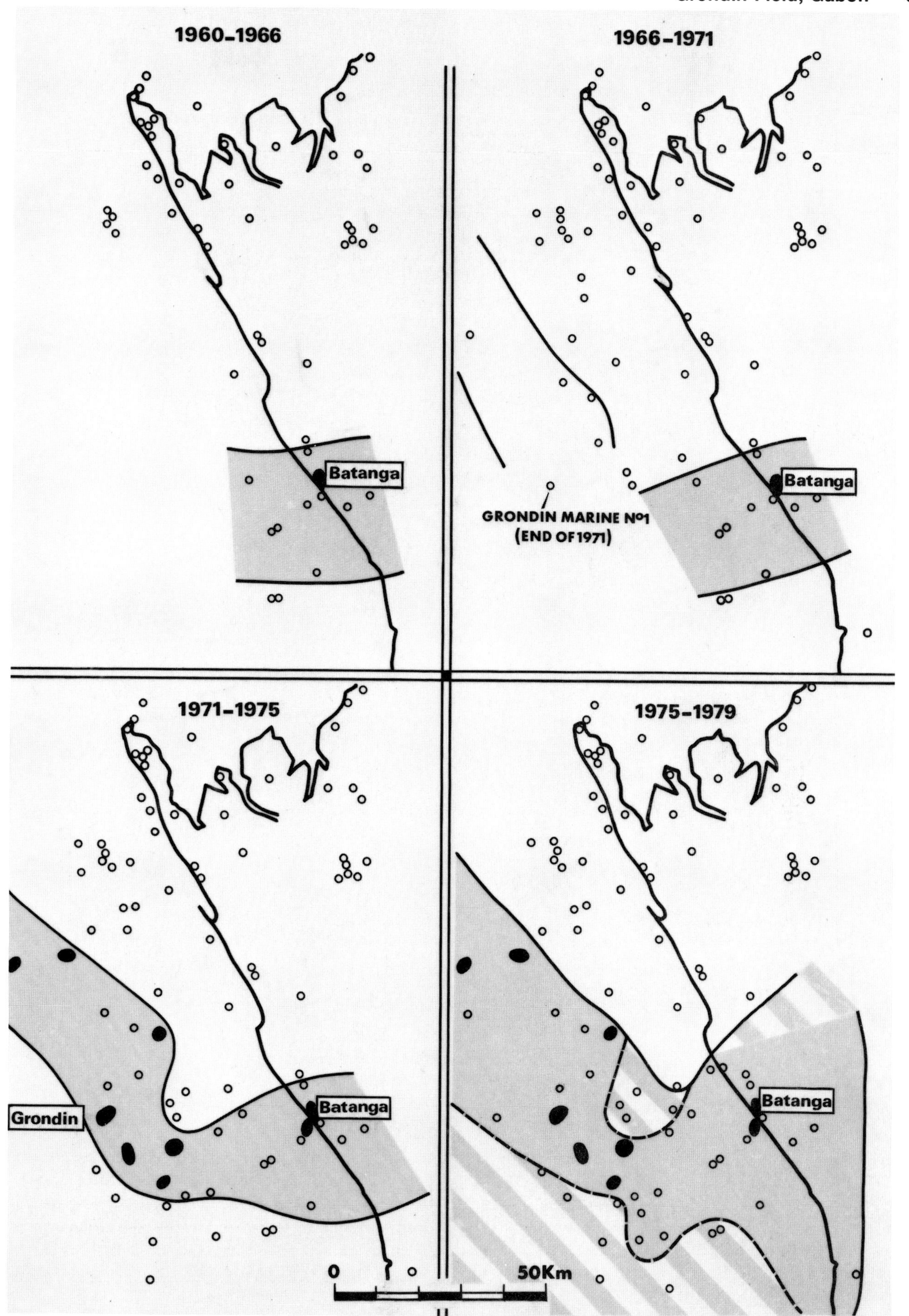

FIG. 4—Map showing the evolution of ideas concerning the extension of Batanga Formations.

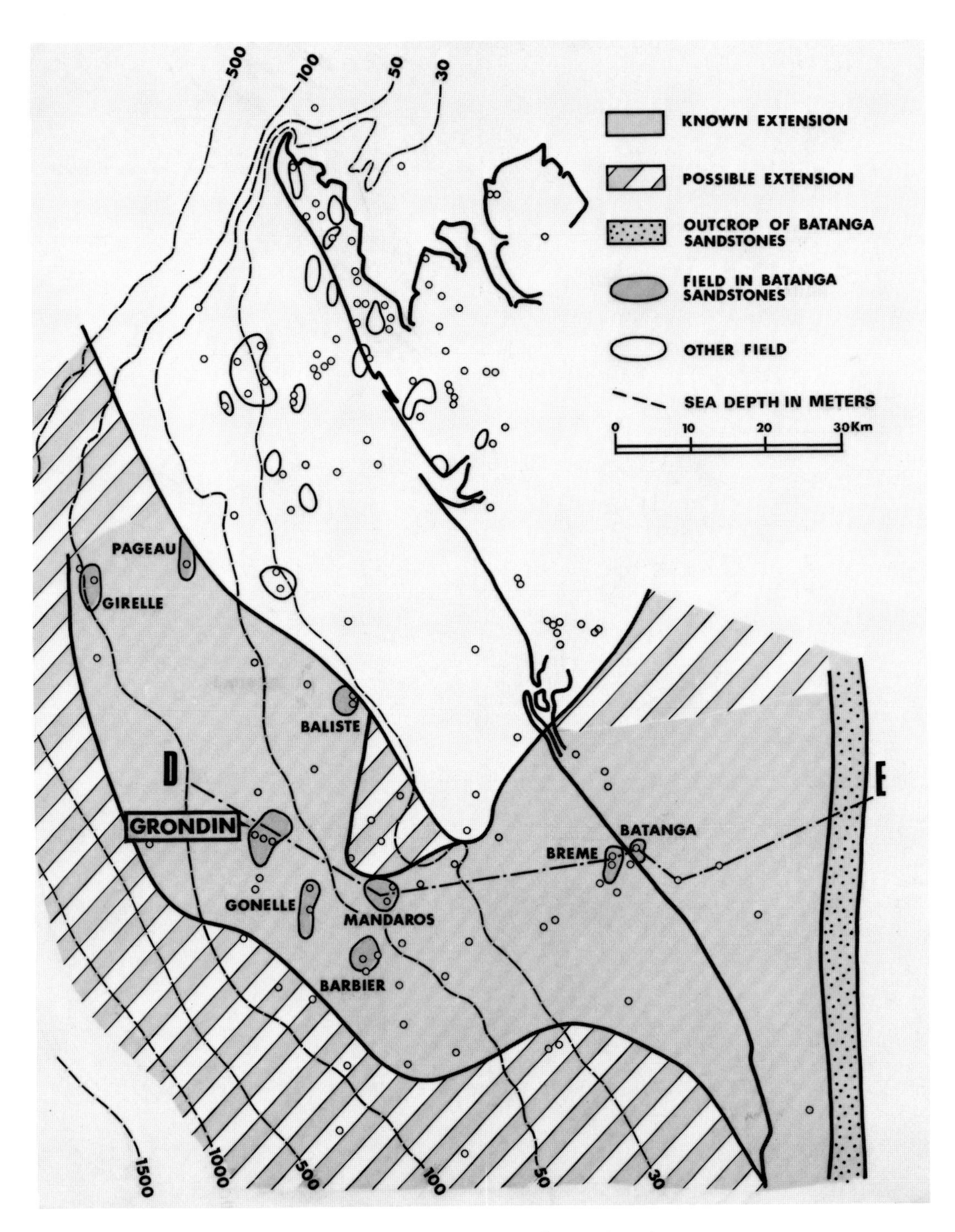

FIG. 5—Extension of Batanga sandstones deposits.

The first field discovered was the Batanga field in 1960. Exploration concentrated on the adjacent offshore region. The Batanga sandstones were identified offshore and though locally exhibited hydrocarbon shows in a relatively narrow zone, no field was discovered.

In 1967, exploration began with the drilling of offshore areas located in water depths exceeding 50 m. The first wells were drilled in the northwest, near the fields producing from the Anguille series. These wells, which were intended to penetrate the Anguille sandstones, revealed an extension zone of the Batanga sandstones. This zone produced only water, however, with some hydrocarbon shows. Exploration then concentrated on the intermediate region where closed structures had been identified. The first well drilled in this zone at the end of 1970 discovered Grondin field. Three other smaller fields were later found.

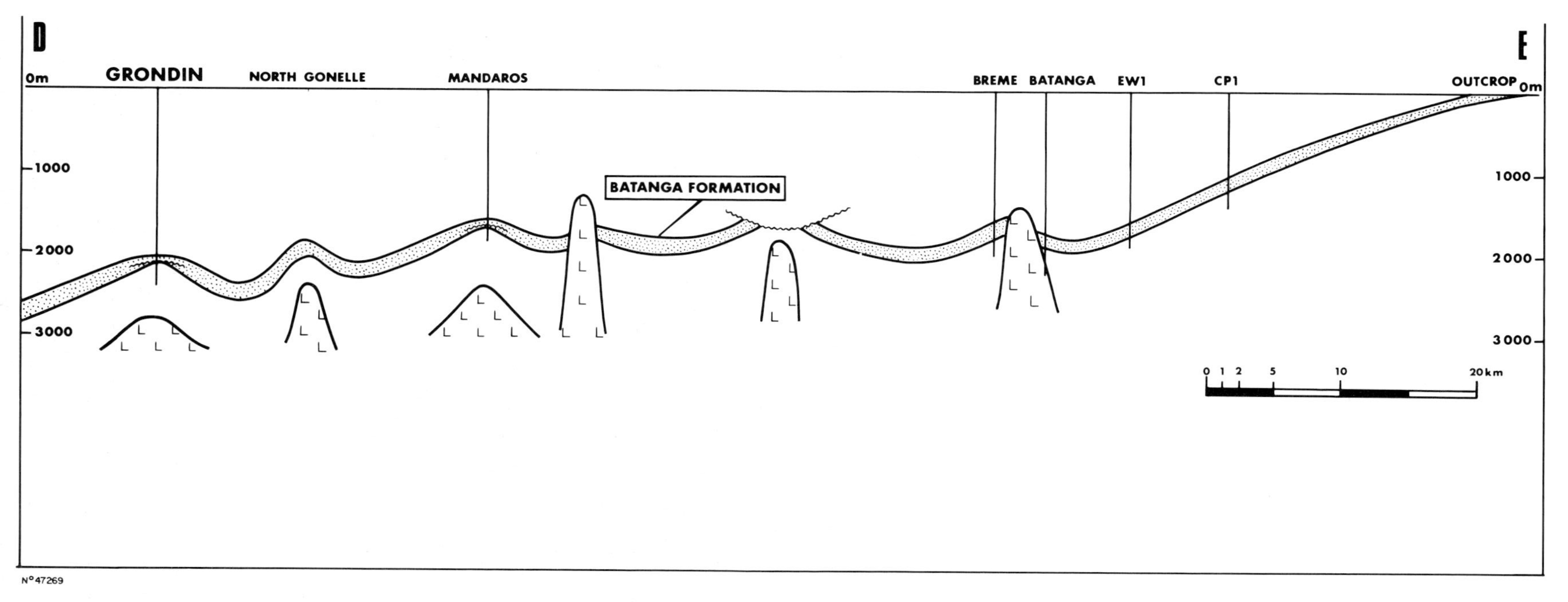

FIG. 6—Schematic regional cross section, line D-E, at Batanga Formation level (for location see Fig. 5).

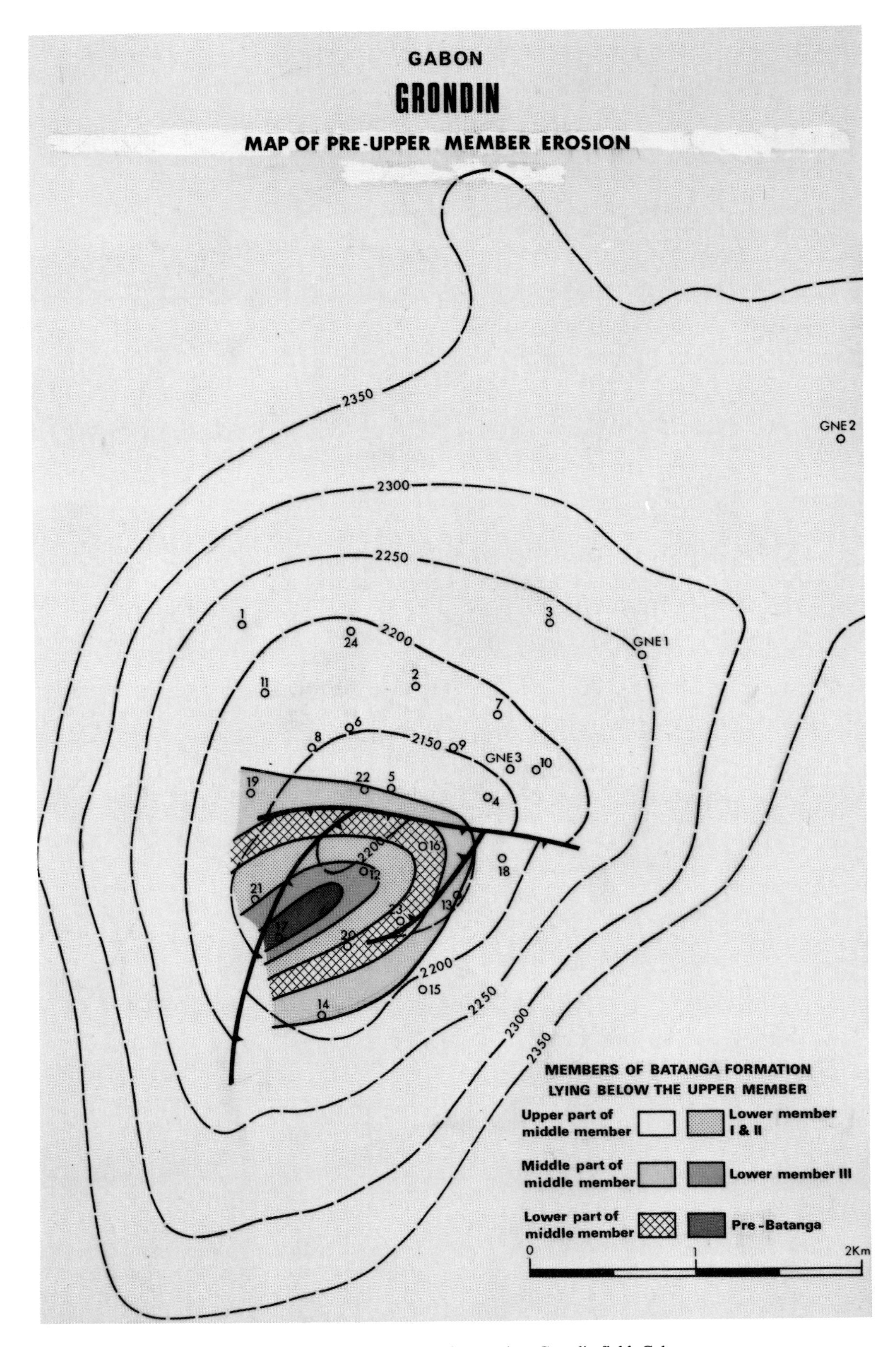

FIG. 7—Map of pre-Upper member erosion, Grondin field, Gabon.

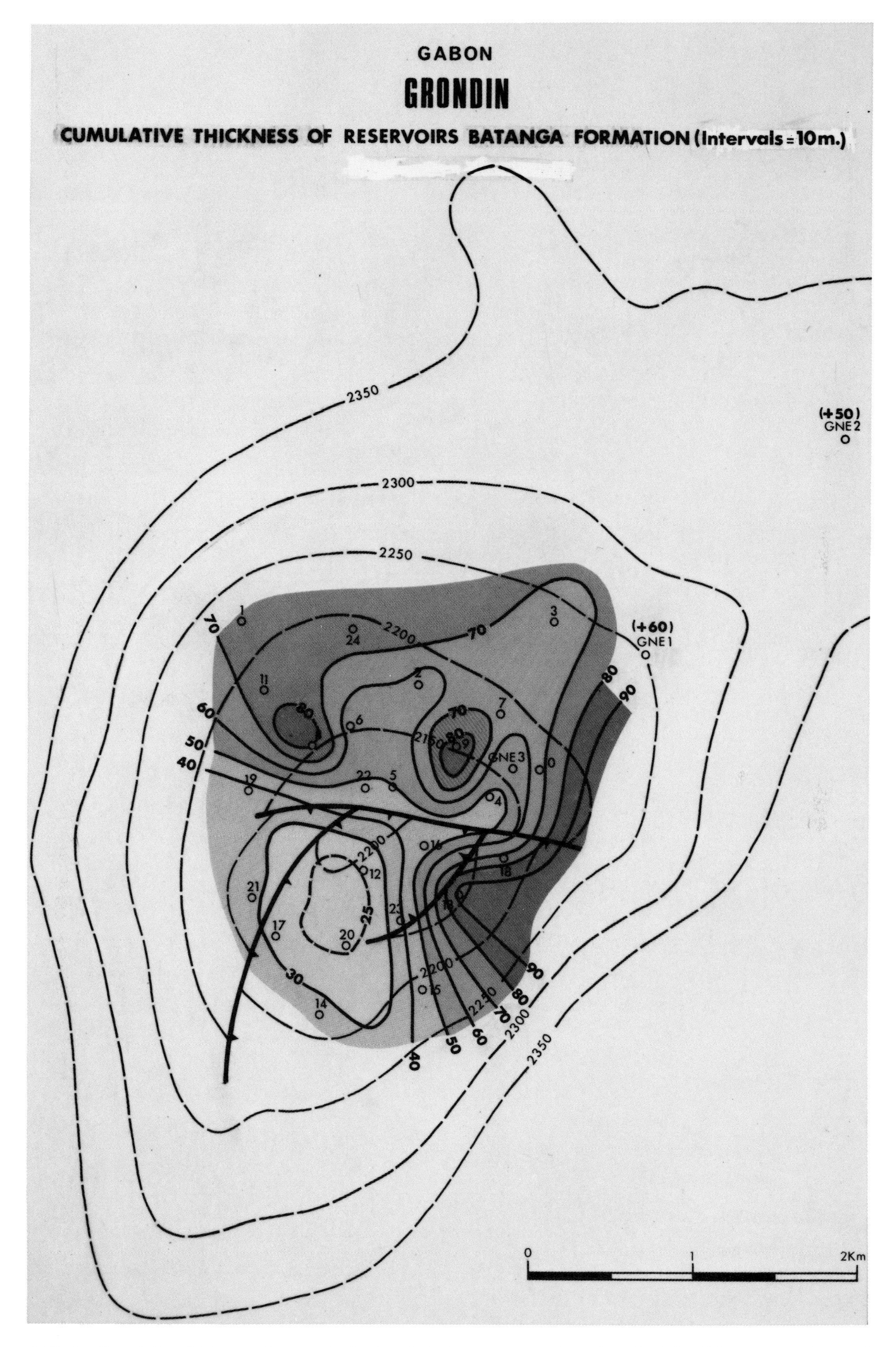

FIG. 8—Cumulative thickness of reservoirs in the whole Batanga Formation, Grondin field, Gabon (contour interval equals 10 m).

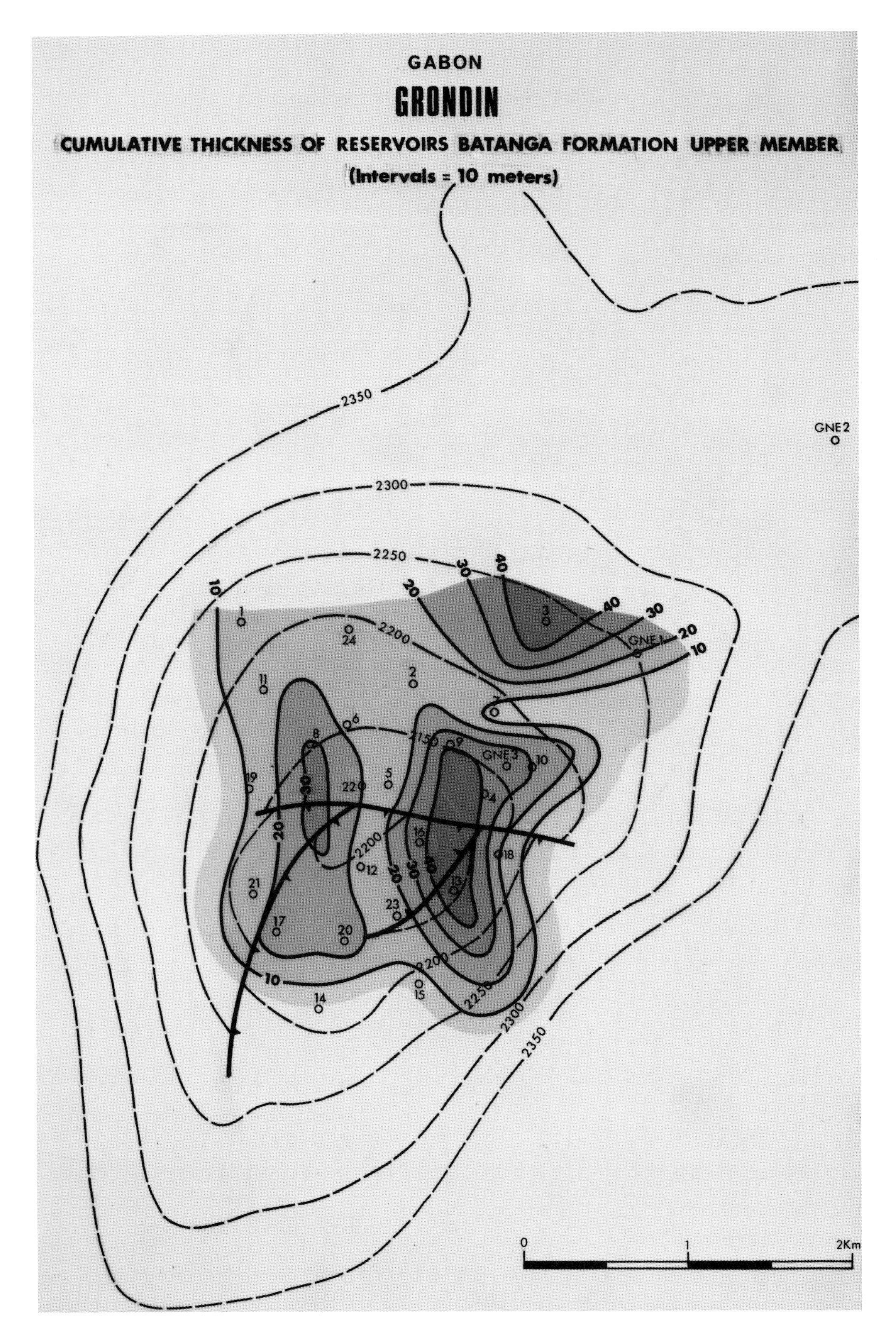

FIG. 9—Cumulative thickness of reservoirs in the Upper member of the Batanga Formation, Grondin field, Gabon.

GRONDIN

STRUCTURE MAP TOP OF BATANGA FORMATION

Depth below sea level

Intervals = 25 meters

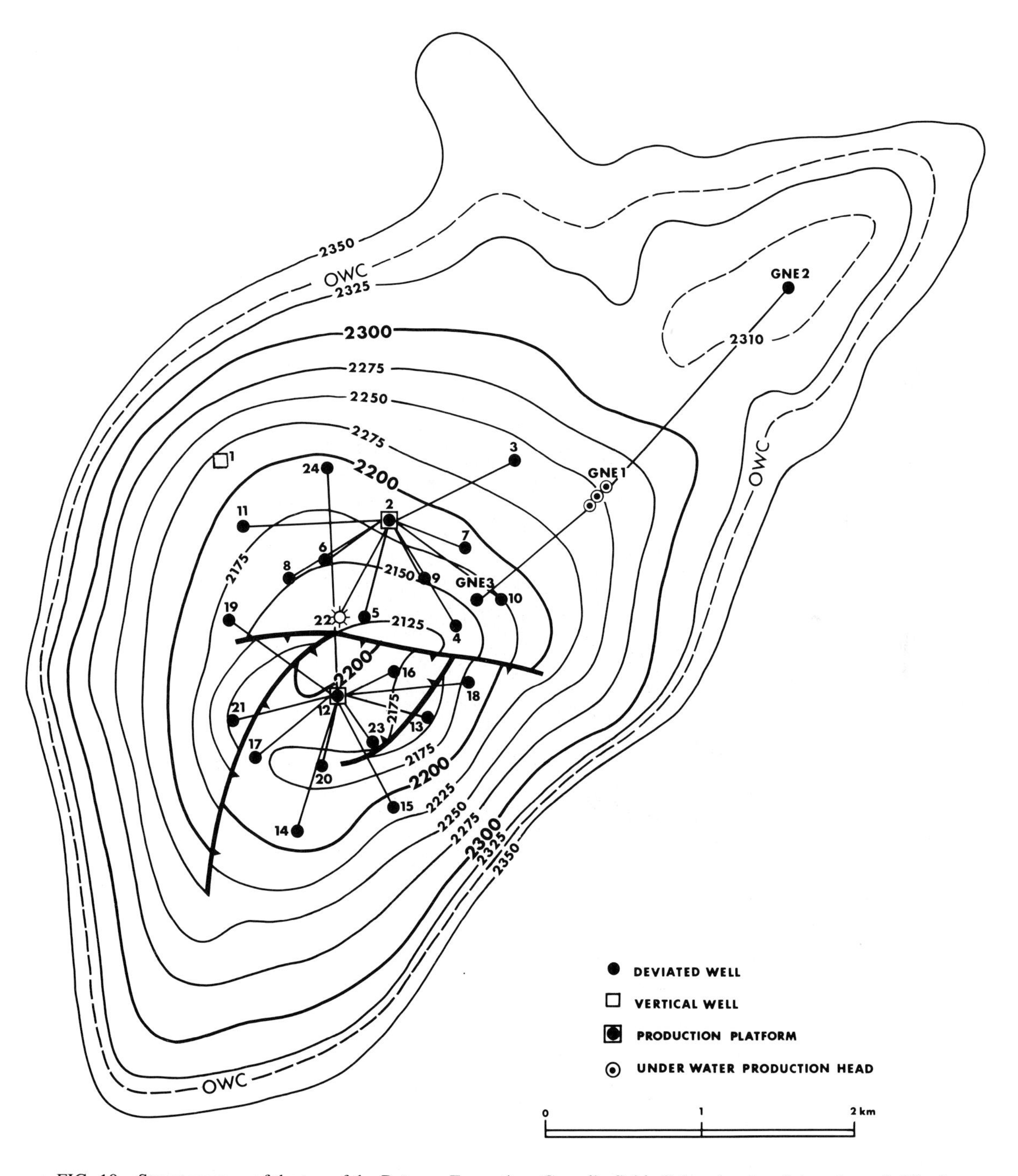

FIG. 10—Structure map of the top of the Batanga Formation, Grondin field, Gabon (contour interval equals 25 m).

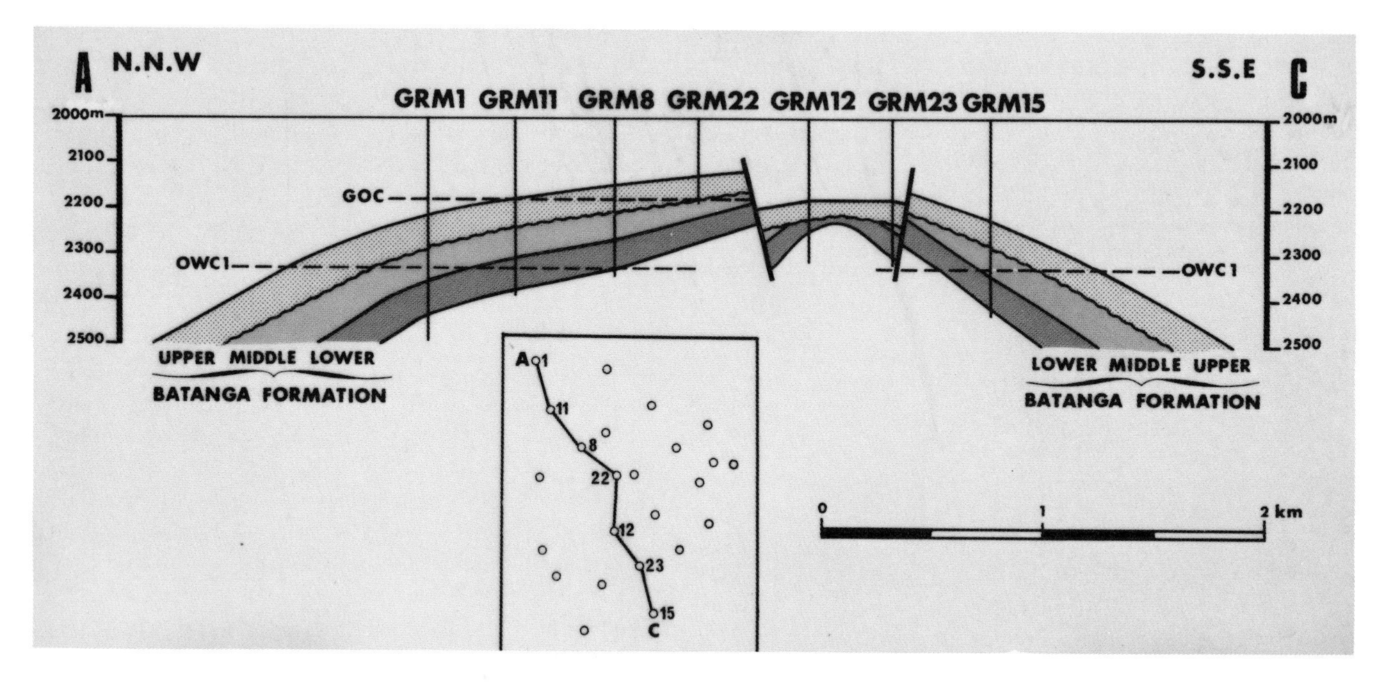

FIG. 11—Structural cross section of line A-C, Grondin field, Gabon.

GRONDIN FIELD

Nearly the entire reserves of Grondin field were found in sandstones of the Batanga Formation, and a very small part (less than 1%) in sandstones of the base of Pointe Clairette and the top of Anguille formations.

The Batanga Formation

Stratigraphy—Figure 3 shows a log section of the Batanga Formation positioned approximately north to south in Grondin field. The Batanga is obvious between the low resistivity shales of Pointe Clairette below and the Ewongué shales above. Good log markers are present in these two shales, making it possible to pinpoint the top and base of the Batanga Formation. The formation is divided into three members (Fig. 3).

The Lower member is separated from the rest of the formation by palynology and subsurface log data. The latter makes it possible to divide the member into three more or less clear "fining and shaling up" sedimenatry cycles. Level I is easily separated from Level II by a radioactive marker observed in eleven wells (only two on the section). The sand beds developed at the base of these cycles have a limited extension and pass laterally to shaly silts or to carbonates.

The Middle member is more sandy and difficult to subdivide. It lies unconformably on the lower member. This unconformity between the Ewongué and Pointe Clairette is recognized regionally and is locally extensive. However at Grondin field its effect is slight and is manifested exclusively by the erosion of Level I of the lower member.

The Upper member is separated from the two other members only by geometric subsurface considerations. Although quite varied in character, this upper member appears definable on logs as a general "fining and shaling up" sequence—the coarsest and cleanest facies meet at its base and it gradually becomes finer and more shaly before passing into the Ewongué shales. The upper member lies unconformably above the other members which are extensively eroded on the top of the structure. This erosion phase has been observed in several places and is regional. Thus, the upper member is considered a shallow-marine deposit, transgressive after regional erosion, and subsequently subjected to progressive subsidence. Irregularities in sand distribution are explained by local depth variations or by irregularities in the wave and tidal currents which spread the sands.

Another geologic interpretation of Grondin field recently suggested is based on different correlations between the wells. The lower member remains approximately the same; however, above it the subdivision of the formation is different and features two sandstone units separated by a shaly zone. The reduced thickness of the formation is interpreted not as the effect of erosion, but as a great irregularity in the deposition of the sand due to gravity deposition. The sand of the Upper member disappears completely without a lateral shaly equivalent throughout the southern part of the field.

According to this theory the fining-up sequences are interpreted as deep channel deposits. This interpretation, which considers the existence of a continuous shaly level within the formation, does not appear to reflect accurately the perfect communication of the reservoirs (Upper and Middle members) observed in producing the field. According to the first hypothesis, the sand level of the Upper member is in contact with all levels of the middle member because it overlies them unconformably (Fig. 7).

Reservoirs—The reservoirs (Fig. 3) consist essentially of two facies: (1) very fine to fine sandstones, well sorted, with subangular grains and local shale; and (2) heterogeneous fine to medium sands, with subangular to subrounded grains, some feldspars and phosphate debris, sparce shaly cement with local silicious or dolomitic cement. In the lower member, dolomitic cement is generally far more abundant. The sandstones sometimes grade into crystalline dolomites with different percentages of quartz.

Porosities are high and can exceed 30%, the mean being about 20%. Permeabilities up to several hundred milidarcys are common.

The cumulative thickness of the reservoirs of the Batanga Formation (Fig. 8) ranges from 25 to 80 m. The distribution of these sandstones is not a function of the deposition conditions, because of the extensive erosion which has occurred. The reservoir thickness map of the Upper member (Fig. 9) is more significant. It appears to show that the thickness of the sandstones is not related to the structure but rather is governed by more regional conditions with irregular thickening toward the east and northeast.

The reservoirs of the Batanga Formation are well sealed by Cretaceous and Tertiary marine shales (over 1,500 m thick).

The productive Anguille series in Gabon, in the region located north and northeast of Grondin, has been penetrated in several wells. It is generally shaly except in two wells (GRM 17 and GRM 21) where fine to coarse sands occur along with a small oil accumulation.

Source Rocks—The source rocks of Grondin field are the organic-rich shales of the Banc-du-Prince Formation (Turonian). It is believed that oil generation and migration occurred fairly late in the Miocene, when burial conditions were satisfactory for maturation.

Structure—The trap of the main Grondin pool is a "salt dome" anticline (Figs. 10, 11). No drilling has penetrated the salt and according to seismic data it is assumed that the salt has produced a "non-pierced" structure.

The structure is unaffected by major faulting. The only observations show a limited collapse of the top zone of the structure with faults exceeding 75 m of throw.

Because of the erosion observed at the top of the structure, two phases of structural formation occurred during the deposition of the Batanga Formation. However, the formation of the structure may have taken place in pulses, that is very short-lived salt thrusts giving rise to erosion. During most of the depositional period the influence of the structure was not significant. This explains the absence of depositional thinning other than that caused by erosion at the top of the structure. This effect is present not only in the Batanga Formation but also in the underlying and overlying formations.

Distribution of Fluids

Batanga—Grondin field has a small gas cap north of the collapse zone. The gas cap is confined within the Upper member.

The field has a single oil-water contact which is encountered in all three members of the formation, depending on the position of the well on the structure. The maximum height of the oil pool is 150 m, while pressure of the pool is hydrostatic. A water-drive was observed during production.

The density of oil is 28.20 API, with a viscosity of 1 centipoise, in bottom-hole conditions. The bubble point of the oil varies with depth.

Anguillle—The Anguille sandstones feature an oil-water contact which is slightly higher than the Batanga.

Development

Grondin field is located in water 62 m deep. It was developed by means of two multi-well platforms and three underwater production heads. It has a total of 25 producing wells distributed as follows: 10 wells including one vertical on platform 1, plus a gas cap control well; 12 wells including one vertical on platform 2; and 3 wells with an underwater production head including one vertical and two directional.

The oil is pumped from a central production platform by means of a 20-in. sealine which conveys the oil to the Cape Lopez terminal near Port Gentil.

Production

Grondin field, discovered in September 1971, began production with six wells less than 18 months after discovery. Maximum delivery per well is 800 cu m/day of oil. Annual production reached a maximum of 4 million cu m/day in 1974 (25,000,000 bbl), and is now about 2,600,000 cu m (16 million bbl).

Production history has shown that good pressure communication exists between the reservoirs of the Upper member and the Middle member. Furthermore, the aquifer exhibits very good participation and the interpretation of rising water in the field confirms the good communication between the producing two members.

Due to the fact that the reservoir is generally unconsolidated, different sand control methods have been used including gravel packing and sand screening.

Reserves

Initial recoverable reserves of the field were estimated at 30 million cu m, or about 190 million bbl. They are distributed as follows: Batanga, Upper members, 56%; Batanga, Middle member, 39%; Batanga, Lower member, 4%; and Anguille, 1%.

As of December 31, 1978, about 60% of these reserves had been produced. The total initial reserves of the fields in the Batanga Formation are about 70,000,000 cu m (440 million bbl).

REFERENCES CITED

Bosio, J., 1974, les recents developpements de l'Offshore du Gabon, le champ de Grondin: Revue de l'Association Francaise des Techniciens du Petrole, July-August 1974, p. 41-45.

Brink, A. H., 1974, Petroleum Geology of Gabon basin: AAPG Bull., v. 58, p. 216-235.

Chateau, G., and Falkner, C., 1977, Deepwater sea-floor production stations being developed off Africa: Oil and Gas Journal, v. 75, p.

148-164.
Hirtz, P., 1964, d'exploration executes par la Spafe au large des cotes du Gabon: Publication of Institut Francais du Petrole, Collection "colloques et seminaires:" Les recherches et la production du Petrole en mer. p. 232-250.
Reyre, D., Belmonte, Y., Derumaux, F., Wenger, R, 1966, evolution geologique du bassin gabonais, *in* Bassins sedimentaires du littoral africain; lere partie: littoral atlantique. Association des services geologiques africains, p. 171-191.
Vidal, J., Joyes R., Van Veen, J., 1975, l'exploration petroliere au Gabon et au Congo, 9th World Petroleum Conference, Tokyo 1975, v. 3, p. 149-165.
Wenger, R., 1975, le bassin sedimentaire gabonais et la derive ees Continents, Anais do XXVII Congress. Brazileiro de Geologia, Revista Brazileira de Geosciencias.

Explanation of Indexing

A reference is indexed according to its important, or "key," words.

Three columns are to the left of a keyword entry. The first column, a letter entry, represents the AAPG book series from which the reference originated. In this case, M stands for Memoir Series. Every five years, AAPG will merge all its indexes together, and the letter M will differentiate this reference from those of the AAPG Studies in Geology Series (S) or from the AAPG Bulletin (B).

The following number is the series number. In this case, 30 represents a reference from Memoir 30.

The last column entry is the page number in this volume where this reference will be found.

Note: This index is set up for single-line entry. Where entries exceed one line of type, the line is terminated. (This is especially evident with manuscript titles, which tend to be long and descriptive.) The reader must sometimes be able to realize keywords, although commonly taken out of context.